T0325913

Analog Circuit Design

A Tutorial Guide to Applications and Solutions

Analog Circuit Design

A Tutorial Guide to Applications and Solutions

Edited by

Bob Dobkin

Jim Williams

ELSEVIER

Amsterdam • Boston • Heidelberg • London • New York • Oxford
Paris • San Diego • San Francisco • Singapore • Sydney • Tokyo
Newnes is an imprint of Elsevier

Newnes

Analog Circuit Design

A Tutorial Guide to Applications and Solutions

Analog Circuit Design

A Tutorial Guide to Applications and Solutions

Edited by

Bob Dobkin

Jim Williams

ELSEVIER

Amsterdam • Boston • Heidelberg • London • New York • Oxford
Paris • San Diego • San Francisco • Singapore • Sydney • Tokyo
Newnes is an imprint of Elsevier

Newnes

Newnes is an imprint of Elsevier
The Boulevard, Langford Lane, Kidlington, Oxford OX5 1GB, UK
225 Wyman Street, Waltham, MA 02451, USA

First edition 2011

British Library Cataloguing in Publication Data
A catalogue record for this book is available from the British Library

Library of Congress Cataloging-in-Publication Data
A catalog record for this book is availabe from the Library of Congress

ISBN: 978-0-12-385185-7

For information on all Newnes publications
visit our web site at books.elsevier.com

Printed and bound in The United States of America
11 12 13 14 15   10 9 8 7 6 5 4 3 2 1

Working together to grow
libraries in developing countries

www.elsevier.com | www.bookaid.org | www.sabre.org

ELSEVIER   BOOK AID International   Sabre Foundation

For Jerrold R. Zacharias, who gave me the sun, the moon and the stars.
For Siu, who is the sun, the moon and the stars.

In memory of Jim Williams, a poet who wrote in electronics.

This book was compiled from Linear Technology Corporation's original *Application Notes*.

These *Application Notes* have been re-named as chapters for the purpose of this book. However, throughout the text there is a lot of cross referencing to different *Application Notes*, not all of which have made it into the book. For reference, this conversion table has been included; it shows the book chapter numbers and the original *Application Note* numbers.

CHAPTER NUMBER	APPLICATION NOTE
1	88
2	101
3	51
4	126
5	19
6	25
7	35
8	70
9	119a
10	199b
11	122
12	83
13	39
14	77
15	81
16	89
17	90
18	92
19	112
20	7
21	71

CHAPTER NUMBER	APPLICATION NOTE
22	86
23	96
24	120
25	3
26	9
27	11
28	20
29	23
30	28
31	40
32	43
33	47
34	72
35	82
36	93
37	94
38	106
39	124
40	99
41	102

Trademarks

These Trademarks all belong to Linear Technology Corporation. They have been listed here to avoid endless repetition within the text. Trademark acknowledgment and protection applies regardless. Please forgive us if we have missed any.

Linear Express, Linear Technology, LT, LTC, LTM, Burst Mode, FilterCAD, LTspice, OPTI-LOOP, Over-The-Top, PolyPhase, SwitcherCAD, TimerBlox, µModule and the Linear logo are registered trademarks of Linear Technology Corporation. Adaptive Power, Bat-Track, BodeCAD, C-Load, Direct Flux Limit, DirectSense, Easy Drive, FilterView, Hot Swap, LinearView, LTBiCMOS, LTCMOS, LTPoE++, LTpowerCAD, LTpowerPlanner, LTpowerPlay, MicropowerSwitcherCAD, Multimode Dimming, No Latency $\Delta\Sigma$, No Latency Delta-Sigma, No R_{SENSE} Operational Filter, PanelProtect, PLLWizard, PowerPath, PowerSOT, PScope, QuikEval, RH DICE Inside, RH MILDICE Inside, SafeSlot, SmartStart, SNEAK-A-BIT, SoftSpan, Stage Shedding, Super Burst, ThinSOT, Triple Mode, True Color PWM, UltraFast, Virtual Remote Sense, Virtual Remote Sensing, VLDO and VRS are trademarks of Linear Technology Corporation. All other trademarks are the property of their respective owners.

Acknowledgments

Spanning three decades of analog technology, this volume represents the hard work of many individuals—too many to name. The lion's share of the credit goes to Linear's dedicated engineer/authors, whose work fills these pages. Jim Williams and Bob Dobkin have given generously of their time and support. I would be remiss not to also acknowledge the contributions of our dedicated publications team of Susan Cooper and Gary Alexander, who put in the extra hours to get the Application Notes ready for publication. Finally, a word of thanks to our publisher, Jonathan Simpson, who helped pave the road from idea to book, Naomi Robertson and Pauline Wilkinson, who smoothed the book's production.

John Hamburger
Linear Technology Corporation

Introduction

Why write applications?

This is seemingly an odd and unlikely way to begin an applications publication, but it is a valid question. As such, the components of the decision to produce this book are worth reviewing.

Producing analog application material requires an intensive, extended effort. Development costs for worthwhile material are extraordinarily high, absorbing substantial amounts of engineering time and money. Further, these same resources could be directed towards product development, the contribution of which is much more easily measured at the corporate coffers.

A commitment to a concerted applications effort must be made despite these concerns. Specifically, the nature of analog circuit design is so diverse, the devices so sophisticated, and user requirements so demanding that designers require (or at least welcome) assistance. Ultimately, the use of analog ICs is tied to the user's ability to solve the problems confronting them. Anything that enhances this ability, in both specific and general cases, obviously benefits all concerned.

This is a very simple but powerful argument, and is the basis of any commitment to applications. Additional benefits include occasional new product concepts and a way to test products under "real world" conditions, but the basic justification is as described.

Traditionally, application work has involved reviewing considerations for successful use of a specific product. Additionally, basic circuit suggestions or concepts are sometimes offered. Although this approach is useful and necessary, some expansion is possible. The applications selected for inclusion in this book are centered on detailed, systems-oriented circuits, (hopefully) similar to users' actual designs. There is broad tutorial content, reflected in the form of frequent text digressions and liberal use of graphics. Discussions of trade-offs, options and techniques are emphasized, as opposed to brief descriptions of circuit operation. Many of the application notes include appended sections which examine related or pertinent topics in detail. Ideally, this treatment provides enough background to allow readers to modify the circuits presented into solutions to their specific problems.

Some comment about the circuit examples is appropriate. They range from relatively simple to quite complex and sophisticated. Emphasis is on high performance, in keeping with the capabilities of contemporary products and users' needs. The circuit's primary function is to serve as a catalyst—once the reader has started thinking, the material has accomplished its mission.

Substantial effort has been expended in working out and documenting these circuits, but they are not necessarily finessed to the highest possible degree. All of the circuits have been breadboarded and bench-tested at the prototype level. Specifications and performance levels quoted in the text represent measured and extrapolated data derived from the breadboard prototype. The volume of material generated prohibits formal worst-case review or tolerance analysis for production.

The content in this volume, while substantial, represents only a portion of the available material. The resultant winnowing process was attended by tears and tantrums. The topics presented are survivors of a selection process involving a number of disparate considerations. These include reader interest, suitability for publication, time and space constraints and lasting tutorial value. Additionally, a minimum 10 year useful lifetime for application notes is desired. This generally precludes narrowly focused efforts. Topics are broad, with a tutorial and design emphasis that (ideally) reflects the reader's long term interest. While the circuits presented utilize existing products, they must be conceptually applicable to succeeding generations of devices. In this regard, it is significant that some of the material presented is still in high demand years after initial publication.

The material should represent a relatively complete and interdisciplinary approach to solving the problem at hand. Solving a problem is usually the reader's overwhelming motivation. The selection and integration of tools and methods towards this end is the priority. For this reason the examples and accompanying text are as complete and practical as possible. This may necessitate effort in areas where we have no direct stake, e.g., the software presented in Chapter 22 or the magnetics developed for Chapters 6 and 7.

Quality, in particular good quality, is obviously desirable in any publication. A high quality application note requires attentive circuit design, thorough laboratory technique, and completeness in its description. Text and figures should be thoughtfully organized and presented, visually pleasing, and easy to read. The artwork and printing should maintain this care in the form of clean text appearance and easily readable graphics.

Application notes should also be efficient. An efficiently written note permits the reader to access desired information quickly, and in readily understandable form. There should be enough depth to satisfy intellectual rigor, but the reader should not need an academic bathyscaphe to get to the bottom of things. Above all, the purpose is to communicate useful information clearly and quickly.

Finally, style should always show. Quite simply, the publication should be enjoyable to read. Style provides psychological lubrication, helping the mind to run smoothly. Clearly, style must only assist the serious purposes of publication and should not be abused; the authors have done their best to maintain the appropriate balance.

This book's many authors deserve any and all forthcoming applause; the named editors accept sole responsibility for philosophical direction, content choice, errors, omissions, and other sins.

Jim Williams
Staff Scientist
Linear Technology Corporation

Foreword

The fundamental difference between analog and digital is "information." With digital information the output is always the same: a set of ones and zeros that represents the information. This information is independent of the supply voltages or the circuitry that is used to generate it. With analog, the output information is basic electrical values—volts, current, charge—and is always related to some real world parameters. With analog, the methodology used to arrive at the answers is intrinsic to the quality of those answers. Errors such as temperature, noise, delay and time stability can all affect the analog output and all are a function of the circuitry that generates the output. It is this analog output that is difficult to derive and requires experience and circuit design talent.

With integrated circuits (ICs) so prevalent, combined with application-specific integrated circuits (ASICs) in most systems, it is becoming increasingly difficult to find good analog examples for teaching engineers analog design. Engineering schools provide the basics of device terminal characteristics and some circuit hookup information, but this is not adequate for designing finished circuits or applying modern IC design techniques. The analog circuitry in today's systems is often difficult to decipher without help from the original designer. The ability to design complex analog systems relies on the ability of engineers to learn from what has gone before.

One of the best avenues for learning analog design is to use the application notes and information from companies who supply analog integrated circuits. These application notes include circuitry, test results, and the basic reasoning for some of the choices made in the design of these analog circuits. They provide a good starting point for new designs.

Since the applications are aimed at solving problems, the application notes, combined with the capability to simulate circuits on Spice, provide a key learning pathway for engineers. The analog information in most of these application notes is timeless and will be as valid twenty years from now as it is today. It's my hope that anyone reading this book is helped through the science and art of good analog design.

Robert Dobkin
Co-Founder, Vice President, Engineering,
and Chief Technical Officer
Linear Technology Corporation

Contents

Section 1

Power Management Tutorials

Ceramic input capacitors can cause overvoltage transients (1)

When it comes to input filtering, ceramic capacitors are a great choice. They offer high ripple current rating and low ESR and ESL. Also, ceramic capacitors are not very sensitive to overvoltage and can be used without derating the operating voltage. However, designers must be aware of a potential overvoltage condition that is generated when input voltage is applied abruptly. After applying an input voltage step, typical input filter circuits with ceramic capacitors can generate voltage transients twice as high as the input voltage. This note describes how to efficiently use ceramic capacitors for input filters and how to avoid potential problems due to input voltage transients.

Minimizing switching regulator residue in linear regulator outputs (2)

Linear regulators are commonly employed to post-regulate switching regulator outputs. Benefits include improved stability, accuracy, transient response and lowered output impedance. Ideally, these performance gains would be accompanied by markedly reduced switching regulator generated ripple and spikes. In practice, all linear regulators encounter some difficulty with ripple and spikes, particularly as frequency rises. This publication explains the causes of linear regulators' dynamic limitations and presents board level techniques for improving ripple and spike rejection. A hardware based ripple/spike simulator is presented, enabling rapid breadboard testing under various conditions. Three appendices review ferrite beads, inductor based filters and probing practice for wideband, submillivolt signals.

Power conditioning for notebook and palmtop systems (3)

Notebook and palmtop systems need a number of voltages developed from a battery. Competitive solutions require small size, high efficiency and light weight. This publication includes circuits for high efficiency 5V and 3.3V switching and linear regulators, backlight display drivers and battery chargers. All the circuits are specifically tailored for the requirements outlined above.

Two wire virtual remote sensing for voltage regulators (4)

Wires and connectors have resistance. This simple, unavoidable truth dictates that a power source's remote load voltage will be less than the source's output voltage. The classical approach to mitigating this utilizes "4-wire" remote sensing to eliminate line drop effects. The power supply's high impedance sense inputs are fed from separate, load-referred sense wires. This scheme works well, but requires dedicated sense wires, a significant disadvantage in many applications. A new approach, utilizing carrier modulation techniques, eliminates sense wires while maintaining load regulation.

Ceramic input capacitors can cause overvoltage transients

1

Goran Perica

A recent trend in the design of portable devices has been to use ceramic capacitors to filter DC/DC converter inputs. Ceramic capacitors are often chosen because of their small size, low equivalent series resistance (ESR) and high RMS current capability. Also, recently, designers have been looking to ceramic capacitors due to shortages of tantalum capacitors.

Unfortunately, using ceramic capacitors for input filtering can cause problems. Applying a voltage step to a ceramic capacitor causes a large current surge that stores energy in the inductances of the power leads. A large voltage spike is created when the stored energy is transferred from these inductances into the ceramic capacitor. These voltage spikes can easily be twice the amplitude of the input voltage step.

Plug in the wall adapter at your own risk

The input voltage transient problem is related to the power-up sequence. If the wall adapter is plugged into an AC outlet and powered up first, plugging the wall adapter output into a portable device can cause input voltage transients that could damage the DC/DC converters inside the device.

Building the test circuit

To illustrate the problem, a typical 24V wall adapter used in notebook computer applications was connected to the input of a typical notebook computer DC/DC converter. The DC/DC converter used was a synchronous buck converter that generates 3.3V from a 24V input.

The block diagram of the test setup is shown in Figure 1.1. The inductor L_{OUT} represents the lumped equivalent inductance of the lead inductance and the output EMI filter inductor found in some wall adapters. The output capacitor in the wall adapter is usually on the order of 1000 μF; for our purposes, we can assume that it has low ESR—in the 10mΩ to 30mΩ range. The equivalent circuit of the wall adapter and DC/DC converter interface is actually a series resonant tank, with the dominant components being L_{OUT}, C_{IN} and the lumped ESR (the lumped ESR must include the ESR of C_{IN}, the lead resistance and the resistance of L_{OUT}).

The input capacitor, C_{IN}, must be a low ESR device, capable of carrying the input ripple current. In a typical notebook computer application, this capacitor is in the range of 10 μF to 100 μF. The exact capacitor value depends on a number of factors but the main requirement is that it must handle the input ripple current produced by

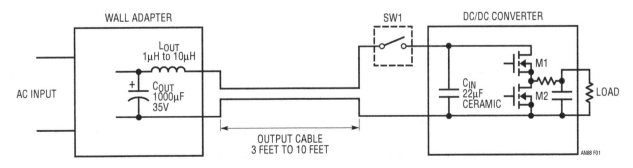

Figure 1.1 • Block Diagram of Wall Adapter and Portable Device Connection

Analog Circuit and System Design: A Tutorial Guide to Applications and Solutions. DOI: 10.1016/B978-0-12-385185-7.00001-9

the DC/DC converter. The input ripple current is usually in the range of 1A to 2A. Therefore, the required capacitors would be either one $10\,\mu F$ to $22\,\mu F$ ceramic capacitor, two to three $22\,\mu F$ tantalum capacitors or one to two $22\,\mu F$ OS-CON capacitors.

Turning on the switch

When switch SW1 in Figure 1.1 is turned on, the mayhem starts. Since the wall adapter is already plugged in, there is 24V across its low impedance output capacitor. On the other hand, the input capacitor C_{IN} is at 0V potential. What happens from t = 0s is pretty basic. The applied input voltage will cause current to flow through L_{OUT}. C_{IN} will begin charging and the voltage across C_{IN} will ramp up toward the 24V input voltage. Once the voltage across C_{IN} has reached the output voltage of the wall adapter, the energy stored in L_{OUT} will raise the voltage across C_{IN} further above 24V. The voltage across C_{IN} will eventually reach its peak and will then fall back to 24V. The voltage across C_{IN} may ring for some time around the 24V value. The actual waveform will depend on the circuit elements.

If you intend to run this circuit simulation, keep in mind that the real-life circuit elements are very seldom linear under transient conditions. For example, the capacitors may undergo a change of capacitance (Y5V ceramic capacitors will lose 80% of the initial capacitance under rated input voltage). Also, the ESR of input capacitors will depend on the rise time of the waveform. The inductance of EMI-suppressing inductors may also drop during transients due to the saturation of the magnetic material.

Testing a portable application

Input voltage transients with typical values of C_{IN} and L_{OUT} used in notebook computer applications are shown in Figure 1.2. Figure 1.2 shows input voltage transients for C_{IN} values of $10\,\mu F$ and $22\,\mu F$ with L_{OUT} values of $1\,\mu H$ and $10\,\mu H$.

Table 1.1 Peak Voltages of Waveforms In Figure 1.2

TRACE	L_{IN} (μH)	C_{IN} (μF)	V_{IN} PEAK (V)
CH1	1	10	57.2
R2	10	10	50
R3	1	22	41
R4	10	22	41

The top waveform shows the worst-case transient, with a $10\,\mu F$ capacitor and $1\,\mu H$ inductor. The voltage across C_{IN} peaks at 57.2V with a 24V DC input. The DC/DC converter may not survive repeated exposure to 57.2V.

The waveform with $10\,\mu F$ and $10\,\mu H$ (trace R2) looks a bit better. The peak is still around 50V. The flat part of the waveform R2 following the peak indicates that the synchronous MOSFET M1, inside of the DC/DC converter in Figure 1.1, is avalanching and taking the energy hit. Traces R3 and R4 peak at around 41V and are for a $22\,\mu F$ capacitor with $1\,\mu H$ and $10\,\mu H$ inductors, respectively.

Input voltage transients with different input elements

Different types of input capacitors will result in different transient voltage waveforms, as shown in Figure 1.3. The reference waveform for $22\,\mu F$ capacitor and $1\,\mu H$ inductor is shown in the top trace (R1); it peaks at 40.8V.

The waveform R2 in Figure 1.3 shows what happens when a transient voltage suppressor is added across the input. The input voltage transient is clamped but not eliminated. It is very hard to set the voltage transient's breakdown voltage low enough to protect the DC/DC converter and far enough from the operating DC level of the input source (24V). The transient voltage suppressor P6KE30A that was used was too close to starting to conduct at 24V.

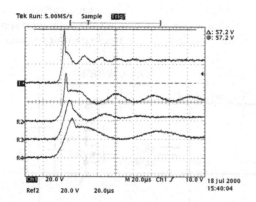

Figure 1.2 • Input Voltage Transients Across Ceramic Capacitors

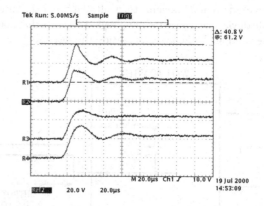

Figure 1.3 • Input Transients with Different Input Components

Unfortunately, using a transient voltage suppressor with a higher voltage rating would not provide a sufficiently low clamping voltage.

The waveforms R3 and R4 are with a 22 µF, 35V AVX TPS type tantalum capacitor and a 22 µF, 30V Sanyo OS-CON capacitor, respectively. With these two capacitors, the transients have been brought to manageable levels. However, these capacitors are bigger than the ceramic capacitors and more than one capacitor is required in order to meet the input ripple current requirements.

Table 1.2 Peak Voltages of Waveforms In Figure 1.3

TRACE	C_{IN} (µF)	CAPACITOR TYPE	V_{IN} PEAK (V)
R1	22	Ceramic	40.8
R2	22	Ceramic with 30V TVS	32
R3	22	AVX, TPS Tantalum	33
R4	22	Sanyo OS-CON	35

Optimizing input capacitors

Waveforms in Figure 1.3 show how input transients vary with the type of input capacitors used.

Optimizing the input capacitors requires clear understanding of what is happening during transients. Just as in an ordinary resonant RLC circuit, the circuit in Figure 1.1 may have an underdamped, critically damped or overdamped transient response.

Because of the objective to minimize the size of input filter circuit, the resulting circuit is usually an underdamped resonant tank. However, a critically damped circuit is actually required. A critically damped circuit will rise nicely to the input voltage without voltage overshoots or ringing.

To keep the input filter design small, it is desirable to use ceramic capacitors because of their high ripple current ratings and low ESR. To start the design, the minimum value of the input capacitor must first be determined. In the example, it has been determined that a 22 µF, 35V ceramic capacitor should be sufficient. The input transients generated with this capacitor are shown in the top trace of Figure 1.4. Clearly, there will be a problem if components that are rated for 30V are used.

To obtain optimum transient characteristic, the input circuit has to be damped. The waveform R2 shows what happens when another 22 µF ceramic capacitor with a 0.5 Ω resistor in series is added. The input voltage transient is now nicely leveled off at 30V.

Critical damping can also be achieved by adding a capacitor of a type that already has high ESR (on the order of 0.5 Ω). The waveform R3 shows the transient response when a 22 µF, 35V TPS type tantalum capacitor from AVX is added across the input.

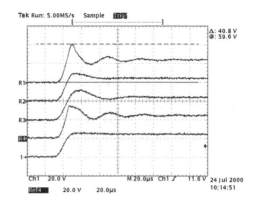

Figure 1.4 • Optimizing Input Circuit Waveforms for Reduced Peak Voltage

Table 1.3 Peak Voltages of Waveforms In Figure 1.4 with 22 µF Input Ceramic Capacitor and Added Snubber

TRACE	SNUBBER TYPE	Vin PEAK (V)
R1	None	40.8
R2	22 µF Ceramic + 0.5 Ω In Series	30
R3	22 µF Tantalum AVX, TPS Series	33
R4	30V TVS, PGKE30A	35
Ch1	47 µF, 35V Aluminum Electrolytic Capacitor	25

The waveform R4 shows the input voltage transient with a 30V transient voltage suppressor for comparison.

Finally, an ideal waveform shown in Figure 1.4, bottom trace (Ch1) is achieved. It also turns out that this is the least expensive solution. The circuit uses a 47 µF, 35V aluminum electrolytic capacitor from Sanyo (35CV47AXA). This capacitor has just the right value of capacitance and ESR to provide critical damping of the 22 µF ceramic capacitor in conjunction with the 1 µH of input inductance. The 35CV47AXA has an ESR value of 0.44 Ω and an RMS current rating of 230mA. Clearly, this capacitor could not be used alone in an application with 1A to 2A of RMS ripple current without the 22 µF ceramic capacitor. An additional benefit is that this capacitor is very small, measuring just 6.3mm by 6mm.

Conclusion

Input voltage transients are a design issue that should not be ignored. Design solutions for preventing input voltage transients can be very simple and effective. If the solution is properly applied, input capacitors can be minimized and both cost and size minimized without sacrificing performance.

Minimizing switching regulator residue in linear regulator outputs

Banishing those accursed spikes

2

Jim Williams

Introduction

Linear regulators are commonly employed to post-regulate switching regulator outputs. Benefits include improved stability, accuracy, transient response and lowered output impedance. Ideally, these performance gains would be accompanied by markedly reduced switching regulator generated ripple and spikes. In practice, all linear regulators encounter some difficulty with ripple and spikes, particularly as frequency rises. This effect is magnified at small regulator V_{IN} to V_{OUT} differential voltages; unfortunate, because such small differentials are desirable to maintain efficiency. Figure 2.1 shows a conceptual linear regulator and associated components driven from a switching regulator output.

The input filter capacitor is intended to smooth the ripple and spikes before they reach the regulator. The output capacitor maintains low output impedance at higher frequencies, improves load transient response and supplies frequency compensation for some regulators. Ancillary purposes include noise reduction and

minimization of residual input-derived artifacts appearing at the regulators output. It is this last category–residual input-derived artifacts—that is of concern. These high frequency components, even though small amplitude, can cause problems in noise-sensitive video, communication and other types of circuitry. Large numbers of capacitors and aspirin have been expended in attempts to eliminate these undesired signals and their resultant effects. Although they are stubborn and sometimes seemingly immune to any treatment, understanding their origin and nature is the key to containing them.

Switching regulator AC output content

Figure 2.2 details switching regulator dynamic (AC) output content. It consists of relatively low frequency ripple at the switching regulator's clock frequency, typically 100kHz to 3MHz, and very high frequency content "spikes" associated with power switch transition times. The switching regulator's pulsed energy delivery creates the ripple. Filter capacitors smooth the output, but not

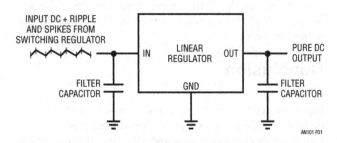

Figure 2.1 • Conceptual Linear Regulator and Its Filter Capacitors Theoretically Reject Switching Regulator Ripple and Spikes

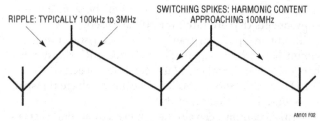

Figure 2.2 • Switching Regulator Output Contains Relatively Low Frequency Ripple and High Frequency "Spikes" Derived From Regulator's Pulsed Energy Delivery and Fast Transition Times

Analog Circuit and System Design: A Tutorial Guide to Applications and Solutions. DOI: 10.1016/B978-0-12-385185-7.00002-0

completely. The spikes, which often have harmonic content approaching 100MHz, result from high energy, rapidly switching power elements within the switching regulator. The filter capacitor is intended to reduce these spikes but in practice cannot entirely eliminate them. Slowing the regulator's repetition rate and transition times can greatly reduce ripple and spike amplitude, but magnetics size increases and efficiency falls[1]. The same rapid clocking and fast switching that allows small magnetics size and high efficiency results in high frequency ripple and spikes presented to the linear regulator.

Ripple and spike rejection

The regulator is better at rejecting the ripple than the very wideband spikes. Figure 2.3 shows rejection performance for an LT1763 low dropout linear regulator. There is 40db attenuation at 100kHz, rolling off to about 25db at 1MHz. The much more wideband spikes pass directly through the regulator. The output filter capacitor, intended to absorb the spikes, also has high frequency performance limitations. The regulator's and filter capacitor's imperfect response, due to high frequency parasitics, reveals Figure 2.1 to be overly simplistic. Figure 2.4 restates Figure 2.1 and includes the parasitic terms as well as some new components.

The figure considers the regulation path with emphasis on high frequency parasitics. It is important to identify these parasitic terms because they allow ripple and spikes to propagate into the nominally regulated output. Additionally, understanding the parasitic elements permits a measurement strategy, facilitating reduction of high frequency output content. The regulator includes high frequency parasitic paths, primarily capacitive, across its pass transistor and into its reference and regulation amplifier. These terms combine with finite regulator gain-bandwidth to limit high frequency rejection. The input and output filter capacitors include parasitic inductance and resistance, degrading their effectiveness as frequency rises. Stray layout capacitance provides additional unwanted feedthrough paths. Ground potential differences, promoted by ground path resistance and inductance, add additional error and also complicate measurement. Some new components, not normally associated with linear regulators, also appear. These additions include ferrite beads or inductors in the regulator input and output lines. These components have their own high frequency parasitic paths but can considerably improve overall regulator high frequency rejection and will be addressed in following text.

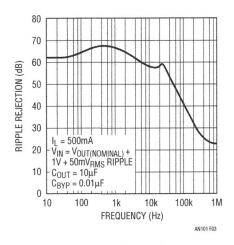

Figure 2.3 • Ripple Rejection Characteristics for an LT1763 Low Dropout Linear Regulator Show 40dB Attenuation at 100kHz, Rolling Off Towards 1MHz. Switching Spike Harmonic Content Approaches 100MHz; Passes Directly From Input to Output

---

Note 1: Circuitry employing this approach has achieved significant harmonic content reduction at some sacrifice in magnetics size and efficiency. See Reference 1.

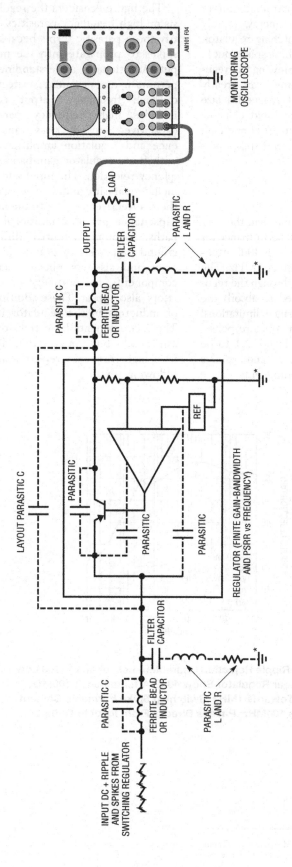

Figure 2.4 • Conceptual Linear Regulator Showing High Frequency Rejection Parasitics. Finite GBW and PSRR vs Frequency Limit Regulator's High Frequency Rejection. Passive Components Attenuate Ripple and Spikes, But Parasitics Degrade Effectiveness. Layout Capacitance and Ground Potential Differences Add Errors, Complicate Measurement

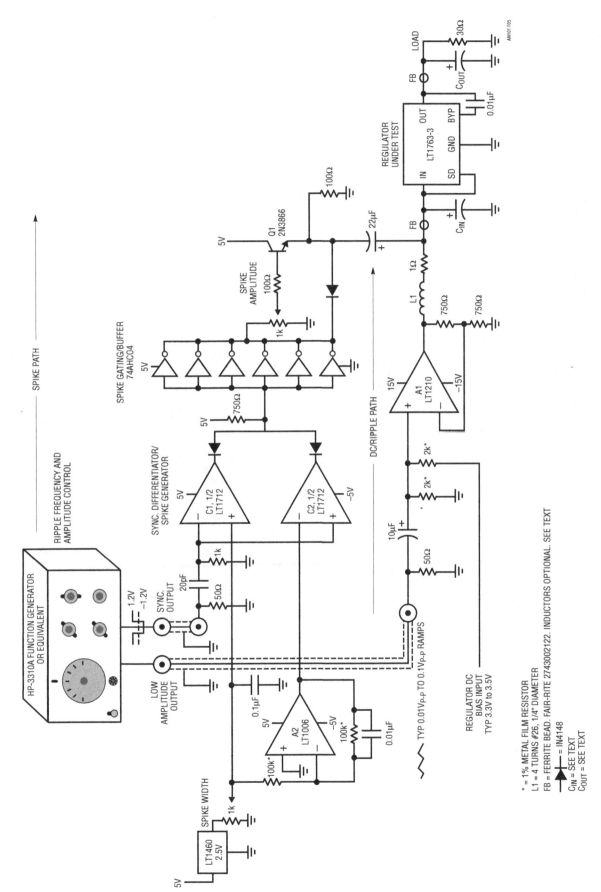

Figure 2.5 • Circuit Simulates Switching Regulator Output. DC, Ripple Amplitude, Frequency and Spike Duration/Height are Independently Settable. Split Path Scheme Sums Wideband Spikes with DC and Ripple, Presenting Linear Regulator with Simulated Switching Regulator Output. Function Generator Sources Waveforms to Both Paths

Ripple/spike simulator

Gaining understanding of the problem requires observing regulator response to ripple and spikes under a variety of conditions. It is desirable to be able to independently vary ripple and spike parameters, including frequency, harmonic content, amplitude, duration and DC level. This is a very versatile capability, permitting real time optimization and sensitivity analysis to various circuit variations. Although there is no substitute for observing linear regulator performance under actual switching regulator driven conditions, a hardware simulator makes surprises less likely. Figure 2.5 provides this capability. It simulates a switching regulator's output with independently settable DC, ripple and spike parameters.

A commercially available function generator combines with two parallel signal paths to form the circuit. DC and ripple are transmitted on a relatively slow path while wideband spike information is processed via a fast path. The two paths are combined at the linear regulator input. The function generator's settable ramp output (trace A, Figure 2.6) feeds the DC/ripple path made up of power amplifier A1 and associated components. A1 receives the ramp input and DC bias information and drives the regulator under test. L1 and the 1Ω resistor allow A1 to drive the regulator at ripple frequencies without instability. The wideband spike path is sourced from the function generator's pulsed "sync" output (trace B). This output's edges are differentiated (trace C) and fed to bipolar comparator C1-C2. The comparator outputs (traces D and E) are spikes synchronized to the ramp's inflection points. Spike width is controlled by complementary DC threshold potentials applied to C1 and C2 with the 1k potentiometer and A2. Diode gating and the paralleled logic inverters present trace F to the spike amplitude control. Follower Q1 sums the spikes with A1's DC/ripple path, forming the linear regulator's input (trace G).

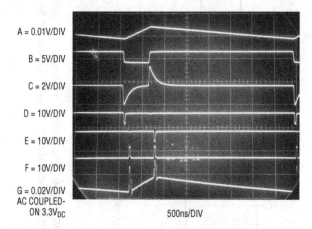

A = 0.01V/DIV
B = 5V/DIV
C = 2V/DIV
D = 10V/DIV
E = 10V/DIV
F = 10V/DIV
G = 0.02V/DIV AC COUPLED-ON 3.3V_{DC}
500ns/DIV

Figure 2.6 • Switching Regulator Output Simulator Waveforms. Function Generator Supplies Ripple (Trace A) and Spike (Trace B) Path Information. Differentiated Spike Information's Bipolar Excursion (Trace C) is Compared by C1-C2, Resulting in Trace D and E Synchronized Spikes. Diode Gating/Inverters Present Trace F to Spike Amplitude Control. Q1 Sums Spikes with DC-Ripple Path From Power Amplifier A1, Forming Linear Regulator Input (Trace G). Spike Width Set Abnormally Wide for Photographic Clarity

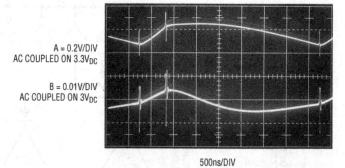

A = 0.2V/DIV AC COUPLED ON 3.3V_{DC}
B = 0.01V/DIV AC COUPLED ON 3V_{DC}
500ns/DIV

Figure 2.7 • Linear Regulator Input (Trace A) and Output (Trace B) Ripple and Switching Spike Content for C_{IN} = 1μF, C_{OUT} = 10μF. Output Spikes, Driving 10μF, Have Lower Amplitude, But Risetime Remains Fast

Linear regulator high frequency rejection evaluation/optimization

The circuit described above facilitates evaluation and optimization of linear regulator high frequency rejection. The following photographs show results for one typical set of conditions, but DC bias, ripple and spike characteristics may be varied to suit desired test parameters. Figure 2.7 shows Figure 2.5's LT1763 3V regulator response to a 3.3V DC input with trace A's ripple/spike contents, $C_{IN} = 1\mu F$ and $C_{OUT} = 10\mu F$. Regulator output (trace B) shows ripple attenuated by a factor of ≈ 20. Output spikes see somewhat less reduction and their harmonic content remains high. The regulator offers no rejection at the spike rise time. The capacitors must do the job. Unfortunately, the capacitors are limited by inherent high frequency loss terms from completely filtering the wideband spikes; trace B's remaining spike shows no risetime reduction. Increasing capacitor value has no benefit at these rise times. Figure 2.8 (same trace assignments as Figure 2.7) taken with $C_{OUT} = 33\mu F$, shows $5\times$ ripple reduction but little spike amplitude attenuation.

Figure 2.9's time and amplitude expansion of Figure 2.8's trace B permits high resolution study of spike characteristics, allowing the following evaluation and optimization. Figure 2.10 shows dramatic results when a ferrite bead immediately precedes C_{IN}[2]. Spike amplitude drops about $5\times$. The bead presents loss at high frequency, severely limiting spike passage[3]. DC and low frequency pass unattenuated to the regulator. Placing a second ferrite bead at the regulator output before C_{OUT} produces Figure 2.11's trace. The bead's high frequency loss characteristic further reduces spike amplitude below 1mV without introducing DC resistance into the regulator's output path[4].

Figure 2.12, a higher gain version of the previous figure, measures $900\mu V$ spike amplitude - almost $20\times$ lower than without the ferrite beads. The measurement is completed by verifying that indicated results are not corrupted by common mode components or ground loops. This is done by grounding the oscilloscope input near the measurement point. Ideally, no signal should appear. Figure 2.13 shows this to be nearly so, indicating that Figure 2.12's display is realistic[5].

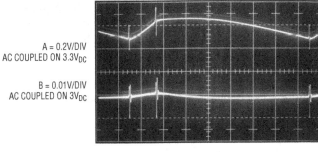

A = 0.2V/DIV
AC COUPLED ON 3.3V_{DC}

B = 0.01V/DIV
AC COUPLED ON 3V_{DC}

500ns/DIV

Figure 2.8 • Same Trace Assignments as Figure 2.7 with C_{OUT} Increased to 33μF. Output Ripple Decreases By 5$\times$, But Spikes Remain. Spike Risetime Appears Unchanged

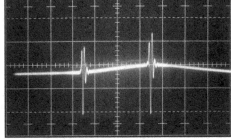

0.005V/DIV
AC COUPLED ON 3V_{DC}

200ns/DIV

Figure 2.9 • Time and Amplitude Expansion of Figure 2.8's Output Trace Permits Higher Resolution Study of Spike Characteristics. Trace Center-Screen Area Intensified for Photographic Clarity in This and Succeeding Figures

Note 2: "Dramatic" is perhaps a theatrical descriptive, but certain types find drama in these things.
Note 3: See Appendix A for information on ferrite beads.
Note 4: Inductors can sometimes be used in place of beads but their limitations should be understood. See Appendix B.
Note 5: Faithful wideband measurement at sub-millivolt levels requires special considerations. See Appendix C.

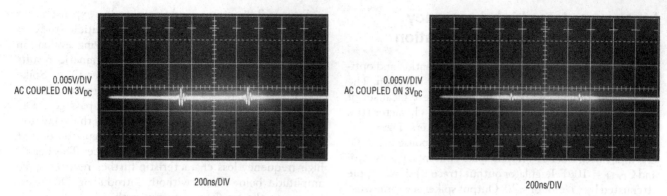

Figure 2.10 • Adding Ferrite Bead to Regulator Input Increases High Frequency Losses, Dramatically Attenuating Spikes

Figure 2.11 • Ferrite Bead in Regulator Output Further Reduces Spike Amplitude

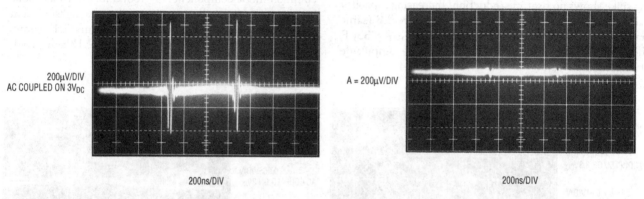

Figure 2.12 • Higher Gain Version of Previous Figure Measures 900µV Spike Amplitude – Almost 20× Lower Than Without Ferrite Beads. Instrumentation Noise Floor Causes Trace Baseline Thickening

Figure 2.13 • Grounding Oscilloscope Input Near Measurement Point Verifies Figure 12's Results Are Nearly Free of Common Mode Corruption

References

1. Williams, Jim, "A Monolithic Switching Regulator with 100µV Output Noise," Linear Technology Corporation, Application Note 70, October 1997 (See Appendices B, C,D,H,I and J)

2. Williams, Jim, "Low Noise Varactor Biasing with Switching Regulators," Linear Technology Corporation, Application Note 85, August 2000 (See pp 4-6 and Appendix C)

3. Williams, Jim, "Component and Measurement Advances Ensure 16-Bit Settling Time," Linear Technology Corporation, Application Note 74, July 1998 (See Appendix G)

4. LT1763 Low Dropout Regulator Datasheet, Linear Technology Corporation

5. Hurlock, Les, "ABCs of Probes," Tektronix Inc., 1990

6. McAbel, W.E., "Probe Measurements," Tektronix Inc., Concept Series, 1971

7. Morrison, Ralph, "Noise and Other Interfering Signals," John Wiley and Sons, 1992

8. Morrison, Ralph, "Grounding and Shielding Techniques in Instrumentation," Wiley-Interscience, 1986

9. Fair-Rite Corporation, "Fair-Rite Soft Ferrites," Fair-Rite Corporation, 1998

Appendix A
About ferrite beads

A ferrite bead enclosed conductor provides the highly desirable property of increasing impedance as frequency rises. This effect is ideally suited to high frequency noise filtering of DC and low frequency signal carrying conductors. The bead is essentially lossless within a linear regulator's passband. At higher frequencies the bead's ferrite material interacts with the conductor's magnetic field, creating the loss characteristic. Various ferrite materials and geometries result in different loss factors versus frequency and power level. Figure A1's plot shows this. Impedance rises from 0.01Ω at DC to 50Ω at 100MHz. As DC current, and hence constant magnetic field bias, rises, the ferrite becomes less effective in offering loss. Note that beads can be "stacked" in series along a conductor, proportionally increasing their loss contribution. A wide variety of bead materials and physical configurations are available to suit requirements in standard and custom products.

Appendix B
Inductors as high frequency filters

Inductors can sometimes be used for high frequency filtering instead of beads. Typically, values of 2µH to10µH are appropriate. Advantages include wide availability and better effectiveness at lower frequencies, e.g., ≤100kHz. Figure B1 shows disadvantages are increased DC resistance in the regulator path due to copper losses, parasitic shunt capacitance and potential susceptibility to stray switching regulator radiation. The copper loss appears at DC, reducing

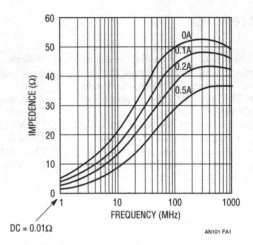

Figure A1 • Impedance vs Frequency at Various DC Bias Currents for a Surface Mounted Ferrite Bead (Fair-Rite 2518065007Y6). Impedance is Essentially Zero at DC and Low Frequency, Rising Above 50Ω Depending on Frequency and DC Current
Source: Fair-Rite 2518065007Y6 Datasheet.

efficiency; parasitic shunt capacitance allows unwanted high frequency feedthrough. The inductor's circuit board position may allow stray magnetic fields to impinge its winding, effectively turning it into a transformer secondary. The resulting observed spike and ripple related artifacts masquerade as conducted components, degrading performance.

Figure B2 shows a form of inductance based filter constructed from PC board trace. Such extended length traces, formed in spiral or serpentine patterns, look inductive at high frequency. They can be surprisingly effective in some circumstances, although introducing much less loss per unit area than ferrite beads.

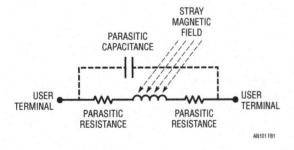

Figure B1 • Some Parasitic Terms of an Inductor. Parasitic Resistance Drops Voltage, Degrading Efficiency. Unwanted Capacitance Permits High Frequency Feedthrough. Stray Magnetic Field Induces Erroneous Inductor Current

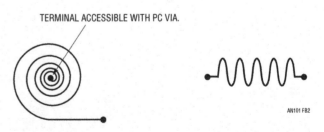

Figure B2 • Spiral and Serpentine PC Patterns are Sometimes Used as High Frequency Filters, Although Less Effective Than Ferrite Beads

Appendix C
Probing technique for sub-millivolt, wideband signal integrity

Obtaining reliable, wideband, sub-millivolt measurements requires attention to critical issues before measuring anything. A circuit board layout designed for low noise is essential. Consider current flow and interactions in power distribution, ground lines and planes. Examine the effects of component choice and placement. Plan radiation management and disposition of load return currents. If the circuit is sound, the board layout proper and appropriate components used, then, and only then, may meaningful measurement proceed.

The most carefully prepared breadboard cannot fulfill its mission if signal connections introduce distortion. Connections to the circuit are crucial for accurate information extraction. Low level, wideband measurements demand care in routing signals to test instrumentation. Issues to consider include ground loops between pieces of test equipment (including the power supply) connected to the breadboard and noise pickup due to excessive test lead or trace length. Minimize the number of connections to the circuit board and keep leads short. Wideband signals to or from the breadboard must be routed in a coaxial environment with attention to where the coaxial shields tie into the ground system. A strictly maintained coaxial environment is particularly critical for reliable measurements and is treated here[1].

Figure C1 shows a believable presentation of a typical switching regulator spike measured within a continuous coaxial signal path. The spike's main body is reasonably well defined and disturbances after it are contained. Figure C2 depicts the same event with a 3 inch ground lead connecting the coaxial shield to the circuit board ground plane. Pronounced signal distortion and ringing occur. The photographs were taken at 0.01V/division sensitivity. More sensitive measurement requires proportionately more care.

Figure C3 details use of a wideband 40dB gain pre-amplifier permitting text Figure 2.12's 200µV/division measurement. Note the purely coaxial path, including the AC coupling capacitor, from the regulator, through the pre-amplifier and to the oscilloscope. The coaxial coupling capacitor's shield is directly connected to the regulator board's ground plane with the capacitor center conductor going to the regulator output. There are no non-coaxial measurement connections. Figure C4, repeating text Figure 2.12, shows a cleanly detailed rendition of the 900µV output spikes. In Figure C5 two inches of ground lead has been deliberately introduced at the measurement site, violating the coaxial regime. The result is complete corruption of the waveform presentation. As a final test to verify measurement integrity, it is useful to repeat Figure C4's measurement with the signal path input (e.g., the coaxial coupling capacitor's center conductor) grounded near the measurement point as in text Figure 2.13. Ideally, no signal should appear. Practically, some **small** residue, primarily due to common mode effects, is permissible.

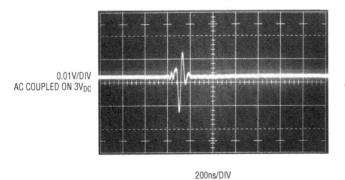

0.01V/DIV
AC COUPLED ON 3V$_{DC}$

200ns/DIV

Figure C1 • Spike Measured Within Continuous Coaxial Signal Path Displays Moderate Disturbance and Ringing After Main Event

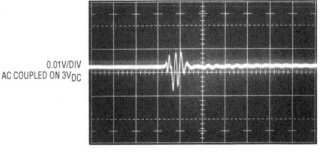

0.01V/DIV
AC COUPLED ON 3V$_{DC}$

200ns/DIV

Figure C2 • Introducing 3″ Non-Coaxial Ground Connection Causes Pronounced Signal Distortion and Post-Event Ringing

Note 1: More extensive treatment of these and related issues appears in the appended sections of References 1 and 2. Board layout considerations for low level, wideband signal integrity appear in Appendix G of Reference 3.

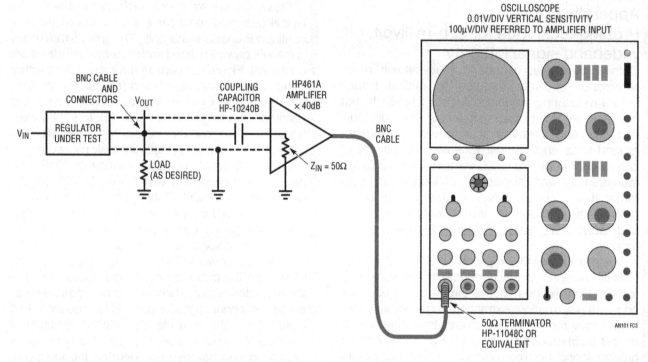

Figure C3 • Wideband, Low Noise Pre-Amplifier Permits Sub-Millivolt Spike Observation. Coaxial Connections Must be Maintained to Preserve Measurement Integrity

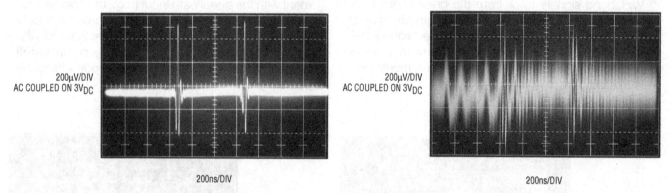

Figure C4 • Low Noise Pre-Amplifier and Strictly Enforced Coaxial Signal Path Yield Text Figure 2.12's 900mV$_{P-P}$ Presentation. Trace Baseline Thickening Represents Pre-Amplifier Noise Floor

Figure C5 • 2 Inch Non-Coaxial Ground Connection at Measurement Site Completely Corrupts Waveform Presentation

Power conditioning for notebook and palmtop systems

3

Robert Dobkin   Carl Nelson   Dennis O'Neill   Steve Pietkiewicz
Tim Skovmand   Milt Wilcox

Introduction

Notebook and palmtop systems need a multiplicity of regulated voltages developed from a single battery. Small size, light weight, and high efficiency are mandatory for competitive solutions in this area. Small increases in efficiency extend battery life, making the final product much more usable with no increase in weight. Additionally, high efficiency minimizes the heat sinks needed on the power regulating components, further reducing system weight and size.

Battery systems include NiCad, nickel-hydride, lead acid, and rechargeable lithium, as well as throw-away alkaline batteries. The ability to power condition a wide range of batteries makes the ultimate product much more attractive because power sources can be interchanged, increasing overall system versatility.

A main rechargeable battery may be any of the four secondary type cells, with a back-up or emergency ability to operate off alkaline batteries. The higher energy density available in non-rechargeable alkaline batteries allows the systems to operate for extended time without battery replacement.

The systems shown here provide power conditioning with high efficiency and low parts count. Trade-offs between complexity and efficiency have been made to maximize manufacturability and minimize cost. All the supplies operate over a wide range of input voltage allowing great flexibility in the choice of battery configuration.

LT1432 driver for high efficiency 5V and 3.3V buck regulator

The LT1432 is a control chip designed to operate with the LT1170 or LT1270 family of switching regulators to make a very high efficiency (Figure 3.1) 5V or 3.3V step-down (buck) switching regulator. These regulators feature a low-loss saturating NPN switch that is normally configured with the negative terminal (emitter) at ground. The LT1432 allows the switch to be floated as required in a step-down converter, yet still provides full switch saturation for highest efficiency.

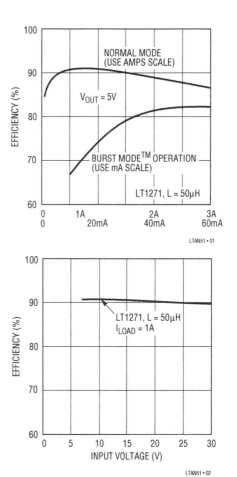

Figure 3.1 • LT1432 5V Efficiency

Analog Circuit and System Design: A Tutorial Guide to Applications and Solutions. DOI: 10.1016/B978-0-12-385185-7.00003-2

Many other features have been incorporated into the LT1432 to enhance operation in battery powered applications. An accurate current limit uses only 60mV sense voltage, allows for foldback, and uses "free" PC board trace material for the sense resistor. Logic controlled shutdown mode draws only 15μA battery current to allow for extremely long shutdown periods. The switching IC is powered from the regulator output to enhance efficiency and to allow input voltages as low as 6.5V.

The LT1432 has optional Burst Mode operation to achieve high efficiency at very light load currents (0mA to 100mA). In normal switching mode, the standby power loss is about 60mW, limiting efficiency at light loads. In burst mode, standby loss is reduced to approximately 15mW. Output ripple is $150mV_{P-P}$ in this mode, but this is normally well within the requirements for digital logic supplies. Burst Mode operation would typically be used for "sleep" conditions where IC memory chips remain powered for data retention, but the remainder of the system is powered down. Load current in this mode is typically in the 5mA–100mA range. The operating mode is under logic control.

The LT1432 is available in 8-pin surface mount and DIP packages. The LT1170 and LT1270 families are available in a surface mount version of the 5-pin TO-220 package.

Circuit description

The circuit shown in Figure 3.2 is a basic 5V positive buck converter which can operate with input voltages from 6.5V to 25V. The power switch is located between the V_{SW} pin and GND pin on the LT1271. Its current and duty cycle are controlled by the voltage on the V_C pin with respect to the GND pin. This voltage ranges from 1V to 2V as switch currents increase from zero to full scale. Correct output voltage is maintained by the LT1432 which has an internal reference and error amplifier. The amplifier output is level shifted with an internal open collector NPN to drive the V_C pin of the switcher. The normal resistor divider feedback to the switcher feedback pin cannot be used because the feedback pin is referenced to the GND pin, which is switching many volts. The feedback pin (FB) is simply bypassed with a capacitor. This forces the switcher V_C pin to swing high with about 200μA sourcing capability. The LT1432 V_C pin then sinks this current to control the loop. C4 forms the dominant loop pole with a loop zero added by R1. C5 forms a higher frequency loop pole to control switching ripple at the V_C pin.

A floating 5V power supply for the switcher is generated by D2 and C3 which peak detect the output voltage during switch "off" time. This is a very efficient way of powering the switcher because power drain does not increase with regulator input voltage. However, the circuit is not self-starting, so some means must be used to start the regulator. This is performed by an internal current path in the LT1432 which allows current to flow from the input supply to the V^+ pin during start-up.

In both the 5V and 3.3V regulators, D1, L1, and C2 act as the conventional catch diode and output filter of the buck converter. These components should be selected carefully to maintain high efficiency and acceptable output ripple.

Current limiting is performed by R2. Sense voltage is only 60mV to maintain high efficiency. This also reduces

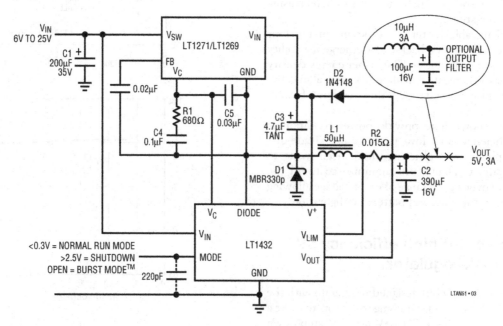

Figure 3.2 • High Efficiency 5V Regulator with Manual Burst Mode Operation

the value of the sense resistor enough to utilize a printed circuit board trace as the sense resistor. The sense voltage has a positive temperature coefficient to match the temperature coefficient of copper.

The basic regulator has three different operating modes, defined by the mode pin drive. Normal operation occurs when the mode pin is grounded. A low quiescent current Burst Mode operation can be initiated by floating the mode pin. Input supply current is typically 1.3mA in this mode, and output ripple voltage is 100mV$_{P-P}$. Pulling the mode pin above 2.5V forces the entire regulator into micropower shutdown where it typically draws less than 20μA.

What are the benefits of using an active (synchronous) switch to replace the catch diode? This is the trendy thing to do, but calculations and actual breadboards show that the improvement in efficiency is only a few percent at best. This can be shown with the following simplified formulas:

$$\text{Diode loss} = V_f(V_{IN} - V_{OUT})(I_{OUT})/V_{IN}$$
$$\text{FETswitch loss} = (V_{IN} - V_{OUT})(R_{SW})(I_{OUT})^2/V_{IN}$$

The change in efficiency is:

$$(\text{Diode loss} - \text{FET loss})(\text{Efficiency})^2/(V_{OUT})(I_{OUT})$$

This is equal to:

$$(V_{IN} - V_{OUT})(V_f - R_{FET} \times I_{OUT})(E)^2/(V_{IN})(V_{OUT})$$

If V_f (diode forward voltage)=0.45V, V_{IN}=10V, V_{OUT}=5V, R_{FET}=0.1Ω, I_{OUT}=1A, and efficiency=90%, the improvement in efficiency is only:

$$(10V - 5V)(0.45V - 0.1\,\Omega \times 1A)$$
$$(0.9)2/(10V)(5V) = 2.8\%$$

This does not take FET gate drive losses into account, which can easily reduce this figure to less than 2%. The added cost, size, and complexity of a synchronous switch configuration would be warranted only in the most extreme circumstances.

Burst Mode efficiency is limited by quiescent current drain in the LT1432 and the switching IC. The typical Burst Mode zero-load input power is 17mW. This gives about one month battery life for a 12V, 1.2AHr battery pack. Increasing load power reduces discharge time proportionately. Full shutdown current is only about 15μA, which is considerably less than the self-discharge rate of typical batteries.

BICMOS switching regulator family provides highest step-down efficiencies

The LTC1148 family of single and dual step-down switching regulator controllers features automatic Burst Mode operation to maintain high efficiencies at low output currents. All members of the family use a constant offtime, current mode architecture. This results in excellent line and load transient response, constant inductor ripple current, and well controlled start-up and short circuit currents. The LTC1147/LTC1143 drive a single external P-channel MOSFET, while the LTC1148/LTC1142 and LTC1149 drive synchronous external power MOSFETs at switching frequencies up to 250kHz.

Table 3.1 gives an overview of the family with applicability to common notebook DC to DC converter requirements. The LTC1147 is available in an 8-pin SOIC and drives only a single power MOSFET, giving it the smallest PC board footprint at a slight penalty in efficiency. The LTC1148HV/LTC1142HV offer synchronous switching capability at input voltages from 4V to 18V (20V abs max) with a low 200μA quiescent current. The LTC1149 extends synchronous switching operation up to input voltages of 48V (60V abs max) with a slight penalty in quiescent current.

The rated current level for all device types is set by the external sense resistor according to the formula I_{OUT}=100mV/R_{SENSE}. The maximum peak inductor current and Burst Mode current are also linked to R_{SENSE}. The peak current is limited to 150mV/R_{SENSE}, while Burst Mode operation automatically begins when the

Table 3.1 LTC1148 Family Applications							
	LTC1147	**LTC1148**	**LTC1143**	**LTC1142**	**LTC1148HV**	**LTC1142HV**	**LTC1149**
Continuous V$_{IN}$ < 48V							X
Continuous V$_{IN}$ < 18V					X	X	
Continuous V$_{IN}$ < 13.5V	X	X	X	X			
Low Dropout 5V	X	X	X	X	X	X	X
Dual 5V and 3.3V			X	X		X	
Adjustable		X			X		X

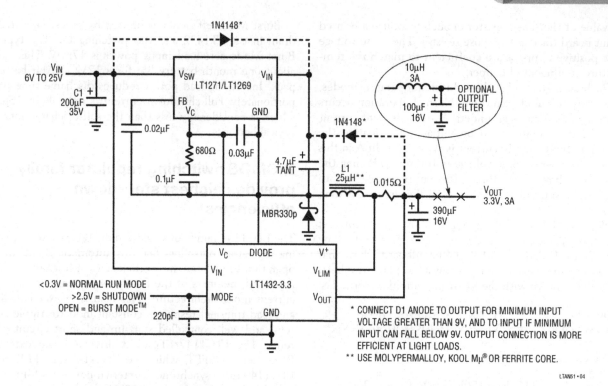

Figure 3.3 • High Efficiency 3.3V Regulator with Manual Burst Mode Operation

output current drops below approximately 15mV/ R_{SENSE}. In this mode, the external MOSFET(s) are held off to reduce switching losses and the controller sleeps at 200μA supply current (600μA for the LTC1149), while the output capacitor supports the load. When the output capacitor discharges 50mV, the controller briefly turns back on, or "bursts," to recharge the capacitor. Complete shutdown reduces the supply current to only 10μA (150μA for the LTC1149).

The first application shown in Figure 3.4 converts 5V to 3.3V at 1.5A output current. By choosing the LTC1147-3.3, a minimum board space solution is achieved at a slight penalty in peak efficiency (the LTC1148-3.3 driving synchronous MOSFETs in this application would add

approximately 2.5% to the high current efficiency). Figure 3.5 shows how Burst Mode operation maintains high operating efficiencies at low output currents.

In the second application (Figure 3.6), an LTC1148HV-5 is used as the controller for a 10W high efficiency regulator. This circuit can be used with as few as 5 NiCad or NiMH cells thanks to its excellent low dropout performance. Like other members of the family, the LTC1148HV goes to 100% duty cycle (P-channel MOSFET turned on DC) in

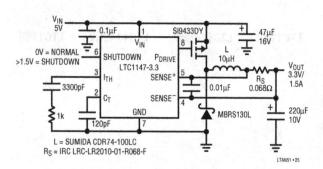

Figure 3.4 • High Efficiency Surface Mount 5V to 3.3V Converter Delivers 1.5A in Minimum Board Area

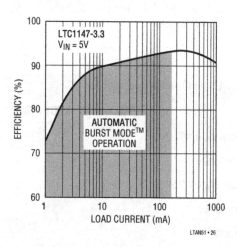

Figure 3.5 • High Operating Efficiency for Figure 3.4 Circuit Spans Three Decades of Output Current

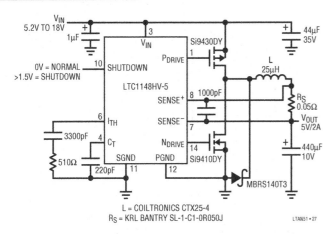

Figure 3.6 ● **High Efficiency Low Dropout 5V Switching Regulator Needs Only 200mV Headroom at 1A Output**

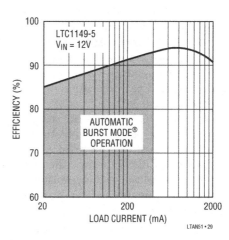

Figure 3.8 ● **Operating Efficiency for LTC1149-5 High Efficiency Converter**

dropout. The input to output voltage differential required to maintain regulation then simply becomes the product of the load current and total resistance of the MOSFET, inductor, and current sense resistor. In the Figure 3.6 circuit, this total resistance is less than 0.2Ω. For operation at low input voltages, logic-level MOSFETs must be used.

While the 18V input voltage rating of the LTC1148HV and LTC1142HV can generally accommodate most battery packs, the AC wall adapters used in conjunction with notebook systems often dictate significantly higher input voltages. This is the primary home for the LTC1149, shown in the Figure 3.7 application. This 2.5A regulator can operate at input voltages from 8V (limited by the standard MOSFET threshold voltages) to 30V, while still providing excellent efficiency as shown in Figure 3.8. The

synchronous switch plays an increasing role at high input voltages due to the low duty cycle of the main switch.

Board layout of Figures 3.4, 3.6 and 3.7's circuits is critical for proper transition between Burst Mode operation and continuous operation. The timing capacitor pin and inductor current are the two most important waveforms to monitor while checking an LTC1148 family regulator. The timing capacitor pin only goes to 0V during sleep intervals, which should only happen when the load current is less than approximately 20% of the rated output current. Consult the appropriate data sheet for information on proper component location and ground routing.

Surface mount capacitors for switching regulator applications

A good rule of thumb for the output capacitor selection in all LTC1148 family circuits is that it must have an ESR less than or equal to the sense resistor value (for example, 0.05Ω for the Figure 3.6 circuit). In surface mount applications multiple capacitors may have to be paralleled to meet the capacitance, ESR, or RMS current handling requirements of the application. Aluminum electrolytic and dry tantalum capacitors are both available in surface mount configurations.

In the case of tantalum, it is critical that the capacitors are surge tested for use in switching power supplies. An excellent choice is the AVX TPS series of surface mount tantalum capacitors, available in case heights ranging from 2mm to 4mm. For example, if $440\mu F/10V$ is called for in an application, $2\times$AVX $220\mu F/10V$ (P/N TPSE227K010) could be used. Consult the manufacturer for other specific recommendations.

High efficiency linear supplies

The switching supplies operate over a wide input range while maintaining high efficiency. Alternative notebook systems have been developed for narrow supply operation

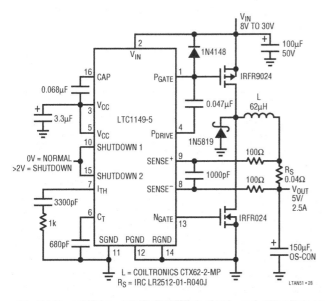

Figure 3.7 ● **High Efficiency 5V/2.5A Regulator Operates from AC Wall Adapters as High as 30V**

using for example, four NiCad batteries and a linear regulator to provide the 5V output. At full charge, four NiCad batteries can be as high as 6V and are allowed to discharge down to 4.5V while directly powering the system. A high efficiency low dropout linear regulator suited for this technique is shown in Figure 3.9.

This is a complete IC in a very low cost TO-92 3-pin package driving a low saturation PNP transistor. Many power PNP transistors can be used. The Motorola MJE1123 and Zetex ZBD949 are specified for this application. The dropout voltage of this regulator depends on the PNP transistor saturation and can be in the range of 0.25V at 3A output current — lower at lower current. The simplicity of this system is attractive for notebook applications and efficiency is good since little power is lost across the linear regulator at low input voltages.

For input voltages of 5.2V* and above, the output is regulated at 5V. As the battery voltage decreases below 5.2V*, the transistor saturates and the output voltage follows the input voltage down with the saturation voltage of the transistor subtracted from the input voltage.

The LT1123 low dropout driver can supply up to 125mA of base current to the pass transistor. At dropout this current is supplied continuously into the base of the pass transistor as the transistor remains in saturation. If

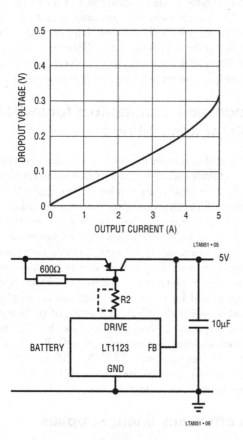

Figure 3.9 • LT1123 Dropout Voltage

*Actual voltage depends on load current.

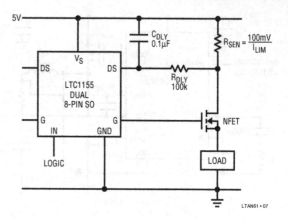

Figure 3.10 • LTC1155 Dual Micropower N-Channel MOSFET Driver

lower drive current is desired an optional resistor (R2) can be inserted in series with the base of the transistor to minimize the drive current and decrease the power dissipation in the IC. An N-channel FET can be inserted in series with the drive lead of the LT1123 to electrically shutdown the system.

Power switching with dual high side micropower N-channel MOSFET drivers

The LT1155 dual high side N-channel FET gate driver allows using low cost N-channel FETs for high side switching applications. No external components are needed since an internal charge pump boosts the gate above the positive rail, fully enhancing an N-channel MOSFET. Micropower operation, with 8µA standby current and 85µA operating current, allows use in virtually all battery powered systems even for main power switching.

Included on the chip is over-current sensing to provide automatic shutdown in case of short circuits. A time delay can be added in series with the current sense to prevent false triggering on high in-rush loads such as capacitors or lamps.

The LTC1155 operates off a 4.5V to 18V supply input and safely drives the gates of virtually all FETs. It is particularly well-suited for portable applications where micropower operation is critical. The device is available in 8-pin SO and DIP packages.

The LTC1157 is a dual driver for 3.3V supplies. The LTC1157 internal charge pump boosts the gate drive voltage 5.4V above the positive rail (8.7V above ground), fully enhancing a logic-level N-channel MOSFET for 3.3V high side switching applications. The charge pump is completely on-chip and therefore requires no external components to generate the higher gate voltage. The charge pump has been designed to be very efficient, requiring only 3µA in the standby mode and 80µA while delivering 8.7V to the power MOSFET gate.

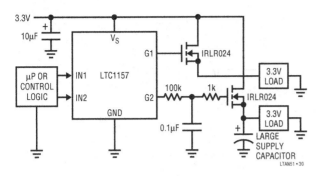

Figure 3.11 • LTC1157 Dual 3.3V MOSFET Driver

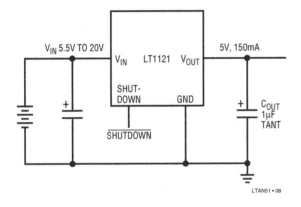

Figure 3.12 • LT1121 Micropower Low Dropout Regulator

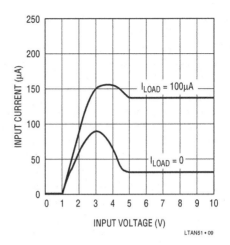

Figure 3.13 • LT1121 Input Current

Figure 3.11 demonstrates how two surface mount MOSFETs and the LTC1157 can be used to switch two 3.3V loads. The gate rise and fall time is typically in the tens of microseconds, but can slowed by adding two resistors and a capacitor as shown on the second channel. Slower rise and fall times are sometimes required to reduce the start-up current demands of large supply capacitors.

LT1121 micropower 150mA regulator with shutdown

The LT1121 is a low dropout regulator designed for applications where quiescent current must be very low when output current is low. It draws only 30μA input current at zero load current. Ground pin current increases with load current, but the ratio is about 1:25, so the efficiency of the regulator is only about 4% below theoretical maximum for a linear regulator. More importantly, the ground pin current does not increase significantly when the input voltage falls below the minimum required to maintain a regulated output.

These characteristics allow the LT1121 to be used in situations where it is desirable to have the output track the input when the input falls below its normal range. Previous regulators drew such high input current in this condition that micropower operation was not possible.

Extra effort was taken to make the LT1121 stable with small output capacitors that have high ESR. A 1μF

tantalum output capacitor is suggested, as compared to 10μF for previous designs. Larger output capacitors can be used without fear of instabilities.

The LT1121 is ideal for the backup and/or suspend mode power supply in notebook computers. A shutdown pin allows the regulator to be fully turned off, with input current dropping to only 16μA. Careful design of the IC circuitry connected to the input and output pins allows the output to be held high while the input is pulled to ground or reversed, without current flowing from the output back to the input. The input pin can be reversed up to 20V.

The LT1121 is available with a fixed output voltage of 3.3V or 5V and as an adjustable device with an output voltage range of 3.75V to 30V. Fixed voltage devices are available in 3-pin SOT-223, and 8-pin SO packages. Adjustable devices are available in an 8-pin SO package.

The LT1129, a 700mA version of the LT1121 is also available. The LT1129 includes all of the protection features of the LT1121. No load quiescent current is slightly higher at 50μA and the LT1129 requires a minimum of 3.3μF of output capacitance. The LT1129 is also available with fixed output voltages of 3.3V and 5V and in an adjustable version with an output range of 3.75V to 5V. The device is available in a 5-pin DD package.

Cold cathode fluorescent display driver

New backlight systems seem universally to use cold cathode fluorescent tubes. Electroluminescent backlights have limited light output and limited life for notebook systems, and have limited usage among notebook and notebook manufacturers. The cold cathode fluorescent, on the other hand, has high efficiency, long life, and high light output. Typically the cold cathode fluorescent wants to be driven with 1mA to 5mA at 30kHz to 50kHz. The driving voltage and current are a function of the manufacturer and tube geometry.

Optimally the current through the tube should be regulated to control its brightness.

To understand the operation of the cold cathode fluorescent display driver in Figure 3.14, the circuit should be looked at as two sections; 1. The regulating loop, 2. The high voltage oscillator/driver.

The regulating loop consists of an LT1172 switching regulator in a buck mode configuration driving constant current into a self-oscillating converter coupled to a high voltage transformer. The architecture of the driver allows a wide input range of battery voltage while maintaining fluorescent tube current constant. In negative buck mode, the LT1172 periodically connects inductor L1 to ground via the switch pin. This creates a flow of current in L1 which is steered by self-oscillating transistors Q1 and Q2 to the primary of transformer L2. The output of L2 is a high voltage AC waveform that is partially ballasted by the 15pF capacitor. To achieve the desired regulation of actual bulb current, D1 and D2 rectify bulb current and pass one

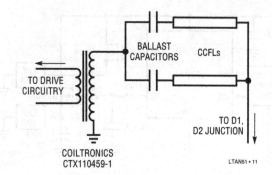

Figure 3.15 • Two Bulb Adaption for Color

phase through R1. This rectified current is converted to a voltage by R1 and filtered by R3 and C6. The filtered signal becomes a feedback signal to the LT1172, which maintains it at 1.25V.

Enclosing the cold cathode fluorescent bulb in a feedback loop allows precise control of its operating current and allows microprocessor control of its brightness. Voltage fed through a resistor to the top of C6, either from a D/A converter or from logic, will control the current through the fluorescent tube, allowing brightness to be varied from a keyboard input.

This architecture of a buck converter driving a self-oscillating inverter was chosen because it allows a wide range of input voltages. It is also tolerant of winding ratios on the cold cathode fluorescent transformer. One caution with this circuit is the voltage applied to the bulb terminals is not limited if the feedback loop is broken, so care must be taken to minimize the possibility of power being applied to this circuit with the fluorescent tube removed. Cures for this problem and much more detail on backlight engineering and circuits appear in LTC Application Note 55, "Techniques for 92% Efficient LCD Illumination."

Battery charging

Lead acid battery charger

Though not as popular as NiCad, lead acid rechargeable gel cells are attractive because of their high energy density per unit volume. These cells have a long life expectancy when treated properly, but often suffer premature failure because of improper charging. The circuit shown in Figure 3.16 provides a near ideal charging system for lead-acid cells. It has precise nonlinear temperature compensation, constant voltage charging with constant current override, and high efficiency over a wide range of input and battery voltages.

The basic charger is a flyback design to allow operation with input voltages above or below battery voltage. The LT1171 IC switcher operates at 100kHz and can deliver up to 15W into the battery. A dual op amp is used to control constant voltage and constant current modes. A1 activates as a current limiter when charging current through R7

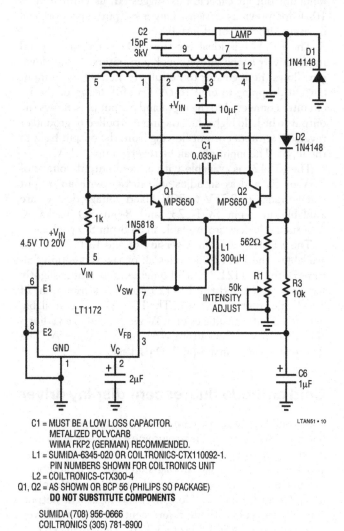

C1 = MUST BE A LOW LOSS CAPACITOR.
    METALIZED POLYCARB
    WIMA FKP2 (GERMAN) RECOMMENDED.
L1 = SUMIDA-6345-020 OR COILTRONICS-CTX110092-1.
    PIN NUMBERS SHOWN FOR COILTRONICS UNIT
L2 = COILTRONICS-CTX300-4
Q1, Q2 = AS SHOWN OR BCP 56 (PHILIPS SO PACKAGE)
    DO NOT SUBSTITUTE COMPONENTS

SUMIDA (708) 956-0666
COILTRONICS (305) 781-8900

Figure 3.14 • CCFL Inverter

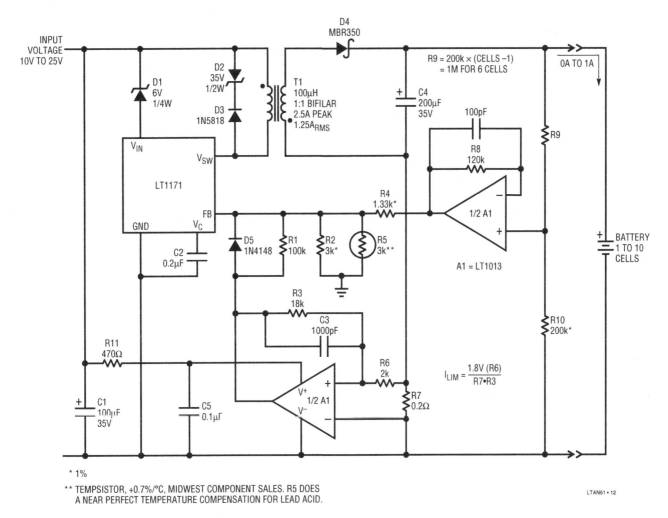

Figure 3.16 • Lead Acid Battery Charger

exceeds a preset limit determined by R3, R6, and R7. This current limit is included to prevent excess charge current for heavily discharged batteries. Losses in R7 are kept low because the voltage drop across R7 is kept to several hundred millivolts.

Lead acid batteries have a nonlinear negative temperature coefficient which must be accurately compensated to ensure long battery life and full charge capacity. R5 is a positive temperature coefficient thermistor (tempsistor) whose +0.7%/°C linear TC is converted to the required nonlinear characteristic by the parallel connection with R2. The combination of R2, R3, and R4 multiply the 1.244V feedback level of the LT1171 to the proper 2.35V level required by one cell at 25°C. A2 is used as a buffer to drive the resistor network. This allows large resistors to be used for the cell multiplier string, R9 and R10. R9 is set at 200k for each series cell over one. R9 current is only 12μA, so it can be left permanently connected to the battery. R1 is added to give the charger a finite output resistance (≈0.025Ω/cell) in constant voltage mode to prevent low frequency hunting.

NiCAD charging

Battery charging is a very important section of any notebook system. The battery charging circuits shown here for nickel cadmium or nickel metal hydride batteries control the current into the battery but do not detect when full battery charge is reached.

The first circuit, Constant Current Battery Charger (Figure 3.17), is built around a flyback configuration. This allows the battery voltage to be lower or higher than the input voltage. For example, a 16V battery stack may be charged off of a 12V automobile battery. The charge current is sensed by R4, a 1.2Ω resistor and set at approximately 600mA. Resistors R5 and R6 limit the peak output voltage when no battery is connected. Diode D3 prevents the battery discharging through the divider network when the charger is off, while transistor Q1 allows electronic shutdown of the charger.

The next two chargers are a high efficiency buck charger configuration. The input voltage must be higher than the battery voltage for charging to occur. These chargers are

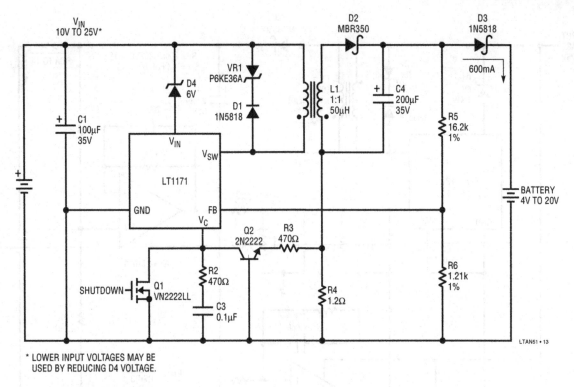

Figure 3.17 • Constant Current Battery Charger

90% efficient when charging at maximum output current. No heat sinks are needed on either the switching regulator or diodes because the efficiency is so high.

The dual rate battery charger in Figure 3.18 uses a logic signal to toggle between a high charge rate, up to 2A, or a trickle rate for keep alive. An LT1006 amplifier senses the current into the battery and drives the feedback pin of an LT1171 switching regulator. The entire control circuit is bootstrapped to the LT1171 and floats at the switching frequency, so stray capacitance must be minimized.

A gain setting transistor changes the gain on the LT1006 by shorting or opening resistor R1. This changes the charge rate, for the value shown, between 0.1A and 1A.

The charger in Figure 3.19 is programmable with a voltage from D-A converters. The charging current is directly proportional to the program voltage. A small sense resistor in the bottom side of the battery senses the battery charging current. This is compared with the program voltage and a feedback signal is developed to drive the LT1171 V_C pin. This controls the charging current from the LT1171 and with appropriate control circuits any battery current may be programmed. Efficiency during high charge currents is 90%.

LCD display contrast power supply

LCD display typically requires between −18V and −24V to set the contrast of the display. Usually, a switching regulator is needed in the system to generate this voltage although it runs at low power. The LT1172 generates the voltage with a minimum parts count.

The circuit in Figure 3.20 works by generating +18V to +24V in a boost configuration and then inverting the voltage by charge pumping. This allows the use of a small inductor for the converter rather than a transformer.

A 4-cell NiCad regulator/charger

The new LTC1155 Dual Power MOSFET Driver delivers 12V of gate drive to two N-channel power MOSFETs when powered from a 5V supply with no external components required. This ability, coupled with its micropower current demands and protection features, makes it an excellent choice for high side switching applications which previously required more expensive P-channel MOSFETs.

A notebook computer power supply system in Figure 3.21 is a good example of an application which benefits directly from this high side driving scheme. A 4-cell, NiCad battery pack can be used to power a 5V notebook computer system. Inexpensive N-channel power MOSFETs have very low ON resistance and can be used to switch power with low voltage drop between the battery pack and the 5V logic circuits.

Figure 3.21 shows how a battery charger and an extremely low voltage drop 5V regulator can be built using the LTC1155 and three inexpensive power MOSFETs. One half of the LTC1155 Dual MOSFET Driver controls

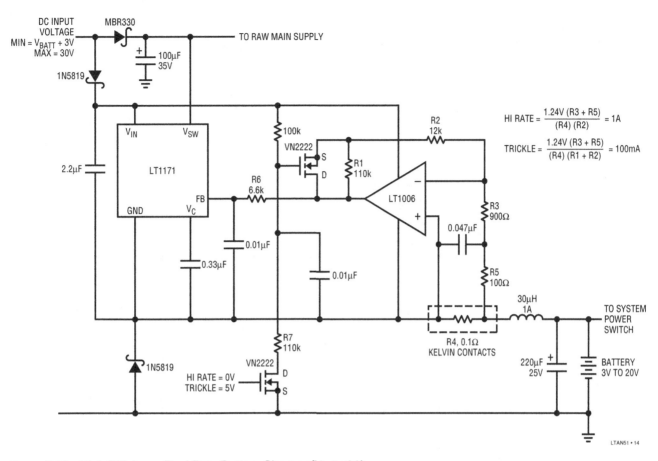

Figure 3.18 ● High Efficiency Dual Rate Battery Charger (Up to 2A)

the charging of the battery pack. The 9V, 2A current limited wall unit is switched directly into the battery pack through an extremely low resistance MOSFET switch, Q2. The gate drive output, pin 2, generates about 13V of gate drive to fully enhance Q1 and Q2. The voltage drop across Q2 is only 0.17V at 2A and, therefore, can be surface mounted to save board space.

An inexpensive thermistor, RT1, measures the battery temperature and latches the LTC1155 off when the temperature rises to 40°C by pulling low on pin 1, the Drain Sense Input. The window comparator also ensures that battery packs which are very cold (<10°C) are not quick charged.

Q1 drives an indicator lamp during quick charge to let the computer operator know that the battery pack is being charged properly. When the battery temperature rises to 40°C, the LTC1155 latches off and the battery charge current flowing through R9 drops to 150mA.

A 4-cell NiCad battery pack produces about 6V when fully charged. This voltage will drop to about 4.5V when the batteries are nearly discharged. The second half of the LTC1155 provides gate voltage drive, pin 7, for an extremely low voltage drop MOSFET regulator. The LT1431 controls the gate of Q4 and provides a regulated

5V output when the battery is above 5V. When the battery voltage drops below 5V, Q4 acts as a low resistance switch between the battery and the regulator output.

A second power MOSFET, Q3, connected between the 9V supply and the regulator output "bypasses" the main regulator when the 9V supply is connected. This means that the computer power is taken directly from the AC line while the charger wall unit is connected. The LT1431 provides regulation for both Q3 and Q4, and maintains a constant 5V at the regulator output. The diode string made up of diodes D2-D4 ensure that Q3 conducts all the regulator current when the wall unit is plugged in by separating the two gate voltages by about 2V.

R14 acts as a current sense for the regulator. The regulator latches off at 3A when the voltage drop between the second Drain Sense Input, pin-8, and the supply, pin-6, rises above 100mV. R10 and C3 provide a short delay. The μP can restart the regulator by turning the second input, pin-5, off and then back on.

The regulator is switched off by the μP when the battery voltage drops below 4.6V. The standby current for the 5V, 2A regulator is less than 10μA. The regulator is switched on again when the battery voltage rises during charging.

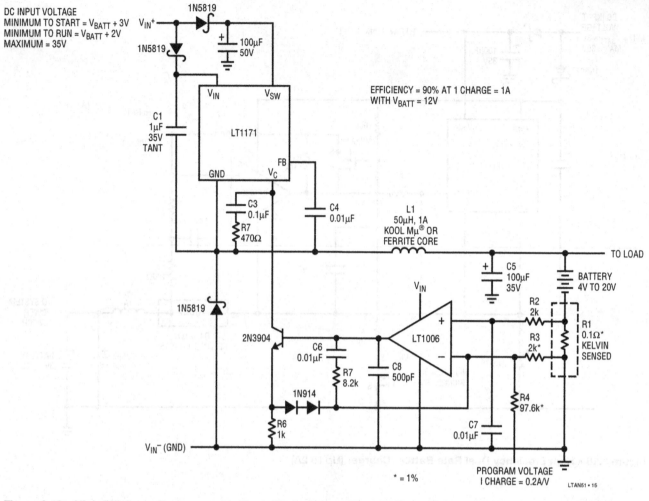

Figure 3.19 • High Efficiency Programmable Buck-Mode Battery Charger (Input Voltage Must be Higher than Battery Voltage)

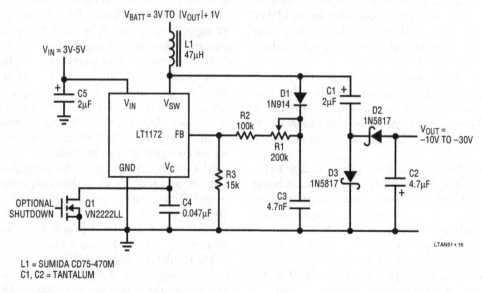

L1 = SUMIDA CD75-470M
C1, C2 = TANTALUM

Figure 3.20 • LCD Bias Circuit Generates −24V

*Kool Mμ is a registered trademark of Magnetics, Inc.

29

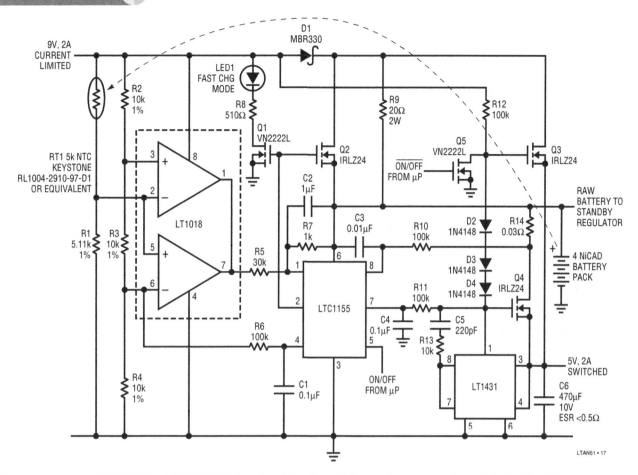

Figure 3.21 • The LTC1155 Dual MOSFET Driver Provides Gate Drive and Protection for a 4-Cell NiCad Charger and Regulator

Power dissipation in the notebook computer itself is generally quite low. The current limited wall unit dissipates the bulk of the power created by quick charging the battery pack. Q2 dissipates less than 0.5W. R9 dissipates about 0.7W. Q4 dissipates about 2W for a very short period of time when the batteries are fully charged and dissipates less than 0.5W as soon as the battery voltage drops to 5V. The three integrated circuits shown are micropower and dissipate virtually no power. Q3, however, can dissipate as much as 7W if the full 2A output current is required while powered from the wall unit.

The circuit shown in Figure 3.21 consumes very little board space. The LTC1155 is available in an 8-pin SO package and the three power MOSFETs can also be housed in SO packaging. Q3 and Q4 must be heat sinked properly however. (Consult the MOSFET manufacturer data sheet for surface mount heat sink recommendations.)

The LTC1155 allows the use of inexpensive N-channel MOSFET switches to directly connect power from a 4-cell NiCad battery pack to the charger and the load. This technique is very cost effective and is also very efficient. Nearly all the battery power is delivered directly to the load to ensure maximum operating time from the batteries.

Power supplies for palmtop computers

Palmtop computer power supply designs present an entirely separate set of problems from notebook computers. Notebook machines typically use a 9V to 15V NiCad stack for the power source. Palmtop machines, due to their extremely small size, have room for only two or four AA cells. The palmtop machines require much longer operating time in sleep mode, since they presently do not have disk drives. A typical palmtop system may have several hours of operating life with the processor at full activity, tens of hours of quiescent operation with the processor shutdown but the display active, and up to two months life in sleep mode where all memory is retained but no computation takes place. Palmtop machines also use a lithium battery for backup power when the AA cells are dead or being replaced.

The power source for palmtops are usually disposable AA alkaline cells. The use of these disposable batteries generates a separate set of problems from notebook computers. Unlike power supply systems powered by rechargeable NiCad or NiMH batteries, high efficiency power converter circuits are not necessarily optimum for use with disposable batteries. Since rechargeable batteries have very low output impedance, the most efficient converter circuits result in maximum operating time.

Disposable cells, on the other hand, have relatively high internal impedance, so maximum battery life results when the battery load is low and relatively constant. Power supply converters that minimize both the loss in the converter circuit and minimize the effect of battery internal resistance will give longest system operating life. Some of the 4-cell designs presented here are optimized for low peak battery current to lengthen the disposable battery life. Other configurations, while they may have higher efficiency, require higher peak energy demands on the battery and consequently shorten the battery life. The converter circuits shown here have been tested using alkaline AA cells and provide long battery life.

2-Cell input palmtop power supply circuits

A regulated 5V supply can be generated from two AA cells using the circuit shown in Figure 3.22. U1, an LT1108-5 micropower DC to DC converter, is arranged as a step-up, or "boost" converter. The 5V output, monitored by U1's SENSE pin, is internally divided down and compared to a 1.25V reference voltage inside the device. U1's oscillator turns on when the output drops below 5V, cycling the switching transistor at a 19kHz rate. This action alternately causes current to build up in L1, then dump into C1 through D1, increasing the output voltage. When the output reaches 5V, the oscillator turns off.

The gated oscillator provides the mechanism to keep the output at a constant 5V. R1 invokes the current limit feature of the LT1108, limiting peak switch current to approximately 1A. U1 limits switch current by turning off the switch when the current reaches the programmed limit set by R1. Switch "on" time, therefore, decreases as V_{IN} is

increased. Switch "off" time is not affected. This scheme keeps peak switch current constant over the entire input voltage range, allowing minimum energy transfer to occur at low battery voltage without exceeding L1's maximum current rating at high battery voltage. Maximum current demands should be carefully considered, with R1 tailored to the individual application to obtain longest possible battery life. For example, if only 75mA maximum is required, R1 can be increased to 100Ω. This will limit switch current to approximately 650mA which has the effect of increasing converter efficiency and lowering peak current demands, considerably extending battery life.

The circuit delivers 5V at up to 150mA from an input range of 3.5V to 2.0V. Efficiency measures 80% at 3.0V, decreasing to 70% at 2.0V for load currents in the 15mA to 150mA range. Output ripple measures $75mV_{P-P}$ and no-load quiescent current is just 135μA.

LCD bias from 2 AA cells

A −24V LCD bias generator is shown in Figure 3.23. In this circuit U1 is an LT1173 micropower DC to DC converter. The 3V input is converted to +24V by U1's switch, L1, D1, and C1. The switch pin (SW1) then drives a charge pump consisting of C2, C3, D2, and D3 to level shift the +24V output to −24V. Line regulation is less than 0.2% from 3.3V to 2.0V inputs. Load regulation measures 2% from a 1mA to 7mA load. The circuit delivers up to 7mA from a 2.0V input at 73% efficiency.

4-Cell input palmtop power supply circuits

Newer, more powerful palmtop machines using 386SX processors require more power than two AA cells can

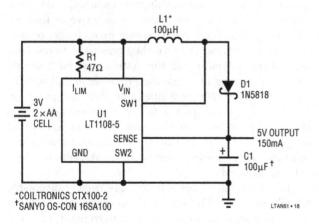

*COILTRONICS CTX100-2
†SANYO OS-CON 16SA100

LTAN51 • 18

Figure 3.22 • 2 AA Cells to 5V Deliver 150mA

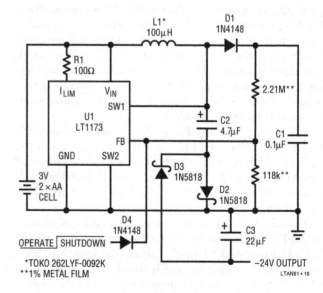

*TOKO 262LYF-0092K
**1% METAL FILM

LTAN51 • 19

Figure 3.23 • 2-Cell LCD Supply Generates −24 at 7mA

31

deliver for reasonable operating life. The circuits shown here provide a switchable 3.6V/5V output for main logic, a −23V output for LCD display bias, a +12V output for Flash memory V_{P-P} generation, and an automatic backup supply using a 3V lithium cell. Under no-load conditions, the quiescent current required by the entire system is 380µA.

The main converter circuit shown in Figure 3.24 is a combination step-up/step-down converter. When the 4 AA cells are fresh, the circuit behaves as a linear regulator. While this may seem to be inefficient, note that the battery voltage normally quickly drops from 6V to 5V. At 5V input, the efficiency is 3.6V/5V or 72%. As battery voltage drops further, efficiency increases, reaching over 90% at 4.2V input. When the battery drops below 4V, the circuit switches over to step-up mode, squeezing every bit of available energy out of the battery.

The converter delivers 200mA at 3.6V with as little as 2.5V input. In step-up mode, efficiency runs between 83% and 73% (at 2.5V_{IN}). The linear regulator has no current spikes. AA alkaline cells have a fairly high internal impedance, and the current spikes that switching regulators demand from the battery reduces battery life. A 4-cell AA alkaline battery has an impedance of about 0.5Ω when fresh, increasing to 2Ω at end-of-life. This topology delivers over 9.3 hours of 3.6V, 200mA output power, compared to just 7 hours using a flyback topology.

A backup function is implemented with another LT1173 circuit also shown in Figure 3.24. Power for the LT1173 comes from the main logic output. The lithium battery sees a load consisting of the 10µF capacitor leakage, switch leakage, and about 1.5µA due to the 910k/1M resistor divider. The total load is less than 5µA. The LT1173 requires 110µA quiescent current, taken from the main logic supply line.

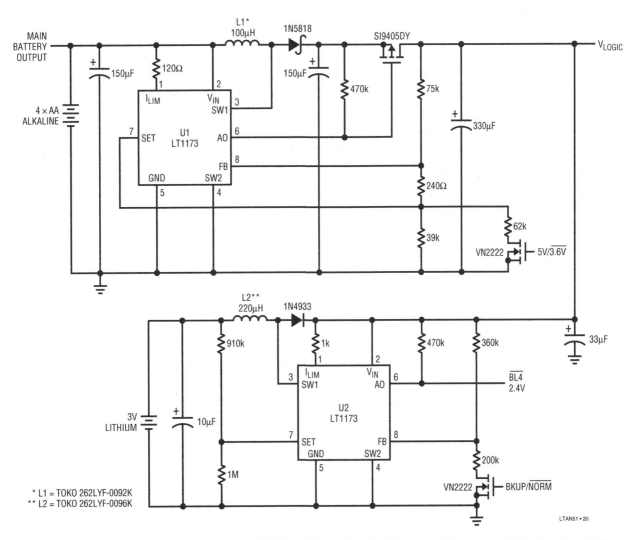

Figure 3.24 • Main Logic Converter Generates 3.6V/5V at 200mA; Backup Converter Generates 3.4V when Main Battery is Dead or Removed

When the BACKUP/NORMAL input goes high, the feedback string is connected, but the converter does not cycle until the main logic supply voltage drops to 3.4V. This converter is capable of supplying 3.6V at 10mA. If the BACKUP/NORMAL signal is driven from Figure 3.27's circuit, the backup converter will automatically kick in when the main AA cells are removed or dead. A low-battery detector function is provided using the gain block inside the LT1173. The 910k/1M divider set the BL4 output to go low when V_{BATT} equals 2.4V.

The −24V LCD bias generator, shown in Figure 3.25, uses the LT1173 as a controller driving the FZT749, a 2A PNP in a SOT-223 package. The LT1173 maintains 1.25V between its FB pin and its GND pin. Current must flow through the 3M resistor to force 1.25V across R1. This forces the "GND" pin negative. The 220μH inductor limits switch current to 500mA from a fresh battery and 300mA from a dead (3.6V) battery. Efficiency of this converter is in the 70% range. Higher efficiency can be obtained merely by decreasing the value of the inductor; however, this will actually DECREASE battery life due to the higher current spikes drawn from the battery.

A Flash memory V_{P-P} generator is shown in Figure 3.26. Up to 40mA at 12V is available from the output. The converter is switched on and off via the small N-channel MOSFET connected to the 124k feedback resistor. When the MOSFET is turned on, the resistor is connected to ground and the converter generates 12V. When the MOSFET is off, the 124k resistor is disconnected and the feedback pin floats high, turning off the converter. When off, output voltage sits at battery voltage minus a diode drop. This condition is approved for Flash memory. Inadvertent programming cannot occur as the Flash chip contains a level detector. When the V_{P-P} pin voltage is less than 11.4V, the Flash chip itself will not

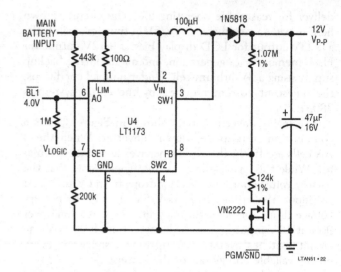

Figure 3.26 • Flash Memory V_{P-P} Generator Delivers 12V, 40mA from 4AA Input

allow programming to take place. Another low-battery detect function is provided using the LT1173's gain block. The main alkaline battery is being sensed here, and the AO pin goes low when the battery voltage falls below 4.0V.

Finally, a micropower two-terminal reference and dual comparator form a pair of battery detectors. The upper comparator in Figure 3.27 senses the main battery directly. When the battery voltage falls below 2.5V (a very dead battery!) or the battery is removed, BL3 will go high. If connected to the BACKUP/NORMAL signal of the lithium backup converter, the backup will take over the main logic supply line automatically. The other comparator goes low when the battery voltage falls below 3.6V.

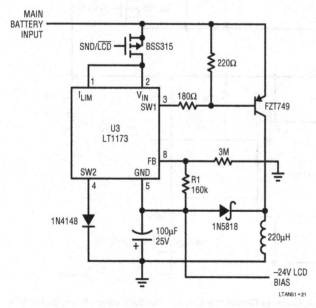

Figure 3.25 • LCD Bias Generator Delivers −24V at 10mA

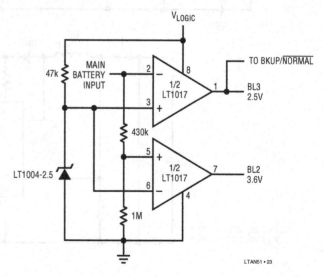

Figure 3.27 • Battery Detectors Sense Removal of Main Battery, Indicate V_{BATT} <3.6V

A CCFL backlight driver for palmtop machines

Backlit displays have greatly enhanced user acceptance of portable computers. Palmtop machines have not used backlit displays because of the high power required by the inverter circuit used to drive the bulb. Figure 3.28's circuit, a micropower CCFL supply, overcomes this problem. A typical notebook CCFL supply drives the bulb at 5mA. This circuit, using an LT1173 micropower DC to DC converter, operates over an input range of 2.0V to 6V. Maximum bulb current is limited to 1mA. Control over bulb current is maintained down to 1μA, a very dim light! It is intended for palmtop applications where the longest possible battery life is desired.

L1, Q1, and Q2 comprise a current driven Royer class converter which oscillates at a frequency primarily set by L1's characteristics (including its load) and the 0.01μF capacitor. This entire converter is gated on and off by the burst mode operation of the LT1173. The 1M/0.01μF RC at the LT1173 feedback pin filters the half-sine appearing at the 3.3k-1M potentiometer chain. This signal represents 1/2 the lamp current. The LT1173 servos the energy in the lamp to maintain 1.25V at its feedback pin, closing a loop. For low bulb currents, the LT1173 idles most of the time, drawing only 110μA quiescent current. At the 1mA maximum bulb current, the circuit draws less than 100mA. A substantial amount of light is emitted by the bulb at an input current drain of less than 5mA.

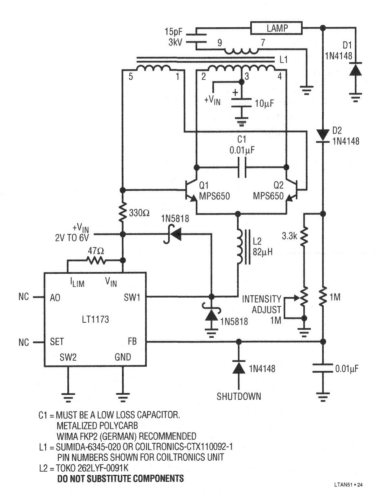

C1 = MUST BE A LOW LOSS CAPACITOR.
METALIZED POLYCARB
WIMA FKP2 (GERMAN) RECOMMENDED
L1 = SUMIDA-6345-020 OR COILTRONICS-CTX110092-1
PIN NUMBERS SHOWN FOR COILTRONICS UNIT
L2 = TOKO 262LYF-0091K
DO NOT SUBSTITUTE COMPONENTS

LTAN51 • 24

Figure 3.28 • Micropower CCFL Driver Delivers Up to 1mA of Bulb Current from 2 AA Cells

2-Wire virtual remote sensing for voltage regulators

Clairvoyance marries remote sensing

4

Jim Williams   Jesus Rosales   Kurk Mathews   Tom Hack

Introduction

Wires and connectors have resistance. This simple, unavoidable truth dictates that a power source's remote load voltage will be less than the source's output voltage. Figure 4.1 shows this, and implies that intended load voltage can be maintained by raising regulator output. Unfortunately, line resistance and load variations introduce uncertainties, limiting achievable performance.

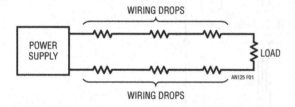

Figure 4.1 • Unavoidable Wiring Drops Cause Low Load Voltage. Line and Load Resistance Variations Introduce Additional Load Voltage Uncertainty, Mitigating Against Compensation by Raising Supply Voltage

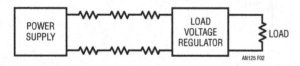

Figure 4.2 • Local Regulation Stabilizes Load Voltage But is Inefficient

Figure 4.2 illustrates one compensatory approach. Locally positioned regulation stabilizes load voltage against line drops but is inefficient due to regulator losses. Figure 4.3, the classical approach, utilizes "4-wire" remote sensing to eliminate line drop effects. The power supply sense inputs are fed from load referred sense wires. The

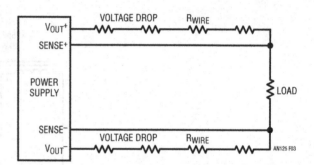

Figure 4.3 • Classical "4-Wire" Remote Sensing. V_{OUT} Line Voltage Drops Are Compensated by Regulator Sensing at Load. High Impedance Sense Inputs Negate Sense Wire Resistance. Approach Requires Four Wires

sense inputs are high impedance, negating sense line resistance effects. This scheme works well, but requires dedicated sense wires, a significant disadvantage in many applications.

"Virtual" remote sensing

Figure 4.4 retains the advantages of classical 4-wire remote sensing while eliminating the sense leads. Here, the LT4180 Virtual Remote Sense™ (VRS) IC alternates output current between 95% and 105% of the nominal required output current. The LT4180 forces the power supply to provide a DC current plus a small square wave current with peak-to-peak amplitude equal to 10% of the DC current. Decoupling capacitor C_{LOAD}, normally required for low impedance under transient conditions in non-VRS systems, takes an additional role by filtering out the VRS square wave excursions.

Because C is sized to produce an "AC short" at the square wave frequency, a square wave voltage is produced at the power supply equal to $V_{OUTAC} = 0.1 \cdot I_{DC} \cdot R_{WIRE}V_{P-P}$ The

Analog Circuit and System Design: A Tutorial Guide to Applications and Solutions. DOI: 10.1016/B978-0-12-385185-7.00004-4

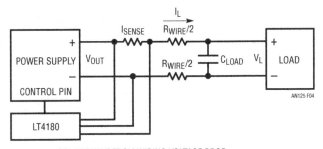

V_{OUT} = DC + SQUAREWAVE FROM WIRING VOLTAGE DROP
C_{LOAD} REMOVES SQUAREWAVE, SO V_L CONTAINS ONLY DC
I_L = DC + SQUAREWAVE

Figure 4.4 • LT4180 2-Wire Virtual Remote Sense Estimates Wiring Voltage Drops, Compensates by Adjusting Supply Output Voltage. Wiring Loss Is Determined by Measuring Small Signal Square Wave Carrier Induced Voltage Drop. Load Capacitor Absorbs Square Wave; Load Is at DC

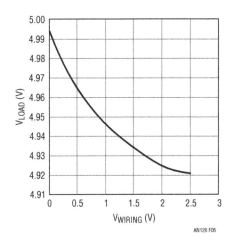

Figure 4.5 • Typical LT4180 Virtual Remote Sense Performance Shows 1.6% Regulation vs 0V → 2.5V Wiring Drop

square wave voltage at the power supply has a peak-to-peak amplitude equal to one tenth the DC wiring drop. This is a direct measurement of wiring drop, not an estimate, accurate over all load currents. Signal processing produces a DC voltage from this AC signal which is introduced into the supply feedback loop to provide accurate load regulation[1]. Note that the "power supply" may be an IC linear or switching regulator, a module or any other power source capable of variable output. Power supplies can be synchronized to the LT4180 and VRS operating frequency is adjustable over more than three decades. Optional spread spectrum operation provides partial immunity from single-tone interference and a 3V to 50V input range simplifies design. Because this technique is based on an estimate of load voltage, not a direct measurement, the resultant correction is an approximation, but a very good one.

Typical LT4180 load regulation is plotted in Figure 4.5. In this example, load current increases from zero until it produces a 2.5V wiring drop. Load voltage drops only 73 mV at maximum current. A voltage drop equivalent to 50% of load voltage results in only a 1.5% shift in load voltage value. Smaller wiring drops produce even better results.

Applications

The following applications are all VRS augmented voltage regulators of various descriptions. The power regulation stages employed are, with one exception, generic LTC designs and are spared exhaustive commentary, permitting emphasis on the LT4180 VRS role. Additionally, the similarity of the VRS associated circuitry across the broad array of applications shown should be noted, and is indicative of the relative ease of implementation. Surprisingly little change is needed to use the VRS in the different situations presented.

VRS linear regulators

Figure 4.6 adds a simple stage to the LT4180 to implement a complete VRS aided linear regulator. The LT4180 senses current via the 0.2 Ω shunt and feedback controls Q1 with Q2, completing a control loop. Cascoded Q2 permits the ICs 5V capable open drain output to control a high voltage at Q1's gate. Components at the compensation pin furnish loop stability, promoting good transient response[2]. Figure 4.7 shows Figure 4.6's load step waveforms. They include V_{SENSE} (trace A), V_{LOAD} (B) and I_{LOAD} (C). Transient response is determined by loop compensation, load capacitance and remote sense sample rate. Figure 4.8 shows response with C_{LOAD} increased to 1100 μF Load voltage transient excursion reduces and duration increases.

Figure 4.9, employing a monolithic regulator, adds current limiting and simplifies loop compensation. Transient response approximates Figure 4.6's. As before, the LT4180's low voltage drain pin requires a cascode transistor to control the high voltage at the LT3080 set pin.

Note 1. Readers finding their intellectual prowess unsatiated by this admittedly cursory description will find more studious coverage in Appendix A, "A Primer on LT4180 VRS Operation."

Note 2. Value selection procedure for LT1480 VRS circuits is detailed in Appendix B, "Design Guidelines for LT4180 VRS Circuits."

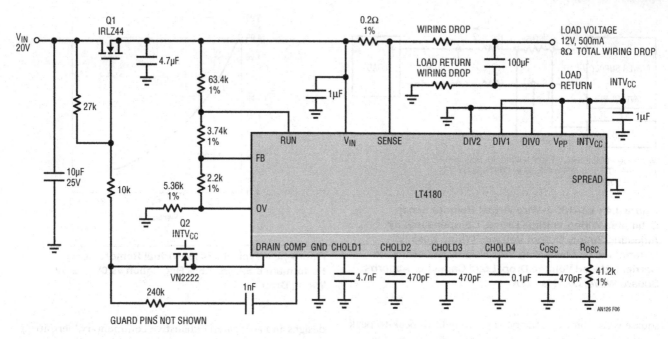

Figure 4.6 • Virtual Remote Sense Controls Discrete Linear Regulator. Q2 Cascodes Drain Output, Buffering High Voltage Q1 Gate Drive. COMP Pin Associated Components Stabilize Loop

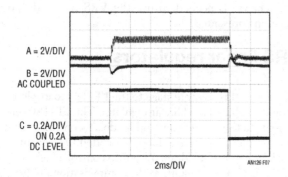

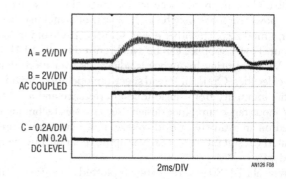

Figure 4.7 • Figure 4.6's Load Step Waveforms with 100µF Load Capacitor Include V_{SENSE} (Trace A), V_{LOAD} (B) and I_{LOAD} (C). Transient Response is Determined by Loop Compensation, Load Capacitance and Remote Sense Sample Rate

Figure 4.8 • Same Conditions as Figure 4.7 with C_{LOAD} Increased to 1100µF. V_{LOAD} Transient Excursion Reduces, Duration Extends

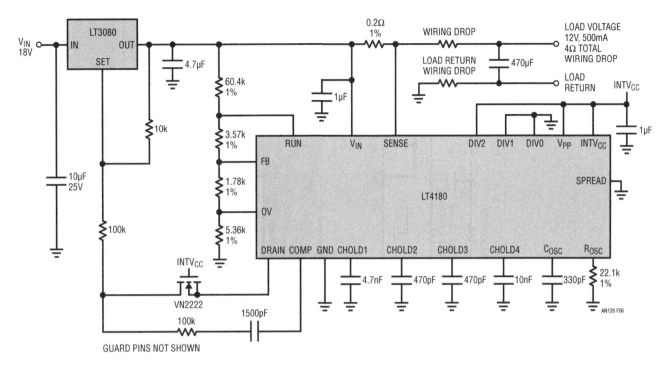

Figure 4.9 • Figure 4.6's Approach Utilizing IC Regulator Adds Current Limiting, Simplifies Loop Compensation. Transient Response Approximates Figure 4.6's

VRS equipped switching regulators

VRS based switching regulators are readily constructed. Figure 4.10's flyback voltage boost configuration has similar architecture to the linear examples although output voltage is above the input. In this case, the LT4180 open drain output is directly compatible with the LT3581 boost regulator low voltage V_C pin—no cascode stage is necessary.

Step down ("Buck") VRS equipped switching regulators are similarly easily achieved. Figure 4.11's scheme, reminiscent of the previously described linear regulators, substitutes an LT3685 step down regulator which is directly controlled from the LT4180 open drain output. A single pole roll-off stabilizes the loop and a 12V, 1.5A output is maintained from a 22V to 36V input despite a 0Ω to 2.5Ω wiring drop loss. Figure 4.11A is similar, except that it provides a 5V, 3A output from a 12V to 36V input.

VRS based isolated switching supplies

The VRS approach is adaptable to isolated output supplies. Figure 4.12's 24V output converter utilizes an approach similar to the previous examples except that it supplies a fully isolated output. The virtual remote sense feature accommodates a 10Ω wire resistance. The LT3825 and T1 form a transformer coupled power stage. Opto-coupled feedback maintains output isolation.

Figure 4.13's 48V → 3.3V, 3A design also has a fully isolated output, facilitated by power delivery through a transformer and optically coupled feedback loop closure. The LT3758 drives T1 via Q1. T1's rectified and filtered secondary supplies output power which is corrected for line drops by the LT4180. Isolation is maintained by transmitting the feedback signal with an opto-isolator. The opto-isolators output collector ties back to the LT3578 V_C pin, closing the control loop.

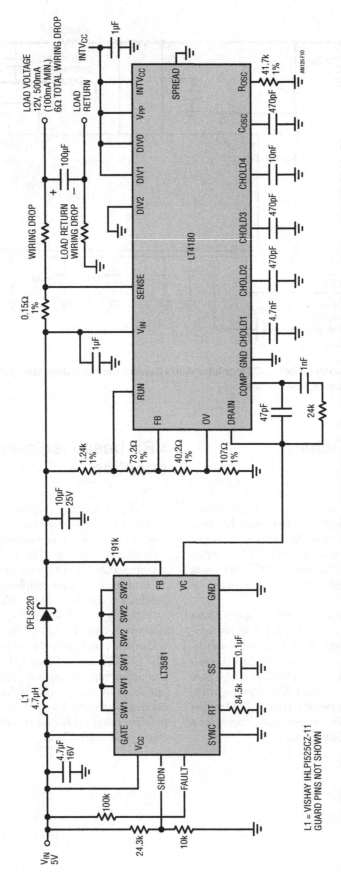

Figure 4.10 • Virtual Remote Sensed Voltage Boost Configuration. LT4180 Drain Output Controls Flyback Regulator via LT3581 V_C Pin

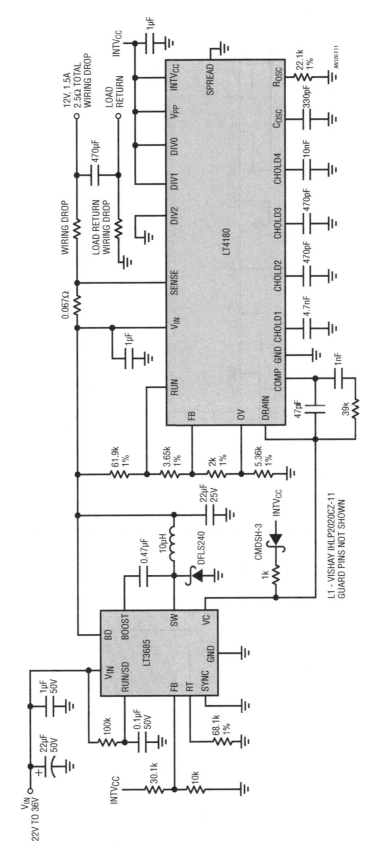

Figure 4.11 • Remote Sense Corrected 22V$_{IN}$ to 36V$_{IN}$ Step-Down Regulator Maintains 12V Output Despite Wiring Losses

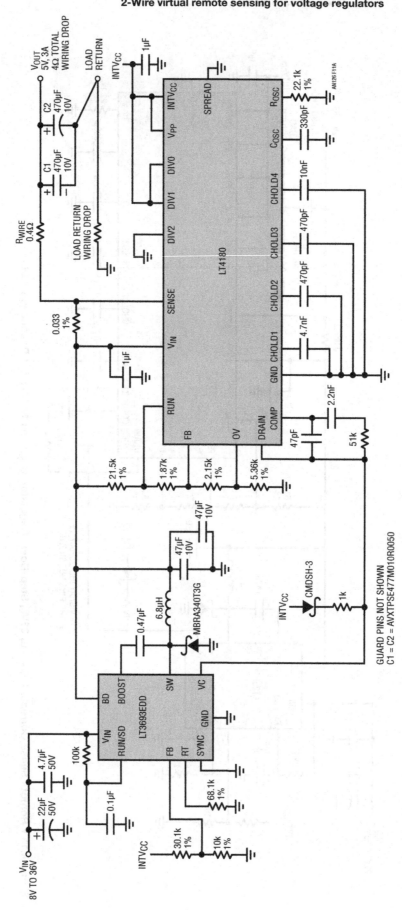

Figure 4.11A • 12V$_{IN}$ → 36V$_{IN}$ to 5V$_{OUT}$ Step-Down Remote Sensed Regulator Has Similar Architecture to Figure 4.11

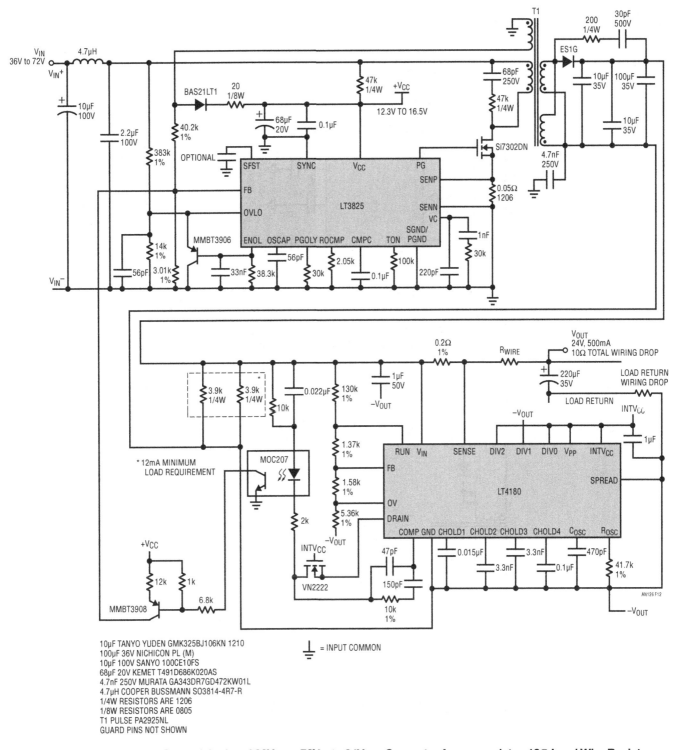

Figure 4.12 • Virtual Remote Sensed, Isolated 36V$_{IN}$ → 72V$_{IN}$ to 24V$_{OUT}$ Converter Accommodates 10Ω Lead Wire Resistance. LT3825/T1 Form Transformer Coupled Power Stage. LT4180 Provides Virtual Remote Sense, Opto-Coupled Feedback Maintains Output Isolation

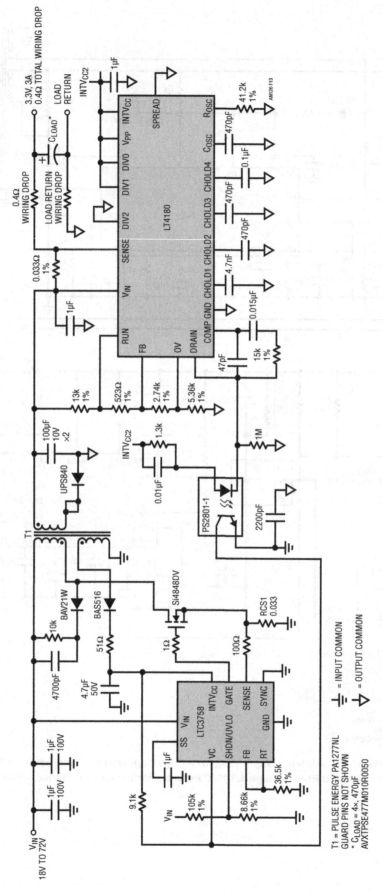

Figure 4.13 • 48V → 3.3V Isolated Step-Down, Remote Sensed Regulator. T1 Delivers Isolated Power, LT4180 Remotely Senses Output, Supplies Feedback via Opto-Isolator

Figure 4.14, also a VRS isolated step-down supply, uses a commercially produced 48V isolated input module augmented with virtual remote sensing. The module sense terminals are unused. The LT4180 wiring drop correction is introduced at the module trim pin. Component values are shown for 3.3 and 5V outputs. The "black box" Vicor module trim pin transient response defines available control bandwidth. Figure 4.15, trace A, is the trim pin input step (see test circuit A), trace B, the module output. The trim pin directed dynamics set practical expectations for VRS equipped loop response around the module. Figures 4.16 and 4.17 do not disappoint. Figure 4.14's load step response appears in Figure 4.16. Trace A is load step current, trace B, the resultant output voltage transient. The response envelope, bounded by module trim pin dynamics, is clean and well controlled. Figure 4.17 shows Figure 4.14's turn-on into a 2.5 Amp load. LT4180 activation arrests the initial abrupt rise at the 3rd vertical division. The ascent's conclusion is controlled to the regulation point in damped fashion.

LT4180 sampling square wave residue is just discernible in the waveform's settled portion.

BEFORE PROCEEDING ANY FURTHER, THE READER IS WARNED THAT CAUTION MUST BE USED IN THE CONSTRUCTION, TESTING AND USE OF THIS CIRCUIT. HIGH VOLTAGE, AC LINE CONNECTED POTENTIALS ARE PRESENT IN THIS CIRCUIT. EXTREME CAUTION MUST BE USED IN WORKING WITH AND MAKING CONNECTIONS TO THIS CIRCUIT. REPEAT: THIS CIRCUIT CONTAINS DANGEROUS, AC LINE CONNECTED HIGH VOLTAGE POTENTIALS. USE CAUTION.

Figure 4.18's VRS aided "Off-Line" isolated output supply has a 5V output with 2 A capacity. The schematic appears complex, but inspection reveals it to be essentially an AC line powered variant of Figure 4.13's isolated approach. The LT4180 provides remote sensing and closes an isolated feedback loop with optical transmission.

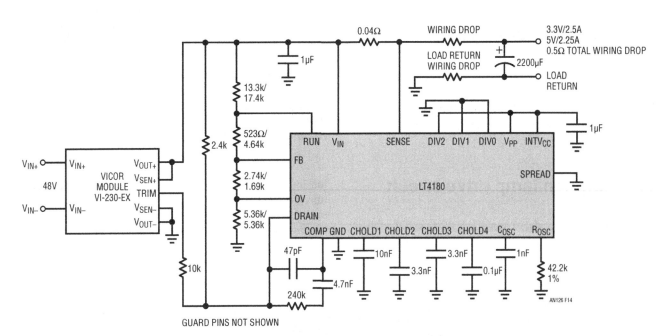

Figure 4.14 • Commercially Produced, Isolated 48V Input Module Augmented with Virtual Remote Sense. Module Sense Terminals Are Unused. Wiring Drop Correction Introduced at Module Trim Pin. Component Values Shown for 3.3V/5V Outputs

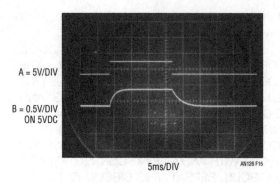

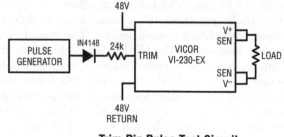

Trim Pin Pulse Test Circuit

Figure 4.15 • Vicor Module Trim Pin Transient Response Defines Available Control Bandwidth. Trace A is Trim Pin Input Step (See Test Circuit), Trace B, Module Output

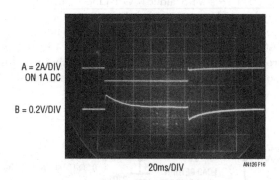

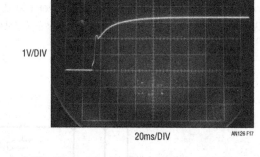

Figure 4.16 • Figure 4.14's **Load Step Response. Trace A is Load Step Current, Trace B Resultant Output Voltage Transient. Response Envelope, Bounded by Module Trim Pin Dynamics, is Well Controlled**

Figure 4.17 • Figure 4.14's **Turn-On into a 2.5A Load. LT4180 Activation Arrests Initial Abrupt Rise at Third Vertical Division. Ascent Conclusion is Controlled to Regulation Point. LT4180 Sampling Square Wave Residue is Discernible**

VRS halogen lamp drive circuit

A final circuit, Figure 4.19, uses the VRS to stabilize drive to a halogen lamp, in this case a 12V 30W automotive type. Lamp output power remains constant despite 9V to 15V input variation and line resistance/connection uncertainties. Additional benefits include constant color output and extended lamp life. The circuit, a step up/down ("SEPIC") converter, maintains 12V at the lamp despite the 9V to 15V input range[3]. The VRS functions in the manner previously described. Line resistance losses due to switches, wiring and connectors are obviated by VRS action. Figure 4.20 plots unaided vs remote sensed and regulated halogen lamp light output. VRS equipped

luminosity is flat over the 9 to 15V input range while unregulated performance suffers dramatically. The regulation also benefits lamp life by greatly reducing lamp turn-on current. Figure 4.21 shows unregulated lamp turn-on exceeding 20A without regulation. In Figure 4.22, regulation cuts current peaking to 7A, a 3× reduction. This soft turn-on and constant 12V drive under high/low line conditions optimizes illumination and improves lamp life.

References

1. LT4180 Data Sheet, Linear Technology Corporation, 2010.
2. Ridley, R. "Analyzing the Sepic Converter", Power Systems Design Europe, November, 2006.

Note 3. SEPIC operation is described in Reference 2.

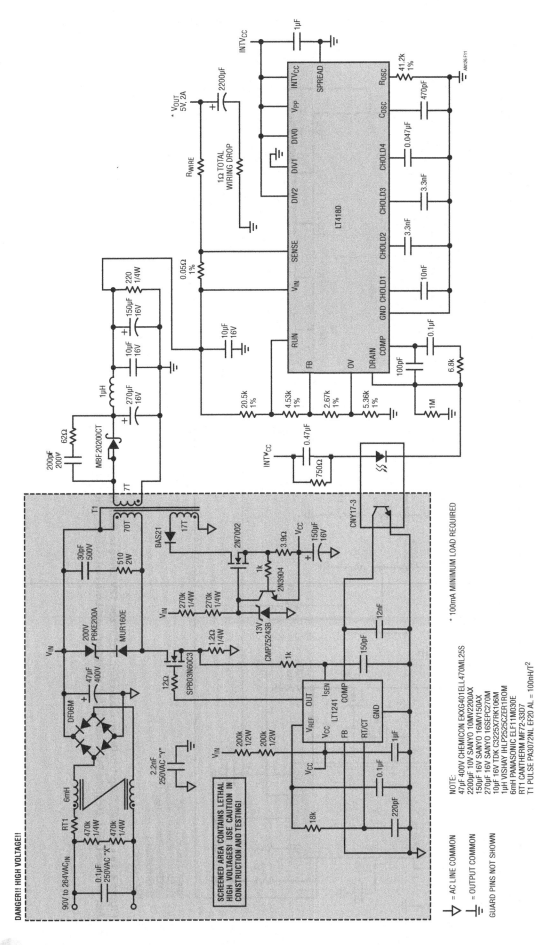

Figure 4.18 • A 5V Output "Off-Line" Converter Equipped with Virtual Remote Sense. LT4180 Provides Remote Sensing, Closes Isolated Feedback Loop via Opto-Isolator

WARNING! SCREENED AREA CONTAINS LETHAL AC LINE CONNECTED HIGH VOLTAGES. USE CAUTION IN CONSTRUCTION AND TESTING

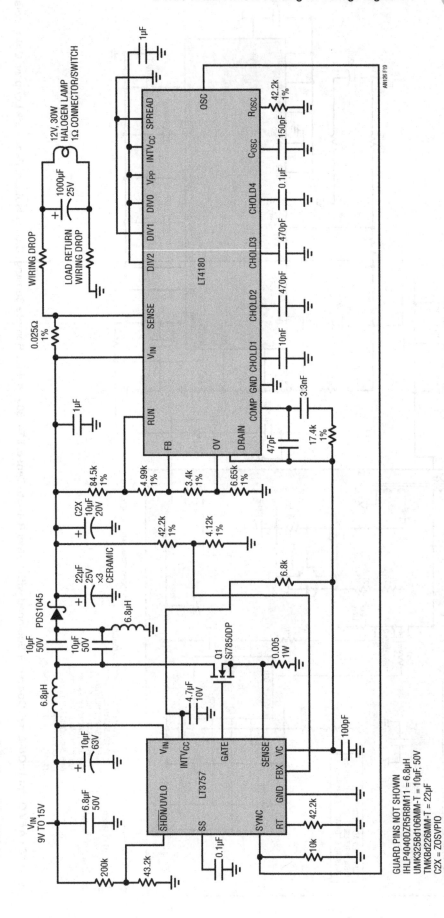

Figure 4.19 • LT4180 Step Up/Down Converter Stabilizes 12V Drive to 30W Halogen Automotive Lamp Despite 9V → 15V Input Variation and Line Resistance Uncertainties

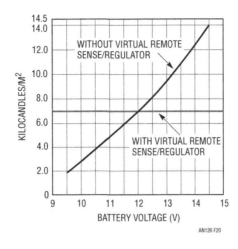

Figure 4.20 • Unaided vs Remote Sensed/Regulated Halogen Lamp Light Output. Regulation Benefits Include Stable Illumination, Constant Color Output and Extended Lamp Life

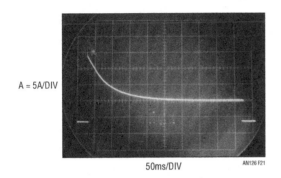

Figure 4.21 • Lamp Turn-On Current Exceeds 20A without Regulation, Degrading Lifetime

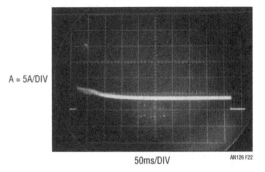

Figure 4.22 • Regulation Promotes Soft Turn-On, 12V Drive Under High/Low Line Conditions, Optimizing Illumination and Improving Lamp Life

Appendix A
A primer on LT4180 VRS operation

Voltage drops in wiring can produce considerable load regulation errors in electrical systems (Figure A1). As load current I_L increases, voltage drop in the wiring ($I_L \cdot RW$) increases and the voltage delivered to the system (V_L) drops. The traditional approach to solving this problem, remote sensing, regulates the voltage at the load, increasing the power supply voltage (V_{OUT}) to compensate for voltage drops in the wiring. While remote sensing works well, it does require an additional pair of wires to measure at the load, which may not always be practical.

The LT4180 eliminates the need for a pair of remote sense wires by creating a virtual remote sense. Virtual remote sensing is achieved by measuring the incremental change in voltage that occurs with an incremental change in current in the wiring (Figure A2). This measurement can be used to infer the total DC voltage drop in the wiring, which can then be compensated for. The

Virtual Remote Sense takes over control of the power supply via its feedback pin (V_{FB}), maintaining tight regulation of load voltage V_L.

Figure A3 shows the timing diagram for Virtual Remote Sensing (VRS). A new cycle begins when the power supply and VRS close the loop around V_{OUT} (Regulate V_{OUT} = H). Both V_{OUT} and I_{OUT} slew and settle to a new value, and these values are stored in the Virtual Remote Sense (Track $V_{OUTHIGH}$ = L and Track I_{OUT} = L). The V_{OUT} feedback loop is opened and a new feedback loop is set up commanding the power supply to deliver 90% of the previously measured current (0.9 I_{OUT}). V_{OUT} drops to a new value as the power supply reaches a new steady state, and this information is also stored in the Virtual Remote Sense. At this point, the change in the output voltage (ΔV_{OUT}) for a −10% change in output current has been measured and is stored in the Virtual Remote Sense. This voltage is used during the next VRS cycle to compensate for voltage drops due to wiring resistance.

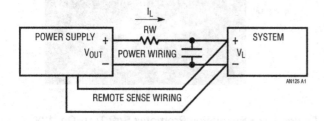

Figure A1 • Traditional Remote Sensing Works Well But Requires Two Sense Wires

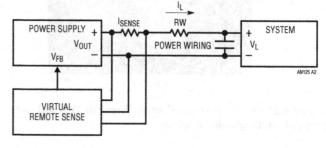

Figure A2 • Virtual Remote Sensing Eliminates Sense Wires

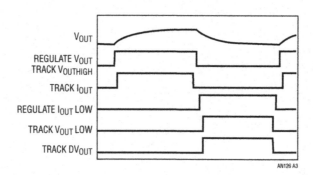

Figure A3 • Simplified Virtual Remote Sense Timing Diagram. State Machine Driven Sequence Samples and Stores Information Necessary to Set Appropriate Power Supply Voltage to Correct for Wiring Losses

Appendix B
Design guidelines for LT4180 VRS circuits

Introduction

The LT4180 is designed to interface with a variety of power supplies and regulators having either an external feedback or control pin. In Figure B1, the regulator error amplifier (which is a g_m amplifier) is disabled by tying its inverting input to ground. This converts the error amplifier into a constant-current source which is then controlled by the drain pin of the LT4180. This is the preferred method of interfacing because it eliminates the regulator error amplifier from the control loop which simplifies compensation and provides best control loop response.

For proper operation, increasing control voltage should correspond to increasing regulator output. For example, in the case of a current mode switching power supply, the control pin ITH should produce higher peak currents as the ITH pin voltage is made more positive.

Isolated power supplies and regulators may also be used by adding an opto-coupler (Figure B2). LT4180 output voltage INTV$_{CC}$ supplies power to the opto-coupler LED. In situations where the control pin V$_C$ of the regulator may exceed 5V a cascode may be added to keep the DRAIN pin of the LT4180 below 5V (Figure B3). Use a Low VT MOSFET for the cascode transistor.

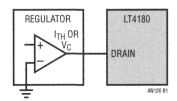

Figure B1 • Nonisolated Regulator Interface

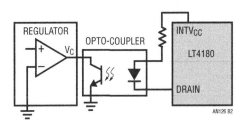

Figure B2 • Isolated Power Supply Interface

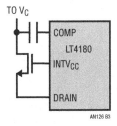

Figure B3 • Cascoded DRAIN Pin for Isolated Supplies

Design procedure

The first step in the design procedure (Figure B4) is to determine whether the LT4180 will control a linear or switching supply/regulator. If using a switching power supply or regulator, it is recommended that the supply be synchronized to the LT4180 by connecting the OSC pin to the SYNC pin (or equivalent) of the supply.

If the power supply is synchronized to the LT4180, the power supply switching frequency is determined by:

$$f_{OSC} = \frac{4}{R_{OSC} \cdot C_{OSC}}$$

Recommended values for R_{OSC} are between 20k and 100k (with 30.1k the optimum for best accuracy) and greater than 100pF for C_{OSC}. C_{OSC} may be reduced to as low as 50pF, but oscillator frequency accuracy will be somewhat degraded.

The following example synchronizes a 250kHz switching power supply to the LT4180. In this example, start with $R_{OSC} = 30.1k$:

$$C_{OSC} = \frac{4}{250kHz \cdot 30.1k} = 531pF$$

This example uses 470 pF. For 250 kHz:

$$R_{OSC} = \frac{4}{250kHz \cdot 470pF} = 34.04k$$

The closest standard 1% value is 34 k.

The next step is to determine the highest practical dither frequency. This may be limited either by the response time of the power supply or regulator, or by the propagation time of the wiring connecting the load to the power supply or regulator.

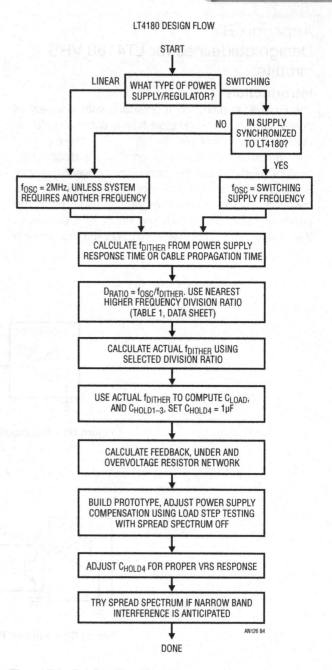

Figure B4 • Design Flow Chart

First determine the settling time (to 1% of final value) of the power supply. The settling time should be the worst-case value (over the whole operating envelope: V_{IN}, I_{LOAD}, etc.).

$$F1 = \frac{1}{2 \cdot t_{SETTLING}} Hz$$

For example, if the power supply takes 1 ms to settle (worst-case) to within 1% of final value:

$$F1 = \frac{1}{2 \cdot 1e-3} = 500Hz$$

Next, determine the propagation time of the wiring. In order to ignore transmission line effects, the dither period should be approximately twenty times longer than this. This will limit dither frequency to:

$$F2 = \frac{V_F}{20 \cdot 1.017ns/ft \cdot L} Hz$$

where V_F is the velocity factor (or velocity of propagation), and L is the length of the wiring (in feet).

For example, assume the load is connected to a power supply with 1000 ft of CAT5 cable. Nominal velocity of propagation is approximately 70%.

$$F2 = \frac{0.7}{20 \cdot 1.017e-9 \cdot 1000} = 34.4kHz$$

The maximum dither frequency should not exceed F1 or F2 (whichever is less):

$$f_{DITHER} < min\ (F1, F2).$$

Continuing this example, the dither frequency should be less than 500Hz (limited by the power supply).

With the dither frequency known, the division ratio can be determined:

$$D_{RATIO} = \frac{f_{OSC}}{f_{DITHER}} = \frac{250,000}{500} = 500$$

The nearest division ratio is 512 (set DIV0 = L, DIV1 = DIV2 = H). Based on this division ratio, nominal dither frequency will be:

$$f_{DITHER} = \frac{f_{OSC}}{D_{RATIO}} = \frac{250,000}{512} = 488Hz$$

After the dither frequency is determined, the minimum load decoupling capacitor can be determined. This load capacitor must be sufficiently large to filter out the dither signal at the load.

$$C_{LOAD} = \frac{2.2}{R_{WIRE} \cdot 2 \cdot f_{DITHER}}$$

where C_{LOAD} is the minimum load decoupling capacitance, R_{WIRE} is the minimum wiring resistance of one conductor of the wiring pair, and f_{DITHER} is the minimum dither frequency.

Continuing the example, our CAT5 cable has a maximum $9.38\Omega/100m$ conductor resistance.

Maximum wiring resistance is:

$$R_{WIRE} = 2 \cdot 1000\ ft \cdot 0.305\ m/ft \cdot 0.0938\Omega/m$$
$$R_{WIRE} = 57.2\Omega$$

With an oscillator tolerance of ±15%, the minimum dither frequency is 414.8 Hz, so the minimum decoupling capacitance is:

$$C_{LOAD} = \frac{2.2}{57.2\Omega \cdot 2 \cdot 414.8Hz} = 46.36\mu F$$

This is the minimum value. Select a nominal value to account for all factors which could reduce the nominal, such as initial tolerance, voltage and temperature coefficients and aging.

CHOLD capacitor selection and compensation

With dither frequency determined, use the following equations to determine CHOLD values:

$$C_{HOLD1} = \frac{11.9nF}{f_{DITHER}\ (kHz)}$$

and

$$C_{HOLD2} = C_{HOLD3} = \frac{2.5nF}{f_{DITHER}\ (kHz)}$$

So, with a dither frequency of 488 Hz:

$$C_{HOLD1} = \frac{11.9\,nF}{0.488kHz} = 24.4nF$$

and

$$C_{HOLD2} = C_{HOLD3} = \frac{2.5nF}{0.488(kHz)} = 5.12nF$$

NPO ceramic or other capacitors with low leakage and dielectric absorption should be used for all hold capacitors.

Set C_{HOLD4} to 1μF

Start with a 47pF capacitor between the COMP and DRAIN pins of the LT4180. Add an RC network in parallel with the 47pF capacitor. 10k and 10nF are good starting values. Connect a DC load corresponding to full-scale load current and verify that V_{OUT} produces a rounded squarewave without any noticeable overshoot or ringing (similar to the V_{OUT} waveform in Figure 4.16). If overshoot or ringing is observed, decrease the value of the resistor until it just disappears. If overshoot or ringing is not observed, increase the value of the resistor until it is observed, then slightly decrease the value of the resistor so that overshoot and ringing disappear. Check for proper voltage drop correction and converter behavior (start-up, regulation etc.), over the load range, and repeat the above procedure with a smaller value of the compensation capacitor, if necessary. Decrease C_{HOLD4} capacitance until V_{OUT} exhibits slight low frequency instability, then increase C_{HOLD4} slightly from this value.

Setting output voltage, undervoltage and overvoltage thresholds

The RUN pin has accurate rising and falling thresholds which may be used to determine when Virtual Remote Sense operation begins. Undervoltage threshold should never be set lower than the minimum operating voltage of the LT4180 (3.1V).

The overvoltage threshold should be set slightly greater than the highest voltage which will be produced by the power supply or regulator:

$$V_{OUT(MAX)} = V_{LOAD(MAX)} + V_{WIRE(MAX)}$$

$V_{OUT(MAX)}$ should never exceed $1.5 \cdot V_{LOAD}$

Since the RUN and OV pins connect to MOSFET input comparators, input bias currents are negligible and a common voltage divider can be used to set both thresholds (Figure B5).

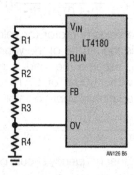

Figure B5 • Voltage Divider for UVL and OVL

The voltage divider resistors can be calculated from the following equations:

$$R_T = \frac{V_{OV}}{200\mu A}, \quad R4 = \frac{1.22\,V}{200\mu A}$$

where R_T is the total divider resistance and V_{OV} is the overvoltage set point.

Find the equivalent series resistance for R2 and R3 (R_{SERIES}). This resistance will determine the RUN voltage level.

$$R_{SERIES} = \left(\frac{1.22 \cdot R_T}{V_{UVL}}\right) - R4$$

$$R1 = R_T - R_{SERIES} - R4$$

$$R3 = \frac{1.22V - \left(V_{OUT(NOM)} \cdot \dfrac{R4}{R_T}\right)}{\dfrac{V_{OUT(NOM)}}{R_T}}$$

$$R2 = R_{SERIES} - R3$$

where V_{UVL} is the RUN voltage and $V_{OUT(NOM)}$ is the nominal output voltage desired.

For example, with $V_{UVL} = 4V$, $V_{OV} = 7.5V$ and $V_{OUT(NOM)} = 5V$,

$$R_T = \frac{7.5V}{200\mu A} = 37.5k$$

$$R4 = \frac{1.22V}{200\mu A} = 6.1k$$

$$R_{SERIES} = \left(\frac{1.22V \cdot 37.5k}{4V}\right) - 6.1k = 5.34k$$

$$R1 = 37.5k - 5.34k - 6.1k = 26.6k$$

$$R3 = \frac{1.22V - \left(\dfrac{5V \cdot 6.1k}{37.5k}\right)}{\dfrac{5V}{37.5k}} = 3.05k$$

$$R2 = R_{SERIES} - R3 = 2.29k$$

R_{SENSE} selection

Select the value of R_{SENSE} so that it produces a 100 mV voltage drop at maximum load current. For best accuracy, V_{IN} and SENSE should be Kelvin connected to this resistor.

Soft-correct operation

The LT4180 has a soft-correct function which insures orderly start-up (Figure B6). When the RUN pin rising threshold is first exceeded (indicating V_{IN} has crossed its undervoltage lockout threshold), power supply output voltage is set to a value corresponding to zero wiring voltage drop (no correction for wiring). Over a period of time (determined by C_{HOLD4}), the power supply output voltage ramps up to account for wiring voltage drops, providing best load-end voltage regulation. A new soft-correct cycle is also initiated whenever an overvoltage condition occurs.

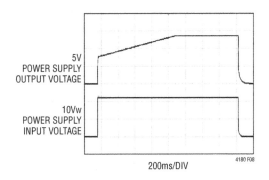

5V
POWER SUPPLY
OUTPUT VOLTAGE

10Vw
POWER SUPPLY
INPUT VOLTAGE

200ms/DIV

4180 F08

Figure B6 • Soft-Correct Operation, $C_{HOLD4} = 1\mu F$

Using guard rings

The LT4180 includes a total of four track/holds in the Virtual Remote Sense path. For best accuracy, all leakage sources on the CHOLD pins should be minimized.

At very low dither frequencies, the circuit board layout may include guard rings which should be tied to their respective guard ring drivers.

To better understand the purpose of guard rings, a simplified model of hold capacitor leakage (with and without guard rings) is shown in Figure B7. Without guard rings, a large difference voltage may exist between the hold capacitor (Pin 1) node and adjacent conductors (Pin 2) producing substantial leakage current through the leakage resistance (R_{LKG}). By adding a guard ring driver with approximately the same voltage as the voltage on the hold capacitor node, the difference voltage across R_{LKG1} is reduced substantially thereby reducing leakage current on the hold capacitor.

Synchronization

Linear and switching power supplies and regulators may be used with the LT4180. In most applications regulator interference should be negligible. For those applications where accurate control of interference spectrum is desirable, an oscillator output has been provided so that switching supplies may be synchronized to the LT4180 (Figure B8). The OSC pin was

designed so that it may directly connect to most regulators, or drive opto-isolators (for isolated power supplies).

Spread spectrum operation

Virtual remote sensing relies on sampling techniques. Because switching power supplies are commonly used, the LT4180 uses a variety of techniques to minimize potential interference (in the form of beat notes which may occur between the dither frequency and power supply switching frequency). Besides several types of internal filtering, and the option for VRS/power supply synchronization, the LT4180 also provides spread spectrum operation.

By enabling spread spectrum operation, low modulation index pseudo-random phasing is applied to Virtual Remote Sense timing. This has the effect of converting any remaining narrow-band interference into broadband noise, reducing its effect.

Increasing voltage correction range

Correction range may be slightly improved by regulating $INTV_{CC}$ to 5V. This may be done by placing an LDO between V_{IN} and $INTV_{CC}$. Contact Linear Technology Applications for more information.

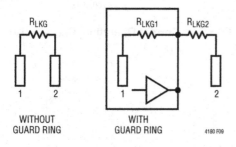

Figure B7 • Simplified Leakage Models (with and without Guard Rings)

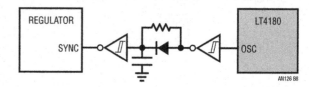

Figure B8 • Clock Interface for Synchronization

Section 2

Switching Regulator Design

Switching regulators: theory and practice (5)

This design manual is an extensive discussion of all standard switching configurations for the LT1070, including buck, boost, flyback, forward, inverting and "Cuk." The manual includes comprehensive information on the LT1070, the external components used with it, and complete formulas for calculating component values.

Switching regulators for poets (6)

Subtitled "A gentle guide for the trepidatious," this is a tutorial on switching regulator design. The text assumes no switching regulator design experience, contains no equations, and requires no inductor construction to build the circuits described. Designs detailed include flyback, isolated telecom, off-line, and others. Appended sections cover component considerations, measurement techniques and steps involved in developing a working circuit.

Step-down switching regulators (7)

Discusses the LT1074, an easily applied step-down regulator IC. Basic concepts and circuits are described along with more sophisticated applications. Six appended sections cover LT1074 circuitry detail, inductor and discrete component selection, current measuring techniques, efficiency considerations and other topics.

A monolithic switching regulator with 100µV output noise (8)

This publication details circuitry and applications considerations for the LT1533 low noise switching regulator. Eleven DC/DC converter circuits are presented, some offering <100µV output noise in a 100MHz bandwidth. Tutorial sections detail low noise DC/DC design, measurement, probing and layout techniques, and magnetics selection.

Powering complex FPGA-based systems using highly integrated DC/DC µModule regulator systems (9)

In a recent discussion with a system designer, the requirement for his power supply was to regulate 1.5V and deliver up to 40A of current to a load that consisted of four FPGAs. This is up to 60W of power that must be delivered in a small area with the lowest profile (height) possible to allow a steady flow of air for cooling. The power supply had to be surface mountable and operate at high enough efficiency to minimize heat dissipation. He also demanded the simplest possible solution so his time could be dedicated to the more complex tasks. Aside from precise electrical performance, this solution had to remove the heat generated during DC to DC conversion quickly so that the circuit and the ICs in the vicinity do not overheat. Such a solution requires an innovative design to meet these criteria:
1. Very low profile to allow efficient air flow and to prevent thermal shadow on surrounding ICs
2. High efficiency to minimize heat dissipation
3. Current sharing capability to spread the heat evenly to eliminate hot spots and minimize or eliminate the need for heat sinks
4. Complete DC/DC circuit in a surface mount package that includes the DC/DC controller, MOSFETs, inductor, capacitors and compensation circuitry for a quick and easy solution

Powering complex FPGA-based systems using highly integrated DC/DC µModule regulator systems (10)

In a previous application note, we discussed the circuit and electrical performance of a compact and low profile 48A,

1.5V DC/DC regulator solution for a four-FPGA design. The new approach uses four DC/DC μModule® regulators in parallel to increase output current while sharing the current equally among each device. This solution relies on the accurate current sharing of these μModule regulators to prevent hot spots by dissipating the heat evenly over a compact surface area. Each DC/DC μModule is a complete power supply with onboard inductor, DC/DC controller, MOSFETs, compensation circuitry and input/output bypass capacitors. It occupies only 15mm×15mm of board area and has a low profile (height) of only 2.8mm. This low profile allows air to flow smoothly over the entire circuit. Moreover, this solution casts no thermal shadow on its surrounding components, further assisting in optimizing thermal performance of the entire system.

Diode turn-on time induced failures in switching regulators (11)

Most circuit designers are familiar with diode dynamic characteristics such as charge storage, voltage dependent capacitance and reverse recovery time. Less commonly acknowledged and manufacturer specified is diode forward turn-on time. This parameter describes the time required for a diode to turn on and clamp at its forward voltage drop. Historically, this extremely short time, units of nanoseconds, has been so small that user and vendor alike have essentially ignored it. It is rarely discussed and almost never specified. Recently, switching regulator clock rate and transition time have become faster, making diode turn-on time a critical issue.

5

LT1070 design manual

Carl Nelson

Introduction

Three terminal monolithic linear voltage regulators appeared almost 20 years ago, and were almost immediately successful for a variety of reasons. In particular, there were relatively few engineers capable of designing a good linear voltage regulator. The new devices were also easy to use, and inexpensive. In currently popular parlance they were "expert systems," containing a good deal of their designer's knowledge in silicon form. Because of these advantages, the regulators quickly eclipsed discrete and earlier monolithic building blocks and dominated the market.

More recently, there has been increasing interest in switching-based regulators. Switching regulators, with their high efficiency and small size, are increasingly desirable as overall package sizes have shrunk. Unfortunately, switching regulators are also one of the most difficult linear circuits to design. Mysterious modes, sudden failures, peculiar regulation characteristics and just plain explosions are common occurrences during the design of a switching regulator.

Most switching regulator ICs are building blocks. Many discrete components are required, and substantial expertise is assumed on the part of the user. Some newer devices include the power switch on the die, but still require a significant amount of engineering to apply.

Finally, there has been a notable lack of comprehensive and practical application literature support from manufacturers.

These considerations are reminiscent of the state of linear regulator design when the first three terminal monolithic regulators appeared. Given this historical lesson, the LT®1070 five terminal switching regulator has been designed for ease of use and economy. It does not require the user to be well-schooled in switching regulator design, and is versatile enough to be used in all the popular switching regulator configurations. To obtain maximum user benefit, a significant applications effort has been associated with this part. This chapter covers both ancillary tutorial material as well as direct operating considerations for the part. It is intended to be used "as required." For those in a mission-oriented hurry, much of the discussion can be ignored, and breadboards constructed with a high probability of success. The more academically inclined reader may choose to peruse the material more carefully. Either approach is valid and the chapter is intended to satisfy both.

— Jim Williams

Analog Circuit and System Design: A Tutorial Guide to Applications and Solutions. DOI: 10.1016/B978-0-12-385185-7.00005-6

Preface

Smaller versions of the LT1070

Since this application note was written, several new versions of the LT1070 have been developed. The LT1071 and LT1072 are identical to the LT1070 except for switch current ratings, 2.5A and 1.25A, respectively. Designs which result in lower switch currents can take advantage of the cost savings of these smaller chips. Design equations for the LT1071 and LT1072 are identical to the LT1070 with the following exceptions:

Peak Switch Current (I_P)	= 5A	LT1070
	= 2.5A	LT1071
	= 1.25A	LT1072
Switch "On" Resistance (R)	≈0.2Ω	LT1070
	≈0.4Ω	LT1071
	≈0.8Ω	LT1072
V_C Pin to Switch Current Transconductance	≈8A/V	LT1070
	≈4A/V	LT1071
	≈2A/V	LT1072

Also available are 100kHz versions of the LT1070/ LT1071/LT1072.

Inductance calculations

Feedback from readers of this chapter shows that there is confusion about the use of ΔI to calculate inductance values. ΔI is the *change* in inductor or primary current during switch "on" time, and the suggested value is approximately 20% of the peak current rating of the LT1070 switch (5A), or in some cases, 20% of the average inductor current. This 20% rule-of-thumb is designed to give near maximum output power for a given switch current rating. If maximum output power is not needed, much smaller inductors/transformers may be used by allowing ΔI to increase. The design approach is to calculate peak inductor/switch current (I_P) using the formulas provided in this chapter, with $L = \infty$.

Then compare this current to the peak switch current. The difference is the "room" allowable for ΔI:

$$\Delta I_{MAX} = 2\left(I_{SWITCH\,(PEAK)} - I_P\right)$$

This formula assumes continuous mode operation. If ΔI, as calculated by this formula, exceeds I_P, it may be possible to go to discontinuous mode operation, with further reductions in inductance. Discontinuous mode requires higher switch currents and not all these chapter topologies show design equations for this mode, but it should definitely be considered for very low output powers or where inductor/ transformer size is critical. All topologies work well in discontinuous mode with the exception of fully isolated flyback. Drawbacks of discontinuous mode include higher output ripple and slightly lower efficiency.

Example 1: Negative buck converter with $V_{IN} = -24V$, $V_{OUT} = -5V$ and $I_{OUT} = 1.5A$,

$$I_P \text{ (Equation 37)} = I_{OUT} + \frac{(V_{IN} - V_{OUT})(V_{OUT})}{2 \bullet V_{IN} \bullet f \bullet (L \approx \infty)}$$
$$= I_{OUT} = 1.5A$$

$$\Delta I_{MAX} = 2\left(I_{SW} - I_P\right) = 2\left(5 - 1.5\right) = 7A \text{ (LT1070)}$$
$$= 2\left(2.5 - 1.5\right) = 2A \text{ 9 (LT1071)}$$
$$= 2\left(1.25 - 1.5\right) = N.A. \text{ (LT1072)}$$

The LT1072 is too small ($I_P > I_{SW}$), so select the LT1071, which yields a maximum ΔI of 2A. A conservative value of actual ΔI is selected at 1A. This allows room for efficiency losses and variations in component values. **Using Equation 37:**

$$L = \frac{(V_{IN} - V_{OUT})(V_{OUT})}{V_{IN}\,(\Delta I) \bullet f} = \frac{(24 - 5)(5)}{24(1) \bullet 40k} = 99\mu H$$

Example 2: Flyback converter with $V_{IN} = 6V$, $V_{OUT} = \pm 15V$ at 35mA and 5V at 0.2A, N = 0.4 (primary to 5V secondary). For calculations, the entire output power of 2.05W is referred to the 5V secondary, yielding one value for N(0.4), V_{OUT} (5V) and $I_{OUT} = 0.41A$.

Using Equation 79:

$$I_P = \frac{I_{OUT}}{E}\left(\frac{V_{OUT}}{V_{IN}} + N\right) + \frac{(V_{IN})(V_{OUT})}{2 \bullet f\,(V_{OUT} + N\,V_{IN})(L = \infty)}$$
$$= \frac{0.41A}{0.75}\left(\frac{5V}{6V} + 0.4\right) = 0.674A$$

The LT1072 is large enough to handle this current, yielding:

$$\Delta I_{MAX} = 2(1.25A - 0.674A) = 1.15A$$

Using a conservative value of 0.7A for ΔI (note that this is 56% of the 1.25A Max LT1072 switch current, not 20%), and Equation 77, yields:

$$L = \frac{(V_{IN})(V_{OUT})}{\Delta I \bullet f(V_{OUT} + N\,V_{IN})} = \frac{(6)(5)}{(0.7)(40k)(5 + 0.4 \bullet 6)} = 145\mu H$$

Protecting the magnetics

A second problem for LT1070 designers has been protection of the magnetics under overload or short-circuit conditions. Physical size restraints often require inductors or transformers which are not specified to handle the full current limit values of the LT1070. This problem can be handled in several ways.

1. Use an LT1071 or LT1072 if full load current requirements allow it.
2. Take advantage of the fact that the LT1070 current limit *drops* at higher temperatures. The worst-case current limit values shown on the old data sheets allow for both temperature extremes with one specification. New data sheets specify a maximum of 10A for the LT1070, 5A for the LT1071 and 2.5A for the LT1072 at temperatures of 25°C or higher. Be aware that the temperature dependence of current limit has been improved considerably on the LT1070 since the original data sheet was printed. The old value was greater than $-0.3\%/°C$, while the new figure is under $-0.1\%/°C$. The current limit graphs on the new data sheets reflect this improved characteristic.
3. Reconsider the necessity of limiting the inductor/transformer current to the manufacturers' specification. Maximum current ratings in many cases are determined by core saturation considerations. Allowing the core to saturate does *not* harm the core. Core or winding damage occurs only if temperatures rise so far that material properties are permanently altered. Core saturation used to be considered a "fatal" condition for conventional switchers because currents would "run away" and destroy switches or diodes. The LT1070 limits current on an instantaneous cycle-by-cycle basis, preventing current "run away" even with grossly overdriven cores. The major consideration then is the heating effect of the winding current (I^2R). Under short-circuit conditions, winding currents in inductors are nearly constant at the current limit value of the LT1070. Transformer *secondary* winding currents are nearly constant at $1/N$ times the LT1070 current limit. This assumes that the core is not heavily saturated. If the core saturates significantly below the current limit values, RMS winding current will be significantly *lower* than the current limit. The best way to resolve this complex situation is to actually measure core/winding temperature with a thermocouple under overload conditions. The thermocouple should be "buried" as deeply as possible in the windings and/or core to reflect peak temperatures. The magnetic and electric fields generated by the switching may affect the thermocouple meter. If this occurs, simply check the temperature periodically by turning off power. Consult with the magnetics manufacturer to determine peak allowable temperatures, with permanent damage as

the criteria, not performance specifications. The major failure mode is winding shorts caused by insulation melting. High temperature insulation is available from most manufacturers.

New switch current specification

The LT1070 was specified at 5A peak switch current, for duty cycles of 50% or less. At higher duty cycles the peak current was limited to 4A. This abrupt change in specification at 50% duty cycle was bothersome because many designs operate near 50% duty cycle and require maximum possible output power. To solve this problem, switch current limits are specified as a linearly decreasing function, from 5A at 50% duty cycle to 4A at 80% duty cycle. The LT1071 and LT1072 are also specified this way.

High supply voltages

It has become apparent that many applications for the LT1070 have maximum input voltages which exceed 40V. The straightforward approach is to use the "HV" devices that are specified at 60V, but in some cases the standard part can be used at lower cost simply by dropping supply voltage with a Zener diode as shown. The LT1070 supply pin (V_{IN}) requires only a few volts to operate, so in most cases the unregulated input voltage range is not compromised with this Zener. Zener dissipation can be calculated from $I_Z \approx 6mA + I_{SW} (0.0015 + DC/40)$:

$I_{SW} =$ LT1070 average switch current during "on" time
$DC =$ duty cycle

For $I_{SW} = 4A$, $DC = 30\%$; $I_Z = 42\,mA$

A 20V Zener would dissipate $(20)(42) = 0.84W$. Note that this power would be dissipated anyway in the LT1070, so no loss in efficiency occurs. The resistor, R_Z, is necessary for start-up. Without it, a latch-off condition exists where the V_{IN} pin sits more than 16V negative with respect to the switch pin. If the LT1070 is not switching and the FB pin is below 0.5V, the LT1070 is in the "isolated flyback" mode where it is trying to regulate the V_{IN}-to-V_{SW} voltage. When this voltage exceeds 16V, the regulator thinks it should reduce duty cycle to zero, resulting in a permanent "no-switching" state. R_Z forces the V_{IN} pin to rise enough to initiate start-up. The user need not be concerned that the V_{IN}-to-ground pin voltage exceeds 40V during this state because R_Z is too large to allow harmful currents to flow.

Some attention needs to be paid to C_Z. The LT1070 is very tolerant of noise and ripple on the V_{IN} pin, but C_Z

may be necessary in some applications. The problem is that D1 must charge C_Z when power is applied. If power comes up very rapidly, D1 might exceed its one cycle surge rating.

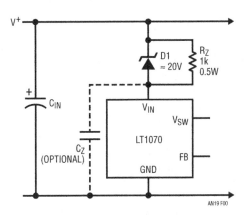

AN19 F00

Discontinuous "oscillations" (ringing)

Many customers have called about oscillations occurring on the switch pin during a portion of the switch "off" time. These are not oscillations. They are a damped ringing caused by the transition to a zero-current state in the inductor or transformer primary. At light loads, or with low inductance values, inductor current will drop to zero during switch off time. This causes the inductor voltage to collapse toward zero. In doing so, however, energy is transferred back to the inductor from the parasitic capacitance of the switch, inductor, and catch diode. The inductor and capacitance form a parallel resonant tank which "rings." This ringing is not harmful as long as its peak amplitude does not result in a negative voltage on the switch pin. It can be damped, if desired, by paralleling the inductor/primary with a series R/C damper, typically 100Ω to $1k\Omega$, and 500pF to 5000pF. Typical undamped ringing frequency is 100kHz to 1MHz.

LT1070 operation

The LT1070 is a current mode switcher. This means that switch duty cycle is directly controlled by switch current rather than by output voltage. Referring to the block diagram, the switch is turned "on" at the start of each oscillator cycle. It is turned "off" when switch current reaches a predetermined level. Control of output voltage is obtained by using the output of a voltage-sensing error amplifier to set current trip level. This

technique has several advantages. First, it has immediate response to input voltage variations, unlike ordinary switchers which have notoriously poor line transient response. Second, it reduces the 90° phase shift at midfrequencies in the energy storage inductor. This greatly simplifies closed-loop frequency compensation under widely varying input voltage or output load conditions. Finally, it allows simple pulse-by-pulse current limiting to provide maximum switch protection under output overload or short conditions. A low dropout internal regulator provides a 2.3V supply for all internal circuitry on the LT1070. This low dropout design allows input voltage to vary from 3V to 60V with virtually no change in device performance. A 40kHz oscillator is the basic clock for all internal timing. It turns "on" the output switch via the logic and driver circuitry. Special adaptive antisat circuitry detects onset of saturation in the power switch and adjusts driver current instantaneously to limit switch saturation. This minimizes driver dissipation and provides very rapid turn-off of the switch.

A 1.2V bandgap reference biases the positive input of the error amplifier. The negative input is brought out for output voltage sensing. This feedback pin has a second function; when pulled low with an external resistor, it programs the LT1070 to disconnect the main error amplifier output and connects the output of the flyback amplifier to the comparator input. The LT1070 will then regulate the value of the flyback pulse with respect to the supply voltage. This flyback pulse is directly proportional to output voltage in the traditional transformer coupled flyback topology regulator. By regulating the amplitude of the flyback pulse, the output voltage can be regulated with no direct connection between input and output. The output is fully floating up to the breakdown voltage of the transformer windings. Multiple floating outputs are easily obtained with additional windings. A special delay network inside the LT1070 ignores the leakage inductance spike at the leading edge of the flyback pulse to improve output regulation.

The error signal developed at the comparator input is brought out externally. This pin (V_C) has four different functions. It is used for frequency compensation, current limit adjustment, soft starting and total regulator shutdown. During normal regulator operation this pin sits at a voltage between 0.9V (low output current) and 2V (high output current). The error amplifiers are current output (g_m) types, so this voltage can be externally clamped for adjusting current limit. Likewise, a capacitor coupled external clamp will provide soft start. Switch duty cycle goes to zero if the V_C pin is pulled to ground through a diode, placing the LT1070 in an idle mode. Pulling the V_C pin below 0.15V causes total regulator shutdown, with only 50μA supply current for shutdown circuitry biasing.

Block Diagram

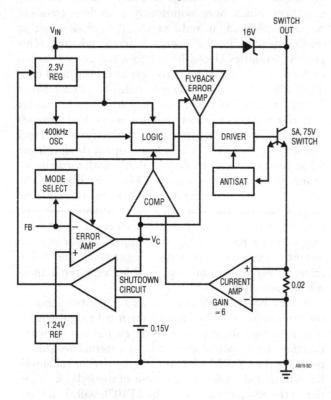

Pin functions

Input supply (V_{IN})

The LT1070 is designed to operate with input voltages from 3V to 40V (standard) or 60V (HV units). Supply current is essentially flat over this range at about 6mA (with zero output current). With increasing switch current, the supply current (during switch on-time) increases at a rate approximately 1/40 of switch current, corresponding to a forced h_{FE} of 40 for the switch.

Undervoltage lockout is incorporated on the LT1070 by sensing saturation of the lateral PNP pass transistor which drives an internal 2.3V regulator. A remote collector on this transistor conducts current and locks out the switch for input voltages below 2.5V. No hysteresis is used to maximize the useful range of input voltage. Operating the regulator right at the 2.5V threshold may result in a "burping" action as the LT1070 turns on and off in response to wobbles in input voltage, but this will not harm the device. External undervoltage lockout can be added if it is desirable to raise the threshold voltage. The circuit shown in Figure 5.1 is one example of how to implement this.

The threshold of this circuit is approximately $V_Z + 1.5V$. Below that voltage, D2 pulls the V_C pin low to shut off the regulator.

Ground pin

The ground pin (case) of the LT1070 is important because it acts as both the negative sense point for the internal error amplifier and as the high current path for the 5A switch. This is not normally good design practice, but was necessary in a 5-pin package configuration. *To avoid degradation of load regulation, Kelvin connections should be made to the ground pin.* This is done on the TO-3 package by tying one end of the package to power ground and the other end to the feedback divider resistor (analog ground). This is illustrated in Figure 5.2.

For best load regulation, the resistance in the switch current path must be kept low. 0.01Ω of wire resistance creates 50mV drop at 5A switch current. This is a 1% change in a 5V output, and actually causes the output to *increase* with increasing load current.

With the TO-220 package, (Figure 5.3) connect the feedback resistor directly to the ground pin with a separate wire if no case connection is made. The case can be used as a second ground pin if desired.

Avoid long wire runs to the ground pin to minimize load regulation effects and inductive voltages created by the high di/dt switch current. A ground plane will keep EMI to a minimum.

Feedback pin

The feedback pin is the inverting input to a single stage error amplifier. The noninverting input to this amplifier is internally tied to a 1.244V reference as shown in Figure 5.4.

Input bias current of the amplifier is typically 350nA with the output of the amplifier in its linear region. The amplifier is a g_m type, meaning that it has high output impedance with controlled voltage-to-current gain ($g_m \approx 4400\mu$mhos). DC voltage gain with no load is ≈ 800.

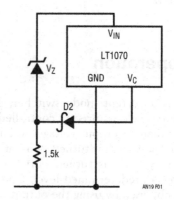

Figure 5.1 • External Undervoltage Lockout

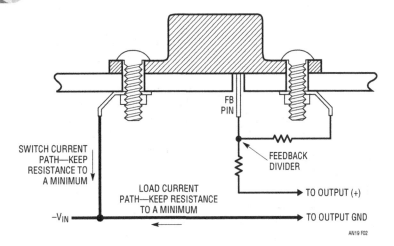

Figure 5.2

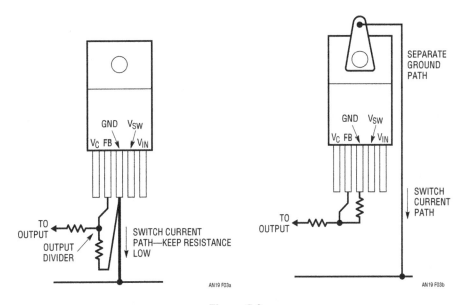

Figure 5.3

The feedback pin has a second function; it is used to program the LT1070 for normal or flyback-regulated operation (see description of block diagram). In Figure 5.4, Q53 is biased with a base voltage approximately 1V. This clamps the feedback pin to about 0.4V when current is drawn out of the pin. A current of ≈10µA or higher through Q53 forces the regulator to switch from normal operation to flyback mode, but this threshold current can vary from 3µA to 30µA. *The LT1070 is in flyback mode during normal start-up until the feedback pin rises above 0.45V.* The resistor divider used to set output voltage will draw current out of the feedback pin until the output voltage is up to about 33% of its regulated value.

If it is desired to run the LT1070 in the *fully isolated* flyback mode, a single resistor is tied from the feedback pin

to ground. The feedback pin then sits at a voltage of ≈0.4V for R = 8.2k. The actual voltage depends on resistor value since the feedback pin has about 200Ω output impedance in this mode. 500µA in the resistor will drop the feedback pin voltage from 0.4V to 0.3V. Minimum current through the resistor to guarantee flyback operation is 50µA. Actual resistor value is chosen to fine-trim flyback regulated voltage. (See discussion of isolated flyback mode operation and graphs of feedback pin characteristics.)

An internal 30Ω resistor and 5.6V Zener protect the feedback pin from overvoltage stress. Maximum transient voltage is ±15V. This high transient condition most commonly occurs during fast fall time output shorts if a feedforward capacitor is used around the feedback divider. If a feedforward capacitor is used for DC output voltages

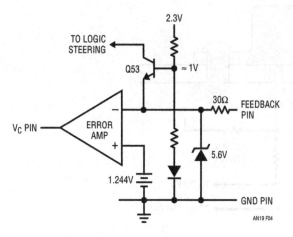

Figure 5.4

greater than 15V, a resistor equal to $V_{OUT}/20mA$ should be used between the divider node and the feedback pin as shown in Figure 5.5.

Keep in mind when using the LT1070 that the feedback pin reference voltage is referred to the ground pin of the regulator, and the ground pin can have switch currents exceeding 5A. Any resistance in the ground pin connection will degrade load regulation. Best regulation is obtained by tying the grounded end of the feedback divider directly to the ground pin of the LT1070, as a separate connection from the power ground. This limits output voltage errors to just the drop across the ground pin resistance instead of multiplying it by the feedback divider ratio. See discussion of ground pin.

Compensation pin (V_C)

The V_C pin is used for frequency compensation, current limiting, soft start and shutdown. It is the output of the error amplifier and the input of the current comparator. The error amplifier circuit is shown in Figure 5.6.

Q57 and Q58 form a differential input stage whose collector currents are inverted and multiplied times four

by Q55 and Q56. Q55 current is further inverted by Q60 and Q61 to generate a current fed balanced output which can swing from the 2.3V rail down to a clamp level of $\approx 0.4V$ as set by R21 and Q62. The $60\mu A$ tail current of the input transistors sets the g_m of the error amplifier at $4400\mu mhos$. Voltage gain with no load is limited by transistor output impedance at ≈ 800. Maximum source and sink current is $\approx 220\mu A$.

The voltage on the V_C pin determines the current level at which the output switch will turn off. For V_C voltage below 0.9V (at 25°C), the output switch will be totally off (duty cycle = 0). Above 0.9V, the switch will turn on at each oscillator cycle, then turn off when switch current reaches a trip level set by V_C voltage. This trip level is zero at $V_C = 0.9V$, and increases to about 9A when V_C reaches its upper clamp level of 2V. These numbers are based on a duty cycle of 10%. Above 10%, switch turn-off is a function of both switch current *and* time. The time dependence is caused by a small ramp fed into the current amplifier input. This ramp starts at $\approx 40\%$ duty cycle, and is the source of the bend in the V_C vs duty cycle graph shown in Figure 5.7. This ramp is used to prevent a phenomenon peculiar to "current mode" switching regulators known as *subharmonic oscillation*. See section on subharmonic oscillations for further details.

A second amplifier output is also tied to the V_C pin. This "flyback mode" amplifier is turned on only when current is drawn *out* of the feedback pin. This condition occurs during start-up in the normal mode until the feedback divider has raised the voltage at the feedback pin above 0.45V.

It is a permanent condition when the LT1070 is programmed for isolated flyback mode by tying a single resistor from the feedback pin to ground.

In the isolated flyback mode, S1 is closed and the feedback pin is low, totally disabling the main amplifier. S2 and

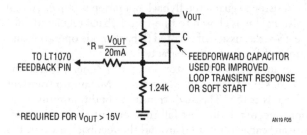

Figure 5.5

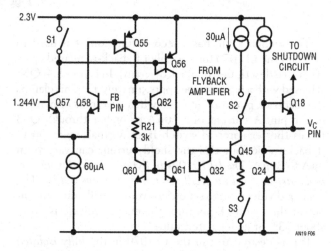

Figure 5.6 • Error Amplifier

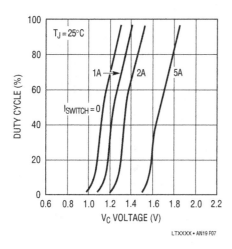

Figure 5.7 • Duty Cycle vs V$_C$ Voltage

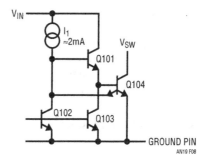

Figure 5.8

S3 are turned on only during the "off" state of the output power transistor and then, only after a 1.5μs delay following output transistor turn-off. This prevents transient flyback spikes from causing poor regulation. S2 current is fixed at 30μA. S3 current can rise to a maximum of ≈70μA, allowing the V$_C$ pin to source 30μA and sink 40μA in the flyback mode. g$_m$ of the flyback amplifier is typically 300μmho.

When the V$_C$ pin is externally pulled below 0.15V, a shutdown circuit is activated. Q24 and Q18 perform this function. Q24 is a special "high V$_{BE}$" diode whose forward voltage is about 150mV higher than Q18 V$_{BE}$. Pulling current out of Q18 activates shutdown and turn off all internal regulator functions except for a 50μA to 100μA trickle current needed to bias Q18 and Q24. See characteristic curves for details of the V/I properties of the V$_C$ pin in shutdown.

Loop frequency compensation can be performed with an RC network connected from the V$_C$ pin to ground. An optional compensation is to connect the RC network between the V$_C$ pin and the feedback pin. See loop frequency compensation section.

Output pin

The V$_{SW}$ pin of the LT1070 is the collector of the internal NPN power switch. This NPN has a typical on-resistance of 0.15Ω and a breakdown voltage (BV$_{CBO}$) of 85V. Very fast switching times and high efficiency are obtained by using a special driver loop which automatically adapts base drive current to the minimum required to keep the switch in a quasi-saturation state. This loop is shown in Figure 5.8.

Q104 is the power switch. Its base is driven by Q101, whose collector is returned to V$_{IN}$. Q101 is turned on and off by Q102. In parallel with Q102 is a second, larger transistor (Q103) which pulls high reverse base current out of Q104 for rapid switch turn-off. The key element

in the loop is the extra emitter on Q104. This emitter carries no current when Q104 collector is high (unsaturated). In this condition, the driver, Q101, can deliver very high base drive to the switch for fast turn-on. When the switch saturates, the extra emitter acts as a collector and pulls base current away from the driver. This linear feedback loop servos itself to keep the switch just at the edge of saturation. Very low switch currents result in near-zero driver current, and high switch currents automatically increase driver current as necessary. The ratio of switch current to driver current is approximately 40:1. This ratio is determined by the sizing of the extra emitter and the value of I$_1$. The quasi-saturation state of the switch permits rapid turn-off without the need for reverse base-emitter voltage drive.

Also tied to the V$_{SW}$ pin is the input circuitry for the flyback mode error amplifier as shown in Figure 5.9. This circuitry draws no current from the V$_{SW}$ pin when the switch pin is *less* than 16V above V$_{IN}$ because the diodes block current. When V$_{SW}$ is more than 16V above V$_{IN}$, ≈500μA is drawn out of the switch pin because the reference diodes (D1 and D2) and Q10 turn on. This 500μA current level is set by the ratio of collector areas on the 2-collector lateral PNP Q10 and the value of I$_2$. Q9 is reverse

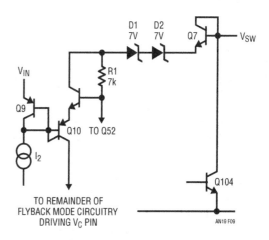

Figure 5.9

biased in this state. The 16V transition point sets the flyback mode reference voltage. The flyback reference voltage can be increased above 16V by drawing additional current through R1 via Q52. The amplitude of this current is determined by the size of the resistor tied to the feedback pin. See discussion in isolated flyback mode operation.

Basic switching regulator topologies

There are many possible switching regulator configurations, or "topologies." In any particular regulator requirement, the possible choices are narrowed somewhat by constraints of polarity, voltage ratio, and fault conditions (simple boost regulators cannot be current limited), but this may still leave the designer with several choices. To convert 28V to 5V, for instance, the list of possible topologies includes buck, flyback, forward and current boosted buck. The following discussion of topologies is limited to those which can be realized with the LT1070, but this covers nearly all the low to medium power DC/DC conversion requirements.

Buck converter

Figure 5.10a shows the basic buck topology. S1 and S2 open and close alternately so that the voltage applied to L1 is either V_{IN} or zero. DC output voltage is then the average voltage applied to L1. If t_1 is the time S1 is closed, and t_2 is the time it is open, V_{OUT} is equal to:

$$V_{OUT} = V_{IN}\frac{t_1}{t_1 + t_2} = (V_{IN})(DC) \qquad (1)$$

where, by convention, duty cycle (DC) is defined as the ratio of t_1 to $t_1 + t_2$:

$$DC = \frac{t_1}{t_1 + t_2} \qquad (2)$$

Note that the definition of duty cycle allows only for values between 0 and 1. The formula for V_{OUT} therefore shows a basic property of buck converters; the *output voltage is always less than the input voltage.*

This simple formula also tells much about switching regulators in general. The most important point is what is *not* in the equation, *and that includes L1, C1, frequency and load current. To a first approximation, the output voltage of a switching regulator depends only on the duty cycle of the switching network and input voltage.* This is a very important point which must be kept firmly in mind when analyzing switching regulators.

Diodes may be used to replace switches when unidirectional current flow exists. In Figures 5.10b and 5.10c, single-switch buck regulators are shown with diodes used to replace S2. Diodes cause some loss in efficiency, but simplify the design and reduce cost. Notice that when S1 is

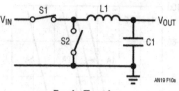

a. Basic Topology

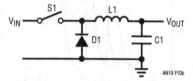

b. Positive Buck Using One Switch

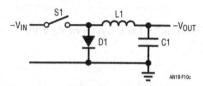

c. Negative Buck Using One Switch

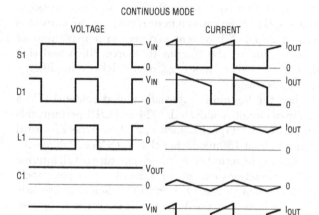

Figure 5.10 • Buck Converter

closed, D1 is reverse biased (off) and that when S1 opens, the current flow through L1 forces the diode to become forward biased (on). This duplicates the alternate switching action of two switches. There is an exception to this condition, however. If the load current is low enough, the current through L1 will drop to zero sometime during S1 off-time. This is known as *discontinuous mode operation.* Buck regulators will be in discontinuous mode for any load current less than:

$$I_{OUT} \leq \frac{V_{OUT}\left(1 - \frac{V_{OUT}}{V_{IN}}\right)}{(2)(f)(L1)} \qquad (3)$$

where f = switching frequency.

Discontinuous mode alters the original statement that output voltage depends only on input voltage and switch duty cycle because a third state of the switches now exists with diodes replacing S2; namely both switches off. Waveforms for voltage and current of S1, D1, L1, C1 and the input source are shown for both continuous and discontinuous modes of operation.

Normally it is not important to avoid discontinuous mode operation at light load currents. A possible exception to this would be when the "on" time of S1 cannot be reduced to a low enough value to prevent the lightly loaded output from drifting unregulated high. If this occurs, most switching regulators will begin "dropping cycles" wherein S1 does not turn on at all for one or more cycles. This mode of operation maintains control of the output, but the subharmonic frequencies generated may be unacceptable in certain situations.

A general property of "perfect" switching regulators is that they do not dissipate power in the process of converting one voltage or current to another; in other words, they are 100% efficient. This is to be expected from an inspection of Figure 5.10a: *there are no components which dissipate power;* only switches, inductors and capacitors. The following formula can then be stated;

$$P_{OUT} = P_{IN} \text{ or, } (I_{OUT})(V_{OUT}) = (I_{IN})(V_{IN}) \qquad (4)$$

and

$$I_{IN} = I_{OUT}\left(\frac{V_{OUT}}{V_{IN}}\right) \qquad (5)$$

This shows that the *average current drawn by the input of a switching regulator can be much higher or lower than the load current,* depending on the ratio of output-to-input voltage. If this simple fact is ignored, the designer may realize too late that his low voltage to high voltage converrter will draw more current from the low voltage supply than it is capable of handling.

Boost regulators

The basic boost regulator shown in Figure 5.11a has an output voltage given by:

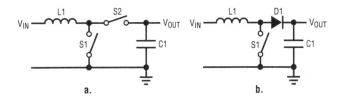

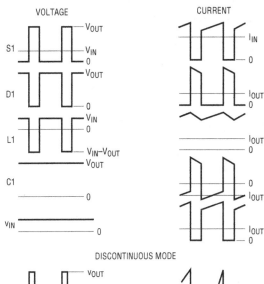

CONTINUOUS MODE

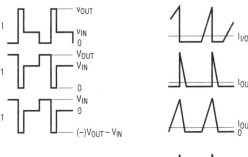

DISCONTINUOUS MODE

Figure 5.11 • Boost Regulators

$$V_{OUT} = \frac{V_{IN}}{1 - DC} \text{(continuous mode)} \qquad (6)$$

DC is duty cycle, the ratio of S1 "on" time to "off" time, assuming that S1 and S2 open and close alternately. Duty cycle can take on values only between 0 and 1; therefore, *the output voltage of a boost regulator is always higher than the input voltage.*

In Figure 5.11b, a diode has replaced S2 to realize a boost regulator with a single switch. The voltage and current waveforms for all the components including the source are shown, both for continuous and discontinuous mode. Note that the current drawn from the input and delivered in pulses to the load is significantly higher than the output load current. The amplitude of input current and peak switch and diode current is equal to:

$$I_P = I_{OUT}\frac{V_{OUT}}{V_{IN}} \text{ (continuous mode)} \qquad (7)$$

Average diode current is equal to I_{OUT} and *average* switch current is $I_{OUT}(V_{OUT} - V_{IN})/V_{IN}$, both of which are significantly less than peak current. *The switch, diode and output capacitor must be specified to handle the peak currents as well as average currents.* Discontinuous mode requires even higher ratios of switch current to output current.

One drawback of boost regulators is that they cannot be current limited for output shorts because the current steering diode, D1, makes a direct connection between input and output.

Combined buck-boost regulator

Buck-boost regulators (Figure 5.12) are used to generate an output with the reverse polarity of the input. They look similar to a boost regulator except that the load is referred to the inductor side of the input instead of the switch side. Buck-boost regulators have an output voltage given by:

$$V_{OUT} = -V_{IN}\left(\frac{DC}{1 - DC}\right) \qquad (8)$$

With duty cycle varying between 0 and 1, the output voltage can vary between zero and an infinitely high value. The current and voltage waveforms show that, like boost regulators, the peak switch, diode, and output capacitor currents can be significantly higher than output currents and these components must be sized accordingly.

$$\begin{aligned}I_{PEAK} &= \frac{I_{OUT}}{1 - DC} \\ &= I_{OUT}\frac{(V_{OUT} + V_{IN})}{V_{IN}} \text{ (continuous mode)} \qquad (9)\end{aligned}$$

Maximum switch voltage is equal to the sum of input plus output voltage. The forward turn-on time of D1 is therefore very important in higher voltage applications to prevent additional switch stress.

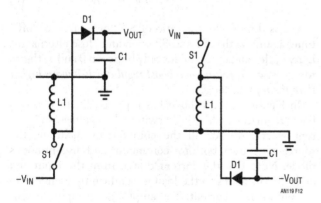

Figure 5.12 • Inverting Topology

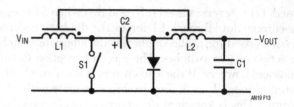

Figure 5.13 • 'Cuk Converter

'Cuk converter

The 'Cuk converter in Figure 5.13 is named after Slobodan 'Cuk, a professor at Cal Tech. It is like a buck-boost converter in that input and output polarities are reversed, but it has the advantage of low ripple current at both input and output. The optimum topology version of the 'Cuk converter eliminates the disadvantage of needing two inductors by winding them both on the same core, with exact 1:1 turns ratio. With slight adjustments to L1 or L2, *either* input ripple current *or* output ripple current can be forced to zero. An impvoved version even exists which results in *both* ripple currents going to zero. This considerably eases the requirements on size and quality of input and output capacitors without requiring filters.

The switch must handle the *sum* of input and output current:

$$I_{PEAK}(S1) = I_{IN} + I_{OUT} = I_{OUT}\left(1 + \frac{V_{OUT}}{V_{IN}}\right) \qquad (10)$$

The ripple current in C2 is equal to I_{OUT}, so this capacitor must be large. It can be electrolytic, however, so physical size is not normally a problem.

Flyback regulator

Flyback regulators (Figure 5.14) use a transformer to transfer energy from input to output. During S1 "on" time, energy builds up in the core due to increasing current in the primary winding. At this time, the polarity of the output winding is such that D1 is reverse biased. When S1 opens, the total stored energy is transferred to the secondary winding and current is delivered to the load. The turns ratio (N) of the transformer can be adjusted for optimum power transfer from input to output.

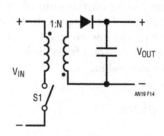

Figure 5.14 • Flyback Converter

Peak switch current in a flyback regulator is equal to:

$$I_{PEAK}(S1) = \frac{I_{OUT}(N V_{IN} + V_{OUT})}{V_{IN}} \text{(continuous mode)}$$

(10)

Notice that peak switch current can be reduced to a minimum by using a very small value for N. This has two negative consequences however; the switch voltage and diode current become very large during switch off time. For a given maximum switch voltage, optimum power transfer occurs at $V_{IN} = 1/2V_{MAX}$.

Both input ripple current *and* output ripple current are high in a flyback regulator, but this disadvantage is more than offset in many cases by the ability to achieve current or voltage gain and the inherent isolation afforded by the transformer. Output voltage is given by:

$$V_{OUT} = V_{IN} \bullet N \bullet \frac{DC}{1 - DC}$$

(11)

With any value of N, a duty cycle between 0 and 1 can be found which generates the required output. *Flyback regulators can have an output voltage which is higher or lower than the input voltage.*

A disadvantage of flyback regulators is the high energy which must be stored in the transformer in the form of DC current in the windings. This requires larger cores than would be necessary with pure AC in the windings.

Forward converter

A forward converter (Figure 5.15) avoids the problem of large stored energy in the transformer core. It does this, however, at the expense of an extra winding on the transformer, two more diodes, and an additional output filter inductor. Power is transferred from input to the load through D1 during switch "on" time. When the switch turns "off," D1 reverse biases and L1 current flows through D2. Output voltage is equal to:

$$V_{OUT} = V_{IN} \bullet N \bullet DC$$

(12)

The additional winding and D3 are required to define switch voltage during switch "off" time. Without this clamp, switch voltage would jump all the way to breakdown at the moment the switch is opened due to the magnetizing current flowing in the primary. This "reset"

winding normally has a 1:1 turns ratio to the primary which limits switch duty cycle to 50% maximum. Above this duty cycle, switch current rises uncontrolled even with no load because the primary winding cannot maintain zero DC voltage. Reducing the number of turns on the reset winding will allow higher switch duty cycles at the expense of higher switch voltage.

Output voltage ripple of forward converters tends to be low because of L1, but input ripple current is high due to the low duty cycles normally used. A smaller core can be used for T1 compared to flyback regulators because there is no net DC current to saturate the core.

Current-boosted boost converter

This topology in Figure 5.16 is an extension of the standard boost converter. A tapped inductor is used to decrease the switch current for a given load current. This allows higher load currents at the expense of higher switch voltage. The increase in maximum output power over a standard boost converter is equal to:

$$\frac{P_{OUT}}{P_{BOOST}} = \frac{(N + 1)(V_{OUT})}{N(V_{OUT} - V_{IN}) + V_{OUT}}$$

(13)

Analysis of this equation shows that significant increases in power are possible when the input-output differential is low. Care must be used, however, to ensure that maximum switch voltage is not exceeded.

Current-boosted buck converter

The current boosted buck converter in Figure 5.17 uses a transformer to increase output current above the maximum current rating of the switch. It accomplishes this at the expense of increased switch voltage during switch "off" time. The increase in maximum output current over a standard buck converter is equal to:

$$\frac{I_{OUT}}{I_{BUCK}} = \frac{V_{IN}}{V_{OUT} + N(V_{IN} - V_{OUT})}$$

(16)

In a 15V to 5V converter, for instance, with N = 1/4,

$$\frac{I_{OUT}}{I_{BUCK}} = \frac{15}{5 + 1/4(15 - 5)} = 2$$

This is a 100% increase in output current.

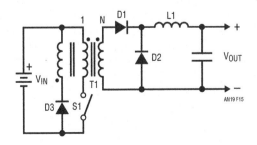

Figure 5.15 • Forward Converter

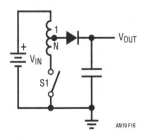

Figure 5.16 • Current-Boosted Boost Converter

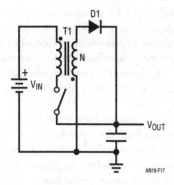

Figure 5.17 • Current-Boosted Buck Converter

Maximum switch voltage for a *current-boosted* buck converter is increased from V_{IN} to:

$$V_{SWITCH} = V_{IN} + V_{OUT}/N \qquad (17)$$

Application circuits

Boost mode (output voltage higher than input)

The LT1070 will operate in the boost mode with input voltages as low as 3V and output voltages over 50V. Figure 5.18 shows the basic boost configuration for positive voltages. This circuit is capable of output power levels that depend mainly on input voltage.

$$P_{OUT(MAX)}^* \approx V_{IN} \bullet I_P \left[1 - I_P \bullet R \left(\frac{1}{V_{IN}} - \frac{1}{V_{OUT}}\right)\right] \qquad (17)$$

*This formula assumes that L1 → ∞
I_P = maximum switch current
R = switch "on" resistance

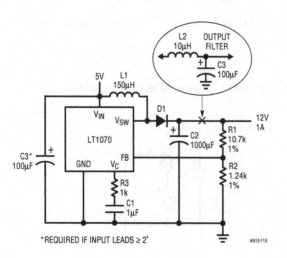

Figure 5.18 • Boost Converter

With V_{IN} = 5V, V_{OUT} = 12V, I_P = 5A, R = 0.2Ω

$$P_{OUT(MAX)} = 5 \bullet 5 \left[1 - 5(0.2)\left(\frac{1}{5} - \frac{1}{12}\right)\right] = 22W$$

With higher input voltages, output power levels can exceed 100W. Power loss internal to the LT1070 in a boost regulator is approximately equal to:

$$P_{IC} \approx (I_{OUT})^2 \bullet R \left[\left(\frac{V_{OUT}}{V_{IN}}\right)^2 - \frac{V_{OUT}}{V_{IN}}\right] + \frac{I_{OUT}(V_{OUT} - V_{IN})}{40} \qquad (18)$$

The first term of this equation is the power loss due to the "on" resistance of the switch (R). The second term is the loss from the switch driver. For the circuit in Figure 5.18, with I_{OUT} = 1A:

$$P_{IC} = (1)^2 \bullet (0.2)\left[\left(\frac{12}{5}\right)^2 - \frac{12}{5}\right] + \frac{(1)(12-5)}{40}$$

$$= 0.672 + 0.175 = 0.85 \text{ W}$$

The only other significant power loss in a boost regulator is in the diode, D1, as given by:

$$P_D = V_F \bullet I_{OUT} \qquad (19)$$

V_F is the forward voltage of the diode at a current equal to $I_{OUT} \bullet V_{OUT}/V_{IN}$. In the example shown, with I_{OUT} = 1A and V_F = 0.8V:

$$P_D = 0.8 \bullet 1 = 0.8W$$

Total power loss in the regulator is the sum of $P_{IC} + P_D$, and this can be used to calculate efficiency (E):

$$E = \frac{P_{OUT}}{P_{IN}} = \frac{P_{OUT}}{P_{OUT} + P_{IC} + P_D}$$

$$E = \frac{(1A)(12\,V)}{(1)(12) + 0.85 + 0.8} = 88\% \qquad (20)$$

With higher input voltages, efficiencies can exceed 90%.

Maximum output voltage in the boost mode is limited by the breakdown of the switch to 65V (standard part) or 75V (HV part). It may also be limited by maximum duty cycle if input voltage is low. The 90% maximum duty cycle of the LT1070 limits output voltage to ten times the input voltage. For the simple boost mode, higher ratios of output to input voltage require a tapped inductor.

Design procedure for a boost regulator is straightforward. R1 and R2 set the regulated output voltage. The feedback pin voltage is internally trimmed to 1.244V, so output voltage is equal to 1.244 (R1 + R2)/R2. R2 is normally set to 1.24k and R1 is found from:

$$R1 = R2\left(\frac{V_{OUT}}{1.244} - 1\right) \qquad (21)$$

The 1.24k value for R2 is chosen to set divider current at 1mA, but this value can vary from 300Ω to 10k with negligible effect on regulator performance. For proper load regulation, R2 must be returned directly to the ground pin of the LT1070, while R1 is connected directly to the load. For further details, see pin description section.

Inductor

Next, L1 is selected. The trade-offs are size, maximum output power, transient response, input filtering, and in some cases, loop stability. Higher inductor values provide maximum output power and low input ripple current, but are physically larger and degrade transient response. Low inductor values have high magnetizing current which reduces maximum output power and increases input current ripple. Low inductance can also cause a subharmonic oscillation problem if duty cycle is above 50%.

With the aforementioned considerations in mind, a simple formula can be derived to calculate L1 based on the maximum ripple current (ΔI) to be allowed in L1.

$$L = \frac{V_{IN}(V_{OUT} - V_{IN})}{\Delta I \bullet f \bullet V_{OUT}} \qquad (22)$$

Example: let $\Delta I = 0.5A$, $V_{IN} = 5V$, $V_{OUT} = 12V$, $f = 40kHz$

$$L = \frac{5(5 - 12)}{(0.5)(40 \bullet 10^3)(12)} = 146\mu H$$

A second formula will allow a calculation of maximum power output with this size inductor:

$$P_{MAX} = V_{IN}\left[I_P - \frac{V_{IN}(V_{OUT} - V_{IN})}{2 \bullet L \bullet f \bullet V_{OUT}}\right]\left[1 - \frac{I_P \bullet R}{V_{IN}} + \frac{I_P \bullet R}{V_{OUT}}\right] \qquad (23)$$

I_P = maximum switch current

Using the values from the previous example, with $I_P = 5A$, $R = 0.2\Omega$,

$$P_{OUT(MAX)} = 5\left[5 - \frac{5(12 - 5)}{2(146 \bullet 10^{-6})(40 \bullet 10^3)(12)}\right] \times$$
$$\left[1 - 5 \bullet (0.2)\left(\frac{1}{5} - \frac{1}{12}\right)\right]$$
$$= 5(5 - 0.25)(0.88) = 21W$$

Note that the second term in the first set of brackets is the only one which contains "L," and that this term drops out of the equation for large values of L. In this example, that term is equal to 0.25A, showing that *maximum effective switch current, and therefore maximum output power is reduced by one-half the inductor ripple current in a boost regulator.* In this example, peak effective switch current is reduced from 5A to 4.75A with 0.5A ripple current, a 5% loss. An additional 12% reduction of maximum available power is caused by switch "on" resistance. At higher input voltages, this switch loss is significantly reduced.

When continuous inductor current is desired, the value of L1 cannot be decreased below a certain limit if duty cycle of the switch exceeds 50%. Duty cycle can be calculated from:

$$DC = \frac{V_{OUT} - V_{IN}}{V_{OUT}} \qquad (24)$$

In this example,

$$DC = \frac{12 - 5}{12} = 58.3\%$$

The reason for a lower limit on the value of L for duty cycles greater than 50% is a *subharmonic* oscillation which can occur in current mode switching regulators. For further details of this phenomenon, see subharmonic oscillation section of this chapter. The minimum value of L1 to ensure no subharmonic oscillations in a boost regulator is:

$$L1_{(MIN)} = \frac{V_{OUT} - 2V_{IN}}{2 \bullet 10^5}$$
$$= \frac{12 - 2(5)}{2 \bullet 10^5} = 10\mu H \qquad (25)$$

Note that for $V_{OUT} \leq 2V_{IN}$, there is no restriction on inductor size. The minimum value of 10µH obtained in this example is below the value which would yield continuous inductor current, so it is an artificial restriction. Subharmonic oscillations do not occur if inductor current is discontinuous. The critical inductor size for continuous inductor current is:

$$L_{CRIT} = \frac{V_{IN}^2(V_{OUT} - V_{IN})}{2 \bullet f \bullet I_{OUT}(V_{OUT})^2}$$
$$= \frac{(5)^2(12 - 5)}{2(40 \bullet 10^3)(1)(12)^2} = 15.2\mu H \qquad (26)$$

Discontinuous mode operation is sometimes chosen because it results in the smallest physical size for the inductor. The maximum power output is considerably reduced, however, and can never exceed $2.5(V_{IN})$ watts with the LT1070. The minimum inductor size required to provide a given output power in the discontinuous mode is given by:

$$L_{MIN}(discontinuous) = \frac{2 \, I_{OUT}(V_{OUT} - V_{IN})}{I_P^2 \bullet f} \qquad (27)$$

Example: let $V_{IN} = 5V$, $V_{OUT} = 12V$, $I_{OUT} = 0.5A$, $I_P = 5A$

$$L_{MIN}(discontinuous) = \frac{(2 \bullet 0.5)(12 - 5)}{(5)^2 \bullet (40 \bullet 10^3)} = 7\mu H$$

This formula does not take into account efficiency losses, so the minimum value of L should probably be increased by at least 50% for worst-case conditions. Efficiency is degraded when using minimum inductor sizes because of higher switch and diode peak currents.

In summation, to choose a value for L1:
1. Decide on continuous or discontinuous mode.
2. If continuous mode, calculate C1 based on ripple current and check maximum power and subharmonic limits.
3. If discontinuous mode, calculate L1 based on power output requirements and check to see that output power does not exceed limit for discontinuous mode ($P_{MAX} = 2.5 V_{IN}$)

L1 must not saturate at the peak operating current. This value of current can be calculated from:

$$I_{L(PEAK)} = I_{OUT} \frac{(V_{OUT} + V_F) - (I_{OUT} \bullet V_{OUT} \bullet R/V_{IN})}{(V_{IN} - I_{OUT} \bullet V_{OUT} \bullet R/V_{IN})} + \frac{V_{IN}(V_{OUT} - V_{IN})}{2L1 \bullet f \bullet V_{OUT}} \quad (27)$$

V_F = forward voltage of D1
R = "on" resistance of LT1070 switch

In this example, with $V_{IN} = 5V$, $V_{OUT} = 12V$, $V_F = 0.8V$, $I_{OUT} = 1A$, $R = 0.2\Omega$, L1 = 150μH, f = 40kHz;

$$I_{L(PEAK)} = \frac{1(12 + 0.8 - 1 \bullet 12 \bullet (0.2)/5)}{5 - 1 \bullet 12 \bullet (0.2)/5}$$
$$+ \frac{5(12 - 5)}{2(150 \bullet 10^{-6})(40 \bullet 10^3)(12)}$$
$$= 2.73 + 0.24 = 3A$$

A core must be selected for L1 which does not saturate with 3A peak inductor current.

Output capacitor

The main criteria for selecting C2 is low ESR (effective series resistance), to minimize output voltage ripple. A reasonable design procedure is to let the *reactance* of the output capacitor contribute no more than 1/3 of the total peak-to-peak output voltage ripple (V_{P-P}), yielding:

$$C2 \geq \frac{V_{OUT} \bullet I_{OUT}}{f(V_{IN} + V_{OUT})(0.33 V_{P-P})}) \quad (28)$$

Using $V_{OUT} = 12V$, $I_{OUT} = 1A$, $V_{IN} = 5V$, f = 40kHz and $V_{P-P} = 200mV$,

$$C2 \geq \frac{12 \bullet 1}{(40 \bullet 10^3)(5 + 12)(0.33 \bullet 0.2)} = 268\mu F$$

This leaves 67% of the ripple attributable to ESR, giving:

$$ESR_{(MAX)} = \frac{0.67 \bullet V_{P-P} \bullet V_{IN}}{I_{OUT}(V_{IN} + V_{OUT})}$$
$$= \frac{0.67 \bullet 0.2 \bullet 5}{1(5 + 12)} = 0.04\Omega \quad (29)$$

After C2 has been selected, output voltage ripple may be calculated from:

$$V_{P-P} = I_{OUT} \left(\frac{V_{IN} + V_{OUT}}{V_{IN}} \bullet ESR + \frac{V_{OUT}}{(V_{IN} + V_{OUT})(f)(C2)} \right) \quad (30)$$

If lower output ripple is required, a larger output capacitor must be used with lower ESR. It is often necessary to use capacitor values much higher than calculated to obtain the required ESR. In the example shown, capacitors with guaranteed ESR less than 0.04Ω with a working voltage of 15V generally fall in the 1000μF to 2000μF range. Higher voltage units have lower capacitance for the same ESR.

A second option to reduce output ripple is to add a small LC output filter. If the LC product of the filter is much smaller than L1 • C2, it will not affect loop phase margin. Dramatic reduction in output ripple can be achieved with this filter, often at lower cost and less board space than simply increasing C2. See section on output filters for details.

Frequency compensation

Loop frequency compensation is performed by R3 and C1. Refer to the frequency compensation part of this chapter for R3 and C1 selection procedure.

Current steering diode

D1 should be a fast turn-off diode. Schottky diodes are best in this regard and offer better efficiency in the forward mode. With higher output voltages, the efficiency aspect is minimal and silicon fast turn-off diodes are a more economical choice. Turn-on time is important also with output voltages above 40V. Diodes with slow turn-on time will have a very high forward voltage for a short time after forward current starts to flow. This transient forward voltage can be anywhere from volts to tens of volts. It must be summed with output voltage to calculate worst-case switch voltage. To minimize switch transient voltage, the wiring of C2 and D1 should be short and close to the LT1070 as shown below.

Short-circuit conditions

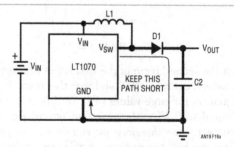

Boost regulators are *not* short-circuit protected because the current steering diode (D1) connects the input to the output. The LT1070 will not be harmed for overloads up to 5A. Beyond that point, D1 can be permanently "on" and the LT1070 switch will be effectively shorted to the output. A fuse in series with the input voltage is the only simple means of protecting the circuit. Fuse sizing can be calculated from:

$$I_{IN} \approx \frac{I_{OUT} \bullet V_{OUT}}{V_{IN}} \qquad (33)$$

The circuit in Figure 5.18 has $I_{OUT} = 1A$, $V_{OUT} = 12V$, $V_{IN} = 5V$, yielding:

$$I_{IN} \approx \frac{1 \bullet 12}{5} = 2.4A$$

A 4A fast-blow fuse would be a reasonable choice in this design.

Negative buck converter

The circuit in Figure 5.19 is a negative "buck" regulator. It converts a higher negative input voltage to a lower negative output voltage. Buck regulators are characterized by low output voltage ripple, but high input current ripple. The feedback path in this design must include a PNP transistor to level shift the output voltage sense signal to the feedback pin of the LT1070, which is referenced to the negative input voltage.

Output divider

R1 and R2 set output voltage;

$$R1 = \frac{(V_{OUT} - V_{BE})(R2)}{V_{REF}} \qquad (34)$$

V_{REF} = LT1070 reference voltage = 1.244V
V_{BE} = base-emitter voltage of Q1

R2 is nominally set to 1.24k. With the 5.2V output shown, and letting $V_{BE} = 0.6V$, R1 is:

$$R1 = \frac{(5.2 - 0.6)(1.24)}{1.244} = 4.585k\Omega$$

The nearest 1% value is 4.64kΩ. It will be apparent to experienced analog designers that the output voltage will have a temperature drift of 2mV/°C caused by the temperature coefficient of V_{BE}. If this drift is too high, it can be compensated by a resistor/diode network in parallel with R2 as shown.

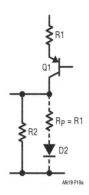

For zero output drift, R_P is made equal to R1 and R1 is now calculated from:

$$R1 = R_P = \left(\frac{V_{OUT}}{V_{REF}} - 1\right)(R2) \qquad (36)$$

Duty cycle

Duty cycle of buck converters in the continuous mode is given by:

$$DC = \frac{V_{OUT} + V_F}{V_{IN}}$$

V_F = forward voltage of D1

Inductor

The inductor, L1, is chosen as a trade-off between maximum output power with minimum output voltage ripple, versus small physical size and faster transient response. A good starting point for higher power designs is to choose a

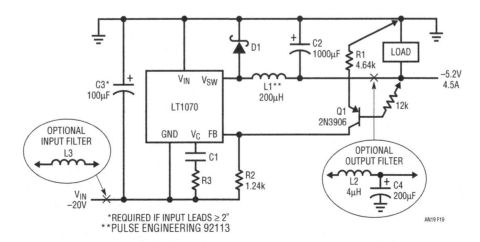

Figure 5.19 • Negative Buck Regulator

ripple current (ΔI). The LT1070 is capable of supplying up to 5A in the buck mode, so a reasonable upper limit on ripple current is 0.5A, or 10% of full load. This sets the value of L1 at:

$$L1 = \frac{(V_{IN} - V_{OUT})(V_{OUT})}{V_{IN}(\Delta I)(f)} \quad (37)$$

With circuit in Figure 5.19, $V_{IN} = 20V$, $V_{OUT} = 5.2V$, f = 40kHz, $\Delta I = 0.5A$, giving:

$$L1 = \frac{(20 - 5.2)(5.2)}{20(0.5)(40 \bullet 10^3)} = 192\mu H$$

The inductor current will go discontinuous (= zero for part of the cycle) when output current is one-half the ripple current. If continuous inductor current is desired for lower load currents, L1 will have to be increased.

Peak inductor and switch current is equal to output current plus one-half the peak-to-peak ripple current;

$$I_{L(PEAK)} = I_{OUT} + \frac{(V_{IN} - V_{OUT})(V_{OUT})}{2(V_{IN})(L)(f)} \quad (37)$$

With the example shown, letting $I_{OUT} = 4.5A$, L1 = 200μH:

$$I_{L(PEAK)} = 4.5 + \frac{(20 - 5)(5)}{2(20)(200 \bullet 10^{-6})(40 \bullet 10^3)}$$
$$= 4.5 + 0.23 = 4.73A$$

The core used for L1 must be sized so that it does not saturate at 4.73A in this example. For lower output current applications, a much smaller core can be used. The core need not be sized for peak current limit conditions (6A to 10A) in most situations because the LT1070 pulse-by-pulse current limit functions even with saturated cores.

Lower values of L1 can be used if maximum output power and low ripple are not as important as physical size or fast transient response. Pure discontinuous mode operation yields the lowest value for L1, and L1 is chosen on the basis of required output current. Maximum output current in the discontinuous mode is one-half maximum switch current and L1 is found from:

$$L1_{(MIN)} \frac{2V_{OUT}(I_{OUT})\left(1 - \frac{V_{OUT}}{V_{IN}}\right)}{(I_P^2)(f)} \quad (38)$$

where I_P = maximum switch current.
Example: let $V_{OUT} = 5.2V$, $I_{OUT} = 2A$, $V_{IN} = 20V$, $I_P = 5A$,

$$L1_{(MIN)} \frac{(2)(5.2)(2)\left(1 - \frac{5.2}{20}\right)}{5^2(40 \bullet 10^3)} = 15.4\mu H$$

It is suggested that, in discontinuous mode, this calculated value be increased by approximately 50% in practice to account for variations in cores, input voltage and frequency.

The core must be sized to not saturate at a peak current of 5A for maximum output in discontinuous mode.

Output capacitor

C2 is chosen for output ripple considerations. ESR of the capacitor may limit ripple voltage, so this parameter should be checked first. Maximum ESR allowed for a given peak-to-peak output ripple (V_{P-P}), assuming C2 $\to \infty$, is given by:

$$ESR_{(MAX)} = \frac{V_{P-P}(L1)(f)}{V_{OUT}\left(1 - \frac{V_{OUT}}{V_{IN}}\right)} \quad (39)$$

with $V_{P-P} = 25mV$, L1 = 200μH, f = 40kHz, $V_{IN} = 20V$, $V_{OUT} = 5.2V$,

$$ESR_{(MAX)} = \frac{0.025(200 \bullet 10^{-6})(40 \bullet 10)^3}{5.2\left(1 - \frac{5.2}{20}\right)} = 0.052\Omega$$

To obtain a reasonable value for C2, actual ESR should be no more than two-thirds of the maximum value. In this example, ESR is selected at 0.035Ω. C2 may now be found:

$$C2 \geq \frac{1/(8Lf^2)}{\left[\frac{V_{P-P}}{V_{DMT}\left(1 - \frac{V_{DMT}}{V_{IN}}\right)} - \frac{ESR}{Lf}\right]}$$

$$\geq \frac{1/[8(200\bullet10^{-6})(40\bullet10^3)^2]}{\left[\frac{0.025}{5.2\left(1-\frac{5.2}{20}\right)} - \frac{0.025}{(200\bullet10^{-6})(40\bullet10^3)}\right]} \geq 184\mu F$$

$$(40)$$

It is very likely that a 184μF capacitor of the right operating voltage cannot be found with an ESR of 0.035Ω maximum. C2 will have to be increased in value significantly to achieve the required ESR.

Output filter

If low output ripple is required, C2 may acquire unreasonably large values. A second option is to add an output filter as shown. Exact calculations for the values of L2 and C4 in this filter are beyond the scope of this note, but a rough approximation can be made by assuming that the ESR of C2 and C4 are the limiting factors. This leads to a value for L2 independent of the actual capacitance of C4:

$$L2 \approx \frac{(ESR2)(ESR4)(V_{IN} - V_{OUT})(V_{OUT})}{(V_{P-P})(2\pi)(f)^2(L1)(V_{IN})} \quad (41)$$

ESR2 = ESR of C2 and ESR4 = ESR of C4 and V_{P-P} = desired output ripple peak-to-peak.

If we assume ESR2 = ESR4 = 0.1Ω, and require V_{P-P} = 5mV_{P-P}:

$$L2 = \frac{(0.1)(0.1)(20 - 5.2)(5.2)}{(0.005)(2\pi)(40 \bullet 10^3)^2(200 \bullet 10^{-6})(20)} = 3.8\mu H$$

L2 may be increased above this value, but the L2 C4 product should be kept at least ten times smaller than L1 C2.

Input filter

Buck regulators have high ripple current fed back into the input voltage supply. Peak-to-peak value of this current is equal to output current. This can cause intolerable EMI conditions in some systems. An input filter formed by L3 and C3 will greatly reduce this ripple current. The major considerations for this filter are its attenuation ratio and the possible effect it has on the regulator loop stability. See discussion of input filters elsewhere in this chapter for more details.

Frequency compensation

R3 and C1 provide frequency compensation. See frequency compensation section for details of selecting these components.

Catch diode

D1 is the current steering diode. During switch off-time, it provides a path for L1 current. This diode should be a high speed switching type with fast turn-on and turn-off. A Schottky type is suggested for lower output voltage applications to improve efficiency. Formulas for average and peak diode current plus diode power dissipation are shown below. These equations assume continuous inductor current with fairly low ripple.

$$I_{PEAK} \approx I_{OUT} \tag{42}$$

$$I_{AV} = I_{OUT}\left(1 - \frac{V_{OUT}}{V_{IN}}\right) \tag{43}$$

$$P_{DIODE} = V_F \bullet I_{OUT}\left(1 - \frac{V_{OUT}}{V_{IN}}\right) \tag{44}$$

where V_F is diode forward voltage at $I = I_{PEAK}$.

Negative-to-positive buck-boost converter

The circuit in Figure 5.20 looks similar to a positive boost regulator except that the output load is referred to the inductor termination (ground) instead of the switch. A

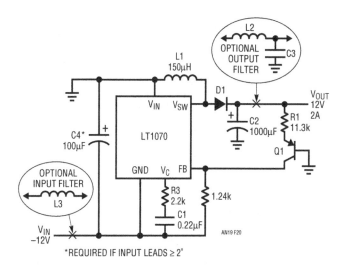

Figure 5.20 • Negative-to-Positive Buck-Boost Converter

transistor (Q1) is used to level shift the output voltage signal down to the feedback pin of the LT1070 which is referred to the negative input voltage.

Unlike buck or boost converters, inverting converters do not have any inherent limitation on input voltage relative to output voltage. Input levels may be either higher or lower than output voltage. The *sum* of input voltage plus output voltage of the LT1070 switch output voltage is given by:

$$V_{OUT} = -V_{IN}\left(\frac{DC}{1 - DC}\right) \tag{45}$$

DC = switch duty cycle (0 to 1)
With DC = 0, output voltage is zero, and as DC Æ 1, output voltage increases without limit.

Duty cycle of an inverting buck-boost converter is given by:

$$DC = \frac{|V_{OUT}|}{|V_{IN}| + |V_{OUT}|}$$

Maximum power output of a buck-boost converter is equal to:

$$P_{OUT(MAX)} = \frac{\dfrac{(I_P)(V_{OUT})(V_{IN})}{V_{OUT} + V_{IN}} - \dfrac{(I_P)^2(R)(V_{OUT})}{V_{OUT} + V_{IN}}}{1 + V_F/V_{OUT}} \tag{46}$$

I_P = peak switch current —1/2 L1 p-p ripple current
R = switch "on" resistance
V_F = forward voltage of D1

The first term on the top of the equation is the theoretical output power with no switch or diode (D1) losses. The second top term is the switch loss. The term on bottom accounts for diode losses.

With the circuit shown, $V_{IN} = -12V$, $V_{OUT} = 12V$, ripple current in L1 = $0.5A_{P-P}$, peak switch current = 5A, $R = 0.2\Omega$, $V_F = 0.8V$,

$$P_{OUT(MAX)} = \frac{\dfrac{(4.75)(12)(12)}{12+12} - \dfrac{(4.75)^2(0.2)(12)}{12+12}}{1 + 0.8/12}$$

$$= 24.6W$$

Setting output voltage

R1 and R2 determine output voltage:

$$R1 = \frac{R2(V_{OUT} - V_{BE})}{V_{REF}} \qquad (52)$$

V_{REF} = LT1070 reference voltage = 1.244V
V_{BE} = base-emitter voltage of Q1

In this example, R2 = 1.24k, $V_{OUT} = 12V$, and the V_{BE} of Q1 is $\approx 0.6V$, giving:

$$R1 = \frac{1.24\,(12 - 0.6)}{1.244} = 11.36k\Omega$$

The output voltage will have a $-2mV/°C$ drift due to the temperature drift of V_{BE}. If this is undesirable, a resistor diode combination can be added in parallel with R2 to correct drift. See section on negative buck converters for details.

Inductor

The inductor is normally calculated on the basis of maximum allowed ripple current, because high ripple currents reduce the maximum available output power and degrade efficiency. For a peak-to-peak ripple current (ΔI_L), L1 is equal to:

$$L1 = \frac{(V_{IN})(V_{OUT})}{(\Delta I_L)(V_{IN} + V_{OUT})(f)} \qquad (53)$$

f = LT1070 operating frequency = 40kHz

In this example, with ΔI chosen at 20% of maximum LT1070 switch current ($\Delta I = 1.0A$),

$$L1 = \frac{(12)(12)}{(1.0)(12 + 12)(40 \bullet 10^3)} = 150\mu H$$

Larger values for L1 will not raise power levels appreciably, will increase size and cost and will degrade transient response. L1 is not acting as a ripple filter for either the input or the output, so large values will not improve ripple either.

If L1 is reduced in value, maximum power output will be degraded. Equation 46 defines I_P as the maximum

allowed switch current minus $1/2\Delta I_L$. Therefore I_P would have to be reduced from 5A to 2.5A if L1 were reduced to the point where the ripple current equaled 5A. This is a 2:1 reduction in maximum output power. Further reductions in L1 result in discontinuous current flow and equation 46 is invalid. The poor efficiency obtained with discontinuous current flow recommends it only for low power outputs when the physical size of L1 is critical. With discontinuous current flow, the minimum recommended size for L1 is:

$$L1_{MIN}(discontinuous) = \frac{2(V_{OUT})(I_{OUT})}{(f)\,(0.7\,I_P)^2} \qquad (54)$$

The (0.7) coefficient in form of I_P is a "fudge" factor to account for variations in f and L1, and switching losses.

Example, $V_{OUT} = 12V$, $I_{OUT} = 0.5A$, f = 40kHz, $I_P = 5A$

$$L1 = \frac{2(12)(0.5)}{(40 \bullet 10^3)(0.7 \bullet 5)^2} = 24.5\mu H$$

Once L1 has been selected, peak inductor current in continuous mode can be calculated from:

$$I_{L(PEAK)} = I_{OUT}\left[1 + \frac{V_{OUT} + V_F}{V_{IN} - (I_{OUT} \bullet R)\dfrac{V_{IN} + V_{OUT}}{V_{IN}}}\right]$$

$$+ \frac{(V_{IN})(V_{OUT})}{2(L1)(V_{IN} + V_{OUT})(f)} \qquad (55)$$

V_F = forward voltage of D1
R = LT1070 switch "on" resistance

With the circuit in Figure 5.20 with L1 = $150\mu H$, $V_F = 0.8V$, $I_{OUT} = 1.5A$ and $R = 0.2\Omega$,

$$I_{L(PEAK)} = 1.5\left[1 + \frac{12 + 0.8}{12 - (1.5 \bullet 0.2)\dfrac{12 + 12}{12}}\right]$$

$$+ \frac{(12)(12)}{2(150 \bullet 10^{-6})(12 + 12)(40 \bullet 10^3)}$$

$$I_{L(PEAK)} = 3.18 + 0.5 = 3.68A$$

3.18A is the *average* current through L1 and 0.5A is the peak AC ripple current. The core used for L1 must be large enough so that it does not saturate at $I_L = 3.68A$.

Peak inductor current for discontinuous mode operation is found from:

$$I_{L(PEAK)} = \sqrt{\frac{(I_{OUT})(V_{OUT} + V_F)(2)}{(L1)(f)}} \qquad (56)$$
(discontinuous mode)

Example, let L1 = 20μH, I_{OUT} = 0.25A, V_F = 0.8V

$$I_{L(PEAK)} = \sqrt{\frac{(0.25)(12+0.8)(2)}{(20 \bullet 10^{-6})(40 \bullet 10^3)}} = 2.83A$$

The core size for this discontinuous application can be considerably smaller than in the previous example. Core volume is approximately proportional to $I_L{}^2 \bullet L$. With L1 = 100μH, and I_L = 3.93A, $I_L{}^2 \bullet L = 1.5 \bullet 10^{-3}$. The 20$\mu$H inductor with I_L = 2.83A has $I_L{}^2 \bullet L = 0.16 \bullet 10^{-3}$. The core can be nearly ten times smaller. This size difference is not free—the discontinuous circuit will supply much less current and have somewhat poorer efficiency.

Output capacitor

C2 must be a high quality (low ESR) switching capacitor because it does all the output filtering. L1 simply functions as an energy transfer element. A reasonable starting point for selecting C2 is to assume that the *ESR* (effective series resistance) of C2 contributes 2/3 of the output ripple and that the *reactance* of C2 contributes 1/3. With this in mind, a formula can be derived for ESR:

$$ESR_{(MAX)} = \frac{(V_{P-P})(V_{IN})(2/3)}{I_{OUT}(V_{IN} + V_{OUT})} \qquad (57)$$

V_{P-P} = peak-to-peak output voltage ripple

With V_{P-P} selected at 100mV, and V_{IN} = 12V, V_{OUT} = 12V, I_{OUT} = 1.5A, ESR is:

$$ESR_{(MAX)} = \frac{(0.1)(12)(2/3)}{1.5(12+12)} = 0.0185\Omega$$

With ESR found, the value of C2 may now be computed:

$$C2 = \frac{(I_{OUT})(V_{OUT})}{\left[V_{P-P} - (I_{OUT})(ESR)\left(\frac{V_{IN}+V_{OUT}}{V_{IN}}\right)\right](V_{OUT}+V_{IN})(f)} \qquad (58)$$

If we specify C2 ESR at 0.015Ω max, C2 is:

$$C2 = \frac{(1.5)(12)}{\left[0.1 - (1.5)(0.015)\left(\frac{12+12}{12}\right)\right](12+12)(40 \bullet 10^3)}$$
$$= 341\mu F \qquad (59)$$

It is most likely that to find a capacitor with a maximum ESR of 0.015Ω, the capacitance will have to be much larger than 341μF. If lower output ripple is desired, the value of C2 may become very large just to meet ESR requirements.

A second solution to the output ripple problem is to add an output filter at the point indicated in Figure 5.20. This filter can provide a large reduction in ripple with almost no effect on loop transient response, phase margin or efficiency. See section on output filters for further details.

Current steering diode

D1 must be a fast recovery diode with an *average* current rating equal to I_{OUT} and peak repetitive rating of I_{OUT} $(V_{OUT} + V_{IN})/V_{IN}$. If continuous output shorts can occur, D1 must be rated for 10A and heat sunk accordingly unless the LT1070 current limit is externally reduced. Power dissipation of D1 under normal load conditions is:

$$P_{(D1)} = (I_{OUT})(V_F)$$
$$V_F \text{ is D1 forward voltage at } I_D = I_{OUT}\left(\frac{V_{OUT} + V_{IN}}{V_{IN}}\right) \qquad (60)$$

Breakdown voltage of D1 must be at least V_{IN} + V_{OUT}. Turn-on time should be short to minimize the voltage spike across the LT1070 switch following switch turn-off.

Positive buck converter

Positive buck converters (Figure 5.21) using the LT1070 require a novel design approach because the negative side of the LT1070 switch is committed to the ground of the chip. This negative switch terminal is the inductor drive point in a positive buck converter. The ground pin of the LT1070 must therefore switch back and forth between the input voltage and converter ground. This is accomplished by tying the positive side of the switch (V_{SW}) to the input supply, and using a peak-detected (C3, D3), bootstrapped supply voltage to operate the chip. As long as the LT1070 is switching, C3 will maintain the chip input-to-ground pin voltage at a voltage equal to the input supply voltage. *It is important to keep the value of C3 to a minimum* to ensure proper start-up of this topology. The 2.2μF value shown should not be increased unless careful tests are done to ensure proper start-up under worst-case *light* loads. If the LT1070 *does not* start, the lightly loaded output will go unregulated high. The minimum recommended load current in any case is 100mA.

The most unusual aspect of this design is the manner in which output voltage information is delivered to the LT1070 feedback pin. This pin is switching along with the LT1070 ground pin to which it is referenced, so the feedback circuit must float on the switching ground pin and at the same time be proportional to the DC value of the output voltage. This is accomplished by peak detecting the output voltage with D2 during the "off" time of the LT1070 switch. The voltage on the ground pin of the chip at this time is one diode drop (D1) negative with respect to system ground, because D1 is forward biased by load

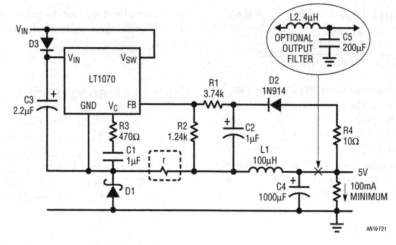

Figure 5.21 • Positive Buck Converter

current flowing through L1. D2 also forward biases, giving a voltage across C2 of:

$$V_{C2} = V_{OUT} - V_{D2} + V_{D1} \qquad (61)$$

V_{D1} = forward voltage of D1
V_{D2} = forward voltage of D2

The feedback network, R1/R2, is therefore biased with a voltage very nearly equal to output voltage, and the LT1070 will regulate output voltage according to:

$$V_{OUT} = V_{C2} + V_{D2} - V_{D1}$$

$$= \frac{V_{REF}(R1 + R2)}{R2} + V_{D2} - V_{D1} \qquad (62)$$

V_{REF} = reference voltage of LT1070 = 1.244V

If V_{D1} is exactly equal to V_{D2}, output regulation will be perfect, but the forward voltage of D1 is load current dependent, while D2 operates at a fixed average current of 1mA. This can cause output voltage variations of 100mV to 400mV if load current varies over a wide range. To minimize this effect, D1 should be conservatively rated with respect to operating current so that the effect of parasitic series resistance is minimized. The unit shown is rated at 10A average current. D1 should also be a fast turn-on type. (See diode discussion elsewhere in this chapter). A long turn-on time for D1 allows C2 to charge to a voltage *higher* than V_{OUT}, creating an abnormally *low* output voltage. R4 is added to minimize this effect. A Schottky diode is recommended for D1 because these diodes have very fast switching times and their low forward voltage improves efficiency, especially for low output voltage.

Load regulation can be significantly improved in this application by inserting a small resistor (r, shown in dashed box) between D1 and L1. The voltage across r will be equal to (r)(I_{OUT}). This voltage *increases* the voltage across R2, forcing the output voltage to *rise* under load. Perfect load regulation will result if the output *rise* created by r just cancels the output *drop* caused by the increased forward voltage of D1. The required value for r is found from:

$$r = r_d \frac{V_{REF}}{V_{OUT}} \qquad (63)$$

r_d = forward series resistance of D1
V_{REF} = LT1070 reference voltage = 1.244V

Load regulation will never be perfect because r_d varies slightly from unit to unit and it is not constant with load current, but regulation better than 2% with V_{OUT} = 5V is easily achieved even with load current varying over a 5:1 range. For higher output voltages, load regulation is even better.

For the circuit shown, with r_d = 0.05Ω, r is:

$$r = \frac{(0.05)(1.244)}{5} = 0.0124 \, \Omega$$

This is most easily obtained by using 9 inches of #22 hookup wire.

Output voltage is determined by R1 and R2:

$$R1 = R2 \frac{V_{OUT} - V_{REF}}{V_{REF}} \qquad (64)$$

R2 is normally fixed at 1.24k to set divider current to 1mA. This equation assumes that $V_{D1} = V_{D2}$. A slight adjustment in R1 will be required if $V_{D1} \neq V_{D2}$.

Duty cycle limitations

Maximum duty cycle for the LT1070 is 90%. This limits the minimum input voltage in buck regulators. Duty cycle can be calculated from:

$$DC = \frac{V_{OUT} + V_F}{V_{IN} - (I_{OUT} \bullet R) + V_F} \qquad (65)$$

V_F = forward voltage of D1
R = "on" resistance of LT1070 switch

Rearranging this formula for V_{IN} yields:

$$V_{IN(MIN)} = \frac{V_{OUT} + V_F}{DC} + (I_{OUT} \bullet R) - V_F \qquad (66)$$

With a maximum duty cycle of 90%, (0.9) and $V_{OUT} = 5V$, $V_F = 0.6V$, $R = 0.2\Omega$, $I_{OUT} = 4A$:

$$V_{IN(MIN)} = \frac{5 + 0.6}{0.9} + (0.2 \bullet 4) - 0.6 = 6.4V$$

Inductor

The energy storage inductor in a buck regulator functions as both an energy conversion element and as an output ripple filter. This double duty often saves the cost of an additional output filter, but it complicates the process of finding a good compromise for the value of the inductor. Large values give maximum power output and low output ripple voltage, but they also can be bulky and give poor transient response. A reasonable starting point is to select a maximum peak-to-peak ripple current, (ΔI). This yields a value for L1 of:

$$L1 = \frac{(V_{IN} - V_{OUT})(V_{OUT})}{(V_{IN})(\Delta I)(f)} \qquad (67)$$

  f = LT1070 operating frequency $\approx$ 40kHz
  ΔI = peak-to-peak inductor ripple current

With the circuit shown, $V_{IN} = 16V$, $V_{OUT} = 5V$ and ΔI set at 20% of 3.5A = 0.7A:

$$L1 = \frac{(16 - 5)(5)}{(16)(0.8)(40 \bullet 10^3)} = 122\mu H$$

The ripple current in L1 reduces the maximum output current by one-half ΔI. For lower output currents this is no problem, but for maximum output power, L1 may be raised by a factor of two to three. For lower output powers, L1 can be reduced to save on size and cost. Discontinuous mode operation will occur even near full load if L1 is reduced far enough. The LT1070 is not affected by discontinuous operation per se, but maximum output power is significantly reduced in discontinuous mode designs:

$$I_{OUT(MAX)} = \frac{(I_P)^2(L)(f)}{2V_{OUT}}\left(\frac{V_{IN}}{V_{IN} - V_{OUT}}\right) \qquad (68)$$
(discontinuous)

  I_P = LT1070 peak switch current

With L1 = 10μH, for instance, and $I_P = 5A$:

$$I_{OUT(MAX)} = \frac{(5)^2(10 \bullet 10^{-6})(40 \bullet 10^3)}{2(5)}\left(\frac{16}{16 - 5}\right) = 1.4A$$

Efficiency is also reduced with discontinuous operation because of increased switch dissipation.

The load current where a buck regulator changes from continuous to discontinuous operation is:

$$I_{CRIT} = \frac{(V_{IN} - V_{OUT})(V_{OUT})}{2(V_{IN})(f)(L1)} \qquad (69)$$

With a 100μH value of L1, inductor current will go discontinuous at:

$$I_{CRIT} = \frac{(16 - 5)(5)}{2(16)(40 \bullet 10^3)(100 \bullet 10^{-6})} = 0.43A \qquad (70)$$

I_{CRIT} can never exceed 2.5A (one half maximum LT1070 switch current).

Peak inductor current in a buck regulator with continuous mode operation is:

$$I_{L(PEAK)} = I_{OUT} + \frac{(V_{IN} - V_{OUT})(V_{OUT})}{2(V_{IN})(L1)(f)} \qquad (71)$$

With $I_{OUT} = 3.5A$ and L1 = 100μH:

$$I_{L(PEAK)} = 3.5 + \frac{(16 - 5)(5)}{2\,(16)(100 \bullet 10^{-6})(40 \bullet 10^3)} = 3.93A$$

The core used for L1 must be able to handle 3.93A peak current without saturating.

Peak inductor currents in discontinuous mode are much higher than output current:

$$\underset{\text{(discontinuous)}}{I_{L(PEAK)}} = \sqrt{\frac{2(V_{OUT})(I_{OUT})(V_{IN} - V_{OUT})}{(V_{IN})(L1)(f)}} \qquad (72)$$

For L1 = 10μH, $I_{OUT} = 1A$:

$$\underset{\text{(discontinuous)}}{I_{L(PEAK)}} = \sqrt{\frac{2(5)(1)(16 - 5)}{(16)(10 \bullet 10^{-6})(40 \bullet 10^3)}} = 4.15A$$

The 10μH inductor, at 1A output current, must be sized to handle 4.14A peak current.

Output voltage ripple

See negative buck regulator section for calculation of output ripple.

Output capacitor

C4 is chosen for output voltage ripple considerations. Its ESR (effective series resistance) is the most important parameter. For details, see negative buck regulators section.

Output filter

For very low output voltage ripple, the value of C4 may become prohibitively high. An output filter, L2 and C5, may be used to reduce output ripple. See Output Filter section for details.

Flyback converter

Flyback converters (Figure 5.22) are able to regulate an output voltage either higher or lower than the input voltage by shuttling stored energy back and forth between the windings of a transformer. During switch "on" time, all energy is stored in the primary winding according to: $E = (I_{PRI})^2(L_{PRI})/2$. When the switch turns off, this energy is transferred to the output winding. The current in the secondary just after switch opening is equal to the reciprocal of turns ratio (1/N) times the current in the primary just prior to switch opening. Output voltage of a flyback converter is not constrained by input voltage as in buck or boost converters.

$$V_{OUT} = \frac{DC}{1 - DC}(N \bullet V_{IN}) \qquad (73)$$

$$DC = \text{switch duty cycle} = \frac{V_{OUT}}{V_{OUT} + (N \bullet V_{IN})} \qquad (74)$$

N = transformer turns ratio

By varying duty cycle between 0 and 1, output voltage can theoretically be set anywhere from 0 to ∞. Practically, however, output voltage is constrained by switch breakdown voltage and the maximum output voltage is limited to:

$$V_{OUT(MAX)} = N(V_M - V_{SNUB} - V_{IN}) \qquad (75)$$

V_{SNUB} = snubber voltage (see snubber details in this section)

V_M = maximum allowed switch voltage

This still allows the LT1070 to regulate output voltages of hundreds or even thousands of volts by using large values of N.

In many applications, N can vary over a wide range without degrading performance. If maximum output power is desired however, N can be optimized:

$$N_{(OPT)} = \frac{V_{OUT} + V_F}{V_M - V_{IN(MAX)} - V_{SNUB}} \qquad (76)$$

V_F = forward voltage of D1

In Figure 5.22, with $V_{OUT} = 5V$, $V_F = 0.7V$ (Schottky), $V_{IN(MAX)} = 30V$, $V_M = 60V$, $V_{SNUB} = 15V$;

$$N_{(OPT)} = \frac{5 + 0.7}{60 - 30 - 15} = 0.38$$

A turns ratio of 1:3 (0.33) was used in this circuit.

A second important transformer parameter which must be determined is primary inductance (L_{PRI}). For maximum output power, L_{PRI} should be high to minimize magnetizing current, but this can lead to unacceptably large core sizes. A reasonable design approach is to reduce the value of L_{PRI} to the point where primary magnetizing current (ΔI) is about 20% of peak switch current. The LT1070 is rated for 5A peak switch current, so for full power applications, ΔI can be set to 1A peak-to-peak. Maximum output current is reduced by one-half of the ratio of ΔI to peak switch current, or =10% in this case.

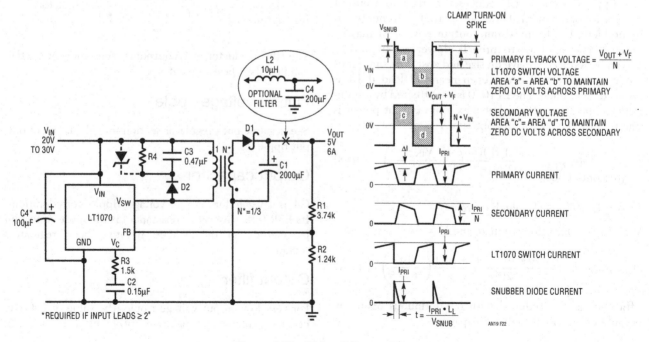

Figure 5.22 • Flyback Converter

With this design approach, L_{PRI} is found from:

$$L_{PRI} = \frac{(V_{IN})(V_{OUT})}{(\Delta I)(f)(V_{OUT} + N \bullet V_{IN})} \qquad (77)$$

With $V_{IN} = 24V$, $V_{OUT} = 5V$, $\Delta I = 1A$, $N = 1/3$:

$$L_{PRI} = \frac{(24)(5)}{(1)(40 \bullet 10^3)(5 + 1/3 \bullet 24)} = 231\mu H$$

Values of L_{PRI} higher than this will raise maximum output current only slightly and will require larger core size. Lower primary inductance may be used for lower output currents to reduce core size.

Maximum output current is a function of peak allowed switch current (I_P):

$$I_{OUT(MAX)} = \frac{E\left(I_P - \frac{\Delta I}{2}\right)(V_{IN})}{(N \bullet V_{IN}) + V_{OUT}} \qquad (78)$$

I_P = maximum LT1070 switch current
E = overall efficiency $\approx 75\%$

With $V_{IN} = 24V$, $V_{OUT} = 5V$, $I_P = 5A$, $\Delta I = 1A$, $N = 1/3$:

$$I_{OUT(MAX)} = \frac{0.75\left(5 - \frac{1}{2}\right)(24)}{(1/3 \bullet 24) + 5} = 6.2A$$

The 75% efficiency number comes from losses in the snubber network ($\approx 6\%$), LT1070 switch ($\approx 4\%$), LT1070 driver ($\approx 3\%$), output diode ($\approx 8\%$) and transformer ($\approx 4\%$). Although this efficiency is not as impressive as the 85% to 95% obtainable with simple buck or boost designs, it is more than justified in many cases by the ability to use the variable N to generate high output currents or high output voltages and the option to add extra windings for multiple outputs.

Peak primary current is used to determine core sizing for the transformer:

$$I_{PRI} = \frac{I_{OUT}}{E}\left(\frac{V_{OUT}}{V_{IN}} + N\right)$$
$$+ \frac{(V_{IN})(V_{OUT})}{2(f)(L_{PRI})(V_{OUT} + N \bullet V_{IN})} \qquad (79)$$

For an output current of 6A, with $V_{IN} = 24V$, $V_{OUT} = 5V$, $E = 75\%$, $L_{PRI} = 231\mu H$, $N = 1/3$:

$$I_{PRI} = \frac{6}{0.75}\left(\frac{5}{24} + 1/3\right)$$
$$+ \frac{(24)(6)}{2(40 \bullet 10^3)(231 \bullet 10^{-6})(5 + 1/3 \bullet 24)}$$
$$= 4.33 + 0.5 = 4.83A$$

The core must be able to handle 4.83A peak current in the $231\mu H$ primary winding without saturating. (See section on inductors and transformers for further details.)

Output divider

R1 and R2 set output voltage:

$$R1 = \frac{V_{OUT} - V_{REF}}{V_{REF}} \bullet R2 \qquad (80)$$

V_{REF} = feedback reference voltage of the LT1070 = 1.244V

R1 and R2 can vary over a wide range, but a convenient value for R2 is 1.24k, a standard 1% value.

For a 5V output,

$$R1 = \frac{(5 - 1.244)(1.24)}{1.244} = 3.756k\Omega$$

Frequency compensation

R3 and C2 provide a pole-zero frequency compensation. For details, see the section on frequency compensation elsewhere in this chapter.

Snubber design

Flyback converters using transformers require a clamp to protect the switch from overvoltage spikes. These spikes are created by leakage inductance in the transformer. Leakage inductance (L_L) is modeled as an inductor in series with the primary winding which is not coupled to the secondary as shown in Figure 5.23.

During switch "on" time, a current is established in L_L equal to peak primary current (I_{PRI}). When the switch turns off, the energy stored in L_L, ($E = I^2 \bullet L_L/2$) will cause the switch voltage to fly up to breakdown if the voltage is not clamped.

If a Zener diode is used for clamping, Zener clamp voltage is selected by assigning a maximum switch voltage and maximum input voltage:

$$V_{ZENER} = V_M - V_{IN(MAX)}$$

V_M = maximum allowed switch voltage

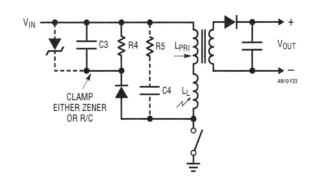

Figure 5.23 • Snubber Clamping

The standard LT1070 maximum switch voltage is 65V, so V_M is typically set at 60V to allow a margin of 5V. If we assume $V_{IN(MAX)} = 30V$ for this circuit:

$$V_{ZENER} = 60 - 30 = 30V$$

Peak Zener current is equal to peak primary current (I_{PRI}) and average power dissipation is equal to:

$$P_{ZENER} = \frac{(V_Z)(I_{PRI})^2(L_L)(f)}{2\left(V_Z - \dfrac{V_{OUT} + V_F}{N}\right)} \qquad (81)$$

An important part of this equation is the term $[V_Z - (V_{OUT} + V_F)/N]$ in the denominator. This voltage is defined as snubber voltage (V_{SNUB}) and is the difference between the Zener voltage and the normal flyback voltage of the primary. (See waveforms with Figure 5.22.) If V_{SNUB} is too low, Zener dissipation rises rapidly. A reasonable minimum for V_{SNUB} is 10V, so this should be checked before proceeding further:

$$V_{SNUB} = V_Z - \frac{V_{OUT} + V_F}{N} = 30 - \frac{5 + 0.7}{1/3} = 12.9V \qquad (82)$$

Leakage inductance in a transformer can be minimized by bifilar winding or by interleaving the primary and secondary. If this is done correctly, leakage inductance is usually less than 1% of primary inductance. If we wind T1 for $L_{PRI} = 230\mu H$, L_L should be less than $2.3\mu H$. Using this value, power dissipation in the Zener at full load current is:

$$P_{ZENER} = \frac{(30)(4.83)^2(2.3 \bullet 10^{-6})(40 \bullet 10^3)}{2\left(30 - \dfrac{5 + 0.7}{1/3}\right)} = 2.5W$$

Zener dissipation under short-circuit conditions is calculated from the same equation (81) by assuming that $V_{OUT} = 0V$ and I_{PRI} is the current limit value of the LT1070. If we let $I_{PRI} = 9A$:

$$P_{ZENER} = \frac{(30)(9)^2(3.5 \bullet 10^{-6})(40 \bullet 10^3)}{2\left(30 - \dfrac{0.7}{1/3}\right)} = 4W$$
(output shorted)

The waveform of LT1070 switch voltage shows a narrow spike extending above the snubber clamp voltage. This spike is caused by the turn-on time of the clamp circuit, in particular the diode in series with the Zener. This diode should be a Schottky or a very fast turn-on type to minimize the height of this spike. It must be rated for peak currents equal to I_{PRI}. The reverse voltage rating of the diode must be at least $V_{IN(MAX)}$.

An alternative to Zener clamping is an R/C clamp. This is less expensive, but has the disadvantage of a less well-defined clamping level. The RC snubber also dissipates power even with no-load conditions. A value for R4 is found from:

$$R_{SNUB} = \frac{2(V_R)^2 - 2(V_R)(V_{OUT}/N)}{(I_{PRI})^2(L_L)(f)} \qquad (83)$$

V_R = voltage across snubber resistor

If we set $V_R = 30V$ (same as V_{ZENER}) and use full load conditions of $I_{PRI} = 4.83A$:

$$R_{SNUB} = \frac{2(30)^2 - 2(30)\left(\dfrac{5}{1/3}\right)}{(4.83)^2(2.3 \bullet 10^{-6})(40 \bullet 10^3)} = 419\Omega$$

Power dissipation in the snubber at full load is equal to:

$$P_R = \frac{(V_R)^2}{R} = \frac{(30)^2}{419} = 2.15W$$

At very light loads, the voltage across the snubber resistor drops to the flyback voltage of the primary, $V_R = (V_{OUT} + V_F)/N$.

In this example, flyback voltage is 16.8V, resulting in a snubber dissipation of $16.8^2/419\Omega = 0.67W$.

This may be a consideration where high efficiency is necessary even with near-zero output loads. *Short circuit power dissipation in the snubber resistor is approximately equal to:*

$$P_R \approx \frac{(I_{PRI})^2(f)(L_L)}{2} \qquad (84)$$
(output shorted)

I_{PRI} in short circuit is the current limit of the LT1070. For $I_{PRI} = 9A$, snubber dissipation with the output shorted is $\approx$ 3.7W in this example.

The value of C3 is not critical, but it should be large enough to keep the ripple voltage across the snubber to only a few volts. This yields a capacitor value of:

$$C3 = \frac{V_R}{(R)(f)(V_S)} \qquad (85)$$

V_S = voltage ripple across C3

For $V_S = 3V$, $V_R = 30V$, $R = 419\Omega$:

$$C3 = \frac{30}{(419)(40 \bullet 10^3)(3)} = 0.6\mu F$$

C3 should be a very low ESR (effective series resistance) film or ceramic type to keep spike voltage to a minimum.

C4 and R5 (shown in dashed lines) form an optional damper, which eliminates primary ringing for light output load conditions when secondary current drops to zero during switch-off time (discontinuous operation). Typical values are $R = 300\Omega$ to 1.5k, C = 500pF to 5000pF.

Output diode (D1)

The output diode has an *average* forward current equal to output current, but the current flows in pulses with an amplitude equal to:

$$I_{D1(PEAK)} = I_{OUT}\left(1 + \frac{V_{OUT} + V_F}{N(V_{IN})}\right) \qquad (86)$$

For the circuit in Figure 5.22, with $I_{OUT} = 6A$:

$$I_{D1(PEAK)} = 6\left(1 + \frac{5 + 0.7}{1/3(24)}\right) = 10.3A$$

To calculate diode power dissipation, use the forward voltage at this peak current multiplied times output current:

$P_{D1} = (V_F)(I_{OUT})$
$V_F = D1$ forward voltage at peak current

With $V_F = 0.55V$ and $I_{OUT} = 6A$, D1 power dissipation is 3.3W.

During start-up and overload conditions, D1 current will increase significantly. *Average* diode current through D1 when the LT1070 is in current limit is equal to:

$$I_{D1} = \frac{\alpha(I_{LIM})(V_{IN})}{N(V_{IN}) + V_{OUT} + V_F} \tag{87}$$

(during LT1070 current limit)

α is an empirical multiplier slightly less than unity. It is very complex to calculate, but it takes into account such things as switch resistance, leakage inductance, snubber losses, and transformer losses. If we assume $\alpha = 0.8$, $I_{LIM} = 9A$, $V_{IN} = 24V$, $N = 1/3$, $V_F = 0.55V$ and a shorted output ($V_{OUT} = 0V$):

$$I_{D1} = \frac{0.8(9)(24)}{1/3(24) + 0 + 0.8} = 20A$$

Peak diode current will be only slightly higher because the duty cycle of the diode is approaching 100% with $V_{OUT} = 0V$.

Output short-circuit current can be reduced, if desired, by clamping the V_C pin of the LT1070. The best way to do this and still be assured of maximum full-load current is to clamp the V_C pin to a portion of output voltage. This generates a foldback current limit that will reduce short-circuit current without affecting normal load current. The clamp network in Figure 5.24 will reduce shorted output current of the circuit in Figure 5.22 to $\approx 5A$.

The clamp point is generated by splitting R1 into two resistors such that the tap point voltage is $\approx 1.75V$ at

normal output voltage. This ensures that D4 will not turn on until the output voltage begins to drop. When $V_{OUT} = 0V$, the voltage at the FB pin is clamped to approximately 0.35V by the internal mode select circuitry and the voltage at the R1 tap point will be approximately the same. The current through the diodes will be maximum available V_C pin current. This sets the clamp voltage on the V_C pin at $\approx 1.55V$, reducing output short-circuit current to $\approx 5A$. Full-load current can be reduced, if desired, by moving the tap point on R1 down, even to the point where it becomes part of R2.

Output capacitor (C1)

Flyback converters do not use the inductance of the transformer as a filter, so the output capacitor must do all the filtering work. The output peak-to-peak voltage ripple is equal to:

$$V_{P-P} = \frac{I_{OUT}}{(f)(C1)\left(1 + \frac{N(V_{IN})}{V_{OUT}}\right)} + (ESR)(I_{OUT})\left(1 + \frac{V_{OUT}}{N(V_{IN})}\right) \tag{88}$$

ESR = effective series resistance of C1

The first term is the ripple due to the *capacitance* of C1; the second term is ripple due solely to the ESR of the capacitor. As it turns out, commercially available capacitors in the range required for this application (100μF to 10,000μF) have ESR high enough to dominate the ripple voltage. A 2,000μF capacitor for instance, might have a guaranteed ESR of 0.02Ω. For $I_{OUT} = 6A$, $V_{OUT} = 5V$, $V_{IN} = 24V$, $N = 1/3$, this gives:

$$V_{P-P} = \frac{6}{(40 \bullet 10^3)(2000 \bullet 10^{-6})\left(1 + \frac{1/3(24)}{5}\right)}$$
$$+ (0.02)(6)\left(1 + \frac{5}{1/3(24)}\right)$$
$$= 28.8\,mV + 195\,mV = 224\,mV$$

The ESR term dominates and will be the main criteria for selecting the size of the output capacitor.

An alternative to brute force output capacitance (to obtain low ESR) is to add an LC output filter (shown as L1 and C4 in Figure 5.22). A relatively small inductor and capacitor can greatly reduce output ripple. If we assume the ripple across C1 is due solely to ESR, and therefore rectangular, the ratio of filter output ripple to input ripple is:

$$\frac{V_{OUT(P-P)}}{V_{IN(P-P)}} = r = \frac{ESR4(V_{OUT})(N \bullet V_{IN})}{(L1)(f)(V_{OUT} + N \bullet V_{IN})^2} \tag{89}$$

ESR4 = effective series resistance of C4

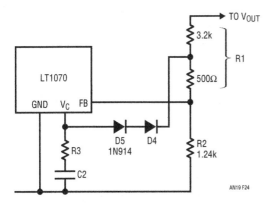

Figure 5.24 • Foldback Current Limiting

This formula again assumes that the ESR of C4 dominates its total impedance. For ESR4 = 0.1Ω, L1 = 10µH, V_{OUT} = 5V, N = 1/3, V_{IN} = 24V,

$$r = \frac{(0.1)(5)(1/3 \bullet 24)}{(10 \bullet 10^{-6})(40 \bullet 10^3)(5 + 1/3 \bullet 24)^2} = 0.059$$

This is a 16:1 reduction in ripple, greatly easing the requirements on C1. Total output ripple, with a filter, is given by:

$$V_{P\text{-}P} = \frac{(ESR1)(ESR4)(V_{OUT})(I_{OUT})}{(L1)(f)(V_{OUT} + N \bullet V_{IN})} \qquad (90)$$

For ESR1 = 0.05Ω, ESR4 = 0.1Ω, V_{OUT} = 5V, V_{IN} = 24V, N = 1/3, I_{OUT} = 6A, L1 = 10µH, output ripple (P-P) is:

$$V_{P\text{-}P} = \frac{(0.05)(0.01)(5)(6)}{[10 \bullet 10^{-6}](40 \bullet 10^3)(5 + 1/3 \bullet 24)} = 28.8 \, mV$$

Totally isolated converter

The LT1070 has a second operating mode called "isolated flyback," as shown in Figure 5.25 (see Note with figure).

While in this mode, it does not use the feedback pin to sense output voltage; instead, it senses and regulates the transformer primary voltage during switch "off" time (t_{OFF}). This voltage is related to V_{OUT} by:

$$V_{OUT} = (N)(V_{PRI}) - V_F \qquad (90)$$
(during t_{OFF})

N = turns ratio of transformer
V_F = forward voltage of output diode
V_{PRI} = primary voltage during switch "off" time

The secondary output voltage will be regulated if V_{PRI} is regulated. The LT1070 switches from normal mode to regulated primary mode when the current *out* of the feedback pin exceeds ≈10µA. An internal clamp holds the voltage (V_{FB}) on this pin at ≈400mV. R2 is used to put the LT1070 in isolated flyback mode. It also doubles as an adjustment in the regulated output. V_{PRI} is regulated to 16V + 7k (V_{FB}/R2), where V_{FB}/R2 is equal to the current through R2, and the 7k is an internal resistor. V_{OUT} is therefore equal to:

$$V_{OUT} = N\left[16 + 7k\left(\frac{V_{FB}}{R2}\right)\right] - V_F \qquad (91)$$

and the required transformer turns ratio is:

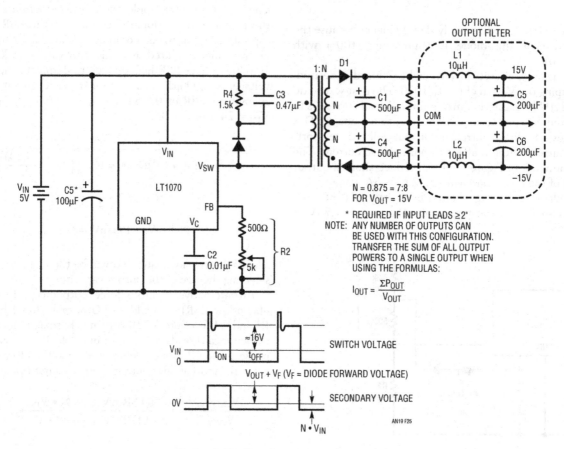

Figure 5.25 • Totally Isolated Converter

$$N = \frac{V_{OUT} + V_F}{16 + 7k\left(\frac{V_{FB}}{R2}\right)} \qquad (92)$$

The term 7k (V_{FB}/R2) is normally set to $\approx$2V to allow some adjustment range in V_{OUT}. Solving for N in Figure 5.25, with V_{OUT} = 15V:

$$N = \frac{15 + 0.7}{16 + 2} = 0.872$$

The smallest integer ratio with N close to 0.872 is 7:8 = 0.875. T1 is to be wound with this turns *ratio* for each output. The *total* number of turns is determined by the required primary inductance (L_{PRI}). This inductance has no optimum value; it is a trade-off between core size, regulation requirements and leakage inductance effects. A reasonable starting value is found by assigning a maximum magnetizing current (ΔI) of 10% of the peak switch current of the LT1070. Magnetizing current is the difference between the primary current at the start of switch "on" time and the current at the end of switch "on" time. This gives a value for L_{PRI} of:

$$L_{PRI} = \frac{V_{IN}}{(\Delta I)(f)\left(1 + \frac{V_{IN}}{V_{PRI}}\right)} \qquad (93)$$

ΔI = primary magnetizing current
V_{PRI} = regulated primary flyback voltage

For V_{IN} = 5V, ΔI = 0.5A, V_{PRI} = 18V:

$$L_{PRI} = \frac{5}{(0.5)(40 \bullet 10^3)(1 + 5/18)} = 196\mu H$$

Again, this value is not an optimum figure, it is simply a compromise between maximum output current and core size.

A second consideration on primary inductance is the transition from continuous mode to discontinuous mode. At light output loads, the flyback pulse across the primary will drop toward zero before the end of switch "off" time. The LT1070 interprets this as a drop in output voltage and raises duty cycle to compensate. This results in an abnormally high output voltage. To avoid this situation, the output should have a minimum load equal to:

$$I_{OUT(MIN)} = \frac{(V_{PRI} \bullet V_{IN})^2}{(V_{PRI} + V_{IN})^2(2V_{OUT})(f)(L_{PRI})} \qquad (94)$$

with V_{PRI} = 18V, V_{IN} = 5V, V_{OUT} = 15V, L_{PRI} = 200μH:

$$I_{OUT(MIN)} = \frac{(18 \bullet 5)^2}{(18 + 5)^2(2 \bullet 15)(40 \bullet 10^3)(200 \bullet 10^{-6})}$$

$$= 64mA$$

This current may be shared equally on each output at 32mA per output. If a lighter minimum load is desired,

primary inductance must be increased. This also increases leakage inductance, so some care must be used.

Leakage inductance is a portion of the primary which is not coupled to the secondary. This leakage inductance will create a flyback spike following switch opening. The height of this spike must be clamped with a snubber (R4, C3, D2) to avoid overvoltage on the switch. (Please read snubber details in the section on normal mode flyback regulators). The *width* of the leakage inductance spike is equal to:

$$t_L = \frac{(I_{PRI})(L_L)}{V_M - V_{PRI} - V_{IN}} \qquad (95)$$

L_L = leakage inductance
I_{PRI} = peak primary current
V_M = peak switch voltage

This spike width is important because it must be less than 1.5μs wide. The LT1070 has internal blanking for $\approx$ 1.5μs following switch turn-off. This blanking time ensures that the flyback error amplifier will not interpret the leakage inductance spike as the actual flyback voltage to be regulated. To avoid poor regulation, the spike must be less than the blanking time.

If transformer T1 is trifilar wound for minimum leakage inductance, L_L may have a typical value of 1.5% of L_{PRI}. Assuming L_{PRI} = 200μH, L_L would be 3μH. To calculate t_L, we still need to assign a value to V_M. In this case, with V_{IN} = 5V, a conservative value for maximum switch voltage would be V_M = 50V. If we assume a maximum primary current of 5A for maximum output current, spike width is:

$$t_L = \frac{5(3 \bullet 10^{-6})}{50 - 18 - 5} = 0.56\mu s$$

This is well within the maximum value of 1.5μs. Note, however, that the pulse width grows rapidly as the sum of $V_{PRI} + V_{IN}$ approaches maximum switch voltage. The following formula will allow one to calculate the maximum ratio of leakage inductance to primary inductance in a given situation.

$$\frac{L_L}{L_P}(MAX) = \frac{t_L(V_M - V_P - V_{IN})(\Delta I)(f)\left(1 + \frac{V_{IN}}{V_P}\right)}{I_{PRI}(V_{IN})} \qquad (96)$$

With a fairly large V_{IN} (36V), even if we use a less conservative value of 60V for V_M, with t_L = 1.5μs, V_p = 18V, ΔI = 0.5A and I_{PRI} = 5A:

$$\frac{L_L}{L_P}(MAX) = \frac{(1.5 \bullet 10^{-6})(60 - 18 - 36)(0.5)(40 \bullet 10^3)\left(1 + \frac{36}{18}\right)}{5(36)}$$

$$= 0.003 = 0.3\%$$

This low ratio of leakage inductance to primary inductance would be nearly impossible to wind, so some compromises must be made. If maximum output current is not required,

I_{PRI} will be less than 5A, (see formula 99). Ripple current (ΔI) can also be increased. Finally, an LT1070HV (high voltage) part can be used, with a switch rating of 75V. Substituting $I_{PRI} = 2.5A$, $\Delta I = 1A$, $V_M = 70V$ into the above calculation yields $L_I/L_{PRI} = 3\%$, which is easily achievable.

Maximum output power with an isolated flyback converter is less than an ordinary flyback converter because transformer turns ratio is fixed by output voltage. This fixes duty cycle at:

$$DC = \frac{V_{PRI}}{V_{PRI} + V_{IN}} \qquad (97)$$

and maximum power is limited to:

$$P_{OUT(MAX)} = \left(\frac{V_{PRI}}{V_{PRI} + V_{IN}}\right)\left[V_{IN}\left(I_P - \frac{\Delta I}{2}\right) - (I_P)^2 R\right](0.8) \qquad (98)$$

R = LT1070 switch "on" resistance
I_P = maximum switch current
0.8 = fudge factor to account for losses in addition to R

With V_{PRI} at a nominal 18V, $V_{IN} = 5V$, $I_p = 5A$, $\Delta I = 0.5A$, duty cycle is 78% and maximum output power is:

$$P_{OUT(MAX)} = \left(\frac{18}{18+5}\right)\left[5\left(5 - \frac{0.5}{2}\right) - (5)^2(0.2)\right](0.8)$$
$$= 11.74W$$

An analysis of the power formula shows that at low V_{IN}, maximum output power is proportional to V_{IN}, and at high V_{IN}, maximum power approaches 50W.

Peak primary current for loads less than the maximum is found from:

$$I_{PRI} = \frac{\left(V_{OUT}\right)\left(I_{OUT}\right)\left(V_{PRI} + V_{IN}\right)}{0.8\left(V_{PRI}\right)\left(V_{IN}\right)} + \frac{\Delta I}{2} + \frac{\left(I_{PRI}\right)^2 R}{V_{IN}} \qquad (99)$$

This formula is actually a quadratic, but rather than solve it explicitly, a much simpler technique, for the range of I_{PRI} involved, is to calculate the first two terms on the right, then use this value of I_{PRI} to calculate the last term. For the circuit in Figure 5.25 with $I_{OUT} = 0.25A$ on each output, $V_{PRI} = 18V$, $V_{IN} = 5V$, $\Delta I = 0.5A$, $R = 0.2\Omega$:

$$I_{PRI} = \underbrace{\frac{(15)(0.5)(18+5)}{0.8(18)(5)}}_{2.64A} + \frac{0.5}{2} + \frac{(2.64)^2(0.2)}{5} = 2.92A$$

The transformer must be sized so that the core does not saturate with 2.92A in the primary winding. Note that there is plenty of margin on 5A maximum switch current. A smaller core could be used if ΔI were increased to 1A,

cutting primary inductance in half. (See section on inductors and transformers.)

Output capacitors

Flyback regulators do not utilize the inductance of the transformer as a filter, so all filtering must be done by the output capacitors, C1 and C4. They should be low ESR types to minimize output ripple. In general, output ripple is limited by the ESR of the capacitor, not the actual capacitance. Output ripple in peak-to-peak volts is given by:

$$V_{P-P} = \frac{I_{PRI}}{2^*N}(ESR) \qquad (100)$$

*This factor of 2 is used because of dual outputs

With $I_{PRI} = 2.92A$, $N = 0.872$ and assigning an ESR of 0.1Ω, output ripple is:

$$V_{P-P} = \frac{(2.92)(0.1)}{(2)(0.872)} = 167mV_{P-P} \text{ at full load}$$

Had we based the output ripple formula on the actual output *capacitance*, rather than its ESR, the result would have been $\approx 10mV$, showing that ESR effects do dominate. The 0.1Ω value chosen for ESR is probably *higher than typical for* a good $500\mu F$ capacitor, but *less than guaranteed maximum*. Note that one reason for high output ripple in this circuit is that the converter is operating at a rather high duty cycle of 78% because of the low input voltage. This leaves only 22% of the time for the secondary to be delivering current to the load. As a consequence, secondary peak currents, and therefore output ripple, are high.

If low output ripple is required, an output filter may be a better choice than simply using huge output capacitors. See output filters section.

Load and line regulation

Load and line regulation are affected by many "open loop" factors in this circuit because the actual output voltage is not sensed—only the primary. Some of these factors are core nonlinearities, diode resistance, leakage inductance, winding resistance, (including skin effect) capacitor ESR and secondary inductance. A typical load regulation for this circuit with a load variation from 20% to 100% is $\approx 3\%$. Line regulation at light loads is better than 0.3% for $V_{IN} = 4.5V$ to 5.5V, but degrades to $\approx 1\%$ for full loads.

With multiple output supplies obtained from a single switching loop, the problem of cross regulation appears. In this supply, and *increase in* load current from 50mA to 200mA on one output, with a constant 50mA load on the second output, will cause the loaded output to *drop* 280mV and the constant load output to *rise* 100mV.

If improved line and load regulation are necessary, a modification can be made to the basic circuit as shown overleaf:

Load Current Compensation

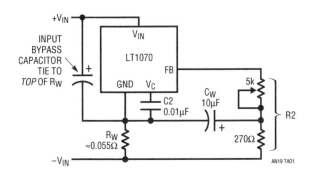

AN19 TA01

R2 is split into two resistors with the center tap coupled to the ground pin of the LT1070 through C_W. A small resistor R_W is inserted in series with the ground pin. When a load is applied to the output, input current flowing through R_W causes the voltage drop across R2 to increase. This increases regulated primary voltage and thereby output voltage, cancelling the open-loop load regulation effects mentioned earlier. Line regulation is also significantly improved at full load.

The value of R_W is found from:

$$R_W = \frac{(R_O)(V_{IN})(E)(R2)}{(V_{OUT})(7k)(N)^*} \quad (101)$$

R_O = output resistance without compensation
$\quad = \Delta V_{OUT}/\Delta I_{OUT}$
E = efficiency ≈ 0.75

*Multiply N by two for dual outputs

For the circuit in Figure 5.25, R_O is found by loading both outputs simultaneously and summing the changes of the two outputs. With 3% load regulation, at $\Delta I_{OUT} = 200mA$, this is a total output change of 900mV. R_O is then 900mV divided by a current change of 200mA, or 4.5Ω. V_{OUT} is the sum of the two outputs, =30V, N is (0.875)(2) = 1.75 and R2 is ≈1.2k:

$$R_W = \frac{(4.5)(5)(0.75)(1,200)}{(30)(7k)(1.75)} = 0.055\Omega$$

This low value of resistance preserves the efficiency of the converter, but is sometimes hard to find "off the shelf." A 15″ length of #26 hookup wire was used for the breadboard. To minimize inductance, the wire is folded in half before winding around a form.

C_W must be made large enough to prevent loop oscillation problems. The product of C_W times the parallel resistance of the two halves of R2 should be several times larger than the basic regulator settling time constant.

With load regulation compensation, the effects of cross regulation are worse than with no compensation. Multiple output supplies should be carefully evaluated for all expected conditions of output loading.

Frequency compensation

The frequency compensation capacitor C2 is much lower in this design than in others because the g_m of the LT1070 is much lower in the isolated mode than in the normal mode. See frequency compensation section for details.

Positive current-boosted buck converter

A current-boosted buck converter is shown in Figure 5.26. It can supply more output current than a standard buck converter or a flyback converter for larger input-output differentials because current flows to the output both when the switch is on *and* when it is off. The "on" cycle can supply up to 5A to the load. The off cycle will deliver 1/N times that much current. With N = 1/3, current delivered to the load during switch off time will be 15A. Total available load current will depend on switch duty cycle, which in turn is fixed by input voltage.

An operational amplifier must be added to generate a feedback signal which floats on top of the regulated output because that is where the ground pin of the LT1070 is tied. A1 is an LM308 selected because its output goes *low* when both its inputs are equal to the op amp negative supply voltage. This condition occurs at $V_{OUT} = 0V$ during start-up. If the op amp output went *high* during this condition, the LT1070 would never start up. R1 and R2 set output voltage, with the bottom of R1 returned directly to the load for "low" sensing. R4 and R5 force Kelvin sensing between the output and the ground pin of the LT1070. It appears that these resistors are shorted out, but the voltage drop across the wire from the ground pin of the LT1070 to the output will cause load regulation problems unless it is "sensed" by R4 and R5. These resistors can be eliminated if that wire is heavy gauge and less than 2″ long.

The following equations should be helpful in designing variations of this circuit.

$$R5 = \frac{(V_{OUT})(R1)}{V_{REF}} = \frac{(V_{OUT})(1.24k)}{1.244V} \quad (102)$$

$$\frac{R5}{R4} = \frac{R1}{R2} \quad (103)$$

$$N_{(MIN)} = \frac{V_{OUT} + V_F}{V_M - V_{IN} - V_{SNUB}} \quad (104)$$

$$DC = \frac{V_{OUT} + V_F}{V_{OUT} + V_F + N(V_{IN} - V_{OUT})} \quad (105)$$

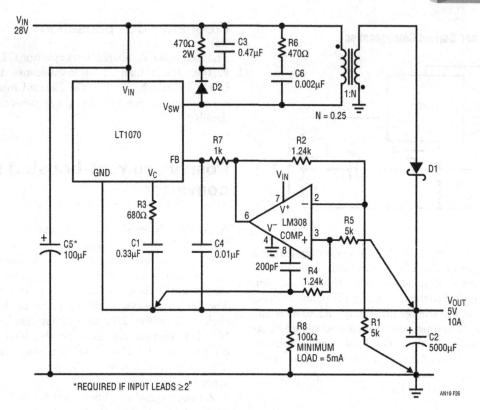

Figure 5.26 • Positive Current Boosted Buck Converter

$$L_{PRI} = \frac{V_{OUT}}{(\Delta I)(f)\left(N + \dfrac{V_{OUT}}{V_{IN} - V_{OUT}}\right)} \quad (106)$$

$$V_{P-P} = (ESR)(I_{OUT}) \frac{\left(\dfrac{V_{OUT}}{N} + V_{IN} - V_{OUT}\right)(1 - N)}{V_{IN}} \quad (107)$$

$$I_{OUT(MAX)} = \left(I_P - \frac{\Delta I}{2}\right)\left[\frac{V_{IN}}{V_{OUT} + V_F + N(V_{IN} - V_{OUT})}\right](0.8) \quad (108)$$

$$I_{PRI} = \frac{I_{OUT}}{V_{IN}}[V_{OUT} + N(V_{IN} - V_{OUT})] \quad (109)$$

(Add $\Delta I/2$ for peak primary current)

N = turns ratio
V_M = LT1070 maximum switch voltage
V_{SNUB} = snubber voltage (see flyback section)
V_F = forward voltage of D1

DC = switch duty cycle
ΔI = peak-to-peak primary ripple current
ESR = effective series resistance of C2
I_{PRI} = average primary current during switch-on time
V_{P-P} = peak-to-peak output ripple voltage
I_P = maximum rated switch current for LT1070

The value for N_{MIN} is based on switch breakdown. Low values give higher output current, but also higher switch voltage. ΔI is normally chosen at 20% to 40% of I_{PRI}. Note that the ripple equation contains the term $(1 - N)$ in the numerator, implying that output ripple current and voltage will be zero for $N = 1$. This is because of the simplifying assumption that ripple current into the output capacitor is the *difference* between primary and secondary current. This difference is zero for $N = 1$ and the equation is no longer valid.

Negative current-boosted buck converter

The negative buck converter in Figure 5.27 is capable of much higher output current than the standard buck converter upper limit of 5A. For design details, see positive current-boosted buck converter and standard negative buck converter sections.

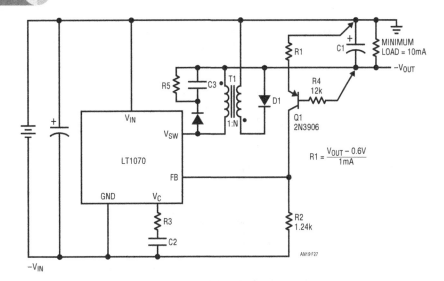

Figure 5.27 • Negative Current Boosted Buck Converter

Negative input/negative output flyback converter

This circuit in Figure 5.28 is normally used for negative output voltages *higher* than the negative input. If voltages *lower* than the input are required, see negative buck converter or negative current-boosted buck converter and standard negative buck converter sections.

The voltage divider, R1 and R2, is required to prevent forward bias on Q1. Connect R1, R2 and R3 exactly as shown for proper output sensing. Further design details may be taken from positive flyback converter section.

Positive-to-negative flyback converter

The positive input-negative output flyback converter in Figure 5.29 requires an external op amp to generate the feedback signal for the LT1070. R1 and R2 set output voltage with R1 scaled at 1kΩ/V. The bottom of R1 goes directly to the output for sensing. R3 and R4 provide the ground (low) sense. Any voltage drop between the ground pin of the LT1070 and the actual ground (+) output can cause load regulation problems. These are eliminated if R3 and R4 are connected exactly as shown. R3 and R4 can be

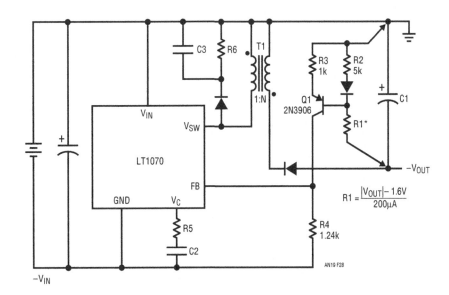

Figure 5.28 • Negative Input-Negative Output Flyback Converter

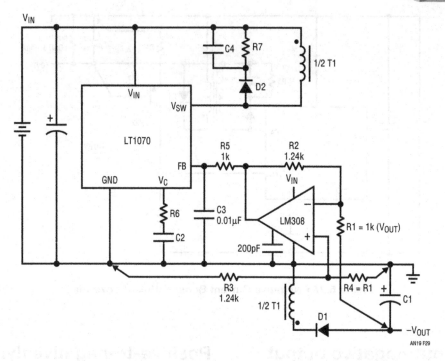

Figure 5.29 • Positive Input-Negative Output Flyback Converter

eliminated if the LT1070 ground pin is connected directly to output ground with a very short heavy wire. For design details, see positive flyback converter.

Voltage-boosted boost converter

The standard boost converter has a maximum output voltage slightly less than the maximum switch voltage of the LT1070. If higher voltages are desired, the inductor can be tapped as shown in Figure 5.30. The effect of the tap is to reduce peak switch voltage by:

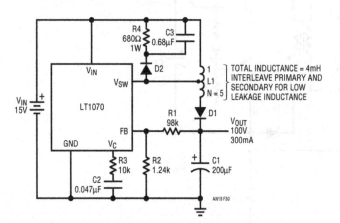

Figure 5.30 • Voltage-Boosted Boost Converter

$$(V_{OUT} - V_{IN})\left(\frac{N}{1+N}\right) \text{volts}$$

A large value for N will allow high output voltages to be regulated without exceeding maximum switch voltage.

A snubber is needed now to handle leakage inductance of the tap point. The following formulas will be helpful for variations on this design.

$$N_{(MIN)} = \frac{V_{OUT} - V_M + V_{SNUB}}{V_M - V_{IN} - V_{SNUB}} \text{ (use max } V_{IN}) \quad (110)$$

$$DC = \frac{V_{OUT} - V_{IN}}{V_{OUT} + N(V_{IN})} \quad (111)$$

$$I_{OUT(MAX)} = \frac{I_P \frac{\Delta I}{2}(V_{IN})}{V_{OUT} + N(V_{IN})} \quad (112)$$

$$I_{PRI} = \frac{I_{OUT}[V_{OUT} + N(V_{IN})]}{V_{IN}} \quad (113)$$

Average during switch-on time. For peak, add $\Delta I/2$.

$$L_{PRI} = \frac{V_{IN}(V_{OUT} - V_{IN})}{(\Delta I)(f)[V_{OUT} + N(V_{IN})]} \quad (114)$$

$$\Delta I \approx 20\% \text{ to } 40\% \text{ of } I_{PRI}$$

$$V_{P-P} = \frac{(I_{OUT})(ESR)[V_{OUT} + N(V_{IN})]}{V_{IN}(N + 1)} \qquad (115)$$

DC = switch duty cycle
V_{SNUB} = snubber voltage (see flyback section for details)
V_M = maximum allowed LT1070 switch voltage
I_P = maximum LT1070 switch current
ΔI = peak-to-peak primary current ripple
ESR = effective series resistance of C
V_{P-P} = peak-to-peak output voltage ripple

L1 should be wound for minimum leakage inductance by using bifilar winding or interleaved windings. R3 and C2 are selected using the technique described in the frequency compensation section. For snubber details see flyback description section. This regulator is not short-circuit proof because L1 and D1 short input to ground when output is shorted.

Negative boost converter

The LT1070 can be used as a negative boost regulator as shown in Figure 5.31 by using the same diode-coupled feedback technique used in the positive buck mode. Basically, D2 and C3 create a peak detector which gives a voltage across C3 equal to the output voltage. R1 and R2 act as a voltage divider to set output voltage at:

$$(V_{REF})(R1 + R2)/R2$$

C3 also acts as a floating power supply for the LT1070. The ground pin of the LT1070 switches back and forth between the output voltage and ground to drive the inductor, L1. For proper circuit operation, a minimum preload of 10mA is required on the output (shown as R_0).

For further design information, see positive boost converter section for details on L1, C1, D1 and output filters. The feedback scheme used here is discussed in more detail in the positive buck section. A refinement in the feedback is that the power transistor driver current flowing into the V_{IN} pin must come from D2 and C3. This tends to

compensate for the series resistance of D1 as it affects load regulation.

Positive-to-negative buck boost converter

This positive-to-negative converter uses the same feedback technique as the positive buck converter. Normal feedback cannot be used because the ground pin of LT1070 is switching back and forth between $+V_{IN}$ and $-V_{OUT}$. To generate a floating feedback signal, D2 peak detects the output voltage during the LT1070 switch-off time. This voltage appears across C3 as a floating DC level which is used as feedback to the LT1070. Output voltage is set by the ratio of R1 to R2. R4 is used to limit the effect of turn-on spikes across the main catch diode, D1. Without this resistor, D1 turn-on spikes would cause C3 to charge to an abnormally high voltage and the output voltage would sag down at high load currents.

D3 and C4 are used to generate a floating supply for the LT1070. The voltage across C4 will peak detect to (V_{OUT}) volts. R5 is added to ensure start-up. R6 is a preload, required only if the normal load can drop to zero current.

For further design details on this circuit, the basic formulas from the negative-to-positive buck/boost converter may be used, along with the feedback explanation from the positive buck converter.

Positive-to-Negative Buck Boost Converter

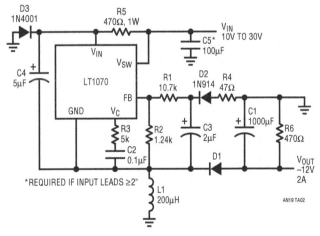

Current-boosted boost converter

This tapped inductor version of the boost converter can offer significant increases in output power when the input-output voltage differential is not too high. The ratio of output current for this converter compared to a standard boost converter is:

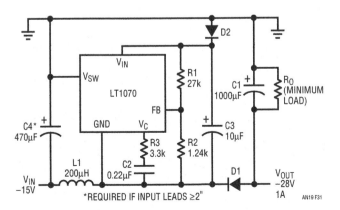

Figure 5.31 ● Negative Boost Regulator

$$\frac{I_{OUT}}{I_{BOOST}} = \frac{N+1}{N\left(1 - \frac{V_{IN}}{V_{OUT}}\right) + 1}$$

If $V_{OUT} \rightarrow V_{IN}$, the increase in output current approaches $N + 1$. Maximum N, however, is limited by switch breakdown voltage:

$$N_{(MAX)} = \frac{V_M - V_{OUT} - V_{SNUB}}{V_{OUT} - V_{MIN}}$$

V_M = maximum LT1070 switch voltage
V_{SNUB} = snubber voltage (see flyback section)
V_{MIN} = minimum input voltage

For V_M = 60V, V_{OUT} = 28V, V_{SNUB} = 8V, V_{MIN} = 16V:

$$N_{(MAX)} = \frac{60 - 28 - 8}{28 - 16} = 2$$

The increase in output current is:

$$\frac{I_{OUT}}{I_{BOOST}} = \frac{2+1}{2\left(1 - \frac{16}{28}\right) + 1} = 1.62 = 62\%$$

Actual maximum output current is:

$$I_{OUT(MAX)} = \frac{I_P - \Delta I/2}{\frac{V_{OUT}}{V_{MIN}} - \frac{N}{N+1}} = \frac{5 - 0.5}{\frac{28}{16} - \frac{2}{3}} = 4.15A$$

I_P = maximum LT1070 switch current
ΔI = increase in inductor current during switch-on time

$$\Delta I = \frac{V_{OUT} - V_{IN}}{(L)(f)\left(\frac{V_{OUT}}{V_{IN}} - \frac{N}{N+1}\right)}$$

L = total inductance
Operating duty cycle is given by:

$$DC = \frac{V_{OUT} - V_{IN}}{V_{OUT} - \left(\frac{N}{N+1}\right)(V_{IN})}$$

A reasonable value for total inductance is found by assuming that this circuit is used near peak switch current of 5A and allowing a 20% increase in switch current during switch-on time→ΔI = 1A:

$$L_{TOTAL} = \frac{V_{OUT} - V_{MIN}}{(f)(\Delta I)\left(\frac{V_{OUT}}{V_{MIN}} - \frac{N}{N+1}\right)}$$

$$= \frac{28 - 16}{(40 \bullet 10^3)(1)\left(\frac{28}{16} - \frac{2}{3}\right)} = 277 \mu H$$

Snubber values are empirically selected to limit snubber voltage to the value chosen ($\approx$8V). For lowest snubber losses, the "1" and "N" sections of the inductor should be wound for maximum coupling (consult manufacturers).

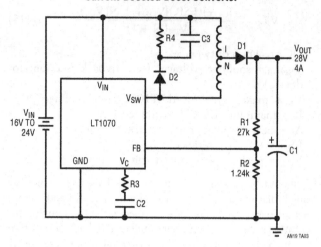

Current-Boosted Boost Converter

Forward converter

Forward converters can use smaller cores than flyback converters because they do not need to store energy in the core. Energy is transferred directly to the output during switch "on" time. The output secondary (N) is positive and delivering current through D1 when the LT1070 switch is on (V_{SW} low). At switch turn-off, the output winding goes negative and output current flows through D2 as in a buck regulator. A third winding (M) is needed in a single switch forward converter to define switch voltage during switch-off time. This "reset" winding, however, limits the maximum duty cycle allowed for the switch. The voltage across the switch during its off state is:

$$V_{SW} = V_{IN} + \frac{V_{IN}}{M} + V_{SNUB}$$

V_{SNUB} = snubber voltage spike caused by leakage inductance

By rearranging this formula, a minimum value for M can be found:

$$M_{(MIN)} = \frac{V_{IN(MAX)}}{V_M - V_{IN(MAX)} - V_{SNUB}}$$

V_M = maximum LT1070 switch voltage
$V_{IN(MAX)}$ = maximum input voltage

For the circuit shown, with $V_{IN(MAX)}$ = 30V, and selecting V_{SNUB} = 5V and V_M = 60V:

$$M_{(MIN)} = \frac{30}{60 - 30 - 5} = 1.2$$

The value of M will define maximum switch duty cycle. If the LT1070 attempts to operate at a duty cycle above this limit, the core will saturate because the volt-second product across the primary in the switch-off state will not be enough to keep flux balance. Duty cycle is limited to:

$$DC_{(MAX)} \frac{1}{1+M} = \frac{1}{1+1.2} = 45\%$$

For maximum output current, N should be as small as possible. Smaller values of N, however, require larger duty cycles, so N is limited to a minimum of:

$$N_{(MIN)} = \frac{(M + 1)(V_{OUT} + V_F)}{V_{IN(LOW)}}$$

V_F = D1 and D2 forward voltage
$V_{IN(LOW)}$ = minimum input voltage

For the circuit shown, with $V_F = 0.6V$, $V_{IN(LOW)} = 20V$:

$$N_{(MIN)} = \frac{(1.2 + 1)(5 + 0.6)}{20} = 0.62$$

To avoid core saturation during normal operation, primary inductance must be a minimum value determined by core volume and core flux density:

$$L_{PRI} \geq \left[\frac{V_{OUT} + V_F}{(N)(B_M)(f)}\right]^2 \left[\frac{(0.4\pi)(\mu_e)}{(V_e)(10^{-8})}\right]$$

B_M = maximum operating flux density
f = LT1070 operating frequency (40kHz)
V_e = core volume
μ_e = effective core permeability

For B_M = 2000 gauss (typical for ferrites), $V_e = 6cm^3$, and μ_e = 1500, $V_{OUT} + V_F = 5.6V$, N = 0.62:

$$L_{PRI} \geq \left[\frac{5.6}{(0.62)(2000)(40 \bullet 10^3)}\right]^2 \left[\frac{(0.4\,\pi)(1500)}{(6)(10^{-8})}\right]$$

$$= 400\mu H$$

The operating flux density of forward converters is often limited by temperature rise rather than saturation. At 2000 gauss, the core loss of a typical ferrite is $0.25W/cm^3$. Total core loss at $V_e = 6cm^3 \approx 1.5W$. When this is combined with copper winding losses, there may be excess core temperatures. Larger cores will allow more space for copper, or can be operated at lower flux density with the same copper loss. See transformer design guide for further details.

Conventional forward converters use a flip-flop to limit maximum duty cycle to 50% and set M = 1. The LT1070 will let duty cycle go to $\approx$ 95% during start-up and low input voltage conditions. This would cause core saturation and subsequent primary and switch currents of up to 10A. To avoid this, Q1 and R5 have been added. Onset of core saturation will cause a voltage drop across R5 high enough to turn on Q1 at each cycle. This pulls down on the V_C pin, reducing duty cycle and maintaining normal switch currents. R6 and C4 filter out spikes.

Operating duty cycle is given by:

$$DC = \frac{V_{OUT} + V_F}{(N)(V_N)}$$

The output filter inductor (L1) is chosen as a trade-off between maximum output power, output ripple, physical size and loop transient response. A reasonable value is one which gives a peak-to-peak inductor ripple current (ΔI_L) of $\approx$ 20% of I_{OUT}. This leads to a value for L1 of:

$$L1 = \frac{V_{OUT}[(N)(V_{IN}) - V_{OUT}]}{[(0.2)(I_{OUT})][(N)(V_{IN})(f)]}$$

For $I_{OUT} = 6A$, $V_{IN} = 25V$, $V_{OUT} = 5V$, N = 0.62:

$$L1 = \frac{5[(0.62)(25) - 5]}{[(0.2)(6)][(0.62)(25)(40 \bullet 10^3)]} = 70\mu H$$

Larger values of L1 will increase maximum output current only slightly. Output ripple voltage will go down inversely with larger L1, but physical size will quickly become a problem for large values because the inductor must handle large DC currents. Peak inductor current is equal to $I_{OUT} + \Delta I_L/2$.

Forward Converter

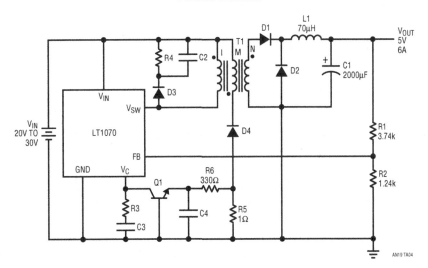

Maximum output current for this forward converter is given by:

$$I_{OUT(MAX)} = \left(\frac{I_P}{N} - \frac{\Delta I_L}{2} - \frac{\Delta I_{PRI}}{N}\right)(0.9)$$

I_P = maximum LT1070 switch current
ΔI_{PFI} = peak primary magnetizing current
$\quad = V_{OUT}/(f)(N)(L_{PRI})$
ΔI_L = peak-to-peak output inductor current
0.9 = fudge factor for losses

For $I_P = 5A$, $N = 0.62$, $\Delta I_L = 1.2A$, $\Delta I_{PFI} = 0.5A$:

$$I_{OUT(MAX)} = \left(\frac{5}{0.62} - \frac{1.2}{2} - \frac{0.5}{0.62}\right)(0.9) = 6A$$

Output voltage ripple (P-P) is assumed to be set by L1 and the ESF of C1:

$$V_{P\text{-}P} = (\Delta I_L)(ESR1) = \frac{ESR1(V_{OUT})[N(V_{IN}) - V_{OUT}]}{(L1)(f)(N)(V_{IN})}$$

$ESR1$ = effective series resistance of C1
If we assume 0.02Ω for ESF1, and $V_{IN} = 25V$,

$$V_{P\text{-}P} = \frac{(0.02)(5)[(0.62)(25) - 5]}{(70 \bullet 10^{-6})(40 \bullet 10^3)(0.62)(25)}$$

$$= 24mV_{P\text{-}P}$$

If less output ripple is desired, the most effective method may be to add an LC filter. See section on output filters.

Frequency compensation

Although the architecture of the LT1070 is simple enough to lend itself to a mathematical approach to frequency compensation, the added complication of input and/or output filters, unknown capacitor ESR, and gross operating point changes with input voltage and load current variations all suggest a more practical empirical method. Many hours spent on breadboards have shown that the simplest way to optimize the frequency compensation of the LT1070 is to use transient response techniques and an "R/C box" to quickly iterate toward the final compensation network.

There are many ways to inject a transient signal into a switching regulator, but the suggested method is to use an AC coupled output load variation. This technique avoids problems of injection point loading and is general to all switching topologies. The only variation required may be an amplitude adjustment to maintain small signal conditions with adequate signal strength. Figure 5.32 shows the setup.

A function generator with 50Ω output impedance is coupled through a $50\Omega/1000\mu F$ series RC network to the regulator output. Generator frequency is noncritical. A good starting point is $\approx 50Hz$. Lower frequencies may cause a blinking scope display which is annoying to work with. Higher frequencies may not allow sufficient settling time for the output transient. Amplitude of the generator output is typically set at $5V_{P\text{-}P}$ to generate a $100mA_{P\text{-}P}$ load variation. For lightly loaded outputs ($I_{OUT} < 100mA$), this level may be too high for small signal response. If the positive and negative transition settling waveforms are significantly different, amplitude should be reduced. Actual *amplitude* is not particularly important because it is the *shape* of the resulting regulator output waveform which indicates loop stability.

A 2-pole oscilloscope filter with $f = 100kHz$ is used to block switching frequencies. Regulators without added LC output filters have switching frequency signals at their outputs which may be much higher amplitude than the low frequency settling waveform to be studied. The filter frequency is high enough to pass the settling waveform with no distortion.

Oscilloscope and generator connections should be made exactly as shown to prevent ground loop errors. The oscilloscope is sync'd by connecting channel "B" probe to the generator output, with the ground clip of the second probe

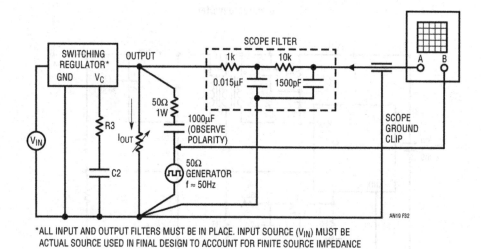

*ALL INPUT AND OUTPUT FILTERS MUST BE IN PLACE. INPUT SOURCE (V_{IN}) MUST BE ACTUAL SOURCE USED IN FINAL DESIGN TO ACCOUNT FOR FINITE SOURCE IMPEDANCE

Figure 5.32 • Testing Loop Stability

connected to exactly the same place as channel "A" ground. The standard 50Ω BNC sync output of the generator should *not* be used because of ground loop errors. It may also be necessary to isolate *either* the generator or oscilloscope from its third wire (earth ground) connection in the power plug to prevent ground loop errors in the 'scope display. These ground loop errors are checked by connecting channel "A" probe tip to exactly the same point as the probe ground clip. Any reading on channel "A" indicates a ground loop problem.

Once the proper setup is made, finding the optimum values for the frequency compensation network is fairly straightforward. Initially, C2 is made large ($\geq 2\mu F$) and R3 is made small ($\approx 1k$). This nearly always ensures that the regulator will be stable enough to start iteration. Now, if the regulator output waveform is single-pole overdamped, (see the waveforms in Figure 5.33) the value of C2 is *reduced* in steps of about 2:1 until the response becomes slightly underdamped. Next, R3 is *increased* in steps of 2:1 to introduce a loop "zero." This will normally improve damping and allow the value of C2 to be further reduced. Shifting back and forth between R3 and C2 variations will now allow one to quickly find optimum values.

If the regulator response is underdamped with the initial large value of C, R should be increased immediately before larger values of C are tried. This will normally bring about the overdamped starting condition for further iteration.

Just what is meant by "optimum values" for R3 and C2? This normally means the smallest value for C2 and the largest value for R3 which still guarantee no loop oscillations, and which result in loop settling that is as rapid as possible. The reason for this approach is that it minimizes the variations in output voltage caused by *input ripple voltage* and *output load transients*. A switching regulator which is grossly overdamped will never oscillate, but it may have unacceptably large output transients following sudden changes in input voltage or output loading. It may also suffer from excessive overshoot problems on startup or short circuit recovery.

To guarantee acceptable loop stability under all conditions, the initial values chosen for R3 and C2 should be checked under all conditions of input voltage and load current. The simplest way of accomplishing this is to apply load currents of minimum, maximum and several points in between. At each load current, input voltage is varied from minimum to maximum while observing the settling waveform. The additional time spent "worst-casing" in this manner is definitely necessary. Switching regulators, unlike linear regulators, have large shifts in loop gain and phase with operating conditions.

If large temperature variations are expected for the regulator, stability checks should also be done at the temperature extremes. There can be significant temperature variations in several key component parameters which affect stability; in particular, input and output capacitor value and their ESR and inductor permeability. LT1070 parametric variations also need some consideration. Those which affect loop stability are error amplifier g_m and the transfer function of V_C pin voltage *versus* switch current (listed as a transconductance under electrical specifications). For modest temperature variations, conservative overdamping under worst-case room temperature conditions is usually sufficient to guarantee adequate stability at all temperatures.

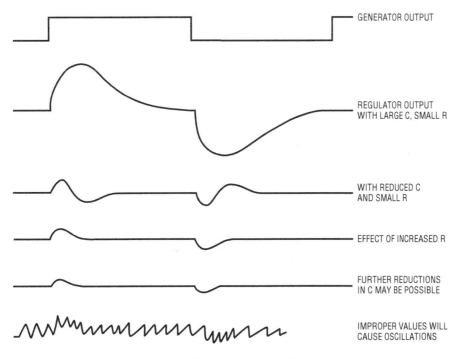

GENERATOR OUTPUT

REGULATOR OUTPUT
WITH LARGE C, SMALL R

WITH REDUCED C
AND SMALL R

EFFECT OF INCREASED R

FURTHER REDUCTIONS
IN C MAY BE POSSIBLE

IMPROPER VALUES WILL
CAUSE OSCILLATIONS

Figure 5.33 • Output Transient Response

Check margins

One measure of stability "margin" is to vary the selected values of both R and C by 2:1 in all possible combinations. If the regulator response remains reasonably well damped under all line and load conditions, the regulator can be considered fairly tolerant of parametric variations. Any tendency towards an underdamped (ringing) response indicates that a more conservative compensation may be needed.

There are several large signal dynamic tests which should also be done on a completed regulator design. The first is to check response to the worst-case large amplitude load variation. A sudden change from light load to full load current may cause the regulator to have an unacceptably large transient dip in output voltage. The simplest cure for this is to increase the size of the output capacitor. Lower inductor values and less conservative frequency compensation also help. A second consideration is the output overshoot created when a large load is suddenly *removed*.

This is potentially more dangerous than a dip because a large overshoot may destroy loads still connected to the regulator output.

Eliminating start-up overshoot

Another transient condition to be checked is *start-up overshoot*. When input voltage is first applied to a switching regulator, the regulator dumps full short-circuit current into the output capacitor attempting to bring the output up to its regulated value. The output can then overshoot well beyond its design value before the control loop is able to idle back the output current. The amplitude of the overshoot can be anywhere from millivolts to tens of volts depending on topology, line and load conditions and component values. This same overshoot possibility exists for

output recovery from output shorts. Again, larger output capacitors, smaller inductors and faster loop response help reduce overshoot. There are also several ways to force slow start-up to eliminate the overshoot. The first is to put a capacitor across the output voltage divider. This creates a time-dependent output voltage setting during start-up and usually eliminates overshoot. This capacitor also has an effect on feedback loop characteristics during normal operation and it can create unacceptably large negative transients on the feedback pin if the output voltage is high and a sudden output short occurs. The transient problem is eliminated by inserting a resistor in series with the feedback pin (see feedback pin part of pin description section). If undesirable loop characteristics are created by the capacitor, they can be eliminated by diode coupling the capacitor as shown in Figure 5.34

Another general technique for forcing slow start-up is to clamp the V_C pin to a capacitor, C4. The value of R4 is chosen to give a voltage across R_Z of 2V at worst-case *low* input voltage ($I_{R4} = 100\mu A$). C4 is then selected to ramp V_C slow enough to eliminate start-up overshoot. C4 should be made no larger than necessary to prevent long reset times. A momentary drop to zero volts at the input may not allow enough time for C4 to discharge fully. If input dropouts of less than 5R4C4 seconds are anticipated, R4 should be paralleled with a diode (cathode to input) for fast reset.

External current limiting

The LT1070 has internal *switch* current limiting which operates on a cycle-by-cycle basis and limits peak switch current to ≈ 9A at low duty cycles and ≈ 6A at high duty cycles. The actual *output* current limit value may be much *higher or lower* depending on topology, input voltage and

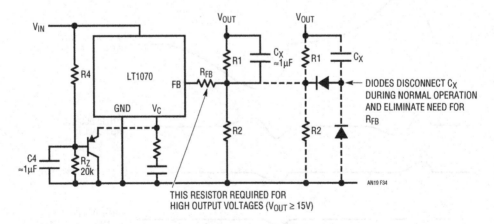

Figure 5.34 • Eliminating Start-Up Overshoot

output voltage. The following formulas give an approximate value for output current limit under short-circuit output conditions and at the point where output voltage just begins to fall below its regulated value.

	OVERLOAD CURRENT (AMPS)	SHORT CIRCUIT CURRENT (AMPS)
Buck	5 to 8	≈ 8
Boost	$(5\,to\,8)(V_{IN}/V_{OUT})$	Not Allowed
Buck-Boost (Inverting)	$\dfrac{5\ to\ 8}{1+V_{OUT}/V_{IN}}$	≈ 8
Current-Boosted Buck	$(5\ to\ 8)\left(\dfrac{V_{IN}}{V_{OUT}+N(V_{IN}-V_{OUT})}\right)$	$\approx 8/N$
Voltage-Boosted Boost	$(5\ to\ 8)\left(\dfrac{V_{IN}}{V_{OUT}+N(V_{IN})}\right)$	Not Allowed
Flyback (Continuous)	$\dfrac{5\ to\ 8}{(V_{OUT}/V_{IN})+N}$	$\approx 8/N$
Flyback (Discontinuous)	Depends on L	$\approx 8/N$
Current-Boosted Boost	$\dfrac{5\ to\ 8}{(V_{OUT}/V_{IN})-(N/N+1)}$	Not Allowed
Forward	5 to 8/N	$\approx 8/N$

These formulas show that short-circuit current can be much higher than full load current for some topologies. If either full load current or short circuit is much higher than is required for a specific application, external current limiting can be added. This has the advantage of reducing stress on external components, avoiding overload on the input supply and reducing heat sink requirements on the LT1070 itself.

The LT1070 is externally current limited by clamping the V_C pin. The techniques shown in Figures 5.35 to 5.39 are just a few of the ways this can be accomplished.

The relationship between *switch* current limit point and V_C clamp voltage is *approximately*:

$$I_{SW(MAX)} = 9(V_C - 1) - 3 \bullet (DC)\ amps$$

DC = switch duty cycle

This relationship is somewhat temperature dependent. The current limit point falls at about 0.3%/°C, so the value set at room temperature should be factored to allow for adequate current limit at higher temperatures. Also, the factor "9" and "3" vary ± 30% in production, so a conservative design will normally clamp switch current to about twice the value needed for maximum load current. This

can result in rather high short-circuit currents, so the current limit scheme may want to include "foldback," wherein the peak switch current is clamped to a lower value with $V_{OUT} = 0V$. By varying the amount of foldback, the short-circuit current can be made greater than, equal to, or less than full load current.

Simple current limiting is shown in Figure 5.35. V_X is an external voltage which could be a separate regulated voltage or the unregulated input voltage. R2 is selected to give approximately 2V across R1. The value of R1 is kept to 500Ω or below to keep the knee of the current limit as sharp as possible. If individual adjustment is not necessary, R1 can be replaced with a fixed resistor. (Note that in some topologies the ground, V_C and FB pins of the LT1070 are switching at high voltage levels. This will require V_X to be referenced to the LT1070 ground pin, not system ground.)

In Figure 5.36, D1 has been replaced with a PNP transistor to reduce the current drain through R1 to 100μA. This is helpful in situations where the LT1070 is used in the total shutdown mode.

In Figure 5.37, *foldback* current limiting is generated by clamping the V_C pin to an output voltage divider. This will reduce short-circuit current by an amount which depends on the relative values of R3, R4 and R5. R5 is needed to prevent "latch off," wherein the output current drops to zero during short circuit and stays at zero even if the short is removed. If this latch-off action is desirable, R5 can be eliminated. A normally closed "start" switch can then be placed in series with D1. If *nonzero* short-circuit current

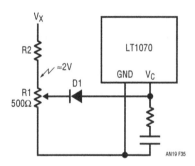

Figure 5.35 • External Current Limit

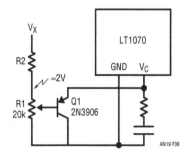

Figure 5.36 • External Current Limit

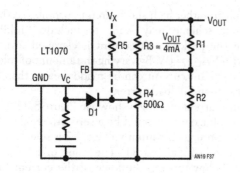

Figure 5.37 • Foldback Current Limit

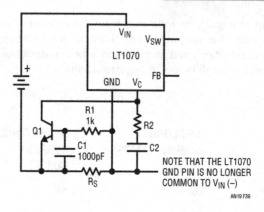

Figure 5.39 • External Current Limit

is desired, R5 is selected to give desired short-circuit current and R4 is adjusted for full load current limit. There is some interaction, so R4 should be set to about midspan for initial selection of R5. If less interaction is desired between R4 and R5 adjustments, a 470Ω resistor can be inserted in series with the wiper on R4 to form a voltage divider with R5.

A current transformer (T1) is used in Figure 5.38 to generate a more precise current limit. The primary is placed in series with the output switching diode for buck, flyback and buck/ boost configurations. Output diode peakcurrent is limited to:

$$I_{PEAK} = \frac{N}{R5}\left(V_{BE} + \frac{V_{OUT} \bullet R4}{R3 + R4}\right)$$

V_{BE} = base-emitter voltage of Q1

The R3/R4 divider provides foldback as shown in the formula, with short-circuit diode current limited to $N(V_{BE}/R5)$. In a typical application, R3 is selected to set the voltage across R4 to ≈1V at normal output voltage. Then R5 is calculated from:

$$R5 = \frac{N(V_{BE} + V_{R4})}{I_{PEAK(PRI)}}$$

The effective *secondary* current limit sense voltage is $V_{BE} + V_{R4}$ at full output voltage and just V_{BE} during short circuit, giving ≈ 2.7:1 foldback ratio. The diode in T1 secondary allows the secondary to "reset" between

current pulses, so that true peak-to-peak diode current is controlled. C1 is used to filter out spikes and noise.

In Figure 5.39 a current limit sense resistor (R_S) is placed in series with the ground pin of the LT1070. Peak *switch* current is limited to $V_{BE}(Q1)/R_S$. This circuit is useful only in situations where the negative input line and the negative output line do not have to be common. Power dissipation in R_S will be fairly high; P ≈ (0.6V) (I_{PEAK})(DC), where DC is the duty cycle of the switch. R1 and C1 filter out noise spikes and catch diode reverse turn-off current spikes.

Driving external transistors

High input voltage applications using the LT1070 require an external high voltage transistor. The transistor is connected in a common gate or common base mode as shown in Figures 5.40 and 5.41. This allows the LT1070 internal current sensing to continue functioning and operates the external transistor in a mode which maximizes both operating voltage and switching speed capability.

In Figure 5.40, the LT1070 drives an N-channel power MOSFET. A separate low voltage supply is used to power the LT1070 and to establish forward gate drive to the MOSFET. Typical gate drive requirement is 10V, with 20V as a typical maximum. The forward gate drive applied to the MOSFET is equal to the supply voltage minus the saturation voltage of the LT1070 switch (saturation voltage

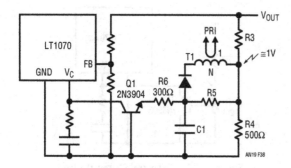

Figure 5.38 • Transformer Current Limit

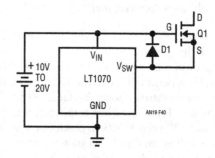

Figure 5.40 • Driving External MOSFET

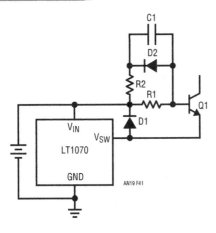

Figure 5.41 • Driving External NPN

provides a forward base current surge at turn-on. Typical values are in the range of $0.005\mu F$ to $0.05\mu F$. D1 clamps the emitter voltage at turn-off. It prevents full collector current from flowing out the base lead during the turn-off delay time (0.5μ to $2\mu s$). D2 and R1 establish the reverse base turn-off current. The voltage across R2 during turn-off delay time is approximately one diode drop. With R2 = 3Ω and a diode drop of 800mV, this would create $\approx 270mA$ reverse base current during turn-off. Reverse leakage in the "off" state is not a problem with this circuit because D1 and D2 force the emitter-base voltage to zero bias when the LT1070 switch is off. D1 should be selected for fast turn-on. It must handle current equal to collector current for times equal to the turn-off time of the transistor. D2 can be any medium speed diode rated for several hundred milli-amps forward current spikes (1N914, etc.).

is typically under 1V). D1 is used to clamp the source during turn-off; it does not slow down turn-off. Diode requirements are that it withstand narrow (100ns) current spikes equal to drain current and that it turn "on" rapidly to provide proper clamping.

In Figure 5.41, the LT1070 drives an NPN bipolar transistor. These devices require high surge base currents at turn-on and turn-off to ensure fast switching times. R1 establishes DC base drive which might be $\approx 1/5$ of collector current. C1

Output rectifying diode

The output diode is often the major source of power loss in switching regulators, especially with output voltages below 10V. It is therefore very important to be able to calculate diode peak current and average power dissipation to ensure adequate diode ratings. The chart in Figure 5.42 lists average diode power dissipation and peak diode current for

TOPOLOGY	AVERAGE DIODE DISSIPATION P_D (WATTS)	PEAK DIODE CURRENT		PEAK DIODE VOLTAGE
		AT FULL LOAD (AMPS)	SHORT CIRCUIT (AMPS)	
Buck	$(I_{OUT})(V_F)(1 - V_{OUT}/V_{IN})$	$I_{OUT} + \dfrac{\Delta I}{2}$	≈ 8	V_{IN}
Current-Boosted Buck	$(I_{OUT})(V_F)(1 - V_{OUT}/V_{IN})$	$\dfrac{I_{OUT}}{V_{IN}}\left(\dfrac{V_{OUT}}{N} - V_{OUT} + V_{IN}\right)$	$\approx 8/N$	$V_{OUT} + N(V_{IN})$
Boost	$(I_{OUT})(V_F)$	$\dfrac{I_{OUT}(V_{OUT})}{V_{IN}} + \dfrac{\Delta I}{2}$	Not Allowed	V_{OUT}
Current-Boosted Boost	$(I_{OUT})(V_F)$	$I_{OUT}\left[\dfrac{V_{OUT}}{V_{IN}} + N\left(\dfrac{V_{OUT}}{V_{IN}} - 1\right)\right] + \dfrac{\Delta I_{PRI}(N+1)}{2}$	Not Allowed	$V_{OUT} - V_{IN}\left(\dfrac{N}{N+1}\right)$
Voltage-Boosted Boost	$(I_{OUT})(V_F)$	$\dfrac{I_{OUT}(N \bullet V_{IN} + V_{OUT})}{V_{IN}(N+1)}$	Not Allowed	$V_{OUT} + N(V_{IN})$
Inverting (Buck Boost)	$(I_{OUT})(V_F)$	$\dfrac{I_{OUT}(V_{IN} + V_{OUT})}{V_{IN}} + \dfrac{\Delta I}{2}$	≈ 8	$V_{OUT} + V_{IN}$
Flyback (Continuous)	$(I_{OUT})(V_F)$	$I_{OUT}\left(1 + \dfrac{V_{OUT}}{N(V_{IN})}\right) + \dfrac{\Delta I_{PRI}}{2N}$	$\approx 7/N$	$V_{OUT} + N(V_{IN})$
Flyback (Discontinuous)	$(I_{OUT})(V_F)$	$\dfrac{1}{N}\sqrt{\dfrac{2(I_{OUT})(V_{OUT})}{f(L_{PRI})}}$	$\approx 7/N$	$V_{OUT} + N(V_{IN})$

AN19 F42

Figure 5.42

normal loads. It also lists diode current under shorted output conditions where the diode duty cycle approaches 100%, and peak and average currents are essentially the same.

The value for diode forward voltage (V_F) used in the average power formulas is the voltage specified for the diode under peak current conditions listed in the next column. The peak current formulas assume no ripple current in the inductor or transformer, but average power calculations will be reasonably close even with fairly high ripple. Boost converters in particular are hard on output diodes when the output voltage is significantly higher than the input voltage. This gives peak diode currents much higher than the average, and manufacturers current ratings must be used with caution.

The most stressful condition for output diodes is overload or short-circuit conditions. The internal current limit of the LT1070 is typically 9A at low switch duty cycles. This is almost a factor of two higher than the 5A rated switch current, so that even if the regulator is used near its limit at full load, the output diode current may double under current limit conditions. If full load output current requires only a fraction of the 5A rated switch current, the ratio of diode short-circuit current to full load current may be much higher than two to one. A regulator designed to withstand continuous short conditions must either use diodes rated for the full short-circuit current shown in the fourth column, or it must incorporate some form of external current limiting. See current limit section for more details.

The last column in Figure 5.42 shows maximum reverse diode voltage. When calculating this number, be sure to use worst-case high input voltage. Transformer or tapped inductor designs may have an additional damped "ringing" waveform which adds to peak diode voltage. This can be reduced with a series R/C damper network in parallel with the diode.

Switching diodes have two important transient characteristics: reverse recovery time and forward turn-on time. Reverse recovery time occurs because the diode "stores" charge during its forward conducting cycle. This stored charge causes the diode to act like a low impedance conductive element for a short period of time after reverse drive is applied. Reverse recovery time is measured by forward biasing the diode with a specified current, then forcing a second specified current *backwards* through the diode. The time required for the diode to change from a reverse *conducting* state to its normal reverse *nonconducting* state, is reverse recovery time. Hard turn-off diodes switch abruptly from one state to the other following reverse recovery time. They, therefore, dissipate very little power even with moderate reverse recovery times. Soft turn-off diodes have a gradual turn-off characteristic that can cause considerable diode dissipation during the turn-off interval. Figure 5.43 shows typical current and voltage waveforms for several commercial diode types used in an LT1070 boost converter with V_{IN} = 10V, V_{OUT} = 20V, 2A.

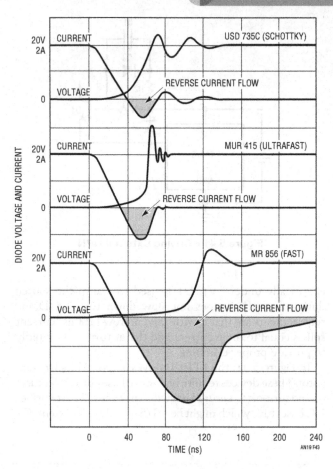

Figure 5.43 • Diode Turn-Off Characteristics

Long reverse recovery times can cause significant extra heating in the diode or the LT1070 switch. Total power dissipated is given by:

$$P_{tRR} = (V)(f)(t_{RR})(I_F)$$

V = reverse diode voltage
f = LT1070 switching frequency
t_{RR} = reverse recovery time
I_F = diode forward current just prior to turn-off

With the circuit mentioned, I_F is 4A, V = 20V and f = 40kHz. Note that diode "on" current is twice output current for this particular boost configuration. A diode with t_{RR} = 300ns creates a power loss of:

$$Pt_{RR} = (20)(40 \bullet 10^3)(300 \bullet 10^{-6})(4) = 0.96\,W$$

If this same diode had a forward voltage of 0.8V at 4A, its forward loss would be 2A *(average* current) times 0.8V equals 1.6W. Reverse recovery losses in this example are nearly as large as forward losses. It is important to realize, however, that reverse losses may not necessarily increase *diode* dissipation significantly. A hard turn-off diode will shift much of the power dissipation to the LT1070 switch, which will undergo a high current *and* high voltage condition during the duration of reverse recovery time. This has not shown to be harmful to the LT1070, but the power loss remains.

Diode *turn-on* time can potentially be more harmful than reverse turn-off. It is normally assumed that the output diode clamps to the output voltage and prevents the inductor or transformer connection from rising higher than the output. A diode that turns "on" slowly can have a very high forward voltage for the duration of turn-on time. The problem is that this increased voltage appears across the LT1070 switch. A 20V turn-on spike superimposed on a 40V boost mode output pushes switch voltage perilously close to the 65V limit. The graphs in Figure 5.44 show diode turn-on spikes for three common diode types, fast, ultrafast and Schottky. The height of the spike will be dependent on rate of rise of current and the final current value, but these graphs emphasize the need for fast turn-on characteristics in applications which push the limits of switch voltage.

Fast diodes can be useless if the stray inductance is high in the diode, output capacitor or LT1070 loop. 20-gauge hookup wire has ≈ 30nH/inch inductance. The current fall time of the LT1070 switch is ≈10^8A/sec. This generates a voltage of $(10^8)(30 • 10^{-9}) = 3V$ *per inch* in stray wiring.

Keep the diode, capacitor and LT1070 ground/switch lead lengths *short*.

Input filters

Most switching regulator designs draw current from the input supply in pulses. The peak-to-peak amplitude of these current pulses is often equal to or higher than the load current. There is significant high frequency energy in the pulses which can cause EMI problems in some systems. The addition of a simple LC filter between the supply and the switching regulator can reduce the amplitude of this EMI by more than an order of magnitude at the switching frequency and several orders of magnitude at higher harmonic frequencies. The basic filter shown in Figure 5.45 can be added to any switching regulator.

The two major design considerations for the filter are the *reverse current transfer function* which determines ripple attenuation and the *filter output impedance function* which must satisfy regulator stability criteria. The stability problem occurs because switching regulators have a *negative input impedance* at low frequencies:

$$Z_{IN}(DC) = -\frac{(V_{IN})^2}{(V_{OUT})(I_{OUT})}$$

The output impedance of the filter has a sharp peak at the LC resonant frequency. If the output impedance is not well below the negative input impedance of the regulator at frequencies up to the bandwidth of the regulator control loop, the possibility for oscillation exists.

There is a basic conflict in the two filter requirements. High ripple attenuation is obtained with a large LC *product* with high Q, but this also tends to aggravate oscillation problems. This conflict is minimized by using large C with smaller L to get the required LC product, but size requirements may also limit this approach. An additional "fix" is to lower the Q of the filter by paralleling L with a small resistor (R_F). This has the disadvantage of limiting the filter attenuation at high frequencies. Filter Q is also reduced by the ESR (R_S) of the capacitor, but deliberately increasing ESR exacts a heavy penalty in ripple attenuation and power loss.

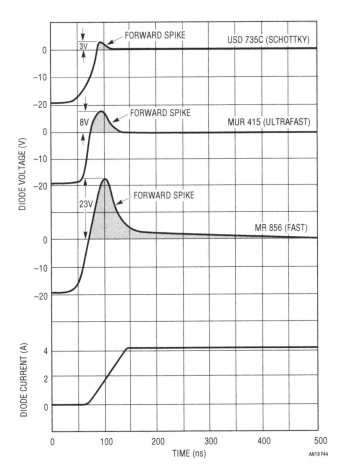

Figure 5.44 • Diode Turn-On Spike

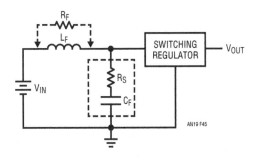

Figure 5.45

Ripple attenuation of an input filter may be calculated from:

$$\frac{I_{OUT(P-P)}}{I_{IN(P-P)}} = \frac{R_S}{R_F} + \frac{R_S(DC)(1-DC)}{(L)(f)}$$

R_S = effective series resistance of C
DC = switching regulator duty cycle

Note that this formula does not contain the value of C. This is because large electrolytic capacitors have a total impedance at 20kHz and above which is essentially equal to ESR. For ripple attenuation, therefore, the value of C is not important; the capacitor is selected on the basis of its ESR.

A typical filter might consist of a 10μH inductor and a 500μF capacitor with $R_S = 0.05\Omega$. Filter attenuation is *least effective* at 50% duty cycle (DC = 0.5), so we will use this number now for worst-case purposes. Ripple attenuation of this filter with $R_F = \infty$ is:

$$\frac{I_{OUT}}{I_{IN}} = \frac{(0.05)(0.5)(1-0.5)}{(10 \bullet 10^{-6})(40 \bullet 10^{3})} = 0.031 = 32:1$$

The formula assumes rectangular wave inputs with triangular outputs and yields the ratio of peak-to-peak values. Higher frequency components of the square wave current are attenuated much more than the overall attenuation figure.

Output impedance of the filter given by:

$$Z_{OUT} = \cfrac{1}{\cfrac{1}{R_F} - \cfrac{J}{\omega L} + \cfrac{J\omega C}{1 + (\omega R_S C)^2} + \cfrac{R_S(\omega C)^2}{1 + (\omega R_S C)^2}}$$

ω = radian frequency = $2\pi(f)$

This formula has a DC ($\omega = 0$) value of zero and a high frequency value equal to R_S in parallel with R_F. If R_S is simply the ESR of the capacitor, both high and low frequency output impedance of the filter is very low. Unfortunately, the output impedance of the filter at its resonant frequency can be significantly higher, and this resonant frequency is typically in the range where switching regulators have negative input impedance. Resonant frequency and peak output impedance formulas are shown below:

$$f = \frac{1}{2\pi\sqrt{LC - (R_S)^2(C)^2}}$$

If R_S is simply the ESR of C, the filter resonant frequency is usually closely approximated by:

$$f = \frac{1}{2\pi\sqrt{LC}} \left[\frac{(R_S)^2(C)}{L} \ll 1 \right]$$

$$Z_{OUT(PEAK)} = \frac{R_F(LC)}{LC + (R_S)(R_F)(C)^2}$$

Resonant frequency for a 500μF, 10μH filter is ≈2kHz. Peak output impedance with $R_F = \infty$ and $R_S = 0.05\Omega$ is ≈0.4Ω.

The criterion for regulator stability is that the filter impedance be much lower than the input impedance of the regulator:

$$\frac{R_F(LC)}{LC + (R_S)(R_F)(C)^2} \ll Z_{IN}$$

The worst case occurs with switching regulators that have low input voltage. If we let $V_{IN} = 5V$, and $V_{OUT} = 20V$, $I_{OUT} = 1A$, regulator input impedance at low frequencies is $(5^2)/(20)(1) = 1.25\Omega$. The peak filter impedance was calculated at 0.4Ω, so it seems that stability criterion is met. There is the problem, however, of the *too good* a capacitor in the filter. If the ESR of C drops to 0.02Ω, peak filter impedance rises to 1Ω and stability becomes questionable. To bring peak filter impedance down, R_F may have to be added. If R_F is set at 1Ω, peak filter impedance drops to 0.5Ω. The penalty in ripple attenuation is a reduction from 32:1 to 12:1 for $R_S = 0.05\Omega$.

In all this discussion, the output impedance of the actual input *source* was assumed to be zero. This is not the actual case obviously, and source impedance may have a significant effect on stability.

The point of all this is that input filters tend to have resonant frequencies and impedances which fall into the range where they can cause stability problems in switching regulators. It is important, therefore, to include the filter design into the overall regulator design from the beginning. The selected filter must be *in place* when the regulator is checked for closed-loop stability and the actual source should be used.

Efficiency calculations

The primary reason for using switching regulators is efficiency, so it is important to be able to estimate that factor with some degree of accuracy. In many cases, the overall efficiency is not as critical as the power loss in the individual components. For reliable operation, each power dissipating component must be properly sized or heat sunk to ensure that maximum operating temperature is not exceeded. Overall efficiency is then found by dividing output power by the sum of all losses plus output power:

$$E = \frac{(I_{OUT})(V_{OUT})}{\Sigma P_L + (I_{OUT})(V_{OUT})}$$

Sources of power loss include the LT1070 quiescent current, switch driver current and switch "on" resistance; the output diode; inductor/transformer winding and core losses; and snubber dissipation.

LT1070 operating current

The LT1070 draws only 6mA quiescent current in its idle state, but this is specified with a voltage on the V_C pin such that the output switch never turns on—duty cycle equals zero. When the V_C pin is servoed by the feedback loop to initiate switching, supply current at the input pin increases in two ways. First, there is a DC increase proportional to V_C pin voltage. This is the result of increasing bias current for the switch driver to ensure adequate switch drive at high switch currents. Second, there is driver current which is "on" only when the output switch is on. The ratio of switch driver current to switch current is $\approx 1{:}40$. Total *average* current into the LT1070 V_{IN} pin is then:

$$I_{IN} \approx 6\,\text{mA} + I_{SW}(0.0015 + DC/40)$$

I_{SW} = switch current
DC = switch duty cycle

Use of this formula requires knowledge of switch duty cycle and switch current. This information is available in the sections that deal with each particular switching configuration. A typical example is a buck converter with 28V input and 5V, 4A output. Duty cycle is $\approx 20\%$ and switch current is 4A. This yields a total supply current of:

$$I_{IN} = 6\text{mA} + 4(0.0015 + 0.2/40) = 32\text{mA}$$

Total power loss due to bias and driver current is equal to input voltage times current:

$$P_{BD} = (I_{IN})(V_{IN}) = (32\text{mA})(28\text{V}) = 0.9\text{W}$$

LT1070 switch losses

Switch "on" resistance losses are proportional to the square of switch current multiplied times duty cycle:

$$P_{SW} = (I_{SW})^2(R_{SW})(DC)$$
$$R_{SW} = \text{LT1070 switch "on" resistance}$$

The maximum specified value for R_{SW} is 0.24Ω at maximum rated junction temperature, with 0.15Ω typical value at room temperature. If we use the worst-case number of 0.24Ω, this yields a switch loss in this example of:

$$P_{SW} = (4)^2(0.24)(0.2) = 0.77\text{W}$$

It is pure coincidence that switch and driver losses are nearly equal in this example. At low switch currents and high input voltages, P_{BD} dominates, whereas switch losses dominate at low input voltages and high switch currents.

AC switching losses in the LT1070 are minimal. Rate of switch current rise and fall is $\approx 10^8$A/sec. This reduces switching times to under 50ns and makes the AC losses small compared to DC losses. An exception to this is the AC switch loss attributable to output diode reverse recovery time. See output diode section.

Output diode losses

For low to moderate output voltages, the output diode is often the major source of power loss. For this reason,

Schottky switching diodes are recommended for minimum forward voltage and reverse recovery time. Diode losses for most topologies can be approximated by the following formula, but please consult the output diode section for further details:

$$P_D \approx (I_{OUT})(V_F)(K) + (V)(f)(t_{RR})(I_F)$$

V_F = diode forward voltage at *peak* diode current
V = diode reverse voltage
t_{RR} = diode reverse recovery time
I_F = diode forward current at turn-off
$K = 1 - (V_{OUT}/V_{IN})$ for buck converters and 1 for most other topologies

In the buck regulator example, with $I_{OUT} = 4A$ and letting $V_F = 0.7V$, $t_{RR} = 100$ns:

$$P_D = (4)(0.7)(1 - 5/28) + (28)(40 \bullet 10^3)(10^{-7})(4)$$
$$= 2.3 + 0.45 = 2.75\text{W}$$

Inductor and transformer losses

See section on inductors and transformers.

Snubber losses

See section on flyback design.

Total losses

In this example of a buck regulator, inductor losses might be ≈ 1W and snubber losses are zero. Total losses therefore are:

$$\Sigma P_L = P_{BD} + P_{SW} + P_D + P_L + P_{SNUB}$$
$$= 0.9 + 0.77 + 2.75 + 1 + 0 = 5.42\text{ W}$$

Efficiency is equal to:

$$E = \frac{(V_{OUT})(I_{OUT})}{\Sigma P_L + (V_{OUT})(I_{OUT})} = \frac{(5)(4)}{5.42 + (5)(4)} = 78.7\%$$

This number is typical of a fairly high efficiency 5V buck regulator. The efficiency of 5V switching supplies is lower than higher voltage outputs because of the high diode losses. A 15V output, for instance, might have $E \approx 86\%$.

Output filters

Output voltage ripple of switching regulators is typically in the range of tens to hundreds of millivolts if no additional output filter is used. A simple output filter can reduce this ripple by a factor of ten to one hundred at little additional cost. The high frequency "spikes" which may be superimposed on the ripple are attenuated even more.

The presence of high amplitude spikes at the output of switching regulators is often puzzling to first time designers. These spikes occur in switching regulators which, by their topology, cannot use the energy storage inductor as an output filter. These include boost, flyback and buck/boost designs. The output of these converters can be modeled as a switched current source driving the output capacitor as shown in Figure 5.46.

The output capacitor is shown as C_{OUT}. Its model includes parasitic resistance (R_S) and inductance (L_S). It is the inductance which creates the output voltage spike. The amplitude of this spike can be calculated if the slew rate (dI/dT) of the switch is known. For simple inductor designs operating at full switch current, dI/dT for the LT1070 switch is approximately 10^8A/sec. Voltage across L_S is:

$$V = L_S \left(\frac{dI}{dT}\right) = L_S\left(10^8\right)$$

Straight wire has an inductance of about 0.02μH per inch. If we assume one inch of wire on each end of the output capacitor, including board trace length, this represents 0.04μH. Allowing an additional 0.02μH *internal* inductance, L_S has a total value of 0.06μH, yielding:

$$V = (0.06)\left(10^{-6}\right)\left(10^8\right) = 6V$$

These spikes are very narrow (<100ns) and are usually attenuated significantly in the wire runs and load bypass capacitors, but these calculations point out the importance of *short lead lengths* on the output capacitor.

Output voltage ripple at the regulator switching frequency is usually of two types. With boost, flyback and inverting (buck/boost) designs, ripple is determined almost totally by the ESR of the output capacitor (R_S).

The *reactance* $1/(2\pi f_C)$, of the capacitor at 40kHz is normally so low compared to R_S that it can be ignored. The output ripple is therefore a square wave with amplitude V_{P-P} and duty cycle DC. A formula for V_{P-P} and DC is given in the discussions of these topologies.

The second type of output ripple is triangular. It occurs in switching regulators which utilize the storage inductor as an output filter. These include buck converters, forward converters and 'Cuk converters. Again, the amplitude of the ripple is determined by R_S, not C, but the waveform is triangular with amplitude V_{P-P} and duty cycle DC.

The attenuation of an output filter with *rectangular inputs* is:

$$\frac{V_{OUT(P-P)}}{V_{P-P}} = \frac{DC(1-DC)(R_F)}{(f)(L)}$$

DC = duty cycle of rectangular inputs (50% = 0.5)

Notice that attenuation is the same for complementary duty cycles, that is 10% and 90% are equal, and 40% and 60% are equal, 50% is the point of poorest attenuation. A converter running at 40% duty cycle with an output filter consisting of a 10μH inductor and a 200μF capacitor with $R_F = 0.05\Omega$ would have a filter attenuation of:

$$\frac{V_{OUT(P-P)}}{V_{P-P}} = \frac{(0.4)(0.6)(0.05)}{(4\bullet10^3)(10\bullet10^{-6})} = 0.03 = 33:1$$

The rectangular input is converted to a triangular output whose peak-to-peak amplitude is 1/33 of the peak-to-peak input. Harmonics of the switching frequency are reduced much more; the third harmonic for instance is attenuated 112:1 with $L_F = 0.06\mu$H. There are no second harmonics.

With buck, forward and 'Cuk converters, the ripple voltage into the filter is already triangular. The output ripple of the filter is of the form $V(t) = mt^2$. Attenuation ratio is given by:

$$\frac{V_{OUT(P-P)}}{V_{P-P}} = \frac{R_F}{(8)(L)(f)}$$

For the same conditions of $R_F = 0.05\Omega$, L = 10μH:

$$\frac{V_{OUT(P-P)}}{V_{P-P}} = \frac{0.05}{(8)(10\bullet10^{-6})(40\bullet10^3)} = 0.0156 = 64:1$$

The ripple voltage of these converters is already lower because of the main inductor filtering, so the output filter inductor can often be only a few μH to obtain adequate filtering. The inductor can even be an air core type. A 1/2" diameter, 3/4" long air-wound coil with 13 turns of #16 wire will have an inductance of 1μH, giving a 6:1 attenuation with $R_F = 0.05\Omega$.

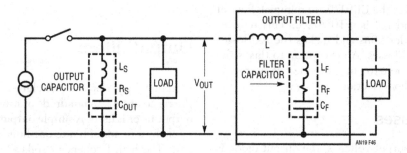

Figure 5.46 • Output Filter

Input and output capacitors

Large electrolytic capacitors used on switching regulators have several important design considerations. The most important is usually effective series resistance (ESR). This is simply the equivalent parasitic resistance in series with the capacitor leads. At frequencies of 10kHz and above, the total impedance of the capacitor is almost identically equal to ESR, and this parasitic resistance limits the filtering effectiveness of the capacitor. The design equations for capacitors used with the LT1070 most often deal simply with ESR; the actual capacitance value is of secondary importance. The following formulas are a very rough guide to maximum ESR vs capacitance for several types of commercially available switching supply capacitors. ESR changes over temperature are shown in Figure 5.47.

Sprague types 673D or 674D

$$ESR = \frac{(400)(10^{-6})}{(C)(V)^{0.6}}\Omega$$

Mallory type VPR

$$ESR = \frac{(200)(10^{-6})}{(C)(V)^{0.6}}\Omega$$

Cornell Dubilier type UFT

$$ESR = \frac{(430)(10^{-6})}{(C)(V)^{0.25}}\Omega$$

C = capacitance value
V = rated working voltage

Note that higher voltage ratings yield lower ESR. This is because higher voltage capacitors are physically larger! Nothing's free, folks. Common design practice is to parallel several capacitors to achieve low ESR and acceptable component height.

A second consideration in capacitor selection is ripple current rating. After a capacitor has been selected, its ripple current rating should be checked to verify that operating ripple is less than the maximum allowed by the manufacturer. Keep in mind, however, that ripple current ratings are normally selected to limit temperature rise in the capacitor. Power dissipation is given by $(I_{RMS})^2(ESR)$. For ambient temperatures below the capacitor's maximum rating, it may be possible to increase ripple current. Consult the capacitor manufacturer. RMS ripple current in the output capacitor for boost, buck-boost and flyback designs can be calculated from output current and switch duty cycle:

$$I_{RMS} = I_{OUT}\sqrt{\frac{DC}{1-DC}}$$

For buck converters, RMS current in the output capacitor is approximately equal to $0.3\Delta I$, where ΔI is the peak-to- peak ripple current in the inductor (continuous mode).

Ripple current in the input capacitor for flyback and buck-boost designs is:

$$I_{RMS} = \frac{(I_{OUT})(V_{OUT})}{V_{IN}}\sqrt{\frac{1-DC}{DC}}$$

For buck designs it is:

$$I_{RMS} = I_{OUT}\sqrt{DC - (DC)^2}$$

and for boost designs, input capacitor ripple current is:

$$I_{RMS} = 0.3\Delta I$$

Inductor and transformer basics

The inductors and transformers used with the LT1070 are very important to the overall performance of the converter, especially with respect to parameters such as efficiency, maximum output power and overall physical size. The many trade-offs associated with the inductance values and the volume of the core require the designer to have a sound basis for selecting the optimum inductor or transformer for each application. Specific guidelines for inductance values are given in the discussion of suggested applications elsewhere in this section, but a general understanding of inductor theory is also needed.

The three important characteristics of a simple 2-terminal inductor used in switching regulators are: inductance value (L, in henries), maximum energy storage ($I^2 \bullet L/2$, in ergs) and power loss (watts). Basic definitions of the parameters which determine these characteristics are shown below.

μ = core permeability. This is basically the *increase* in inductance which is obtained when the inductor is wound on a core instead of just air. A μ of 2000, for instance, will increase inductance by 2000:1.

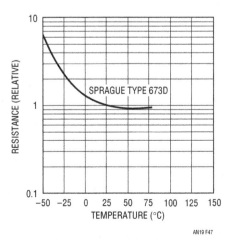

AN19 F47

Figure 5.47 • Typical Capacitor ESR vs Temperature

ℓ = magnetic path length. In a simple toroid this is the average circumference of the core (see sketch).

A = cross-sectional area of the core (see sketch).

g = thickness of air gap (if any) used to increase the energy storage capability of a core (see sketch).

B = magnetic flux density in the core. If B rises too high, the core will "saturate," allowing μ and therefore L, to drop drastically.

N = number of turns in the winding.

I = instantaneous winding current.

V_C = volume of actual core material.

In most converter applications, the required inductance is determined by constraints such as maximum output power, ripple requirements, input voltage and transient response. I is determined by load current. For purposes of this discussion, therefore, it is assumed that L and I are known quantities, and the quantities to be determined are N, A, ℓ, V_C and g.

Inductance is determined by core permeability, path length, cross sectional area and number of turns:

$$L = \frac{(\mu)(A)(N^2)}{\ell}(0.4\,\pi)(10^{-8}) \quad \text{(no gap)}$$

Magnetic flux density is a function of winding current, number of turns and path length:

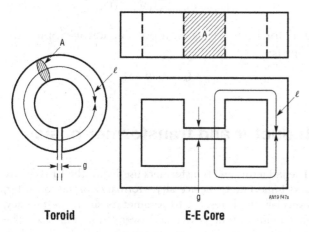

Toroid   **E-E Core**

$$B = \frac{(I)(N)(\mu)}{\ell}(0.4\,\pi) \quad \text{(no gap)}$$

A properly selected inductor must provide the right value of L without exceeding the maximum limit on flux density, (B_M). In other words, the core must not "saturate" under conditions of peak winding current (I_P). By combining the formulas for inductance and flux density, it can be shown that core *volume* (V_C) required is a direct function of the energy to be stored by the inductor:

$$\text{Stored energy} = E = \frac{(I_P{}^2)(L)}{2}$$

$$V_C = (A)(\ell) = \frac{(I_P{}^2)(L)(\mu)(0.4\pi)}{(B^2)(10^{-8})} = (E)\frac{(2\mu)(0.4\pi)}{(B^2)(10^{-8})}$$

In any given application, the value of I_P can be determined from maximum load current and duty cycle. Formulas for maximum I_P are provided in the individual sections on each topology.

In many cases, the maximum load current is much less than the LT1070 is capable of providing. A core designed to handle only full load current may saturate under overload or short-circuit conditions. The *cycle-by-cycle current limiting of the LT1070 protects the regulator against damage even with saturated cores*. This considerably improves the reliability of converters using the LT1070 and eases the design complexity.

Although core volume is the main criterion for selecting a given core, the volume still consists of two variables, A and ℓ. For minimum overall size of the inductor it is generally best to increase A as much as possible at the expense of ℓ, thereby minimizing the number of turns required to obtain the desired inductance. This process can be taken only so far before the "window" in the core becomes too small to accommodate the windings.

Cores with gaps

The energy storage capability of a core can be increased by "gapping" the core. A significant portion of the total energy is stored in the air gap. The drawback of a gapped core is that the effective permeability drops, requiring many more turns to achieve the required inductance. More turns require a larger winding window. The overall size of the inductor, however, can be considerably less with a properly gapped core, especially with high permeability core material. The formula for inductance with a gapped core is:

$$L = \frac{(\mu)(A)(N^2)(0.4\,\pi)(10^{-8})}{\ell\left(1 + \frac{\mu g}{\ell}\right)}$$

Inductance drops by the factor $\left(1 + \frac{\mu \bullet g}{\ell}\right)$

With a μ of 2000, $\ell = 2''$ and g = 0.02'', inductance will drop by 22:1, requiring that N be increased by $\sqrt{22}$ to maintain constant inductance. Increase in energy storage is equal to the decrease in permeability:

$$\frac{E_{MAX}(\text{with gap})}{E_{MAX}(\text{no gap})} = 1 + \frac{\mu \bullet g}{\ell}$$

There are several practical limits on the amount which gap size may be increased. First, large gaps require many more turns to achieve the same inductance. This requires smaller diameter wire which increases copper losses from I^2R heating. Secondly, with large gaps the *effective* gap size is considerably less than the actual gap because of fringing fields around the gap.

When using commercially available cores, data sheet information on ℓ, A and μ is usually given in *effective* values. The theoretical value of μ, for instance, is the bulk value for the core material. The *effective* value for a single piece core may approach the bulk value, but with 2-piece cores, the tiny air spaces left in the mating surfaces can reduce the *effective* permeability by as much as 2:1. This may sound unreasonably pessimistic, but a core with bulk $\mu = 3000$ and $\ell = 1.5''$, will lose half its permeability for $g = 0.0005''$. Data sheets for gapped cores list *effective* values of μ for each gap size to make calculations simple. They may also list a parameter, "inductance per (turn)^2" for each gap to further simplify inductance calculations.

There are two types of core material which are effectively self-gapped: iron powder and permalloy. These materials distribute the gap evenly throughout the core, allowing gapless core to be constructed with much higher energy storage capability. The permeability of this material is much reduced, but if the winding window will accommodate the extra turns, the current handling capability of the inductor will be much higher for the same inductance compared to a high-μ formulation.

Iron powder cores are cheaper than ferrite and can be custom tailored quickly, but high core loss limits their application to low AC flux density applications such as inductors. A significant advantage of iron powder is that it saturates very "softly," preventing catastrophic total loss of inductance for large overcurrent conditions. Note that commercially available powdered iron inductors are typically "optimized" so that core losses and winding (I^2R) losses are the same order of magnitude. Core loss is dependent on peak-to-peak ripple current which depends on the voltage-time product applied to the inductor. The inductors are therefore specified for a maximum DC current and a maximum vol•microsecond product to limit heating. For applications which require highest possible efficiency, consider using oversized cores or permalloy, which is more expensive, but has much lower core loss. Consult with inductor manufacturers about trading off DC current for ripple current, or vice versa.

Inductor selection process

The simplest way to select an inductor is to find an off-the-shelf unit that meets the minimum inductance and current requirements. This may not be cost effective, however, if the standard types are not fairly close to your requirements. The next best approach is to have the unit custom wound by one of the many companies in the business. They will select the best core and winding combination for your particular application. A third approach is to scan the literature for standard core types which you can custom wind to meet your particular requirement. This is a quick way to get a prototype up and running. It can also be very cost effective for some production situations. At the end of this application note is a list of core and inductor/transformer manufacturers.

The procedure for selecting a do-it-yourself core starts with defining the values of peak winding current and inductance. If the LT1070 is to be used at or near full output power, peak winding current will approach 5A, so a conservative value of 5A should be used for core calculations.

If external current limiting is used or if output power levels are lower, peak winding currents can be calculated from the equations supplied in the discussions of each topology. Likewise, inductance values are calculated from specific equations in these sections. Actual values for L generally fall into the range of $50\mu H$ to $1000\mu H$, with $200\mu H$ to $500\mu H$ being most typical.

For ferrite cores, the next step is to calculate the core volume required to prevent saturation:

$$V_e = \frac{(I_P)^2(L)(\mu_e)(0.4\,\pi)}{(B_0)^2(10^{-8})} \qquad \text{(ferrite cores)}$$

L = required inductance (henries)
I_P = peak inductor current (amps)
μ_e = effective relative permeability $\Big\}$ supplied on core
B_0 = maximum operating flux density $\Big\}$ data sheets
    (gauss)
V_e = effective core volume

Example: let $L = 200\mu H$, $I_P = 5A$, $\mu_e = 100$, $B_0 = 2500$ gauss,

$$V_e = \frac{(5)^2(200 \bullet 10^{-6})(100)(0.4\,\pi)}{(2500)^2(10^{-8})} = 10\ cm^3$$

The values chosen for μ_e and B_0 are typical for a gapped ferrite core. Some cores come with several standard gaps.

Others are left ungapped with the user supplying spacers for setting gap length. Custom gapped cores are also available. A reasonable place to start is with a gap length of 0.02 inches. A core with $\mu = 3000$ and path length (ℓ_e) of 2 inches would have an effective permeability of $\mu_e = \mu/(1 + \mu_g/\ell_e) = 3000/(1 + 3000 \bullet 0.02/2) = 97$. Notice that by simply selecting a large gap we can arbitrarily reduce the required core volume. The problem with attempting to use a large gap is that the effective permeability drops so low that a large number of turns are required to achieve the desired inductance. This forces the use of small diameter wire where the copper losses get high enough to cause overheating of the core.

Powdered iron cores, because of their high core loss and ability to operate at very high *DC* flux densities, generally have a different design procedure based on temperature rise due to core loss and winding loss. AC flux densities generally need to be kept below 400 gauss. This leads to a volume formula based on AC flux density:

$$V_C = \frac{(\Delta I)^2(L)(\mu)(0.4\,\pi)}{(4)(B_{AC})^2(10^{-8})}$$

ΔI = peak-to-peak ripple current

For $\Delta I = 1A$, $L = 200\mu H$, $\mu = 75$ and $B_{AC} = 300$ gauss,

$$V_C = \frac{(1)^2(200 \bullet 10^{-6})(75)(0.4\pi)}{(4)(300)^2(10^{-8})} = 5.25 cm^3$$

To reduce core size, inductance (L) must be *increased*. This seems backwards according to the formula, but ΔI is inversely proportional to L, so the $(\Delta I)^2$ term drops rapidly as L is increased, reducing required core volume. The penalty is increased wire (copper) loss due to the increased turns required.

After a tentative core is selected based on volume, a check must be done to see if the power losses from the winding(s) and the core itself are within the allowed limits.

The first step is to calculate the number of turns required:

$$N = \sqrt{\frac{(L)(\ell_e)}{(\mu_e)(A_e)(0.4\pi \bullet 10^{-8})}}$$

N = turns
ℓ_e = effective magnetic path length (cm) ⎫
A_e = effective core area (cm²)   ⎬ supplied on core data sheets
μ_e = effective permeability (with gap)   ⎭

Using the ferrite example, and assigning $\ell_e = 9$cm, $A_e = 1.2$cm², $\mu_e = 100$, a 200μH inductor would require:

$$N = \sqrt{\frac{(200 \bullet 10^{-6})(9)}{(100)(1.2)(0.4\,\pi \bullet 10^{-8})}} = 34.6 \text{ turns (use 35)}$$

To calculate wire size, the usable winding window area (Aw) must be ascertained from the core dimensions. Many data sheets list this parameter directly. The usable window area must allow for bobbin thickness and clearances. Total copper area is only about 60% of window area due to air gaps around the wire. We can now express the required wire gauge in terms of N and Aw:

$$\text{Wire gauge (AWG)} = 10\left(\log\frac{(0.08)(N)}{(0.6)(Aw)}\right)$$

0.08 factor = area of #1 gauge wire
0.6 factor = air space loss around wire

If we assume a value for Aw of 0.2in² and use N = 35:

$$\text{AWG} = 10\log\frac{(0.08)(35)}{(0.6)(0.2)} = 13.68 \text{ (use #14)}$$

The next step is to determine the number of winding layers. This is determined by bobbin length, or toroid inside circumference:

$$\text{Layers} = \frac{N(D + 0.01)}{L_B} = \frac{N\left[(0.32)\left(10^{\frac{-AWG}{20}}\right) + 0.01\right]}{L_B}$$

D = wire diameter in inches
L_B = bobbin length or toroid inside circumference
0.01 = allowance for enamel and spacing

For N = 35, AWG = #14 and $L_B = 0.9''$:

$$\text{Layers} = \frac{35\left[(0.32)\left(10^{\frac{-14}{20}}\right) + 0.01\right]}{0.9} = 2.87$$

The reason for calculating layers is that the *AC copper losses* are very dependent on the number of layers in a winding. To calculate AC losses, a table is used (Figure 5.48) which requires a factor K:

$$K = D\sqrt{(f)(F_P)}$$

D = wire diameter or foil thickness

For foil conductors, F_P is 1. For round wires it is equal to:

$$F_P = \frac{(T_L + 1)(N_C)(D)}{L_W}$$

T_L = turns per layer
N_C = number of paralleled conductors (bifilar → $N_C = 2$)
D = wire diameter
L_W = length of winding ($\approx L_B$)

For 35 turns and 3 layers, $T_L \approx 12$. #14 wire has D = 0.064. N_C for a single wire is 1. With $L_W = 0.9''$:

$$F_P = \frac{(12 + 1)(1)(0.064)}{0.9} = 0.92$$

K is now equal to:

$$K = D\sqrt{(f)(F_P)} = 0.064\sqrt{(40 \bullet 10^3)(0.92)} = 12.3$$

This is a very high K factor; in fact it is slightly off the graph in Figure 5.48, but for now it illustrates the importance of AC resistance calculations. The various lines on the graph represent the number of layers. With three layers, the AC resistance factor is off scale at approximately 23. This means that AC resistance is *23 times* DC resistance. Now we can calculate winding losses. DC winding resistance is found from:

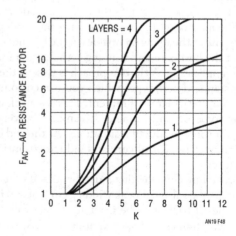

Figure 5.48 • AC Resistance Factor

$$R_{DC} = \frac{(N)(\ell_m)}{12}\left(10\frac{AWG}{10} - 4\right)$$

ℓ_m = mean turn length (core specification)

For N = 35, ℓ_m = 2.4″, AWG = #14:

$$R_{DC} = \frac{(35)(2.4)}{12}\left(10^{\frac{AWG}{10}-4}\right) = 0.0176\Omega$$

AC resistance is then DC resistance multiplied by AC resistance factor (F_{AC}):

$$R_{AC} = (R_{DC})(F_{AC}) = (0.0176)(23) = 0.404\Omega$$

To calculate total losses, DC and AC losses are summed:

$$P_W = (I_{DC})^2(R_{DC}) + (I_{AC})^2(R_{AC})$$

Formulas for I_{DC} and I_{AC} are shown in Figure 5.50. If we assume I_{DC} = 5A and I_{AC} = 1A, total winding losses are:

$$P_w = (5)^2(0.0176) + (1)^2(0.404) = 0.44 + 0.4 = 0.94\text{ W}$$

In this example, AC losses are about equal to DC losses. Simple inductors used in buck, boost and buck/boost designs may have the ratio of AC to DC losses in the range of 0.25 to 4.0. Transformer designs like flyback usually have AC losses *much* higher than DC losses. Losses in the primary and secondary are calculated separately. In many cases, multiple strands of smaller wire or copper foil must be used to reduce the AC resistance factor to acceptable limits.

After winding losses are found, core loss must be calculated. The first step is to find peak *AC* flux density:

$$B_{AC} = \frac{L(\Delta I)}{(2N)(A_e)(10^{-8})}$$

ΔI = peak-to-peak winding ripple current

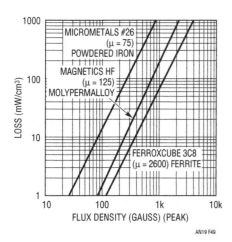

Figure 5.49 • Core Loss vs Flux Density

ΔI is the *ripple* current in the winding. It is the *change* in winding current during the time current is flowing in the winding. For L = 200μH, ΔI = 2A, N = 35 and A_e = 1.2cm²:

$$B_{AC} = \frac{(200 \bullet 10^{-6})(2)}{(35)(1.2)(10^{-8})} = 476\text{ gauss}$$

Core loss *per unit volume* (F_{fe}) is found from the manufacturers tables (see Figure 5.49) of F_{fe} vs flux density and frequency or from the following formula for typical $M_N Z_N$ ferrite material (ferroxcube type 3C8):

$$F_{fe} = \left(1.3 - 10^{-14}\right)\left(B_{AC}\right)^2\left(f^{1.45}\right)$$

For B_{AC} = 476 gauss, f = 40kHz:

$$F_{fe} = \left(1.3 \bullet 10^{-14}\right)(476)^2\left(40 \bullet 10^3\right)^{1.45} = 0.014\text{ W/cm}^3$$

Total core loss is F_{fe} times core volume:

$$P_C = (F_{fe})(V_e) = (0.014)(10) = 0.14\text{ W}$$
$$V_e = \text{effective core volume (cm}^3)$$

Core loss for a powdered iron core is approximately 25 times higher than for ferrite. At a lower flux density of 150 gauss, a powdered iron core would still have core losses 2.5 times that of ferrite. Copper losses would also be higher because of the higher inductance required to reduce AC flux density. Powdered iron cores must be carefully designed to avoid overheating.

Overall losses in the ferrite core are the sum of winding losses plus core losses:

$$P = P_W + P_C = 0.94 + 0.14 = 1.08\text{W}$$

This loss reflects on regulator efficiency, and more importantly, core temperature rise. A 10cm³ core might have a typical thermal resistance of 20°C/W. Temperature rise in this core with P = 1.08W = (1.08)(20) = 21.6°C. 40°C rise is considered a typical design criterion, so this core is being under utilized.

Transformer design example

Requirements: A flyback converter with V_{IN} = 28V_{DC}, V_{OUT} = 5V, I_{OUT} = 6A. From previous calculations it is found that N = 1/3, L_{PRI} = 200μH and $I_{PRI(PEAK)}$ = 4.5A, with ΔI = 1A.

1. Calculate volume of core required with a gapped core. First assume an effective permeability of ≅150 and B_0 = 2500 gauss:

$$V_e = \frac{(I_{PRI})^2(L)(\mu_e)(0.4\,\pi)}{(B_0)^2(10^{-8})}$$
$$= \frac{(4.5)^2(200 \bullet 10^{-6})(150)(0.4\,\pi)}{(2500)^2(10^{-8})} = 12\text{cm}^3$$

A Pulse Engineering core #0128.005 has V_e = 13.3cm³, A_e = 1.61cm², ℓ_e = 8.26cm, μ = 2000.

2. Calculate required gap:

$$g = \frac{\ell_e \left(\frac{\mu}{\mu_e} - 1 \right)}{\mu}$$

$$= \frac{8.26 \left(\frac{2000}{150} - 1 \right)}{2000} = 0.051 \text{cm} = 0.02''$$

If an ungapped core is used with spacers, spacer thickness should be $0.02/2 = 0.01''$.

3. Calculate required turns:

$$N = \sqrt{\frac{(L)(\ell_e)}{(\mu_e)(A_e)(0.4\,\pi \bullet 10^{-8})}}$$

$$= \sqrt{\frac{(200 \bullet 10^{-6})(8.26)}{(150)(1.61)(0.4\,\pi \bullet 10^{-8})}} = 23.3$$

4. Calculate wire size. Allocate 1/2 the window space for the primary winding. Window height (build) for the 0128.005 core is 0.25″ and coil length is 0.782″, giving a window area = (0.25) (0.782) = 0.196in^2:

$$AWG = 10 \log \frac{0.08N}{(0.6)(Aw)} = 10 \log \frac{(0.08)(23)}{(0.6)\left(\frac{0.196}{2}\right)}$$

$$= 14.95 \text{ (use #16)}$$

5. Calculate layers:

$$\text{Layers} = \frac{N \left[(0.32)\left(10^{\frac{-AWG}{20}}\right) + 0.01 \right]}{L_B}$$

$$= \frac{23 \left[(0.32)\left(10^{\frac{-16}{20}}\right) + 0.01 \right]}{0.782}$$

$$= 1.79 \text{ (assume 2 layers)}$$

6. Calculate K factor (#16 wire has D = 0.05):

$$F_P = \frac{(T_L + 1)(N_C)(D)}{L_W} = \frac{\left(\frac{23}{2} + 1\right)(1)(0.05)}{0.782} = 0.8$$

$$K = D\sqrt{(f)(F_P)} = 0.05\sqrt{(40 \bullet 10^3)(0.8)} = 8.94$$

7. Calculate DC winding resistance:

$$R_{DC} = \frac{(N)(\ell_m)}{12}\left(10^{\frac{AWG}{10} - 4}\right)$$

$$= \frac{(23)(3)\left(10^{\frac{16}{10} - 4}\right)}{12} = 0.023\,\Omega$$

$\left(\ell_m \text{ for this core is} \approx 3'' \right)$

8. Use graph to find AC resistance factor. Interleaving of primary and secondary reduces effective layers by two only if primary and secondary conduct simultaneously, which they *do not* in a flyback design. Use layers = 2 line:

$$F_{AC} = 8.3 \text{ (from graph, for K = 8.95)}$$

9. Calculate AC winding resistance:

$$R_{AC} = (R_{DC})(F_{AC}) = (0.023)(8.3) = 0.19\,\Omega$$

10. Calculate primary winding losses.
First, primary AC RMS current must be calculated. From the chart in Figure 5.50:

$$I_{AC} = \frac{I_{OUT}}{E}\sqrt{\frac{(N)(V_{OUT})}{V_{IN}}}$$

$$= \frac{6}{0.75}\sqrt{\frac{(1/3)(5)}{28}} = 1.95A$$

$$I_{DC} = \frac{I_{OUT}}{E}\sqrt{\frac{V_{OUT}(V_{OUT} + N \bullet V_{IN})}{(V_{IN})^2}}$$

$$= \frac{6}{0.75}\sqrt{\frac{5(5 + 1/3 \bullet 28)}{(28)^2}} = 2.4A$$

Power loss in the primary winding is:

$$P_W = (I_{AC})^2 R_{AC} + (I_{DC})^2 R_{DC}$$

$$= (1.95)^2(0.19) + (2.4)^2(0.023) = 0.85W$$

11. Calculate secondary winding loss.
Turns ratio is 1/3, so the secondary will have 23/3 = 7.67 turns. Use 8 turns:

$$AWG = 10 \log \frac{0.08\,N}{0.6\,Aw} = 10 \log \frac{(0.08)(8)}{(0.6)\left(\frac{0.196}{2}\right)} = 10.4$$

TOPOLOGY	DC PRIMARY I	AC PRIMARY I	DC SECONDARY I	AC SECONDARY I
Flyback	$\dfrac{I_{OUT}}{E}\sqrt{\dfrac{V_{OUT}[V_{OUT}+N(V_{IN})]}{(V_{IN})^2}}$	$\dfrac{I_{OUT}}{E}\sqrt{\dfrac{N(V_{OUT})}{V_{IN}}}$	$I_{OUT}\sqrt{\dfrac{V_{OUT}+N(V_{IN})}{N(V_{IN})}}$	$I_{OUT}\sqrt{\dfrac{V_{OUT}}{N(V_{IN})}}$
Buck	I_{OUT}	$(0.29)(\Delta I)$	NA	NA
Current-Boosted Buck	$I_{OUT}\sqrt{\dfrac{V_{OUT}[V_{OUT}+N(V_{IN}-V_{OUT})]}{(V_{IN})^2}}$	$I_{OUT}\sqrt{\dfrac{N[V_{OUT}(V_{IN}-V_{OUT})]}{(V_{IN})^2}}$	$I_{OUT}\sqrt{\dfrac{(V_{IN}-V_{OUT})[V_{OUT}+N(V_{IN}-V_{OUT})]}{N(V_{IN})^2}}$	$I_{OUT}\sqrt{\dfrac{V_{OUT}(V_{IN}-V_{OUT})}{N(V_{IN})^2}}$
Boost	$I_{OUT}\left(\dfrac{V_{OUT}}{V_{IN}}\right)$	$(0.29)(\Delta I)$	NA	NA
Voltage-Boosted Boost	$I_{OUT}\left(\dfrac{V_{OUT}}{V_{IN}}\right)$	$(I_{OUT})(N)\sqrt{\dfrac{V_{OUT}-V_{IN}}{V_{IN}(N+1)}}$	$I_{OUT}\sqrt{\dfrac{V_{OUT}+N(V_{IN})}{V_{IN}(N+1)}}$	$I_{OUT}\sqrt{\dfrac{V_{OUT}-V_{IN}}{V_{IN}(N+1)}}$
Current-Boosted Boost	$I_{OUT}\sqrt{\dfrac{(V_{OUT}-V_{IN})[V_{OUT}+V_{IN}(N+1)]}{(V_{IN})^2}}$	$I_{OUT}\sqrt{\dfrac{N(V_{OUT}-V_{IN})}{V_{IN}}}$	$I_{OUT}\sqrt{\dfrac{V_{OUT}+V_{IN}(N+1)}{N(V_{IN})}}$	$I_{OUT}\sqrt{\dfrac{V_{OUT}-V_{IN}}{N(V_{IN})}}$
Buck-Boost (Inverting)	$I_{OUT}\left(1+\dfrac{V_{OUT}}{V_{IN}}\right)$	$(0.29)(\Delta I)$	NA	NA
Forward	$I_{OUT}\sqrt{\dfrac{N(V_{OUT})}{V_{IN}}}$	$I_{OUT}\sqrt{\dfrac{V_{OUT}[N(V_{IN}-V_{OUT})]}{(V_{IN})^2}}$	$I_{OUT}\sqrt{\dfrac{V_{OUT}}{N(V_{IN})}}$	$I_{OUT}\sqrt{\dfrac{V_{OUT}[N(V_{IN}-V_{OUT})]}{N(V_{IN})^2}}$
'CUK	$I_{OUT}\left(\dfrac{V_{OUT}}{V_{IN}}\right)$	"0" or $(0.29\Delta I)$	I_{OUT}	"0" or $(0.29\Delta I)$

I_{OUT} = DC output current V_{OUT} = DC output voltage V_{IN} = DC input voltage

AN19 F50

Figure 5.50 • AC and DC Winding Currents (RMS Equivalent)

This is rather large, stiff wire and the large diameter will lead to large AC winding losses. A good solution might be to use multiple smaller diameter wire wound in parallel. If we use the length of the coil divided by 2N, it will tell us what diameter wire can be bifilar wound to just fill one layer:

$$D = \frac{L_B}{2N} = \frac{0.782}{(2)(8)} = 0.049''$$

The next smallest standard wire diameter is 18. Two #18 wires have three times the DC resistance of a single #10 wire, but AC resistance will not increase nearly that much. Assume one layer bifilar wound #18 secondary interleaved between the two primary layers (to reduce leakage inductance):

$$R_{DC}=\frac{(N)(\ell_m)}{12}\left(10^{\frac{AWG}{10}}-4\right)=\frac{(8)(3)}{12}\left(10^{\frac{18}{10}}-4\right)$$
$$=0.013\ \Omega\ per\ wire$$

With two wires, total R_{DC} = 0.013/2 = 0.0065Ω.

$$F_P=\frac{(T_L+1)(N_C)(D)}{L_W}=\frac{(8+1)(2)(0.04)}{0.782}=0.92$$

$$K=D\sqrt{(f)(F_P)}=0.04\sqrt{(40\bullet10^3)(0.92)}=7.7$$

From graph, with layers = 1, F_{AC} = 2.3:

$$R_{AC}=(R_{DC})(F_{AC})=(0.0065)(2.3)=0.015\ \Omega$$

From the chart in Figure 5.50:

$$I_{AC}=I_{OUT}\sqrt{\frac{(V_{OUT})}{N(V_{IN})}}=6\sqrt{\frac{5}{1/3(28)}}=4.4A$$

$$I_{DC}=I_{OUT}\sqrt{\frac{V_{OUT}+N(V_{IN})}{N(V_{IN})}}=6\sqrt{\frac{5+1/3(28)}{1/3(28)}}=7.4A$$

$$P_W=(4.4)^2(0.015)+(7.4)^2(0.0065)=0.65W$$

12. Calculate core loss. Core loss is proportional to AC flux density which is determined by *change* in

primary current (ΔI) during primary current flow period. For $\Delta I = 1A$:

$$B_{AC} = \frac{L(\Delta I)}{2(N)(A_e)(10^{-8})} = \frac{(200 \bullet 10^{-6})(1)}{2(23)(1.61)(10^{-8})}$$

$$= 270 \text{ gauss}$$

$$F_{fe} = (1.3 \bullet 10^{-14})(B_{AC})^2(f^{1.45}) = 0.0045 \text{W/cm}^3$$

$$P_C = (F_{fe})(V_e) = (0.0045)(13.3) = 0.06 \text{W}$$

Total power loss with this core is:

$$P = P_W + P_C = 0.85 + 0.65 + 0.06 = 1.56 \text{W}$$

The 0128.005 core is specified at 2.78W for a 40°C temperature rise, yielding $\theta = 40/2.78 = 14.4°\text{C/W}$.

$$\Delta T(\text{core}) = (P)(\theta) = (1.56)(14.4) = 22°\text{C}$$

This is a very conservative design. If minimum core size is required, the procedure now is to go back to step 1 and assume a lower effective permeability (μ_e), perhaps 100. This would reduce core volume and require a larger gap. More turns would be required and the available space for copper would go down, so copper losses would go up. Flux density remains constant, so core loss drops. Thermal resistance goes up however, so the smaller core gets hotter. In addition, the increased number of turns will increase leakage inductance, which will increase snubber losses. It isn't easy, folks!

Heat sinking information

The efficiency of the LT1070 allows it to be used without a heat sink in many applications, but for full-power output a heat sink is required. The equations contained in the efficiency section of this application note will allow the user to estimate fairly accurately the total power dissipation of the chip under full load conditions. Short-circuit power dissipation can be either *more or less* than full load, depending on the topology. Calculation of short-circuit power dissipation in the LT1070 is very complicated because the "on" time of the switch is strongly dependent on parasitic effects such as diode and inductor series resistance, wiring losses and leakage inductance. If continuous output shorts must be tolerated, it is strongly suggested that a temperature probe be used to ensure that maximum junction is not exceeded. Thermal resistance from junction to case is 2°C/W maximum, and short-circuit power dissipation almost never exceeds 10W, so a case temperature of 100°C for commercial units and 130°C for military units will ensure that maximum junction temperature is not exceeded.

Heat sink size for the LT1070 can be calculated if maximum power dissipation and maximum ambient temperature are known.

$$\theta_{HS} = \frac{T_J - T_A - (P)(\theta_{JC})}{P}$$

θ_{HS} = heat sink thermal resistance
P = LT1070 power dissipation
θ_{JC} = LT1070 junction-to-case thermal resistance (2°C/W)
T_J = LT1070 maximum junction temperature
T_A = maximum ambient temperature

For $T_J = 100°C$, $T_A = 60°C$, $P = 5W$:

$$\theta_{HS} = \frac{100 - 60 - (5)(2)}{5} = 6°\text{C/W}$$

Troubleshooting hints

The following is a list of "gotchas" we've put together to help you avoid some of the pitfalls of switching power supply design. They range from obvious to subtle and serious to hilarious. The LT1070 was specifically designed to eliminate many of the problems commonly found in power supply design and be easy to use. The problem is that there are a significant number of easily overlooked mistakes in breadboarding switching regulators which result in either instant death of the IC or electrical characteristics which are puzzling to even highly experienced power supply designers. So here's the list we've collected so far. We hope your problem is on it to save you time and frustration. If not, give us a call and we'll help fix the problem.

Warning

Before reading this section, be aware that the intent of the author is not to insult, but rather to relate in an attention-getting manner a list of goofs that, in many cases, he personally has had to own up to.

1. **Transformer wired backwards**
   Those dots indicate polarity, not smashed flies.

2. **Electrolytic capacitors installed backwards**
   This is no problem until you bend over to see what is wrong—then "bang," a personal demonstration of explosive venting.

3. **LT1070 input and switch pins reversed**
   The catalog and some preliminary data sheets got out with the wrong pinout for the plastic T0-220 package. Our apologies. *Pin 5 is input on T0-220 packages.*

4. **No input bypass capacitor**
   Switching regulators draw current from the input supply in pulses. Long input wires can cause dips in the input voltage at the switching frequency.

Breadboards should have a large ($\geq 100\mu F$) input capacitor up close to the regulator.

5. **Fred's inductor (or transformer)**
   Inductors are not like lawn mowers. If you want to borrow the one out of Fred's drawer, make sure it's the right value for your application.

   A 50µH inductor with 50V applied will have a current increasing with time at the rate of 1A per microsecond. It doesn't take a calculator to see that things can get out of hand quickly during the 25µs period of a 40kHz switcher. Likewise, if "Fred's inductor" is 50 *millihenries*, it will probably saturate at such low current levels that it is useless, not to mention the fact that the transient response can be measured on a Simpson VOM. *Use the formulas in the application note to get a ball park inductance value before starting a breadboard.*

6. **Wimpy magnetic cores**
   Core sizes for the LT1070 will vary from 3cm^3 to 20cm^3 of core material for properly designed inductors or transformers. A thumbnail size core will simply saturate and get hot when asked to operate at ampere current levels. *Breadboard with man-sized cores, then optimize the core size for production.*

7. **Rat's nest wiring**
   The LT1070 is not a jelly bean op amp that can be wired up with 2-foot clip leads. It achieves its high efficiency by switching current at very high speeds. Long wires will cause every component connected to them to look like an inductor at these speeds. This not only causes totally unpredictable operation; it can generate fatal (to components) transient voltages. *Use very short wires to interconnect power components on the breadboard, including bypass capacitors, catch diodes, LT1070 pins, transformer leads, etc.*

8. **No snubber network**
   The LT1070 will tolerate a lot of abuse, but it cannot be overvoltaged on the switch pin and survive to tell the tale. The 65V maximum switch voltage must be observed. Any design using a transformer or tapped inductor will have enough leakage inductance to cause transients well above 65V if no snubber network is used. Load currents and input voltages should be increased slowly while monitoring switch voltage to ensure that the initial snubber design is adequate.

9. **60Hz diodes**
   The LT1070 will *eat* 1N914 and 1N4001 diodes and not even burp. Diode currents, especially during start-up, can exceed 5A. This takes care of the 1N914s. The 1N4001s will last for a little while, until the heat generated by their horribly slow turn-off characteristics causes them to self-destruct. *Use diodes designed for switching applications, with adequate current ratings. Turn-on time is also important*

to avoid overvoltage stress on other components (see diode section).

10. **Something from nothing**
    The *first* step in designing with the LT1070 is to see if it will provide the required power level. Each topology has a different maximum output power that it can provide, depending on things such as input voltage, output voltage and transformer turns ratio. Secondary effects such as inductance values and switch resistance may also limit power. *The power graph on the next page is a rough guide to maximum power levels. Use it as a quick guide only.* More exact formulas are contained in the application section. Oh, by the way, if you thought about paralleling LT1070s for more power—sorry, it won't work. You cannot get to the internal 40kHz oscillator to get them in sync.

11. **Input supply gets clobbered**
    The LT1070 can draw input currents of up to 6A during start-up. It has to charge up the large output capacitor and it does this at a rate set by the internal current limit unless optional soft start is added. *The start-up surge may trip overcurrent latches on some supplies, causing them to stay off until power is recycled.*

    Steady-state problems can also occur. Switching regulators try to deliver constant load voltage. With a given load, this means constant load power. For a high efficiency system, input power also remains constant, *so input current increases as input voltage decreases. Low input voltage conditions may require such high input currents that the input supply current limits. This causes the supply voltage to drop further, forcing a permanent latch condition.* See current limit and soft start sections.

12. **Didn't read the data sheet**
    Then you shall have no pie.

13. **Stray coupling to the V_C or F_B pins**
    Voltages on the FB and V_C pins are referenced to the LT1070 ground pin. In some topologies the ground pin is switching between input voltage and system ground. *Stray capacitance between V_C or FB pins and system ground* will act like coupling to a switching source. *Minimize this capacitance.* The problem is particularly acute when using an RC box to iterate frequency compensation on the V_C pin. Even configurations which have the LT1070 ground pin "grounded" may have problems if the RC box picks up switching energy.

Subharmonic oscillations

Current mode switching regulators which operate with a duty cycle greater than 50% and have continuous inductor current can exhibit a duty cycle instability known as subharmonic oscillations. This effect is not harmful to the regulator and in many cases it does not even affect the output regulation. Its most annoying effect is to produce a high pitched squeal from power components which effectively have their 40kHz operating frequency *modulated* by submultiple frequencies; 20kHz, 10kHz, etc. Subharmonic oscillations do *not* depend on the closed-loop characteristics of the regulator. They can occur even when zero feedback is used. Ordinary closed-loop instabilities can also cause audible sounds from switching regulators, but they tend to be in the range of hundreds of hertz to several kilohertz.

The source of subharmonic oscillations is the simultaneous conditions of fixed frequency and fixed peak amplitude of inductor current as shown in part a of the accompanying figure.

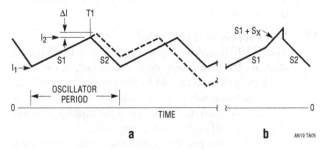

The inductor current starts at l_1, at the beginning of each switch on cycle. Current increases at a rate (S1) equal to input voltage divided by inductor value. When current reaches the trip level, I_2, the current mode loop shuts off the switch and current begins to fall at a rate S2 until the switch is again turned on by the oscillator. Now watch what happens when the point T1 is perturbed so that the current exceeds I_2 by ΔI. The time left for the current to fall is

reduced so that the *minimum* current point is *increased* by $\Delta I + \Delta I S2/S1$. This will cause the minimum current on the *next* cycle to *decrease* by $(\Delta I + \Delta I\, S2/S1)(S2/S1)$. On each succeeding cycle the current perturbation is multiplied by S2/S1. If S2/S1 is greater than 1, the system is unstable. The condition $S2/S1 \geq 1$ occurs at a duty cycle of 50% or higher.

Subharmonic oscillations can be eliminated if an artificial ramp is superimposed on the inductor current waveform as shown in part b of the figure. If this ramp has a slope of S_X, the requirement for stability is that $S_X + S1$ be larger than S2. This leads to the following equation:

$$S_X \geq \frac{S1(2DC - 1)}{1 - DC}$$

DC = duty cycle

For duty cycles less than 50% (DC = 0.5), S_X is a negative number and is not required. For larger duty cycles, S_X takes on values dependent on S1 and duty cycle. S1 is simply V_{IN}/L. This yields an equation for the minimum value of inductance for a fixed value of S_X:

$$L_{MIN} \geq \frac{V_{IN}(2DC - 1)}{S_X(1 - DC)}$$

The LT1070 has an internal S_X *voltage* ramp fed into the current amplifier whose equivalent current referred value is $2\,(10^5 \text{A/sec})$. A sample calculation for minimum inductance with $V_{IN} = 15V$, DC = 60% is shown:

$$L_{MIN} = \frac{(15)(2 \bullet 0.6 - 1)}{(2 \bullet 10^5)(1 - 0.6)} = 37.5\mu H$$

Remember that *for discontinuous operation, no subharmonic oscillations can occur. Likewise, with duty cycle less than 50%, there is no restriction on inductor size.*

ABSOLUTE MAXIMUM RATINGS (Note 1)

Supply Voltage
LT1070/LT1071 (Note 2) 40V
LT1070HV/LT1071HV (Note 2) 60V
Switch Output Voltage
LT1070/LT1071 .... 65V
LT1070HV/LT1071HV 75V
Feedback Pin Voltage (Transient, 1ms) .... ±15V

Operating Junction Temperature Range
Commercial (Operating) ....0°C to 100°C
Commercial (Short Circuit) ....0°C to 125°C
Industrial ....– 40°C to 125°C
Military ....– 55°C to 150°C
Storage Temperature Range ...... – 65°C to 150°C
Lead Temperature (Soldering, 10 sec) ....300°C

PACKAGE/ORDER INFORMATION

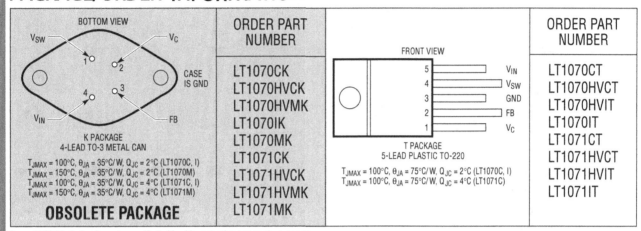

	ORDER PART NUMBER
BOTTOM VIEW K PACKAGE 4-LEAD TO-3 METAL CAN CASE IS GND	LT1070CK LT1070HVCK LT1070HVMK LT1070IK LT1070MK LT1071CK LT1071HVCK LT1071HVMK LT1071MK

T_JMAX = 100°C, θ_JA = 35°C/W, Q_JC = 2°C (LT1070C, I)
T_JMAX = 150°C, θ_JA = 35°C/W, Q_JC = 2°C (LT1070M)
T_JMAX = 100°C, θ_JA = 35°C/W, Q_JC = 4°C (LT1071C, I)
T_JMAX = 150°C, θ_JA = 35°C/W, Q_JC = 4°C (LT1071M)

OBSOLETE PACKAGE

FRONT VIEW (T PACKAGE 5-LEAD PLASTIC TO-220) 5 V_IN, 4 V_SW, 3 GND, 2 FB, 1 V_C	ORDER PART NUMBER
	LT1070CT LT1070HVCT LT1070HVIT LT1070IT LT1071CT LT1071HVCT LT1071HVIT LT1071IT

T_JMAX = 100°C, θ_JA = 75°C/W, Q_JC = 2°C (LT1070C, I)
T_JMAX = 100°C, θ_JA = 75°C/W, Q_JC = 4°C (LT1071C)

Consult LTC Marketing for parts specified with wider operating temperature ranges.

ELECTRICAL CHARACTERISTICS
V_{IN} = 15V, V_C = 0.5V, V_{FB} = V_{REF}, output pin open unless otherwise specified

SYMBOL	PARAMETER	CONDITIONS		MIN	TYP	MAX	UNITS
V_{REF}	Reference Voltage	Measured at Feedback		1.224	1.244	1.264	V
		Pin, V_C = 0.8V	•	1.214	1.244	1.274	V
I_B	Feedback Input Current	V_{FB} = V_{REF}			350	750	nA
			•			1100	nA
g_m	Error Amplifier Transconductance	ΔI_C = ± 25µA		3000	4400	6000	µmho
			•	2400		7000	µmho
	Error Amplifier Source or Sink Current	V_C = 1.5V		150	200	350	µA
			•	120		400	µA
	Error Amplifier Clamp Voltage	Hi Clamp, V_{FB} = 1V		1.80		2.30	V
		Lo Clamp, V_{FB} = 1.5V		0.25	0.38	0.52	V
	Reference Voltage Line Regulation	3V ≤ V_{IN} ≤ V_{MAX}, V_C = 0.8V	•			0.03	%/V

(continued)

SYMBOL	PARAMETER	CONDITIONS		MIN	TYP	MAX	UNITS
A_V	Error Amplifier Voltage Gain	$0.9V \leq V_C \leq 1.4V$		500	800		V/V
	Minimum Input Voltage		•		2.6	3.0	V
I_Q	Supply Current	$3V \leq V_{IN} \leq V_{MAX}$, $V_C = 0.6V$			6	9	mA
	Control Pin Threshold	Duty Cycle = 0		0.8	0.9	1.08	V
			•	0.6		1.25	V
	Normal/Flyback Threshold on Feedback Pin			0.4	0.45	0.54	V
V_{FB}	Flyback Reference Voltage	$I_{FB} = 50\mu A$		15	16.3	17.6	V
			•	14		18.0	V
	Change in Flyback Reference Voltage	$0.05 \leq I_{FB} \leq 1mA$		4.5	6.8	8.5	V
	Flyback Reference Voltage Line Regulation	$I_{FB} = 50\mu A$, $3V \leq V_{IN} \leq V_{MAX}$ (Note 3)			0.01	0.03	%/V
	Flyback Amplifier Transconductance (g_m)	$\Delta I_C = \pm 10\mu A$		150	300	650	μmho
	Flyback Amplifier Source and Sink Current	$V_C = 0.6V$, $I_{FB} = 50\mu A$ (Source)	•	15	32	70	μA
		$V_C = 0.6V$, $I_{FB} = 50\mu A$ (Sink)	•	25	40	70	μA
B_V	Output Switch Breakdown Voltage	$3V \leq V_{IN} \leq V_{MAX}$, $I_{SW} = 1.5mA$ (LT1070/LT1071)	•	65	90		V
		(LT1070HV/LT1071HV)	•	75	90		V
V_{SAT}	Output Switch "On" Resistance (Note 4)	LT1070	•		0.15	0.24	Ω
		LT1071	•		0.30	0.50	Ω
	Control Voltage to Switch Current Transconductance	LT1070			8		A/V
		LT1071			4		A/V
I_{LIM}	Switch Current Limit (LT1070)	Duty Cycle < 50%, $T_j > 25°C$	•	5		10	A
		Duty Cycle < 50%, $T_J < 25°C$	•	5		11	A
		Duty Cycle = 80% (Note 5)	•	4		10	A
I_{LIM}	Switch Current Limit (LT1071)	Duty Cycle < 50%, $T_J > 25°C$	•	2.5		5.0	A
		Duty Cycle < 50%, $T_J < 25°C$	•	2.5		5.5	A
		Duty Cycle = 80% (Note 5)	•	2.0		5.0	A

(continued)

SYMBOL	PARAMETER	CONDITIONS	MIN	TYP	MAX	UNITS
$\dfrac{\Delta I_{IN}}{\Delta I_{SW}}$	Supply Current Increase During Switch "On" Time			25	35	mA/A
f	Switching Frequency		35	40	45	kHz
		•	33		47	kHz
DC (Max)	Maximum Switch Duty Cycle		90	92	97	%
	Flyback Sense Delay Time			1.5		μs
	Shutdown Mode Supply Current	$3V < V_{IN} < V_{MAX}$, $V_C = 0.05V$		100	250	μA
	Shutdown Mode Threshold Voltage	$3V < V_{IN} < V_{MAX}$		150	250	mV
		•	50		300	mV

The • denotes the specifications which apply over the full operating temperature range.

Note 1: Absolute Maximum Ratings are those values beyond which the life of a device may be impaired.

Note 2: Minimum switch "on" time for the LT1070/LT1071 in current limit is ≈1μs. This limits the maximum input voltage during short-circuit conditions, *in the buck and inverting modes only*, to ≈35V. Normal (unshorted) conditions are not affected. Mask changes are being implemented which will reduce minimum "on" time to ≤ 1μs, increasing maximum short-circuit input voltage above 40V. If the present LT1070/LT1071 (contact factory for package date code) is being operated in the buck or inverting mode at high input voltages and short-circuit conditions are expected, a resistor must be placed in series with the inductor, as follows:

The value of the resistor is given by:

$$R = \frac{t \bullet f \bullet V_{IN} - V_F}{I_{LIMIT}} - R_L$$

t = Minimum "on" time of LT1070/LT1071 in current limit, =1μs
f = Operating frequency (40kHz)
V_F = Forward voltage of external catch diode at I_{LIMIT}
I_{LIMIT} = Current limit of LT1070 (≈8A), LT1071 (≈4A)
R_L = Internal series resistance of inductor

Note 3: V_{MAX} = 55V for LT1070HV and LT1071HV to avoid switch breakdown.

Note 4: Measured with V_C in hi clamp, V_{FB} = 0.8V. I_{SW} = 4A for LT1070 and 2A for LT1071.

Note 5: For duty cycles (DC) between 50% and 80%, minimum guaranteed switch current is given by I_{LIM} = 3.33 (2 − DC) for the LT1070 and I_{LIM} = 1.67 (2 − DC) for the LT1071.

TYPICAL PERFORMANCE CHARACTERISTICS

Switch Current Limit vs Duty Cycle

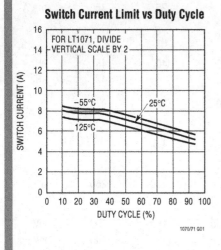

1070/71 G01

Maximum Duty Cycle

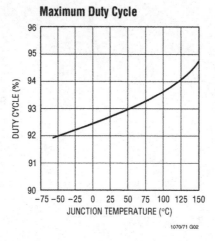

1070/71 G02

Flyback Blanking Time

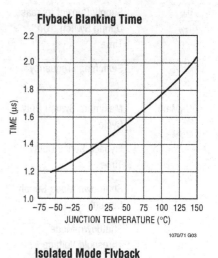

1070/71 G03

Minimum Input Voltage

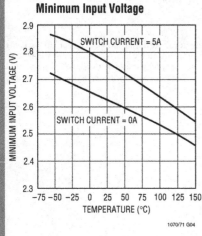

1070/71 G04

Switch Saturation Voltage

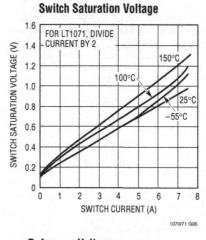

1070/71 G05

Isolated Mode Flyback Reference Voltage

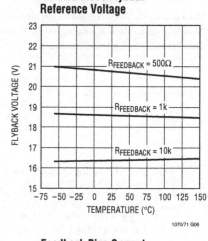

1070/71 G06

Line Regulation

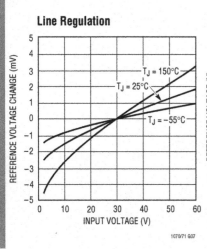

1070/71 G07

Reference Voltage vs Temperature

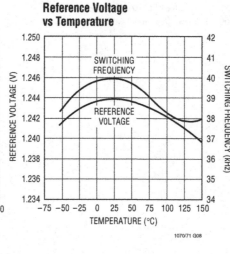

1070/71 G08

Feedback Bias Current vs Temperature

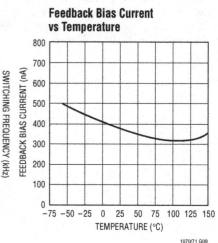

1070/71 G09

TYPICAL PERFORMANCE CHARACTERISTICS

Driver Current* vs Switch Current

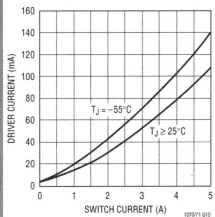

1070/71 G10

*AVERAGE LT1070 POWER SUPPLY CURRENT IS
FOUND BY MULTIPLYING DRIVER CURRENT BY
DUTY CYCLE, THEN ADDING QUIESCENT CURRENT

Supply Current vs Input Voltage*

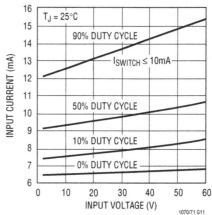

1070/71 G11

*UNDER VERY LOW OUTPUT CURRENT
CONDITIONS, DUTY CYCLE FOR MOST
CIRCUITS WILL APPROACH 10% OR LESS

Supply Current vs Supply Voltage
(Shutdown Mode)

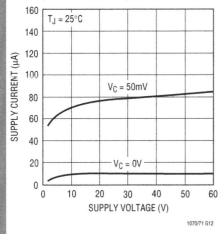

1070/71 G12

Normal/Flyback Mode Threshold
on Feedback Pin

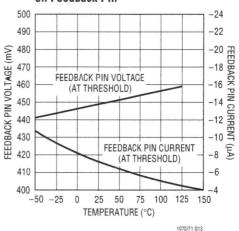

1070/71 G13

Shutdown Mode Supply Current

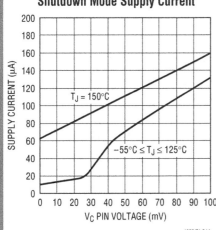

1070/71 G14

Error Amplifier Transconductance

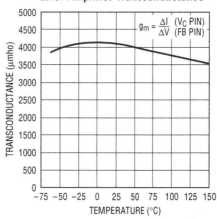

1070/71 G14

TYPICAL PERFORMANCE CHARACTERISTICS

Shutdown Thresholds

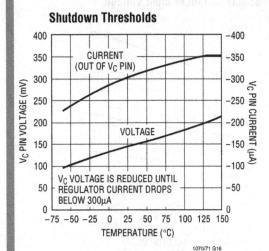

1070/71 G16

Feedback Pin Clamp Voltage

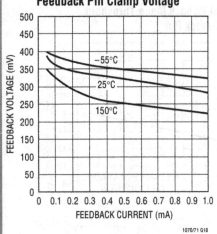

1070/71 G18

V$_C$ Pin Characteristics

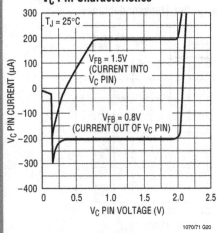

1070/71 G20

Idle Supply Current vs Temperature

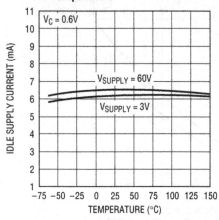

1070/71 G14

Switch "Off" Characteristics

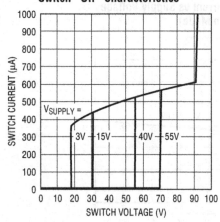

1070/71 G19

Transconductance of Error Amplifier

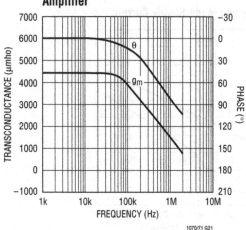

1070/71 G21

Inductor/transformer manufacturers

Pulse Engineering Inc. (619/268-2400)
P.O. Box 12235, San Diego, CA 92112
Hurricane Electronics Lab (801/635-2003)
P.O. Box 1280, Hurricane, UT 84737
Coilcraft Inc. (312/639-2361)
1102 Silver Lake Rd., Cary, IL 60013
Renco Electronics, Inc. (516/586-5566)
60 Jefryn Blvd. East, Deer Park, NY 11729

Core manufacturers

Ferroxcube (ferrites)(914/246-2811)
5083 Kings Highway, Saugerties, NY 12477
Micrometals (powdered iron)(714/630-7420)
1190 N. Hawk Circle, Anaheim, CA 92807
Pyroferric International Inc. (powdered iron)
(217/849-3300)
200G Madison St., Toledo, IL 62468
Fair-Rite Products Corp. (ferrites)(914/895-2055)
P.O. Box J, Wallkill, NY 12589

Stackpole Corp., Ferrite Products Group (814/781-1234)
Stackpole St., St. Mary's, PA 15857
Magnetics Division–Spang & Co.(ferrites)(412/282-8282)
P.O. Box 391, Butler, PA 16003
TDK Corp. of America, Industrial Ferrite Products (312/679-8200)
4709 W. Golf Rd., Skokie, IL 60076

Bibliography

Pressman, A.I., "Switching and Linear Power Supply, Power Converter Design," Hayden Book Co., Hasbrouck Heights, New Jersey, 1977, ISBN 0-8104-5847-0.

Chryssis, G., "High Frequency Switching Power Supplies, Theory and Design," McGraw Hill, New York, 1984, ISBN 0-07-010949-4.

Grossner, N.R., "Transformers for Electronic Circuits," McGraw Hill, New York, 1983, ISBN 0-07-024979-2.

Middlebrook, R.D., and 'Cuk, S., "Advances in Switched Mode Power Conversion," Volumes I, II, III, TESLA Co., Pasadena, CA, 1983.

Proceedings of Powercon, Power Concepts, Inc. Box 5226, Ventura, CA.

"Linear Ferrite Magnetic Design Manual," Ferroxcube Inc., Saugerties, NY.

"Design Manual for SMPS Power Transformers," Pulse Engineering Inc., San Diego, CA.

PACKAGE DESCRIPTION

T Package
5-Lead Plastic TO-220 (Standard)
(LTC DWG # 05-08-1421)

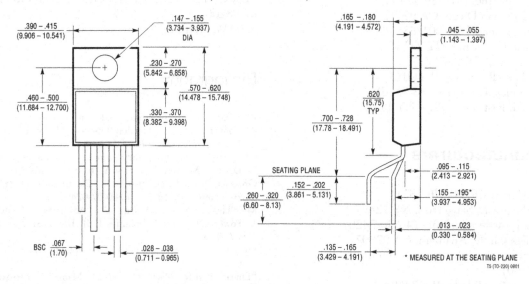

K Package
4-Lead TO-3 Metal Can
(LTC DWG # 05-08-1311)

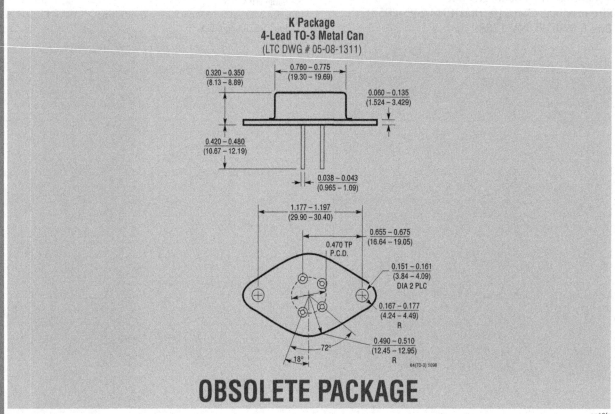

OBSOLETE PACKAGE

an19fc

Switching regulators for poets

A gentle guide for the trepidatious

6

Jim Williams

The above title is not happenstance and was arrived at after considerable deliberation. As a linear IC manufacturer, it is our goal to encourage users to design and build switching regulators. A problem is that while everyone agrees that *working* switching regulators are a good thing, everyone also agrees that they are difficult to get working. Switching regulators, with their high efficiency and small size, are increasingly desirable as overall package sizes shrink. Unfortunately, switching regulators are also one of the most difficult linear circuits to design. Mysterious modes, sudden, seemingly inexplicable failures, peculiar regulation characteristics and just plain explosions are common occurrences. Diodes conduct the wrong way. Things get hot that shouldn't. Capacitors act like resistors, fuses don't blow and transistors do. The output is at ground, and the ground terminal shows volts of noise.

Added to this poisonous brew is the regulator's feedback loop, sampled in nature and replete with uncertain phase shifts. Everything, of course, varies with line and load conditions—and the time of day, or so it seems. In the face of such menace, what are Everyman and the poets to do?

The classic approach is to seek wisdom. Substantial expertise exists but is concentrated in a small number of corporate and academic areas. These resources are not readily accessed by Everyman and some cynics might suggest that they are deliberately protected by a self-serving priesthood. A glance through conference proceedings and published literature yields either a storm of mathematics or absurdly coy and simple little block diagrams that make everything look just so easy. Either way, Everyman loses. And the poets don't even get to try.

Something to think about is that most people who want switching regulators don't need 98.2% efficiency or 100W/cubic inch. They aren't trying to get tenure and don't care about inventing a new type of circuit. What they want are concepts directly applicable to construction of working circuits with readily-available parts. Thus equipped, Everyman can build and sell useful products,

presumably buy more components and everyone's interests (not incidentally, including ours) are served.

As author, I must confess that I am more poet than switching regulator designer, and my poetry ain't very good. Before this effort, my enthusiasm level for switchers resided somewhere between trepidation and terror. This position has changed to one of cautiously respectful optimism. Several things aided this transformation and significantly influenced this publication. The "encouragement" of the Captains of this corporation, emphasized over the last year at increasingly insistent levels, constituted one form of inspiration. Conversations with users (or people who wanted to be) provided more valuable perspective and strength in the knowledge that I was not alone in my difficulties with switchers.

At the circuit level, a significant decision was to employ standard, off-the-shelf magnetics exclusively.[1] This policy was driven by the observation that the majority of problems encountered with switchers centered around inductive components. This approach almost certainly prevents precisely-optimized performance and may horrify some veteran switcher designers. It also eliminates inductor construction uncertainties, saves time and greatly increases the likelihood of getting a design running. It's much easier to work with, and get enthusiastic about, a functional circuit than the smoking carcass of a devastated breadboard. If standard inductor characteristics aren't optimal, it's easier to see the evidence on a 'scope than to guess why you don't see anything.

Additionally, once the circuit is running, an optimized version of the standard product can be supplied by the inductor manufacturer. It's generally easier for the inductor manufacturer to modify its standard product than to start from scratch. The process of communicating and translating circuit performance requirements into inductor construction details is tricky. Using standard product as

---

Note 1: For recommended magnetics supplier, see page 137.

Analog Circuit and System Design: A Tutorial Guide to Applications and Solutions. DOI: 10.1016/B978-0-12-385185-7.00006-8

a starting point accelerates the dialogue, minimizing the number of iterations required for satisfactory results. Often, the standard product suffices for the purpose and no further effort is required.

Strictly speaking, it makes more sense to design the inductor to meet circuit requirements than to fashion a circuit around a standard inductor. Deliberately ignoring this consideration considerably complicated the author's work, but hopefully will simplify the reader's (such is the lot of an application note writer's life). Those interested in inductor design theory are commended to LTC Application Note appropriate chapter-19, "LT®1070 Design Manual."

A final aid in achieving my new outlook on switchers was the LT1070 family. In terms of circuit construction and ease of use they really are superior switching regulator ICs. A 75V, 5A (LT1070HV) on-chip power switch, complete control loop, oscillator and only 5 pins eliminate a lot of the ambiguity of other devices. Internal details and operating features of the LT1070 family are detailed in Appendix A, "Physiology of the LT1070."

Basic flyback regulator

Figure 6.1 shows a basic flyback regulator using the LT1070. It converts a 5V input to a 12V output. Figure 6.2 shows the voltage (Trace A) and the current (Trace B) waveforms at the V_{SWITCH} pin. The V_{SW} output is the collector of a common emitter NPN, so current flows when it is low. Current is pulled through the 100μH inductor and controlled to a value of which forces the 12V output to be constant. The circuit's 40kHz repetition rate is set by the LT1070's internal oscillator. During the time V_{SW} is low, current flow through the inductor causes a magnetic field to be induced into the area around the inductor. The amount of energy stored in this field is a function of the current level, how long current flows, the characteristics of the inductor and its core material. It is often useful to think of the inductor as a bucket and analogize current flow as

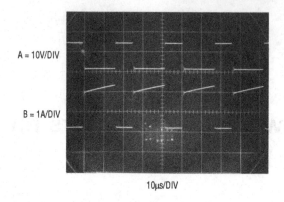

A = 10V/DIV

B = 1A/DIV

10μs/DIV

Figure 6.2 • Flyback Regulator's Waveforms at 7W Loading

water pouring into it. The ultimate limit on energy storage is set by the bucket's capacity, corresponding to the inductor's saturation limitations. The amount of energy that can be put into an inductor in a given time is limited by the applied voltage and the inductance. The amount of energy that can be stored without saturating the inductor is limited by the core characteristics. Size, core material, operating frequency, voltage and current influence inductor design.

If the inductor is enclosed in a feedback-enforced loop, such as Figure 6.1, the energy put into it will be controlled to meet circuit output demands. Figure 6.3 shows what happens when output demand doubles. In this case duty cycle doesn't change much but current doubles. This requires the inductor to store more energy. If it couldn't meet the storage requirement, e.g., it saturated and could not hold any more magnetic flux, it would cease to look inductive. If this point is reached, current flow is limited only by the resistance of the wire and rapidly builds to excessive and destructive values. This behavior is exactly the opposite of a capacitor, where current diminishes upon entering saturation. Capacitors can maintain energy storage with no current flowing; inductors cannot. See Appendix C, "A Checklist for Switching Regulator Designs," for details.

At the end of each inductor charge cycle, current flow in the inductor decays, and the magnetic field around it abruptly collapses. The V_{SW} pin is seen to rise rapidly to a voltage higher than the 5V input. This flyback action gives the regulator its voltage boost characteristics and its name. The boost characteristic is caused by the collapsing magnetic field's lines of flux cutting across the inductor's conductive wire turns. This satisfies the basic requirement for generation of a current in (and hence, a voltage across) a conductor. This moving magnetic field deposits energy into the wire in proportion to how much was stored in the core during the current charge cycle. It is worth noting that the operating characteristics shown here are similar to the Kettering ignition system used in automobiles, explaining why spark occurs when the points open.[2]

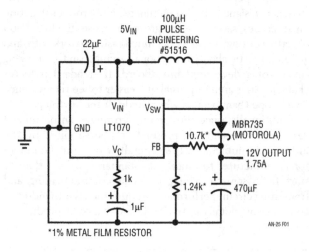

22μF   5V$_{IN}$

100μH
PULSE
ENGINEERING
#51516

V$_{IN}$ V$_{SW}$

GND   LT1070

V$_C$ FB

10.7k*

MBR735
(MOTOROLA)

12V OUTPUT
1.75A

1k

1.24k*

470μF

1μF

AN-25 F01

*1% METAL FILM RESISTOR

Figure 6.1 • Flyback-Type Regulator

Note 2: Back when giants walked the earth, Real Cars used ignition points.

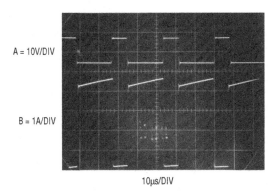

A = 10V/DIV

B = 1A/DIV

10μs/DIV

Figure 6.3 • Flyback Regulator's Waveforms at 14W Loading

In this circuit the flyback is seen to clamp to a level just above the output voltage. This is so because the flyback pulse is steered through the Schottky diode to the output. The 470μF capacitor integrates the repetitive flyback events to DC, providing the circuit's output. The feedback pin (FB) samples this output via the 10.7k to 1.24k divider. The LT1070 compares the feedback pin voltage to its internal 1.24V reference and controls the V_{SW} pin's duty cycle and current, closing a loop. Since the LT1070 is trying to force its feedback pin to 1.24V, output voltage may be set by varying the 10.7k or 1.24k values.

All feedback loops require some form of stability compensation (see the appended section of LTC Application Note appropriate chapter-18, "The Oscillation Problem—Frequency Compensation Without Tears," for general discussion). The LT1070 is no exception. Its voltage gain characteristics, combined with the substantial phase shift of the circuit's sampled energy delivery, ensure oscillation if uncompensated. While the large output capacitor smooths the output to DC, it also teams up with the sampled energy coming into it to create phase shift. To complicate matters,

the load, which may vary, also influences phase characteristics. The regulator can only source into the output capacitor. The load determines the sink time constant, influencing phase performance and overall stability.

The LT1070's internals have been designed with all this in mind and compensation is usually fairly simple. In this case the 1k to 1μF combination at the compensation pin (V_C) rolls off the circuit, providing stable compensation for all operating conditions (see Appendix B, "Frequency Compensation," for details and suggestions on achieving stability in switching regulator loops).

As innocent as Figure 6.1 appears, it's not too difficult to get into odd and seemingly inexplicable problems. Note that the ground connection appears at the ground pin, as opposed to its customary location at the bottom of the diagram. This is deliberate and the supply and load return connections should be made there. The high speed, high current returns from the output transistor's emitter (the "other end" of the V_{SW} pin) should not be allowed to mix with the small currents of the output divider or the V_C pin.

Such mixing can promote poor regulation, unstable operation or outright oscillation. Similarly, the 22μF bypass capacitor ensures clean local power at the LT1070, even during the fast, high current drain periods when V_{SW} comes on. It should have good high frequency characteristics (tantalum or aluminum paralleled by a disc ceramic type). More discussion of these considerations appears in Appendix C.

−48V to 5V telecom flyback regulator

Figure 6.4's circuit is operationally similar to Figure 6.1 but is intended for telecom applications. The raw telecom supply is nominally −48V but can vary from −40V to −60V. This range of voltages is acceptable to the V_{SW} pin but

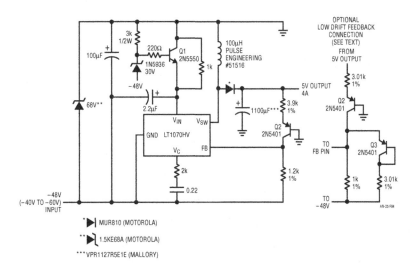

* ▷| MUR810 (MOTOROLA)

** ▷| 1.5KE68A (MOTOROLA)

*** VPR1127R5E1E (MALLORY)

Figure 6.4 • Nonisolated −48V to 5V Regulator

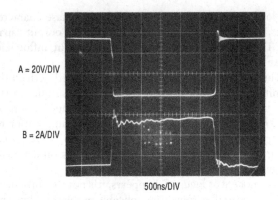

A = 20V/DIV

B = 2A/DIV

500ns/DIV

Figure 6.5 • Nonisolated Regulator's Waveforms

protection is required for the V_{IN} pin ($V_{MAX} = 60V$). Q1 and the 30V zener diode serve this purpose, dropping V_{IN}'s voltage to acceptable levels under all line conditions.

Here, the "top" of the inductor is at ground and the LT1070's ground pin at −48V. The feedback pin senses with respect to the ground pin, so a level shift is required for the 5V output. Q2 serves this purpose, introducing only −2mV/°C drift. This is normally not objectionable in a logic supply, but can be compensated with the optional appropriately scaled diode-resistor shown.

Frequency compensation is similar to Figure 6.1, although a low ESR (equivalent series resistance) capacitor gives less phase shift, permitting faster loop response with the reduced compensation time constant. The 68V zener is a type designed to clamp and absorb excessive line transients which might otherwise damage the LT1070 (V_{SW} maximum voltage is 75V).

Figure 6.5 shows operating waveforms at the V_{SW} pin. Trace A is the voltage and Trace B the current. Switching characteristics are fast and clean. The ripples in the current trace are due to nonoptimal breadboard layout (ground as I say, not as I do). Inductor ringing on turn-off (Trace A) is characteristic of flyback configurations.

Fully-isolated telecom flyback regulator

Figure 6.6's circuit is another telecom regulator. Although it looks more complex, it's really a closely related extension of the previous flyback circuits. The fundamental difference is that the output is fully galvanically isolated from the input, often a requirement in equipment. This necessitates a transformer instead of a simple 2-terminal inductor. It also requires output feedback information to be transmitted to the regulator across a nonconducting path. The transformer complicates the circuit's start-up and switching characteristics while the isolated feedback requires attention to frequency compensation.

In this circuit the V_{IN} pin receives power from a transformer winding. This winding cannot supply power at start-up because the circuit is nonfunctional. Q1 through Q4 address this issue. When power is applied, Q5 cannot conduct because the LT1071 is unpowered. Q1 zener-connected Q2 and Q3 are off. Under these conditions Q4 is on, pulling the V_C pin down and strobing off the LT1071. The potential at Q1's emitter slowly rises as the 10k–100µF combination charges. When Q1's emitter rises high enough, it turns on. Zener-connected Q2 conducts when the voltage across it is about 7V, biasing Q3 on. Q1 sees regenerative feedback, turning Q3 on harder. Q3's turn-on cuts off Q4, allowing the V_C pin to rise and biasing up the LT1071. The rate of rise is limited by the 10µF diode combination at the V_C pin. This network forces the V_C pin to come up slowly, providing a soft-start characteristic (the 100Ω diode string discharges the 10µF capacitor when circuit input power is removed). Because of this sequence, the LT1071 cannot start up the circuit until the V_{IN} potential is well established. This prevents start-up at "starved" or unstable V_{IN} voltages which could cause erratic or destructive modes. When start-up does occur, the transformer feeds the V_{IN} pin with DC via the MUR120 diode.

The 50Ω resistor combines with the 100µF capacitor to give good ripple and transient filtering. This voltage is ample to run the LT1071 and reduces the current through the 10k resistor, saving power. Q1, Q2 and Q3 remain on, biasing Q4 to allow LT1071 operation.

In the previous flyback circuits, the V_{SW} pin drove the inductor directly. Here, a power MOSFET is interposed between the V_{SW} pin and the inductor. In this arrangement the inductor is a transformer and its flyback characteristics are different from a simple 2-terminal inductor. For the simple inductor, the flyback energy was clamped by and dumped directly into the output capacitor. Excessive voltages did not occur. In the transformer case, all the flyback energy does not end up in the output capacitor. Substantial flyback voltage spikes (>100V) appear across the transformer primary when the LT1071 driven MOSFET turns off.

Several measures prevent these spikes from destroying the circuit. The 0.47µF-2k-diode combination, a damper network, conducts during the flyback event. This loads the transformer primary, minimizing flyback amplitude. The damper values are selected empirically, with the trade-off being power dissipation in them. Very low values markedly reduce flyback potentials but cause excessive dissipation. High values permit low dissipation but allow excessive flyback voltages. The damper values should be selected under fully-loaded output conditions because flyback energy is proportionate to transformer power levels. Appendix C contains additional information on damper network considerations.

Even with the damper network, the flyback voltage is too high for the LT1071 output transistor. Q5 prevents the LT1071 from seeing the high voltage. It is connected in series with the LT1071's output transistor. This

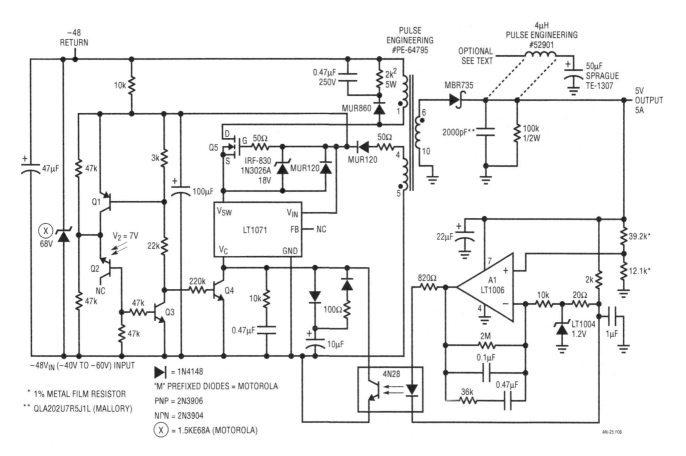

Figure 6.6 • Fully-Isolated −48V to 5V Regulator

connection, sometimes called a cascode, lets Q5 stand off the high voltage and the LT1071 operates well within its breakdown limits. Development and testing of this configuration is detailed in Appendix D. Q5 has large parasitic capacitances associated with all terminals. During switching, these capacitances can cause excessive transient voltages to appear. The 18V zener diode insures against gate-source breakdown ($V_{GSMAX} = 20V$) and the diode clamps the V_{SW} pin to the V_{IN} potential. Mention of these considerations appears in Appendix C.

The transformer's rectified and filtered secondary produces the 5V output. This output is galvanically isolated from the circuit's input. To preserve this desired feature, the feedback path must also be galvanically isolated. A1, the optoisolator and their associated components serve this function. A1, powered by the 5V output, compares a resistively-sampled portion of the output with the LT1004 1.2V reference. Operating at a gain of 200, it drives the optoisolator's LED. The optoisolator's output transistor biases the LT1071's V_C pin, closing a regulation loop. The feedback amplifier inside the LT1071 is essentially bypassed by the A1 optoisolator combination and is not used. Normally, the optoisolator's drifty transmission characteristics over time and temperature would result in unstable feedback.

Here, A1's gain is placed ahead of the optoisolator. This attenuates these uncertainties, providing a stable loop. This approach is not too different from inside-the-loop booster transistors and buffers used with op amps. Both schemes rely on the op amp's gain to eliminate uncertainties and drifts. Returning the optoisolator to V_{REF} instead of ground forces the op amp to bias well above ground, minimizing saturation effects during output transients.

Frequency compensation is somewhat more involved in this circuit than the previous examples. A1 is rolled off by the 0.1µF unit. This keeps gain low at high frequency, preventing amplified ripple and noise from being fed back to the LT1071. Local compensation at the LT1071 V_C pin stabilizes the loop. The 100Ω resistor at the 5V output, a deliberate sink path, allows loop stability at light or no load. Appendix B discusses frequency compensation.

Additional transformer secondary windings could be added if desired. The input zener clips transient voltages.

Circuit waveforms appear in Figure 6.7. Trace A is Q5's drain voltage and Trace B the drain current. Trace A shows that the MOSFET sees about 100V due to flyback effects, but this is well within its rating. The ringing on turn-off is normal and is similar to the waveform observed in Figure 6.4's circuit. Trace B shows that the current flow

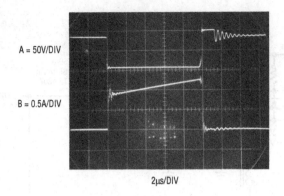

A = 50V/DIV

B = 0.5A/DIV

2µs/DIV

Figure 6.7 • Fully-Isolated Regulator's Waveforms

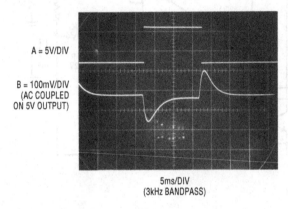

A = 5V/DIV

B = 100mV/DIV
(AC COUPLED
ON 5V OUTPUT)

5ms/DIV
(3kHz BANDPASS)

Figure 6.8 • Fully-Isolated Regulator's Transient Response for a 1A Change on a 2.5A Load

is fast, clean and controlled. Figure 6.8 shows transient response for a 1A step on a 2.5A output. When Trace A goes high the step occurs. Trace B shows that output sag is corrected in about 8ms. When Trace A returns low the 1A load is removed and recovery is similar to the positive step. Broadband output noise, about 75mV$_{P-P}$, may be reduced with the optional filter shown.

100W off-line switching regulator

One of the most desirable switching regulator circuits is also one of the most difficult to design. Figure 6.9's circuit has many similarities to the previous design but is powered directly from the 115V AC line. This off-line operation is desirable because it eliminates large, heavy and inefficient 60Hz magnetics and filter capacitors. The circuit provides an isolated 5V, 20A output as well as isolated ±12V, 1A outputs. Additional features include operation over a 90V AC to 140V AC input range, AC line surge suppression, soft-starting and loop stability under all conditions. Efficiency exceeds 75%.

AC line power is rectified and filtered by the diode bridge–470µF combination. The MOV device provides surge suppression and the thermistor limits turn-on in-rush current. Start-up and soft-start circuitry is similar to Figure 6.6's circuit, with some changes necessitated by the higher input voltage. Erratic operation at extremely low AC line voltages (70V AC) is prevented by the 220k-1.24k divider. At very low AC line inputs, this divider forces the LT1071 feedback pin to a low state, shutting down the circuit. The high input voltage, typically 160V DC, means that the LT1071's internal current limit is set too high to protect the regulator if the circuit's output is shorted. Q6 and its associated components provide about 2A limiting. The LT1071's GND pin current flows through the 0.3Ω resistor, turning on Q6 if current is too high. The 22k-50pF RC filters noise, preventing erratic Q6 operation.

Q5, a power MOSFET, is cascoded with the LT1071 for high voltage switching. Circuit topology is similar to Figure 6.6, with Q5's voltage breakdown increased to 500V.

Additionally, the 50Ω resistor combines with the gate capacitance to slightly slow Q5's transitions, reducing high frequency harmonics. This measure eases layout considerations. The transformer's damper network borrows from Figure 6.6, with values reestablished for this circuit.

The A1-optocoupler-enforced feedback loop preserves the transformer's galvanic isolation, allowing the regulator output to be ground referenced. The feedback loop is also similar to Figure 6.6. Compensation values at A1 and the LT1071 have changed, reflecting this circuit's different gain-phase characteristics.

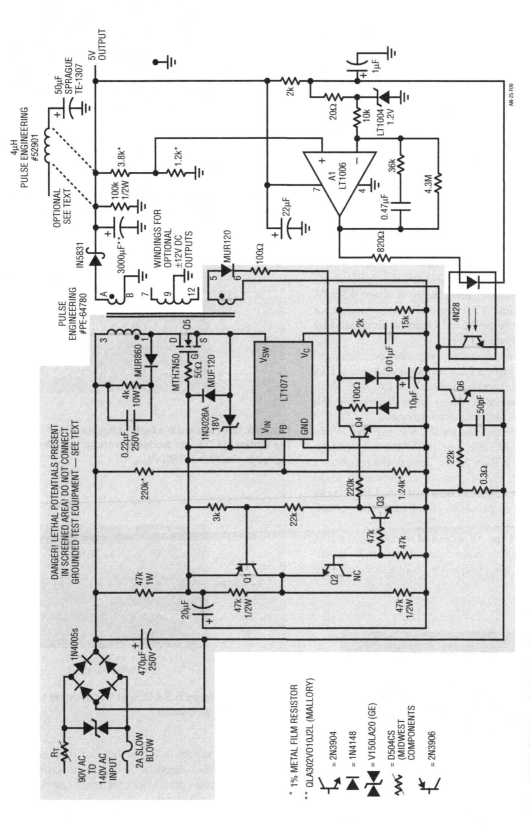

Figure 6.9 • 100W Off-Line Switching Regulator

DANGER! LETHAL POTENTIALS PRESENT—SEE TEXT

Figure 6.10 shows circuit waveforms at 15A output. Trace A, Q5's drain, shows the flyback pulse being damped below 300V (for a discussion of the procedures used to design the damper network and other design techniques in this circuit, see Appendix D, "Evolution of a Switching Regulator Design"). Trace B, the LT1071's V_{SW} pin, stays well within its voltage rating, despite Q5's high voltage switching. Trace C, Q5's drain current, shows that transformer current is well controlled with no saturation effects. Trace D, damper network current, is active when Q5 goes off.

Figure 6.11 is a time and amplitude expansion of Q5's drain (Trace A) and transformer primary current (Trace B). Switching is clean, with residual noise due to nonideal transformer behavior. The damper network clamps the flyback pulse well below Q5's 500V rating and the transformer rings off after the flyback interval. The noise on the current pulse, due to resonances in the transformer, has no significant effect on circuit operation.

Figure 6.12 shows output noise with the optional LC filter in use. Without the filter, noise is about 150mV. Superimposed, residual 120Hz modulation accounts for trace thickening at the peaks and could be eliminated by increasing the 470µF value.

Figure 6.13 shows transient response performance. When Trace A goes high, a 5A transient is added to a 10A steady-state load. Recovery amplitude is low and clean with a first order response. When Trace A goes low, the transient load is removed with similar results.

Figure 6.14 shows response for shifts in the line. When Trace A is high, the AC line is at 140V AC. Line voltage drops to 90V AC with Trace A low. Trace B, the regulator's AC-coupled output, shows a clean recovery with small amplitude error. The ripples in the waveform, 120Hz input residue, could be reduced by increasing the 470µF capacitor.

Figure 6.15 shows the 5V output at start-up into a 20A load. Response is slightly underdamped and can be modified by adjusting the frequency compensation. The

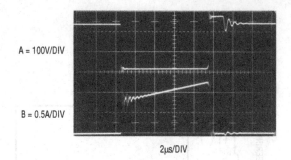

Figure 6.11 • Detail of Off-Line Switcher's Transformer Primary Voltage and Current Waveforms

DANGER! TAKE THIS MEASUREMENT *ONLY* WITH AN ISOLATION TRANSFORMER IN USE—SEE TEXT

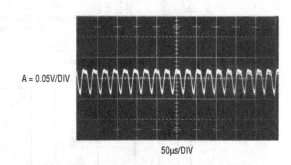

Figure 6.12 • Figure 6.9's Output Ripple at 10A Output with the Optional LC Filter Added—Without the Filter, Ripple Increases to About 150mV$_{P-P}$

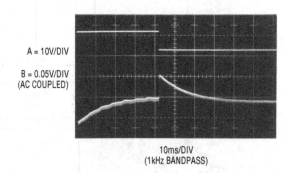

Figure 6.13 • Figure 6.9's Circuit Responding to a 5A Change on a 10A Output

compensation shown in Figure 6.9 is a good compromise between transient response and turn-on characteristics. The delay on turn-on and the controlled rise time are due to the slow-start circuitry.

Figure 6.16 plots regulator efficiency. As would be expected, efficiency is best at high currents, where static losses are a small percentage of output power.

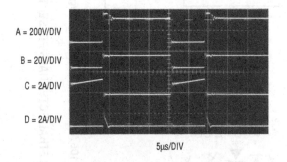

Figure 6.10 • Off-Line Switcher's Waveforms

DANGER! TAKE THIS MEASUREMENT *ONLY* WITH AN ISOLATION TRANSFORMER IN USE—SEE TEXT

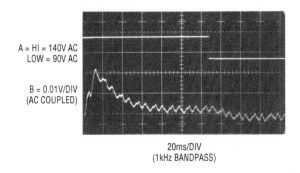

A = HI = 140V AC
LOW = 90V AC

B = 0.01V/DIV
(AC COUPLED)

20ms/DIV
(1kHz BANDPASS)

Figure 6.14 • Figure 6.9 Responds to a 90V AC—140V AC Line Change—Loading is 10A—120Hz Residue in Output Could be Reduced by Increasing the 470µF Input Filter

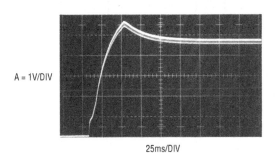

A = 1V/DIV

25ms/DIV

Figure 6.15 • Start-Up for Figure 6.9 at 20A Loading—The 10µF Capacitor at the LT1070's V_C Pin Produces the Slow-Start Characteristic. If the Small Overshoot is Objectionable, Modified Frequency Compensation Can Eliminate it at Some Cost to Transient Response

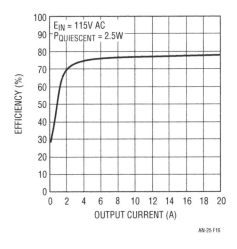

Figure 6.16 • Figure 6.9's Efficiency vs Operating Point

Switch-controlled motor speed controller

Voltage regulators are not the only switching power circuits. Figure 6.17 shows a motor speed regulator. The LT1070 provides simplicity and switch-mode control

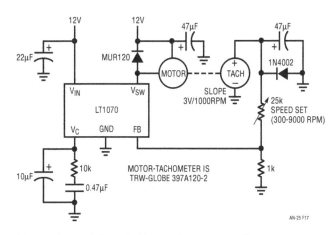

Figure 6.17 • A Simple Motor-Tachometer Servo Loop

efficiency. Although this circuit controls a motor, it shares many considerations common to voltage regulators. When power is applied, the tachometer output is zero and the feedback pin (FB) is also at zero. This causes the LT1070 to begin pulsing its V_{SW} pin at maximum duty cycle. The motor turns, forcing tachometer output. When the FB pin arrives at the LT1070's internal voltage reference value (1.24V), the loop stabilizes. Speed is adjustable with the 25k potentiometer in the feedback string. The MUR120 damps the motor's flyback spike. The characteristics of the motor specified permit no current limiting in series with the diode. Other motors might require this and damper network optimization should be done for any specific unit. Similarly, frequency compensation values will vary with different motor types. The diode at the tachometer output prevents transient reverse voltages due to tachometer commutator switching.

Switch-controlled peltier 0°C reference

Figure 6.18 is another switch-mode control circuit. Here, the LT1070 controls power to a Peltier cooler, providing a 0°C temperature reference for transducer calibration.

A platinum RTD is thermally mated to the Peltier cooler. The RTD combines with a bridge network to give a differential output. A1 provides maximum bridge drive without introducing significant heating in the RTD. The LTC®1043 switched capacitor network converts this output to a single-ended signal at A2. A2, operating at a gain of 400, biases the LT1070's V_C pin. This closes a control loop around the Peltier cooler, forcing its temperature low enough to balance the bridge. The 0°C trim adjusts the servo point to precisely 0°C. A standard RTD should monitor Peltier temperature when making this trim. Alternately, the sensor specified should be supplied

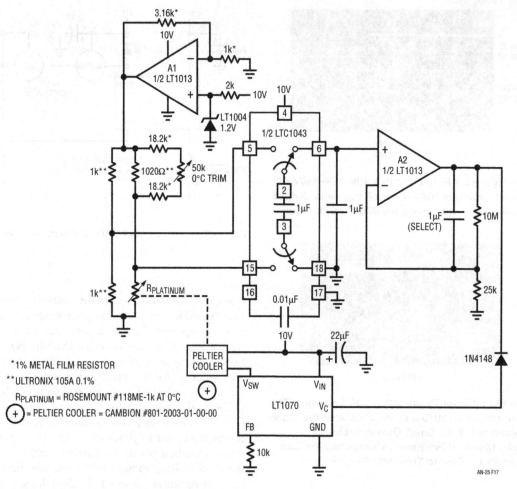

Figure 6.18 • A Peltier-Cooled Switched-Mode 0°C Reference

with a certified 0°C resistance. With the RTD and Peltier cooler tightly mated, stability is excellent. Figure 6.19, a plot of stability over hours in a 25°C ± 3°C ambient, shows a 0.15°C baseline.

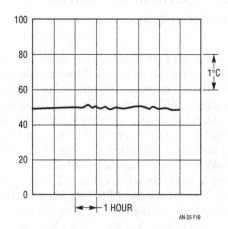

Figure 6.19 • Stability of Figure 6.18's Circuit Over Many Hours with a 25°C ± 3°C Ambient

Acknowledgments

The author acknowledges Carl Nelson's abundance of commentary, some of which was useful, during preparation of this work. Bob Dobkin's thoughts and patience are also appreciated. Ron Young made significant contributions towards Figure 6.6's circuit. Bill McColey and other members of the Engineering staff of Pulse Engineering, Inc.[3], supplied invaluable insight and assistance on magnetics issues. As usual, our customers' requests and requirements provided the most valuable source of guidance, and they are due a special thanks.

Appendix A
Physiology of the LT1070

The LT1070 is a current-mode switcher. This means that switch duty cycle is directly controlled by switch

Note 3: P.O. Box 12235, San Diego, CA 92112 (619/268-2400).

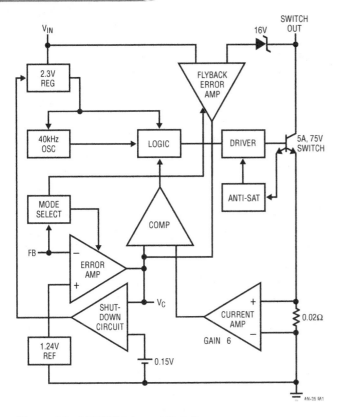

Figure A1 • LT1070 Internal Details

current rather than by output voltage. Referring to Figure A1, the switch is turned on at the start of each oscillator cycle. It is turned off when switch current reaches a predetermined level. Control of output voltage is obtained by using the output of a voltage-sensing error amplifier to set current trip level. This technique has several advantages. First, it has immediate response to input voltage variations, unlike ordinary switchers which have notoriously poor line transient response. Second, it reduces the 90° phase shift at mid-frequencies in the energy storage inductor. This greatly simplifies closed loop frequency compensation under widely varying input voltage or output load conditions. Finally, it allows simple pulse-by-pulse current limiting to provide maximum switch protection under output overload or short conditions. A low dropout internal regulator provides a 2.3V supply for all internal circuitry on the LT1070. This low dropout design allows input voltage to vary from 3V to 6V with virtually no change in device performance. A 40kHz oscillator is the basic clock for all internal timing. It turns on the output switch via the logic and driver circuitry. Special adaptive antisat circuitry detects onset of saturation in the power switch and adjusts driver current instantaneously to limit switch saturation. This minimizes driver dissipation and provides very rapid turn-off of the switch.

A 1.2V bandgap reference biases the positive input of the error amplifier. The negative input is brought out for output voltage sensing. This feedback pin has a second function; when pulled low with an external resistor, it programs the LT1070 to disconnect the main error amplifier output and connects the output of the flyback amplifier to the comparator input. The LT1070 will then regulate the value of the flyback pulse with respect to the supply voltage. This flyback pulse is directly proportional to output voltage in the traditional transformer-coupled flyback topology regulator. By regulating the amplitude of the flyback pulse the output voltage can be regulated with no direct connection between input and output. The output is fully floating up to the breakdown voltage of the transformer windings. Multiple floating outputs are easily obtained with additional windings. A special delay network inside the LT1070 ignores the leakage inductance spike at the leading edge of the flyback pulse to improve output regulation.

The error signal developed at the comparator input is brought out externally. This pin (V_C) has four different functions. It is used for frequency compensation, current limit adjustment, soft-starting and total regulator shutdown. During normal regulator operation this pin sits at a voltage between 0.9V (low output current) and 2.0V (high output current). The error amplifiers are current output (gm) types, so this voltage can be externally clamped for adjusting current limit. Likewise, a capacitor-coupled external clamp will provide soft-start. Switch duty cycle goes to zero if the V_C pin is pulled to ground through a diode, placing the LT1070 in an idle mode. Pulling the V_C pin below 0.15V causes total regulator shutdown with only 50µA supply current for shutdown circuitry biasing. For more details, see Linear Technology Application Note appropriate chapter-5-19, pages 4-8.

Appendix B
Frequency compensation

Although the architecture of the LT1070 is simple enough to lend itself to a mathematical approach to frequency compensation, the added complications of input and/or output filters, unknown capacitor ESR, and gross operating point changes with input voltage and load current variations all suggest a more practical empirical method. Many hours spent on breadboards have shown that the simplest way to optimize the frequency compensation of the LT1070 is to use transient response techniques and an R/C box to quickly iterate toward the final compensation network.

There are many ways to inject a transient signal into a switching regulator, but the suggested method is to use an AC-coupled output load variation. This technique avoids problems of injection point loading and is general to all switching topologies. The only variation required may be an amplitude adjustment to maintain small signal conditions with adequate signal strength. Figure B1 shows the setup.

A function of generator with 50Ω output impedance is coupled through a 50Ω/1000µF series RC network to the regulator output. Generator frequency is non-critical. A good starting point is 50Hz. Lower frequencies may cause a blinking scope display which is annoying to work with. Higher frequencies may not allow sufficient settling time for the output transient. Amplitude of the generator output is typically set at 5V$_{P-P}$ to generate a 100mA$_{P-P}$ load variation.

For lightly loaded output (I$_{OUT}$ <100mA), this level may be too high for small signal response. If the positive and negative transition settling waveforms are significantly different, amplitude should be reduced. Actual amplitude is not particularly important because it is the shape of the resulting regulator output waveform that indicates loop stability.

A 2-pole oscilloscope filter with f = 10kHz is used to block switching frequencies. Regulators without added LC output filters have switching frequency signals at their outputs which may have much higher amplitude than the low frequency settling waveform to be studied. The filter frequency is high enough to pass the settling waveform with no distortion.

Oscilloscope and generator connections should be made exactly as shown to prevent ground loop errors. The oscilloscope is synced by connecting the channel B probe to the generator output, with the ground clip of the second probe connected to exactly

the same place as the channel A ground. The standard 50Ω BNC sync output of the generator should not be used because of ground loop errors. It may also be necessary to isolate either the generator or oscilloscope from its third wire (earth ground) connection in the power plug to prevent ground loop errors in the 'scope display. These ground loop errors are checked by connecting the channel A probe tip to exactly the same point as the probe ground clip. Any reading on channel A indicates a ground loop problem.

Once the proper setup is made, finding the optimum values for the frequency compensation network is fairly straightforward. Initially, C2 is made large (≥2µf), and R3 is made small (≈1kΩ). This nearly always ensures that the regulator will be stable enough to start iteration. Now, if the regulator output waveform is single-pole overdamped (see the waveforms in Figure B2), the value of C2 is reduced in steps of about 2:1 until the response becomes slightly underdamped. Next, R3 is increased in steps of 2:1 to introduce a loop "zero." This will normally improve damping and allow the value of C2 to be further reduced. Shifting back and forth between R3 and C2 variations will now allow one to quickly find optimum values.

If the regulator response is underdamped with the initial large value of C, R should be increased immediately before larger values of C are tried. This will normally bring about the overdamped starting condition for further iteration.

Just what is meant by "optimum values" for R3 and C2? This normally means the smallest value for C2 and the largest value for R3, which still guarantee no loop oscillations, and which result in loop settling that is as rapid as possible. The reason for this

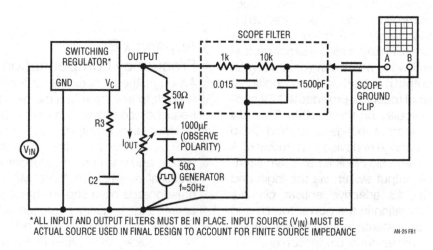

*ALL INPUT AND OUTPUT FILTERS MUST BE IN PLACE. INPUT SOURCE (V$_{IN}$) MUST BE ACTUAL SOURCE USED IN FINAL DESIGN TO ACCOUNT FOR FINITE SOURCE IMPEDANCE   AN-25 FB1

Figure B1 • Testing Loop Stability

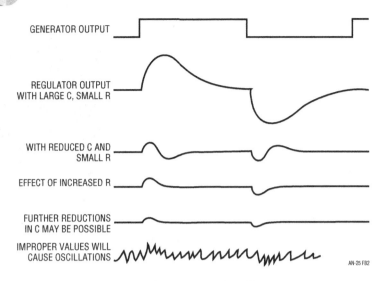

GENERATOR OUTPUT

REGULATOR OUTPUT
WITH LARGE C, SMALL R

WITH REDUCED C AND
SMALL R

EFFECT OF INCREASED R

FURTHER REDUCTIONS
IN C MAY BE POSSIBLE

IMPROPER VALUES WILL
CAUSE OSCILLATIONS

AN-25 FB2

Figure B2 • Output Transient Response

approach is that it minimizes the variations in output voltage caused by input ripple voltage and output load transients. A switching regulator which is grossly overdamped will never oscillate but it may have unacceptably large output transients following sudden changes in input voltage or output loading. It may also suffer from excessive overshoot problems on start-up or short-circuit recovery.

To guarantee acceptable loop stability under all conditions, the initial values chosen for R3 and C2 should be checked under all conditions of input voltage and load current. The simplest way to accomplish this is to apply load currents of minimum, maximum, and several points in between. At each load current, input voltage is varied from minimum to maximum while observing the settling waveform. The additional time spent "worst-casing" in this manner is definitely necessary. Switching regulators, unlike linear regulators, have large shifts in loop gain and phase with operating conditions. If large temperature variations are expected for the regulator, stability checks should also be done at the temperature extremes. There can be significant temperature variations in several key component parameters which affect stability—in particular, input and output capacitor values and their ESRs and inductor permeability. The LT1070 parametric variations also need some consideration. Those which affect loop stability are error amplifier gm, and the transfer function of V_C pin voltage versus switch current (listed as a transconductance under electrical specifications.) For modest temperature variations, conservative overdamping under worst-case temperature conditions is usually sufficient to guarantee adequate stability at all temperatures.

If external amplifiers or other active devices are included in the loop (e.g., Figures 6.6 and 6.9), their effects must be included in stabilizing the loop. LTC Application Note appropriate chapter-18, pages 12-15, provides commentary that may be useful in these situations.

Appendix C
A checklist for switching regulator designs

1. The most common problem area in switching designs is the inductor and the most common difficulty is saturation. An inductor is saturated when it cannot hold any more magnetic flux. As an inductor arrives at saturation it begins to look more resistive and less inductive. Under these conditions the current flow through it is limited only by its DC copper resistance and the source capacity. This is why saturation often results in destructive failures. Figure C1 demonstrates saturation effects. The pulse generator drives Q1, forcing current into the inductor. The diode and RC combination form a typical load. Figure C2 shows results. The voltage at Q1's collector falls when it turns on (Trace A is pulse generator output, Trace B is Q1's collector). Trace C, the inductor current, ramps in controlled fashion. When Q1 goes off, current falls and the inductor rings off. In Figure C3, drive pulse width is longer, allowing more inductor current buildup. This requires the inductor to store more magnetic flux. Its ramp waveform is clean and controlled,

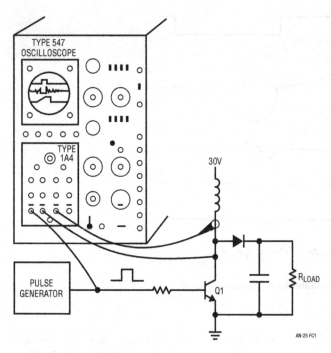

Figure C1 • Inductor Saturation Test Circuit

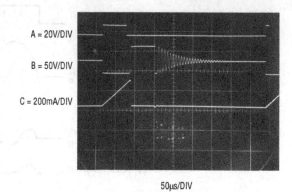

50µs/DIV

Figure C3 • Normal Inductor Operation at Increased Current

indicating that it has the necessary capacity. Figure C4 brings some unpleasant surprises. Drive pulse width has been increased. Now, the inductor current departs from its ramp characteristic into a nonlinear slope. The nonlinear behavior starts between the third and fourth vertical divisions. This curve shows a rapidly increasing current characteristic. These conditions indicate that the inductor is entering saturation. If pulse width is increased much more, the current will rise to destructive levels. It is worth noting that some inductors saturate much more abruptly than this case.

2. Always consider inductive flyback effects. Are semiconductor breakdown ratings adequate to withstand them? Is a snubber (damper) network required? Consider all possible voltages and current paths, including the transient ones via semiconductor junction capacitances, to avoid evil problems.

3. Think about requirements in capacitors. All operating conditions should be accounted for. Voltage rating is the most obvious consideration, but remember to plan for the effects of equivalent series resistance (ESR) and inductance. These specifications can have significant impact on circuit performance. In particular, an output capacitor with high ESR can make loop compensation difficult.

4. Layout is vital. Don't mix signal, frequency compensation, and feedback returns with high current returns. Arrange the grounding scheme for the best compromise between AC and DC performance. In many cases, a ground plane may help. Account for possible effects of stray inductor-generated flux on other components and plan layout accordingly.

5. Semiconductor breakdown ratings must be thought through. Account for all conditions. Transient events usually cause the most trouble,

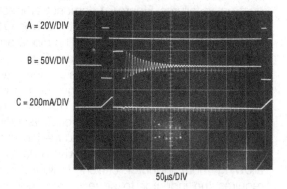

50µs/DIV

Figure C2 • Normal Inductor Operation

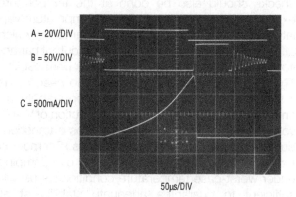

50µs/DIV

Figure C4 • Inductor Being Driven into Saturation

introducing stresses that are often hard to predict. Things to watch for include effects of feedthrough via semiconductor junction capacitances (note the clamping of Q5's gate in Figures 6.6 and 6.9). Such capacitances can allow excessive voltages to occur for brief durations at what is normally a low voltage node. Study the data sheet breakdown, current capacity, and switching speed ratings carefully. Were these specifications written under the same conditions that your circuit is using the device in? If in doubt, consult the manufacturer.

"Simple" diodes furnish a good example of how carefully semiconductor operating conditions must be considered in switching regulators. Switching diodes have two important transient characteristics—reverse recovery time and forward turn-on time. Reverse recovery time occurs because the diode stores charge during its forward conducting cycle. This stored charge causes the diode to act as a low impedance conductive element for a short period of time after reverse drive is applied. Reverse recovery time is measured by forward biasing the diode with a specified current, then forcing a second specified current backwards through the diode. The time required for the diode to change from a reverse conducting state to its normal reverse nonconducting state is reverse recovery time. Hard turn-off diodes switch abruptly from one state to the other following reverse recovery time. They therefore dissipate very little power even with moderate reverse recovery times. Soft turn-off diodes have a gradual turn-off characteristic that can cause considerable diode dissipation during the turn-off interval. Figure C5 shows typical current and voltage waveforms for several commercial diode types used in an LT1070 flyback converter with $V_{IN} = 10V$, $V_{OUT} = 20V$, 2A.

Long reverse recovery times can cause significant extra heating in the diode or the LT1070 switch. Total power dissipated is given by:

$$P_{trr} = V \cdot f \cdot t_{RR} \cdot I_F$$

V = reverse diode voltage

f = LT1070 switching frequency

t_{RR} = reverse recovery time

I_F = diode forward current just prior to turn-off

With the circuit mentioned, I_F is 4A, $V = 20V$, and $f = 40kHz$. Note that diode on current is twice output current for this particular boost configuration. A diode with $t_{rr} = 300ns$ creates a power loss of:

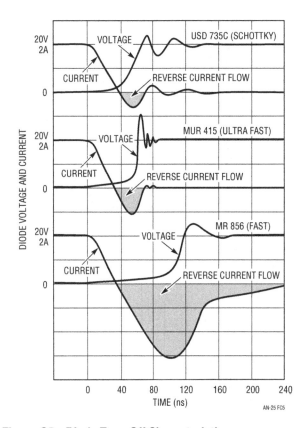

Figure C5 • Diode Turn-Off Characteristics

$$P_{trr} = (20) (40 \cdot 10^3) (300 \cdot 10^{-6}) (4) = 0.96W$$

If this same diode had a forward voltage of 0.8V at 4A, its forward loss would be 2A (average current) times 0.8V equals 1.6W. Reverse recovery losses in this example are nearly as large as forward losses. It is important to realize however, that reverse losses may not necessarily increase diode dissipation significantly. A hard turn-off diode will shift much of the power dissipation to the LT1070 switch, which will undergo a high current and high voltage condition during the duration of reverse recovery time. This has not been shown to be harmful to the LT1070, but the power loss remains.

Diode turn-on time can potentially be more harmful than reverse turn-off. It is normally assumed that the output diode clamps to the output voltage and prevents the inductor or transformer connection from rising higher than the output. A diode that turns on slowly can have a very high forward voltage for the duration of turn-on time. The problem is that this increased voltage appears across the LT1070 switch. A 20V turn-on spike superimposed on a 40V flyback mode output pushes switch voltage perilously close to the 65V limit. The graphs in Figure C6 show diode turn-on spikes for three common diode

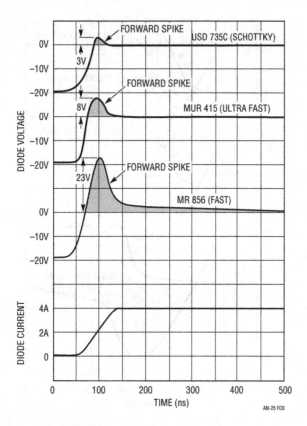

Figure C6 • Diode Turn-On Spike

types—fast, ultrafast, and Schottky. The height of the spike will be dependent on rate of rise of current and the final current value, but these graphs emphasize the need for fast turn-on characteristics in applications which push the limits of switch voltage.

Fast diodes can be useless if the stray inductance is high in the diode, output capacitor or LT1070 loop. 20-gauge hook-up wire has 30nH/inch inductance. The current fall time of the LT1070 switch is 10^8A/sec. This generates a voltage of $(10^8)(30 \cdot 10^{-9}) = 3$V per inch in stray wiring. Keep the diode, capacitor and LT1070 ground/switch lead lengths *short!*

Appendix D
Evolution of a switching regulator design

A good way to approach designing a switching regulator is to break the problem into small tasks and then integrate everything. The combination of inductors, a sampled feedback loop, and high speed currents and voltages leaves much room for confusion. The approach used in Figure 6.9's design is illustrated as an example of an iterative approach in switching regulator design. This off-line circuit

features high power, an isolated feedback loop and the aforementioned complexities. Any attempt to get everything working on the first try is beyond risky.

The transformer drive is the most critical part of Figure 6.9's design. Fast switching of over 100W at high voltage requires care. In particular, two issues must be addressed. Will the high voltage FET-LT1071 cascode connection really work? What amplitudes of flyback voltage are going to occur and what will their effects be?

Figure D1 begins the investigation. This test circuit allows checking of the high voltage cascode. To start, a resistive load is used, eliminating the possible (certain!) complications of the inductive load. Figure D2 shows waveforms. Switching is clean. Trace A is the FET drain, while Trace B is the LT1071 V_{SW} pin. Drain current appears in Trace C. Pulse width is kept deliberately low, minimizing load power dissipation. Everything appears well ordered, and the LT1071 V_{SW} pin does not see any high voltage excursions. Artifacts of the MOSFET's high voltage switching do, however, appear at the LT1071 V_{SW} pin. On the falling edge, the ringing appears, albeit at lower amplitude. The rising edge shows a slight peaking. These effects are due to the high voltage coupling through the MOSFET's junction capacitances. The diode clamps the source to 10V, but the effects of the high voltage slewing are still noticeable. This doesn't cause much trouble with the resistive load, but what will happen with the inductor's higher flyback voltages?

Figure D3 shows the test circuit rearranged to accommodate the transformer load. The transformer replaces the resistor. Its terminated secondary allows

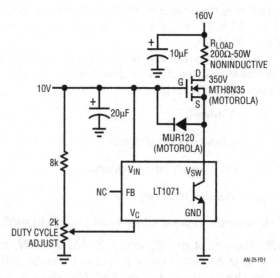

Figure D1 • Test Circuit for MOSFET-LT1071 Cascode with Resistive Load

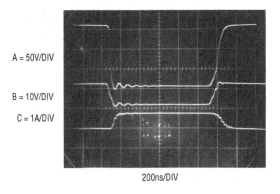

A = 50V/DIV

B = 10V/DIV

C = 1A/DIV

200ns/DIV

Figure D2 • Testing the MOSFET-LT1071 Cascode Switch with a Resistive Load

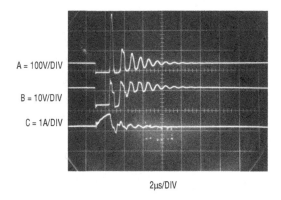

A = 100V/DIV

B = 10V/DIV

C = 1A/DIV

2µs/DIV

Figure D4 • MOSFET-LT1071 Cascode Switching the Transformer Primary—Secondary Load is 0.2Ω

it to present a significant load. The fixed 160V supply has been replaced with a 0V to 200V unit, permitting voltage to be slowly and cautiously increased[1]. The 350V transistor is replaced with a 1000V unit, in preparation for inductive events. Figure D4 shows waveforms. As expected, the inductive flyback (Trace A) is significant, even at low supply voltage (V_{SUPPLY} = 60V in this photo).

I race C, the drain current, rises with a characteristic indicating the inductive load. Trace B, the source

voltage, is of greater concern. The flyback event, feeding through the MOSFET's capacitances, causes the source (and gate) to rise above nominal clamped value. At the higher supply voltages planned, this could cause excessive gate-source voltages with resultant device destruction. Because of this, the zener diode in dashed lines is installed, clamping gate-source voltages to safe values. This component appears in Figure 6.9's final design. With this correction, behavior at higher supply voltages may be investigated.

Figure D5 shows the MOSFET drain at V_{SUPPLY} = 160V. The load draws about 2.5A. Flyback voltage rises to 400V. At 5A loading this voltage approaches 500V (Figure D6), while a 10A load (Figure D7) forces almost 900V flyback. In actual regulator operation, supply voltages, switch on-time and output current can go higher, meaning flyback potentials will exceed 1000V. This graphically mandates the need for a damper network. A simple reverse-biased diode or zener clipper will work, but will suffer from excessive dissipation. The network shown in Figure 6.9 is a good compromise between dissipation and reasonable flyback voltages.

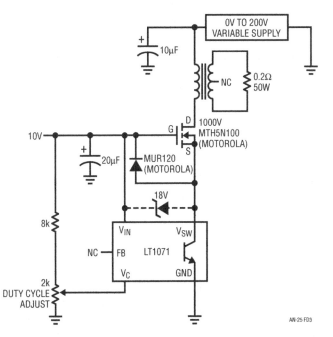

Figure D3 • Test Circuit for MOSFET-LT1071 Cascode with Transformer Load

Note 1: "For fools rush in where angels fear to tread"—An Essay on Criticism, A. Pope.

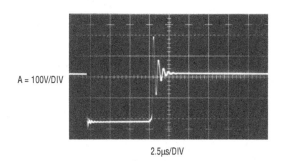

A = 100V/DIV

2.5µs/DIV

Figure D5 • Undamped Regulator Flyback Pulse at 2.5A Output

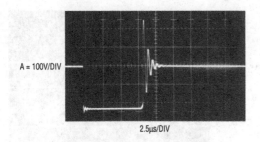

Figure D6 • Undamped Regulator Flyback Pulse at 5A Output

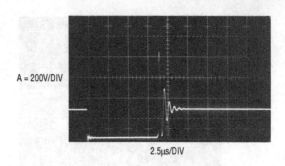

Figure D7 • Undamped Regulator Flyback Pulse at 10A Output

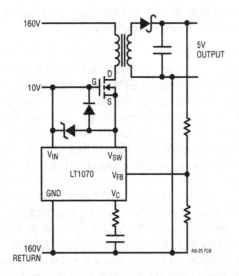

Figure D8 • Developmental Version of Off-Line Switching Regulator—No Isolation is Included and the Scheme is Solely Intended to Verify that a Loop Can be Closed Around the Transformer

Once the drive-flyback issues are settled, a feedback loop is closed around the transformer. This allows checking to see that loop stabilization is possible. Figure D8 diagrams the loop. In this configuration the regulator will function, but is unusable. The output is not galvanically isolated from the input, which ultimately must be directly AC line driven. After this loop has been successfully closed, the isolated version is tried (Figure D9). This introduces more phase shift, but is also found to be stable with appropriate frequency compensation. Finally, the connection between the input and output common potentials is broken, achieving the desired galvanic isolation. The start-up, soft-start and current limit features are then added and optimized. Testing involves checking performance under various line and load conditions. Details on circuit operation are covered in the text associated with Figure 6.9.

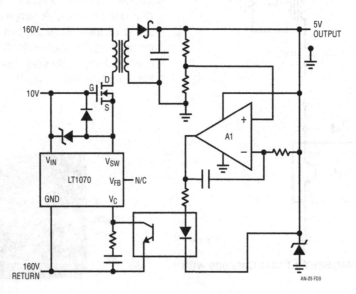

Figure D9 • Developmental Version of Off-Line Regulator with Isolation— the Circuit Verifies That Loop Stability is Achievable with the Added Phase Shift of A1 and the Opto-Isolator—Start-Up, Current Limit and Soft-Start Features Must be Added to Complete the Design

Step-down switching regulators

7

Jim Williams

A substantial percentage of regulator requirements involve stepping down the primary voltage. Although linear regulators can do this, they cannot achieve the efficiency of switching based approaches[1]. The theory supporting step-down ("buck") switching regulation is well established, and has been exploited for some time. Convenient, easily applied ICs allowing implementation of practical circuits are, however, relatively new. These devices permit broad application of step-down regulation with minimal complexity and low cost. Additionally, more complex functions incorporating step-down regulation become realizable.

Basic step down circuit

Figure 7.1 is a conceptual voltage step-down or "buck" circuit. When the switch closes the input voltage appears at the inductor. Current flowing through the inductor-capacitor combination builds over time. When the switch opens

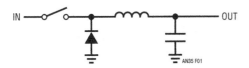

Figure 7.1 • Conceptual Voltage Step-Down ("Buck") Circuit

current flow ceases and the magnetic field around the inductor collapses. Faraday teaches that the voltage induced by the collapsing magnetic field is opposite to the originally applied voltage. As such, the inductor's left side heads negative and is clamped by the diode. The capacitor's accumulated charge has no discharge path, and a DC potential appears at the output. This DC potential is lower than the input because the inductor limits current during the switch's on-time.

Note 1: While linear regulators cannot compete with switchers, they can achieve significantly better efficiencies than generally supposed. See LTC Application Note 32, "High Efficiency Linear Regulators," for details.

Ideally, there are no dissipative elements in this voltage step-down conversion. Although the output *voltage* is lower than the input, there is no *energy* lost in this voltage-to-current-to-magnetic   field-to-current-to-charge-to-voltage conversion. In practice, the circuit elements have losses, but step-down efficiency is still higher than with inherently dissipative (e.g., voltage divider) approaches. Figure 7.2 feedback controls the basic circuit to regulate output voltage. In this case switch on-time (e.g., inductor charge time) is varied to maintain the output against changes in input or loading.

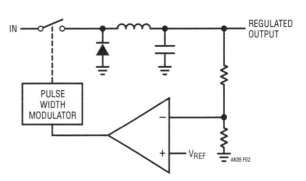

Figure 7.2 • Conceptual Feedback Controlled Step-Down Regulator

Practical step-down switching regulator

Figure 7.3, a practical circuit using the LT®1074[2] IC regulator, shows similarities to the conceptual regulator. Some new elements have also appeared. Components at the LT1074's "V_{COMP}" pin control the IC's frequency compensation, stabilizing the feedback loop. The feedback resistors are selected to force the "feedback" pin to the device's internal 2.5V reference value. Figure 7.4 shows

Note 2: See Appendix A for details on this device.

Analog Circuit and System Design: A Tutorial Guide to Applications and Solutions. DOI: 10.1016/B978-0-12-385185-7.00007-X

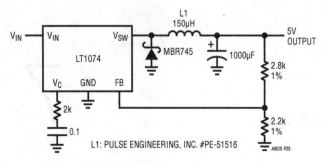

Figure 7.3 • A Practical Step-Down Regulator Using the LT1074

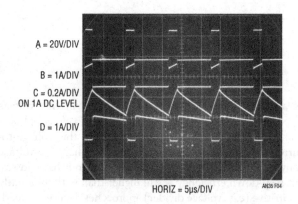

HORIZ = 5µs/DIV   AN35 F04

Figure 7.4 • Waveforms for the Step-Down Regulator at V_{IN} = 28V and V_{OUT} = 5V at 1A

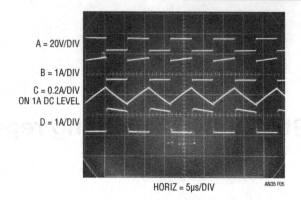

HORIZ = 5µs/DIV   AN35 F05

Figure 7.5 • Waveforms for the Step-Down Regulator at V_{IN} = 12V and V_{OUT} = 5V at 1A

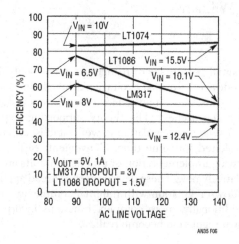

Figure 7.6 • Efficiency vs AC Line Voltage for the LT1074. LT1086 and LM317 Linear Regulators are Shown for Comparison

operating waveforms for the regulator at V_{IN} = 28V with a 5V, 1A load.

Trace A is the V_{SW} pin voltage and Trace B is its current. Inductor current[3] appears in Trace C and diode current is Trace D. Examination of the current waveforms allows determination of the V_{SW} and diode path contributions to inductor current. Note that the inductor current's waveform occurs on top of a 1A DC level. Figure 7.5 shows significant duty cycle changes when V_{IN} is reduced to 12V. The lower input voltage requires longer inductor charge times to maintain the output. The LT1074 controls inductor charge characteristics (see Appendix A for operating details), with resulting waveform shape and time proportioning changes.

Figure 7.6 compares this circuit's efficiency with linear regulators in a common and important situation. Efficient regulation under varying AC line conditions is a frequent requirement. The figure assumes the AC line has been transformed down to acceptable input voltages. The input voltages shown correspond to the AC line voltages given on the horizontal axis. Efficiency for the LM317 and LT1086 linear regulators suffers over the wide input range.

The LT1086 is notably better because its lower dropout voltage cuts dissipation over the range. Switching pre-regulation[4] can reduce these losses, but cannot equal the

LT1074's performance. The plot shows minimum efficiency of 83%, with some improvement over the full AC line excursion. Figure 7.7 details performance. Efficiency

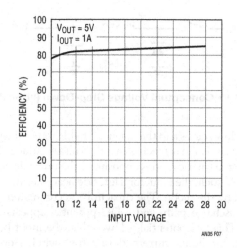

Figure 7.7 • Efficiency Plot for Figure 7.3. Higher Input Voltages Minimize Effects of Saturation Losses, Resulting in Increased Efficiency

---

Note 3: Methods for selecting appropriate inductors are discussed in Appendix B.
Note 4: See Reference 1.

approaches 90% as input voltage rises. This is due to minimization of the effects of fixed diode and LT1074 junction losses as input increases. At low inputs these losses are a higher percentage of available supply, degrading efficiency. Higher inputs make the fixed losses a smaller percentage, improving efficiency. Appendix D presents detail on optimizing circuitry for efficiency.

Dual output step-down regulator

Figure 7.8, a logical extension of the basic step-down converter, provides positive and negative outputs. The circuit is essentially identical to Figure 7.3's basic converter with the addition of a coupled winding to L1. This floating winding's output is rectified, filtered and regulated to a −5V output. The floating bias to the LT1086 positive voltage regulator permits negative outputs by assigning the regulator's output terminal to ground. Negative output power is set by flux pick-up from L1's driven winding. With a 2A

load at the +15 output the −5V output can supply over 500mA. Because L1's secondary winding is floating its output may be referred to any point within the breakdown capability of the device. Hence, the secondary output could be 5V or, if stacked on the +15 output, 20V

Negative output regulators

Negative outputs can also be obtained with a simple 2-terminal inductor. Figure 7.9 demonstrates this by essentially grounding the inductor and steering the catch diodes negative current to the output. A1 facilitates loop closure by providing a scaled inversion of the negative output to the LT1074's feedback pin. The 1% resistors set the scale factor (e.g., output voltage) and the RC network around A1 gives frequency compensation. Waveforms for this circuit are reminiscent of Figure 7.5, with the exception that diode current (Trace D) is negative. Traces A, B and C are V_{SW} voltage, inductor current and V_{SW} current respectively.

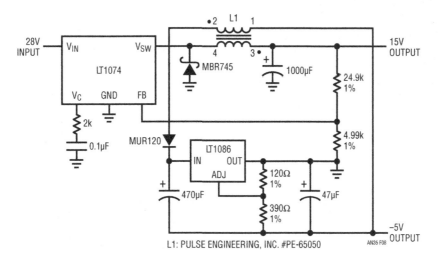

Figure 7.8 • Coupled Inductor Provides Positive and Negative Outputs

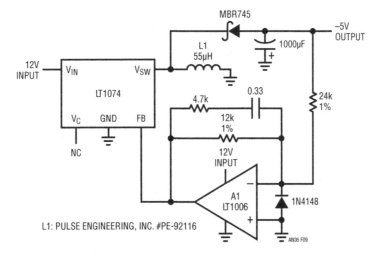

Figure 7.9 • A Negative Output Step-Down Regulator

Figure 7.11, commonly referred to as "Nelson's Circuit," provides the same function as the previous circuit, but eliminates the level-shifting op amp. This design accomplishes the level shift by connecting the LT1074's "ground" pin to the negative output. Feedback is sensed from circuit ground, and the regulator forces its feedback pin 2.5V above its "ground" pin. Circuit ground is common to input and output, making system use easy. Operating waveforms are essentially identical to Figure 7.10. Advantages of the previous circuit compared to this one are that the LT1074 package can directly contact a grounded heat sink and that control signals may be directly interfaced to the ground referred pins.

The inductor values in both negative output designs are notably lower than in the positive case. This is necessitated

by the reduced loop phase margin of these circuits. Higher inductance values, while preferable for limiting peak current, will cause loop instability or outright oscillation.

Current-boosted step-down regulator

Figure 7.12 shows a way to obtain significantly higher output currents by utilizing efficient energy storage in the LT1074 output inductor. This technique increases the duty cycle over the standard step-down regulator allowing more energy to be stored in the inductor. The increased output current is achieved at the expense of higher output voltage ripple.

The operating waveforms for this circuit are shown in Figure 7.13. The circuit operating characteristics are similar to that of the step-down regulator (Figure 7.3). During the V_{SW} (Trace A) "on" time the input voltage is applied to one end of the coupled inductor. Current through the V_{SW} pin (Trace B) ramps up almost instantaneously (since inductor current (Trace F) is present) and then slows as energy is stored in the core. The current proceeds into the inductor (Trace D) and finally is delivered to the load. When the V_{SW} pin goes off, current is no longer available to charge the inductor. The magnetic field collapses, causing the V_{SW} pin voltage to go negative. At this point similarity with the basic regulator vanishes. In this modified version the output current (Trace F) receives a boost as the magnetic field collapses. This results when the energy

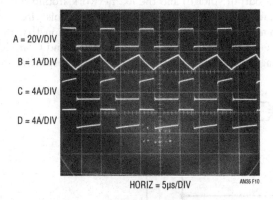

HORIZ = 5µs/DIV   AN35 F10

Figure 7.10 • Figure 7.9's Waveforms

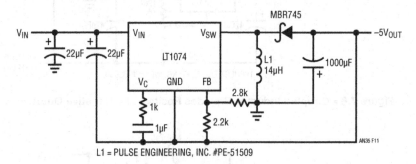

L1 = PULSE ENGINEERING, INC. #PE-51509

Figure 7.11 • Nelson's Circuit ··· A (Better) Negative Output Step-Down Regulator

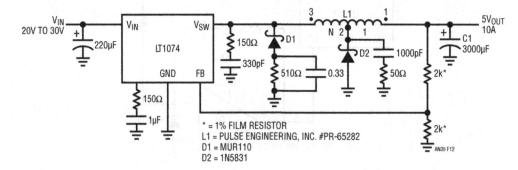

* = 1% FILM RESISTOR
L1 = PULSE ENGINEERING, INC. #PR-65282
D1 = MUR110
D2 = 1N5831

AN35 F12

Figure 7.12 • "Current Boosted" Step-Down Regulator. Boost Current is Supplied By Energy Stored in the Tapped Inductor

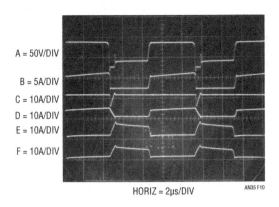

Figure 7.13 • AC Current Flow for the Boosted Regulator

1.2V to 28V output. The remainder of the circuit forms a switched mode pre-regulator which maintains a small, fixed voltage across the LT1085 regardless of its output voltage. A1 biases the LT1074 to produce whatever voltage is necessary to maintain the "E diodes" potential across the LT1085. A1's inputs are balanced when the LT1085 output is "E diodes" above its input. A1 maintains this condition regardless of line, load or output voltage conditions. Thus, good efficiency is maintained over the full range of output voltages. The RC network at A1 compensates the loop. Loop start-up is assured by deliberately introducing a positive offset to A1. This is done by grounding A1's appropriate balance pin (5), resulting in a positive 6mV offset. This increases amplifier drift, and is normally considered

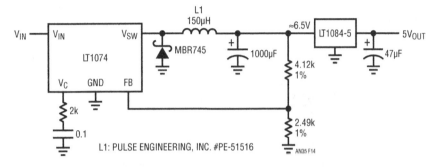

Figure 7.14 • Linear Post-Regulator Improves Noise and Transient Response

stored in the core is transferred to the output. This current step circulates through C1 and D2 (Trace E), somewhat increasing output voltage ripple. Not all the energy is transferred to the "1" winding. Current (Trace C) will continue to flow in the "N" winding due to leakage inductance. A snubber network suppresses the effects of this leakage inductance. For lowest snubber losses the specified tapped inductor is bifilar wound for maximum coupling.

Post regulation-fixed case

In most instances the LT1074 output will be applied directly to the load. Those cases requiring faster transient response or reduced noise will benefit from linear post regulation. In Figure 7.14 a 3-terminal regulator follows the LT1074 output. The LT1074 output is set to provide just enough voltage to the LT1084 to maintain regulation. The LT1084's low dropout characteristics combined with a high circuit input voltage minimizes the overall efficiency penalty.

Post regulation-variable case

Some situations require variable linear post regulation. Figure 7.15 does this with little efficiency sacrifice. The LT1085 operates in normal fashion, supplying a variable

poor practice, but causes no measurable error in this application.

As shown, the circuit cannot produce outputs below the LT1085's 1.2V reference. Applications requiring output adjustability down to 0V will benefit from option "A" shown on the schematic. This arrangement replaces L1 with L2. L2's primary performs the same function as L1 and its coupled secondary winding produces a negative bias output (−V). The full-wave bridge rectification is necessitated by widely varying duty cycles. A2 and its attendant circuitry replace all components associated with the LT1085 V_{ADJ} pin. The LT1004 reference terminates the 10k to 250k feedback string at −1.2V with A2 providing buffered drive to the LT1085 V_{ADJ} pin. The negative bias allows regulated LT1085 outputs down to 0V The −V potential derived from L2's secondary varies considerably with operating conditions. The high feedback string values and A2's buffering ensure stable circuit operation for "starved" values of −V

Low quiescent current regulators

Many applications require very wide ranges of power supply output current. Normal conditions require currents in the ampere range, while standby or "sleep" modes draw

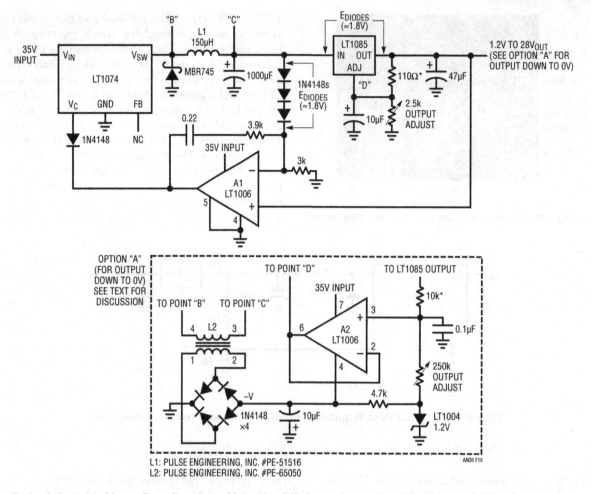

Figure 7.15 • Adjustable Linear Post-Regulator Maintains Efficiency Over Widely Varying Operating Conditions

only microamperes. A typical laptop computer may draw 1 to 2 amperes running while needing only a few hundred microamps for memory when turned off. In theory, any regulator designed for loop stability under no-load conditions will work. In practice, a converter's relatively large quiescent current may cause unacceptable battery drain during low output current intervals. Figure 7.16's simple loop effectively reduces circuit quiescent current from 6mA to only 150µA. It does this by utilizing the LT1074's shutdown pin. When this pin is pulled within 350mV of ground the IC shuts down, pulling only 100µA. Comparator C1 combines with the LT1004 reference and Q1 to form a "bang-bang" control loop around the LT1074. The LT1074's internal feedback amplifier and voltage reference are bypassed by this loop's operation. When the circuit output (Trace C, Figure 7.17) falls

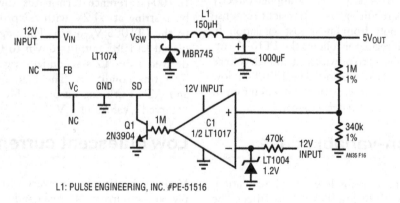

Figure 7.16 • A Simple Loop Reduces Quiescent Current to 150µA

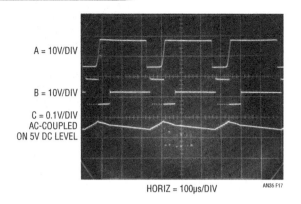

Figure 7.17 • **The Low Quiescent Current Loop's Waveforms**

produce such noise. Additionally, the control loop's operation causes about 50mV of ripple on the output. Ripple frequency ranges from 0.2Hz to 10kHz depending upon input voltage and output current.

Figure 7.18's more sophisticated circuit eliminates these problems with some increase in complexity. Quiescent current is maintained at 150µA. The technique shown is particularly significant, with broad implication in battery powered systems. It is easily applied to a wide variety of regulator requirements, meeting an acknowledged need across a wide spectrum of applications.

Figure 7.18's signal flow is similar to Figure 7.16, but additional circuitry appears between the feedback divider and the LT1074. The LT1074's internal feedback amplifier

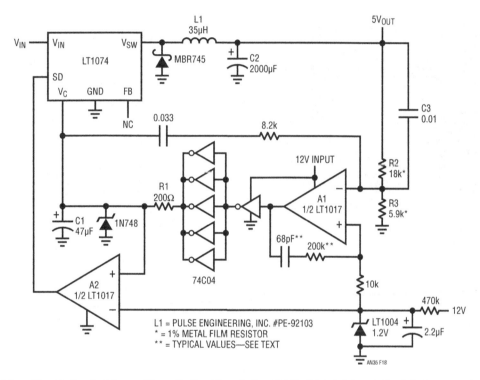

Figure 7.18 • **A More Sophisticated Loop Gives Better Regulation While Maintaining 150µA Quiescent Current**

slightly below 5V C1's output (Trace A) switches low, turning off Q1 and enabling the LT1074. The V_{SW} pin (Trace B) pulses at full duty cycle, forcing the output back above 5V. C1 then biases Q1 again, the LT1074 goes into shutdown, and loop action repeats. The frequency of this on-off control action is directly load dependent, with typical repetition rates of 0.2Hz at no load. Short on-times keep duty cycle low, resulting in the small effective quiescent current noted. The on-off operation combines with the LC filtering action in the regulator's V_{SW} line to generate an output hysteresis of about 50mV (again, see Figure 7.17, Trace C).

The loop performs well, but has two potential drawbacks. At higher output currents the loop oscillates in the 1kHz to 10kHz range, causing audible noise which may be objectionable. This is characteristic of this type of loop, and is the reason that ICs employing gated oscillators invariably

and reference are not used. Figure 7.19 shows operating waveforms under no-load conditions. The output (Trace A) ramps down over a period of seconds. During this time comparator A1's output (Trace B) is low, as are the 74C04 paralleled inverters. This pulls the V_C pin (Trace D) low, forcing the regulator to zero duty cycle. Simultaneously, A2 (Trace C) is low, putting the LT1074 in its 100µA shutdown mode. The V_{SW} pin (Trace E) is off, and no inductor current flows. When the output drops about 60mV, A1 triggers and the inverters go high, pulling the V_C pin up and biasing the regulator. The Zener diode prevents V_C pin overdrive. A2 also rises, taking the IC out of shutdown mode. The V_{SW} pin pulses the inductor at the 100kHz clock rate, causing the output to abruptly rise. This action trips A1 low, forcing the V_C pin back low and shutting off V_{SW} pulsing. A2 also goes low, putting the LT1074 into shutdown.

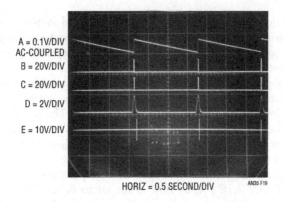

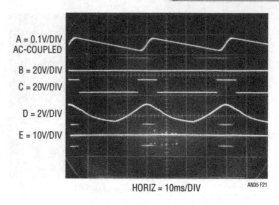

Figure 7.19 • Low Quiescent Current Regulator's Waveforms with No Load (Traces B, C and E Retouched for Clarity)

Figure 7.21 • Low Quiescent Current Regulator's Waveforms at 7mA Loading

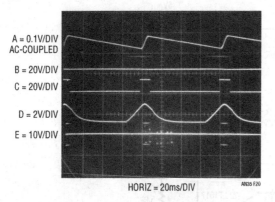

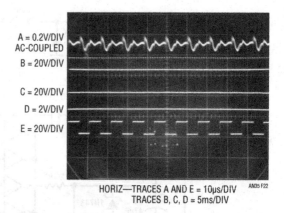

Figure 7.20 • Low Quiescent Current Regulator's Waveforms at 2mA Loading

Figure 7.22 • Low Quiescent Current Regulator's Waveforms at 2A Loading

This "bang-bang" control loop keeps the 5V output within the 60mV ramp hysteresis window set by the loop. Note that the loop oscillation period of seconds means the R1-C1 time constant at V_C is not a significant term. Because the LT1074 spends almost all of the time in shutdown, very little quiescent current (150μA) is drawn.

Figure 7.20 shows the same waveforms with the load increased to 2mA. Loop oscillation frequency increases to keep up with the load's sink current demand. Now, the V_C pin waveform (Trace D) begins to take on a filtered appearance. This is due to R1-C1's 10ms time constant. If the load continues to increase, loop oscillation frequency will also increase. The R1-C1 time constant, however, is fixed. Beyond some frequency, R1-C1 must average loop oscillations to DC. At 7mA loading (Figure 7.21) loop frequency further increases, and the V_C waveform (Trace D) appears heavily filtered.

Figure 7.22 shows the same circuit points at 2A loading. Note that the V_C pin is at DC, as is the shutdown pin. Repetition rate has increased to the LT1074's 100kHz clock frequency. Figure 7.23 plots what is occurring, with a pleasant surprise. As output current rises, loop oscillation frequency also rises until about 23Hz. At this point the R1-C1 time constant filters the V_C pin to DC and the LT1074 transitions into "normal" PWM operation. With the V_C pin at DC it is convenient to think of A1 and the inverters as

a linear error amplifier with a closed-loop gain set by the R2-R3 feedback divider. In fact, A1 is still duty cycle modulating, but at a rate far above R1-C1's break frequency. The phase error contributed by C2 (which was selected for low loop frequency at low output currents) is dominated by the R1-C1 roll off and the C3 lead into A1. The loop is stable and responds linearly for all loads beyond 10mA. In this high current region the LT1074 is desirably "fooled" into behaving like a conventional step-down regulator.

A formal stability analysis for this circuit is quite complex, but some simplifications lend insight into loop operation. At 250μA loading (20kΩ) C2 and the load form a decay time constant exceeding 30 seconds. This is orders of magnitude larger than R2-C3, R1-C1, or the LT1074's 100kHz commutation rate. As a result, C2 dominates the loop. Wideband A1 sees phase shifted feedback, and very low frequency oscillations similar to Figure 7.19's occur[5]. Although C2's *decay* time constant is long, its *charge* time constant is short because the circuit has low sourcing impedance. This accounts for the ramp nature of the oscillations.

Note 5: Some layouts may require substantial trace area to A1's inputs. In such cases the optional RC network around A1 ensures clean transitions at A1's output.

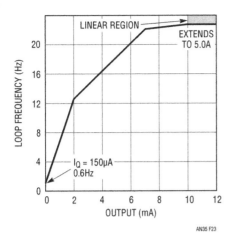

AN35 F23

Figure 7.23 • Figure 7.18's Loop Frequency vs Output Current. Note Linear Loop Operation Above 10mA

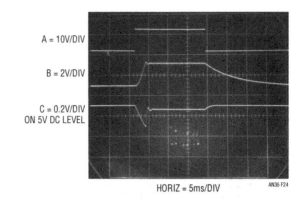

HORIZ = 5ms/DIV   AN35 F24

Figure 7.24 • Load Transient Response for Figure 7.18

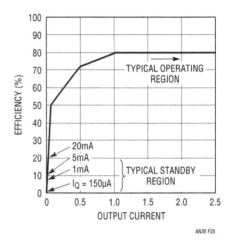

AN35 F25

Figure 7.25 • Efficiency vs Output Current for Figure 7.18. Standby Efficiency is Poor, But Power Loss Approaches Battery Self-Discharge

Increased loading reduces the C2-load decay time constant. Figure 7.23's plot reflects this. As loading increases, the loop oscillates at a higher frequency due to C2's decreased decay time. When the load impedance becomes low enough C2's decay time constant ceases to dominate the loop. This point is almost entirely determined by R1 and C1. Once R1 and C1 "take over" as the dominant time constant the loop begins to behave like a linear system. In this region (e.g., above about 10mA, per Figure 7.23) the LT1074 runs continuously at its 100kHz rate. Now, C3 becomes significant, performing as a simple feedback lead[6] to smooth output response. There is a fundamental trade-off in the selection of the C3 lead value. When the converter is running in its linear region it must dominate the loop's time lag generated hysteretic characteristic. As such, it has been chosen for the best compromise between output ripple at high load and loop transient response.

Despite the complex dynamics transient response is quite good. Figure 7.24 shows performance for a step from no load to 1A. When Trace A goes high a 1A load appears across the output (Trace C). Initially, the output sags almost 200mV due to slow loop response time (the R1-C1 pair delay V_C pin (Trace B) response). When the LT1074 comes on response is reasonably quick and surprisingly well behaved considering circuit dynamics. The multi-time constant recovery[7] ("rattling" is perhaps more appropriate) is visible in Trace C's response.

Figure 7.25 plots efficiency versus output current. High power efficiency is similar to standard converters. Low power efficiency is somewhat better, although poor in the lowest ranges. This is not particularly bothersome, as *power* loss is very small.

The loop provides a controlled, conditional instability instead of the usually more desirable (and often elusive) unconditional stability. This deliberately introduced

characteristic dramatically lowers converter quiescent current without sacrificing high power performance.

Wide range, high power, high voltage regulator

BEFORE PROCEEDING ANY FURTHER, THE READER IS WARNED THAT CAUTION MUST BE USED IN THE CONSTRUCTION, TESTING AND USE OF THIS CIRCUIT. HIGH VOLTAGE, LETHAL POTENTIALS ARE PRESENT IN THIS CIRCUIT. EXTREME CAUTION MUST BE USED IN WORKING WITH AND MAKING CONNECTIONS TO THIS CIRCUIT. REPEAT: THIS CIRCUIT CONTAINS DANGEROUS, HIGH VOLTAGE POTENTIALS. USE CAUTION.

Figure 7.26 is an example of the LT1074 making a complex function practical. This regulator provides outputs

---

Note 6: "Zero Compensation" for all you technosnobs out there.
Note 7: Once again, "multi-pole settling" for those who adore jargon.

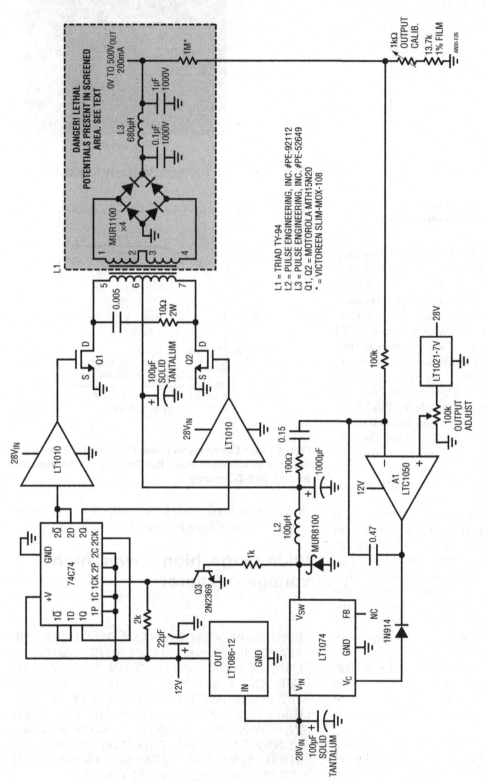

**Figure 7.26 • LT1074 Permits High Voltage Output Over 100dB Range with Power and Efficiency
DANGER! Lethal Potentials Present—See Text**

L1 = TRIAD TY-94
L2 = PULSE ENGINEERING, INC. #PE-92112
L3 = PULSE ENGINEERING, INC. #PE-52649
Q1, Q2 = MOTOROLA MTH15N20
* = VICTOREEN SLIM-MOX-108

from millivolts to 500V at 100W with 80% efficiency. A1 compares a variable reference voltage with a resistively scaled version of the circuit's output and biases the LT1074 switching regulator configuration. The switcher's DC output drives a toroidal DC/DC converter comprised of L1, Q1 and Q2. Q1 and Q2 receive out of phase square wave drive from the 74C7÷4/4 flip-flop stage and the LT1010 buffers. The flip-flop is clocked from the LT1074 V_{SW} output via the Q3 level shifter. The LT1086 provides 12V power for A1 and the 74C74. A1 biases the LT1074 regulator to produce the DC input at the DC/DC converter required to balance to loop. The converter has a voltage gain of about 20, resulting in high voltage output. This output is resistively divided down, closing the loop at A1's negative input. Frequency compensation for this loop must accommodate the significant phase errors generated by the LT1074 configuration, the DC/DC converter and the output LC filter. The 0.47µF roll-off term at A1 and the 100 Ω-0.15µF RC lead network provide the compensation, which is stable for all loads.

Figure 7.27 gives circuit waveforms at 500V output into a 100W load. Trace A is the LT1074 V_{SW} pin while Trace B is its current. Traces C and D are Q1 and Q2's drain waveforms. The disturbance at the leading edges is due to cross-current conduction, which lasts about 300ns—a small percentage of the cycle. Transistor currents during this interval remain within reasonable values, and no overstress or dissipation problems occur. This effect could be eliminated with non-overlapping drive to Q1 and Q2[8], although there would be no reliability or significant efficiency gain. The 500kHz ringing on the same waveforms is due to excitation of transformer resonances. These phenomena are not deleterious, although L1's primary RC damper is included to minimize them.

All waveforms are synchronous because the flip-flop drive stage is clocked from the LT1074 V_{SW} output. The LT1074's maximum 95% duty cycle means that the Q1-Q2 switches can never see destructive DC drive. The only

condition allowing DC drive occurs when the LT1074 is at zero duty cycle. This case is clearly non-destructive, because L1 receives no power.

Figure 7.28 shows the same circuit points as Figure 7.27, but at only 5mV output. Here, the loop restricts drive to the DC/DC converter to small levels. Q1 and Q2 chop just 70mV into L1. At this level L1's output diode drops look large, but loop action forces the desired 0.005V output.

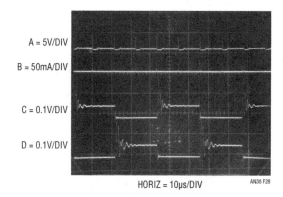

A = 5V/DIV

B = 50mA/DIV

C = 0.1V/DIV

D = 0.1V/DIV

HORIZ = 10µs/DIV   AN35 F28

Figure 7.28 • Figure 7.26's Operating Waveforms at 0.005V Output

The LT1074's switched mode drive to L1 maintains high efficiency at high power; despite the circuits wide output range[9].

Figure 7.29 shows output noise at 500V into a 100W load. Q1-Q2 chopping artifacts and transformer related ringing are clearly visible, although limited to about 80mV The coherent noise characteristic is traceable to the synchronous clocking of Q1 and Q2 by the LT1074.

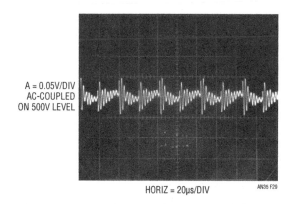

A = 0.05V/DIV
AC-COUPLED
ON 500V LEVEL

HORIZ = 20µs/DIV   AN35 F29

Figure 7.29 • Figure 7.26's Output Noise at 500V into a 100W Load. Residue is Composed of Q1-Q2 Chopping Artifacts and Transformer Related Ringing
DANGER! Lethal Potentials Present—See Text

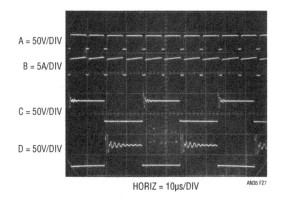

A = 50V/DIV

B = 5A/DIV

C = 50V/DIV

D = 50V/DIV

HORIZ = 10µs/DIV   AN35 F27

Figure 7.27 • Figure 7.26's Operating Waveforms at 500V Output into a 100W Load

---

Note 8: For an example of this technique see LTC Application Note 29, Figure 7.1.

Note 9: A circuit related to the one presented here appears in the LTC Application Note 18 (Figure 7.13). Its linear drive to the step-up DC/DC converter forces dissipation, limiting output power to about 15W. Similar restrictions apply to Figure 7.7 in Application Note 6.

A 50V to 500V step command into a 100W load produces the response of Figure 7.30. Loop response on both edges is clean, with the falling edge slightly underdamped. This slew asymmetry is typical of switching configurations, because the load and output capacitor determine negative slew rate. The wide range of possible loads mandates a compromise when setting frequency compensation. The falling edge could be made critically or even over damped, but response time for other conditions would suffer. The compensation used seems a reasonable compromise.

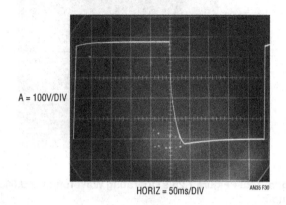

A = 100V/DIV

HORIZ = 50ms/DIV   AN35 F30

Figure 7.30 • 500V Step Response with 100W Load (Photo Retouched for Clarity)
DANGER! Lethal Potentials Present—See Text

Regulated sinewave output DC/AC converter

BEFORE PROCEEDING ANY FURTHER, THE READER IS WARNED THAT CAUTION MUST BE USED IN THE CONSTRUCTION, TESTING AND USE OF THIS CIRCUIT. HIGH VOLTAGE, LETHAL POTENTIALS ARE PRESENT IN THIS CIRCUIT, EXTREME CAUTION MUST BE USED IN WORKING WITH AND MAKING CONNECTIONS TO THIS CIRCUIT. REPEAT: THIS CIRCUIT CONTAINS DANGEROUS HIGH VOLTAGE POTENTIALS. USE CAUTION.

Figure 7.31 is another example of the LT1074 permitting the practical implementation of a complex function. It converts a 28V DC input to a regulated $115V_{AC}$ 400Hz sinewave output with 80% efficiency. Waveform distortion is below 1.6% at 50W output. This design shares similarities with the previous circuit. The LT1074 supplies efficient drive to a high voltage converter despite large line and load variations. An amplifier (A1) controls the input to the high voltage converter via A2 and the LT1074 switching regulator. The high voltage output is divided down and fed back to the amplifier where it is compared to a reference to

close a loop. In the previous circuit the output is DC; here the output is AC. As such, A1's reference (Trace A, Figure 7.32) is an amplitude and frequency stabilized 800Hz half-sine[10]. The high voltage converter is driven from a flip-flop clocked by a reference synchronized pulse (negative going excursions just visible in Trace B) via level shift transistor Q3. The reference synchronized pulse occurs at the zero voltage point of the half-sine. The flip-flop outputs (Traces C and D, respectively) drive the Q1 and Q2 gates. RC filters in the gate line retard the drive's slew rate.

A1 biases the LT1074's V_C pin via A2 to produce an 800Hz half-sine signal at L2's center tap (Trace E). Because Q1 and Q2 are synchronously driven with the reference half-sine their drain waveforms (Traces F and G) reveal alternate chopping of complete half cycles. L2 receives balanced drive and its secondary recombines the chopped half-sines into a $115V_{AC}$ 400Hz sinewave output (Trace H). The diode bridge rectifies L2's output back to an 800Hz half-sine which is fed to A1 via the resistor divider. A1 balances this signal against the reference half-sine to close a loop. Transmitting the 800Hz waveform around the loop requires attention to available bandwidth. The LT1074's 100kHz switching frequency is theoretically high enough to permit this, but the bandwidth attenuation of its output LC filter must be considered. The unusually low output filter capacitor value allows the necessary frequency response. A1's 330k-0.01μF components combine with the RC lead network across the 16k feedback resistor to stabilize the loop.

A2 closes a local loop around the LT1074 configuration. This is necessary because L2 blocks DC information from being conducted around A1's loop. This is a concern because the waveform presented to L2's primary center tap must have no DC component. DC content at this point will cause waveform distortion, transformer power dissipation or both. The LT1074's V_C pin operates with substantial and uncertain DC bias, making A1's inability to control DC errors unacceptable. A2 corrects this by biasing the LT1074 V_C pin at its DC threshold so that no DC component is presented to L2. A1's output represents the difference between the AC-coupled circuit output and the half-sine reference. A2's output contains this information in addition to DC restoration information. L2 and A1 contribute essentially no DC error, so A2's loop may be closed at the LT1074 configuration's output. A2's feedback capacitor stabilizes this local loop.

The drive to L2 cannot sink current. This means that any residual energy stored in L2 when the drive waveform goes to zero sees no exit path. This is a relatively small effect, but can cause output crossover distortion. The synchronous switch option shown on the schematic provides such a path, and is recommended for lowest output distortion. This optional circuitry is detailed in Appendix E.

Note 10: Complete operating details of the half-sine reference generator appear in Appendix E.

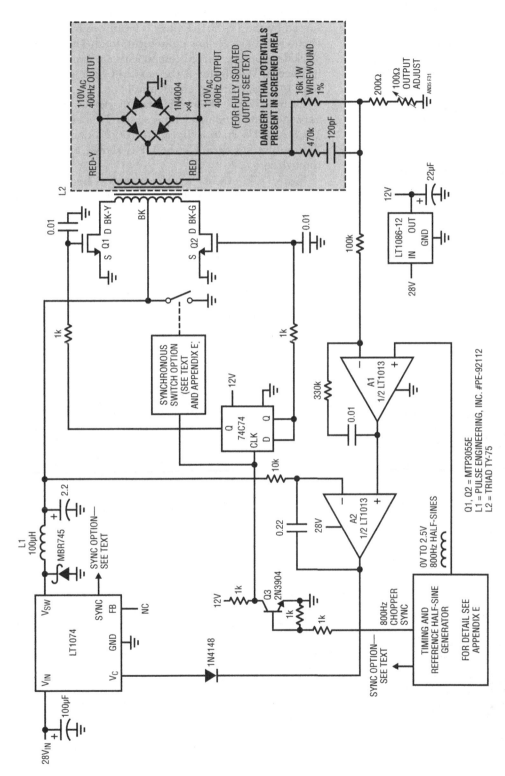

Figure 7.31 • LT1074 Based +28 to 110V_AC 400Hz Converter. Sinewave Output Shows Only 1.6% Distortion
DANGER! Lethal Potentials Present—See Text

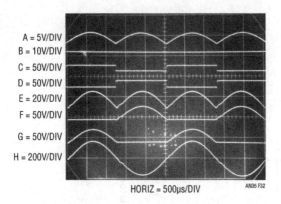

Figure 7.32 • +28 to 110V_AC, 400Hz Converter's Waveforms. The Optional Synchronous Switch is Disabled in this Photo, Resulting in Relatively High Crossover Distortion (Trace H)
DANGER! Lethal Potentials Present—See Text

Figures 7.33a and 7.33b show waveforms in the "turnaround" region of circuit operation. This is the most critical part of the converter, and its characteristics directly determine output waveform purity. Figure 33a (Trace A), a highly amplitude and time expanded version of L2's center tap drive, arrives at 0V (upper cross-etched horizontal line) and turns around cleanly. This action is just slightly time skewed from the reference synchronized pulse (Trace B). The aberration on the rising edge is due to the optional

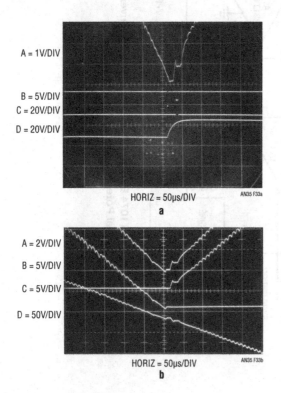

Figure 33a and 33b • Details of "Turnaround" Sequence. Switching Characteristics Directly Determine Output Crossover Distortion
DANGER! Lethal Potentials Present—See Text

synchronous switch's operation. This switch is shorted during the on-time of Trace C's pulse (see Appendix E for operating details of this option). Trace D, Q2's gate drive, aligns with Trace B's pulse. The slew reduction caused by the 1k-0.01μF filter is clearly visible, and contributes to Trace A's low noise turnaround. The LT1074's 100kHz chopping related components are easily observed in Trace A. Waveforms at the next half cycle's zero point (e.g., Q1's gate driven) are identical.

Figure 7.33b shows additional details at highly expanded amplitude and time scales. L2's center tap is Trace A, Q1's drain is Trace B and Q2's drain Trace C. The output sinewave (Trace D) is shown as it crosses through zero.

Figure 7.34 studies waveform purity. Trace A is the sinewave output at 50W loading. Trace B shows distortion products, which are dominated by turnaround related crossover aberrations and LT1074 100kHz chopping residue. Although not strictly necessary, the LT1074's switching can be synchronized to the reference half-sine for coherent noise characteristics. This option is discussed in Appendix E, along with other reference generator details. Trace C is a spectrum analysis centered at 400Hz[11]. In this photo the optional synchronous switch is used, accounting for improved crossover characteristics over Figure 7.32.

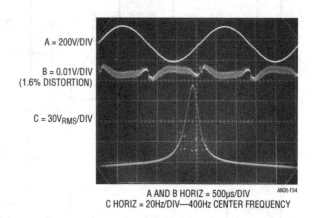

Figure 7.34 • Distortion and Spectral Characteristics for the Sinewave Output Converter. Distortion (Trace B) Shows Crossover Aberrations and the LT1074 Wideband Chopping Residue. The Synchronous Switch Option is Employed in this Photo for Lowest Distortion
DANGER! Lethal Potentials Present—See Text

If a fully floating output is desired the output diode bridge can be isolated by a simple 1:1 ratio transformer. To calibrate this circuit trim the "output adjust" potentiometer for a 115V_AC output. Regulation remains within 1% over wide variations of input and load.

Note 11: Test equipment aficionados may wish to consider how this picture was taken. Hint: Double exposure techniques were not used. This photograph is a real time, simultaneous display of frequency and time domain information.

References

1. Williams, J., "High Efficiency Linear Regulators," Linear Technology Corporation, Application Note 32.
2. Williams, J. and Huffman, B., "Some Thoughts on DC/DC Converters," Linear Technology Corporation, Application Note 29.
3. 1074LT Data Sheet, Linear Technology Corporation.
4. Nelson C., "LT1070 Design Manual," Linear Technology Corporation, Application Note 19.
5. Williams, J., "Switching Regulators for Poets," Linear Technology Corporation, Application Note 25.
6. Williams, J., "Applications of New Precision Op Amps," Linear Technology Corporation, Application Note 6.
7. Williams, J., "Power Conditioning Techniques for Batteries," Linear Technology Corporation, Application Note 8.
8. Pressman, A.I., "Switching and Linear Power Supply, Power Converter Design," Hayden Book Co., Hasbrouck Heights, New Jersey, 1977, ISBN 0-8104-5847-0.
9. Chryssis, G., "High Frequency Switching Power Supplies, Theory and Design," McGraw Hill, New York, 1984, ISBN 0-07-010949-4

Note: This application note was derived from a manuscript originally prepared for publication in EDN magazine.

Appendix A
Physiology of the LT1074

The LT1074 uses standard (as opposed to current mode) pulse width modulation, with two important differences. First, it is a clocked system with a maximum duty cycle of approximately 95%. This allows a controlled start-up when it is used as a positive-to-negative converter or a negative boost converter. Second, duty cycle is an inverse function of input voltage (DC ≈ $1/V_{IN}$), without any change in error amplifier output. This greatly improves line transient response and ripple rejection, especially for designs which have the control loop over-damped.

Referring to the block diagram, the heart of the LT1074 consists of the oscillator, the error amplifier A1, an analog multiplier, comparator C6, and an RS flip-flop. A complete switching cycle begins with the reset (down ramp) period of the oscillator. During this time (≈0.7µs), the RS flip-flop is set and the switch driver Q104 is kept off via the "and" gate G1. At the end of the reset time, Q104 turns on and drives the output switch Q111, Q112 and Q113. The oscillator ramp starts upward, and when it is equal to the output voltage of the analog multiplier, C6 resets the RS flip-flop, turning off the output switch. Duty cycle is therefore controlled by the output of the multiplier which in turn is controlled by the output of the error amplifier, A1.

A multiplier is used in the LT1074 to provide a perfect "feedforward" function. Conventional switching regulators sometimes use a simple form of feedforward to adjust duty cycle immediately when input voltage changes. This reduces the requirement for voltage swing at the output of the error amplifier as it tries to correct for line variations. Bandwidth of switching regulator error amplifiers must be fairly low to maintain loop stability, so rather large output perturbations occur when the output of the error amplifier must move quickly to correct for

line variations. Conventional feedforward schemes typically operate well over a restricted input voltage range or are effective only at certain frequencies. The multiplier technique is very effective over the full range of input voltage and at all frequencies. The basic function is to compensate for the generalized buck regulator transfer function; $V_{OUT} = (V_{IN})(DC)$, where DC = switch duty cycle. This transfer function has two implications. First, it is obvious that to maintain a constant output, duty cycle must change inversely with input voltage. Second, input voltage appears in the loop transfer function, i.e., a fixed variation in duty cycle gives different variations in output voltage depending on input voltage. Loop gain is directly proportional to input voltage, and this can cause loop instability or slow loop response if input voltage varies over a wide range. The multiplier takes out all input voltage effects by automatically adjusting loop gain inversely with input voltage. The multiplier output (V_O) is equal to error amplifier output (V_E) divided by input voltage; $V_O = (V_R) \cdot (V_E)/(V_{IN})$. V_R is a fixed voltage required by all analog multipliers to define multiplier gain. It has an effective value of approximately 20V in the LT1074.

The error amplifier used in the LT1074 is a transconductance type. It has high output impedance (≈500kΩ), so that its AC voltage gain is defined by the impedance of external shunt frequency compensation components (Z_C) and the transconductance (g_m) of the amplifier, $A_V = (g_m)(A_C)$. g_m is ≈3500µmho. The error amplifier has its noninverting input committed to an internal 2.3V reference. The inverting input (fb) is brought out for connection to an external voltage divider that establishes regulator output voltage.

Two other connections are made internally to the fb pin. A window comparator consisting of C4, C5, and some logic provides an "output status" function. It monitors the voltage on the fb pin and gives a "high" output only when the fb voltage is within ±5% of the internal reference voltage. This status

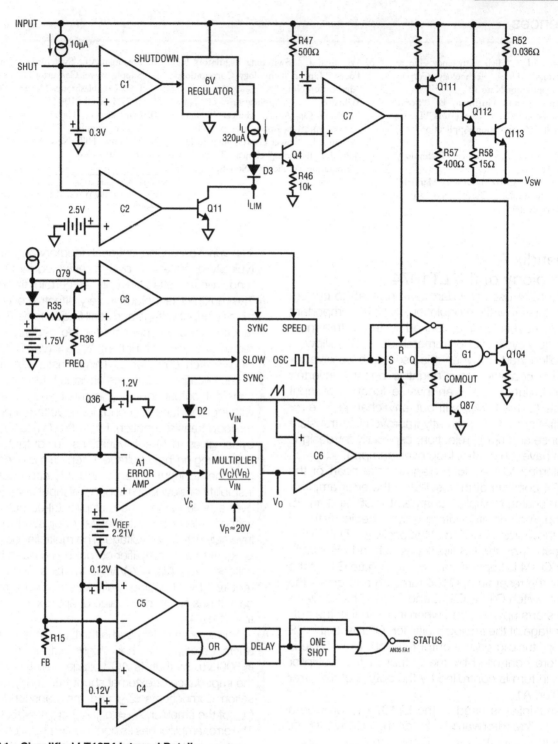

Figure A1 • Simplified LT1074 Internal Details

output can be used to alert external circuitry that the regulator output is "in" or "out" of regulation. The delay and one-shot circuits ensure that switching EMI will not cause spurious outputs, and that the minimum time for an "out-of-bounds" (low) status output is ≅20μs. Also tied to the fb pin is a frequency shift circuit consisting of R15 and Q36. The base of Q36 is biased at ≅1V so that Q36 turns on when the fb pin drops below ≈0.6V Current through Q36 smoothly decreases oscillator frequency. This is necessary for maintaining control of current limit at high input voltages. A "dead short" on the output of a switching regulator requires that switch "on" time reduce to $(V_D)/(V_{IN})(f)$, where V_D is the forward voltage of the output catch diode and f is switching frequency. V_D is typically 0.5V for a Schottky catch diode, forcing switch "on" time to shrink to a theoretical $0.1\mu s$ for a 50V input and 100kHz switching frequency with a shorted output. In practical circuits, effective "on" time can stretch to $0.3\mu s$ under these conditions due to losses in the inductor wire resistance and switch rise and fall time. The LT1074 cannot reduce switch "on" time to less than $\cong 0.6\mu s$ in current limit because it has true pulse-by-pulse switch current limiting. The current limit circuitry must sense switch current *after* the switch turns on, and then send a signal to turn the switch off. Minimum time for this signal path is $0.6\mu s$. Full control of current limit is maintained by reducing switching frequency when the output falls to less than approximately 15% of its regulated value. This has no effect on normal operation and does not change the selection of external components such as the inductor or output capacitor.

True pulse-by-pulse current limiting is performed by comparator C7. It monitors the voltage across sense resistor R52 and resets the RS flip-flop. Current limit threshold is set by the voltage across R47 which in turn is set by the voltage on the I_{LIM} pin. The I_{LIM} voltage is determined by an external resistor or by an internal clamp of 5V if no external resistor is used. To compensate for the temperature coefficient of R47 ($\approx +0.25\%/$°C), the internal current source I_L has a matching positive temperature coefficient. Its nominal value is $300\mu A$ at 25 °C. Current limit can be set from 1A to 6A with one external resistor between I_{LIM} and ground. If no resistor is used, the I_{LIM} pin will self clamp at $\cong 5V$ and current limit will be $\approx 6.5A$. A small pre-bias is added to the negative input of C7 to ensure that current limit will go to zero (no switching) when the I_{LIM} pin is pulled to 0V either by an external short or via Q11 during undervoltage lockout. Soft-start can be achieved by connecting a capacitor to the I_{LIM} pin. Foldback current limiting can also be implemented by connecting a resistor from I_{LIM} to the regulated output.

Switching frequency of the LT1074 is internally set at 100kHz, but can be increased by connecting a resistor from the frequency pin to ground. This resistor biases on Q79 and feeds extra current into the oscillator. Maximum suggested frequency is 200kHz. A comparator, C3, is also connected to the frequency pin and allows this pin to be used for synchronizing the oscillator to an external clock, even when the pin is also being used to boost oscillator frequency. R35 keeps the frequency pin biased

correctly in a no-function state when it is left open and R36 limits Q79 current if the frequency pin is accidentally shorted to ground.

The shutdown pin on the LT1074 can be used as a logic control of output switching, as an undervoltage lockout, or to put the regulator into complete shutdown with $I_{SUPPLY} \approx 100\mu A$. Comparator C2 has a threshold of 2.5V. It forces the output switch to a 100% "off" condition by pulling the I_{LIM} pin low via Q11. Undervoltage lockout is implemented through C2 by connecting the tap of an input voltage divider to the shutdown pin. Full micropower shutdown of the regulator is achieved by pulling the shutdown pin below the 0.30V threshold of C1. This turns off the chip by shutting down the internal 6V bias supply. An internal $10\mu A$ current source pulls the shutdown pin high (inactive) if it is left open.

The Comout pin is an open-collector NPN whose collector voltage is the complement of the switch output (V_{SW}). Q87 is specified to drive up to 10mA and 30V. It is intended to drive the gate of an external N-channel MOS switch which is in parallel with the catch diode. The MOS switch then acts as a synchronous rectifier, which can significantly improve converter efficiency in low output voltage applications. The Comout pin can also be used to drive the gate of an external P-channel MOS switch in parallel with the internal bipolar switch to provide ultrahigh efficiency switching at lower input voltages. A slight time-shift in the Comout signal prevents switch overlap problems.

The combination of these features produces a DC/DC converter with the electrical characteristics shown in Figure A2.

PARAMETER	CONDITIONS	UNITS
Input Voltage Range		4.5V to 60V
Output Voltage Range		2.5V to 50V
Output Current Range	Standard Buck Tapped Buck	0A to 5A 0A to 10A
Quiescent Input Current		7mA
Switching Frequency		100kHz to 200kHz
Switch Rise/Fall Times		50ns
Switch Voltage Loss	1A 5A	1.6V 2V
Reference Voltage		2.35V ±1.5%
Line/Load Regulation		0.05%
Efficiency	$V_{OUT} = 15V$ $V_{OUT} = 5V$	90% 80%

Figure A2 • LT1074 Electrical Characteristics

Appendix B
General considerations for switching regulator design

Inductor selection

Magnetic considerations are easily the most common problem area in switching regulator design. Ninety percent of the difficulties encountered are traceable to the inductive components in the circuit. The overwhelming level of difficulty associated with magnetics mandates a judicious selection process. The most common difficulty is saturation. An inductor is saturated when it cannot hold any more magnetic flux. As an inductor arrives at saturation it begins to look more resistive and less inductive. Under these conditions current flow is limited only by the inductor's DC copper resistance and the source capacity. This is why saturation often results in destructive failures.

While saturation is a prime concern, cost, heating, size, availability and desired performance are also significant. Electromagnetic theory, although applicable to these issues, can be confusing, particularly to the non-specialist.

Practically speaking, an empirical approach is often a good way to address inductor selection. It permits realtime analysis under actual circuit operat-

ing conditions using the ultimate simulator—a breadboard. If desired, inductor design theory can be used to augment or confirm experimental results.

Figure B1 shows a typical step-down converter utilizing the LT1074 switching regulator. A simple approach may be employed to determine the appropriate inductor. A very useful tool is the #845 inductor kit[12] shown in Figure B2. This kit provides a broad range of inductors for evaluation in test circuits such as Figure B1.

Figure B3 was taken with a 450µH value, high core capacity inductor installed. Circuit operating conditions such as input voltage and loading are set at levels appropriate to the intended application. Trace A is the LT1074's V_{SW} pin voltage while Trace B shows its current. When V_{SW} pin voltage is high, inductor current flows. The high inductance means current rises relatively slowly, resulting in the shallow slope observed. Behavior is linear, indicating no saturation problems. In Figure B4, a lower value

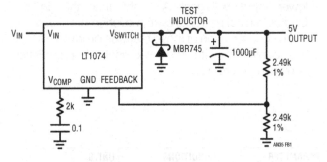

Figure B1 • Basic LT1074 Test Circuit

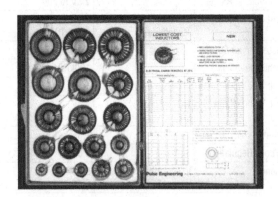

Figure B2 • Model 845 Inductor Selection Kit from Pulse Engineering, Inc. Includes 18 Fully Specified Devices

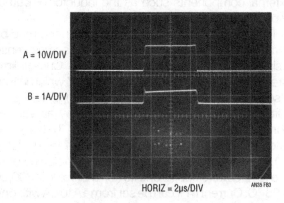

HORIZ = 2µs/DIV

Figure B3 • Waveforms for 450µH, High Capacity Core Unit

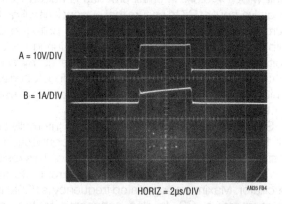

HORIZ = 2µs/DIV

Figure B4 • Waveforms for 170µH, High Capacity Core Unit

Note 12: Available from Pulse Engineering, Inc., PO. Box 12235, San Diego, California 92112, 619-268-2400.

unit with equivalent core characteristics is tried. Current rise is steeper; but saturation is not encountered. Figure B5's selected inductance is still lower, although core characteristics are similar. Here, the current ramp is quite pronounced, but well controlled. Figure B6 brings some informative surprises. This high value unit, wound on a low capacity core, starts out well but heads rapidly into saturation, and is clearly unsuitable.

The described procedure narrows the inductor choice within a range of devices. Several were seen to produce acceptable electrical results, and the "best" unit can be further selected on the basis of cost, size, heating and other parameters. A standard device in the kit may suffice, or a derived version can be supplied by the manufacturer

Using the standard products in the kit minimizes specification uncertainties, accelerating the dialogue between user and inductor vendor.

Inductor selection—alternate method

There are alternate inductor selection methods to the one described. One of the most popular is utilized by those devoid of the recommended inductor kit, time or adequate instrumentation. What is usually desired is to get a prototype LT1074 circuit running *NOW*. What is often available is limited to a drawer of inductors (see Figure B7) of unknown or questionable lineage. Selection of an appropriate inductor is (hopefully) made by simply inserting one of these drawer dwellers into an unsuspecting LT1074 circuit. Although this method's theoretical premise is perhaps questionable, its seemingly limitless popularity compels us to address it. We have developed a 2-step procedure for screening inductors of unknown characteristics. Inductors passing both stages of the test have an excellent chance (75%—based on our sample of randomly selected inductors) of performing adequately in a prototype LT1074 circuit. The only instrumentation required is an ohmmeter and a scale.

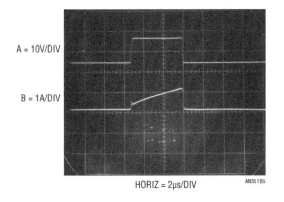

A = 10V/DIV

B = 1A/DIV

HORIZ = 2µs/DIV   AN35 FB5

Figure B5 • Waveforms for 55µH, High Capacity Core Unit

Figure B7 • "Yeah, We Got Some Inductors in a Drawer. I'm Sure They'll Work..."

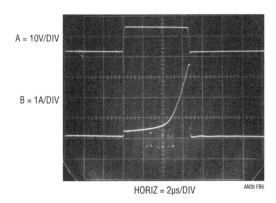

A = 10V/DIV

B = 1A/DIV

HORIZ = 2µs/DIV   AN35 FB6

Figure B6 • Waveforms for 500µH, Low Capacity Core Inductor (Note Saturation Effects)

Test 1 consists of weighing the candidate inductor. Acceptable limits range between 0.01 and 0.25 pounds. This test is best performed at an Inductor Test Facility (see Figure B8), where precision scales are readily available. To save time the quick checkout line is recommended (but only if you have nine[13] inductors or less—no cheating).

Figure B9 shows an inductor under test. The 0.13 pound weight indicated by the scale places this unit well within acceptable limits.

The second test involves measuring the inductor's DC resistance. Acceptable limits are usually between 0.01Ω and 0.25Ω. Inductors passing both tests will probably function in a prototype LT1074 circuit.

Note 13: The maximum permitted number of items in the quick checkout line varies from facility to facility. Please be familiar with and respect local regulations.

Figure B8 • Typical Inductor Test Facility

Figure B10 • Typical Acceptable Inductors

Figure B9 • Inductor Under Test (Don't Forget to Pick Up a Loaf of Bread and a Dozen Eggs)

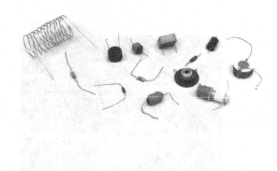

Figure B11 • Typical Unacceptable Inductors

Figures B10 and B11 show typical acceptable and unacceptable inductors. Graduates tend to be relatively dense, with (where visible) thick wire. Flunkers are usually less dense, with small (again, where visible) wire sizes.

When using an inductor selected with this method try low power first, then gradually increase loading. Observe inductor and LT1074 heating, making sure their dissipation is reasonable (warm to the touch). Disproportionate increases in heating as load is increased probably indicate inductor saturation. Either reduce the load, or go back to the drawer and try again.

While these two tests are somewhat lacking in rigor they do increase the chances of quickly getting a circuit to run with available components. In the longer term, the appropriate inductor can be decided upon and specified.

For the theoretically minded, test 1 grades out inductors which are unlikely to have enough flux storage capability (core mass) to avoid saturation. Test 2 eliminates units with too high a resistance to efficiently support typical LT1074 operating currents. Expanded discussion and design considerations for inductors will be found in Reference 4.

Capacitors

Think about requirements in capacitors. All operating conditions should be accounted for. Voltage rating is the most obvious consideration, but remember to plan for the effects of equivalent series resistance (ESR) and inductance. These specifications can have significant impact on circuit performance. In particular, an output capacitor with high ESR can make loop compensation difficult or decrease efficiency.

Layout

Layout is vital. Don't mix signal, frequency compensation, and feedback returns with high current returns. Arrange the grounding scheme for the best compromise between AC and DC performance.

In many cases, a ground plane may help. Account for possible effects of stray inductor-generated flux on other components and plan layout accordingly.

Diodes

Diode breakdown and switching ratings must be thought through. Account for all conditions. Transient events usually cause the most trouble, introducing stresses that are often hard to predict. Study the data sheet breakdown, current capacity, and switching speed ratings carefully. Were these specifications written under the same conditions that your circuit is using the device in? If in doubt, consult the manufacturer.

Switching diodes have two important transient characteristics—reverse recovery time and forward turn-on time.

Reverse recovery time occurs because the diode stores charge during its forward conducting cycle. This stored charge causes the diode to act as a low impedance conductive element for a short period of time after reverse drive is applied. Reverse recovery time is measured by forward biasing the diode with a specified current, then forcing a second specified current backwards through the diode. The time required for the diode to change from a reverse conducting state to its normal reverse non-conducting state is reverse recovery time. Hard turn-off diodes switch abruptly from one state to the other following reverse recovery time. They therefore dissipate very little power even with moderate reverse recovery times. Soft turn-off diodes have a gradual turn-off characteristic that can cause considerable diode dissipation during the turn-off interval.

Fast diodes can be useless if stray inductance is high in the diode, output capacitor or LT1074 loop. 20-gauge hook-up wire has 30nH/inch inductance. Switching currents on the order of 10^8A/sec are typical in regulator circuits. They can easily generate volts per inch in wiring. Keep the diode, capacitor and LT1074 input/switch lead lengths *SHORT!*

Frequency compensation

The basic LT1074 step-down configuration is relatively free of frequency compensation difficulties. The simple RC damper networks shown from the V_C pin to ground will usually suffice. Things become more complex when gain and phase contributing elements are added to the basic loop. In these cases it is often useful to look at the LT1074 as a low bandwidth power stage. The delays are due to the sampled data nature of power delivery (100kHz switching frequency) and the output LC filter. In general, complex loops can be stabilized by limiting the gain bandwidth of the LT1074 below that of the added elements. This

is in accordance with well know feedback theory. A discussion of practical techniques for stabilizing such loops, "The Oscillation Problem (Frequency Compensation Without Tears)," appears at the end of LTC Application Note 18. Other pertinent comments appear in the "Frequency Compensation" sections of LTC Application Notes 19 and 25.

Appendix C
Techniques and equipment for current measurement

Accurate measurement of current flow under rapidly changing circuit conditions is essential to switching regulator design. In many cases current waveforms contain more valuable information than voltage measurements. The most powerful and convenient current measuring tool is the clip-on current probe. Several types appear in Figure C1. The Tektronix P-6042, shown bottom left, is a Hall effect stabilized current transformer which responds from DC to 50MHz. The more recent Tektronix AM-503 (not shown) has similar specifications. The combination of convenience, broad bandwidth and DC response make Hall effect stabilized current probes the instrument of choice for converter design. The DC response allows determination of DC content in high speed current waveforms. The clip-on probe contains a current transformer and a Hall effect device. The Hall device senses at DC and low frequency while the transformer simultaneously processes high frequency content. Careful roll-off matching allows a composite output with no peaking or response aberrations at the two sensors bandwidth crossover.

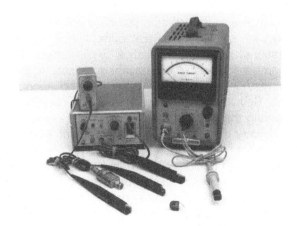

Figure C1 • Various Current Probe Types Provide Different Capabilities. Selection Criteria is Application Dependent

Sensitivity ranges from fractions of a milliampere to amperes and is switch selectable.

Transformer based clip-on current probes are also available. These types lack the DC response of their Hall effect augmented cousins, but are still quite useful. The Tektronix type 131 (and the more modern 134) responds from hundreds of hertz to about 40MHz. AC current probes (type 131 appears in C1, upper left) are as convenient to use as Hall types, but cannot respond at low frequency. AC current probes are also available with a simple termination (left foreground, Figure C1). These types are more difficult to use than the actively terminated models (e.g., type 131 shown) because of complex gain switching. Their low frequency limitations are also poorer, although their high frequency response exceeds 100MHz.

A final form of AC current probe is the simple transformer shown in Figure C1's foreground. These are not clip-on devices, and usually have significant performance limitations compared to the instruments discussed. However they are inexpensive and can provide meaningful measurement results when used according to manufacturers' recommendations. In use, the conductor is threaded through the opening provided and the signal monitored at the output pins.

Figure C1 also shows a wide ranging DC clip-on current probe. The Hewlett-Packard 428B (upper right) responds from DC to only 400Hz, but features 3% accuracy over a 100µA to 10A range. This instrument obviously cannot discern high speed events, but is invaluable for determining overall efficiency and quiescent current.

A great strength of the probes described is that they take a fully floating measurement. The extraction of current information by magnetic connection eliminates common mode voltage considerations. Additionally, the clip-on convenience makes the probes as easy to use as a standard voltage mode probe. As good as they are, current probes have limitations and characteristics which must be remembered to avoid unpleasant surprises. At high currents, probe saturation limits may be encountered. Resultant CRT waveforms will be corrupted, rendering the measurement useless and confusing the unwary. For Hall types, measurement below a few hundred microamperes is limited by noise, which is much more obvious on the display. Keep in mind that current probes have different signal transit delay times than voltage probes or dissimilar current probes. At high sweep speeds this effect shows up in multi-trace displays as time skewing between individual channels. The current probes transit time delay

can be mentally factored in to reduce error when interpreting the display. Note that active probes have the longest signal transit times, on the order of 25ns.

The AC probe's low frequency bandwidth restriction must be kept in mind when interpreting CRT displays. Figure C2 clearly demonstrates this by showing the AC probe's inability to follow low frequency. Similarly, remember that the probe's stated bandpass is a −3dB figure, meaning signal information is not entirely present in the display at this frequency. When working in regions approaching either end of the probe bandpass consider that displayed information may be distorted or incomplete.

There are other ways, albeit less convenient and desirable than clip-on current probes, to measure wideband current signals. Ohm explains that measuring voltage across a resistor gives current. Current shunts (Figure C3 foreground) are low value (for LT1074 circuits 0.1Ω to 0.01Ω is typical) resistors with four terminal connections for accurate measurement. In theory, measuring voltage across a current shunt should yield accurate information. In practice, common mode voltages introduce measurement difficulties, particularly at speed. Making this

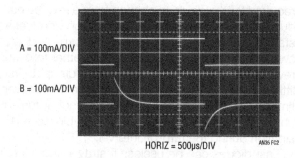

A = 100mA/DIV

B = 100mA/DIV

HORIZ = 500µs/DIV   AN35 FC2

Figure C2 • Hall Stabilized (Trace A) and Transformer

Figure C3 • Typical Current Shunts and an "Isolated" Probe

measurement requires an isolated probe or a high speed differential plug-in. The Signal Acquisition Technologies SL-10 (Figure C3) has 10MHz bandwidth, a galvanically floating input and 600V common mode capability. This probe allows any oscilloscope to take a floating measurement across a shunt.

Differential oscilloscope plug-ins, while not galvanically floating, can measure across a shunt. Tektronix types W, 1A5 and 7A13 have 1mV sensitivity with up to 100MHz bandwidth and excellent common mode rejection. Types 1A7 and 7A22 have 10μV sensitivity, although bandwidth is limited to 1MHz. All differential plug-ins have bandwidth and/or common mode voltage restrictions that vary with sensitivity. These trade-offs must be reviewed when selecting the optimal shunt value for a particular measurement. In general the smallest practical shunt value is desirable. This minimizes the inserted resistances parasitic effects on circuit operation.

Appendix D
Optimizing switching regulators for efficiency

Squeezing the utmost efficiency out of a switching regulator is a complex, demanding design task. Efficiency exceeding 80% to 85% requires some combination of finesse, witchcraft and just plain luck. Interaction of electrical and magnetic terms produces subtle effects which influence efficiency. A detailed, generalized method for obtaining maximum converter efficiency is not readily described but some guidelines are possible.

Losses fall into several loose categories including junction, ohmic, drive, switching, and magnetic losses.

Semiconductor junctions produce losses. Diode drops increase with operating current and can be quite costly in low voltage output converters. A 700mV drop in a 5V output converter introduces more than 10% loss. Schottky devices will cut this nearly in half, but loss is still appreciable. Germanium (rarely used) is lower still, but switching losses negate the low DC drop at high speeds. In very low power converters germanium's reverse leakage may be equally oppressive. Synchronously switched rectification is more complex, but can sometimes simulate a more efficient diode (see LTC Application Note 29, Figure 32). The LT1074's "Comout" pin is intended to drive external synchronous switches. See Appendix A for details.

When evaluating synchronous rectification schemes remember to include both AC and DC drive losses in efficiency estimates. DC losses include base or gate current in addition to DC consumption in any driver stage. AC losses might include the effects of gate (or base) capacitance, transition region dissipation (the switch spends some time in its linear region) and power lost due to timing skew between drive and actual switch action.

The LT1074's output switch is composed of a PNP driving a power NPN (Figure D1). The switch drop can reach 2V at high currents. This will usually be the major loss in the circuit. Its effect on efficiency can be mitigated by using the highest possible input voltage. Text Figure 7.7 shows 5V regulator efficiency improving almost 10% for higher input voltages. Higher output voltages will further minimize the switch losses.

Actual losses caused by switch saturation effects and diode drops are sometimes difficult to ascertain. Changing duty cycles and time variant currents make determination tricky. One simple way to make relative loss judgments is to measure device temperature rise. Appropriate tools here include thermal probes and (at low voltages) the perhaps more readily available human finger. At lower power (e.g., less dissipation, even though loss percentage may be as great) this technique is less effective. Sometimes deliberately adding a known loss to the component in question and noting efficiency change allows loss determination.

Ohmic losses in conductors are usually only significant at higher currents. "Hidden" ohmic losses include socket and connector contact resistance and equivalent series resistance (ESR) in capacitors. ESR generally drops with capacitor value and rises with operating frequency, and should be specified on the capacitor data sheet. Consider the copper resistance of inductive components. It is often necessary to evaluate trade-offs of an inductor's copper resistance versus magnetic characteristics.

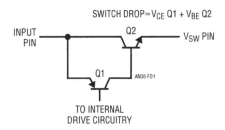

Figure D1 • Simplified LT1074 Output Switch

Switching losses occur when the LT1074 spends significant amounts of time in its linear region relative to operating frequency. At higher switching frequencies

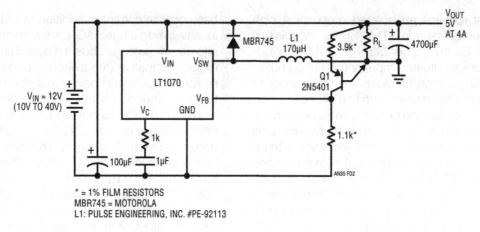

* = 1% FILM RESISTORS
MBR745 = MOTOROLA
L1: PULSE ENGINEERING, INC. #PE-92113

Figure D2 • Floating Input Buck Regulator

transition times can become a substantial loss source. The LT1074's 100kHz preset switching frequency is a good compromise (for this device) and changes should be carefully considered. Raising the switching frequency to gain some desired benefit necessitates consideration of increased LT1074 losses. 200kHz is the maximum practical operating frequency.

Magnetics design also influences efficiency. Design of inductive components is well beyond the scope of this appended section, but issues include core material selection, wire type, winding techniques, size, operating frequency current levels, temperature and other issues. Some of these topics are discussed in LTC Application Note 19, but there is no substitute for access to a skilled magnetics specialist. Fortunately, the other categories mentioned usually dominate losses, allowing good efficiencies to be obtained with standard magnetics. Custom magnetics are usually only employed after circuit losses have been reduced to lowest practical levels.

A special circuit

In cases where input voltage must be low, but may float, Figure D2's circuit may be preferable to an LT1074 based approach. This circuit uses the LT1070, a common emitter output device. With the emitter connected to the ground pin this device's (LT1070 operating details are available in its data sheet, and in LTC Application Notes 19, 25 and 29) switch loss is significantly lower than the LT1074's. Although intended for voltage step-up in flyback configurations the LT1070 can be arranged to perform the step-down function. The advantage is the efficiency gain due to the reduced switch loss. The circuit's primary restriction is that the input must float with respect to the output. Q1 performs a level shift to get the feedback information referenced to the LT1070 "ground" pin, which floats with the input.

The LT1070 is effectively "fooled" and behaves like a flyback regulator. It is oblivious to the fact that the overall function is step down, because the floating input is driven to the output potential. The negative side of the output filter capacitor is connected to the ground (⏚) of the powered system, and the LT1070 input rail becomes the 5V output. Other voltages are obtainable by altering the 3.9k–1.1k feedback ratio. Efficiency approaches 85%.

Appendix E
A half-sine reference generator

Text Figure 7.31's half-sine reference must be amplitude and frequency stabilized to a fairly high degree. It is not unreasonable to expect a 115V_{AC} 400Hz source to be within 1V and 0.1Hz. Additionally, Figure 7.31's reference requires a half-sine, as opposed to the more normal full-sine. These requirements are achievable by classical analog techniques, but a digital approach eases complexity with no performance trade-off[14]. Figure E3 shows such an approach. C1 forms a 1.024MHz crystal oscillator which is divided down by the 7490. The 7490's differentiated ÷10 output becomes the LT1074's 102.4kHz sync. option output. The 7490's ÷5 output (204.8kHz) is fed to the 74191 counters. These counters parallel load a 2716 EPROM which is programmed to produce an 8-bit (256 states) digitally coded half-sine. The program, developed by Sean Gold and Guy M. Hoover, appears in Figure E1.

Note 14: The sinewave is probably the paramount expression of the analog world. The Old Man Himself, George A. Philbrick, once elegantly discussed analog functions as "those which are continuous in excursion and time." He might have viewed digital production of a sinewave with considerable suspicion, or simply labeled it blasphemous. Such are the wages of progress.

The 2716's parallel output is fed to an 8-bit DAC, which produces 800Hz 2.5V (peak) half-sines.

Those wishing to utilize this reference for full-sines will find the appropriate software in Figure E2.

Figure E3 also shows the synchronous switch option discussed in the text. The 74C122 mono-stable forms a simple delayed pulse generator which drives the Q4 switch. The 20μs delay and 6μs pulse width set at the 74C122 were empirically determined to produce lowest overall crossover distortion in Figure 7.31's output.

```
Line 10801   Column   Wrap              APL2/PC
     GENCODES
00 03 06 09 0D 10 13 16 19 1C 1F 22 26 29 2C 2F
32 35 38 3B 3E 41 44 47 4A 4D 50 53 56 59 5C 5F
62 65 68 6B 6D 70 73 76 79 7B 7E 81 84 86 89 8C
8E 91 93 96 98 9B 9D A0 A2 A5 A7 A9 AC AE B0 B3
B5 B7 B9 BB BE C0 C2 C4 C6 C8 CA CB CD CF D1 D3
D5 D6 D8 DA DB DD DE E0 E1 E3 E4 E6 E7 E8 EA EB
EC ED EE EF F1 F2 F3 F3 F4 F5 F6 F7 F8 F8 F9 FA
FA FB FB FC FC FD FD FE FE FE FE FF FF FF FF FF
FF FF FF FF FF FE FE FE FE FD FD FC FC FB FB FA
FA F9 F8 F8 F7 F6 F5 F4 F3 F3 F2 F1 EF EE ED EC
EB EA E8 E7 E6 E4 E3 E1 E0 DE DD DB DA D8 D6 D5
D3 D1 CF CD CB CA C8 C6 C4 C2 C0 BE BB B9 B7 B5
B3 B0 AE AC A9 A7 A5 A2 A0 9D 9B 98 96 93 91 8E
8C 89 86 84 81 7E 7B 79 76 73 70 6D 6B 68 65 62
5F 5C 59 56 53 50 4D 4A 47 44 41 3E 3B 38 35 32
2F 2C 29 26 22 1F 1C 19 16 13 10 0D 09 06 03 00
```

Figure E1 • Half-Sine Software Code for the 2716 EPROM

```
Line 10736   Column   Wrap              APL2/PC
     GENCODES
FF FF FF FF FE FE FE FD FD FC FB FA F9 F9 F7 F6
F5 F4 F3 F1 F0 EE ED EB E9 E8 E6 E4 E2 E0 DE DC
D9 D7 D5 D2 D0 CE CB C9 C6 C3 C1 BE BB B8 B6 B3
B0 AD AA A7 A4 A1 9E 9B 98 95 92 8E 8B 88 85 82
7F 7C 78 75 72 6F 6C 69 66 63 60 5D 5A 57 54 51
4E 4B 48 45 42 40 3D 3A 38 35 33 30 2E 2B 29 27
25 22 20 1E 1C 1A 18 17 15 13 11 10 0E 0D 0C 0A
09 08 07 06 05 04 03 03 02 02 01 01 00 00 00 00
00 00 00 00 01 01 02 02 03 03 04 05 06 07 08 09
0A 0C 0D 0E 10 11 13 15 17 18 1A 1C 1E 20 22 25
27 29 2B 2E 30 33 35 38 3A 3D 40 42 45 48 4B 4E
51 54 57 5A 5D 60 63 66 69 6C 6F 72 75 78 7C 7F
82 85 88 8B 8E 92 95 98 9B 9E A1 A4 A7 AA AD B0
B3 B6 B8 BB BE C1 C3 C6 C9 CB CE D0 D2 D5 D7 D9
DC DE E0 E2 E4 E6 E8 E9 EB ED EE F0 F1 F3 F4 F5
F6 F7 F9 F9 FA FB FC FD FD FE FE FE FF FF FF FF
```

Figure E2 • Full-Sine Software Code for the 2716 EPROM (Bonus)

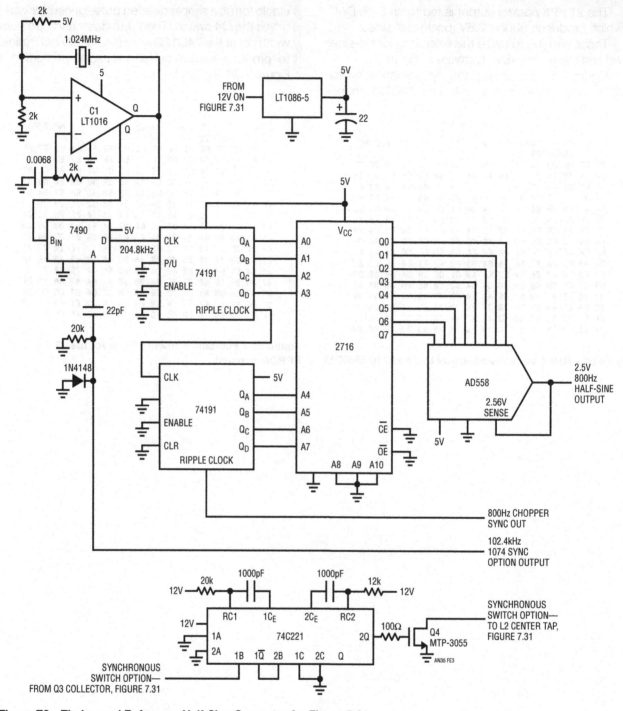

Figure E3 • Timing and Reference Half-Sine Generator for Figure 7.31

Appendix F
The magnetics issue

Magnetics is probably the most formidable issue in converter design. Design and construction of suitable magnetics is a difficult task, particularly for the non-specialist. It is our experience that the majority of converter design problems are associated with magnetics requirements. This consideration is accented by the fact that most converters are employed by non-specialists. As a purveyor of switching power ICs we incur responsibility towards addressing the magnetics issue (our publicly spirited attitude is, admittedly, capitalistically polluted). As such, it is LTC's policy to use off-the-shelf magnetics in our circuits. In some cases, available magnetics serve a particular design. In other situations the magnetics have been specially designed, assigned a part number and made available as standard product.

In many circumstances a standard product is suitable for production. Other cases may require modifications or changes which the manufacturer can provide or advise on. Hopefully, this approach serves the needs of all concerned.

Recommended magnetics manufacturers include the following;

Pulse Engineering, Inc
PO. Box 12235
7250 Convoy Court
San Diego, California 92112
619-268-2400

Coiltronics
984 Southwest 13th Court
Pompano Beach, FL 33069
305-781-8900

A monolithic switching regulator with 100µV output noise

"Silence is the perfectest herald of joy ..."

Jim Williams

8

Introduction

Size, output flexibility and efficiency advantages have made switching regulators common in electronic apparatus. The continued emphasis on these attributes has resulted in circuitry with 95% efficiency that requires minimal board area. Although these advantages are welcome, they necessitate compromising other parameters.

Switching regulator "noise"

Something commonly referred to as "noise" is a primary concern. The switched mode power delivery that permits the aforementioned advantages also creates wideband harmonic energy. This undesirable energy appears as radiated and conducted components commonly labeled as "noise." Actually, switching regulator output "noise" is not really noise at all, but coherent, high frequency residue directly related to the regulator's switching.[1] Figure 8.1 shows typical switching regulator output noise. Two distinct characteristics are present. The slow, ramping output ripple is caused by finite storage capacity of the regulator's output filter components. The quickly rising spikes are associated with the switching transitions. Figure 8.2 shows another switching regulator output. In this case the ripple has been eliminated by adequate filtering and linear postregulation, but the wideband spikes remain. It is these fast spikes that cause so much difficulty in systems. Their high frequency content often corrupts associated circuitry, degrading performance or even disabling operation. Noise gets into adjacent circuitry via three paths. It is conducted out of the regulator output lead, it is conducted back to the driving source ("reflected" noise) and it is radiated. The multiple transmission paths combine with the high frequency content to make noise suppression difficult. Unconscionable

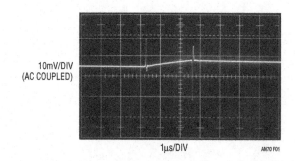

10mV/DIV
(AC COUPLED)

1µs/DIV   AN70 F01

Figure 8.1 • Typical Switching Regulator Output "Noise." Wideband Spikes Are Difficult to Suppress, Causing System Interference Problems. Ripple Component Has Low Harmonic Content, Is Relatively Easily Filtered

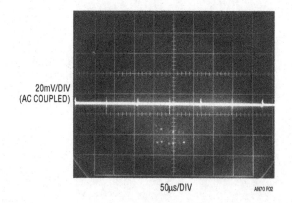

20mV/DIV
(AC COUPLED)

50µs/DIV   AN70 F02

Figure 8.2 • Linear Regulator Eliminates Ripple, but Wideband Spikes Remain. Peak-to-Peak Amplitude Exceeds 30mV (Just Visible Near 2nd, 5th and 8th Vertical Graticule Divisions)

Note 1: Noise contains no regularly occurring or coherent components. As such, switching regulator output "noise" is a misnomer. Unfortunately, undesired switching related components in the regulated output are almost always referred to as "noise." As such, although technically incorrect, this publication treats all undesired output signals as "noise." See Appendix B, "Specifying and measuring something called noise."

Analog Circuit and System Design: A Tutorial Guide to Applications and Solutions. DOI: 10.1016/B978-0-12-385185-7.00008-1

amounts of bypass capacitors, ferrite beads, shields, Mu-metal and aspirin have been expended in attempts to ameliorate noise-induced effects.

Alternate approaches involve synchronizing switching regulator operation to the host system or turning off switching during critical system operation (an "interrupt driven" power supply). Another approach places critical system operations between switch cycles, literally running between electronic rain drops.[2]

The difficulty of debugging a noise-laden system and the compromises involved in synchronized approaches could be eliminated with a low noise switching regulator. An inherently low noise switching regulator is the most attractive approach because it eliminates noise concerns while maintaining system flexibility.

A noiseless switching regulator approach

The key to an inherently low noise regulator is to minimize harmonic content in the switching transitions. Slowing down the switching interval does this, although power dissipated during the transition causes some efficiency loss. Reducing switch repetition rate can largely offset the losses, resulting in a reasonably efficient design with small

magnetics and the desired low noise. Noise reduction by restricting harmonic generation has been employed before, although the implementations were complex and narrowly applicable.[3] A monolithic approach, broadly usable over a range of magnetics and applications, is described here.

A practical, low noise monolithic regulator

Figure 8.3 describes the LT®1533, a monolithic regulator designed for low noise switching supplies. Figure 8.4 details the pin functions. Figure 8.3's functional blocks show a fairly conventional push-pull architecture with a major exception. The push-pull approach has good magnetics utilization (power transfer is always occurring in the transformer; the core does not store energy) and pulls current continuously from the source. The even, continuous current drain from the source eliminates the fast, high peak currents required by flyback and other approaches. The source sees a benign load and is not corrupted. The switches also receive nonoverlapping drive, ensuring they do not conduct simultaneously. Simultaneous conduction would cause excessive, quickly rising currents, degrading efficiency and generating noise.

Note 2: See References 2 and 3 for details and practical examples of these techniques.

Note 3: See Appendix A, "A history of low noise DC/DC conversion." See also References 4 through 10.

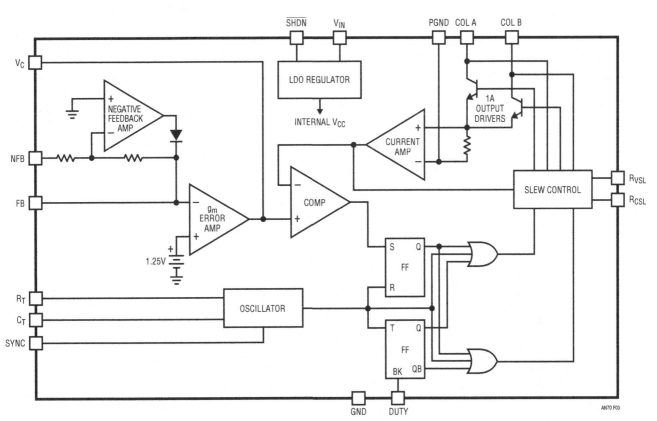

Figure 8.3 • LT1533 Simplified Block Diagram. 1A Slew-Controlled Output Stages Provide Low Noise Switching

COL A, COL B:Output transistor collectors which switch out-of-phase.

DUTY: Grounding this pin forces the outputs to switch at a 50% duty cycle. This pin must float if not used.

SYNC: Used to synchronize to an external clock. Float or tie to ground if unused.

C_T: Oscillator timing capacitor.

R_T: Oscillator timing resistor.

FB: Used for positive output voltage sensing.

NFB: Used for negative output voltage sensing.

GND: Analog ground pin.

PGND: High current ground return. Should be returned to ground via ≈ 50nH (≅1" of PC trace or wire, or a small ferrite bead). See Appendix F or schematic (Figures 5 and 26) notes for details. In some package options, this pin may be internally connected to "GND" pin.

V_C: Frequency compensation node.

SHDN: Normally high. Grounding this pin shuts the part down. I_{SHDN} = 20μA.

R_{CSL}: Current slew control resistor.

R_{VSL}: Voltage slew control resistor.

V_{IN}: Input supply pin. 2.7V to 30V range. Undervoltage lockout at 2.55V.

Figure 8.4 • LT1533 Short Form Pin Function Descriptions

The design's most significant aspect is the output stage. Each 1A power transistor operates inside a broadband control loop. The voltage across each transistor and the current through it are sensed and the loop control's slew rate of each parameter. The voltage and current slew rates are independently settable by external programming resistors. This ability to control the switching's rate-of-change makes low noise switching regulation practical. Operating the switching transistors in a local loop permits predictable, wide range control over a variety of situations.[4] Figure 8.5 is a 40kHz, 5V to 12V converter using the LT1533 in a push-pull, "forward" configuration. The feedback resistor's ratio produces a 12V output. A two-section LC filter provides high ripple attenuation, although a single section will give good performance. It is particularly noteworthy that high frequency noise content (as opposed to the 40kHz fundamental related ripple) is unaffected by output filter characteristics. This is so simply because there is so little high frequency energy developed in this circuit. If there's nothing there, it doesn't need to be filtered!

L2 provides compensation for the output current control loop. In practice, L2 may be a length of PC trace, a small inductor, a coiled section of wire or a ferrite bead. See Appendix F, "Magnetics considerations" for complete discussion.

Measuring output noise

Measuring the LT1533's unprecedented low noise levels requires care.[5] Figure 8.6 shows a test setup for taking the measurement. Good connection and signal handling technique combined with judicious instrumentation choice should yield a 100μV noise floor in a 100MHz bandwidth. This corresponds to the noise of a 50Ω resistor in a 100MHz bandwidth.

Before measuring regulator output noise, it is good practice to verify test setup performance. This is done by running the test setup with no input. Figure 8.7 shows a noise base line of 100μV in a 100MHz bandwidth, indicating the

---

Note 4: Patent pending.
Note 5: Equipment selection and measurement techniques are detailed in Appendix B, "Specifying and measuring something called noise." See also Appendix C, "Probing and connection techniques for low level, wideband signal integrity."

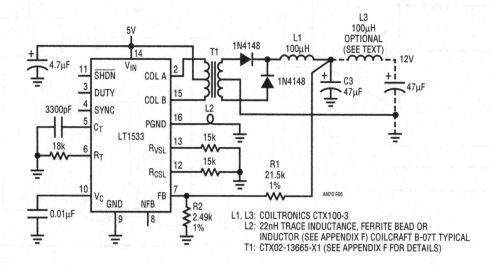

L1, L3: COILTRONICS CTX100-3
L2: 22nH TRACE INDUCTANCE, FERRITE BEAD OR INDUCTOR (SEE APPENDIX F) COILCRAFT B-07T TYPICAL
T1: CTX02-13665-X1 (SEE APPENDIX F FOR DETAILS)

Figure 8.5 • 100μV Noise 5V-to-12V Converter. Output LC Section May Be Deleted If Low Frequency Ripple Is Acceptable

171

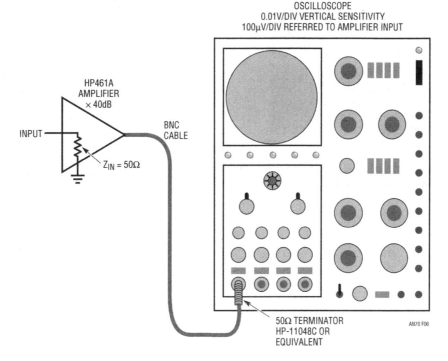

OSCILLOSCOPE
0.01V/DIV VERTICAL SENSITIVITY
100μV/DIV REFERRED TO AMPLIFIER INPUT

HP461A
AMPLIFIER
× 40dB

INPUT

$Z_{IN} = 50\Omega$

BNC
CABLE

50Ω TERMINATOR
HP-11048C OR
EQUIVALENT

AN70 F06

Figure 8.6 • Test Setup Noise Baseline Is 100μV$_{P-P}$ in 100MHz Bandwidth. Performance Is 50Ω Resistor Noise Limited. BNC Cable Connections and Terminations Provide Coaxial Environment, Ensuring Wideband, Low Noise Characteristics

instrumentation is operating properly. Measuring Figure 8.5's noise involves AC coupling the circuit's output into the test setup's input. Figure 8.8 shows this. Coaxial connections must be maintained to preserve measurement integrity.[6] Figure 8.9's waveforms detail circuit operation. Traces A and C are switching transistor collector voltages, B and D are the respective transistor currents. The test setup's output, representing circuit output noise, is Trace E. Wideband spiking and ripple, just visible in the noise floor, is inside 100μV, even in a 100MHz bandpass.[7]

This is spectacularly good performance and is, in fact, actually better than the photo shows. Removing all probes from the breadboard leaves only Trace E's coaxial connection. This eliminates any possible ground loop-induced error.[8] Figure 8.10's trace shows 40kHz ripple with about the same amplitude as in Figure 8.9. Switching related spikes, just faintly outlined in the noise, are reduced.

Measurement bandwidth is reduced to 10MHz in Figure 8.11, attenuating test fixture amplifier noise. Switching and ripple residue amplitude and shape do not change, indicating no signal activity beyond this frequency.

Figure 8.12's horizontal expansion of Figure 8.10 returns to 100MHz bandpass. The switching spike appears in the center screen region. At 2μs/division sweep, there is no wideband activity observable. Figure 8.13, a 10MHz bandpass version of Figure 8.12, retains all signal information, further suggesting no signal power beyond 10MHz.

Figure 8.14 is the noise floor of an HP4195A spectrum analyzer in a 500MHz sweep. When Figure 8.5's circuit is AC coupled into the analyzer, the output (Figure 8.15a) is essentially identical. The analyzer is unable to detect switching-induced noise in a 500MHz bandpass. Some 40kHz fundamental-related components are detectable in Figure 8.15b's 1MHz wide plot, although the rest of the sweep is analyzer noise limited. Additional filtering or a linear postregulator could eliminate the 40kHz ripple-related residue if desired.

The preamplified oscilloscope is a more sensitive tool for these measurements because its triggered operation has the advantage of synchronous detection. This is demonstrable by free running the preamplified oscilloscope sweep; the switching-related components are indistinguishable in the noise background.

Figure 8.16 studies ripple at the first LC filter section output. The ripple's 40kHz fundamental is clearly seen, although no wideband spikes are visible. Figure 8.17 horizontally expands Figure 8.14's time scale, but high frequency harmonics and spikes are not observable.

Low frequency noise is rarely a concern, although Figure 8.18 shows it is inside 50μV in a 10Hz to 10kHz

Note 6: Again, see Appendices B and C for extended treatment of these and related issues.
Note 7: It is common industry practice to specify switching regulator noise in a 20MHz bandpass. There can be only one reason for this, and it is a disservice to users. See Appendix B for tutorial on observed noise versus measurement bandwidth.
Note 8: See Appendix C for related discussion and techniques for triggering oscilloscopes without invasively probing the circuit.

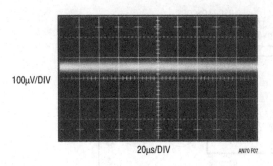

100µV/DIV

20µs/DIV   AN70 F07

Figure 8.7 • Oscilloscope Verifies Test Setup 100µV Noise Floor in 100MHz Bandwidth. Indicated Noise Is That of a 50Ω Resistor

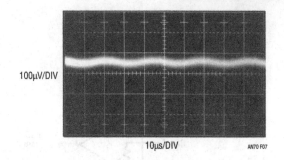

100µV/DIV

10µs/DIV   AN70 F07

Figure 8.10 • Removing Probes from Figure 8.9's Test Eliminates Ground Loops, Slightly Reducing Observed Noise. Switching Artifacts Are Just Discernible Above Noise Floor

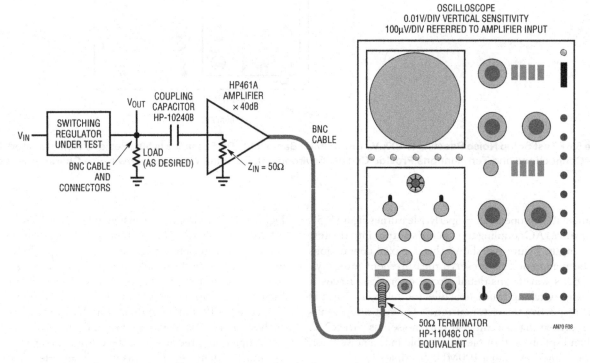

OSCILLOSCOPE
0.01V/DIV VERTICAL SENSITIVITY
100µV/DIV REFERRED TO AMPLIFIER INPUT

V_{IN}

SWITCHING
REGULATOR
UNDER TEST

V_{OUT}

COUPLING
CAPACITOR
HP-10240B

HP461A
AMPLIFIER
× 40dB

BNC
CABLE

BNC CABLE
AND
CONNECTORS

LOAD
(AS DESIRED)

$Z_{IN} = 50Ω$

50Ω TERMINATOR
HP-11048C OR
EQUIVALENT

AN70 F08

Figure 8.8 • Connecting Figure 8.5's Circuit to the Test Setup. Coaxial Connections Must Be Maintained to Preserve Measurement Integrity

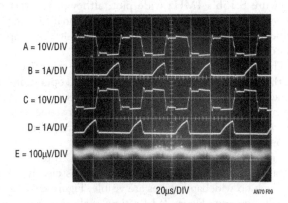

A = 10V/DIV

B = 1A/DIV

C = 10V/DIV

D = 1A/DIV

E = 100µV/DIV

20µs/DIV   AN70 F09

Figure 8.9 • Waveforms for Figure 8.5 at 100mA Loading. Traces A and C Are Voltage; B and D are Current, Respectively. Switching Transistion's Noise Signature Appears in Trace E, the Circuit's Output Noise

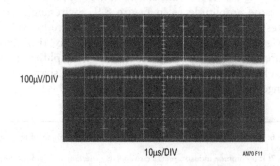

100µV/DIV

10µs/DIV   AN70 F11

Figure 8.11 • Reducing Measurement Bandwidth to 10MHz Attenuates Amplifier Noise. Switching Residue Characteristics Remain Unchanged, Indicating No Signal Activity Beyond This Frequency

bandpass. Input current noise is usually of more interest. Excessive "reflected" noise can corrupt the regulator's driving source, causing system level interference. Figure 8.19 shows Figure 8.5's input current as DC with a small, 40kHz fundamental-related sinusoidal component. There is no high frequency content, and the sinusoidal variations are easily handled by the driving source.

System-based noise "measurement"

In the final analysis, the effect of switching regulator output noise on the system it is powering is the ultimate test. Appendix K, "System-based noise measurement," presents results when the LT1533 is used to power a 16-bit A/D converter.

Transition rate effects on noise and efficiency

In theory, simply setting transition rate to low values will achieve low noise. Practically, such an approach, while workable, wastes power during transitions, lowering efficiency. A good compromise sets transition time at the fastest rate permitting desired noise performance. The LT1533's slew adjustments allow easy determination of this point. Figure 8.20's photographs dramatically demon-

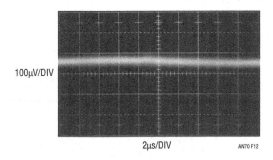

100μV/DIV

2μs/DIV   AN70 F12

Figure 8.12 • Horizontal Expansion of Figure 8.10 Shows No Wideband Components. Switching Originated Noise Appears in Center Screen Region

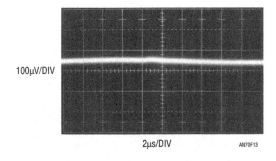

100μV/DIV

2μs/DIV   AN70 F13

Figure 8.13 • A 10MHz Band Limited Version of Figure 8.12. As Before, Signal Information Is Retained, Although Amplifier Noise Is Reduced. Results Indicate No Signal Power Beyond 10MHz

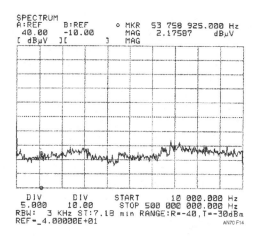

Figure 8.14 • Noise Floor of Test Fixture and HP-4195A Spectrum Analyzer in a 500MHz Sweep

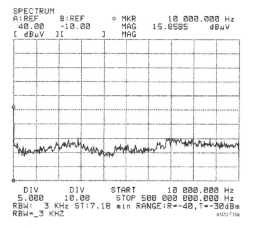

Figure 8.15a • Figure 8.5's Circuit Connected to the Spectrum Analyzer Produces Essentially Identical Results to Figure 8.14. Circuit's Noise Is Undetectable

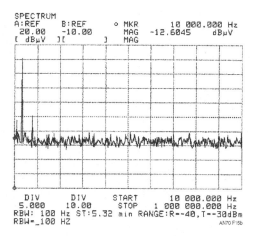

Figure 8.15b • Reducing Analyzer Sweep to 1MHz Width Reveals 40kHz Related Components. Remainder of Plot Is Analyzer Noise Floor Limited, Even in Sensitive 455kHz Band

strate the relationship between transition time and output noise for Figure 8.5's circuit. The sequence shows >5:1 noise reduction as switch transition time slows from 100ns (20a) to 1µs (20d). Figure 8.20d's displayed noise is actually lower, as the probing-induced error caused by monitoring the switch corrupts the measurement.[9]

Figure 8.21 graphically summarizes Figure 8.20's information. Significant noise reduction coincides with descending transition slew time until about 1.3µs. Little additional noise benefit occurs beyond this point. Figure 8.22 shows efficiency fall-off with slew time. There is a 6% penalty between 100ns and 1.3µs, the same region where noise performance improves by a factor of 5 (per previous figure). There is an additional 6% penalty beyond 1.3µs, although no significant noise reduction occurs (again, per Figure 8.21). As such, operation in this region is undesirable. Figure 8.23 clearly shows the inflection point in the efficiency versus noise trade-off.[10]

Negative output regulator

The LT1533 has a separate feedback input that directly accepts negative inputs.[11] This permits negative outputs without the usual discrete level shifting stage. Figure 8.24's 5V to −12V converter is similar to Figure 8.5's circuit, except that the negative output is fed back to the negative feedback input. The feedback scale factor change is necessitated by the higher effective reference voltage. In all other respects, the circuit (and its performance) is similar to Figure 8.5.

Floating output regulator

Figure 8.25's isolation stage permits a fully floating, regulated output. The LT1431 shunt regulator compares a portion of the output to its internal reference and drives the optoisolator with the error signal. The optoisolator's collector output biases the LT1431's V_C pin, closing a feedback loop to regulate circuit output. The 0.22µF capacitor stabilizes the loop and the 240kΩ resistor biases the optoisolator into a favorable operating region. This circuit's operation and characteristics are similar to Figure 8.5 with the added benefit of the isolated output.

Floating bipolar output converter

Grounding the LT1533's "DUTY" pin and biasing FB forces the device into its 50% duty cycle mode. Figure 8.26's output is full wave rectified with respect to T1's secondary center tap, producing bipolar outputs. The forced 50% duty cycle combined with no feedback means

---

Note 9: See Appendix C, "Probing and connection techniques for low level, wideband signal integrity" for guidance.
Note 10: The noise and efficiency characteristics appearing in Figures 8.20 to 8.23 were generated at the bench in about ten minutes. All you CAD modeling types out there might want to think about that.
Note 11: See Figure 8.3's block diagram.

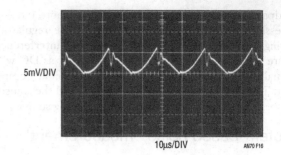

Figure 8.16 • Ripple at Figure 8.5's First LC Output Has No Wideband Spikes

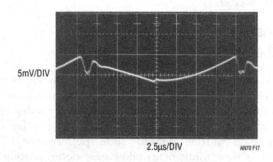

Figure 8.17 • Time Expansion of Previous Figure. No High Frequency Content Is Visible

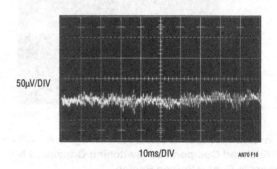

Figure 8.18 • Low Frequency Noise in a 10Hz to 10kHz Bandpass

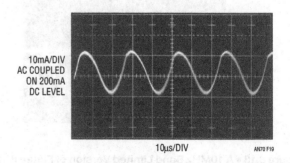

Figure 8.19 • Figure 8.5's Small Sinusoidal Input Current Variations Contain No High Frequency Content and Are Easily Absorbed by Input Supply

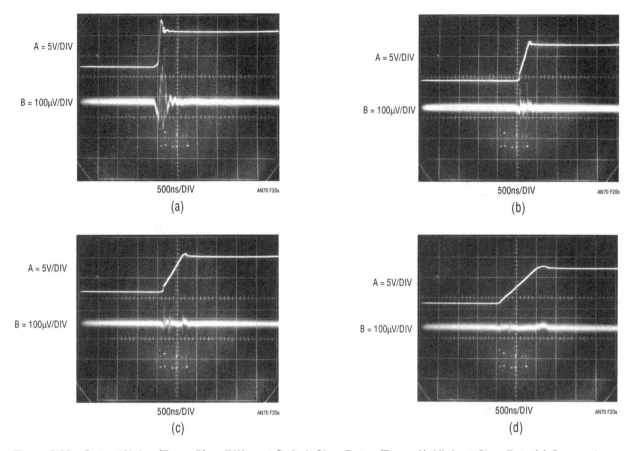

Figure 8.20 • Output Noise (Trace B) vs Different Switch Slew Rates (Trace A). Highest Slew Rate (a) Causes Largest Noise. Retarding Slew Rate (b and c) Decreases Noise Until Lowest Noise Performance Is Achieved (d)

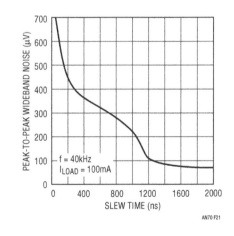

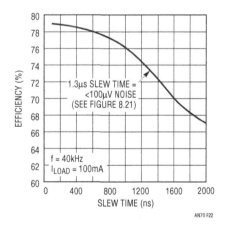

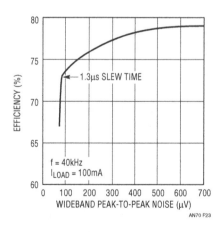

Figure 8.21 • Figure 8.5's Noise vs Slew Time at 40kHz Switching Frequency. Noise Reduction Beyond 1.3μs Is Minimal

Figure 8.22 • Figure 8.5's Efficiency Drops 6% as Slew Time Extends to 1.3μs. Operation Beyond This Point Gains Little Noise Performance (See Previous Curve) with 6% Efficiency Penalty

Figure 8.23 • Efficiency vs Noise for Figure 8.5. Data Shows Significant Efficiency Fall-Off for Noise Below 80μV

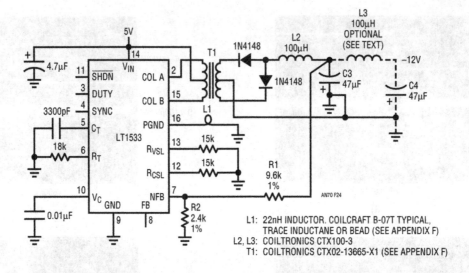

Figure 8.24 • A Negative Output Version of Figure 8.5. LT1533's Negative Feedback Input Requires Minimal Configuration Changes. Noise Performance Is Identical to Positive Output Version

the outputs are unregulated, proportioning to T1's drive voltage. An output inductor is usually not required, as in Figure 8.5's "forward" converter. At the very highest output currents, some inductance may be necessary to limit inrush current. If this is not done, the circuit may not start. Typically, linear regulators provide regulation.[12]

Figure 8.26's waveforms appear in Figure 8.27. Collector voltage (Traces A and C) and current (Traces B and D) are shown, along with the indicated output noise (Trace E). In this case linear regulators and an output filter are in use. In Figure 8.28 all probes except the coaxial output connection are removed. This eliminates probing induced parasitics,[13] allowing a higher fidelity signal presentation. Here, the switching residuals are barely detectable in the noise floor. Removing the optional output filter (Figure 8.29) allows linear regulator contributed noise and switching spikes to rise, but noise is still below 300µV$_{P-P}$.

As in Figure 8.5's case, spectrum analyzer measurements are instrument limited. Figure 8.30 shows the analyzer's noise floor in a 500MHz sweep when monitoring the unpowered Figure 8.26's breadboard. In Figure 8.31, the breadboard is powered, but analyzer output is noise limited and essentially indistinguishable from the unpowered case. Similarly, Figure 8.32's 1MHz wide "power-on" plot is identical to Figure 8.33's noise floor limited "power-off" sweep. Note that linear postregulation is in use and the 40kHz fundamental components are not detectable. Figure 8.5's circuit did not have linear postregulation and 40kHz fundamental residue appeared in Figure 8.15b.

Note 12: See Appendix E, "Selection criteria for linear regulators."
Note 13: See Appendix C, "Probing and connection techniques for low level, wideband signal integrity," for relevant discussion.

Battery-powered circuits

The basic configurations may be battery-powered for use in portable apparatus. Figure 8.34, similar to Figure 8.5, runs from 2.7V$_{MIN}$ (e.g., three NiCd batteries), producing 12V output. This design induces no noise-based error when powering a fast 16-bit A/D converter, something almost no DC/DC converter can do. Appendix K contributes compelling testimony to this somewhat boastful claim.

Figure 8.35 also operates from three NiCd cells, producing a 9V output. This design achieves 100µV output noise, qualifying it as the electronic equivalent of a 9V battery.

Performance augmentation

In some cases it may be desirable to augment LT1533 performance characteristics. Usually, this involves additional circuitry, and may necessitate trading off performance in one area to gain the desired benefit.

Low quiescent current regulator

The LT1533 has a quiescent current of about 6mA. Figure 8.36's circuit reduces this figure to 100µA by running an on-off control loop around the device. The control loop replaces the normal error amplifier, achieving regulation by switching the IC in and out of shutdown in accordance with loop demands.

Comparator C1 compares a scaled version of the output with its internal reference and biases the regulators shutdown pin. Loop hysteresis is obtained by utilizing the phase shift (e.g., time delay) of the output LC components. In a

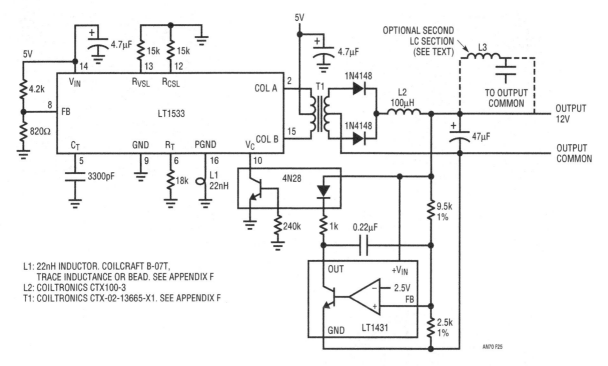

Figure 8.25 • An Optoisolated Output Variant of Figure 8.5. Loop Closure to V_C Pin Bypasses LT1533 Error Amplifier, Enhancing Loop Stability. Noise Performance is Maintained

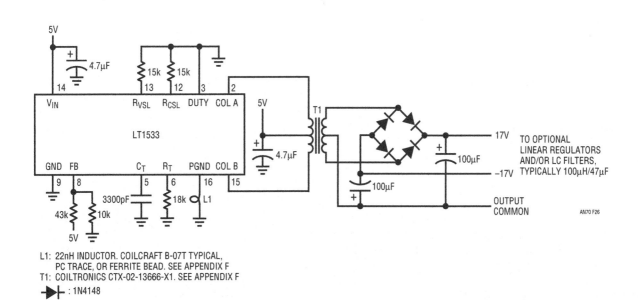

Figure 8.26 • A Bipolar, Floating Output Converter. Grounding "DUTY" Pin and Biasing FB Puts Regulator into 50% Duty Cycle Mode. Floating, Unregulated Outputs Proportion to T1's Center Tap Voltage. Linear Regulators Are Optional

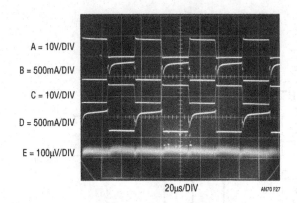

Figure 8.27 • **Waveforms for the Floating Output Converter at 100mA Loading. Linear Postregulator and Optional LC Filter Are Employed. Slew-Controlled Collector Voltage (Traces A and C) and Current (Traces B and D) Produce Output (Trace E) with Under 100μV Noise**

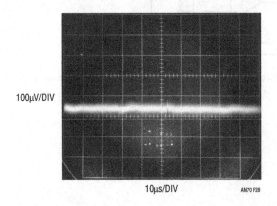

Figure 8.28 • **Removing All Probes Except Coaxial Output Connection Reveals Figure 8.27's True Noise Figure. Switching Residue Is Just Detectable in Amplifier Noise**

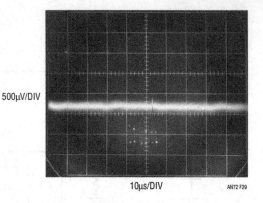

Figure 8.29 • **Removing Optional LC Filter Causes Linear Regulator-Contributed Noise and Switching Spikes to Rise. Peak-to-Peak Noise Is Still <300μV**

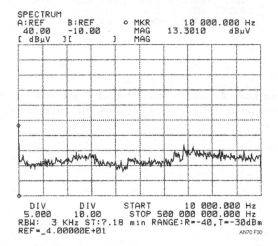

Figure 8.30 • **HP-4195A Analyzer's Noise Floor in a 500MHz Sweep When Connected to Unpowered Figure 8.26**

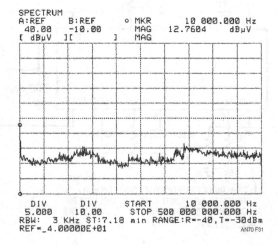

Figure 8.31 • **Figure 8.26's "Power-On" Output Noise Is Undetectable in Analyzer's Noise Floor Limited 500MHz Sweep**

normal continuously closed loop this phase shift must be minimized and compensated. In this case it promotes the desired hysteretic control characteristic. Local AC positive feedback at C1 ensures clean transitions. Figure 8.37 shows the loop at work. When circuit output drops below the regulation point, C1's output (Trace A) goes high. This enables the regulator and it responds with a burst of drive (Trace B) to the transformer. The output is restored and C1 goes low until the next cycle. During C1's low time the regulator is shut down, resulting in the extremely low quiescent current noted. The loop's on-off control characteristic causes low frequency output noise related to LC tank ring. Trace C shows 600μV peaks, although no wideband components are observable.

High voltage input regulator

The LT1533's IC process limits collector breakdown to 30V. A complicating factor is that the transformer swings to 2× supply. Thus, 15V represents the maximum

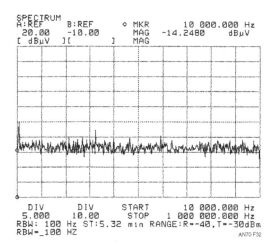

Figure 8.32 • Linear Postregulation Eliminates 40kHz Fundamental-Related Components in 1MHz Sweep

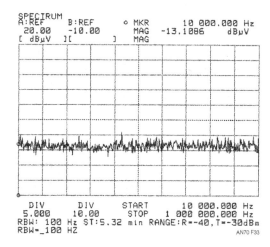

Figure 8.33 • Turning Circuit Power Off Verifies Figure 8.32's Plot Is Analyzer Noise Floor Limited. Sweep Results Are Identical to Figure 8.32's "Power-On" Data

allowable input supply. Many applications require higher voltage inputs and Figure 8.38 uses a cascoded[14] output stage to achieve such high voltage capability. This 24V-to-50V converter is reminiscent of previous circuits, except that Q1 and Q2 appear. These devices, interposed between the IC and the transformer, constitute a cascoded high voltage stage. They provide voltage gain while isolating the IC from their large collector voltage savings.

Note 14: The term "cascode," derived from "cascade to cathode," is applied to a configuration that places active devices in series. The benefit may be higher breakdown voltage, decreased input capacitance, bandwidth improvement, etc. Cascoding has been employed in op amps, power supplies, oscilloscopes and other areas to obtain performance enhancement.

Normally, high voltage cascodes are designed to simply supply voltage isolation. Cascoding the LT1533 presents special considerations because the transformer's instantaneous voltage and current information must be accurately transmitted, albeit at lower amplitude, to the LT1533. If this is not done, the regulator's slew control loops will not function, causing a dramatic output noise increase. The AC compensated resistor dividers associated with the Q1-Q2 base collector biasing serve this purpose. Q3 and associated components provide a stable DC termination for the dividers. Figure 8.39 shows waveforms for Q1's operation (Q2 is identical, although of opposing phase). Trace A is Q1's emitter, Trace B its base and Trace C the collector. T1's ring-off obscures the fact that waveform fidelity is maintained through the cascode, although inspection reveals

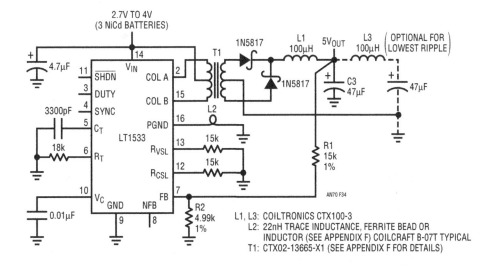

Figure 8.34 • Circuit Delivers 5V from Three NiCd Batteries, Has 100μV Wideband Output Noise. This Design Contributes No Noise-Based Error When Powering a 16-Bit A/D Converter (See Appendix K)

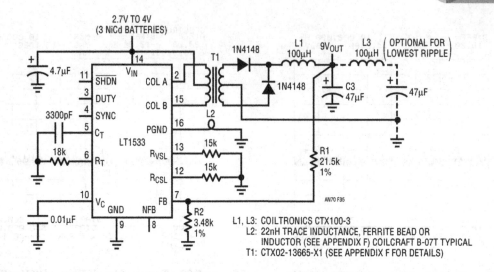

Figure 8.35 • Electronic Equivalent of 9V Battery Operates from Three NiCd Cells. Output Noise Is Below 100µV

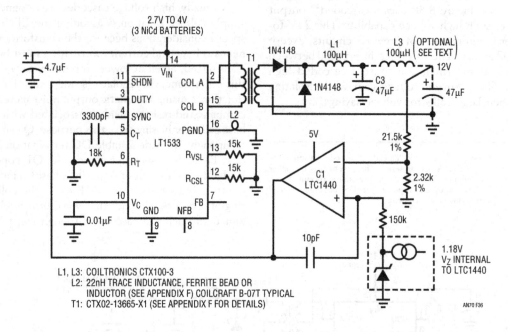

Figure 8.36 • Hysteretic "Burst Mode"™ Loop Lowers Quiescent Current to 100µA While Maintaining Low Output Noise

this to be the case. Additional testimony is given by circuit output noise (Trace D), which measures about 100µV peak.

24V-to-5V low noise regulator

Figure 8.40 extends Figure 8.38's cascoding technique in a step-down design.[15] Inputs from 20V to 50V are converted

to a 5V/2A capacity output. Q3 and Q4 protect the regulator's V_{IN} pin from the high input voltages. The cascode must accommodate 100V transformer swings. In this instance MOSFETs (Q1-Q2) are utilized, although the divider technique is necessarily retained. RC gate damper networks prevent transformer swings coupled via gate-channel capacitance from corrupting the cascode's waveform transfer fidelity. Figure 8.41 shows that resultant cascode response is faithful, even with 100V swings. Trace A is Q1's source, with Traces B and C its gate and drain, respectively. Under these conditions, at 2A output,

---

Note 15: This circuit was developed from a design by Jeff Witt of Linear Technology Corporation.

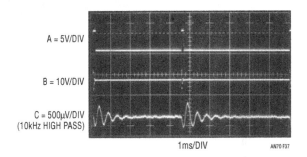

A = 5V/DIV

B = 10V/DIV

C = 500μV/DIV
(10kHz HIGH PASS)

1ms/DIV   AN70 F37

Figure 8.37 • Operating Waveforms for the Low Quiescent Current Converter. Comparator Output (Trace A) Restores Output Voltage by Turning LT1533 On (Trace B). Output Noise Shows LC Ringing (Trace C), Although High Frequency Content Is Negligible

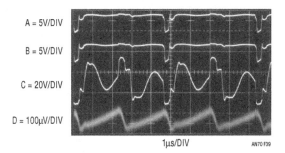

A = 5V/DIV

B = 5V/DIV

C = 20V/DIV

D = 100μV/DIV

1μs/DIV   AN70 F39

Figure 8.39 • Cascode Transmits Instantaneous Voltage and Slew Information, Permitting LT1533 to Maintain Low Noise Output. Trace A is Q1 Emitter, Trace B Is Its Base and Trace C the Collector. Transformer Ring-Off Obscures Cascode Action, but Study Reveals Faithful Transmission. Output (Trace D) Has 100μV Noise

noise is inside 400μV peak. Note that Q3 and Q4 protect the regulator from excessive input voltages.

10W, 5V to 12V low noise regulator

Figure 8.42 boosts the regulator's 1A output capability to over 5A. It does this with simple emitter followers (Q1-Q2). Theoretically, the followers preserve T1's voltage and current waveform information, permitting the LT1533's slew control circuitry to function. In practice, the transistors must be relatively low beta types. At 3A collector current their beta of 20 sources ≈150mA via the Q1-Q2

base paths, adequate for proper slew loop operation.[16] The follower loss limits efficiency to about 68%. Higher input voltages minimize follower-induced loss, permitting efficiencies in the low 70% range.

Figure 8.43 shows noise performance. Ripple measures 4mV (Trace A) using a single LC section, with high frequency content just discernible. Adding the optional sec-

---

Note 16: Operating the slew loops from follower base current was suggested by Bob Dobkin of Linear Technology Corporation.

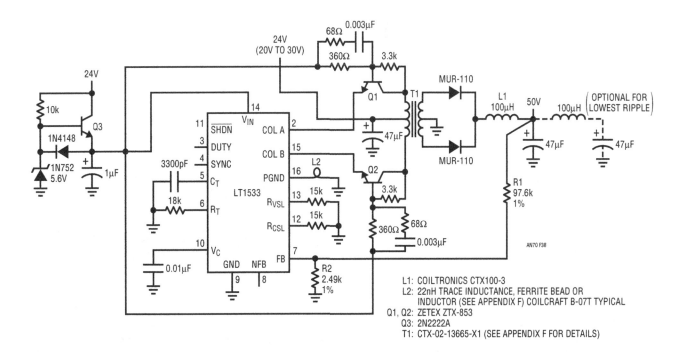

L1: COILTRONICS CTX100-3
L2: 22nH TRACE INDUCTANCE, FERRITE BEAD OR
    INDUCTOR (SEE APPENDIX F) COILCRAFT B-07T TYPICAL
Q1, Q2: ZETEX ZTX-853
Q3: 2N2222A
T1: CTX-02-13665-X1 (SEE APPENDIX F FOR DETAILS)

AN70 F38

Figure 8.38 • A 50V Output Low Noise Regulator. Cascoded Bipolar Transistors Accommodate 60V Transformer Swings, Permitting 24V (20V$_{IN}$ to 30V$_{IN}$) Powered Operation

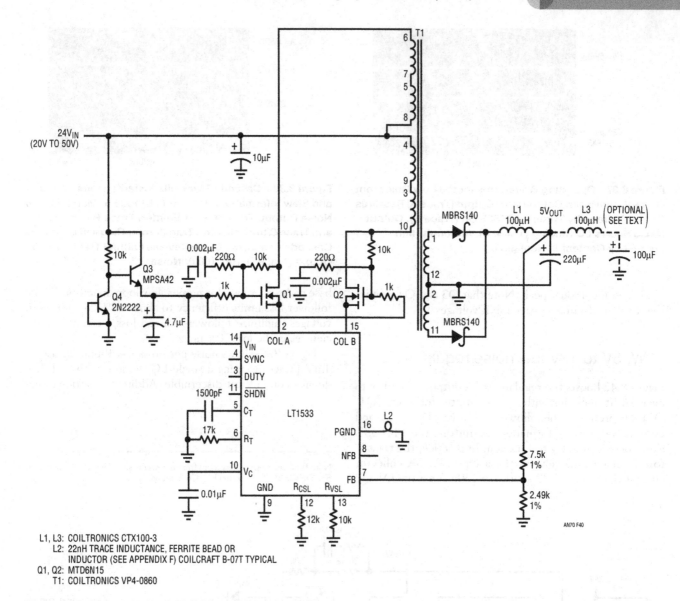

L1, L3: COILTRONICS CTX100-3
L2: 22nH TRACE INDUCTANCE, FERRITE BEAD OR
        INDUCTOR (SEE APPENDIX F) COILCRAFT B-07T TYPICAL
Q1, Q2: MTD6N15
T1: COILTRONICS VP4-0860

Figure 8.40 • A Low Noise 24V-(20V$_{IN}$ to 50V$_{IN}$)-to-5V Converter. Cascoded MOSFETs Withstand 100V Transformer Swings, Permitting LT1533 to Control 5V/2A Output

ond LC section drops ripple below 100µV (Trace B), and high frequency content is seen (note ×50 vertical scale factor change) to be inside 180µV.

7500V isolated low noise supply

A final form of performance augmentation is extremely high voltage isolation. This is often required in situations where circuitry must withstand high common mode voltage effects. Figure 8.44 is similar to Figure 8.25's isolated supply, except that it has 7500V (peak) breakdown capability. Transformer and optoisolator changes permit this. The remaining operating and performance characteristics are identical to Figure 8.25.

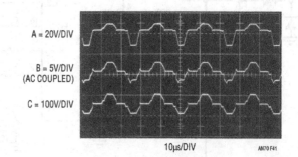

A = 20V/DIV

B = 5V/DIV
(AC COUPLED)

C = 100V/DIV

10µs/DIV   AN70 F41

Figure 8.41 • MOSFET-Based Cascode Permits Regulator to Control 100V Transformer Swings While Maintaining Low Noise 5V Output. Trace A Is Q1's Source, Trace B Q1's Gate and Trace C the Drain. Waveform Fidelity Through Cascode Permits Proper Slew Control Operation

183

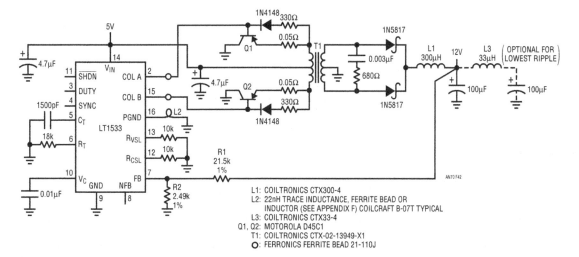

L1: COILTRONICS CTX300-4
L2: 22nH TRACE INDUCTANCE, FERRITE BEAD OR
      INDUCTOR (SEE APPENDIX F) COILCRAFT B-07T TYPICAL
L3: COILTRONICS CTX33-4
Q1, Q2: MOTOROLA D45C1
T1: COILTRONICS CTX-02-13949-X1
O: FERRONICS FERRITE BEAD 21-110J

Figure 8.42 • A 10W Low Noise 5V-to-12V Converter. Q1-Q2 Provide 5A Output Capacity While Preserving LT1533's Voltage/ Current Slew Control. Efficiency Is 68%. Higher Input Voltages Minimize Follower Loss, Boosting Efficiency Above 71%

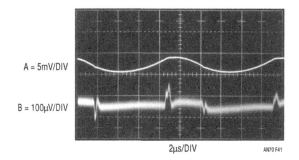

A = 5mV/DIV

B = 100µV/DIV

2µs/DIV   AN70 F41

Figure 8.43 • Waveforms for Figure 8.42 at 10W Output. Trace A Shows Fundamental Ripple with Higher Frequency Residue Just Discernible. Optional LC Section Produces Trace B's 180µV$_{P-P}$ Wideband Noise Performance

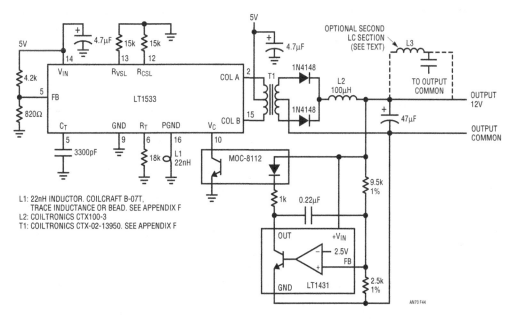

L1: 22nH INDUCTOR. COILCRAFT B-07T,
      TRACE INDUCTANCE OR BEAD. SEE APPENDIX F
L2: COILTRONICS CTX100-3
T1: COILTRONICS CTX-02-13950. SEE APPENDIX F

Figure 8.44 • A 7500V Isolation Version of Figure 8.25. Transformer and Optoisolator Are Changed to Achieve Isolation and Noise Immunity. Circuit Operation Is as Before

Note: This Application Note was derived from a manuscript originally prepared for publication in EDN magazine.

References

1. Shakespeare, William, "Much Ado About Nothing," II, i, 319, 1598-1600.
2. Williams, Jim, "Design DC/DC Converters to Catch Noise at the Source," Electronic Design, October 15, 1981, page 229.
3. Williams, Jim, "Conversion Techniques Adapt Voltages to Your Needs," EDN, November 10, 1982, page 155.
4. Tektronix, Inc. "Type 535 Operating and Service Manual," CRT Circuit, 1954.
5. Tektronix, Inc. "Type 454 Operating and Service Manual," CRT Circuit, 1967.
6. Tektronix, Inc. "7904 Oscilloscope Operating and Service Manual," Converter-Rectifiers, 1972.
7. Hewlett-Packard Co. "1725A Oscilloscope Operating and Service Manual," High Voltage Power Supply, 1980.
8. Arthur, Ken, "Power Supply Circuits," High Voltage Power Supplies, Tektronix Concept Series, 1967.
9. Williams, Jim and Huffman, Brian, "Some Thoughts on DC/DC Converters," Low Noise 5V to ± 15V Converter and Ultralow Noise 5V to ± 15V Converter, pages 1 to 5, Linear Technology Corporation Application Note 29, 1988.
10. Williams, Jim and Huffman, Brian, "Precise Converter Designs Enhance System Performance," EDN, October 13, 1988, pages 175 to 185.
11. Tektronix, Inc. "Type 1A7A Differential Amplifier Instruction Manual," Check Overall Noise Level Tangentially, pages 5-36 and 5-37, 1968.
12. Williams, Jim, "High Speed Amplifier Techniques," Linear Technology Corporation Application Note 47, 1991.
13. Witt, Jeff, "The LT1533 Heralds a New Class of Low Noise Switching Regulators," Linear Technology, Vol. VII, No. 3, August 1997, Linear Technology Corporation.
14. Morrison, Ralph, "Noise and Other Interfering Signals," John Wiley and Sons, 1992.
15. Morrison, Ralph, "Grounding and Shielding Techniques in Instrumentation," Wiley-Interscience, 1986.
16. Sheehan, Dan, "Determine Noise of DC/DC Converters," Electronic Design, September 27, 1973.
17. Hewlett-Packard Co. "HP-11941A Close Field Probe Operation Note," 1987.
18. Terrien, Mark, "The HP-11940A Close Field Probe: Characteristics and Application to EMI Troubleshooting," RF and Microwave Symposium, available from Hewlett-Packard Co.
19. Pressman, A.I., "Switching and Linear Power Supply, Power Converter Design," Hayden Book Co., Hasbrouck Heights, New Jersey, 1977.
20. Chryssis, G., "High Frequency Switching Power Supplies, Theory and Design," McGraw Hill, New York, 1984.

Appendix A
A history of low noise DC/DC conversion

Why are batteries low noise power sources? Why do 60Hz AC power line derived linear regulators have low output noise? As with most innocent questions, thoughtful answers provide surprising insights. These sources have low output noise because they have low harmonic energy content. A 60Hz fundamental driven supply produces some harmonic activity, but power becomes very small well inside 1kHz. A battery is even better.

These conclusions set a direction towards designing low noise DC/DC converters. If the goal is low noise, the key is reduction of harmonic energy, in particular, wideband harmonics. This simple guideline is central to LT1533 operation, although refinements are necessary for a generally applicable IC.

History

The notion of minimizing harmonics in DC/DC conversion to get low output noise is not new. Oscilloscopes have used this technique to generate high voltage CRT accelerating potentials without degrading instrument operation.[1] Designing a 10,000V output DC/DC converter that does not disrupt a 500MHz, high sensitivity vertical amplifier is challenging.

Figure A1 shows the CRT DC/DC converter from a Tektronix 454 oscilloscope. Q1430, configured as a modified Hartley power oscillator, drives T1430. T1430's output is multiplied by the diode-capacitor tripler, producing 12,000V. Feedback to Q1414 is summed against a 75V derived reference, closing a regulation loop around the power oscillator.

The sine wave transformer drive (see waveforms in the figure) has low harmonic content, resulting in the desired low conducted and radiated noise. This approach is not very efficient—Q1430 operates in its linear region—but the power loss is acceptable in a 125W instrument.

Tektronix 7000 series oscilloscopes used a resonant, offline converter to power the entire instrument. As before, CRT high voltage was generated separately (see Footnote 1). Figure A2, a partial schematic of a Tektronix 7904 power converter, shows a series resonant network, L1237-C1237 in the Q1234-Q1241 drive path. This results in sine wave drive to output transformer T1310, despite Q1234-Q1241's rectagular waveshape. Feedback (not shown) closes a loop around this stage, stabilizing its operating point. The resonant, sine wave transformer drive provides the desired low noise characteristics with good efficiency.

Note 1: Ancillary benefits include eliminating a complex and expensive high voltage winding in the main power transformer, avoidance of long, high voltage wire runs and space and weight savings.

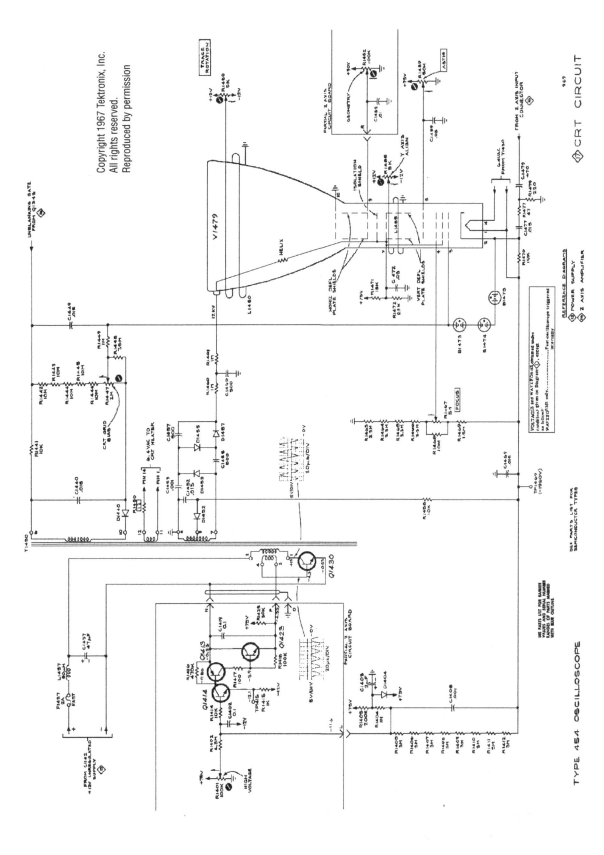

TYPE 454 OSCILLOSCOPE

Figure A1 • Tektronix 454 CRT Circuit Uses Sine Wave Drive for Low Noise DC/DC Conversion. Efficiency Is Poor, Because Q1430 Remains in Linear Region

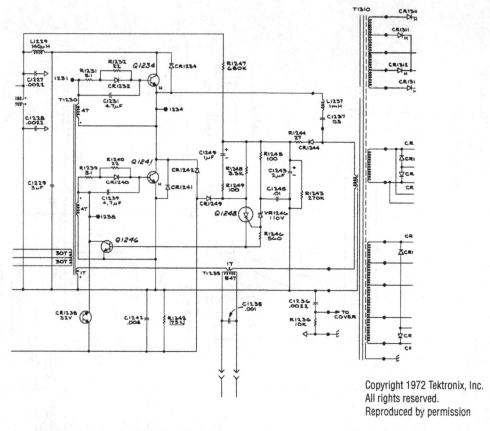

Figure A2 • Tektronix 7904 Main Inverter Obtains Low Noise by Converting Q1234-Q1241 Rectangular Drive to Sine Wave via L1237-C1237 Resonating Network. Output Transformer Produces Low Noise Power with Good Efficiency. Approach Is Application Specific and Inflexible

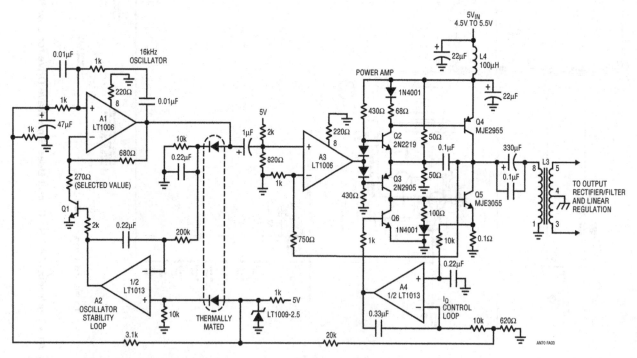

Figure A3 • Sine Wave-Based DC/DC Converter Appeared in LTC Application Note 29. Output Noise Is Low, but Circuit Is Complex and Inefficient

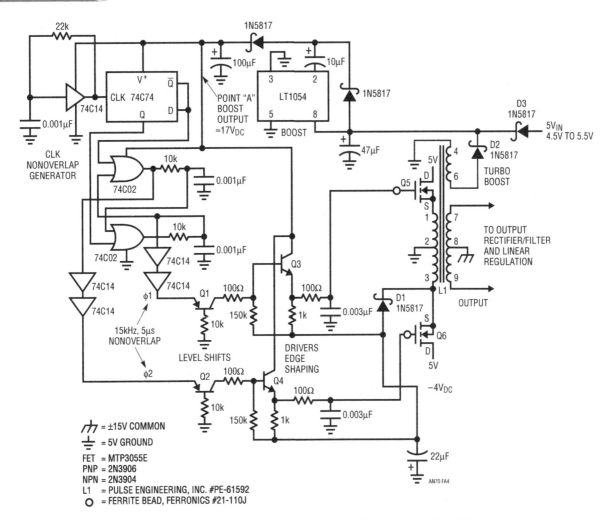

Figure A4 • LTC Application Note 29 Circuit Slopes Edge Drive for Low Noise and Better Efficiency. Gate Drive Circuitry Is Complex and Poorly Controlled, Making Circuit Inflexible

A less specific example appears in LTC Application Note 29. Figure A3, a partial schematic of Application Note 29's Figure 8.4, shows a sine wave oscillator (A1 based) driving a power amplifier (A3 and Q2 to Q6). L3, the output transformer, provides voltage boosted secondary drive to linear regulators (not shown). This brute force approach provides a converter with extraordinarily low noise, but is complex and inefficient. Q4 and Q5, operating in their linear regions, dissipate considerable power, and efficiency is 30%.

Figure A4's approach, also from AN29's Figure 8.1, achieves better efficiency. The partial schematic shows source followers driven from 100Ω-0.003µF edge slow-down networks. This slows down the transistor's transitions, resulting in harmonic reduction and low noise. Unfortunately, the

drive scheme is complex and somewhat inflexible, requiring bootstrapped voltages to fully switch the transistors on and off. Additionally, a transformer change would require drive rework to maintain efficiency and low noise characteristics. Finally, the dynamic voltage and current control in the transistors is passively determined and not very well controlled.

The LT1533 uses closed-loop control[2] around its output stages to tightly control voltage and current slewing. This allows a variety of circuits and magnetics to be easily accommodated, resulting in a true general purpose solution. Text Figure 8.3 and the associated discussion provide more LT1533 operating details.

_____

Note 2: Patent pending.

Appendix B
Specifying and measuring something called noise

Undesired output components in switching regulators are commonly referred to as "noise." The rapid, switched mode power delivery that permits high efficiency conversion also creates wideband harmonic energy. This undesirable energy appears as radiated and conducted components, or "noise." Actually switching regulator output "noise" isn't really noise at all, but coherent, high frequency residue directly related to the regulator's switching. Unfortunately, it is almost universal practice to refer to these parasitics as "noise," and this publication maintains this common, albeit inaccurate, terminology.[1]

Measuring noise

There are an almost uncountable number of ways to specify noise in a switching regulator's output. It is common industrial practice to specify peak-to-peak noise in a 20MHz bandpass.[2] Realistically, electronic systems are readily upset by spectral energy beyond 20MHz, and this specification restriction benefits no one.[3] Considering all this, it seems appropriate to specify peak-to-peak noise in a verified 100MHz bandwidth. Reliable low level measurements in this bandpass require careful instrumentation choice and connection practices.

Our study begins by selecting test instrumentation and verifying its bandwidth and noise. This necessitates the arrangement shown in Figure B1. Figure B2 diagrams signal flow. The pulse generator supplies a subnanosecond rise time step to the attenuator, which produces a <1mV version of the step. The amplifier takes 40dB of gain (A = 100) and the oscilloscope displays the result. The "front-to-back" cascaded bandwidth of this system should be about 100MHz (t_{rise} = 3.5ns) and Figure B3 reveals this to be so. Figure B3's trace shows 3.5ns rise time and about 100µV of noise. The noise is limited by the amplifier's 50W noise floor.[4]

Figure B4's presentation of text Figure 8.5's output noise shows barely visible switching artifacts (at vertical graticule lines 4, 6 and 8) in the 100MHz bandpass. Fundamental ripple is seen more clearly, although similarly noise floor dominated. Restricting

measurement bandwidth to 10MHz (Figure B5) reduces noise floor amplitude, although switching noise and ripple amplitudes are preserved. This indicates that there is no signal power beyond 10MHz. Further measurements as bandwidth is successively reduced can determine the highest frequency content present.

The importance of measurement bandwidth is further illustrated by Figures B6 to B8. Figure B6 measures a commercially available DC/DC converter in a 1MHz bandpass. The unit appears to meet its claimed 5mV$_{P-P}$ noise specification. In Figure B7, bandwidth is increased to 10MHz. Spike amplitude enlarges to 6mV$_{P-P}$, about 1mV outside the specification limit. Figure B8's 50MHz viewpoint brings an unpleasant surprise. Spikes measure 30mV$_{P-P}$—six times the specified limit![5]

Low frequency noise

Low frequency noise is rarely a concern, because it almost never affects system operation. Text Figure 8.5's low frequency noise is shown in Figure B9. It is possible to reduce low frequency noise by rolling off control loop bandwidth (e.g., via a 0.68µF feedback capacitor across R1 and V_C value of 2000pF in text Figure 8.5). Figure B10 shows about a five times improvement when this is done, even with greater measurement bandwidth. A possible disadvantage is loss of loop bandwidth and slower transient response.

Preamplifier and oscilloscope selection

The low level measurements described require some form of preamplification for the oscilloscope. Current generation oscilloscopes rarely have greater than 2mV/DIV sensitivity, although older instruments offer more capability. Figure B11 lists representative preamplifiers and oscilloscope plug-ins suitable for noise measurement. These units feature wideband, low noise performance. It is particularly significant that the majority of these instruments are no longer produced. This is in keeping with current instrumentation trends, which emphasize digital signal acquisition as opposed to analog measurement capability.

The monitoring oscilloscope should have adequate bandwidth and exceptional trace clarity. In the latter regard high quality analog oscilloscopes are unmatched. The exceptionally small spot size of these instruments is well-suited to low level noise

---

Note 1: Less genteelly, "If you can't beat 'em, join 'em."
Note 2: One DC/DC converter manufacturer specifies *RMS* noise in a 20MHz bandwidth. This is beyond deviousness and unworthy of comment.
Note 3: Except, of course, eager purveyors of power sources who specify them in this manner.
Note 4: Observed peak-to-peak noise is somewhat affected by the oscilloscope's "intensity" setting. Reference 11 describes a method for normalizing the measurement.

Note 5: *Caveat emptor.*

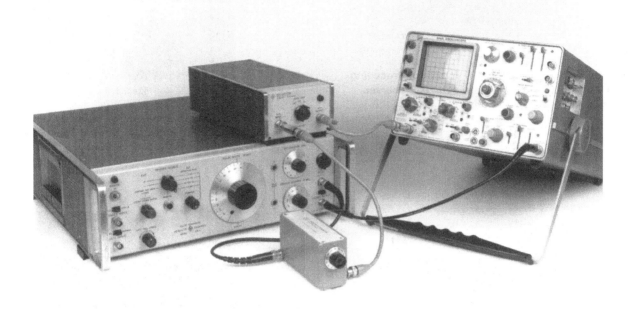

Figure B1 • 100MHz Bandwidth Verification Test Setup. Note Coaxial Connections for Wideband Signal Integrity

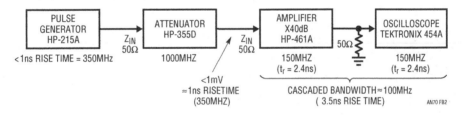

Figure B2 • Subnanosecond Pulse Generator and Wideband Attenuator Provide Fast Step to Verify Test Setup Bandwidth

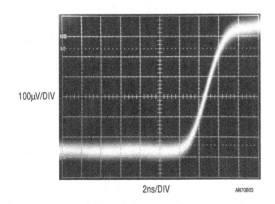

Figure B3 • Oscilloscope Display Verifies Test Setup's 100MHz (3.5ns Rise Time) Bandwidth. Baseline Noise Derives from Amplifier's 50Ω Input Noise Floor

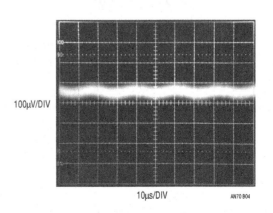

Figure B4 • Text Figure 8.5's Output Switching Noise Is Just Discernible in A 100MHz Bandpass

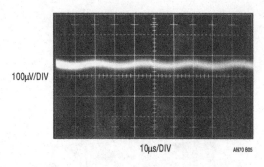

Figure B5 • 10MHz Band Limited Version of Preceding Photo. All Switching Noise Information Is Preserved, Indicating Adequate Bandwidth

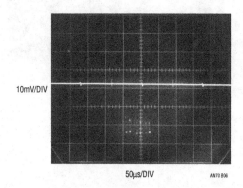

Figure B6 • Commercially Available Switching Regulator's Output Noise in a 1MHz Bandpass. Unit Appears to Meet Its 5mV$_{P-P}$ Noise Specification

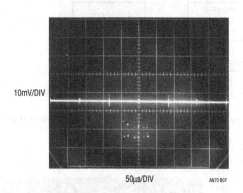

Figure B7 • Figure B6's Regulator Noise in a 10MHz Bandpass. 6m$_{VP-P}$ Noise Exceeds Regulator's Claimed 5mV Specification

measurement.[6] The digitizing uncertainties and raster scan limitations of DSOs impose display resolution penalties. Many DSO displays will not even register the small levels of switching-based noise.

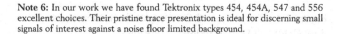

Note 6: In our work we have found Tektronix types 454, 454A, 547 and 556 excellent choices. Their pristine trace presentation is ideal for discerning small signals of interest against a noise floor limited background.

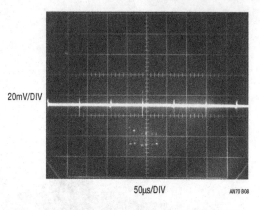

Figure B8 • Wideband Observation of Figure B7 Shows 30mV$_{P-P}$ Noise—Six Times the Regulator's Specification!

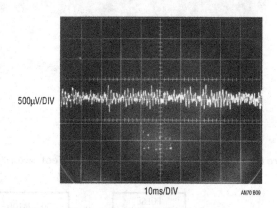

Figure B9 • 1Hz to 3kHz Noise Using Standard Frequency Compensation. Almost All Noise Power Is Below 1kHz

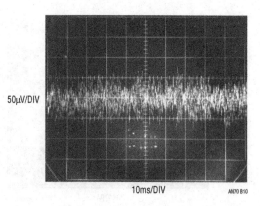

Figure B10 • Feedback Lead Network Decreases Low Frequency Noise, Even as Measurement Bandwidth Expands to 100kHz

INSTRUMENT TYPE	MANUFACTURER	MODEL NUMBER	BANDWIDTH	MAXIMUM SENSITIVITY/GAIN	AVAILABILITY	COMMENTS
Amplifier	Hewlett-Packard	461A	150MHz	Gain = 100	Secondary Market	50Ω Input, Stand-Alone
Differential Amplifier	Tektronix	1A5	50MHz	1mV/DIV	Secondary Market	Requires 500 Series Mainframe
Differential Amplifier	Tektronix	7A13	100MHz	1mV/DIV	Secondary Market	Requires 7000 Series Mainframe
Differential Amplifier	Tektronix	11A33	150MHz	1mV/DIV	Secondary Market	Requires 11000 Series Mainframe
Differential Amplifier	Tektronix	P6046	100MHz	1mV/DIV	Secondary Market	Stand-Alone
Differential Amplifier	Preamble	1855	100MHz	Gain = 10	Current Production	Stand-Alone, Settable Bandstops
Differential Amplifier	Tektronix	1A7/1A7A	1MHz	10μV/DIV	Secondary Market	Requires 500 Series Mainframe, Settable Bandstops
Differential Amplifier	Tektronix	7A22	1MHz	10μV/DIV	Secondary Market	Requires 7000 Series Mainframe, Settable Bandstops
Differential Amplifier	Tektronix	5A22	1MHz	10μV/DIV	Secondary Market	Requires 5000 Series Mainframe, Settable Bandstops
Differential Amplifier	Tektronix	ADA-400A	1MHz	10μV/DIV	Current Production	Stand-Alone with Optional Power Supply, Settable Bandstops
Differential Amplifier	Preamble	1822	10MHz	Gain = 1000	Current Production	Stand-Alone, Settable Bandstops
Differential Amplifier	Stanford Research Systems	SR-560	1MHz	Gain = 50000	Current Production	Stand-Alone, Settable Bandstops, Battery or Line Operation

Figure B11 ● Some Applicable High Sensitivity, Low Noise Amplifiers. Trade-Offs Include Bandwidth, Sensitivity and Availability

Appendix C
Probing and connection techniques for low level, wideband signal integrity

The most carefully prepared breadboard cannot fulfill its mission if signal connections introduce distortion. Connections to the circuit are crucial for accurate information extraction. The low level, wideband measurements demand care in routing signals to test instrumentation.

Ground loops

Figure C1 shows the effects of a ground loop between pieces of line-powered test equipment. Small current flow between test equipment's nominally grounded chassis creates 60Hz modulation in the measured circuit output.

This problem can be avoided by grounding all line powered test equipment at the same outlet strip or otherwise ensuring that all chassis are at the same ground potential. Similarly, any test arrangement that permits circuit current flow in chassis interconnects must be avoided.

Pickup

Figure C2 also shows 60Hz modulation of the noise measurement. In this case, a 4-inch voltmeter probe at the feedback input is the culprit. *Minimize the number of test connections to the circuit and keep leads short.*

Poor probing technique

Figure C3's photograph shows a short ground strap affixed to a scope probe. The probe connects to a point which provides a trigger signal for the

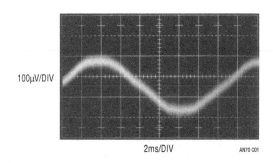

100μV/DIV

2ms/DIV   AN70 C01

Figure C1 ● Ground Loop Between Pieces of Test Equipment Induces 60Hz Display Modulation

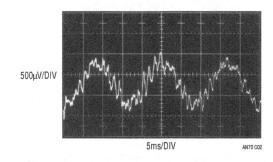

500μV/DIV

5ms/DIV   AN70 C02

Figure C2 ● 60Hz Pickup Due to Excessive Probe Length at Feedback Node

Figure C3 • Poor Probing Technique. Trigger Probe Ground Lead Can Cause Ground Loop-Induced Artifacts to Appear in Display

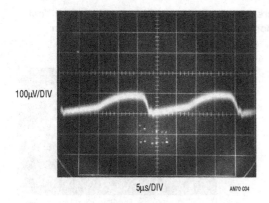

100µV/DIV

5µs/DIV   AN70 C04

Figure C4 • Apparent Excessive Ripple Results from Figure C3's Probe Misuse. Ground Loop on Board Introduces Serious Measurement Error

oscilloscope. Circuit output noise is monitored on the oscilloscope via the coaxial cable shown in the photo.

Figure C4 shows results. A ground loop on the board between the probe ground strap and the ground referred cable shield causes apparent excessive ripple in the display. *Minimize the number of test connections to the circuit and avoid ground loops.*

Violating coaxial signal transmission— felony case

In Figure C5, the coaxial cable used to transmit the circuit output noise to the amplifier-oscilloscope has been replaced with a probe. A short ground strap is employed as the probe's return. The error inducing

Figure C5 • Floating Trigger Probe Eliminates Ground Loop, but Output Probe Ground Lead (Photo Upper Right) Violates Coaxial Signal Transmission

trigger channel probe in the previous case has been eliminated; the 'scope is triggered by a noninvasive, isolated probe.[1] Figure C6 shows excessive display noise due to breakup of the coaxial signal environment. The probe's ground strap violates coaxial transmission and the signal is corrupted by RF. *Maintain coaxial connections in the noise signal monitoring path.*

Violating coaxial signal transmission—misdemeanor case

Figure C7's probe connection also violates coaxial signal flow, but to a less offensive extent. The probe's ground strap is eliminated, replaced by a tip grounding attachment. Figure C8 shows better results over the preceding case, although signal corruption is still evident. *Maintain coaxial connections in the noise signal monitoring path.*

Proper coaxial connection path

In Figure C9, a coaxial cable transmits the noise signal to the amplifier-oscilloscope combination. In theory, this affords the highest integrity cable signal transmission. Figure C10's trace shows this to be true. The former example's aberrations and

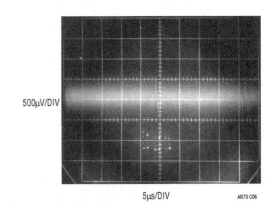

500μV/DIV

5μs/DIV   AN70 C06

Figure C6 • Signal Corruption Due to Figure C5's Noncoaxial Probe Connection

Note 1: To be discussed. Read on.

Figure C7 • Probe with Tip Grounding Attachment Approximates Coaxial Connection

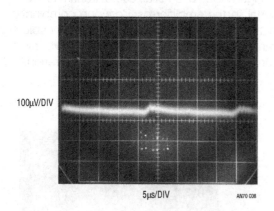

100µV/DIV

5µs/DIV

AN70 C08

Figure C8 • Probe with Tip Grounding Attachment Improves Results. Some Corruption Is Still Evident

excessive noise have disappeared. The switching residuals are now faintly outlined in the amplifier noise floor. *Maintain coaxial connections in the noise signal monitoring path.*

Direct connection path

A good way to verify there are no cable-based errors is to eliminate the cable. Figure C11's approach eliminates all cable between breadboard, amplifier and oscilloscope. Figure C12's presentation is indistinguishable from Figure C10, indicating no cable-introduced infidelity. *When results seem optimal, design an experiment to test them. When results seem poor, design an experiment to test them. When results are as expected, design an experiment to test them. When results are unexpected, design an experiment to test them.*

Test lead connections

In theory, attaching a voltmeter lead to the regulator's output should not introduce noise. Figure C13's increased noise reading contradicts the theory. The regulator's output impedance, albeit low, is not zero, especially as frequency scales up. The RF noise injected by the test lead works against the finite output impedance, producing the 200µV of noise

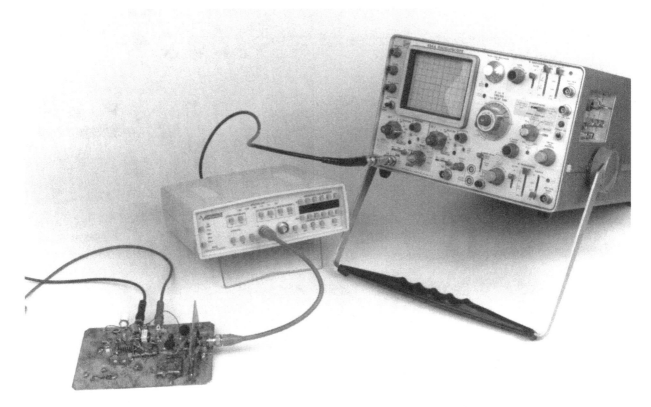

Figure C9 • Coaxial Connection Theoretically Affords Highest Fidelity Signal Transmission

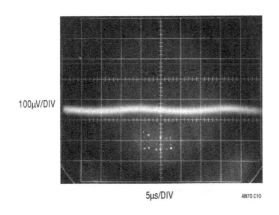

100µV/DIV

5µs/DIV   AN70 C10

Figure C10 • Life Agrees with Theory. Coaxial Signal Transmission Maintains Signal Integrity. Switching Residuals Are Faintly Outlined in Amplifier Noise

indicated in the figure. If a voltmeter lead must be connected to the output during testing, it should be done through a 10kΩ-10µF filter. Such a network eliminates Figure C13's problem while introducing minimal error in the monitoring DVM. *Minimize the number of test lead connections to the circuit while*

checking noise. Prevent test leads from injecting RF into the test circuit

Isolated trigger probe
The text associated with Figure C5 somewhat cryptically alluded to an "isolated trigger probe." Figure C14 reveals this to be simply an RF choke terminated against ringing. The choke picks up residual radiated field, generating an isolated trigger signal. This arrangement furnishes a 'scope trigger signal with essentially no measurement corruption. The probe's physical form appears in Figure C15. For good results the termination should be adjusted for minimum ringing while preserving the highest possible amplitude output. Light compensatory damping produces Figure C16's output, which will cause poor 'scope triggering. Proper adjustment results in a more favorable output (Figure C17), characterized by minimal ringing and well-defined edges.

Trigger probe amplifier
The field around the switching magnetics is small and may not be adequate to reliably trigger some oscilloscopes. In such cases, Figure C18's

196

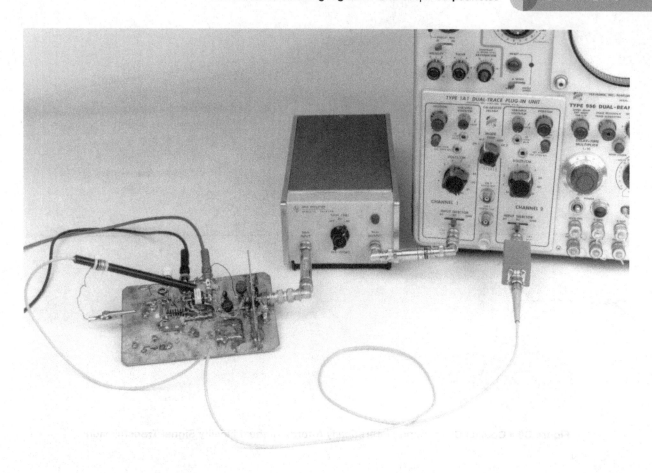

Figure C11 • Direct Connection to Equipment Eliminates Possible Cable-Termination Parasitics, Providing Best Possible Signal Transmission

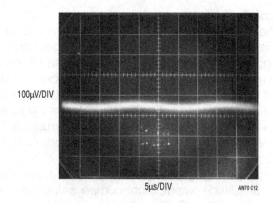

100μV/DIV

5μs/DIV   AN70 C12

Figure C12 • Direct Connection to Equipment Provides Identical Results to Cable-Termination Approach. Cable and Termination Are Therefore Acceptable

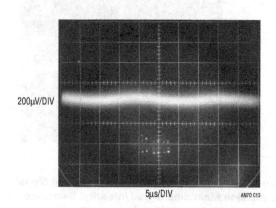

200μV/DIV

5μs/DIV   AN70 C13

Figure C13 • Voltmeter Lead Attached to Regulator Output Introduces RF Pickup, Multiplying Apparent Noise Floor

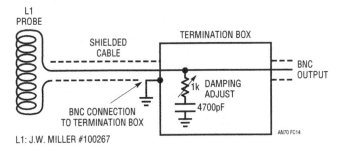

Figure C14 • Simple Trigger Probe Eliminates Board Level Ground Loops. Termination Box Components Damp L1's Ringing Response

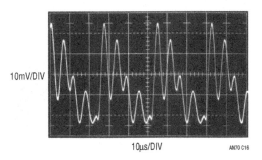

Figure C16 • Misadjusted Termination Causes Inadequate Damping. Unstable Oscilloscope Triggering May Result

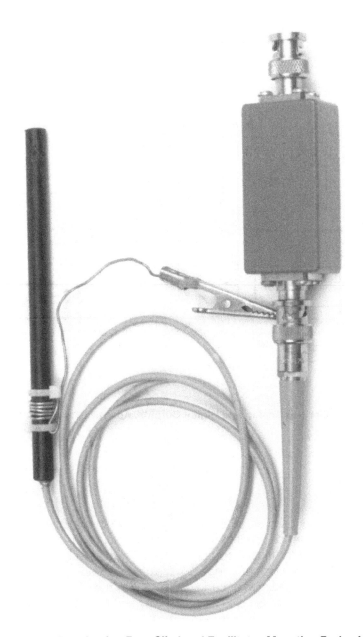

Figure C15 • The Trigger Probe and Termination Box. Clip Lead Facilitates Mounting Probe, Is Electrically Neutral

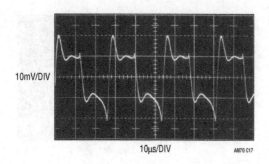

10mV/DIV

10µs/DIV AN70 C17

Figure C17 • Properly Adjusted Termination Minimizes Ringing with Small Amplitude Penalty

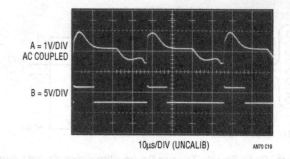

A = 1V/DIV
AC COUPLED

B = 5V/DIV

10µs/DIV (UNCALIB) AN70 C19

Figure C19 • Trigger Probe Amplifier Analog (Trace A) and Digital (Trace B) Outputs

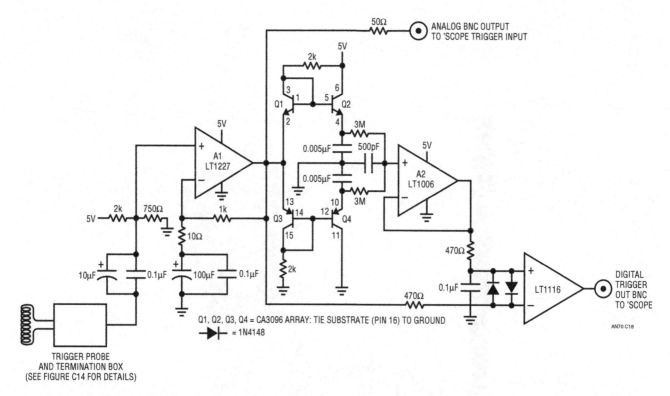

Figure C18 • Trigger Probe Amplifier Has Analog and Digital Outputs. Adaptive Threshold Maintains Digital Output over 50:1 Probe Signal Variations

trigger probe amplifier is useful. It uses an adaptive triggering scheme to compensate for variations in probe output amplitude. A stable 5V trigger output is maintained over a 50:1 probe output range. A1, operating at a gain of 100, provides wideband AC gain. The output of this stage biases a 2-way peak detector (Q1 through Q4). The maximum peak

is stored in Q2's emitter capacitor, while the minimum excursion is retained in Q4's emitter capacitor. The DC value of the midpoint of A1's output signal appears at the junction of the 500pF capacitor and the 3MΩ units. This point always sits midway between the signal's excursions, regardless of absolute amplitude. This signal-adaptive voltage is

199

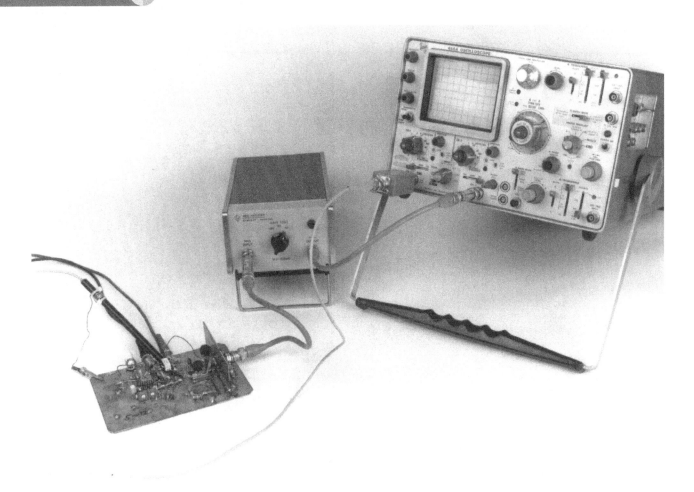

Figure C20 ● Typical Noise Test Setup Includes Trigger Probe, Amplifier, Oscilloscope and Coaxial Components

buffered by A2 to set the trigger voltage at the LT1116's positive input. The LT1116's negative input is biased directly from A1's output. The LT1116's output, the circuit's trigger output, is unaffected by >50:1 signal amplitude variations. An ×100 analog output is available at A1.

Figure C19 shows the circuit's digital output (Trace B) responding to the amplified probe signal at A1 (Trace A).

Figure C20 is a typical noise testing setup. It includes the breadboard, trigger probe, amplifier, oscilloscope and coaxial components.

Appendix D
Breadboarding and Layout Considerations
LT1533-based circuit's low harmonic content allows their noise performance to be less layout sensitive than other switching regulators. However, some degree of prudence is in order. As in all things,

cavalierness is a direct route to disappointment. Obtaining the absolute lowest noise figure requires care, but performance below 500µV is readily achieved. In general, lowest noise is obtained by preventing mixing of ground currents in the return path. Indiscriminate disposition of ground currents into a bus or ground plane will cause such mixing, raising observed output noise. The LT1533's restricted edge rates mitigate against corrupted ground path-induced problems, but best noise performance occurs in a "single-point" ground scheme. Single-point return schemes may be impractical in production PC boards. In such cases, provide the lowest possible impedance path to the power entry point from the inductor associated with the LT1533's power ground pin. (Pin 16). Locate the output component ground returns as close to the circuit load point as possible. Minimize return current mixing between input and output sections by restricting such mixing to the smallest possible common conductive area.

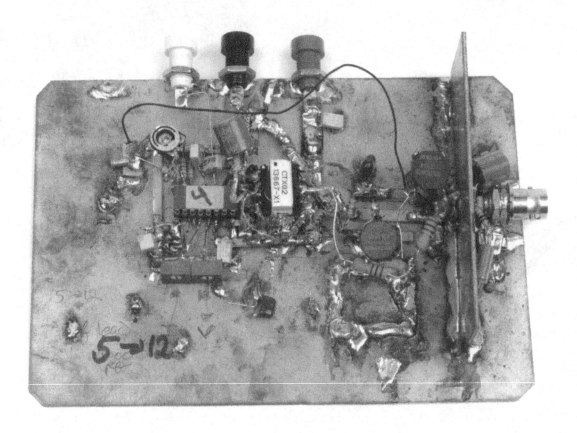

Figure D1 • Text Figure 8.5's 5V-to-12V Converter Breadboard. Construction Is Easy to Change, Facilitating Experiments. Note Single-Point Ground Returns. Ground Plane Carries No Current, Is Tied to Input Common at Board Entry Point (Middle Banana Jack)

5V to 12V Breadboard

Figure D1 shows text Figure 8.5's breadboard. In keeping with a breadboard's purpose, it is constructed to be fast and easy to modify. Single-point returns arrive separately from the output area (right side of photo) and Pin 16 of the LT1533 (center left photo). The ground plane carries no current. The dummy load resistors are *not* terminated to the plane, but returned to the transformer's center tap. The center tap and plane are separately tied into the ground system at the power input common jack.

5V to ±15V breadboard

Text Figure 8.24's breadboard appears in Figure D2. Layout considerations are similar to Figure D1, although the design's floating output mandates

changes. The output load (photo's right, above BNC connector) returns directly to the transformer secondary, which floats from input (and plane) ground potential. The main ground plane is tied to input common at the power entry port (left banana jack). The floating output potentials are referred to a separate, smaller planed area (photo lower right) which is tied to the transformer secondary center tap.

Demonstration board

Figure D3 enticingly portrays an LT1533 demonstration board. The board's practical layout is readily adaptable to production versions. This board is useful for observing LT1533 performance and as an example of a practical layout. Noise performance is similar to the text's breadboards.

Figure D2 • Text Figure 8.26's Breadboard with Linear Output Regulators. Construction Encourages Changes and Measurement. Layout Is Similar to Figure D1, Although Floating Output Necessitates Changes. Separate Planed Area (Photo Center Right) Maintains Low Impedance Between Output-Related Returns

Figure D3 • The Very Civilized LT1533 Demonstration Board in All Its Comely Splendor

Appendix E
Selection criteria for linear regulators

Some applications, particularly floating output circuits, may require linear postregulators. Selection criteria include regulator output accuracy, dropout, ripple rejection and line regulation. Often, short-circuit protection is not needed because drive circuit output impedance and current limiting prevent destructive overload. In such cases, if relatively poor output load regulation and accuracy are acceptable, simple Zener diode-emitter follower-based regulation may suffice. LM78L/79L type devices offer 5% output accuracy and improved line regulation, although dropout is about 2V—significantly higher than a simple Zener-emitter follower regulator. Ripple rejection in LM78L/79L types degrades as they approach dropout, which is the desirable operating region for best efficiency. High performance regulators such as the LT1575 (negative) and LT1521 (positive) offer dropout voltages below

0.5V, tight line regulation, 1% accuracy and fully specified ripple rejection close to dropout.

It is usually desirable to operate close to dropout to maintain good overall efficiency. Because of this, regulator ripple rejection should be tested in this intended operating region. Additionally, cost, size and performance trade-offs of various filter components and regulators should be evaluated to determine the best solution for a particular application.

Testing ripple rejection
Ripple rejection may be tested with Figure E1's arrangement. The generator should operate over the frequency range of interest and be capable of supplying the required output drive. In practice, the generator is set to supply the regulator input operating voltage at the expected LT1533 switching frequency. Comparison of different regulators and filter components under varying operating conditions is easily carried out.

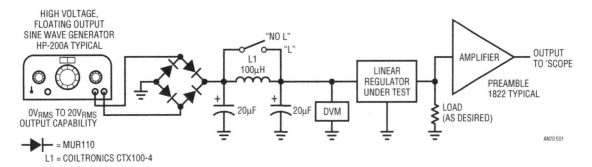

Figure E1 • Ripple Rejection Test Setup for Linear Regulators. LC Combinations and Regulators May Be Evaluated

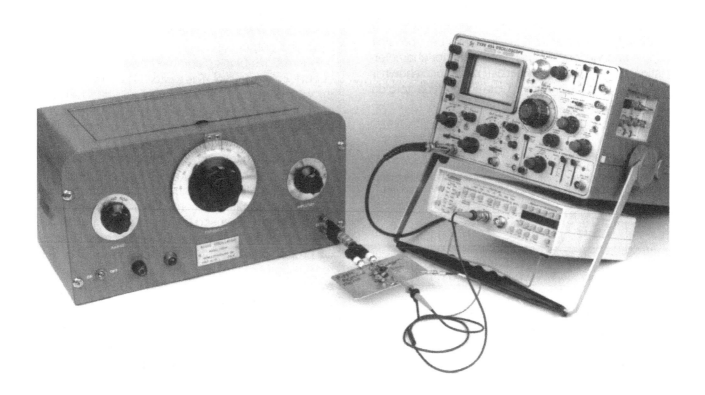

Figure E2 • Ripple Rejection Test Setup Includes Sine Wave Generator, Breadboard, Amplifier and Oscilloscope

Appendix F
Magnetics considerations

Transformers

The LT1533's symmetrical "push-pull" drive makes transformer behavior quite predictable. As such, transformers may usually be specified by indicating the operating frequency, power and desired input/output voltages. Figure F1 lists the transformers used in the text circuits along with some of their characteristics. These components, and variations on them, are available from Coiltronics, telephone #561-241-7876.

Inductors

Inductors in LT1533 circuits do not have special characteristics. Text Figure 8.5's circuit, a "forward" type converter,[1] *requires* an inductor ahead of its filter capacitor, although additional LC filtering is optional. Figure 8.26's "50%" mode circuit has no output inductor requirement unless heavily loaded (see text), although LC sections may be used for best possible ripple attenuation. In either case, inductor characteristics are not particularly critical. All circuits shown in the text use Coiltronics "Octa-Pak" type toroidal core-based inductors.

NOMINAL INPUT VOLTAGE	NOMINAL OUTPUT VOLTAGE AFTER LINEAR REGULATOR	OUTPUT POWER	COILTRONICS PART NUMBER	CONNECTION DIAGRAM
5V	12V	1.5W	CTX-02-13716-X1	A
5V	12V	3.0W	CTX-02-13665-X1	A
5V	±15V	1.5W	CTX-02-13713-X1	B
5V	±15V	3.0W	CTX-02-13664-X1	B
5V	12V	1.5W	CTX-02-13834-X3*	A
5V	12V	10W	CTX-02-13949-X1	A

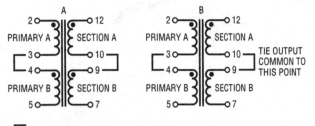

⌐ = TIED TOGETHER

* = HIGH TURNS RATIO VERSION OF CTX-02-13716-X1. ACCOMMODATES LOW SUPPLY VOLTAGES OR HIGH DROPOUT REGULATORS

AN70 FF01

Figure F1 • Transformer Types Used in Text Circuits. Variations for Specific Requirements Are Available from Coiltronics, 561-241-7876

Note 1: See References 16 and 17 for basic forward converter theory.

The 22nH inductor used in the LT1533's power ground return (Pin 16) is mandatory. It may take several forms, including trace inductance, a small coil of wire, a ferrite bead or the packaged inductor specified in the schematics. If coiled wire is employed, five turns of #28 is sufficient. An equivalent length of PC trace gives similar results. A ferrite bead (e.g., Ferronics #21-110J or equivalent) with one or two turns of wire also works well. An example of a packaged 22nH inductor is the Coilcraft B-07T which is specified in the test circuits.

Appendix G
Why voltage and current slew control?
Carl T. Nelson

The LT1533 gives dramatic reduction of high frequency noise by controlling both voltage and current slew rates in the switch. This technique also has the advantage of controlling noise in the other switching regulator components, namely, the catch diode and input and output capacitors.

Figure G1's block diagram shows the basic concepts for slew control. The switch Q1 is driven on and off with currents I_1 and I_2 switched via S1. These currents are large enough to drive the switch at very high slew rates. Actual slew rates are set by I_3 for voltage slew and I_4 for current slew.

During switch turn-on, the collector of Q1 is initially high and current is zero. Inductor current holds the switch high until switch current equals inductor current. The first limiting action occurs as current builds in Q1. Current is sensed by a fixed gain amplifier A3. The increasing current generates a current through C2 proportional to switch current slew rate. This current is compared to I_4, and the difference is amplified by A2, which shunts away all excess I_1 current to control switch current slew rate.

When switch current exceeds inductor current, Q1's collector would normally fall low at a speed limited only by diode and switch parasitic capacitance. To control voltage slew, the current through C1 is compared to I_3, and the difference is amplified by A1 to clamp the base current of Q1. This stops any further rise in switch current and forces switch voltage to fall at a controlled rate.

At switch turn-off, current and voltage must be controlled in the reverse order. Switch S1 is flipped to provide reverse base drive, and the polarity of I_3 and I_4 is reversed. Almost immediately, switch current falls slightly below inductor current. This would normally cause the switch voltage to slew up, limited only by diode and switch capacitance. Here, C1 senses voltage slew and A1 controls switch base

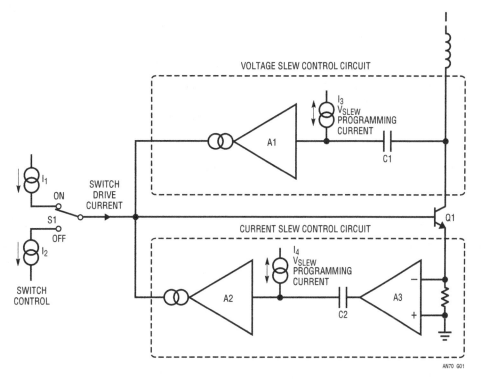

Figure G1 • Slew Control Conceptual Block Diagram

drive to limit switch rise time. Switch current remains essentially constant during voltage slew.

When switch voltage reaches the level where the catch diode turns on, switch current would normally drop rapidly, creating fast B field transients around the switch, diode and output capacitor lines. A3 and C2 come into play here, sensing the decreasing switch current and controlling the base drive via A2 to force a controlled decrease in switch, diode and capacitor current.

Figure G2 shows switch, diode and output capacitor waveforms with controlled switch drive in operation. Note that current and voltage slew limiting do not occur simultaneously. One must take over when

the first is complete. This requires very fast control circuitry to avoid crossover glitches that would create noise spikes.

Appendix H
Hints for lowest noise performance

The LT1533's controlled switching times allow extraordinarily low noise DC/DC conversion with surprisingly little design effort. Wideband output noise well below 500µV is easily achieved. In most situations this level of performance is entirely adequate. Applications requiring the lowest possible output noise will benefit from special attention to several areas.

Noise tweaking

The slew time versus efficiency trace-off discussed in the text should be weighted towards lowest noise to the extent tolerable. Typically, slew times beyond 1.3µs result in "expensive" noise reduction in terms of lost efficiency, but the benefit is available. The issue is how much power is expendable to obtain incremental decreases in output noise. Similarly, the layout techniques discussed in Appendix D should be reviewed. Rigid adherence to these guidelines will result in correspondingly lower noise

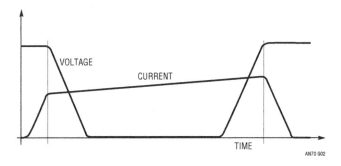

Figure G2 • Switch Voltage and Current During Turn-On and Turn-Off

performance. The text's breadboards were originally constructed to provide the lowest possible noise levels, and then systematically degraded to test layout sensitivity. This approach allows experimentation to determine the best layout without expending fanatical attention to details that provide essentially no benefit.

The slow edge times greatly minimize radiated EMI, but experimentation with the component's physical orientation can sometimes improve things. Look at the components (yes, literally!) and try and imagine just what their residual radiated field impinges on. In particular, the optional output inductor may pick up field radiated by other magnetics, resulting in increased output noise. Appropriate physical layout will eliminate this effect, and experimentation is useful. The EMI probe described in Appendix J is a useful tool in this pursuit and highly recommended. Appendix I contributes hints on magnetics-based noise and is similarly recommended.

Capacitors

The filter capacitors used should have low parasitic impedance. Sanyo OS-CON types are excellent in this regard and contributed to the performance levels quoted in the text. Tantalum types are nearly as good. The input supply bypass capacitor, which should be located directly at the transformer center tap, needs similarly good characteristics. Aluminum electrolytics are not suitable for any service in LT1533 circuits.

Damper network

Some circuits may benefit from a small (e.g., 330Ω-1000pF) damper network across the transformer secondary if the absolutely lowest noise is needed. Extremely small (20µV to 30µV) excursions can briefly appear during the switching interval when no energy is coming through the transformer. These events are so minuscule that they are barely measurable in the noise floor, but the damper will eliminate them.

Measurement technique

Strictly speaking, measurement technique is not a way to obtain lowest noise performance. Realistically, it is essential that measurement technique be trustworthy. Uncountable hours have been lost chasing "circuit problems" that in reality are manifestations of poor measurement technique. Please read Appendices B and C before pursuing solutions to circuit noise that isn't really there.[1]

Appendix I
Protection against magnetics noise is knowledge and good common sense
Jon Roman—Coiltronics, Inc.

Noise test data

For this test I chose four of the most common magnetics geometries that are currently in production use today. They are as follows: Pot Core, ER Core, E Core and the Toroid. The following test data was taken using the methods as described in the following paragraphs. The test circuit used to determine the amount of noise radiation is as shown in Figure I1. The push-pull configuration, power ratings and turns ratios were chosen to align with the Jim Williams' low noise designs presently under study. The distance chosen for the Sniffer Noise Probe[1] was set at 0.250 inches from the surface of the core structure. This distance was chosen as a result of preliminary testing to allow for a measurable reading on the smallest amount of flux lines coming from the quietest core structure. The measured worst-case full load noise is shown in Figure I2, for each of the four geometries chosen for this test. Note that the noise is shown in millivolts rather than gauss, the conversion for gauss is:

$$V_{PROBE}(mV_{P-P}) = 2.88mGauss/\mu s^2$$

After taking the noise reading for each of the UUT's under their full load conditions, the load resistor was removed to allow for observation of magnetizing flux noise. Then, using the same measurement techniques, the noise was measured a second time to determine the difference between the load noise

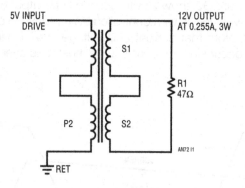

Figure I1 • Test Circuit

GEOMETRY	FULL LOAD MAGNETIZING
Pot Core	20mV
ER Core	63mV
E Core	488mV
Toroid	860mV

Figure I2 • Worst-Case Full Load Noise

GEOMETRY	FULL LOAD MAGNETIZING
Pot Core	16mV
ER Core	49mV
E Core	95mV
Toroid	91mV

Figure I3 • Worst-Case Magnetizing Load Noise

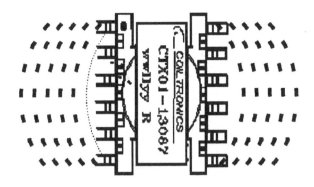

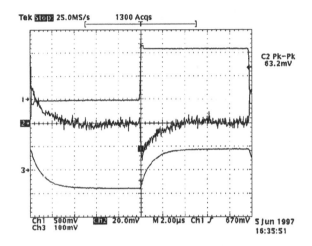

Figure I5

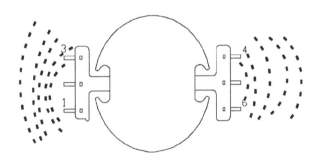

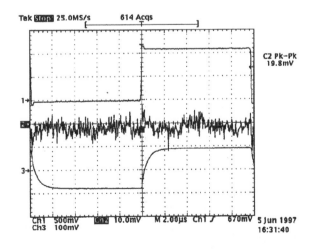

Figure I4

and the magnetizing noise. The measured worst-case magnetizing noise is shown in Figure I3.

Pot core

The pot core tested was as predicted the quietest geometry of the ones tested. Just as expected the "Hot Spot" for noise was located at the window of the gap where the leads exit the core. Reference the waveform shown in Figure I4. Note the top waveform shows the voltage input, the middle waveform shows the noise as recorded using an amplifier and the bottom waveform is the current through the UUT.

ER core

The ER core tested was the surprise of the group with a much lower noise reading than one would have originally thought possible. Reference the waveform shown in Figure I5. Note the top waveform shows the voltage input, the middle waveform shows the noise as recorded using an amplifier and the bottom waveform is the current through the UUT.

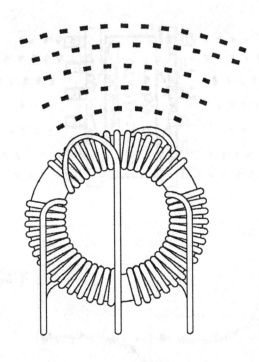

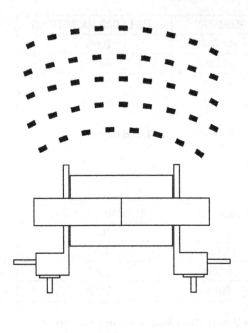

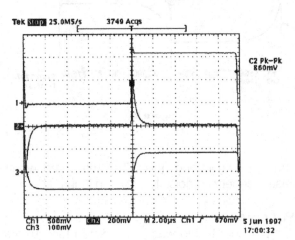

Figure I6

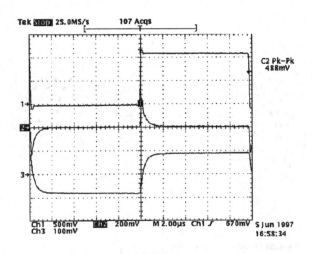

Figure I7

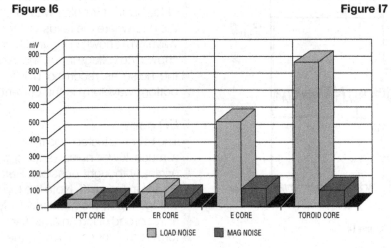

Figure I8

Toroid

The toroid core (Figure 16) tested showed a much higher electronic noise than was originally expected from a closed-path geometry. The worst-case noise came from the top of the core, with the winding placed as evenly on the core as possible. Note the top waveform shows the voltage input, the middle waveform shows the noise as recorded using an amplifier and the bottom waveform is the current through the UUT.

E core

The Ecore (Figure 17) showed the highest concentration of noise just above the winding on the top center of the device. The noise on the surrounding sides was measurable but far below the field that was found directly above and below the part. Note the top waveform shows the voltage input, middle waveform shows the noise as recorded using an amplifier and the bottom waveform is the current through the UUT.

Summary

Figure 18 is a graph showing the relative difference comparing the full load noise against the magnetizing load noise. It is recognized that closed core structures such as toroids in inductors produce less stray (leakage) flux than open structures like rod cores or bobbin cores. Some recent products offer "magnetic shields" or tubes of magnetic material around a bobbin-type core in an attempt to provide a "magnetic shunt" for the flux to follow. These structures offer very little reduction in noise because of the high reluctance in the air gap between the "shield" and the inner bobbin core. The reluctance in this gap is much higher than that of the magnetic material. The resultant leakage flux from the gap can defeat the purpose of the shield almost entirely! The best approach for reducing noise in inductors is to use true closed-field geometries, such as the toroid.

When designing for the lowest possible noise in transformer applications, it is important to observe the effects of the load current, as opposed to the magnetizing current. The preceding test demonstrates that the traditional low noise structures (toroid) can radiate relatively high amounts of leakage flux due to the coupling characteristics between windings. The reflected load currents in both the primary and the secondary do not affect the magnetizing flux, but create a magnetic leakage field around the wire, if the coupling is less than perfect. This is, by definition, leakage flux. The size and shape of the window area can have an effect on coupling between windings, as well as the shape of the flux field emanating from the transformer.

Winding technique also has an effect on coupling and noise. Multifilar winding, as opposed to layer winding, can offer better coupling characteristics, which in turn, lowers noise by lessening leakage flux.

Conclusion

Every *millivolt* counts.

Appendix J
Measuring EMI radiation

EMI (electromagnetic interference) is a form of switching regulator noise. It is a radiated, as opposed to conducted, phenomenon. LT1533-based circuits produce low amounts of EMI for the same reason they minimize conducted noise—controlled switching times. This appendix, guest written by Bruce Carsten, describes an excellent tool for relative EMI measurement and how to use it.[1] Carsten's methods not only show how to measure relative EMI, but how to identify and silence its source.

APPLICATION NOTE E101: EMI SNIFFER PROBE
Bruce Carsten Associates, Inc.
6410 NW Sisters Place, Corvallis, Oregon 97330
541-745-3935

The EMI Sniffer Probe[2] is used with an oscilloscope to locate and identify magnetic field sources of electromagnetic interference (EMI) in electronic equipment. The probe consists of a miniature 10 turn pickup coil located in the end of a small shielded tube, with a BNC connector provided for connection to a coaxial cable (Figure J1). The Sniffer Probe output voltage is essentially proportional to the rate of change of the ambient magnetic field, and thus to the rate of change of nearby currents.

The principal advantages of the Sniffer Probe over simple pickup loops are:
1. Spatial resolution of about a millimeter.
2. Relatively high sensitivity for a small coil.
3. A 50Ω source termination to minimize cable reflections with unterminated scope inputs.
4. Faraday shielding to minimize sensitivity to electric fields.

The EMI Sniffer Probe was developed to diagnose sources of EMI in switch mode power converters, but

---

Note 1: Calibrated measurements are discussed in References 14 and 15.
Note 2: The EMI Sniffer Probe is available from Bruce Carsten Associates at the address noted in the title of this appendix.

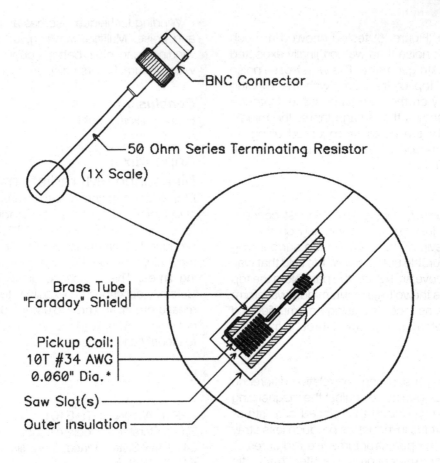

BNC Connector

50 Ohm Series Terminating Resistor

(1X Scale)

Brass Tube "Faraday" Shield

Pickup Coil: 10T #34 AWG 0.060" Dia. *

Saw Slot(s)

Outer Insulation

© 1997, Bruce Carsten Associates, Inc.

* Approx. 160 µ Wire, 1.5mm Coil Dia.

Figure J1 • Construction of the EMI Sniffer Probe for Locating and Identifying Magnetic Field Sources of EMI

it can also be used in high speed logic systems and other electronic equipment.

Sources of EMI

Rapidly changing voltages and currents in electrical and electronic equipment can easily result in radiated and conducted noise. Most EMI in switch mode power converters is thus generated during switching transients when power transistors are turned on or off.

Conventional scope probes can readily be used to see dynamic voltages, which are the principal sources of common mode conducted EMI. (High dV/dt can also feed through poorly designed filters as normal mode voltage spikes and may radiate fields from a circuit without a conductive enclosure.)

Dynamic currents produce rapidly changing magnetic fields which radiate far more easily than electric fields as they are more difficult to shield. These changing magnetic fields can also induce low impedance voltage transients in other circuits, resulting in unexpected normal and common mode conducted EMI.

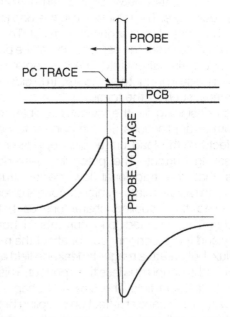

Figure J2 • Sniffer Probe Response to Current in a Physically "Isolated" Conductor

211

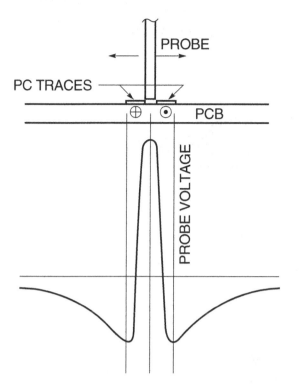

PROBE

PC TRACES

PROBE VOLTAGE

PCB

⊕ ⊙

Figure J3 • Sniffer Probe Response with Return Current in a Parallel Conductor

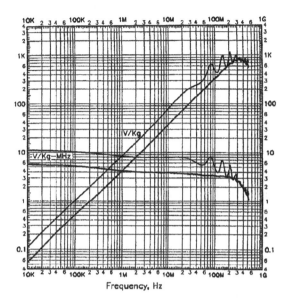

©1996, Bruce Carsten Associates, Inc.

Figure J4 • Typical EMI Sniffer Probe Frequency Response Measured with 1.3m (51″) of 50Ω Coax to Scope. Upper Traces: 1Meg Scope Input Impedance. Lower Traces: 50Ω Scope Input Impedance

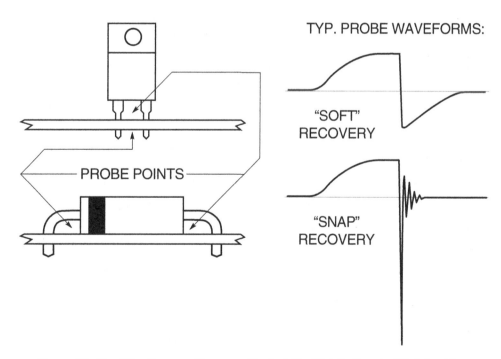

PROBE POINTS

TYP. PROBE WAVEFORMS:

"SOFT" RECOVERY

"SNAP" RECOVERY

Figure J5 • Rectifier Reverse Recovery Typical Fix: Tightly Coupled R-C Snubber

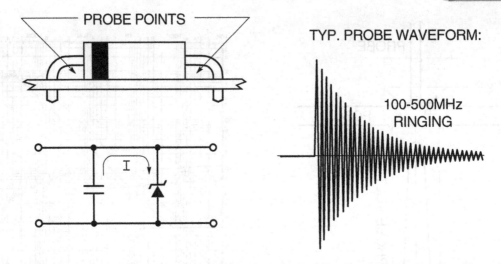

Figure J6 • Ringing Between Clamp Zener and Capacitor. Typical Fix: Small Ferrite Bead on Zener Lead(s)

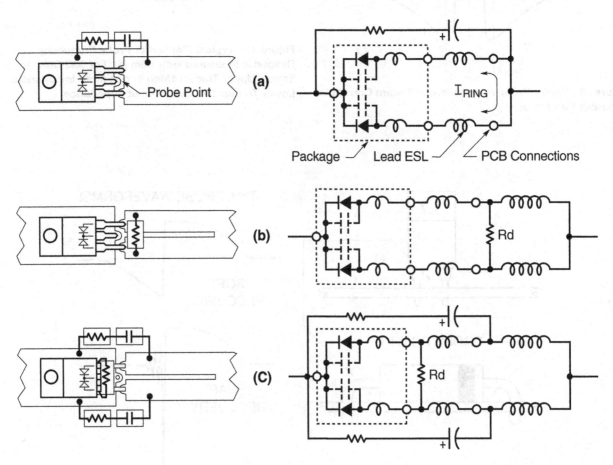

Figure J7 • Ringing in Paralleled Dual Rectifiers

These high dI/dt currents and resultant fields can not be directly sensed by voltage probes, but are readily detected and located with the Sniffer Probe.

While current probes can sense currents in discrete conductors and wires, they are of little use with printed circuit traces or in detecting dynamic magnetic fields.

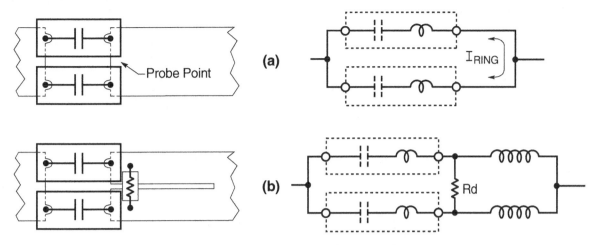

Figure J8 • Ringing in Paralleled "Snubber" Capacitors

Probe response characteristics

The Sniffer Probe is sensitive to magnetic fields only along the probe axis. This directionality is useful in locating the paths and sources of high dI/dt currents. The resolution is usually sufficient to locate which trace on a printed circuit board, or which lead on a component package, is conducting the EMI generating current.

For "isolated" single conductors or PC traces, the Probe response is greatest just to either side of the conductor where the magnetic flux is along with probe axis. (Probe response may be a little greater with the axis tilted towards the center of the conductor.) As shown in Figure J2, there is a sharp response null in the middle of the conductor, with a 180° phase shift to either side and a decreasing response with distance. The response will increase on the inside of a bend where the flux lines are crowded together, and is reduced on the outside of a bend where the flux lines spread apart.

When the return current is in an adjacent parallel conductor, the Probe response is greatest between the two conductors as shown in Figure J3. There will be a sharp null and phase shift over each conductor, with a lower peak response outside the conductor pair, again decreasing with distance.

The response to a trace with a return current on the opposite side of the board is similar to that of a single isolated trace, except that the probe response may be greater with the Probe axis tilted away from the trace. A "ground plane" below a trace will have a similar effect, as there will be a counter-flowing "image" current in the ground plane.

The Probe frequency response to a uniform magnetic field is shown in Figure J4. Due to large variations in field strength around a conductor, the Probe should be considered as a qualitative indicator only, with no attempt made to "calibrate" it. The response fall-off near 300MHz is due to the pickup coil inductance driving the coax cable impedance, and the mild resonant peaks (with a 1MΩ scope termination) at multiples of 80MHz are due to transmission line reflections.

Principles of probe use

The Sniffer Probe is used with at least a 2-channel scope. One channel is used to view the noise whose source is to be located (which may also provide the scope trigger) and the other channel is used for the Sniffer Probe. The probe response nulls make it inadvisable to use this scope channel for triggering.

A third scope trigger channel can be very useful, particularly if it is difficult to trigger on the noise. Transistor drive waveforms (or their predecessors in the upstream logic) are ideal for triggering; they are usually stable, and allow immediate precursors of the noise to be viewed.

Start with the Probe at some distance from the circuit with the Probe channel at maximum sensitivity. Move the probe around the circuit, looking for "something happening" in the circuit's magnetic fields at the same time as the noise problem. A precise "time domain" correlation between EMI noise transients and internal circuit fields is fundamental to the diagnostic approach.

As a candidate noise source is located, the Probe is moved closer while the scope sensitivity is decreased to keep the probe waveform on-screen. It should be possible to quickly bring the probe down to the PC board trace (or wiring) where the probe signal seems to be a maximum. This may not be near the point of EMI generation, but it should be near a PC

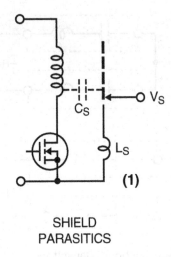

(1)

SHIELD
PARASITICS

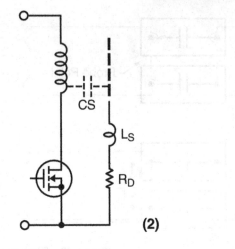

(2)

SHIELD RESONANCE DAMPING

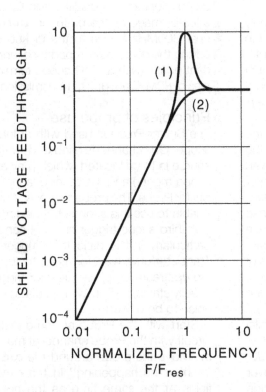

SHIELD RESONANCE
CAN BE DAMPED WITH
A RESISTOR "R_D" OR A
SMALL FERRITE BEAD:

$$R_D \cong \sqrt{\frac{L_S}{C_S}}$$

Figure J9 • Shield Effectiveness at High Frequencies Is Limited by Shield Capacity and Lead Inductance

trace or other conductor carrying the current from the EMI source. This can be verified by moving the probe back and forth in several directions; when the appropriate PC trace is crossed at roughly right angles, the probe output will go through a sharp null over the trace, with an evident phase reversal in probe voltage on each side of the trace (as noted above).

This EMI "hot" trace can be followed (like a bloodhound on the scent trail) to find all or much of the EMI generating current loop. If the trace is hidden on the back side (or inside) of the board, mark its path with a felt pen and locate the trace on disassembly, on another board or on the artwork. From the current path and the timing of the noise

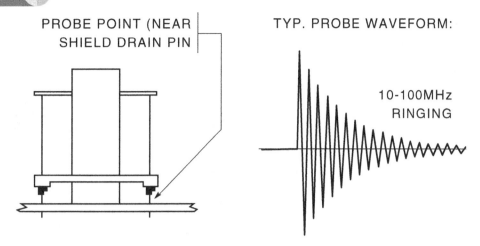

PROBE POINT (NEAR SHIELD DRAIN PIN

TYP. PROBE WAVEFORM:

10-100MHz RINGING

Figure J10 ● Transformer Shield Ringing Typical Fix: 10Ω to 100Ω Resistor (or Ferrite Bead in Drain Wire)

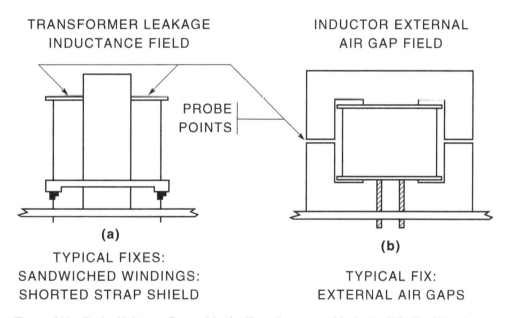

TRANSFORMER LEAKAGE INDUCTANCE FIELD

INDUCTOR EXTERNAL AIR GAP FIELD

PROBE POINTS

(a)

(b)

TYPICAL FIXES: SANDWICHED WINDINGS: SHORTED STRAP SHIELD

TYPICAL FIX: EXTERNAL AIR GAPS

Figure J11 ● Probe Voltages Resemble the Transformer and Inductor Winding Waveforms

transient, the source of the problem usually becomes almost self-evident.

Several not-uncommon problems (all of which have been diagnosed with various versions of the Sniffer Probe) are discussed here with suggested solutions or fixes.

Typical Di/Dt EMI problems

Rectifier reverse recovery
Reverse recovery of rectifiers is the most common source of dI/dt-related EMI in power converters; the

charge stored in P-N junction diodes during conduction causes a momentary reverse current flow when the voltage reverses. This reverse current may stop very quickly (<1ns) in diodes with a "snap" recovery (more likely in devices with a PIV rating of less than 200V), or the reverse current may decay more gradually with a "soft" recovery. Typical Sniffer Probe waveforms for each type of recovery are shown in Figure J5.

The sudden change in current creates a rapidly changing magnetic field, which will both radiate external fields and induce low impedance voltage spikes in

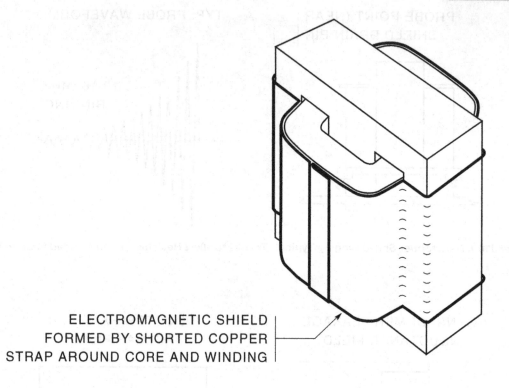

ELECTROMAGNETIC SHIELD
FORMED BY SHORTED COPPER
STRAP AROUND CORE AND WINDING

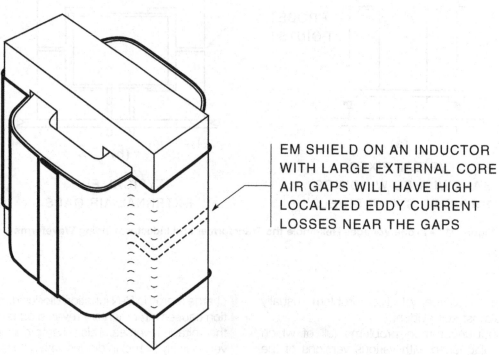

EM SHIELD ON AN INDUCTOR
WITH LARGE EXTERNAL CORE
AIR GAPS WILL HAVE HIGH
LOCALIZED EDDY CURRENT
LOSSES NEAR THE GAPS

Figure J12 • A "Sandwiched" PRI-SEC Transformer Winding Construction Reduces Electromagnetic Shield Eddy Current Losses

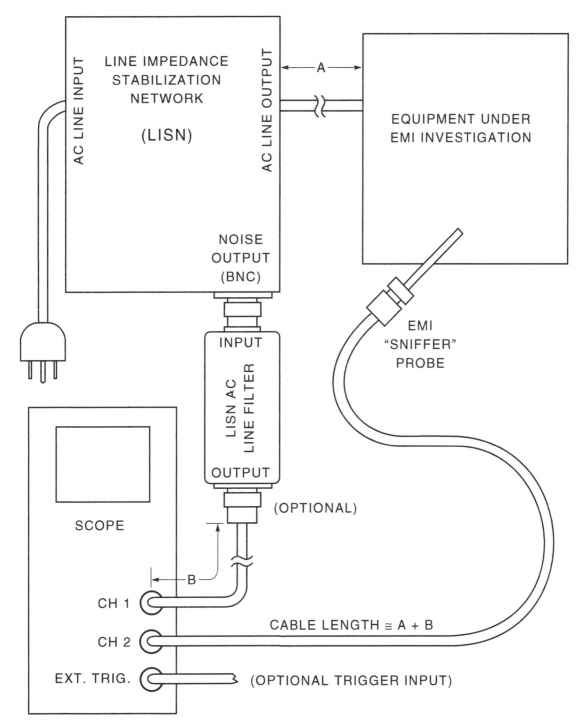

Figure J13 • Using the Probe with a "LISN"

other circuits. This reverse recovery may "shock" parasitic L-C circuits into ringing, which will result in oscillatory waveforms with varying degrees of damping when the diode recovers. A series R-C damper circuit in parallel with the diode is the usual solution.

Output rectifiers generally carry the highest currents and are thus the most prone to this problem,

but this is often recognized and they may be well-snubbed. It is not uncommon for unsnubbed catch or clamp diodes to be more of an EMI problem. (The fact that a diode in an R-C-D snubber may need its *own* R-C snubber is not always self-evident, for example).

The problem can usually be identified by placing the Sniffer Probe near a rectifier lead. The signal will

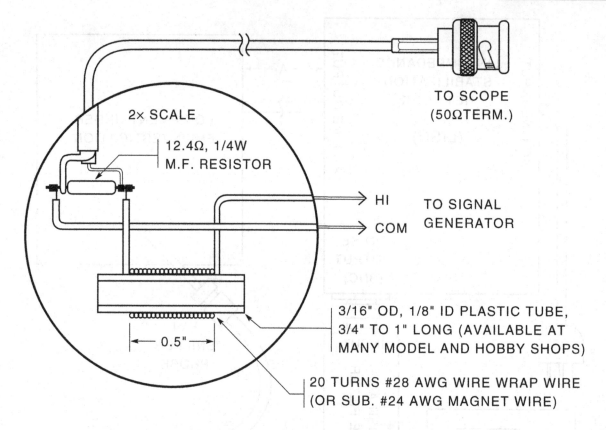

2× SCALE

12.4Ω, 1/4W
M.F. RESISTOR

TO SCOPE
(50ΩTERM.)

HI   TO SIGNAL
COM   GENERATOR

3/16" OD, 1/8" ID PLASTIC TUBE,
3/4" TO 1" LONG (AVAILABLE AT
MANY MODEL AND HOBBY SHOPS)

20 TURNS #28 AWG WIRE WRAP WIRE
(OR SUB. #24 AWG MAGNET WIRE)

← 0.5" →

The Sniffer Probe Tip is centered inside the test coil where the Probe voltage
is greatest. The approximate flux density in the middle of a coil can be calculated
from the formula:

$$B = H = 1.257\, NI/I \quad \text{(CGS Units)}$$

For the 1.27cm long, 20-turn test coil, the flux density is about 20 Gauss per amp.
At 1MHz, the Sniffer Probe voltage is 19mV$_{P-P}$ (±10%) per 100mA$_{P-P}$ for a 1MΩ
load impedance, and half that for a 50Ω load.

Figure J14 • EMI "Sniffer" Probe Test Coil

be strongest on the inside of a lead bend in an axial package, or between the anode and cathode leads in a TO-220, TO-247 or similar type of package, as shown in Figure J5.

Using "softer" recovery diodes is a possible solution and Schottky diodes are ideal in low voltage applications. However, it must be recognized that a P-N diode with soft recovery is also inherently lossy (while a "snap" recovery is not), as the diode simultaneously develops a reverse voltage while still conducting current: the fastest possible diode (lowest recovered charge) with a moderately soft recovery is usually the best choice. Sometimes a faster, slightly "snappy" diode with a tightly coupled R-C snubber works as well or better than a soft but excessively slow recovery diode.

If significant ringing occurs, a "quick-and-dirty" R-C snubber design approach works fairly well: increasingly large damper capacitors are placed across the diode until the ringing frequency is halved. We know that the total ringing capacity is now quadrupled or that the original ringing capacity is 1/3 of the added capacity. The damper resistance required is about equal to the capacitive reactance of the original ringing capacity at the original ringing frequency. The "frequency halving" capacity is then connected in series with the damping resistance and placed across the diode, as tightly coupled as possible.

1) Use a 2-channel scope, preferably one with an external trigger.

2) One scope channel is used for the Sniffer Probe, which is not to be used for triggering.

3) The second channel is used to view the noise transient whose source is to be located, which may also be used for triggering if practical.

4) More stable and reliable triggering is achieved with an "external trigger" (or a 3rd channel) on a transistor drive waveform (or preceding logic transition), allowing immediate precursors to the transient to be viewed. (Nearly all noise transients occur during, or just after, a power transistor turn-on or turn-off.

5) Start with the Probe at some distance from the circuit with maximum sensitivity and "sniff around" for something happening in *precise sync* with the noise transient. The Probe waveform will not be identical to the noise transient, but will usually have a strong resemblance.

6) Move the Probe closer to the suspected source while decreasing sensitivity. The conductor carrying the responsible current is located by the sharp response null on top of the conductor with inverted polarity on each side.

7) Trace out the noise current path as much as possible. Identify the current path on the schematic.

8) The source of the noise transient is usually evident from the current path and the timing information.

Figure J15 • EMI "Sniffer" Probe Procedure Outline

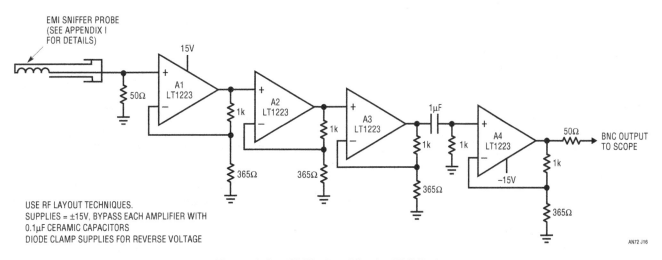

Figure J16 • 40MHz Amplifier for EMI Probe

Snubber capacitors must have a high pulse current capability and low dielectric loss. Temperature stable (disc or multilayer) ceramic, silvered mica and some plastic film-foil capacitors are suitable. Snubber resistors should be noninductive; metal film, carbon film and carbon composition resistors are good, but wirewound resistors must be avoided. The maximum snubber resistor dissipation can be estimated from the product of the damper capacity, switching frequency and the square of the peak snubber capacitor voltage.

Snubbers on passive switches (diodes) or active switches (transistors) should always be coupled as closely as physically possible, with minimal loop inductance. This minimizes the radiated field from the change in current path from the switch to the snubber. It also minimizes the turn-off voltage overshoot "required" to force the current to change path through the switch-snubber loop inductance.

Ringing in clamp Zeners

A capacitor-to-capacitor ringing problem can occur when a voltage clamping Zener or TransZorb® is placed across the output of a converter for overvoltage protection (OVP). Power Zeners have a large junction capacity, and this can ring in series with the lead ESL and the output capacitors, with some of the ringing voltage showing up on the output. This ringing current can be most easily detected near the Zener leads, particularly on the inside of a bend as shown in Figure J6.

R-C snubbers have not been found to work well in this case as the ringing loop inductance is often as low or lower than the obtainable parasitic inductance in the snubber. Increasing the external loop inductance to allow damping is not advisable as this would limit dynamic clamping capability. In this case, it was found that a small ferrite bead on one or both of the Zener leads dampened the HF oscillations with minimal adverse side effects (a high permeability ferrite bead quickly saturates as soon as the Zener begins to conduct significant current).

Paralleled rectifiers

A less evident problem can occur when dual rectifier diodes in a package are paralleled for increased current capability, even with a tightly coupled R-C snubber. The two diodes seldom recover at exactly the same time, which can cause a very high frequency oscillation (hundreds of MHz) to occur between the capacities of the two diodes in series with the anode lead inductances, as shown in Figure J7. This effect can really only be observed by placing the probe

between the two anode leads, as the ringing current exists almost nowhere else (the ringing is nearly "invisible" to a conventional voltage probe, like many other EMI effects that can be easily found with a magnetic field Sniffer Probe).

This "teeter-totter" oscillation has a voltage "null" about where the R-C snubber is connected, so it provides little or no damping (see Figure J7a). It is actually very difficult to insert a suitable damping resistance into this circuit.

The easiest way to dampen the oscillation is to "slit" the anode PC trace for an inch or so and place a damping resistor at the anode leads as shown in Figure J7b. This increases the inductance in series with the diode-diode loop external to the package and leads, while having minimal effect on the effective series inductance. Even better damping is obtained by placing the resistor across the anode leads at the entry point to the case, as shown in Figure J7c, but this violates the mindset of many production engineers.

It is also preferable to split the original R-C damper into two (2R) - (C/2) dampers, one on each side of the dual rectifier (also shown in Figure J7c). In practice, it is always preferable to use dual R-C dampers, one each side of the diode; loop inductance is cut about in half, and the external dI/dt field is reduced even further due to the oppositely "handed" currents in the two snubber networks.

Paralleled snubber or damper caps

A problem similar to that with the paralleled diodes occurs when two or more low loss capacitors are paralleled and driven with a sudden current change. There is a tendency for a current to ring between the two capacitors in series with their lead inductances (or ESL), as shown in Figure J8a. This type of oscillation can usually be detected by placing the Sniffer Probe between the leads of the paralleled capacitors. The ringing frequency is much lower than with the paralleled diodes (due to the larger capacity), and the effect *may* be benign if the capacitors are sufficiently close together.

If the resultant ringing *is* picked up externally, it can be damped in a similar way as with the parallel diodes as shown in Figure J8b. In either case, the dissipation in the damping resistor tends to be relatively small.

Ringing in transformer shield leads

The capacity of a transformer shield to other shields or windings (C_S in Figure J9) forms a series resonant circuit with its "drain wire" inductance (L_S) to the bypass point. This resonant circuit is readily excited by typical square wave voltages on windings, and a poorly damped oscillatory current may flow in the

---

®TransZorb is a registered trademark of General Instruments, GSI.

drain wire. The shield current may radiate noise into other circuits, and the shield voltage will often show up as common mode conducted noise. The shield voltage is very difficult to detect with a voltage probe in most transformers, but the ringing shield current can be observed by holding the Sniffer Probe near the shield drain wire (Figure J10), or the shield current's return path in the circuit.

This ringing can be dampened by placing a resistor R_D in series with the shield drain wire, whose value is approximately equal to the surge impedance of the resonant circuit, which may be calculated from the formula in Figure J9.

The shield capacitance (C_S) can readily be measured with a bridge (as the capacity from the shield to all facing shields and/or windings), but L_S is usually best calculated from C_S and the ringing frequency (as sensed by the Sniffer Probe). This resistance is typically on the order of tens of ohms.

One or more small ferrite beads can also be placed on the drain wire instead to provide damping. This option may be preferable as a late "fix" when the PC board has already been laid out.

In either case, the damper losses are typically quite small. The damper resistor has a moderately adverse impact on shield effectiveness below the shield and drain wire resonant frequency; damper beads are superior in this respect as their impedance is less at lower frequencies. The drain wire connection should also be as short as possible to the circuit bypass point, both to minimize EMI and to raise the shield's maximum effective (i.e., resonant) frequency.

Leakage inductance fields

Transformer leakage inductance fields emanate from between primary and secondary windings. With a single primary and secondary, a significant dipole field is created, which may be seen by placing the Sniffer Probe near the winding ends as shown in Figure J11a. If this field is generating EMI, there are two principal fixes:
1. Split the Primary or secondary in two, to "sandwich" the other winding, and/or:
2. Place a shorted copper strap "electromagnetic shield" around the complete-core and winding assembly as shown in Figure J12. Eddy currents in the shorted strap largely cancel the external magnetic field.

The first approach creates a "quadrupole" instead of a dipole leakage field, which significantly reduces the distant field intensity. It also reduces the eddy current losses in any shorted strap electromagnetic shield used, which may or may not be an important consideration.

External air gap fields

External air gaps in an inductor, such as those in open "bobbin core" inductors or with "E" cores spaced apart (Figure J11b), can be a major source of external magnetic fields when significant ripple or AC currents are present. These fields can also be easily located with the Sniffer Probe; response will be a maximum near an air gap or near the end of an open inductor winding.

"Open" inductor fields are not readily shielded and if they present an EMI problem the inductor must usually be redesigned to reduce external fields. The external field around spaced E cores can be virtually eliminated by placing all of the air gap in the center leg. Fields due to a (possibly intentional) residual or minor outside air gap can be minimized with the shorted strap electromagnetic shield of Figure J12, if eddy current losses prove not to be too high.

A less obvious problem may occur when inductors with "open" cores are used as second stage filter chokes. The minimal ripple current may not create a significant field, but such an inductor can "pick up" external magnetic fields and convert them to noise voltages or be an EMI susceptibility problem.[3]

Poorly bypassed high speed logic

Ideally, all high speed logic should have a tightly coupled bypass capacitor for each IC and/or have power and ground distribution planes in a multilayer PCB.

At the other extreme, I have seen one bypass capacitor used at the power entrance to a logic board, with power and ground led to the ICs from opposite sides of the board. This created large spikes on the logic supply voltage and produced significant electromagnetic fields around the board.

With a Sniffer Probe, I was able to show which pins of which ICs had the larger current transients in synchronism with the supply voltage transients. (The logic design engineers were accusing the power supply vendor of creating the noise. I found that the supplies were fairly quiet; it was the poorly designed logic power distribution system that was the problem.)

Probe use with a "LISN"

A test setup using the Sniffer Probe with a Line Impedance Stabilization Network (LISN) is shown in Figure J13. The optional "LISN AC LINE FILTER" reduces AC line voltage feedthrough from a few 100mV to microvolt levels, simplifying EMI diagnosis when a suitable DC voltage source is not available or cannot be used.

Note 3: Ed. Note. See Appendix H for additional commentary.

Testing the sniffer probe

The Sniffer Probe can be functionally tested with a jig similar to that shown in Figure J14, which is used to test probes in production.

Conclusion

The Sniffer Probe is a simple, but very fast and effective means to locate dI/dt sources of EMI. These EMI sources are very difficult to locate with conventional voltage or current probes.

Summary

A summarized procedure for using the EMI "Sniffer" Probe appears in Figure J15.

Sniffer Probe amplifier

Figure J16 shows a 40MHz amplifier for the Sniffer Probe. A gain of 200 allows an oscilloscope to display probe output over a wide range of sensed inputs. The amplifier is built into a small aluminum box. The probe should connect to the amplifier via BNC cable, although the 50Ω termination does not have to be a high quality coaxial type. The Probe's uncalibrated, relative output means high frequency termination aberrations are irrelevant. A simple film resistor, contained in the amplifier box, is adequate. Figure J17 shows the Sniffer Probe and the amplifier.

Appendix K
System-based noise "measurement"

The ultimate test of switching regulator noise is its effect on the system being powered. The data below was taken using an LT1533 powering an LT1605 16-bit A/D converter. Crossplots for integral and differential nonlinearity are shown for bench supply vs LT1533 supply powered operation. The difference is within the test system's limit-of-error.

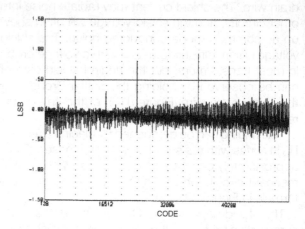

Figure K2 • Differential Nonlinearity Using LT1533 Supply

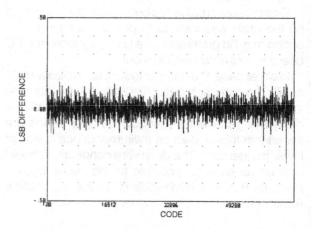

Figure K3 • Subtraction of Above Plots. Residual Error Is Test System Limited

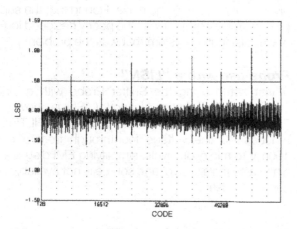

Figure K1 • Differential Nonlinearity Using Bench Supply

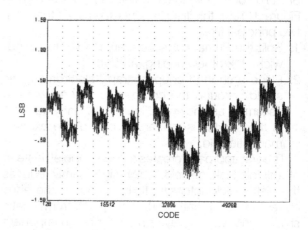

Figure K4 • Integral Nonlinearity Using Bench Supply

223

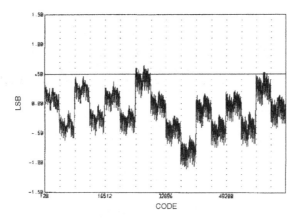

Figure K5 • Integral Nonlinearity Using LT1533 Supply

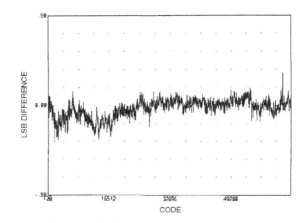

Figure K6 • Subtraction of Above Plots. Residual Error Is Test System Limited

Powering complex FPGA-based systems using highly integrated DC/DC µModule regulator systems

Part 1 of 2 Circuit and electrical performance

Alan Chern   Afshin Odabaee

In a recent discussion with a system designer, the requirement for his power supply was to regulate 1.5V and deliver up to 40A of current to a load that consisted of four FPGAs. This is up to 60W of power that must be delivered in a small area with the lowest profile (height) possible to allow a steady flow of air for cooling. The power supply had to be surface mountable and operate at high enough efficiency to minimize heat dissipation. He also demanded the simplest possible solution so his time could be dedicated to the more complex tasks. Aside from precise electrical performance, this solution had to remove the heat generated during DC to DC conversion quickly so that the circuit and the ICs in the vicinity do not overheat. Such a solution requires an innovative design to meet these criteria:

1. Very low profile to allow efficient air flow and to prevent thermal shadow on surrounding ICs
2. High efficiency to minimize heat dissipation
3. Current sharing capability to spread the heat evenly to eliminate hot spots and minimize or eliminate the need for heat sinks
4. Complete DC/DC circuit in a surface mount package that includes the DC/DC controller, MOSFETs, inductor capacitors and compensation circuitry for a quick and easy solution

Innovation in DC/DC design

The innovation is a modular but surface mount approach that uses efficient DC/DC conversion, precise current sharing and low thermal impedance packaging to deliver the output power while requiring minimal cooling. Because of the low profile and power sharing among four devices, a system using this solution depends on fewer fans or a slower fan speed as well as few or no heat sinks. (These contribute to lower system cost, consuming less power to remove heat).

Figure 9.1 shows a test board for such a circuit. The design regulates 1.5V output while delivering 40A (up to 48A) of load current. Each "black square" is a complete DC/DC circuit and is housed in a 15mm × 15mm × 2.8mm surface mount package. With a few input and output capacitors and resistors, the design using these DC/DC µModule™ systems is as simple as it's shown in the photo.

Figure 9.1 • Four DC/DC µModule Regulator Systems Current Share to Regulate 1.5V at 48A with Only 2.8mm Profile and 15mm × 15mm of Board Area. Each µModule Regulator Weighs Only 1.7g and Has an IC Form-Factor That Can Easily Be Used with Any Pick-and-Place Machine During Board Assembly

DC/DC µModule regulators: complete systems in an LGA package

The LTM4601 µModule DC/DC regulator is a high performance power module shrunk down to an IC form factor. It is a completely integrated solution—including the PWM controller, inductor, input and output capacitors, ultralow

Analog Circuit and System Design: A Tutorial Guide to Applications and Solutions. DOI: 10.1016/B978-0-12-385185-7.00009-3

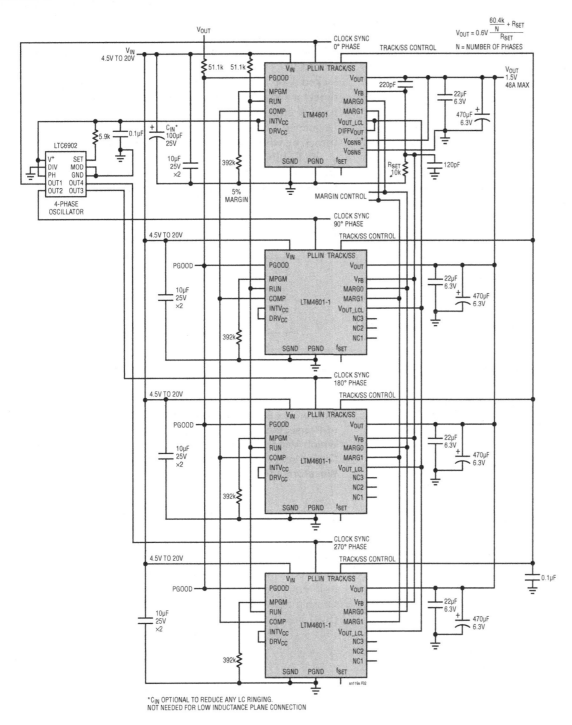

$$V_{OUT} = 0.6V \cdot \frac{\frac{60.4k}{N} + R_{SET}}{R_{SET}}$$

N = NUMBER OF PHASES

*C_{IN} OPTIONAL TO REDUCE ANY LC RINGING.
NOT NEEDED FOR LOW INDUCTANCE PLANE CONNECTION.

Figure 9.2 • Simply Parallel Multiple DC/DC μModule Regulator Systems to Achieve Higher Output Current. Board Layout Is as Easy as Copying and Pasting Each μModule Regulator's Layout With Very Few External Components Required

$R_{DS(ON)}$ FETs, Schottky diodes and compensation circuitry. Only external bulk input and output capacitors and one resistor are needed to set the output from 0.6V to 5V. The supply can produce 12A (more, if paralleled) from a wide input range of 4.5V to 20V, making it extremely versatile. The pin-compatible LTM4601HV extends the input range to 28V.

Another significant advantage of the LTM4601 over power-module- or IC-based systems is its ability to easily scale up as loads increase. If load requirements are greater than one µModule regulator can produce, simply add more modules in parallel. The design of a parallel system involves little more than copying and pasting the layout of each 15mm × 15mm µModule regulator. Electrical layout issues are taken care of within the µModule package—there are no external inductors, switches or other components to worry about.

Output features include output voltage tracking and margining. The high switching frequency, typically 850kHz at full load, constant on-time, zero latency controller delivers fast transient response to line and load changes while maintaining stability. Should frequency harmonics be a concern, an external clock can control synchronization via an on-chip phase-locked loop.

48A from four parallel DC/DC µModule regulators

Figure 9.2 shows a regulator comprising four parallel LTM4601s, which can produce a 48A (4 × 12A) output. The regulators are synchronized but operate 90° out-of-phase with respect to each other, thereby reducing the amplitude of input and output ripple currents through cancellation (Figure 9.3).

Synchronization and phase shifting is implemented via the LTC6902 oscillator, which provides four clock outputs, each 90° phase shifted (for 2- or 3-phase relationships, the LTC6902 can be adjusted via a resistor). By operating the µModule regulators out-of-phase, peak input and output current is reduced by approximately 20% depending on the duty cycle (see the LTM4601 data sheet). The attenuated ripple, in turn, decreases the external capacitor RMS

current rating and size requirements, further reducing solution cost and board space.

The clock signals serve as input to the PLLIN (phase-locked loop in) pins of the four LTM4601s. The phase-locked loop of the LTM4601 is comprised of a phase detector and a voltage controlled oscillator, which combine to lock onto the rising edge of an external clock with a frequency range of 850kHz. The phase-locked loop is turned on when a pulse of at least 400ns and 2V amplitude at the PLLIN pin is detected, though it is disabled during start-up. Figure 9.3 shows the switching waveforms of four LTM4601 µModule regulators in parallel.

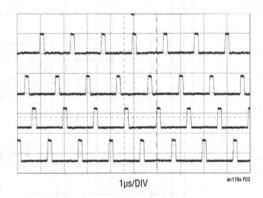

1µs/DIV   an119a F03

Figure 9.3 • By Operating Each DC/DC µModule 90° Out-of-Phase, the Input and Output Ripples Are Reduced, Which Also Reduces the Requirement for Input and Output Capacitors. Photo Shows Individual µModule Switching Waveforms for Figure 9.2

Only one resistor is required to set the output voltage. In a parallel set-up, the value of the resistor depends on the number of LTM4601s used. This is because the effective value of the top (internal) feedback resistor changes as you parallel LTM4601s. The LTM4601's reference voltage is 0.6V and its internal top feedback resistor value is 60.4kΩ, so the relationship between V_{OUT}, the output voltage setting resistor (R_{FB}), and the number of modules (n) placed in parallel is:

$$V_{OUT} = 0.6V \frac{\frac{60.4k}{n} + R_{FB}}{R_{FB}}$$

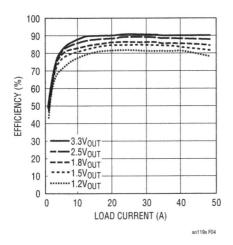

an119a F04

Figure 9.4 • Efficiency of the Four DC/DC µModules in Parallel Remains High Over a Wide Range of Output Voltages (12V Input)

Figure 9.4 illustrates the system's high efficiency over the vast output current range up to 48A. The system performs impressively with no dipping in the efficiency curve for a broad range of output voltages.

Start-up, soft-start and current sharing

The soft-start feature of the LTM4601 prevents large inrush currents at start-up by slowly ramping the output

voltage to its nominal value. The relation of start-up time to V_{OUT} and the soft-start capacitor (C_{SS}) is:

$$t_{SOFTSTART} = 0.8 \bullet \left(0.6V - V_{OUT(MARGIN)}\right) \bullet \frac{C_{SS}}{1.5\mu A}$$

where

$$V_{OUT(MARGIN)} = \frac{\%V_{OUT}}{100} \bullet V_{OUT}$$

For example, a $0.1\mu F$ soft-start capacitor yields a nominal 8ms ramp (see Figure 9.5) with no margining.

Current sharing among parallel regulators is well balanced through start-up to full load. Figure 9.6 shows an evenly distributed output current curve for a 2-parallel LTM4601 system, as each rises to a nominal 10A each, 20A total.

Conclusion

The DC/DC µModule regulators are self-contained and complete systems in an IC form factor. The low profile, high efficiency and current sharing capability allow practical high power solutions for the new generation of digital systems. Thermal performance is impressive at 48A of output current with balanced current sharing and smooth uniform start-up. The ease and simplicity of this design minimizes development time while saving board space. In part two of this discussion, the focus will be on thermal performance and layout of this circuit.

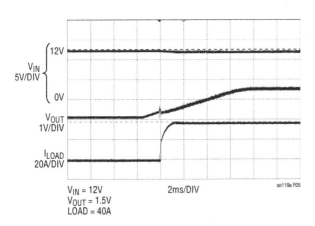

an119a F05

Figure 9.5 • Controlled Soft-Start Is Important in Proper Start-Up of the FPGA or the System as a Whole; Soft-Start Current and Voltage Ramp for Four DC/DC µModule Regulators in Parallel

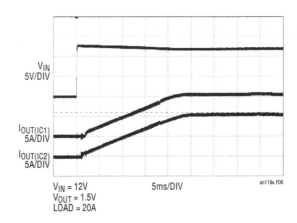

an119a F06

Figure 9.6 • Each DC/DC µModule Regulator Starts and Ends By Sharing the Load Current Evenly and Balanced, a Crucial Feature to Prevent One Regulator from Overheating; Two Parallel LTM4601s, as Each Rises to a Nominal 10A Each, 20A Total

Powering complex FPGA-based systems using highly integrated DC/DC μModule regulator systems

Part 2 of 2 Thermal performance and layout

Alan Chern   Afshin Odabaee

60W by paralleling four DC/DC μModule regulators

In part one of this article, we discussed the circuit and electrical performance of a compact and low profile 48A, 1.5V DC/DC regulator solution for a four-FPGA design. The new approach uses four DC/DC μModule™ regulators in parallel (Figure 10.1) to increase output current while sharing the current equally among each device. This solution relies on the accurate current sharing of these μModule regulators to prevent hot-spots by dissipating the heat evenly over a compact surface area. Each DC/DC μModule is a complete power supply with on-board inductor, DC/DC controller MOSFETs, compensation circuitry and input/output bypass capacitors. It occupies only 15mm x 15mm of board area and has a low profile (height) of only 2.8mm. This low profile allows air to flow smoothly over the entire circuit. Moreover, this solution casts no thermal shadow on its surrounding components, further assisting in optimizing thermal performance of the entire system.

Thermal performance

Figure 10.2 is a thermal image of the board shown in Figure 10.1 with readings of the temperatures at specific locations. Cursors 1 to 4 show an estimation of the surface temperature on each module. Cursors 5 to 7 indicate the surface temperature of the PCB. Notice the difference in temperature between the inner two regulators, cursors

Figure 10.1 • Four DC/DC μModule Regulator Systems Current Share to Regulate 1.5V at 48A With Only 2.8mm Profile and 15mm × 15mm of Board Area for Each Device. Each μModule Regulator Weighs Only 1.7g and Has an IC Form-Factor that Can Easily be Used With Any Pick-and-Place Machine During Board Assembly

Analog Circuit and System Design: A Tutorial Guide to Applications and Solutions. DOI: 10.1016/B978-0-12-385185-7.00010-X

1 and 2, and the outside, cursors 3 and 4. The LTM4601 μModule regulators placed on the outside have large planes to the left and right promoting heat sinking to cool the part down a few degrees. The inner two only have small top and bottom planes to draw heat away, thus becoming slightly warmer than the outside two.

Airflow also has a substantial effect on the thermal balance of the system. Note the difference in temperature between Figures 10.2 and 10.3. In Figure 10.3, a 200LFM airflow travels evenly from the bottom to the top of the demo board, causing a 20°C drop across the board compared to the no airflow case in Figure 10.2.

The direction of airflow is also important. In Figure 10.4 the system is placed inside a 50°C ambient chamber with airflow traveling from right to left, pushing the heat from one μModule regulator to the next, creating a stacking

effect. The μModule device on the right, the closest to the airflow source, is the coolest. The leftmost μModule regulator has a slightly higher temperature because of spillover heat from the other μModule regulators.

Heat transfer to the PCB also changes with airflow. In Figure 10.2, heat transfers evenly to both left and right sides of the PCB. In Figure 10.4, most of the heat moves to the left side. Figure 10.5 shows an extreme case of heat stacking from one μModule device to the next. Each of the four μModule regulators is fitted with a BGA heat sink and the entire board is operated in a chamber with an ambient temperature of 75°C.

Simple copy and paste layout

Layout of the parallel μModule regulators is relatively simple, in that there are few electrical design considerations. Nevertheless, if the intent of a design is to minimize the required PCB area, thermal considerations become paramount, so the important parameters are spacing, vias, airflow and planes.

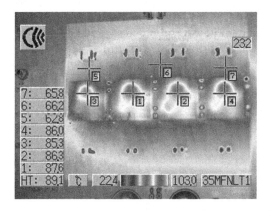

Figure 10.2 • Thermograph of 48A, 1.5V Circuit of Figure 10.1 Shows Balanced Current Sharing Among Each DC/DC μModule Regulator and Low Temperature Rise Even Without Airflow (V$_{IN}$ = 20V to 1.5V$_{OUT}$ at 40A)

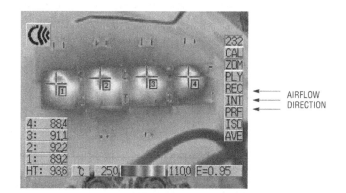

Figure 10.4 • Thermograph of Four Parallel LTM4601 with 400LFM Right-to-Left Airflow in 50°C Ambient Chamber (12V$_{IN}$ to 1V$_{OUT}$ at 40A)

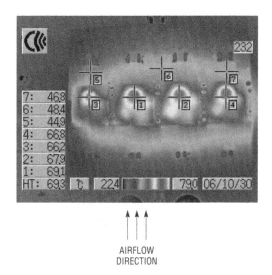

Figure 10.3 • Thermograph of Four Parallel LTM4601 with 200LFM Bottom-to-Top Airflow (20V$_{IN}$ to 1.5V$_{OUT}$ at 40A)

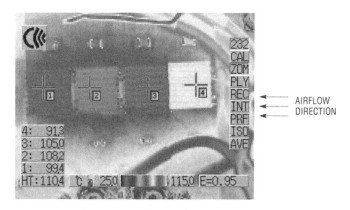

Figure 10.5 • Thermograph of Four Parallel LTM4601 With BGA Heatsinks and 400LFM Right-to-Left Airflow in a 75°C Ambient Chamber (12V$_{IN}$ to 1V$_{OUT}$ at 40A)

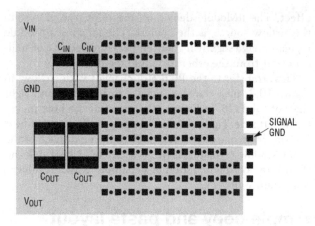

Figure 10.6 • The LTM4601's Pin Layout Provides Simple Power Plane Placement and Easy Paralleling Capability (Copy and Paste Approach)

The LTM4601 µModule regulator has a unique LGA package footprint, which allows solid attachment to the PCB while enhancing thermal heat sinking. The footprint itself simplifies layout of the power and ground planes, as shown in Figure 10.6. Laying out four parallel µModule

regulators is just as easy, as shown in Figures 10.7 and 10.8.

If laid out properly, the LGA packaging and the power planes alone can provide enough heat sinking to keep the LTM4601 cool.

Conclusion

Delivering 60W of power in a compact space without efficient means to remove the heat from the power supply exacerbates the already difficult task of system heat management and cooling. The DC/DC µModule family is designed with careful attention to the layout of its internal components, package type, and electrical operation, which ease thermal management of a very dense power supply circuit. The LGA package and simple layout allow 100% surface mountable and low profile design for maximum efficiency in air flow. This new approach in power supply design takes advantage of paralleling multiple DC/DC µModule regulators and following a copy and paste approach in layout design, to provide a 60W power supply with minimum components while operating efficiently in a compact and low profile space.

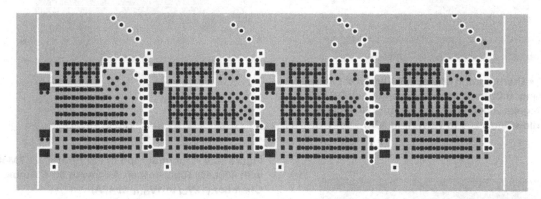

Figure 10.7 • Top Layer Planes for Figure 10.1 Circuit

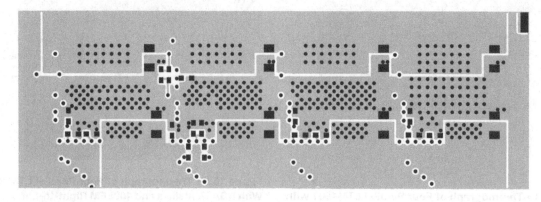

Figure 10.8 • Bottom Layer Planes for Figure 10.1 Circuit

Diode turn-on time induced failures in switching regulators

Never has so much trouble been had by so many with so few terminals

11

Jim Williams David Beebe

Introduction

Most circuit designers are familiar with diode dynamic characteristics such as charge storage, voltage dependent capacitance and reverse recovery time. Less commonly acknowledged and manufacturer specified is diode forward turn-on time. This parameter describes the time required for a diode to turn on and clamp at its forward voltage drop. Historically, this extremely short time, units of nanoseconds, has been so small that user and vendor alike have essentially ignored it. It is rarely discussed and almost never specified. Recently, switching regulator clock rate and transition time have become faster, making diode turn-on time a critical issue. Increased clock rates are mandated to achieve smaller magnetics size; decreased transition times somewhat aid overall efficiency but are principally needed to minimize IC heat rise. At clock speeds beyond about 1MHz, transition time losses are the primary source of die heating.

A potential difficulty due to diode turn-on time is that the resultant transitory "overshoot" voltage across the diode, even when restricted to nanoseconds, can induce overvoltage stress, causing switching regulator IC failure. As such, careful testing is required to qualify a given diode for a particular application to insure reliability. This testing, *which assumes low loss surrounding components and layout in the final application*, measures turn-on overshoot voltage due to diode parasitics only. Improper associated component selection and layout will contribute additional overstress terms.

Diode turn-on time perspectives

Figure 11.1 shows typical step-up and step-down voltage converters. In both cases, the assumption is that the diode clamps switch pin voltage excursions to safe limits. In the step-up case, this limit is defined by the switch pin's maximum allowable forward voltage. The step-down case limit is set by the switch pin's maximum allowable reverse voltage.

Figure 11.2 indicates the diode requires a finite length of time to clamp at its forward voltage. This *forward turn-on time* permits transient excursions above the nominal diode clamp voltage, potentially exceeding the IC's breakdown limit. The turn-on time is typically measured in nanoseconds, making observation difficult. A further complication is that the turn-on overshoot occurs at the amplitude extreme of a pulse waveform, precluding high resolution amplitude measurement. These factors must be considered when designing a diode turn-on test method.

Figure 11.3 shows a conceptual method for testing diode turn-on time. Here, the test is performed at 1A although other currents could be used. A pulse steps 1A into the diode under test via the 5Ω resistor. Turn-on time voltage excursion is measured directly at the diode under test. The figure is deceptively simple in appearance. In particular, the current step must have an exceptionally fast, high-fidelity transition and faithful turn-on time determination requires substantial measurement bandwidth.

Detailed measurement scheme

A more detailed measurement scheme appears in Figure 11.4. Necessary performance parameters for various elements are called out. A sub-nanosecond rise time pulse generator 1A, 2ns rise time amplifier and a 1GHz oscilloscope are required. These specifications represent realistic operating conditions; other currents and rise times can be selected by altering appropriate parameters.

The pulse amplifier necessitates careful attention to circuit configuration and layout. Figure 11.5 shows the amplifier includes a paralleled, Darlington driven RF transistor output stage. The collector voltage adjustment ("rise time

Analog Circuit and System Design: A Tutorial Guide to Applications and Solutions. DOI: 10.1016/B978-0-12-385185-7.00011-1

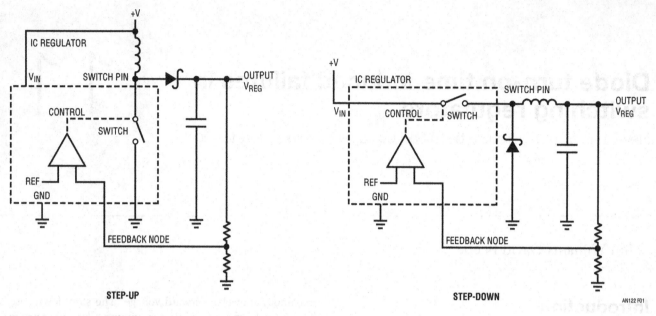

Figure 11.1 • Typical Voltage Step-Up/Step-Down Converters. Assumption is Diode Clamps Switch Pin Voltage Excursion to Safe Limits

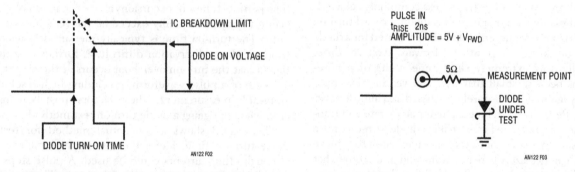

Figure 11.2 • Diode Forward Turn-On Time Permits Transient Excursion Above Nominal Diode Clamp Voltage, Potentially Exceeding IC Breakdown Limit

Figure 11.3 • Conceptual Method Tests Diode Turn-On Time at 1A. Input Step Must Have Exceptionally Fast, High Fidelity Transition

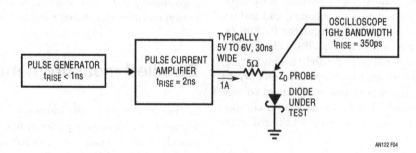

Figure 11.4 • Detailed Measurement Scheme Indicates Necessary Performance Parameters for Various Elements. Sub-Nanosecond Rise Time Pulse Generator, 1A, 2ns Rise Time Amplifi er and 1GHz Oscilloscope are Required

trim") peaks Q4 to Q6 F_T; an input RC network optimizes output pulse purity by slightly retarding input pulse rise time to within amplifier passband. Paralleling allows Q4 to Q6 to operate at favorable individual currents, maintaining bandwidth. When the (mildly interactive) edge purity and rise time trims are optimized, Figure 11.6 indicates the amplifier produces a transcendently clean 2ns rise time output pulse devoid of ringing, alien components or post-transition excursions. Such performance makes diode turn-on time testing practical.[1]

Figure 11.7 depicts the complete diode forward turn-on time measurement arrangement. The pulse amplifier,

driven by a sub-nanosecond pulse generator, drives the diode under test. A Z_0 probe monitors the measurement point and feeds a 1GHz oscilloscope.[2,3,4]

Note 1: An alternate pulse generation approach appears in Appendix F "Another Way to Do It."
Note 2: Z_0 probes are described in Appendix C, "About Z_0 Probes". See also References 27 thru 34.
Note 3: The sub-nanosecond pulse generator requirement is not trivial. See Appendix B, "Subnanosecond Rise Time Pulse Generators For The Rich and Poor."
Note 4: See Appendix E, "Connections, Cables, Adapters, Attenuators, Probes and Picoseconds" for relevant commentary.

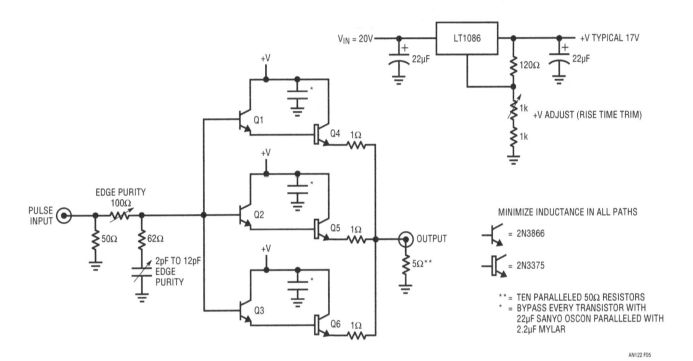

Figure 11.5 • Pulse Amplifier Includes Paralleled, Darlington Driven RF Transistor Output Stage. Collector Voltage Adjustment ("Rise Time Trim") Peaks Q4 to Q6 F_T, Input RC Network Optimizes Output Pulse Purity. Low Inductance Layout is Mandatory

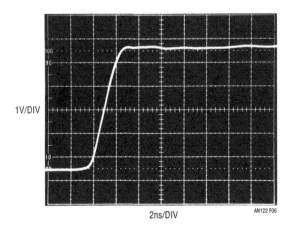

Figure 11.6 • Pulse Amplifier Output into 5Ω. Rise Time is 2ns with Minimal Pulse-Top Aberrations

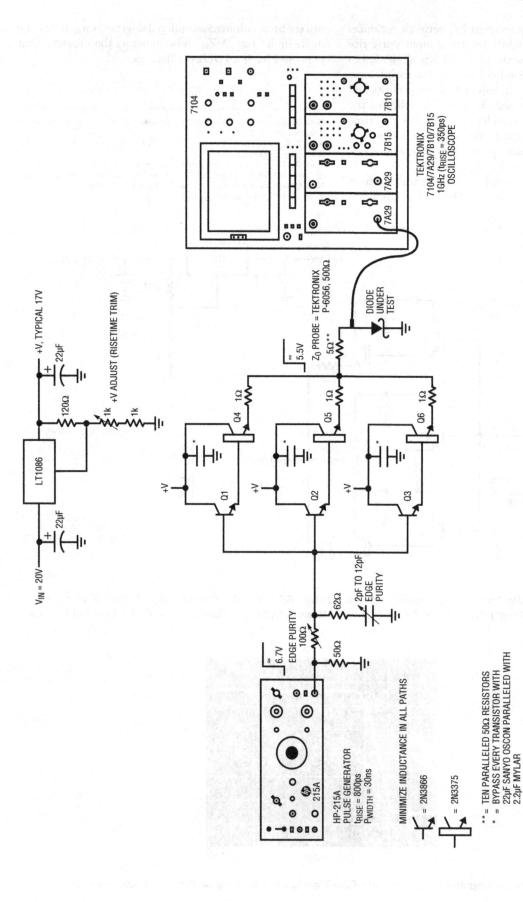

Figure 11.7 • Complete Diode Forward Turn-On Time Measurement Arrangement Includes Sub-Nanosecond Rise Time Pulse Generator, Pulse Amplifier, Z_0 Probe and 1GHz Oscilloscope

Diode testing and interpreting results

The measurement test fixture, properly equipped and constructed, permits diode turn-on time testing with excellent time and amplitude resolution.[5] Figures 11.8 through 11.12 show results for five different diodes from various manufacturers. Figure 11.8 (Diode Number 1) overshoots steady state forward voltage for 3.6ns, peaking 200mV. This is the best performance of the five. Figures 11.9 through 11.12 show increasing turn-on amplitude and time which are detailed in the figure captions. In the worst cases, turn-on amplitudes exceed nominal clamp voltage by more than 1V while turn-on times extend for tens of nanoseconds. Figure 11.12 culminates this unfortunate parade with huge time and amplitude errors. Such errant excursions can and will cause IC regulator breakdown and failure. The lesson here is clear. Diode turn-on time must be characterized and measured in any given application to insure reliability.

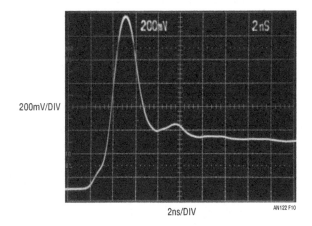

Figure 11.10 • "Diode Number 3" Peaks 1V Above Nominal 400mV V_FWD, a 2.5x Error

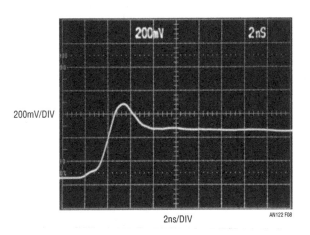

Figure 11.8 • "Diode Number 1" Overshoots Steady State Forward Voltage for ≈3.6ns, Peaking 200mV

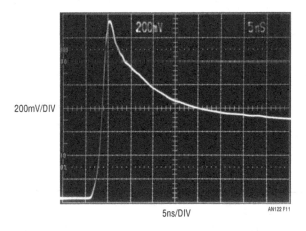

Figure 11.11 • "Diode Number 4" Peaks ≈750mV with Lengthy (Note Horizontal 2.5x Scale Change) Tailing Towards VFWD Value

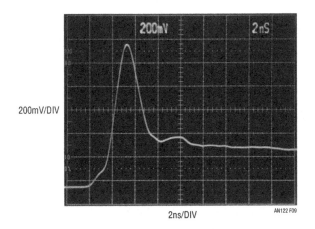

Figure 11.9 • "Diode Number 2" Peaks ≈750mV Before Settling in 6ns... > 2x Steady State Forward Voltage

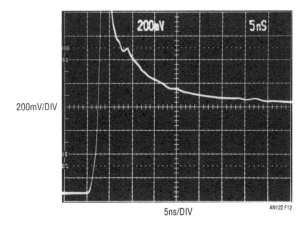

Figure 11.12 • "Diode Number 5" Peaks Offscale with Extended Tailing (Note Horizontal Slower Scale Compared to Figures 11.8 thru 11.10)

Note 5: See Appendix A, "How Much Bandwidth is Enough?" for discussion on determining necessary measurement bandwidth.

References

1. Churchill, Winston S., "Never in the field of human conflict was so much owed by so many to so few." Speech, "The Few", Tribute to the Royal Airforce, House of Commons, August 20th 1940.

2. Zettler, R., and Cowley, A. M., "Hybrid Hot Carrier Diodes," Hewlett-Packard Journal, February 1969.

3. Motorola Inc., "Motorola Rectifier Applications Handbook", Motorola, Inc., 1993.

4. RCA RF/Microwave Devices, RCA, 1975.

5. Chessman, M., and Sokol, N., "Prevent Emitter-Follower Oscillation", Electronic Design, 13, pp. 110–113, 21 June 1976.

6. DeBella, G. B., "Stability of Capacitively-Loaded Emitter Followers– a Simplified Approach", Hewlett-Packard Journal, 17, pp. 15–16, April 1966.

7. Hamilton, D. J., Shaver, F. H., and Griffith, P. G., "Avalanche Transistor Circuits for Generating Rectangular Pulses," Electronic Engineering, December 1962.

8. Seeds, R. B., "Triggering of Avalanche Transistor Pulse Circuits," Technical Report No. 1653-1, August 5, 1960, Solid-State Electronics Laboratory, Stanford Electronics Laboratories, Stanford University, Stanford, California.

9. Beale, J. R. A., et al., "A Study of High Speed Avalanche Transistors", Proc. I.E.E., 104, Part B, July 1957 pp. 394–402.

10. Braatz, Dennis., "Avalanche Pulse Generators", Private Communication, Tektronix, Inc., 2003.

11. Tektronix Inc., Type 111 Pretrigger Pulse Generator Operating and Service Manual, Tektronix, Inc., 1960.

12. Haas, Isy., "Millimicrosecond Avalanche Switching Circuit Utilizing Double-Diffused Silicon Transistors," Fairchild Semiconductor, Application Note 8/2, December 1961.

13. Beeson, R. H., Haas, I., and Grinich, V. H., "Thermal Response of Transistors in Avalanche Mode," Fairchild Semiconductor, Technical Paper 6, October 1959.

14. Chaplin, G. B. B, "A Method of Designing Transistor Avalanche Circuits with Applications to a Sensitive Transistor Oscilloscope," paper presented at the 1958 IRE-AIEE Solid State Circuits Conference, Philadelphia, PA., February 1958.

15. Motorola Inc., "Avalanche Mode Switching", Motorola Transistor Handbook, 1963. Chapter 9 pp. 285-304.

16. Williams, Jim., "A Seven-Nanosecond Comparator for Single Supply Operation," "Programmable, Subnanosecond Delayed Pulse Generator", Linear Technology Corporation, May 1998, Application Note 72, pp. 32–34.

17. Williams, Jim., "Power Conversion, Measurement and Pulse Circuits", Linear Technology Corporation, Application Note 113, August 2007.

18. Moll, J.L., "Avalanche Transistors as Fast Pulse Generators". Proc. I.E.E., Vol 106, Part B, Supplement 17, 1959, pp. 1082 to 1084.

19. Williams, Jim., "Circuitry for Signal Conditioning and Power Conversion", Linear Technology Corporation, March 1999 Application Note 75.

20. Williams, Jim., "Signal Sources, Conditioners and Power Circuitry", Linear Technology Corporation, November 2004, Application Note 98, pp. 20–21.

21. Williams, Jim., "Practical Circuitry for Measurement and Control Problems", Linear Technology Corporation, August 1994. Application Note 61.

22. Williams, Jim., "Measurement and Control Circuit Collection", Linear Technology Corporation, June 1991 Application Note 45.

23. Williams, Jim., "Slew Rate Verification for Wideband Amplifiers", Linear Technology Corporation, May 2003 Application Note 94.

24. Williams, Jim., "30 Nanosecond Settling Time Measurement for a Precision Wideband Amplifier", Linear Technology Corporation, Application Note 79, September 1999.

25. Williams, Jim., "A Monolithic Switching Regulator with 100–uV Output Noise", Linear Technology Corporation, Application Note 70, October 1997.

26. Andrews, James R., "Pulse Measurements in the Picosecond Domain", Picosecond Pulse Labs, Application Note AN-3a, 1988.

27. Williams, Jim., "High Speed Amplifier Techniques", Linear Technology Corporation, Application Note 47, August 1991.

28. Williams, Jim., "About Probes and Oscilloscopes," Appendix B, in "High Speed Comparator Techniques", Linear Technology Corporation, Application Note 13, April 1985.

29. Weber, Joe., "Oscilloscope Probe Circuits", Tektronix, Inc., Concept Series, 1969.

30. McAbel, W. E., "Probe Measurements", Tektronix, Inc., Concept Series, 1969.

31. Hurlock, L., "ABC's of Probes", Tektronix, Inc., 1991.

32. Bunze, V., "Matching Oscilloscope and Probe for Better Measurements", Electronics, pp. 88–93, March 1 1973.

33. Tektronix Inc., "P6056/P6057 Probe Instruction Manual", Tektronix, Inc., December 1981.

34. Tektronix Inc., "P6034 Probe Instruction Manual", Tektronix, Inc., 1963.

35. Hewlett-Packard., "HP215A Pulse Generator Operating and Service Manual", Hewlett-Packard, 1962.

36. Tektronix Inc., "Type 109 Pulse Generator Operating and Service Manual", Tektronix, Inc., 1963.

Appendix A
How much bandwidth is enough?

Accurate wideband oscilloscope measurements require bandwidth. A good question is just how much is needed. A classic guideline is that "end-to-end" measurement system rise time is equal to the root-sum-square of the system's individual components' rise times. The simplest case is two components; a signal source and an oscilloscope.

Figure A1's plot of $\sqrt{Signal^2 + Oscilloscope^2}$ rise time versus error is illuminating. The figure plots signal-to-oscilloscope rise time ratio versus observed rise time (rise time is bandwidth restated in the time domain, where:

$$Rise\ Time\ (ns) = \frac{350}{Bandwith(MHz)})$$

The curve shows that an oscilloscope 3 to 4 times faster than the input signal rise time is required for measurement accuracy inside about 5%. This is why trying to measure a 1ns rise time pulse with a 350MHz oscilloscope ($t_{RISE} = 1ns$) leads to erroneous conclusions. The curve indicates a monstrous 41% error. Note that this curve does not include the effects of passive probes or cables connecting the signal to the oscilloscope. Probes do not necessarily follow root-sum-square law and must be carefully chosen and applied for a given measurement. Figure A2, included for reference, gives 10 cardinal points of rise time/bandwidth equivalency between 1MHz and 5GHz.

Figures A3 through A10 illustrate pertinent effects of these considerations by viewing the text's

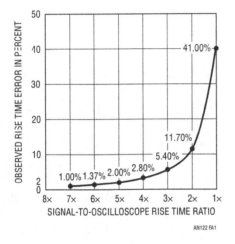

AN122 FA1

Figure A1 • Oscilloscope Rise Time Effect on Rise Time Measurement Accuracy. Measurement Error Rises Rapidly as Signal-to-Oscilloscope Rise Time Ratio Approaches Unity. Data, Based on Root-Sum-Square Relationship, Does Not Include Probe, Which May Not Follow Root-Sum-Square Law

RISE TIME	BANDWIDTH
70ps	5GHz
350ps	1GHz
700ps	500MHz
1ns	350MHz
2.33ns	150MHz
3.5ns	100MHz
7ns	50MHz
35ns	10MHz
70ns	5MHz
350ns	1MHz

Figure A2 • Some Cardinal Points of Rise Time/ Bandwidth Equivalency. Data is Based on Rise Time/ Bandwidth Formula in Text

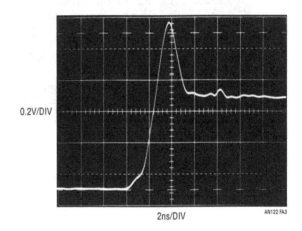

0.2V/DIV

2ns/DIV

AN122 FA3

Figure A3 • Typical Diode Turn-On Viewed in 2.5GHz Sampled Bandpass Displays 500mV Turn-On Peak

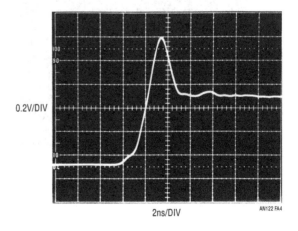

0.2V/DIV

2ns/DIV

AN122 FA4

Figure A4 • Figure A3's Diode Turn-On Observed in 1GHz Real Time Bandwidth Has Nearly Identical Characteristics, Indicating Adequate Oscilloscope Bandwidth

diode turn-on time measurement at various bandwidths.[1] Figure A3 displays a typical diode turn-on in a 2.5GHz sampled bandpass, showing 500mV turn-on amplitude.[2] Figure A4's 1GHz bandwidth measurement has nearly identical characteristics, indicating adequate oscilloscope bandwidth. The dramatic error in observed turn-on overshoot amplitude as bandwidth decreases in succeeding figures is readily apparent and should not be lost to the experimenter.

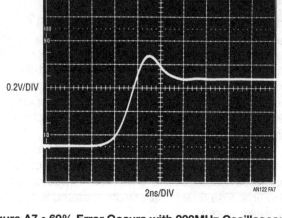

Figure A7 • 60% Error Occurs with 200MHz Oscilloscope Bandwidth

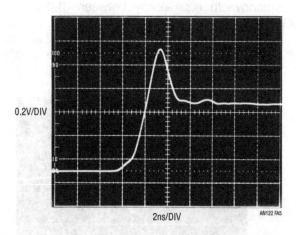

Figure A5 • 600MHz Oscilloscope Bandwidth Results in ≈440mV Observed Peak, an 12% Amplitude Error

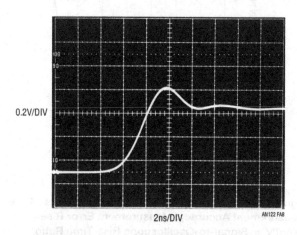

Figure A8 • 65% Error (!) in 75MHz Bandwidth

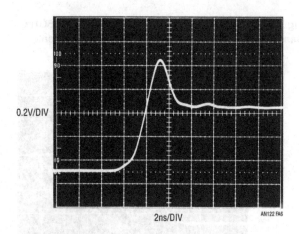

Figure A6 • 400MHz Measurement Bandwidth Causes 20% Error

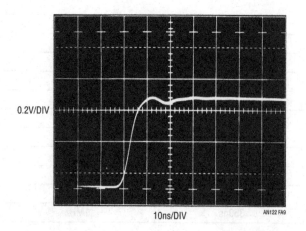

Figure A9 • 50MHz Oscilloscope Just Hints at Peaking. Note 5x Horizontal Scale Change vs Figures A3 through A8

Note 1: Prudent investigation requires verifying bandwidth of all elements in the signal path. See Appendix D, "Verifying Rise Time Measurement Integrity."
Note 2: 3.9GHz oscilloscope + 3.5GHz probe = 2.5GHz probe tip bandwidth.

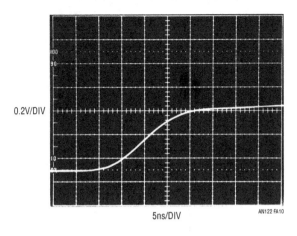

5ns/DIV   AN122 FA10

Figure A10 • 20MHz Oscilloscope Bandwidth Presentation is Smooth...and Worthless. Note 2.5x Horizontal Scale Change vs Figures A3 through A8

Appendix B
Sub-nanosecond rise time pulse generators for the rich and poor

The pulse amplifier requires a sub-nanosecond input rise time pulse to cleanly switch current to the diode under test. The majority of general purpose pulse generators have rise times in the 2.5ns to 10ns range. Instrument rise times below 2.5ns are relatively rare, with only a select few types getting down to 1ns. The ranks of sub-nanosecond rise time generators are even thinner, and costs are, in this author's view, excessive. Sub-nanosecond rise time generation, particularly if relatively large swings (e.g. 5V to 10V) are desired, employs arcane technologies and exotic construction techniques. Available instruments in this class work well, but can easily cost $10,000 with prices rising towards $30,000 depending on features. For bench work, or even production testing, there are substantially less expensive approaches.

The secondary market offers sub-nanosecond rise time pulse generators at attractive cost. The Hewlett-Packard HP-8082A transitions in under 1ns, has a full complement of controls and costs about $500. The Tektronix type 111 has edge times of 500ps, with fully variable repetition rate and external trigger capabilities. Pulse width is set by external charge line length. Price is usually about $25. The HP-215A, long out of manufacture, has

800ps edge times and is a clear bargain, with typical price below $50.[1] This instrument also has a very versatile trigger output, permitting continuous trigger time phase adjustment from before to after the main output. External trigger impedance, polarity and sensitivity are also variable. The output, controlled by a stepped attenuator, will put a clean ±10V pulse into 50Ω in 800ps.[2]

400ps rise time avalanche pulse generator

A potential problem with older instruments is availability.[3] As such, Figure B1 shows a circuit for producing sub-nanosecond rise time pulses. Rise time is 400ps, with adjustable pulse amplitude. Output pulse occurrence is settable from before-to-after a trigger output. This circuit uses an avalanche pulse generator to create extremely fast rise time pulses.[4]

Q1 and Q2 form a current source that charges the 1000pF capacitor. When the LTC1799 clock is high (trace A, Figure B2) both Q3 and Q4 are on. The current source is off and Q2's collector (trace B) is at ground. C1's latch input prevents it from responding and its output remains high. When the clock goes low, C1's latch input is disabled and its output drops low. The Q3 and Q4 collectors lift and Q2 comes on, delivering constant current to the 1000pF capacitor (trace B). The resulting linear ramp is applied to C1 and C2's positive inputs. C2, biased from a potential derived from the 5V supply, goes high 30ns after the ramp begins, providing the "trigger output" (trace C) via its output network. C1 goes high when the ramp crosses the potentiometer programmed delay at its negative input, in this case about 170ns. C1 going high triggers the avalanche-based output pulse (trace D), which will be described. This arrangement permits the delay programming control to vary output pulse occurrence from 30ns before to 300ns after the trigger output. Figure B3 shows the output pulse (trace D) occurring 25ns before the trigger output. All other waveforms are identical to Figure B2.

When C1's output pulse is applied to Q5's base, it avalanches. The result is a quickly rising pulse across Q5's emitter termination resistor. The collector capacitors and the charge line discharge, Q5 collector voltage falls and breakdown ceases. The collector capacitors and the charge line then recharge. At C1's next pulse, this action repeats. The capacitors supply initial pulse response, with the charge lines

---

Note 1: The absurdly low valuation may be due to the instrument's front panel controls and markings, which only subtly hint at its capabilities.
Note 2: Instrument afficionados would do well to study this instrument's elegant step-recovery diode based output stage, a thing of exotic beauty. See Reference 35.

Note 3: Residents of Silicon Valley tend towards inbred techno-provincialism. Citizens of other locales cannot simply go to a flea market, junk store or garage sale and buy a sub-nanosecond pulse generator.
Note 4: The circuit's operation essentially duplicates the aforementioned Tektronix type 111 pulse generator (see Reference 11). Information on avalanche operation appears in References 7 through 25.

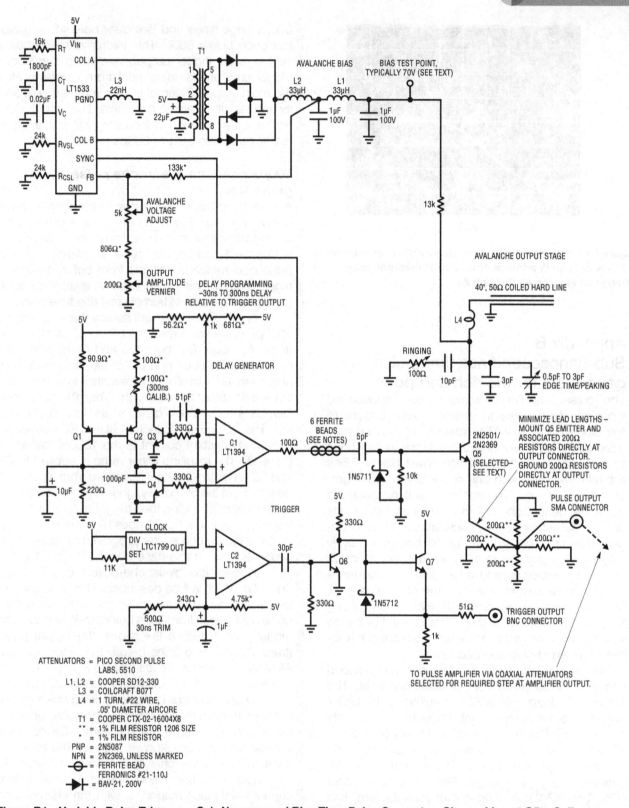

Figure B1 • Variable Delay Triggers a Sub-Nanosecond Rise Time Pulse Generator. Charge Line at Q5's Collector Determines ≈10ns Output Width. Output Pulse Occurrence is Settable from Before-to-After Trigger Output

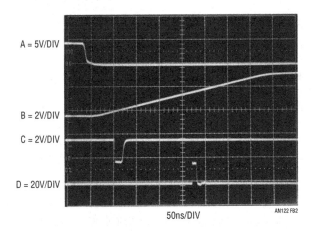

A = 5V/DIV

B = 2V/DIV

C = 2V/DIV

D = 20V/DIV

50ns/DIV

AN122 FB2

Figure B2 • Pulse Generator's Waveforms Include Clock (Trace A), Q2's Collector Ramp (Trace B), Trigger Output (Trace C) and Pulse Output (Trace D). Delay Sets Output Pulse ≈170ns After Trigger Output

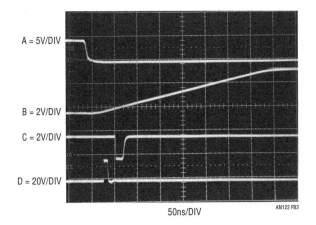

A = 5V/DIV

B = 2V/DIV

C = 2V/DIV

D = 20V/DIV

50ns/DIV

AN122 FB3

Figure B3 • Pulse Generator's Waveforms with Delay Adjusted for Output Pulse Occurrence (Trace D) 25ns Before Trigger Output (Trace C). All Other Activity is Identical to Previous Figure

prolonged discharge contributing the pulse body. The 40" charge line length forms an output pulse width about 12ns in duration.

Avalanche operation requires high voltage bias. The LT1533 low noise switching regulator and associated components supply this high voltage. The LT1533 is a "push-pull" output switching regulator with controllable transition times.

Output harmonic content ("noise") is notably reduced with slower switch transition times.[5] Switch current and voltage transition times are controlled by resistors at the R_{CSL} and R_{VSL} pins,

respectively. In all other respects the circuit behaves as a classical push-pull, step-up converter

Circuit optimization

Circuit optimization begins by setting the "Output Amplitude Vernier" to maximum and grounding Q4's collector. Next, set the "Avalanche Voltage Adjust" so free running pulses just appear at Q5's emitter, noting the bias test points voltage. Readjust the "Avalanche Voltage Adjust" 5V below this voltage and unground Q4's collector. Set the "30ns Trim" so the trigger output goes low 30ns after the clock goes low. Adjust the delay programming control to maximum and set the "300ns Calib." so C1 goes high 300ns after the clock goes low. Slight interaction between the 30ns and 300ns trims may require repeating their adjustments until both points are calibrated.

Q5 requires selection for optimal avalanche behavior Such behavior, while characteristic of the device specified, is not guaranteed by the manufacturer. A sample of 30 2N2501s, spread over a 17-year date code span, yielded ≈90%. All "good" devices switched in less than 475ps with some below 300ps.[6] In practice, Q5 should be selected for "in-circuit" rise time under 400ps. Once this is done, output pulse shape is optimized by adjusting Q5's collector damping trims ("edge time/peaking" and "ringing").

The trims are somewhat interactive, but not unduly so, and optimal adjustment converges nicely. The pulse edge is carefully adjusted so that maximum transition speed is attained with minimal sacrifice of

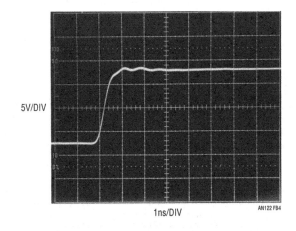

5V/DIV

1ns/DIV

AN122 FB4

Figure B4 • Excessive Damping is Characterized by Front Corner Rounding and Minimal Pulse-Top Aberrations. Trade Off is Relatively Slow Rise Time

---

Note 5: The LT1533's low noise performance and its measurement are discussed in Reference 25.

Note 6: 2N2501s are available from Semelab plc. Sales@semelab.co.uk; Tel. 44-0-1455-556565. A more common transistor, the 2N2369, may also be used but switching times are rarely less than 450ps.

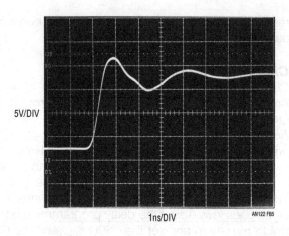

5V/DIV

1ns/DIV   AN122 FB5

Figure B5 • Minimal Damping Accentuates Rise Time, Although Pulse-Top Ringing is Excessive

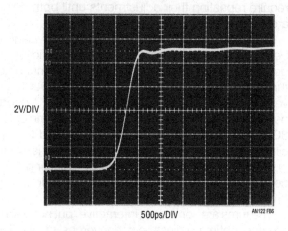

2V/DIV

500ps/DIV   AN122 FB6

Figure B6 • Optimal Damping Retards Pulse-Top Ringing While Preserving Rise Time

pulse purity.[7] Figures B4 through B6 detail the optimization procedure. In Figure B4, the trims are set for significant effect, resulting in a reasonably clean pulse but sacrificing rise time.[8] Figure B5 represents the opposite extreme. Minimal trim effect accentuates rise time, but promotes post-transition ring. Figure B6's compromise trimming is more desirable. Edge rate is only slightly reduced, but post-transition ring is significantly retarded, resulting in a 400ps rise time with high pulse purity.[9,10]

Note 7: Optimization procedures for obtaining high degrees of pulse purity while preserving rise time appear in Reference 11.
Note 8: The strata is becoming rarefied when a sub-nanosecond rise time is described as "sacrificed".
Note 9: Faster rise times are possible, although considerable finesse is required in Q5's selection, layout, mounting, terminal impedance choice and triggering. The 400ps rise time quoted represents readily reproducible results. Rise times below 300ps have been achieved, but require tedious effort.
Note 10: Accurate rise time determination at these speeds mandates verifying measurement signal path (cables, attenuators, probes, oscilloscope) integrity. See Appendix D, "Verifying Rise Time Measurement Integrity" and Appendix E, "Connections, Cables, Adapters, Attenuators, Probes and Picoseconds."

Appendix C
About Z_0 probes

When to roll your own and when to pay the money

Z_0 (e.g. "low impedance") probes provide the most faithful high speed probing mechanism available for low source impedances. Their sub-picofarad input capacitance and near ideal transmission characteristic make them the first choice for high bandwidth oscilloscope measurement. Their deceptively simple operation invites "do-it-yourself" construction but numerous subtleties mandate difficulty for prospective constructors. Arcane parasitic effects introduce errors as speed increases beyond about 100MHz (t_{RISE} 3.5ns). The selection and integration of probe materials and the probes physical incarnation require extreme care to obtain high fidelity at high speed. Additionally, the probe must include some form of adjustment to compensate small, residual parasitics. Finally, true coaxiality must be maintained when fixturing the probe at the measurement point, implying a high grade, readily disconnectable, coaxial connection capability.

Figure C1 shows that a Z_0 probe is basically a voltage divided input 50Ω transmission line. If R1 equals 450Ω, 10x attenuation and 500Ω input resistance result. R1 of 4950Ω causes a 100x attenuation with 5k input resistance. The 50Ω line theoretically constitutes a distortioness transmission environment. The apparent simplicity seemingly permits "do-it-yourself" construction but this section's remaining figures demonstrate a need for caution.

Figure C2 establishes a fidelity reference by measuring a clean 700ps rise time pulse using a 50Ω line terminated via a coaxial attenuator – no probe is employed. The waveform is singularly clean and crisp with minimal edge and post-transition aberrations. Figure C3 depicts the same pulse with a commercially produced 10x Z_0 probe in use. The probe is faithful and there is barely discernible error in the presentation. Photos C4 and C5, taken with two separately constructed "do-it-yourself" Z_0 probes, show errors.

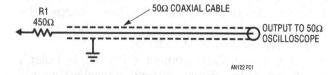

R1
450Ω   50Ω COAXIAL CABLE

OUTPUT TO 50Ω
OSCILLOSCOPE

AN122 FC1

Figure C1 • Conceptual 500Ω, "Z_0", 10x Oscilloscope Probe. If R1 = 4950Ω, 5k Input Resistance with 100x Signal Attenuation Results. Terminated Into 50Ω, Probe Theoretically Constitutes a Distortionless Transmission Line. "Do-It-Yourself" Probes Suffer Uncompensated Parasitics, Causing Unfaithful Response Above ≈100MHz (t_{RISE} = 3.5ns)

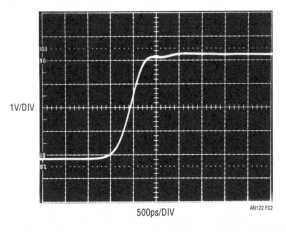

Figure C2 • 700ps Rise Time Pulse Observed Via 50Ω Line and Coaxial Attenuator Has Good Pulse Edge Fidelity With Controlled Post-Transition Events

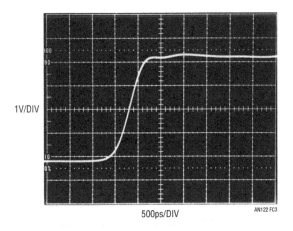

Figure C3 • Figure C2's Pulse Viewed With Tektronix Z_0 500Ω Probe (P-6056) Introduces Barely Discernible Error

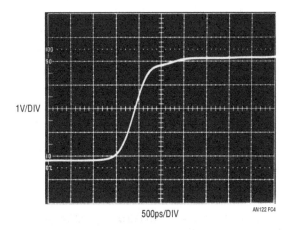

Figure C4 • "Do-It-Yourself" Z_0 Probe #1 Introduces Pulse Corner Rounding, Likely Due to Resistor/Cable Parasitic Terms or Incomplete Coaxiality. "Do-It-Yourself" Z_0 Probes Typically Manifest This Type Error at Rise Times ≤ 2ns

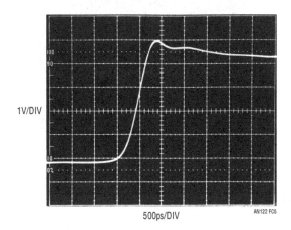

Figure C5 • "Do-It-Yourself" Z_0 Probe #2 Has Overshoot, Again Likely Due to Resistor/Cable Parasitic Terms or Incomplete Coaxiality. Lesson: At These Speeds, Don't "Do It Yourself"

In C4, "Probe #1" introduces pulse front corner rounding; "Probe #2" in C5 causes pronounced corner peaking. In each case, some combination of resistor/cable parasitics and incomplete coaxiality are likely responsible for the errors. In general, "do-it-yourself" Z_0 probes cause these types of errors beyond about 100MHz (t_{RISE} 3.5ns). At higher speeds, if waveform fidelity is critical, it's best to pay the money.

Appendix D
Verifying rise time measurement integrity

Any measurement requires the experimenter to insure measurement confidence. Some form of calibration check is always in order. High speed time domain measurement is particularly prone to error and various techniques can promote measurement integrity.

Figure D1's battery-powered 200MHz crystal oscillator produces 5ns markers, useful for verifying oscilloscope time base accuracy. A single 1.5 AA cell supplies the LTC3400 boost regulator, which produces 5V to run the oscillator. Oscillator output is delivered to the 50Ω load via a peaked attenuation network. This provides well defined 5ns markers (Figure D2) and prevents overdriving low level sampling oscilloscope inputs.

Once time base accuracy is confirmed it is necessary to check rise time. The lumped signal path rise time, including attenuators, connections, cables, probes, oscilloscope and anything else, should be included in this measurement. Such "end-to-end" rise time checking is an effective way to promote

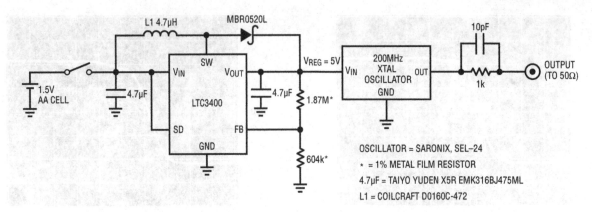

Figure D1 • 1.5V Powered, 200MHz Crystal Oscillator Provides 5ns Time Markers. Switching Regulator Converts 1.5V to 5V to Power Oscillator

OSCILLATOR = SARONIX, SEL–24
* = 1% METAL FILM RESISTOR
4.7µF = TAIYO YUDEN X5R EMK316BJ475ML
L1 = COILCRAFT D0160C-472

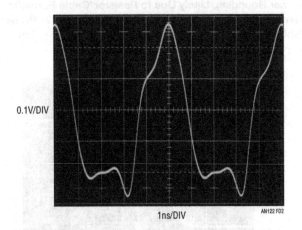

0.1V/DIV

1ns/DIV

AN122 FD2

Figure D2 • Time Mark Generator Output Terminated into 50Ω. Peaked Waveform is Optimal for Verifying Time Base Calibration

meaningful results. A guideline for insuring accuracy is to have 4x faster measurement path rise time than the rise time of interest. Thus, Appendix Figure B6's 400ps rise time measurement requires a verified 100ps measurement path rise time to support it. Verifying the 100ps measurement path rise time, in turn, necessitates a 25ps rise time test step. Figure D3 lists some very fast edge generators for rise time checking.[1]

The Hewlett-Packard 1105A/1106A, specified at 20ps rise time, was used to verify Appendix Figure A3's measurement signal path. Figure D4 indicates a 140ps rise time, promoting measurement confidence.

Note 1: This is a fairly exotic group, but equipment of this caliber really is necessary for rise time verification.

MANUFACTURER	MODEL NUMBER	RISE TIME	AMPLITUDE	AVAILABILITY	COMMENTS
Avtech	AVP2S	40ps	0V to 2V	Current Production	Free Running or Triggered Operation, 0MHz to 1MHz
Hewlett-Packard	213B	100ps	≈175m	Secondary Market	Free Running or Triggered Operation to 100kHz
Hewlett-Packard	1105A/1108A	60ps	≈200m	Secondary Market	Free Running or Triggered Operation to 100kHz
Hewlett-Packard	1105A/1106A	20ps	≈200m	Secondary Market	Free Running or Triggered Operation to 100kHz
Picosecond Pulse Labs	TD1110C/TD1107C	20ps	≈230m	Current Production	Similar to Discontinued HP1105/1106/8A. See above
Stanford Research Systems	DG535 OPT 04A	100ps	0.5V to 2	Current Production	Must be Driven with Stand-alone Pulse Generator
Tektronix	284	70ps	≈200Vm	Secondary Market	50kHz Repetition Rate. Pre-trigger 75ns to 150ns Before Main Output. Calibrated 100MHz and 1GHz Sine Wave Auxiliary Outputs
Tektronix	111	500ps	≈±10V	Secondary Market	10kHz to 100kHz Repetition Rate. Positive or Negative Outputs. 30ns to 250ns Pre-trigger Output. External Trigger Input. Pulse Width Set with Charge Lines
Tektronix	067-0513-00	30ps	≈400m	Secondary Market	60ns Pre-trigger Output. 100kHz Repetition Rate
Tektronix	109	250ps	0V to ±55V	Secondary Market	≈600Hz Repetition Rate (High Pressure Hg Reed Relay Based). Positive or Negative Outputs. Pulse Width Set by Charge Lines

Figure D3 • Picosecond Edge Generators Suitable for Rise Time Verification. Considerations Include Speeds, Features and Availability

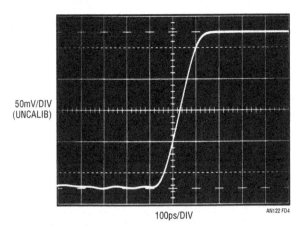

50mV/DIV
(UNCALIB)

100ps/DIV

AN122 FD4

Figure D4 • 20ps Step Produces ≈140ps Probe/ Oscilloscope Rise Time, Verifying Appendix Figure A3's Signal Path Rise Time

connection mechanism and high frequency compensation. Passive "Z_0" types, commercially available in 500Ω (10x) and 5kΩ (100x) impedances, have input capacitance below 1pF.[1] Any such probe must be carefully frequency compensated before use or misrepresented measurement will result. Inserting the probe into the signal path necessitates some form of signal pick-off which nominally does not influence signal transmission. In practice, some amount of disturbance must be tolerated and its effect on measurement results evaluated. High quality signal pick-offs always specify insertion loss, corruption factors and probe output scale factor.

The preceding emphasizes vigilance in designing and maintaining a signal path. Skepticism, tempered by enlightenment, is a useful tool when constructing a signal path and no amount of hope is as effective as preparation and directed experimentation.

Appendix E
Connections, cables, adapters, attenuators, probes and picoseconds

Sub-nanosecond rise time signal paths must be considered as transmission lines. Connections, cables, adapters, attenuators and probes represent discontinuities in this transmission line, deleteriously affecting its ability to faithfully transmit desired signal. The degree of signal corruption contributed by a given element varies with its deviation from the transmission line's nominal impedance. The practical result of such introduced aberrations is degradation of pulse rise time, fidelity, or both. Accordingly, introduction of elements or connections to the signal path should be minimized and necessary connections and elements must be high grade components. Any form of connector, cable, attenuator or probe must be fully specified for high frequency use. Familiar BNC hardware becomes lossy at rise times much faster than 350ps. SMA components are preferred for the rise times described in the text. Additionally, to minimize inductance and cable induced mismatch and distortion, the text's pulse amplifier output should be connected *directly* (no cable) to the diode under test. Mixing signal path hardware types via adapters (e.g. BNC/SMA) should be avoided. Adapters introduce significant parasitics, resulting in reflections, rise time degradation, resonances and other degrading behavior.

Similarly, oscilloscope connections should be made directly to the instrument's 50Ω inputs, avoiding probes. If probes must be used, their introduction to the signal path mandates attention to their

Appendix F
Another way to do it

If some restrictions are tolerable, an elegantly simple alternative method for generating the fast rise 1A pulse is available. The Tektronix type 109 mercury wetted reed relay based pulse generator will put a 50V pulse into 50Ω (1A) in 250ps.[1] Pulse width is set by an externally connected charge line with an approximate scale factor of 2ns/ft. Figure F1, a

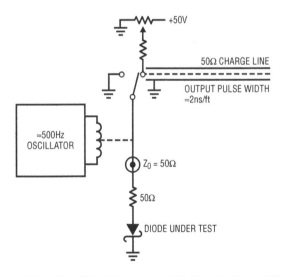

Figure F1 • Simplified Operation of Tektronix Type 109 Mercury Wetted Reed Relay Based Pulse Generator. When Right Side Contacts Close, Charge Line Discharges Into 50Ω-Diode Load. Strict Attention to Construction Allows Wideband, 50Ω Characteristics, Permitting 250ps Rise Time, High Purity Output Pulse

Note 1: See Appendix C, "About Z0 Probes"

Note 1: See Reference 36.

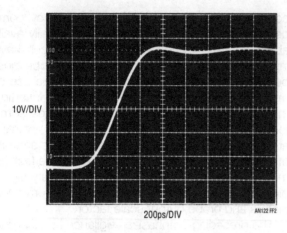

10V/DIV

200ps/DIV   AN122 FF2

Figure F2 • Tektronix Type 109 Produces High Purity, 50V, 1A Pulse, Driving Monitoring 1GHz Oscilloscope to its 350ps Rise Time Limit

simplified schematic, shows type 109 operation. When the relay contacts close, the charge line discharges via the 50Ω-diode path. The pulse extends until the line depletes; depletion time depends on line length. The relay structure is very carefully arranged to assume wideband, 50Ω characteristics. Figure F2 shows the result. The 109 drives the monitoring 1GHz oscilloscope to its 350ps rise time limit with a 50V high fidelity pulse.

Operating restrictions include finite relay life (-≈200 hours), obtaining the instrument (out of production for 20+ years), difficulty in observing its low frequency output on some oscilloscopes and test fixture layout sensitivity due to the 250ps rise time. Additionally, the faster rise time may not approximate actual circuit operating conditions as closely as the text's 2ns circuit.

Section 3

Linear Regulator Design

Performance verification of low noise, low dropout regulators (12)

In an increasing trend, telecommunications, networking, audio and instrumentation require low noise power supplies. In particular, there is interest in low noise, low dropout linear regulators (LDO). Establishing and specifying LDO dropout performance is relatively easy to do. Verifying that a regulator meets dropout specification is similarly straightforward. Accomplishing the same missions for noise and noise testing is considerably more involved. The noise bandwidth of interest must be called out, along with operating conditions. Low noise performance is affected by numerous subtleties; changes in operating conditions can cause unwelcome surprises. Because of this, LDO noise must be quoted under specified operating and bandwidth conditions to be meaningful. Failure to observe this precaution results in misleading data and erroneous conclusions. This Application Note suggests a noise testing method, details its implementation and presents results.

Performance verification of low noise, low dropout regulators

Silence of the amps

12

Jim Williams Todd Owen

Introduction

In an increasing trend, telecommunications, networking, audio and instrumentation require low noise power supplies. In particular, there is interest in low noise, low dropout linear regulators (LDOs). These components power noise-sensitive circuitry, circuitry that contains noise-sensitive elements or both. Additionally, to conserve power, particularly in battery driven apparatus such as cellular telephones, the regulators must operate with low input-to-output voltages.[1] Devices presently becoming available meet these requirements (see separate section, "A family of $20\mu V_{RMS}$ noise, low dropout regulators").

Noise and noise testing

Establishing and specifying LDO dropout performance is relatively easy to do. Verifying that a regulator meets dropout specification is similarly straightforward. Accomplishing the same missions for noise and noise testing is considerably more involved. The noise bandwidth of interest must be called out, along with operating conditions. Operating conditions can include regulator input and output voltage, load, assorted discrete components, etc. Low noise performance is effected by numerous subtleties; changes in operating conditions can cause unwelcome surprises.[2] Because of this, LDO noise must be quoted under specified operating and bandwidth conditions to be meaningful. Failure to observe this precaution will result in misleading data and erroneous conclusions.

Noise testing considerations

What noise bandwidth is of interest and why is it interesting? In most systems, the range of 10Hz to 100kHz is the information signal processing area of concern. Additionally, linear regulators produce little noise energy outside this region.[3] These considerations suggest a measurement bandpass of 10Hz to 100kHz, with steep slopes at the band limits. Figure 12.1 shows a conceptual filter for LDO noise testing. The Butterworth sections are the key to steep slopes and flatness in the passband. The small input level requires 60dB of low noise gain to provide adequate signal for the Butterworth filters. Figure 12.2 details the filter scheme. The regulator under test is at the diagram's center.[4] A1-A3 make up a 60dB gain highpass section. A1 and A2, extremely low noise devices ($<1nV\sqrt{Hz}$), comprise a 60dB gain stage with a 5Hz highpass input. A3 provides a 10Hz, 2nd order Butterworth highpass characteristic. The LTC1562 filter block is arranged as a 4th order Butterworth lowpass. Its output is delivered via the $330\mu F$-100Ω highpass network. The circuit's output drives a thermally responding RMS voltmeter.[5] Note that all circuit power is furnished by batteries, precluding ground loops from corrupting the measurement.

Instrumentation performance verification

Good measurement technique dictates verifying the noise test instrumentation's performance. Figure 12.3's spectral

(Continued on page 252)

Note 3: Switching regulators are an entirely different proposition, requiring very broadband noise measurement. See Reference 1.
Note 4: Component choice for the regulator, more critical than might be supposed, is discussed in Appendix B, "Capacitor selection considerations."
Note 5: The choice of RMS voltmeter is absolutely crucial to obtain meaningful measurements. See Appendix C, "Understanding and selecting RMS voltmeters."

Note 1: See Appendix A, "Architecture of a low noise LDO," for design considerations of these devices.
Note 2: See Appendix D, "Practical considerations for selecting a low noise LDO."

Analog Circuit and System Design: A Tutorial Guide to Applications and Solutions. DOI: 10.1016/B978-0-12-385185-7.00012-3

A family of 20μV$_{RMS}$ noise, low dropout regulators

Telecom and instrumentation applications often require a low noise voltage regulator. Frequently this requirement coincides with the need for low regulator dropout and small quiescent current. A recently introduced family of devices addresses this problem. Figure 12.A shows a variety of packages, power ranges and features in three basic regulator types. The SOT-23 packaged LT1761 has only 20μV$_{RMS}$ noise with 300mV dropout at 100mA. Quiescent current is only 20μA.

Applying the regulators

Applying the regulators is simple. Figure 12.B shows a minimum parts count, 3.3V output design. This circuit appears similar to conventional approaches with a notable exception: a bypass pin (BYP) is returned to the output via a 0.01μF capacitor. This path filters the internal reference's output, minimizing regulator output noise. It is the key to the 20μV$_{RMS}$ noise performance. A shutdown pin ($\overline{SHDN}$), when pulled low, turns off the regulator output while keeping current drain inside 1μA. Dropout characteristics appear in Figure 12.C. Dropout scales with output current, falling to less than 100mV at low currents.

These devices provide the lowest available output noise in a low dropout regulator without compromising other parameters. Their performance, ease of use and versatility allow use in a variety of noise-sensitive applications.

REGULATOR TYPE	OUTPUT CURRENT	RMS NOISE (10Hz to 100kHz) C$_{BYP}$ = 0.01μF	PACKAGE OPTIONS	FEATURES	QUIESCENT CURRENT	SHUTDOWN CURRENT
LT1761	100mA	20μV$_{RMS}$	SOT23-5	Shutdown, Reference Bypass, Adjustable Output; SOT23 Package Mandates Selecting Any Two Features	20μA	<1μA
LT1762	150mA	20μV$_{RMS}$	MSOP-8	Shutdown, Reference Bypass, Adjustable Output	25μA	<1μA
LT1962	300mA	20μV$_{RMS}$	MSOP-8	Shutdown, Reference Bypass, Adjustable Output	30μA	<1μA
LT1763	500mA	20μV$_{RMS}$	SO-8	Shutdown, Reference Bypass, Adjustable Output	30μA	<1μA
LT1963	1.5A	40μV$_{RMS}$	SO-8, SOT223-3, DD-5, TO220	Shutdown, Adjustable Output, Fast Transient Response	1mA	<1μA
LT1764	3A	40μV$_{RMS}$	DD-5, TO220	Shutdown, Adjustable Output, Fast Transient Response	1mA	<1μA

Figure 12.A • Low Noise LDO Family Short-Form Specifications. Quiescent Current Scales with Output Current Capability, Although Noise Performance Remains Nearly Constant

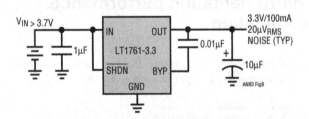

Figure 12.B • Applying the Low Noise, Low Dropout, Micropower Regulator. Bypass Pin and Associated Capacitor are Key to Low Noise Performance

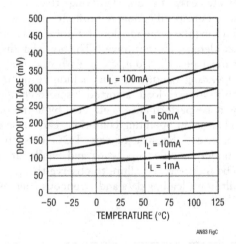

Figure 12.C • Figure 12.B's Dropout Voltage at Various Currents

plot of the filter section shows essentially flat response in the 10Hz to 100kHz passband with abrupt slopes at the band extremes. Figure 12.4, expanding the vertical scale to 1dB/division, reveals some flatness deviation but well within 1dB throughout nearly the entire passband. Grounding the filter's input determines the tester's noise floor. Figure 12.5 shows less than $4\mu V_{P-P}$, corresponding to a $0.5\mu V_{RMS}$ voltmeter reading. This is only about 0.5% of full scale ($100\mu V_{RMS}$), contributing negligible error. These results assure the confidence necessary to proceed with regulator noise measurement.

Regulator noise measurement

Regulator noise measurement begins with attention to test setup details. The extremely low signal levels require attention to shielding, cable management, layout and component choice.[6] Figure 12.6a is the bench arrangement. The photo shows the completely shielded environment required to obtain faithful noise measurements. The metal can[7] encloses the regulator under test and its internal battery power supply. A BNC fitting (photo lower center) connects the regulator output to the noise filter test circuit (black box). Note that the monitoring oscilloscope and voltmeter are not simultaneously connected to the output, precluding ground loops which would corrupt the measurement.

Figure 12.6b details the regulator enclosure with its cover removed. The battery supply is visible; the regulator occupies the can center. The BNC fitting connecting the noise filter box (lower left) eliminates triboelectric disturbances a cable might contribute.

Figure 12.6c is the noise test circuit box. Functions are as labeled in the photo. The two capped BNC connectors (box lower) are unused box entries.

Figure 12.7's oscilloscope photo shows an LT1761 regulator's noise measured at the filter output. Monitoring this point with the RMS voltmeter shows a $20\mu V_{RMS}$ reading. Figure 12.8's spectral plot of this noise indicates diminished power above 1kHz, in accordance with expected regulator noise density. Figure 12.9 shows more complete spectral noise density data for three regulator types. Noise power decays uniformly with increasing frequency, although the three regulators show some dispersion below 200Hz.

Bypass capacitor (C_{BYP}) influence

The regulator's internal voltage reference contributes most of the device's noise. The reference bypass capacitor filters reference noise, precluding it from appearing, in amplified form, at the output.[8] Figure 12.10 is a study of regulator noise vs various values of C_{BYP}. Figure 12.10a shows substantial noise for $C_{BYP}=0\mu F$, while 10d displays nearly $9\times$ improvement with $C_{BYP}=0.01\mu F$; intermediate values of C_{BYP} (10b and 10c) produce commensurate results.

Interpreting comparative results

Figure 12.11's photos compare an LT1761-5's output noise (12.11d) with three other regulators (12.11a, 12.11b and 12.11c). These three devices are manufacturer specified for low noise performance, but the photos do not indicate this. The seeming contradiction is probably due to ambiguity in testing methods or specifications. For example, inappropriate choice of test equipment (see Appendix C) or measurement bandwidth can easily cause huge ($5\times$) errors. This uncertainty mandates the noise testing described to insure realistic conclusions.

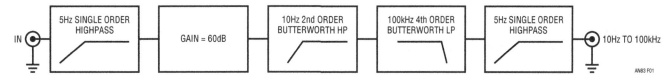

Figure 12.1 • Filter Structure for Noise Testing LDOs. Butterworth Sections Provide Appropriate Response in Desired Frequency Range

---

Note 6: Capacitor choice is discussed in Appendix B, "Capacitor selection considerations."
Note 7: The cookies were excellent, particularly the thin ones with sugar on top.

Note 8: See Appendix A, "Architecture of a low noise LDO," for details.

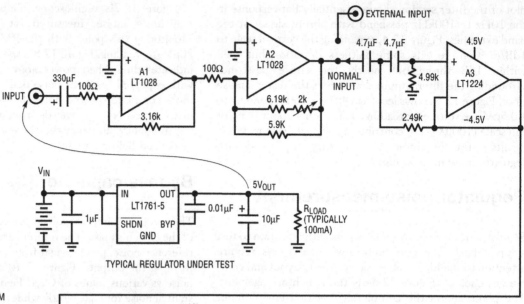

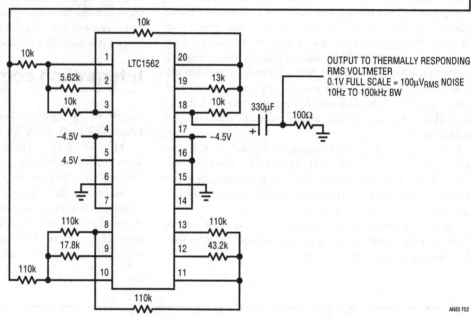

ALL RESISTORS 1% METAL FILM
4.7µF CAPACITORS = MYLAR, WIMA MKS-2
330µF CAPACITORS = SANYO OSCON
±4.5V DERIVED FROM 6AA CELLS
POWER REGULATOR FROM APPROPRIATE
NUMBER OF D SIZE BATTERIES

Figure 12.2 • Implementation of Figure 12.1. Low Noise Amplifiers Provide Gain and Initial Highpass Shaping. LTC1562 Filter Supplies 4th Order Butterworth Lowpass Characteristic

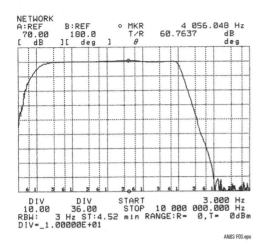

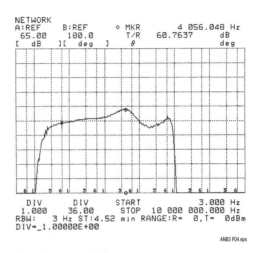

Figure 12.3 • HP-4195A Spectrum Analyzer Plot of Filter Characteristics. Filter Performance Is Nearly Flat Over Desired 10Hz to 100kHz Range with Steep Rolloff Outside Bandpass Region

Figure 12.4 • Expanded Scale Examination of Passband Shows Flatness within 1dB Over Almost Entire Measurement Range

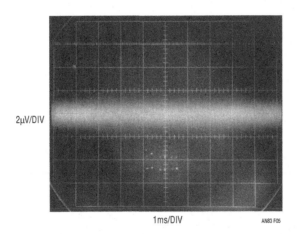

2μV/DIV

1ms/DIV

Figure 12.5 • <4μV$_{P-P}$ Test Setup Noise Residue Corresponds to About 0.5μV$_{RMS}$ Measurement Noise Floor

Figure 12.6a • LDO Noise Measurement Bench Setup. Shielded Can Contains Regulator; Noise Filter Circuitry Occupies Box at Photo Lower Center. Oscilloscope and RMS Voltmeter Are Not Simultaneously Connected, Precluding Ground Loop from Corrupting Measurement

Figure 12.6b • Shielded Can with Cover Removed. LDO Under Test Occupies Center. D-Cells Provide Power, Eliminate Potential Ground Loop. BNC Fitting (Photo Lower Left) Connects Output to Filter Circuit Test Box, Minimizing Triboelectric Based Errors

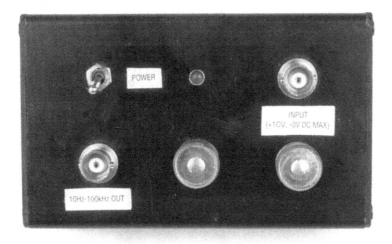

AN83 F06b

Figure 12.6c • Noise Filter Test Box Is Fully Shielded. Connections Are As Indicated. Capped BNC Fittings (Photo Low Center and Right) Are Unused Entries

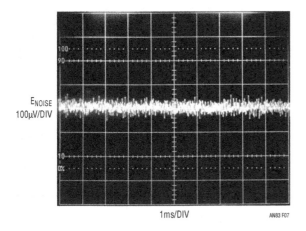

1ms/DIV   AN83 F07

Figure 12.7 • LT1761 Output Voltage Noise in a 10Hz to 100kHz Bandwidth. RMS Noise Measures 20μV$_{RMS}$

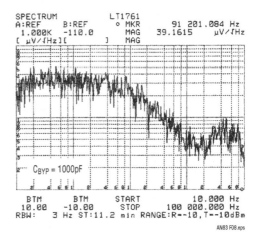

AN83 F08.eps

Figure 12.8 • Noise Spectrum Plot Shows Diminishing Power Above 1kHz

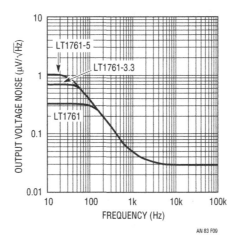

AN 83 F09

Figure 12.9 • Output Noise Spectral Density Data Curves for Three Regulators Show Dispersion Below 200Hz

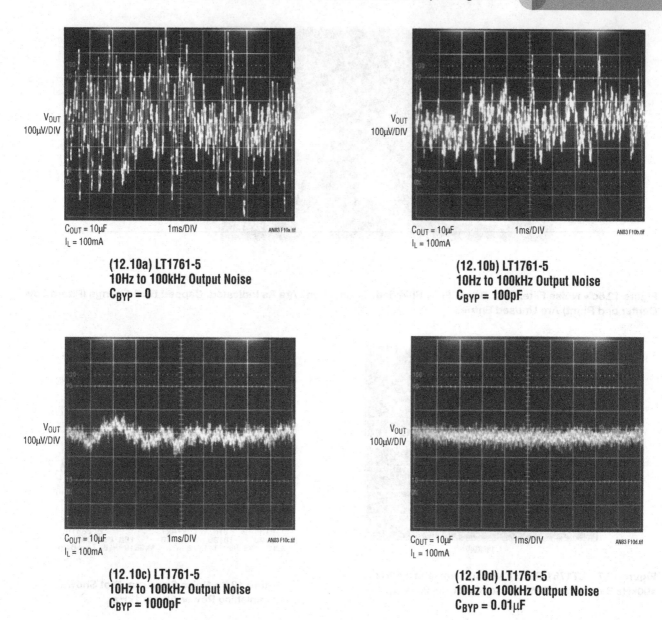

$C_{OUT} = 10\mu F$ 1ms/DIV
$I_L = 100mA$

(12.10a) LT1761-5
10Hz to 100kHz Output Noise
$C_{BYP} = 0$

$C_{OUT} = 10\mu F$ 1ms/DIV
$I_L = 100mA$

(12.10b) LT1761-5
10Hz to 100kHz Output Noise
$C_{BYP} = 100pF$

$C_{OUT} = 10\mu F$ 1ms/DIV
$I_L = 100mA$

(12.10c) LT1761-5
10Hz to 100kHz Output Noise
$C_{BYP} = 1000pF$

$C_{OUT} = 10\mu F$ 1ms/DIV
$I_L = 100mA$

(12.10d) LT1761-5
10Hz to 100kHz Output Noise
$C_{BYP} = 0.01\mu F$

Figure 12.10 • Regulator Noise for Various Bypass Capacitor (C_{BYP}) Values. Noise Decreases with Increasing C_{BYP}

257

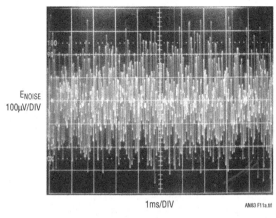

(12.11a) Manufacturer "MI" Output Voltage Noise (5V Output)

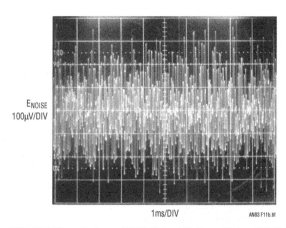

(12.11b) Manufacturer "NS" Output Voltage Noise (5V Output)

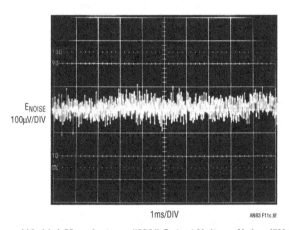

(12.11c) Manufacturer "MA" Output Voltage Noise (5V Output)

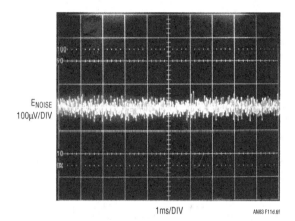

(12.11d) LT1761-5 Output Voltage Noise (5V Output)

Figure 12.11 • Noise for LT1761-5 vs Three Other Devices. "C" Is Specified for RMS Noise Figure Approaching the LT1761-5, but in a Restricted Noise Measurement Bandwidth. Caveat emptor!

References

1. Williams, J., "A Monolithic Switching Regulator with 100μV Output Noise," Linear Technology Corporation Application Note 70, October 1997.
2. Sheingold, D. H. (editor), "Nonlinear Circuits Handbook," 2nd Edition, Analog Devices, Inc., 1976.
3. Kitchen, C., and Counts, L., "RMS-to-DC Conversion Guide," Analog Devices, Inc. 1986.
4. Williams, J., "Practical Circuitry for Measurement and Control Problems," "Broadband Random Noise Generator," "Symmetrical White Gaussian Noise" Appendix B, Linear Technology Corporation, Application Note 61, August 1994, pp. 24–26, 38–39.
5. General Radio Company, Type 1390B Random Noise Generator Operating Instructions, October 1961.
6. Hewlett-Packard Company, "1968 Instrumentation. Electronic—Analytical—Medical", AC Voltage Measurement. Hewlett-Packard Company, 1968, pp. 197–198.
7. Justice, G., "An RMS-Responding Voltmeter with High Crest Factor Rating," Hewlett-Packard Journal, Hewlett-Packard Company, January 1964.
8. Hewlett-Packard Company. "Model HP3400A RMS Voltmeter Operating and Service Manual," Hewlett-Packard Company, 1965.
9. Williams, J., "A Monolithic IC for 100MHz RMS/DC Conversion," Linear Technology Corporation, Application Note 22, September 1987.
10. Ott, W.E., "A New Technique of Thermal RMS Measurement," IEEE Journal of Solid State Circuits, December 1974.
11. Williams, J.M. and Longman, T.L., " A 25MHz Thermally Based RMS-DC Converter," 1986 IEEE ISSCC Digest of Technical Papers.
12. O'Neill, P.M., "A Monolithic Thermal Converter," H.P. Journal, May 1980.
13. Williams, J., "A Fourth Generation of LCD Backlight Technology," "RMS Voltmeters," Linear Technology Corporation, Application Note 65, November 1995, pp. 82–83.
14. Tektronix, Inc., "Type 1A7A Differential Amplifier Instruction Manual," Check Overall Noise Level Tangentially, 1968, pp. 5-36 and 5-37.

Appendix A
Architecture of a low noise LDO

Noise minimization

The low noise LDOs use Figure A1's scheme, with special attention to minimizing noise transmission within the loop and from unregulated input. The internal voltage reference's noise is filtered by C_{BYP}. Additionally, the error amplifier's frequency response is shaped to minimize noise contribution while preserving transient response and PSRR. Regulators which do not do this have poor noise rejection and transient performance.

Pass element considerations

Extremely low dropout voltage requires considering the pass element. Dropout limitations are set by the pass elements on-impedance limits. The ideal pass element has zero impedance capability between input and output and consumes no drive energy.

A number of design and technology options offer various trade-offs and advantages. Figure A2 lists some pass element candidates. Followers offer current gain, ease of loop compensation (voltage gain is below unity) and the drive current ends up going to the load. Unfortunately, saturating a follower requires voltage overdriving the input (e.g. base, gate). Since drive is usually derived directly from V_{IN}, this is difficult. Practical circuits must either generate the overdrive or obtain it elsewhere. Without voltage

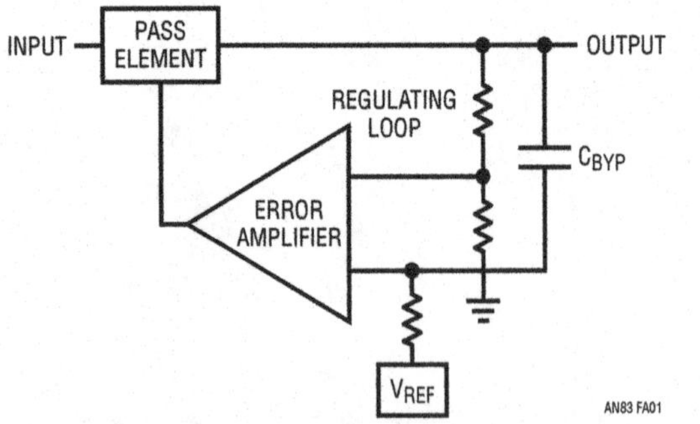

Figure A1 • Simplified Low Noise LDO Regulator. Voltage Reference Is Filtered by C_{BYP}, Isolating Noise from Regulating Loop. Error Amplifier's Frequency Compensation Prevents Transmission of Input Noise While Preserving Transient Response

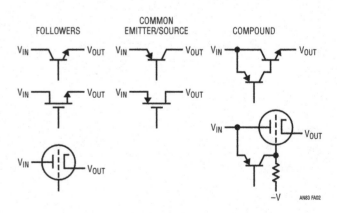

Figure A2 • Linear Regulator Pass Element Candidates

overdrive, the saturation loss is set by V_{BE} in the bipolar case and channel on-resistance for MOS. MOS channel on-resistance varies considerably under these conditions, although bipolar losses are more predictable. Note that voltage losses in driver stages (Darlington, etc.) add directly to the dropout voltage. The follower output used in conventional 3-terminal IC regulators combines with drive stage losses to set dropout at 3V.

Common emitter/source is another pass element option. This configuration removes the V_{BE} loss in the bipolar case. The PNP version is easily fully saturated, even in IC form. The trade-off is that the base current never arrives at the load, wasting power. At higher currents, base drive losses can negate a common emitter's saturation advantage. As in the follower example, Darlington connections exacerbate the problem. Achieving low dropout in a monolithic PNP regulator requires a PNP structure that attains low dropout while minimizing base drive loss. This is particularly the case at higher pass currents. Considerable effort was expended in this direction in the LT176X through LT196X designs.

Common source connected P-channel MOSFETs are also candidates. They do not suffer the drive losses of bipolars, but typically require volts of gate-channel bias to fully saturate. In low voltage applications this may require generation of negative potentials. Additionally, P-channel devices have poorer saturation than equivalent size N-channel devices.

The voltage gain of common emitter and source configurations is a loop stability concern but is manageable.

Compound connections using a PNP driven NPN are a reasonable compromise, particularly for high power (beyond 250mA) IC construction. The trade-off between the PNP V_{CE} saturation term and reduced drive losses over a conventional PNP structure is favorable. Also, the major current flow is through a power NPN, easily realized in monolithic form. The connection has voltage gain, necessitating attention to loop frequency compensation. Regulators utilizing this pass scheme can supply up to 7.5A with dropouts below 1.5V (LT1083 through LT1086 series).

Readers are invited to submit results obtained with our emeritus thermionic friends, shown out of respectful courtesy.

Dynamic characteristics

The LT176X through LT196X's low quiescent currents do not preclude good dynamics. Usually, low quiescent power devices are associated with slow dynamics and instability. The devices are stable (no output oscillation) even with low ESR ceramic output capacitors. This contrasts with conventional LDO regulators that often oscillate with ceramic capacitors.

The internal architecture provides an added bonus in transient performance when the 0.01µF noise capacitor is added. Transient response for a 10mA to 100mA load step with a 10µF output capacitor appears in Figure A3 with the capacitor deleted. Figure A4 shows the same situation with the 0.01µF bypass capacitor in place. Settling time and amplitude are markedly reduced.

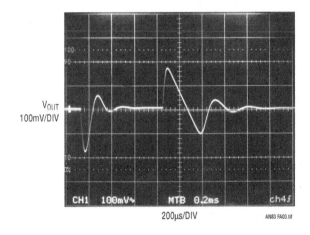

Figure A3 • Transient Response without Noise Bypass Capacitor

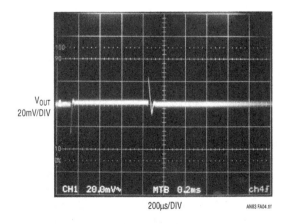

Figure A4 • Noise Bypass Capacitor Improves Transient Response. Note Voltage Scale Change

Appendix B
Capacitor selection considerations

Bypass capacitance and low noise performance

Adding a capacitor from the regulator's V_{OUT} to BYP pin lowers output noise. A good quality low leakage capacitor is recommended. This capacitor bypasses the regulator's reference, providing a low frequency noise pole. A 0.01μF capacitor lowers the output voltage noise to 20μV$_{RMS}$. Using a bypass capacitor also improves transient response. With no bypassing and a 10μF output capacitor, a 10mA to 500mA load step settles within 1% of final value in under 100μs. With a 0.01μF bypass capacitor, the output settles within 1% for the same load step in under 10μs; total output deviation is inside 2.5%. Regulator start-up time is inversely proportional to bypass capacitor size, slowing to 15ms with a 0.01μF bypass capacitor and 10μF at the output.

Output capacitance and transient response

The regulators are designed to be stable with a wide range of output capacitors. Output capacitor ESR affects stability, most notably with small capacitors. A 3.3μF minimum output value with ESR of 3Ω or less is recommended to prevent oscillation. Transient response is a function of output capacitance. Larger values of output capacitance decrease peak deviations, providing improved transient response for large load current changes. Bypass capacitors, used to decouple individual components powered by the regulator, increase effective output capacitor value. Larger values of reference bypass capacitance dictate larger output capacitors. For 100pF of bypass capacitance, 4.7μF of output capacitor is recommended. With 1000pF of bypass capacitor or larger, a 6.8μF output capacitor is required.

Figure B1's shaded region defines the regulator's stability range. Minimum ESR needed is set by the amount of bypass capacitance used, while maximum ESR is 3Ω.

Ceramic capacitors

Ceramic capacitors require extra consideration. They are manufactured with a variety of dielectrics, each with different behavior across temperature and applied voltage. The most common dielectrics are Z5U, Y5V, X5R and X7R. The Z5U and Y5V dielectrics provide high capacitance in a small package, but exhibit strong voltage and temperature coefficients as shown in Figures B2 and B3. Used with a 5V regulator, a 10μF Y5V capacitor shows values as low as 1μF to 2μF over the operating temperature range. The X5R and X7R

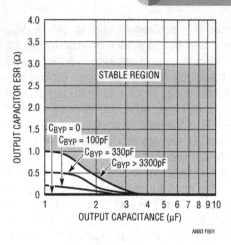

Figure B1 • Regulator Stability for Various Output and Bypass (C_{BYP}) Capacitor Characteristics

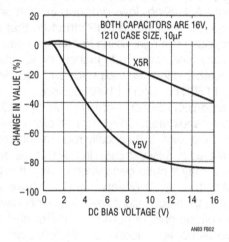

Figure B2 • Ceramic Capacitor DC Bias Characteristics Indicate Pronounced Voltage Dependence. Device Must Provide Desired Capacitance Value at Operating Voltage

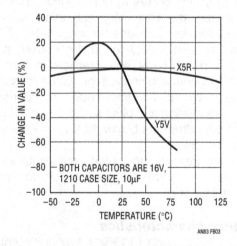

Figure B3 • Ceramic Capacitor Temperature Characteristics Show Large Capacitance Shift. Effect Should Be Considered When Determining Circuit Error Budget

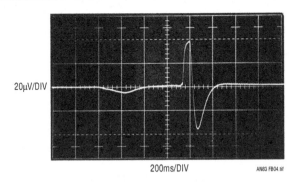

20µV/DIV

200ms/DIV   AN83 FB04.tif

Figure B4 • A Ceramic Capacitor Responds to Light Pencil Tapping. Piezoelectric Based Response Approaches 80µV$_{P-P}$

dielectrics have more stable characteristics and are more suitable for output capacitor use. The X7R type has better stability over temperature, while the X5R is less expensive and available in higher values.

Voltage and temperature coefficients are not the only problem sources. Some ceramic capacitors have a piezoelectric response. A piezoelectric device generates voltage across its terminals due to mechanical stress, similar to the way a piezoelectric accelerometer or microphone works. For a ceramic capacitor the stress can be induced by vibrations in the system or thermal transients. The resulting voltages produced can cause appreciable amounts of noise, especially when a ceramic capacitor is used for noise bypassing. A ceramic capacitor produced Figure B4's trace in response to light tapping from a pencil. Similar vibration induced behavior can masquerade as increased output voltage noise.

Appendix C
Understanding and selecting RMS voltmeters

The choice of AC voltmeter is absolutely crucial for meaningful measurements. The AC voltmeter must respond faithfully to the RMS value of the measured noise. The majority of AC voltmeters (including DVMs with AC ranges) are *not* capable of doing this. This includes instruments with "true RMS" AC scales. As such, selecting an appropriate instrument requires care. The selection process begins with a basic understanding of AC voltmeter types.[1]

AC voltmeter types

There are three basic AC voltmeter types. They include rectify and average, analog computing and thermal. The thermal approach is the only one that is inherently accurate, regardless of input waveshape. This feature is particularly relevant to noise RMS amplitude determination.

Rectify and average

The rectify and average scheme (Figure C1) applies the AC input to a precision rectifier. The rectifier output feeds a simple RC averager which is gain scaled, providing the output. In practice, gain is set so the DC output equals the RMS value of a sine wave input. If the input remains a pure sine wave, accuracy can be quite good. However, nonsinusoidal inputs cause large errors. This type of voltmeter is only accurate for sine wave inputs, with increasing error as the input departs from sinosoidal.

Analog computation

Figure C2 shows a more sophisticated AC voltmeter method. Here, the instantaneous value is (ideally) continuously computed by an analog computational loop. The DC output follows the equation noted, resulting in much better accuracy as input waveshape varies. Almost all commercial implementations of this approach utilize logarithmically based analog computing techniques. Unfortunately, dynamic limitations in the ZY/X block dictate bandwidth restrictions. These circuits typically develop significant errors beyond 20kHz to 200kHz.

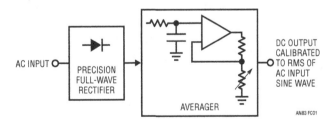

AC INPUT — PRECISION FULL-WAVE RECTIFIER — AVERAGER — DC OUTPUT CALIBRATED TO RMS OF AC INPUT SINE WAVE

AN83 FC01

Figure C1 • Rectify-and-Average Based AC-DC Converter. Gain Is Set So DC Output Equals RMS Value of Sine Wave Input. Nonsine Inputs Produce Errors

Note 1: Another way to approximately measure RMS AC noise using an oscilloscope is the tangential method. See Reference 14.

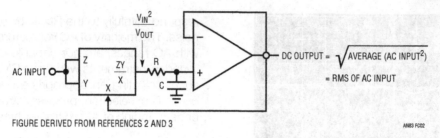

FIGURE DERIVED FROM REFERENCES 2 AND 3

AN83 FC02

Figure C2 • Analog Computer Based AC-DC Converter. Loop Continuously Computes Input's RMS Value. Bandwidth Limitations Produce High Frequency Errors

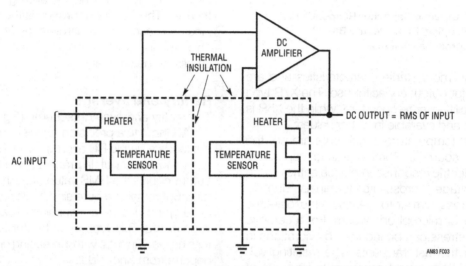

AN83 FC03

Figure C3 • Thermally Based AC-DC Converter Converts AC Input to Heat. Remaining Circuitry Determines DC Value (Output) Required to Produce Identical Heating. Error Is Extraordinarily Low, Even at Waveshape and Bandwidth Extremes

Thermal

The thermally based AC voltmeter is inherently insensitive to input waveshape, particularly suiting it to noise RMS amplitude measurement. Additionally, high accuracy at bandwidths exceeding 100MHz is achievable. Figure C3 diagrams the classic thermal scheme.[2] It is composed of matched heater-temperature sensor pairs and an amplifier. The AC input drives a heater, warming it. The temperature sensor associated with this heater responds, biasing the amplifier. The amplifier closes its feedback loop by driving the output heater to warm its associated temperature sensor. When the loop closes, the heaters are at the same temperature. This "force-balance" action results in the DC output equalling the input heater's RMS heating value—the fundamental definition of RMS. Changes in waveshape have no effect, as they are effectively down-converted to heat. This "first principles" nature of operation makes thermally based AC voltmeters ideal for quantitative RMS noise measurement.

Performance comparison of noise driven AC voltmeters

The wide performance variation of the above methods, and even within a method, mandates caution in selecting an AC voltmeter. Comparing AC voltmeters intended for use in RMS noise measurements is illuminating. Figure C4 shows a simple evaluation arrangement. The noise generator drives the external input of text Figure 12.2, producing a suitably band passed input at the voltmeter under test.[3]

Figure C5 shows results for 20 voltmeters. Four of the voltmeters are thermal types; the remainder utilize logarithmic analog computing or rectify and average AC-DC conversion. The four thermal types agreed well within 1%. In fact, three of the thermal types were within 0.2%, while the fourth (HP-3400A), a metered instrument, is only readable to 1%. The other 16 voltmeters showed errors up to 48% relative to the thermal group! Note that the errors cause lower readings than are actually warranted; a poorly chosen voltmeter will give unfairly optimistic readings.

Note 2: See the References section for publications covering thermal AC–DC conversion.

Note 3: Noise generators are worthy of study and considered in References 4 and 5.

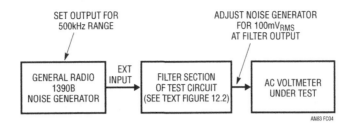

Figure C4 • Arrangement for Evaluating AC Voltmeters. Noise Source, Filtered by Figure 12.2's Circuit, Drives the Voltmeter

VOLTMETER TYPE OR SAMPLE NUMBER		READING IN MILLIVOLTS	ERROR IN %	AC-DC CONVERSION METHOD
	HP-3403C	100	0	Thermal
	HP-3400A	100	0	Thermal
	Fluke 8920A	100	0	Thermal
	LTC Special (See Figure C6)	100	0	Thermal
1		84	−16	Log
2		85	−15	Rect-Avg
3		84	−16	Rect-Avg
4	Fluke 8800A	90	−10	Rect-Avg
5	HP-3455	100	0	Log
6	HP-334	92	−8	Rect-Avg
7	Handheld	52	−48	Rect-Avg
8	HP-3478	100	0	Log
9	Inexpensive Handheld	56	−44	Rect-Avg
10	HP-403B	93	−7	Rect-Avg
11	HP-3468B	93	−7	Log
12		80	−20	Rect-Avg
13		72	−28	Rect-Avg
14		62	−38	Rect-Avg
15	Fluke 87	95	−5	Log
16	HP-34401A	93	−7	Log

Figure C5 • Results of AC Voltmeter Evaluation. Four Thermally Based Types Agree within 1%. Other Instruments Show Relative Error As Large As 48%

The lesson here is clear. It is essential to verify AC voltmeter accuracy before proceeding with RMS noise measurement. Failure to do so may cause highly misleading "results."

Thermal voltmeter circuit

It is sometimes desirable to construct, rather than purchase, a thermal voltmeter. Figure C6's circuit is applicable to noise measurement. Text Figure 12.2's filter feeds A1. A1's output biases A2, which provides additional AC gain. The LT1088 based RMS/DC converter is made up of matched pairs of heaters and diodes and a control amplifier. The LT1206 drives R1, producing heat which lowers D1's voltage. Differentially connected A3 responds by driving R2, via Q3, to heat D2, closing a loop around the amplifier. Because the diodes and heater resistors are matched, A3's DC output is related to the RMS value of the input, regardless of input frequency or waveshape. In practice,

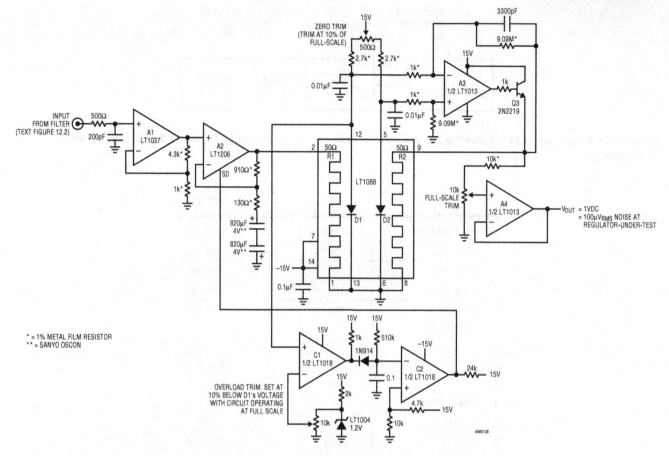

Figure C6 • Inexpensive Thermally Based RMS Voltmeter Suitable for LDO Noise Measurement

residual LT1088 mismatches necessitate a gain trim, which is implemented at A4. A4's output is the circuit output.

Start-up or input overdrive can cause A2 to deliver excessive current to the LT1088 with resultant damage. C1 and C2 prevent this. Overdrive forces D1's voltage to an abnormally low potential. C1 triggers low under these conditions, pulling C2's negative input low. This causes C2's output to go high, putting A2 into shutdown and terminating the overload. After a time determined by the RC at C2's input, A2 will be enabled. If the overload condition still exists, the loop will almost immediately shut A2 down again. This oscillatory action will continue, protecting the LT1088 until the overload condition is removed.

To trim this circuit, connect its input to a 10mV$_{RMS}$, 100kHz signal. Set the 500Ω adjustment for exactly 100mV DC out. Next, apply a 100kHz, 100mV$_{RMS}$ input and trim the 10k potentiometer for 1V DC out. Repeat this sequence until the adjustments do not interact. Two passes should be sufficient.

Appendix D
Practical considerations for selecting a low noise LDO

Any design has particular requirements for a low noise LDO; every individual situation should be carefully examined for specific needs. However, some general guidelines apply in selecting a low noise LDO. Significant issues are summarized below.

Current capacity

Insure the regulator has adequate output current capacity for the application, including worst-case transient loads.

Power dissipation

The device must be able to dissipate whatever power is required. This affects package choice. Usually, the V$_{IN}$-V$_{OUT}$ differential is low in LDO applications, obviating this issue. Prudence dictates checking to be sure.

Package size

Package size is important in limited space applications. Package size is dictated by current capacity and power dissipation constraints. See the preceding paragraphs.

Noise bandwidth

Insure that the LDO meets the system's noise requirement over the entire bandwidth of interest. 10Hz to 100kHz is realistic, as information usually occupies this range.

Input noise rejection

Insure that the regulator can reject input related disturbances originating from clocks, switching regulators and other power bus users. If the regulator's power supply rejection is poor, its low noise characteristics are useless.

Load profile

Know the load characteristics. Steady state drain is obviously important, but transient loads must also be evaluated. The regulator must maintain stability and low noise characteristics under all such transient loads.

Discrete components

The choice of discrete components, particularly capacitors, is important. The wrong capacitor dielectric can adversely effect stability, noise performance, or both. See Appendix B, "Capacitor selection considerations."

Section 4

High Voltage and High Current Applications

Parasitic capacitance effects in step-up transformer design (13)

This chapter explores the causes of the large resonating current spikes on the leading edge of the switch current waveform. These anomalies are exacerbated in very high voltage designs.

High efficiency, high density, PolyPhase converters for high current applications (14)

This application note addresses the following questions. How much do I gain by using a PolyPhase™ architecture? How many phases do I need for my application? How do I design a PolyPhase converter? The design example of an LTC1629-based, 6-phase 90A power converter is presented. The mathematical equations and graphical curves for calculating the ripple currents are included.

Parasitic capacitance effects in step-up transformer design

13

Brian Huffman

One of the most critical components in a step-up design like Figure 13.1 is the transformer. Transformers have parasitic components that can cause them to deviate from their ideal characteristics, and the parasitic capacitance associated with the secondary can cause large resonating current spikes on the leading edge of the switch current waveform. These spikes can cause the regulator to exhibit erratic operating conditions that manifests itself as duty cycle instability. This effect is exacerbated in very high voltage designs. Attention to transformer design will cure this problem.

Figure 13.2 shows the high frequency current paths of the parasitic capacitors. In the analysis of operation assume the input and output voltages are at AC ground. Thus, the parasitic capacitors are all in parallel. The transformer's secondary provides the AC current path for these capacitors. The current flowing through the secondary produces N times the current in the primary. As the parasitic capacitance and turns ratio increase, the primary current becomes progressively larger.

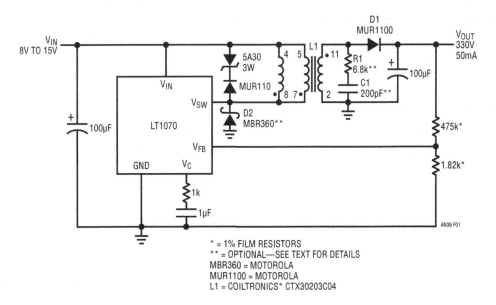

Figure 13.1 • High Voltage Power Supply

Analog Circuit and System Design: A Tutorial Guide to Applications and Solutions. DOI: 10.1016/B978-0-12-385185-7.00013-5

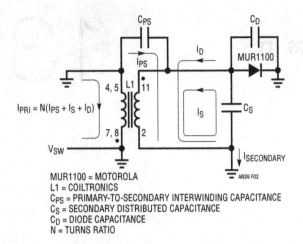

Figure 13.2 • AC Current Paths for Parasitic Capacitors

MUR1100 = MOTOROLA
L1 = COILTRONICS
C_{PS} = PRIMARY-TO-SECONDARY INTERWINDING CAPACITANCE
C_S = SECONDARY DISTRIBUTED CAPACITANCE
C_D = DIODE CAPACITANCE
N = TURNS RATIO

The operation waveforms for this circuit are shown in Figure 13.3. When the switch (V_{SW} pin—Trace A) is turned "on" the primary is pulled to ground. The secondary cannot instantly follow this action because of the loading effects of its parasitic capacitors. The effects of the parasitic capacitance can easily be seen in the leading edge of the switch current waveform (Trace B). This current spike is caused by the loading effect of the secondary parasitic capacitances. The secondary output (Trace D) swing can exceed 600V. As such, a large amount of charge

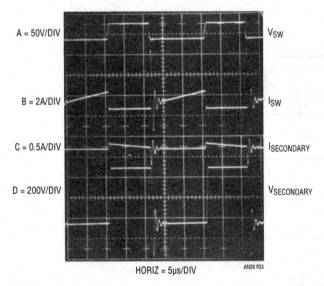

Figure 13.3 • Operating Waveforms for High Voltage Converter

$(Q = C \cdot V)$ is needed to swing this node during the switching transients.

The secondary current is amplified by the turns ratio and produces a primary current; $I_{PRI} = N(I_{PS} + I_S + I_D)$. This amplifying effect can be observed by comparing the secondary current (Trace C), which does not include the effects of I_S, with the switch current (Trace B). The result is a rather large current spike.

The oscillatory nature of the response is formed by a series combination of the leakage inductance and reflected secondary capacitance (Figure 13.4). Any impedance placed across the secondary appears at the primary terminals reduced in magnitude by a factor of the turns ratio squared. For example, if $N = 10$, then a 200Ω resistor looks like 2Ω and a 100pF capacitor looks like $0.01\mu F$ so even a small secondary capacitance can heavily load the primary. The series LC forms a self-resonating circuit that rings at the resonant frequency of the transformer.

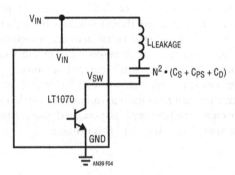

Figure 13.4 • Simplified Primary Model During Transient Conditions

The output switching diode, D1, can also cause narrow spikes on the current waveform. In this case, the reverse recovery time of the diode is the important parameter. Reverse recovery time occurs because the diode "stores" charge during its forward conducting cycle. This stored charge causes the diode to act like a low impedance conductive element for a short period of time after reverse drive is applied.

There is a short period of time (blanking time) following switch turn "on" during which the LT1070 ignores the switch current waveform. This blanking time eliminates premature termination of the "on" pulse caused by the leading edge current spike in the current waveform. Once this blanking time has elapsed the output switch will turn "off" when the peak switch current reaches the threshold level established by the error amplifier output (V_C pin).

This internally fixed blanking time, 400ns, is appropriate for typical applications. However, for high voltage applications the blanking time becomes a critical parameter. The effects of the transformer's parasitic capacitance is to extend the "spike" width past the blanking time causing the LT1070 to mistrigger, as shown in Figure 13.5. The duty cycle variations in the switch pin (Trace A) and switch current (Trace B) waveforms are the result of this problem. Figure 13.6 details the interaction

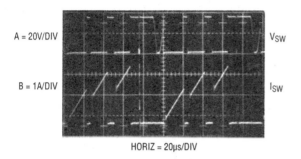

Figure 13.5 • **Duty Cycle Instability Due to Current Spike**

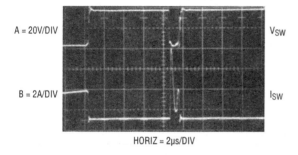

Figure 13.6 • **Detail Waveforms of Switch Current Spike**

between blanking time and peak switch current. Notice that the switch is turned "off" as soon as the LT1070 samples the switch current. For the LT1070 to function properly, the spike current must be below the normal peak switch current before the 400ns blanking period is over.

Another problem induced by the parasitic capacitance can also be seen in Figure 13.3. High **reverse** current can flow in the V_{SW} pin (Trace B) due to the oscillatory nature of the transformer This will forward bias the LT1070's substrate diode, which is inherent in the output transistor (see Appendix A). Unwanted current can flow almost anywhere within the IC's circuitry when the substrate diode is forward biased, causing unpredictable duty cycle behavior.

The substrate diode current can be eliminated by placing a Schottky diode, D2, between the V_{sw} pin and ground (Figure 13.1). The reverse primary current will flow through the Schottky diode instead of the LT1070. Another way to prevent the substrate diode from conducting is to place an RC snubber R1-C1, on the secondary. This will attenuate the ringing.

Well known transformer winding techniques can be used to minimize parasitic capacitance in step-up transformers. The basic technique is to wind the secondary layers in a manner that minimizes the voltage difference between adjacent layers. A standard way of accomplishing this is to wind several separate secondary "stacks" on a split bobbin. This and other techniques will cause a dramatic reduction in secondary capacitance.

For transformer information contact:
Coiltronics
984 Southwest 13th Court
Pompano Beach, FL 33069
305-781-8900

Appendix A

In a junction isolated IC the monolithic transistors are surrounded by an isolating P-N junction, as illustrated in Figure A1. When this junction is reverse biased, it electrically isolates one device from another on the chip. However, this device structure also forms parasitic lateral NPN transistors (see Figure A2). The P substrate is the base, the N-epi region is the emitter and the collector is any other N-epi pocket. A simplified schematic diagram of the NPN cell is shown in Figure A3.

If the vertical NPN output switch transistor is operating in its normal mode, either "on" or "off," the parasitic transistor is off and will have no effect. The parasitic becomes active when the vertical collector is pulled to a potential below that of the substrate by the reverse switch current. This forward biases the collector substrate junction, which is commonly called the substrate diode. As a result, the parasitic NPN draws current from other portions of the circuit. This current will add to the base drive of lateral PNPs and to the collector current of NPNs, causing the IC to behave in mysterious and unpredictable ways (see Figure A2).

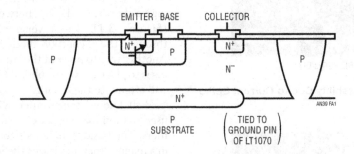

Figure A1 • Junction Isolated NPN Structure

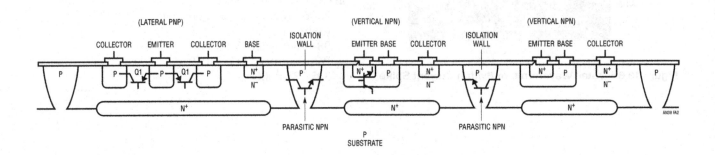

Figure A2 • Locations Where the Lateral NPN Occur

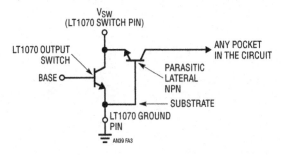

Figure A3 • Schematic Diagram of LT1070's Output Switch with Parasitic

High efficiency, high density, PolyPhase converters for high current applications

14

Wei Chen

Introduction

As logic systems get larger and more complex, their supply current requirements continue to rise. Systems requiring 100A are fairly common. A high current power supply to meet such requirements usually requires paralleling several power regulators to alleviate the thermal stress on the individual power components. A power supply designer is left with the choice of how to drive these paralleled regulators: brute-force single-phase or smart PolyPhase.

A PolyPhase converter interleaves the clock signals of the paralleled power stages, reducing input and output ripple current without increasing the switching frequency. The decreased power loss from the ESR of the input capacitor and the low switching losses associated with MOSFETs at relatively low switching frequencies help achieve high power conversion efficiency. The size and cost of the input capacitors are also greatly reduced as a result of input ripple current cancellation. Since output ripple current cancellation also occurs, lower value inductors can be used. This results in improved dynamic response to load transients. A combination of lower current rating and decreased inductance also allows the use of smaller-sized, low profile, surface mounted inductors. For multi-output applications, PolyPhase converters may also provide the benefit of smaller input capacitors.

Previously, the implementation of multiphase designs was difficult and expensive because of complex timing and current-sharing requirements. The newly developed LTC1629 solves these problems for high current, single output designs, while the LTC1628 addresses dual-output applications. Both ICs are dual, current mode, PolyPhase controllers that can drive two synchronous buck stages simultaneously. The features of the LTC1629 include a unity-gain differential amplifier for true remote sensing, low impedance gate drives, current-sharing, overvoltage protection, optional overcurrent latch-off and foldback current limit. Additionally, the LTC1629 can be configured for 2-, 3-, 4-, 6- and 12-phase operation with a simple phase selection signal (high, low or open). Optimizing the number of phases can help achieve the smallest and the most cost-effective power supply design.

This chapter analyzes the performance of PolyPhase converters and provides guidelines for selecting the phase number and designing a PolyPhase converter using the LTC1629. The following questions will be answered as the discussion goes on:

- How much do I gain by using a PolyPhase architecture?
- How many phases do I need for my application?
- How do I design a PolyPhase converter?

How do PolyPhase techniques affect circuit performance?

In general, PolyPhase operation improves the large signal performance of a switched mode power converter, by such means as reducing ripple current and ripple voltage. A synchronous buck converter is used as an example in this chapter to analyze the effects of PolyPhase techniques on circuit performance.

High current outputs usually require paralleling several regulators. The single-regulator approach is not feasible because of the unacceptable thermal stress on the individual power components. Paralleled regulators are synchronized to have the same switching frequency to eliminate beat frequency noise at both the input and output terminals. Based on the phase relationship between the paralleled regulators, these converters can be divided into two types: single-phase and PolyPhase. To balance the thermal stress in each component, paralleled regulators must also share the load current.

Analog Circuit and System Design: A Tutorial Guide to Applications and Solutions. DOI: 10.1016/B978-0-12-385185-7.00014-7

In this chapter, the number of channels refers to the number of the paralleled regulators in one supply. The following symbols are defined to facilitate reference:

- V_0: DC output voltage
- I_0: DC output current
- V_{IN}: DC input voltage
- T: switching period
- mc: number of paralleled channels
- m: number of phases. The possible phase numbers are usually determined by the channel number, mc. For example, if mc = 6, the possible phase numbers are m = 1, 2, 3, 6
- C_0: output capacitor
- ESR: equivalent series resistance of C_0
- L_f: output inductor
- D: duty cycle, approximated by V_0/V_{IN} in buck circuits

Current-sharing

The current-sharing can be easily achieved by implementing peak current mode control. In a current mode control regulator, the load current is proportional to the error voltage in the voltage feedback loop. If the paralleled regulators see the same error voltage, they will source equal currents. A 2-channel circuit is used as the example to explain this current-sharing mechanism.

As shown in Figure 14.1, peak current mode control requires that the high side switch turn off when the peak inductor current (I_{L1}, I_{L2}) intersects the error voltage, V_{ER}, resulting in the same peak inductor currents. If the inductors are identical, the peak-to-peak ripple currents of the inductors will be the same. The DC currents

of two inductors, which are the peak current less half of the peak-to-peak ripple current, will be equivalent. Two modules therefore share the load current equally. The same current-sharing mechanism can be extended to any number of channels in parallel. This current-sharing scheme will prevent an individual module from suffering excessive current stress in steady state operation and during line/load transient conditions. Note that the sharing mechanism is open loop, so no oscillations will occur due to current-sharing.

Output ripple current cancellation and reduced output ripple voltage

The phase relationship of Figure 14.1(b) shows how ripple current cancellation at the output works. Because of the 180 degree phase difference between the two converters, the two inductor ripple currents in the two-phase converter tend to cancel each other, resulting in a smaller ripple current flowing into the output capacitor. The frequency of the output ripple current is doubled as well. All of these factors contribute to a smaller output capacitor for the same ripple voltage requirement.

Figure 14.2 shows the measured waveforms of the inductor currents and output ripple currents in a 2-channel converter. The output ripple cancellation reduces the output ripple current from $14A_{P-P}$ (single-phase) to $6A_{P-P}$ (dual-phase). The ripple frequency in the dual-phase circuit is twice the switching frequency.

To quantify the output ripple current amplitude in an m-phase circuit, a closed-form expression was developed. The derivation starts with the 2-phase circuit shown in

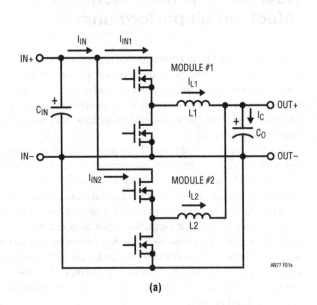

(a)

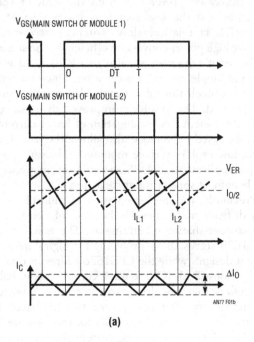

(a)

Figure 14.1 • 2-Channel Converter: (a) Schematic and (b) Typical Waveforms

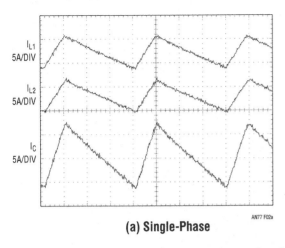

(a) Single-Phase

AN77 F02a

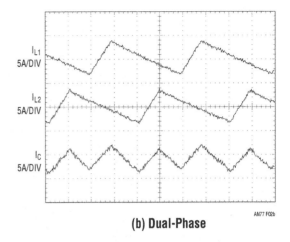

(b) Dual-Phase

AN77 F02b

Figure 14.2 • Output Ripple Current Waveforms In a 2-Channel Circuit. I_{L1} and I_{L2} Are the Inductor Currents In Two Channels and I_C Is the Net Ripple Current Flowing Into Output Capacitor. Test Conditions: $V_{IN} = 12V$, $V_0 = 2V$, $I_0 = 20A$

Figure 14.1. During the interval [from DT, T] when the high side switch in module 1 is off and the high side switch of module 2 is on, the inductor current in module 1 decreases and the inductor current in module 2 increases. The net ripple current flowing into the output capacitors is smaller. The output ripple current for the 2-phase circuit is derived as:

$$\Delta I_0 = \frac{2V_0(1-D)T}{L_f} \cdot \frac{|1-2D|}{|1-2D|+1} \tag{1}$$

See Appendix A for the detailed derivation procedure. By extending the same derivation procedure to an m-phase configuration, the output ripple current for an m-phase circuit is obtained.

Output ripple current peak-to-peak amplitude in *m*-phase circuit:

$$\Delta I_0 = \begin{cases} \dfrac{V_0 T(1-D)}{L_f}, & m = 1 \\[4mm] \dfrac{mc \bullet V_0 T}{L_f} \bullet \dfrac{\prod\limits_{i=1}^{m}\left|\dfrac{i}{m}-D\right|}{\prod\limits_{i=1}^{m-1}\left(\left|\dfrac{i}{m}-D\right|+\dfrac{1}{m}\right)}, & m = 2,3\ldots \end{cases} \tag{2}$$

The output ripple voltage is estimated to be:

$$\Delta V_{0,PP} < \frac{\Delta I_0 T}{8mC_0} + \Delta I_0 \bullet ESR \tag{3}$$

The first term in equation (3) represents the ripple voltage on the pure capacitive components of C_0 and the second term represents the ripple voltage generated on the ESR of C_0. Intuitively, a higher phase number helps reduce the ripple component in the first term, and therefore, the overall ripple voltage amplitude at the output. Another interesting fact that can be observed is that the

output ripple current and voltage will reach zero if the duty cycle is equal to one of the following critical points:

$$D_{crit} = \frac{i}{m}, \quad i = 1, 2, \ldots, m-1 \tag{4}$$

In a buck converter, the duty cycle is the ratio of the output voltage and input voltage. By interpreting equation (4) in terms of V_{IN} and V_0, the zero output ripple conditions can be rewritten as:

$$\frac{V_0}{V_{IN}} = \frac{i}{m}, \quad i = 1, 2 \ldots, m-1 \tag{5}$$

The plots in Figure 14.3 demonstrate the influence of phase number and duty cycle on the output ripple current. In this plot, the output ripple current is normalized against the inductor ripple current at zero duty cycle (DIr = V_0T/ L_f). The following assumptions are made: the number of channels equals the phase number, the output voltage is fixed and the power conversion efficiency is assumed to be 100%. This plot can be used to estimate the output ripple current without tedious calculations.

The output ripple current approaches zero when the duty cycle is near the critical points for the selected phase number. For a buck circuit, the duty cycle is approximately the ratio of V_0/V_{IN}. Therefore, if the input and output voltages are relatively fixed, there exists an optimum phase number to minimize the output ripple voltage.

Assuming that the maximum available phase number is six and the efficiency is 100%, the optimum phase numbers for some common input and output voltages are shown in Table 14.1.

For high step-down ratio or low duty cycle applications (for example, $V_{IN} = 12V$, $V_0 = 1.2V$, $D = 0.1$), a high phase number helps reduce the maximum ripple current. For wide duty cycle range applications, high phase number

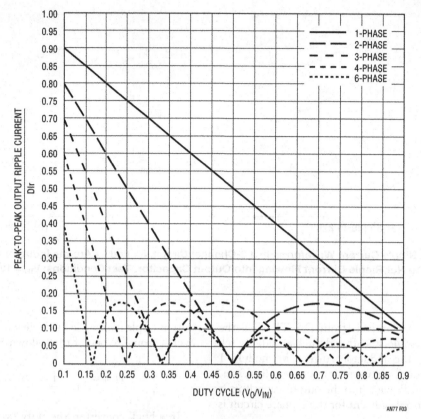

AN77 F03

Figure 14.3 • Normalized Output Ripple Current vs Duty Cycle, $\text{Dir} = \dfrac{V_0 T}{L_f}$

tends to, but does not necessarily, yield lower output ripple current. The optimum phase number needs to be evaluated over the complete operating duty cycle range. The reduction in the ripple current by increasing phase number is not significant above four phases in most duty cycle ranges.

Figure 14.4 shows the measured output ripple current near the critical duty cycle point, which is $D_{crit} = 0.5$ for a 2-phase circuit. The test conditions were $V_{IN} = 5V$, $V_0 = 2V$, $I_0 = 20A$, fs $= 250kHz$. Because of the voltage drop across the MOSFET switches, the operating duty cycle was very close to 50%. The dual-phase technique was able to reduce the output ripple current considerably, from $10A_{P-P}$ (in the single-phase circuit) to $2.5A_{P-P}$ As

a result, the output ripple voltage near the critical duty cycle point is negligible, as shown in Figure 14.5.

Improved load transient response

The influences of PolyPhase techniques on the load transient performance are numerous. First, the reduced output ripple voltage allows more room for voltage variations during the load transient because the ripple voltage will consume a smaller portion of the total error budget. With the same number of capacitors on the output terminals of the power supply, the sum of the overshoot and undershoot can be reduced dramatically. Second, the reduced ripple current allows the use of lower value inductors. This speeds up the output current slew rate of the power supply. Consequently, PolyPhase helps improve the load transient performance of the power supply. Figure 14.6 shows the output voltages during a load transient. It is noted that the two circuits have the same electrical design. The dual-phase technique reduces the voltage variation from $69mV_{P-P}$ to $58mV_{P-P}$, a 16% reduction with no changes in component values. The inductor values could be reduced while still achieving lower output ripple voltage than the single-phase design, and further improvement in the peak-to-peak voltage variations for the load transient response could be realized.

Table 14.1 Optimum Phase Number for Minimizing the Ripple Currents (Assuming That the Maximum Phase Number Is 6 and the Efficiency Is 100%)

	$V_0 = 1.2V$	$V_0 = 1.5V$	$V_0 = 2.0V$	$V_0 = 2.5V$
$V_{IN} = 5V$	4	6	5	2, 4, 6[1]
$V_{IN} = 12V$	6	6	6	5

[1] 6 is the optimum phase number for minimum input ripple current

275

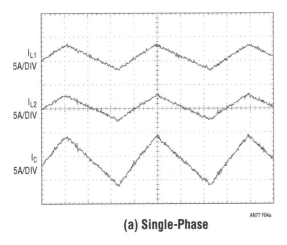

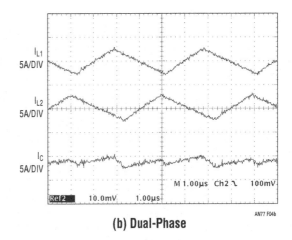

(a) Single-Phase

(b) Dual-Phase

Figure 14.4 • Experimental Waveforms of Output Ripple Current Near Critical Duty Cycle Point (V_{IN} = 5V, V_0 = 2V, I_0 = 20A, f_s = 250kHz): Top Trace, Inductor 1 Current; Middle Trace, Inductor 2 Current; Bottom Trace, Output Ripple Current

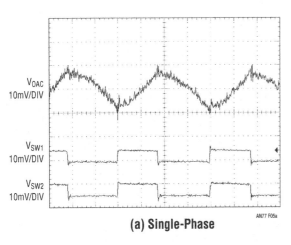

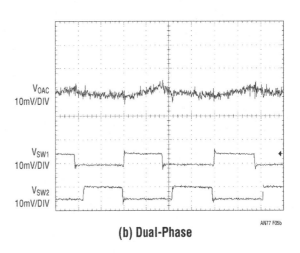

(a) Single-Phase

(b) Dual-Phase

Figure 14.5 • Measured Output Ripple Voltage (Top Trace) near Critical Duty Cycle Point (V_{IN} = 5V, V_0 = 2V, I_0 = 20A, f_s = 250kHz, V_{SW1} and V_{SW2} are the Switch Node Voltages Across the Bottom FETs)

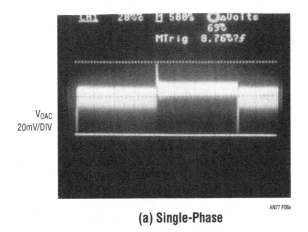

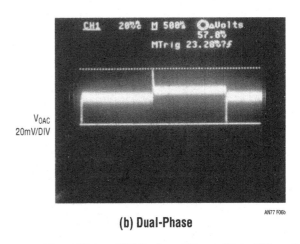

(a) Single-Phase

(b) Dual-Phase

Figure 14.6 • Measured Output Voltage During Load Transients (V_{IN} = 12V, V_0 = 2V, f_s = 250kHz. Load Steps: 5A to 20A and 20A to 5A, 50μs Rise and Fall Times. Time Scale: 500μs/DIV)

Input ripple current cancellation

The input current of a buck converter is discontinuous. With the input supply mainly sourcing DC current, the input capacitor supplies a pulsating current to the buck converter. In a single-phase circuit, the high side switches of the paralleled buck modules turn on simultaneously. The input capacitor needs to provide the sum of the pulsed currents. In a PolyPhase circuit, however, the paralleled buck stages switch at different times and the pulsating current flowing through the input capacitor is reduced dramatically.

Figure 14.7 shows the measured input ripple current in a 2-channel converter. The PolyPhase converter reduces the peak amplitude of the input ripple current by half and doubles the ripple frequency. The reduced ripple current amplitude results in a much smaller RMS current in the input capacitor. Because the power loss on the ESR of the input capacitor is proportional to the square of the RMS current, the loss reduction can be significant. The size of the input capacitor is reduced and the life of the capacitor will likely be improved. The increased ripple frequency and the reduced ripple amplitude also facilitate EMI filtering.

In order to quantitatively evaluate the input ripple current in an m-phase circuit, a close-form expression is derived by using some mathematical manipulations on the input ripple current waveforms.

Input ripple current RMS value:

$$I_{irms} = \sqrt{\left(D - \frac{k}{m}\right)\left(\frac{k+1}{m} - D\right)\Big|_0^2 + \frac{mc^2}{12mD^2}\left(\frac{V_0(1-D)T}{L_f}\right)^2 \cdot \left[(k+1)^2\left(D - \frac{k}{m}\right)^3 + k^2\left(\frac{k+1}{m} - D\right)^3\right]}$$

(6)

where k = FLOOR(m•D), m = 1,2,...

The variable k is determined by the phase number (m) and the duty cycle (D). For example, in a five-phase converter, at 45% duty cycle, k = FLOOR(5 • 0.45) = 2. The FLOOR(x) function provides a greatest integer that is smaller than or equal to x.

As indicated in equation (6), the input ripple current of a PolyPhase converter consists of two major factors: the DC load current (first term) and the inductor ripple current (second term). Since the inductor ripple current is almost unaffected by the load conditions, the maximum RMS input ripple current is reached at full load.

Usually, the size of the input capacitor is determined by the power dissipation on its ESR. The full load condition contributes to a maximum RMS input ripple current, and therefore, determines the size of the input capacitor.

Figure 14.8 plots the RMS input ripple current against the duty cycle for different phase configurations. In this plot, the RMS input ripple current is normalized against the DC load current. The output voltage is assumed fixed at 5V and the input voltage is varied, resulting in a duty cycle range from 0.1 to 0.9. Several facts can be observed from the curves.

When the duty cycles are close to the critical duty cycle points (determined in equation (4)), the first term in equation (6) is zero. The RMS input ripple current reaches the local minimum values. These values are not zero due to the output inductor ripple current. Consequently, there exists an optimum phase number to achieve the minimum RMS input ripple current for a fixed input and output application. For some common input and output voltages, the optimum phase numbers for minimizing the input ripple current are shown in Table 14.1. Note that these are the same as the values for the minimum output ripple voltage. For a wide duty cycle range application, higher phase number helps reduce the maximum input ripple current. But the reduction in the input ripple current by increasing phase number may not be significant at higher phase

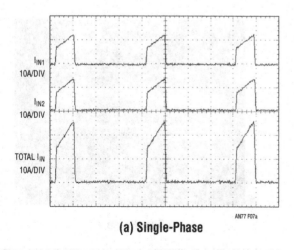

(a) Single-Phase

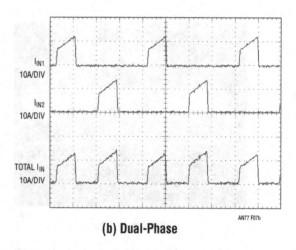

(b) Dual-Phase

Figure 14.7 • Measured Input Ripple Current: I_{in1} and I_{in2} are the Ripple Currents into the Paralleled Modules. Total I_{in} is the net Ripple Current into the Input Capacitor. (V_{IN} = 12V, V_0 = 2V, I_0 = 20A, f_s = 250kHz)

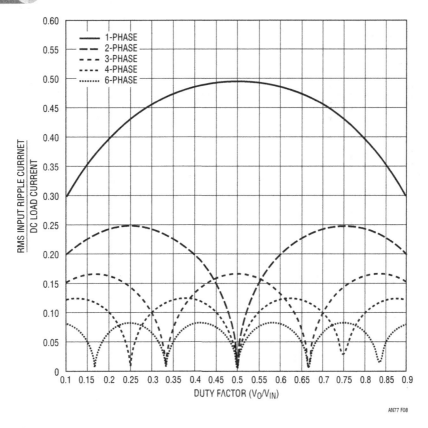

AN77 F08

Figure 14.8 • Normalized RMS Input Ripple Current

numbers in certain duty cycle ranges. The optimum phase number needs to be evaluated over the complete operating duty cycle range.

Figure 14.9 shows the experimental waveform of the input ripple currents in a 2-channel circuit. The circuit operated at close to 50% duty cycle, which is the critical duty cycle point for the two-phase circuit. Compared to the single-phase technique, the PolyPhase technique reduced the ripple current in the input capacitor dramatically.

Design considerations

Similar to the design of conventional paralleled regulators, the design of a PolyPhase converter involves the choice of the number of paralleled channels and the selection of the power components (MOSFETs, inductors, capacitors, etc.).

Usually, the number of phases is set to be equal to the number of channels. However, the number of channels and the number of phases may be different. The number of

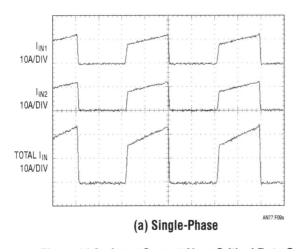

(a) Single-Phase

AN77 F09a

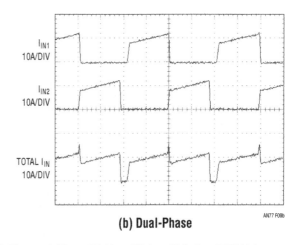

(b) Dual-Phase

AN77 F09b

Figure 14.9 • Input Current Near Critical Duty Cycle In a 2-Channel (V_{IN} = 5V, V_O = 2V, I_O = 20A, f_s = 250kHz)

TOP VIEW

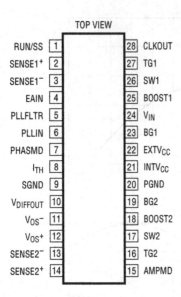

RUN/SS	1		28	CLKOUT
SENSE1$^+$	2		27	TG1
SENSE1$^-$	3		26	SW1
EAIN	4		25	BOOST1
PLLFLTR	5		24	V$_{IN}$
PLLIN	6		23	BG1
PHASMD	7		22	EXTV$_{CC}$
I$_{TH}$	8		21	INTV$_{CC}$
SGND	9		20	PGND
V$_{DIFFOUT}$	10		19	BG2
V$_{OS}^-$	11		18	BOOST2
V$_{OS}^+$	12		17	SW2
SENSE2$^-$	13		16	TG2
SENSE2$^+$	14		15	AMPMD

Figure 14.10 • Pinouts of LTC1629

Table 14.2 Phase Function Table for LTC1629

PHASMD	0V	OPEN	INTV$_{CC}$
PLLIN	0°	0°	0°
CONTROLLER 1	0°	0°	0°
CONTROLLER 2	180°	180°	240°
CLKOUT	60°	90°	120°

channels is usually determined by the total load current and the acceptable current stress in each channel. For example, if the required load current is 60A and the maximum current stress per channel is 15A, 4 channels need to be paralleled. The number of phases, on the other hand, can be selected to minimize the input and output filter capacitors. Note that each phase must have an equal number of channels. In this example, a 4-channel configuration, one, two or four phases may be used.

Selection of phase number

As discussed in the previous section, the selection of a different number of phases will greatly affect the input and output ripple current.

For a narrow input and output range, the duty cycle range is relatively narrow. The optimum phase number should be chosen such that the circuit operates at or near one of the critical duty cycle points (determined by equation 4). For some practical input voltages and output voltages, the optimum phase numbers for the minimum input ripple currents and the lowest output ripple voltage are listed in Table 14.1. For wide input or output voltage range, the phase number should be chosen such that the worst-case RMS input ripple current and the worst-case output ripple voltage are minimized for the complete operating duty cycle range.

PolyPhase converters using the LTC1629

The LTC1629 integrates proprietary phase-locked-loop-based phasing circuitry. Each IC can be synchronized to an external signal at the PLLIN pin and produce a CLKOUT signal to synchronize other ICs. Table 14.2 shows

the phase function table of the LTC1629. By applying the command signal (INTV$_{CC}$, open or SGND) to the PHASMD pin and connecting the CLKOUT pin of one IC to the PLLIN pin of the next one, different numbers of phases can be achieved. Figure 14.11 shows 2-phase, 3-phase, 4-phase, 6-phase and 12-phase configurations using the LTC1629.

A higher number of phases is usually needed for very high output current or multioutput applications. For example, in a 2-output system, 3.3V/90A and 5V/60A, if each output is provided by a 6-phase power supply, the two power supplies can be interleaved by using a 12-phase configuration. As shown in Figure 14.12, U1, U2 and U3 are used to produce the 3.3V output, and U4, U5, and U6 are used for the 5V output. The resulting input ripple current frequency is twelve times the switching frequency and the ripple current amplitude is reduced.

The LTC1629 includes a unity gain differential amplifier, enabling true remote sensing of the output voltage. This is particularly useful for maintaining tight output voltage regulation for high current applications. Each LTC1629 based regulator consists of two synchronous buck stages and two or more power regulators can be paralleled directly. The inherent peak current mode control permits automatic current-sharing. When several LTC1629-based regulators are in parallel, the LTC1629 of the master regulator senses the output voltage (V$_{0+}$, V$_{0-}$) via its on-chip differential amplifier and divides this voltage (V$_{DIFFOUT}$) down through the resistor divider to utilize the built-in 0.8V reference for output voltage regulation. This control voltage is then fed to the EAIN pins (error amplifier input) of each LTC1629. Since the error amplifier inside the LTC1629 is a g$_m$ transconductance amplifier, directly paralleling the I$_{TH}$ pins (error amplifier outputs) and the EAIN pins is allowed. The paralleled regulators now share the same error voltage. Because the load current of a current mode regulator is proportional to the error voltage, the paralleled regulators must source equal currents.

Layout considerations

To take full advantage of the ripple cancellation of the PolyPhase technique, the input capacitors and output capacitors are ideally placed at the summation points of all the input ripple currents and all the output ripple currents, respectively. Figure 14.13 shows the layout for

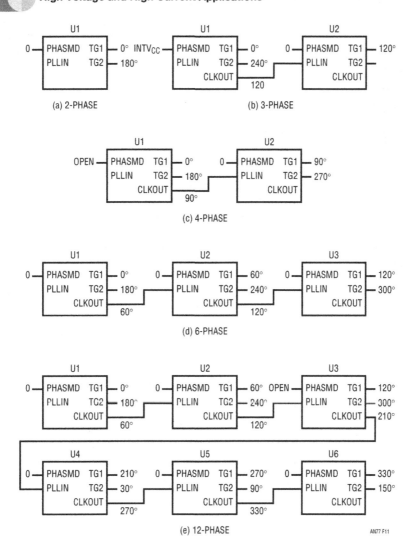

Figure 14.11 • Configuration of Different Phases Using the LTC1629

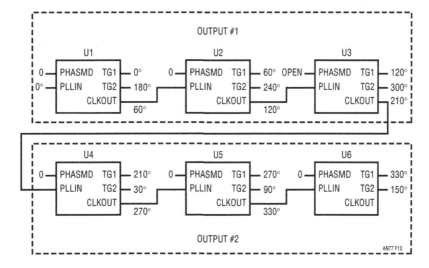

Figure 14.12 • 2-Output System Using 12-Phase Configuration

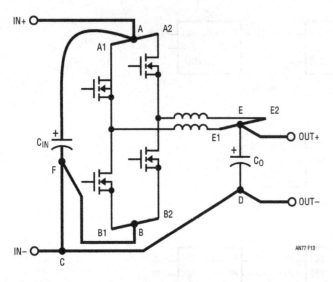

Figure 14.13 • Layout Diagram of Power Stage for 2-Phase Converter

a 2-phase converter. In practice, the filter capacitors may be placed across the individual modules' inputs and outputs (such as A1B1, A2B2, etc.). The traces (AA1, AA2, BB1, BB2, etc.) between modules should be the shortest and widest possible to balance the current stress in each capacitor. The impedance of the traces highlighted in Figure 14.13 should be minimized. It is preferable that these traces be large copper planes. It is also important that the sources of bottom MOSFETs (B1, B2, etc.) be connected to the input filter capacitors before joining the ground plane (CD). Otherwise, the ground noise generated by the pulsating current through the trace inductance will be seen on the output terminals as spikes.

Design example: 100A PolyPhase power supply

The specifications for a high current PolyPhase power supply are as follows:
- Input: 12V (±10%)
- Outputs: 3.3V at 90A nominal, 100A maximum
- Load regulation: <20mV from 0A to full load
- Switching Noise: peak-to-peak voltage <1% of DC voltage
- Efficiency: >89% at V_{IN} = 12V, V_0 = 3.3V, I_0 = 90A

Design details

To utilize off-the-shelf, surface mount inductors and to avoid the use of very thick PCB copper traces, it is desirable to limit the individual module current to about 16A. Six

channels are needed for this application. The design now becomes only a 15A regulator which gets repeated six times.

MOSFETs

The selection of MOSFETs is determined by the current requirement and switching frequency. Low $R_{DS(ON)}$ MOSFETs usually drive down the conduction losses but tend to introduce high switching related losses at high switching frequencies because of the large gate charge and parasitic capacitances. For a given current requirement and selected switching frequency, both $R_{DS(ON)}$ and gate charge (Q_g) should be evaluated to minimize the sum of the conduction losses, driving losses and the switching loss. For the specified application, a number of MOSFETs can be good choices: Si4420 (Siliconix), FDS6670A (Fairchild), FDS7760A (Fairchild) and IRF7811 or IRF7805 (International Rectifier). In this application, we need two MOSFETs for each high side switch and three MOSFETs for each low side switch. The power dissipation in the MOSFETs includes the conduction loss, switching loss and reverse recovery loss of the body diodes in the low side MOSFETs. The gate driving loss is seen by the controller IC. In this design, if Si4420s are used, each top MOSFET dissipates about 0.5W and each low side MOSFET consumes about 0.9W. Based on the thermal resistance provided in the datasheet, 30°C/W junction-to-ambient, the maximum junction temperature of the MOSFETs is about 30°C above the ambient temperature. Refer to the LTC1629 data sheet and the literature from the MOSFET vendors for more information on estimating the power loss of MOSFETs.

Inductors

The selection of inductors is driven by the load current amplitude and the switching frequency. The LTC1629 senses the inductor current with a current sense resistor. The inductor ripple current must be large enough to produce a reasonable AC sense voltage on the small value sense resistor required for a high current application. A reasonable starting point is to choose an inductor such that its ripple current amplitude is about 40% of the maximum channel current. It is estimated that an inductor value between 1.0µH and 1.6µH will be suitable for 3.3V output at a frequency of 200kHz. A number of off-the-shelf, surface mount inductors are available for the specified application: P1608 (Pulse), PE53691 (Pulse), ETQP6F1R3L (Panasonic) and CEPH149-1R6MC (Sumida). Any inductor of similar inductance value and current capability should work correctly.

The maximum available phase number is six and the possible phase number options are 1, 2, 3 and 6. By using equations (1–6), the ripple currents at the input and output for different-phase configuration are estimated as in Table 14.3.

Table 14.3 Input and Output Ripple Current for Different Phase Configurations

Channels	6	6	6	6
Phases	1	2	3	6
Input Ripple Current (A_{RMS})	46.8	25.7	15.2	8.5
Output Ripple Current (A_{P-P})	57.1	19.0	6.3	2.1

Assuming that the efficiency is 100%, the inductance is 1.3μH, frequency is 200kHz

A six-phase configuration minimizes both the input capacitor size and output ripple voltage. Compared to the single-phase technique, the six-phase technique reduces the input ripple current by more than 81%, and the output ripple by more than 96%. Therefore, the six-phase configuration was adopted for this design.

Capacitors

The selection of the input capacitors is driven by the RMS value of the input ripple current. Large capacitor ripple currents introduce high power losses due to the ESR of the capacitor. The resulting internal heating tends to reduce the capacitor life. Low ESR capacitors must be used. This design uses Sanyo OS-CON capacitors (16SA150M 150μF/16V) whose maximum allowable ripple current is rated at about 3.26A_{RMS}. The input RMS ripple current for the six-phase configuration is estimated to be about 8.5A_{RMS}. Therefore, a minimum of three OS-CON capacitors are needed. If a conventional single-phase technique were used instead, the input RMS ripple current would be about 46.8A_{RMS}. A minimum of fifteen OS-CON

capacitors would then be needed. Therefore, the use of an LTC1629 based PolyPhase design saves at least twelve (15 − 3 = 12) OS-CON capacitors.

The output capacitors are KEMET (T510X477 M006AS 470μF/6.3V), ultralow ESR (30mohm), surface mount tantulum capacitors. The peak-to-peak ripple current is estimated to be 2.1A_{P-P} It would be 57.1A_{P-P} if a conventional single-phase approach were taken instead.

Test results

The complete schematic diagram is shown in Figure 14.14. The power supply consists of six buck channels and three LTC1629s. Figure 14.15 shows the measured waveforms of gate voltages and switch-node voltages in a typical six-phase converter. The gate voltages and switch-node voltages of six buck stages are interleaved 60 degrees output of phase. Since the inductor current is driven by the difference between the switch-node voltage and the DC output voltage, the ripple currents in the six inductors are also 60 degrees out of phase. As a result, the amplitude of the net ripple current flowing into the output capacitor decreases substantially and the ripple frequency increases to six times the switching frequency. The output switching noise and the power loss from the ESR are greatly attenuated.

Figure 14.16 shows the output voltage, which was measured at the output capacitor (C14 in the schematics). The output ripple voltage is less than 10mV$_{P-P}$ at 90A output current and the ripple frequency is six times the switching frequency.

Efficiency was measured under different loading conditions. Figure 14.17 shows the efficiency curve. For most of the load range, the efficiency was about 90%. At 100A, the efficiency was measured 89.4%.

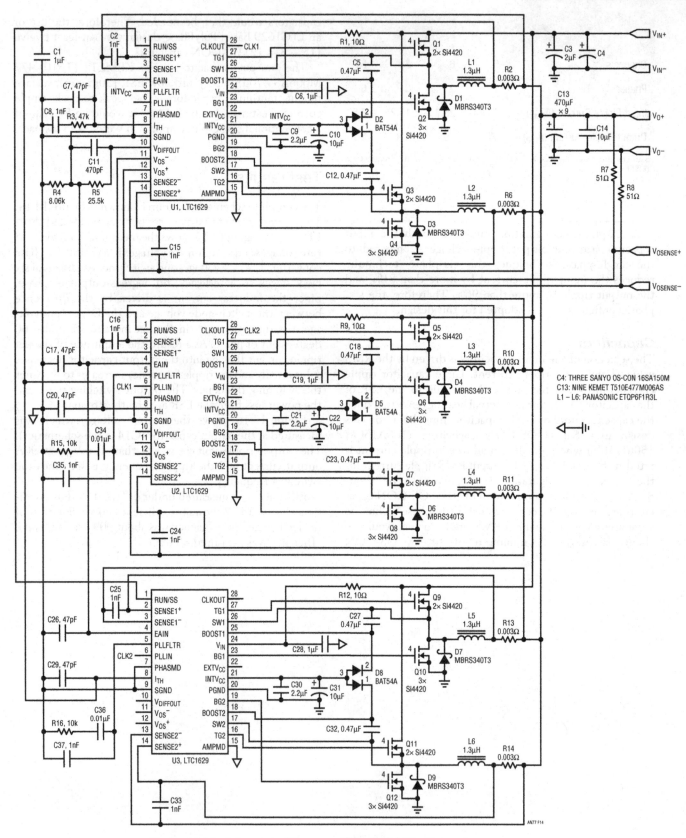

C4: THREE SANYO OS-CON 16SA150M
C13: NINE KEMET T510E477M006AS
L1 – L6: PANASONIC ETQP6F1R3L

Figure 14.14 • Schematic Diagram of 3.3V/100A Six-Phase Converter

AN77 F14

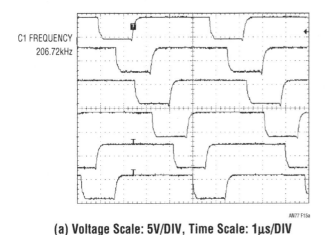

AN77 F15a

(a) Voltage Scale: 5V/DIV, Time Scale: 1µs/DIV

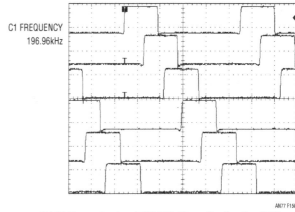

AN77 F15b

(b) Voltage Scale: 10V/DIV, Time Scale: 1µs/DIV

Figure 14.15 • Typical Waveforms of 6-phase Converter: (a) Gate Voltages, (b) Switch-Node Voltages (Drain-to-Source Voltage of the Synchronous Switch)

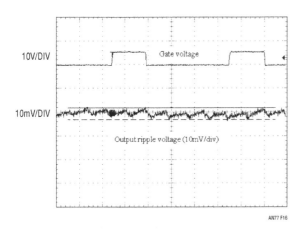

AN77 F16

Figure 14.16 • Output Ripple Voltage Waveforms (Time Scale: 1µs/DIV): V_{IN} = 12V, V_0 = 3.3V, I_0 = 90A

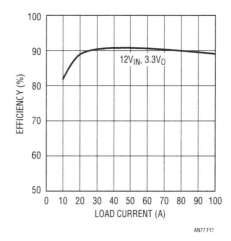

AN77 F17

Figure 14.17 • Measured Efficiency at Different Load

Summary

PolyPhase converters reduce the input and output ripple currents by interleaving the clock signals of paralleled power stages. With a proper choice of phase number, the output ripple voltage and the input capacitor size can be minimized without increasing the switching frequency. Lower output ripple voltage and smaller output inductors help improve the circuit's dynamic performance during load transients. The low switching losses and driving losses of MOSFETs at relatively low switching frequency, and the reduced power losses due to ESRs of the capacitors, help achieve high efficiency.

The LTC1629 dual, PolyPhase current mode controller is able to achieve the benefits of the PolyPhase technique without complicating the control circuit. The LTC1629 helps minimize the external component count and simplify the complete power supply design by integrating two PWM current mode controllers, true remote sensing, selectable phasing control, current-sharing capability, high current MOSFET drivers and protection features (such as overvoltage protection, optional overcurrent latch-off and foldback current limit) into one IC. The resulting manufacturing simplicity helps improve power supply reliability. The high current MOSFET drivers allow the use of low $R_{DS(ON)}$ MOSFETs to minimize the conduction losses for high current applications. Lower current ratings on the individual inductors and MOSFETs also make it possible to use low profile, surface mount components. Therefore, an LTC1629-based PolyPhase high current converter can achieve high efficiency, small size and low profile simultaneously. The savings on the input and output capacitors, inductors and heat sinks minimize the overall cost and size of the complete power supply.

Appendix A
Derivation of output ripple current in a 2-phase circuit

During the interval from DT to T as shown in Figure 14.1, the high side switch in module 1 is off and the high side switch of module 2 is on. The inductor current in module 1 decreases and the inductor current in module 2 increases. The variations of these inductors are estimated to be:

$$\Delta I_{L1} = \frac{-V_0(1-D)T}{L_f} \qquad (A1)$$

$$\Delta I_{L2} = \frac{(V_{IN}-V_0)(1-D)T}{L_f} \qquad (A2)$$

$$\text{where} \qquad D = \frac{V_0}{V_{IN}} \qquad (A3)$$

The net output ripple current is the sum of these inductor ripple currents:

$$\Delta I_0 = |\Delta I_{L1} + \Delta I_{L2}| = \frac{V_0(1-D)T}{L_f} \frac{|1-2D|}{D} \qquad (A4)$$

Equation (A4) is derived based on the waveform shown in Figure 14.1, where D is greater than 0.5. When D is smaller than 0.5, one can easily derive the output ripple current as:

$$\Delta I_0 = |\Delta I_{L1} + \Delta I_{L2}| = \frac{V_0(1-D)T}{L_f} \frac{|1-2D|}{1-D} \qquad (A5)$$

By combining equations (A4) and (A5), we can derive the output ripple current for the 2-phase configuration as:

$$\Delta I_0 = |\Delta I_{L1} + \Delta I_{L2}| = \frac{2V_0(1-D)T}{L_f} \frac{|1-2D|}{|1-2D|+1} \qquad (A6)$$

Section 5

Powering Lasers and Illumination Devices

Ultracompact LCD backlight inverters (15)

It has become desirable to fashion laptop computers with large area screens, leaving little room for the display's backlight inverter electronics. Miniaturization limitations of high voltage magnetic transformers impose limits on achievable space reduction. Another voltage step-up technology, piezoelectric transformers, permits the desired size reduction and provides additional benefits. This publication describes practical piezoceramic transformers and support circuitry. Ancillary benefits of the piezoelectric approach are also described. Appended sections detail transformer operation and feedback loop considerations.

A thermoelectric cooler temperature controller for fiber optic lasers (16)

This application note presents circuitry for maintaining 0.01°C temperature control of fiber optic lasers over wide ambient range variations. The circuitry also features high efficiency power delivery, compact size and low noise. Detailed descriptions of circuitry and results are given with special emphasis on thermal loop optimization. An appended section covers practical considerations for thermoelectric cooler-based control loops.

Current sources for fiber optic lasers (17)

A large group of fiber optic lasers is powered by DC current. Laser drive is supplied by a current source with modulation added to the signal. The current source, although conceptually simple, constitutes an extraordinarily tricky design problem. There are a number of practical requirements for a fiber optic current source and failure to consider them can cause laser and/or optical component destruction. This application note describes ten laser current source circuits with a range of capabilities. High and low current types are presented, along with designs for grounded anode, cathode or floating operation. Each circuit also includes laser protection features. Appended sections cover laser load simulation and current source noise measurement techniques.

Bias voltage and current sense circuits for avalanche photodiodes (18)

Avalanche photodiodes, used in laser based fiber optic systems, require high voltage bias and accurate, wide range current monitoring. Bias voltage varies from 15V–90V and current ranges from 100nA to 1mA, a 10,000:1 dynamic range. This publication presents various 5 volt powered circuits which meet these requirements. Appended sections detail specific circuit techniques and cover measurement practice.

Ultracompact LCD backlight inverters

A svelte beast cuts high voltage down to size

Jim Williams, *Linear Technology Corporation*

Jim Phillips, Gary Vaughn, *CTS Wireless Components*

Introduction

The liquid crystal display (LCD) has become ubiquitous. It is in use everywhere, from personal computers of all sizes to point-of-sale terminals as well as instruments, autos and medical apparatus. The LCD utilizes a cold cathode fluorescent lamp (CCFL) as a light source to back light the display. The CCFL requires a high voltage AC supply for operation. Typically, over 1000 volts RMS is required to initiate lamp operation, with sustaining voltages ranging from 200VAC to 800VAC. To date, the high voltage section of backlight "inverters" has been designed around magnetic transformers. A great deal of effort has been directed towards these ends, accompanied by a large volume of descriptive material.[1] Unfortunately, as available circuit board space continues to shrink, magnetic transformer based approaches begin to encounter difficulty. In particular, it is highly desirable to fashion laptop computers with large area screens, leaving little room for the backlight inverter board. In many cases there is so little space available that building the inverter function inside the LCD panel has become attractive, although to date impractical.

Limitations and problems of magnetic CCFL transformers

Construction and high voltage breakdown characteristics of magnetic transformers present barriers to implementing them in these forthcoming space intensive designs. Additionally, as refined as magnetic technology is, other inverter problems associated with it also exist. Such problems include the necessity to optimize and calibrate the inverter for best performance with a given display type. Practically, this means that the manufacturer must, via either hardware or software, adjust inverter parameters to achieve optimum performance with a given display type. Changes in display choice must be accompanied by commensurate adjustments in inverter characteristics. Another problem area is fail-safe protection due to self-destructive transformer malfunctions. Finally, the magnetic field provided by conventional transformers can interfere with operation of adjacent circuitry. With the exception of size, all of these problems are addressable, although incurring economic and circuit/system penalties.[2] What is really needed is high voltage generating capability that is inherently better suited to coming generations of backlight inverters. Piezoceramic transformers, an arcane and little known technology, have been tamed and made available for CCFL inverter use. This publication summarizes results of an extensive collaborative development effort between LTC and CTS Wireless Components (formerly Motorola Ceramic Products).

Piezoelectric transformers

Piezoelectric transformers (PZT), like magnetic devices, are basically energy converters. A magnetic transformer operates by converting an electrical input to magnetic energy and then reconverting the magnetic energy back to an electrical output. A PZT has an analogous operating mechanism. It converts an electrical input into mechanical energy and subsequently reconverts this mechanical energy back to an electrical output. The mechanical transport causes the PZT to vibrate, similar to quartz crystal operation, although at acoustic frequencies. The resonance associated with this acoustic activity is extraordinarily high; Q factors over 1000 are typical. This transformer action is accomplished by utilizing properties of certain ceramic materials and structures. A PZT's voltage gain is set by its physical configuration and the number of layers in its construction. This is obviously very different from a

---

Note 1: See References 1 through 3 for examples.

Note 2: Again, see References 1 through 3.

Analog Circuit and System Design: A Tutorial Guide to Applications and Solutions. DOI: 10.1016/B978-0-12-385185-7.00015-9

Figure 15.1 • Piezoceramic Transformers Compared to a Dime for Size. 1.5 Watt (Photo Upper) and 10 Watt (Photo Lower) Units Are Shown. Devices Are Narrower Than Magnetic Transformers

magnetic transformer, although some (very rough) magnetic analogs are turns ratio and core configuration. Also very different, and central to any serious drive scheme attempt, is that a PZT has a large input capacitance, as opposed to a magnetic transformer's input inductance.[3]

Figures 15.1 and 15.2 show PZTs; the surprisingly small size is readily apparent. The form factor is ideal for constructing space efficient CCFL backlight inverters. A complete, practical inverter appears in Figure 15.3.

Piezoelectric transformer technology is not new. It has been employed before, in various forms.[4] More familiar examples of piezoelectrics are barbecue grill ignitors (direct mechanical input to the PZT produces an electrical discharge) and marine sonar transducers (electrical input produces a pronounced sonic output). Piezoelectrics are also used in speakers (tweeters), medical ultrasound transducers, mechanical actuators and fans. Piezoelectric based backlight inverters have been attempted, but previous transformer and circuit approaches could not provide power, efficiency and wide dynamic range of operation.

Transformer operating regions were restricted, with complex and ill-performing electronic control schemes. Additionally, the PZT mounting schemes employed enlarged size, negating the PZT's size advantage.

Developing a PZT transformer control scheme

It is instructive to review the path to a practical circuit. Figure 15.4 treats the PZT like a quartz crystal, placing it in a Pierce type oscillator.[5] Self-resonance occurs and a sinosoidal AC high voltage drives the lamp. This circuit has a number of unpleasant features. Very little power is available, due to the circuit's high output impedance. Additionally, the PZT has a number of other spurious modes besides its desired 60kHz fundamental. Changes in drive level or load characteristics induce "mode-hopping," manifested by PZT resonance jumping to subharmonics or harmonics. Sometimes, several modes occur simultaneously! Operation in these modes is characterized by low efficiency and instability. Practically, this circuit was never intended as a serious candidate, only an exploratory exercise. Its contribution is demonstrating that PZT self-resonance is a potentially viable path. Figure 15.5,

---

Note 3: To call this description of PZT operation abbreviated is the kindest of verbiage. Those interested in piezoceramic theory, whether savant or scholar, will find tutorial in Appendices A and B. Appendix A is a brief treatment; B is considerably more detailed. Both sections were written by Jim Phillips of CTS Wireless Components.
Note 4: For examples, see References 4 through 11.

Note 5: See References 12 through 14.

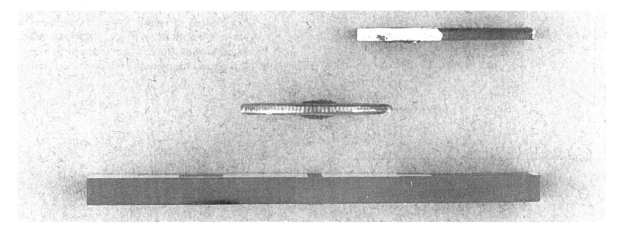

Figure 15.2 • Figure 15.1's Piezoceramic Transformers Rotated to Show Height. Dime Is Photo Center. Height Is Smaller Than Magnetic Types

Figure 15.3 • Complete LCD Backlight Inverter Compared to Dime for Size. Piezoceramic Transformer (Photo Left) Permits Significantly Thinner Board Size

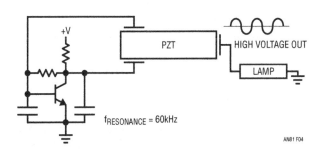

Figure 15.4 • Pierce Type Circuit Sustains Resonance, But Cannot Deliver Power Efficiently. Circuit Also "Mode-Hops" Due to Transformer's Parasitic Resonances

a feedback oscillator, addresses the high output impedance problem with a totem style pair. This is partially successful, although efficient totem drive devoid of simultaneous conduction requires care. Mode-hopping persists, in this case aggravated by the long acoustic transit time through the PZT and the wideband feedback. The sonic transit time produces enormous feedback phase error. Even worse, this phase error varies with line and load conditions. The

alternate feedback scheme shown senses current, as opposed to voltage. This eliminates the voltage divider induced loading but does nothing to address the phase uncertainties and mode-hopping. A final problem, common to all resonant oscillators, concerns start-up. Gentle tapping of the PZT will usually start a reticent circuit but this is hardly reassuring.[6] Figure 15.6 is similar but uses a ground-referred push-pull power stage, simplifying the drive scheme. This is a better approach but phase, mode-hopping and start-up problems are as before.

Figure 15.7 retains the drive scheme and solves the remaining problems. Central to circuit operation is a new transformer terminal labeled "feedback." This connection, precisely positioned on the transformer, provides constant-phase resonance information regardless of operating conditions.

When power is applied, the RC oscillator drives Q1 and Q2 at a frequency outside resonance. The PZT, excited off-resonance, at first responds very inefficiently, although voltage-amplified resonant waveforms appear at the feedback and output terminals. The resonant information present at the feedback terminal injection-locks the RC

Note 6: At the low voltage end, please!

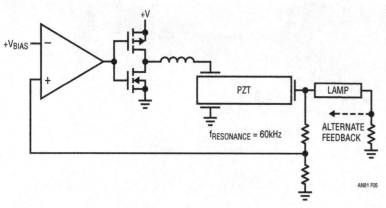

Figure 15.5 • Feedback Based Oscillator Has More Efficient Drive Stage. Poorly Defined Transformer Phase Characteristics Cause Spurious Modes with Line and Load Variations

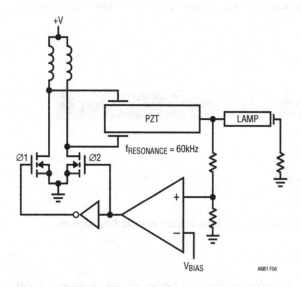

Figure 15.6 • Push-Pull Version of Figure 5 Retains Efficiency While Permitting Simple All N-Channel Drive. Poor Phase Characteristics Still Preclude Stable Loop Operation

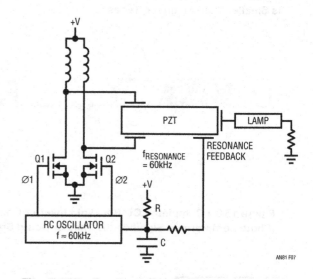

Figure 15.7 • Feedback Tap On Transformer Synchronizes RC Oscillator, Providing Stable Phase Characteristics. Loop Maintains Fundamental Resonance Under All Conditions

oscillator, pulling it to the PZT's resonance. Now, the PZT is supplied with on-resonance drive and efficient operation commences.[7] The feedback terminal's constant-phase characteristic is maintained over all line and load conditions, and the loop enforces resonance.

Figure 15.8, retaining the resonance loop, adds an amplitude control loop to stabilize lamp intensity. Lamp current is sensed and fed back to a voltage regulator to control PZT drive power. The regulator's reference point is variable, permitting lamp intensity to be set at any desired level. The amplitude and resonance loops operate simultaneously, although fully independent of each other. This two loop operation is the key to high power, wide range, reliable control.

Figure 15.9 is a detailed schematic of Figure 15.8's concept. The resonance loop is comprised of Q4 and the CMOS inverter based oscillator. The amplitude

loop centers around the LT1375 switching regulator. Figure 15.10 shows waveforms. Traces A and B are Q2 and Q1 gate drives, respectively. Resultant Q1 and Q2 drain responses are traces C and D. The LT1375 step-down switching regulator, responding to the rectified and averaged lamp current, closes the amplitude loop by driving the L1-L2 junction (trace E). The 4.7μF capacitor at the V_C pin stabilizes the loop.[8] Note that no filtering is used—the raw LT1375 500kHz PWM output directly drives the L1-L2-PZT network. This is permissible because the PZT Q factor is so high that it responds only at resonance (again, see traces C and D).

Note 7: This is the heart and soul of a bootstrapped start-up.

Note 8: This is a deceptively innocent sentence. The PZT's acoustic transport speed furnishes an almost pure delay in the loop, making compensation an interesting exercise. See Appendix C, "A Really Interesting Feedback Loop."

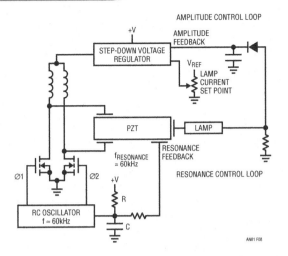

Figure 15.8 • Previous Circuit with Amplitude Control Loop Added. New Circuitry Senses Lamp Current, Accordingly Controls PZT Drive Power. Resonance and Amplitude Control Loops Do Not Interact

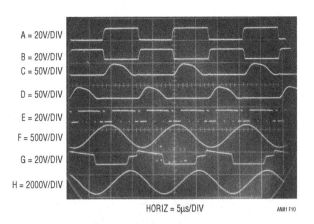

HORIZ = 5µs/DIV   AN81 F10

Figure 15.10 • Waveforms for Figure 15.9. Traces A and B Are Q2 and Q1 Drive, Respectively. Resultant Q1 and Q2 Drain Responses Are Traces C and D. LT1375 V_{SW} Output Is Trace E. PZT Resonance Tap Is Trace F; Q4's Collector, Trace G. PZT High Voltage Output Is Trace H. PZT Acts As a Mechanical Filter, Producing Low Distortion Sine Waves

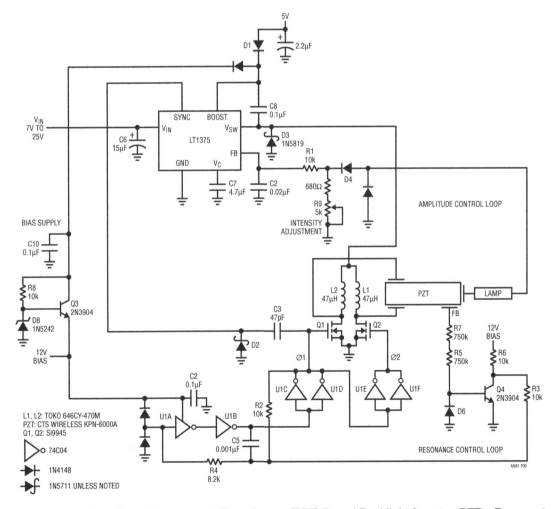

Figure 15.9 • Complete Piezoceramic Transformer (PZT) Based Backlight Inverter. PZT's Resonant Feedback Synchronizes Inverter Based RC Oscillator Via Q4. Amplitude Control Loop Powers PZT Via LT1375 Switching Regulator

The feedback tap (trace F), supplying phase coherent information, looks like a current source to Q4 under all conditions (note trace F's vertical scale factor). The 750k resistors in series minimize parasitic capacitance at the transformer feedback terminal. Q4's collector (trace G) clamps this information to a lower voltage and injection-locks the CMOS inverter based oscillator, closing the resonance loop. The oscillator insures start-up (refer to text associated with Figure 15.7) and effectively filters the already narrow band resonant feedback, further insuring resonance loop fidelity under all conditions. Trace H is the PZT's high voltage output delivered to the lamp.

In this example dimming is set with a potentiometer, although simple current summing to the LT1375 feedback pin allows electronic control.[9]

Additional considerations and benefits

As previously mentioned, the PZT has other benefits besides small size. One of these is safety. A PZT cannot fail due to output shorts or opens. Short circuits knock the PZT off-resonance and it simply stops, absorbing no energy. Open circuits do not cause arc-induced PZT failures because the PZT cannot "arc turns" like a magnetic transformer. However, it is always wise to sense and arrest an overvoltage condition. The PZT is capable of large outputs, despite its small size. Powered by a 10 volt supply, it can easily produce $3000V_{RMS}$ if uncontrolled. This mandates some form of overvoltage protection in a production circuit. Another significant attribute is that amplitude control loop scale factor is almost completely independent of load, including parasitic capacitance. The practical advantage is that a wide range of displays may be used with no recalibration of any kind. This is in direct opposition to magnetically based inverters which require some form (either hardware or software based) of scale factor recalibration

when displays are changed. Understanding why this is so requires some study.

Display parasitic capacitance and its effects

Almost all displays introduce some amount of parasitic capacitance between the lamp, its leads and electrically conductive elements within the display. Such elements may include the display enclosure, the lamp reflector or both. Figure 15.11 diagrams this situation. The parasitic capacitance to ground has two major impacts. It absorbs energy, causing lost power. This raises overall inverter input power because the inverter must supply both parasitic and intended load paths. Some techniques can minimize the effects of parasitic capacitance loss paths but they cannot be completely compensated.[10]

A second effect of parasitic capacitance, manifested in magnetically based inverters, is much more subtle. Magnetically based inverters have finite source impedance at frequency, corrupting the produced sinosoid. The amount of parasitic capacitance influences the degree of corruption. Different displays have varying amounts of parasitic capacitance, resulting in varying degrees of waveform distortion with different displays. The RC averaging time constant is *not* an RMS to DC converter and produces different outputs as distortion content in its input waveform changes. The amplitude loop acts on the DC output of the RC averager and the assumption is that the input waveform distortion content is constant. In a well designed magnetically based inverter this is essentially true, even as operating conditions vary. The averager's output error is consistent and can be "calibrated away" by scale factor adjustments. However, if the display type is changed, the averager is subjected to a differently distorted waveform and scale factor adjustments are required. This necessitates

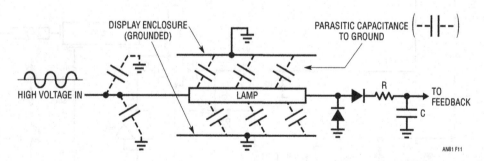

Figure 15.11 • Parasitic Capacitance Absorbs Energy and Corrupts Drive Waveform Due to Finite Source Impedance vs Frequency in Magnetic Based Inverters. Differing Amounts of Capacitance with Various Displays Cause RC Averaging Errors, Necessitating Calibration for Each Display Type. PZT's Highly Resonant Characteristics Eliminate Calibration Requirement

---

Note 9: See Reference 1 for information on various dimming control schemes.

Note 10: Complete treatment of this issue is sacrificed here to maintain focus. A more thorough investigation appears in Reference 1.

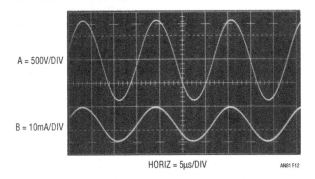

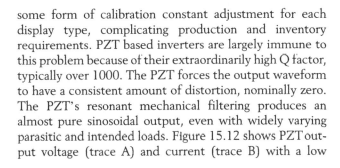

Figure 15.12 • PZT Inverter Driving a Low Parasitic Capacitance Display. Trace A Is Lamp Voltage, Trace B Lamp Current. Waveforms Are Nearly Ideal Sinosoids

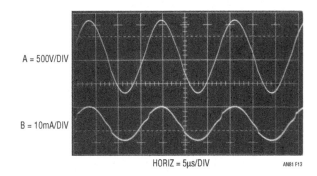

Figure 15.13 • Display with Higher Capacitive Loss Causes Minor Distortion. Lamp RMS Current Varies Only 0.5% vs Figure 15.12

some form of calibration constant adjustment for each display type, complicating production and inventory requirements. PZT based inverters are largely immune to this problem because of their extraordinarily high Q factor, typically over 1000. The PZT forces the output waveform to have a consistent amount of distortion, nominally zero. The PZT's resonant mechanical filtering produces an almost pure sinosoidal output, even with widely varying parasitic and intended loads. Figure 15.12 shows PZT output voltage (trace A) and current (trace B) with a low

parasitic loss display. Waveshapes are essentially ideal sinosoids. In Figure 15.13 a display with much higher parasitic losses has been substituted. Minor waveform distortion, particularly in the current trace (B), is evident, although minimal. The RC averager produces little error vs Figure 15.12 and less than 0.5% lamp current difference occurs between the two cases. In contrast, a magnetically based inverter can easily suffer 10% to 15% lamp current differences, impacting display luminosity and/or lamp lifetime.

References

1. Williams, J.,"A Fourth Generation of LCD Backlight Technology," Linear Technology Corporation, Application Note 65, November 1995.

2. Williams, J., "Techniques for 92% Efficient LCD Illumination," Linear Technology Corporation, Application Note 55, August 1993.

3. Williams, J., "Illumination Circuitry for Liquid Crystal Displays," Linear Technology Corporation, Application Note 49, August 1992.

4. Rosen, C. A., "Ceramic Transformers and Filters," Proc. Electronic Components Symposium, 1956, pp. 205–211.

5. Mason, W. P., "Electromagnetic Transducers and Wave Filters," D. Van Nostrand Company, Inc., 1948.

6. Ohnishi, O., Kishie, H., Iwamoto, A., Sasaki, Y., Zaitsu, T., Inoue, T., "Piezoelectric Ceramic Transformer Operating in Thickness Extensional Vibration Mode for Power Supply," Ultrasonic Symposium Proc., 1992, pp. 483–488.

7. Zaitsu, T., Ohnishi, O., Inoue, T., Shoyama, M., Ninomiya, T., Hua, G., Lee, F. C., "Piezoelectric Transformer Operating in Thickness Extensional Vibration and Its Application to Switching Converter," PESC '94 Record, June 1994, pp. 585–589.

8. Zaitsu, T., Shigehisa, T., Inoue, T., Shoyama, M., Ninomiya, T., "Piezoelectric Transformer Converter with Frequency Control," INTELEC '95, October 1995.

9. Williams, J., "Piezoceramics Plus Fiber Optics Boost Isolation Voltages," EDN, June 24, 1981.

10. Williams, J., Huffman, B., "DC-DC Converters, Part III. Design DC-DC Converters for Power Conservation and Efficiency," "Ceramic Power Converter," EDN, November 10, 1988, pp. 209–224, pp. 222–224.

11. Williams, J., Huffman, B., "Some Thoughts On DC-DC Converters," "20,000V Breakdown Converter," Linear Technology Corporaton, Application Note 29, October 1988, pp. 27–28.

12. Mattheys, R. L., "Crystal Oscillator Circuits," Wiley, New York, 1983.

13. Frerking, M. E., "Crystal Oscillator Design and Temperature Compensation," Van Nostrand Reinhold, New York, 1978.

14. Williams, J., "Circuit Techniques for Clock Sources," Linear Technology Corporation, Application Note 12, October 1985.

Appendix A
Piezoelectric transformers

"Good Vibrations"

James R. Phillips
Sr. Member of Technical Staff
CTS Wireless Components
4800 Alameda Blvd. N.E.
Albuquerque, New Mexico 87113
(505) 348-4286
Jim.Phillips@CTSWireless1.com

Piezoelectric transformers are, in fact, not transformers. There are no wires or magnetic fields. A better analogy would actually be that of a dynamo. The piezoelectric "transformer" works exactly like a motor mechanically coupled to a generator. In order to understand this concept, one must start by understanding the basics of piezoelectricity.

Piezowhat?

Many materials exhibit some form of piezoelectric effect. The ones most commonly used are Quartz, Lithium Niobate, and PZT or Lead Zirconate Titanate with the transformer using the latter. There are two piezoelectric effects, the direct effect and the inverse effect. In the case of the direct effect, placing a force or vibration (*stress*) on the piezoelectric element will result in the generation of a charge. (Figure A1) The polarity of this charge depends upon the orientation of the stress when compared to the direction of polarization in the piezoelectric element. The polarization direction, for PZT, is set by poling or applying a high D.C. field in the range of 45KV/cm to the element during the manufacturing process.

The inverse piezoelectric effect is, as the name implies, the opposite of the direct effect. (Figure A2) Applying an electric field (voltage) to the piezoelectric element results in a dimensional change (*strain*). The direction of the change is likewise linked to the polarization direction. Fields applied at the same polarity of the element result in a dimensional increase, while those of opposite polarity result in a decrease. It should be noted that an increase in one dimension in a structure would result in a decrease in the other two through Poisson's coupling. This phenomenon is an important factor in the operation of the transformer.

The piezoelectric transformer uses both the direct and inverse effects in concert to create high voltage step-up ratios (Figure A3). The input portion of the transformer is driven by a sine wave voltage, which causes it to vibrate (*inverse effect-motor*). The vibration is coupled through the structure to the output, which results in the generation of an output voltage (*direct effect-generator*).

Alchemy and black magic

The piezoelectric transformer is constructed of PZT ceramic, but more exactly, a multilayer ceramic. The manufacture of the transformer is similar to the manufacture of ceramic chip capacitors. Layers of flexible, unfired PZT ceramic tape are printed with metallic patterns and then aligned and stacked to form the required structure. The stacks are then pressed, diced and fired to create the final ceramic device.

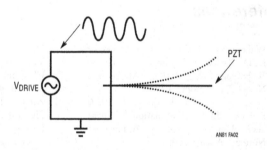

Figure A2 • Inverse Piezoelectric Effect: Applied Voltage Results in Vibration or Movement

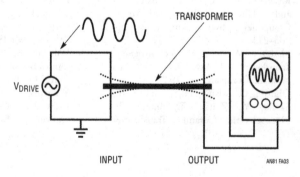

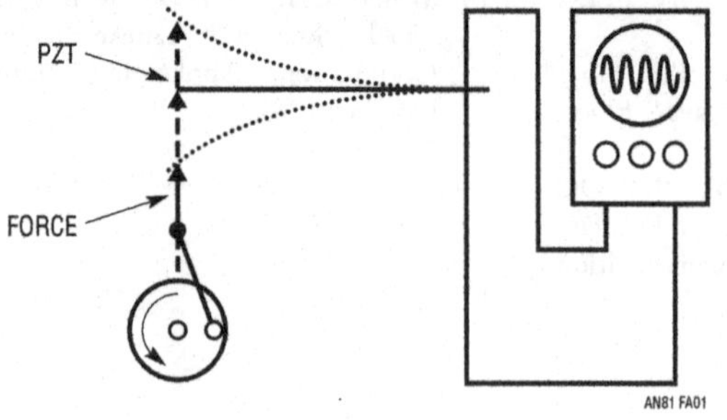

Figure A3 • Transformer: Applied Voltage Results in Vibration Which Causes Output Voltage

Figure A1 • Direct Piezoelectric Effect: Force or Vibration Results in Output Voltage

The input section of the transformer has, in fact, a multilayer ceramic capacitor structure (Figure A4). The metal electrodes are patterned in such a way as to create an inter-digitated plate configuration (*section A-A*). The output section of the transformer has no electrode plates between the ceramic layers and, as a result, fires into a single ceramic structure. The end of the output section is coated with a conductive material, which forms the output electrode for the transformer.

The next step in the construction is to establish the polarization directions in the two halves of the transformer. The input section of the transformer is poled across the inter-digitated electrodes resulting in a polarization direction aligned vertically to the thickness. The output section is poled to create a horizontal or length oriented polarization direction. During operation, the input is driven in thickness mode. This means that a voltage is applied between the parallel plates of the input causing it to become thicker and thinner on alternate halves of the sine wave. The change in input thickness couples through to the output section causing it to become longer and shorter, generating the output voltage. The resulting voltage step-up ratio is then proportional to the ratio of the output length, which generates the voltage, to the thickness of the input layers, across which the drive voltage is applied.

The fun part

The equivalent circuit model for the piezoelectric transformer outwardly looks identical to that of its series resonant magnetic counterpart (Figure A5). The differences, however, extend past the nominal values to what the various components represent.

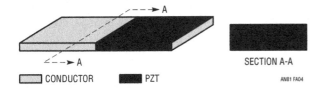

Figure A4 • Piezoelectric Transformer: Multilayer PZT Construction

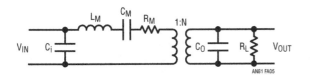

Figure A5 • Piezoelectric Transformer: Equivalent Circuit Model

The input and output capacitance are simply the result of having a dielectric between two metal plates. The effective dielectric constant of PZT runs between 400 and 5,000, depending upon composition. This, unfortunately, is where basic electronics end. The rest of the components are more complicated. The inductance, L_M, is actually the mass of the transformer. The capacitance, C_M, is the compliance of the material or the inverse of spring rate. The compliance is calculated from the applicable generalized beam equation and the Young's modulus. The resistor, R_M, represents the combination of dielectric loss and the mechanical Q of the transformer. It is already obvious to most that truly understanding this device requires background in electronics, mechanics and materials, but we're not quite done yet.

The resonant frequency is related to the product of the capacitance, C_M, and inductance, L_M. This, however, represents the acoustic, not electrical, resonate frequency. The transformer is designed to operate in length resonance. The associated motions are identical to that of a vibrating string. The major difference is that the frequencies are in the ultrasonic range and vary, by design, between 50kHz and 2MHz. Like the string, the transformer has displacement nodes and anti-nodes. Mechanically clamping a node will prevent vibration. This will reduce efficiency in the best case and prevent operation in the worst. Mounting the transformer is crucial. It can not be simply reflowed to a PCB.

The final element in the model is the "ideal" transformer with ratio N. This transformer actually represents three separate transformations. The first is the transformation of electrical energy into mechanical vibration. This is a function of the piezoelectric constant, which is electric field divided by stress, the stress area and the electric field length. The second transformation is the transfer of the mechanical energy from the input section to the output section and is a function of the Poisson's ratio for the material. The final transformation is the transfer of mechanical energy back into electrical energy. This is calculated in a similar fashion to the input side.

A resonant personality

Resonant magnetic high voltage transformers have an electrical Q of between 20 and 30. The equivalent for the piezoelectric transformer is its mechanical Q, which approaches 1,000. This is both good and bad. The ultimate efficiency can be higher, but the usable bandwidth of the transformer is only 2.5% of that of the magnetic. In addition, as shown earlier, the resonant frequency is dependent upon the compliance of the material, which, in turn, is a function of the

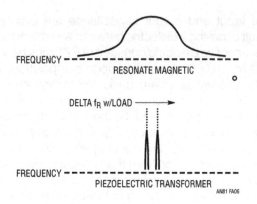

Figure A6 • Transformer Bandwidth Comparison: Magnetic vs Piezoelectric

Young's modulus. Piezoelectric materials have the unusual effect that Young's modulus changes with electrical load. In most, if not all, cases, the shift in resonant frequency over rated load is greater than the usable bandwidth (Figure A6). The piezoelectric transformer must be run at resonance to maintain efficiency and stability. The near-resonance designs used with magnetic transformers work poorly, if at all with piezoelectrics. Tracking oscillators are a requirement.

Rosen[1] first proposed the concept of the piezoelectric transformer in 1956. It is now evident why it took 43 years to get it right.

Appendix B
Piezoelectric technology primer

James R. Phillips
Sr. Member of Technical Staff
CTS Wireless Components
4800 Alameda Blvd. N.E.
Albuquerque, New Mexico 87113
(505) 348-4286
Jim.Phillips@CTSWireless1.com

Piezoelectricity
The piezoelectric effect is a property that exists in many materials. The name is made up of two parts; piezo, which is derived from the Greek word for pressure, and electric from electricity. The rough translation is, therefore, pressure-electric effect. In a piezoelectric material, the application of a force or stress results in the development of a charge in the material. This is known as the direct piezoelectric effect. Conversely, the application of a charge to the same material will result in a change in mechanical

---

Note 1: See Reference 4.

dimensions or strain. This is known as the indirect piezoelectric effect.

Several ceramic materials have been described as exhibiting a piezoelectric effect. These include lead-zirconate-titanate (PZT), lead-titanate (PbTiO$_2$), lead-zirconate (PbZrO$_3$), and barium-titanate (BaTiO$_3$). These ceramics are not actually piezoelectric but rather exhibit a polarized electrostrictive effect. A material must be formed as a single crystal to be truly piezoelectric. Ceramic is a multi crystalline structure made up of large numbers of randomly orientated crystal grains. The random orientation of the grains results in a net cancelation of the effect. The ceramic must be polarized to align a majority of the individual grain effects. The term piezoelectric has become interchangeable with polarized electrostrictive effect in most literature.

Piezoelectric effect
It is best to start with an understanding of common dielectric materials in order to understand the piezoelectric effect. The defining equations for high permittivity dielectrics are:

$$C = \frac{K\varepsilon_r A}{t} = \frac{\varepsilon_0 \varepsilon_r A}{t} = \frac{\varepsilon A}{t} \quad \text{and} \quad Q = CV \rightarrow Q = \frac{\varepsilon A V}{t}$$

where:
- C = Capacitance
- A = Capacitor plate area
- ε_r = Relative dielectric constant
- ε_0 = Dielectric constant of air = 8.85×10^{-12} farads/meter
- ε = Dielectric constant
- V = Voltage
- t = Thickness or plate separation
- Q = Charge

In addition, we can define electric displacement, D, as charge density or the ratio of charge to the area of the capacitor:

$$D = \frac{Q}{A} = \frac{\varepsilon V}{t}$$

and further define the electric field as:

$$E = \frac{V}{t} \quad \text{or} \quad D = \varepsilon E$$

These equations are true for all isotropic dielectrics. Piezoelectric ceramic materials are isotropic in the unpolarized state, but they become anisotropic in the poled state. In anisotropic materials, both the electric field and electric displacement must be represented as vectors with three dimensions in a fashion similar to the mechanical force vector. This is a direct result of the dependency of the ratio of dielectric displacement, D, to electric field, E, upon the orientation of the capacitor plate to the crystal (or poled ceramic) axes. This means that the general equation for electric displacement can be written as a state variable equation:

$$D_i = \varepsilon_{ij} E_j$$

The electric displacement is always parallel to the electric field, thus each electric displacement vector, D_i, is equal to the sum of the field vector, E_j, multiplied by its corresponding dielectric constant, ε_{ij}:

$$D_1 = \varepsilon_{11} E_1 + \varepsilon_{12} E_2 + \varepsilon_{13} E_3$$
$$D_2 = \varepsilon_{21} E_1 + \varepsilon_{22} E_2 + \varepsilon_{23} E_3$$
$$D_3 = \varepsilon_{31} E_1 + \varepsilon_{32} E_2 + \varepsilon_{33} E_3$$

Fortunately, the majority of the dielectric constants for piezoelectric ceramics (as opposed to single crystal piezoelectric materials) are zero. The only non-zero terms are:

$$\varepsilon_{11} = \varepsilon_{22}, \ \varepsilon_{33}$$

Axis nomenclature

The piezoelectric effect, as stated previously, relates mechanical effects to electrical effects. These effects, as shown above, are highly dependent upon their orientation to the poled axis. It is, therefore, essential to maintain a constant axis numbering scheme (Figure B1).

For electro-mechanical constants:

$$d_{ab}, \ a = \text{Electrical direction}$$
$$b = \text{Mechanical direction}$$

Electrical-mechanical analogies

Piezoelectric devices work as both electrical and mechanical elements. There are several electrical-mechanical analogies that are used in designing modeling the devices.

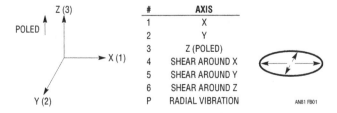

#	AXIS
1	X
2	Y
3	Z (POLED)
4	SHEAR AROUND X
5	SHEAR AROUND Y
6	SHEAR AROUND Z
P	RADIAL VIBRATION

AN81 FB01

Figure B1

ELECTRICAL UNIT		MECHANICAL UNIT	
e	Voltage (Volts)	f	Force (Newtons)
i	Current (Amps)	v	Velocity (Meter/Second)
Q	Charge (Coulombs)	s	Displacement (Meters)
C	Capacitance (Farads)	C_M	Compliance (Meters/Newton)
L	Inductance (Henrys)	M	Mass (Kg)
Z	Impedance	Z_M	Mechanical Impedance

$$i = \frac{dQ}{dt} \qquad v = \frac{ds}{dt}$$
$$e = L\frac{di}{dt} = L\frac{d^2Q}{dt^2} \quad f = M\frac{dv}{dt} = M\frac{d^2s}{dt^2}$$

Coupling

Coupling is a key constant used to evaluate the "quality" of an electro-mechanical material. This constant represents the efficiency of energy conversion from electrical to mechanical or mechanical to electrical.

$$k^2 = \frac{\text{Mechanical Energy Converted to Electrical Charge}}{\text{Mechanical Energy Input}}$$

or

$$k^2 = \frac{\text{Mechanical Energy Converted to Mechanical Displacement}}{\text{Electrical Energy Input}}$$

Electrical, mechanical property changes with load

Piezoelectric materials exhibit the somewhat unique effect that the dielectric constant varies with mechanical load and the Young's modulus varies with electrical load.

Dielectric constant:

$$\varepsilon_{r \ FREE}(1 - k^2) = \varepsilon_{r \ CLAMPED}$$

This means that the dielectric "constant" of the material reduces with mechanical load. Here "Free" stands for a state when the material is able to change dimensions with applied field. "Clamped" refers to either a condition where the material is physically clamped or is driven at a frequency high enough above mechanical resonance that the device can't respond to the changing E field.

Elastic modulus (Young's modulus):

$$Y_{OPEN}(1 - k^2) = Y_{SHORT}$$

This means that the mechanical "stiffness" of the material reduces when the output is electrically shorted. This is important in that both the mechanical Q_M and resonate frequency will change with load. This is also the property that is used in the variable dampening applications.

Elasticity

All materials, regardless of their relative hardness, follow the fundamental law of elasticity (Figure B2). The elastic properties of the piezoelectric material control how well it will work in a particular application. The first concepts, which need to be defined, are stress and strain.

For any given bar of any material:

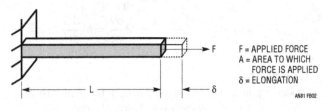

F = APPLIED FORCE
A = AREA TO WHICH
    FORCE IS APPLIED
δ = ELONGATION

AN81 FB02

Figure B2

$$Stress = \sigma = F/A$$

$$Strain = \lambda = \delta/L$$

The relationship between stress and strain is Hooke's Law which states that, within the elastic limits of the material, strain is proportional to stress.

$$\lambda = S\sigma$$

or, for an anisotropic material

$$\lambda i = Sij\sigma j$$

Note: The constant relating stress and strain is the modulus of elasticity or Young's modulus and is often represented by S, E or Y.

Piezoelectric equation

It has been previously shown that when a voltage is applied across a capacitor made of normal dielectric material, a charge results on the plates or electrodes of the capacitor. Charge can also be produced on the electrodes of a capacitor made of a piezoelectric material by the application of stress. This is known as the Direct Piezoelectric Effect. Conversely, the application of a field to the material will result in strain. This is known as the Inverse Piezoelectric Effect. The equation, which defines this relationship, is the piezo-electric equation.

$$Di = dij\sigma j$$

where:

D_i = Electrical displacement (or charge density)

d_ij = Piezoelectric modulus, the ratio of strain to applied field or charge density to applied mechanical stress

Stated differently, d measures charge caused by a given force or deflection caused by a given voltage. We can, therefore, also use this to define the piezoelectric equation in terms of field and strain.

$$Di = \frac{\sigma_j \lambda_i}{E_j}$$

Earlier, electric displacement was defined as

$$Di = \varepsilon_{ij}E_j$$

therefore,

$$e_{ij} E_j = d_{ij}\sigma$$

and

$$E_j = \frac{d_{ij}\sigma_j}{E_{ij}}$$

which results in a new constant

$$g_{ij} = \frac{d_{ij}}{E_{ij}}$$

This constant is known as the piezoelectric constant and is equal to the open circuit field developed per unit of applied stress or as the strain developed per unit of applied charge density or electric displacement. The constant can then be written as:

$$g = \frac{field}{Stress} = \frac{volts/meter}{newtons/meter^2} = \frac{\Delta L/L}{\varepsilon V/t}$$

Fortunately, many of the constants in the formulas above are equal to zero for PZT piezoelectric ceramics. The nonzero constants are:

$s_{11} = s_{22}, s_{33}, s_{12}, s_{13} = s_{23}, s_{44}, s_{66}, = 2(s_{11}-s_{12})$
$d_{31} = d_{32}, d_{33}, d_{15} = d_{24}$

Basic piezoelectric modes
See Figure B3.

Poling
Piezoelectric ceramic materials, as stated earlier, are not piezoelectric until the random ferroelectric domains are aligned. This alignment is accomplished through a process known as "poling." Poling consists of inducing a D.C. voltage across the material. The ferroelectric domains align to the induced field resulting in a net piezoelectric effect. It should be noted that not all the domains become exactly aligned. Some of the domains only partially align and some do not align at all. The number of domains

THICKNESS EXPANSION

THICKNESS SHEAR

FACE SHEAR

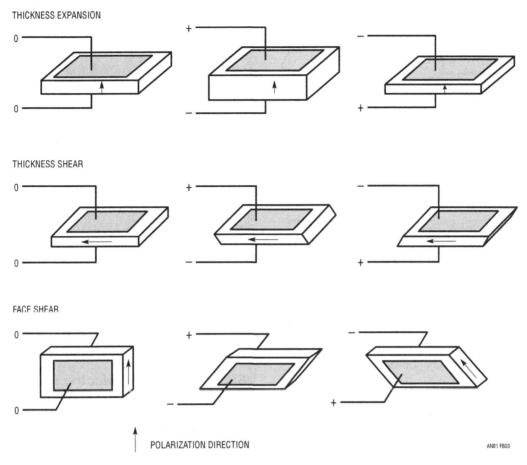

POLARIZATION DIRECTION   AN81 FB03

Figure B3

that align depends upon the poling voltage, temperature, and the time the voltage is held on the material. During poling the material permanently increases in dimension between the poling electrodes and decreases in dimensions parallel to the electrodes. The material can be de-poled by reversing the poling voltage, increasing the temperature beyond the material's Currie point, or by inducing a large mechanical stress.

Post Poling

Applied voltage
Voltage applied to the electrodes at the same polarity as the original poling voltage results in a further increase in dimension between the electrodes and decreases the dimensions parallel to the electrodes. Applying a voltage to the electrodes in an opposite direction decreases the dimension between the electrodes and increases the dimensions parallel to the electrodes.

Applied force
Applying a compressive force in the direction of poling (perpendicular to the poling electrodes) or a tensile force parallel to the poling direction results in a voltage generated on the electrodes which has the same polarity as the original poling voltage. A tensile force applied perpendicular to the electrodes or a compressive force applied parallel to the electrodes results in a voltage of opposite polarity.

Shear
Removing the poling electrodes and applying a field perpendicular to the poling direction on a new set of electrodes will result in mechanical shear. Physically shearing the ceramic will produce a voltage on the new electrodes.

Piezoelectric benders

Piezoelectric benders are often used to create actuators with large displacement capabilities (Figure B4). The bender works in a mode which is very similar to the action of a bimetallic spring. Two separate bars or wafers of piezoelectric material are metallized and poled in the thickness expansion mode. They are then assembled in a $+ - + -$ stack and mechanically bonded. In some cases, a thin membrane is placed between the two wafers. The outer electrodes

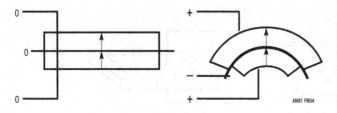

Figure B4

are connected together and a field is applied between the inner and outer electrodes. The result is that for one wafer the field is in the same direction as the poling voltage while the other is opposite to the poling direction, This means that one wafer is increasing in thickness and decreasing in length while the other wafer is decreasing in thickness and increasing in length, resulting in a bending moment.

Loss

There are two sources for loss in a piezoelectric device. One is mechanical, the other is electrical.

$$\text{Mechanical Loss}: \quad Q_M = \frac{\text{Mechanical Stiffness or Mass Resistance}}{\text{Mechanical Resistance}}$$

$$\text{Electrical Loss}: \quad \tan\delta = \frac{\text{Effective Series Resistance}}{\text{Effective Series Reactance}}$$

Simplified Piezoelectric Element Equivalent Circuit

R_i = Electrical resistance
C_i = Input capacitance = $\frac{\varepsilon_0 \varepsilon_r A}{t}$
$\varepsilon_0 = 8.85 \times 10\text{-}12$ farads/meter
A = Electrode area
t = Dielectric thickness
L_M = Mass (Kg)
C_M = Mechanical compliance = 1/Spring Rate (M/N)
N = Electro-mechanical Linear Transducer Ratio (newtons/volts or coulomb/meter)

This model (Figure B5) has been simplified and it is missing several factors. It is only valid up to and slightly beyond resonance. The first major problem

with the model is related to the mechanical compliance (C_M). Compliance is a function of mounting, shape, deformation mode (thickness, free bend, cantilever, etc.) and modulus of elasticity. The modulus of elasticity is, however, anisotropic and it varies with electrical load. The second issue is that the resistance due to mechanical Q_M has been left out. Finally, there are many resonant modes in the transformers, each of which has its own C_M as shown in Figure B6.

Mechanical compliance
Mechanical compliance, which is the inverse of spring constant, is a function of the shape, mounting method, modulus and type of load. Some simple examples are shown in Figure B7.

The various elements that have been explained can now be combined into the design of a complete piezoelectric device. The simple piezoelectric stack transformer will be used to demonstrate the way they are combined to create a functional model.

Simple stack piezoelectric transformer

The piezoelectric transformer acts as an ideal tool to explain the modeling of piezoelectric devices in that it utilizes both the direct and indirect piezoelectric effects. The transformer operates by first converting electrical energy into mechanical energy in one half of the transformer. This energy is in the form of a vibration at the acoustic resonance of the device. The mechanical energy produced is then mechanically coupled into the second half of the transformer. The second half of the transformer then reconverts the mechanical energy into electrical energy. Figure B8 shows the basic layout of a stack transformer. The transformer is driven across the lower half (dimension d_1) resulting in a thickness mode

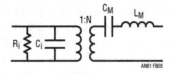

Figure B5

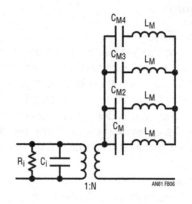

Figure B6

SIMPLE BEAM—UNIFORM END LOAD

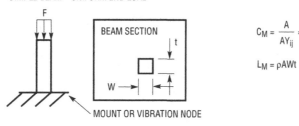

$$C_M = \frac{A}{AY_{ij}} = \frac{A}{Wt}$$

$$L_M = \rho AWt$$

MOUNT OR VIBRATION NODE

SIMPLE BEAM—UNIFORM LOAD—END MOUNTS

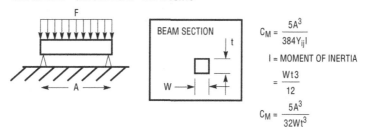

$$C_M = \frac{5A^3}{384Y_{ij}I}$$

$$I = \text{MOMENT OF INERTIA}$$

$$= \frac{Wt3}{12}$$

$$C_M = \frac{5A^3}{32Wt^3}$$

SIMPLE BEAM—UNIFORM LOAD—CANTILEVER MOUNTS

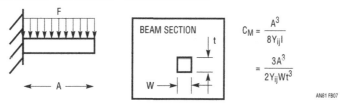

$$C_M = \frac{A^3}{8Y_{ij}I}$$

$$= \frac{3A^3}{2Y_{ij}Wt^3}$$

AN81 FB07

Figure B7

vibration. This vibration is coupled into the upper half and the output voltage is taken across the thinner dimension d_2.

Equivalent circuit
The equivalent circuit model for the transformer (shown in Figure B9) can be thought of as two piezoelectric elements that are assembled back to back.

These devices are connected together by an ideal transformer representing the mechanical coupling between the upper and lower halves. The input resistance, R_i, and the output resistance, R_0, are generally very large and have been left out in this model. The resistor R_L represents the applied load. Determining the values of the various components can be calculated as shown previously.

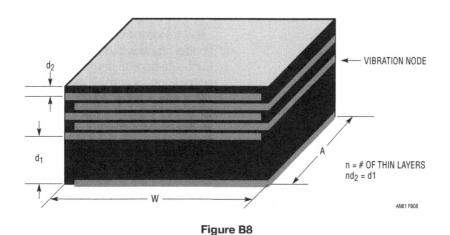

VIBRATION NODE

n = # OF THIN LAYERS
$nd_2 = d1$

AN81 FB08

Figure B8

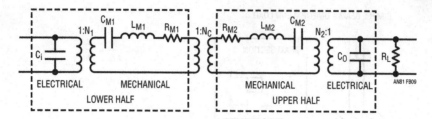

Figure B9

Input/output capacitance

$$Ci = \varepsilon_0 \varepsilon_r \frac{\text{Input Area}}{\text{Input Thickness}} = \varepsilon_0 \varepsilon_r \frac{AW}{d_1}$$

similarly,

$$C_0 = \varepsilon_0 \varepsilon_r \frac{\text{Output Area}}{\text{Output Thickness}} = \varepsilon_0 \varepsilon_r \frac{nAW}{d_2}$$

Mechanical compliance

The mechanical compliance, C_M, can be represented by a simple beam subjected to a uniform axial load. This is because the thickness expansion mode will apply uniform stress across the surface. It should be noted that the beam length is measured with respect to the vibration node. The vibration node is used as this is the surface which does not move at resonance and can, therefore, be thought of as a fixed mounting surface.

$$C_M = \frac{\text{Beam Length}}{\text{Beam Area } Y_{33}}$$

$$C_{M1} = \frac{d_1}{AWY_{33}}$$

$$C_{M2} = \frac{d_2}{AWY_{33}}$$

Note: Even if $nd_2 \neq d_1$ the vibration node will still be located in the mechanical center of the transformer.

Mass

$$L_{M1} = \rho AWd_1$$

$$L_{M2} = \rho AWnd_2 = \rho AWd_1$$

Resistance

The resistances in the model are a function of the mechanical Q_M and Q of the material at resonance and will be calculated later.

Ideal transformer ratio

The transformer ratio, N_1, can be thought of as the ratio of electrical energy input to the resulting mechanical energy output. This term will then take the form of newtons per volt and can be derived from the piezoelectric constant, g.

As before:

$$g = \text{Electrical Field Stress} = \frac{\text{Volts/Meter}}{\text{Newtons/Meter}^2}$$

therefore:

$$\frac{1}{g} = \frac{n/m}{V/m}$$

$$N_1 = \frac{1}{g} = \frac{\text{Area of Applied Force}}{\text{Length of Generated Field}}$$

or

$$N_1 = \frac{AW}{g_{33}d_1}$$

The output section converts mechanical energy back to electrical energy and the ratio would normally be calculated in an inverse fashion to N_1. In the model, however, the transformer ratio is shown as $N_2 : 1$. This results in a calculation for N_2 that is identical to the calculation of N_1.

$$N_2 = \frac{1}{g} = \frac{\text{Area of Applied Force}}{\text{Length of Generated Field}}$$

or

$$N_2 = \frac{AW}{g_{33}d_2}$$

The transformer $1 : N_C$, represents the mechanical coupling between the two halves of the transformer. The stack transformer is tightly coupled and the directions of stress are the same in both halves. This results in $N_C \cong 1$.

Model simplification

The response of the transformer can be calculated from this model, but it is possible to simplify the model through a series of simple network conversion and end up in an equivalent circuit whose form is the same as that of a standard magnetic transformer (Figure B10), where, due to translation through the transformer,

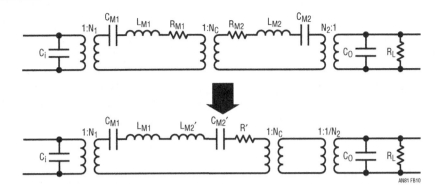

Figure B10

$$C_{M2}' = N_C^2 C_{M2} \quad \text{and} \quad L_{M2}' = L_{M2}/N_C^2$$

but $N_C^2 \cong 1$, therefore

$$C_{M2}' = C_{M2} = C_{M1} \quad \text{and} \quad L_{M2}' = L_{M2} = L_{M1}$$

which allows the next level of simplification (Figure B11). Here

$$L' = L_{M1} + L_{M2}' = 2L_1 = 2\rho AWd1$$

$$C' = \frac{\left(C_{M1} C_{M2}'\right)}{\left(C_{M1} + C_{M2}'\right)} = \frac{C_{M1}^2}{2C_{M1}} = \frac{C_{M1}}{2} = \frac{d_1}{2AWY_{33}}$$

Final simplification (Figure B12) where

$$C = C' N_1^2 \quad \text{and} \quad L = L'/N_1^2$$

and, from before

$$N_1 = \frac{AW}{g_{33}d_1}$$

therefore

$$C = \frac{d1}{2WLY_{33}} \frac{A^2 W^2}{g_{33}^2 d_1^2} = \frac{AW}{2Y_{33}g_{33}^2 d_1}$$

$$L = 2\rho AWd_1 \frac{g_{33}^2 d_1^2}{A^2 W^2} = \frac{2\rho g_{33}^2 d_1^2}{AW}$$

$$N = \frac{N_1 N_C}{N_2} = \frac{AW}{g_{33}d_1} \frac{g_{33}d_2}{AW} = \frac{d_2}{d_1}$$

The last value we need to calculate is the motional resistance. This value is based upon the mechanical QM of the material and the acoustic resonant frequency.

Resonant frequency

$$\omega_0 = 1/\sqrt{LC}$$

$$= \frac{1}{\sqrt{\dfrac{2\rho d_1 g_{33}^2}{AW} \dfrac{AW}{2Y_{33}g_{33}^2 d1}}}$$

$$= \frac{1}{\sqrt{\dfrac{\rho d_1^2}{Y_{33}}}} = \frac{1}{d_1 \sqrt{\dfrac{\rho}{Y_{33}}}}$$

$$c_{PZT} \equiv \text{Speed of Sound in PZT} = \sqrt{Y/\rho}$$

therefore

$$\omega_0 = c_{PZT}/d_1$$

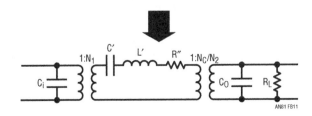

Figure B11

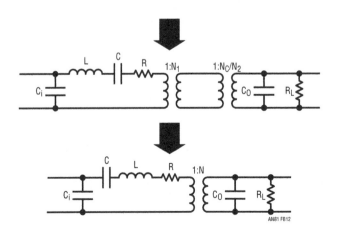

Figure B12

The equation above states that the resonant frequency is equal to the speed of sound in the material divided by the acoustic length of the device. This is the definition of acoustic resonance and acts as a good check of the model. The final derivation is the value of resistance.

$$Q_M \equiv 1/\omega_0 RC$$

or

$$R = 1/\omega_0 Q_M C$$

$$R = \frac{d_1 \sqrt{\rho Y_{33}}}{Q_M} \frac{2Y_{33}g_{33}^2 d_1}{AW}$$

$$= \frac{2d_1^2 g_{33}^2 \sqrt{\rho Y_{33}}}{Q_M AW}$$

Note: C_M and R are both functions of Y_{33} and Y_{33} is a function of R_L.

It should be noted that the model is only valid for transformers driven at or near their fundamental resonate frequencies. This is because the initial mechanical model assumed a single vibration node located at the center of the stack which is only true when the transformer is driven at fundamental resonance. There are more nodes when the transformer is driven at harmonic frequencies (Figure B13).

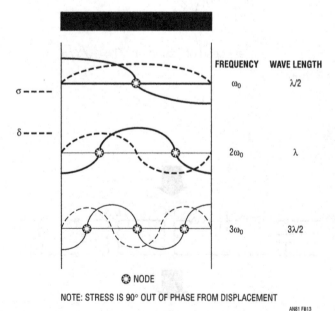

	FREQUENCY	WAVE LENGTH
	ω_0	$\lambda/2$
	$2\omega_0$	λ
	$3\omega_0$	$3\lambda/2$

✿ NODE

NOTE: STRESS IS 90° OUT OF PHASE FROM DISPLACEMENT

AN81 FB13

Figure B13

There are no fixed nodes at frequencies other than resonance. This means that the transformer must be designed with the resonate mode in mind or phase cancellations will occur and there will be little or no voltage gain. It is often difficult to understand the concept of nodes and phase cancellation, so a simple analogy can be used. In this case, waves created in a waterbed will be used to explain the effect.

Pressing on the end of a waterbed creates a "wave" of displacement that travels down the length of the bed until it reaches the opposite end and bounces back. The water pressure (stress) is the lowest, or negative with respect to the water at rest, at a point just in front of the wave and highest at a point just behind the wave. The pressures at the crest and in the trough are at the same pressure as the bed at rest. The wave will reflect back and forth until resistance to flow causes it to dampen out. The average pressure over time at any point in the bed will be exactly the same as the pressure at rest. Similarly, the average stress in a transformer off resonance will approach zero and there will be no net output.

Pressing on the end of the same bed repeatedly just after the wave has traveled down the length, reflected off the end, returned and reflected off the "driven" end will result in a standing wave. This means that one half of the bed is getting thicker as the other half is getting thinner and the center of the bed will be stationary. The center is the node and the thickness plotted over time of either end will form a sine wave. There will be no net pressure difference in the center, but the ends will have a pressure wave which form a sine wave 90° out of phase with the displacement. The transformer again works in the same manner with no voltage at the node and an AC voltage at the ends. It is fairly simple to expand this concept to harmonics and to other resonate shapes.

Conclusion

The number of different applications for piezoelectric ceramic, and in particular PZT ceramic, is too great to address in a single paper. The basic principals that have been set forth in this primer can, however, be used to both understand and design piezoelectric structures and devices. The ability to create devices of varying applications and shapes is greatly enhanced by the use of multilayer PZT ceramics.

APPENDIX C
A really interesting feedback loop

The almost pure mechanical delay presented by the PZT's acoustic transport presents a fascinating exercise in loop compensation.[1] Veterans of feedback loop compensation battles will exercise immediate caution when confronted with a pure and lengthy delay in a loop. Neophyte designers will gain a lesson they will not easily forget.

Figure C1 diagrams Figure 15.9's amplitude control loop, with significant contributions to loop transmission represented. The PZT delivers phase delayed information at about 60kHz to the lamp. This information is smoothed to DC by the RC averaging time constant and delivered to the LT1375's feedback terminal. The LT1375 controls PZT power with its 500kHz PWM output, closing the control loop. The capacitor at the LT1375's V_C pin rolls off gain, nominally stabilizing the loop. This compensation capacitor must roll off gain-bandwidth at a low enough value to prevent loop delays from causing oscillation. Which of these delays is the most significant? From a stability viewpoint, the LT1375's output repetition rate and the PZT's oscillation frequency are sampled data systems. Their information delivery rate is far above the PZT's 200μs delay and the averaging time constants, and is not significant. The PZT delay and the RC time constant are the major contributors to loop delay. The RC time constant must be large enough to turn the half wave rectified waveform into DC. The lumped delay of the PZT and the RC thus dominates loop transmission. It must be compensated by the capacitor at the LT1375 V_C pin. A large enough value for this capacitor rolls off loop gain at low enough frequency to provide stability. The loop simply does not have enough gain to oscillate at a frequency commensurate with the RC and PZT delays.[2]

A good way to begin to establish a value for loop roll-off is to let the loop oscillate. This is facilitated by initially deleting the compensation capacitor and turning the circuit on. Figure C2 shows the Vc pin (trace A) and FB node (trace B). The frequency of oscillation and the phase relationship between these two signals provide valuable insight into achievable closed-loop bandwidth.[3] Loop delay sets oscillation frequency at about 2.3kHz. Selecting the Vc value to roll off bandwidth well below this frequency is appropriate. Figure C3 shows results with Figure 15.9's 4.7μF capacitor installed. Trace A, a step input, is applied to the LT1375's shutdown pin. When the shutdown pin is enabled, V_C (trace B) slews up and the feedback pin (trace C) moves toward its control point as PZT output voltage (trace D) rises. The V_C pin damping causes extended slew time but settling is completed in 25 milliseconds with minimal overshoot. The small overshoot (past mid screen) derives from the lamp's negative resistance characteristic and is easily handled by loop dynamics.

Some situations may require significantly faster loop response than simple first order compensation can provide. An example is wide range dimming via

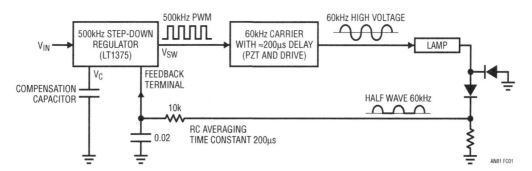

Figure C1 • Delay Terms in Figure 15.9's Amplitude Control Feedback Path. PZT's 200μs Mechanical Delay and RC Time Constant Dominate Loop Transmission and Must Be Compensated for Stable Operation

---

Note 1: Perhaps this verbiage indulges drama. Conversely, nerds like me find arcana such as this fascinating.

Note 2: The high priests of feedback refer to this as "Dominant Pole Compensation." The rest of us are reduced to more pedestrian descriptives. As such, this technique is sometimes loosely referred to as "glop comp."

Note 3: Deliberately sustaining loop oscillation is a valuable investigative tool but may encounter problems in some applications. Consider aircraft flap control servos or power plant generator stabilizing loops.

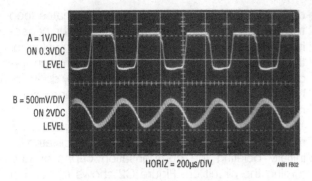

A = 1V/DIV
ON 0.3VDC
LEVEL

B = 500mV/DIV
ON 2VDC
LEVEL

HORIZ = 200μs/DIV   AN81 FB02

Figure C2 • Allowing Loop to Oscillate Hints at Compensation Requirements. 2.3kHz Oscillation Frequency Appears to Derive from Figure C1's Delay Terms. Trace A Is LT1375 V_C Pin, Trace B Its Feedback Terminal. High Frequency Content in Trace B's Outline, PZT Carrier Residue, Is Not Pertinent

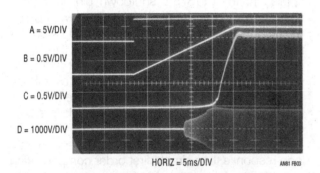

A = 5V/DIV

B = 0.5V/DIV

C = 0.5V/DIV

D = 1000V/DIV

HORIZ = 5ms/DIV   AN81 FB03

Figure C3 • Figure 9's Loop Compensation Is Stable Despite PZT's Mechanically Induced Delay. First Order Roll-Off from Large V_C Pin Capacitor Stabilizes Loop. Trace A Is Step Input at LT1375 Shutdown Pin. Traces B, C and D are V_C, Feedback and PZT Output Nodes, Respectively

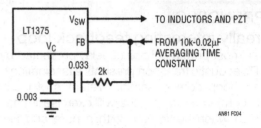

LT1375 V_{SW} ──→ TO INDUCTORS AND PZT

          FB ←── FROM 10k-0.02μF
V_C AVERAGING TIME
               CONSTANT
     0.033 2k

0.003

AN81 FC04

Figure C4 • Very Light V_C Damping Augmented by a Feedback Lead Provides 20× Faster Loop Response. Technique Allows Wide Range Lamp Dimming Via PWM without Sacrificing Line Regulation

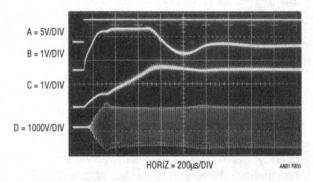

A = 5V/DIV

B = 1V/DIV

C = 1V/DIV

D = 1000V/DIV

HORIZ = 200μs/DIV   AN81 FB05

Figure C5 • Feedback Based Compensation Waveforms. Trace A Is Step Input at LT1375 Shutdown Pin. Traces B, C and D Are V_C, Feedback and PZT Output Nodes, Respectively. Note 25× Sweep Speed Increase Over Figure C3

PWM methods. Such methods rely on rapid on-off loop cycling to achieve wide range dimming. Typically, the cycling occurs well above the flicker rate, in the 100Hz to 200Hz region. Loop settling must be very quick with respect to this frequency or line regulation will be poor. This is so because during slew the loop is out of control. If slew time approaches on-time, control is, by definition, poor. Figure C4 uses a feedback lead, allowing very light V_C damping with resultant faster slew time.[4] Figure C5 shows results. When the shutdown pin (trace A) is enabled, V_C (trace B) slews rapidly, followed by the FB node (trace C). The PZT high voltage envelope is trace D. Loop capture occurs in 1.2 milliseconds—about 20x faster than Figure C3's simple first order compensation. Note that this performance begins to approach the limits implied by Figure C2's information.[5]

Note 4: Proper Bodese for this compensation technique is a "feedback zero."

Note 5: Score one for lightning empiricism.

A thermoelectric cooler temperature controller for fiber optic lasers

Climatic pampering for temperamental lasers

16

Jim Williams

Introduction

Continued demands for increased bandwidth have resulted in deployment of fiber optic-based networks. The fiber optic lines, driven by solid state lasers, are capable of very high information density. Highly packed data schemes such as DWDM (dense wavelength division multiplexing) utilize multiple lasers driving a fiber to obtain large multichannel data streams. The narrow channel spacing relies on laser wavelength being controlled within 0.1nm (nanometer). Lasers are capable of this but temperature variation influences operation. Figure 16.1 shows that laser output peaks sharply vs wavelength, implying that laser wavelength must be controlled well within 0.1nm to maintain performance. Figure 16.2 plots typical laser wavelength vs temperature. The 0.1nm/°C slope means that although temperature facilitates tuning laser wavelength, it must not vary once the laser has been peaked. Typically, temperature control of 0.1°C is required to maintain laser operation well within 0.1nm.

Temperature controller requirements

The temperature controller must meet some unusual requirements. Most notably, because of ambient temperature variation and laser operation uncertainties, the controller must be capable of either sourcing or removing heat to maintain control. Peltier-based thermoelectric coolers (TEC) permit this but the controller must be truly bidirectional. Its heat flow control must not have dead zone or untoward dynamics in the "hot-to-cold" transition region. Additionally, the temperature controller must be a precision device capable of maintaining control well inside 0.1°C over time and temperature variations.

Laser based systems packaging is compact, necessitating small solution size with efficient operation to avoid excessive heat dissipation. Finally, the controller must operate from a single, low voltage source and its (presumably

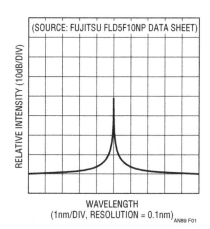

Figure 16.1 • Laser Intensity Peak Approaches 40dB within a 1nm Window

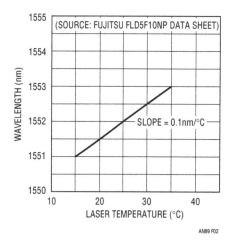

Figure 16.2 • Laser Wavelength Varies ≈0.1nm/°C. Typical Application Requires Wavelength Stability within 0.1nm, Mandating Temperature Control

Analog Circuit and System Design: A Tutorial Guide to Applications and Solutions. DOI: 10.1016/B978-0-12-385185-7.00016-0

switched mode) operation must not corrupt the supply with noise.

Temperature controller details

Figure 16.3, a schematic of the thermoelectric cooler (TEC) temperature controller, includes three basic sections. The DAC and the thermistor form a bridge, the output of which is amplified by A1. The LTC1923 controller is a pulse width modulator which provides appropriately modulated and phased drive to the power output stage. The laser is an electrically delicate and very expensive load. As such, the controller provides a variety of monitoring, limiting and over-load protection capabilities. These include soft-start and overcurrent protection, TEC voltage and current sense and "out-of-bounds" temperature sensing. Aberrant operation results in circuit shutdown, preventing laser module damage. Two other features promote system level compatibility. A phase-locked loop based oscillator permits reliable clock synchronization of multiple LTC1923s in multilaser systems. Finally, the switched mode power delivery to the TEC is efficient but special considerations are required to ensure that switching related noise is not introduced ("reflected") into the host power supply. The LTC1923 includes edge slew limiting which minimizes switching related harmonics by slowing down the power stages' transition times. This greatly reduces high frequency harmonic content, preventing excessive switching related noise from corrupting the power supply or the laser.[1] The switched mode power output stage, an "H-bridge" type, permits efficient bidirectional drive to the

TEC, allowing either heating or cooling of the laser. The thermistor, TEC and laser, packaged at manufacture within the laser module, are tightly thermally coupled.

The DAC permits adjusting temperature setpoint to any individual laser's optimum operating point, normally specified for each laser. Controller gain and bandwidth adjustments optimize thermal loop response for best temperature stability.

Thermal loop considerations

The key to high performance temperature control is matching the controller's gain bandwidth to the thermal feedback path. Theoretically, it is a simple matter to do this using conventional servo-feedback techniques. Practically, the long time constants and uncertain delays inherent in thermal systems present a challenge. The unfortunate relationship between servo systems and oscillators is very apparent in thermal control systems.

The thermal control loop can be very simply modeled as a network of resistors and capacitors. The resistors are equivalent to the thermal resistance and the capacitors to thermal capacity. In Figure 16.4 the TEC, TEC-sensor interface and sensor all have RC factors that contribute to a lumped delay in the system's ability to respond. To prevent oscillation, gain bandwidth must be limited to account for this delay. Since

Note 1: This technique derives from earlier efforts. See Reference 1 for detailed discussion and related topics.

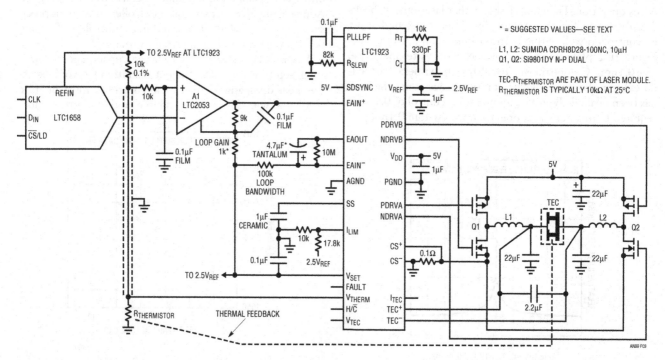

Figure 16.3 • Detailed Schematic of TEC Temperature Controller Includes A1 Thermistor Bridge Amplifier, LTC1923 Switched Mode Controller and Power Output H-Bridge. DAC Establishes Temperature Setpoint. Gain Adjust and Compensation Capacitor Optimize Loop Gain Bandwidth. Various LTC1923 Outputs Permit Monitoring TEC Operating Conditions

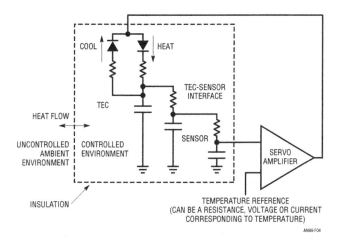

Figure 16.4 • Simplified TEC Control Loop Model Showing Thermal Terms. Resistors and Capacitors Represent Thermal Resistance and Capacity, Respectively. Servo Amplifier Gain Bandwidth Must Accommodate Lumped Delay Presented by Thermal Terms to Avoid Instability

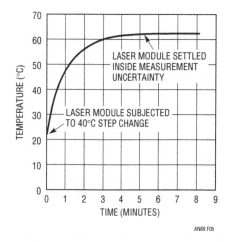

Figure 16.5 • Ambient-to-Sensor Lag Characteristic for a Typical Laser Module Is Set by Package Thermal Resistance and Capacity

high gain bandwidth is desirable for good control, the delays should be minimized. This is presumably addressed by the laser module's purveyor at manufacture

The model also includes insulation between the controlled environment and the uncontrolled ambient. The function of insulation is to keep the loss rate down so the temperature control device can keep up with the losses. For any given system, the higher the ratio between the TEC-sensor time constants and the insulation time constants, the better the performance of the control loop.[2]

Temperature control loop optimization

Temperature control loop optimization begins with thermal characterization of the laser module. The previous section emphasized the importance of the ratio between the TEC-sensor and insulation time constants. Determination of this information places realistic bounds on achievable controller gain bandwidth. Figure 16.5 shows results when a typical laser module is subjected to a 40°C step change in ambient temperature. The laser module's internal temperature, monitored by its thermistor, is plotted vs time with the TEC unpowered. An ambient-to-sensor lag measured in minutes shows a classic first order response.

The TEC-sensor lumped delay is characterized by operating the laser module in Figure 16.3's circuit with gain set at maximum and no compensation capacitor installed. Figure 16.6 shows large-signal oscillation due to thermal lag dominating the loop. A great deal of valuable information is

contained in this presentation.[3] The frequency, primarily determined by TEC-sensor lag, implies limits on how much loop bandwidth is achievable. The high ratio of this frequency to the laser module's thermal time constant (Figure 16.5) means a simple, dominant pole loop compensation will be effective. The saturation limited waveshape suggests excessive gain is driving the loop into full cooling and heating states. Finally, the asymmetric duty cycle reflects the TEC's differing thermal efficiency in the cooling and heating modes.

Controller gain bandwidth reduction from Figure 16.6's extremes produced Figure 16.7's display. The waveform results from a small step (≈0.1°C) change in temperature setpoint. Gain bandwidth is still excessively high, producing a damped, ringing response over 2 minutes in duration! The loop is just marginally stable. Figure 16.8's test conditions are identical but gain bandwidth has been significantly reduced. Response is still not optimal but settling occurs in ≈ 4.5 seconds, about 25× faster than the previous case. Figure 16.9's response, taken at further reduced gain bandwidth settings, is nearly critically damped and settles cleanly in about 2 seconds. A laser module optimized in this fashion will easily attenuate external temperature shifts by a factor of thousands without overshoots or excessive lags. Further, although there are substantial thermal differences between various laser modules, some generalized guidelines on gain bandwidth values are possible.[4] A DC gain of 1000 is sufficient for required temperature control, with bandwidth below 1Hz providing adequate loop stability. Figure 16.3's suggested gain and bandwidth values reflect these conclusions, although stability testing for any specific case is mandatory.

Note 2: For the sake of text flow, this somewhat academic discussion must suffer brevity. However, additional thermodynamic gossip appears in Appendix A, "Practical considerations in thermoelectric cooler based control loops."

Note 3: When a circuit "doesn't work" because "it oscillates," whether at millihertz or gigahertz, four burning questions should immediately dominate the pending investigation. What frequency does it oscillate at, what is the amplitude, duty cycle and waveshape? The solution invariably resides in the answers to these queries. Just stare thoughtfully at the waveform and the truth will bloom.

Note 4: See Appendix A, "Practical considerations in thermoelectric cooler based control loops," for additional comment.

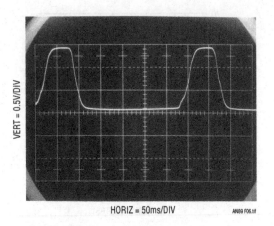

VERT = 0.5V/DIV

HORIZ = 50ms/DIV   AN89 F06.tif

Figure 16.6 • Deliberate Excess of Loop Gain Bandwidth Introduces Large-Signal Oscillation. Oscillation Frequency Provides Guidance for Achievable Closed-Loop Bandwidth. Duty Cycle Reveals Asymmetric Heating-Cooling Mode Gains

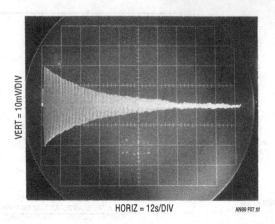

VERT = 10mV/DIV

HORIZ = 12s/DIV   AN89 F07.tif

Figure 16.7 • Loop Response to Small Step in Temperature Setpoint. Gain Bandwidth Is Excessively High, Resulting in Damped, Ringing Response Over 2 Minutes in Duration

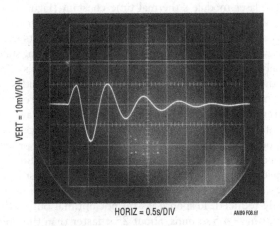

VERT = 10mV/DIV

HORIZ = 0.5s/DIV   AN89 F08.tif

Figure 16.8 • Same Test Conditions as Figure 16.7 but at Reduced Loop Gain Bandwidth. Loop Response Is Still Not Optimal but Settling Occurs in 4.5 Seconds—Over 25× Faster than Previous Case

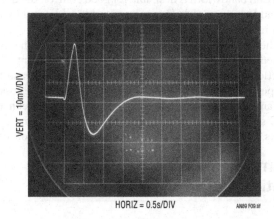

VERT = 10mV/DIV

HORIZ = 0.5s/DIV   AN89 F09.tif

Figure 16.9 • Gain Bandwidth Optimization Results in Nearly Critically Damped Response with Settling in 2 Seconds

Temperature stability verification

Once the loop has been optimized, temperature stability can be measured. Stability is verified by monitoring thermistor bridge offset with a stable, calibrated differential amplifier.[5] Figure 16.10 records ±1 millidegree baseline stability over 50 seconds in the cooling mode. A more stringent test measures longer term stability with significant variations in ambient temperature. Figure 16.11's strip-chart recording measures cooling mode stability against an environment that steps 20°C above ambient every hour over 9hours.[6] The data shows 0.008°C resulting variation, indicating a thermal gain of 2500.[7] The 0.0025°C baseline tilt over the 9 hour plot length derives from varying ambient temperature. Figure 16.12 utilizes identical test conditions, except that the controller operates in the heating mode. The TEC's higher heating mode efficiency furnishes greater thermal gain, resulting in a 4× stability improvement to about 0.002°C variation. Baseline tilt, just detectable, shows a similar 4× improvement vs Figure 16.11.

This level of performance ensures the desired stable laser characteristics. Long-term (years) temperature stability is primarily determined by thermistor aging characteristics.[8]

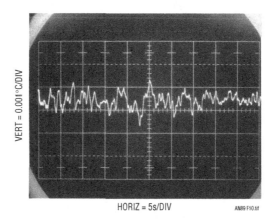

VERT = 0.001°C/DIV

HORIZ = 5s/DIV   AN89 F10.tif

Figure 16.10 ● Short-Term Monitoring in Room Environment Indicates 0.001°C Cooling Mode Baseline Stability

---

Note 5: This measurement monitors thermistor stability. Laser temperature stability will be somewhat different due to slight thermal decoupling and variations in laser power dissipation. See Appendix A.
Note 6: That's right, a *strip-chart recording*. Stubborn, locally based aberrants persist in their use of such archaic devices, forsaking more modern alternatives. Technical advantage could account for this choice, although deeply seated cultural bias may be a factor.

Note 7: Thermal gain is temperature control aficionado jargon for the ratio of ambient-to-controlled temperature variation.
Note 8: See Appendix A for additional information.

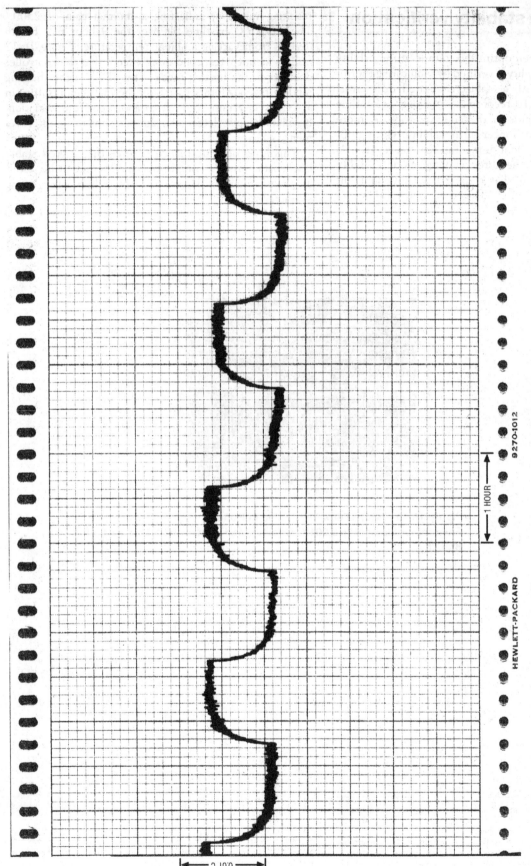

HEWLETT-PACKARD

9270-1012

Figure 16.11 • Long-Term Cooling Mode Stability Measured in Environment that Steps 20 Degrees Above Ambient Every Hour. Data Shows Resulting 0.008°C Peak-to-Peak Variation, Indicating Thermal Gain of 2500. 0.0025°C Baseline Tilt Over Plot Length Derives From Varying Ambient Temperature

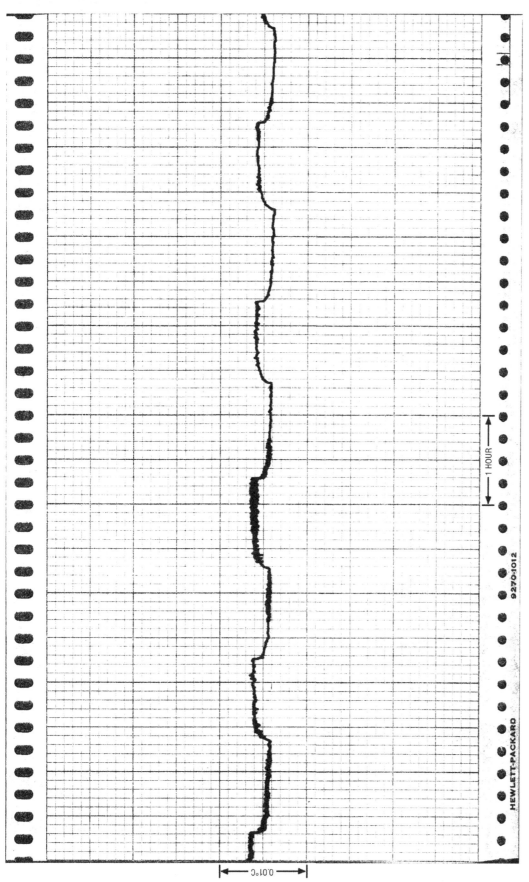

Figure 16.12 • Identical Test Conditions as Figure 16.11, Except in Heating Mode. TEC's Higher Heating Mode Efficiency Results in Higher Thermal Gain. 0.002°C Peak-to-Peak Variation Is 4× Stability Improvement. Baseline Tilt, Just Detectable, Shows Similar 4× Improvement vs Figure 16.11

Reflected noise performance

The switched mode power delivery to the TEC provides efficient operation but raises concerns about noise injected back into the host system via the power supply. In particular, the switching edge's high frequency harmonic content can corrupt the power supply, causing system level problems. Such "reflected" noise can be troublesome to deal with. The LTC1923 avoids these issues by controlling the slew of its switching edges, minimizing high frequency harmonic content.[9] This slowing down of switching transients typically reduces efficiency by only 1% to 2%, a small penalty for the greatly improved noise performance. Figure 16.13 shows noise and ripple at the 5V supply with slew control in use. Low frequency ripple, 12mV in

amplitude, is usually not a concern, as opposed to the high frequency transition related-components, which are much lower. Figure 16.14, a time and amplitude expansion of Figure 16.13, more clearly studies high frequency residue. High frequency amplitude, measured at center screen, is about 1mV. The slew limiting's effectiveness is measured in Figure 16.15 by disabling it. High frequency content jumps to nearly 10mV, almost 10× worse performance. Leave that slew limiting in there!

Most applications are well served by this level of noise reduction. Some special cases may require even lower reflected noise. Figure 16.16's simple LC filter may be employed in these cases. Combined with the LTC1923's slew limiting, it provides vanishingly small reflected ripple and high frequency harmonics.

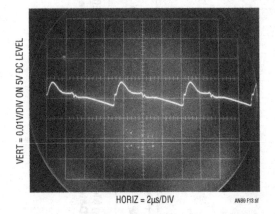

Figure 16.13 • "Reflected" Noise at 5V DC Input Supply Due to Switching Regulator Operation with Edge Slew Rate Limiting in Use. Ripple Is 12mV$_{P-P}$, High Frequency Edge Related Harmonic Is Much Lower

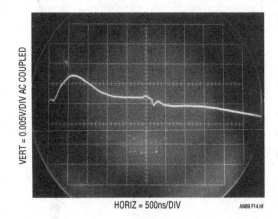

Figure 16.14. • Time and Amplitude Expansion of Figure 16.13 More Clearly Shows Residual High Frequency Content with Slew Limiting Employed

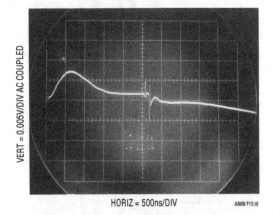

Figure 16.15 • Same Test Conditions as Previous Figure, Except Slew Limiting Is Disabled. High Frequency Harmonic Content Rises ≈×10. Leave that Slew Limiting in There!

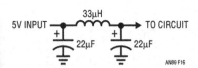

Figure 16.16 • LC Filter Produces 1mV Reflected Ripple and 500μV High Frequency Harmonic Noise Residue

---

Note 9: This technique derives from previous work. See Reference 1

Figure 16.17, taken using this filter, shows only about 1mV of ripple, with submillivolt levels of high frequency content. Figure 16.18 expands the time scale to examine the high frequency remnants. Amplitude is 500μV, about 1/3 Figure 16.14's reading. As before, slew limiting effectiveness is measurable by disabling it. This is done in Figure 16.19, with a resulting 4.4× increase in high frequency content to about 2.2mV. As in Figure 16.15, if lowest reflected noise is required, leave that slew limiting in there!

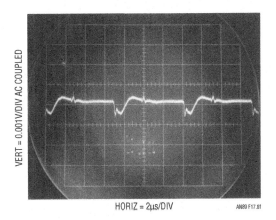

Figure 16.17 • **5V Supply Reflected Ripple Measures 1mV with Figure 16.16's LC Filter in Use, a 10× Reduction Over Figure 16.13. Switching Edge Related Harmonic Content Is Small Due to Slew Limiting Action**

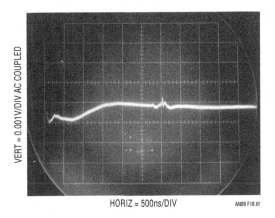

Figure 16.18 • **Horizontal Expansion Permits Study of High Frequency Harmonic with Slew Limiting Enabled. Amplitude Is 500μV, About 1/3 Figure 16.14's Reading**

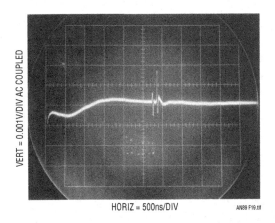

Figure 16.19 • **Same Conditions as Previous Figure, Except Slew Limiting Is Disabled. Harmonic Content Amplitude Rises to 2.2mV, a 4.4× Degradation. As in Figure 16.15, Leave that Slew Limiting in There!**

References

1. Williams, J., "A Monolithic Switching Regulator with 100µV Output Noise," Linear Technology Corporation, Application Note 70, October 1997.
2. Williams, J., "Thermal Techniques in Measurement and Control Circuitry," Linear Technology Corporation, Application Note 5, December 1984.
3. Williams, J., "Temperature Controlling to Microdegrees," Massachusetts Institute of Technology Education Research Center, October 1971.
4. Fulton, S. P., "The Thermal Enzyme Probe," Thesis, Massachusetts Institute of Technology, 1975.
5. Olson, J. V., "A High Stability Temperature Controlled Oven," Thesis, Massachusetts Institute of Technology, 1974.
6. Harvey, M. E., "Precision Temperature Controlled Water Bath," *Review of Scientific Instruments*, p. 39–1, 1968.
7. Williams, J. "Designer's Guide to Temperature Control,". *EDN* Magazine, June 20, 1977.
8. Trolander, Harruff and Case, "Reproducibility, Stability and Linearization of Thermistor Resistance Thermometers," ISA, Fifth International Symposium on Temperature, Washington, D. C., 1971.

Note. This Application Note was derived from a manuscript originally prepared for publication in *EDN* magazine.

Appendix A
Practical considerations in thermoelectric cooler based control loops

There are a number of practical issues involved in implementing thermoelectric cooler (TEC) based control loops. They fall within three loosely defined categories. These include temperature setpoint, loop compensation and loop gain. Brief commentary on each category is provided below.

Temperature setpoint

It is important to differentiate between temperature accuracy and stability requirements. The exact temperature setpoint is not really important, so long as it is stable. Each individual laser's output maximizes at some temperature (see text Figures 16.1 and 16.2). Temperature setpoint is typically incremented until this peak is achieved. After this, only temperature setpoint stability is required. This is why thermistor tolerances on laser module data sheets are relatively loose (5%). Long-term (years) temperature setpoint stability is primarily determined by thermistor stability over time. Thermistor time stability is a function of operating temperatures, temperature cycling, moisture contamination and packaging. The laser modules' relatively mild operating conditions are very benign, promoting good long-term stability. Typically, assuming good grade thermistors are used at module fabrication, thermistor stability comfortably inside 0.1°C over years may be expected.

Also related to temperature setpoint is that the servo loop controls *sensor* temperature. The laser operates at a somewhat different temperature, although laser temperature stability depends upon the stable loop controlled environment. The assumption is that laser dissipation constant remains fixed, which is largely true.[1]

Loop compensation

Figure 16.3's "dominant pole" compensation scheme takes advantage of the long time constant from ambient into the laser module (see text Figure 16.5). Loop gain is rolled off at a frequency low enough to accommodate the TEC-thermistor lag (see text Figure 16.6) but high enough to smooth transients arriving from the outside ambient. The relatively high TEC-thermistor to module insulation time constant ratio (<1 second to minutes) makes this approach viable. Attempts at improving loop response with more sophisticated compensation schemes encounter difficulty due to laser module thermal term uncertainties. Thermal terms can vary significantly between laser module brands, rendering tailored compensation schemes impractical or even deleterious. Note that this restriction still applies, although less severely, even for modules of "identical" manufacture. It is very difficult to maintain tight thermal term tolerances in production.

Note 1: Academics will be quick to note that this phenomenon also occurs in the sensor's operation. Strictly speaking, the sensor operates at a slightly elevated temperature from its nominally isothermal environment. The assumption is that its dissipation constant remains fixed, which is essentially the case. Because of this, its temperature is stable.

The simple dominant pole compensation scheme provides good loop response over a wide range of laser module types. It's the way to go.

Loop gain

Loop gain is set by both electrical and thermal gain terms. The most unusual aspect of this is different TEC gain in heating and cooling modes. Significantly more gain is available in heating mode, accounting for the higher stabilities noted in the text (Figures 16.11 and 16.12). This higher gain means that loop gain bandwidth limits should be determined in heating mode to avoid unpleasant surprises. Figure 16.3's suggested loop gain and compensation values reflect this. It is certainly possible to get cute by changing loop gain bandwidth with mode but performance improvement is probably not worth the ruckus.[2]

It is important to remember that the TEC is a heat pump, the efficiency of which depends on the temperature across it. Gain varies with efficiency, degrading temperature stability as efficiency decreases. The laser module should be well coupled to some form of heat sink.[3] The small amount of power involved does not require large sink capability but adequate thermal flow must be maintained. Usually, coupling the module to the circuit's copper ground plane is sufficient, assuming the plane is not already thermally biased.

Note 2: The LTC1923's "heat-cool" status pin beckons alluringly.

Note 3: Yes, this means you should use that messy white goop. A less obnoxious alternative is the thermally conductive gaskets, which are nearly as good.

Current sources for fiber optic lasers

A compendium of pleasant current events

17

Jim Williams

Introduction

A large group of fiber optic lasers are powered by DC current. Laser drive is supplied by a current source with modulation added further along the signal path. The current source, although conceptually simple, constitutes an extraordinarily tricky design problem. There are a number of practical requirements for a fiber optic current source and failure to consider them can cause laser and/or optical component destruction.

Design criteria for fiber optic laser current sources

Figure 17.1 shows a conceptual laser current source. Inputs include a current output programming port, an output current clamp and an enable command. Laser current is the sole output. This block diagram is deceptively simple. In practice, a laser current source must meet a number of practical requirements, some quite subtle. The key to a successful design is a thorough understanding of individual system requirements. Various approaches suit different sets of freedoms and constraints, although all must address some basic concerns.

There are two basic sets of concerns for laser current sources: performance and protection. Performance issues include the current source's magnitude and stability under all conditions, output connection restrictions, voltage compliance, efficiency, programming interface and power requirements.

Protection features must be included to prevent laser and optical component damage. The laser, an expensive and delicate device, must be protected under all conditions, including supply ramp up and down, improper control input commands, open or intermittent load connections and "hot plugging."

Detailed discussion of performance issues

It is useful to expand on the above cursory discussion to clarify design goals. As such, each previously called out issue is treated in greater detail below.

Required power supply
The available power supply should be defined. A single rail 5V supply is presently the most common and desirable. Supply tolerances, typically ±5%, must be accounted for. System distribution voltage drops may result in surprisingly low rail voltages at the point of load. Occasionally, split rails are available, although this is relatively rare. Additionally, split rail operation can complicate laser protection, particularly during supply sequencing. See discussion below for additional comment.

Output current capability
Low power lasers operate on less than 250mA. Higher power types can require up to 2.5A.

Output voltage compliance
Current source output voltage compliance must be able to accommodate the laser's forward junction drop and any additional drops in the drive path. Typically, voltage compliance of 2.5V is adequate.

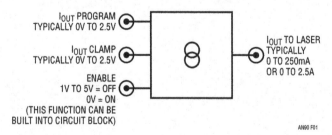

IOUT PROGRAM
TYPICALLY 0V TO 2.5V

IOUT CLAMP
TYPICALLY 0V TO 2.5V

ENABLE
1V TO 5V = OFF
0V = ON
(THIS FUNCTION CAN BE
BUILT INTO CIRCUIT BLOCK)

IOUT TO LASER
TYPICALLY
0 TO 250mA
OR 0 TO 2.5A

AN90 F01

Figure 17.1 • Conceptual Laser Current Source is Deceptively Simple. Practical System Issues and Laser Vulnerability Necessitate Careful Design

Analog Circuit and System Design: A Tutorial Guide to Applications and Solutions. DOI: 10.1016/B978-0-12-385185-7.00017-2

Efficiency

Heat build up in fiber optic systems is often a concern due to space limitations. Accordingly, current source efficiency can be an issue. At low current, linear regulation is often adequate. Switching regulator based approaches may be necessary at higher current.

Laser connection

In some cases, the laser may float off ground; other applications require grounded anode or cathode operation. Grounding the anode seemingly mandates a negative supply but single rail operation can be retained if switching regulator techniques are employed.

Output current programming

Output current is set by a programming port voltage. The voltage may be derived from a potentiometer, DAC or filtered PWM. Typically, a range of 0V to 2.5V corresponds to 0A to 250mA or 0A to 2.5A. Set point accuracy is usually within 0.5%, although better tolerances are readily achievable. Output current stability, discussed below, is considerably tighter.

Stability

The current source should be well regulated against line, load and temperature changes. Line and load induced variations should be held well within 0.05%, with typical temperature drifts of 0.01%. Judicious component choice can considerably improve these figures.

Noise

Current source noise, which can modulate laser output, must be minimized. Typically, noise bandwidth to 100MHz is of interest. A linearly regulated current source has inherently low noise and usually presents no problems. Switching regulator based current sources require special techniques to maintain low noise.

Transient response

The current source does not need fast transient response but it cannot overshoot the programmed current under any circumstances. Such overshoots can damage the laser or associated optical components.

Detailed discussion of laser protection issues

Overshoot

As noted above, outputs overshooting the nominal programmed current can be destructive. Any possible combination of improper control input or power supply turn on/off characteristics must be accounted for. Also, any spurious laser current under any condition is impermissible. Note that portions of the current source circuitry may have undesired and unpredictable responses during supply ramp up/down, complicating design.

Enable

An enable line allows shutting the current source output off. The enable line can also be used to hold current output off during supply ramp up, preventing undesired outputs. This can be tricky because the enable signal circuitry may be powered by the same supply that runs the laser. The enable signal must reliably operate independent of power supply turn-on profile. Optionally, the enable function can be self-contained within the current source, eliminating the necessity to generate this signal.

Output current clamp

The output current clamp sets maximum output current, overriding the output current programming command. This voltage controlled input can be set by a potentiometer, DAC or filtered PWM.

Open laser protection

An unprotected current source's output rises to maximum voltage if the load is disconnected. This circumstance can lead to "hot plugging" the laser, a potentially destructive event. Intermittent laser connections can produce similar undesirable results. The current source output should latch off if the load disconnects. Recycling power clears the latch but only if the load has been established.

The preceding discussion dictates considerable care when designing laser current sources. The delicate, expensive load, combined with the uncertainties noted, should promote an aura of thoughtful caution.[1] The following circuit examples (hopefully) maintain this outlook while simultaneously presenting practical, usable circuits. A variety of approaches are shown, in keeping with the broad area of application. The designs can be directly utilized or serve as starting points for specific cases.

Basic current source

Figure 17.2, a basic laser current source, supplies up to 250mA via Q1. This circuit requires both laser terminals to float. The amplifier controls laser current by maintaining the 1Ω shunt voltage at a potential dictated by the programming input. Local compensation at the amplifier stabilizes the loop and the 0.1μF capacitor filters input commands, assuring the loop never slew limits. This precaution prevents overshoot due to programming input dynamics. The enable input turns off the current source by simultaneously grounding Q1's base and starving the

---

Note 1. "For fools rush in where angels fear to tread." An essay on criticism, A. Pope. 1711.

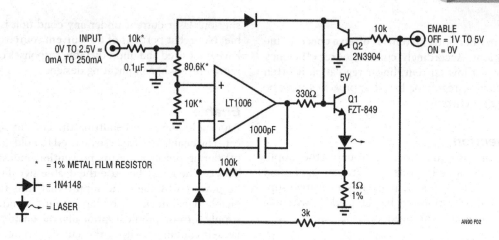

Figure 17.2 • Basic Current Source Requires Off-Ground Operation of Laser Terminals. Amplifier Controls Current by Comparing 0.1Ω Shunt to Input. Biasing Enable Until Supply is Verified Prevents Spurious Outputs

amplifier's "+" input while biasing the "−" input high. This combination also insures the amplifier smoothly ramps to the desired output current when enable switches low. The enable input must be addressed by an external "watchdog" which switches after the power supply has been verified to be within operating limits. Because the external circuitry may operate from the same supply as the current source, the enable threshold is set at 1V. The 1V threshold assures the enable input will dominate the current source output at low supply voltages during power turn on. This prevents spurious outputs due to unpredictable amplifier behavior below minimum supply voltage.

High efficiency basic current source

The preceding circuit uses Q1's linear regulation to close the feedback loop. This approach offers simplicity at the expense of efficiency. Q1's power dissipation can approach 1W under some conditions. Many applications permit this but some situations require heating minimization. Figure 17.3 minimizes heating by replacing Q1 with a step-down switching regulator. The switched mode power delivery eliminates almost all of the transistor's heat.

The figure shows similarities to Figure 17.2's linear approach, except for the LTC1504 switching regulator's

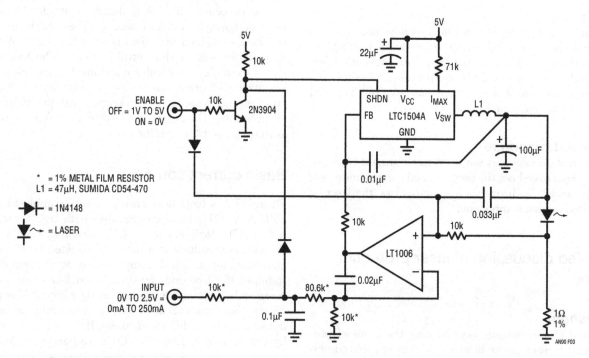

Figure 17.3 • Switching Regulator Replaces Figure 17.2's Q1, Providing Higher Efficiency. Feedback Control and Enable Input Considerations are as Before

addition. It is useful to liken the switching regulator's input (V_{CC}), feedback (FB) and output (V_{SW}) to the transistor's collector, base and emitter. This analogy reveals the two circuits to have very similar operating characteristics, with the switched mode version enhancing efficiency. The LTC1504's output LC filter introduces phase shift, necessitating attention to loop compensation. The amplifier's local rolloff is similar to Figure 17.2, although phase leading AC feedback elements (0.01µF and 0.033µF capacitors) are required for good loop damping. In all other respects, including enable and programming input considerations, this circuit's operation is identical to Figure 17.2.

Grounded cathode current source

Figure 17.4 allows the laser's cathode to be grounded, as is sometimes required, by sensing anode current. It utilizes A1, a device with 500mA output capability and programmable output current limit. A1 senses output current across the 1Ω shunt, with limiting controlled by the circuit's current programming input. A1 is set up as a unity-gain follower with respect to the laser, allowing its positive input to serve as a laser voltage clamp input. At laser voltages below the V_{CLAMP} input, A1 appears as a current source, controlled by the current programming input's setting. At laser voltages equaling or above the V_{CLAMP} input, A1 is a voltage source, controlled by V_{CLAMP}'s value.

This permits the V_{CLAMP} input to limit maximum voltage across the laser terminals.

The enable function operates similarly to previous descriptions and the 1µF capacitors restrict output movement to safe, well damped speeds. The diode shunting the laser prevents reverse bias during power supply sequencing. The 1N5817 protects against uncontrolled positive outputs if the negative supply drops out or sequences too slowly. This circuit's simplicity and laser connection versatility (appropriate modifications permit grounded anode operation) are attractive but A1's negative supply requirement may be detrimental. The negative supply complicates the external "watchdog" circuitry required for the enable input. In the worst case, it simply may not be available in the host system.

Single supply, grounded cathode current source

Figure 17.5 preserves Figure 17.4's grounded cathode operation while operating from a single supply. This circuit is reminiscent of Figure 17.2, with a notable exception. Here, differential amplifier A2 senses across a shunt in the laser anode, permitting cathode grounding. A2's gain-scaled output feeds back to A1 for loop closure. Loop compensation and enable input considerations are related to previous examples and, as before, Q1 could be replaced with a switching regulator.

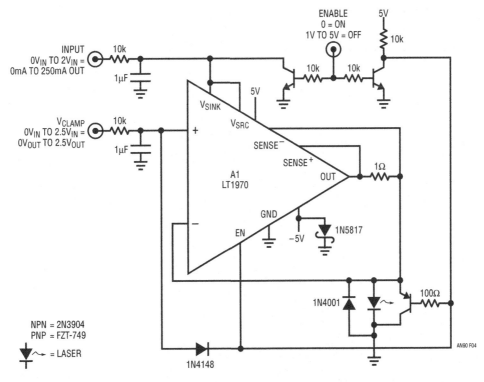

Figure 17.4 • LT1970 Power Ampifier/Current Source Permits Grounding Laser Cathode, Although Requiring Split Rails. Appropriate Modifications would Allow Grounded Anode Operation. Enable Input Must be Biased Until Supplies are Verified

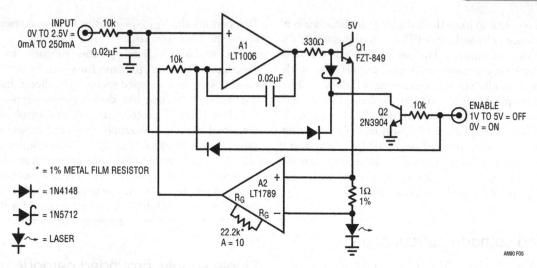

Figure 17.5 • Differentially Sensed Shunt Voltage Allows Grounded Cathode Laser Drive with Single Supply. Loop and Enable Input Considerations Derive from Previous Figures

Fully protected, self-enabled, grounded cathode current source

All of Figure 17.5's elements are repeated in Figure 17.6; no additional comment on them is necessary. However, three new features appear, allowing this circuit to operate

in a fully protected and self-contained fashion. The circuit monitors its power supply and "self-enables" when the supply is within limits, eliminating the "enable" port and the external "watchdog" previously required. A settable current clamp and open laser protection prevent laser damage.

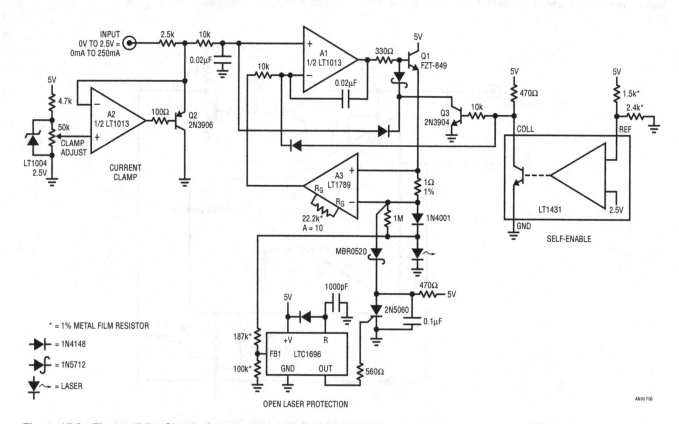

Figure 17.6 • Figure 17.5's Circuit, Augmented with "Self-Enable," Monitors Power Supply, Operates when V_{SUPPLY} = 4V. Current and Open Laser Clamps Protect Laser

The self-enable is designed around an LT1431 shunt regulator. It has the highly desirable property of maintaining a predictable open collector output when operating below its minimum supply voltage. At initial turn on, supply voltage is very low (e.g., 1V), the LT1431's output does not switch and current flows to Q3's base. Q3 turns on, preventing Q1's base from receiving bias. Additionally, the circuit's current programming input is pulled down and A1's "−" input is driven. This arrangement ensures that the laser cannot receive current until Q3 turns off. Also, when Q3 does go off, A1's output will cleanly ramp up to the desired programmed current. The resistor values at the LT1431's "REF" input dictate the device will go low when V_{SUPPLY} passes through 4V. This potential ensures proper circuit operation. Supply start-up waveforms appear in Figure 17.7. Trace A, the nominal 5V rail, ramps for 3ms before arriving at 5V. During this interval, the LT1431 (trace B) follows the ramp, biasing Q3 on. A1's output (trace C) is uncontrolled during this period, Q1's emitter (trace D), however, is cut off due to Q3's conduction and cannot pass the disturbance. As a result, the laser conducts no current (trace E) during this time. When the supply (trace A) ramps beyond 4V (just before the photo's 4th vertical division), the LT1431 switches low (trace B), Q3 switches off and the circuit "self-enables." A1's output (trace C) ramps up, with Q1's emitter (trace D) and laser current (trace E) slaved to its movement. This action prevents any undesired current in the laser during supply turn on, regardless of unpredictable circuit behavior at low supply voltages.

Supply turn on is not the only time laser current must be controlled. Response to programming input changes must be similarly well behaved. Figure 17.8 shows laser current response (trace B) to a programming input step (trace A). Damping is clean, with no hint of overshoot.

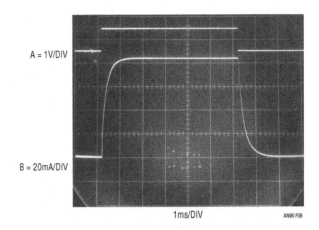

Figure 17.8 • Figure 17.6's Output (Trace B) Responding to Trace A's Input Step. Trace B's Laser Current Has Controlled Damping, No Overshoot

The circuit also includes open laser protection. If the current source operates into an open load (no laser), it will produce maximum voltage at the laser output terminals. This circumstance can lead to "hot plugging" the laser, a potentially destructive event. Intermittent laser connections can produce similar undesirable results. The LTC1696 overvoltage protection controller guards against open laser operation. This device's output latches high when its feedback input (FB) exceeds 0.88V. Here, the FB pin is biased so that laser output voltage exceeding 2.5V forces the LTC1696 high, triggering the SCR to shunt current away from the laser. The 470Ω resistor supplies SCR holding current and the diodes insure no current flows in the output.

Figure 17.9 details events with a properly connected laser at supply turn on. Trace A is the supply, trace B the laser voltage, trace C the LTC1696 output and trace D the laser current. The waveforms show laser voltage (trace B)

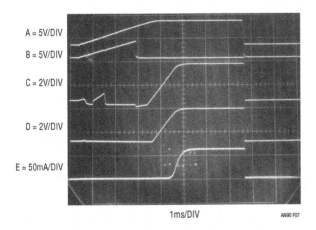

Figure 17.7 • Figure 17.6's Waveforms During Power Supply Application (Trace A). Traces B and C are LT1431 and A1 Outputs, Respectively. Q1's Emitter (Trace D) Provides Power Gain. Feedback Sets Laser Current (Trace E). Self-Enable Circuit Prevents Spurious A1 Outputs (Trace C) During Supply Ramp Up from Corrupting Laser Current (Trace E)

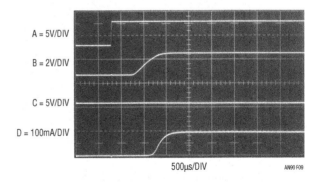

Figure 17.9 • Figure 17.6's Open Laser Protection Does Not Act During Normal Turn On. Trace A is Supply, Trace B Laser Voltage, Trace C LTC1696 Output and Trace D Laser Current. LTC1696 Overvoltage Threshold is Not Exceeded, SCR is Unbiased (Trace C) and Laser Conducts Current (Trace D)

rising to about 2V at supply turn on (trace A). Under these normal conditions, the LTC1696 output (trace C) stays low and laser current (trace D) rises to the programmed value.

Figure 17.10 shows what happens when the circuit is turned on into an open laser connection. Trace assignments are identical to the previous photo. At supply turn on (trace A), the laser voltage (trace B) transitions beyond the 2.5V open laser threshold. The LTC1696 output (trace C) goes high, the SCR latches and no current flows in the shunted laser line (trace D). Once this occurs, power must be recycled to reset the LTC1696-SCR latch. If the laser has not been properly connected, the circuit will repeat its protective action. Open laser protection is not restricted to turn on. It will also act if laser connection is lost at any time during normal circuit operation.

A final protection feature in Figure 17.6 is a current clamp. It prevents uncontrolled programming inputs from being transmitted by clamping them to a settable level. A2, Q2 and associated components form the clamp. Normally, A2's "+" input is above the circuit's programming input (Q2's emitter voltage), A2's output is high and Q2 is off. If the programming input exceeds A2's "+" input level, A2 swings low, Q2 comes on and the amplifier feedback controls Q2's emitter to the "clamp adjust" wiper potential. This clamps A1's input to the "clamp adjust" setting, preventing laser current overdrive. Clamp action need not be particularly fast to be effective, because of A1's 10k-0.02µF input filter. Figure 17.11's traces show clamp response to programming input overdrive. When the programming input (trace A) exceeds the clamp's preset level, Q2's emitter (trace B) does the same, causing A2's output (trace C) to swing down. A2 feedback controls Q2's emitter to the clamp level, arresting the voltage applied to the 10k-0.02µF filter. The filter band limits the abrupt clamp operation, resulting in a smooth corner at A1's positive

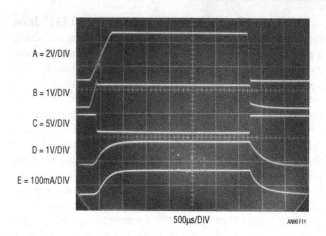

A = 2V/DIV

B = 1V/DIV

C = 5V/DIV

D = 1V/DIV

E = 100mA/DIV

500µs/DIV   AN90 F11

Figure 17.11 • Figure 17.6's Current Clamp Reacting to Programming Input Overdrive. Waveforms Include Programming Input (Trace A), Q2 Emitter (Trace B), A2 Output (Trace C), A1 + Input (Trace D) and Laser Current (Trace E). When Programming Input Exceeds Clamp Threshold, A2 Swings Abruptly (Trace C), Causing Q2's Emitter (Trace B) to Clamp. A1's + Input (Trace D) Remains at Clamp Level, Maintaining Safe Laser Current (Trace E)

input (trace D). A1's clamped input dictates a similarly shaped and clamped laser current (trace E). The clamp remains active until the programming input falls below the "clamp adjust" setting.

2.5A, grounded cathode current source

Figure 17.12, derived from Figures 17.3 and 17.6, provides up to 2.5A to a grounded cathode laser. A1 is the control amplifier, output current is efficiently delivered by the LT1506 switching regulator and A2 senses laser current via a 0.1Ω shunt. Loop operation is similar to the descriptions given for Figures 17.3 and 17.6 with DC feedback to A1 coming from A2. Frequency compensation differs from the previous figures. Stable loop operation is achieved by local roll off at A1, augmented by two lead networks associated with L1. Midband lead is provided by feeding back a lightly filtered (1k-0.47µF) version of LT1506 V_SW output activity. High frequency lead, arriving via the 330Ω-0.05µF pair, optimizes edge response. Figure 17.13's waveforms detail dynamic response. Trace A's input step arrives in filtered form at A1's positive input (trace B). The loop produces trace C's faithfully profiled laser current output.

As shown, the circuit has the externally controlled enable function, although Figure 17.6's "self-enable" feature may be used. Similarly, Figure 17.6's current clamp and open laser protection may be employed in this circuit.

This circuit's switched mode energy delivery provides high efficiency at high power but output noise may be an issue. Residual harmonic content related to switching regulator operation appears in the laser current. The resultant low

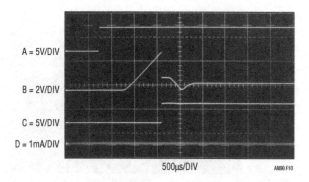

Figure 17.10 • Open Laser Protection Circuit Responding to Open Laser Turn On. Trace Assignments Identical to Previous Figure. Laser Line (Trace B) Excursion Beyond Overvoltage Threshold Causes LTC1696 Output (Trace C), Biasing SCR to Clamp Open Laser Line. No Current Flows in Laser Line, Trace D (Note 100× Increase in Measurement Sensitivity vs Figure 17.9)

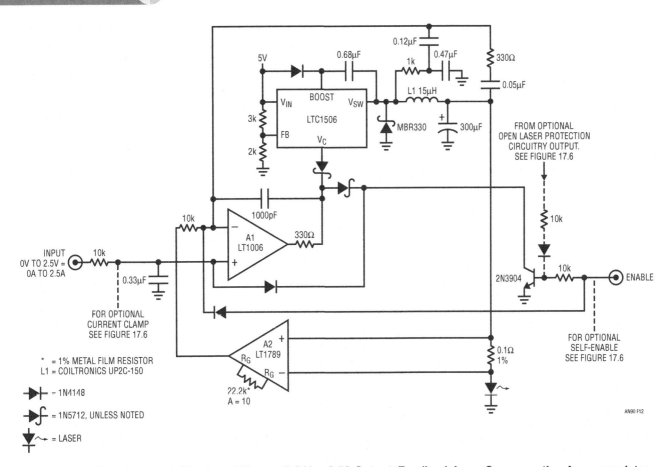

Figure 17.12 • Switched Mode Version of Figure 17.6 Has 2.5A Output. Feedback Loop Compensation Accommodates Switching Regulator Delay. Clamp, Protection and Self-Enable Circuits are Optional

level modulation of laser output may be troublesome in some applications. Figure 17.14 shows about 800µA$_{P-P}$ switching regulator related noise in the 2A laser current output.[2] This disturbance is composed of fundamental ripple and switching transition related harmonic. This 0.05% noise is below most optical system requirements, although the following circuit achieves substantially lower noise figures.

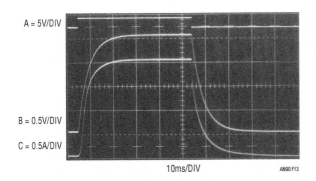

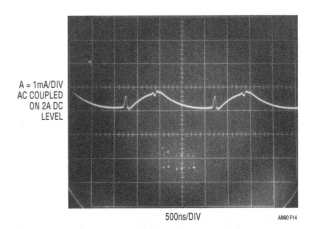

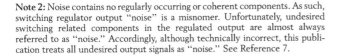

Figure 17.13 • 2.5A Current Source Waveforms for an Input Step (Trace A). A1's Input Filter (Trace B) Smooths Step, Resulting in Trace C's Similarly Shaped Laser Current

Note 2: Noise contains no regularly occurring or coherent components. As such, switching regulator output "noise" is a misnomer. Unfortunately, undesired switching related components in the regulated output are almost always referred to as "noise." Accordingly, although technically incorrect, this publication treats all undesired output signals as "noise." See Reference 7.

Figure 17.14 • High Power Current Source Noise Includes Switching Regulator Fundamental Ripple and Harmonic Content. 800µA$_{P-P}$ Noise is About 0.05% of 2A DC Output

0.001% noise, 2A, grounded cathode current source

The previous circuit's 0.05% noise content suits many optical system applications. More stringent requirements will benefit from Figure 17.15's extremely low noise content. This grounded cathode, 2A circuit has only 20μA$_{P-P}$ noise, about 0.001%. Special switching regulator techniques are used to attain this performance. Substantial noise reduction is achieved by limiting edge switching speed in the regulator's power stage.[3] Voltage and current rise times in switches Q1 and Q2 are controlled by the LT1683 pulse width modulator. The LT1683's output stage operates Q1 and Q2 in local loops which sense and control their edge times. Transistor voltage information is fed back via the 4.7pF capacitors; current status is derived from the 0.033Ω shunt and also fed back. This arrangement permits the PWM control chip to fix transistor switching times, regardless of power supply or load changes. The transition rates are set by resistors (R$_{VSL}$ and R$_{CSL}$) associated with

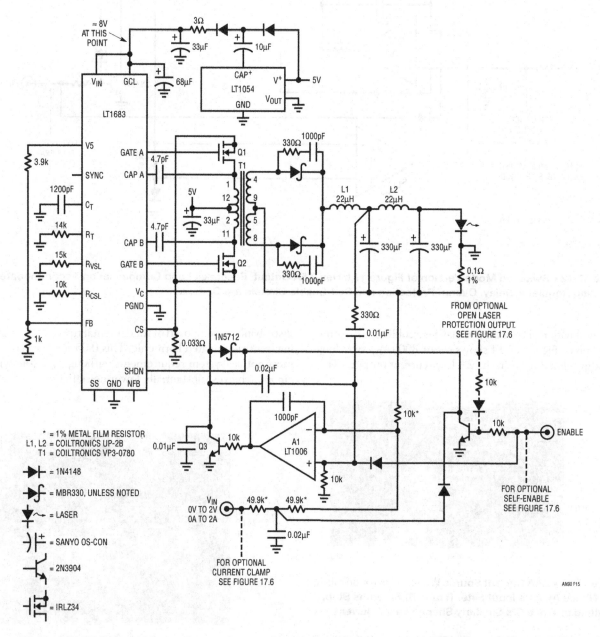

Figure 17.15 • 0.001% Noise, 2A Laser Current Source Has Grounded Cathode Output. Clamp, Protection and Self-Enable Circuits May be Added

Note 3: See Reference 7 for details on this technique.

the LT1683 controller. In practice, these resistor values are set by adjusting them to minimize output noise. The remainder of the circuit forms a grounded cathode laser current source.

Q1 and Q2 drive T1, whose rectified output is filtered by LC sections. Because T1's secondary floats, the laser cathode and the 0.1Ω shunt may be declared at circuit ground. The shunt is returned to T1's secondary center tap, completing a laser current flow path. This arrangement produces a negative voltage corresponding to laser current at the shunt's ungrounded end. This potential is resistively summed at A1 with the positive voltage current programming input information. A1's output feedback controls the LT1683's pulse width drives to Q1 and Q2 via Q3, closing a loop to set laser current. Loop compensation is set by band limiting at A1 and Q3's collector, aided by a single lead network arriving from the L1-L2 junction.

Some circuit details merit attention. The LT1683's supply input pins are fed from an LT1054 based voltage multiplier. This boosted voltage provides enough gate drive to ensure Q1-Q2 saturation. Damper networks across T1's rectifiers minimize diode switching related events in the output current. Finally, this circuit is compatible with the "self-enable"

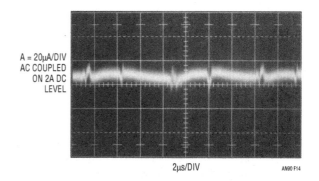

A = 20μA/DIV
AC COUPLED
ON 2A DC
LEVEL

2μs/DIV   AN90 F14

Figure 17.16 • Figure 17.15's Output Noise Measures ≈20μA$_{P-P}$, About 0.001%. Coherent, Identifiable Components Include Fundamental Ripple Residue and Switching Artifacts

and laser protection features previously described. Appropriate connection points appear in the figure.

The speed controlled switching times result in a spectacular decrease in noise. Figure 17.16 shows just 20μA$_{P-P}$ noise, about 0.001% of the 2A DC laser current. Fundamental ripple residue and switching artifacts are visible against the measurement noise floor.[4]

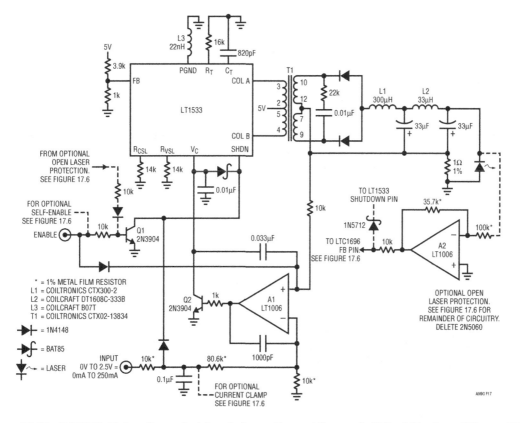

Figure 17.17 • 0.0025% Noise, Grounded Anode Laser Current Source is 250mA Version of Figure 17.15

Note 4: Reliable wideband current noise measurement at these levels requires special techniques. See Appendix B, "Verifying switching regulator related noise" and Appendix C, "Notes on current probes and noise measurement," for details.

0.0025% noise, 250mA, grounded anode current source

This circuit, similar to the previous one, uses edge time control to achieve an exceptionally low noise output. It is intended for lower power lasers requiring grounded anode operation. The LT1533, a version of the previous circuit's LT1683, has internal power switches. These switches drive T1. T1's rectified and filtered secondary produces a negative output, biasing the laser. The laser's anode is grounded and its current path to T1's secondary completed via the 1Ω shunt. This configuration makes T1's center tap voltage positive and proportional to laser current. This voltage is compared by A1 to the current programming input. A1 biases Q2, closing a loop around the LT1533. Loop compensation is provided by local bandwidth limiting at A1 and Q2's collector damping and feedback capacitors.

This circuit's 2.5μA$_{P-P}$ noise qualifies it for the most demanding applications. Figure 17.18 shows residual switching related noise approaching the measurement noise floor.

The enable function operates as previously described. Additionally, this circuit is compatible with Figure 17.6's

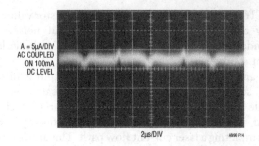

Figure 17.18 • Figure 17.17's 2.5μA$_{P-P}$ Switching Related Noise is Detectable in Measurement Noise Floor

"self-enable" and laser protection accessory circuits. Changes necessitated by the grounded anode operation appear on the schematic.

Low noise, fully floating output current source

Figure 17.19 retains the preceding example's low noise but also has a fully floating output. Either laser terminal may be

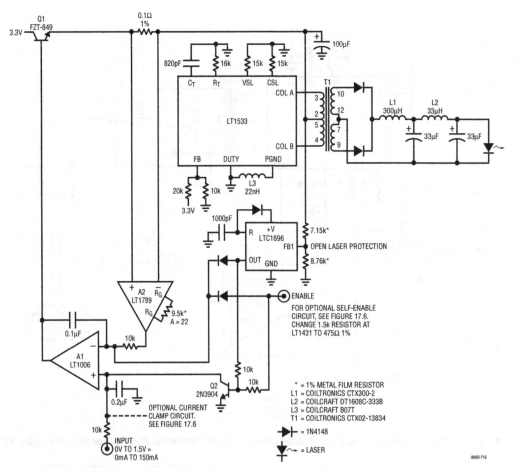

Figure 17.19 • Switched Mode, Low Noise Current Source Has Floating Output, Permitting Grounding Laser Anode or Cathode. Open Laser Protection is Included; Circuit is Compatible with Current Clamp and Self-Enable Options

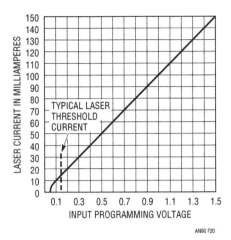

Figure 17.20 • Laser Current vs Input Programming Voltage for Floating Current Source. Conformance is within 1% over Nearly Entire Range. Error Below 10mA, Due to Nonideal Transformer Behavior, is Below Typical Laser Threshold Current

grounded without effecting circuit operation. This feature is realized by feedback controlling transformer primary current and relying on interwinding coupling to maintain regulation.[5] This coupling varies slightly with operating point, limiting output current regulation to about 1%.

The schematic shows the LT1533 low noise switching regulator driving T1. The LT1533, while retaining its controlled edge time characteristics, is forced to run at 50% duty cycle by grounding its "duty" pin. Current flows through Q1 and the 0.1Ω shunt into T1's primary. The LT1533 open collector power outputs alternately chop primary current to ground. Primary current magnitude, and hence the 0.1Ω shunt voltage, is set by Q1's bias. Q1's bias, in turn, is set by A1's output, which represents the difference between the output current programming input and A2's amplified version of the shunt voltage. This loop enforces a shunt voltage proportionate to current programming input value. In this way, the current programming input sets T1 primary current, determining T2 secondary current through the laser. Current programming input scaling is calibrated by differential amplifier A2's gain setting resistor.

The primary side feedback's lack of global feedback mandates current regulation compromise. Figure 17.20's plot of laser current vs programming input voltage shows 1% conformance over nearly the entire range. The error below 10mA, due to nonideal transformer behavior, is normally insignificant because it is below typical laser threshold current. Line regulation, also degraded by the sensing scheme, still maintains about 0.05%/V. Similarly, load regulation, over a 1V to 1.8V compliance voltage, is typically 2%.

This circuit's floating output complicates inclusion of the laser protection and "self-enable" features described

in Figure 17.6's text but they are accommodated. Open laser protection, shown in Figure 17.20, is accomplished by biasing the LTC1696 from T1's center tap. If the laser opens, the loop forces a marked rise at T1's center tap, latching the LTC1696's output high. This skews A1's inputs, sending its output low and shutting off Q1. All T1 drive ceases. Because the LTC1696 output latches, power must be recycled to reset the circuit. If the laser has not been connected, the latch will act again, protecting the laser from "hot plugging" or intermittent connections. The "self-enable" and current clamp options may be added in accordance with the notations on the schematic.

Anode-at-supply current source

Figure 17.21's current source is useful where the laser anode is committed to the power supply. A1, sensing Q1's emitter, closes a loop which forces constant current in the laser. Local compensation at A1 and input band limiting stabilize the loop.

This circuit also includes an inherent "self-enable" feature. The LT1635 operates at supply voltages down to 1.2V. Above 1.2V the LT1635's comparator configured section (C1) holds off circuit output until supply voltage reaches 2V. Below 1.2V supply, Q1's base biasing prevents unwanted outputs. Figure 17.22 details operation during supply turn on. At supply ramp up (trace A), output current (trace D) is disabled. When the supply reaches 2V, C1 (trace B) goes low, permitting A1's output (trace C) to rise. This biases Q1 and laser current flows (trace D). The LT1635 operates on supply voltages as low as 1.2V. Below this level, spurious outputs are prevented by junction stacking and band limiting at Q1's base. Q1's base components also prevent unwanted outputs when the supply rises rapidly. Such rapid rise could cause uncontrolled A1 outputs before the amplifier and its feedback loop are established. Figure 17.23 shows circuit events during a rapid supply rise. Trace A shows the supply's quick ascent. Trace B, C1's output, responds briefly but goes low some time after the supply moves past 2V. A1 (trace C) produces an uncontrolled output for about 100µs. The RC combination in Q1's base line filters this response to insignificant levels and no laser current (trace D) flows.

The slew retarded input and loop compensation yield clean dynamic response with no overshoot. Figure 17.24, trace A, is an input step. This step, filtered at A1's input (trace B), is represented as a well controlled laser current output in trace C.

Current clamping and open laser protection options are annotated in the schematic. Additionally, higher output current is possible at increased supply voltages, although Q1's dissipation limits must be respected.

Note: This application note was derived from a manuscript originally prepared for publication in EDN magazine.

Note 5: We have engaged this stunt before to serve a variety of purposes. See References 2, 3 and 4.

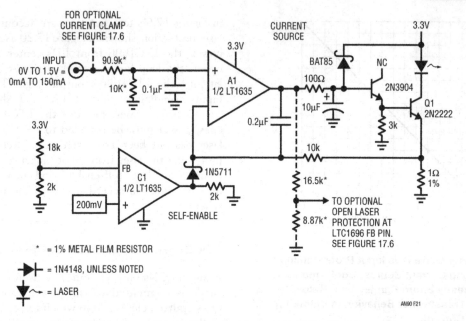

Figure 17.21 • Circuit Has Laser Anode Committed to Supply, Inherent Self-Enabled Operation. LT1635 Functions at 1.2V, Although Self-Enable Feature Holds Off Output Until Power Supply Exceeds 2V. Current Clamp and Open Laser Protection are Optional

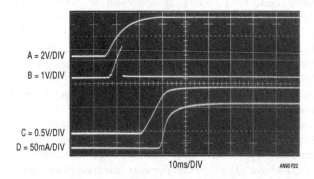

Figure 17.22 • Output Current (Trace D) is Held Off Until Supply (Trace A) Ramps Past 2V. Self-Enable Comparator (Trace B) Operates Above 1.2V; Q1 Base (Trace C) Biasing Prevents Output Below 1.2V

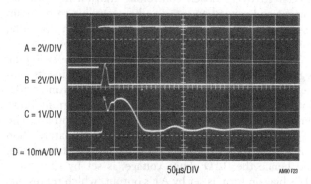

Figure 17.23 • Rapidly Rising Supply (Trace A) Produces No Current Output (Trace D) Despite A1's Transient Uncontrolled Output (Trace C). C1 (Trace B) Reacts Properly but A1's Inactive Loop Cannot Respond. Q1's Base Line Components Preclude Spurious Current Output (Trace D)

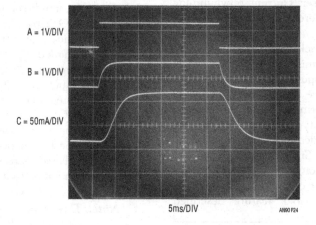

Figure 17.24 • Output Current (Trace C) Responds Cleanly to Filtered Version (Trace B) of Trace A's Input Step

References

1. Hewlett-Packard Company, "Model 6181C Current Source Operating and Service Manual," 1975.
2. Williams, J., "Designing Linear Circuits for 5V Single Supply Operation," "Floating Output Current Loop Transmitter," Linear Technology Corporation, Application Note 11, September 1985, p. 10.
3. Williams, J., "A Fourth Generation of LCD Backlight Technology," "Floating Lamp Circuits," Linear Technology Corporation, Application Note 65, November 1995, pp. 40-42.
4. Linear Technology Corporation, "LT1182/LT1183/LT1184/LT1184F CCFL/LCD Contrast Switching Regulators," Data Sheet, 1995.
5. Grafham, D. R., "Using Low Current SCRs," General Electric Co, AN-200.19, January 1967.
6. General Electric Co., "SCR Manual," 1967.
7. Williams, J., "A Monolithic Switching Regulator with 100µV Output Noise," Linear Technology Corporation, Application Note 70, October 1997.
8. Pope, A., "An essay on criticism," 1711.

Appendix A
Simulating the laser load

Fiber optic lasers are a delicate, unforgiving and expensive load. This is a poisonous brew when breadboarding with high likelihood of catastrophe. A much wiser alternative is to simulate the laser load using either diodes or electronic equivalents. Lasers look like junctions with typical forward voltages ranging from 1.2V to 2.5V. The simplest way to simulate a laser is to stack appropriate numbers of diodes in series. Figure A1 lists typical junction voltages at various currents for two popular diode types. The MR750 is suitable for currents in the ampere range, while the 1N4148 serves well at lower currents. Typically, stacking two to three diodes allows simulating the laser in a given current range. Diode voltage tolerances and variations due to temperature and current changes limit accuracy, although results are generally satisfactory.

Electronic laser load simulator

Figure A2 is a laser load simulator powered by a 9V battery. It eliminates diode load junction voltage drop uncertainty. Additionally, any desired "junction drop" voltage may be conveniently set with the indicated potentiometer. Electronic feedback enforces establishment and maintenance of calibrated junction drop equivalents.

The potentiometer sets a voltage at A1's negative input. A1 responds by biasing Q1. Q1's drain voltage controls Q2's base and, hence, Q2's emitter potential. Q2's emitter is fed back to A1, closing a loop around the amplifier. This forces the voltage across Q2 to equal the potentiometer's output voltage under all conditions. The capacitors at A1 and Q1 stabilize the loop and Q2's base resistor and ferrite bead suppress parasitic oscillation. The 1N5400 prevents Q1-Q2 reverse biasing if the load terminals are reversed.

MR750 (25°C)	
TYPICAL JUNCTION CURRENT	TYPICAL JUNCTION VOLTAGE
0.5A	0.68V
1.0A	0.76V
1.5A	0.84V
2.0A	0.90V
2.5A	0.95V

1N4148 (25°C)	
TYPICAL JUNCTION CURRENT	TYPICAL JUNCTION VOLTAGE
0.1A	0.83V
0.2A	0.96V
0.3A	1.08V

Figure A1 • Characteristics of Diodes Suitable for Simulating Lasers. Appropriate Series Connections Approximate Laser Forward Voltage

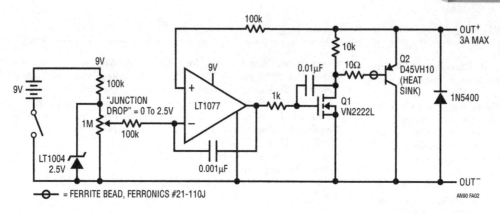

Figure A2 • Floating, Battery-Powered Laser Simulator Sets Desired "Junction Drop" Across Output Terminals. Amplifier Feedback Controls Q2's V_CE to Potentiometer Voltage

Appendix B
Verifying switching regulator related noise

Measuring the switching regulator related current noise levels discussed in the text requires care. The microamp amplitudes and wide bandwidth of interest (100MHz) mandates strict attention to measurement technique. In theory, simply measuring voltage drop across a shunt resistor permits current to be determined. In practice, the resultant small voltages and required high frequency fidelity pose problems. Coaxial probing techniques are applicable but probe grounding requirements become severe. The slightest incidence of multiple ground paths ("ground loops") will corrupt the measurement, rendering observed "results" meaningless. Differentially configured coaxial probes offer some relief from ground loop based difficulties but there is an inherently better approach.[1]

Current transformers offer an attractive way to measure noise while eliminating probe grounding concerns. Two types of current probes are available: split core and closed core. The split core "clip on" types are convenient to use but have relatively low gain and a higher noise floor than closed core types.[2] The closed core transformer's gain and noise floor advantages are particularly attractive for wideband, low current measurement.

Figure B1's test setup allows investigation of the closed core transformer's capabilities. The transformer specified has flat gain over a wide bandwidth,

a well shielded enclosure and a coaxial 50Ω output connection. Its 5mV/mA output feeds a low noise x100, 50Ω input amplifier. The amplifier's terminated output is monitored by an oscilloscope with a high sensitivity plug-in. A 1V pulse driving a known resistor value ("R") provides a simple way to source calibrated current into the transformer.

If R = 10k, resultant pulsed current is 100μA. Figure B2's oscilloscope photo shows test setup response. The waveform is crisp, essentially noise free and agrees with predicted amplitude. More sensitive measurement involves determining the test setup's noise floor. Figure B3, taken with no current flowing in the transformer, indicates a noise limit of about 10μA_P-P. Most of this noise is due to the x100 amplifier.

The preceding exercise determines the test setup's gain and noise performance. This information provides the confidence necessary to make a meaningful low level current measurement. Figure B4, taken with Figure B1's R = 100k, sources only 10μA to the transformer. This is comparable to the previously determined noise floor but the trace, clearly delineated against the noise limit, indicates a 10μA amplitude. This level of agreement qualified this test method to obtain the text's quoted noise figures.

Isolated trigger probe
The performance limits noted above were determined with a well defined, pulsed input test signal. Residual switching regulator noise has a much less specific profile. The oscilloscope may encounter problems triggering on an ill-defined, noise laden waveform. Externally triggering the 'scope from the switching regulator's clocking solves this problem but introduces ground loops, corrupting the

Note 1: This is not to denigrate low level voltage probing methods. Their practice is well refined and directly applicable in appropriate circumstances. See Appendix C in Reference 7 for tutorial.
Note 2: See Appendix C, "Notes on current probes and noise measurement," for a detailed comparison.

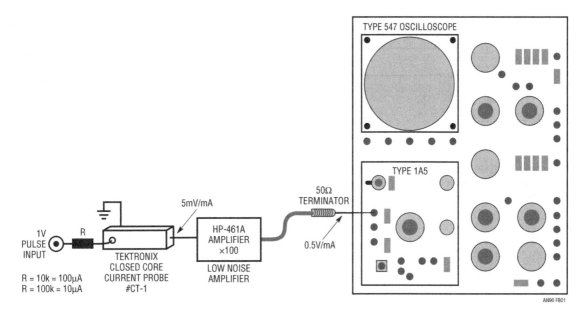

Figure B1 • Noise Measurement Instrumentation Includes Resistors, Closed Core Current Probe, Low Noise Wideband Amplifier and Oscilloscope

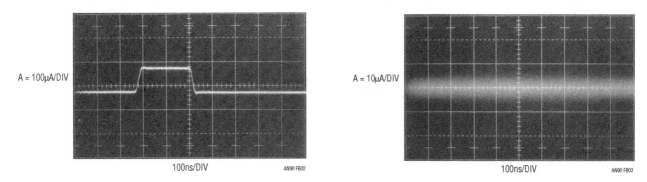

Figure B2 • Response to a 100μA Input is Clean. Displayed Amplitude Agrees with Input Stimulus, Indicating Calibrated Measurement

Figure B3 • 10μA Noise Floor is Determined by Removing Current Loop from Transformer. Remaining Noise is Primarily Due to ×100 Amplifier

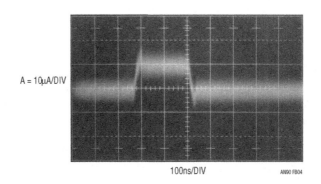

Figure B4 • Verifying Gain Near Noise Floor. 10μA Input Pulse Produces Calibrated, Readily Discernible Output

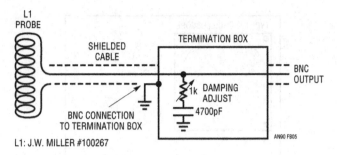

L1: J.W. MILLER #100267

AN90 FB05

Figure B5 • Simple Trigger Probe Eliminates Board Level Ground Loops. Termination Box Components Damp L1's Ringing Response

measurement.[3] It is possible, however, to externally trigger the 'scope without making any galvanic connections to the circuit, eliminating ground loop concerns. This is accomplished by coupling to the field produced by the switching regulator magnetics. A probe which does this is simply an RF choke terminated against ringing (Figure B5). The choke, appropriately positioned, picks up residual switching frequency related magnetic field, generating an isolated trigger signal.[4] This arrangement furnishes a 'scope trigger signal with essentially no measurement corruption. The probe's physical form appears in Figure B6. For good results, the termination should be adjusted for minimum ringing while preserving the highest possible amplitude output. Light compensatory damping produces Figure B7's output, which will cause poor 'scope triggering. Proper adjustment results in a more favorable output (Figure B8), characterized by minimal ringing and well defined edges.

Trigger probe amplifier
The field around the switching magnetics is small and may not be adequate to reliably trigger some oscilloscopes. In such cases, Figure B9's trigger probe amplifier is useful. It uses an adaptive triggering scheme to compensate for variations in probe output amplitude. A stable 5V trigger output is maintained over a 50:1 probe output range. A1, operating at a gain of 100, provides wideband AC gain. The output of this stage biases a 2-way peak detector (Q1 through Q4). The maximum peak is stored in Q2's emitter capacitor, while the minimum excursion is

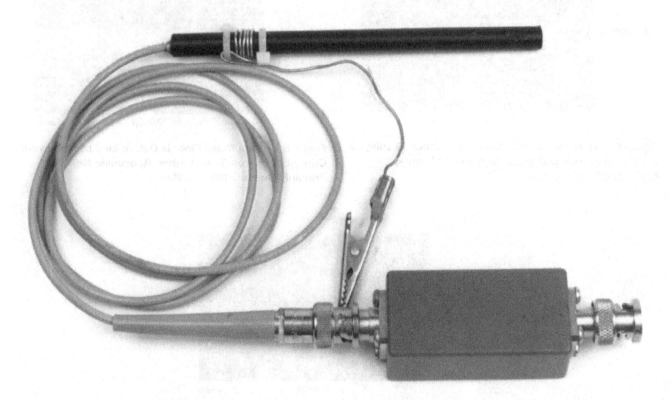

Figure B6 • The Trigger Probe and Termination Box. Clip Lead Facilitates Positioning Probe, is Electrically Neutral

Note 3: See previous comments at the beginning of this appendix.

Note 4: Veterans of LTC application notes, a hardened crew, will recognize this probe's description from LTC Application Note 70 (Reference 7). It directly applies to this topic and is reproduced here for reader convenience.

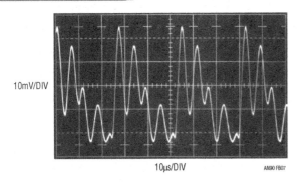

Figure B7 • Misadjusted Termination Causes Inadequate Damping. Unstable Oscilloscope Triggering May Result

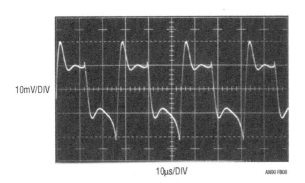

Figure B8 • Properly Adjusted Termination Minimizes Ringing with Small Amplitude Penalty

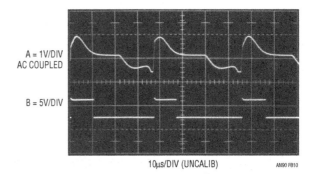

Figure B10 • Trigger Probe Amplifier Analog (Trace A) and Digital (Trace B) Outputs

retained in Q4's emitter capacitor. The DC value of the midpoint of A1's output signal appears at the junction of the 500pF capacitor and the 3MΩ units. This point always sits midway between the signal's excursions, regardless of absolute amplitude. This signal-adaptive voltage is buffered by A2 to set the trigger voltage at the LT1394's positive input. The LT1394's negative input is biased directly from A1's output. The LT1394's output, the circuit's trigger output, is unaffected by >50:1 signal amplitude variations. An x100 analog output is available at A1.

Figure B10 shows the circuit's digital output (trace B) responding to the amplified probe signal at A1 (trace A).

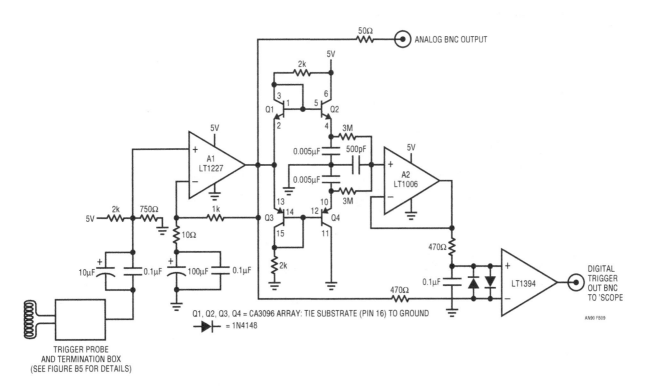

Figure 17.B9 • Trigger Probe Amplifier Has Analog and Digital Outputs. Adaptive Threshold Maintains Digital Output over 50:1 Probe Signal Variations

Appendix C
Notes on current probes and noise measurement

Appendix B explained current probes advantages in switching regulator related current noise measurement. Their minimally invasive nature eases connection parasitics, enhancing measurement fidelity. Different combinations of current probes and amplifiers provide varying degrees of performance and convenience. Figure C1 summarizes characteristics for two probes and applicable amplifiers. In general, the noise floor uncertainties of the convenient split core types are compromised by their construction. The closed core probes are less noisy and some types have inherently higher gain, a dis-

tinct advantage. A laboratory based comparison is revealing.

Figure C2 shows the CT-1 (closed core) HP-461A combination responding to a 100µA pulsed input. The waveform is clearly outlined, with pulse top and bottom trace thickening deriving from the noise floor.[1] Figure C3, taken with the same input, is degraded. The split core P6022/Preamble 1855 combination used has much greater noise. The decreased performance is almost entirely due to the split core probe's construction.

In closing, it is worthwhile noting that Hall element stabilized current probes (e.g., Tektronix AM503, P6042) are not suitable for low level measurement. The Hall device based flux nulling loop extends probe response to DC but introduces ≈300µA of noise.

CURRENT PROBE	AMPLIFIER	NOISE FLOOR (100 MHz BW)	COMMENTS
Tektronix P6022 (1mV/mA)	Preamble 1855 (1MΩ)	100µA	Split Core is Convenient to Use but Sensitivity is Low, Resulting in Relatively High Overall Noise Floor
Tektronix CT-1 (5mV/mA)	Hewlett-Packard 461A (50Ω)	15µA	Probe's Higher Gain Accounts for Most Noise Floor Reduction— 50Ω Input Amplifier Provides Some Additional Benefit. Closed Core Probe Requires Breaking Conductor to Make Measurement

Figure C1 • Recommended Instrumentation for Current Noise Measurement. Split Core "Current Probe" is Convenient; Closed Core Provides Higher Gain and Lower Noise

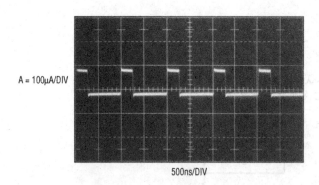

A = 100µA/DIV

500ns/DIV

Figure C2 • CT-1/HP-461A Combination Clearly Displays a 100µA Pulse Train. Noise Floor Causes Slight Pulse Top and Bottom Trace Thickening

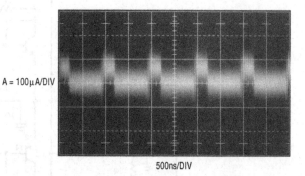

A = 100µA/DIV

500ns/DIV

Figure C3 • P6022/Preamble 1855 Presentation of Previous Figure's Waveform Has Degraded Signal-to-Noise Performance. Split Core "Current Probe" Convenience Necessitates Measurement Fidelity Compromise

Note 1: Diehard curmudgeons still using high quality analog oscillscopes routinely discern noise presence due to trace thickening. Those stuck with modern instruments routinely view thick, noisy traces.

Bias voltage and current sense circuits for avalanche photodiodes

Feeding and reading the APD

Jim Williams

18

Introduction

Avalanche photodiodes (APDs) are widely utilized in laser based fiberoptic systems to convert optical data into electrical form. The APD is usually packaged with a signal conditioning amplifier in a small module. An APD receiver module and attendant circuitry appears in Figure 18.1. The APD module (figure right) contains the APD and a transimpedance (e.g., current-to-voltage) amplifier. An optical port permits interfacing fiberoptic cable to the APD's photosensitive portion. The module's compact construction facilitates a direct, low loss connection between the APD and the amplifier, necessary because of the extremely high speed data rates involved.

The receiver module needs support circuitry. The APD requires a relatively high voltage bias (figure left) to operate, typically 20V to 90V. This voltage is set by the bias supply's programming port. This programming voltage may also include corrections for the APD's temperature dependent response. Additionally, it is desirable to monitor the APD's average current (figure center), which indicates optical signal strength. This information can be combined with feedback techniques to maintain optical signal strength at an optimal level. The feedback loop's operating characteristics can also determine if deleterious degradation of optical components has occurred, permitting corrective measures to be taken. APD current is typically between 100nA and 1mA, a dynamic range of 10,000:1. This measurement, which should be taken with an accuracy inside 1%, normally must occur in the APD's "high side," complicating circuit design. This restriction applies because the APD's anode is committed to the receiver amplifier's summing point.

The APD module, an expensive and electrically delicate device, must be protected from damage under all conditions. The support circuitry must never produce spurious outputs which could destroy the APD module. Particular

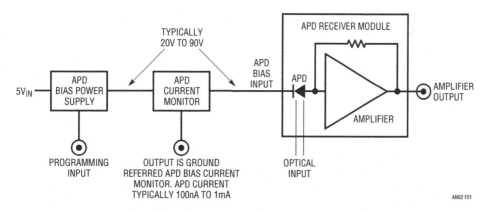

Figure 18.1 • Avalanche Photodiode (APD) Module (Figure Right) Contains APD, Amplifier and Optical Port. Power Supply (Figure Left) Provides APD Bias Voltage. APD Current Monitor (Figure Center) Operates at High Common Mode Voltage, Complicating Signal Conditioning

Analog Circuit and System Design: A Tutorial Guide to Applications and Solutions. DOI: 10.1016/B978-0-12-385185-7.00018-4

attention must be devoted to the bias supply's dynamic response under programming and power-up/down conditions. Finally, it is desirable to power the support circuitry from a single 5V rail.

The bias voltage and current measurement requirements described above constitute a significant design challenge and are addressed in the following text.

Simple current monitor circuits (with problems)

Figure 18.2's straightforward approaches attempt to address the current monitor problem. Figure 18.2a uses an instrumentation amplifier powered by a separate 35V rail to measure across the 1kΩ current shunt. Figure 18.2b is similar but derives its power supply from the APD bias line. Although both approaches function, they do not meet APD current sensing requirements. APD bias voltages can range to 90V, exceeding the amplifier's supply and common mode voltage limits. Additionally, the measurement's wide dynamic range requires the single rail powered amplifier to swing within 100µV of zero, which is impractical. Finally, it is desirable for the amplifiers to operate from a single, low voltage rail.

Figure 18.3's circuit divides down the high common mode current shunt voltage, theoretically permitting the 5V powered amplifier to extract the current measurement over a 20V to 90V APD bias range. In practice, this arrangement introduces prohibitive errors, primarily because the desired signal is also divided down. The current measurement information is buried in the divider resistor's tolerance, even with 0.01% components. The desired 1% accuracy over a 100mA to 1nA range cannot be achieved. Finally, although the amplifier operates from a single 5V supply, it cannot swing all the way to zero.

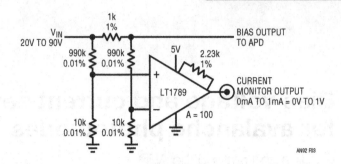

Figure 18.3 • Dividing Down High Common Mode Voltage Introduces Huge Errors, Even with Precision Components. Desired 1% Accuracy Over 100nA to 1mA Current Monitor Range Is Buried by Resistor Mismatch, Even with 0.01% Resistors. Single Rail Powered Amplifier Cannot Swing Close Enough to Zero. Approach Is Impractical

It is clear from the preceding circuits that common circuit approaches will not meet APD signal conditioning requirements. More sophisticated techniques are necessary.

Carrier based current monitor

Figure 18.4 utilizes AC carrier modulation techniques to meet APD current monitor requirements. It features 0.4% accuracy over the sensed current range, runs from a 5V supply and has the high noise rejection characteristics of carrier based "lock in" measurements.

The LTC1043 switch array is clocked by its internal oscillator. Oscillator frequency, set by the capacitor at Pin 16, is about 150Hz. S1 clocking biases Q1 via level shifter Q2. Q1 chops the DC voltage across the 1kΩ current shunt, modulating it into a differential square wave signal which feeds A1 through 0.2µF AC coupling

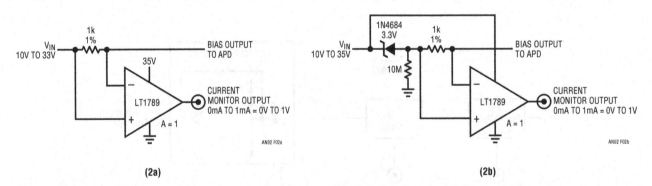

(2a)   (2b)

Figure 18.2 • Instrumentation Amplifiers Extract Current Measurement from Modest Common Mode Voltages. Figure 18.2a Requires Separate Amplifier Power and Bias Supply Connections; Figure 18.2b Derives Both Connections from Single Point. Zener Level Shift Accommodates Amplifier Input Common Mode Range. Circuits Cannot Operate from Single, Low Voltage Rail, Swing Close to Zero or Accommodate High Bias Voltages

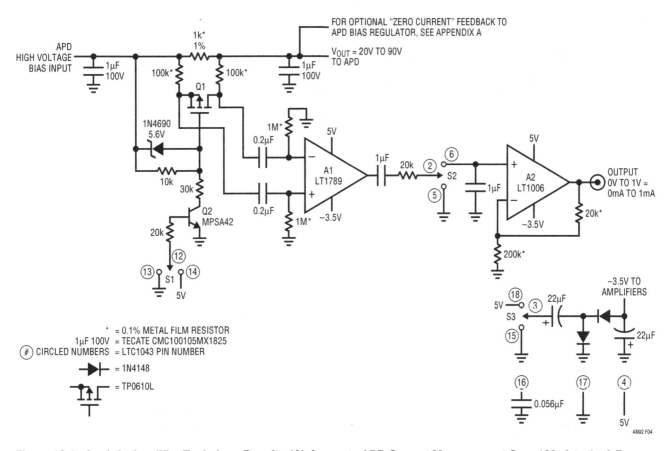

Figure 18.4 • Lock-In Amplifier Technique Permits 1% Accurate APD Current Measurement Over 100nA to 1mA Range. APD Current Is AC Modulated by Q1, Single-Ended at A1 and Demodulated to DC by S2-A2

capacitors. A1's single-ended output biases demodulator S2, which presents a DC output to buffer amplifier A2. A2's output is the circuit output.

Switch S3 clocks a negative output charge pump which supplies the amplifier's V⁻ pins, permitting output swing to (and below) zero volts. The 100k resistors at Q1 minimize its on-resistance error contribution and prevent destructive potentials from reaching A1 (and the 5V rail) if either 0.2µF capacitor fails. A2's gain of 1.1 corrects for the slight attenuation introduced by A1's input resistors. In practice, it may be desirable to derive the APD bias voltage regulator's feedback signal from the indicated point, eliminating the 1kΩ shunt resistor's voltage drop.[1] Verifying accuracy involves loading the APD bias line with 100nA to 1mA and noting output agreement.[2]

Note 1: See Appendix A, "Low error feedback signal derivation techniques," for details.
Note 2: Appropriate high value load resistors, perhaps augmented with a monitoring current meter, are available from Victoreen and other suppliers. Tight resistor tolerance, while convenient, is not strictly required, as output target value is set by current meter indication.

DC coupled current monitor

Figure 18.5's DC coupled current monitor eliminates the previous circuit's trim but pulls more current from the APD bias supply. A1 floats, powered by the APD bias rail. The 15V Zener diode and current source Q2 ensure A1 never is exposed to destructive voltages. The 1kΩ current shunt's voltage drop sets A1's positive input potential. A1 balances its inputs by feedback controlling its negative input via Q1. As such, Q1's source voltage equals A1's positive input voltage and its drain current sets the voltage across its source resistor. Q1's drain current produces a voltage drop across the ground referred 1k resistor identical to the drop across the 1kΩ current shunt and, hence, APD current. This relationship holds across the 20V to 90V APD bias voltage range. The 5.6V Zener assures A1's inputs are always within their common mode operating range and the 10M resistor maintains adequate Zener current when APD current is at very low levels.

Two output options are shown. A2, a chopper stabilized amplifier, provides an analog output. Its output is able to swing to (and below) zero because its V⁻ pin is supplied

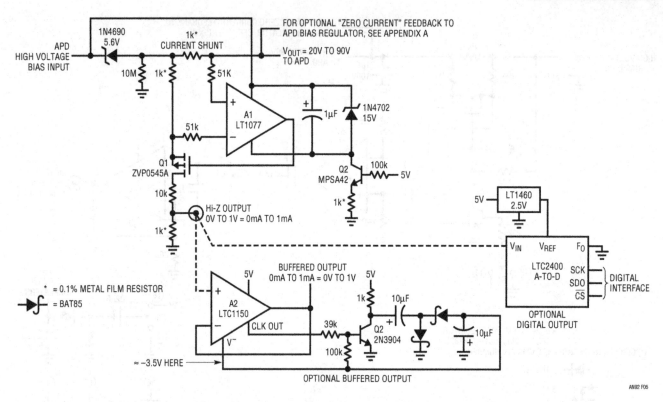

Figure 18.5 • A1-Q1 Float at High Voltage Rail, Measuring APD Current Via 1kΩ Shunt. Q1's Ground Referred Drain Current Provides Hi-Z Output. Buffer Options Include Analog (Figure Bottom Left) and Digital (Figure Bottom Right)

with a negative voltage. This potential is generated by using A2's internal clock to activate a charge pump which, in turn, biases A2's V⁻ pin.[3]

A second output option substitutes an A-to-D converter, providing a serial format digital output. No V⁻ supply is required, as the LTC2400 A-to-D will convert inputs to (and slightly below) zero volts.

Resistors at strategic locations prevent destructive failures. The 51kΩ unit protects A1 if the APD bias line shorts to ground. The 10k resistor limits current to a safe value if Q1 fails and the 100k resistor serves a similar purpose if Q2 malfunctions. As in the previous figure, APD voltage regulator feedback may be taken at the current shunt's output to maintain optimal regulation.[4] As stated, this circuit does not require trimming and maintains 0.5% accuracy. It does, however, pull current approximately equalling the current delivered to the APD, in addition to Q2's collector current. This can be an issue if the APD bias supply has restricted current capability.

APD bias supply

All previous examples have been current monitors. Figure 18.6, developed by Michael Negrete, is a high voltage APD bias supply.[5] The LT1930A switching regulator and L1 form a flyback based boost stage. The flyback events pump a diode-capacitor network tripler, producing a high voltage DC output. Feedback from the output via the R1-R2 combination stabilizes the regulator's operating point. D6 and D7 protect the switch and feedback pins, respectively, from parasitic negative excursions and the 10Ω resistors prevent excessive switch current. C8 and C9, series connected for high voltage capability, minimize output noise. A 0V to 4.5V programming voltage results in a corresponding 90V to 30V output (3% accuracy) with about 2mA of current capacity.

Circuit output noise is quite low. Figure 18.7, taken with 500μA loading at V_OUT = 50V, shows about 200μV ripple and harmonic residue in a 10MHz bandwidth. This is adequate for most APD receivers.[6]

Note 3: Circuit veterans will exercise extreme wariness when confronted with a bootstrapped biasing scheme such as this. Appendix D, "A single rail amplifier with true zero volt output swing," should soothe anxieties.
Note 4: See Appendix A, "Low error feedback signal derivation techniques."

Note 5: See Reference 7.
Note 6: Faithful noise measurements at these low levels require considerable care. See Appendices B and C for practical details.

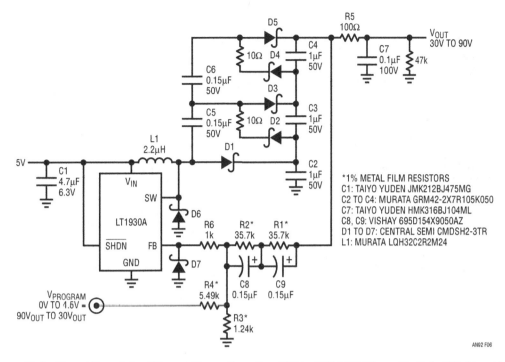

Figure 18.6 ● Boost Regulator/Charge Pump Supplies 30V to 90V APD Bias with Only 200μV$_{P-P}$ Noise

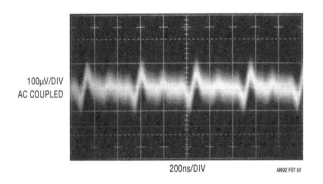

200ns/DIV

Figure 18.7 ● Figure 18.6's APD Bias Supply Shows 200μV$_{P-P}$ Ripple and Harmonic Residue in 10MHz Bandwidth

APD bias supply and current monitor

Figure 18.8, the Martin Configuration, combines the previous circuit with Figure 18.5's current monitor, providing a complete APD signal conditioner.[7] The programmable APD bias supply is as before, except that feedback comes via A2. A2, sensing after the 1kΩ current shunt, isolates the R1-R2 path loading, preventing it from influencing the

shunt's voltage drop. A2's action also insures tight output regulation, despite the current shunt's presence.[8]

The current monitor, borrowing from Figure 18.5, measures across the 1kΩ current shunt, presenting its output in Q1's drain line. As shown, the output has about 1kΩ output impedance, although either of Figure 18.5's output options may be employed.

When considering circuit operation, note that both amplifiers are powered by the charge pump's high voltage output, with their V⁻ pin returned to the "2/3 V$_{OUT}$" point. This biasing permits the amplifiers to process high voltage signals, although the voltage across them never exceeds 30V.

Transformer based APD bias supply and current monitor

Figure 18.9's circuit, another complete APD bias supply and current monitor, uses different techniques than the previous example. Advantages include 0.25% bias voltage and current monitoring accuracy, small size and fewer high voltage components for greater reliability. The LT1946 switching regulator and T1 form a flyback type boost configuration. T1's turn ratio provides voltage gain and the high voltage flyback events are rectified and smoothed to

Note 7: This circuit is based on work by Alan Martin.

Note 8: See Appendix A, "Low error feedback signal derivation techniques," for further discussion.

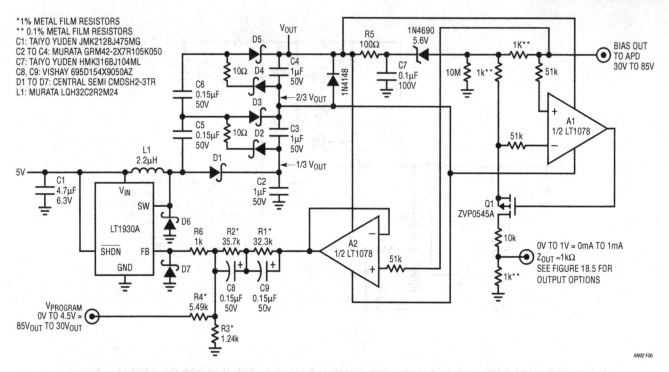

Figure 18.8 • Figure 18.5's Current Monitor Combines with Figure 18.6's Bias Supply, Providing APD Bias and Current Measurement. A2 Buffers LT1930A's Feedback Path Loading from Bias Supply Output, Eliminating Current Error. Amplifiers Process 85V Signals, Although Voltage Across Them Never Exceeds 30V

DC by the diode and capacitor in T1's secondary. This DC potential is divided down and fed back to A1. A1 compares this signal to the APD bias programming input and sets the LT1946's operating point, closing a control loop. Loop compensation is furnished by local rolloff at A1 and a lead network across the 10M feedback resistor. This loop establishes and maintains the APD bias output in accordance with the programming input's value. C1, active at $V_{SUPPLY} = 1.2V$, prevents output overshoot at power turn on by grounding the programming input command while simultaneously forcing A1's output low. This shuts off the switching regulator and no high voltage is produced. When power at turn on reaches ≈4V, C1 changes state and A1's positive input ramps to the programming voltage. The switching regulator's output follows this turn-on profile and no overshoot occurs. The LT1004 clamps spurious programming inputs beyond 2.5V, preventing excessive high voltage outputs.[9]

The circuit's current monitor portion takes full advantage of T1's floating secondary. Here, the 1kΩ current shunt resides in T1's secondary return path (Pin 3), eliminating the high common mode voltages encountered in the previous "high side" sensed examples. Circuit ground is declared at the shunt's uncommitted terminal, meaning

T1's Pin 3 will undergo increasing negative excursion with greater APD current. Inverter A2 converts the shunt's negative voltage to a buffered positive output. Its gain, scaled 1% above unity, compensates its input resistor's shunt loading error. Swing to zero is facilitated by returning A2's V⁻ pin to a small negative potential derived from the LT1946's V_{SW} pin switching. The 10M-287k divider's current loading error is prevented from appearing in A2's output by a compensatory current from the APD bias programming input. This compensating current, arriving at A2 via the 100k-3.65k-1M network, is scaled to precisely balance out the shunt's output portion due to the 10M-287k path's loading error. See Appendix A for detailed discussion of this technique.

Output noise for this circuit, shown in Figure 18.10, is about 1mV$_{P-P}$ in a 10MHz bandwidth. This is characteristic of flyback regulators and somewhat higher than Figure 18.8's charge pump based arrangement. It is still acceptable for most APD receivers, although special switching regulator techniques (read on!) can considerably reduce this figure.

Inductor based APD bias supply

Figure 18.11 borrows from Figure 18.9's flyback technique to form a simple, small area APD bias supply. Figure 18.9's current monitor function has been deleted—this circuit provides only the bias supply. Additionally, Figure 18.9's

---

Note 9: Optional circuitry allows input clamping at any desired voltage. See Appendix E, "APD protection circuits."

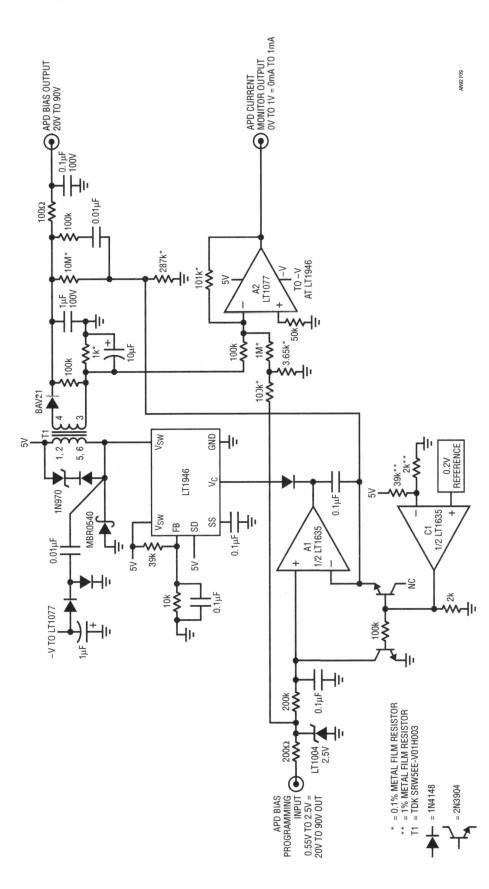

Figure 18.9 • A1 Controls LT1946 Boost Regulator to Supply 20V to 90V Bias. C1 Prevents Output Overshoot at Power Turn-On. A2 Senses APD Current Across 1kΩ Shunt in T1's Output Return. Programming Input Feedforward to A2 Cancels 10M–287k Feedback Divider's Loading Error, Preserving Current Monitor Accuracy

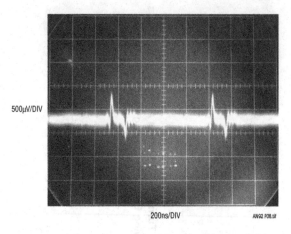

Figure 18.10 • Figure 18.9's Output Noise Measures 1mV$_{P-P}$ in 10MHz Bandwidth

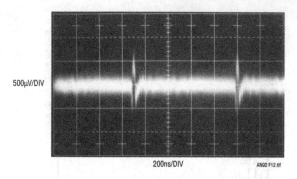

Figure 18.12 • Cascode Based Bias Supply Noise in 10MHz Bandwidth Is About 1.3mV$_{P-P}$

capacitance. The high voltage is rectified and filtered to DC, forming the circuit's output. Feedback to the regulator stabilizes the output, which may be varied by appropriate biasing at the V$_{PROGRAM}$ input. Components at the LT1946 V$_C$ pin compensate the loop. Over a 20V to 90V output range, the circuit remains within 2% of the V$_{PROGRAM}$ input dictated output voltage. Figure 18.12 shows switching related output noise is about 1.3 millivolts peak-to-peak in a 10MHz bandwidth.

200μV output noise APD bias supply

Some APD receiver applications require extremely low noise in an extended bandwidth. Figure 18.13's APD bias supply uses special switching regulator techniques to achieve 200μV noise in a 100MHz bandwidth. The LT1533 is a "push-pull" output switching regulator with controllable switch transition times. Output harmonic content ("noise") is notably reduced with slower switch transition times.[11] Switch current and voltage transition times are controlled by resistors at the R$_{CSL}$ and R$_{VSL}$ pins, respectively. In all other respects, the circuit behaves as a classical push-pull, transformer based, step-up converter. The V$_{PROGAM}$ input biases a feedback loop, setting the output anywhere between 20V and 90V.

The controlled transition times result in a dramatic decrease in output noise. Figure 18.14 shows ripple and switching related residue of 200μV in a 100MHz bandwidth. This is far below conventional regulators, meeting the most stringent noise requirement.

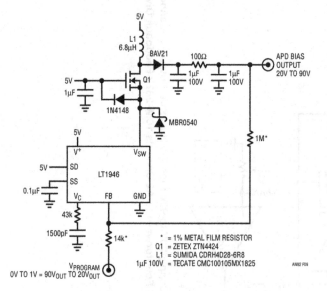

Figure 18.11 • Q1 Cascoded with LT1946 Switches L1, Providing 20V to 90V APD Bias Output. Q1's Source Diodes Clamp Parasitic Conducted Spikes to Safe Levels

transformer has been replaced with a 2-terminal inductor. The circuit is a basic inductor flyback boost regulator with a single important deviation. Q1, a high voltage device, has been interposed between the LT1946 switching regulator and the inductor. This permits the regulator to control Q1's high voltage switching without undergoing high voltage stress. Q1, operating as a "cascode" with the LT1946's internal switch, withstands L1's high voltage flyback events.[10] Diodes associated with Q1's source terminal clamp L1 originated spikes arriving via Q1's junction

---

Note 10: See Reference 11.

Note 11: Noise contains no regularly occurring or coherent components. As such, switching regulator output "noise" is a misnomer. Unfortunately, undesired switching related components in the regulated output are almost always referred to as "noise." Accordingly, although technically incorrect, this publication treats all undesired output signals as "noise." See Reference 2.

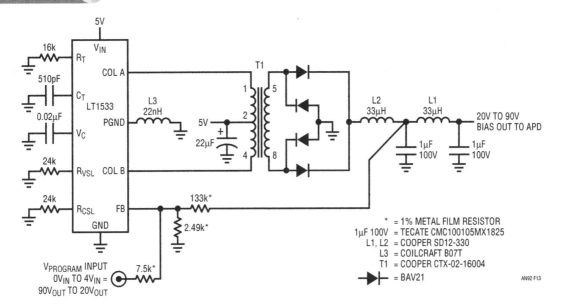

Figure 18.13 • Transformer Coupled 20V to 90V APD Bias Supply Controls Switch Transition Time for Extremely Low Output Noise

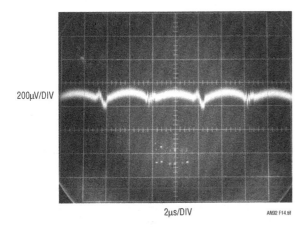

200μV/DIV

2μs/DIV   AN92 F14.tif

Figure 18.14 • LT1533's Controlled Transition Times Achieve Spectacularly Low Output Harmonic Residue. Switching Related Noise Is Below 100μV, Fundamental Ripple About 200μV. Measurement Bandwidth is 100MHz

Low noise APD bias supply and current monitor

Figure 18.15 builds on the previous circuit's performance, forming a complete, high performance APD signal conditioner. The bias supply is identical to Figure 18.13's low noise example, with the addition of the A1 based feedback buffer. This stage, similar to the one in Figure 18.8, isolates the regulator's feedback path current from the 1kΩ shunt, preserving current monitor accuracy. A1's Zener-current source power biasing scheme permits it

to process high voltage signals even though it is a low voltage device.[12] The current monitor, shown in block form, may be selected from the choices indicated depending upon requirements.

0.02% accuracy current monitor

Some APD current monitor applications call for high accuracy and stability. Figure 18.16's unusual optical switching based approach achieves 0.02% accuracy over a sensed 100nA to 1μA range. This scheme measures shunt current by switching (S1A, S1B) a capacitor across the shunt ("ACQUIRE"). After a time the capacitor charges to the voltage across the shunt. S1A and S1B open and S2A and S2B close ("READ"). This grounds one capacitor plate and the capacitor discharges into the grounded 1μF unit at S2B. This switching cycle is continuously repeated, resulting in A1's ground referred positive input assuming the same voltage that is across the floating 1kΩ shunt. The LED driven MOSFET switches specified do not have junction potentials and the optical drive contributes no charge injection error. A nonoverlapping clock prevents simultaneous conduction in S1 and S2, which would result in charge loss, causing errors and possible circuit damage. The 5.1V Zener prevents switched capacitor failure if the bias output is shorted to ground.

A1, a chopper stabilized amplifier, has a clock output. This clock, level shifted and buffered by Q3, drives a logic divider chain. The first flip-flop activates a charge pump,

---

Note 12: The feedback buffer is considered in detail in Appendix A, "Low error feedback signal derivation techniques."

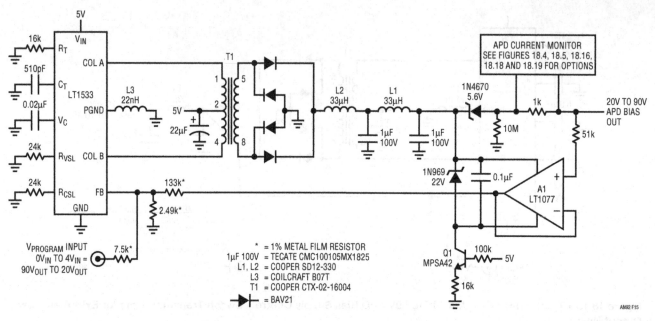

Figure 18.15 • Figure 18.13 Augmented with Feedback Divider Buffer and Current Monitor Provides Complete 100μV Noise APD Signal Conditioner

pulling A1's V⁻ pin negative, permitting amplifier swing to (and below) zero volts.[13] The divider chain terminates into a logic network. This network provides phase opposed charging of the 0.02μF capacitors (Traces A and B, Figure 18.17). The gating associated with these capacitors is arranged so the logic provides nonoverlapping, complementary biasing to Q1 and Q2. These transistors supply this nonoverlapping drive to the S1 and S2 actuating LEDs (Traces C and D).

The extremely small parasitic error terms in the LED driven MOSFET switches results in nearly theoretical circuit performance. However, residual error (=0.1%) is caused by SIA's high voltage switching pumping S2B's 3pF to 4pF junction capacitance. This results in a slight quantity of unwanted charge being transferred to the 1μF capacitor at S2B. The amount of charge transferred varies with the APD bias voltage (20V to 90V) and, to a lesser extent, the varactor-like response of S2B's off-state capacitance.

These terms are partially cancelled by DC feedforward to A1's negative input and AC feedforward from Q1's gate to S2B. The corrections compensate error by a factor of five, resulting in 0.02% accuracy.

Optical switch failure could expose A1 to high voltage, destroying it and possibly presenting destructive voltages to the 5V rail. This most unwelcome state of affairs is prevented by the 47k resistors in A1's positive input.

Digital output 0.09% accuracy current monitor

Figure 18.18 modifies the optically based current monitor to supply a digital output. The schematic is essentially identical to Figure 18.16's, with two significant differences. Here, a digital output is supplied via the LTC2431 A-to-D converter. The converter's differential inputs allow the same feedforward based error correction used in the previous example. The divider chain countdown ratio has changed to accommodate a higher speed clock, sourced by the LTC1799 oscillator. This higher speed clock, which times A-to-D operation, centers the A-to-D's internal notch filter at the optical switches commutation frequency, maximizing rejection.[14]

This circuit's 0.09% accuracy does not equal the previous analog ouput's version because of the LT1460 reference's 0.075% tolerance, which is not trimmable. The circuit can be adjusted to 0.02% accuracy by trimming the 1kΩ shunt so measured output current directly corresponds to A-to-D output.

Digital output current monitor

Previous current monitor examples furnish digital output from ground referenced A-to-D converters fed from analog

---

Note 13: This scheme, a variant of the one described back in Figure 18.5, is detailed in Appendix D, "A single rail amplifier with true zero volt output swing."

Note 14: The LTC2431's internal digital filter's first null occurs at 1/2560 of the frequency applied to its F₀ pin. For details, see the LTC2431 data sheet.

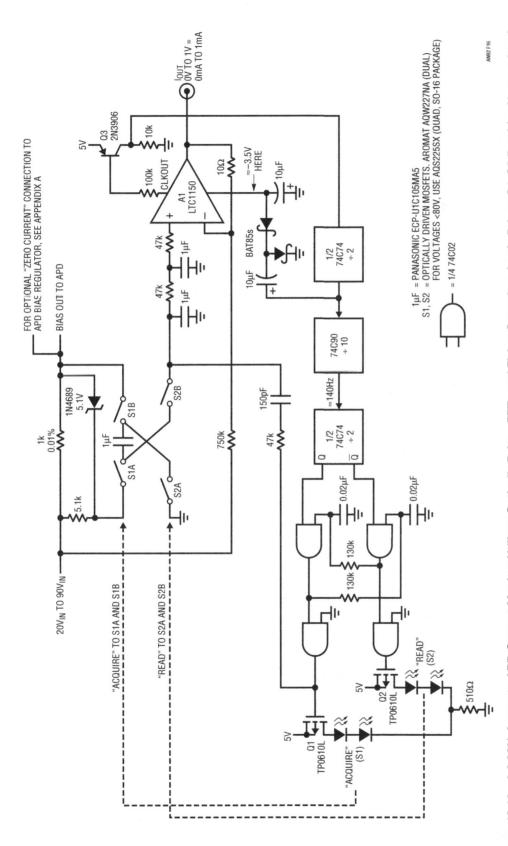

Figure 18.16 • A 0.02% Accurate APD Current Monitor Utilizes Optically Driven FETs and Flying Capacitor. Logic Driven Q1-Q2 Provides Nonoverlapping Clocking to S1-S2 LEDs. Clock Derives from A1's Internal Oscillator

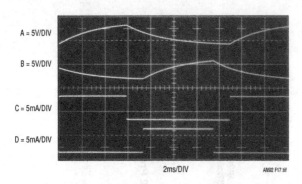

Figure 18.17 • Clocked, Cross Coupled Capacitors (Traces A and B) in 74C02 Based Network Result in Nonoverlapping Drive (Traces C and D) to S1-S2 Actuating LEDs

level shifting stages. Figure 18.19 directly digitizes shunt current by floating the A-to-D converter in the APD bias line. The A-to-D output is level shifted in the digital domain, presenting ground referred digital data. This simple approach is attractive, although the available APD bias

supply must supply about 3mA to the A-to-D and its attendant circuitry.

The LTC2410 and its LT1029 reference are powered directly from the high voltage APD bias supply input. Current sink Q3 and the LT1029 bias the LTC2410 V^- pin, maintaining 5V across the A-to-D over the 20V to 90V bias rail range. The A-to-D's differential inputs measure across the $1k\Omega$ current shunt. Resistors and a Zener clamp protect the A-to-D from excessive voltages if the APD bias line is shorted to ground. The A-to-D's digital outputs, floating at high voltage, drive level shifts which provide ground referred data. One of the identical stages is shown; the other indicated in conceptual form. The stage is designed for low quiescent and dynamic current consumption while maintaining data fidelity. This is necessary to minimize current drain from the APD bias supply and to avoid modulating it with transient loading artifacts. High voltage common emitter Q1 sources current to Q2, which provides a ground referred logic compatible output. Capacitive feedforwards maintain data edge speed while minimizing standing current requirements.

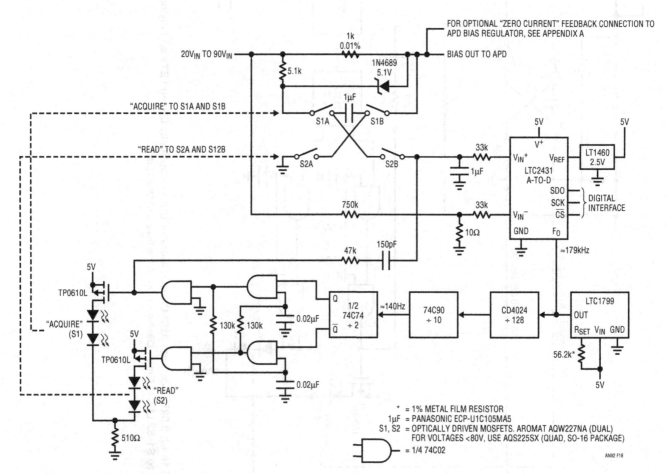

Figure 18.18 • Figure 18.16's Optically Driven FET Based Current Monitor Modified for Digital Output. LTC1799 Clocks A-to-D and Optical Switch LEDs. 0.09% Accuracy, Trimmable to 0.02%

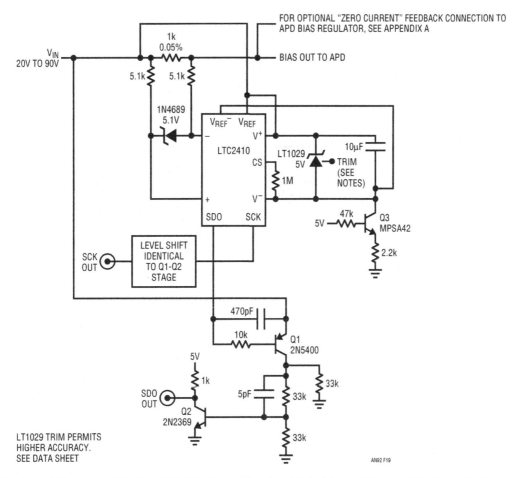

Figure 18.19 • A-to-D Converter Floats at High Voltage, Forming Digital Output Current Monitor. Q1-Q2 Level Shift Provides Ground Referenced Digital Output. 0.25% Accuracy Is Trimmable to 0.05%

This circuit's 0.25% untrimmed accuracy is due to shunt and LT1029 tolerances. Trimming the LT1029 (see schematic note) permits 0.05% accuracy.

Digital output current monitor and APD bias supply

Figure 18.20 also floats an A-to-D converter across the shunt, while including an APD bias supply. The bias supply is derived from the LT1946 switching regulator and Q1, operating in nearly identical fashion to Figure 18.11's circuit. The primary difference is that Figure 18.11's inductor is replaced here with a transformer. The transformer's primary winding furnishes high voltage step-up, similar to Figure 18.11. The floating secondary drives an isolated LT1120 based 3.75V regulator. This floating regulator's output, stacked on top of the APD bias line, powers the LTC2400 A-to-D converter. The isolated 3.75V supply permits the A-to-D to measure across the 1kΩ shunt without pulling operating power from the APD supply. Resistive current limiting and the 5.1V Zener protect the A-to-D from high voltage if the APD bias output is shorted to ground. Low

power optoisolators provide ground referred digital output while eliminating floating supply "starve out" due to cross regulation interaction with the APD regulation loop. Specifically, very low power APD bias outputs could result in insufficient transformer flux to furnish floating supply loading requirements. Common optoisolators require significant current, mandating the low power types specified. The previous circuit's discrete level shift stage would draw even less power but the optoisolators are simple and adequate.

The LT1120 2.5V reference and 1kΩ shunt tolerances dictate 2% circuit accuracy. If the tighter tolerance components noted in the schematic are used, 0.1% accuracy is practical.

Summary

Figure 18.21's chart is an attempt to summarize the circuits presented, although such brevity breeds oversimplification. As such, although the chart reviews salient features, there is no substitute for a thorough investigation of any particular application's requirements.

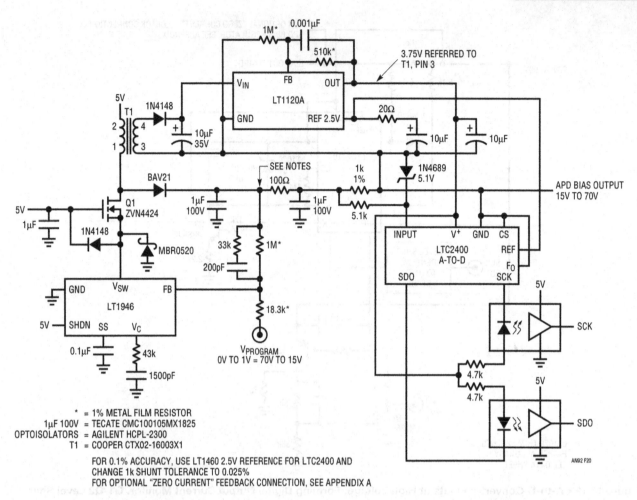

Figure 18.20 • APD Bias Supply with Digital Output Current Monitor. T1's Primary Supplies APD High Voltage Source, Similar to Figure 18.11; Secondary Furnishes Power to Floating Circuitry. 1kΩ Shunt Voltage Drop Is Compensatible Using Optional Feedback Circuitry. Optoisolators Provide Ground Referred Digital Output. Current Monitor Accuracy is 2%, Trimmable to 0.1%

FIGURE NUMBER	BIAS SUPPLY CAPABILITY	ANALOG OUTPUT CURRENT MONITOR (100nA to 1mA)	DIGITAL OUTPUT CURRENT MONITOR (100nA to 1mA)	COMMENTS
18.4	No	Yes	No	0.4% Accuracy. High Noise Rejection
18.5	No	Yes	Yes	0.5% Accuracy. Draws Current from APD Bias Supply Approximately Equalling Current Delivered to the APD in Addition to Housekeeping Current
18.6	Yes 30V to 90V	No	No	200µV Noise in 10MHz Bandwidth. 3% Accuracy
18.8	Yes 30V to 85V	Yes	No	3% Bias Voltage Accuracy. 0.5% Current Monitor Accuracy. Current Monitor Has 1kΩ Output Impedance
18.9	Yes 20V to 90V	Yes	No	0.25% Bias Voltage Accuracy. 1mV Output Noise in 10MHz Bandwidth. 0.25% Current Monitor Accuracy. Small Size. Few Large Value, High Voltage Capacitors Improves Reliability. Low Current Drain from APD Rail Permits Smaller High Voltage Capacitors for a Given Ripple Level
18.11	Yes 20V to 90V	No	No	2% Bias Voltage Accuracy. 1.5mV Output Noise in 10MHz Bandwidth. Small Size, Simple
18.13	Yes 20V to 90V	No	No	2% Bias Voltage Accuracy. 200µV Ripple and Noise in 100MHz Bandwidth. Relatively Large Solution Size Due to 250kHz Oscillator Frequency
18.15	Yes 20V to 90V	Yes	Yes	2% Bias Voltage Accuracy. 200µV Ripple and Noise in 100MHz Bandwidth. Current Monitor Accuracy Depends on Option Selected. Relatively Large Solution Size Due to 250kHz Oscillator Frequency
18.16	No	Yes	No	0.02% Accuracy. Low Current Drain from APD Rail Permits Smaller High Voltage Capacitors for a Given Ripple Level
18.18	No	No	Yes	0.09% Accuracy. 0.02% Achievable with Shunt Trimming. Low Current Drain from APD Rail Permits Smaller High Voltage Capacitors for a Given Ripple Level
18.19	No	No	Yes	0.25% Accuracy. Trimmable to 0.05% by Adjusting Reference
18.20	Yes 15V to 70V	No	Yes	2% Bias Voltage Accuracy. 2% Current Monitor Accuracy. 0.1% Accuracy Obtainable with Optional LT1460 Reference. Low Current Drain from APD Rail Permits Smaller High Voltage Capacitors for a Given Ripple Level

Figure 18.21 • Summarized Characteristics of Techniques Presented. Applicable Circuit Depends on Application Specifics

Note: This chapter was derived from a manuscript originally prepared for publication in EDN magazine.

References

1. Meade, M.L. "Lock-In Amplifiers and Applications," London, P. Peregrinus, Ltd.

2. Williams, J, "A Monolithic Switching Regulator with 100μV Output Noise," Linear Technology Corporation, Application Note 70, October 1997.

3. Williams, J., "Measurement and Control Circuit Collection," "ΔV_BE Based Thermometer," Linear Technology Corporation, Application Note 45, June 1991, pp. 7–8.

4. Williams, J., "Applications for a Switched Capacitor Instrumentation Building Block," Linear Technology Corporation, Application Note 3. July 1985.

5. Williams, J., "Monolithic CMOS-Switch IC Suits Diverse Applications," EDN Magazine, October 4, 1984.

6. Williams, J., "A Fourth Generation of LCD Backlight Technology," "Floating Lamp Circuits," Linear Technology Corporation, Application Note 65, November 1995, pp. 40–43, Figure 18.48.

7. Negrete, M., "Fiberoptic Communication Systems Benefit from Tiny, Low Noise Avalanche Photodiode Bias Supply," Linear Technology Corporation, Design Note 273, December 2001.

8. Martin, A., "Charge Pump Based APD Circuits," Private Communication, May 2002.

9. Williams, J., "Applications of New Precision Op Amps," "Instrumentation Amplifier with V_CM = 300V and 160dB CMRR," Linear Technology Corporation, Application Note 6, January 1985, pp. 1–2.

10. Williams, J., "Bridge Circuits," "Optically Coupled. Switched Capacitor Instrumentation Amplifier," Linear Technology Corporation, Application Note 43, June 1990, pp. 9–10.

11. Hickman, R. W. and Hunt, F. V., "On Electronic Voltage Stabilizers," "Cascode," Review of Scientific Instruments, January 1939, pp. 6–21, p. 16.

Appendix A
Low error feedback signal derivation techniques

Various text circuits either detail or make reference to counteracting loading effects of the APD bias supply's output feedback divider. If the divider is located before the 1kΩ current shunt, its current drain is not included in the current monitor's output and no error is incurred. A potential difficulty with this approach is that the 1kΩ shunt appears in series with the bias supply output, degrading load regulation. The maximum 1mA shunt current produces a 1V output regulation drop. In some cases this is permissible and no further consideration is required. Circumstances dictating tighter load regulation require compensation techniques.

Divider current error compensation—"low side" shunt case

When the shunt is in a transformer's return path ("low side shunt"), divider error is cancelled by introducing a compensatory term into the APD current monitor circuitry.

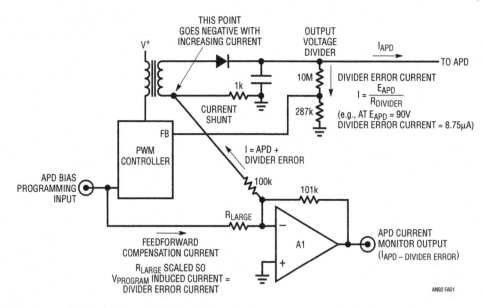

Figure A1 • Output Voltage Divider Current Loading Error Is Compensated with Feedforward from Programming Input. A1 Algebraically Sums Feedforward Term and Current Shunt Information, Presents Corrected Output

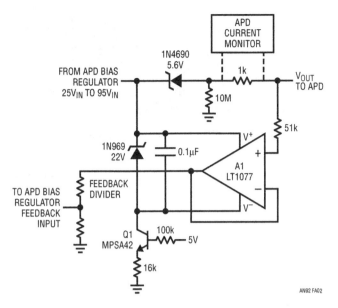

AN92 FA02

Figure A2 • A1 Follower Floats from High Voltage Rail, Eliminates Feedback Divider Current Loading Error. Q1 Current Source and 22V Zener Maintain Low Voltage Across Amplifier; 5.6V Zener Accommodates A1's Input Range

Figure A1 shows details. The output voltage divider's current loading error is prevented from appearing in A1's output by feeding forward a compensatory current from the APD bias programming input. This compensating current, arriving at A1 via R_{LARGE}, is scaled to precisely balance out the portion of shunt output contributed by the voltage divider's loading error.

Divider current error compensation — "high side" shunt case

Figure A2 addresses situations where the shunt resides in the "high side." Such arrangements involve high common mode voltages, seemingly mandating a high voltage buffer amplifier to isolate the divider's current loading. Figure A2 shows a way around this, using standard low voltage amplifiers to process high voltage signals. A1, sensing after the 1kΩ shunt, isolates the feedback divider's loading while permitting the APD bias regulator to include the shunt within its feedback loop. A1 is powered directly from the bias regulator's high voltage output but its V⁻ pin is Zener clamped with respect to its V⁺ pin. Current sink Q1 maintains this bias over the wide range of possible APD regulator outputs. Although A1 processes high voltage signals, the voltage across it is held to safe levels. The 5.6V Zener in the APD bias line ensures A1's inputs are always inside their common mode operating range. The 10M resistor maintains adequate Zener bias when APD currents are extremely low. A 51k resistor protects A1 from destructive high voltage if the APD bias output is shorted to ground. Similarly, the 100k resistor prevents high voltage from appearing on the 5V supply if Q1 fails.

Appendix B
Preamplifier and oscilloscope selection

The low level measurements described require some form of preamplification for the oscilloscope. Current generation oscilloscopes rarely have greater than 2mV/DIV sensitivity, although older instruments offer more capability. Figure B1 lists representative preamplifiers and oscilloscope plug-ins suitable for noise measurement. These units feature wideband, low noise performance. It is particularly significant that many of these instruments are no longer produced.

INSTRUMENT TYPE	MANUFACTURER	MODEL NUMBER	BANDWIDTH	MAXIMUM SENSITIVITY/GAIN	AVAILABILITY	COMMENTS
Amplifier	Hewlett-Packard	461A	150MHz	Gain = 100	Secondary Market	50Ω Input, Standalone
Differential Amplifier	Tektronix	1A5	50MHz	1mV/DIV	Secondary Market	Requires 500 Series Mainframe
Differential Amplifier	Tektronix	7A13	100MHz	1mV/DIV	Secondary Market	Requires 7000 Series Mainframe
Differential Amplifier	Tektronix	11A33	150MHz	1mV/DIV	Secondary Market	Requires 11000 Series Mainframe
Differential Amplifier	Tektronix	P6046	100MHz	1mV/DIV	Secondary Market	Standalone
Differential Amplifier	Preamble	1855	100MHz	Gain = 10	Current Production	Standalone, Settable Bandstops
Differential Amplifier	Preamble	1822	10MHz	Gain = 1000	Current Production	Standalone, Settable Bandstops

Figure B1 • Some Applicable High Sensitivity, Low Noise Amplifiers. Trade-Offs Include Bandwidth, Sensitivity and Availability

This is in keeping with current instrumentation trends, which emphasize digital signal acquisition as opposed to analog measurement capability.

The monitoring oscilloscope should have adequate bandwidth and exceptional trace clarity. In the latter regard high quality analog oscilloscopes are unmatched. The exceptionally small spot size of these instruments is well-suited to low level noise measurement.[1] The digitizing uncertainties and raster scan limitations of DSOs impose display resolution penalties. Many DSO displays will not even register the small levels of switching-based noise.

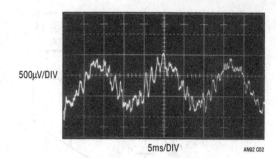

500μV/DIV

5ms/DIV   AN92 C02

Figure C2 • 60Hz Pickup Due to Excessive Probe Length at Feedback Node

---

Note 1: In our work we have found Tektronix types 453, 453A, 454, 454A, 547 and 556 excellent choices. Their pristine trace presentation is ideal for discerning small signals of interest against a noise floor limited background.

Appendix C
Probing and connection techniques for low level, wideband signal integrity[1]

The most carefully prepared breadboard cannot fulfill its mission if signal connections introduce distortion. Connections to the circuit are crucial for accurate information extraction. The low level, wideband measurements demand care in routing signals to test instrumentation.

Ground loops

Figure C1 shows the effects of a ground loop between pieces of line-powered test equipment. Small current flow between test equipment's nominally grounded chassis creates 60Hz modulation in the measured circuit output. *This problem can be avoided by grounding all line powered test equipment at the same outlet strip or otherwise ensuring that all chassis are at the same ground potential. Similarly, any test*

arrangement that permits circuit current flow in chassis interconnects must be avoided.

Pickup

Figure C2 also shows 60Hz modulation of the noise measurement. In this case, a 4-inch voltmeter probe at the feedback input is the culprit. *Minimize the number of test connections to the circuit and keep leads short.*

Poor probing technique

Figure C3's photograph shows a short ground strap affixed to a scope probe. The probe connects to a point which provides a trigger signal for the oscilloscope. Circuit output noise is monitored on the oscilloscope via the coaxial cable shown in the photo.

Figure C4 shows results. A ground loop on the board between the probe ground strap and the ground referred cable shield causes apparent excessive ripple in the display. *Minimize the number of test connections to the circuit and avoid ground loops.*

Violating coaxial signal transmission— felony case

In Figure C5, the coaxial cable used to transmit the circuit output noise to the amplifier-oscilloscope has been replaced with a probe. A short ground strap is employed as the probe's return. The error inducing trigger channel probe in the previous case has been eliminated; the 'scope is triggered by a noninvasive, isolated probe.[2] Figure C6 shows excessive display noise due to breakup of the coaxial signal environment. The probe's ground strap violates coaxial

---

Note 1: Veterans of LTC Application Notes, a hardened crew, will recognize this Appendix from AN70 (see Reference 2). Although that publication concerned considerably more wideband noise measurement, the material is directly applicable to this effort. As such, it is reproduced here for reader convenience.
Note 2: To be discussed. Read on.

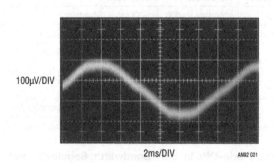

100μV/DIV

2ms/DIV   AN92 C01

Figure C1 • Ground Loop Between Pieces of Test Equipment Induces 60Hz Display Modulation

Figure C3 • Poor Probing Technique. Trigger Probe Ground Lead Can Cause Ground Loop-Induced Artifacts to Appear in Display

transmission and the signal is corrupted by RF. *Maintain coaxial connections in the noise signal monitoring path.*

Violating coaxial signal transmission — misdemeanor case

Figure C7's probe connection also violates coaxial signal flow, but to a less offensive extent. The probe's ground strap is eliminated, replaced by a tip grounding attachment. Figure C8 shows better results over the preceding case, although signal corruption is still evident. *Maintain coaxial connections in the noise signal monitoring path.*

Proper coaxial connection path

In Figure C9, a coaxial cable transmits the noise signal to the amplifier-oscilloscope combination. In theory, this affords the highest integrity cable signal transmission.

Figure C10's trace shows this to be true. The former example's aberrations and excessive noise have disappeared. The switching residuals are now faintly outlined in the amplifier noise floor. *Maintain coaxial connections in the noise signal monitoring path.*

Direct connection path

A good way to verify there are no cable-based errors is to eliminate the cable. Figure C11's approach

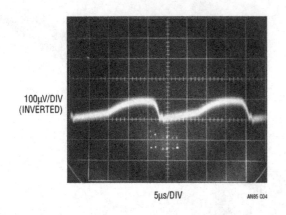

100µV/DIV
(INVERTED)

5µs/DIV   AN85 C04

Figure C4 • Apparent Excessive Ripple Results from Figure C3's Probe Misuse. Ground Loop on Board Introduces Serious Measurement Error

eliminates all cable between breadboard, amplifier and oscilloscope. Figure C12's presentation is indistinguishable from Figure C10, indicating no cable-introduced infidelity. *When results seem optimal, design an experiment to test them. When results*

seem poor, design an experiment to test them. When results are as expected, design an experiment to test them. When results are unexpected, design an experiment to test them.

Test lead connections

In theory, attaching a voltmeter lead to the regulator's output should not introduce noise. Figure C13's increased noise reading contradicts the theory. The regulator's output impedance, albeit low, is not zero, especially as frequency scales up. The RF noise injected by the test lead works against the finite output impedance, producing the 200µV of noise indicated in the figure. If a voltmeter lead must be connected to the output during testing, it should be done through a 10kΩ-10µF filter. Such a network eliminates Figure C13's problem while introducing minimal error in the monitoring DVM. *Minimize the number of test lead connections to the circuit while checking noise. Prevent test leads from injecting RF into the test circuit.*

Figure C5 • Floating Trigger Probe Eliminates Ground Loop, But Output Probe Ground Lead (Photo Upper Right) Violates Coaxial Signal Transmission

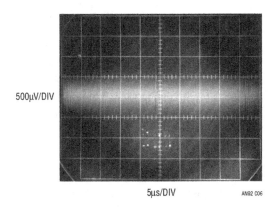

500µV/DIV

5µs/DIV   AN92 C06

Figure C6 • Signal Corruption Due to Figure C5's Noncoaxial Probe Connection

Isolated trigger probe

The text associated with Figure C5 somewhat cryptically alluded to an "isolated trigger probe." Figure C14 reveals this to be simply an RF choke terminated against ringing. The choke picks up

residual radiated field, generating an isolated trigger signal. This arrangement furnishes a 'scope trigger signal with essentially no measurement corruption. The probe's physical form appears in Figure C15. For good results, the termination should be adjusted for minimum ringing while preserving the highest possible amplitude output. Light compensatory damping produces Figure C16's output, which will cause poor 'scope triggering. Proper adjustment results in a more favorable output (Figure C17), characterized by minimal ringing and well-defined edges.

Trigger probe amplifier

The field around the switching magnetics is small and may not be adequate to reliably trigger some oscilloscopes. In such cases, Figure C18's trigger probe amplifier is useful. It uses an adaptive triggering scheme to compensate for variations in probe output amplitude. A stable 5V trigger output is maintained over a 50:1 probe output range. A1, operating at a gain of 100, provides wideband AC gain. The output of this stage biases a 2-way peak detector (Q1

Figure C7 • Probe with Tip Grounding Attachment Approximates Coaxial Connection

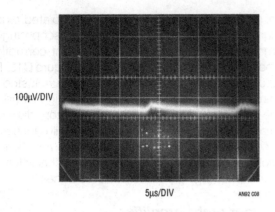

100µV/DIV

5µs/DIV   AN92 C08

Figure C8 • Probe with Tip Grounding Attachment Improves Results. Some Corruption Is Still Evident

through Q4). The maximum peak is stored in Q2's emitter capacitor, while the minimum excursion is retained in Q4's emitter capacitor. The DC value of the midpoint of A1's output signal appears at the junction of the 500pF capacitor and the 3MΩ units. This point always sits midway between the signal's excursions, regardless of absolute amplitude. This signal-adaptive voltage is buffered by A2 to set the trigger voltage at the LT1394's positive input. The LT1394's negative input is biased directly from A1's output. The LT1394's output, the circuit's trigger output, is unaffected by >50:1 signal amplitude variations. An ×100 analog output is available at A1.

Figure C19 shows the circuit's digital output (Trace B) responding to the amplified probe signal at A1 (Trace A).

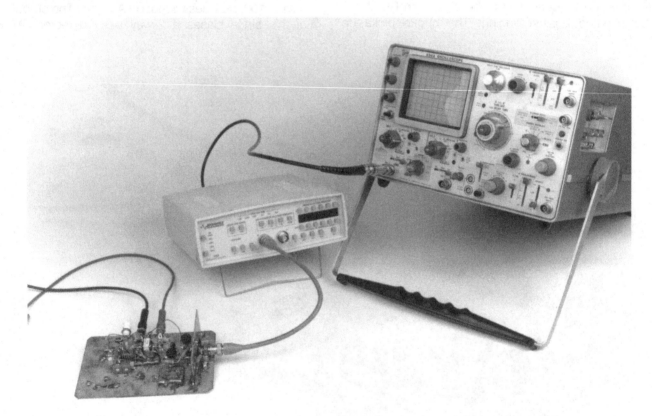

Figure C9 • Coaxial Connection Theoretically Affords Highest Fidelity Signal Transmission

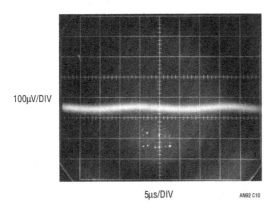

100µV/DIV

5µs/DIV   AN92 C10

Figure C10 • Life Agrees with Theory. Coaxial Signal Transmission Maintains Signal Integrity. Switching Residuals Are Faintly Outlined in Amplifier Noise

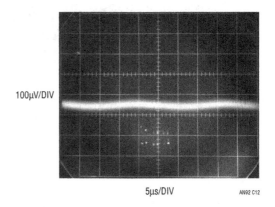

100µV/DIV

5µs/DIV   AN92 C12

Figure C12 • Direct Connection to Equipment Provides Identical Results to Cable-Termination Approach. Cable and Termination Are Therefore Acceptable

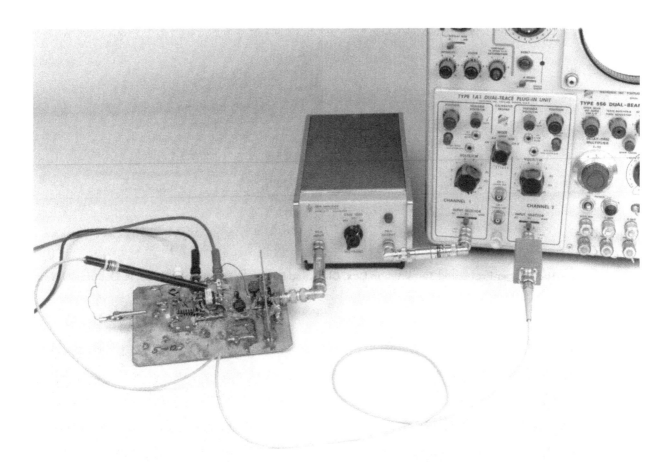

Figure C11 • Direct Connection to Equipment Eliminates Possible Cable-Termination Parasitics, Providing Best Possible Signal Transmission

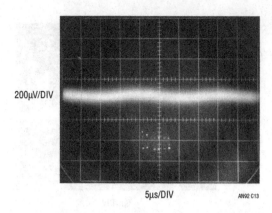

200µV/DIV

5µs/DIV   AN92 C13

Figure C13 • Voltmeter Lead Attached to Regulator Output Introduces RF Pickup, Multiplying Apparent Noise Floor

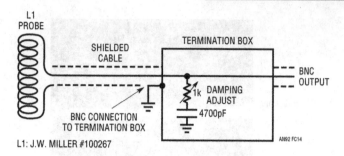

L1: J.W. MILLER #100267

Figure C14 • Simple Trigger Probe Eliminates Board Level Ground Loops. Termination Box Components Damp L1's Ringing Response

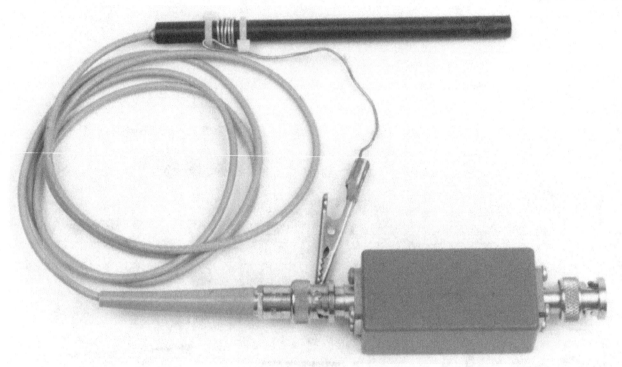

Figure C15 • The Trigger Probe and Termination Box. Clip Lead Facilitates Mounting Probe, Is Electrically Neutral

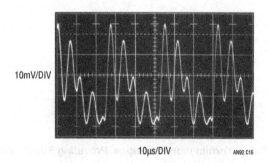

10mV/DIV

10µs/DIV   AN92 C16

Figure C16 • Misadjusted Termination Causes Inadequate Damping. Unstable Oscilloscope Triggering May Result

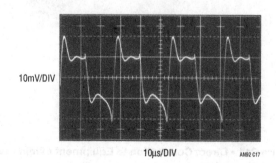

10mV/DIV

10µs/DIV   AN92 C17

Figure C17 • Properly Adjusted Termination Minimizes Ringing with Small Amplitude Penalty

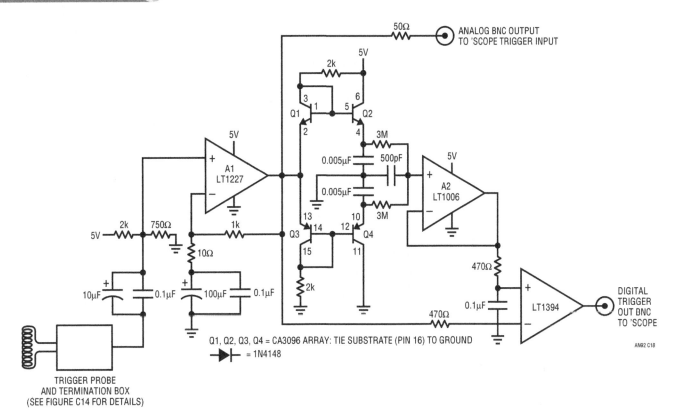

Figure C18 • Trigger Probe Amplifier Has Analog and Digital Outputs. Adaptive Threshold Maintains Digital Output Over 50:1 Probe Signal Variations

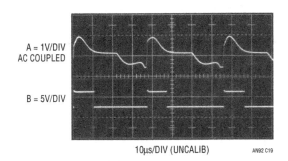

A = 1V/DIV
AC COUPLED

B = 5V/DIV

10μs/DIV (UNCALIB)   AN92 C19

Figure C19 • Trigger Probe Amplifier Analog (Trace A) and Digital (Trace B) Outputs

Figure C20 is a typical noise testing setup. It includes the breadboard, trigger probe, amplifier, oscilloscope and coaxial components.

Appendix D
A single rail amplifier with true zero volt output swing

Performance requirements necessitate analog output current monitors to swing within 100μV of zero. This is difficult because the circuits run from a single, positive rail. No single rail amplifier can swing this close to zero while maintaining accurate outputs. Figure D1's power supply bootstrapping scheme achieves the desired characteristics with minimal component addition.

A1, a chopper stabilized amplifier, has a clock output. This output switches Q1, providing drive to the diode-capacitor charge pump. The charge pump output feeds A1's V⁻/terminal, pulling it below zero, permitting output swing to (and below) ground.

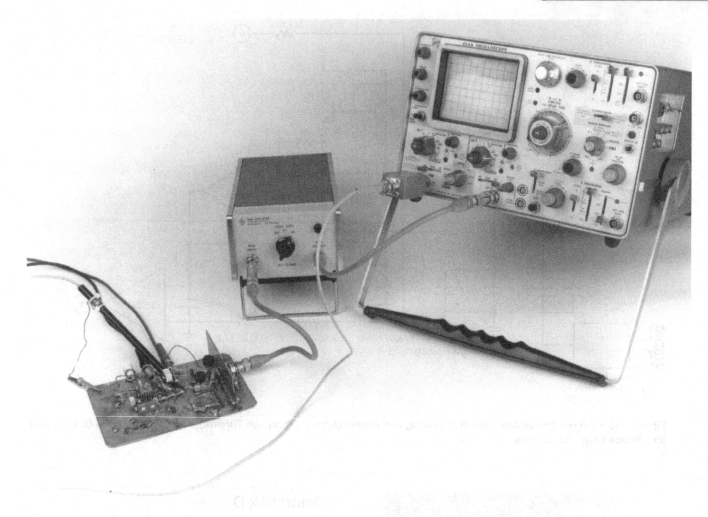

Figure C20 • Typical Noise Test Setup Includes Trigger Probe, Amplifier, Oscilloscope and Coaxial Components

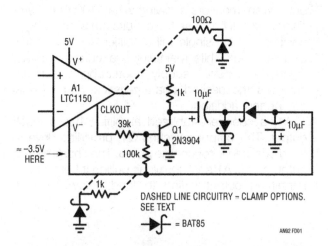

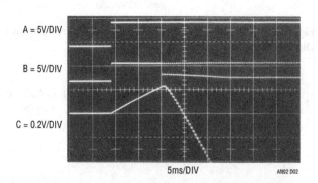

Figure D1 • Single Rail Powered Amplifier Has True Zero Volt Output Swing. A1's Clock Output Switches Q1, Driving Diode-Capacitor Charge Pump. A1's V⁻ Pin Assumes Negative Voltage, Permitting Zero (and Below) Volt Output Swing

Figure D2 • Amplifier Bootstrapped Supply Start-Up. Amplifier V⁻ Pin (Trace C) Initially Rises Positive at 5V Supply (Trace A) Turn-On. When Amplifier Internal Clock Starts (Trace B, 5th Vertical Division), Charge Pump Activates, Pulling V⁻ Pin Negative

If desired, negative output excursion can be limited by either clamp option shown.

Reliable start-up of this bootstrapped power supply scheme is a valid concern, warranting investigation. In Figure D2, the amplifier's V⁻ pin (Trace C) initially rises at supply turn-on (Trace A) but heads negative when amplifier clocking (Trace B) commences at about midscreen.

The circuit provides a simple way to obtain output swing to zero volts, permitting a true "live at zero" output.

Appendix E
APD protection circuits

APD receiver modules are electrically delicate and expensive devices. Because of this, Figure E1's protection circuits may be of interest. They are designed to protect the APD module from bias programming overvoltage error (Figure E1a), excessive current

(E1b) or destructive voltage (Figure E1c). In Figure E1a, Q1 is normally off and programming voltage passes to the bias regulator voltage programming input. Abnormally high inputs, defined by the potentiometer's setting, cause A1 to swing low, biasing Q1 and closing A1's feedback loop. This causes Q1's emitter to clamp at the potentiometer wiper's voltage, safely limiting the bias regulator's programming input.

Figure E1b is an APD current limiter. This particular circuit is designed for use with "low side" shunts in transformer coupled APD supplies, such as text Figure 18.9, although the technique is generally applicable. As long as the shunt current's absolute value is below the current limit point, A2 is saturated high and the associated APD bias regulator functions normally. Shunt overcurrent forces A2's output lower, pulling the regulator's control pin (V_C) lower and limiting current. The 100pF 1MΩ combination stabilizes A2 and the bias regulator assumes the characteristics of a current source.

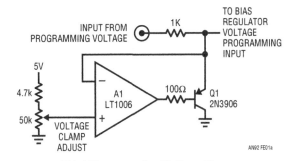

(E1a) Programming Voltage Clamp

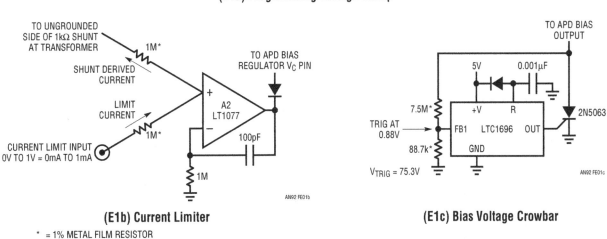

(E1b) Current Limiter

(E1c) Bias Voltage Crowbar

* = 1% METAL FILM RESISTOR
= 1N4148

Figure E1 • Protection Circuits Prevent APD Destruction Due to Hardware or Software Failures. Options Include Programming Voltage Clamp (E1a), Current Limiter (E1b) and Bias Voltage Crowbar (E1c)

Figure E1c is an overvoltage crowbar. It is intended as the last line of defense against uncontrolled APD bias supply high voltage outputs. Normally, the LTC1696 crowbar IC is below its 0.88V trigger threshold and the SCR is off. If the APD bias rises too high the LTC1696 triggers, firing the SCR. SCR turn-on "crowbars" the APD bias line, arresting the high voltage and maintaining a short across the line via its latch characteristic. If the APD bias supply has significant output impedance, prolonged SCR loading is not deleterious; if not, the bias supply should be fused.

Section 6

Automotive and Industrial Power Design

Developments in battery stack voltage measurement (19)

Automobiles, aircraft, marine vehicles, uninterruptible power supplies and telecom hardware represent areas utilizing series-connected battery stacks. These stacks of individual cells may contain many units, reaching potentials of hundreds of volts. In such systems it is often desirable to accurately determine each individual cell's voltage. Obtaining this information in the presence of the high "common mode" voltage generated by the battery stack is more difficult than might be supposed.

Developments in battery stack voltage measurement

A simple solution to a not so simple problem

19

Jim Williams   Mark Thoren

Automobiles, aircraft, marine vehicles, uninterruptible power supplies and telecom hardware represent areas utilizing series connected battery stacks. These stacks of individual cells may contain many units, reaching potentials of hundreds of volts. In such systems it is often desirable to accurately determine each individual cell's voltage. Obtaining this information in the presence of the high "common mode" voltage generated by the battery stack is more difficult than might be supposed.

The battery stack problem

The "battery stack problem" has been around for a long time. Its deceptively simple appearance masks a stubbornly resistant problem. Various approaches have been tried, with varying degrees of success.[1]

Figure 19.1's voltmeter measures a single cell battery. Beyond the obvious, the arrangement works because there are no voltages in the measurement path other than the measurand. The ground referred voltmeter only encounters the voltage to be measured.

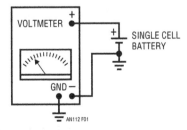

Figure 19.1 • Voltmeter Measuring Ground Referred Single Cell is Not Subjected to Common Mode Voltage

Note 1: See Appendix A, "A Lot of Cut Off Ears and No Van Goghs" for detail and commentary on some typical approaches.

Figure 19.2's "stack" of series connected cells is more complex and presents problems. The voltmeter must be switched between the cells to determine each individual cell's voltage. Additionally, the voltmeter, normally

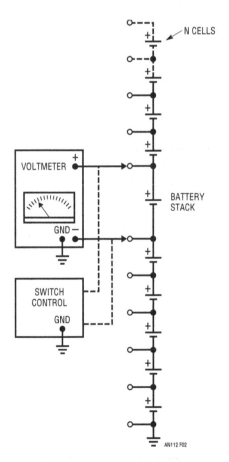

Figure 19.2 • Voltmeter Measuring Cell in Stack Undergoes Increasing Common Mode Voltage as Measurement Proceeds Up Stack. Switches and Switch Control Also Encounter High Voltages

Analog Circuit and System Design: A Tutorial Guide to Applications and Solutions. DOI: 10.1016/B978-0-12-385185-7.00019-6

composed of relatively low voltage breakdown components, must withstand input voltage relative to its ground terminal. This "common mode" voltage may reach hundreds of volts in a large series connected battery stack such as is used in an automobile. Such high voltage operation is beyond the voltage breakdown capabilities of most practical semiconductor components, particularly if accurate measurement is required. The switches present similar problems. Attempts at implementing semiconductor based switches encounter difficulty due to voltage breakdown and leakage limitations. What is really needed is a practical method that accurately extracts individual cell voltages while rejecting common mode voltages. This method cannot draw any battery current and should be simple and economically implemented.

Transformer based sampling voltmeter

Figure 19.3's concept addresses these issues. Battery voltage ($V_{BATTERY}$) is determined by pulse exciting a transformer (T1) and recording transformer primary clamp voltage after settling occurs. This clamp voltage is predominately set by the diode and $V_{BATTERY}$ shunting and similarly clamping T1's secondary. The diode and a small transformer term constitute predictable

errors and are subtracted out, leaving $V_{BATTERY}$ as the output.

Detailed circuit operation

Figure 19.4 is a detailed version of the transformer based sampling voltmeter. It closely follows Figure 19.3 with some minor differences which are described at this section's conclusion. The pulse generator produces a 10μs wide event (Trace A, Figure 19.5) at a 1kHz repetition rate. The pulse generator's low impedance output drives T1 via a 10k resistor and also triggers the delayed pulse generator.

T1's primary (Trace B) responds by rising to a value representing the sum of $V_{DIODE} + V_{BATTERY}$ along with a small fixed error contributed by the transformer. T1's primary clamps at this value. After a time (Trace C) dictated by the delayed pulse generator a pulse (Trace D) closes S1, allowing C1 to charge towards T1's clamped value. After a number of pulses C1 assumes a DC level identical to T1's primary clamp voltage. A1 buffers this potential and feeds differential amplifier A2. A2, operating at a gain near unity, subtracts the diode and transformer error terms, resulting in a direct reading $V_{BATTERY}$ output.

Accuracy is critically dependent on transformer clamping fidelity over temperature and clamp voltage range. The carefully designed transformer specified yields Figure 19.6's waveforms. Primary (Trace A) and secondary

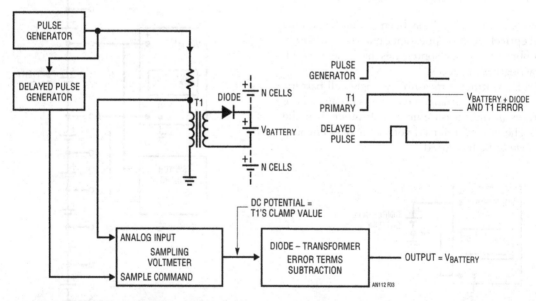

Figure 19.3 • Transformer-Based Sampling Voltmeter Operates Independently of High Common Mode Voltages. Pulse Generator Periodically Activates T1. Delayed Pulse Triggers Sampling Voltmeter, Capturing T1's Clamped Value. Residual Error Terms are Corrected in Following Stage

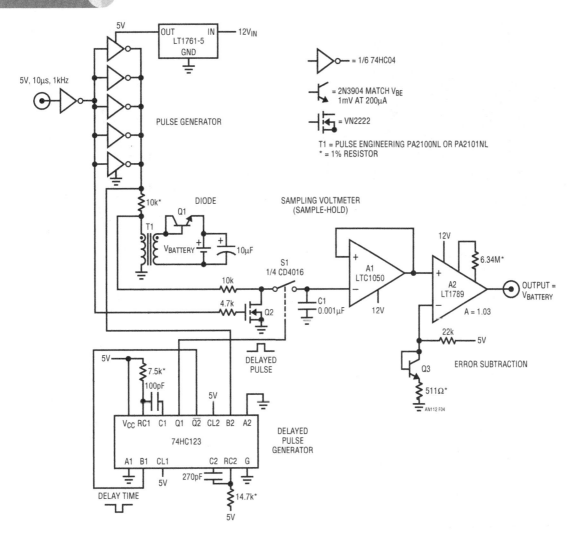

Figure 19.4 • Transformer Fed Sampling Voltmeter Schematic Closely Follows Figure 19.3's Concept. Error Subtraction Terms Include Q3 Compensating Q1 and Resistor/Gain Corrections for Errors in T1's Clamping Action. Q1-Q3 Transistors Replace Diodes for More Consistent Matching. Q2 Prevents T1's Negative Recovery Excursion from Influencing S1

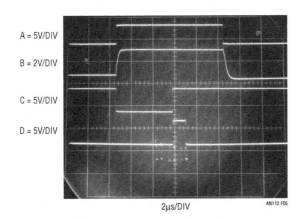

Figure 19.5 • Figure 19.4's Waveforms Include Pulse Generator Input (Trace A), T1 Primary (Trace B), 74HC123 $\overline{Q2}$ Delay Time Output (Trace C) and S1 Control Input (Trace D). Timing Ensures Sampling Occurs When T1 is Settled in Clamped State

(Trace B) clamping detail appear at highly expanded vertical scale. Clamping flatness is within millivolts; trace center aberrations derive from S1 gate feedthrough. Tight transformer clamp coupling promotes good performance. Circuit accuracy at 25°C is 0.05% over a 0V to 2V battery range with 120ppm/°C drift, degrading to 0.25% at $V_{BATTERY} = 3V$.[2]

Several details aid circuit operation. Transistor V_{BE}'s, substituted for diodes, provide more consistent initial matching and temperature tracking. The 10μF capacitor at Q1 maintains low impedance at frequency, minimizing cell voltage movement during the sampling interval. Finally, synchronously switched Q2 prevents T1's negative recovery excursion from deleteriously influencing S1's operation.

This approach's advantage is that its circuitry does not encounter high common mode voltages—T1 galvanically

---

Note 2: Battery stack voltage monitor development is aided by the floating, variable potential battery simulator described in Appendix B.

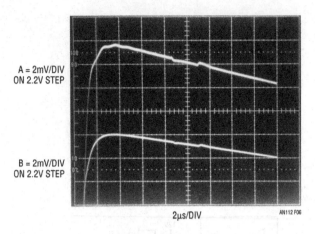

A = 2mV/DIV
ON 2.2V STEP

B = 2mV/DIV
ON 2.2V STEP

2μs/DIV   AN112 F06

Figure 19.6 • T1 Primary (Trace A) and Secondary (Trace B) Clamping Detail. Highly Expanded Vertical Scale Shows Primary and Secondary Clamping Flatness Within Millivolts. Trace Center Aberrations Derive from S1 Gate Feedthrough

isolates the circuit from common mode potentials associated with $V_{BATTERY}$. Thus, conventional low voltage techniques and semiconductors may be employed.

Multi-cell version

The transformer-based method is inherently adaptable to the multi-cell battery stack measurement problem previously described. Figure 19.7's conceptual schematic shows a multi-cell monitoring version. Each channel monitors one cell. Any individual channel may be read by biasing its appropriate enable line to turn on a FET switch, enabling that particular channel's transformer. The hardware required for each channel is typically limited to a transformer, a diode connected transistor and a FET switch.

Automatic control and calibration

This scheme is suited to digitally based techniques for automatic calibration. Figure 19.8 uses a PIC16F876A microcontroller, fed from an LTC1867 analog to digital converter, to control the pulse generators and channel selection. As before, even though the cell stack may reach hundreds of volts, the transformer galvanic isolation allows the signal path components to operate at low voltage.

A further benefit of processor driven operation is elimination of Figure 19.4's V_{BE} diode matching requirement. In practice, a processor-based board is tested at room

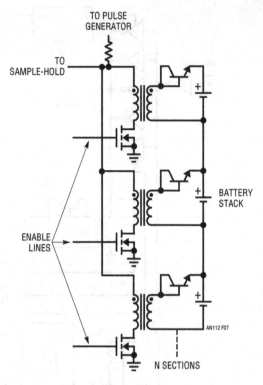

TO PULSE
GENERATOR

TO
SAMPLE-HOLD

ENABLE
LINES

BATTERY
STACK

AN112 F07

N SECTIONS

Figure 19.7 • Multiple Channels are Facilitated by Adding Enable Lines and Transistor Switches

temperature with known voltages applied to all input terminals. The channels are then read, furnishing the information necessary for the processor to determine each channel's initial V_{BE} and gain. These parameters are then stored in nonvolatile memory, permitting a one-time calibration that eliminates both V_{BE} mismatch and gain mismatch induced errors.

Channels 6 and 7 provide zero and 1.25V reference voltages representing cell voltage extremes. The room-temperature values are stored to nonvolatile memory. As temperature changes occur, readings from channel 6 and 7 are used to calculate a change in offset and a change in gain that are applied to the six measurement channels. The calibration is maintained as temperature varies because each channel's −2mV/°C V_{BE} drift slopes are nearly identical. Similarly, gain errors from channel to channel are nearly identical.

Since the gain and offset are continuously calibrated, the gain and offset of the LTC1867 drop out of the equation. The only points that must be accurate are the 0V measurement (easy, just short the channel 6 inputs together) and the 1.25V reference voltage, provided by an LT®1790-1.25. The LTC®1867 internally amplifies its internal 2.5V reference to 4.096V at the REFCOMP pin, which sets the full scale of the ADC (4.096V when it is

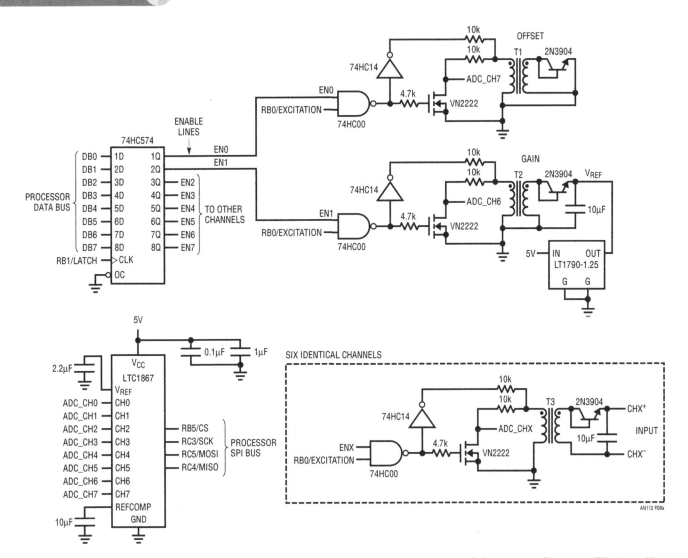

Figure 19.8a • Pulse Generators, Calibration Channels, Measurement Channels. ADC Calibration Channels Eliminate V$_{BE}$ Matching Requirement and Compensate for Temperature Dependent Errors

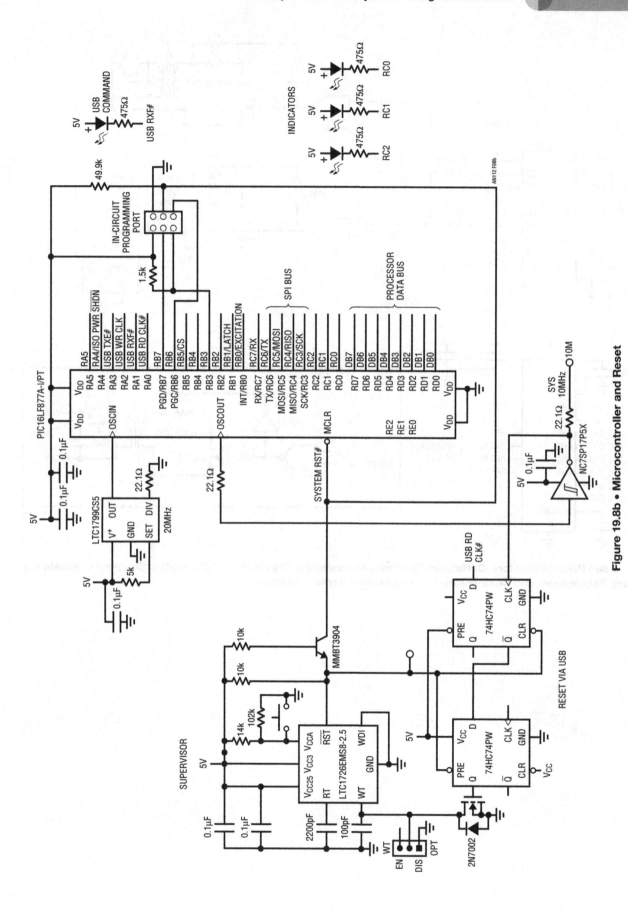

Figure 19.8b • Microcontroller and Reset

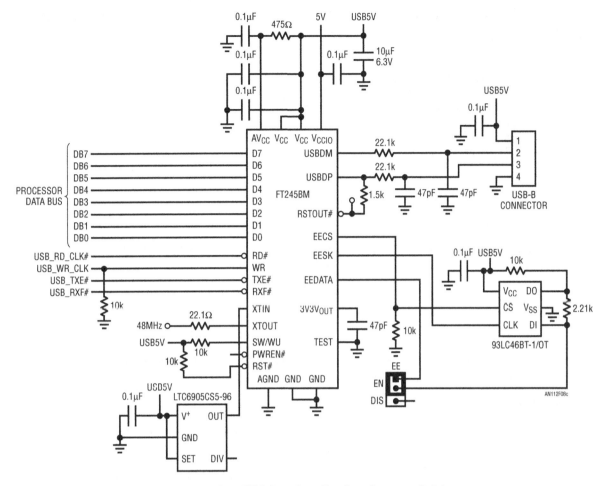

Figure 19.8c • USB Interface (for Development Only)

configured for unipolar mode, ±2.048V in bipolar mode). Thus the absolute maximum cell voltage that can be measured is 3.396V. And since the offset measurement is nominally 0.7V at the ADC input it is never in danger of clamping at zero. (A zero reading will result if a given LTC1867 has a negative offset and the input voltage is any positive voltage less than or equal to the offset.)

Accuracy of the processor-driven circuit is 1mV over a 0V to 2V input range at 25°C. Drift drops to less than 50ppm/°C—almost 3x lower than Figure 19.4.

Firmware description

The complete firmware code listing is in Appendix C. The code for this circuit is designed to be a good starting point for an actual product. Data is displayed to a PC via an FTDI FT242B USB interface IC. The PC has FTDI's Virtual Com Port drivers installed, allowing control through any terminal program. Data for all channels is continuously displayed to the terminal, and simple text commands control program operation.

A timer interrupt is called 1000 times per second. It controls the pulse generators and ADC, and stores the ADC readings to an array that can be read at any time. Thus if the main program is reading the buffer, the most out-of-date any reading will be is 1ms.

Automatic calibration routines are also included. Two functions store a zero reading and a full-scale reading for all channels, including the calibration voltages applied to channels 6 and 7, to nonvolatile memory. These are subsequently used to calibrate out the initial gain and offset errors as well as temperature dependent errors. The entire procedure is to apply zero volts to all inputs and issue a command to store the zero calibration, then apply 1.25V to all inputs and issue a command to store the full-scale calibration. Note that this is no more complicated than a basic functionality test that would be part of any manufacturing process. The 1.25V factory calibration source can be from a voltage calibrator, or from a selected "golden" LT1790-1.25 that is kept at a stable temperature.

A digital filter is also included for testing purposes. The filter is a simple exponential IIR (infinite impulse response) filter with a constant of 0.1. This reduces the noise seen in the readings by a factor of $\sqrt{10}$.

Measurement details

To take a reading from a given channel, the processor must apply the excitation to the transformer, wait for the voltage signal to settle out, take a reading with the ADC, and then

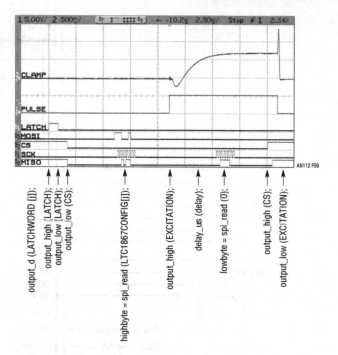

Figure 19.9 • Pulse Generator and ADC Sequencing

remove the excitation. This is driven by an interrupt service routine that is called once every millisecond. Refer to Appendix C for the code listing. Figure 19.9 shows the digital signals, excitation pulse, and clamp voltage at the ADC input along with the C code that performs these operations.[3]

Individual channels are enabled by loading an 8 bit byte with one bit set high into the 74HC574 latch.

Note that the excitation is applied after 8 bits of the LTC1867 data are read out. This is perfectly acceptable, since there is no conversion taking place and all of the data in the LTC1867 output register is static. Depending on the specific timing of the processor being used, excitation may be applied before reading any data, in the middle of reading data, or after reading the data but before initiating a conversion. If the serial clock is very slow—1MHz for instance, applying excitation before reading any data would result in the excitation being applied for 16µs which is too long. The only constraint is that the voltage at the ADC input must

---

Note 3: Sometimes a jack-of-all-trades is exactly what you need. A high speed digital designer would never dream of trading a good logic analyzer for a mixed-signal oscilloscope to test signal integrity across a complicated backplane. And its 100MHz analog channels pale in comparison to a good four channel, half-gig scope. But for testing a circuit with a microcontroller and data converters up to a few megasamples per second, a good mixed signal oscilloscope is the master of the trade.

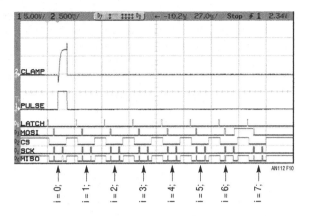

Figure 19.10 • ISR Scanning 8 Channels

have enough time to settle properly and that the excitation is not left on for too long. Figure 19.10 shows the same signals over the entire interrupt service routine. There are similar analog signals at each transformer and the other LTC1867 inputs.

Adding more channels

There are lots of ways to add more channels to this circuit. Figure 19.11 shows a 64 channel concept. Figure 19.11 decodes the 64 channels into eight banks of eight channels using 74HC138 address decoders. The selected bank corresponds to one LTC1867 input that is programmed through the SPI interface. The additional analog multiplexing

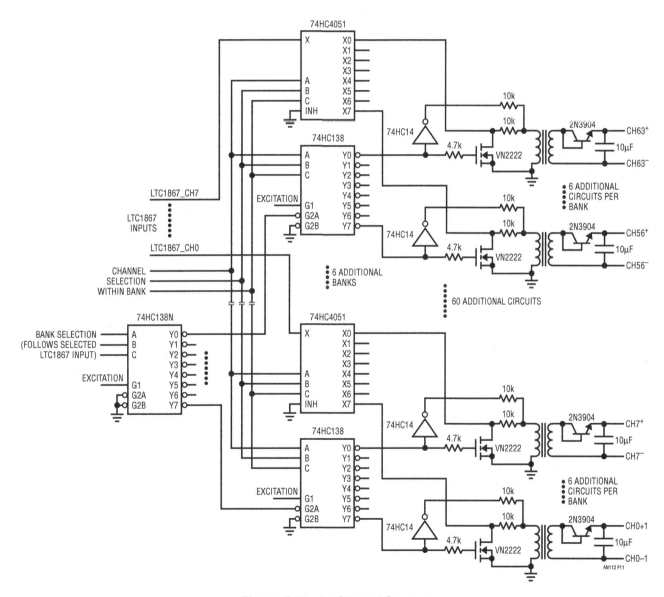

Figure 19.11 • 64-Channel Concept

is done with 74HC4051 8:1 analog switches. A single 74HC4051 feeding each LTC1867 input gives 64 inputs. The LTC1867 is still a great choice in high channel count applications, rather than a single channel ADC, because it is good idea to break up multiplexer trees into several stages to minimize total channel capacitance. The LTC1867 takes care of the last stage. And with a maximum sample rate of 200ksps, it can digitize up to 200 channels at the maximum 1ksps limitation of the sense transformer. That's a lot of batteries.

References

1 Williams, Jim, "Transformers and Opto-couplers Implement Isolation Techniques," "Isolated Temperature Measurement," pp. 116-117. EDN Magazine (January 1982).
2 Williams, Jim, "Isolated Temperature Sensor," LT198A Data Sheet. Linear Technology Corporation (1983).
3 Dobkin, R. C., "Isolated Temperature Sensor," LM135 Data Sheet. National Semiconductor Corporation (1978).

4 Williams, Jim, "Isolation Techniques for Signal Conditioning," "Isolated Temperature Measurement," pp. 1-2. National Semiconductor Corporation, Application Note 298 (May 1982).
5 Sheingold, D. H., "Transducer Interfacing Handbook," "Isolation Amplifiers," pp. 81-85. Analog Devices Inc. (1980).
6 Williams, Jim, "Signal Sources, Conditioners and Power Circuitry," "0.02%

Accurate Instrumentation Amplifier with 125 V_{CM} and 120dB CMRR," pp. 11-13. Linear Technology Corporation, Application Note 98 (November 2004).

Appendix A
A lot of cut off ears and no Van Goghs

Things that don't work

The "battery stack problem" has been around a long time. Various approaches have been tried, with varying degrees of success. The problem appears deceptively simple; technically and economically qualified solutions are notably elusive. Typical candidates and their difficulties are presented here.

Figure A1 presumably solves the problem by converting cell potentials to current, obviating the high common mode voltages. Op amps feed a multiplexed input A/D; the decoded A/D output presents individual cell voltages. This approach is seriously flawed. Required resistor precision and values are unrealistic, becoming progressively more unrealistic as the number of cells in the stack increases. Additionally, the resistors drain current from the cells, a distinct and often unallowable disadvantage.

An isolation amplifier based approach appears in Figure A2. Isolation amplifiers feature galvanically floating inputs, fully isolated from their output terminals. Typically, the device contains modulation-demodulation circuitry and a floating supply which powers the signal input section[4]. The amplifier inputs monitor the cell; its isolation barrier prevents battery stack common mode voltage from corrupting output referred

measurement results. This approach works quite well but, requiring an isolation amplifier per cell, is complex and quite expensive. Some simplification is possible; e.g., a single power driver servicing many amplifiers, but the method remains costly and involved.

Figure A3 employs a switched capacitor technique to measure individual cell voltage while rejecting common mode voltage. The clocked switches alternately connect the capacitor across its associated cell and discharge it into an output common referred capacitor.[5] After a number of such cycles the output capacitor assumes the cell voltage. A buffer amplifier provides the output. This arrangement rejects common mode voltages but requires many expensive high voltage switches, a high voltage level shift and non-overlapping switch drive. More subtly, switch leakage degrades accuracy, particularly as temperature rises. Optically driven switches, particularly those available as conveniently packaged LED driven MOSFETS, can simplify the level shift but expense, voltage breakdown and leakage concerns remain[6].

Switch related disadvantages are eliminated by Figure A4's approach. Each cell's potential is digitized by a dedicated A/D converter. A/D output is transmitted across an isolation barrier via a data isolator (optical, transformer). In its most elementary form, each A/D is powered by a separate, isolated power supply. This isolated supply population is

---

Note 4: See Reference 5 for details on isolation amplifiers.

Note 5: Old timers amongst the readership will recognize this configuration as a derivative of the venerable reed switched "flying capacitor" multiplexer.
Note 6: An optically coupled variant of this approach is given in Reference 6.

reducible, but cannot be eliminated. Constraints include cell voltage and the A/D's maximum permissible supply and input common mode voltages. Within these limitations, several A/D channels are serviceable by one isolated supply. Further refinement is possible through employment of multiplexed input A/Ds. Even with these improvements, numerous isolated supplies are still mandated by large battery stacks. Although this scheme is technologically sound, it is complex and expensive.

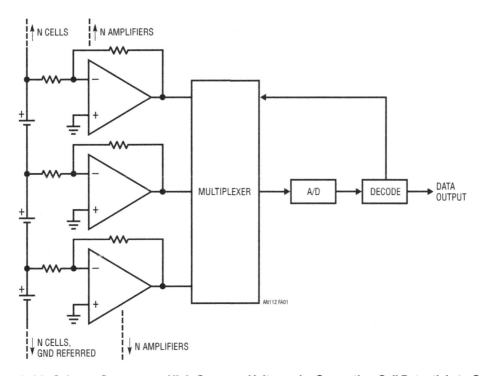

Figure A1 • Unworkable Scheme Suppresses High Common Voltages by Converting Cell Potentials to Current. Circuit Decodes Amplifier Outputs to Derive Individual Cell Voltages. Required Resistor Precision and Values are Unrealistic. Resistors Draw Current from Cells

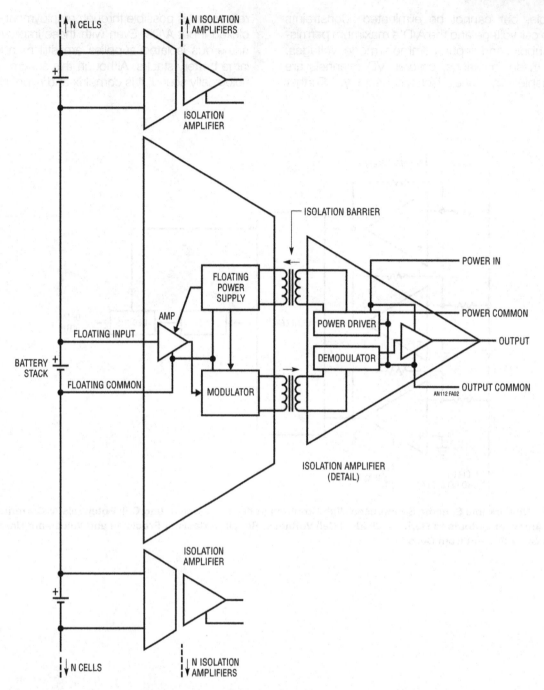

Figure A2 • Isolation Amplifier's Galvanically Floating Input Eliminates Common Mode Voltage Effects. Approach Works, but is Complex and Expensive Requiring Isolation Amplifier per Cell

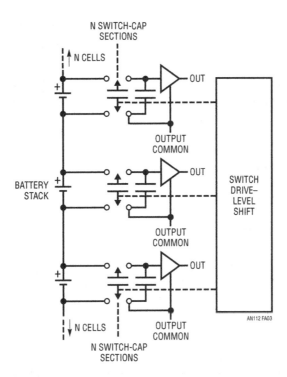

Figure A3 • Switched Capacitor Scheme Rejects Common Mode Voltage but Requires High Voltage Switches, Nonoverlapping Drive and Level Shift. Switch Leakage Degrades Accuracy. Optically Driven Switches Can Simplify Level Shift but Breakdown and Leakage Issues Remain

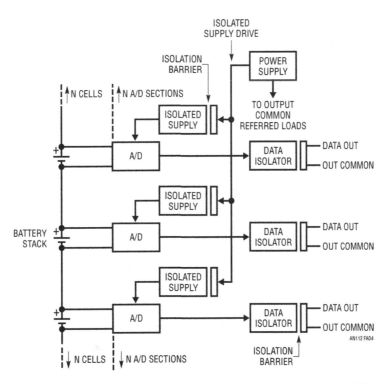

Figure A4 • A/D per Cell Requires Isolated Supplies and Data Isolators. Multiplexed Input A/Ds can Minimize A/D Usage. Isolated Supply Population is Reducible, but Cannot be Eliminated

Appendix B
A floating output, variable potential battery simulator

Battery stack voltage monitor development is aided by a floating, variable potential battery simulator. This capability permits accuracy verification over a wide range of battery voltage. The floating battery simulator is substituted for a cell in the stack and any desired voltage directly dialed out. Figure B1's circuit is simply a battery-powered follower (A1) with current boosted (A2) output. The LT1021 reference and high resolution potentiometric divider specified permits accurate output settling within 1mV. The composite amplifier unloads the divider and drives a 680µF capacitor to approximate a battery. Diodes preclude reverse biasing the output capacitor during supply sequencing and the 1µF-150k combination provides stable loop compensation. Figure B2 depicts loop response to an input step; no overshoot or untoward dynamics occur despite A2's huge capacitive load. Figure B3 shows battery simulator response (trace B) to trace A's transformer clamp pulse. Closed-loop control and the 680µF capacitor limit simulator output excursion within 30µV. This error is so small that noise averaging techniques and a high gain oscilloscope preamplifier are required to resolve it.

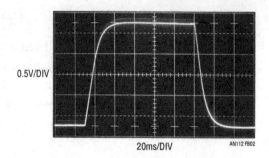

0.5V/DIV

20ms/DIV   AN112 FB02

Figure B2 • 150k-1µF Compensation Network Provides Clean Response Despite 680µF Output Capacitor

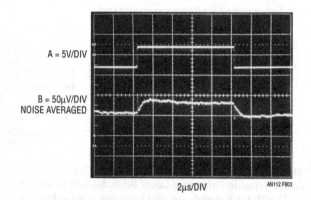

A = 5V/DIV

B = 50µV/DIV
NOISE AVERAGED

2µs/DIV   AN112 FB03

Figure B3 • Battery Simulator Output (Trace B) Responds to Trace A's Transformer Clamp Pulse. Closed-Loop Control and 680µF Capacitor Maintain Simulator Output Within 30µV. Noise Averaged, 50µV/Division Sensitivity is Required to Resolve Response

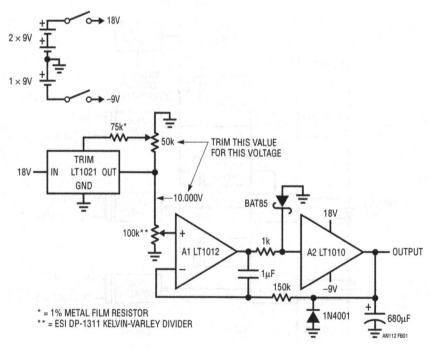

* = 1% METAL FILM RESISTOR
** = ESI DP-1311 KELVIN-VARLEY DIVIDER

Figure B1 • Battery Simulator Has Floating Output Settable Within 1mV. A1 Unloads Kelvin-Varley Divider; A2 Buffers Capacitive Load

Appendix C
Microcontroller code listing

The microcontroller code consists of three files:

Battery_monitor.c contains the main program loop, including calibration and temperature correction, and support functions.

Interrupts.c is the code for the timer2 interrupt that drives the transformer excitation and controls the LTC1867 ADC.

Battery monitor.h contains various defines, global variable declarations and function prototypes.

```
/*****************************************************************************
battery_monitor.c

Six Channel Battery Monitor with continuous gain and offset
correction. Includes a "factory calibration" feature. On first power up,
apply zero volts to all inputs, allow data to settle, and type 'o'.

Next apply 1.25V to all inputs, allow data to settle, and type 'p'.

This calibrates the circuit, and it is ready to run.

Offset correction technique:
Present offset correction = init_offset[7] - voltage[7]
Hotter = less counts on voltage[7] so correction goes POSITIVE,
so ADD this to voltage[i]

voltage[i] = voltage[i] - init_offset[i] + present_offset

Slope correction Technique:
Initial slope = init_fs[6] - init_offset[7] counts per 1.25V
Present slope = voltage[6] - voltage[7] counts per 1.25V

Keyboard command summary:
'a': increment conversion period (default is 1ms)
'z': decrement conversion period
's': increment by 10
'x': decrement by 10
'd': increment pulse-convert delay (default is 2us)
'c': decrement pulse-convert delay
'f': increment pulse-convert delay by 10
'v': decrement pulse-convert delay by 10
'n': Calculate voltages for display
'm': Display raw ADC values
't': Echo text to terminal so you can insert comments into
     data that is being captured. Terminate with '!'
'k': Disable digital filter
'l': Enable digital filter
'o': Store offsets to nonvolatile memory
'p': Store full-scale readings to nonvolatile memory

Written for CCS Compiler Version 3.242
Mark Thoren
Linear Technology Corporation
January 15, 2007
*****************************************************************************/

#include "battery_monitor.h"
#include "interrupts.c"

void main(void)
    {
    int8 i;
    unsigned int16 adccode;
    float temp=0.0, offset_correction, slope, slope_correction;

    initialize();              // Initialize hardware
    rx_usb();                  // Wait until any character is received
    print_cal_constants();     // display calibration constants before starting.

    while(1)
        {
        if(usb_hit()) parse();  // get keyboard command if necessary
        for(i=0; i<=7; ++i)     // Read raw data first
            {
            readflag[i] = 1;    // Tell interrupt that we're reading!!
```

an112f

```
        adccode = data[i];
        readflag[i] = 0;
        temp = (float) adccode; // convert to floating point
        if(filter)              // Simple exponential IIR filter
            {
            voltage[i] = 0.9 * voltage[i];
            voltage[i] += 0.1* temp;
            }
        else
            {
            voltage[i] = temp;
            }
        }

    if(calculate) // Display temperature corrected voltages
        {
        // Calculate Corrections.
        // offset correction is stored CH7 reading minus the present reading
        offset_correction = read_offset_cal(7) - voltage[7];
        // Slope correction is the stored slope based on initial CH6 and CH7
        // readings divided by the present slope. Units are (dimensionless)
        slope_correction = (float) read_fs_cal(6) -
                        (float) read_offset_cal(7); // Initial counts/1.25V
        slope_correction = slope_correction / (voltage[6] - voltage[7]);

        for(i=0; i<=5; ++i)       // Print Measurement Channels
            {
            // Units on slope are "volts per ADC count"
            slope = 1.25000 / ((float) read_fs_cal(i) -    // Inefficient but
                            (float) read_offset_cal(i)); // we are RAM limited
            // Correct for initial offset and temperature dependent offset.
            // units on temp are "ADC counts"
            temp = voltage[i] - (float) read_offset_cal(i) + offset_correction;
            // Correct for initial slope
            temp = temp * slope;
            // Units on temp is now "volts"
            // Correct for temperature dependent slope
            temp = temp * slope_correction;
            busbusy = 1;
            printf(tx_usb, "%1.5f, ", temp);
            busbusy = 0;
            }
        busbusy = 1;        // Print to terminal
        printf(tx_usb, "%1.6f, %1.1f, ", slope_correction, offset_correction);
        busbusy = 0;
        }

    else // Display raw ADC counts
        {
        for(i=0; i<=7; ++i)
            {
            busbusy = 1;      // Print to terminal
            printf(tx_usb, "%1.0f, ", voltage[i]);
            busbusy = 0;
            }
        }

    busbusy = 1;
    printf(tx_usb, "D:%d, P:%d\r\n", delay, period); // print period and delay
    busbusy = 0;
    // Delay and blink light
    delay_ms(100); output_high(PIN_C0); delay_ms(100); output_low(PIN_C0);
    } //end of loop
} //end of main
```

an112f

```
/*****************************************************************************
Parse keyboard commands
arguments: none
returns: void
*****************************************************************************/

void parse(void)
   {
   char ch;
   switch(rx_usb())
      {
      case 'a': period += 1; break;     // increment period
      case 'z': period -= 1; break;     // decrement period
      case 's': period += 10; break;    // increment by 10
      case 'x': period -= 10; break;    // decrement by 10
      case 'd': delay += 1; break;      // increment pulse-convert delay
      case 'c': delay -= 1; break;      // decrement pulse-convert delay
      case 'f': delay += 10; break;     //     "    by 10
      case 'v': delay -= 10; break;     //     "    by 10
      case 'n': calculate = 1; break;   // Calculate voltages
      case 'm': calculate = 0; break;   // Display raw values
      case 't':    // Echoes text to terminal so you can insert comments into
         {         // data that is being captured. Terminate with '!'
         busbusy = 1;
         printf(tx_usb, "enter comment\r\n");
         while((ch=rx_usb())!='!') tx_usb(ch);
         tx_usb('\r');
         tx_usb('\n');
         busbusy = 0;
         } break;
      case 'k': filter = 0; break;      // Disable filter
      case 'l': filter = 1; break;      // Enable Filter
      case 'o': write_offset_cal(); break;  // Store offset to nonvolatile mem.
      case 'p': write_fs_cal(); break;      // Store FS to nonvolatile mem.
      }
   setup_timer_2(T2_DIV_BY_16,period,8);     // Update period if necessary
   }

/*****************************************************************************
write offset and full-scale calibration constants to non-volatile memory
arguments: none
returns: void
*****************************************************************************/
void write_offset_cal(void)
   {
   int i;
   unsigned int16 intvoltage;
   for(i=0; i<=7; ++i)
      {
      intvoltage = (unsigned int16) voltage[i];    // Cast as unsigned int16
      write_eeprom (init_offset_base+(2*i), intvoltage >> 8); // Write high byte
      delay_ms(20);
      write_eeprom (init_offset_base+(2*i)+1, intvoltage);  // Write low byte
      delay_ms(20);
      }
   }

void write_fs_cal(void)
   {
   int i;
   unsigned int16 intvoltage;
   for(i=0; i<=7; ++i)
      {
```

```
        intvoltage = (unsigned int16) voltage[i];     // Cast as unsigned int16
        write_eeprom (init_fs_base+(2*i), intvoltage >> 8);  // Write high byte
        delay_ms(20);
        write_eeprom (init_fs_base+(2*i)+1, intvoltage);   // Write low byte
        delay_ms(20);
        }
    }

/***********************************************************************************
read offset and full-scale calibration constants from non-volatile memory
arguments: none
returns: void
***********************************************************************************/
unsigned int16 read_offset_cal(int channel)
    {
    return make16(read_eeprom(init_offset_base+(2*channel)),
                  read_eeprom(init_offset_base+(2*channel)+1));
    }

unsigned int16 read_fs_cal(int channel)
    {
    return make16(read_eeprom(init_fs_base+(2*channel)),
                  read_eeprom(init_fs_base+(2*channel)+1));
    }


/***********************************************************************************
Print calibration constants (raw ADC counts)
arguments: none
returns: void
***********************************************************************************/
void print_cal_constants(void)
    {
    int i;
    for(i=0; i<=7; ++i)
        {
        printf(tx_usb, "ch%d offset: %05Lu, fs: %05Lu\r\n"
        , i, read_offset_cal(i),read_fs_cal(i));
        }
    }

/***********************************************************************************
Interface to the FT24BM USB controller

usb_hit()    arguments: none  returns: 1 if data is ready to read, zero otherwise
rx_usb() arguments: none returns: character from USB controller
tx_usb() argments: data to send to PC, returns: void
***********************************************************************************/
char usb_hit(void)
    {
    return !input(RXF_);
    }

char rx_usb(void)
    {
    char buf;
    while(input(RXF_)) {} // Low when data is available, wait around
    output_low(RD_);
    delay_cycles(1);
    buf=input_d();
    output_high(RD_);
    return(buf);
    }
```

```
void tx_usb(int8 value)
   {
   while(input(TXE_))     //Low when FULL, wait around
      {
      }
   output_d(value);
   output_high(WR);
   delay_cycles(1);
   output_low(WR);
   input_d();
   }

/******************************************************************************
Hardware initialization
arguments: none
returns: void
******************************************************************************/
void initialize(void)
   {
   output_high(ISO_PWR_SD_);   //turn on power
   setup_adc_ports(NO_ANALOGS);
   setup_adc(ADC_OFF);
   setup_psp(PSP_DISABLED);
   setup_spi(SPI_CONFIG);
   CKP = 0; // Set up clock edges - clock idles low, data changes on
   CKE = 1; // falling edges, valid on rising edges.
   output_low(I2C_SPI_);
   output_low(AUX_MAIN_); // SPI is only MAIN
   setup_counters(RTCC_INTERNAL,RTCC_DIV_1);
   setup_timer_0(RTCC_INTERNAL|RTCC_DIV_1);
   setup_timer_1(T1_DISABLED);
   setup_timer_2(T2_DIV_BY_16,period,8);
   setup_comparator(NC_NC_NC_NC);
   setup_vref(FALSE);

   output_low(PIN_C0);
   delay_ms(100);
   output_high(PIN_C0); // Turn off LEDs
   output_high(PIN_C1);
   output_high(PIN_C2);

// I/O Initialization
   input(RXF_);
   input(TXE_);
   output_high(RD_);
   output_low(WR);
   delay_ms(100);
   output_low(CS);
   delay_us(5);
   output_high(CS);
// Turn on interrupts  (only one)
   enable_interrupts(INT_TIMER2);
   enable_interrupts(GLOBAL);
   }
```

```
/*********************************************************************
Timer 2 Interrupt
This interrupt service routine does all of the work of controlling transformer
excitation and controlling the LTC1867.
*********************************************************************/
#int_TIMER2        // Tell compiler that this is the Timer 2 ISR
TIMER2_isr()
{
    static int8 ledstatus;
    int8 j, highbyte, lowbyte;
    if(++ledstatus == 0x80) output_low(LED);       // Blink light every 256 calls
    if(ledstatus == 0x00) output_high(LED);

    if(!busbusy) // If main() is using the bus, do nothing.
    {
        for(j=0; j<=7; ++j)
        {
            output_d(LATCHWORD[j]);              // Place excitation data on the bus
            output_high(LATCH);                  // Latch in data
            output_low(LATCH);
            output_low(CS);                      // Enable LTC1867 serial interface
            highbyte = spi_read(LTC1867CONFIG[j]); // Read out high byte.
                                                 // Acquisition begins on 6th falling clock edge
            output_high(EXCITATION);             // Apply transformer excitation
            delay_us(delay);                     // Wait for analog signal to settle
            lowbyte = spi_read(0);               // Finish reading data. Input is also settling
                                                 // During this time.
            output_high(CS);                     // Start conversion!!!
            output_low(EXCITATION);              // Remove excitation. One instruction cycle is plenty
                                                 // of "analog hold time"
            if(!readflag[j]) data[j] = make16(highbyte, lowbyte); // Don't write if main() is reading!!
                                                 // This is a simple anti-collision technique. The worst
                                                 // case latency is a single reading, or 1ms.

        }//end of for loop
    }//end of if(!busbusy)
}// end of ISR
```

an112f

```
/******************************************************************************
battery_monitor.h
defines, global variables, function prototypes
******************************************************************************/


#include <16F877A.h>     // Standard header
#device adc=8
#use delay(clock=20000000) // Clock frequency is 20MHz
#use rs232(baud=9600,parity=N,xmit=PIN_C6,rcv=PIN_C7,bits=9)
#define SPI_CONFIG SPI_MASTER|SPI_L_TO_H|SPI_CLK_DIV_4   // 5MHz SPI clk when
                                                         // master clk = 20MHz


//#fuses NOWDT,RC, NOPUT, NOPROTECT, NODEBUG, BROWNOUT, LVP, NOCPD, NOWRT
// This is less confusing - set up configuration word with #rom statement
//   Bit     13 12   11   10    9  8  7   6  5 4   3    2     1      0
// Function  CP -- DEBUG WRT1 WRT0 CPD LVP BOREN - - PWRTEN# WDTEN FOSC1 FOSC0
//
#rom 0x2007 = {0x3F3A}


/////////////////////////////////////
// Battery Monitor Project Defines //
/////////////////////////////////////

// Global variables
int16 data[8];              // Raw data from the LTC1867
int8 readflag[8];           // Tells ISR that main is reading data, do not write
int1 busbusy = 0;           // Tells ISR that main is talking on the bus
int1 calculate = 1;         // Send calculated voltages to terminal when asserted
int1 filter = 1;            // Enables digital filter when asserted
unsigned int8 period = 40;  // Period between reads
unsigned int8 delay = 2;    // Additional settling time after applying excitation
float voltage[8];           // Holds floating point calculated voltages

// Non-volatile memory base addresses for calibration constants
#define init_offset_base 0
#define init_fs_base 16

// First, define the SDI words to be sent to the LTC1867
// All are Single ended, unipolar, 4.096V range.
#define LTC1867CH0     0x84
#define LTC1867CH1     0xC4
#define LTC1867CH2     0x94
#define LTC1867CH3     0xD4
#define LTC1867CH4     0xA4
#define LTC1867CH5     0xE4
#define LTC1867CH6     0xB4
#define LTC1867CH7     0xF4


// Excitation enable lines. Write this to the '574 register
// before enabling excitation pulse.
#define EXC0     0x01
#define EXC1     0x02
#define EXC2     0x04
#define EXC3     0x08
#define EXC4     0x10
#define EXC5     0x20
#define EXC6     0x40
#define EXC7     0x80
```

```
// Now define two lookup tables such that the excitation signal lines up with
// the selected LTC1867 input.
byte CONST LTC1867CONFIG [8] = {LTC1867CH1, LTC1867CH2, LTC1867CH3, LTC1867CH4,
                                LTC1867CH5, LTC1867CH6, LTC1867CH7, LTC1867CH0};
byte CONST LATCHWORD [8] = {EXC6, EXC5, EXC4, EXC3, EXC2, EXC1, EXC0, EXC7};

//Pin Definitions
#define EXCITATION    PIN_B0    // Enables excitation to the selected channel
#define LATCH         PIN_B1    // 74HC573 latch pin
#define LED           PIN_C1    // Spare blinky light

#define RD_           PIN_A0
#define RXF_          PIN_A1
#define WR_           PIN_A2
#define TXE_          PIN_A3
#define ISO_PWR_SD_   PIN_A4
#define LCD_EN        PIN_A5
#define CS_           PIN_B5

#define AUX_MAIN_     PIN_E1
#define I2C_SPI_      PIN_E2

#byte SSPCON   = 0x14
#byte SSPSTAT  = 0x94
#bit CKP            = SSPCON.4
#bit CKE            = SSPSTAT.6

// Function Prototypes
void parse(void);
void write_offset_cal(void);
void write_fs_cal(void);
unsigned int16 read_offset_cal(int channel);
unsigned int16 read_fs_cal(int channel);
void print_cal_constants(void);
char usb_hit(void);
void initialize(void);
void tx_usb(int8 value);
char rx_usb(void);
```

Section 1

Data Conversion

Some techniques for direct digitization of transducer outputs (20)

Analog-to-digital conversion circuits which directly digitize low level transducer outputs, without DC preamplification, are presented. Covered are circuits that operate with thermocouples, strain gauges, humidity sensors, level transducers and other sensors.

The care and feeding of high performance ADCs: get all the bits you paid for (21)

This application note describes proper techniques for applying high performance ADCs. It describes the problems designers encounter, how to recognize their symptoms and how to avoid them. Topics include ground planes and grounding, supply and reference bypassing, analog input signal conditioning, sampling clock generation, signal jitter and proper handling of the data outputs. A sample board layout is provided as well as performance curves showing the effects of correct and incorrect application.

A standards lab grade 20-bit DAC with 0.1ppm/°C drift (22)

This publication details a true 1ppm D-to-A converter. Total DC error of this processor-corrected DAC remains within 1ppm from 18–32 °C, including reference drift. DAC error exclusive of reference drift is substantially better. Construction details and performance verification techniques are included, along with a complete software listing.

Delta sigma ADC bridge measurement techniques (23)

This chapter features several applications that demonstrate how to take full advantage of Linear Technology's delta sigma ADCs when interfacing to sensors. In many cases, signal conditioning can be greatly simplified or eliminated completely. This note explains where it is appropriate to use amplifiers and how to optimize amplifier gain. Also included are discussions on measuring effective number of bits (ENOB) and the relationship to instrument performance, frequency response of delta sigma ADCs, and test techniques. C source code for all of the applications is included to aid firmware development.

1ppm settling time measurement for a monolithic 18-bit dac (24)

DAC DC specifications are relatively easy to verify. Measurement techniques are well understood, albeit often tedious. AC specifications require more sophisticated approaches to produce reliable information. In particular, the settling time of a DAC and its output amplifier is extraordinarily difficult to determine to 18-bit (4ppm) resolution. DAC settling time is the elapsed time from input code application until the output arrives at, and remains within, a specified error band around the final value. To measure an 18-bit DAC, a settling time measurement technique has been developed with 20-bit (1ppm) resolution for times as short as 265ns. The new method will work with any DAC. Detailed descriptions of the method are given, along with performance characterization. Nine appendices cover related topics, including a technique for verifying the settling time methods specification.

Some techniques for direct digitization of transducer outputs

20

Jim Williams

Almost all transducers produce low level signals. Normally, high accuracy signal conditioning amplifiers are used to boost these outputs to levels which can easily drive cables, additional circuitry, or data converters. This practice raises the signal processing range well above the error floor permitting high resolution over a wide dynamic range.

Some emerging trends in transducer-based systems are causing the use of signal conditioning amplifiers to be reevaluated. While these amplifiers will always be useful, their utilization may not be as universal as it once was. In particular, many industrial transducer-fed systems are employing digital transmission of signals to eliminate noise-induced inaccuracies in long cable runs. Additionally, the increasing digital content of systems, along with pressures on board space and cost, make it desirable to digitize transducer outputs as far forward in the signal chain as possible. These trends point toward direct digitization of transducer outputs—a difficult task.

Classical A/D conversion techniques emphasize high level input ranges. This allows LSB step size to be as large as possible, minimizing offset and noise-caused errors. For this reason, A/D LSB size is almost always above a millivolt, with $100\mu V$ to $200\mu V$ per LSB available in a few 10V full-scale devices. The requirements to directly A/D convert the output of a typical strain gauge transducer are illuminating. The transducer's full-scale output is 30mV, meaning a 10-bit A/D converter must have an LSB increment of only $30\mu V$. Performing a 10-bit conversion on a type K thermocouple monitoring a 0°C to 60°C environment proves even more stringent. The type K thermocouple puts out $41.4\mu V/°C$ over the 0°C to 60°C range. The LSB increment is found by:

$$\frac{60°C \bullet 41.4\mu V/°C}{1024} = 2.42\mu V/LSB$$

These examples furnish extraordinarily small step sizes, far below commercially available A/D units and seemingly impossible to digitize without DC preamplification. In fact, both transducers' outputs may be directly digitized to stable 10-bit resolution using circuitry specifically designed for the function.

This chapter details circuit techniques which directly digitize the low level outputs of a variety of transducers. The approaches described are unique in that they do not utilize any DC gain stage. The transducer outputs receive no DC signal conditioning; A/D conversion is directly performed at low level. The circuits produce a serial data output which may be transmitted over a single wire with the characteristic noise immunity of digital systems. By eliminating the traditional DC gain stage, these circuits furnish a direct, economical way to digitize low level transducer outputs without sacrificing performance.

Figure 20.1 shows a simple way to convert the current output of an LM334 temperature sensor to a corresponding output frequency. The sensor pulls a temperature-dependent current (0.33%/°C) from A1's positive input node. This point, biased from the LM329-driven resistor string, responds with a varying, temperature-dependent voltage. The voltage varies the operating point of A1, configured as a self-resetting integrator. A1 integrates the LM329 referenced current into its summing point, producing a negative-going ramp at its output. When the ramp amplitude becomes large enough, the transistors turn on, resetting the feedback capacitor and forcing A1's output to zero. When the capacitor's reset current goes to zero, the transistors go off and A1 begins to integrate negatively again. The frequency of this oscillation action is dependent on A1's DC operating point, which varies with the LM334's temperature. The circuit's DC biasing values are arranged so that a 0°C to 100°C sensor temperature excursion produces 0kHz to 1kHz at the output. Additionally, only 2V appear across the LM334, minimizing sensor power dissipation related errors. The differentiator-transistor network at A1's output provides a TTL compatible output. To calibrate this circuit, place the LM334 in a 0°C environment and trim the "0°C adjust" for 0Hz.

Analog Circuit and System Design: A Tutorial Guide to Applications and Solutions. DOI: 10.1016/B978-0-12-385185-7.00020-2

Next, put the LM334 in a 100°C environment and set the "100°C adjust" for 1kHz output.

Repeat this procedure until both points are fixed. This circuit has a stable 0.1°C resolution with ±1.0°C accuracy.

Figure 20.2 shows another temperature-to-frequency converter but this circuit uses the popular type K thermocouple as a sensor. The design includes cold junction compensation for the thermocouple over a 0°C to 60°C range. Accuracy is ±1°C and resolution is 0.1°C.

The thermocouple's extremely low output (41.4µV/°C) and the requirement for cold junction compensation make it one of the most difficult transducers to directly digitize. The approach used is based on the 50nV/°C input offset drift performance of the LTC1052 chopper-stabilized amplifier.

In this circuit, A1's positive input is biased by the thermocouple. A1's output drives a crude V→F converter comprised of the 74C04 inverters and associated components. Each V→F output pulse causes a fixed quantity of charge to be dispensed into a 1µF capacitor from the 100pF capacitor via the LTC1043 switch. The larger capacitor integrates the packets of charge, producing a DC voltage at A1's negative input. A1's output forces the V→F converter to run at whatever frequency is required to balance the amplifier's inputs. This feedback action eliminates drift and nonlinearities in the V→F converter as an error term and the output frequency is solely a function of the DC conditions at A1's inputs. The 3300pF capacitor forms a dominant response pole at A1, stabilizing the loop.

A1's low drift eliminates offset errors in the circuit, despite an LSB value of only 4.14µV (0.1°C)!

RT, a thermistor, and the 1.8k, 187Ω, 487Ω and 301k values form a cold junction compensation network which is biased from the LT1004 1.2V reference. In addition to cold junction compensation, the network provides offset-ting, permitting a 0°C sensor temperature to yield 0Hz at the output.

Figure 20.3 details circuit operation. A1's output drives the 33k-0.68µF combination, producing a ramp (Trace A, Figure 20.3) across the capacitor. When the ramp crosses inverter A's threshold, the cascaded inverter chain switches, producing a low output at E (Trace B). This causes the 0.68µF capacitor to discharge through the diode, resetting the capacitor to 0V. The 820pF unit provides positive AC feedback to inverter B's input (Trace C), assuring a clean reset. The frequency of this ramp-and-reset sequence varies with A1's output. Inverter F's output controls the LTC1043 switch. When the inverter output is high, Pins 2 and 6 are connected, allowing the 100pF capacitor to charge to a potential derived from the LT1004 1.2V reference. When the inverter goes low, Pin 2 is connected to Pin 5. During this interval, the 100pF capacitor completely discharges (Trace D) into the 1µF unit. The amount of charge delivered is constant over each cycle (Q = CV), so the voltage the 1µF capacitor charges to is a function of frequency and discharge path resistance. This voltage is summed with the LT1004-derived offset-ting potential at A1's negative input, closing a loop around A1. The −120ppm/°C drift of the 100pF charge-dispensing polystyrene capacitor is compensated by the opposing tempco of the specified resistors used in the 1µF's discharge path. Typical circuit gain is 20ppm/°C, allowing less than 1LSB (0.1°C) output drift over a 0°C to 70°C ambient operating range.

The thermocouple's known characteristics, combined with A1's low offset and the cold junction/offsetting network components specified, eliminate zero trimming. Calibration is accomplished by placing the thermocouple in a 60°C environment and adjusting the 50kΩ potentiometer for a 600Hz output. Beyond 60°C the cold junction

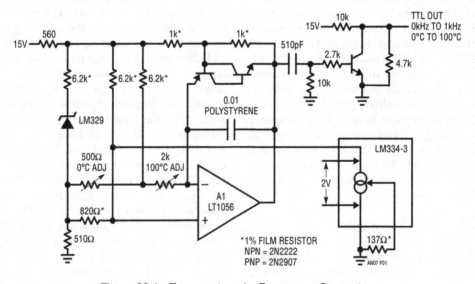

Figure 20.1 • Temperature-to-Frequency Converter

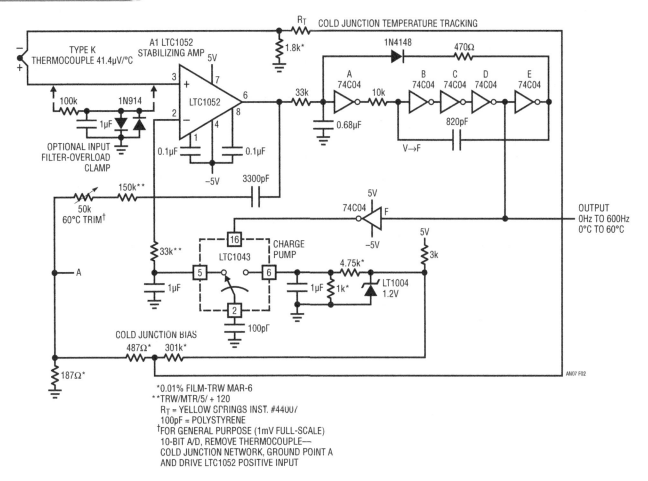

Figure 20.2 ● Thermocouple-to-Frequency Converter

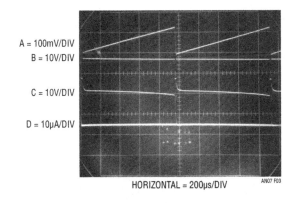

A = 100mV/DIV
B = 10V/DIV
C = 10V/DIV
D = 10µA/DIV

HORIZONTAL = 200µs/DIV   AN07 F03

Figure 20.3 ● Thermocouple Digitizer Waveforms

network departs from the thermocouple's response and output error increases rapidly. Although the digital output will be a function of the thermocouple's temperature over hundreds of degrees, linearization by a monitoring processor is required.

It is worth noting that this circuit can directly convert any low level, single-ended signal. If the offsetting/cold junction network is removed and the 50kΩ potentiometer

returned directly to ground, inputs may be applied to A1's positive terminal. The circuit produces a 10-bit accurate output with a full-scale range of only 1mV (1µV per LSB)! The high impedance of A1's input allows filtering or overload clamping of the input signal without introducing error.

Figure 20.4 is another temperature measuring circuit, but the transducer used is unusual. The circuit measures temperature by utilizing the relationship between the speed of sound and temperature in a medium. In dry air the relationship is governed by:

$$C = 331,5\sqrt{\frac{T}{273}}\,\text{meters/second}$$

where C = speed of sound.

Acoustic thermometry is used where extremes in operating temperature are encountered, such as cryogenics and nuclear reactors. Additionally, acoustic temperature standards have been built by operating the acoustic transducer inside a sealed, known medium.

The inherent time domain operation of acoustic thermometers allows a direct conversion into a digital output. Figure 20.4 shows a circuit that does this. A1, the inductor and their associated components for a simple flyback type

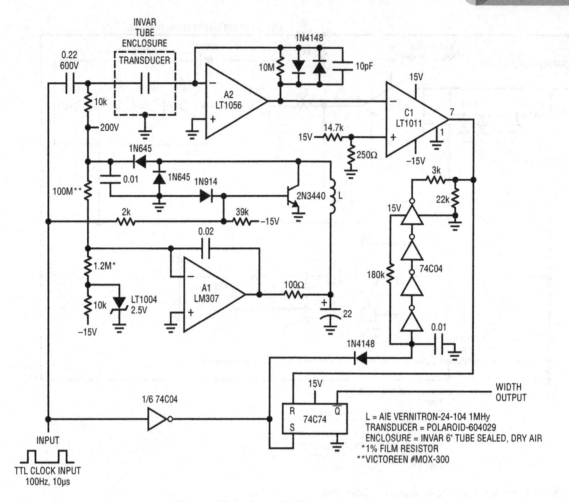

Figure 20.4 • Acoustic Thermocouple

regulated 200V supply which biases the transducer. The transducer is composed of the Polaroid ultrasonic element noted, mounted at one end of a sealed, 6-inch long Invar tube. The Invar material minimizes mechanical tube deformation with temperature. The medium inside the tube is dry air. The transducer may be thought of as a capacitor, composed of an insulating disc with a conductive coating on each side.

Each time the TTL clock (Trace A, Figure 20.5) goes high, the transducer receives AC drive via the 0.22µF capacitor This drive causes mechanical movement of the disc and ultrasonic energy is emitted. The clock input simultaneously sets the 74C74 flip-flop output (Trace E) low and pulls the 0.01µF capacitor to ground. This cuts off drive to C1's 3k output pull-up resistor (Trace C), forcing C1's output (Trace D) to zero. During the clock pulse's period, A2's output (Trace B) is saturated due to excessive signal at its input. When the clock pulse ceases, A2 comes out of bound and amplifies in its linear region. The ultrasonic transducer now acts like a capacitance microphone, with the 200V supply providing bias. Residual disc ringing is picked up and appears at A2's output. This signal cannot trigger C1, however, because the 0.01µF capacitor has not

charged high enough to allow the inverter to chain output to bias C1's output pull-up resistor.

The ultrasonic energy emitted by the transducer travels down the tube, bounces off the far end and heads toward the transducer. Before it returns, the 0.01µF capacitor crosses the inverter's threshold and C1's 3k resistor

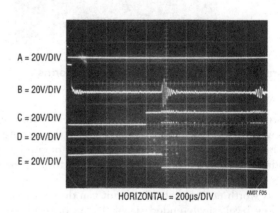

Figure 20.5 • Acoustic Thermocouple Waveforms

(Trace C) receives bias. Upon returning, the sonic energy causes a mechanical displacement of the transducer; forcing a shift in capacitance. This capacitance shift causes charge to be displaced into C2's summing point, and the output responds with an amplified version of this signal (Trace B). C1's output (Trace D) triggers, resetting the flip-flop. The flip-flop's output pulse (Trace E) represents the transit time down the tube and will vary with temperature according to the equation given. A monitoring processor can convert this pulse width into the desired temperature information.

In the photograph another received signal, lower in amplitude, is visible at the extreme right-hand side of Trace B. Its position in time identifies it as a second bounce return from the tube's far end. Also, note the increased detected noise level after the return of the first bounce. This is due to sonic energy dispersion inside the tube. The transducer picks up energy deflected from the tube walls, which is phase shifted from the desired signal. C1 is seen to respond to these unwanted signal sources, but the circuit's final pulse output is unaffected. Additionally, the time window gating supplied to C1's pull-up resistor greatly

reduces the likelihood of false triggering due to noise coming from outside the tube.

Temperature sensors are not the only transducers which can be directly digitized. Strain gauge transducers account for a large class of pressure and force measurements. Typically, a strain gauge bridge-based transducer produces 3mV of full-scale output per volt of bridge drive. Figure 20.6 shows a way to directly digitize a strain gauge bridge's output to 10-bit accuracy. For a 7.5V bridge drive, an LSB increment is 25μV considerably larger than the thermocouple example but still far below conventional A/D converters. The bridge's differential output complicates the required converter input structure, but is accommodated.

A1 and the transistor provide bridge excitation. One signal output of the bridge is connected to A1's negative input. A1's positive input is at ground. A1 drives the transistor to bias the bridge at whatever voltage is required to bring its negative input to ground potential. The diode drops in the bridge's −5V return line allow the transistor to force the bridge's positive end far enough to servo A1's inputs. This arrangement allows the bridge's other output

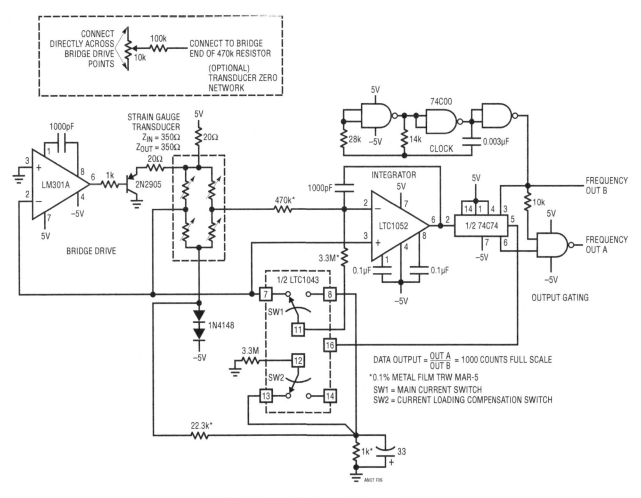

Figure 20.6 • Strain Gauge Digitizer

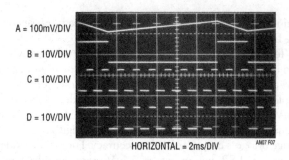

A = 100mV/DIV

B = 10V/DIV

C = 10V/DIV

D = 10V/DIV

HORIZONTAL = 2ms/DIV   AN07 F07

Figure 20.7 • Strain Gauge Digitizer Waveforms

to be sensed in a single-ended, ground-referred fashion. In practice, a slight error exists due to A1's offset voltage. This error is eliminated by referring the A/D converter input to A1's negative input instead of ground.

The A/D converter is made up of A2, a flip-flop and some gates. It is based on a current balancing technique. Once again, the chopper-stabilized LTC7652's 50nV/°C input drift is required to implement the low level input A/D. Figure 20.7 details key A/D waveforms. Assume the flip-flop's Q output (Trace B) is low, connecting LTC1043 Pins 11 and 12 to Pins 7 and 13, respectively. The main current switch passes no current, as the 3.3M resistor is placed across A2's inputs. The current loading compensation switch puts a 3.3M value across the 1k divider resistor, lowering the voltage across it by 0.03%.

Under these conditions the only current into A2's summing point is from the bridge via the 470k resistor. This positive current forces A2's output (Trace A, Figure 20.7) to integrate in a negative direction. The negative ramp continues and finally passes the 74C74 flip-flop's switching threshold. At the next clock pulse (clock is Trace C), the flip-flop changes state (Trace B), causing the LTC1043 switch positions to reverse. Pin 12 connects to Pin 14 and Pin 11 to Pin 8. In this case, the 3.3M resistor, controlled by the current loading compensation switch, is disconnected from the 1k unit, but the 3.3M value, controlled by the main current switch, replaces it. The 0.03% loading of the 3.3M resistor, combined with this switching scheme, eliminate any sag or loading effects across the 1k resistor during switching. The result is a quickly rising, precise current flow out of A2's summing point.

This current, scaled to be greater than the bridge's maximum output, forces A2's output movement to reverse and integrate in the positive direction. At the first clock pulse after A2's output has crossed the flip-flop's triggering threshold, switching occurs and the entire cycle repeats. Because the reference current is fixed, the flip-flop's duty cycle is solely a function of the bridge signal current into A2's summing point. Additionally, the reference current is supplied from the 22.3k-1k divider, which is derived from the bridge drive. Thus, the A/D's reference current varies ratiometrically with the bridge output, eliminating bridge drive variations as an error source. The flip-flop's output gates the clock, producing the "frequency output A"

waveform (Trace D). The 10k resistor combines with the output gate's input capacitance to slightly delay the clock signal, eliminating spurious output pulses due to flip-flop delay. The circuit's data output, the ratio of output A to the clock frequency, may be extracted with counters. Because the output is expressed as a ratio, clock frequency stability is unimportant.

Several subtle factors are critical in setting up and using this circuit. The 470k input resistor at A2 has been selected to produce less than 1LSB loading error on the strain gauge bridge. The bridge receives only about 7.5V of drive due to the deliberate resistor and diode drops in its supply lines. At 3mV output per volt of bridge drive, full-scale signal is 22.5mV This produces a signal current of only:

$$I = \frac{0.0225V}{470k} = 48nA$$

To maintain 10-bit accuracy, leakage and amplifier bias current into A2's summing point must be less than 0.1% of this figure or:

$$I = \frac{48nV}{1000} = 48pA$$

Although A2's bias current is much lower than this, board leakage can cause trouble. At a minimum, careful layout and a clean PC board are required. The best practice is to use a Teflon stand-off for all summing point connections. The 470k and 3.3M resistors associated with A2's negative input should be placed as close as possible to the IC pin. Note also that the 3.3M current summing resistor is switched to A2's positive input when it is not sourcing current to the summing point. This seemingly unnecessary connection prevents minute stray 60Hz and noise currents from being coupled to A2's summing point when the current reference is off. Failure to utilize this connection will cause jitter in the LSB. Gain trimming of this circuit may be accomplished by varying the 22.3k value. If the particular strain gauge transducer used requires zero trimming, use the optional network shown. Over a 0°C to 70°C range the circuit will typically maintain its 10-bit output within 1LSB accuracy. The tracking errors of the starred resistors are the primary contributors to this small error.

Because of their extremely wide dynamic range, photo diodes present a difficult challenge for signal conditioning circuitry. A high quality device furnishes a linear current output over a 100dB range, requiring a 17-bit A/D converter as well as a current-to-voltage input amplifier. A common approach employs a logarithmically responding current-to-voltage input amplifier to nonlinearly compress the photodiode's output, allowing a much lower resolution A/D converter to be used. Although this scheme saves the cost of the 17-bit A/D, it has the inconvenience of a nonlinear output. Also, logarithmic amplifiers respond relatively slowly, which may be detrimental in some photometric measurements. Figure 20.8's circuit directly converts a photodiode's current output into an output

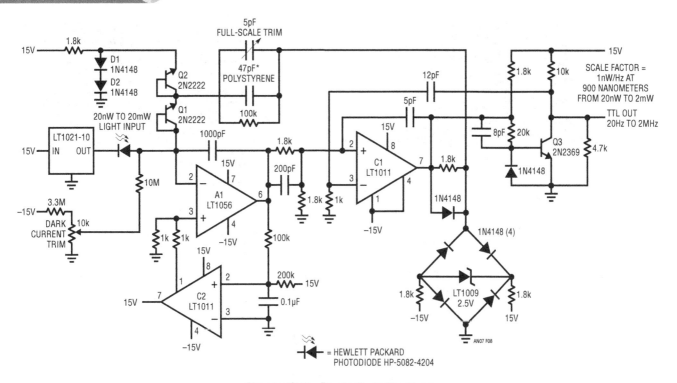

Figure 20. 8 • Photodiode Digitizer

frequency with 100dB of dynamic range. Optical input power of 20nW to 2mW produces a linear, calibrated 20Hz-to-2MHz output. Output response to input light steps is fast and cost is low.

The photodiode's output current feeds a highly modified, high frequency version of a Pease type charge pump I→F converter. Diode output current biases A1's negative input, causing its output (Trace A, Figure 20.9) to ramp in a negative direction. When A1's output crosses zero, C1's output (Trace B) goes low, causing the LT1009 diode bridge to bound at −3.7V The 200pF-1.8k lead network at C1's positive input aids comparator high frequency response. C1's output going low also provides AC positive feedback to its positive input (Trace D). Additional AC positive feedback is supplied by output transistor Q3's collector

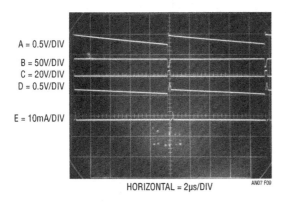

A = 0.5V/DIV
B = 50V/DIV
C = 20V/DIV
D = 0.5V/DIV

E = 10mA/DIV

HORIZONTAL = 2μs/DIV   AN07 F09

Figure 20.9 • Photodiode Digitizer Waveforms

(Trace C). During this interval, charge is pulled from A1's summing point via the 47pF-5pF capacitors (Trace E). This causes A1's output to move quickly positive, switching C1 after the positive feedback around it has decayed. The LT1009 diode bridge now bounds at 3.7V. The 47pF-5pF pair receives charge, A2's summing junction recovers and the entire cycle repeats at a frequency linearly related to photodiode output current. D1 and D2 compensate the bridge diodes. Diode connected Q1 compensates steering diode Q2. The diode connected transistors provide lower leakage than simple diodes. C2 provides circuit latch-up protection, necessary because of the circuit's AC-coupled feedback loop. If latch-up occurs, A1's output saturates low, causing C2's emitter-follower connected output to go high. This forces A1's output positive, initiating normal circuit action.

The LT1021-10 reference biases the photodiode, providing optimum optical current response characteristics. To trim this circuit, place the photodiode in a *completely* dark environment. Trim the "dark current" adjustment so the circuit oscillates at the lowest possible frequency, typically 1Hz to 2Hz. Next, apply or electrically simulate (see manufacturer's data sheet for light input versus output current data) a 2mW optical input. Trim the 5pF adjustment for an output frequency of 2MHz. If the adjustment is outside the range of the trimmer, alter the 47pF capacitor's value appropriately. Once calibrated, this circuit will maintain 1% accuracy over the photodiodes's entire 100dB range. The accuracy obtained is limited by photodiode characteristics and not the circuit. Figure 20.10 shows

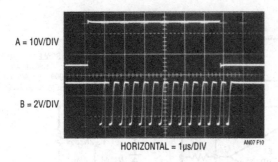

A = 10V/DIV

B = 2V/DIV

HORIZONTAL = 1µs/DIV   AN07 F10

Figure 20.10 • Step Response of Photodiode Digitizer

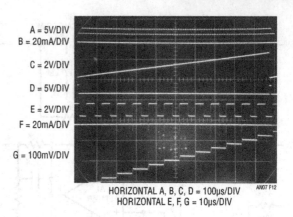

A = 5V/DIV
B = 20mA/DIV

C = 2V/DIV

D = 5V/DIV

E = 2V/DIV
F = 20mA/DIV

G = 100mV/DIV

HORIZONTAL A, B, C, D = 100µs/DIV   AN07 F12
HORIZONTAL E, F, G = 10µs/DIV

Figure 20.12 • Humidity-to-Frequency Converter Waveforms

dynamic response of the circuit to a fast light pulse (Trace A, Figure 20.10). The frequency output settles within 1µs on both edges.

One of the most difficult physical parameters to transduce is relative humidity. A recently introduced humidity transducer, based on a capacitance shift versus relative humidity (RH), offers good accuracy, fast response, wide range and linear response. The transducer features a nominal 1.7pF per percent RH capacitance shift with a 500pF value at RH = 76%. It does not require temperature compensation. A significant consideration in signal conditioning this transducer is that the average voltage across the device must be zero. No net DC may pass through the transducer. Figure 20.11's circuit converts the RH transducer's capacitive shifts directly into a calibrated frequency output. The LTC1043 switched-capacitor instrumentation building block IC free runs at 150kHz. Pin 2 (Figure 20.12, Trace A) is alternately connected between the LT1004 negative reference and A1's summing junction. The 1µF-22MΩ

combination associated with the RH transducer ensures the device's required pure AC biasing.

When Pin 2 is connected to Pin 6, the transducer receives a negative charge. When the LTC1043's internal clock switches, Pin 2 is tied to Pin 5, depositing all of the transducer's charge into A1's summing point. A1's input (Trace B), just faintly visible, shows transducer current, while Trace C is A1's output. A1, an integrator, ramps up in stepped fashion as successive discrete packets of charge are deposited into its summing point. Concurrent with this action, a second set of LTC1043 switches (Pins 7, 8, 11,12, 13 and 14) works to synchronously transfer a fixed amount of charge of opposing polarity into A1's summing junction. The amount of fixed charge is set to cancel the sensor offset (e.g., 0% RH does not extrapolate to 0pF sensor capacitance). Thus, the slope of the stepped ramp at A1's output

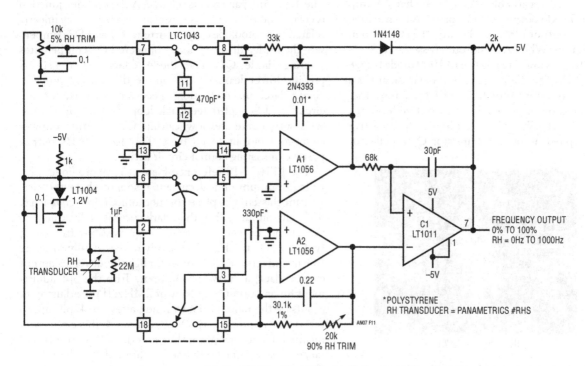

Figure 20.11 • Humidity-to-Frequency Converter

is a function of the sensor's value minus its offset term. A1 continues to ramp positive until it equals the voltage at C1's negative input. This triggers C1's output high (Trace D). AC positive feedback holds C1's output high long enough for the 2N4393 FET to completely discharge A1's feedback capacitor. A1's output drops to zero and the entire cycle repeats. The frequency of repetition is a function of the RH transducer's capacitance. C1's input voltage is derived from the LT1004 reference. LTC1044 Pins 3, 18 and 15 and the 330pF value form a simple charge pump which biases A2's summing point. A2's output assumes whatever value is required to maintain its summing point at zero. The 0.22μF capacitor integrates A2's response to DC, while the feedback resistors establish the operating point. Because A2's output voltage determines ramp height, its feedback resistor's value sets the circuit's gain slope. Traces E, F and G, time and amplitude expansions of Traces A, B and C, permit a detailed look at the effects of the transducer's charge dumping on A1's output ramp.

Circuit temperature dependence is low because the 330pF and 0.01μF polystyrene capacitors' (both gain terms) −120ppm drifts ratiometrically cancel. Further ratiometric error cancellation occurs because the transducer's charge source and A2's output voltage are both derived from the LT1004 reference. The sole uncompensated term in the circuit is the 470pF capacitor which supplies the offsetting charge. Its −120ppm/°C drift is well below the transducer's 2% accuracy specification, and circuit temperature independence is assured.

To calibrate this circuit, place the transducer in a 5% RH environment and adjust the 5% trim for 50Hz output. Next place the transducer in a 90% RH environment and adjust the 90% trim for a 900Hz output. Repeat this procedure until both points are fixed. Relative humidity accuracy will be 2% over the 5% to 90% RH range. If RH standards are not available, the circuit may be approximately calibrated against using fixed capacitors in place of the sensor. Ideal values are 5% RH = 379.3pF and 90% = 523.8pF Note that these values assume an ideal sensor. An actual device may depart from them by as much as 10%.

Another frequently required physical parameter is level. Level transducers which measure angle from ideal level are employed in road construction, machine tools, inertial navigation systems and other applications requiring a gravity reference. One of the most elegantly simple level transducers is a small tube nearly filled with a partially conductive liquid. Figure 20.13 shows such a device. If the tube is level with respect to gravity, the bubble resides in the tube's center and the electrode resistances to common are identical. As the tube shifts away from level, the resistances increase and decrease proportionally. By controlling the tube's shape at manufacture, it is possible to obtain a linear output signal when the transducer is incorporated into a bridge circuit.

Transducers of this type must be excited with an AC waveform to avoid damage to the partially conductive liquid inside the tube. Signal conditioning involves generating this excitation as well as extracting angle information and polarity determination (e.g., which side of level the tube is on). Figure 20.14 shows a circuit which does this, directly producing a calibrated frequency output corresponding to level. A sign bit, also supplied at the output, gives polarity information.

The level transducer is configured with a pair of 2k resistors to form a bridge. The required AC bridge excitation is developed at C1A, which is configured as a multivibrator. C1 biases Q1, which switches the LT1009's 2.5V potential through the 100μF capacitor to provide the AC bridge drive. The bridge differential output AC signal is converted to a current by A1, operating as a Howland current pump. This current, whose polarity reverses as bridge drive polarity switches, is rectified by the diode bridge. Thus, the 0.03μF capacitor receives unipolar charge. A2, running at a differential gain of 2, senses the voltage across the capacitor and presents its single-ended output to C1B. When the voltage across the 0.03μF capacitor becomes high enough, C1B's output goes high, turning on the paralleled sections of the LTC1043 switch. This discharges the capacitor. The 47pF capacitor provides enough AC feedback around C1B to allow a complete zero reset for the capacitor. When the AC feedback ceases, C1B's output goes low and the LTC1043 switch goes off. The 0.03μF unit again receives constant-current charging and the entire cycle repeats. The frequency of this oscillation is determined by the magnitude of the constant current delivered to the bridge-capacitor configuration. This current's magnitude is determined by the transducer bridge's offset, which is level related.

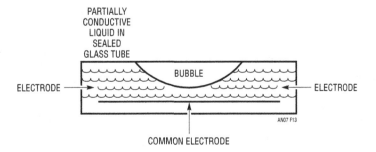

Figure 20.13 ● Bubble-Based Level Transducer

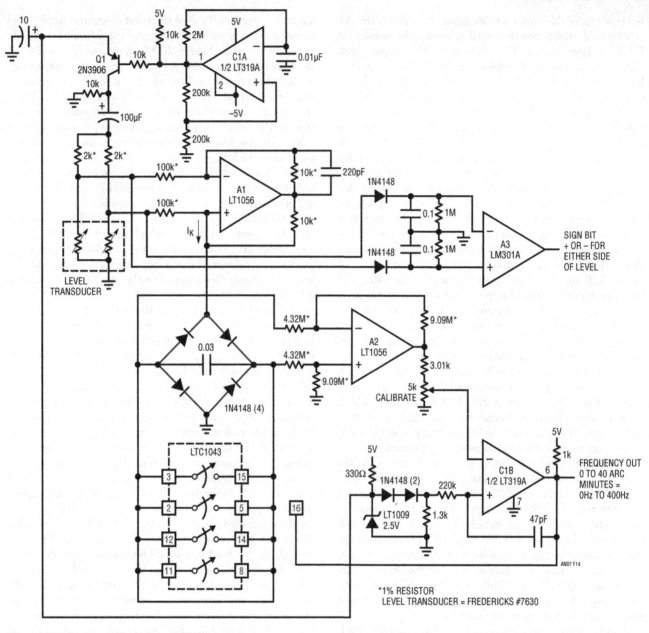

Figure 20.14 • Level Transducer Digitizer

Figure 20.15 shows circuit waveforms. Trace A is the AC bridge drive, while Trace B is A1's output. Observe that when the bridge drive changes polarity, A1's output flips sign rapidly to maintain a constant current into the bridge-capacitor configuration. A2's output (Trace C) is a unipolar ground-referred ramp. Trace D is C1B's output pulse and the circuit's output. The diodes at C1B's positive input provide temperature compensation for the sensor's positive tempco, allowing C1B's trip voltage to ratiometrically track bridge output over temperature.

A3, operating open loop, determines polarity by comparing the rectified and filtered bridge output signals with respect to ground.

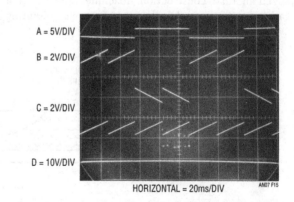

A = 5V/DIV

B = 2V/DIV

C = 2V/DIV

D = 10V/DIV

HORIZONTAL = 20ms/DIV

Figure 20.15 • Level Transducer Digitizer Waveforms

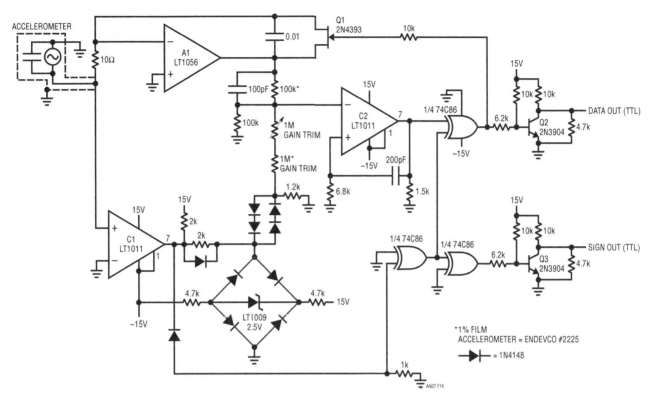

Figure 20.16 • Accelerometer Digitizer

To calibrate this circuit, place the level transducer at a known 40 arc-minute angle and adjust the 5k trimmer at C1B for a 400Hz output. Circuit accuracy is limited by the transducer to about 2.5%.

The final example concerns direct digitization of a piezoelectric accelerometer. These transducers rely on the property of ceramic materials to produce charge when mechanically excited. In this device a mass is coupled to the ceramic element. An acceleration acting on the mass causes charge to be dispensed from the ceramic element. Sensitivity and frequency response are related to the characteristics of the ceramic used and the mechanical design of the transducer. The best way to signal condition a piezoelectric output is to unload it directly into the virtual ground of an op amp's summing point. This method provides no voltage difference between the center conductor and the shield of the coaxial cable connecting the accelerometer and the single conditioning amplifier This eliminates cable capacitance as a parasitic term, an important consideration in any charge output transducer Because the accelerometer produces AC outputs, a direct digitization of its output must produce a sign bit as well as amplitude data.

Figure 20.16's circuit accomplishes a complete, direct A/D conversion on the piezoelectric accelerometer noted and is generally applicable to other devices in this class. To understand the circuit it is convenient to replace the accelerometer with a square wave source coming through

a resistor. When the square wave is positive, the A1 integrator responds with a negative-going ramp output (Trace A, Figure 20.17). C1, detecting the square wave polarity, goes high and the LT1009 diode bridge (Trace B) limits at 3.7V. A1's ramp output is summed with the bridge's output at C2's negative input. The series diodes temperature-compensate the bridge diodes. When A1's output goes far enough negative, C2's (Trace C) output goes high. The output gating is arranged so that with C1's output low and C2 high, Q1's gate (Trace D) receives turn-on bias. Q1 comes on, discharging A1's feedback capacitor and resetting A1's output to zero. Local AC positive feedback at C2 ensures adequate time for a complete zero reset of A1's feedback capacitor. The 100pF capacitor at C2's

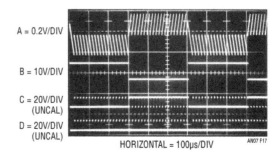

A = 0.2V/DIV

B = 10V/DIV

C = 20V/DIV (UNCAL)

D = 20V/DIV (UNCAL)

HORIZONTAL = 100μs/DIV

Figure 20.17 • Accelerometer Digitizer Waveforms with Square Wave Test Drive

input aids high frequency response. When the AC feedback decays away, Q1 goes off, A1 begins to ramp negative again and the cycle repeats as long as the input square wave is positive. The frequency of oscillation is directly proportional to the current into A1's summing point. When the input square wave goes negative, A1 abruptly begins to ramp in the positive direction. Simultaneously, the C1 input polarity detector output goes negative, forcing the LT1009 bridge output negative. C2's output now switches when A2's output exceeds a positive limit. The output gating, directed by C1's polarity signal, inverts C2's output to supply proper drive to Q1's gate. Q1 turns on and resetting occurs. Thus, the loop maintains oscillation, but with all signs reversed. The Q2 and Q3 level shifters supply TTL data outputs for data and sign.

This circuit constitutes an I→F converter which responds to AC inputs. If the square wave source is replaced with a piezoelectric accelerometer, direct digitization results. Figure 20.18 shows circuit response when an acceleration (Trace A), in this case a damped sinusoid, is applied to the transducer. The sign bit (Trace B) keeps track of acceleration

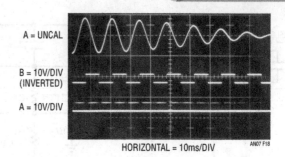

Figure 20.18 • Accelerometer Digitizer Response

polarity, while the frequency output supplies amplitude data. Observe the drop in output frequency as the input waveform damps. A monitoring processor, sampling the sign and frequency waveforms faster than twice the highest acceleration frequency of interest, can extract desired acceleration waveform data. To trim the circuit, apply a known amplitude acceleration and adjust the 1MΩ gain trim at C2. Alternately, the accelerometer may be electrically simulated (see manufacturer's data sheet for scale factors).

The care and feeding of high performance ADCs: get all the bits you paid for

21

William C. Rempfer

Introduction

A new generation of ADCs currently appearing on the scene brings higher performance and lower cost to new markets. Figure 21.1 shows an example of how high speed 12-bit converters are becoming affordable for the first time to a new range of applications. At the same time, the new converters achieve better dynamic performance with high frequency input signals. This means that more system designers are facing the challenge of using high

performance ADCs. In this chapter, we will talk about some of the problems designers encounter, how to recognize their symptoms and how to avoid them. We will focus on the particular case of the LTC1410, a 1.25Msps, 12-bit ADC. The same considerations become important in higher resolution ADCs at lower speeds. Conversely, lower resolution ADCs will need this same attention at higher speeds.

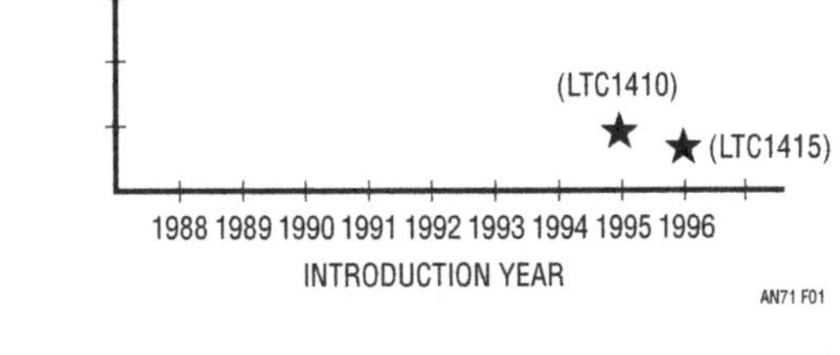

An ADC has many "inputs"

Providing a clean analog input signal to an ADC doesn't always guarantee a clean digital output signal. This is because an ADC has not just one input, but many. Ground pins, supply pins and reference pins also act as "inputs" and must be given special care to prevent noise and unwanted signals from corrupting the ADC output. Grounding, bypassing of the supplies and the reference and driving the analog and clock inputs are the major weapons in this battle against corruption.

Ground planes and grounding

Designing a high speed ADC system without using a proper ground is like trying to play basketball on a huge trampoline. No matter how well you mount the baskets to the court, the whole court will bounce and wobble as the players jump and try to shoot. To play the game, you must have a solid floor. Similarly, to give a solid ground for your data converter circuit, you must use an analog

Figure 21.1 • High Performance 1.25Msps, 12-Bit ADCs Are Becoming Affordable to a New Range of Applications. More System Designers Will Need to Know How to Use Them Effectively

Analog Circuit and System Design: A Tutorial Guide to Applications and Solutions. DOI: 10.1016/B978-0-12-385185-7.00021-4

ground plane. This will put your circuit on a solid foundation.

Figure 21.2 shows grounding techniques for the LTC1410, a 1.25Msps, 12-bit ADC. This provides an example that can be modified for the particular high performance converter used. All bypass caps, reference caps and ground connections for the ADC should be tied to the analog ground plane. Tie them as close together as possible to reduce the sensitivity to currents that may flow in the ground plane. The input signal circuitry, filter caps and op amp bypass caps (not shown) should also be grounded to the ground plane near the ADC.

Noise from digital components in the system must be kept out of the analog ground. To do this, boards should be designed with separate analog and digital ground planes, as shown in Figure 21.2. (The figure shows a 2-layer board layout. If more layers are available, separate layers may be used for analog and digital ground planes.) All noisy digital logic devices must be on the digital ground plane. All the grounds and bypass caps of the ADC (even the digital ones) should tie to the analog ground plane. Tie the two ground planes together at only one point to keep digital currents from taking shortcuts through the analog ground. In single ADC systems this connection can be made at the ADC's

digital output driver ground pin (or the digital ground pin). In systems with more than one ADC, this cannot be done without creating more than one tie point between the ground planes. In this case, a different connection point can be used (for example, at the power supply). In any case, be sure to use only a single connection point.

Supply bypassing

The high conversion rates of high performance converters require proper bypassing on the supply pins. The key to good bypassing is low lead inductance between the ADC and the bypass capacitors. The goal is to force AC currents to flow in the shortest possible loop from the supply pin through the bypass cap and back through ground to the ground pin.

In Figure 21.2, the first components placed around the ADC are the bypass caps, which are located as close as possible to the supply pins. The capacitors must have low inductance and low equivalent series resistance (ESR). Tantalum 10μF surface mount devices are good if they are used in conjunction with 0.1μF ceramics. Even better are the new surface mount ceramic capacitors

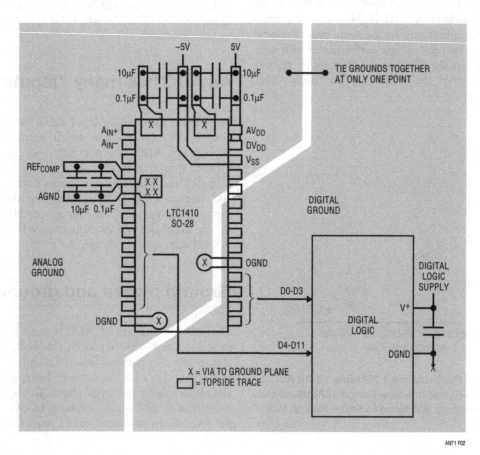

Figure 21.2 • High Performance ADC Layout Must Have Separate Analog and Digital Ground Planes, Bypass Caps with Short Connections and Digital Outputs Routed Away from the Inputs

(e.g., Murata, 1210 sized, 10μF/16V units), which can be used alone. They come in values of 10μF or more and have ESR values as low as 20m.

Figure 21.3a shows the differential nonlinearity (DNL) of the LTC1410 with good supply bypassing. Figure 21.3b shows the effects of 2 inches of lead length (corresponding to roughly 60nH of inductance) in series with the supply bypass caps. This is an exaggerated case of poor bypassing layout, which causes the DNL to degrade beyond 1LSB, reducing the accuracy to 11 bits. For best performance, use supply bypass leads of less than one-half inch. A little care pays off with excellent performance (Figure 21.3a).

Reference bypassing

The ADC's analog reference input provides the scale factor for the conversion. For a clean data output the reference must be noise free. Dynamic currents pulled from the reference by the ADC as it converts perturb the reference unless it is properly bypassed. Surface mount tantalum or ceramic capacitors provide good results. They should be located near the reference pin and should be grounded very near the ADC analog ground pin, as shown in Figure 21.2.

Figure 21.3c shows the easily recognizable signature of a reference bypassing problem—a bow-tie shape to the error curve. This occurs because reference perturbations feed in with full strength for inputs near plus or minus full scale but have less effect for inputs near zero scale. This degradation in DNL results from several inches of lead length in series with the reference bypass caps. Once again, this is an exaggerated case to make the consequences of poor bypassing more visible. To maintain high accuracy, keep the lead lengths less than half an inch (Figure 21.3a).

Driving the analog input

Switched capacitor inputs

The inputs to switched capacitor ADCs are easy to drive if you allow for the fact that they draw a small input-current transient at the end of each conversion. This happens when the internal sampling capacitors switch back onto the input to acquire the next sample. For accurate results, the circuitry driving the analog input must settle from this transient before the next conversion is started.

There are two ways to accomplish this. One is to drive the ADC with an op amp that settles from a load transient in less than the acquisition time of the ADC. Fortunately, most op amps settle much more quickly from a load transient than from an input step, so meeting this requirement

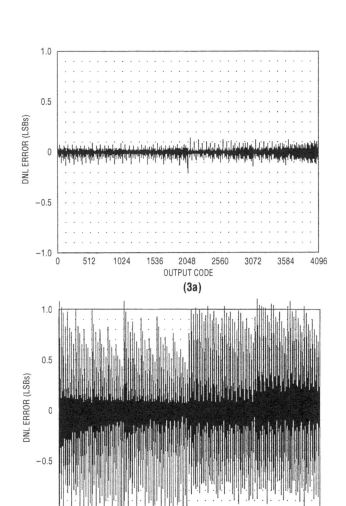

(3a)

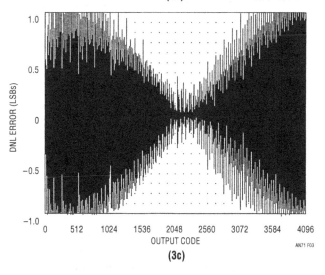

(3b)

(3c)

Figure 21.3 • Poor Layout Will Degrade the Differential Nonlinearity (DNL) of Fast ADCs: (a) a Clean LTC1410 Layout with Bypass Cap Wires of Less than 0.5 Inch; (b) 2-Inch Wires to Supply-Bypass Caps; (c) a Wire of More than 2 Inches to the Reference-Bypass Cap

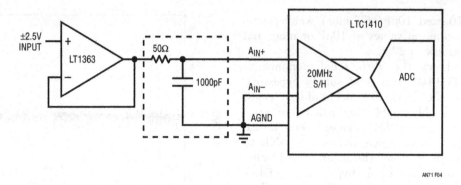

AN71 F04

Figure 21.4 • Many New ADCs Have Wide Bandwidth Sample-and-Holds. In Lower Bandwidth Applications, a Simple RC Filter Will Remove Wideband Noise that May Be Present in the Input Signal

is not too difficult. The LT1363, for example, is a good choice for driving the LTC1410 input.

A second solution to handling the input transient is to use an input RC filter with a capacitor much larger than the ADC input capacitance. This larger capacitor provides the charge for the sampling capacitor, which eliminates the voltage transient altogether. Figure 21.4 shows such a filter for the LTC1410. The 1000pF capacitor provides the input charge for the ADC's sampling capacitor. The LT1363's capacitive load driving capability makes it a good choice for use with this filter.

Filtering wideband noise from the input signal

Many new converters have wide S/H input bandwidths. This is great for capturing high frequency input signals, but for lower input bandwidth applications the converter will pick up any wideband noise that may be in the input signal. To avoid this, use a filter at the ADC input to pass only your desired signal bandwidth.

The simple filter in Figure 21.4 bandlimits the input signal to 3MHz and still allows clean sampling up to the Nyquist frequency (625kHz). Figure 21.7a shows the Nyquist performance of the LTC1410 using this filter. The signal to noise and distortion ratio (SINAD) is 71.5dB and total harmonic distortion (THD) is −84dB.

Choosing an op amp

To drive high performance ADCs, you will need a high performance op amp.

The noise and distortion of good ADCs are now so low that they no longer mask the performance of the op amp. This adds another tradeoff to op amp selection.

High speed, current feedback op amps have lower DC precision and don't settle as well to high accuracy (for example, 0.01%) as the voltage feedback types. However, they have the best distortion and drive for high speed AC

frequency domain applications. Figure 21.5a shows the FFT result of an LT1227 current feedback amp driving a 172kHz signal into the LTC1410. The distortion (THD) of −82dB is about 3dB worse than the −85dB of the ADC alone.

High speed voltage feedback amplifiers have better precision and settling. They work well in frequency domain applications but are best suited for high speed, time domain or multiplexed applications where their DC precision and settling are required. Figure 21.5b shows the voltage feedback LT1363's 2dB further degradation in distortion (to −80dB) under the same conditions.

Slower op amps like the OP-27/OP-37 are excellent in noise and precision but are simply not fast enough for high frequency applications. They distort as they are pushed beyond their slewing capabilities (as shown in the FFT plot of Figure 21.5c).

Driving the convert-start input

An improperly driven conversion-start input can create conversion errors in a couple of ways. First, if an ADC has internal timing, the returning edge of the convert signal (the opposite edge from the one that starts the conversion) can couple noise into the converter if it occurs during the conversion time. To avoid this, use a narrow pulse for convert-start instead of a square wave. This ensures that it either returns quickly (after the sample is taken but before the conversion gets underway), or returns after the conversion is over. (This does not apply to those ADCs that draw all their timing from a clock input and require precise 50% duty-cycle clock inputs.)

A convert-start signal that overshoots or rings can also degrade performance. If it overshoots beyond the supply rails it can turn on the ADC's input protection diodes and couple noise into the converter. If it rings, it may still be bouncing around as the ADC's sample-and-hold captures the input signal, which can affect the conversion result. Normally, overshoot and ringing are not a problem with CMOS logic on a well designed board but they are still things to watch out for.

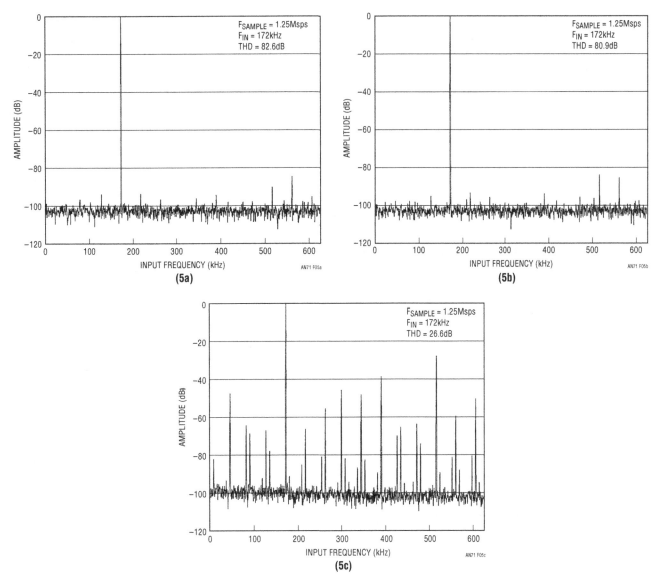

Figure 21.5 • Op Amp Selection is Important When an ADC Has Low Distortion Levels. (a) Current Feedback Op Amps Such as the LT1227 (Seen Here Driving the LTC1410) Provide the Lowest THD in the FFT Output; (b) Fast Voltage Feedback Op Amps Do Nearly as Well in THD as Current Feedback Amps and Offer Better Precision; (c) Slower Op Amps Pushed Beyond Their Slew Limits Will Severely Distort Fast Signals

Effects of jitter

High frequency or high slew rate input signals impose another requirement on the ADC: low aperture jitter. Aperture jitter is the variation in the ADC's aperture delay from conversion to conversion and results in an uncertainty in the time when the input sample is taken. Figure 21.6 shows how this jitter causes an equivalent input noise by working against the slew rate of the analog input signal. The faster the input signal slew rate, the worse the noise for a given jitter. The best possible SINAD for an ADC is limited by the jitter according to the formula:

$$SINAD(dB) = 20\log[1/(2 \bullet t_{JITTER(RMS)} \bullet f_{INPUT})]$$

where:

$t_{JITTER(RMS)}$ = the RMS jitter in seconds
f_{INPUT} = the analog input frequency in Hz

The LTC1410's 5ps (RMS) aperture jitter allows clean sampling of inputs far beyond the Nyquist frequency.

However, to achieve this performance, the convert-start input signal applied to the ADC must also have low jitter. Figure 21.7a shows the ADC, driven from a low jitter source, capturing a 600kHz input with 71.5dB SINAD.

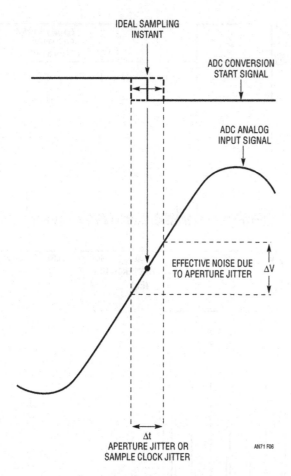

Figure 21.6 • Aperture Jitter in a Sampling ADC or Jitter in the Conversion-Start Signal Applied to the ADC Can Degrade Its Noise Performance. The Time Jitter Works Against the Slope of the Analog Input Signal to Generate an Effective Noise Voltage that Appears in the ADC's Output Spectrum

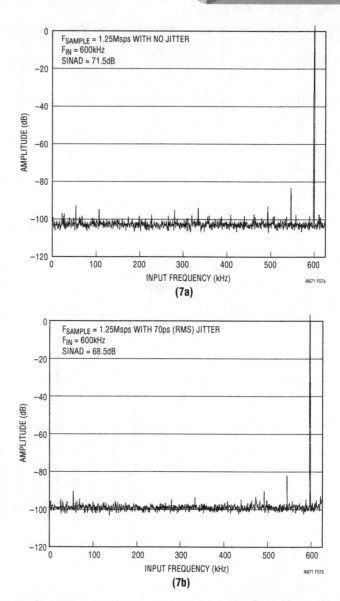

Figure 21.7 • Jitter in the Conversion Start Signal Creates Noise: (a) With a Low Jitter Source, the LTC1410 Will Give 71.5dB SINAD when Sampling a Nyquist Input Signal; (b) Adding 70ps of Jitter to the Convert-Start Signal Will Raise the Noise Floor by 3dB to 68.5dB

As Figure 21.7b shows, adding 70ps of jitter to the conversion-start input signal will raise the noise floor, and reduce the SINAD, by 3dB.

If generating a lower jitter signal is a problem, one trick is to start with a higher frequency clock, which will usually have lower jitter, and then divide the frequency down with fast logic (which retains the lower jitter) to get the desired sample clock frequency.

Routing the data outputs

One of the worst potential sources of digital noise and coupling in an ADC is its output data bus. Fortunately, the user can control this with proper board layout. First, to prevent the data outputs of the ADC from capacitively coupling to the analog input circuitry, they should be routed in the opposite direction. This will tend to occur naturally if separate digital and analog ground plane layouts are used, as in Figure 21.2. Second, the digital output drivers in the ADC switch quickly and will create large current transients if they are loaded with too much capacitance. Locating the receiving buffers or latches close to the ADC will minimize this loading.

Although reduced, some capacitive currents still flow, and it is important to control their return path to the driver of the ADC. Starting from the output drivers of the ADC, the current goes through the output lines, charges the input capacitance of the receiving latches or buffers, and returns through the digital ground plane to the ADC's output driver. For a falling edge, this current returns into the output driver ground pin. For a rising edge, it returns to the ground point of the output driver's supply-bypass cap. Tying the digital and analog grounds together at the ADC output driver ground pin (as in Figure 21.2) helps prevent this current from flowing across the analog ground plane. If the grounds must be tied at the power supply instead of at the ADC, the return currents will flow through the analog ground plane. In this case, it is especially important to

minimize these currents by minimizing the capacitive loading on the digital outputs.

Conclusion

With attention to these principles of layout and design, the new generation of high speed ADCs will provide excellent results. The new devices provide cleaner signal capture than older devices at affordable prices. It makes sense to get all the performance you can.

High speed A/D converters — world's best power/speed ratio

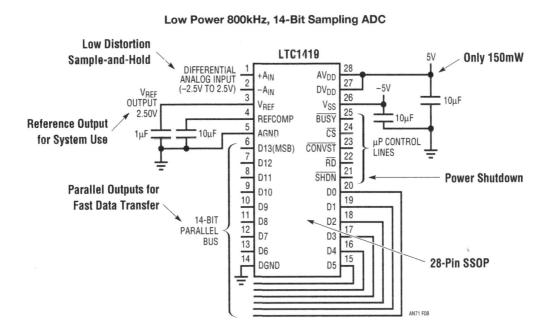

Low Power 800kHz, 14-Bit Sampling ADC

Family features

- Up to 3Msps (LTC1412)
- 12-Bit to 16-Bit Resolution
- ±5V or 5V Supply Operation
- High Bandwidth Sample-and-Holds

- Excellent SINAD at Nyquist
- Surface Mount Packages (SSOP)
- Nap and Sleep Modes
- Internal References

PART NUMBER	RESOLUTION	SPEED	COMMENTS	DATABOOK
16-Bit				
LTC1604	16	333ksps	±2.5V Input Range, ±5V Supply	New
LTC1605	16	100ksps	±10V Input Range, Single 5V Supply	New
14-Bit				
LTC1414	14	2.2Msps	150mW, 81.5dB SINAD and 95dB SFDR	New
LTC1419	14	800ksps	150mW, 81.5dB SINAD and 95dB SFDR	New
LTC1416	14	400ksps	75mW, Low Power with Excellent AC Specs	New
LTC1418	14	200ksps	15mW, Single 5V, Serial/Parallel I/O	New
12-Bit				
LTC1412	12	3Msps	150mW, 71.5dB SINAD and 84dB THD	New
LTC1410	12	1.25Msps	150mW, 71.5dB SINAD and 84dB THD	**96** 6-58
LTC1415	12	1.25Msps	55mW, Single 5V Supply	**96** 6-73
LTC1409	12	800ksps	80mW, 71.5dB SINAD and 84dB THD	**96** 6-47
LTC1279	12	600ksps	60mW, Single 5V or ±5V Supply	**95** 6-8
LTC1278-5	12	500ksps	75mW, Single 5V or ±5V Supply	**94** 6-80
LTC1278-4	12	400ksps	75mW, Single 5V or ±5V Supply	**94** 6-80
LTC1400	12	400ksps	High Speed Serial I/O in SO-8 Package	**96** 6-36

A standards lab grade 20-bit DAC with 0.1ppm/°C drift

The dedicated art of digitizing one part per million

22

Jim Williams   J. Brubaker   P. Copley   J. Guerrero   F. Oprescu

Introduction

Significant progress in high precision, instrumentation grade D-to-A conversion has recently occurred. Ten years ago 12-bit D-to-A converters (DACs) were considered premium devices. Today, 16-bit DACs are available and increasingly common in system design. These are true precision devices with less than 1LSB linearity error and 1ppm/°C drift.[1] Nonetheless, there are DAC applications that require even higher performance. Automatic test equipment, instruments, calibration apparatus, laser trimmers, medical electronics and other applications often require DAC accuracy beyond 16 bits. 18-bit DACs have been produced in circuit assembly form, although they are expensive and require frequent calibration. 20 and even 23+ (0.1ppm!) bit DACs are represented by manually switched Kelvin-Varley dividers. These devices, although amazingly accurate, are large, slow and extremely costly. Their use is normally restricted to standards labs.[2] A useful development would be a practical, 20-bit (1ppm) DAC that is easily constructed and does not require frequent calibration.

20-bit DAC architecture

Figure 22.1 diagrams the architecture of a 20-bit (1ppm) DAC. This scheme is based on the availability of a true 1ppm analog-to-digital converter with scale and zero drifts below 0.02ppm/°C. This device, the LTC® 2400, is used as a feedback element in a digitally corrected loop to realize a 20-bit DAC.[3]

In practice, the "slave" 20-bit DAC's output is monitored by the "master" LTC2400 A-to-D, which feeds digital information to a code comparator. The code comparator differences the user input word with the LTC2400 output, presenting a corrected code to the slave DAC. In this fashion, the slave DAC's drifts and nonlinearity are continuously corrected by the loop to an accuracy determined by the A-to-D converter and V_{REF}.[4] The sole DAC requirement is that it be monotonic. No other components in the loop need to be stable.

This loop has a number of desirable attributes. As mentioned, accuracy limitations are set by the A-to-D converter and its reference. No other components need be stable. Additionally, loop behavior averages low order bit indexing and jitter, obviating the loop's inherent small-

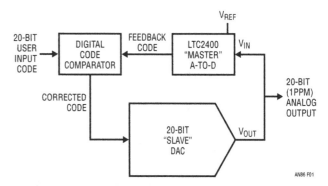

Figure 22.1 • Conceptual Loop-Based 20-Bit DAC. Digital Comparison Allows A-to-D to Correct DAC Errors. LTC2400 A-to-D's Low Uncertainty Characteristics Permit 1ppm Output Accuracy

---

Note 1: See Appendix A, "A history of high accuracy digital-to-analog conversion," for a review of high accuracy digital-to-analog conversion.
Note 2: Consult Appendix C, "Verifying data converter linearity to 1ppm," for discussion on Kelvin-Varley dividers. Also, see Appendix A, "A history of high accuracy digital-to-analog conversion."
Note 3: The LTC2400 analog-to-digital converter is profiled in Appendix B, "The LTC2400—a monolithic 24-bit analog-to-digital converter."

Note 4: D-to-A converters have been placed in loops to make A-to-D converters for a long time. Here, an A-to-D converter feeds back a loop to form a D-to-A converter. There seems a certain justified symmetry to this development. Turnabout is indeed fair play.

Analog Circuit and System Design: A Tutorial Guide to Applications and Solutions. DOI: 10.1016/B978-0-12-385185-7.00022-6

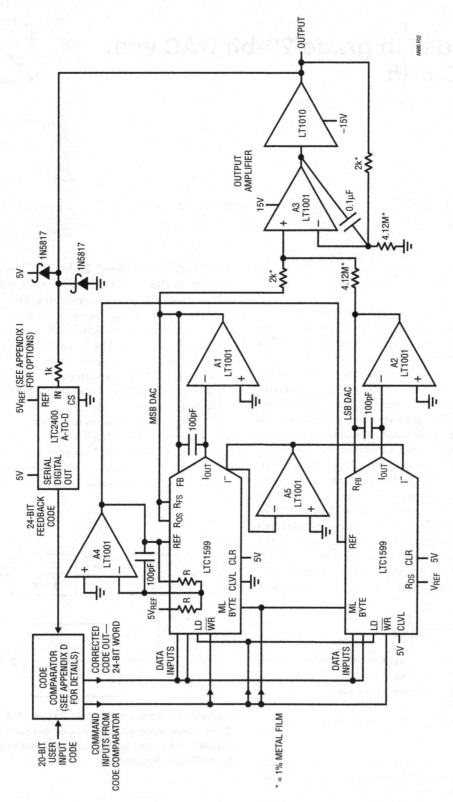

Figure 22.2 • Detail of 1ppm DAC. Composite DAC Is Comprised of Two DAC Values Summed at Output Amplifier. LTC2400 A-to-D and Code Comparator Furnish Stabilizing Feedback

signal instability. Finally, classical remote sensing may be used or digitally based sensing is possible by placing the A-to-D converter at the load. The A-to-D's SO-8 package and lack of external components makes this digitally incarnated Kelvin sensing scheme practical.[5]

Circuitry details

Figure 22.2 is a detailed schematic of the 1ppm DAC. The slave DAC is comprised of two DACs. The upper 16 bits of the code comparator's output are fed to a 16-bit DAC ("MSB DAC"), while the lower bits are converted by a separate DAC ("LSB DAC"). Although a total of 32 bits are presented to the two DACs, there are 8 bits of overlap, assuring loop capture under all conditions. The composite DACs' resultant 24-bit resolution provides 4 bits of index-ing range below the 20th bit, ensuring a stable LSB of 1ppm of scale. A1 and A2 transform the DAC's output currents into voltages, which are summed at A3. A3's scaling is arranged so that the correction loop can always capture and correct any combination of zero- and full-scale errors. A3's output, the circuit output, feeds the LTC2400 A-to-D. The LT® 1010 provides buffering to drive loads and cables. The A-to-D's digital output is differenced against the input word by the code comparator, which produces a corrected code. This corrected code is applied to the MSB and LSB DACs, closing a feedback loop.[6] The loop's integrity is determined by A-to-D converter and voltage refer-ence errors.[7] The resistor and diodes at the 5V powered A-to-D protect it from inadvertent A3 outputs (power up, transient, lost supply, etc.). A4 is a reference inverter and A5 provides a clean ground potential to both DACs.

Linearity considerations

A-to-D linearity determines overall DAC linearity. The A-to-D has about ±2ppm nonlinearity. In applications where this error is permissible, it may be ignored. If 1ppm linear-ity is required, it is obtainable by correcting the residual linearity error with software techniques. Details on LTC2400 linearity and this feature are presented in Appendices D and E.

DC performance characteristics

Figure 22.3 is a plot of linearity vs output code. The data shows linearity is within 1ppm over all codes.[8] Output noise, measured in a 0.1Hz to 10Hz bandpass, is seen in

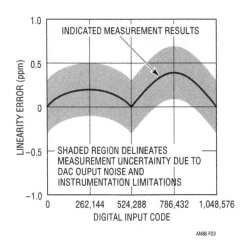

Figure 22.3 • Linearity Plot Shows No Error Outside 1ppm for All Codes

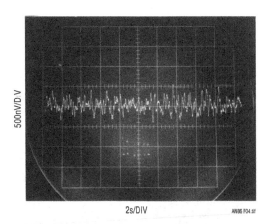

Figure 22.4 • Output Noise Indicates Less Than 1µV, About 0.2LSB. Measurement Noise Floor, Due to Equipment Limitations, Is 0.2µV

Figure 22.4 to be about 0.2LSB.[9] This measurement is somewhat corrupted by equipment limitations, which set a noise floor of about 0.2µV.

Dynamic performance

The A-to-D's conversion rate combines with the loop's sampled data characteristic and slow amplifiers to dictate relatively slow DAC response. Figure 22.5's slew response requires about 150 microseconds.

Figure 22.6 shows full-scale DAC settling time to within 1ppm (±5µV) requiring about 1400 milliseconds. A smaller step (Figure 22.7) of 500µV needs only 100 millise-conds to settle within 1ppm.[10]

Note 5: One wonders what Lord Kelvin's response would be to the digizatation of his progeny. Such uncertainties are the residue of progress.
Note 6: The code comparator is detailed in Appendix D, "A processor based code comparator."
Note 7: Voltage reference options are discussed in Appendix I, "Voltage refer-ences." For tutorial on the LTC2400, refer to Appendix B.
Note 8: Establishing and maintaining confidence in a 1ppm linearity measure-ment is uncomfortably close to the state of the art. The technique used is shown in Appendix C, "Verifying data converter linearity to 1ppm."

Note 9: Noise measurement considerations appear in Appendix H, "Microvolt level noise measurement."
Note 10: Measuring DAC settling time to 1ppm is by no means straightforward, even at the relatively slow speed involved here. See Appendix G, "Measuring DAC settling time."

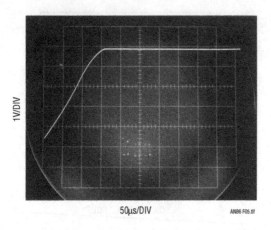

Figure 22.5 • DAC Output Full-Scale Slew Characteristics

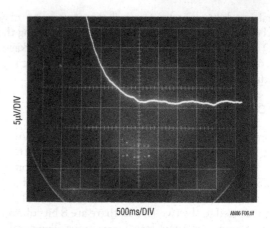

Figure 22.6 • High Resolution Settling Detail After a Full-Scale Step. Settling Time Is 1400 Milliseconds to Within 1ppm (±5µV)

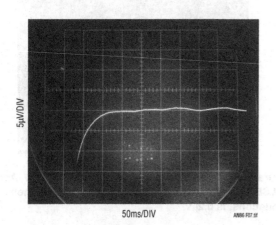

Figure 22.7 • Small Step Settling Time Measures 100 Milliseconds to Within 1ppm (±5µV) for a 500µV Transition

PARAMETER	SPECIFICATION
Resolution	1ppm
Full-Scale Error	4ppm of V_{REF} (Trimmable to 1ppm by V_{REF} Adjustment)
Full-Scale Error Drift	0.04ppm/°C Exclusive of Reference (0.1ppm/°C with LTZ1000A Reference[1])
Offset Error	0.5ppm
Offset Error Drift	0.01ppm/°C
Nonlinearity	±2ppm, Trimmable to Less Than 1ppm[2]
Output Noise	0.2ppm (≈0.9µV, 0.1Hz to 10Hz BW)
Slew Rate	0.033V/µs
Settling Time—Full-Scale Step	1400 Milliseconds
Settling Time—500µV Step	100 Milliseconds
Output Voltage Range	0V to 5V. For Other Ranges See Note 3

Note 1: See Appendix I
Note 2: See Appendix E
Note 3: See Appendices E and F

Figure 22.8 • Summarized Specifications for the 20-Bit DAC

Conclusion

Summarized 1ppm DAC specifications appear in Figure 22.9 These specifications should be considered guidelines, as the options and variations noted will affect performance.

Consult the appropriate appendices for design specifics and trade-offs.

Note: This Application Note was derived from a manuscript originally prepared for publication in EDN magazine.

References

1. Linear Technology Corporation, "LTC2400 Data Sheet," Linear Technology Corporation, January 1999.

2. Linear Technology Corporation, "LTC2410 Data Sheet," Linear Technology Corporation, April 2000.

3. Keithley Instruments, "Low Level Measurements," Keithley Instruments, 1984.

4. Williams, J., "Testing Linearity of the LTC2400 24-Bit A/D Converter," Linear Technology. Corporation, Design Solution 11, November 1999.

5. Seebeck, T. Dr., "Magnetische Polarisation der Metalle und Erze durch Temperatur-Differenz," Abhaandlungen der Preussischen Akademic der Wissenschaften (1822-1823), pp. 265–373.

6. Williams, J., "Component and Measurement Advances Ensure 16-Bit DAC Settling Time," Linear Technology Corporation, Application Note 74, July 1998.

7. Lee, M., "Understanding and Applying Voltage References," Linear Technology Corporation, Application Note 82, November 1999.

8. Williams, J., "Applications Considerations and Circuits for a New Chopper-Stabilized Op Amp," Linear Technology Corporation, Application Note 9, August 1987.

9. Huffman, B., "Voltage Reference Circuit Collection," Linear Technology Corporation, Application Note 42, June 1991.

10. Spreadbury, P. J., "The Ultra-Zener—A Portable Replacement for the Weston Cell?" IEEE Transactions on Instrumentation and Measurement, Vol. 40, No. 2, April 1991, pp. 343–346.

11. Williams, J., "Thermocouple Measurement," Linear Technology Corporation, Application Note 28, February 1988.

12. Hueckel, J. H., "Input Connection Practices for Differential Amplifiers," Neff Inst. Corporation, Duarte, California.

13. Gould Inc., "Elimination of Noise in Low Level Circuits," Gould Inc., Instrument Systems Division, Cleveland, Ohio.

14. Williams, J., "Prevent Low Level Amplifier Problems,". *Electronic Desig*, February 15, *1975*, 62.

15. Pascoe, G., "The Choice of Solders for High Gain Devices," *New Electronics* (Great Britain), February 6, 1977.

16. Pascoe, G., "The Thermo-E.M.F. of Tin-Lead Alloys,". *Journal Phys.*, December 1976.

17. Brokaw, A. P., "Designing Sensitive Circuits? Don't Take Grounds for Granted,". *EDN.*, October 5, 1975, p. 44.

18. Morrison, R., "Noise and Other Interfering Signals,". John Wiley and Sons 1992.

19. Morrison, R., "Grounding and Shielding Techniques in Instrumentation," Wiley-Interscience, 1986.

20. Vishay Inc., "Vishay Foil Resistors," Vishay Inc., 1999.

Appendix A
A history of high accuracy digital-to-analog conversion

People have been converting digital-to-analog quantities for a long time. Probably among the earliest uses was the summing of calibrated weights (Figure A1, left center) in weighing applications. Early electrical digital-to-analog conversion inevitably involved switches and resistors of different values, usually arranged in decades. The application was often the calibrated balancing of a bridge or reading, via null detection, some unknown voltage. The most accurate resistor-based DAC of this type is Lord Kelvin's Kelvin-Varley divider (Figure, large box). Based on switched resistor ratios, it can achieve ratio accuracies of 0.1ppm (23+ bits) and is still widely employed in standards laboratories.[1] High speed digital-to-analog conversion resorts to electronically switching the resistor network. Early electronic DACs

were built at the board level using discrete precision resistors and germanium transistors (Figure, center foreground, is a 12-bit DAC from a Minuteman missile D-17B inertial navigation system, circa 1962). The first electronically switched DACs available as standard product were probably those produced by Pastoriza Electronics in the mid 1960s. Other manufacturers followed and discrete- and monolithically-based modular DACs (Figure, right and left) became popular by the 1970s. The units were often potted (Figure, left) for ruggedness, performance or to (hopefully) preserve proprietary knowledge. Hybrid technology produced smaller package size (Figure, left foreground). The development of Si-Chrome resistors permitted precision monolithic DACs such as the LTC1595 (Figure, immediate foreground). In keeping with all things monolithic, the cost-performance trade off of modern high resolution IC DACs is a bargain. Think of it! A 16-bit DAC in an 8-pin IC package. What Lord Kelvin would have given for a credit card and LTC's phone number.

Note 1: See Appendix C, "Verifying data converter linearity to 1ppm," for details on Kelvin-Varley Dividers.

Figure A1 • Historically Significant Digital-to-Analog Converters Include: Weight Set (Center Left), 23+ Bit Kelvin-Varley Divider (Large Box), Hybrid, Board and Modular Types, and the LTC1595 IC (Foreground). Where Will It All End?

Appendix B
The LTC2400—a monolithic 24-bit analog-to-digital converter

The LTC2400 is a micropower 24-bit A-to-D converter with an integrated oscillator, 4ppm nonlinearity and 0.3ppm RMS noise. It uses delta-sigma technology to provide extremely high stability. The device can be configured for better than 110dB rejection at 50Hz or 60Hz ±2%, or it can be driven by an external oscillator for a user defined rejection frequency in the range 1Hz to 120Hz.

This ultraprecision A-to-D converter in an SO-8 pin package forms the heart of the 20-bit DAC described in the text. It is significant that the device is used here as a circuit *component* rather than in the traditional standalone role accorded precision A-to-D converters. This freedom, in keeping with the IC's economy and ease of use, is a noteworthy opportunity. Alert designers will recognize this development and capitalize on it. Key specifications for the A-to-D are given in Figure B1.

PARAMETER	CONDITIONS	
Resolution (No Missing Codes)	$0.1V \leq V_{REF} \leq V_{CC}$	24 Bits
Integral Nonlinearity	$V_{REF} = 2.5V$ $V_{REF} = 5V$	2ppm of V_{REF} 4ppm of V_{REF}
Offset Error	$2.5V \leq V_{REF} \leq V_{CC}$	0.5ppm of V_{REF}
Offset Error Drift	$2.5V \leq V_{REF} \leq V_{CC}$	0.01ppm of V_{REF}/°C
Full-Scale Error	$2.5V \leq V_{REF} \leq V_{CC}$	4ppm of V_{REF}
Full-Scale Error Drift	$2.5V \leq V_{REF} \leq V_{CC}$	0.02ppm of V_{REF}/°C
Total Unadjusted Error	$V_{REF} = 2.5V$ $V_{REF} = 5V$	5ppm of V_{REF} 1ppm of V_{REF}
Output Noise		$1.5\mu V_{RMS}$
Normal Mode Rejection 60Hz ±2%		110dB (Min)
Normal Mode Rejection 50Hz ±2		110dB (Min)
Input Voltage Range		$0.125V \cdot V_{REF}$ to $1.125V \cdot V_{REF}$
Reference Voltage Range		$0.1V \leq V_{REF} \leq V_{CC}$
Supply Voltage		$2.7V \leq V_{CC} \leq 5.5V$

Supply Current
  Conversion Mode
  Sleep Mode
	$\overline{CS} = 0V$
	$\overline{CS} = V_{CC}$
		200μA
		20μA

Figure B1 • Key Specifications for LTC2400 A-to-D Converter. High Linearity and Extreme Stability Allow Realization of 1ppm DAC

Appendix C
Verifying data converter linearity to 1ppm—help from the nineteenth century

Introduction

Verifying 1ppm linearity of the DAC and the analog-to-digital converter used to construct it requires special considerations. Testing necessitates some form of voltage source that produces equal amplitude output steps for incremental digital inputs. Additionally, for measurement confidence, it is desirable that the source be substantially more linear than the 1ppm requirement. This is, of course, a stringent demand and painfully close to the state of the art.

The most linear "D to A" converter is also one of the oldest. Lord Kelvin's Kelvin-Varley divider (KVD), in its most developed form, is linear to 0.1ppm. This manually switched device features ten million individual dial settings arranged in seven decades. It may be thought of as a 3-terminal potentiometer with fixed "end-to-end" resistance and a 7-decade switched wiper position (Figure C1).

The actual construction of a 0.1ppm KVD is more artistry and witchcraft than science. The market is relatively small, the number of vendors few and resultant price high. If $13,000 for a bunch of switches and resistors seems offensive, try building and certifying your own KVD. Figure C2 shows a detailed schematic.

The KVD shown has a 100kΩ input impedance. A consequence of this is that wiper output resistance is high and varies with setting. As such, a very low bias current follower is required to unload the KVD without introducing significant error. Now, our KVD looks like Figure C3. The LT1010 output buffer allows driving cables and loads and, more subtly, maintains the amplifier's high open-loop gain.

Approach and error considerations

This schematic is deceptively simple. In practice, construction details are crucial. Parasitic thermocouples (Seebeck effect), layout, grounding, shielding, guarding, cable choice and other issues affect

Figure C1 • Conceptual Kelvin-Varley Divider

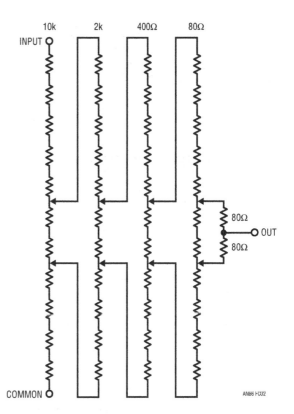

Figure C2 • A 4-Decade Kelvin-Varley Divider. Additional Decades Are Implemented By Opening Last Switch, Deleting Two Associated 80Ω Values and Continuing ÷ 5 Resistor Chains

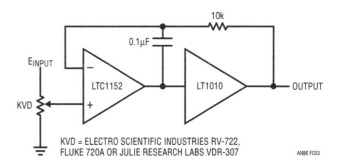

Figure C3 • KVD with Buffer Gives Output Drive Capability

achievable performance.[1] In fact, as good as the chopper-stabilized LTC1152 is with respect to drift, offset, bias current and CMRR, selection is required if we seek sub-ppm nonlinearity performance. Figure C4, an error budget analysis, details some of the selection criteria.

The buffer is tested with Figure C5's circuit. As the KVD is run through its entire range, the floating null

---

Note 1: See Appendix J, "Cables, connections, solder, layout, component choice, terror and arcana," for relevant tutorial.

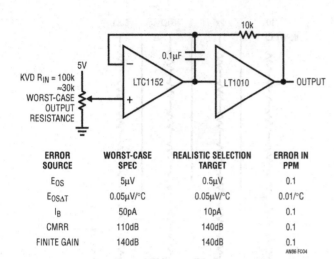

ERROR SOURCE	WORST-CASE SPEC	REALISTIC SELECTION TARGET	ERROR IN PPM
E_{OS}	5µV	0.5µV	0.1
$E_{OS\Delta T}$	0.05µV/°C	0.05µV/°C	0.01/°C
I_B	50pA	10pA	0.1
CMRR	110dB	140dB	0.1
FINITE GAIN	140dB	140dB	0.1

AN86 FC04

Figure C4 • Error Budget Analysis for the KVD Buffer. Selection Permits ≈0.4ppm Predicted Linearity Error

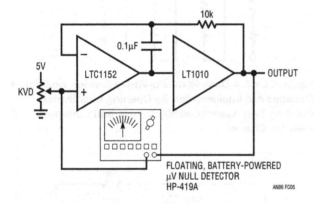

Figure C5 • Determining Buffer Error By Measuring Input-Output Deviation with Floating Microvolt Null Detector. Technique Permits Evaluation of Fixed and Operating Point Induced Errors

detector must remain well within 1ppm (5µV), preferably below 0.5ppm. This test ensures that all error sources, particularly I_B and CMRR, whose effects vary with operating point, are accounted for. Measured performance indicates the sum of all errors called out in Figure C4 is well within desired limits.

In Figure C6, we add offset trim, a stable voltage source and a second KVD to drive the main KVD. Additionally, an ensemble of three HP-3458A voltmeters monitor the output.

The offset trim bleeds a small current into the main KVD ground return, producing a few microvolts of offset-trim range. This functionally trims out all sources of zero error (amplifier offsets, parasitic thermocouple mismatches and the like), permitting a true zero volt output when the main KVD is set to all zeros.

The voltmeters, specified for <0.1ppm nonlinearity on the 10V range, "vote" on the source's output.

Circuitry details

Figure C7 is a more detailed schematic. It is similar to Figure C6 but highlights issues and concerns. The grounding scheme is single point, preventing mixing of return currents and the attendant errors. The shielded cables used for interconnections between the KVDs and voltmeters should be specified for low thermal activity. Keithley type SC-93 and Guildline #SCW are suitable. Crush type copper lugs (as opposed to soldered types) provide lower parasitic thermocouple activity at KVD and DVM connection points. However, they must be kept clean to prevent oxidation, thus avoiding excessive thermal voltages.[2] A copper deoxidant (Caig Labs "Deoxit" D100L) is quite effective for maintaining such cleanliness. Low thermal lugs and jacks, preterminated to cables, are also available (Hewlett-Packard 11053, 11174A) and convenient.

Thermal baffles enclosing KVD and DVM connections tend to thermally equilibrate their associated banana jack terminals, minimizing residual parasitic thermocouple activity. Additionally, restrict the number of connections in the signal path. Necessary connections should be matched in number and material so that differential cancellation occurs. Complying with this guideline may necessitate deliberate introduction of solder-copper junctions (marked "X" on Figure C7) to obtain optimum differential cancellation.[3] This is normally facilitated by simply breaking the appropriate wire or PC trace and soldering it. Ensure that the introduced thermocouples temperature track the junctions they are supposed to cancel. This is usually accomplished by locating all junctions within close physical proximity.

The noise filtering capacitor at the main KVD is a low leakage type, with its metal case driven by the output buffer to guard out surface leakage.

When studying the approach used, it is essential to differentiate between linearity and absolute accuracy. This eliminates concerns with absolute standards, permitting certain freedoms in the measurement scheme. In particular, although single-point grounding was used, remote sensing was not. This is a deliberate choice, made to minimize the number of potential error-causing parasitic thermocouples in the signal path. Similarly, a ratiometric reference connection between the KVD LTZ1000A voltage source and the voltmeters was not utilized for the same

---

Note 2: See Note 1.
Note 3: See Appendix J, "Cables, connections, solder, layout, component choice, terror and arcana," for further discussion.

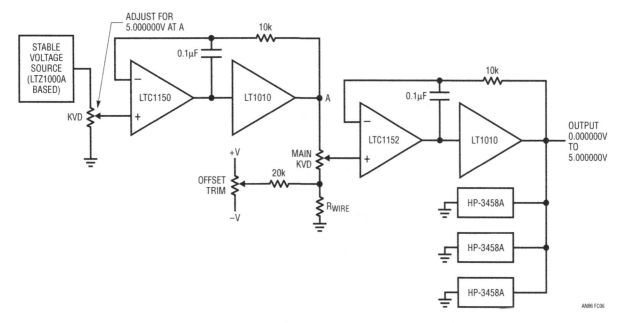

Figure C6 • Simplified Sub-ppm Linearity Voltage Source

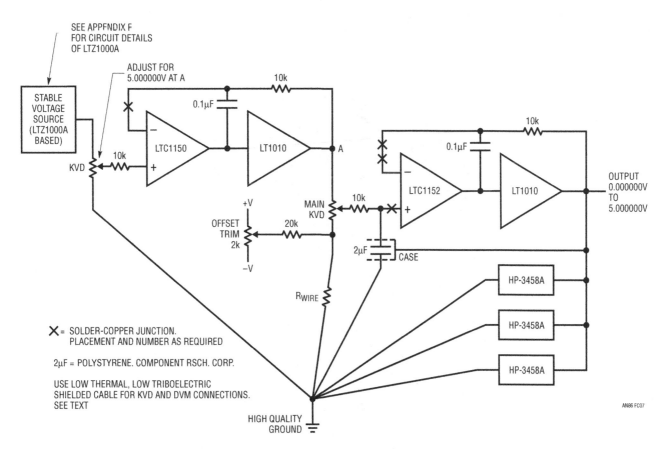

X = SOLDER-COPPER JUNCTION.
PLACEMENT AND NUMBER AS REQUIRED

2µF = POLYSTYRENE. COMPONENT RSCH. CORP.

USE LOW THERMAL, LOW TRIBOELECTRIC
SHIELDED CABLE FOR KVD AND DVM CONNECTIONS.
SEE TEXT

Figure C7 • Complete Sub-ppm Linearity Voltage Source

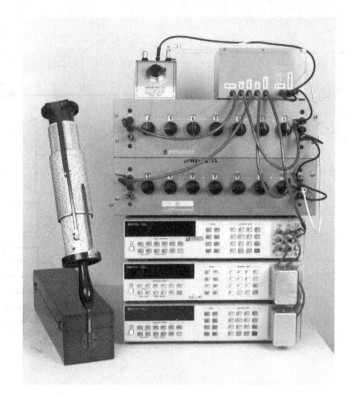

Figure C8 • The Sub-ppm Linearity Voltage Source. Box Upper Right Is LTZ1000A Based Reference and Buffers. Upper Left Is Offset Trim. Reference and Main Kelvin-Varley Dividers Are Photo Center—Upper and Center-Middle, Respectively. Three HP-3458 DVMs (Photo Lower) Monitor Output. Computer (Left Foreground) Aids Linearity Calculations

Figure C9 • Reference-Buffer Box Construction. LTZ1000A Reference Circuitry Is Photo Lower Left, Buffer Amplifiers Photo Center. Note Capacitor Case Bootstrap Connection (Photo Center—Right). Single Point Ground Mecca Appears Photo Upper Left. Power Supply (Photo Top) Mounts Outside Box, Minimizing Magnetic Field Disturbance

1. VERIFY KVD LINEARITY BY INTERCOMPARISON AND INDEPENDENT CAL. LAB.
2. TAKE WORST-CASE VOLTMETER ENSEMBLE DEVIATIONS OVER 0V TO 5V, EVERY 0.5V
3. 100 RUNS (10 PER DAY, ONCE PER HOUR)
4. INDICATED RESULT IS 0.3ppm NONLINEARITY

AN86 FC10

Figure C10 • Testing Regime for the High Linearity Voltage Source

reason. In theory, a ratiometric connection affords lower drift. In practice, the resultant introduced parasitic thermocouples obviate the desired advantage. Additionally, the aggregate stability of the LTZ1000A reference and the voltmeter references (also, incidentally, LTZ1000A based) is comfortably inside 0.1ppm for periods of 10 minutes.[4] This is more than enough time for a 10-point linearity measurement.

Construction
Figures C8 and C9 are photographs of the voltage source and the reference-buffer box internal construction. The figure captions annotate some significant features.

Results
This KVD-based, high linearity voltage source has been in use in our lab for years. During this period, the total linearity uncertainty defined by the source and its monitoring voltmeters has been just 0.3ppm (see Figure C10's measurement regime). This is more than 3 times better than the desired 1ppm performance, promoting confidence in our measurements.[5]

Acknowledgments
The author is indebted to Lord Kelvin and to Warren Little of the C. S. Draper Laboratory (née M. I. T.

Instrumentation Laboratory) standards lab. Warren taught me, with great patience, the wonders of KVDs some thirty years ago and I am still trading on his efforts.

Appendix D
A processor-based code comparator
The code comparator enforces the loop by setting the slave DAC inputs to the code that equalizes the user input and the LTC2400 A-to-D output. This action is more fully described on page one of the text.

Figure D1 is the code comparator's digital hardware. It is composed of three input data latches and a PIC-16C5X processor. Inputs include user data (e.g., DAC inputs), linearity curvature correction (via DIP switches), convert command ("DA $\overline{WR}$") and a selectable filter time constant. An output ("DAC RDY") indicates when the DAC output is settled to the user input value. Additional outputs and an input control and monitor the analog section (text Figure 22.2) to effect loop closure. Note that although a total of 32 bits are presented to the two 16-bit slave DACs, there are 8 bits of overlap, allowing a total dynamic range of 24 bits. This provides 4 bits of indexing range below the 20th bit, ensuring a stable LSB of 1ppm of scale. The 8-bit overlap assures the loop will always be able to capture the correct output value.

The processor is driven by software code, authored by Florin Oprescu, which is described below.

Note 4: The LTZ1000A reference is detailed in Appendix I, "Voltage references."

Note 5: The author, wholly unenthralled by web surfing, has spent many delightful hours "surfing the Kelvin." This activity consists of dialing various Kelvin-Varley divider settings and noting monitoring A-to-D agreement within 1ppm. This is astonishingly nerdy behavior, but thrills certain types.

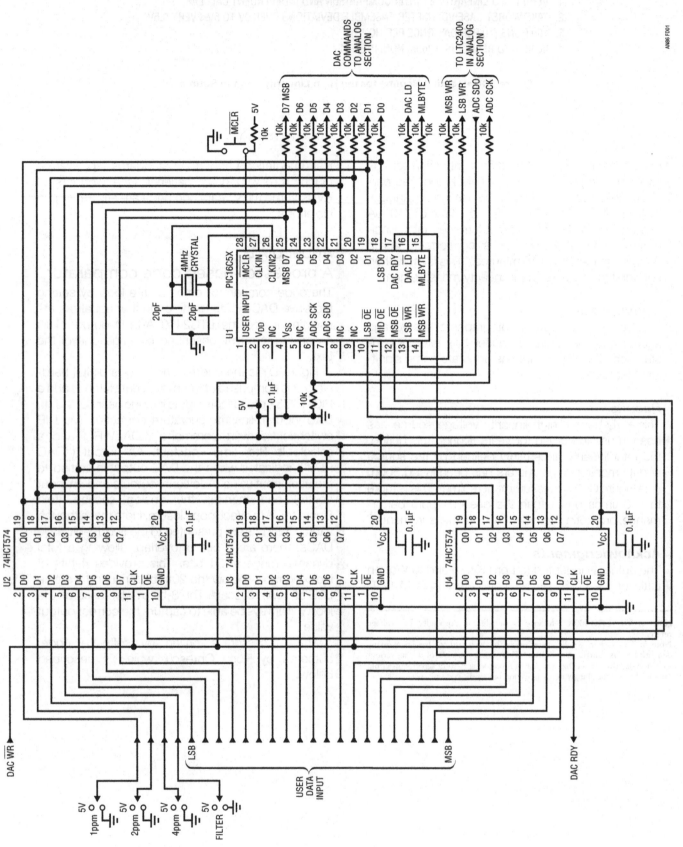

Figure D1 • Code Comparator Hardware. User Control Lines Are at Left, Analog Section Connections Appear at Right Side

```
;20bit DAC code comparator
;
;****************************************************
;                                                    *
;      Filename:            dac20.asm                *
;      Date         12/4/2000                        *
;      File Version:        1.1                      *
;                                                    *
;      Author:              Florin Oprescu           *
;      Company:             Linear Technology Corp.  *
;                                                    *
;                                                    *
;****************************************************
;
; Variables
;============
; uses 17 bytes of RAM as follows:
;
; {UB2, UB1, UB0} user input word buffer
;────────────────────────────────────────
; 24 bits unsigned integer (3 bytes):
;
; The information is transferred from the external input register
; into {UB2, UB1, UB0} whenever a "user input update" event
; is detected by testing the timer0 content. Following the data
; transfer, the UIU ("user input update") flag is set and the DAC
; ready flags RDY and RDY2 are cleared. UB0 uses the same physical
; location as U0. The user input double buffering is necessary
; because the loop error corresponding to the current ADC reading
; must be calculated using the previous user input value.
; The old user input value can be replaced by the new user input
; value only after the loop error calculation.
; The worst case minimum response time to an UIU event must be
; calculated. The user shall not update the external input register
; at intervals shorter than this response time. For the moment the
; program can not block the user access to the external input
; register during a read operation. In such a situation the result
; of the read operation can be very wrong.
;
;      UB0 - least significant byte. Same physical location as U0
;
;      UB1 - second byte.
;
;      UB2 - most significant byte.
;
;
; {U2, U1, U0} user input word
;────────────────────────────────────────
; 24 bits unsigned integer (3 bytes):
;
; The information is transferred from {UB2, UB1, UB0[7:4], [0000]}
; into {U2, U1, U0} whenever the UIU flag is found set within the
; CComp ("code comparator") procedure. The UIU flag is reset
; following the data transfer.
```

```
;
;       U0 -   least significant byte of current DAC input
;              The 4 least significant bits U0[3:0] are set
;              to zero.
;
;       U1 -   second byte of current DAC input
;
;       U2 -   most significant byte of current DAC input
;
;
; {CON} control byte
;───────────────────────
; (1 byte):
;
; The 3 least significant bits CON[2:0] represent the ADC linearity
; correction factor transferred from UB[2:0] when the UIU flag
; is found set within the CComp procedure - at the same time as the
; {U2, U1, U0} variable is updated.
;
; The effect of CON[2:0] is additive and its weight is as follows:
;
;       CON[0] = 1 linearity correction effect is about 1ppm
;       CON[1] = 1 linearity correction effect is about 2ppm
;       CON[2] = 1 linearity correction effect is about 4ppm
;
; The LTC2400 has a typical 4ppm INL error therefore the default
; curvature correction value can be set at CON[2:0] = 100
;
; CON[3] is the control loop integration factor transferred from
; UB[3] when the UIU flag is found set within the CComp procedure.
; If CON[3]=0, after the control loop error becomes less than 4ppm
; the error correction gain is reduced from 1 to 1/4
; If CON[3]=1, after the control loop error becomes less than 4ppm
; the error correction gain is reduced from 1 to 1/16
;
; CON[7] is used as the UIU ("user input update") flag. It is set
; when {UB2, UB1, UB0} is updated and it is reset when {U2, U1, U0}
; and CON[3:0] are updated.
;
; CON[6] is used as the RDY ("DAC ready") flag. It is set when
; the DAC loop error becomes less than 4ppm and it is reset when
; the UIU flag is set.
;
; CON[5] is used as the RDY2 ("DAC ready twice") flag. It is set
; whenever the DAC loop error becomes less than 4ppm and the RDY
; flag has been previously set. It is reset when the UIU flag is set.
;
;
; The bit CON[4] is not used and is always set to 0.
;
;
```

```
; {ADC3, ADC2, ADC1, ADC0} formatted ADC conversion result
;────────────────────────────────────────────────────────────
; 32 bits signed integer (4 bytes).
;
; The ADC reading is necessary only for the calculation of the control
; loop error and in order to save RAM space, it can share the same
; physical space as the loop error variable.
;
;
; The LTC2400 output format is offset binary. It must be converted
; to 2's complement before any arithmetic operation. A number of
; possible codes are not valid LTC2400 output codes. If such codes
; are detected it can be inferred that a serial transfer error has
; occurred, the data must be discarded and a new conversion must
; be started. For all LTC2400 devices B31=0 and B30=0 always. In
; addition, with the exception of some early samples of the device
; the sequence B[29:28]=00 should not occur in a valid transaction.
;
;      ADC0 -   least significant byte
;               contains ADC output bits B11(MSBIT) to B4 (LSBIT)
;
;      ADC1 -   second byte
;               contains ADC output bits B19(MSBIT) to B12 (LSBIT)
;
;      ADC2 -   third byte
;               contains ADC output bits B27(MSBIT) to B20 (LSBIT)
;
;      ADC3 -   most significant byte
;               contains ADC output bits ~B29(as 7 MSBITS for
;               2's complement sign extension) and B28 (LSBIT)
;
;
; {ADCC} ADC curvature correction
;────────────────────────────────────────────
; 8 bits unsigned integer (1 byte)
;
; The LTC2400 transfer characteristic has a typical INL of about
; 4ppm and a parabolic shape symmetric with respect to mid-scale.
; This error can be corrected to a first and second order and
; ADDC contains the magnitude of this correction.
;
;
; {ER3, ER2, ER1, ER0} control loop error value
;────────────────────────────────────────────────────
; signed integer (4 bytes)
;
; Contains the value of the current control loop error calculated
; as the difference between the previous user input and the last
; ADC reading. It is used to adjust the Low_DAC setting. Uses the
; same physical location as {ADC3, ADC2, ADC1, ADC0}:
;
;      ER0 -    least significant byte, same location as ADC0
;
;      ER1 -    second byte, same location as ADC1
```

```
;
;      ER2 -    third byte, same location as ADC2
;
;      ER3 -    most significant byte, same location as ADC3
;
;
; {DL3, DL2, DL1, DL0} Low DAC control value
;─────────────────────────────────────────────────────────
; signed integer (4 bytes):
;
; Contains the Low_DAC setting in a 2's complement, 32 bit
; format. Must be initialized to 0!
;
;      DL0 -    least significant byte - used for Low_DAC
;               control
;
;      DL1 -    second byte - used for Low_DAC control after
;               conversion to offset binary format {DL1, DL0}
;
;      DL2 -    third byte - may be only 00 or FF
;
;      DL3 -    most significant byte - may be only 00 or FF
;
;
; {INDX} Index variable for various program functions
;─────────────────────────────────────────────────────────
; 1 byte.
;
;
; {TMPV} Temporary variable for various program functions
;─────────────────────────────────────────────────────────
; 1 byte.
;
;
;
; Algorithm
;===========
;
; After each ADC conversion cycle the processor calculates the control
; loop error value as the difference between the desired output and
; the latest conversion result. Than it updates the DACs command
; such as to reduce the error magnitude. A new ADC conversion cycle
; is started following the DACs update operation.
;
; In order to maintain adequate control loop stability it is necessary
; for the DACs and the associated amplifiers to settle to better than
; 20 bits accuracy before the ADC starts sampling the system output. For
; an LTC2400 based system this settling time is 66ms.
;
; Initialization:
;  Initializes the PIC controller and the hardware interface
;  and starts the Scan procedure.
;
```

```
;    1. Load ADC control port with default values
;         SCKAD = 0
;         SDOAD = 1
;    2. Set ADC control port I and O pins
;         SCKAD = output
;         SDOAD = input
;    3. Load register control port with default values
;         NCSR[2:0] = 111
;         NCSD[1:0] = 11
;         ADDAC     = 1
;         NLDAC     = 1
;         DACRDY    = 0
;    4. Set register control port in output mode
;    5. Set data bus to default value DBUS[7:0]=0x00
;    6. Set data bus in output mode
;    7. Initialize internal registers and variables:
;       OPTION    = 0x2F
;         Timer0 used as counter is incremented by low-to-high edge
;         Prescaler works with watch dog timer in div128 mode
;       CON       = 0x80
;          Simulate a UIU "user interface update" event to force
;          the update of both Low_DAC and High_DAC
;       {DL3, DL2, DL1, DL0} = 0
;       {     U2,  U1,  U0} = 0
;    8. Update hardware using the initialized variables
;    9. Start new ADC conversion by reading and discarding
;       32 serial bits.
;    10. Start the Scan procedure
;
; Scan:
;  Continuously looks for "user input update" events. When
;  a "user input update" event is detected updates the
;  input buffer {UB2, UB1, UB0}, resets timer, sets UIU flag
;  and resets RDY and RDY2 flags.
;
;  The active low write signal for the external input register
;  (which is the same as the user interface NWR input signal)
;  is driven by the user and it is connected to the counter
;  input of Timer0. The Timer0 is used in counter mode without a
;  prescaler and it increments whenever a low-to-high transition
;  is detected at its input. This is the same transition which
;  latches in the input register a new user command.
;  Because of the PIC controller timing constraints, this write
;  signal must be maintained low for at least 2*Tosc + 20ns
;  where Tosc is the maximum PIC clock period. When a 4 MHz
;  clock is used for the PIC processor, the low time must be
;  longer than about 520ns.
;
;  1. Test for "user input update" events by testing the Timer0
;     value.
;          If Timer0>0 an UIU event has occurred. Reset the timer
;          and answer Yes.
;          If Timer0=0 answer No.
```

```
;     1.1 If Yes, read input latch into {UB2, UB1, UB0},
;         reset DACRDY output line, set UIU flag and
;         and reset RDY and RDY2 flags (CON[7:5]=100)
;         Than continue
;     1.2 If No continue
;
; Continuously looks for the ADC end of conversion event. When
; the "end of conversion" is detected it reads the 28 most
; significant bits from the ADC and it constructs the ADC
; result {ADC3, ADC2, ADC1, ADC0} in 2's complement format
;         If ADC3[1] == 0 => ADC3[7:1]=1111 111
;         If ADC3[1] == 1 => ADC3[7:1]=0000 000
;         For very early LTC2400 samples only, it is possible
;         to obtain as a valid 0 conversion result ADC3[1:0]=00
;         In this case:
;         If ADC3[1:0] == 0 => ADC3=0
; It also calculates the first (x1) and second (x2) order ADC
; curvature correction ADCC as follows:
;         x1 = {0x00, 0x80} -
;             -abs({ADC3, ADC2, ADC1, ADC0}/(2^16)-{0x00, 0x80})
;         x2 = {0x00, 0x40} -
;             -abs({0x00,{0,ADC2[6:0]},ADC1,ADC0}/(2^16)-{0x00,0x40})
;         ADCC = floor((x1 + x2/2) * {00000 CON[2:0]} / 4 )
; The actual implementation uses only the least significant
; byte of x without any substantial additional error.
; Thus the above relation can be modified as follows:
;         ADCC = floor((abs(ADC2) + abs({ADC2[6],ADC2[6:0]})/2) *
;             * {00000 CON[2:0]} / 4 )
; The maximum correction range is about 7ppm INL at mid
; scale for CON[2:0] = 111.
;
; 2. Test for ADC "end of conversion" event by testing the
;    value of the ADC_SDO signal.
;         If ADC_SDO = LOW answer Yes.
;         If ADC_SDO = HIGH answer No.
;   2.1 If Yes read 28 most significant bits from the ADC,
;       update {ADC3, ADC2, ADC1, ADC0} and calculate the
;       curvature correction byte ADCC. Than start the CComp
;       procedure.
;       It should be noticed that while reading the first 28
;       most significant bits from the ADC the controller
;       generates 27 serial clock pulses. An additional 5 serial
;       clock pulses (for a total of 32) are necessary to restart
;       the conversion.
;   2.2 If No restart the Scan procedure.
;
;
; CComp:
;  Calculates the current control loop error as:
;
;  error = current_user_input - ADC_reading +
;        + new_user_input_LSB - current_user_input_LSB
;
```

```
;   The curvature correction is included in the ADC
;   conversion result and is always positive therefore:
;
;   ADC_reading = {ADC3, ADC2, ADC1, ADC0} +
;               + {   0,    0,    0, ADCC}
;
;   The term "new_user_input_LSB - current_user_input_LSB"
;   represents the residue of the new user command which
;   is added to the Low_DAC.
;
;   {ER3, ER2, ER1, ER0} =
;       = {0, U2, U1,  U0} - {ADC3, ADC2, ADC1, ADC0} -
;       - {   0,    0,    0, ADCC} +
;       + {0,  0,  0, UB0} - {   0,    0,    0,    U0} =
;
;       = {0, U2, U1, UB0} - {ADC3, ADC2, ADC1, ADC0} -
;       - {   0,    0,    0, ADCC}.
;
;   The loop error {ER3, ER2, ER1, ER0} is a 32 bit signed number
;   and the weight of the least significant bit is 1/16ppm of
;   the ADC reference voltage. A 4ppm error value is represented
;   as {0, 0, 0, 0x40}.
;
;   The ADC output noise is dominated by thermal noise and has a
;   white distribution. The control loop noise can be reduced by
;   the square root of N by averaging N successive ADC readings.
;   The obvious penalty is a slow settling time. Due to the
;   limited amount of RAM available a direct implementation
;   of this noise reduction strategy is difficult. In an equivalent
;   implementation, when the absolute value of the loop error
;   {ER3, ER2, ER1, ER0} decreases below a certain threshold, the
;   gain of the error correction loop can be decreased. The default
;   threshold is set at a very conservative 4ppm. This value must
;   always be larger than the peak noise level of the ADC. A very
;   quiet implementation can probably operate with a threshold of
;   2ppm. If CON[3]=0 the gain of the error correction loop is
;   decreased from 1 to 1/4. If CON[3]=1 the gain of the error
;   correction loop is decreased from 1 to 1/16.
;
;   The High_DAC is always controlled by the 16 most significant
;   bits of the most recent user command {UB2, UB1}
;
;   The Low_DAC is controlled by the {DL3, DL2, DL1, DL0}
;   variable which integrates the control loop error. Under
;   correct operating condition abs({DL3, DL2, DL1, DL0})<2^15.
;   In order to avoid roll-overs during large transients the
;   {DL3, DL2, DL1, DL0} must be clamped within the +/- 2^15 range.
;   The 16 bit Low_DAC can than be controlled by {DL1, DL0}
;   after conversion to offset binary format.
;
```

```
;   The DACRDY output line reflects the state of the
;   internal RDY2 flag.
;
;   After the updates are completed we must start a new ADC
;   conversion by completing the serial transfer.
;
;   1. Test if UIU flag is set
;       1.1 If Yes, move UB[3:0] into CON[3:0]
;           and {UB0[7:4], 0000} into U0. The last ADC result
;           is curvature corrected using the previous CON[3:0] value!.
;   2. Calculate {ER3, ER2, ER1, ER0}.
;   3. Test if UIU flag is set
;       3.1 If Yes, move {UB2, UB1} into {U2, U1} and
;           clear UIU, RDY and RDY2 flags (CON[7:5]=000 )
;       3.2 If No, test if abs({ER3, ER2, ER1, ER0}) < 4ppm
;           3.2.1 If Yes, test if CON[6]=1 (RDY flag)
;                       3.2.1.1 If Yes, set RDY2 flag (CON[5]=1 )
;                       3.2.1.2 If No,  set RDY  flag (CON[6]=1 )
;                   and test if CON[3]=0 (filter flag)
;                       3.2.1.3 If Yes, {ER3, ER2, ER1, ER0} =
;                                     = {ER3, ER2, ER1, ER0}/4
;                       3.2.1.4 If No,  {ER3, ER2, ER1, ER0} =
;                                     = {ER3, ER2, ER1, ER0}/16
;           3.2.2 If No, clear UIU, RDY and RDY2
;                   flags (CON[7:5]=000 )
;   4   {DL3, DL2, DL1, DL0} = {DL3, DL2, DL1, DL0} +
;                             +{ER3, ER2, ER1, ER0}.
;   5. Update High_DAC, Low_DAC and DACRDY output line
;   6. Read the 4 least significant bits from ADC and start
;      a new conversion
;   7. Restart the Scan procedure
;
;
; Hardware resources
;=====================
;
; Uses 8 input/output pins, 9 output pins, 1 input pin and 1
; counter input pin
;
; DBUS[7:0] data bus
;———————————————————
; 8 bit bi-directional data bus is used to read the 20 bit input
; command IC[19:0], the one bit filter selection FS[0] and the 3 bit
; curvature correction selection CC[2:0]. It is also used to write
; the 16 bit Low_DAC command LDAC[15:0] and the 16 bit High_DAC
; command HDAC[15:0].
;
; assigned to PIC port C[7:0]
;
```

```
; The data format for the read and write operations is as follows:
;
; DBUS[ 7:0] = IC[19:12] when NCSR[2] = 0
; DBUS[ 7:0] = IC[11: 4] when NCSR[1] = 0
; DBUS[ 7:0] = {IC[3:0], FS[0], CC[2:0]} when NCSR[0] = 0
; LDAC[ 7:0] = DBUS[7:0] when NCSD[0] = 0 and ADDAC = 0
; LDAC[15:8] = DBUS[7:0] when NCSD[0] = 0 and ADDAC = 1
; HDAC[ 7:0] = DBUS[7:0] when NCSD[1] = 0 and ADDAC = 0
; HDAC[15:7] = DBUS[7:0] when NCSD[1] = 0 and ADDAC = 1
;
;
; NCSR[2:0] active low output enable controls for input registers
;------------------------------------------------------------------
; 3 output lines used to selectively enable the three 8-bit input
; registers in order to read the user updated DAC command, the 3
; curvature correction bits and the one filter control bit.
;
; NCSR[0] enables the low input byte (LSB) and is assigned to port B[0]
;
; NCSR[1] enables the second input byte and is assigned to port B[1]
;
; NCSR[2] enables the high input byte (MSB) and is assigned to port B[2]
;
;
; NCSD[1:0] active low input enable controls for the DACs
;------------------------------------------------------------------
; 2 output lines used to selectively enable the two DACs
;
; NCSD[0] enables the Low_DAC and is assigned to port B[3]
;
; NCSD[1] enables the High_DAC and is assigned to port B[4]
;
;
; ADDAC DAC address control
;--------------------------------
; output line. A low enables a write operation to the low byte of
; Low_DAC or High_DAC. A high enables a write operation to the high
; byte of Low_DAC or High_DAC.
;
; ADDAC is assigned to port B[5]
;
;
```

```
; NLDAC active low DAC load control
;————————————————————————————————————————
; output line. A high to low transition on this line updates the
; Low_DAC and High_DAC output values
;
; NLDAC is assigned to port B[6]
;
;
; DACRDY active high ready output signal
;————————————————————————————————————————
; output line. Indicates that the control loop error has been
; within a +/- 4ppm range for at least 250 ms
;
; DACRDY is assigned to port B[7]
;
;
; SCKAD external serial clock line for the ADC
;————————————————————————————————————————
; output line. ADC external serial clock. An external 10Kohm
; pull-down resistor is necessary on this line for correct
; power-up configuration.
;
; SCKAD is assigned to port A[0]
;
;
; SDOAD serial data line from ADC
;————————————————————————————————————————
; input line. Used to read ADC serial data.
;
; SDOAD is assigned to port A[1]
;
;
;
; NWRUI active low user interface write control
;————————————————————————————————————————
; input line. The user must bring this line low in order to update
; the DAC input value. A minimum low and high time is required !
;
; NWRUI is assigned to TOCKI counter input pin
;
;
;
```

```
; The spare I/O pins A[3:2] are configured as outputs and maintained LOW.
;
;;;;;;;;;;;;;;;;;;;;;;;;;;;;;;;;;;;;;;;;;;;;;;;;;;;;;;;;;;;;;;;;;;

        list        p=16c55A            ; list directive to define processor
        #include    <p16c5x.inc>        ; processor specific variable definitions

        __CONFIG    _CP_OFF & _WDT_ON & _XT_OSC




;VARIABLE DEFINITIONS
;====================

UB0         EQU     H'0008'             ; user input word buffer LSB
UB1         EQU     H'0009'             ; user input word buffer second byte
UB2         EQU     H'000A'             ; user input word buffer MSB

U0          EQU     H'0008'             ; user input word LSB
U1          EQU     H'000B'             ; user input word second byte
U2          EQU     H'000C'             ; user input word MSB

CON         EQU     H'000D'             ; control byte

ADC0        EQU     H'000E'             ; ADC conversion result LSB
ADC1        EQU     H'000F'             ; ADC conversion result second byte
ADC2        EQU     H'0010'             ; ADC conversion result third byte
ADC3        EQU     H'0011'             ; ADC conversion result MSB

ADCC        EQU     H'0012'             ; ADC curvature correction byte

ER0         EQU     H'000E'             ; control loop error LSB
ER1         EQU     H'000F'             ; control loop error second byte
ER2         EQU     H'0010'             ; control loop error third byte
ER3         EQU     H'0011'             ; control loop error MSB

DL0         EQU     H'0013'             ; Low_DAC LSB
DL1         EQU     H'0014'             ; Low_DAC second byte
DL2         EQU     H'0015'             ; Low_DAC third byte
DL3         EQU     H'0016'             ; Low_DAC MSB

INDX        EQU     H'0017'             ; index variable

TMPV        EQU     H'0018'             ; temporary variable


    #define OPRDF   0x2F                ; OPTION register default value
    #define CONDF   0x80                ; CON register default value
```

```
;HARDWARE ASSIGNMENT DEFINITIONS
;================================

    #define  DBUS    PORTC          ; 8bit I/O data bus


    #define  REGCN   PORTB          ; register control port
    #define  REGDF   0x7F           ; register control port default value
    #define  NCSR0   PORTB,0        ; LSB input register active low output enable
    #define  NCSR1   PORTB,1        ; second byte input register active low output enable
    #define  NCSR2   PORTB,2        ; MSB input register active low output enable

    #define  NCSD0   PORTB,3        ; Low_DAC active low write enable
    #define  NCSD1   PORTB,4        ; High_DAC active low write enable
    #define  ADDAC   PORTB,5        ; address bit for Low_DAC and High_DAC
    #define  NLDAC   PORTB,6        ; active low load control for Low_DAC and High_DAC

    #define  DACRDY  PORTB,7        ; 20bit_DAC ready indicator


    #define  ADCCN   PORTA          ; ADC control port
    #define  ADCTR   0x02           ; ADC control port configuration
                                    ; SDOAD input, the rest outputs
    #define  ADCDF   0x02           ; ADC control port default value
    #define  SCKAD   PORTA,0        ; ADC external serial clock
    #define  SDOAD   PORTA,1        ; ADC serial data output




;THE CODE
;================================

RESET      ORG     0x1FF           ; processor reset vector
           goto    start

           ORG     0x000

                                   ;Initialization procedure
                                   ;────────────────────────
start      movlw   ADCDF           ;write ADC control port default value
           movwf   ADCCN           ;
           movlw   ADCTR           ;set the I and O pin states for the
           tris    ADCCN           ;ADC control port
                                   ;
           movlw   REGDF           ;write register control port default value
           movwf   REGCN           ;
           clrw                    ;set register control port pins as
           tris    REGCN           ;output only
                                   ;
           movwf   DBUS            ;set DBUS default value of 0
           tris    DBUS            ;set DBUS as output
                                   ;
           movlw   OPRDF           ;set OPTION register default value
           option                  ;
                                   ;
```

```
          clrf    TMR0            ;
          btfss   STATUS,NOT_TO   ;if this is not a power-on reset
          movwf   TMR0            ;load Timer0 with a nonzero value
                                  ;to force an initial read of the
                                  ;external input register
                                  ;
          clrf    DL3             ;initialize {DL3, DL2, DL1, DL0}=0
          clrf    DL2             ;
          clrf    DL1             ;
          clrf    DL0             ;
          clrf    U2              ;initialize {U2, U1, U0}=0
          clrf    U1              ;
          clrf    U0              ;
                                  ;
          movlw   CONDF           ;set CON variable default value
          movwf   CON             ;
                                  ;prepare to trigger a new ADC conversion
                                  ;after completing a hardware update
          movlw   0x20            ;read and discard 32 serial bits from
          movwf   INDX            ;the ADC
                                  ;
          goto    iupdt           ;go to the hardware update function

                                  ;ADC output buffer flush function
                                  ;--------------------------------------
fladc     movlw   0x20            ;reads and discards 32 serial bits from
          movwf   INDX            ;the ADC

                                  ;ADC dummy serial read function
                                  ;--------------------------------------
                                  ;reads and discards the number of serial
                                  ;bits indicated by the INDX variable
rddmy     bsf     SCKAD           ;low-to-high ADC serial clock edge
          bcf     SCKAD           ;high-to-low ADC serial clock edge
          decfsz  INDX,1          ;test if we read enough bits
          goto    rddmy           ;if No, read one more bit
          btfss   SDOAD           ;if Yes test that a new conversion has started
          goto    fladc           ;if No, there is an interface problem. Flush the
                                  ;ADC output buffer and start a new conversion
          goto    scan            ;if Yes restart the scan procedure

                                  ;external input register read function
                                  ;--------------------------------------
rduiu     movlw   0xFF            ;input register read routine
          tris    DBUS            ;set data bus in read mode (input)
          bcf     NCSR0           ;output enable for input reg. LSB
          nop                     ;wait for data bus to settle
          movf    DBUS,0          ;read input reg. LSB
          bsf     NCSR0           ;output disable for input reg. LSB

          bcf     NCSR1           ;output enable for input reg. second byte
          movwf   UB0             ;store input reg. LSB into input buffer
          movf    DBUS,0          ;read input reg. second byte
          bsf     NCSR1           ;output disable for input reg. second byte
```

```
        bcf     NCSR2           ;output enable for input reg. MSB
        movwf   UB1             ;store input reg. second byte into input buffer
        movf    DBUS,0          ;read input reg. MSB
        movwf   UB2             ;store input reg. MSB into input buffer

        clrw                    ;terminate input reg. read operation
        bsf     NCSR2           ;output disable for input reg. MSB
        tris    DBUS            ;return data bus to write mode

        clrf    TMR0            ;clear Timer0 to continue wait for a UIU event
        bcf     DACRDY          ;signal user that input data has been read

        bsf     CON,7           ;set UIU flag
        bcf     CON,6           ;clear RDY flag
        bcf     CON,5           ;clear RDY2 flag


                                ;scan procedure
                                ;--------------
                                ;monitors UIU and end-of-conversion events
scan    movf    TMR0,1          ;test if Timer0 = 0
        btfss   STATUS,Z        ;if Timer0=0 no UIU has occurred, skip next
        goto    rduiu           ;a user interface update has occurred
                                ;go and read the new DAC input data

        btfsc   SDOAD           ;test ADC end of conversion signal
        goto    scan            ;conversion not ready, rescan


                                ;ADC serial read function
                                ;------------------------
rdadc   movlw   0x1B            ;ADC conversion is done, read first 28 bits
        movwf   INDX            ;the first bit must be "0" to get here
                                ;so do not bother with it
rdbit   bsf     SCKAD           ;low-to-high ADC serial clock edge
        bcf     SCKAD           ;high-to-low ADC serial clock edge
        bcf     STATUS,C        ;move ADC output bit to carry. First clear carry
        btfsc   SDOAD           ;read ADC output bit
        bsf     STATUS,C        ;if ADC output is "1" set carry
        rlf     ADC0,1          ;load carry as msb of ADC result
        rlf     ADC1,1          ;and shift left all 4 bytes of the ADC result
        rlf     ADC2,1          ;
        rlf     ADC3,1          ;
        decfsz  INDX,1          ;test if all 28 bits have been read
        goto    rdbit           ;if not, continue
                                ;
                                ;we have skipped the first ADC bit (ADC bit31=0)
                                ;which has been tested as =0 when we detected the
                                ;end of conversion.
                                ;we have read 27 additional bits and have generated
                                ;27 clock pulses. To restart the conversion we must
                                ;produce the 5 remaining clock pulses
```

```
                                    ;verify validity of ADC serial data and format it
                                    ;————————————————————————————————————————
            btfsc  ADC3,2           ;test if the ADC bit30 is "0"
            goto   fladc            ;if not there is an interface problem. Flush the
                                    ;ADC output buffer and start a new conversion
                                    ;if yes, put the ADC result in 2's complement form

            movlw  0x03             ;first clear the 6 most significant bits in ADC3
            andwf  ADC3,1           ;

            btfsc  STATUS,Z         ;tests for the [ADC_B29,ADC_B28]=00 ADC output

            goto   rdend            ;if Yes the formatting is completed.
                                    ;in very early LTC2400 samples the ADC output code
                                    ;[ADC_B29,ADC_B28]=00 is valid

;           goto   fladc            ;for current LTC2400 devices improved error
                                    ;detection capability is obtained if the
                                    ;previous line is replaced with this line.
                                    ;The replacement is not mandatory.
                                    ;For current LTC2400 parts the output code
                                    ;[ADC_B29,ADC_B28]=00 is not valid thus it may
                                    ;be assumed that an ADC interface error has
                                    ;occurred. Flush the ADC output buffer and start
                                    ;a new conversion

            movlw  0x02             ;if No, convert ADC3 in 2's complement form
            btfss  ADC3,1           ;
            movlw  0xFE             ;
            xorwf  ADC3,1           ;

                                    ;curvature correction calculator
                                    ;————————————————————————————————
                                    ;first order curvature correction multiplier
                                    ;use ADC2[7:0] as a 2's complement number
rdend       movf   ADC2,0           ;calculate abs(ADC2)
            btfsc  ADC2,7           ;if ADC2[7]=0  w =  ADC2
            comf   ADC2,0           ;else          w = !ADC2
            movwf  ADCC             ;ADCC=w=abs(ADC2)
                                    ;second order curvature correction multiplier
                                    ;use ADC2[6:0] as a 2's complement number
            movf   ADC2,0           ;calculate abs(ADC2[6:0])
            btfsc  ADC2,6           ;if ADC2[6]=0  w =  ADC2
            comf   ADC2,0           ;else          w = !ADC2
            movwf  TMPV             ;TMPV=w=abs(ADC2[6:0])
            rrf    TMPV,0           ;w=TMPV/2 in order to scale the second order
                                    ;curvature correction
            andlw  0x1f             ;clear 3 MSB of w to complete calculation
            addwf  ADCC,0           ;w=abs(ADC2)+abs(ADC2[6:0])/2
            movwf  TMPV             ;TMPV contains the curvature correction multiplier
                                    ;
            clrf   ADCC             ;
            bcf    STATUS,C         ;clear carry for div-by-2 operation
            btfsc  CON,2            ;if CON[2]=1
```

```
          movwf   ADCC            ;ADCC=ADCC+abs(ADC2)
          rrf     TMPV,1          ;TMPV=TMPV/2
          movf    TMPV,0          ;
          bcf     STATUS,C        ;clear carry for div-by-2 operation
          btfsc   CON,1           ;if CON[1]=1
          addwf   ADCC,1          ;ADCC=ADCC+abs(ADC2)/2
          rrf     TMPV,1          ;TMPV=TMPV/2
          movf    TMPV,0          ;
          btfsc   CON,0           ;if CON[0]=1
          addwf   ADCC,1          ;ADCC=ADCC+abs(ADC2)/4

                                  ;code comparator procedure
                                  ;————————————————————————————
ccomp     btfss   CON,7           ;if the UIU flag is clear
          goto    ercalc          ;skip CON[3:0] and U0 update
          movlw   0xF0            ;else update CON[3:0]
          andwf   CON,1           ;clear CON[3:0]
          movlw   0x0F            ;extract new CON[3:0]
          andwf   UB0,0           ;from input buffer
          iorwf   CON,1           ;and load it

          movlw   0xF0            ;move UB[7:4] to U0[7:4]
          andwf   UB0,1           ;UB0 and U0 use the same
                                  ;physical location

                                  ;calculate control loop error
                                  ;————————————————————————————
ercalc    comf    ADCC,1          ;ADCC 1's complement
          comf    ADC0,1          ;ADC0 1's complement
          movlw   0x02            ;add carry-in for ADCC and for ADC0
                                  ;2's complement conversion
          clrf    TMPV            ;prepare carry-out accumulator
          addwf   UB0,0           ;w=carry-in + UB0
          btfsc   STATUS,C        ;if there is a carry-out
          incf    TMPV,1          ;accumulate it
          addwf   ADCC,0          ;w=carry-in + UB0 - ADCC
          btfsc   STATUS,C        ;if there is a carry-out
          incf    TMPV,1          ;accumulate it
          addwf   ADC0,1          ;ER0=UB0 - ADC0 - ADCC
                                  ;has same location as ADC0
          btfsc   STATUS,C        ;if there is a carry-out
          incf    TMPV,1          ;accumulate it
```

```
comf    ADC1,1        ;ADC1 1's complement
movlw   0xff          ;w=0xff (1's complement of ADCC second byte)
addwf   TMPV,0        ;w=0xff + carry-in
clrf    TMPV          ;prepare carry-out accumulator
btfsc   STATUS,C      ;if there is a carry-out
incf    TMPV,1        ;accumulate it
addwf   U1,0          ;w=0xff + carry-in + UB1
btfsc   STATUS,C      ;if there is a carry-out
incf    TMPV,1        ;accumulate it
addwf   ADC1,1        ;ER1=U1 - ADC1 - 0 + carry-in
                      ;has same location as ADC1
btfsc   STATUS,C      ;if there is a carry-out
incf    TMPV,1        ;accumulate it

comf    ADC2,1        ;ADC2 1's complement
movlw   0xff          ;w=0xff (1's complement of ADCC third byte)
addwf   TMPV,0        ;w=0xff + carry-in
clrf    TMPV          ;prepare carry-out accumulator
btfsc   STATUS,C      ;if there is a carry-out
incf    TMPV,1        ;accumulate it
addwf   U2,0          ;w=0xff + carry-in + UB2
btfsc   STATUS,C      ;if there is a carry-out
incf    TMPV,1        ;accumulate it
addwf   ADC2,1        ;ER2=U2 - ADC2 - 0 + carry-in
                      ;has same location as ADC2
btfsc   STATUS,C      ;if there is a carry-out
incf    TMPV,1        ;accumulate it

comf    ADC3,1        ;ADC3 1's complement
decf    TMPV,1        ;ADCC 2's complement term. The next
                      ;carry-in is not useful - discard
movf    TMPV,0        ;w=carry-in
addwf   ADC3,1        ;ER3= 0 - ADC3 - 0 + carry-in
                      ;has same location as ADC3

btfsc   CON,7         ;test if the UIU flag is set
goto    lduiu         ;go to U1, U2 update
```

```
                 ;error comparator
                 ;————————————
                 ;calculate absolute value of loop error and
                 ;compare loop error magnitude with the 4ppm
                 ;threshold
movf   ER3,0     ;W = ER3
btfsc  ER3,7     ;test if {ER3, ER2, ER1, ER0} < 0
comf   ER3,0     ;if yes  W = -ER3
btfss  STATUS,Z  ;test if W=0
goto   nrdy      ;if not absolute error >= 4ppm

movf   ER2,0     ;W = ER2
btfsc  ER3,7     ;test if {ER3, ER2, ER1, ER0} < 0
comf   ER2,0     ;if yes  W = -ER2
btfss  STATUS,Z  ;test if W=0
goto   nrdy      ;if not absolute error >= 4ppm

movf   ER1,0     ;W = ER1
btfsc  ER3,7     ;test if {ER3, ER2, ER1, ER0} < 0
comf   ER1,0     ;if yes  W = -ER1
btfss  STATUS,Z  ;test if W=0
goto   nrdy      ;if not absolute error >= 4ppm

movf   ER0,0     ;W = ER0
btfsc  ER3,7     ;test if {ER3, ER2, ER1, ER0} < 0
comf   ER0,0     ;if yes  W = -ER0
andlw  0xC0      ;keep only W[7:6] which are bits >= 4ppm
btfss  STATUS,Z  ;test if W[7:6]=0
goto   nrdy      ;if not absolute error >= 4ppm

                 ;if we are here the absolute loop error is
                 ;less than 4 ppm. Set the flags and
                 ;scale the loop error.
btfsc  CON,6     ;test if RDY flag is already set
bsf    CON,5     ;if Yes, set RDY2 flag
bsf    CON,6     ;set RDY flag in any case
```

```
                              ;error scaling
                              ;————————————
                              ;reduce error correction value for loop
                              ;damping and ADC noise reduction
          btfsc  CON,3        ;test if CON[3]=0
          goto   div4         ;if Yes ER0=ER0/4
                              ;if No  ER0=ER0/16
          rrf    ER0,1        ;*1/2
          rrf    ER0,0        ;*1/2
          andlw  0x3F         ;clear 2 most significant bits
          btfsc  ER3,7        ;if {ER3, ER2, ER1, ER0} < 0
          iorlw  0xC0         ;set 2 most significant bits
          movwf  ER0          ;ER0=ER0/4
div4      rrf    ER0,1        ;*1/2
          rrf    ER0,0        ;*1/2
          andlw  0x3F         ;clear 2 most significant bits
          btfsc  ER3,7        ;if {ER3, ER2, ER1, ER0} < 0
          iorlw  0xC0         ;set 2 most significant bits
          movwf  ER0          ;ER0=ER0/4
          goto   eracc        ;go to error accumulator

                              ;load latest user input
                              ;————————————————————
lduiu     movf   UB1,0        ;
          movwf  U1           ;U1=UB1
          movf   UB2,0        ;
          movwf  U2           ;U2=UB2
nrdy      movlw  0x1F         ;
          andwf  CON,1        ;clear UIU, RDY and RDY2 flags

                              ;error accumulator
                              ;————————————————
                              ;adds the current loop error to the
                              ;previous Low_DAC control value
                              ;{DL3, DL2, DL1, DL0}={DL3, DL2, DL1, DL0}+
                              ;             +{ER3, ER2, ER1, ER0}
eracc     movf   ER0,0        ;the carry-in is 0
          clrf   TMPV         ;clear carry-in accumulator
          addwf  DL0,1        ;DL0=DL0+ER0
          btfsc  STATUS,C     ;if there is a carryout
          incf   TMPV,1       ;accumulate in carry-in
          movf   TMPV,0       ;load carry-in
          clrf   TMPV         ;clear carry-in accumulator
          addwf  ER1,0        ;W=ER1+carry-in
          btfsc  STATUS,C     ;if there is a carryout
          incf   TMPV,1       ;accumulate in carry-in
          addwf  DL1,1        ;DL1=DL1+ER1
          btfsc  STATUS,C     ;if there is a carryout
          incf   TMPV,1       ;accumulate in carry-in
          movf   TMPV,0       ;load carry-in
          clrf   TMPV         ;clear carry-in accumulator
          addwf  ER2,0        ;W=ER2+carry-in
```

```
        btfsc   STATUS,C        ;if there is a carryout
        incf    TMPV,1          ;accumulate in carry-in
        addwf   DL2,1           ;DL2=DL2+ER2
        btfsc   STATUS,C        ;if there is a carryout
        incf    TMPV,1          ;accumulate in carry-in
        movf    TMPV,0          ;load carry-in
        addwf   ER3,0           ;W=ER3+carry-in
        addwf   DL3,1           ;DL3=DL3+ER3


                                ;Low_DAC control truncation
                                ;————————————————————————
                                ;limits the {DL3, DL2, DL1, DL0} range to
                                ; abs({DL3, DL2, DL1, DL0}) < 2^15 by
                                ;truncation
        btfss   STATUS,Z        ;test if DL3=0
        goto    negpot          ;if No, DL may be negative
                                ;if Yes, DL is positive
                                ;test for overflow (>= 2^15)
        movf    DL2,1           ;
        btfss   STATUS,Z        ;test if DL2=0
        goto    ovflow          ;if No, DL >= 2^15, must truncate
                                ;if Yes continue testing for overflow
        btfsc   DL1,7           ;test if DL1[7]=1
        goto    ovflow          ;if No, DL >= 2^15, must truncate
        goto    updt            ;if Yes we are done with DL range control

ovflow  clrf    DL3             ;if we arrive here DL >= 2^15. Must
        clrf    DL2             ;truncate to DL=2^15-1 => DL3=DL2=0
        movlw   0xFF            ;and DL1=0xEF, DL0=0xFF
        movwf   DL0             ;
        movwf   DL1             ;
        bcf     DL1,7           ;
        goto    updt            ;done with overflow correction
```

```
udflow      clrf    DL1             ;if we arrive here DL < -2^15. Must
            bsf     DL1,7           ;truncate to DL=-2^15-1 => DL3=DL2=0xFF
            clrf    DL0             ;and DL1=0x80, DL0=0
            movlw   0xFF            ;
            movwf   DL3             ;
            movwf   DL2             ;
            goto    updt            ;done with underflow correction

negpot      btfss   DL3,7           ;DL may be negative. Test if DL3[7]=1
            goto    ovflow          ;if No, DL > 2^15, must truncate
            incf    DL3,0           ;if Yes, DL <0.
            btfss   STATUS,Z        ;test if DL3=FF
            goto    udflow          ;if No, DL < -2^15, must truncate
            incf    DL2,0           ;if Yes continue testing for underflow
            btfss   STATUS,Z        ;test if DL2=FF
            goto    udflow          ;if No, DL < -2^15, must truncate
                                    ;if Yes continue testing for underflow
            btfss   DL1,7           ;test if DL1[7]=0
            goto    udflow          ;if No, DL < -2^15, must truncate
                                    ;if Yes we are done with DL range control

                                    ;Hardware update function
                                    ;————————————————————————
                                    ;Low_DAC and High_DAC update
                                    ;
                                    ;This is the hardware update function
                                    ;entry point for normal operation.
                                    ;
updt        movlw   0x05            ;prepare to generate the last 5 ADC external
            movwf   INDX            ;serial clock pulses
                                    ;when going to restart the scan procedure
                                    ;at the end of the hardware update function
                                    ;This will trigger a new ADC conversion.
                                    ;
                                    ;This is the hardware update function
                                    ;entry point for initial operation.
                                    ;The INDX variable has been initialized
                                    ;before to 0x2F which will generate
                                    ;32 serial clock pulses to the ADC thus
                                    ;starting a new conversion
                                    ;
```

```
iupdt       clrw                     ;set the data bus in write mode
            tris    DBUS             ;
            bcf     ADDAC            ;set DAC address for LSB
            movf    U1,0             ;load High_DAC LSB
            movwf   DBUS             ;
            bcf     NCSD1            ;write to High_DAC
            bsf     NCSD1            ;
            movf    DL0,0            ;load Low_DAC LSB
            movwf   DBUS             ;
            bcf     NCSD0            ;write to Low_DAC
            bsf     NCSD0            ;
            bsf     ADDAC            ;set DAC address for MSB
            movf    U2,0             ;load High_DAC MSB
            movwf   DBUS             ;
            bcf     NCSD1            ;write to High_DAC
            bsf     NCSD1            ;
            movlw   0x80             ;change DL1 to offset binary
            xorwf   DL1,0            ;load Low_DAC MSB
            movwf   DBUS             ;
            bcf     NCSD0            ;write to Low_DAC
            bsf     NCSD0            ;
            bcf     NLDAC            ;load Low_DAC and High_DAC
            bsf     NLDAC            ;
                                     ;DACRDY output update
            btfsc   CON,5            ;test if RDY2 flag is set
            bsf     DACRDY           ;if Yes, set DACRDY output
            btfss   CON,5            ;if No
            bcf     DACRDY           ;and only if No, clear DACRDY output
                                     ;
            clrwdt                   ;clear watch dog timer
                                     ;
            goto    rddmy            ;generate the necessary number
                                     ;of ADC serial clock pulses in order
                                     ;to start a new conversion

            END                      ; directive 'end of program'
```

Appendix E
Linearity and output range options

The LTC2400 used as the feedback A-to-D element in the DAC has a typical ± 2ppm residual nonlinearity. Figure E1's lower curve shows this, along with the first order correction necessary (upper curve) to get nonlinearity inside 1ppm (center curve). If true 1ppm performance is necessary, the software based correction described in Appendix D can be utilized. The software generates the desired "inverted bowl" correction characteristic. The correction may be set to complement the residual nonlinearity characteristics

of any individual LTC2400 via DIP switches at the code comparator.

The LTC2410 offers another approach to improved linearity. This LTC2400 variant has improved linearity but specifies a maximum 2.5V input range. Figure E2 divides the DAC output with a precision resistor ratio set, allowing LTC2410 use while maintaining the 5V full-scale output. The disadvantage of this approach is the ratio set's additional 0.1ppm/°C and 5ppm/year error contribution.[1] Figure E3 is similar, although the ratio set's new value permits a 10V full-scale output.

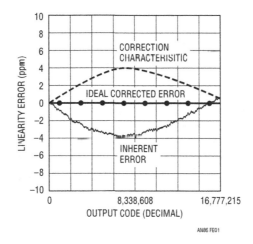

Figure E1 • LTC2400 A-to-D Inherent Residual Linearity Error (Lower), Correction Characteristic (Upper) and Resultant Corrected Linearity (Center). Correction Ensures <1ppm Nonlinearity

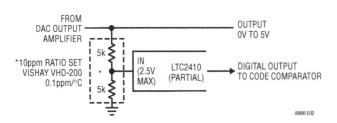

Figure E2 • Precision Resistor Ratio Set Divides DAC Output, Permitting Higher Inherent Linearity LTC2410 Utilization. Disadvantage Is 5ppm/Yr and 0.1ppm/°C Additional Drift Terms

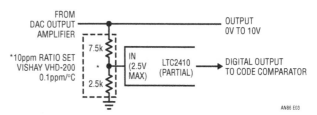

Figure E3 • Similar to Figure E5, Except 3:1 Ratio Set Permits 10V Output While Accommodating LTC2410's 2.5V Input

Note 1: The strata are becoming rarified when "error contribution" is delineated in fractional parts-per-million and the yearly drift rate noted.

Appendix F
Output stages

Some applications may require outputs other than the text circuit's 0V to 5V range. The simplest variation is a bipolar output, shown in Figure F1. The circuit, a summing inverter, subtracts the DAC output from a reference to obtain a bipolar output. Resistor and reference values may be varied to obtain different output excursions. The LT1010 output buffer provides drive capability and the chopper stabilized amplifier maintains 0.05μV/°C stability. The resistors introduce a 0.3ppm/°C error contribution[1]

Figure F2 yields voltage gain by dividing the DAC output prior to its application to the feedback A-to-D. In this case, the 1:1 divider ratio sets a 10V output, assuming an A-to-D reference of 5V. As in Figure F1, the resistors add a slight temperature error, about 0.1ppm/°C for the ratio set specified.[2]

Figure F3 uses active devices for voltage outputs as high as ±100V. The discrete high voltage stage is driven in closed-loop fashion by a chopper stabilized amplifier. Q1 and Q2 furnish voltage gain, and feed the Q3-Q4 emitter follower outputs. Q5 and Q6 set current limit at 25mA by diverting output drive when voltages across the 27Ω shunts become too high. The local 1M-50k feedback pairs set stage gain at 20, allowing LTC1152 drives to cause full ± 120V output

swing. The local feedback reduces stage gain-bandwidth, making dynamic control easier. This stage is relatively simple to frequency compensate because only Q1 and Q2 contribute voltage gain. Additionally, the high voltage transistors have large junctions, resulting in low f_ts, and no special high frequency roll-off precautions are needed. Because the stage inverts, feedback is returned to the amplifier's positive input. Frequency compensation is achieved by rolling off the amplifier with the local 0.005μF-10k pair.

Heating and voltage coefficient errors are minimized in the feedback term by using four individual resistors. Trimming involves selecting the indicated resistor for exactly 100.0000V output with the DAC at full scale.

Figure F4 increases output current capability with a current gain stage inside the DAC output amplifier's feedback loop. This stage replaces the LT1010 150mA buffer shown in the text. The figure shows two options, differing in output capacity. It is worth noting that as output current rises, wiring resistance becomes a large potential error term. For example, at only 10mA output, 0.001Ω of wiring resistance introduces 10μV drop—a 2ppm error. Because of this, heavy loads should be supplied via short, highly conductive paths and remote sensing employed.

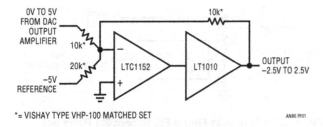

Figure F1 • Precision Resistors and Chopper Stabilized Output Amplifier Allow Bipolar DAC Output. Trade-Off Is ≈0.3ppm/°C Additional Resistor Based Error

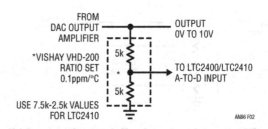

Figure F2 • ×2 Voltage Gain Obtained By Feedback Division at A-to-D. Slight Increase in Overall Temperature Coefficient Results

---

Note 1: See Note 1 in Appendix E.
Note 1: See Note 1 in Appendix E.

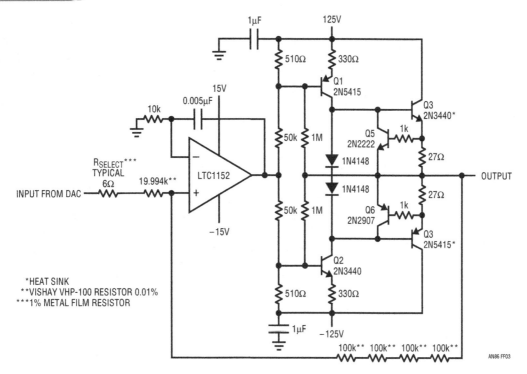

Figure F3 • High Voltage Output Stage Delivers ±100V at 25mA. Multiple Feedback Resistors Minimize Dissipation and Voltage Coefficient Effects

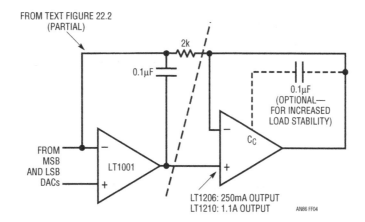

Figure F4 • LT1206/LT1210 Output Stages Supply 250mA and 1.1A Loads, Respectively. Remote Sensing Is Usually Necessary to Compensate IR Drops

Appendix G
Measuring DAC settling time

Measuring the 20-bit DAC's output settling time is a challenging task. Although the time scale involved is relatively slow, the 5µV LSB step size presents problems. The issue reduces to obtaining a great deal of gain without inducing overdrive in the monitoring oscilloscope. Such overdrive will corrupt the measurement, rendering displayed results meaningless.

Figure G1 is a solution. The DAC output is resistively balanced against a precision variable reference supply, adjustable in 0.5µV steps.[1] The circuit's remainder constitutes a clamped, distributed gain of 2000 amplifier. Diode clamping, placed at each gain stage input, prevents saturation from occurring even with large DAC-reference supply imbalances. The distributed gain allows 10kHz bandwidth while maintaining clamping effectiveness. The monitoring oscilloscope, operating at 5mV or 10mV/DIV (5µV to 10µV at the DAC output), can readily discern 5µV settling without incurring deleterious overdrive.

Layout and construction of this circuit requires care. Figure G2 shows construction details. A linear layout minimizes parasitic feedback paths, preventing oscillation. The DAC input step is fully shielded, preventing feedthrough to various sensitive points within the amplifier. Finally, the entire circuit is built into a shielded enclosure to minimize effects of stray RF and pick up.

The circuit is tested by applying a test step that settles much faster than the DAC. Figure G3 uses a mercury wetted reed relay based pulse generator to supply the step. The unit noted is commercially produced, although similar results are obtainable with standard mercury based reed relays. When the relay opens the circuit's output settles essentially instantaneously (Figure G4) relative to DAC speed and settling time amplifier bandwidth.

Figure G1's response is tested by grounding one of its inputs and driving the other with the pulse generator. Figure G5 shows settling to within 1ppm (±5µV) in 2ms. This is much faster than the DAC settles, lending confidence to text Figures 22.6 and 22.7 indicated results.

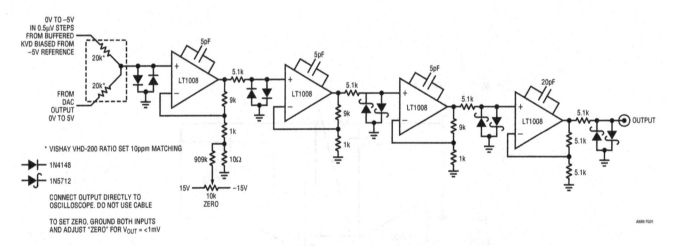

Figure G1 • Clamped, Distributed Gain-of-2000 Amplifier Permits DAC Settling Time Measurement Without Saturation Effects

---

Note 1: See Appendix C for details on such a supply.

AN86 FG02.tif

Figure G2 ● Settling Time Amplifier Construction. Bandwidth Is Only 10kHz, Although Gain of 2000 Necessitates Layout Care to Avoid Parasitic Feedback Induced Oscillation. Input (Photo Lower Left) Is Fully Shielded, Preventing Radiative Feedthrough to Amplifier. Enclosure Shields Circuit from Stray RF and Pickup

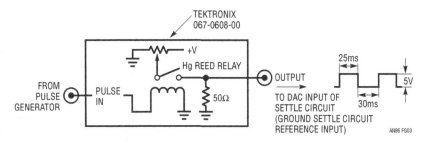

Figure G3 ● Reed Relay Based Pulser Supplies Clean Step to Test Settling Time Circuit

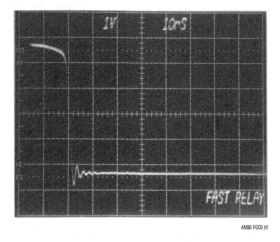

AN86 FG03.tif

Figure G4 ● Mercury Wetted Reed Relay Opens in 5 Nanoseconds, Settles Quickly to Zero. 500MHz Ring-Off Derives from Source-Termination Impedance Mismatch

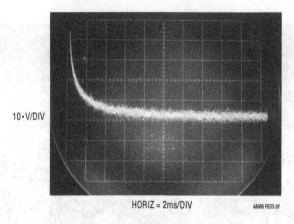

10•V/DIV

HORIZ = 2ms/DIV   AN86 FE03.tif

Figure G5 • Settling Time Circuit Responds to Test Step with 2ms Settling to ±1ppm (±5μV)

Appendix H
Microvolt level noise measurement

Verifying DAC output noise requires a quiet, high gain amplifier at the oscilloscope. Figure H1 shows one way to take the measurement. The input preamplifier, operating at a gain of 1000, supplies a high pass cutoff at 0.1Hz. It drives the oscilloscope via a 10Hz discrete low pass filter. The oscilloscope, set to 1mV/DIV, indicates 1 LLV/DIV referred to the preamplifier input. Figure H2 indicates DAC output noise well below an LSB, about 0.9μV. Equipment limitations set measurement noise floor at 0.2μV.

Figure H3 shows the noise measurement test setup. Note that the signal levels involved dictate a completely shielded, coaxial path from breadboard to oscilloscope.

Figure H4 lists some applicable high sensitivity amplifiers suitable for the noise measurement. Current generation oscilloscopes rarely have greater than 2mV/DIV sensitivity, although older instruments offer more capability. The figure lists representative preamplifiers and oscilloscope plug-ins suitable for noise measurement. These units feature wideband, low noise performance. It is particularly significant that many of these instruments are no longer

produced. This is in keeping with current instrumentation trends, which emphasize digital signal acquisition as opposed to analog measurement capability.

The monitoring oscilloscope should have exceptional trace clarity. In the latter regard high quality analog oscilloscopes are unmatched. The exceptionally small spot size of these instruments is well-suited to low level noise measurement.[1] The digitizing uncertainties and raster scan limitations of DSOs impose display resolution penalties. Many DSO displays will not even register the fine structure of the noise waveform.

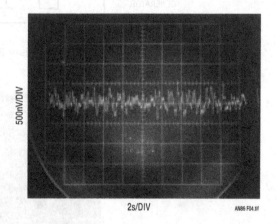

500nV/DIV

2s/DIV   AN86 F04.tif

Figure H2 • Indicated DAC Output Noise in a 0.1Hz to 10Hz Bandpass Is Below 1μV, About 0.2LSB. Equipment Limitations Set Measurement Noise Floor at 0.2μV

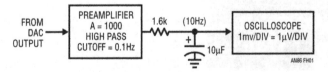

FROM DAC OUTPUT → PREAMPLIFIER A = 1000 HIGH PASS CUTOFF = 0.1Hz — 1.6k — (10Hz) + 10μF → OSCILLOSCOPE 1mv/DIV = 1μV/DIV   AN86 FH01

Figure H1 • Microvolt Noise Measurement Necessitates High Gain Preamplifier for Oscilloscope. Preamplifier and Discrete Filter Set 0.1Hz to 10Hz Measurement Bandpass

Note 1: In our work we have found Tektronix types 453, 453A, 454, 454A, 547 and 556 excellent choices. Their pristine trace presentation is ideal for discerning small signals of interest against a noise floor limited background.

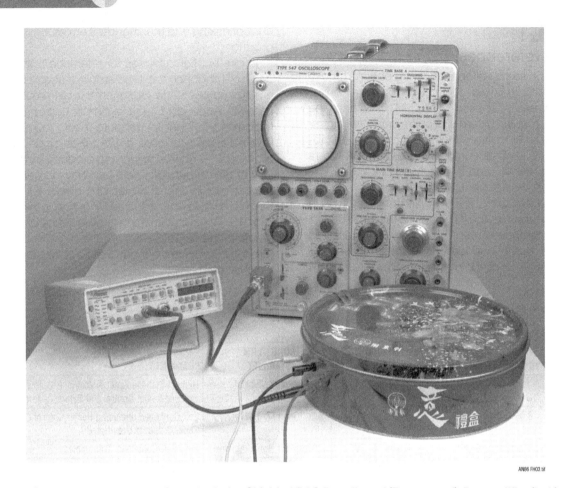

AN86 FH03.tif

Figure H3 • Noise Measurement Test Setup Includes Shielded DAC Breadboard (Foreground), Preamplifier (Left) and Low Pass Filter Attached to Oscilloscope (Center). Measurement Path Is Fully Coaxial

INSTRUMENT TYPE	MANUFACTURER	MODEL NUMBER	MAXIMUM BANDWIDTH	SENSITIVITY/GAIN	AVAILABILITY	COMMENTS
Differential Amplifier	Tektronix	1A7/1A7A	500kHz/1MHz	10µV/DIV	Secondary Market	Requires 500 Series Mainframe, Settable Bandstops
Differential Amplifier	Tektronix	7A22	1MHz	10µV/DIV	Secondary Market	Requires 7000 Series Mainframe, Settable Bandstops
Differential Amplifier	Tektronix	5A22	1MHz	10µV/DIV	Secondary Market	Requires 5000 Series Mainframe, Settable Bandstops
Differential Amplifier	Tektronix	ADA-400A	1MHz	10µV/DIV	Current Production	Standalone with Optional Power Supply, Settable Bandstops
Differential Amplifier	Tektronix	AM-502	1MHz	100,000	Secondary Market	Standalone with Optional Power Supply, Settable Bandstops
Differential Amplifier	Preamble	1822	10MHz	Gain = 1000	Current Production	Standalone, Settable Bandstops
Differential Amplifier	Stanford Research Systems	SR-560	1MHz	Gain = 50000	Current Production	Standalone, Settable Bandstops, Battery or Line Operation

Figure H4 • Some Applicable High Sensitivity, Low Noise Amplifiers. Trade-Offs Include Compatibility, Sensitivity and Availability

Appendix I
Voltage references

Figure I1 lists some voltage reference options for use with the DAC. The self-contained types are convenient and easily applied. The LM199A and the LTZ1000A require external circuitry but offer higher performance. All choices must be trimmed to establish absolute DAC accuracy. The LTZ1000A offers the highest stability and is discussed below.

Figure I2 shows the LTZ1000A and its support circuitry. A1 senses LTZ1000A die temperature and accordingly controls the IC heater via the 2N3904. A2 controls reference current. The Zener reference is sensed via Kelvin connections, minimizing voltage drop effects. A single point ground eliminates return current mixing and the attendant errors that would be produced.

Figure I3 offers choices for reference buffering. All employ a chopper stabilized amplifier augmented with a buffer output stage. Buffer error is extremely low, as noted in Appendix C. I3a, a simple unity-gain stage, transmits the input to the output with low error and minimal reference loading. I3b takes moderate gain, allowing a 7V reference input to produce (in this case) 10V at the output. I3c offers two ways to get 5V from the nominal 7V input. A precision divider lightly loads the reference in one case; the 5V output is taken at the LT1010. Reference loading is avoided by placing the divider at the output (optional case shown) and driving the A-to-D reference input from the divider output, which is permissible.

TYPE	VOLTAGE	INITIAL ACCURACY	TEMPERATURE DRIFT	LONG-TERM STABILITY	COMMENTS
LTZ1000A	7.2V	Minimum 7V Maximum 7.5V	0.05ppm/°C	4ppm/Yr Typical	Highest Stability Zener Available. Requires External Heater Control and Reference Buffer Circuitry
LM199A	6.95V	2%	0.5ppm/°C	10ppm/Yr Typical	Self-Contained, Including Heater Control Circuitry. Zener Output Is Unbuffered
LT1021	5V, 7V, 10V	0.05V (7V)	2ppm/°C (7V)	20ppm/kHr Noncumulative	Fully Self-Contained. Trimmable
LT1027	5V	0.02%	2ppm/°C	20ppm/kHr Noncumulative	Fully Self-Contained. Trimmable
LT1236	5V, 10V	0.05%	5ppm/°C	20ppm/kHr Noncumulative	Fully Self-Contained. Trimmable

Figure I1 • Reference Choices Compared for Output Voltage, Accuracy and Stability. Highest Stability Types Require External Circuitry

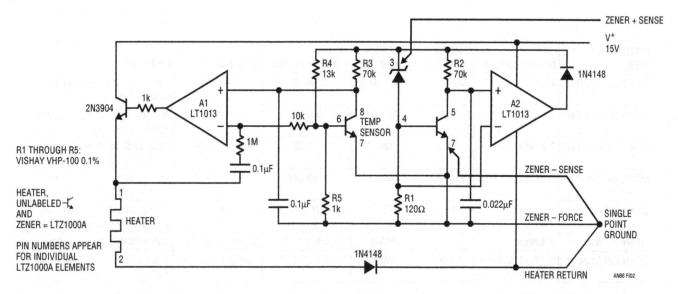

Figure I2 • 7V Reference Includes A1 Heater Control Amplifier, A2 Zener Current Regulator and LTZ1000A Zener. Note Zener Kelvin Connections and Single Point Ground

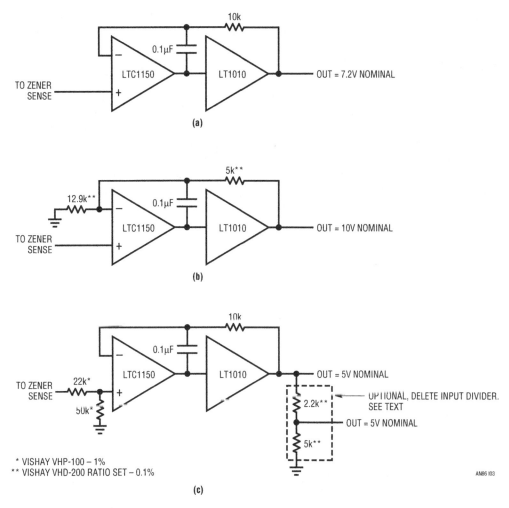

Figure I3 • Chopper Stabilized Reference Buffer Options Include Unity Gain (a), 10V (b) and 5V (c) Output. Trimming Is Required for Absolute Accuracy

Appendix J
Cables, connections, solder, component choice, terror and arcana

Subtle parasitic effects can have pronounced and seemingly inexplicable effects on low level circuit performance. Perhaps the most prevalent detractor to microvolt level circuitry is unintended thermocouples. Considerable discussion for dealing with thermocouples appeared in Appendix C and should be considered preliminary to this appendix.

In 1822, Thomas Seebeck, an Estonian physician, accidentally joined semicircular pieces of bismuth and copper (Figure J1) while studying thermal effects on galvanic arrangements. A nearby compass indicated a magnetic disturbance. Seebeck

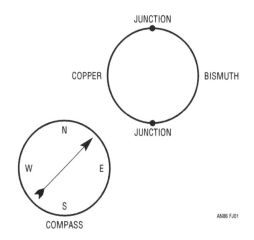

Figure J1 • The Arrangement for Dr. Seebeck's Accidental Discovery of "Thermomagnetism"

experimented repeatedly with different metal combinations at various temperatures, noting relative magnetic field strengths. Curiously, he did not believe that electric current was flowing and preferred to describe the effect as "thermomagnetism." He published his results in a paper, "Magnetische Polarisation der Metalle und Erze durch Temperatur-Differenz" (see references).

Subsequent investigation has shown the "Seebeck Effect" to be fundamentally electrical in nature, repeatable and quite useful. Thermocouples, by far the most common transducer, are Seebeck's descendants. Unfortunately, unintended and unwanted thermocouples are also Seebeck's progeny.

In low drift circuits, unwanted thermocouples are probably the primary source of error. Connectors, switches, relay contacts, sockets, wire and even solder are all candidates for thermal EMF generation. It is relatively clear that connectors and sockets can form thermal junctions. However, it is not at all obvious that junctions of wire from different manufacturers can easily generate 200nV/°C— four times a precision amplifier's drift specification! Figure J2 shows a plot obtained for such a wire junction. Even solder can become an error term at low levels, creating a junction with copper or Kovar wires or PC traces (see Figure J3). Figure J4 lists thermocouple potentials for some common materials found in electronic assemblies.

The unusually energetic response of Cu-CuO necessitated the treatment described in Appendix C (Figure C7 and associated text) for cleaning DVM and Kelvin-Varley divider connections. Readers finding this figure's information seemingly academic should be awakened by Figure J5. This chart lists thermoelectric potentials for commonly employed laboratory connectors. Thermocouple activity of some types is more than 20 times greater than others. Be careful!

Minimizing thermal EMF induced errors is possible if judicious attention is given to circuit board layout. In general, it is good practice to limit the number of junctions in the signal path. Avoid connectors, sockets, switches and other potential error sources to the extent possible. In some cases this will not be possible. In these instances, attempt to balance the number and type of junctions in the signal path so that differential cancellation occurs. Doing this may involve deliberately creating and introducing

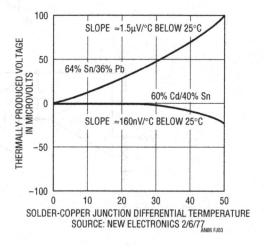

Figure J3 • Solder-Copper Thermal EMFs. Cd/Sn Has Notably Lower Activity but Is Toxic, Not Available and Not Recommended

MATERIALS	POTENTIAL (µV/°C)
Cu-Cu	< 0.2
Cu-Ag	0.3
Cu-Au	0.3
Cu-Cd/Sn	0.3
Cu-Pb/Sn	1 to 3
Cu-Kovar	40
Cu-Si	400
Cu-CuO	1000

Source: Low Level Measurements, Keithley Instruments, 1984 (see References)

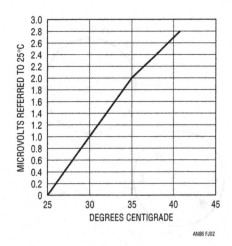

Figure J2 • Thermal EMF Generated by Two "Identical" Copper Wires Due to Oxidation and Impurities

Figure J4 • Thermoelectric Potentials for Various Materials Indicates Inadvisability of Mixing Materials in Signal Path. Cu-Cu Connections (Chart Top) Must Be Kept Clean or 5000:1 Degradation Occurs As They Oxidize (Chart Bottom)

CONNECTION TYPE	DESCRIPTION	THERMOELECTRIC POTENTIAL (μV/°C)
BNC-BNC Mate		0.4
BNC-Banana Adapter		0.35
BNC-BNC "Barrel" Adapter		0.4
Male/Female Banana Mate Sample #1		0.35
Male/Female Banana Mate Sample #2		1.1
Male/Female Banana Mate (Type Specified for Low Thermal Activity) Sample #3		0.07
Copper Lug-Copper Banana Binding Post		0.08
Copper Lug-Standard Banana Binding Post		0.5
Plated Lug-Copper Banana Binding Post		1.7

Figure J5 ● Measured Thermoelectric Potentials for Some Common Laboratory Connectors. Pronounced Difference Between "Banana" Samples Is Due to Manufacturer's Materials Choice. Note That Copper Lug/Copper Banana Post Has 20× Lower Activity Than Plated Lug/Copper Banana Post

junctions to offset unavoidable junctions. This can be a tricky procedure. Repeated deliberate temperature excursions may be necessary to determine the optimal number and placement of added junctions. Experimentation, tempered by a healthy reserve of patience and abundance of time, is required. This practice, borrowed from standards lab procedures, can be quite effective in reducing thermal EMF originated drifts. Figure J6 shows a simple example where a nominally unnecessary resistor is included to promote such thermal balancing. For remote signal sources connectors may be unavoidable. In these cases, choose a connector specified for relatively low thermal EMF activity and ensure a similarly balanced approach in routing signals through the connector along the circuit board and to circuitry. If some imbalance is unavoidable, deliberately introduce an intentional counterbalancing junction. In all cases maintain the differencing junctions in close physical proximity, which will keep them at the same temperature. Avoid drafts and temperature gradients, which can introduce thermal imbalances and cause problems. Figure J7 shows the LTC1150 set up in a test circuit to measure its temperature stability. The lead lengths

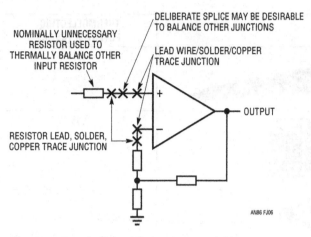

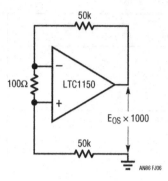

Figure J7 • Amplifier Drift Test Circuit. Thermal EMFs and Thermal Capacity at Each Input Must Be Similar for Cancellation to Occur

Figure J6 • Typical Thermal Layout Considerations Emphasize Minimizing and Differencing Parasitic Thermocouples. Thermal Mass at Amplifier Inputs Should Be Equal, Allowing Differenced Parasitic Thermocouple Outputs to Arrive Matched In Phase and Amplitude

of the resistors connected to the amplifier's inputs are identical. The thermal capacity each input sees is also balanced because of the symmetrical connection of the resistors and their identical size. Thus, thermal EMF induced shifts are equal in phase and amplitude and cancellation occurs. Very slight air currents can still affect even this arrangement. Figure J8 shows a strip chart of output noise with the circuit covered by a small styrofoam cup (HANDI-KUP™ Company Model H8-S) and with no cover in "still" air. This data illustrates why it is often prudent to enclose low level circuitry inside some form of thermal baffle.

Thermal EMFs are the most likely, but not the only, potential low level error source. Electrostatic and electromagnetic shielding may be required. Power supply transformer fields are notorious sources of errors often mistakenly attributed to amplifier DC drift and noise. A transformer's magnetic field impinging on a PC trace can easily generate microvolts across that conductor in accordance with well known magnetic theory. The circuit cannot distinguish between this spurious signal and the desired input. Attempts to eliminate the problem by rolling off circuit response may work, but often the filtered version of the undesired pickup masquerades as an unstable DC term. The most direct approach is to use shielded transformers but careful layout may be equally effective and less costly. A circuit that requires the transformer to be close by to achieve a good quality grounding scheme may be disturbed by the transformer's magnetic field. An RF choke connected across a

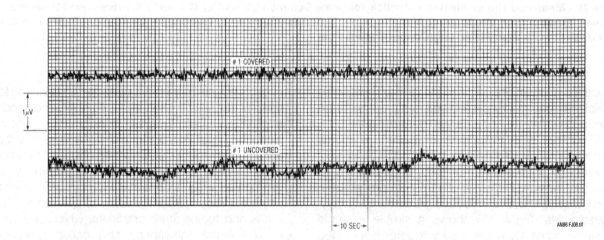

Figure J8 • Effect of Thermal Baffle on Low Frequency Amplifier Noise in "Still" Air. Amplifier Is Covered By Small Cup in Upper Trace, Uncovered in Lower Trace. Instability Worsens If Air Movement Increases

scope probe can determine the presence and relative intensity of transformer fields, aiding layout experimentation.

Another source of parasitic error is stray leakage current. Such leakage currents must be prevented from influencing circuit operation. The simplest way to do this is to connect leakage sensitive points via Teflon standoffs. Because the points never contact the PC board, stray leakage currents do not affect them. Although this approach is effective, its implementation may not be acceptable in production. Guarding is another technique for minimizing board leakage effects. The guard is a PC trace completely encircling the leakage sensitive points. This trace is driven at a potential equal to that of the point, preventing leakage to the "guarded" point. On PC boards, the guard should enclose the node(s) to be protected. Guarding was used to eliminate the effects of capacitor surface leakage in Appendix C's Figure C7.

Delta sigma ADC bridge measurement techniques

23

Mark Thoren

Introduction

Sensors for pressure, load, temperature, acceleration and many other physical quantities often take the form of a Wheatstone bridge. These sensors can be extremely linear and stable over time and temperature. However, most things in nature are only linear if you don't bend them too much. In the case of a load cell, Hooke's law states that the strain in a material is proportional to the applied stress—as long as the stress is nowhere near the material's yield point (the "point of no return" where the material is permanently deformed). The consequence is that a load cell based on a resistance strain gauge will have a very small electrical output—often only a few millivolts. In spite of this, instruments with 100,000 counts of resolution and more are possible. This Application Note presents some new approaches to high resolution measurement made possible by Linear Technology's family of 20- and 24-bit delta-sigma analog-to-digital converters.

Often the biggest hurdle in implementing a 20- or 24-bit bridge measurement circuit is moving beyond the conventional signal conditioning circuitry required for 12- to 16-bit ADCs. Simply substituting a 24-bit ADC in a circuit designed for a 12-bit ADC does not guarantee a 4096 fold increase in resolution. While performance may be improved, the 24-bit ADC may simply reveal the limitations of the analog front-end, and the full benefit of the 24-bit ADC will not be realized.

In a typical 12-bit measurement system, the signal conditioning amplifier requires a very high gain in order to make use of the full ADC input range. A sensor with a 10mV full-scale output requires a gain of 500 to use the full input range of a typical 5V input ADC. A filter may be required to reduce noise at the expense of settling time. Additional difficulties arise when dealing with the large common mode voltage typical of a bridge sensor. Most instrumentation amplifiers have a large discrepancy between the typical CMRR and guaranteed minimum, which may require an additional trim.

Linear Technology's differential input delta sigma ADCs address these concerns and make it possible to extract maximum resolution from a variety of bridge measurement devices. The outstanding resolution of these ADCs eliminates the need for amplification in many applications. Even in extreme cases where amplification is required, high resolution ADCs such as the LTC®2440 allow very modest amplifier gains. A fully differential amplifier topology eliminates the differential-to-single-ended stage of a typical

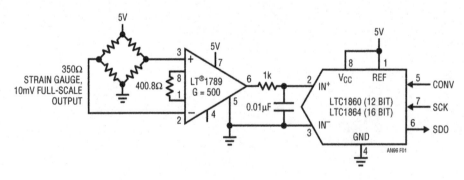

Figure 23.1 • Typical 12-Bit to 16-Bit System

Analog Circuit and System Design: A Tutorial Guide to Applications and Solutions. DOI: 10.1016/B978-0-12-385185-7.00023-8

3-amplifier instrumentation amplifier along with the CMRR limitations of this stage.

Linear Technology's delta sigma ADCs also simplify noise filtering. The internal SINC^4 FIR digital filter results in a very small noise bandwidth (6Hz at 7.5 samples/second) without the long settling time associated with an analog filter of equivalent bandwidth. Thus no filter is required to prevent an amplifier's noise from reducing resolution, or to reject 50Hz-60Hz line pickup.

The 120dB *minimum* CMRR of the ADC itself eliminates concern about the mid-supply common mode voltage of a typical bridge sensor. The ADC also lends itself readily to floating measurements with little additional circuitry. See Design Note 341.

Understanding the capability of the sensor is a critical aspect of measurement circuit design. Often a requirement is stated as a number of "bits of resolution" without taking into account the sensor's limitations, or even the sensor's output voltage. Lack of familiarity with the sensor can result in an improperly designed circuit (bad for product performance) or an overdesigned circuit (bad for financial performance). The applications that follow are developed starting at the sensor and considering its performance in terms of physical units; i.e., kgs, PSI, etc., disregarding the number of "bits" as a figure of merit until the conclusion.

Low cost, precision altimeter uses direct digitization

The availability of small, low cost, piezoresistive barometric pressure sensors has led to a plethora of portable devices with barometer and altimeter functions. The LTC2411 is a perfect companion to these sensors, as it is capable of outstanding resolution with no analog signal conditioning at all.

The NPP301-100 absolute pressure sensor from G.E. Lucas Novasensor is a piezoresistive device in an SO-8 package. Full-scale output is 20mV per volt of excitation voltage at one atmosphere (the sensor's full-scale input). Temperature coefficient of span is nominally −0.2%/°C (typical of piezoresistive sensors) and the temperature coefficient of offset is ±0.04%/°C.

The formula for altitude vs pressure, based on the 1976 U.S. standard atmosphere is:

$$\text{Altitude (Ft)} = \frac{10^{\left(\frac{\log_{10}\left(\frac{P}{P_0}\right)}{5.2558797}\right)}}{-6.8755856 \bullet 10^{-6}}$$

where P is the pressure at a given altitude and P_0 is the pressure at sea level.

The input resolution of the LTC2411 is $1.45\mu V_{RMS}$ (independent of reference voltage), providing a pressure resolution of:

$$1000\text{mB} \bullet \frac{1.45\mu V}{100\text{mV}} = 0.0145\text{mB}$$

(0.00043 inches of mercury)

when the bridge is excited with 5V. 0.0145mB corresponds to an altitude resolution of 5 inches at sea level, everything else being perfect.

The circuit shown in Figure 23.2 records pressure and temperature data once every 15 seconds to a 32k EEPROM. An FM75 pre-calibrated temperature sensor measures temperature. Each data point contains the entire 32-bit output word from the LTC2411 and temperature to the nearest 1/16°C. Three AA cells provide both power and a quiet reference voltage. The ratiometric bridge measurement cancels the effect of long-term drift of the battery voltage. The overall accuracy of the measurement is not affected by reduced battery voltage, however the noise floor increases proportionally.

Figure 23.3 shows altitude data from a steep hike to the top of Mission Peak (summit at 767m (2517ft) in Fremont, CA) on a cold evening. The raw altitude data clearly shows the barometric sensor's temperature sensitivity—the indicated altitude at the summit is more than 152m (500ft) too low due to the drop in temperature. Correcting for temperature effects based on the nominal −1900ppm/°C bridge temperature coefficient brings the maximum altitude to 780m (2560ft), much closer to the true value of 767m (2517ft). More accurate results are possible with further calibration.

How many bits?

The effective number of bits based on the ADC resolution and 100mV output of the sensor is

$$\text{LOG}_2\left(\frac{100\text{mV}}{1.45\mu V}\right) = 16.1 \text{ Bits}$$

The Marketing department will raise their eyebrows and devote whole ad campaigns around the extra 0.1 bit and five inch resolution. A skydiver or mountaineer, though, is more concerned with the overall absolute accuracy in a wide range of conditions than the impressive sounding number of bits.

Increasing resolution with amplifiers

Linear Technology's delta sigma ADCs eliminate the need for instrumentation amplifiers in many instances. However, if the ADC by itself is not adequate, a properly designed amplifier can increase the overall resolution.

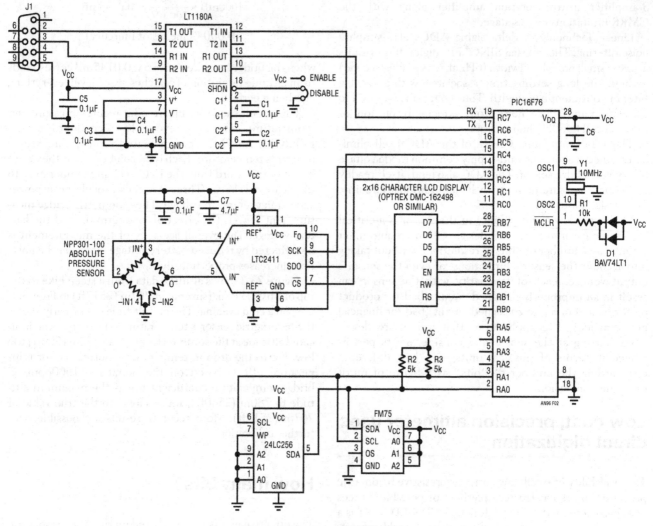

Figure 23.2 • Logging Altimeter

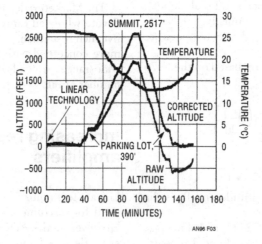

Figure 23.3 • Example Altitude Data

Figure 23.4 shows a differential amplifier with an input resolution of $375nV_{RMS}$ over the bandwidth of the LTC2431. The LTC2051 is an autozero amplifier with an input offset of $3\mu V$ and an offset drift of $30nV/°C$, ideal for applications in which long-term drift and accuracy are critical.

How much gain?

With high resolution ADCs, it is often unnecessary (and sometimes detrimental) to match the amplifier's output range to the input range of the ADC. Instead, limit the gain to a value such that that the overall resolution is limited by the amplifier's input noise.

Table 23.1 shows the effects of various gains on the performance of the circuit in Figure 23.4. At a gain of 1, the $5.6\mu V_{RMS}$ input noise of the LTC2431 ADC dominates. At a gain of 2, the noise of the ADC still dominates, but the input resolution is doubled. Around a gain of 30, the amplifier's input noise starts to dominate. Increasing gain to more than 40 only provides an incremental improvement in resolution. Increasing the gain further increases the measurement errors due to the LTC2051's bias current and finite open-loop gain. Even at a gain of 80, less than 20% of the ADC's input range is used if the input is a standard 10mV output strain gauge. Don't be tempted to bump the gain to 250 to use the whole $\pm 2.5V$ input range; the overall resolution will not be improved.

ADC response to amplifier noise

When properly designed, the resolution of this circuit is limited almost entirely by the amplifier's input noise. This is the case whenever the amplifier's input noise is lower than the ADC's input noise (there is no point in using an amplifier that is noisier than the ADC). So how does one determine the total noise from the ADC and amplifier data sheets?

This is an easy exercise with low resolution sampling converters. The ADC inputs typically have a flat frequency response out to the full-power bandwidth given in the data

Table 23.1

R_G	GAIN	FRACTION OF ADC SPAN USED	NOISE AT AMPLIFIER OUTPUT (μV_{RMS})	NOISE REFERRED TO INPUT (μV_{RMS})
(None)	1	1/500	6	6
82.5k	2	1/250	6	3
20.6k	5	1/100	6	1.2
9.15k	10	1/50	7	0.7
4.33k	20	1/25	9.5	0.475
2.11k	40	1/12	15	0.375
1.04k	80	1/6	26	0.325

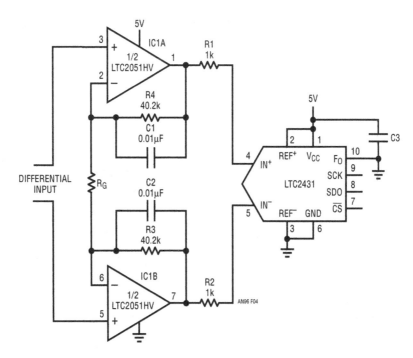

Figure 23.4 • Low Noise Differential Amplifier

sheet (which may be significantly higher than 1/2 of the ADC's maximum sample rate). The bandwidth of the amplifier-ADC combination is limited by either the amplifier's bandwidth or the ADC's full-power bandwidth. As long as the total noise that gets through this system (calculated from the amplifier's wideband noise and the square root of the smaller of the amplifier bandwidth or the ADC bandwidth) is a small fraction of an LSB, the problem is solved. Common noise reduction techniques are simple RC filters directly in front of the ADC or a 2nd order Sallen-Key filter.

Once again, Linear Technology's delta sigma ADCs change the approach to noise reduction. The full-power bandwidth of these devices is approximately 3Hz when operating in the 50Hz or 60Hz rejection mode (7.5sps or 6.25sps) and the response to wideband noise is equivalent to a 6Hz brickwall filter. It is exceedingly difficult to build an analog filter that will improve upon this without degrading some other parameter.

The 6Hz bandwidth figure predicts total noise fairly accurately for autozero amplifiers that have flat noise power densities from DC to some frequency higher than 6Hz. (Note that this bandwidth must be scaled appropriately if the conversion rate is changed by applying an external conversion clock. For instance, applying a 300.7kHz conversion clock to an LTC2411 will result in a 12Hz noise bandwidth.)

For most amplifiers, a good first estimate of the $\Delta\Sigma$ ADC's response to amplifier noise is the 0.1Hz to 10Hz peak-to-peak noise specification divided by a crest factor of six.[1] It is only fair to calculate noise from 10 seconds worth of ADC data for non-autozero amplifiers; longer sets of data contain 1/f noise components below 0.1Hz as well as thermal drift components.

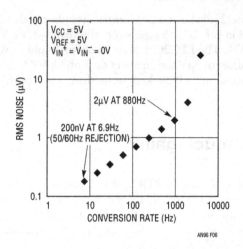

Figure 23.6 • LTC2440 Speed vs RMS Noise

The 0.1Hz to 10Hz noise for the LTC2051HV is $1.5\mu V_{P-P}$ or $1.5\mu V/6 = 250nV_{RMS}$. Since two amplifiers are used, the total noise is $250nV_{RMS} \cdot \sqrt{2} = 353nV_{RMS}$, very close to the measured value. (This rule has proved fairly accurate in a large number of experiments with both autozero and low-noise bipolar amplifiers.)

How many bits?

Table 23.2 extends Table 23.1 with another column for the effective number of bits based on the input-referred noise and a 10mV full-scale input signal at 7.5 sample/sec.

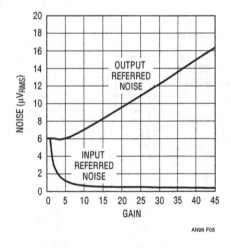

Figure 23.5 • Noise vs Gain

Note 1: Based on a 99.97% probability.

Table 23.2					
R_G	GAIN	FRACTION OF ADC SPAN USED	NOISE AT AMPLIFIER OUTPUT	NOISE REFERRED TO INPUT	EFFECTIVE NUMBER OF BITS
(none)	1	1/500	6	6	10.7
82.5k	2	1/250	6	3	11.7
20.6k	5	1/100	6	1.2	13
9.15k	10	1/50	7	0.7	13.8
4.33k	20	1/25	9.5	0.475	14.4
2.11k	40	1/12	15	0.375	14.7
1.04k	80	1/6	26	0.325	14.9

Faster *or* more resolution with the LTC2440

The LTC2440 has an input noise of $200nV_{RMS}$ at the base data output rate of 6.8sps—seven times lower than the LTC2411. Also, the variable oversample ratio allows speed and resolution to be optimized for a given application.

Decreasing the oversample ratio by a factor of 2 increases the data rate and effective bandwidth by a factor of 2, the total noise increases by a factor of $\sqrt{2}$, and the effective number of bits is reduced by 1/2 bit. (Increasing data rate by 4 reduces ENOB by 1 bit.) This relationship holds for data output rates from 6.8Hz to 880Hz. The highest two output rates include additional quantization noise, but the linearity, offset, and full-scale error do not change.

This increase in performance does not come entirely for free. The LTC2411 family devices are relatively easy to drive. Source impedances of up to 10k do not degrade performance if parasitic capacitance at the inputs is kept to a minimum, and the effective input impedance is on the order of 3MΩ to 5MΩ. On the other hand, the LTC2440 samples the inputs at 1.8Msps, or 11.5 times faster than the LTC2411. This means that any disturbance at the inputs must settle more than 11.5 times faster and the effective input impedance is approximately 110kΩ. There are two ways to meet the settling time requirement: lower the source impedance or keep any sampling disturbance from happening at all. Resistive sources located very close to the LTC2440 can meet these criteria; current shunts and 350Ω strain gauges may not require any signal conditioning at all. Most signals, though, require buffering of some sort. The LTC2440 data sheet contains a further discussion of these effects.

Sampling glitches from the LTC2440 contain very high frequency components (>250MHz). Loading the inputs to the LTC2440 with large (1μF) capacitors reduces the sampling glitches to an insignificant level. Of course, most amplifiers do not behave very well when driving this sort of load, so additional compensation is required.

Figure 23.7 shows a basic high impedance buffer for applications that require DC accuracy. During a conversion, the disturbances emanating from the ADC's inputs are attenuated by the ratio of $C_{SAMPLE}/1\mu F$, or ~100dB for the LTC2440, and the amplifier simply supplies the average sampling current.

The input noise of two LTC2051s is approximately equal to the input noise of the LTC2440 by itself, which means that the overall resolution is reduced by a factor of 2 (one effective bit) when both inputs are buffered.

Low noise bipolar amplifiers can be orders of magnitude quieter than chopper or autozero amplifiers at frequencies above the amplifier's 1/f corner. Low frequency noise and drift are not a problem for AC applications, but may require canceling for high accuracy DC

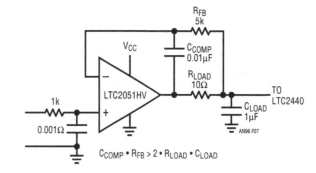

Figure 23.7 ● Basic Buffer for the LTC2440

measurements. AC excitation cancels any DC error sources and allows for extremely high-resolution bridge measurements.

The circuit in Figure 23.8 uses the LT1678 low noise amplifier and correlated double sampling to achieve a single reading resolution of $14nV_{RMS}$. The amplifiers are compensated to drive 1μF capacitors at the inputs to the LTC2440. The 200nV input noise of the LTC2440 allows for a conservative gain of 21, thus reducing errors associated with finite amplifier gain and large output swings. Even rail-to-rail amplifiers tend to perform better when their outputs are far away from the rails; this circuit only produces a 210mV differential output for a full scale 10mV input, keeping both outputs happily close to mid-supply. This is in contrast to circuits that use an instrumentation amplifier with a gain of 500 in order to fill the entire input range of a lower resolution ADC—another major advantage of the LTC2440.

Reversing the excitation to the bridge produces several benefits. First of all, autozero amplifiers are not required to maintain DC accuracy, since all offsets are cancelled. Parasitic thermocouples are cancelled as well since these voltages appear in series with the amplifier offsets. If the signal wires from the sensor are of even slightly different composition, a thermocouple pair will result. Consider the case of two "identical" copper wires that have a thermoelectric potential of 0.2μV/°C. A "temperature noise" of $1°C_{RMS}$ translates directly to $200nV_{RMS}$ measurement noise (Linear Technology Application Note 86, Appendix J).

Bridge excitation switching is accomplished by brute force with two Si9801DY FET half bridges. While this may seem less elegant than driving the excitation with force/sense amplifiers, it eliminates the need for supply voltages outside of 0V and 5V and the errors introduced are far less than those of most very high quality sensors. The total $R_{DS(ON)}$ of 0.13Ω translates to a mere 0.037% full-scale error and the 5000ppm/°C temperature coefficient of $R_{DS(ON)}$ only causes −2ppm/°C drift when used with a 350Ω bridge. These errors will be almost three times less with a 1000Ω bridge.

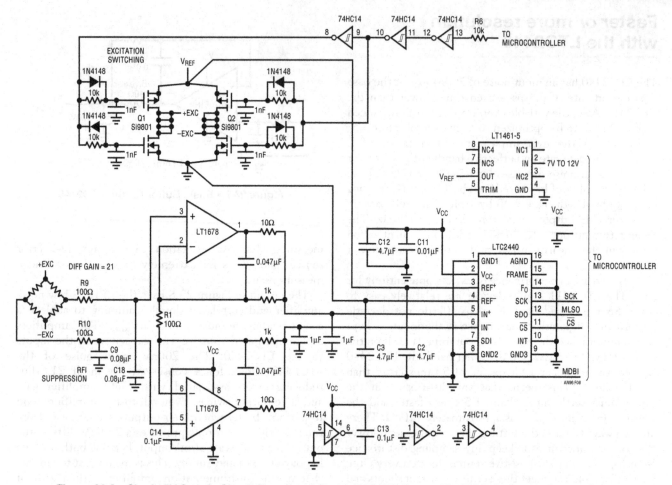

Figure 23.8 • Single 5V Supply Circuit Resolves 1 Part in 200,000 Over a 10mV Strain Gauge Output Span

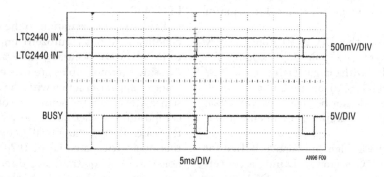

Figure 23.9 • Amplifier Waveforms

Figure 23.9 shows the inputs to the LTC2440 and the state of BUSY pin, indicating when sampling is taking place.

How many bits?

A full-scale input of 10mV and an input noise of $14nV_{RMS}$ gives 19.4 bits, noteworthy to the Marketing department. With this level of precision, the bridge itself is the dominant error source. Consider the Omega LC1001 series calibrated, temperature compensated load cells. This device has a full-scale error of 0.25%, a zero offset of 1% of full scale and a linearity of 0.05% of full scale. Temperature coefficient of both zero and span are 20ppm/°C. All of these errors can be calibrated out, including thermal drift if the temperature of the cell is measured. However, mechanical hysteresis and non-repeatability affect the maximum resolution that can be achieved under all conditions. The hysteresis specification for this load cell is 0.03% of full scale, or 1 part in 3333.

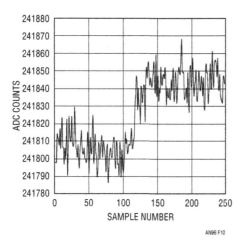

Figure 23.10 • 1 Gram Step

This does not necessarily mean that this will be an "11.7-bit" system—it depends on the application. If this load cell is used in a parts counting scale, then it will be zeroed frequently and the load will be quickly applied and measured. Long term creep effects that lead to hysteresis will be negligible.

To illustrate the real resolution of the circuit in Figure 23.8, half a penny was dropped onto a Tedea model 1240 load cell with a 200kg full-scale range, equivalent to adding 1 part in 200,000. As Figure 23.10 shows, the additional weight is clearly registered.

Appendix A
Frequency response of an AC excited bridge

The AC excitation technique described in the last application circuit is actually a simple finite impulse response (FIR) filter. Any signal not caused by the excitation voltage (which includes all offsets and amplifier input noise) sees a 2-tap FIR filter with coefficients of (0.5, −0.5). This produces a null at DC, f_S, $2f_S$, etc., and a gain of 1 (0dB) at $f_S/2$, $3f_S/2$, $5f_S/2$, etc., and the frequency response is $0.5 - 0.5 \cdot COS(2\pi \cdot f/f_S)$.

Noise at 3.4Hz is not attenuated, but it actually isn't quite so bad. The ADC's own internal filter is down 3dB at the sampling frequency so the net gain is −3dB at 3.4Hz. Although 3.4Hz is below the amplifier's 13Hz 1/f corner, it is still faster than any expected thermal effects.

The autozeroing process also affects the response of the input signal. Since the excitation reversal is calculated out in software, the result is a 2-tap FIR filter with coefficients of (0.5, 0.5). This produces a gain of 1 at DC, f_S, $2f_S$, etc., and a null at $f_S/2$, $3f_S/2$, $5f_S/2$, etc., and the frequency response

is $0.5-0.5 \cdot COS(2\pi \cdot f/f_S)$. This also reduces the bandwidth by a factor of 2, which reduces the noise by a factor of 2.

Since the LTC2440 input noise is Gaussian, averaging more samples further reduces the noise. (Noise is reduced by a factor of the square root of the number of readings averaged.) The optimum number of samples to average depends on the application—where the cost for additional noise reduction is increased settling time.

Appendix B
Measuring resolution, RMS vs peak-to-peak noise and psychological factors

When testing signal conditioning circuits, it is important to isolate error sources and be familiar with the limitations of the sensor. In the case of the circuit in Figure 23.8, the first test is to short the differential amplifier inputs together and bias them to mid-supply. This emulates a perfect load cell with zero offset and zero output impedance. In this configuration, the only output is the noise and offset of the circuit itself. Another test is to connect the inputs to a mid-supply bias through two 350Ω resistors, adding the effects of amplifier bias current and current noise.

The last test, before actually connecting the load cell, is to connect a DJ Instruments model DJ-101ST strain gauge bridge simulator. This instrument is a precision resistor bridge that simulates a strain gauge. Full-scale outputs of 1, 2, 3, 5, 10 and 15mV/V are provided, and for a given range the output can be adjusted from 0% to 120% of full scale in 10% steps. The objective is to emulate all of the properties of a strain gauge *except* measuring strain. This device verifies that Figure 23.8 maintains 19nV$_{RMS}$ resolution across its entire input range. The LT1461 was selected based on its low drift and high output current—there are more quiet references available, but this test shows that its noise performance is adequate. Applying a full-scale input will determine if the reference voltage is sufficiently quiet —a noisy reference will elevate ADC noise when the input is close to full scale. (The output code of an ADC is V_{IN}/V_{REF}. Thus noise in the output data that is due to reference noise is directly proportional to the input voltage, assuming the reference noise is small compared to V_{REF}.)

The last test is to actually connect a load cell. This circuit will detect 1 gram on a Tedea model 1240 load cell with a 200kg capacity, and it will detect one sheet of copier paper dropped on a stack of 20 reams (10,000 sheets). The importance of testing the circuit with a bridge simulator becomes clear when the air conditioner turns on—a single sheet of paper gets buried in noise. This exposes a very real limitation on high resolution weigh scales; a large tare weight turns the scale into a very sensitive seismograph. Other limitations include mechanical hysteresis, creep and linearity of the load cell.

RMS vs peak-to-peak noise

Another topic that often surfaces in high resolution applications is whether to use RMS noise or peak-to-peak noise. Some would eschew RMS noise as useless, asserting that peak-to-peak noise is the true measure of performance. However RMS noise has a distinct advantage—it is standardized and easy to calculate by taking the standard deviation of a set of ADC readings. This is only valid if the output noise is Gaussian, which is the case for all of Linear Technology's delta sigma ADCs.

Consider a high resolution ADC that has $1\mu V_{RMS}$ input noise and zero offset. With the inputs shorted together, it will produce a distribution of output values, 68.27% of which will be within $1\mu V$ of zero. Of course this implies that 31.73% of the readings are more than $1\mu V$ away from zero. It would be nice to know the true "peak-to-peak" noise of an ADC; some bound on the error for any given data point. But electrical noise does not work this way; there is always a nonzero probability that any one reading can have a value very far from the average. Fortunately there is a systematic way to relate the performance requirements of an application to the RMS noise level, but it does require a small dose of realism.

Say a weigh scale application requires 10,000 "counts" of resolution. If the load cell has a 10mV full-scale output, this implies a voltage resolution of $1\mu V$. The LTC2440 has $200nV_{RMS}$ input noise in its

highest resolution mode, so it should work. But one common rule says that the ratio of peak-to-peak to RMS (also called crest factor) is 6 for Gaussian noise, so the peak-to-peak noise is $1.2\mu V$— greater than the $1\mu V$ spec. The piece of information missing is the probability that a reading will be outside of the acceptable limit. If the noise at the ADC inputs is Gaussian, then the answer should lie in the normal error integral table found in the back of any statistics textbook.

Figure 23.12 shows 10,000 readings from an LTC2440 in the lowest noise mode (6.8 samples per second), and Figure 23.13 is a histogram of these readings that appears Gaussian at first glance.

The RMS noise of this particular LTC2440 is 194nV, slightly better than the 200nV typical spec. Table 23.3 compares the textbook probability that a reading will be within a certain crest factor to the actual measured values.

This shows that the noise at the input of the LTC2440 is indeed very close to Gaussian. It also shows that only one reading in 100 will be outside of a $\pm 0.5\mu V$ range. If the display shows 10,000 counts, one would have to stare at it for almost 15 seconds to see more than one count of noise. If this is acceptable, then assuming a crest factor of 5 is adequate and there is no need for additional signal conditioning. If it is not acceptable, follow the design procedure in the section on increasing resolution with amplifiers.

Psychological factors

Linear Technology's 20- and 24-bit ADCs have an LSB size that is smaller than the noise voltage (this is typical of any high resolution ADC). As a result, these devices will have a number of bits that do not hold

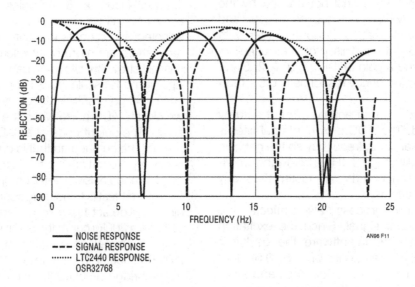

- —— NOISE RESPONSE
- --- SIGNAL RESPONSE
- ······ LTC2440 RESPONSE, OSR32768

AN96 F11

Figure 23.11 • Correlated Double Sampling Frequency Response

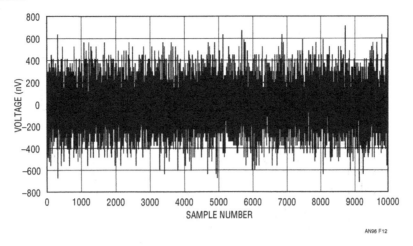

Figure 23.12

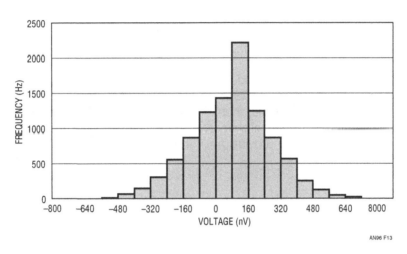

Figure 23.13

CREST FACTOR	TEXTBOOK PROBABILITY (%)	MEASURED PROBABILITY (%)
2	68.27	70.95
3	86.64	85.12
4	95.45	95.66
5	98.76	98.83
6	99.73	99.76
7	99.95	99.98

Table 23.3

steady for a constant input voltage. It is almost impossible to estimate ADC noise by looking at the binary data output. Looking at a decimal representation of the output data is not much better. In fact, the only way to get an objective opinion on the RMS noise level of a given circuit is to collect a set of data points and take the standard deviation.

This is a good lead-in to one last hint for designing circuits with Linear Technology's delta sigma ADCs. *Always* provide an easy way to get data points into a PC. Even in a handheld application that will never be attached to a computer in real life, the ability to analyze data will save an incredible amount of troubleshooting time. A convenient way to do this is to send data to the PC's serial port using the circuit shown in Figure 23.14.

Use a spare I/O pin to send the serial data and write a small test program that only takes data from the ADC and sends it to the PC. If all of the processor's I/O pins are used up, pick one that can be safely used for this purpose in a "test mode," such as a pin that normally reads the state of a pushbutton. It is not even necessary to use a hardware UART to generate the serial data; a simple "bit-bang" routine at 1200

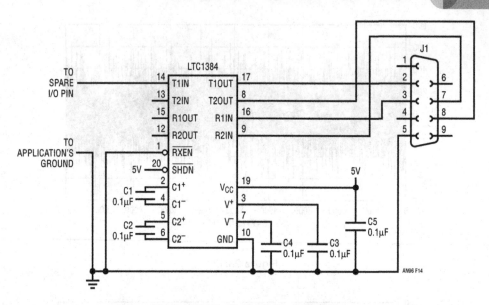

Figure 23.14 • Serial Debug Circuit

baud will work just fine. Most compilers provide a printf or sprintf function to format data as ASCII decimal characters. If these are not available, then sending ADC data as eight ASCII hexadecimal characters will work as most spreadsheets provide a

''HEX2DEC'' function to convert hexadecimal to decimal data. Most operating systems have a terminal program included that is suitable for capturing data to a text file that can then be imported into a spreadsheet.

Appendix C
Altimeter code

Altimeter.c

```
/*
Logging Altimeter based on LTC2411
Mark Thoren
Linear Technology Corporation

On power-up, program searches for first free memory location in 32KB EEPROM,
then starts taking temperature and absolute pressure data at 7.5 samples per
second, the data output rate of the LTC2411. Every 15 seconds a temperature /
pressure data point is stored to nonvolatile memory.

To read back data for processing, connect to a PC terminal set to 9600 8 N 1
and type "D" for Dump. Capture text to a file and
import to your favorite spreadsheet.

To erase data, type "E" for Erase. This zeros memory, necessesary because of the
data point flag in bit 31. (See data format.)

Data Format: 8 bytes, ms bit = 1, indicating data is present.
Next are 31 data bytes, Then the 16 bit temperature word.
*/


#include <16F73.h>
#include "pcm73a.h"
#include "altimeter.h"
#include "lcd.c"
#include "math.h"

int1 new_period, old_period;    // global flags for detecting new period

void main()
   {
   char i;

   // Voltage from LTC2411 and temperature from FM75 are stored as unions
   // allowing byte access to 16 and 32 bit variables.
   union                        // voltage.bits32       the whole thing
      {                         // voltage.by.te0        byte 0
      signed int32 bits32;      // voltage.by.te1        byte 1
      struct fourbytes by;      // voltage.by.te2        byte 2
      } voltage;                // voltage.by.te3        byte 3

   union                        // temperature.bits16       the whole thing
      {                         // temperature.by.te0       byte 0
      signed int16 bits16;      // temperature.by.te1       byte 1
      struct twobytes by;
      } temperature;

   float dec_voltage;           // Floating point for display purposes only
   float dec_temperature;
   int deg_c;
   unsigned int16 address = 0;
```

an96fa

```
    initialize(); // Set up hardware

    lcd_gotoxy(1,1);                    // Print obligatory hello message
    printf(lcd_putc, "hello!");         // First to LCD display
    printf("hello serial port!\r\n");// Then to serial port

// Detect next available address
// To avoid overwriting when cycling power
    while(address <= 0x7FFF)
        {
        if(!(ee_read(address) & 0x80)) break;
        address +=8;
        }

// Main loop
    while(1)
        {
        while(!kbhit())                         // If serial data is in, take action
            {                                   // else start collecting data
            output_low(ADC_CS_);                // Enable LTC2411 serial interface
            while(input(PIN_C4)) {}             // Make sure data is ready
            voltage.by.te3 = spi_read(0x00);    // Read data as four bytes
            voltage.by.te2 = spi_read(0x00);
            voltage.by.te1 = spi_read(0x00);
            voltage.by.te0 = spi_read(0x00);
            output_high(ADC_CS_);

            // Subtract offset (result is in 2's complement)
            voltage.bits32 = voltage.bits32 - 0x20000000;
            dec_voltage = (float) voltage.bits32;              // cast as float
            dec_voltage = dec_voltage * 2.5 / 268435456.0; // normalize to vref

            i2c_start();// Get temperature data from FM75 sensor
            i2c_write(FM75_ADDR | I2WRITE);
            i2c_write(0x00); i2c_start();
            i2c_write(FM75_ADDR | I2READ);
            temperature.by.te1=i2c_read(1);
            temperature.by.te0 = i2c_read(0);
            i2c_stop();

            // Convert from bits to degrees C
            dec_temperature = ((float) temperature.bits16) / 256.0;

            lcd_gotoxy(1,1);
            printf(lcd_putc, "%f", dec_voltage);               // Display sensor voltage
            lcd_gotoxy(1,2);
            printf(lcd_putc, "%3.3f C", dec_temperature);


            // Write data if a new 15 second period has begun
            if(old_period != new_period && address <= 0x7FF0)
                {
```

```
              old_period = new_period;                        // Set flag
              LED = new_period;                               // Blink light
              printf("writing address %ld\r\n", address);     // Serial port display
              ee_write(address, voltage.by.te3 | 0x80);       // Set MS bit as data flag
              ee_write(++address, voltage.by.te2);            // Write voltage data.
              ee_write(++address, voltage.by.te1);
              ee_write(++address, voltage.by.te0);
              ee_write(++address, temperature.by.te1);
              ee_write(++address, temperature.by.te0);
              address+=3;
              }// End of writing data
          }// End of detect character

          // Detect commands at serial port
          printf("got a character\r\n");
          switch(getc())
              {
              case 'D': printf("Dumping Data\r\n"); dump_data(); break;
              case 'E': printf("Erasing Data\r\n"); erase_data(); address = 0; break;
              default: break;
              }
      }// End of while loop
    }// End of main()

void initialize(void)
    {
    output_high(ADC_CS_);
    setup_adc_ports(NO_ANALOGS);
    setup_adc(ADC_CLOCK_DIV_2);
    setup_spi(SPI_MASTER|SPI_L_TO_H|SPI_CLK_DIV_16|SPI_SS_DISABLED);
    CKP = 0; // Set up clock edges - clock idles low, data changes on
    CKE = 1; // falling edges, valid on rising edges.
    setup_counters(RTCC_INTERNAL,RTCC_DIV_2);
    setup_timer_1(T1_INTERNAL|T1_DIV_BY_2); // 16 bit counter overflows every 52.4 ms
    setup_timer_2(T2_DISABLED,0,1);
    setup_ccp1(CCP_OFF);
    setup_ccp2(CCP_OFF);
    enable_interrupts(INT_TIMER1);
    enable_interrupts(global);
    lcd_init();
    delay_ms(6);

    i2c_stop();
    i2c_start();
    i2c_write(FM75_ADDR | I2WRITE);
    i2c_write(0x01);                    // Configuration register
    i2c_write(0x60);                    // Set to 12 bit resolution
    i2c_stop();
    }

#int_TIMER1
TIMER1_isr()
    {
```

an96fa

474

```
// 286.25  interrupt periods at 52.4ms each = 15 seconds
static int16 count;
if(++count == 286)
   {
   new_period = new_period ^ 0x01; // Toggle new_min
   count = 0;
   }
}
```

Altimeter.h

```
#fuses HS,NOWDT, PUT
#use delay(clock=10000000) // 10MHz clock
#use rs232(baud=9600,parity=N,xmit=PIN_C6,rcv=PIN_C7)
#use i2c(master,slow,sda=PIN_A0,scl=PIN_A1)

#define ADC_CS_    PIN_C2
#bit    LED      = PORTC.0

#define MAX_EE_ADDR 0x7FFC
#define EE_ADDR     0b10100000   // Ground all address pins
#define FM75_ADDR   0b10011110
#define I2READ      0x01
#define I2WRITE     0x00

// Prototypes
void initialize(void);

// Structures to allow byte access to word and long word data
struct fourbytes
   {        //Define structure of four consecutive bytes
   int8 te0;
   int8 te1;
   int8 te2;
   int8 te3;
   };

struct twobytes
   {        //Define structure of two consecutive bytes
   int8 te0;
   int8 te1;
   };

// Read data from I2C EEPROM
char ee_read(int16 address)
   {
   char x;
   i2c_stop();
   i2c_start();
   if(i2c_write(EE_ADDR | I2WRITE)) printf("\r\NAK DEVICE ADDRESS!!");
   if(i2c_write(address >> 8)) printf("\r\nNAK MSB!!");
   if(i2c_write(address)) printf("\r\nNAK LSB!!");
   i2c_start();
```

```
    if(i2c_write(EE_ADDR | I2READ)) printf("\r\nNAK WRITE COMMAND!!");
    x=i2c_read(0);
    i2c_stop();
    return x;
    }

// Write data to I2C EEPROM
void ee_write(int16 address, char data)
    {
    int8 x;
    i2c_stop();
    i2c_start();
    if(i2c_write(EE_ADDR | I2WRITE)) printf("\r\nNO ACK!!");
    if(i2c_write(address >> 8)) printf("\r\nNO ACK!!");
    if(i2c_write(address)) printf("\r\nNO ACK!!");
    if(i2c_write(data)) printf("\r\nNO ACK!!");
    i2c_stop();
    do        // Make sure data is written
        {
        i2c_start();
        x=i2c_write(EE_ADDR | I2WRITE);
        i2c_stop();
        } while(x);
    }

// Send data to PC
void dump_data()
    {
    int16 address = 0;
    char x = 0x80;

    union                       // voltage.bits32        the whole thing
        {                       // voltage.by.te0        byte 0
        signed int32 bits32;    // voltage.by.te1        byte 1
        struct fourbytes by;    // voltage.by.te2        byte 2
        } altidata;             // voltage.by.te3        byte 3

    union                       // voltage.bits32        the whole thing
        {                       // voltage.by.te0        byte 0
        signed int16 bits16;    // voltage.by.te1        byte 1
        struct twobytes by;     // voltage.by.te2        byte 2
        } tempdata;             // voltage.by.te3        byte 3

    do
        {
        altidata.by.te3 = x = ee_read(address);
        altidata.by.te2 = ee_read(++address);
        altidata.by.te1 = ee_read(++address);
        altidata.by.te0 = ee_read(++address);
        tempdata.by.te1 = ee_read(++address);
        tempdata.by.te0 = ee_read(++address);
        address +=3;
        altidata.by.te3 &= 0x7F;   // Clear detect bit (would screw up math)
        printf("%ld,%ld\r\n", altidata.bits32, tempdata.bits16);
```

an96fa

476

```
        }while((x & 0x80) && address <= 0x7FF0);
    printf("done\r\n");
    }

// Clear data in I2C EEPROM
void erase_data()
    {
    int16 address;
    int i;
    char x;
    for(address=0; address<=0x7FFF; address+=64)
        {
        i2c_start();
        i2c_write(EE_ADDR | I2WRITE);
        i2c_write(address >> 8);
        i2c_write(address);
        for(i=0; i<64; ++i)                    // 64 byte burst write
            {
            i2c_write(0);
            }
        i2c_stop();
        do          // Make sure data is written
            {
            i2c_start();
            x=i2c_write(EE_ADDR | I2WRITE);
            i2c_stop();
            } while(x);
        }
    }
```

Appendix D–Correlated double sampling driver code

```c
/*
This function takes care of switching the bridge excitation, allowing
the amplifiers to settle, programming the LTC2440 oversample ratio,
and reading the LTC2440 output. The return value is the difference between
the present reading and the previous reading. The first time this function
is called the return value will not be valid. EXCITATION, CS, AND MISO refer to
the I/O pins for the bridge excitation switching, LTC2440 chip select,
and the LTC2440 SDO pin, respectively.
*/

int32 double_sample(void)
   {
#define SETTLING_TIME   2       // Miliseconds, settling time after switching bridge
#define OSR 0xFF                // Hex code to program the LTC2440 oversample ratio
                                // (0xFF = OSR32768)

   static union{                // These are static because they need to be remembered
       struct fourbytes by;     // from one conversion to the next.
       signed int32 bits32;
       } positive;

   static union{
      struct fourbytes by;
      signed int32 bits32;
      } negative;

   union{                       // Doesn't need to be static
      struct fourbytes by;
      signed int32 bits32;
      } voltage;

   static boolean polarity = 0; // 0=negative, 1=positive

      output_low(CS);           // Enable SPI interface
      while(input(MISO)) {}     // Wait for SDO to go low indicating end of conversion

      // The following block takes care of switching bridge polarity and loading
      // the proper
      if(polarity==1)
         {
         polarity = 0;
         output_low(EXCITATION);              // Switch excitation
         delay_ms(SETTLING_TIME);             // Here is the place to insert extra
         // Read out positive result           // settling time for amplifiers
         positive.by.te3 = spi_read(OSR);     // Program OSR, read first byte
         positive.by.te2 = spi_read(0);
         positive.by.te1 = spi_read(0);
         positive.by.te0 = spi_read(0);
         }
      else
         {
```

an96fa

```
        polarity = 1;
        output_high(EXCITATION);                  // Switch excitation
        delay_ms(SETTLING_TIME);
        // Read out negative result
        negative.by.te3 = spi_read(OSR);
        negative.by.te2 = spi_read(0);
        negative.by.te1 = spi_read(0);
        negative.by.te0 = spi_read(0);
        }
    output_high(CS);

    voltage.bits32 = (positive.bits32 - negative.bits32); // Take difference of last
    voltage.bits32 = voltage.bits32 >> 1;                 // two readings, divide by 2

    return voltage.bits32;    // Return result
}
```

1ppm settling time measurement for a monolithic 18-bit DAC

When does the last angel stop dancing on a speeding pinhead?

24

Jim Williams

Introduction

Performance requirements for instrumentation, function generation, inertial navigation systems, trimming, calibrators, ATE, medical apparatus and other precision applications are beginning to eclipse capabilities of 16-bit data converters. More specifically, 16-bit digital-to-analog converters (DACs) have been unable to provide required resolution in an increasing number of ultra-precision applications.

New components (see Components for 18-bit D/A Conversion, page 2) have made 18-bit DACs a practical design alternative[1]. These ICs provide 18-bit performance at reasonable cost compared to previous modular and hybrid technologies. The monolithic DACs DC and AC specifications approach or equal previous converters at significantly lower cost.

DAC settling time

DAC DC specifications are relatively easy to verify. Measurement techniques are well understood, albeit often tedious. AC specifications require more sophisticated approaches to produce reliable information. In particular, the settling time of a DAC and its output amplifier is extraordinarily difficult to determine to 18-bit (4ppm) resolution. DAC settling time is the elapsed time from input code application until the output arrives at, and remains within, a specified error band around the final value. To measure a new 18-bit DAC, a settling time measurement technique has been developed with 20-bit (1ppm) resolution for times as short as 265ns. The new method will work with any DAC. Realizing this measurement capability and its performance verification has

required an unusually intense, extensive and protracted effort. Hopefully, the data converter community will find the results useful[2].

DAC settling time is usually specified for a full-scale 10V transition. Figure 24.1 shows that DAC settling time has three distinct components. The *delay time* is very small and is almost entirely due to propagation delay through the DAC and output amplifier. During this interval, there is no output movement. During *slew time*, the output amplifier moves at its highest possible speed towards the final value. *Ring time* defines the region where the amplifier recovers from slewing and ceases movement within some defined error band. There is normally a trade-off between slew and ring time. Fast slewing amplifiers generally have extended ring times, complicating amplifier choice and frequency compensation.

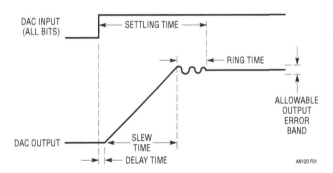

Figure 24.1 • DAC Settling Time Components Include Delay, Slew and Ring Times. Fast Amplifiers Reduce Slew Time, Although Longer Ring Time Usually Results. Delay Time is Normally a Small Term

---

Note 1: See Appendix A, "A History of High Accuracy Digital-to-Analog Conversion".

Note 2: A historical note is in order. In early 1997, LTC's DAC design group tasked the author to measure 16-bit DAC settling time. The result was published in July 1998 as Application Note 74, "Component and Measurement Advances Ensure 16-Bit DAC Settling Time". Almost exactly 10 years later, the DAC group raised the ante, requesting 18-bit DAC settling time measurement. This constitutes 2 bits of resolution improvement per decade of author age. Since it was unclear how many decades the author (born 1948) had left, it was decided to double jump the performance requirement and attempt 20-bit resolution. In this way, even if the author is unavailable in 10 years, the DAC group will still get its remaining 2 bits.

Analog Circuit and System Design: A Tutorial Guide to Applications and Solutions. DOI: 10.1016/B978-0-12-385185-7.00024-X

Components for 18-bit D/A conversion

Components suitable for 18-bit D/A conversion are members of an elite class. 18 binary bits is one part in 262,144—just 0.0004% or 4 parts-per-million. This mandates a vanishingly small error budget and the demands on components are high. The LTC2757 digital-to-analog converter listed in the chart uses Si-Chrome thin-film resistors for high stability and linearity over temperature. Gain drift is typically 0.25ppm/°C or about 4.6LSBs over 0°C to 70°C. Some amplifiers shown contribute less than 1LSB error over 0°C to 70°C with 18-bit DAC driven settling times of 1.8μs available. The references offer drifts as low as 1LSB over 0°C to 70°C with initial trimmed accuracy to 0.05%

COMPONENT TYPE	ERROR CONTRIBUTION OVER 0°C TO 70°C	COMMENTS
Short Form Descriptions of Components Suitable for 18-Bit Digital-to-Analog Conversion		
LTC®2757 DAC	≈4.6LSB Gain Drift 1LSB Linearity	Full Parallel Inputs Current Output
LT®1001 Amplifier	<1LSB	Good Low Speed Choice 10mA Output Capability
LT1012 Amplifier	<1LSB	Good Low Speed Choice Low Power Consumption
LT1468/LT1468-2 Amplifier	<8LSB	1.8μs Settling to 18 Bits Fastest Available
LTC1150 Amplifier	<1LSB	Lowest Error. ≈10ms Settling Time. Requires LT1010 Output Buffer. Special Case. See Appendix E
LTZ1000A Reference	<1LSB	Lowest Drift Reference in This Group. 4ppm (1LSB)/Yr. Time Stability Typical
LM199A Reference-6.95V	≈4LSB	Low Drift. 10ppm (2.5LSB) Yr. Time Stability Typical
LT1021 Reference-10V	≈16LSB	Good General Purpose Choice
LT1027 Reference-5V	≈16LSB	Good General Purpose Choice
LT1236 Reference-10V	≈40LSB	Trimmed to 0.05% Absolute Accuracy

Additionally, the architecture of very fast amplifiers usually dictates trade-offs which degrade DC error terms[3].

Measuring anything at any speed to 20-bit resolution (1ppm) is hard. Dynamic measurement to 20-bit resolution is particularly challenging. Reliable 1ppm DAC settling time measurement constitutes a high order difficulty problem requiring exceptional care in approach and experimental technique. This publication's remaining sections describe a method enabling an oscilloscope to accurately display DAC settling time information for a 10V step with 1ppm resolution (10μV) within 265ns. The approach employed permits observation of small amplitude information at the excursion limits of large waveforms without overdriving the oscilloscope.

Considerations for measuring DAC settling time

Historically, DAC settling time has been measured with circuits similar to that in Figure 24.2. The circuit uses the "false sum node" technique. The resistors and DAC form a bridge type network. Assuming ideal components, the DAC output will step to $-V_{REF}$ when the DAC inputs move to all ones. During slew, the settle node is bounded by the diodes, limiting voltage excursion. When settling occurs, the oscilloscope probe voltage should be zero. Note that the resistor divider's attenuation means the probe's output will be one-half of the DAC's settled voltage.

In theory, this circuit allows settling to be observed to small amplitudes. In practice, it cannot be relied upon to produce useful measurements. The oscilloscope connection presents problems. As probe capacitance rises, AC loading of the resistor junction influences observed settling waveforms.

Note 3: This issue is treated in detail in latter portions of the text. Also see Appendix D, "Practical Considerations for DAC-Amplifier Compensation".

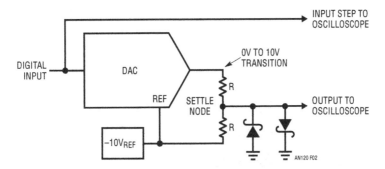

Figure 24.2 • Popular Summing Scheme for DAC Settling Time Measurement Provides Misleading Results. 18-Bit Measurement Causes >800× Oscilloscope Overdrive. Displayed Information is Meaningless

A 10pF probe alleviates this problem but its 10× attenuation sacrifices oscilloscope gain. 1× probes are not suitable because of their excessive input capacitance. An active 1× FET probe will work, but a more significant issue remains.

The clamp diodes at the settle node are intended to reduce swing during amplifier slewing, preventing excessive oscilloscope overdrive. Unfortunately, oscilloscope overdrive recovery characteristics vary widely among different types and are not usually specified. The Schottky diodes' 400mV drop means the oscilloscope may see an unacceptable overload, bringing displayed results into question[4].

At 10-bit resolution (10mV at the DAC output, resulting in 5mV at the oscilloscope), the oscilloscope typically undergoes a 2× overdrive at 50mV/DIV and the desired 5mV baseline is just discernible. At 12-bit or higher resolution, the measurement becomes hopeless with this arrangement. Increasing oscilloscope gain brings commensurate increased vulnerability to overdrive induced errors. At 18 bits, there is clearly no chance of measurement integrity.

The preceding discussion indicates that measuring 18-bit settling time requires a high gain oscilloscope that is somehow immune to overdrive. The gain issue is addressable with an external wideband preamplifier that accurately amplifies the diode-clamped settle node. Getting around the overdrive problem is more difficult.

The only oscilloscope technology that offers inherent overdrive immunity is the classical sampling 'scope[5]. Unfortunately, these instruments are no longer manufactured (although still available on the secondary market). It is possible, however, to construct a circuit that utilizes sampling techniques to avoid the overload problem. Additionally, the circuit can be endowed with features particularly suited for measuring 20-bit DAC settling time.

Sampling based high resolution DAC settling time measurement

Figure 24.3 is a conceptual diagram of the 20-bit DAC settling time measurement circuit. This figure shares attributes with Figure 24.2, although some new features appear. In this case, the preamplified oscilloscope is

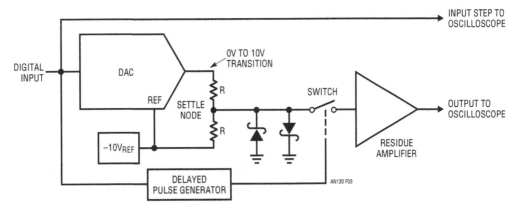

Figure 24.3 • Conceptual Arrangement Eliminates Oscilloscope Overdrive Delayed Pulse Generator Controls Switch, Preventing Oscilloscope from Monitoring Settle Node Until Settling is Nearly Complete

Note 4: For a discussion of oscilloscope overdrive considerations, see Appendix B, "Evaluating Oscilloscope Overdrive Performance".

Note 5: Classical sampling oscilloscopes should not be confused with modern era digital sampling 'scopes that have overdrive restrictions. See Appendix B, "Evaluating Oscilloscope Overload Performance" for comparisons of various type oscilloscopes with respect to overdrive. For detailed discussion of classical sampling oscilloscope operation, see references 17 through 21 and 23 through 25. Reference 18 is noteworthy; it is the most clearly written, concise explanation of classical sampling instruments the author is aware of. A 12-page jewel.

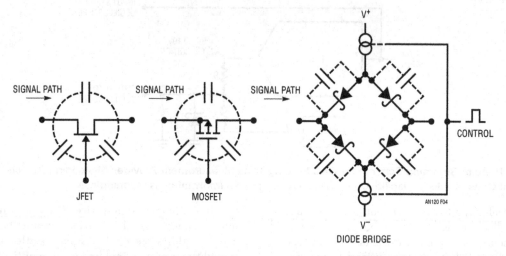

Figure 24.4 • Conventional Choices for the Sampling Switch Include JFET, MOSFET and Diode Bridge. FET Parasitic Capacitances Result in Large Gate Drive Originated Feedthrough to Signal Path. Diode Bridge is Better; Its Small Parasitic Capacitances Tend to Cancel. Bridge Requires DC and AC Trims and Complex Drive Circuitry

connected to the settle point by a switch. The switch state is determined by a delayed pulse generator, which is triggered from the same pulse that controls the DAC. The delayed pulse generator's timing is arranged so that the switch does not close until settling is very nearly complete. In this way, the incoming waveform is sampled in time, as well as amplitude. The oscilloscope is never subjected to overdrive—no off-screen activity ever occurs.

Developing a sampling switch

Requirements for Figure 24.3's sampling switch are stringent. It must faithfully pass signal path information without introducing alien components, particularly those deriving from the switch command channel. Figure 24.4 shows conventional choices for the sampling switch. They include FETs and the diode bridge. The FET's parasitic gate to channel capacitances result in large gate drive originated feedthrough into the signal path. For almost all FETs, this feedthrough is many times larger than the signal to be observed, inducing overload and obviating the switches' purpose. The diode bridge is better; its small parasitic capacitances tend to cancel and the symmetrical, differential structure results in very low feedthrough. Practically, the bridge requires DC and AC trims and complex drive and support circuitry. This approach, incarnated with great care, can reliably measure DAC settling time to 16-bit resolution[6]. Beyond 16 bits, residual feedthrough becomes objectionable and another approach is needed.

Note 6: LTC Application Note 74, "Component and Measurement Advances Ensure 16-bit DAC Settling Time" utilized such a sampling bridge and it is detailed in that text.

Electronic switch equivalents

A low feedthrough, high resolution "switch" can be constructed with wideband active components. The great advantage of this approach is that the switch control channel can be maintained "in-band"; that is, its transition rate is within the circuits' bandpass. The circuit's wide bandwidth means the switch command transition is under control at all times. There are no out-of-band responses, greatly reducing feedthrough. Figure 24.5 lists some candidates for low feedthrough electronic switch equivalents. A and B, while theoretically possible, are cumbersome to implement. C and D are practical. C must be optimized for low feedthrough on rising and falling control pulse edges because of the multiplier's unrestricted wideband response. D's falling edge feedthrough is inherently minimized by the g_m amplifier's transconductance collapse when the control pulse goes low. This allows feedthrough to be optimized for the control pulse's rising edge without regard to falling edge effects. This is a significant advantage in constructing an electronic equivalent switch.

Transconductance amplifier based switch equivalent

Figure 24.6 is a conceptual transconductance amplifier based "switch". The wideband control and signal paths faithfully track 1000:1 transconductance change, resulting in exceptionally pure switch dynamics. The switched current source is carefully optimized for lowest feedthrough on the rising control edge without regard to falling edge characteristics. The g_m amplifier's transconductance collapse on the falling edge ensures low feedthrough for that condition,

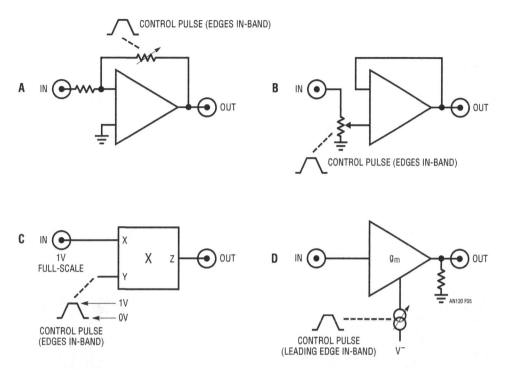

Figure 24.5 • Conceptual Low Feedthrough Electronic Switch Equivalents A and B are Difficult to Implement, C and D are Practical. C Must be Optimized for Low Feedthrough on Rising and Falling Control Pulse Edges. D's Falling Edge Feedthrough is Inherently Minimized by Attendant Bandwidth Reduction

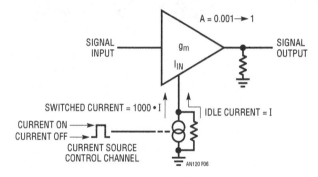

Figure 24.6 • Transconductance Amplifier Based "Switch" Has Minimal Control Channel Feedthrough Wideband Control and Signal Paths Faithfully Track 1000:1 Transconductance Change, Resulting in Exceptionally Pure Switch Dynamics

preventing oscilloscope overdrive. Figure 24.7 details the transconductance amplifier-based switch. This design switches signals over a ±30mV range with peak control channel feedthrough of millivolts and settling times inside 40ns.

The circuit approximates switch action by varying A1A's transconductance; the maximum gain is unity. At low transconductance, A1A's gain is nearly zero, and essentially no signal is passed. At maximum transconductance, signal passes at unity gain. The amplifier and its transconductance control channel are very wideband, permitting them to faithfully track rapid variations in transconductance

setting. This characteristic means the amplifier is never out of control, affording clean response and rapid settling to the "switched" input's value.

A1A, one section of an LT1228, is the wideband transconductance amplifier. Its voltage gain is determined by its output resistor load and the current magnitude into its "I_{SET}" terminal. A1B, the second LT1228 section, unloads A1A's output. As shown, it provides a gain of 2, but when driving a back-terminated 50Ω cable, its effective gain is unity at the cable's receiving end. The back termination enforces a 50Ω environment. Current source Q1, controlled by the "switch control input", sets A1A's transconductance, and, hence, gain. With Q1 gated off (control input at zero), the 10M resistor supplies about 1.5µA into A1A's I_{SET} pin, resulting in a voltage gain of nearly zero, blocking the input signal. When the switch control input goes high, Q1 turns on, sourcing approximately 1.5mA into the I_{SET} pin. This 1000:1 set current change forces maximum transconductance, causing the amplifier to assume unity gain and pass the input signal. Trims for zero and gain ensure accurate input signal replication at the circuit's output. The Q1 associated 50pF variable capacitor purifies turn-on switching. The specified 10k resistor at Q1 has a 3300ppm/°C temperature coefficient, compensating A1A's complementary transconductance temperature dependence to minimize gain drift.

Figure 24.8 shows circuit response for a switched 10mV DC input and $C_{ABERRATION}$ = 35pF. When the control input (trace A) is low, no output (trace B) occurs.

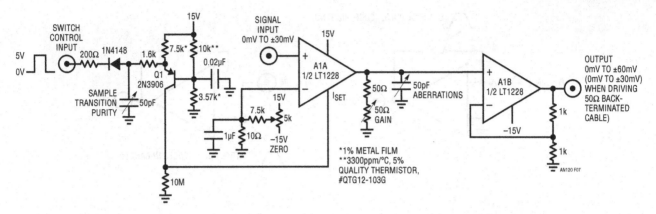

Figure 24.7 • Transconductance Amplifier-Based 100MHz Low Level Switch Has Minimal Control Channel Feedthrough. A1A's Unity-Gain Output is Cleanly Switched by Logic Controlled Q1's Transconductance Bias. A1B Provides Buffering and Signal Path Gain

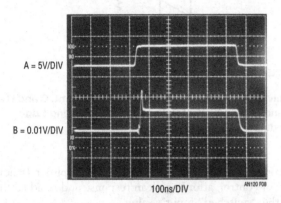

Figure 24.8 • Control Input (Trace A) Dictates Switch Output's (Trace B) Representation of 0.01V DC Input. Control Channel Feedthrough, Evident at Switch Turn-On, Settles in 20ns. Turn-Off Feedthrough is Undetectable Due to Deceased Signal Channel Transconductance and Bandwidth. $C_{ABERRATION} \approx 35pF$ for this Test

When the control input goes high, the output reproduces the input with "switch" feedthrough settling in about 20ns. Note that turn-off feedthrough is undetectable due to the 1000× transconductance reduction and attendant 25× bandwidth drop. Figure 24.9 speeds the sweep up to 10ns/division to examine zero volt settling detail. The output (trace B) settles inside 1mV 40ns after the switch control (trace A) goes high. Peak feedthrough excursion, damped by $C_{ABERRATION}$, is only 5mV Figure 24.10 was taken under identical conditions, except that $C_{ABERRATION} = 0pF$. Feedthrough increases to approximately 20mV, although settling time to 1mV remains at 40ns. Figure 24.11, using double exposure technique, compares signal channel rise times for $C_{ABERRATION} = 0pF$ (leftmost trace) and approximately 35pF (rightmost trace) with the control channel tied high. The larger $C_{ABERRATION}$ value, while minimizing feedthrough amplitude (see Figure 24.9), increases rise time by 7× versus $C_{ABERRATION} = 0pF$.

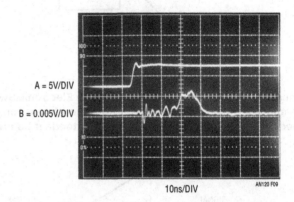

Figure 24.9 • High Speed Delay and Feedthrough for 0V Signal Input. Output (Trace B) Peaks Only 0.005V Before Settling Inside 0.001V 40ns After Switch Control Command (Trace A). $C_{ABERRATION} \approx 35pF$ for This Test

The transconductance switches' small DC and AC errors nicely accommodate the applications' requirements. The low feedthrough, already sufficient, becomes irrelevant because its small time and amplitude error will be buried in the DAC ring time interval. The transconductance amplifier based "switch" points the way towards practical 1ppm DAC settling time measurement.

DAC settling time measurement method

Figure 24.12, a more complete representation of Figure 24.3, utilizes the above described sampling switch. Figure 24.3's blocks appear in greater detail and some new refinements show up. The DAC-amplifier summing area is unchanged. Figure 24.3's delayed pulse generator has been split into two blocks; a delay and a pulse generator, both independently variable. The input step to the oscilloscope runs through a section that compensates settling time-measurement path propagation delay. This path includes settle

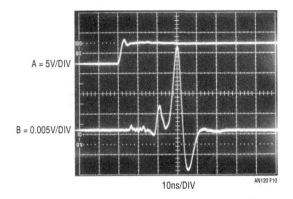

Figure 24.10 • Identical Conditions as Figure 24.9 Except C$_{ABERRATION}$ = 0pF. Feedthrough Related Peaking Increases to ≈0.02V; 0.001V Settling Time Remains at 40ns

A = 5V/DIV

B = 0.005V/DIV

10ns/DIV   AN120 F10

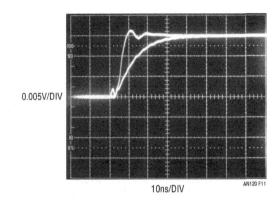

Figure 24.11 • Signal Channel Rise Time for C$_{ABERRATION}$ = 0pF (Leftmost Trace) and ≈35pF (Rightmost Trace) Record 3.5ns and 25ns, Respectively. Switch Control Input High for this Measurement. Photograph Utilizes Double Exposure Technique

0.005V/DIV

10ns/DIV   AN120 F11

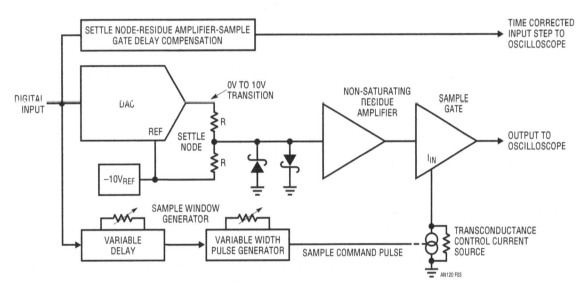

Figure 24.12 • Block Diagram of Sampling-Based DAC Settling Time Measurement Scheme. Placing Transconductance Controlled Sample Gate After Residue Amplifier Minimizes Sample Command Feedthrough Impact, Eliminating Oscilloscope Overdrive. Input Step Time Reference is Compensated for Settle Node, Residue Amplifier and Sample Gate Delays

node, amplifier and sample gate delays. The transconductance sampling switch ("sample gate"), driven from a nonsaturating residue amplifier, feeds the oscilloscope. Placing the sampling switch after the residue amplifier gain further minimizes sample command feedthrough impact.

Detailed settling time circuitry

Figure 24.13 is a detailed schematic of the 20-bit DAC settling time measurement circuitry. The input pulse switches all DAC bits simultaneously and is also routed to the oscilloscope via the delay compensation network. The delay network, composed of CMOS inverters and an adjustable RC network, compensates the oscilloscope's input step signal for the 44ns delay through the circuit

measurement path[7]. The DAC-amplifier output is compared against the LT1021 10V reference via the precision 10k summing resistors. The LT1021 also furnishes the DAC reference, making the measurement ratiometric. The clamped settle node is unloaded by A1, which takes gain. A2 provides additional clamped gain for a total summing node referred amplification of 40. A2's output feeds the sampling switch whose operation is identical to Figure 24.7's description. The A1-A2 amplifier's clamping and gain are arranged so saturation never occurs—the amplifier is always in its active region.

The input pulse triggers the 74HC123 dual one shot. The one shot is arranged to produce a delayed (controllable by

---

Note 7: See Appendix C, "Measuring and Compensating Signal Path Delay and Circuit Trimming Procedures".

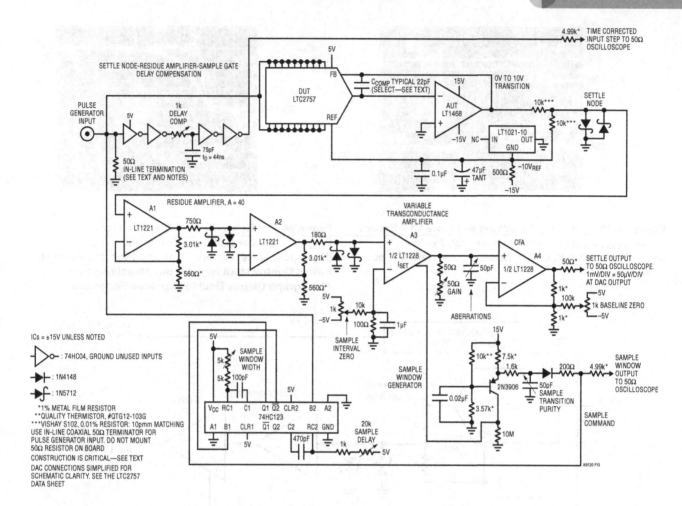

Figure 24.13 • Detailed DAC Settling Time Measurement Circuit Closely Follows Preceding Figure. Optimum Performance Requires Attention to Layout

the 20k potentiometer) pulse whose width (controllable by the 5k potentiometer) sets sampling switch on-time. If the delay is set appropriately, the oscilloscope will not see any input until settling is nearly complete, eliminating overdrive. The sample window width is adjusted so that all remaining activity is observable. In this way, the oscilloscope output is reliable and meaningful data may be taken.

Figure 24.14 shows circuit waveforms. Trace A is the time corrected input pulse, trace B the sample gate, trace C the DAC-amplifier output and trace D the circuit output. When the sample gate goes high, trace D's switching is clean, the last millivolt of ring time is easily observed and the amplifier settles nicely to final value bounded by broadband noise. When the sample gate goes low, the transconductance switch goes off and no feedthrough is discernible. Note that there is no off-screen activity at any time—the oscilloscope is never subjected to overdrive.

The circuit requires trimming to achieve this level of performance[8]. Figure 24.15 shows a typical display resulting

from poor "Sample Interval Zero" adjustment. This adjustment, corrected in Figure 24.16, results in a continuous baseline. Sample command feedthrough is just visible at trace B's leading edge. Figure 24.17 shows output response (trace B) to the sample command (trace A) turn-on before

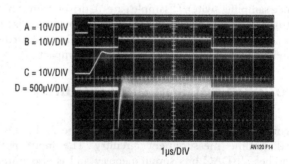

A = 10V/DIV
B = 10V/DIV

C = 10V/DIV
D = 500µV/DIV

1µs/DIV   AN120 F14

Figure 24.14 • Settling Time Circuit Waveforms Include Time Corrected Input Pulse (Trace A), Sample Command (Trace B), DAC Output (Trace C) and Settling Time Output (Trace D). Sample Window Delay and Width are Variable

Note 8. To maintain text flow and focus, trimming procedures are not presented here. Detailed trimming information appears in Appendix C.

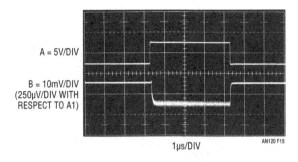

A = 5V/DIV

B = 10mV/DIV
(250µV/DIV WITH
RESPECT TO A1)

1µs/DIV

AN120 F15

Figure 24.15 • Poor Sample Interval Zero Adjustment Causes Shifted Output Baseline (Trace B) During Trace A's Sample Interval

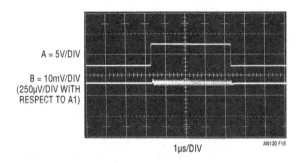

A = 5V/DIV

B = 10mV/DIV
(250µV/DIV WITH
RESPECT TO A1)

1µs/DIV

AN120 F16

Figure 24.16 • Trimmed Sample Interval Zero Has No Output Baseline Deviation (Trace B) During Sample Interval (Trace A). Sample Command Feedthrough is Just Visible at Trace B's Leading Edge

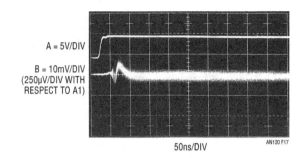

A = 5V/DIV

B = 10mV/DIV
(250µV/DIV WITH
RESPECT TO A1)

50ns/DIV

AN120 F17

Figure 24.17 • Output Response (Trace B) To Sample Command (Trace A) Turn-On Before Trimming Aberrations and Transition Purity. Delay is ≈20ns. Aberrations Peak 350µV, Settle in 50ns. A1's Positive Input Grounded via 5kΩ for This and Succeeding Figures

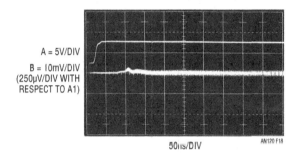

A = 5V/DIV

B = 10mV/DIV
(250µV/DIV WITH
RESPECT TO A1)

50ns/DIV

AN120 F18

Figure 24.18 • Post-Trim Output Response (Trace B) To Sample Command Turn-On, Trace A. Delay Increases to 70ns but Aberrations Peak Only 5µV, Settling in 50ns

trimming "aberrations" and "transition purity"[9]. Delay is approximately 20ns with aberrations peaking 350µV and settling in 50ns. Figure 24.18 shows post trim response to sample command turn-on. Delay increases to 70ns but aberrations peak only 50µV settling in 50ns. Figure 24.19 shows output response (trace B) to sample command (trace A) turn-off. The 1000:1 transconductance drop ensures a clean transition independent of the turn-on optimized trims.

Circuit gain is adjusted with the indicated potentiometer.

Settling time circuit performance

Figure 24.20 summarizes settling time circuit performance. The graph indicates the minimum measurable settling time for a given resolution. Speed limitations are imposed by sample command path delays and sample gate switching residue profile[10]. Minimum measurable settling time below 160ns is available to 16-bit resolution. Beyond this point, the sample gate's switching residue profile dictates increased minimum measurable settling time to about 265ns at 20 bits. Circuit noise limitations are imposed by

the DAC/amplifier, summing resistors, and residue amplifier/sampling switch with about equal weighting. Because of this, resolution beyond approximately 15ppm requires filtering or noise averaging techniques.

Using the sampling-based settling time circuit

It is good practice to "walk" the sampling window backwards in time from the settled region up to the last 100µV or so of amplifier movement so ring time cessation is observable. The sampling-based approach provides this capability and it is a very powerful measurement tool. Additionally, slower amplifiers may require extended delay and/or sampling window times. This may necessitate larger capacitor values in the 74HC123 one-shot timing networks.

Figure 24.21 shows DAC settling in an unfiltered bandpass. The DAC settles (trace B) to 16 bits 1.7µs after trace A's time corrected input step[11]. Sample gate feedthrough is undetectable, indicating higher resolution is possible without overdriving the oscilloscope. Noise is the fundamental measurement limit. Figure 24.22 attenuates noise by reducing

---

Note 9: A1's positive input was grounded via 5kΩ (precision 10k resistors disconnected) for Figure 24.17 and 24.18's tests.
Note 10. Driving the sample command path (74HC123 B2 input) with a phase-advanced version of the pulse generator input largely eliminates sample command path delay induced error, considerably improving minimum measurable settling time. This benefit is not germane to the present effort's purposes and was not implemented.

Note 11: Settling time is significantly affected by the DAC-amplifier compensation capacitor. See Appendix D, "Practical Considerations for DAC-amplifier Compensation" for tutorial.

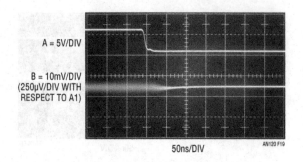

A = 5V/DIV

B = 10mV/DIV
(250µV/DIV WITH
RESPECT TO A1)

50ns/DIV   AN120 F19

Figure 24.19 • Output Response (Trace B) To Sample Command (Trace A) Turn-Off. 1000:1 Transconductance Drop Ensures Clean Transition, Independent of Trim State

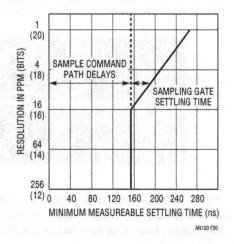

Figure 24.20 axes:
RESOLUTION IN PPM (BITS)
1 (20)
4 (18)
16 (16)
64 (14)
256 (12)

SAMPLE COMMAND PATH DELAYS

SAMPLING GATE SETTLING TIME

MINIMUM MEASUREABLE SETTLING TIME (ns)
0 40 80 120 160 200 240 280

AN120 F20

Figure 24.20 • Minimum Measureable Settling Time vs Resolution. Limits are Imposed by Sample Command Path Delays and Sample Gate Settling Profile. Resolution Beyond ≈15ppm Requires Filtering or Noise Averaging

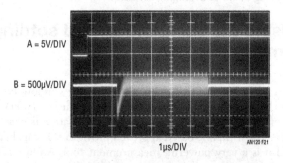

A = 5V/DIV

B = 500µV/DIV

1µs/DIV   AN120 F21

Figure 24.21 • 0V to 10V DAC Settling in Unfiltered Bandpass. DAC Settles (Settle Output, Trace B) to 16 Bits (15ppm) <2µs After Trace A's Time Corrected Input Step. Sample Gate Feedthrough is Well Controlled, Indicating Higher Resolution is Possible Without Overdriving Oscilloscope. Noise Limits Measurement

measurement bandwidth to 250kHz. Trace assignments are as in the previous photo. 18-bit settling (4ppm) requires approximately 5µs. The reduced bandwidth permits higher resolution although the indicated settling time is likely pessimistic due to the filter's lag. Figure 24.23, decreasing

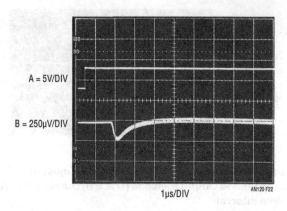

A = 5V/DIV

B = 250µV/DIV

1µs/DIV   AN120 F22

Figure 24.22 • Same Trace Assignments as Previous Photo; Measurement Taken in 250kHz Bandpass Settling to 18 Bits (4ppm) Requires ≈5µs. Filtering Permits Increased Resolution Although Indicated Settling Time Increases

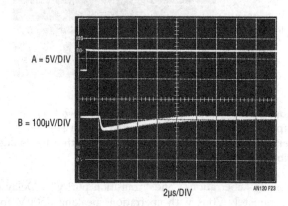

A = 5V/DIV

B = 100µV/DIV

2µs/DIV   AN120 F23

Figure 24.23 • 19 Bit (2ppm) Settling is Discerniable About 9µs After Input Command in 50kHz Bandwidth

bandwidth to 50kHz, permits 19-bit (2ppm) resolution with indicated settling in about 9µs. Again, the same filtering which permits high resolution almost certainly lengthens observed settling time.

Figure 24.24 uses noise averaging techniques to measure settling time to 20 bits (1ppm-10µV) without the band limiting filter's time penalty[12]. Photo A shows the DAC-amplifier adjusted for overdamped response, B and C underdamped and optimum responses, respectively. Averaging eliminates noise, permitting determination of settling time due to DAC dynamics[13]. Settling time ranges from 4µs to 6µs with fractional LSB tailing evident.

Note: *This application note was derived from a manuscript originally prepared for publication in EDN magazine.*

Note 12: Most oscilloscopes require preamplification to resolve Figure 24.24's signal amplitudes. See Appendix I, "Auxiliary Circuits" for an example.
Note 13: More properly, this measurement determines DAC settling time due solely to step input initiated dynamics. For this reason, Figure 24.24's averaged results may be considered somewhat academic. Noise limits measurement certainty at any given instant to approximately 100µV. It is not unreasonable to maintain that this 100µV of noise means the DAC never settles inside this limit. The averaged measurement defines settling time with noise limitations removed. Hopefully, this disclosure will appease technolawyers among the readership.

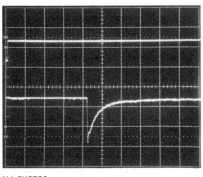 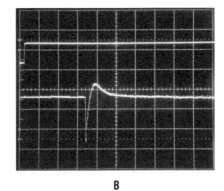 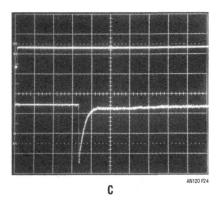

ALL PHOTOS   **A**   **B**   **C**
TRACE A = 5V/DIV
TRACE B = 25µV/DIV (AVERAGED)
HORIZ = 1µs/DIV

AN120 F24

Figure 24.24 • Noise Averaging Oscilloscope Permits 1ppm, (10µV) Settling Time Measurement Without Bandlimiting Filter Time Penalty. Photo A Shows Overdamped Response, B and C Underdamped and Optimum Responses, Respectively. Averaging Eliminates Noise, Permitting Determination of Settling Time Due To DAC Dynamics. Settling Times Range From 4µs to 6µs; Fractional LSB Tailing is Evident

References

1. Williams, Jim, "Component and Measurement Advances Ensure 16-Bit DAC Settling Time," Linear Technology Corporation, Application Note 74, July 1998.

2. Williams, Jim, "Measuring 16-Bit Settling Times: The Art of Timely Accuracy," EDN, November 19, 1998.

3. Williams, Jim, "Methods for Measuring Op Amp Settling Time," Linear Technology Corporation, Application Note 10, July 1985.

4. Demerow, R., "Settling Time of Operational Amplifiers," Analog Dialogue, Volume 4-1, Analog Devices, Inc., 1970.

5. Pease, R.A., "The Subtleties of Settling Time," The New Lightning Empiricist, Teledyne Philbrick, June 1971.

6. Harvey, Barry, "Take the Guesswork Out of Settling Time Measurements," EDN, September 19, 1985.

7. Williams, Jim, "Settling Time Measurement Demands Precise Test Circuitry," EDN, November 15, 1984.

8. Schoenwetter, H.R., "High-Accuracy Settling Time Measurements," IEEE Trans-actions on Instrumentation and Mea-surement, Vol. IM-32, No.1, March 1983.

9. Sheingold, D.H., "DAC Settling Time Measurement," Analog-Digital Conversion Handbook, pg. 312-317. Prentice-Hall, 1986.

10. Williams, Jim, "30 Nanosecond Settling Time Measurement for a Precision Wideband Amplifier," Linear Technology Corporation, Application Note 79, September 1999.

11. Williams, Jim, "Evaluating Oscilloscope Overload Performance," Box Section A, in "Methods for Measuring Op Amp Settling Time," Linear Technology Corporation, Application Note 10, July 1985.

12. Orwiler, Bob, "Oscilloscope Vertical Amplifiers", Tektronix, Inc., Concept Series, 1969.

13. Addis, John, "Fast Vertical Amplifiers and Good Engineering". "Analog Circuit Design; Art, Science and Personalities," Butterworth-Heinemann, 1991.

14. Travis, W, "Settling Time Measurement Using Delayed Switch". Private Communication, 1984.

15. Hewlett-Packard, "Schottky Diodes for High-Volume, Low Cost Applications," Application Note 942, Hewlett-Packard Company, 1973.

16. Korn, G.A. and Korn, T.M., "Electronic Analog and Hybrid Computers," "Diode Switches," pg. 223–226. McGraw-Hill, 1964.

17. Carlson, R., "A Versatile New DC-500 MHz Oscilloscope with High Sensitivity and Dual Channel Display," Hewlett-Packard Journal, Hewlett-Packard Company, January 1960.

18. Tektronix, Inc., "Sampling Note". Tektronix, Inc., 1964.

19. Tektronix, Inc., "Type 1S1 Sampling Plug-In Operating and Service Manual," Tektronix, Inc., 1965.

20. Mulvey, J., "Sampling Oscilloscope Circuits," Tektronix, Inc., Concept Series, 1970.

21. Addis, John, "Sampling Oscilloscopes," Private Communication, February 1991.

22. Williams, Jim, "Bridge Circuits-Marrying Gain and Balance," Linear Technology Corporation, Application Note 43, June 1990.

23. Tektronix, Inc., "Type 661 Sampling Oscilloscope Operating and Service Manual," Tektronix, Inc., 1963.

24. Tektronix, Inc., "Type 4S1 Sampling Plug-In Operating and Service Manual," Tektronix, Inc., 1963.

25. Tektronix, Inc., "Type 5T3 Timing Unit Operating and Service Manual," Tektronix, Inc., 1965.

26. Morrison, Ralph, "Grounding and Shielding Techniques in Instrumentation," 2nd Edition, Wiley Interscience, 1977.

27. Ott, Henry W, "Noise Reduction Techniques in Electronic Systems," Wiley Interscience, 1976.

28. Williams, Jim, "High Speed Amplifier Techniques," Linear Technology Corporation, Application Note 47, 1991.

29. Tektronix, Inc., "Type 109 Pulse Generator Operating and Service Manual,", Tektronix, Inc., 1963.

30. Williams, Jim, "Signal Sources, Conditioners and Power Circuitry," "Wideband, Low Feedthrough, Low Level Switch", pg. 13-15. Appendix A, "How Much Bandwidth is Enough?", pg. 26, Linear Technology Corporation, Application Note 98, November 2004.

31. Williams, Jim, "Applications Considerations and Circuits for a New Chopper-Stabilized Op Amp," Linear Technology Corporation, Application Note 9, March 1985.

Appendix A
A history of high accuracy digital-to-analog conversion

People have been converting digital-to-analog quantities for a long time. Probably among the earliest uses was the summing of calibrated weights (Figure A1, left) in weighing applications. Early electrical digital-to-analog conversion inevitably involved switches and resistors of different values, usually arranged in decades. The application was often the calibrated balancing of a bridge or reading, via null detection, some unknown voltage. The most accurate resistor-based DAC of this type is Lord Kelvin's Kelvin-Varley divider (Figure, large box). Based on switched resistor ratios, it can achieve ratio accuracies of 0.1ppm (23+ bits) and is still widely employed in standards laboratories. High speed digital-to-analog conversion resorts to electronically switching the resistor network. Early electronic DACs were built at the board level using discrete precision resistors and

germanium transistors (Figure, center foreground, is a 12-bit DAC from a Minuteman missile D-17B inertial navigation system, circa 1962). The first electronically switched DACs available as standard product were probably those produced by Pastoriza Electronics in the mid 1960s. Other manufacturers followed and discrete- and monolithically-based modular DACs (Figure, right and left) became popular by the 1970s. The units were often potted (Figure, left) for ruggedness, performance or to (hopefully) preserve proprietary knowledge. Hybrid technology produced smaller package size (Figure, left foreground). The development of Si-Chrome resistors permitted precision monolithic DACs such as the LTC2757 (Figure, immediate foreground). In keeping with all things monolithic, the cost-performance trade-off of modern high resolution IC DACs is a bargain. Think of it! An 18-bit DAC in an IC package. What Lord Kelvin would have given for a credit card and LTC's phone number.

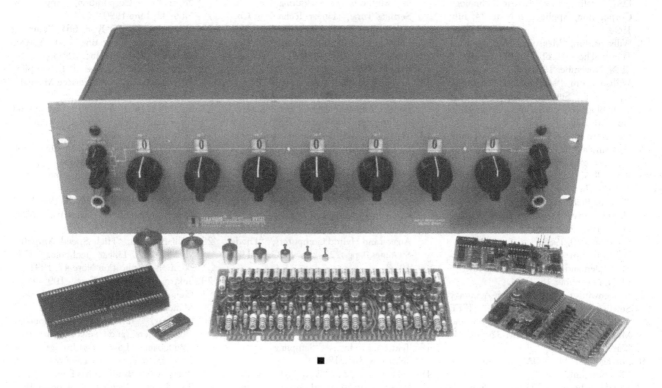

Figure A1 • Historically Significant Digital-to-Analog Converters Include Weight Set (Center Left), 23+ Bit Kelvin-Varley Divider (Large Box), Hybrid, Board and Modular Types, and the LTC2757 IC (Foreground). Where Will It All End?

Appendix B
Evaluating oscilloscope overdrive performance

The settling-time circuit is heavily oriented towards eliminating overdrive at the monitoring oscilloscope. Oscilloscope recovery from overdrive is a murky area and almost never specified. How long must one wait after an overdrive before the display can be taken seriously? The answer to this question is quite complex. Factors involved include the degree of overdrive, its duty cycle, its magnitude in time and amplitude and other considerations. Oscilloscope response to overdrive varies widely between types and markedly different behavior can be observed in any individual instrument. For example, the recovery time for a 100× overload at 0.005V/DIV may be very different than at 0.1V/DIV. The recovery characteristic may also vary with waveform shape, DC content and repetition rate. With so many variables, it is clear that measurements involving oscilloscope overdrive must be approached with caution.

Why do most oscilloscopes have so much trouble recovering from overdrive? The answer to this question requires some study of the three basic oscilloscope types' vertical paths. The types include analog (Figure B1A), digital (Figure B1B) and classical sampling (Figure B1C) oscilloscopes. Analog and digital 'scopes are susceptible to overdrive. The classical sampling 'scope is the only architecture that is inherently immune to overdrive.

An analog oscilloscope (Figure B1A) is a real-time, continuous linear system[1]. The input is applied to an attenuator which is unloaded by a wideband buffer. The vertical preamp provides gain, and drives the trigger pick-off, delay line and the vertical output amplifier. The attenuator and delay line are passive elements and require little comment. The buffer, preamp and vertical output amplifier are complex linear gain blocks, each with dynamic operating range restrictions. Additionally, the operating point of each block may be set by inherent circuit balance, low frequency stabilization paths or both. When the input is overdriven, one or more of these stages may saturate, forcing internal nodes and components to abnormal operating points and temperatures. When the overload ceases, full recovery of the electronic and thermal time constants may require surprising lengths of time[2].

The digital sampling oscilloscope (Figure B1B) eliminates the vertical output amplifier, but has an attenuator buffer and amplifiers ahead of the A/D converter. Because of this, it is similarly susceptible to overdrive recovery problems.

The classical sampling oscilloscope is unique. Its nature of operation makes it inherently immune to overload. Figure B1C shows why. The sampling occurs before any gain is taken in the system. Unlike Figure B1B's digitally sampled 'scope, the input is fully passive to the sampling point. Additionally, the output is fed back to the sampling bridge, maintaining its operating point over a very wide range of inputs. The dynamic swing available to maintain the bridge output is large and easily accommodates a wide range of oscilloscope inputs. Because of all this, the amplifiers in this instrument do not see overload, even at 1000× overdrives, and there is no recovery problem. Additional immunity derives from the instrument's relatively slow sample rate—even if the amplifiers were overloaded, they would have plenty of time to recover between samples[3].

The designers of classical sampling 'scopes capitalized on the overdrive immunity by including variable DC offset generators to bias the feedback loop (see Figure B1C, lower right). This permits the user to offset a large input, so small amplitude activity on top of the signal can be accurately observed. This is ideal for, among other things, settling time measurements. Unfortunately, classical sampling oscilloscopes are no longer manufactured, so if you have one, take care of it!

Although analog and digital oscilloscopes are susceptible to overdrive, many types can tolerate some degree of this abuse. The early portion of this Appendix stressed that measurements involving oscilloscope overdrive must be approached with caution. Nevertheless, a simple test can indicate when the oscilloscope is being deleteriously affected by overdrive.

The waveform to be expanded is placed on the screen at a vertical sensitivity that eliminates all off-screen activity. Figure B2 shows the display. The lower right hand portion is to be expanded. Increasing the vertical sensitivity by a factor of two (Figure B3) drives the waveform off-screen, but the remaining display appears reasonable. Amplitude has doubled and waveshape is consistent with the original display. Looking carefully, it is possible to see small amplitude information presented as a dip in the waveform at about the third vertical division. Some

---

Note 1: Ergo, the Real Thing. Hopelessly bigoted residents of this locale mourn the passing of the analog 'scope era and frantically hoard every instrument they can find.

Note 2: Some discussion of input overdrive effects in analog oscilloscope circuitry is found in Reference 13.

Note 3: Additional information and detailed treatment of classical sampling oscilloscope operation appears in References 17–20 and 23–25.

small disturbances are also visible. This observed expansion of the original waveform is believable. In Figure B4, gain has been further increased, and all the features of Figure B3 are amplified accordingly. The basic waveshape appears clearer and the dip and small disturbances are also easier to see. No new waveform characteristics are observed. Figure B5 brings some unpleasant surprises. This increase in gain causes definite distortion. The initial

negative-going peak, although larger, has a different shape. Its bottom appears less broad than in Figure B4. Additionally, the peak's positive recovery is shaped slightly differently. A new rippling disturbance is visible in the center of the screen. This kind of change indicates that the oscilloscope is having trouble. A further test can confirm that this waveform is being influenced by overloading. In Figure B6, gain remains the same but the vertical position knob has

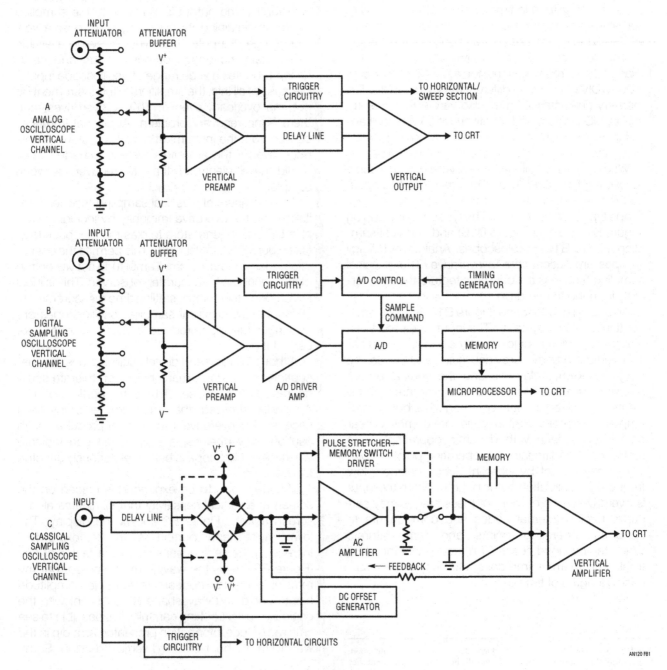

AN120 FB1

Figure B1 • Simplified Vertical Channel Diagrams for Different Type Oscilloscopes. Only the Classical Sampling Scope (C) Has Inherent Overdrive Immunity. Offset Generator Allows Viewing Small Signals Riding on Large Excursions

493

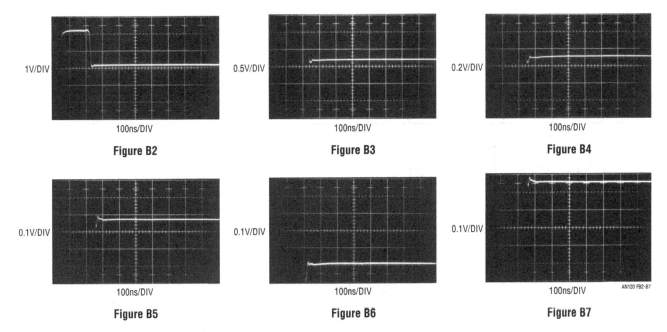

1V/DIV	0.5V/DIV	0.2V/DIV
100ns/DIV	100ns/DIV	100ns/DIV
Figure B2	Figure B3	Figure B4

0.1V/DIV	0.1V/DIV	0.1V/DIV
100ns/DIV	100ns/DIV	100ns/DIV
Figure B5	Figure B6	Figure B7

AN120 FB2-B7

Figures B2-B7 • The Overdrive Limit is Determined by Progressively Increasing Oscilloscope Gain and Watching for Waveform Aberrations

been used to reposition the display at the screen's bottom.[4] This shifts the oscilloscope's DC operating point which, under normal circumstances, should not affect the displayed waveform. Instead, a marked shift in waveform amplitude and outline occurs. Repositioning the waveform to the screen's top produces a differently distorted waveform (Figure B7). It is obvious that for this particular waveform, accurate results cannot be obtained at this gain.

Appendix C
Measuring and compensating signal path delay and circuit trimming procedures

Delay compensation
The settling time circuit utilizes an adjustable delay network to time correct the input pulse for delays in the signal-processing path. Typically, these delays introduce errors of a few percent, so a first-order correction is adequate. Setting the delay trim involves observing the network's input-output delay and adjusting for the appropriate time interval. Determining the "appropriate" time interval is somewhat more complex. Measuring the settling time

circuit's signal path delay involves modifications to Figure 24.13, shown in Figure C1. These changes lock the circuit into its "sample" mode, permitting an input-to-output delay measurement under signal-level conditions similar to normal operation. In Figure C2, trace A is the pulse-generator input at 200µV/DIV (note 10k-1Ω divider feeding the settle node). Trace B shows the circuit output at A4, delayed by about 44ns. This delay is a small error but is readily corrected by adjusting the delay network for the same time lag. If Appendix F's serial interface is utilized, 10ns should be added to the delay correction. Similarly, if Appendix I's post amplifier is used, the delay correction must be increased by 17ns.

Circuit trimming procedure
The following procedure, given in numerical order, trims the settling time circuit for optimum performance. It is advisable to execute trimming in the order given, avoiding out-of-sequence adjustments.
1. Turn off input pulses.
2. Trim "Baseline Zero" for 0V out at oscilloscope at 10mV per division or less.
3. Disconnect precision 10k resistors and ground settle node via 5.1kΩ.
4. Set sample delay to mid-range, sample window width to minimum.
5. Drive pulse generator input with 40kHz square wave.

Note 4: *Knobs* (derived from Middle English, "knobbe", akin to Middle Low German, "knubbe"), cylindrically shaped, finger rotatable panel controls for controlling instrument functions, were utilized by the ancients.

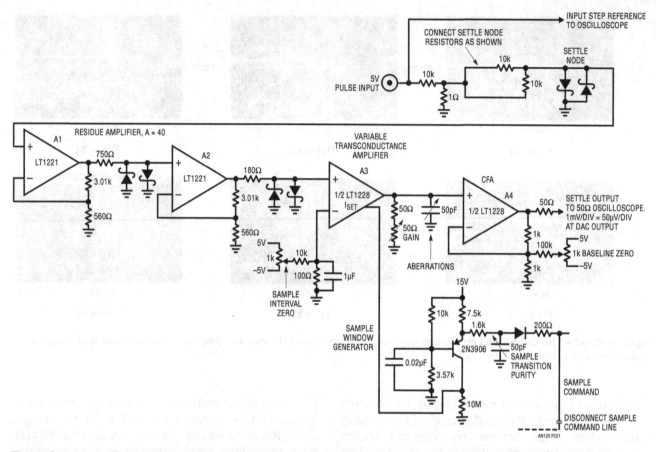

Figure C1 • Partial Text Figure 24.13 Schematic Shows Modifications for Measuring Signal Path Delay. Changes Lock Circuit into Sample Mode, Permitting Input-to-Output Delay Measurement

6. Adjust "Sample Interval Zero" for no offset between the sample interval and the unsampled baseline[1].
7. Adjust "Sample Transition Purity" and "Aberration" trims for minimum amplitude

disturbances when the sample gate opens with oscilloscope horizontal at 50ns per division and vertical sensitivity of 10mV per division.
8. Reconnect precision 10k resistors and remove 5.1kΩ unit from the settle node.

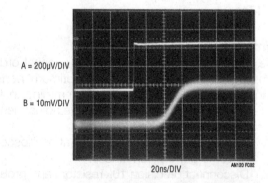
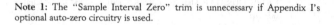

Figure C2 • Sampling Circuit Input-Output Delay Measures About 44ns

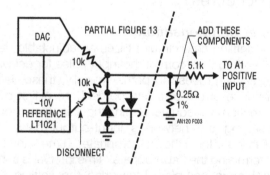

Figure C3 • Added Components Furnish 250μV Gain Calibration Source with Input Pulses Replaced by 5V Level. DAC Output Assumes 10V Reference Potential Under These Conditions; 5.1k Resistor Mimics 10kΩ Divider Output Impedance at A1

Note 1: The "Sample Interval Zero" trim is unnecessary if Appendix I's optional auto-zero circuitry is used.

9. Adjust "Delay Compensation" for 44ns delay from the pulse generator input to the time corrected output pulse.

10. Turn off input pulses. Disconnect the pulse generator and its 50Ω termination. Apply 5V DC to the pulse input.

11. Connect Figure C3's network to the settle node. The added components shown furnish a 250µV DC gain calibration source when the input pulses are replaced by a 5V level. Under the figure's conditions, the DAC assumes a 10V output with the 5.1k resistor mimicking the 10kQ divider output impedance at A1. Figure 24.13's "Gain" trim is adjusted for a 10mV DC deflection at the oscilloscope. This completes the trimming procedure and the circuit is ready for use.

Appendix D
Practical considerations
for DAC-amplifier compensation

There are a number of practical considerations in compensating the DAC-amplifier pair to get fastest settling time. Our study begins by revisiting text Figure 24.1 (repeated here as Figure D1). Settling time components include delay, slew and ring times. Delay is due to propagation time through the DAC-amplifier and is a very small term. Slew time is set by the amplifier's maximum speed. Ring time defines the region where the amplifier recovers from slewing and ceases movement within some defined error band. Once a DAC-amplifier pair have been chosen, only ring time is readily adjustable. Because slew time is usually the dominant lag, it is tempting to select the fastest slewing amplifier available to obtain best settling. Unfortunately, fast slewing amplifiers usually have extended ring times, negating their brute force speed advantage. The penalty for raw speed is, invariably, prolonged ringing, which can only be damped with large compensation capacitors. Such compensation works, but results in protracted settling times. The key to good settling times is to choose an amplifier with the right balance of slew rate and recovery characteristics and compensate it properly. This is harder than it sounds because amplifier settling time cannot be predicted or extrapolated from any combination of data sheet specifications. It must be measured in the intended configuration. In the case of a DAC-amplifier, a number of terms combine to influence settling time. They include amplifier slew rate and AC dynamics, DAC output resistance and capacitance, and the

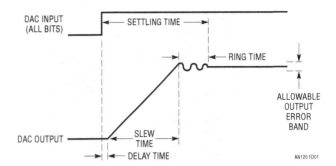

Figure D1 ● DAC-Amplifier Settling Time Components Include Delay, Slew and Ring Times. For Given Components, Only Ring Time is Readily Adjustable

compensation capacitor. These terms interact in a complex manner, making predictions hazardous[1]. If the DAC's parasitics are eliminated and replaced with a pure resistive source, amplifier settling time is still not readily predictable. The DAC's output impedance terms just make a difficult problem more messy. The only real handle available to deal with all this is the feedback compensation capacitor C_F. C_F's purpose is to roll off amplifier gain at the frequency that permits best dynamic response. Normally, the DAC's current output is unloaded directly into the amplifier's summing junction, placing the DAC's parasitic capacitance to ground at the amplifier's input. The capacitance introduces feedback phase shift at high frequencies, forcing the amplifier to "hunt" and ring about the final value before settling. Different DACs have different values of output capacitance. CMOS DACs have the highest output capacitance, typically 100pF and it varies with code.

Best settling results when the compensation capacitor is selected to functionally compensate for all the above parasitics. Figure D2, taken with an LTC2757/LT1468 DAC-amplifier combination, shows results for an optimally selected (in this case, 20pF) feedback capacitor. Trace A is the DAC input pulse and trace B the amplifier's settle signal. The amplifier is seen to come cleanly out of slew and settle very quickly.

In Figure D3, the feedback capacitor is too large (27pF). Settling is smooth, although overdamped, and a 300ns penalty results. Figure D4's feedback capacitor is too small (15pF), causing a somewhat underdamped response with resultant excessive ring time excursions. Settling time goes out to 2.8µs. Note that the above compensation values for 18-bit settling are not necessarily indicative of results for 16 or 20 bits. Optimal compensation values must be

---

Note 1: Spice aficionados take notice.

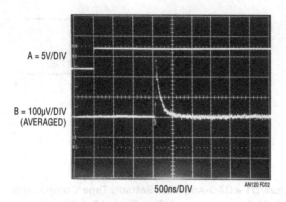

Figure D2 • Optimized Compensation Capacitor Permits Nearly Critically Damped Response, Faster Settling Time. t_{SETTLE} = 1.8µs to 0.0004% (18 Bits)

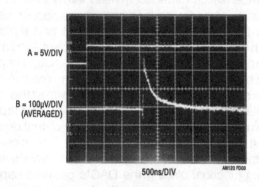

Figure D3 • Overdamped Response Ensures Freedom from Ringing, Even with Production Component Variations. Penalty is Increased Settling Time t_{SETTLE} = 2.1µs to 0.0004% (18 Bits)

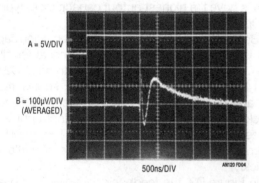

Figure D4 • Underdamped Response Results from Undersized Capacitor. Component Tolerance Budgeting Will Prevent This Behavior. t_{SETTLE} = 2.8µs to 0.0004% (18 Bits)

established for any given desired resolution. Typical values range from 15pF to 39pF.

When feedback capacitors are individually trimmed for optimal response, DAC, amplifier and compensation capacitor tolerances are irrelevant. If individual trimming is not used, these tolerances must be considered to determine the feedback capacitor's production value. Ring time is affected by DAC capacitance and resistance, as well as the feedback capacitor's value. The relationship is nonlinear although some guidelines are possible. The DAC impedance terms can vary by ±50% and the feedback capacitor is typically a ±5% component. Additionally, amplifier slew rate has a significant tolerance, which is stated on the data sheet. To obtain a production feedback capacitor value, determine the optimum value by individual trimming with the production board layout (board layout parasitic capacitance counts too!). Then, factor in the worst-case percentage values for DAC impedance terms, slew rate and feedback capacitor tolerance. Combine this information with the trimmed capacitor's measured value to obtain the production value. This budgeting is perhaps unduly pessimistic (RMS error summing may be a defensible compromise), but will keep you out of trouble[2].

Appendix E
A very special case—measuring settling time of chopper-stabilized amplifiers

The text box section (page 2) lists the LTC1150 chopper-stabilized amplifier The term "special case" appears in the "comments" column. A special case it is! To see why requires some understanding of how these amplifiers work. Figure E1 is a simplified block diagram of the LTC1150 CMOS chopper-stabilized amplifier. There are actually two amplifiers. The "fast

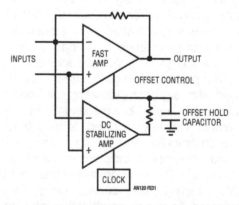

Figure E1 • Highly Simplified Block Diagram of Monolithic Chopper-Stabilized Amplifier. Clocked Stabilizing Amplifier and Hold Capacitor Cause Settling Time Lag

Note 2: The potential problems with RMS error summing become clear when sitting in an airliner that is landing in a snowstorm.

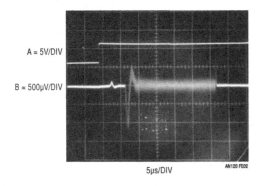

A = 5V/DIV

B = 500μV/DIV

5μs/DIV   AN120 FE02

Figure E2 • Short-Term Settling Profile of Chopper-Stabilized Amplifier Seems Typical. Settling Appears to Occur in 10μs

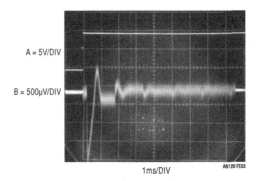

A = 5V/DIV

B = 500μV/DIV

1ms/DIV   AN120 FE03

Figure E3 • Surprise! Actual Settling Requires 700× More Time Than Figure E2 Indicates. Slow Sweep Reveals Monstrous Tailing Error (Note Horizontal Scale Change) Due to Amplifier's Clocked Operation. Stabilizing Loop's Iterative Corrections Progressively Reduce Error Before Finally Disappearing Into Noise

amp" processes input signals directly to the output. This amplifier is relatively quick, but has poor DC offset characteristics. A second, clocked, amplifier is employed to periodically sample the offset of the fast channel and maintain its output "hold" capacitor at whatever value is required to correct the fast amplifier's offset errors. The DC stabilizing amplifier is clocked to permit it to operate (internally) as an AC amplifier, eliminating its DC terms as an error source[1]. The clock chops the stabilizing amplifier at about 500 Hz, providing updates to the hold capacitor-offset control every 2ms[2].

The settling time of this composite amplifier is a function of the fast and stabilizating paths response. Figure E2 shows amplifier short-term settling. Trace A is the DAC input pulse and trace B the settle signal. Damping is reasonable and the 10μs settling time and profile appear typical. Figure E3 brings an unpleasant surprise. If the DAC slewing interval happens to coincide with the amplifier's sampling cycle, serious error is induced. In Figure E3, trace A is the

amplifier output and trace B the settle signal. Note the slow horizontal scale. The amplifier initially settles quickly (settling is visible in the 2nd vertical division region) but generates a huge error 200μs later when its internal clock applies an offset correction. Successive clock cycles progressively chop the error into the noise but 7 *milliseconds* are required for complete recovery. The error occurs because the amplifier sampled offset when its input was driven well outside its bandpass. This caused the stabilizing amplifier to acquire erroneous offset information. When this "correction" was applied, the result was a huge output error.

This is admittedly a worst case. It can only happen if the DAC slewing interval coincides with the amplifier's internal clock cycle, but it can happen[3,4].

Note 1: This AC processing of DC information is the basis of all chopper and chopper-stabilized amplifiers. In this case, if we could build an inherently stable CMOS amplifier for the stabilizing stage, no chopper stabilization would be necessary.
Note 2: Those finding this description intolerably brief are commended to Reference 31.

Note 3: Readers are invited to speculate on the instrumentation requirements for obtaining Figure E3's photo.
Note 4: The spirit of Appendix D's footnote 2 is similarly applicable in this instance.

Appendix F
Settling time measurement of serially loaded DACS

Measuring serially loaded DACs settling time requires additional circuitry. This circuitry must provide a "start" pulse to the settling time measurement circuit after serially loading a full-scale step into the DAC.

Figure F1's processor based circuitry, designed and constructed by LTC's Mark Thoren, does this. The "start" pulse (trace A, Figure F2) initiates the measurement. Traces B, C and D are CS/LD, SCK, and SDI, respectively. Trace E, the resultant DAC output, is measured for settling time in (what should be by now) familiar fashion. Figure F3 is a complete processor software code listing.

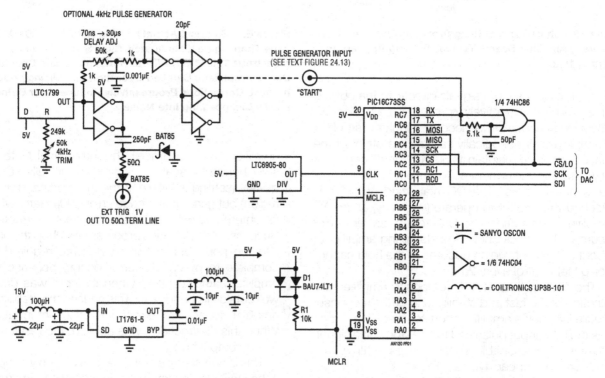

Figure F1 • The Serial Interface. Processor Responds to Input Pulse, Directs DAC to Perform 10V Steps

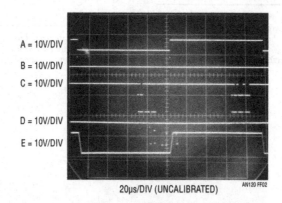

20µs/DIV (UNCALIBRATED)

Figure F2 • Serial Interface Operation Includes Input Start Pulse (Trace A), CS/LD (Trace B), SCK (C), SDI (D) and Resultant DAC Output (E), Digital Data Lines are Static During Measurement Interval, Precluding Crosstalk Induced Corruption

```
/*
Serial DAC step program. Makes controlling the serial DAC as easy as the old way of
tying all the digital lines of a parallel DAC to a pulse generator.

The serial DAC CS/LD signal is the output of an XOR gate edge detector that gives
a 1µs pulse on either the rising edge or falling edge of CONTROL signal.
Program enters main loop when CONTROL is high. When CONTROL goes low, the code
for +5V is sent. When CONTROL goes high, the code for -5V is sent. Thus the
timing of the load pulse accurately follows the input signal by about 20ns.

A delay of 60µs is inserted after the load pulse so that you can look at
settling details without having to worry about digital feedthrough.
*/

#include <16F73.h>
#include "pcm73a.h"
#use delay(clock=20000000)      // 20 meg clock
#fuses HS,NOWDT,PUT//,MCLR

// Defines for DAC addresses
#define DACA 0
#define DACB 2
#define DACC 4
#define DACD 6

#define PM10 0x03
#define PM5 0x02


// Control input
#define CONTROL PIN_C7

void init(void);

void main()
    {
    init();  // set up hardware

    // This just allows the program to sync up to a pulse generator that
    // may not have a clean output on power-up. You need to see at least one rising
    // and one falling edge before continuing.
    while(!input(CONTROL)){} delay_us(2); // wait for rising edge
    while(input(CONTROL)){} delay_us(2); // wait for falling edge
    delay_ms(100);
    while(!input(CONTROL)){} delay_us(2); // wait for rising edge
    while(input(CONTROL)){} delay_us(2); // wait for falling edge
```

an120f

```
// Okay, now we're all synchronized.
// Since program does not have direct access to the CS/LD line, you
// have to rely on the externally applied pulse.
while(!input(CONTROL)){} delay_us(2); // wait for rising edge
while(input(CONTROL)){} delay_us(2); // wait for falling edge
spi_write(0x6F);  // Set all DACs to +/-10V range
spi_write(0x00);
spi_write(PM5);

while(!input(CONTROL)){} delay_us(2); // wait for rising edge
spi_write(0x70 | DACB);        // Set DACB to -10 volts
spi_write(0x00);
spi_write(0x00);

while(input(CONTROL)){} delay_us(2); // wait for falling edge
spi_write(0x70 | DACC);        // Set DACC to 0 volts
spi_write(0x80);
spi_write(0x00);

while(!input(CONTROL)){} delay_us(2); // wait for rising edge
spi_write(0x70 | DACD);        // Set DACD to +10 volts
spi_write(0xFF);
spi_write(0xFF);

while(1)
    {
    while(input(CONTROL)){} delay_us(80); // wait for falling edge
    spi_write(0x70 | DACA);        // Set DACA to 0 volts
    spi_write(0xFF);
    spi_write(0xFF);

    while(!input(CONTROL)){} delay_us(80); // wait for rising edge
    spi_write(0x70 | DACA);        // Set DACA to +10 volts
    spi_write(0x00);
    spi_write(0x00);
    }
}

void init()
    {
    setup_adc_ports(NO_ANALOGS);
    setup_adc(ADC_CLOCK_DIV_2);
    setup_spi(SPI_MASTER|SPI_L_TO_H|SPI_CLK_DIV_4|SPI_SS_DISABLED);
    CKP = 0; // Set up clock edges - clock idles low, data changes on
    CKE = 1; // falling edges, valid on rising edges.
    setup_counters(RTCC_INTERNAL,RTCC_DIV_2);
    setup_timer_1(T1_DISABLED);
    setup_timer_2(T2_DISABLED,0,1);
    setup_ccp1(CCP_OFF);
    setup_ccp2(CCP_OFF);
    set_tris_a(0b00000000);
    set_tris_b(0b00000000);
    set_tris_c(0b10000100); // Make sure control signal is input
    }
```

Figure F3 • Software Listing for PIC16C73SS Processor. Code Directs DAC to Step 10V at Each Input Pulse Transition

Appendix G
Breadboarding, layout and connection techniques

The measurement results presented in this publication required painstaking care in breadboarding, layout and connection techniques. Wideband, 10μV resolution measurement does not tolerate cavalier laboratory attitude. The oscilloscope photographs presented, devoid of ringing, hops, spikes and similar aberrations, are the result of an exhaustive (and frustrating) breadboarding exercise[1]. The breadboard was rebuilt numerous times and required weeks of layout and shielding experimentation before obtaining a noise/uncertainty floor worthy of 20-bit measurement. In particular, extreme measures were required to minimize sample command signal feedthrough. Layout techniques include minimization and restriction of radiative paths, ground plane current management and mounting the LT1228 "switch" upside down, allowing its V-referred substrate to approximate a monolithic shield for the IC's internal circuitry.

Ohm's law

It is worth considering Ohm's law as a key to successful layout[2]. Consider that 1mA running through 0.1Ω generates 100μV—almost 3LSB at 18 bits! Now, run that milliampere at 5ns to 10ns rise times (approximately 75MHz) and the need for layout care becomes clear. A paramount concern is disposal of circuit ground return current and disposition of current in the ground plane. The impedance of the ground plane between any two points is not zero,

Figure G1 • Settling Time Breadboard Overview. Pulse Generator Input Enters Top Left—50Ω Coaxial Terminator Mounted On Extension Tube (Not Visible—See Appendix H, Figure H7) Minimizes Pulse Generator Return Current Mixing Into Signal Ground Plane. DAC-Amplifier and Support Circuitry are at Left. Sampler Circuitry Occupies Board Right. Sampler Digital Support Circuitry is Contained Within Large Shield (Board Lower). Nonsaturating Amplifier Occupies Board Center. Partially Visible X10 Post Amplifier (See Appendix I) is BNC Fixtured, Thin Board at Photo Right. Auto-Zero Circuit (see Appendix I), Mounted on Thin Strut (Lower Right), is Not Visible

---

Note 1: "War" is perhaps a more accurate descriptive.

Note 2: I do not wax pedantic here. My abuse of this postulate runs deep.

Figure G2 • DAC-Amplifier Detail DAC and Output Amplifier are at Photo Center Left. Precision Summing Resistors (Box-Shaped, Just Below Large Round Capacitor Near Photo Center) are Oriented to Restrict Radiative Coupling While Minimizing Summing Node Capacitance. Variable Capacitor (Lower Left Center) Sets Amplifier (Photo Center) Compensation. Non-Saturating Amplifier Appears at Right. Shield (Bottom) Restricts Sampler Digital Section's Radiation. Vertically Mounted Board (Extreme Upper Left) is Optional Serial DAC Interface (See Appendix F). Coaxial Connectors (Center Lower) Facilitate High Purity Signal Extraction

particularly as frequency scales up. This is why the entry point and flow of ''dirty'' ground returns must be carefully placed within the grounding system. In the sampler-based breadboard, the approach was separate ''dirty'' and ''signal'' ground planes, tied together at the supply ground origin.

A good example of the importance of grounding management involves delivering the input pulse to the breadboard. The pulse generator's 50Ω termination must be an in-line coaxial type, and it cannot be directly tied to the signal ground plane. The high speed, high density (5V pulses through the 50Ω termination generate 100mA current spikes) current flow must return directly to the pulse generator. The coaxial terminator's construction ensures this substantial current does this, instead of being dumped into the signal ground plane (100mA termination current flowing through 1mΩ of ground plane produces approximately 3LSB of error!). The 50Ω

termination is physically distanced from the breadboard via a coaxial extension tube (visible in Figure H7)[3]. This further ensures that pulse generator return current circulates in a tight local loop at the terminator and does not mix into the signal plane.

It is worth mentioning that every ground return in the entire circuit must be evaluated with these concerns in mind. A paranoiac mindset is quite useful.

Shielding

The most obvious way to handle radiation-induced errors is shielding. Various following figures show shielding. Determining where shields are required should come *after* considering what layout will minimize their necessity. Often, grounding requirements conflict with minimizing radiation effects, precluding

Note 3: Strictly considered, this technique introduces mis-termination originated transmission line reflections but no appreciable error results at the bandwidth of interest.

Figure G3 ● LT1228 Sampling "Switch" (Photo Center) Is Mounted Upside Down, Permitting V⁻ Referred Die Backside To Shield Residual Radiative Coupling, Reducing Sampling Switch Drive Feedthrough. Switch Signal Channel Is Fed From Non-Saturating Amplifier (Photo Left). Sample Command From Shielded Digital Section (Lower) Arrives Via Coaxial Cable Tunneling Through Shield (Lower Center)

maintaining distance[4] between sensitive points. Shielding[5] is usually an effective compromise in such situations.

A similar approach to ground path integrity should be pursued with radiation management. Consider what points are likely to radiate, and try to lay them out at a distance from sensitive nodes. When in doubt about odd effects, experiment with shield placement and note results, iterating towards favorable performance[6]. *Above all, never rely on filtering or measurement bandwidth limiting to "get rid of" undesired signals whose origin is not fully understood.* This is not only intellectually dishonest, but may produce wholly invalid measurement "results", even if they look pretty on the oscilloscope.

Connections

All signal connections to the breadboard must be coaxial. Ground wires used with oscilloscope probes are forbidden. A one inch ground lead used with a 'scope probe can easily generate several LSBs of observed "noise". Use coaxially mounting probe tip adapters[7]. Figures G1 to G5 restate the above sermon in visual form while annotating the text's circuits.

Note 4: Distance is the physicist's approach to reducing radiation induced effects.
Note 5: Shielding is the engineer's approach to reducing radiation induced effects.
Note 6: After it works, you can figure out why.

Note 7: See Reference 28 for additional nagging along these lines.

Figure G4 • A Dedicated, Serially Interfaced Settling Time Breadboard. Serial Interface Digital Board is Obscured Beneath Visible Analog Board (See Appendix H, Figure H7). Note Insulating Nylon Screws (Right and Left Lower Corners and Upper Edge Between BNC Connectors) Used to Attach Digital and Analog Boards. Digital Board Ground is Single-Point Connected at Analog Board Ground Entry Point (Middle Banana Jack). Serial Signals Access Vertical DAC Board Via Coaxial (Photo Left). Shields Isolate Sample Switch Digital Section (Lower) and Summing Node (Left Center). Non-Saturating Amplifier and Sampling Switch are as in Figure G1. Optional Auto-Zero Board (See Appendix I) Mounts from BNC Fitting at Center Right. Upper Left Corner Components are Input Pulse Time Correction

Figure G5 • Serially Interfaced Settling Time Breadboard Signal Path Detail. DAC Board is at Left. Precision Resistors Feed Summing Node and Non-Saturating Amplifier (Photo Upper Center). Shield Protects Summing Node from Input Pulse Originated Radiation. LT1228 Sampling "Switch" (Far Right) is Mounted Upside Down, Permitting V⁻ Referred Die Backside to Shield Radiative Coupling, Reducing Switch Drive Feedthrough. Switch Drive Level Shift-Current Source (Extreme Right Upper) Receives Sample Command Via Coaxial, Partially Visible at Lower Right. Large Vertical Shield Confines Sample Command Digital Section Radiation (Lower). SMA Connector (Center) Enables Test and Calibration Signal Connection

Appendix H
How do you know it works?

Settling time circuit performance verification

High purity pulse generator

Any prudent investigation requires performance verification of the test method. Strictly considered, it may not be possible to furnish indisputable proof that the circuit in question is functioning at its design limits, particularly in a state-of-the art measurement. However, a reasonable level of confidence is a realistic goal. Performance verification for the settling time test circuit requires a high purity pulse generator that transitions and settles to 1ppm as quickly as possible. This is a high order difficulty requirement and the author is unaware of any electronic means of achieving this capability. Fortunately, electromechanical technology offers a solution.

Figure H1 shows a conceptual mercury wetted reed relay pulse generator. Theoretically, when the contacts open, an infinitely fast falling edge appears across the 50Ω termination with zero settling time to 0V. Figure H2 reveals this to be not the case. This photograph, taken with a typical commercially available relay, shows <5ns transition time with a 500MHz ring off over 10ns[1]. These imperfections are not surprising when Figure H3's parasitic terms are considered. Figure H1's deceptively simple schematic is revealed to have a number of unintentional terms which severely limit performance. These terms include, but are not limited to, parasitic resistance,

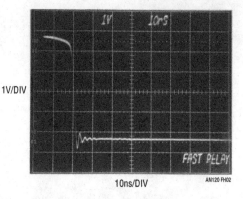

Figure H2 • Mercury Wetted Reed Relay Opens in <5ns, Settles Quickly to Zero. 500MHz Ring Off Derives from Source-Termination Impedance Mismatch

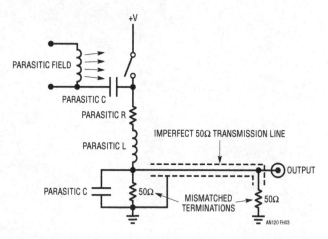

Figure H3 • Parasitic Terms Limit Achievable Performance. Transmission Line, Required to Route Output Pulse, Adds Termination Mismatch Errors and Line Related Infidelities

inductance, and capacitance as well as undesirable field interaction within the relay. Additionally, the connection through the relay to the output terminal constitutes an ill-defined transmission line which promotes additional vagaries. The parasitic terms interact in a haphazard and unpredictable way, resulting in alien terms at the pulse output. What is really needed is a relay specifically designed and constructed for inclusion into a wideband 50Ω system.

A true 50Ω, wideband mercury wetted reed relay

In the 1960s, Tektronix manufactured the type 109 mercury wetted reed relay, intended for use as a pulse generator. In its preferred configuration, energized charge lines are switched by the relay into a 50Ω termination, resulting in a 250ps risetime pulse. Here, charge lines are not employed; rather, the

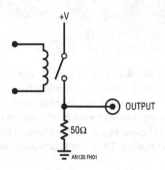

Figure H1 • Conceptual Mercury Wetted Reed Relay Pulse Generator Produces Infinitely Fast Falling Edge Across 50Ω Termination with Zero Settling Time to 0V

Note 1: Tangential to this discussion, but nonetheless interesting, is the corner rounding at the pulse top just before its rapid fall. This may be due to "teasing" of the mercury, causing its resistance to increase just before it fully opens. John Willison of Stanford Research Systems suggests the mechanism may be charge displacement in the capacitor formed as the relay contacts just open. Scott Hamilton, Manchester University (UK), has raised the possibility of quantum tunneling across the brief, small contact gap *a la* scanning tunneling microscope operation. Comments from the readership are welcome.

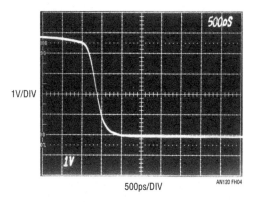

Figure H4 • High Grade Mercury Wetted Reed Relay (Tektronix Type 109) Falls in 800ps Viewed in 1GHz Real Time Bandwidth. Strict Attention To Parasitic Minimization in Relay Structure and Transmission Path Produces High Fidelity Transition Without Alien Components

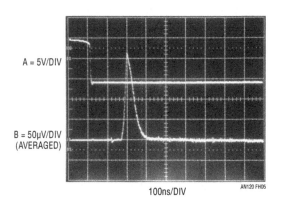

Figure H5 • Sampling Switch Techniques Permit Measuring Type 109 Falling Edge Residue to Microvolts in 20MHz Bandpass. Trace A is 109 Falling Edge, Trace B the Last 220μV of Movement Before Settling to 1ppm in Indicated 265ns

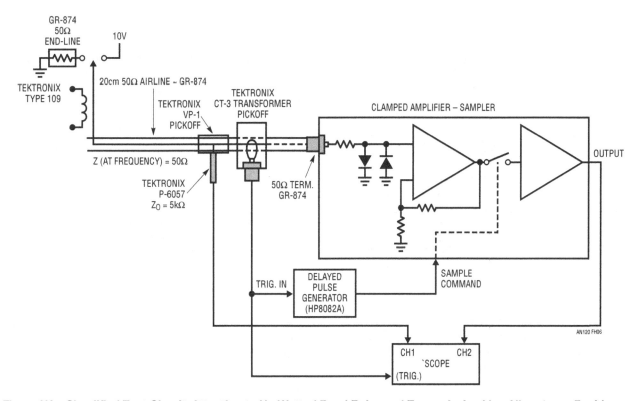

Figure H6 • Simplified Test Circuit. Attention to Hg Wetted Reed Relay and Transmission Line Allow 1ppm Residue ≈265ns After Contacts Open

device is used as a simple switch. Advantage is taken of the exquisite care at manufacture to make the relay transparent in a 50Ω system. Figure H4's large scale transition reveals 800 ps fall time in a 1GHz bandwidth. Actual fall time is probably somewhat faster as the monitoring oscilloscope has a

350ps risetime[2]. The transition is singularly clean and devoid of discontinuities with the exception of the previously noted pre-fall corner rounding. Figure H5's

---

Note 2: A Root-Sum-Square risetime calculation indicates 720 picoseconds. See Reference 30.

Figure H7 • High Purity Step Generator Connected to Figure G4's Settling Time Breadboard. Tektronix Type 109 Mercury Wetted Relay Based Pulse Generator (Photo Left) Delivers Pulse Via General Radio 20 Centimeter Airline. Tektronix CT-3 Coaxial Transformer (Right End of Air Line) Supplies Trigger Pick-Off Via 50Ω BNC Vertical Connector. GR-874 Coaxial 50Ω Load (Right Side of CT-3) Terminates Line And Supplies Pulse to Breadboard. Pulse Amplitude is Set by DC Voltage Applied to Tektronix 109 Via Banana Input Adapter. Unused Type 109 Contact is Terminated with End-Line GR-874 Fitting. All Connections Must Be Polished and Mechanically Secure to Ensure High Fidelity 50Ω Environment. Long Coaxial Extension Tube On Breadboard Isolates Input Pulse Termination (Tube Top) Current from Board Ground Plane (see Text and Figure G1 For Commentary). Lower Board Contains Digital Serial Interface (See Appendix F and Figure G4)

remarkable photograph uses sampling switch techniques similar to those described in the text to measure type 109 fall-time purity to microvolts[3]. The 109 output (trace B) moves the final 220μV settling inside 10μV approximately 265ns after the relay contacts open (trace A). Actual 109 settling time may be faster as the measurement is likely sampling circuit limited[4]. Figure H6 is a simplified version of the test circuit which produced these results. The type 109 drives 20 centimeters of 50Ω GR-874 airline into a high quality GR-874 50Ω termination at the clamped amplifier-sampling switch. The delayed pulse

generator and oscilloscope are set up similarly to the main text description. The Tektronix pickoff components noted allow signal extraction from the airline without degrading transmission line integrity. It is interesting to note that airline is a non-negotiable requirement. The highest quality Teflon 50Ω cable produced impure response, albeit minor.

Figure H7's photograph shows the high purity step generator connected to Figure G4's settling time breadboard. The type 109 (photo left) delivers its pulse via a General Radio 20 centimeter airline. The Tektronix CT-3 coaxial transformer (right end of airline) supplies the trigger pick-off via the vertically mounted 50Ω BNC connector. A GR-874 50Ω load (right side of CT-3) terminates the airline and supplies the pulse to the breadboard. Pulse amplitude is set

---

Note 3: Some may find "remarkable" to be excessively enthusiastic verbiage but these things thrill certain types.
Note 4: Sampling circuit delay ≈70 nanoseconds; 109 1ppm settling time is probably inside 195 nanoseconds.

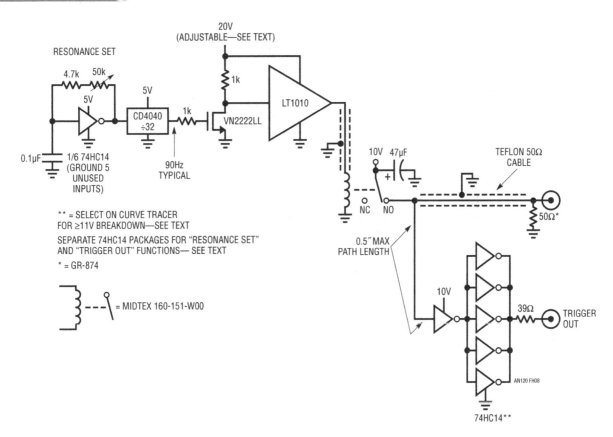

Figure H8 ● "Pretty Good" Substitute for Figure H6's Tektronix 109, Constructed from Commercially Available Components, Settles 10V Step to 1ppm in 950ns

by a DC voltage applied at the 109 via banana inputs. The unused 109 contact is terminated with an end-line GR-874 fitting. All connections must be polished and mechanically secured to ensure a high fidelity 50Ω environment. Any component substitution or mechanical connection imperfection will degrade Figure H5's results.

"Pretty good" mercury wetted reed relay pulse generator

It may be unrealistic for readers to duplicate the Tektronix 109 based results. The specified exotic apparatus is difficult and expensive to obtain and the set up requires arduous labor and almost fanatical attention to detail. In this spirit, Figure H8's "pretty good" mercury wetted reed relay pulse generator is offered. Its performance, while falling well short of Figure H6, still furnishes a 10V step which settles to 1ppm in 950ns[5]. A simple clock ("resonance set") furnishes low frequency drive to the relay via the transistor level shift and the LT1010 power

buffer. Trigger pick-off is provided by the paralleled CMOS inverters. Separate packages for the "resonance set" and trigger functions must be used to avoid output pulse contamination. Figure H9 shows results for the "pretty good" step generator. Its

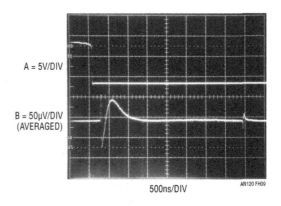

Figure H9 ● "Pretty Good" Step Generator Output (Trace A) Settles (Trace B) to 1ppm (10μV) in 950ns. Ill-Defined Relay Impedance Results in Approximately 3.6× Slower Settling and Alien Reside Components Compared to Figure H5's Tektronix 109 Based Results. Event at 10th Vertical Division is Sample Gate Turn-Off Feedthrough

Note 5: Footnote 4's 70ns timing allowance applies here. Figure H8 likely settles in ≈880ns.

slower settling and alien residue components compared to the Tektronix 109 approach are apparent. The event at the 10th vertical division is sample gate turn-off feedthrough related.

Circuit calibration involves adjusting "resonance set" until the relay emits a reasonably pure audible tone. Next, set the 20V supply to a value which promotes the cleanest settling characteristics.

Appendix I
Auxiliary circuits

Several auxiliary circuits have been found useful in the DAC settling time work described in the text. Figure I1 is a simple, wideband, X10 preamplifier for oscilloscopes lacking the required sensitivity for 1ppm (10μV) settling time resolution. This

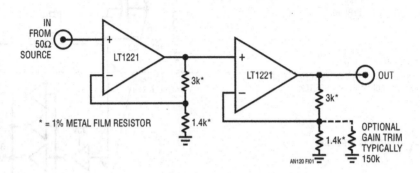

Figure I1 • Simple X10 Pre-Amplifier for Oscilloscopes Lacking Required Sensitivity for 1ppm (10μV) Settling Time Resolution

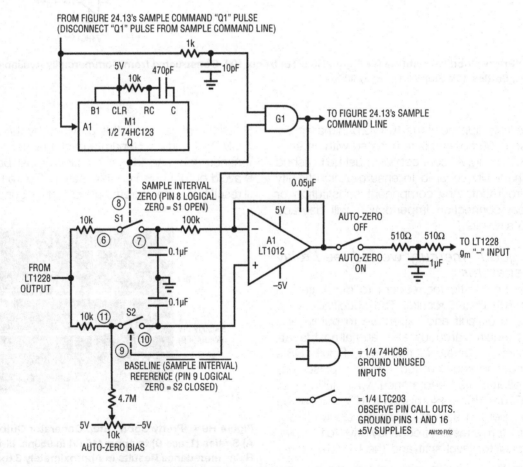

Figure I2 • Auto-Zero Locks Sample Interval Zero Value to Non-Sampled Region Baseline. Synchronously Switched A1 Compares Sample Interval and Non-Sampled Region Values and Applies Appropriate Offset, Closing Correction Loop Around LT1228. M1 Precludes Settling Activity from Influencing Sample Interval Zero Value

preamplifier should be placed directly (no cable) at the oscilloscope input and connected via 50Ω terminated BNC cable to the settling time fixture output.

Figure I2, an auto-zero circuit, locks the sample interval zero value to the non-sampled region baseline. It eliminates the need for periodic readjustment of the "Sample Interval Zero" trim when working at the highest levels of resolution over a protracted time. Synchronously switched A1 compares sample interval and non-sampled region zero values and applies an appropriate offset, closing a correction loop around the LT1228. M1's extended pulse precludes settling activity from influencing the sample interval zero value. The "Auto-Zero Bias" trim corrects for slight errors and should not require readjustment once set to equalize the sample interval zero value and the non-sample region baseline. The commented schematic provides information for the auto-

zero's interconnection to the settling time circuit. Figure I3 shows auto-zero related waveforms. They include the time corrected input pulse (trace A), DAC output (B), sample delay (C), M1 input (D), M1's sample interval zero pulse (E), G1's sample command (F), and settle signal output (G). M1's delayed output maintains the sample interval zero value independent of the settling signature.

Figure I4 is a simple time calibrator used to verify oscilloscope time base accuracy. Q1 and Q2 form a 1MHz quartz oscillator. The 74C90 provides switch selectable output periods of 2μs and 5μs and the attenuator supplies a 50Ω output impedance. The period values have been selected for calibration points appropriate for expected DAC settling times. Other periods are available by varying oscillator frequency, division ratio or both. 9V battery drain is about 10mA.

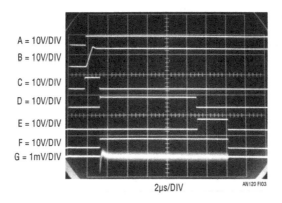

2μs/DIV

AN120 FI03

Figure I3 • Auto-Zero Related Waveforms Include Time Corrected Input Pulse (Trace A), DAC Output (B), Sample Delay (C), M1's Input (D), M1's Sample Interval Zero Pulse (E), G1's Sample Command (F) and Settle Signal Output (G). M1's Delayed Output Maintains Sample Interval Zero Value Independent of Settling Signature

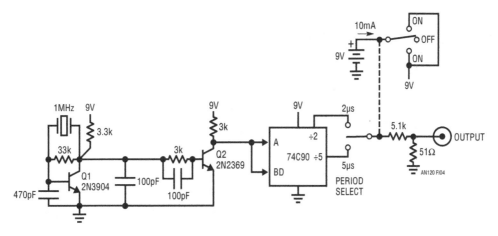

Figure I4 • Battery-Powered Oscilloscope Time Base Verifier Has 2μs and 5μs Period Outputs. Quartz Oscillator Q1 Is Buffered by Q2; Digital Divider Supplies Outputs Via Attenuator

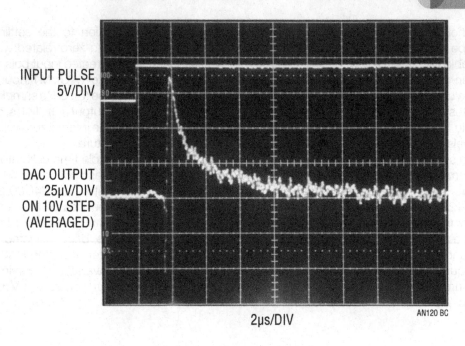

INPUT PULSE
5V/DIV

DAC OUTPUT
25µV/DIV
ON 10V STEP
(AVERAGED)

2µs/DIV

AN120 BC

''A part-per-million is a part-per-million. It's magic. It's the brass ring. It's the holy grail of every measurement artist. It will mesmerize you, it will goad you, it will drive you crazy and, if you're lucky, will reward you. A part-per-million is a part-per-million.''

Jerrold R. Zacharias
M.I.T. physicist, mentoring a young,
very naïve investigator.
— 1971

Applications for a switched-capacitor instrumentation building block

25

Jim Williams

CMOS analog IC design is largely based on manipulation of charge. Switches and capacitors are the elements used to control and distribute the charge. Monolithic filters, data converters and voltage converters rely on the excellent characteristics of IC CMOS switches. Because of the importance of switches in their circuits, CMOS designers have developed techniques to minimize switch induced errors, particularly those associated with stray capacitance and switch timing. Until now, these techniques have been used only in the internal construction of monolithic devices. A new device, the LTC®1043, makes these switches available for board-level use. Multi-pole switching and a self-driven, non-overlapping clock allow the device to be used in circuits which are impractical with other switches.

Conceptually, the LTC1043 is simple. Figure 25.1 details its features. The oscillator, free-running at 200kHz, drives a non-overlapping clock. Placing a capacitor from Pin 16 to ground shifts the oscillator frequency downward to any desired point. The pin may also be driven from an external source, synchronizing the switches to external circuitry. A non-overlapping clock controls both DPDT switch sections. The non-overlapping drive prevents simultaneous conduction in the series connected switch sections.

Charge balancing circuitry cancels the effects of stray capacitance. Pins 1 and 10 may be used as guard points for Pins 3 and 12 in particularly sensitive applications.

Although the device's operation is simple, it permits surprisingly sophisticated circuit functions. Additionally, the careful attention paid to switching characteristics makes implementing such functions relatively easy. Discrete timing and charge-balance compensation networks are eliminated, reducing component count and trimming requirements.

Classical analog circuits work by utilizing continuous functions. Their operation is usually described in terms of voltage and current. Switched-capacitor based circuits are sampled data systems which approximate continuous functions with bandwidth limited by the sampling frequency. Their operation is described in the distribution of charge over time. To best understand the circuits which follow, this distinction should be kept in mind. Analog sampled data and carrier-based systems are less common than true continuous approaches, and developing a working familiarity with them requires some thought.

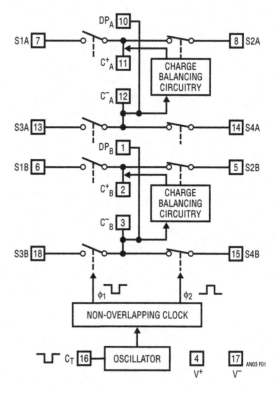

Figure 25.1 • Block Diagram of LTC1043 Showing Individual Switches

Analog Circuit and System Design: A Tutorial Guide to Applications and Solutions. DOI: 10.1016/B978-0-12-385185-7.00025-1

Switched-capacitor approaches have greatly aided analog MOS IC design. The LTC1043 brings many of the freedoms and advantages of CMOS IC switched-capacitor circuits to the board level, providing a valuable addition to available design techniques.

Instrumentation amplifier

Figure 25.2 uses the LTC1043 to build a simple, precise instrumentation amplifier. An LTC1043 and an LT1013 dual op amp are used, allowing a dual instrumentation amplifier using just two packages. A single DPDT section converts the differential input to a ground referred single-ended signal at the LT1013's input. With the input switches closed, C1 acquires the input signal. When the input switches open, C2's switches close and C2 receives charge. Continuous clocking forces C2's voltage to equal the difference between the circuit's inputs. The 0.01μF capacitor at Pin 16 sets the switching frequency at 500Hz. Common mode voltages are rejected by over 120dB and drift is low.

Amplifier gain is set in the conventional manner. This circuit is a simple, economical way to build a high performance instrumentation amplifier. Its DC characteristics rival any IC or hybrid unit and it can operate from a single 5V supply. The common mode range includes the supply rails, allowing the circuit to read across shunts in the supply lines. The performance of the instrumentation amplifier

depends on the output amplifier used. Specifications for an LT1013 appear in the figure. Lower figures for offset, drift and bias current are achievable by employing type LT1001, LT1012, LT1056 or the chopper-stabilized LTC1052.

Ultrahigh performance instrumentation amplifier

Figure 25.3 is similar to Figure 25.2, but utilizes the remaining LTC1043 section to construct a low drift chopper amplifier. This approach maintains the true differential inputs while achieving 0.1μV/°C drift. The differential input is converted to a single-ended potential at Pin 7 of the LTC1043. This voltage is chopped into a 500Hz square wave by the switching action of Pins 7, 11 and 8. A1, AC-coupled, amplifies this signal. A1's square wave output, also AC-coupled, is synchronously demodulated by switches 12, 14 and 13. Because this switch section is synchronously driven with the input chopper, proper amplitude and polarity information is presented to A2, the DC output amplifier. This stage integrates the square wave into a DC voltage to provide the output. The output is divided down and fed back to Pin 8 of the input chopper where it serves as the zero signal reference. Because the main amplifier is AC-coupled, its DC terms do not affect overall circuit offset, resulting in the extremely low offset and drift noted in the specifications. This circuit offers lower offset and drift than any commercially available instrumentation amplifier.

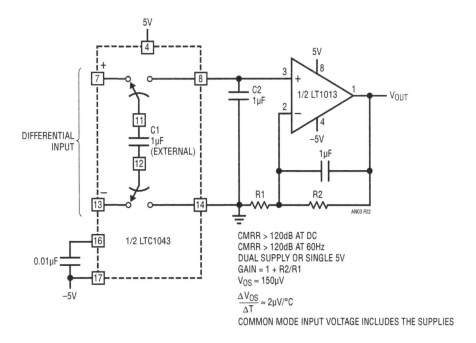

Figure 25.2 • ±5 V Precision Instrumentation Amplifier

Lock-in amplifier

The AC carrier approach used in Figure 25.3 may be extended to form a "lock-in" amplifier. A lock-in amplifier works by synchronously detecting the carrier modulated output of the signal source. Because the desired signal information is contained within the carrier, the system constitutes an extremely narrow-band amplifier. Non-carrier related components are rejected and the amplifier passes only signals which are coherent with the carrier. In practice, lock-in amplifiers can extract a signal 120dB below the noise level.

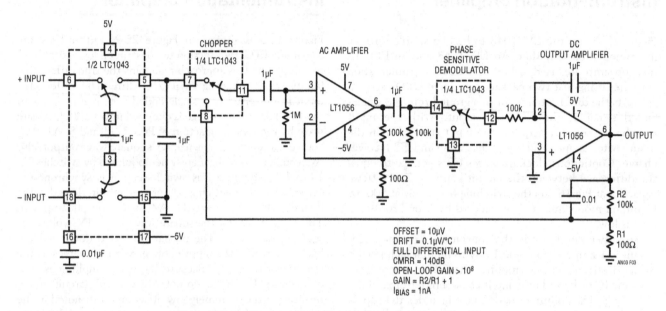

Figure 25.3 • Chopper-Stabilized Instrumentation Amplifier

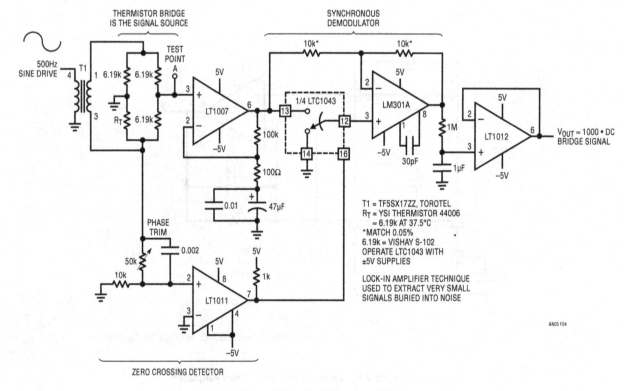

Figure 25.4 • Lock-In Amplifier

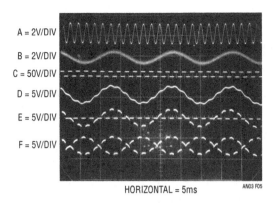

Figure 25.5

Figure 25.4 shows a lock-in amplifier which uses a single LTC1043 section. In this application, the signal source is a thermistor bridge which detects extremely small temperature shifts in a biochemical microcalorimetry reaction chamber.

The 500Hz carrier is applied at T1's input (Trace A, Figure 25.5). T1's floating output drives the thermistor bridge, which presents a single-ended output to A1. A1 operates at an AC gain of 1000. A 60Hz broadband noise source is also deliberately injected into A1's input (Trace B). The carrier's zero crossings are detected by C1. C1's output clocks the LTC1043 (Trace C). A1's output (Trace D) shows the desired 500Hz signal buried within the 60Hz noise source. The LTC1043's zero-cross-synchronized switching at A2's positive input (Trace E) causes A2's gain to alternate between plus and minus one. As a result, A1's output is synchronously demodulated by A2. A2's output (Trace F) consists of demodulated carrier signal and non-coherent components. The desired carrier amplitude and polarity information is discernible in A2's output and is extracted by filter averaging at A3. To trim this circuit, adjust the phase potentiometer so that C1 switches when the carrier crosses through zero.

Wide range, digitally controlled, variable gain amplifier

Aside from low drift and noise rejection, another dimension in amplifier design is variable gain. Designing a wide range, digitally variable gain block with good DC stability is a difficult task. Such configurations usually involve relays or temperature compensated FET networks in expensive and complex arrangements. The circuit shown in Figure 25.6 uses the LTC1043 in a variable gain amplifier which features continuously variable gain from 0 to 1000, gain stability of 20ppm/°C and single-ended or differential input. The circuit uses two separate

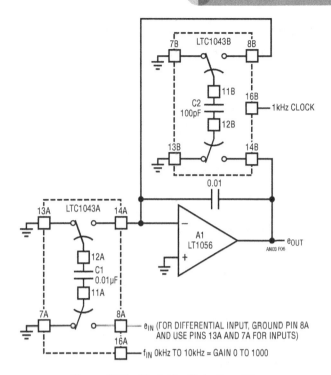

Figure 25.6 • Variable-Gain Amplifier

LTC1043s. Unit A is clocked by a frequency input which could be derived from a host processor. LTC1043B is continuously clocked by a 1kHz source which could also be processor supplied. Both LTC1043s function as the sampled data equivalent of a resistor within the bandwidth set by A1's 0.01µF value and the switched-capacitor equivalent feedback resistor. The time-averaged current delivered to the summing point by LTC1043A is a function of the 0.01µF capacitor's input-derived voltage and the commutation frequency at Pin 16. Low commutation frequencies result in small time-averaged current values, approximating a large input resistor. Higher frequencies produce an equivalent small input resistor. LTC1043B, in A1's feedback path, acts in a similar fashion. For the circuit values given, the gain is simply:

$$G = \frac{f_{IN}}{10} \bullet \frac{0.01\mu F}{100pF}$$

Gain stability depends on the ratiometric stability between the 1kHz and variable clocks (which could be derived from a common source) and the ratio stability of the capacitors. For polystyrene types, this will typically be 20ppm/°C. The circuit input, determined by the pin connections shown in the figure, may be either single-ended or fully differential. Additionally, although A1 is connected as an inverter, the circuit's overall transfer function may be either positive or negative. As shown, with Pins 13A and 7A grounded and the input applied to 8A, it is negative.

Precision, linearized platinum RTD signal conditioner

Figure 25.7 shows a circuit which provides complete, linearized signal conditioning for a platinum RTD. One side of the RTD sensor is grounded, often desirable for noise considerations. This LTC1043 based circuit is considerably simpler than instrumentation or multi-amplifier based designs and will operate from a single 5V supply. A1 serves as a voltage-controlled ground referred current source by differentially sensing the voltage across the 887Ω feedback resistor. The LTC1043 section which does this presents a single-ended signal to A1's negative input, closing a loop. The 2k-0.1µF combination sets amplifier roll-off well below the LTC1043's switching frequency and the configuration is stable. Because A1's loop forces a fixed voltage across the 887Ω resistor, the current through Rp is constant. A1's operating point is primarily fixed by the 2.5V LT1009 voltage reference.

The RTD's constant current forces the voltage across it to vary with its resistance, which has a nearly linear positive temperature coefficient. The nonlinearity could cause several degrees of error over the circuit's 0°C to 400°C

operating range. A2 amplifies Rp's output, while simultaneously supplying nonlinearity correction. The correction is implemented by feeding a portion of A2's output back to A1's input via the 10k to 250k divider. This causes the current supplied to Rp to slightly shift with its operating point, compensating sensor nonlinearity to within ±0.05°C. The remaining LTC1043 section furnishes A2 with a differential input. This allows an offsetting potential, derived from the LT1009 reference, to be subtracted from Rp's output. Scaling is arranged so 0°C equals 0V at A2's output. Circuit gain is set by A2's feedback values and linearity correction is derived from the output.

To calibrate this circuit, substitute a precision decade box (e.g., General Radio 1432k) for Rp. Set the box to the 0°C value (100.00Ω) and adjust the offset trim for a 0.00V output. Next, set the decade box for a 140°C output (154.26Ω) and adjust the gain trim for a 1.400V output reading. Finally, set the box to 249.0°C (400.00°C) and trim the linearity adjustment for a 4.000V output. Repeat this sequence until all three points are fixed. Total error over the entire range will be within ±0.05°C. The resistance values given are for a nominal 100.00Ω (0°C) sensor. Sensors deviating from this nominal value can be used by factoring in the deviation from 100.00Ω. This deviation,

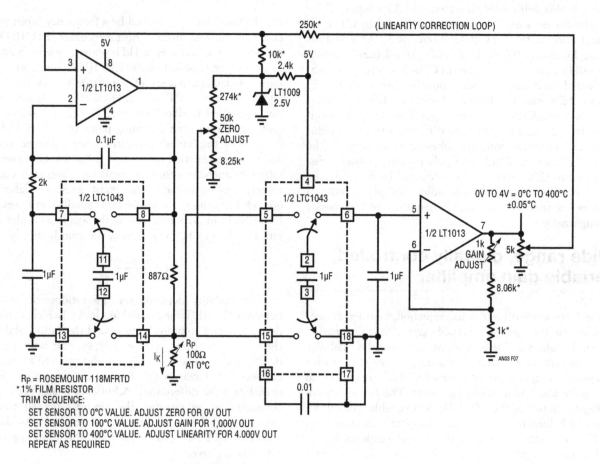

Figure 25.7 • Linearized Platinum Signal Conditioner

which is manufacturer specified for each individual sensor, is an offset term due to winding tolerances during fabrication of the RTD. The gain slope of the platinum is primarily fixed by the purity of the material and is a very small error term.

Note that A1 constitutes a voltage controlled current source with input and output referred to ground. This is a difficult function to achieve and is worthy of separate mention.

Relative humidity sensor signal conditioner

Relative humidity is a difficult physical parameter to transduce, and most transducers available require fairly complex signal conditioning circuity. Figure 25.8 combines two LTC1043s with a recently introduced capacitively based humidity transducer in a simple charge pump based circuit.

The sensor specified has a nominal 500pF capacitance at RH = 76%, with a slope of 1.7pF/% RH. The average voltage across this device must be zero. This provision prevents deleterious electrochemical migration in the sensor. LTC1043A inverts a resistively scaled portion of the LT1009 reference, generating a negative potential at Pin 14A. LTC1043B alternately charges and discharges the humidity sensor via Pins 12B, 13B and 14B. With 14B

and 12B connected, the sensor charges via the 1µF unit to the negative potential at Pin 14A. When the 14B-12B pair opens, 12B is connected to A1's summing point via 13B. The sensor now discharges into the summing point through the 1µF capacitor. Since the charge voltage is fixed, the average current into the summing point is determined by the sensor's humidity related value. The 1µF value AC couples the sensor to the charge-discharge path, maintaining the required zero average voltage across the device. The 22M resistor prevents accumulation of charge, which would stop current flow. The average current into A1's summing point is balanced by packets of charge delivered by the switched-capacitor network in A1's feedback loop. The 0.1µF capacitor gives A1 an integrator-like response, and its output is DC.

To allow 0% RH to equal 0V, offsetting is required. The signal and feedback terms biasing the summing point are expressed in charge form. Because of this, the offset must also be delivered to the summing point as charge, instead of a simple DC current. If this is not done, the circuit will be affected by frequency drift of LTC1034B's oscillator. Section 8B-11B-7B serves this function, delivering LT1009-referenced offsetting charge to A1.

Drift terms in this circuit include the LT1009 and the ratio stability of the sensor and the 100pF capacitors. These terms are well within the sensor's 2% accuracy specification and temperature compensation is not required. To calibrate this circuit, place the sensor in a

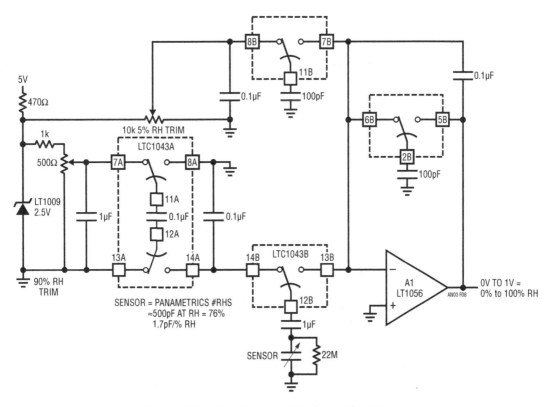

Figure 25.8 • Relative Humidity Signal Conditioner

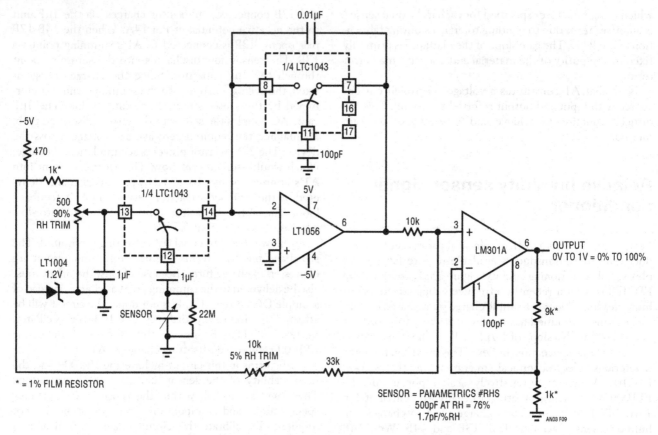

Figure 25.9 ● Relative Humidity Signal Conditioner

known 5% RH environment and adjust the "5% RH trim" for 0.05V output. Next, place the sensor in a 90% RH environment and set the "90% RH trim" for 900mV output. Repeat this procedure until both points are fixed. Once calibrated, this circuit is accurate within 2% in the 5% to 90% RH range.

Figure 25.9 shows an alternate circuit which requires two op amps but needs only one LTC1043 package. This circuit retains insensitivity to clock frequency while permitting a DC offset trim. This is accomplished by summing in the offset current after A1.

LVDT signal conditioner

LVDTs (linear variable differential transformers) are another example of a transducer which the LTC1043 can signal condition. An LVDT is a transformer with a mechanically actuated core. The primary is driven by a sine wave, usually amplitude stabilized. Sine drive eliminates error inducing harmonics in the transformer. The two secondaries are connected in opposed phase. When the core is positioned in the magnetic center of the transformer, the secondary outputs cancel and there is no output. Moving the core away from the center position unbalances the flux

ratio between the secondaries, developing an output. Figure 25.10 shows an LTC1043 based LVDT signal conditioner. A1 and its associated components furnish the amplitude stable sine wave source. A1's positive feedback path is a Wein bridge, tuned for 1.5kHz. Q1, the LT1004 reference, and additional components in A1's negative loop unity-gain stabilize the amplifier. A1's output (Trace A, Figure 25.11), an amplitude stable sine wave, drives the LVDT. C1 detects zero crossings and feeds the LTC1043 clock pin (Trace B). A speed-up network at C1's input compensates LVDT phase shift, synchronizing the LTC1043's clock to the transformer's output zero crossings. The LTC1043 alternately connects each end of the transformer to ground, resulting in positive half-wave rectification at Pins 7 and 14 (Traces C and D, respectively). These points are summed (Trace E) at a lowpass filter which feeds A2. A2 furnishes gain scaling and the circuit's output.

The LTC1043's synchronized clocking means the information presented to the lowpass filter is amplitude and phase sensitive. The circuit output indicates how far the core is from center and on which side.

To calibrate this circuit, center the LVDT core in the transformer and adjust the phase trim for 0V output. Next, move the core to either extreme position and set the gain trim for 2.50V output.

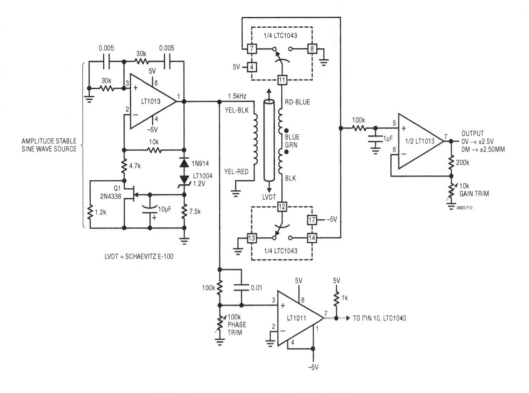

Figure 25.10 • LVDT Signal Conditioner

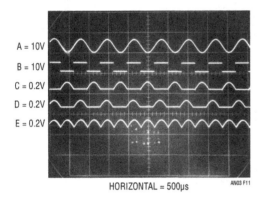

HORIZONTAL = 500µs   AN03 F11

Figure 25.11

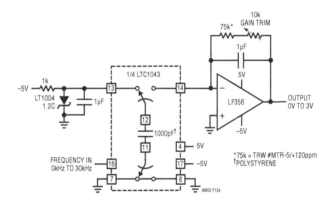

Figure 25.12a • Frequency-to-Voltage Converter

Charge pump F→V and V→F converters

Figure 25.12 shows two related circuits, both of which show how the LTC1043 can simplify a precision circuit function. Charge pump F→V and V→F converters usually require substantial compensation for non-ideal charge gating behavior. These examples equal the performance of such circuits, while requiring no compensations. These circuits are economical, component count is low, and the 0.005% transfer linearity equals that of more complex designs. Figure 25.12a is an F→V converter. The LTC1043's clock pin is driven from the input (Trace A, Figure 25.13). With the input high, Pins 12 and 13 are shorted and 14 is open. The 1000pF capacitor receives charge from the 1µF unit, which is biased by the LT1004. At the input's negative-going edge, Pins 12 and 13 open and 12 and 14 close. The 1000pF capacitor quickly removes current (Trace B) from A1's summing node. Initially, current is transferred through A1's feedback capacitor and the amplifier output goes negative (Trace C). When A1 recovers, it slews positive to a level which resets the summing junction to zero. A1's 1µF feedback capacitor

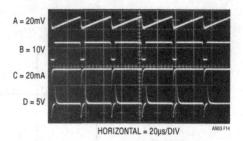

Figure 25.14

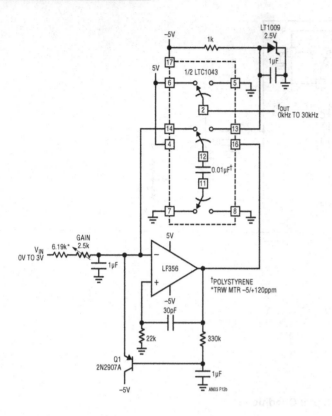

Figure 25.12b • Voltage-to-Frequency Converter

averages this action over many cycles and the circuit output is a DC level linearly related to frequency. A1's feedback resistors set the circuit's DC gain. To trim the circuit, apply 30kHz in and set the 10kΩ gain trim for exactly 3V output. The primary drift term in this circuit is the 120ppm/°C tempco of the 1000pF capacitor, which should be polystyrene. This can be reduced to within 20ppm/°C by using a feedback resistor with an opposing tempco (e.g., TRW # MTR-5/+120ppm). The input pulse width must be low for at least 100ns to allow complete discharge of the 1000pF capacitor.

In Figure 25.12b, the LTC1043 based charge pump is placed in A1's feedback loop, resulting in a V→F converter. The clock pin is driven from A1's output. Assume that A1's

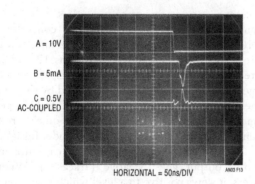

Figure 25.13

negative input is just below 0V. The amplifier output is positive. Under these conditions, LTC1043's Pins 12 and 13 are shorted and 14 is open, allowing the 0.01µF capacitor to charge toward the negative 1.2V LT1004. When the input-voltage-derived current forces A1's summing point (Trace A, Figure 25.13) positive, its output (Trace B) goes negative. This reverses the LT1043's switch states, connecting Pins 12 and 14. Current flows from the summing point into the 0.0µF capacitor (Trace C). The 30pF-22k combination at A1's positive input (Trace D) ensures A1 will remain low long enough for the 0.01µF capacitor to completely reset to zero. When the 30pF-22k positive feedback path decays, A1's output returns positive and the entire cycle repeats. The oscillation frequency of this action is directly related to the input voltage with a transfer linearity of 0.005%.

Start-up or overdrive conditions could force A1 to go to the negative rail and stay there. Q1 prevents this by pulling the summing point negative if A1's output stays low long enough to charge the 1µF-330k RC. Two LTC1043 switch sections provide complementary sink-source outputs. Similar to the F→V circuit, the 0.01µF capacitor is the primary drift term, and the resistor type noted above will provide optimum tempco cancellation. To calibrate this circuit, apply 3V and adjust the gain trim for a 30kHz output.

12-bit A→D converter

Figure 25.15 shows the LTC1043 used to implement an economical 12-bit A→D converter. The circuit is self-clocking, has a serial output, and completes a full-scale conversion in 25ms.

Two LTC1043s are used in this design. Unit A free-runs, alternately charging the 100pF capacitor from the LT1004 reference source and then dumping it into A1's summing point. A1, connected as an integrator, responds with a linear ramp output (Trace B, Figure 25.16). This ramp is compared to the input voltage by C1B. When the crossing occurs, C1B's output goes low (Trace C, just faintly visible in the photograph), setting the flip-flop high (Trace D). This pulls LTC1043's Pin 16 high, resetting A1's integrator capacitor via the paralleled switches. Simultaneously, Pin 14B opens, preventing charge from being delivered to A1's summing point during the reset. The flip-flop's Q output, low during this interval, causes

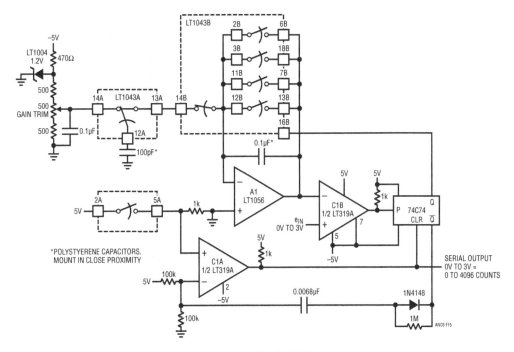

Figure 25.15 • 12-Bit A→D Converter

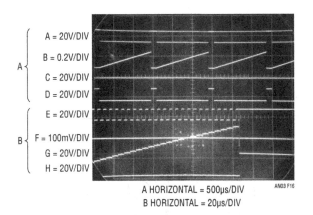

A = 20V/DIV
B = 0.2V/DIV
C = 20V/DIV
D = 20V/DIV
E = 20V/DIV
F = 100mV/DIV
G = 20V/DIV
H = 20V/DIV

A HORIZONTAL = 500µs/DIV
B HORIZONTAL = 20µs/DIV

AN03 F16

Figure 25.16

an AC negative-going spike at C1A. This forces C1A's output high, inserting a gap in the output clock pulse stream (Trace A). The width of this gap, set by the components at C1A's negative input, is sufficient to allow a complete reset of A1's integrating capacitor. The number of pulses between gaps is directly related to the input voltage. The actual conversion begins at the gap's negative edge and ends at its positive edge. The flip-flop output may be used for resetting. Alternately, a processor driven "time-out" routine can determine the end of conversion. Traces E through H offer expanded scale versions of Traces A through D, respectively. The staircase detail of A1's ramp output reflects the charge pumping action at its summing point. Note that drift in the 100pF and 0.1µF capacitors, which should be polystyrene, ratiometrically cancels. Full-scale drift for this circuit is typically 20ppm/°C, allowing it to hold 12-bit accuracy over 25°C + 10°C. To calibrate the circuit, apply 3V in and trim the gain potentiometer for 4096 pulses out between data stream gaps.

Miscellaneous circuits

Figures 25.17 to 25.22 show a group of miscellaneous circuits, most of which are derivations of applications covered in the text. As such, only brief comments are provided.

527

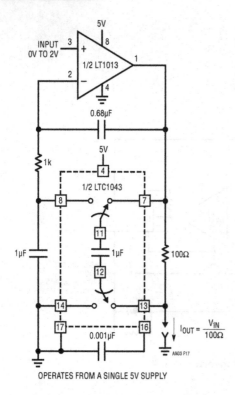

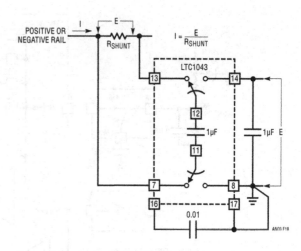

Figure 25.18 • Precision Current Sensing in Supply Rails

Figure 25.17 • Voltage Controlled Current Source with Ground Referred Input and Output

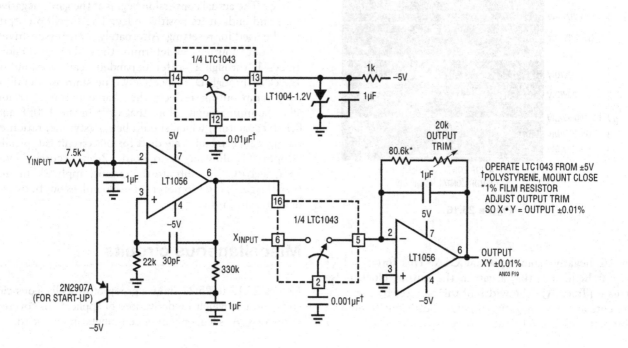

Figure 25.19 • Analog Multiplier with 0.01% Accuracy

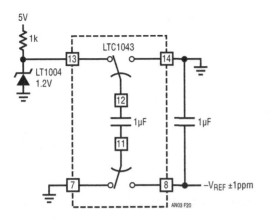

Figure 25.20 • Precision Voltage Inverter

Voltage-controlled current source—grounded source and load

This is a simple, precise voltage-controlled current source. Bipolar supplies will permit bipolar output. Configurations featuring a grounded voltage control source and a grounded load are usually more complex and depend upon several components for stability. In this circuit, accuracy and stability are almost entirely dependent on the 100Ω shunt.

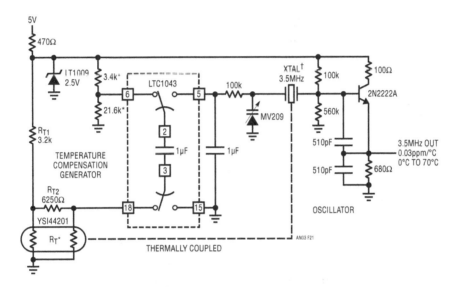

Figure 25.21 • Temperature Compensated Crystal Oscillator

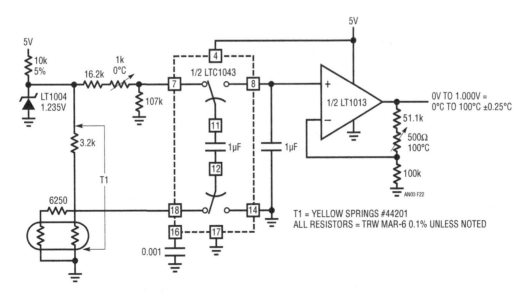

Figure 25.22 • Linear Thermometer

529

Current sensing in supply rails

The LTC1043 can sense current through a shunt in either of its supply rails (Figure 25.18). This capability has wide application in battery and solar-powered systems. If the ground-referred voltage output is unloaded by an amplifier, the shunt can operate with very little voltage drop across it, minimizing losses.

0.01% analog multiplier

Figure 25.19, using the V→F and F→V circuits previously described, forms a high precision analog multiplier. The F→V input frequency is locked to the V→F output because the LTC1043's clock is common to both sections. The F→V reference is used as one input of the multiplier, while the V→F furnishes the other. To calibrate, short the X and Y inputs to 1.7320V and trim for a 3V output.

Inverting a reference

Figure 25.20 allows a reference to be inverted with 1ppm accuracy. This circuit features high input impedance and requires no trimming.

Low power, 5V driven, temperature compensated crystal oscillator

Figure 25.21 uses the LTC1043 to differentiate between a temperature sensing network and a DC reference. The single-ended output biases a varactor-tuned crystal oscillator to compensate drift. The varactor-crystal network has high DC impedance, eliminating the need for an LTC1043 output amplifier.

Simple thermometer

Figure 25.22's circuit is conceptually similar to the platinum RTD example of Figure 25.7. The thermistor network specified eliminates the requirement for a linearity trim, at the expense of accuracy and range of operation.

High current, "inductorless," switching regulator

Figure 25.23 shows a high efficiency battery driven regulator with a 1A output capacity. Additionally, it does not require an inductor, an unusual feature

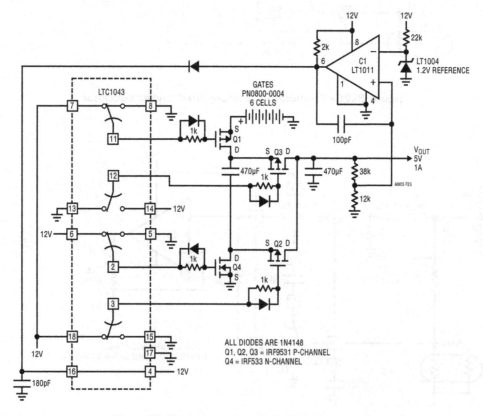

Figure 25.23 • Inductorless Switching Regulator

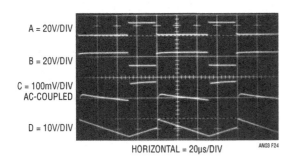

A = 20V/DIV

B = 20V/DIV

C = 100mV/DIV
AC-COUPLED

D = 10V/DIV

HORIZONTAL = 20µs/DIV   AN03 F24

Figure 25.24

for a switching regulator operating at this current level.

The LTC1043 switched-capacitor building block provides non-overlapping complementary drive to the Q1-Q4 power MOSFETs. The MOSFETs are arranged so that C1 and C2 are alternately placed in series and then in parallel. During the series phase, the 12V battery's current flows through both capacitors, charging them and furnishing load current. During the parallel phase, both capacitors deliver current to the load. Traces A and B, Figure 25.24, are the LTC1043-supplied drives to Q3 and Q4, respectively. Q1 and Q2 receive similar drive from Pins 3 and 11. The diode-resistor networks provide additional non-overlapping drive characteristics, preventing simultaneous drive to the series-parallel phase switches. Normally, the output would be one-half of the supply voltage, but C1 and its associated components close a feedback loop, forcing the output to 5V. With the circuit in the series phase, the output (Trace C) heads rapidly positive. When the output exceeds 5V, C1 trips, forcing the LTC1043 oscillator pin (Trace D) high. This truncates the LTC1043's triangle wave oscillator cycle. The circuit is forced into the parallel phase and the output coasts down slowly until the next LTC1043 clock cycle begins. C1's output diode prevents the triangle down-slope from being affected and the 100pF capacitor provides sharp transitions. The loop regulates the output to 5V by feedback controlling the turn-off point of the series phase. The circuit constitutes a large-scale switched-capacitor voltage divider which is never allowed to complete a full cycle. The high transient currents are easily handled by the power MOSFETs and overall efficiency is 83%.

Application considerations and circuits for a new chopper-stabilized op amp

26

Jim Williams

A great deal of progress has been made in op amp DC characteristics. Carefully executed designs currently available provide sub-microvolt V_{OS} ΔT drift, low bias currents and open-loop gains exceeding one million. Considerable design and processing advances were required to achieve these specifications. Because of this, it is interesting to note that amplifiers with even better DC specification were available in 1963 (Philbrick Researches Model SP656). Although these modular amplifiers were large and expensive ($\approx 3'' \times 2'' \times 1.5''$ at $195.00 1963 dollars) by modern standards, their DC performance anticipated today's best monolithic amplifiers while using relatively primitive components. This was accomplished by employing chopper-stabilization techniques (see Box "Choppers, Chopper-Stabilization and the LTC®1052") instead of the more common DC-differential stage approach.

The chopper-stabilized approach, developed by E. A. Goldberg in 1948, uses the amplifier's input to amplitude modulate an AC carrier. This carrier, amplified and synchronously demodulated back to DC, furnishes the amplifier's output. Because the DC input is translated to and amplified as an AC signal, the amplifier's DC terms have no effect on overall drift. This is the reason chopper-stabilized amplifiers are able to achieve significantly lower time and temperature drifts than classic differential types. Additionally, the AC processing of the signal aids low frequency amplifier noise performance and eliminates many of the careful design and layout procedures necessary in a classic differential approach. The most significant trade-off is increased complexity. The chopping circuitry and sampled data operation of these amplifiers require significant attention for good results. Additionally, the AC dynamics of chopper-stabilized amplifiers are complex if bandwidths greater than the chopping carrier frequency are required.

The LTC1052 is a third generation monolithic chopper-stabilized amplifier. As the table in Figure 26.1 shows, it is significantly better than previous monolithic chopper-

PARAMETER	LTC1052 CHOPPER-STABILIZED	ICL7652 CHOPPER-STABILIZED	HA2904/5 CHOPPER-STABILIZED	AD547 FET	LM11 LOW I_B BIPOLAR	LT1012 LOW I_B BIPOLAR
E_{OS} – 25°C	±5μV	±5μV	±50μV	±250μV	±300μV	±35μV
E_{OS} ΔT/°C	0.05μV/°C	0.05μV/°C	0.4μV/°C	1μV/°C	3μV/°C	1.5μV/°C
Noise (1Hz BW)	0.5μV$_{P-P}$ Typ	0.2μV$_{P-P}$ Typ*	Specified as 900nV/√Hz at 10Hz**	4μV$_{P-P}$	6μV$_{P-P}$ $R_S = 100k\Omega$	0.5μV$_{P-P}$
Open-Loop Gain	120dB	120dB (25°C)	5×10^8 Typ	2.5×10^5	2.5×10^5	3×10^5
Bias Current—25°C	30pA	30pA	150pA Typ	25pA	50pA	100pA
CMRR	120dB	110dB (25°C)	120dB	80dB	110dB	114dB
PSRR	120dB	110dB (25°C)	120dB	100dB	100dB	114dB
Input Common Mode Range	V⁺/–2.3V V⁻/+0V	V⁺/–1.5V V⁻/+0.7V	±10V at ±15V Supply	V⁺/–2V V⁻/+3V	V⁺/–0.5V V⁻/+1.5V	V⁺/–1.5V V⁻/+1.5V
Slew Rate	4V/μs	0.5V/μs	2.5V/μs	3V/μs	0.3V/μs	0.1V/μs
GBW	1MHz	0.45MHz	3MHz	1MHz	0.8MHz	0.8MHz

*Unable to verify by laboratory testing. Measured at 0.7μV$_{P-P}$
**Measured at 5μV$_{P-P}$ in a 1Hz bandwidth

Figure 26.1

Analog Circuit and System Design: A Tutorial Guide to Applications and Solutions. DOI: 10.1016/B978-0-12-385185-7.00026-3

stabilized amplifiers in several areas. For comparison purposes, conventional FET input and bias current compensated bipolar types are also listed. Noise has been a particular concern with previous monolithic chopper designs and Figure 26.2 is a strip chart of the LTC1052's performance at two measurement bandwidths. Additionally, the LTC1052's input common mode range includes V⁻, making single supply operation more practical.

Considerable attention to DC parasitics, particularly thermal EMFs, is required if the LTC1052's ultralow drift is to be fully utilized. Any connection of dissimilar metals produces a potential which varies with the junction's temperature (Seeback effect). As temperature sensors, thermocouples exploit this phenomenon to produce useful information. In low drift amplifier circuits the effect is probably the primary source of error. Connectors, switches, relay contacts, sockets, wire and even solder are all candidates for thermal EMF generation. It is relatively clear that connectors and sockets can form thermal junctions. However, it is not at all obvious that junctions of wire from different manufacturers can easily generate

200nV/°C—four times the LTC1052's drift specification! Figure 26.3 shows a plot obtained for such a wire junction. Even solder can become an error term at low levels, creating a junction with copper or Kovar wires or PC traces (see Figure 26.4).

Minimizing thermal EMF induced errors is possible if judicious attention is given to circuit board layout. In general, it is good practice to limit the number of junctions in the amplifier's input signal path. Avoid connectors, sockets, switches and other potential error sources to the extent possible. In some cases this will not be possible. In these instances, attempt to balance the number and type of junctions in the amplifier inputs so that differential cancellation occurs. Doing this may involve deliberately creating and introducing junctions to offset unavoidable junctions. This practice, borrowed from standard lab procedures, can be quite effective in reducing thermal EMF originated drifts. Figure 26.5 shows a simple example where a nominally unnecessary resistor is included to promote such thermal balancing. For remote signal sources such as transducers, connectors may be unavoidable. In these cases choose

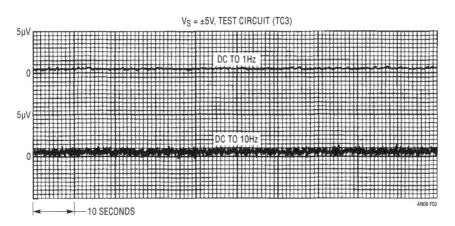

Figure 26.2 • LTC1052 Input Noise Voltage

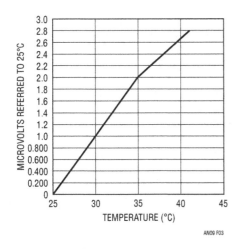

Figure 26.3 • Thermal EMF Generated by Two Wires

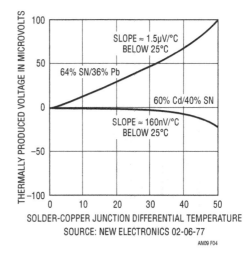

Figure 26.4 • Solder-Copper Thermal EMFs

533

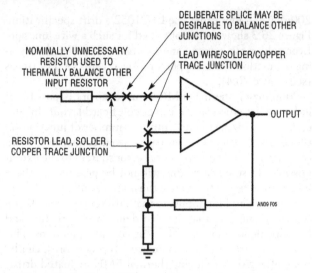

Figure 26.5 • Typical Thermal Layout Considerations

a connector specified for relatively low thermal EMF activity and ensure a similarly balanced approach in routing signals through the connector, along the circuit board and to the amplifier. If some imbalance is unavoidable, deliberately introduce an intentional counterbalancing junction. In all cases maintain the junctions in close physical proximity, which will keep them at the same temperature. Avoid drafts and temperature gradients, which can introduce thermal imbalances and cause problems. Figure 26.6 shows the LTC1052 set up in a test circuit to measure its temperature stability. The lead lengths of the resistors connected to the amplifier's inputs are identical. The thermal capacity each input sees is also balanced because of the symmetrical connection of the resistors and their identical size. Thus, thermal EMF induced shifts are equal in phase and amplitude and cancellation occurs. Very slight air currents can still affect even this arrangement. Figure 26.7 shows strip charts of output noise with the circuit covered by a small Styrofoam cup (HANDI-KUP Company Model H8-S) and with no cover in "still" air.

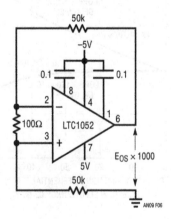

Figure 26.6 • Recommended Drift Test Circuit

This data illustrates why it is often prudent to enclose the LTC1052 and its attendant components inside some form of thermal baffle.

Thermal EMFs are the most likely, but not the only, potential low level error source. Electrostatic and electromagnetic shielding may be required. Power supply transformer fields are notorious sources of errors often mistakenly attributed to amplifier DC drift and noise. A transformer's magnetic field impinging on a PC trace can easily generate microvolts across that conductor in accordance with well-known magnetic theory. The amplifier cannot distinguish between this spurious signal and the desired input. Attempts to eliminate the problem by rolling off amplifier gain with a feedback capacitor may work, but often the filtered version of the undesired pickup masquerades as an unstable DC term in the output. The most direct approach is to use shielded transformers, but careful layout may be equally effective and less costly. A circuit which requires the transformer to be close by to achieve a good quality grounding scheme may be distributed by the transformer's magnetic field. An RF choke connected across a scope probe can determine the presence and relative intensity of transformer fields, aiding layout experimentation.

Another source of parasitic error is stray leakage current. The LTC1052's 30pA bias current allows operation from very high source impedances. In such cases it is desirable to prevent stray leakage currents from reaching the inputs. The simplest way to do this is to connect the amplifier inputs directly to the signal source via a Teflon standoff. Because the amplifier inputs never contact the PC board, stray leakage currents do not affect them. Although this approach is effective, its implementation may not be acceptable in production. Guarding is another technique to minimize board leakage effects. The guard is a PC trace completely encircling the input. This trace is driven (see Figure 26.8) at a potential equal to that of the input, preventing leakage to the amplifier input terminal. On PC boards, the guard should enclose the input(s) to be protected, with signal connections made directly to the amplifier input.

A final form of parasitic is one particular to all carrier-based amplifiers. If the amplifier is operating in a circuit which contains clocking or oscillation with substantial harmonic content at or near its carrier frequency (e.g., from another LTC1052), erratic operation is possible. This is particularly the case if inductors or transformers radiate magnetic fields related to the clocking or oscillation. The undesired interaction between the amplifier's chopping sequence and the externally generated AC signals may cause mixing the beat frequencies to occur, resulting in errors in the output. The LTC1052 is not particularly sensitive in this regard, but synchronizing its internal oscillator with external circuit clocking precludes this problem. The 14-pin version of the LTC1052 features a pin which allows the internal clock to be synchronized to an external signal. Input signals containing substantial

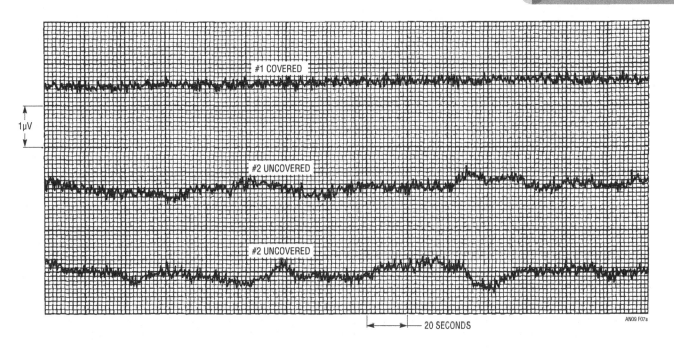

Figure 26.7 • DC-to-1Hz Noise Test Circuit

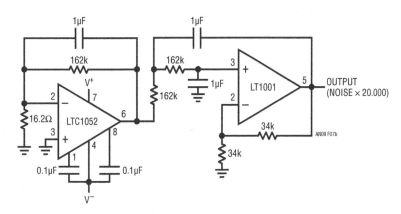

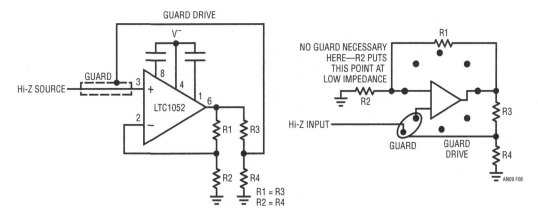

Figure 26.8 • Guarding Technique and Typical Layout

AC content may also cause this problem if the AC signal has strong spectral components related to the chopping frequency. In applications where such AC input components exist, it may be necessary to drive the LTC1052 from an external clock source at a frequency which has no harmonic relationship with the input signal. For example, a 372Hz clock frequency will prevent 60Hz input components from affecting amplifier operation.

Applications

Once alerted to the potential problems previously outlined, the engineer is prepared to design circuits around the LTC1052. The most obvious applications are at low level DC, where the low drift will improve performance over other amplifiers. More subtly, it is possible to exploit the LTC1052's low offset uncertainties to extend the dynamic range of circuit operation. The circuits which follow demonstrate these points, using relatively straightforward examples of improvements in low level, precision performance. Additional, less obvious, circuits use the LTC1052 to stabilize and enhance the performance of a variety of functions including data converters, buffers and comparators.

Standard grade variable voltage reference

Figure 26.9 diagrams a standard lab grade variable voltage reference. This circuit combines a pair of LTC1052s with high grade saturated standard cells and other components to produce an extremely stable reference source. The circuit may be used to calibrate 6 1/2 digit voltmeters, ultrahigh resolution data converters and other apparatus requiring high order traceability to primary standards.

SCO-106 Reference/2ppm/Year		= 2ppm
A1 – 0.05µV/°C × A = 3 × 5°C = 0.75µV		= 0.075ppm
A2 – 0.05µV/°C × A = 1 × 5°C = 0.25µV		= 0.025ppm
A1 + A2 Time Drift/Year = 2µV		= 0.02ppm
KVD – 0.5ppm/°C × 5°C + 1ppm/Year = 3.5ppm		= 3.5ppm
Resistors – 0.1ppm/°C Ratio 5°C + 2ppm/Year = 2.5ppm		= 2.5ppm
A2 CMRR Error	= 1ppm 1.2ppm	= 1.2ppm
A2 Loading Error	= 0.2ppm	9.32ppm

Figure 26.10 • Error Sources for Ultra-Precision Voltage Reference

The SCO-106 saturated cells furnish a reference voltage which is buffered and amplified to precisely 10V by A1. A1's output drives a seven place settable Kelvin-Varley divider with 1ppm accuracy. A2's low bias current and high CMRR allow it to unload the divider without introducing significant error To calibrate this circuit, adjust A1's output for exactly 10V by selecting the feedback resistor and fine trimming the 20MΩ potentiometer. A1's output should be measured with equipment having order traceability to primary NBS standards. Once calibrated, this circuit will provide *worst-case* 0.0014% accuracy over one year's time and ±5°C temperature excursions. Figure 26.10 details error sources. Note that the amplifiers contribute only about 1.3ppm (0.00013%) of the total.

Ultra-precision instrumentation amplifier

An ultra-precision instrumentation amplifier appears in Figure 26.11. This circuit offers greater accuracy and lower drift than any commercially available IC, hybrid or module.

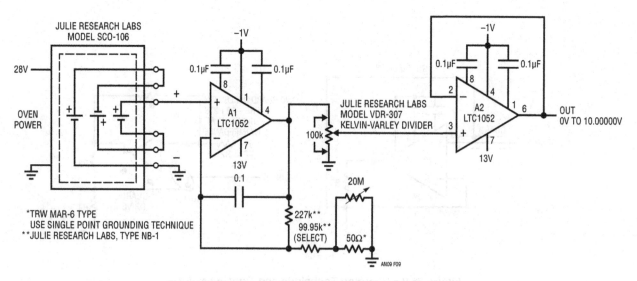

Figure 26.9 • Standard Grade Variable Voltage Reference

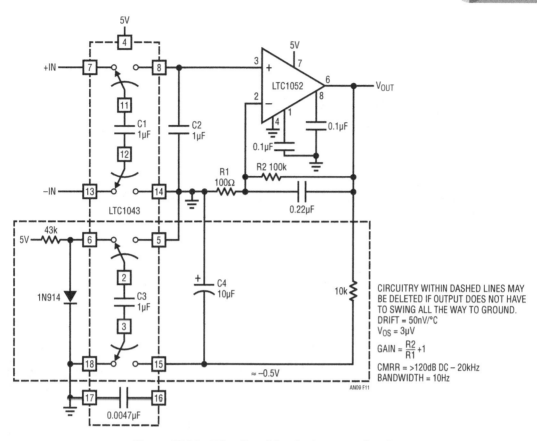

Figure 26.11 • Ultra-Precision Instrumentation Amp

Additionally, it will run from a single 5V power supply. The LTC1043 switched-capacitor instrumentation building block provides a differential-to-single-ended transition using a flying capacitor technique. C1 alternately samples the differential input signal and charges ground referred C2 with this information. The LTC1052 measures the voltage across C2 and provides the circuit's output. Gain is set by the ratio of the amplifier's feedback resistors. Normally, the LTC1052's output stage can swing within 15mV of ground. If operation all the way to zero is required, the circuit shown in dashed lines may be employed. This configuration uses the remaining LTC1043 section to generate a small negative voltage by inverting the diode drop. This potential drives the 10k pull-down resistor, forcing the LTC1052's output into class A operation for voltages near zero. Note that the circuit's switched-capacitor front end forms a sampled data filter allowing common mode rejection ratio to remain high, even with increasing frequency. The 0.0047µF unit sets front-end switching frequency at a few hundred hertz. The chart details circuit performance.

High performance isolation amplifier

Instrumentation amplifiers cannot be used to signal condition all differential signals. In factory and process control environments, severe grounding and common mode voltages often mandate the requirement for isolation amplifiers. Isolation amplifiers feature inputs which are galvanically isolated from their output and power connections. This allows the amplifier to ignore the effects of ground loops and operate at input common mode voltages many times the power supply voltage. Implementing a precise, low drift isolation amplifier is not easy, and commercial units are quite expensive. Figure 26.12 shows a circuit with 0.03% transfer accuracy and the 50nV/°C input drift of the LTC1052. As shown, the circuit provides a gain of 1000 and will operate at 250V input common mode levels.

The circuit works by amplitude modulating the output of a signal conditioning amplifier through a transformer. A synchronous demodulator filter reconstructs the amplifier's original output and furnishes the circuit's output. A separate oscillator and transformer provide power to the amplifier preserving galvanic isolation between the circuit's input and output ports.

Three 74C04 gates and their associated components form an oscillator which provides complementary drive to Q5 and Q6. These devices energize L1, which generates floating power on the input side of the dashed barrier shown. Simultaneously, the oscillator provides slightly delayed complementary drive to the Q1-Q2 FET switches via the 330Ω-100pF network and the additional inverters.

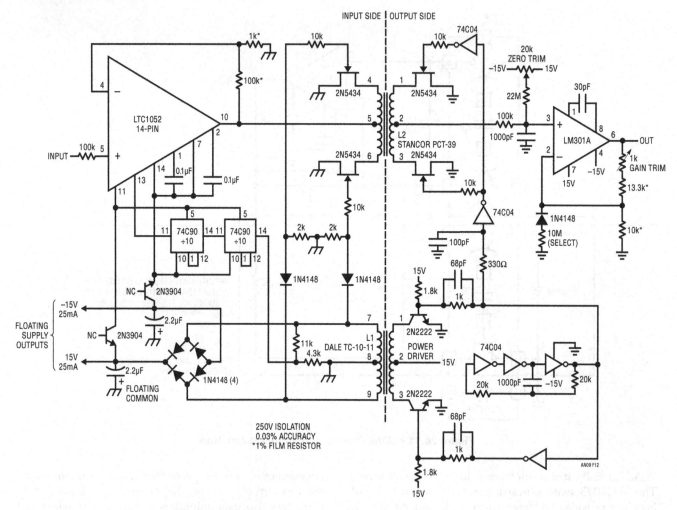

Figure 26.12 ● Precision Isolation Amplifier

The floating power produced by L1 is rectified and filtered and drives the LTC1052 (A1) via the Zener drops of the transistors. The ±15V floating power is brought out so it can be used to power transducers or other loads. Interaction between the transformer's chopping carrier and A1's internal oscillator is avoided by synchronizing the amplifier to the carrier via the two decade counters. Q3 and Q4, driven by opposing phase carrier signals derived from L1, chop A1's output into L2. This modulated signal information is received at L2's other winding. Because Q1 and Q2 are driven synchronously with Q3 and Q4, they demodulate the amplitude and phase (e.g., plus or minus polarity) information in the carrier. The 330Ω-100pF network compensates for the slight skew in switch drive signals on opposing sides of L2, minimizing gain error. L2's output (Pin 2) is RC filtered at A2, which also provides the circuit's output. Slight switching errors in the modulator-demodulator result in very small gain differences between positive and negative outputs at Pin 2 of L2. This effect is compensated by the diode-resistor network in A2's output, which provides a small decrease in gain for negative outputs.

Figure 26.13 shows the response of the isolation amplifier to a sine wave input. For this test, the floating common and circuit grounds are tied together. Trace A is the input applied to A1. Trace B, taken at Pin 4 of L2, shows A1's amplified output being modulated into the transformer. Trace C, obtained at Pin 1 of L2, depicts the received modulated waveform as it is synchronously demodulated. The filtered and final output of A2 appears in Trace D. The 25kHz carrier limits full-power bandwidth of this circuit to about 500Hz, adequate for process control and transducer

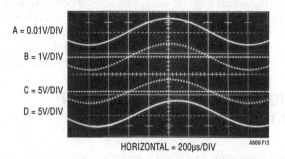

A = 0.01V/DIV

B = 1V/DIV

C = 5V/DIV

D = 5V/DIV

HORIZONTAL = 200μs/DIV

Figure 26.13 ● Waveforms for Isolation Amplifier

applications. The transformers used set a voltage break-down specification of 250V, although higher levels are achievable with different devices. As shown, circuit gain is 1000, allowing amplification of a ±5mV signal riding on 250V of common mode to a ±5V output. Gain accuracy is 0.03% with a gain drift of typically 50ppm/°C. Input referred drift is set by the LTC1052's 50nV/°C specification.

To trim this circuit, tie A1's input to floating common and adjust the zero trim for 0V output. Next, connect A1's input to a 5mV source and adjust the gain trim at A2 for exactly 5.000V$_{OUT}$. Finally, connect A1's input to a −5mV source and select the 10MΩ value in A2's feedback path for a −5.000V output reading. Repeat this procedure until all three points are fixed.

Stabilized, low input capacitance buffer (FET probe)

A recurring requirement in automatic semiconductor test-ing and probing equipment is for a highly stable unity-gain buffer amplifier with low input capacitance. Such an amplifier is also useful for other circuit chores where it is desirable to accurately monitor a point without introducing any significant AC or DC loading terms. Figure 26.14 shows such a circuit. Q1 and Q2 constitute a simple, high speed FET input buffet. Q1 functions as a source follower, with the Q2 current source load setting the drain-source channel current. The LT1010 buffer provides output drive capability for cables or whatever load is required. Normally, this open-loop configuration would be quite drifty because there is no DC feedback. The LTC1052 contributes this function to stabilize the circuit. It does this by comparing the filtered circuit output to a similarly filtered version of the input signal. The amplified

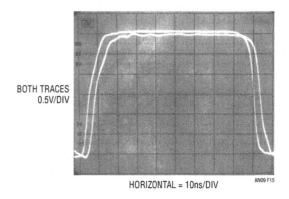

Figure 26.15 • Stabilized Buffer Delay

difference between these signals is used to set Q2's bias, and hence Q1's channel current. This forces Q1's V$_{GS}$ to whatever voltage is required to match the circuit's input and output potentials. The diode in Q1's source line ensures that the gate never forward biases and the 2000pF capacitor at A1 provides stable loop compensation. The RC network in A1's output prevents it from seeing high speed edges coupled through Q2's collector-base junction. A2's output is also fed back to the shield around Q1's gate lead, bootstrapping the circuit's effective input capacitance down to less than 1pF.

The LT1010's 15MHz bandwidth and 100V/μs slew rate, combined with its 150mA output, are fast enough for most circuits. For very fast requirements, the alternate discrete component buffer shown will be useful. Although its output is current limited at 75mA, the GHz range transistors employed provide exceptionally wide band-width, fast slewing and very little delay. Figure 26.15 shows the LTC1052 stabilized buffer circuit's response using the discrete stage. Response is clean and quick, with delay inside 4ns. Note that rise time is limited by the pulse

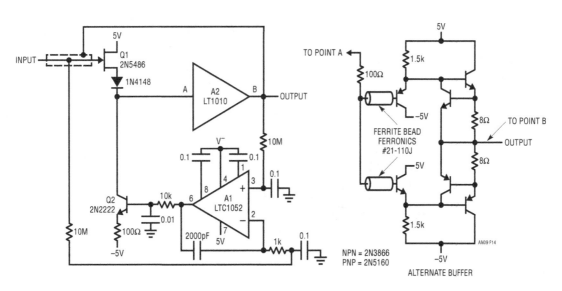

Figure 26.14 • Fast, Stabilized FET Buffer

generator and not the circuit. For either stage, offset is set by the LTC1052 at 3µV with gain about 0.95. It is worth noting that this circuit performs the same function as commercial FET probes in the $1,000.00 range.

Chopper-stabilized comparator

It is often desirable to use a reasonably fast voltage comparator with low input offset drift. Such a device is useful in high resolution A/D converters, crossing detectors and anywhere else a precise, stable, high speed comparison must be made. Unfortunately, obtaining reasonable comparator speed and low input drift in a design is difficult and monolithic comparators must be constructed around this trade-off. Figure 26.16 shows

a way to use the LTC1052 to eliminate the offset and drift in a comparator without sacrificing speed or differential input versatility. This circuit is applicable only in situations where some dead time is available for zeroing action to occur.

The circuit functions by periodically shorting the comparator inputs together and forcing the comparator into its linear region via its offset pins. The voltage at the offset pins required to do this is stored. When the comparator inputs are returned to their normal states, the stored voltage is maintained at the comparator's offset pins, effectively controlling the device's offset. Periodic updating ensures long-term stability of the correction. In this circuit, A1 is the stabilizing amplifier for C1. C1's inputs are controlled by a dual DPDT switch section furnished by the LTC1043. When LTC1043 Pin 16 is high, Pins 12 and 11 are

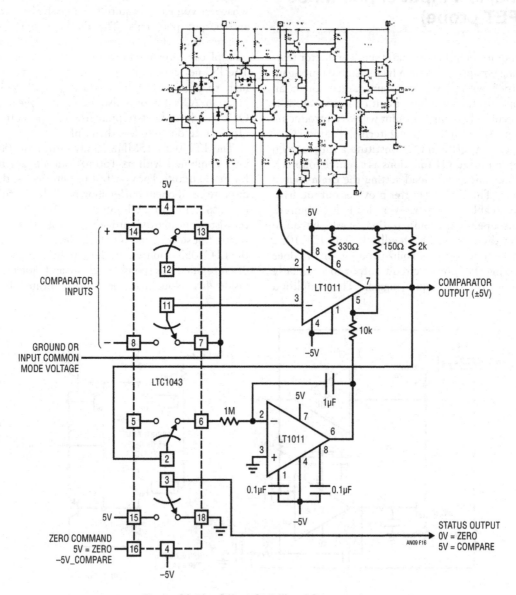

Figure 26.16 • Offset Stabilized Comparator

connected to Pins 13 and 7, respectively. Pin 3 at C1's output, is connected to Pin 18. Under these conditions, A1 is effectively connected in a negative feedback loop between C1's output and its offset Pin 5 (see detail of LT1011 input stage in Figure 26.16). This forces C1 into its linear region and its output oscillates at a high frequency between the rail voltages. A1, connected as a low frequency integrator; filters this action, compares its DC equivalent value to ground (its positive input potential) and drives C1's offsets to zero. When Pin 16 of the LTC1043 goes low, all switch states reverse and C1's inputs are free to compare the signals present at LTC1043 Pins 14 and 8 in the normal fashion. During this interval, A1's output remains fixed at the voltage stored in its feedback capacitor. A1's low bias current allows long durations between correction cycles—periods of seconds are practical—while maintaining effective comparator offset well within 5μV with negligible temperature drift.

Figure 26.17 shows the circuit's response to a sine wave (Trace A) applied to C1's positive input at LTC1043 Pin 14. C1's negative input, LTC1043 Pin 8, is grounded. With the circuit's zero command low (Trace B), C1's output (Trace C) responds in the normal fashion. During this period, the status output (Trace D) is low, indicating C1 is in its normal mode. When the zero command occurs (Trace B, just to the right-center of the screen), C1 is forced into its linear region, where it oscillates. During this time, A1 updates the correction voltage stored in its feedback capacitor. When the zero command pulse falls, normal comparator action is seen to resume. Note that the circuit's status output reflects the true operating mode of the circuit, because its timing includes the 50ns delay of the LTC1043 switch. For this reason the status output, and not the zero command, should be used to indicate the circuit's actual operating state.

Stabilized data converter

Amplifiers and comparators are not the only elements which can benefit from chopper-stabilization by the

LTC1052. Figure 26.18 shows a way to offset-stabilize a data converter, thereby doubling its dynamic range of operation, eliminating the necessity for an offset trim and reducing zero drift to negligible levels. In this circuit, the LTC1052 corrects for offset deficiencies in the AD650 V→F converter. Although specified for 1MHz full-scale operation, this device's 4mV input offset limits untrimmed dynamic range of operation to only 3 1/2 decades of output frequency. Under normal operating conditions, the AD650's positive input is grounded and its negative input is driven via the resistor string shown. Obtaining more than 3 1/2 decades of operation requires an offset trim at Pins 13 and 14. Even after trimming, the input amplifier's 30μV/°C drift contributes a 3Hz/°C zero point error. The LTC1052 corrects these problems by measuring the offset voltage at the circuit's summing node, comparing it to ground and driving the AD650 positive input (normally grounded in the manufacturer's recommended circuit configuration) with the appropriate stabilizing correction voltage. The dual FETs eliminate bias current caused errors. The LTC1052's integrator configuration keeps its gain at low frequency and DC, preserving the AD650's fast dynamic response while eliminating its offset errors. The divider network in the LTC1052's output is scaled to allow enough correction range to zero the AD650's offsets without causing overdrive during start-up and transients. With this scheme in use, the circuit does not require any zero trim to achieve full 6 decade operation. To calibrate, apply 10V and trim the output for exactly 1MHz.

Wide range V→F converter

Figure 26.19 shows another stabilized V→F converter. It features 1Hz to 1.25MHz operation, 0.05% linearity, and a temperature coefficient of typically 20ppm/°C, all substantially better than Figure 26.18's circuit. Additionally, it is less expensive and runs from a single 5V supply. Trade-offs include slower step response and a larger component count. This circuit uses a charge feedback scheme to allow the LTC1052 to close a loop around the entire V→F converter, instead of simply controlling offset. This approach enhances linearity and stability but introduces the loop's settling time into the overall V→F step response characteristic. Figure 26.20 shows waveforms of operation.

A positive input voltage directs A1's output to go negative, biasing the Q1 current source. Q1's collector puts current into the 330pF capacitor, causing it to rise in voltage (Trace A, Figure 26.20). The low input current CMOS inverter changes state when the ramp crosses 1/2 of the supply voltage. This causes all of the inverters to switch. The two paralleled inverters at the end of the chain go low (Trace B), simultaneously supplying positive feedback at the 10k-3.3pF junction (Trace C) and forcing the 330pF capacitor to a lower voltage by removing current

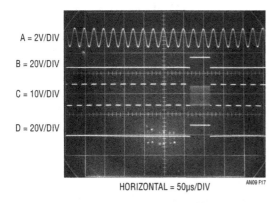

A = 2V/DIV

B = 20V/DIV

C = 10V/DIV

D = 20V/DIV

HORIZONTAL = 50μs/DIV   AN09 F17

Figure 26.17 • Stabilized Comparator Waveforms

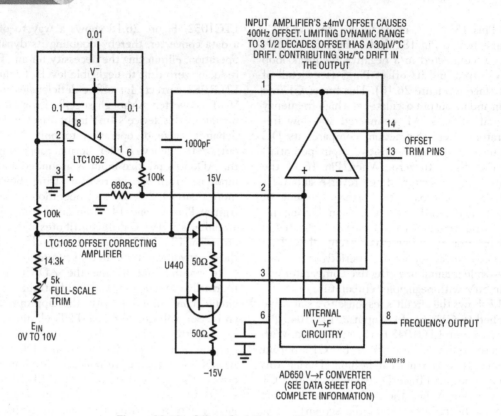

Figure 26.18 • Offset Stabilizing a V→F Converter

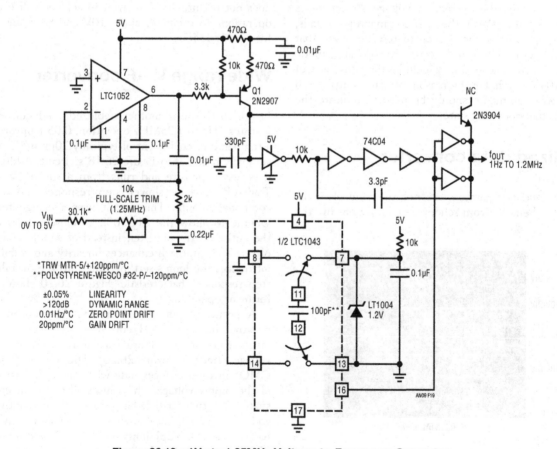

Figure 26.19 • 1Hz to 1.25MHz Voltage-to-Frequency Converter

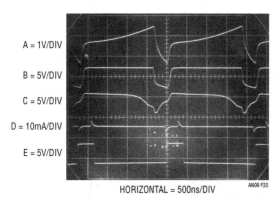

A = 1V/DIV

B = 5V/DIV

C = 5V/DIV

D = 10mA/DIV

E = 5V/DIV

HORIZONTAL = 500ns/DIV   AN09 F20

Figure 26.20 • V→F Waveforms

from it (Trace D) via the diode connected 2N3904. During the ramping interval, LTC1043 switch Pins 11 and 12 are connected to Pins 8 and 14 discharging the 100pF capacitor into Pin 14. When the output inverters go low, the LTC1043's control pin (Pin 16) also switches, placing the charged 100pF capacitor across Pins 7 and 13. Thus, each time the inverters switch, a fixed quantity of charge is dispensed into the 2k-0.22µF-10k potentiometer junction (Q = CV). The LT1043's switching is arranged so that this charge is of opposite polarity to the positive input current. The 0.22µF capacitor integrates the discrete charge dumps to DC. A1 servo controls the Q1 inverter oscillator to run at whatever frequency is required to force its negative input to 0V. In this manner, drift and nonlinear response in the Q1 inverter oscillator are compensated by A1's closed-loop control. The circuit's frequency output is delivered by another LTC1043 section (Trace E).

Several factors contribute to this circuit's performance. The low input current to the CMOS inverter, combined with the low leakage of the 2N3904 base-emitter diode and the circuit's servo action, allows operation to well below 1Hz, despite the small 330pF integrating capacitor. In the lower frequency ranges, currents at this junction are small and board leakage can cause jitter. A clean board will work well, but the best approach is to mount the capacitor Q1's collector, the inverter input and the transistor base connection on a Teflon stand-off, using short connections. The resistor and capacitor specified in the figure, both gain terms, have opposing temperature coefficients, aiding gain drift performance. The LTC1052's low offset eliminates the need for a zero trim while preserving the circuit's >120dB dynamic range of operation. To trim the circuit, apply 5.000V and adjust the 1.25MHz trim for 1.2500MHz out.

1Hz to 30MHz V→F converter

Although Figure 26.19's circuit is impressive, it still does not tax the LTC1052's dynamic range of operation. Figure 26.21 shows a highly modified version of Figure 26.19. It has a 1Hz to 30MHz output (150dB

dynamic range) for a 0V to 3V input. This is by far the widest dynamic range and highest operating frequency of any V→F discussed in the literature at the time of writing*. It is a good application of the extremely wide signal processing range afforded by the LTC1052. The circuit maintains 0.08% linearity over its entire 7 1/3 decade range with a full-scale drift of about 20ppm/°C. Zero point error is 0.3Hz/°C and is directly related to the LTC1052's 50nV/°C drift specification.

To get the additional bandwidth, Figure 26.19's CMOS inverters are replaced with a fast JFET buffer driving a Schottky TTL Schmitt trigger. The Schottky diode prevents the Schmitt trigger from ever seeing negative voltage at its input. The diode connected 2N3904 is retained for resetting the capacitor, which has a smaller value. Figure 26.19's positive AC feedback, with its attendant recovery time constant, is avoided in this circuit. Instead, the Schmitt's input voltage hysteresis provides the limits which the oscillator runs between. The 30MHz full-scale output is much faster than the LTC1043 can accept, so the digital divider stages are used to reduce the feedback frequency signal by a factor of 20. Remaining Schmitt sections furnish complementary outputs. Good high frequency wiring techniques should be used when constructing the current source-buffer-Schmitt trigger sections.

Figure 26.22 shows the key waveforms with the circuit loafing at 20MHz. Trace A is the Schmitt trigger input, which is seen to ramp between two voltage limits, while Trace B is the Schmitt output. The closed-loop approach results in very low output jitter and noise over the entire 150dB operating range. Figure 26.23 plots this, showing frequency jitter versus output frequency. Jitter does not rise above 0.01% until 20kHz, which is only 0.05% of scale. Even at 1ppm of scale (30Hz), jitter is still about 1%, finally rising to 10% at 1Hz (0.000003% of full scale). As V→F operating frequency decreases toward the LTC1052's feedback loop roll-off, the loop dominates the jitter characteristic. In the high frequency ranges the loop poles are not a factor and current source and Schmitt trigger switching noise dominate. As with Figure 26.19's circuit, the feedback loop slows step response. Figure 26.24 shows this, with a full-scale input step requiring almost 50ms to settle. To trim this circuit, ground the input and adjust the 1Hz trim until oscillation just starts. Next, apply 3.000V and set the 30MHz trim for a 30.00MHz output. Repeat this procedure until both points are fixed.

16-bit A/D converter

V→F converters are not the only types of data converters which can benefit from the LTC1052's performance.

---

*1Hz to 100MHz circuit I discussed in AN14, "Designs for High Performance V→F Converters."

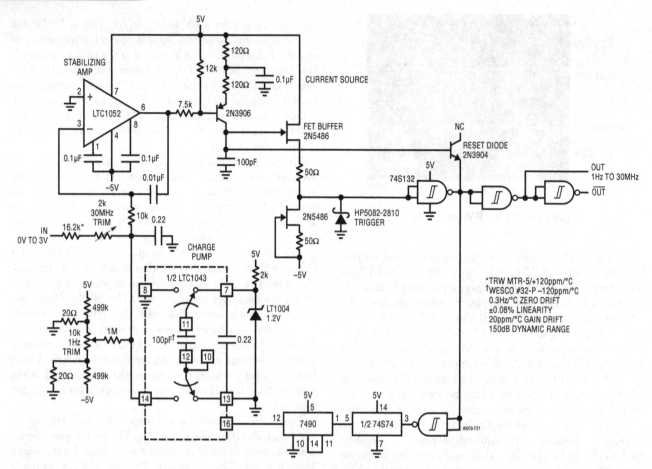

Figure 26.21 • 1Hz to 30MHz V→F Converter

Figure 26.25 shows a 16-bit A/D converter (overrange to 100,000 counts is provided).

The A/D converter, made up of A2, a flip-flop, some gates and a current sink, is based on a current balancing technique. Once again, the chopper-stabilized LTC1052's 50nV/°C input drift is required to eliminate offset errors in the A/D. Figure 26.26 details key A/D waveforms. Assume the flip-flop's Q output (Trace B) is low, connecting LTC1043 Pins 3 and 18. The current sink switch directs its output to ground.

Under these conditions, the only current into A2's summing point is from the input via the 95k resistor. This positive current forces A2's output (Trace A, Figure 26.26) to integrate in a negative direction. The negative ramp continues and finally passes the 74C74 flip-flop's switching threshold. At the next clock pulse

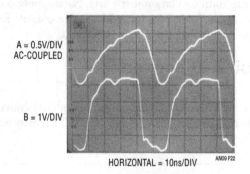

Figure 26.22 • Fast V→F Ramp-Reset Detail

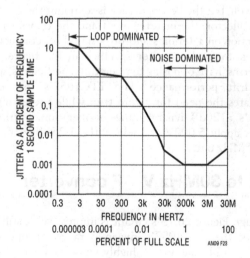

Figure 26.23

544

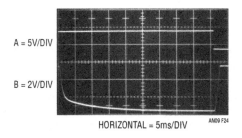

Figure 26.24 • Fast V→F Step Response

A = 5V/DIV

B = 2V/DIV

HORIZONTAL = 5ms/DIV   AN09 F24

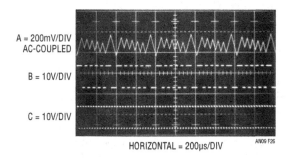

A = 200mV/DIV
AC-COUPLED

B = 10V/DIV

C = 10V/DIV

HORIZONTAL = 200μs/DIV   AN09 F26

Figure 26.26 • 16-Bit A/D Waveforms

(clock is Trace C), the flip-flop changes state (Trace B), causing the LTC1043 switch positions to reverse. Pin 3 connects to Pin 15, allowing the current sink to bias A2's summing point.

This results in a quickly rising, precise current flow out of A2's summing point. This current, scaled to be greater than the maximum input derived current, forces A2's output movement to reverse and integrate in the positive

direction. At the first clock pulse after A2's output has crossed the flip-flop's triggering threshold, switching occurs and the entire cycle repeats. Because the reference current is fixed, the flip-flop's duty cycle is solely a function of the input signal current into A2's summing point. The flip-flop's output gates the clock, producing the "frequency output A" output. The 10k-10pF RC slightly

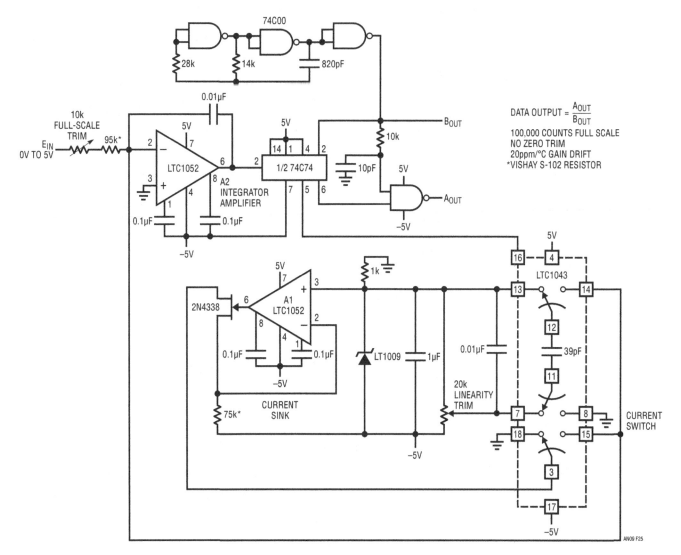

Figure 26.25 • 16-Bit A/D Converter

delays the clock signal, eliminating spurious output pulses due to flip-flop delay. The circuit's data output, the ratio of output A to the clock frequency, may be extracted with counters. Because the output is expressed as a ratio, clock frequency stability is unimportant.

Slight parasitic charge pumping at the current switch introduces an error term which varies with loop operating frequency. This effect will cause a small nonlinearity in the A/D's transfer function unless compensated. The remaining LTC1043 sections accomplish this by inverting the reference and returning a very small, compensatory charge to the current sink output each time circuit switching occurs. The charge delivered is scaled by the linearity trim to cancel the parasitic term. To calibrate this circuit, apply 5.00000V and adjust the full-scale trim for 100,000 counts out. Next, set the input to 1.25000V and adjust the linearity trim for 25,000 counts out. Repeat this procedure until both points are fixed. Converter accuracy is ±1 count with a temperature coefficient of typically 15ppm/°C. Better TC is possible by employing a more stable reference. The high offset stability of the LTC1052 at A2 eliminates zero errors and trimming.

Simple remote thermometer

Although many remote thermometer circuits have appeared, few allow the temperature transducer's output to be directly transmitted over an unshielded wire. The relatively high output impedance of most temperature transducers make their outputs sensitive to noise on the line and shielding is required. The low offset drift of the LTC1052 permits the circuit of Figure 26.27, which offers one solution to this problem. Here, the low output impedance of a closed-loop op amp gives ideal line noise immunity, while the op amp's offset voltage drift provides a temperature sensor. Using the op amp in this way requires no external components and has the additional advantages of a hermetic package and unit-to-unit mechanical uniformity if replacement is ever required.

The op amp's offset drift is amplified to drive the meter by the LTC1052. The diode bridge connection allows either positive or negative op amp temperature sensor offsets to interface directly with the circuit. In this case, the circuit is arranged for a 10°C to 40°C output, although other ranges are easily accommodated. To calibrate this circuit, subject the op amp sensor to a 10°C environment and adjust the 10°C trim for an appropriate meter indication. Next, place the op amp sensor in a 40°C environment and trim the 40°C adjustment for the proper reading. Repeat this procedure until both points are fixed. Once calibrated, this circuit will typically provide accuracy within ±2°C, even in high noise environments.

Output stages

In some circumstances it may be required to obtain more output current or swing from the LTC1052 than it can provide. The CMOS output stage cannot provide the current levels of bipolar op amps. Additionally, it may be necessary to run the device off ±15V supplies and to obtain increased voltage and current outputs. Figure 26.28 parallels a package of CMOS inverters to obtain 10mA to 20mA output current capability. The inversion in the loop requires the feedback connection to go to the amplifier's positive input. The RC damper eliminates oscillation in the inverter stage, which is running in its linear region. The local capacitive feedback at the amplifier gives loop compensation. Figure 26.29 shows a way to run the LTC1052 from 15V supplies while obtaining the increased current and voltage output capabilities of the LT318A amplifier The transistors run in Zener mode, dropping the supply to about ±7V at the LTC1052. The LT318A serves as an output stage with a voltage gain of 4. The output swing is that of the LT318A, typically, ±13V into 2kΩ with a short-circuit current of 20mA. This circuit is dynamically stable at any gain in either the inverting or noninverting configuration, although the LTC1052's input common mode range (−7V to 5V with the ±15V power supply used) must not be exceeded.

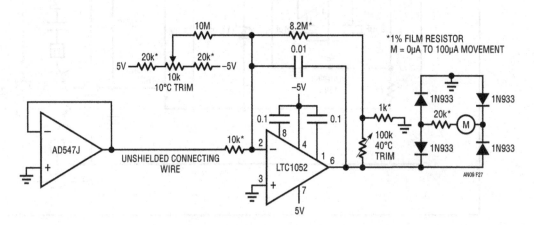

Figure 26.27 • High Noise Rejection Thermometer

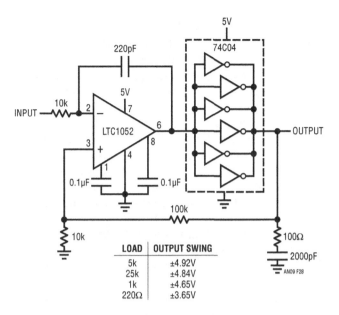

Figure 26.28 • Increasing Output Current

LOAD	OUTPUT SWING
5k	±4.92V
25k	±4.84V
1k	±4.65V
220Ω	±3.65V

of a second switch, synchronously driven with the input switch. The output integrator stage smooths the switch output to DC and presents the final output. Drifts in the output integrator stage are of little consequence because they are preceded by the AC gain stage. The DC drifts in the AC stage are also irrelevant because they are isolated from the rest of the amplifier by the coupling capacitors. Overall DC gain is extremely high, being the product of the gains of the AC stage and the DC gain of the integrator. Although this approach easily yields drifts of 100nV/°C and open-loop gains of 100 million, there are some drawbacks. The amplifier has a single-ended, noninverting input and cannot accept differential signals without additional circuitry added at the front end. Also, the carrier-based approach constitutes a sampled data system and overall amplifier bandwidth is limited to a small fraction of the carrier frequency. Carrier frequency, in turn, is restricted by AC amplifier gain-phase limitations and errors induced by switch response time. Maintaining good DC performance involves keeping the effects of these considerations small and carrier frequencies

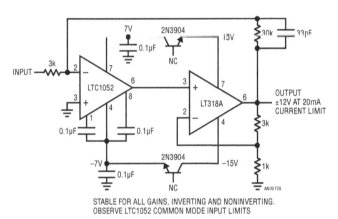

STABLE FOR ALL GAINS, INVERTING AND NONINVERTING.
OBSERVE LTC1052 COMMON MODE INPUT LIMITS

Figure 26.29 • Increasing Output Current and Voltage
($V_{SUPPLY} = ±15V$)

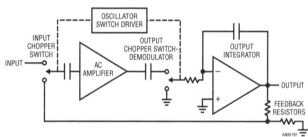

Figure B1 • Chopper Amplifier

Box Section—Choppers, Chopper-Stabilization and the LTC1052

All chopper-stabilized amplifiers achieve high DC stability by converting the DC input into an AC signal. An AC gain stage amplifies this signal. After amplification it is converted back to DC and presented as the amplifier's output. Figure B1 shows a conceptual chopper amplifier:

The AC amplifier's input is alternately switched between the signal input and feedback divider network. The AC amplifier's output amplitude represents the difference between the feedback signal and the circuit's input. This output is converted back to DC by a phase sensitive demodulator composed

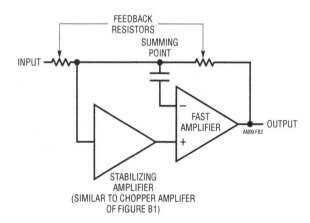

Figure B2 • Classic Chopper-Stabilized Amplifier

are usually in the low kilohertz range, dictating low overall bandwidth.

The classic chopper-stabilized amplifier solves the chopper amplifier's low bandwidth problem. It uses a parallel path approach (Figure B2) to provide wider bandwidth while maintaining good DC characteristics. The stabilizing amplifier, a chopper type, biases the fast amplifier's positive terminal to force the summing point to zero. Fast signals directly drive the AC amplifier, while slow ones are handled by the stabilizing chopper amplifier. The low frequency cut-off of the fast amplifier must coincide with the high frequency roll-off of the stabilizing amplifier to achieve smooth overall gain-frequency characteristics. With proper design, the chopper-stabilized approach yields bandwidths of several megahertz with the low drift characteristic of the chopper amplifier. Unfortunately, because the stabilizing amplifier controls the fast amplifier's positive terminal, the classic chopper-stabilized approach is restricted to inverting operation only.

The LTC1052 uses a different approach which permits full differential input operation, good bandwidth and retains ultralow drift. It relies on an autozero technique.

During the LTC1052's autozero cycle, the inputs are shorted together and a feedback path is closed around the input stage to null its offset. Switch S2 (Figure B3) and capacitor C_{EXTA} act as a sample and hold to store the nulling voltage during the sampling cycle.

In the sampling cycle, the now almost ideal amplifier is used to amplify the differential input voltage. Switch S2 connects the amplified input voltage to C_{EXTB} and the output gain stage. C_{EXTB} and S2 act as a sample and hold to store the amplified input signal during the autozero cycle. By switching between these two states at a frequency much higher than the signal frequency, a continuous output results.

Notice that during the autozero cycle the inputs are not only shorted together, but are also shorted to the negative input. This forces nulling with the common mode voltage present. The same argument applies to power supply variations and accounts for the extremely good CMRR and PSRR specifications on the LTC1052.

The complete amplifier contains stabilizing elements, feedforward for high frequency signals, and antialiasing circuitry, but the superior DC performance is completely described by this simple loop.

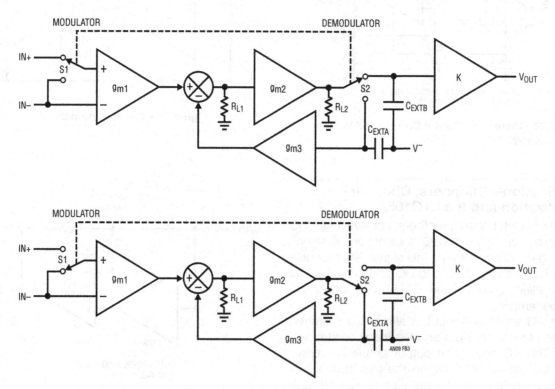

Figure B3 ● LTC1052 Conceptual Amplifier (Simplified)

References

Goldberg, E. A. "Stabilization of Wideband Amplifiers for Zero and Gain", RCA Review, June, 1950, page 298.

SP656 Data Sheet, Philbrick Researches, 1963.

Brokaw, A. P. "Designing Sensitive Circuits? Don't Take Grounds for Granted", EDN, October 5, 1975, page 44.

Morrison, Ralph, "Grounding and Shielding Techniques in Instrumentation", 2nd Edition, Wiley Interscience, 1977.

Ott, Henry W., "Noise Reduction Techniques in Electronic Systems", Wiley Interscience, 1976.

Hueckel, John H., "Input Connection Practices for Differential Amplifiers", Neff Inst. Corp., Duarte, California.

"Elimination of Noise in Low-Level Circuits", Gould, Inc., Instrument Systems Division, Cleveland, Ohio.

Williams, J., "Prevent Low Level Amplifier Problems", Electronic Design, February 15, 1975, page 62.

Pascoe, G. "The Choice of Solders for High-Gain Devices", New Electronics (Great Britain), February 6, 1977.

Pascoe, G., "The Thermo-E.M.F. of Tin-Lead Alloys", Journal Phys. E: December, 1976.

Designing linear circuits for 5V single supply operation

27

Jim Williams

In predominantly digital systems it is often necessary to include linear circuit functions. Traditionally, separate power supplies have been used to run the linear components (see Box, "Linear Power Supplies—Past, Present, and Future").

Recently, there has been increasing interest in powering linear circuits directly from the 5V logic rail. The logic rail is a difficult place for analog components to function. The high amplitude broadband current and voltage noise generated by logic clocking makes analog circuit operation difficult. (See Box, "Using Logic Supplies for Linear Functions".)

Generally speaking, analog circuitry which must achieve very high performance levels should be driven from dedicated supplies. The difficulties encountered in maintaining the lowest possible levels of noise and drift in an analog system are challenging enough without contending with a digitally corrupted power supply.

Many analog applications, however, can be successfully implemented using the logic supply. Combining components intended to provide high performance from the logic rail with good design can give excellent results (see Box, "High Performance, Single Supply Analog Building Blocks"). The examples which follow show this in a variety of precision measurement and control circuits which function from a 5V supply.

Linearized RTD signal conditioner

Figure 27.1 shows a circuit which provides complete, linearized signal conditioning for a platinum RTD. One side of the RTD sensor is grounded, often desirable for noise considerations. The Q1-Q2 current source is referenced to A1's output. A1's operating point is primarily fixed by

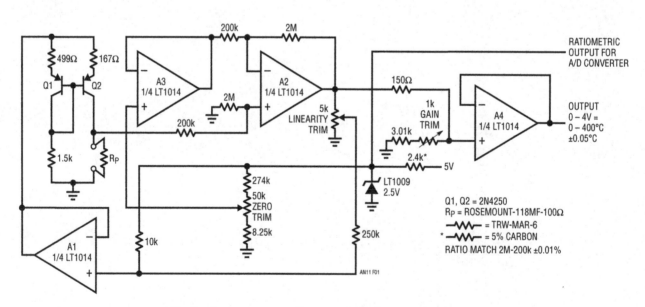

Figure 27.1 • Linearized Platinum RTD Signal Conditioner

Analog Circuit and System Design: A Tutorial Guide to Applications and Solutions. DOI: 10.1016/B978-0-12-385185-7.00027-5

the 2.5V LT1009 voltage reference. The RTD's constant current forces the voltage across it to vary with its resistance, which has a nearly linear positive temperature coefficient. The nonlinearity causes several degrees of error over the circuit's 0°C to 400°C operating range. A2 amplifies R_P's output, while simultaneously supplying nonlinearity correction. The correction is implemented by feeding a portion of A2's output back to A1's input via the 10k-250k divider. This causes the Q1-Q2 current source output to shift with R_P's operating point, compensating sensor nonlinearity to within ±0.05°C. A3, also referenced to the LT1004, voltage sums an offsetting signal at A2's negative input, allowing 0°C to equal 0V at A2's output. The resistive divider in A4's input line sets circuit gain, and the circuit's output is taken at A4.

To calibrate this circuit, substitute a precision decade box (e.g., General Radio 1432K) for R_P. Set the box to 0°C value (100.00Ω) and adjust the zero trim for a 0.0V output. Next, set the decade box for a 140°C output (154.26Ω) and adjust the gain trim for a 1.400V output reading. Finally, set the box to 400°C (249.0Ω) and trim the linearity adjustment for a 4.000V output. Repeat this sequence until all three points are fixed. Total error over the entire range will be within ±0.05°C. The resistance values given

are for a nominal 100.00Ω (0°C) sensor. Sensors deviating from this nominal value can be used by factoring in the deviation from 100.00Ω. This deviation, which is manufacturer-specified for each individual sensor, is an offset term due to winding tolerances during fabrication of the RTD. The gain slope of the platinum is primarily fixed by the purity of the material and is a very small error term.

Linearized output methane detector

Figure 27.2 is another 5V powered transducer circuit. Like the platinum RTD example, this circuit linearizes the transducer's output, but it performs a more complex mathematical operation. The circuit's frequency output is directly proportional to the methane concentration detected by the transducer specified. Figure 27.3 shows that the transducer output varies as:

$$\approx \frac{1}{\sqrt{\text{Concentration}}}$$

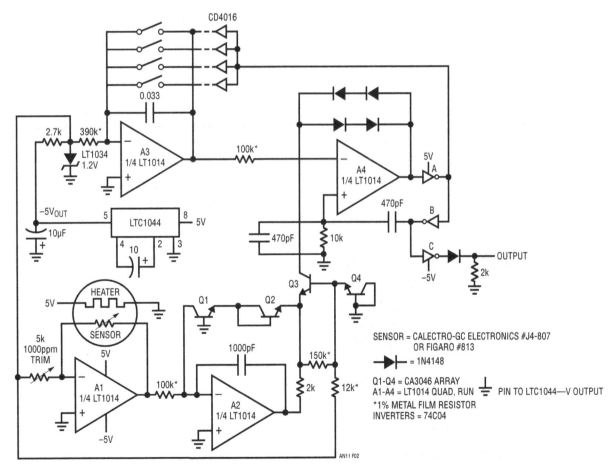

Figure 27.2 • Linearized Methane Transducer Signal Conditioner

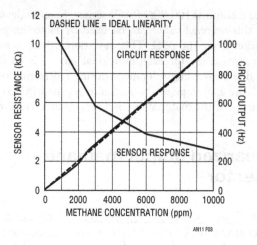

Figure 27.3 • Transducer vs Circuit Response

The circuit linearizes this function and its frequency output is also plotted.

The sensor's resistance vs methane concentration is converted to a voltage by A1. The LT1034 serves as a reference. A1's output feeds A2. The exponential relationship between V_{BE} and collector current in transistors is utilized to generate a current in Q3's collector proportional to the square of A2's input current. This operation compensates the sensor's square root term. Q3's collector current sets the operating point of the A3-A4 oscillator. A3, an integrator, generates a positive going linear ramp (Trace A, Figure 27.4). The ramp is compared with Q3's current at A4's summing point (Trace B). A4 is configured as a current summing comparator. The feedback diode-bound network minimizes delay due to output slew time. When the ramp forces the summing point positive, A4's output (Trace C) swings negative. CMOS inverter A (Trace D) goes high, turning on the CD4016 switch to reset the A3 integrator. Simultaneously, inverter B goes low (Trace E), supplying AC positive feedback to A4's " + " input (Trace F). When the positive feedback decays, A4's output goes

high, A3 begins to integrate, and the entire cycle repeats. Q3's collector current determines how long A3's ramp runs before A4 resets it. The ramp time is directly proportional to Q3's collector current, meaning that oscillation frequency is inversely (1/X) related to the current. The overall circuit transfer function is:

$$\frac{1}{X^2}$$

This linearizes the sensor's output. In practice, the sensor's response slightly deviates from the equation shown, actually being:

$$\frac{1}{\sqrt[1.9]{\text{Concentration}}}$$

The reset time constants at A4's input introduce enough oscillator "dead time" to partially compensate for the deviation. The dead time's frequency retarding characteristic effects the oscillator's high frequency range, providing a first order correction. The overall linearization achieved is within the sensor's manufacturing tolerances. The slight "bump" in the circuit's response curve is due to the mismatch between the sensor's term and the circuit's X^2 function.

The dead time correction in the oscillator smooths this error out above 4000ppm. The LTC®1044 voltage converter generates a negative supply directly from the 5V rail. This approach provides necessary negative circuit potentials while maintaining compatibility with 5V supply only operation. The sensor's heater is powered directly from the 5V rail, as specified by the manufacturer. To calibrate the circuit, place the sensor in a 1000ppm methane environment and adjust the 5k trim for a 100Hz output. Accuracy from 500ppm to 10,000ppm is limited by the sensor's 10% specification.

Cold junction compensated thermocouple signal conditioner

Figure 27.5 shows a 5V powered, complete thermocouple signal conditioner. Cold junction compensation is included, and the circuit allows one leg of the thermocouple to be grounded, desirable for noise considerations. The LTC1043 combines the cold junction network differential output with the grounded thermocouple's signal at the LTC1052. The LTC1052 provides stable, low drift gain. To enable swing all the way to ground, the LTC1043's other switch section generates a small negative potential. This allows the LTC1052 output stage to run Class A for small outputs, permitting swing to 0V. The table gives proper values for R1 for various thermocouples. Output scaling may be set by R_F/R_I to whatever slope is desirable. Cold junction compensation holds within $\pm 1°C$ over 0°C to 60°C.

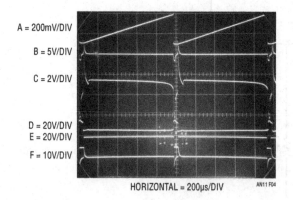

A = 200mV/DIV
B = 5V/DIV
C = 2V/DIV
D = 20V/DIV
E = 20V/DIV
F = 10V/DIV

HORIZONTAL = 200µs/DIV   AN11 F04

Figure 27.4 • Linearized Methane Detector Waveforms

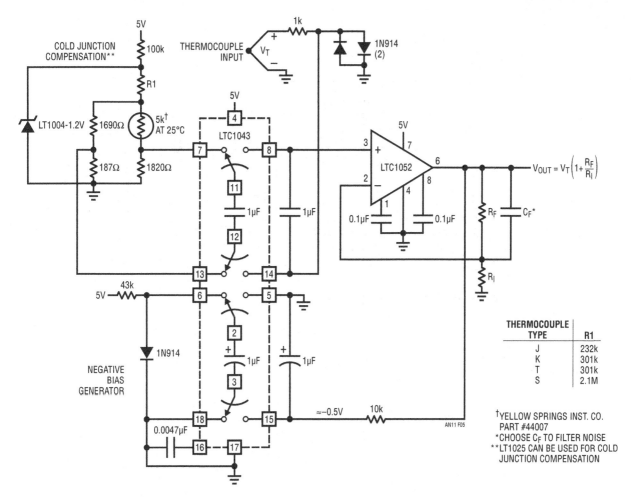

Figure 27.5 • Cold Junction Compensated Thermocouple Signal Conditioner

5V powered precision instrumentation amplifier

Many transducer outputs require a true differential input "instrumentation" type amplifier. Transducers with single-ended outputs do not, in theory, require a differential input amplifier but common mode noise often exceeds the signal of interest. For these reasons, transducer systems often employ these type amplifiers.

No commercially manufactured instrumentation amplifier will function from a 5V supply and achieving good precision in a design is difficult. The circuits in Figure 27.6 meet these requirements. They also feature input protection, filtering capability and a shield driving output.

In Figure 27.6a, A1, A2 and A3 accomplish the differential input-to-single-ended output conversion, with R_G setting gain. The LT1014's high open-loop gain permits accurate circuit gain. Offset and drift performance allows use with low level transducers such as thermocouples and strain gauges. The 100kΩ-1μF combination filters noise and 60Hz pickup; the amplifier is never exposed to high

frequency common mode noise. The transistor-diode clamps combine with the 100k resistors to prevent high voltage spikes or faults (common in industrial environments) from damaging the amplifiers. A4 is used as a shield driver to reduce the effects of input cable capacitance. It drives the shield at the input common mode voltage, which is derived from the input amplifier's output. Performance characteristics are summarized in the table.

Figure 27.6b achieves greater DC precision at the expense of bandwidth. Here, the LTC1043 switched-capacitor building block alternately commutates a 1μF capacitor between the circuit input and the LTC1052's input. This stage, accomplishing a differential-to-single-ended transition, allows the chopper-stabilized LTC1052 to take a ground referred measurement. The LTC1043's other switch stage generates a small negative potential, allowing the LTC1052 output to swing all the way to 0V. DC precision is excellent, surpassing all monolithic ±15V instrumentation amps, although bandwidth is limited to 10Hz. The LTC1043's switching action, set at about 400Hz by the 0.01μF value, forms a lowpass filter. This permits extremely high rejection of noise, allowing the

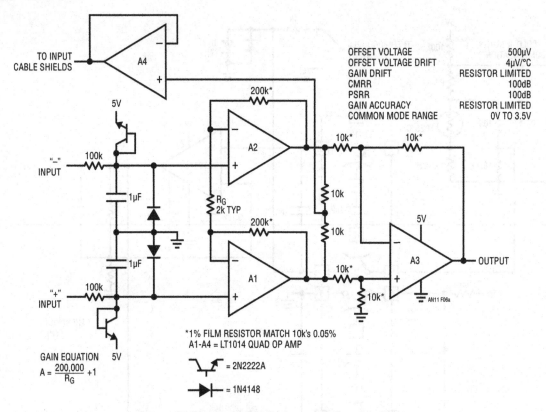

*1% FILM RESISTOR MATCH 10k's 0.05%
A1–A4 = LT1014 QUAD OP AMP

⊣ = 2N2222A

▷⊢ = 1N4148

OFFSET VOLTAGE 500µV
OFFSET VOLTAGE DRIFT .... 4µV/°C
GAIN DRIFT .... RESISTOR LIMITED
CMRR 100dB
PSRR 100dB
GAIN ACCURACY .... RESISTOR LIMITED
COMMON MODE RANGE .... 0V TO 3.5V

GAIN EQUATION
$$A = \frac{200,000}{R_G} + 1$$

Figure 27.6a ● Precision Instrumentation Amplifier

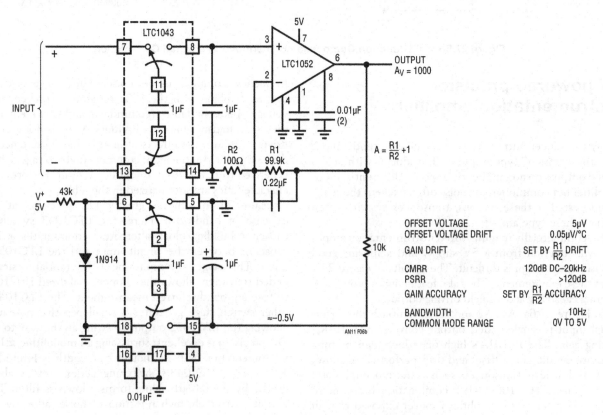

$$A = \frac{R1}{R2} + 1$$

OFFSET VOLTAGE 5µV
OFFSET VOLTAGE DRIFT .... 0.05µV/°C
GAIN DRIFT .... SET BY $\frac{R1}{R2}$ DRIFT
CMRR 120dB DC–20kHz
PSRR .... >120dB
GAIN ACCURACY .... SET BY $\frac{R1}{R2}$ ACCURACY
BANDWIDTH 10Hz
COMMON MODE RANGE .... 0V TO 5V

Figure 27.6b ● Ultra-Precision Instrumentation Amplifier

CMRR to remain above 120dB at 20kHz. Overall performance is summarized in the table.

5V powered strain gauge signal conditioner

Figure 27.7 shows an unusual approach to signal conditioning the bridge output of a strain gauge pressure transducer. The 5V circuit needs only two amplifiers and provides an auxiliary ratio output for a monitoring A/D converter. The design functions by converting the bridge's differential output into a ground-referred single-ended signal which is then amplified. This approach eliminates common mode errors by eliminating the bridge's common mode output component. Additionally, the number of precision resistors required is minimized and no matching is required.

and A2's gain is set to provide the desired output scale factor. Because bridge drive is derived from the LT1034 reference, A2's output is not affected by supply shifts. The LT1034's output is available for ratio operation. To calibrate this circuit, apply or electrically simulate 0psi and trim the zero adjustment for 0V output. Next, apply or electrically simulate 350psi and trim gain for 3.500V out. Repeat these adjustments until both points are fixed.

"Tachless" motor speed controller

Figure 27.8 shows a way to servo control the speed of a DC motor. This circuit is particularly applicable to digitally controlled systems in robotic and X-Y positioning applications. By functioning from the 5V logic supply it eliminates

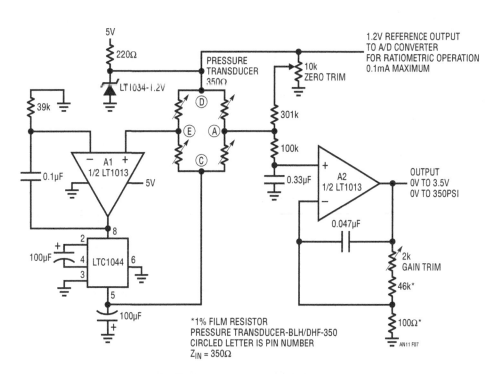

Figure 27.7 • Strain Gauge Signal Conditioner

A1 biases the LTC1044 positive-to-negative converter. The LTC1044's output pulls the bridge's output negative, causing A1's input to balance at 0V. This local loop permits a single-ended amplifier (A2) to extract the bridge's output signal. The 100k-0.33μF RC filter's noise

additional motor drive supplies. The "tachless" feedback saves additional space and cost. The circuit senses the motor's back EMF to determine its speed. The difference between the speed and a setpoint is used to close a sampled loop around the motor. Because no commercially available sample-hold circuit will run from a 5V supply, special techniques are required.

A1 generates a pulse train (Trace A, Figure 27.9). When A1's output is high, Q1 is biased and Q3 drives the motor's ungrounded terminal (Trace B). When A1 goes low, Q3 turns off and the motor's back EMF appears after the inductive flyback ceases. During this period, S1's input (Trace C) is turned on, and the 0.047 capacitor acquires the back EMF's value. A2 compares this value with the setpoint and the amplified difference (Trace D) changes A1's duty cycle, controlling motor speed. A2 has the desirable characteristic of assuming unity gain in the absence of a feedback signal.

Start-up or input overdrive cannot force servo lock-up due to loss of sampling action. The loop is self-restoring and will establish control when abnormal conditions cease.

Drive to S1's control input must be carefully controlled for proper operation. Q2 prevents the switch from closing until the negative going flyback interval is over and the 0.068μF capacitor slows the switches turn-on edge. These measures ensure clean back EMF acquisition in the 0.047μF unit. Q2's collector line diodes compensate the motor's clamp diode drop, preventing destructive negative voltages at S1. The circuit controls from 20rpm to full speed with good transient response

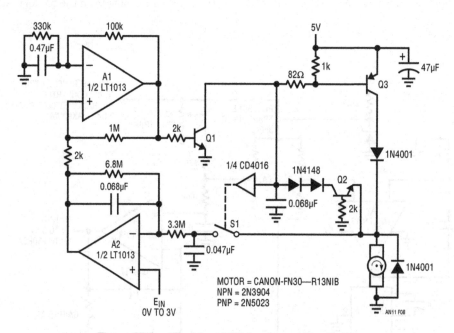

Figure 27.8 • "Tachless" Motor Speed Controller

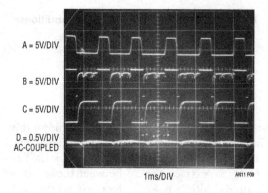

Figure 27.9 • Motor Speed Controller Waveforms

under all shaft loads. The gain and roll-off terms in A2's feedback loop are optimal for the motor listed, and should be reestablished for other types.

4-20mA current loop transmitter

Transmission of industry standard 4-20mA current loop signals to values and other actuators is a common requirement. Resistive line losses and actuator impedances require current transmitters to be able to force a compliance voltage of a least 20V. Because of this, 5V powered systems usually cannot meet current loop transmitter requirements, but Figure 27.10 shows a way to do this. This 5V powered circuit utilizes a servo controlled DC/DC converter to generate the compliance voltage necessary for loop current requirements. It will drive 4-20mA into loads as high as 2200Ω (44V compliance) and is inherently short-circuit protected. The circuit's input is applied to A2, whose output biases A1's " + " input via the offsetting network. A1's output goes high, biasing Q4 to turn on Q3. Q3's collector drives the T1-Q1-Q2 DC/DC converter, which is clocked by the RC gate oscillator. T1 furnishes voltage step-up and the rectified and filtered

secondary current flows through the 100Ω resistor and the load. A1's negative input measures the voltage across the 100Ω resistor, completing a current control loop around T1. The 0.33µF capacitor furnishes stable loop roll-off and the 100pF unit suppresses local oscillation at Q4. Within the compliance limit, A1 maintains constant output current, regardless of load impedance shifts or supply changes. To calibrate this circuit, short the output, apply 0V to the input and adjust the "4mA trim" for 0.3996V across the 100Ω resistor. Next, shift the input to 4.000V and trim the 20mA adjustment for 1.998V across the 100Ω resistor Repeat this procedure until both points are fixed. The gain trim network shunting the 100Ω resistor necessitates the odd voltage trim target values, but output current swings between 4.000mA and 20.00mA.

Figure 27.11 details modifications which permit the circuit's output current to galvanically float. This is often useful in industrial situations where the output lines may be exposed to common mode voltages or high voltage fault conditions. The transformer's primary current, which theoretically reflects current delivered by the secondary, is sensed across a shunt and fed back via A2. In practice, current control precision is limited by non-ideal transformer behavior to about 1%. Common mode voltage is limited by T1's 300V breakdown.

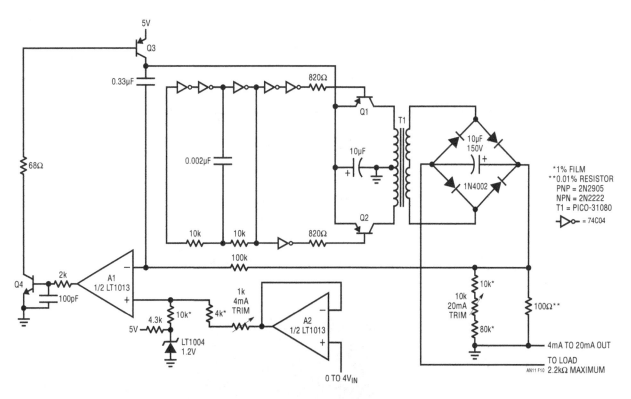

Figure 27.10 • 4-20mA Current Loop Transmitter

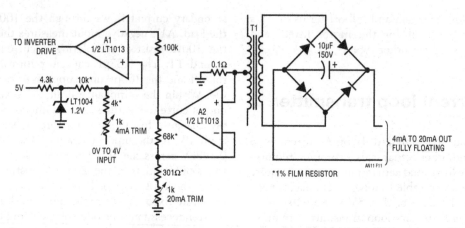

Figure 27.11 • Floating Output Option for Current Loop Transmitter

Fully isolated limit comparator

Figure 27.12's 5V powered circuit performs a fully isolated limit comparison on low level signals. It produces a digital output indicating if the input is above or below a preset limit. This circuit is ideal for process control applications where transducers operate at high common mode voltages or where large ground loops exist. An uncommitted gain of 100 amplifier allow thermocouples and other low level sources to be used with the circuit. The circuit functions by echoing an interrogation pulse if the input is above a preset level. If the input is below this level, no echo pulse occurs. A transformer is used to allow a 2-way, galvanically isolated signal path and the energy contained in the interrogation pulse powers the circuit's floating elements. Figure 27.13 shows operating waveforms for the "above limit" case. When an input interrogation pulse is applied, Q1's collector drives the transformer primary (Figure 27.13, Trace A). Energy is transferred to the transformer's secondary and stored in the

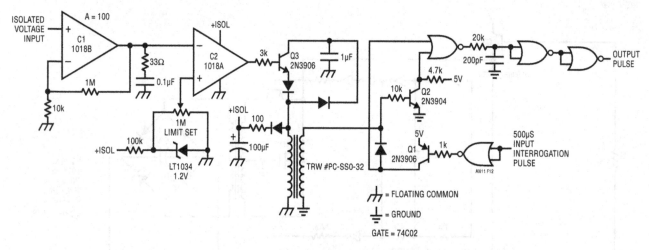

Figure 27.12 • Fully Isolated Limit Comparator with Gain of 100

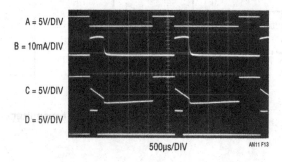

A = 5V/DIV

B = 10mA/DIV

C = 5V/DIV

D = 5V/DIV

500µs/DIV

Figure 27.13 • Isolated Limit Comparator Waveforms

100μF capacitor. The charge in the capacitor powers the iso-lated circuitry ("+ISOL" potential indicated in Figure 27.12). If low power dual comparator C2's output is low, Q3 biases and drives current into the transformer secondary (Trace B). This is reflected in the transformer's primary (Trace C). Q2 and the associated gate circuitry form a demodulator which produces an output pulse (Trace D). If C2's output had been high (below limit set), the transformer would have received no secondary drive and there would be no output pulse. The demodulator is designed so that nothing appears on the output line unless the circuit is above the preset limit.

Comparator C1's output damper network allows it to function as an op amp for low level signals. This circuit easily extracts millivolt signals buried in high common mode voltage or ground noise and delivers its limit decision to the output. The maximum common mode voltage is limited by the transformer's 500V breakdown.

Fully isolated 10-bit A/D converter

Figure 14's 5V powered circuit is a complete 10-bit A/D converter which is fully floating from system ground. It is ideal for performing 10-bit A/D conversions in the face of the high common mode noise characteristics of predominantly digital systems. It is also useful in industrial environments, where noise and high common mode voltages are present in transducer fed systems.

Circuit operation is initiated by applying a pulse to the "convert command" input (Trace A, Figure 27.15). This pulse appears at the transformer secondary and charges the 100μF capacitor. This potential is used to supply power to the floating A/D conversion circuitry. The transformer's secondary pulse biases the inverter-open drain buffer combination to discharge the 4μF capacitor (Trace B).

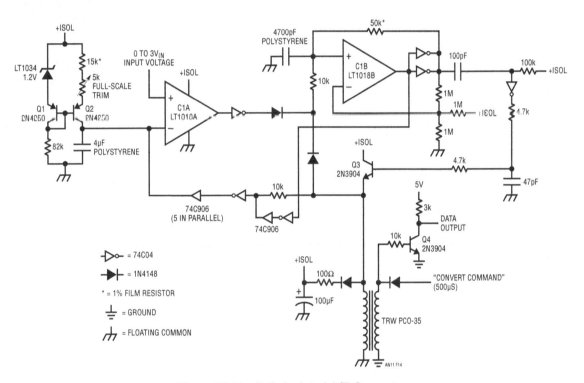

Figure 27.14 • Fully Isolated A/D Converter

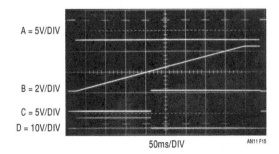

Figure 27.15 • A/D Waveforms—Isolated Section

The secondary pulse also biases a diode to stop the C1B 3kHz oscillator output (Trace D). Concurrently, C1A goes high, forcing the inverter in its output line low (Trace C). When the convert command pulse ceases, the Q1-Q2 current source charges the 4µF capacitor with a linear ramp. The C1B oscillator now runs. When the ramp crosses the input voltage's value, C1A's output switches and its output line inverter (Trace C) goes high, cutting off the oscillator. The number of oscillator pulses occurring during this interval is proportional to the input voltage value. These pulses are differentiated and fed to Q3, which drives the transformer. The differentiation causes narrow spikes to be fed to the transformer, easing power drain on the 100µF energy storage capacitor. Q3's RC base delay and inverter-buffer combination at C1B's output prevent Q3's emitter pulses from triggering a ramp reset. The waveforms appearing at the transformer's input do not reflect the circuit's complex operation, and easily interface to a digital system. Figure 27.16, Trace A shows the "convert command" pulse. Trace B is the transformer primary. The differentiated oscillator pulses coming from the transformer secondary appear as small amplitude spikes. In this case, a small voltage is being converted and the number of pulses is small. Trace C, the "data output", is taken at Q4's collector; and the pulses are TTL compatible.

Several subtile factors contribute to the 10-bit performance of this circuit. The 4700pF and 4µF polystyrene capacitors are both −120ppm/°C gain terms. Because of this, their temperature drift's ratio and overall circuit gain drift is about 25ppm/°C. The five parallel 74C906 open-drain buffers provide an effective 0V reset for the 4µF capacitor, minimizing reset offset errors. Parallel inverters in C1B's output line reduce saturation caused errors, aiding oscillator stability with shifts in supply and temperature. Finally, the diode path at Q3's emitter averts a ±1 count uncertainty error by synchronizing the oscillator to the conversion sequence. The 5k potentiometer in the current source trims calibration to equal 1024 counts out for 3.000V input. The transformer used allows the converter to function at common mode levels up to 175V. The circuit requires 330ms to complete a 10-bit conversion and drifts less than 1LSB over 25°C ±25°.

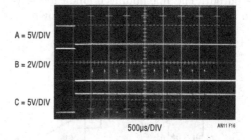

A = 5V/DIV

B = 2V/DIV

C = 5V/DIV

500µs/DIV   AN11 F16

Figure 27.16 • Detail of A/D Waveforms—Grounded Section

High performance single supply analog building blocks

Two new components provide the basic building blocks needed for high performance 5V single supply circuits. The LT1014 quad op amp, also available as the LT1013 dual, features DC specifications nearly as good as the best ±15V op amps. The LT1017/ LT1018 series dual comparators combine low power and high DC precision with speed adequate for most applications. To ease single supply operation, both units' common mode range includes ground, and the op amp's output swings nearly to ground.

LT1014 basic features

E_{OS}—70µV

$E_{OS}\Delta TC$—2µV/°C

I_{BIAS}—15nA

Gain—1.0×10^6

Supply I—310µA

Noise—0.1Hz-10Hz-0.55µV$_{P-P}$

Common Mode Range—Ground to $(V^+) - 1.5V$

Supply Range—3.4V to 40V

Output Swing   No Load $(V^+) - 0.6V$

                 $(V^-) + 0.015V$

         600Ω Load $(V^+) - 1.0V$

                 $(V^-) + 0.005V$

LT1017/LT1018 basic features

E_{OS}—500µV

$E_{OS}\Delta TC$—5µV/°C

I_{BIAS}—15nA

Gain—1 Million

Supply I—LT1017—30µA, LT1018—110µA

Response Time—6µs (LT1018)

Common Mode Range—V^- to $(V^+) - 0.9V$

Supply Range—1.1V to 40V

Output Current—65mA Pull-Down—60µA Pull-Up

Output Stage Can Pull Down Loads Above V_S

Linear power supplies—past, present, and future

The amplitude of linear circuitry's power supplies has been determined by the available technology used for linear functions.

Probably the first standard linear supply was ±300V, used in analog computers. The operating characteristics of vacuum tubes necessitated high voltages. Additionally, because analog computers (from which operational amplifiers descended) were mathematical machines, bipolar supplies were desirable for computational purposes. With the arrival of solid state linear components in the early 1960s, new supply voltages were necessary and desirable. ±15V was adopted because of transistor breakdown limits, as well as the availability of low voltage references (Zener and avalanche diodes). Power and size advantages afforded by the new supply standard were also obviously attractive. It is significant that while the supply drop decreased available dynamic range of operation by over 20dB, the usable signal processing range did not decrease. This was due to the lower noise and drift of the solid state components.

The arrival of monolithic linear circuits in the late 1960s and subsequent design refinements expanded the available territory at the lower end of the signal processing range.

Currently available precision linear components and technology shifts are causing reevaluation of the power supply issue. In particular, several trends argue for linear functions to be able to operate from the low voltage digital rail. The increasing digital content of systems makes 5V compatible linear components attractive. Cost and space considerations in these systems often make separate linear supplies undesirable. This situation is not universal, and never will be, but is increasingly common.

A move to lower voltages for digital circuits, which must occur, underscores the need for low voltage, high performance linear ICs. Drops in digital supply value will be forced by increasing density requirements, which will lower IC breakdown limits.

Lower power consumption in systems goes along with supply voltage and density issues. Increasingly complex systems are being put into smaller physical spaces, necessitating attention to dissipation. In many instances, portable operation is desirable, so circuitry must be directly compatible with battery potentials, as well as lower power. Linear components must not give up performance to function in this low voltage, digitally driven environment. Demands for precision remain high, and low voltage linear circuits, despite their narrowed dynamic operating range, must meet these requirements. This is not easy, considering that a 12-bit ±15V system typically has a 2.5mV/LSB error budget. At 5V, this number shrinks to 1mV. To deal with this, design techniques developed for ±15V components are being used in new, low voltage ICs and new approaches will also be employed.

Using logic supplies for linear functions

The fast clocking and transient high currents characteristic of digital systems make logic supplies an unfriendly place for linear components to operate. A key to achieving good results is considering power bus routing as an integral part of the signal processing chain. The figure shows that supply rail impedances will cause both DC and AC errors at various points in a system. This is true of any power distribution scheme, but is especially troublesome in digitally oriented systems, where fast current spiking and clock harmonics are present. Circuitry located at position "A" will see appreciable positive rail noise and ground potential will be corrupted by fast, relatively high currents returning through conductor impedances. Supply bypassing can reduce positive rail noise, but ground potential uncertainty can cause unacceptable errors. Locating linear circuits as shown in "B" reduces both positive and ground rail problems by eliminating the digitally derived currents. The linear devices lower operating currents allow lower errors due to supply distribution impedances. Appropriate bypassing techniques are also shown. LC filters can be substantially more effective than simple capacitors, especially in cases where it is not practical to route the positive rail directly from the supply. RC filtering forces voltage drop across the resistor, but is often acceptable due to the linear components low power requirements.

In many cases it is not possible to arrange a "clean" supply for the linear components. In such circumstances it may be possible to synchronize noise sensitive linear circuit operations to occur between system clock pulses. This approach utilizes the synchronous nature of most digital systems and the fact that supply bus disturbances are often minimized between clock pulses.

Probably the most effective technique for dealing with digital supply noise is to galvanically isolate the linear circuits from the supply. The most obvious way to do this is provision of separate power supplies for the linear circuits, but this is often unacceptable. Instead, transformer and optical isolation circuit techniques allow logic rail driven, galvanically isolated circuits (see Figures 27.11, 27.12 and 27.14).

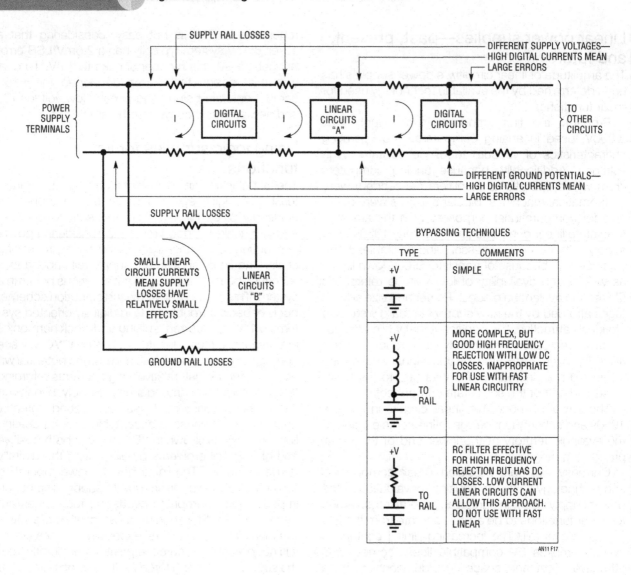

Application considerations for an instrumentation lowpass filter

28

Nello Sevastopoulos

Description

The LTC1062 is a versatile, DC accurate, instrumentation lowpass filter with gain and phase that closely approximate a 5th order Butterworth filter. The LTC1062 is quite different from presently available lowpass switched-capacitor filters because it uses an external (R, C) to isolate the IC from the input signal DC path, thus providing DC accuracy. Figure 28.1 illustrates the architecture of the circuit. The output voltage is sensed through an internal buffer, then applied to an internal switched-capacitor network which drives the bottom plate of an external capacitor to form an input-to-output 5th order lowpass filter. The input and output appear across an external resistor and the IC part of the overall filter handles only the AC path of the signal. A buffered output is also provided (Figure 28.1) and its maximum guaranteed offset voltage over temperature is 20mV. Typically the buffered output offset is 0mV to 5mV and drift is 1μV/°C. As will be explained later, the use of an input (R, C) provides other advantages, namely lower noise and antialiasing.

Tuning the LTC1062

In Figure 28.1, the filter function is formed by using an external (R, C) to place the LTC1062 inside an AC loop. Because of this, the value of the (R • C) product is critically related to the filter passband flatness and to the filter stability. The internal circuitry of the LTC1062 is driven by a clock which also determines the filter cutoff frequency. For a maximally flat amplitude response, the clock should be 100 times the desired cutoff frequency and the (R, C) should be chosen such as:

$$\frac{f_C}{1.62} \leq \frac{1}{2\pi RC} \leq \frac{f_C}{1.63}$$

where:

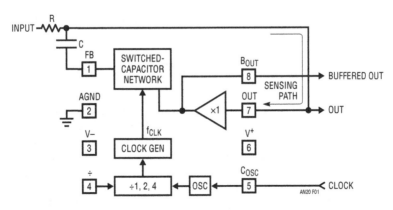

Figure 28.1 • LTC1062 Architecture

Analog Circuit and System Design: A Tutorial Guide to Applications and Solutions. DOI: 10.1016/B978-0-08-089064-7.00028-X

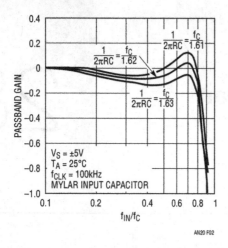

Figure 28.2 • Passband Gain vs Input Frequency

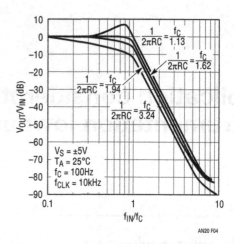

Figure 28.4 • Filter Frequency Response for Various (R, C) Values and Constant Clock

f_C = filter cutoff frequency, (−3dB point).

For instance, to make a maximally flat filter with a −3dB frequency at 10Hz, we need a 100 • 10Hz = 1kHz internal or external clock and an external (R, C) such as:

$$\frac{1}{2\pi RC} = \frac{10Hz}{1.62} = 6.17Hz$$

The minimum value of the resistor, R, depends upon the maximum signal we want to attenuate, and the current sinking capability of Pin 1 which is typically 1mA. The 10Hz filter of the previous example should attenuate a 40Hz signal by 60dB. If the instantaneous amplitude of this signal is 1V peak, the minimum value of the external resistor should be 1kΩ.

Figure 28.2 shows the high accuracy of the passband response for values of $1/2\pi RC$ around $(f_C/1.62)$. If

maximum flatness is required, the (R • C) product should be well controlled. Figures 28.3 and 28.4 are similar to Figure 28.2 but with wider range of $(1/2\pi RC)$ values. When the input (R, C) cutoff frequency approaches the cutoff frequency of the filter, the filter peaks and the circuit may become oscillatory. This can accidentally happen when using input ceramic capacitors with strong negative temperature coefficient. As the temperature increases, the value of the external capacitor decreases and if the clock driving the LTC1062 stays constant, the resulting $(1/2\pi RC)$ approaches the filter cutoff frequency. On the other hand, if the external (R • C) has a strong positive temperature coefficient, the filter passband at high temperatures will become droopy. It is important to note that the filter attenuation slope is mainly set by the internal LTC1062 circuitry and it is quasi-independent from the values of the external (R, C). This is shown in Figure 28.4, where a 100Hz cutoff frequency LTC1062 was tested with an external 10kHz clock and for:

$$\frac{f_C}{3.24} \leq \frac{1}{2\pi RC} \leq \frac{f_C}{1.13}$$

Also, Figure 28.4 shows that the −30dB/octave slope remains constant even though the external (R • C) changes.

LTC1062 clock requirements

Using an external clock: the internal switched-capacitor network is clock driven and the clock frequency should be 100 times the desired filter cutoff frequency. Pin 5 of the LTC1062 is the clock input and an external clock swinging close to the LTC1062 power supplies will provide the clock requirements for the internal circuitry. The typical logic threshold levels of Pin 5 are shown on Table 28.1.

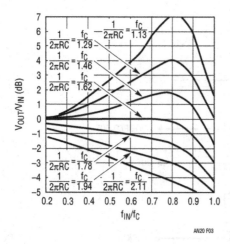

Figure 28.3 • Passband Gain vs Input Frequency

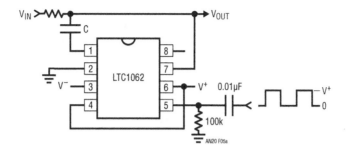

Figure 28.5a • AC Coupling an External CMOS Clock Powered from a Single Positive Supply, V⁺

V_{SUPPLY}	V_{th}^+	V_{th}^-
±2.5V	+0.9V	−1V
±5V	+1.3V	−2.1V
±6V	+1.7V	−2.5V
±7V	+1.75V	−2.9V
±8V	+1.95V	−3.3V
±9V	+2.15V	−3.7V

The temperature coefficient of the threshold levels is $-1mV/°C$.

Because the trip levels of Pin 5 are asymmetrically centered around ground and because $(V_{th}^+ - V_{th}^-)$ is less than the positive supply voltage, V^+, CMOS level clocks operating from V^+ and ground can be AC coupled into Pin 5 and drive the IC, Figure 28.5a.

Internal oscillator

A simple oscillator is internally provided and it is overridden when an external clock is applied to Pin 5. The internal oscillator can be used for applications for clock requirements below 130kHz and where maximum passband flatness over a wide temperature range is not required. The internal oscillator can be tuned for frequencies below 130kHz by connecting an external capacitor C_{OSC}, from Pin 5 to ground (or negative supply). Under this condition, the clock frequency can be calculated by:

$$f_{CLK} \cong 130kHz\left(\frac{33pF}{33pF + C_{OSC}}\right) \quad (1)$$

Due to process tolerances, the internal 130kHz frequency varies and also has a negative temperature coefficient. The LTC1062 data sheet publishes curves characterizing the internal oscillator. To tune the frequency of the internal oscillator to a precise value, it is necessary to adjust the value of the external capacitor, C_{OSC}, or to use a potentiometer in series with the C_{OSC}, Figure 28.5b. The new clock frequency, f'_{CLK}, can be calculated by:

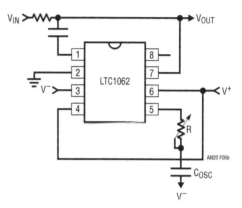

Figure 28.5b • Adding a Resistor in Series with COSC to Adjust the Internal Clock Frequency

$$f'_{CLK} = \frac{f_{CLK}}{(1 - 4RC_{OSC}f_{CLK})} \quad (2)$$

where f_{CLK} is the value of the clock frequency, when $R = 0$, from (1). When an external resistor (potentiometer) is used, the new value of the clock frequency is always higher than the one calculated in (1). To achieve a wide tuning range, calculate from (1) the ideal (f_{CLK}, C_{OSC}) pair, then double the value of C_{OSC} and use a 50k potentiometer to adjust f'_{CLK}.

Example: To obtain a 1kHz clock frequency, from (1) C_{OSC} typically should be 4250pF. By using 8500pF for C_{OSC} and a 50k potentiometer, the clock frequency can be adjusted from 500Hz to 3.3kHz as calculated by (2).

The internal oscillator frequency can be measured directly at Pin 5 by using a low capacitance probe.

Clock feedthrough

Clock feedthrough is defined as the amount of clock frequency appearing at the output of the filter. With ±5V supplies the measured clock feedthrough was $420\mu V_{RMS}$, while with ±7.5V supplies it increased to $520\mu V_{RMS}$. The clock feedthrough can be eliminated by using an (R, C) at the buffered output, Pin 8, provided that this pin is used as an output. If an external op amp is used to buffer the DC accurate output of the LTC1062, an input (R, C) can be

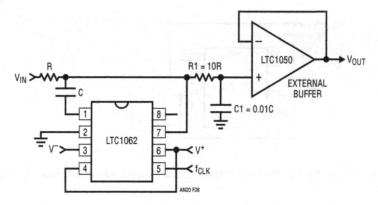

Figure 28.6 ● Adding an External (R1, C1) to Eliminate the Clock Feedthrough and to Improve the High Frequency Attenuation Floor

used to eliminate the clock feedthrough, Figure 28.6, and to further increase the attenuation floor of the filter. Note that this (R, C) does not really improve the noise floor of the circuit since the major noise components are located near the filter cutoff frequency.

Single 5V supply operation

Figure 28.7 shows the LTC1062 operating with a single supply. The analog ground, Pin 2, as well as the buffer input, Pin 7, should be biased at 1/2 supply. The value of resistor R1 should conduct 100µA or more. In Figure 28.7, the resistor R'DC biases the buffer and the capacitor C' isolates the buffer bias from the DC value of the output. Under these conditions the external (R, C) should be adjusted such that $(1/2\pi RC) = f_C/1.84$. This accounts for the extra AC loading of the (R', C') combination.

The resistor and capacitor (R', C') are not needed if the input voltage has a DC value around 1/2 supply.

If an external capacitor is used to activate the internal oscillator, its bottom plate should be tied to system ground.

Dynamic range and signal/noise ratio

There is some confusion with these two terminologies. Because monolithic switched-capacitor filters are inherently

more noisy than (R, C) active filters, it is necessary to take a hard look at the way some IC manufacturers describe the S/N ratio of their circuit. For instance, when dividing the filter's typical RMS swing by its wideband noise, the result is called "best case" S/N ratio. But this is definitely not "dynamic range". Under max swing conditions, many monolithic filters exhibit high harmonic distortion. This indicates poor dynamic range even though the S/N ratio looks great on paper.

With ±5V supplies and higher, the LTC1062 has a typical $100\mu V_{RMS}$ wideband noise. With $1V_{RMS}$ output levels, the signal/noise ratio is 80dB. The test circuit of Figure 28.8 is used to illustrate the harmonic distortion of the device. The worst-case occurs when the input

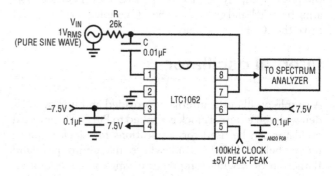

Figure 28.8 ● 1kHz Cutoff Frequency, 5th Order LP Filter, Test Circuit for Observing Distortion

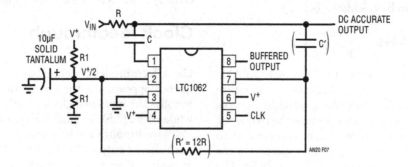

Figure 28.7 ● Single Supply Operation

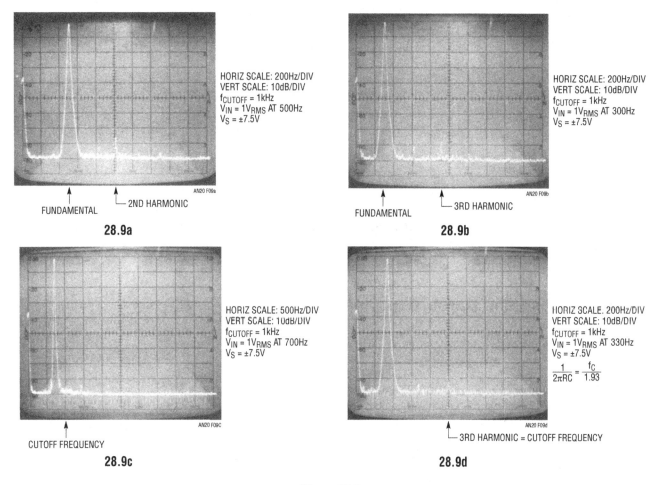

HORIZ SCALE: 200Hz/DIV
VERT SCALE: 10dB/DIV
f_{CUTOFF} = 1kHz
V_{IN} = 1V_{RMS} AT 500Hz
V_S = ±7.5V

FUNDAMENTAL   2ND HARMONIC

28.9a

HORIZ SCALE: 200Hz/DIV
VERT SCALE: 10dB/DIV
f_{CUTOFF} = 1kHz
V_{IN} = 1V_{RMS} AT 300Hz
V_S = ±7.5V

FUNDAMENTAL   3RD HARMONIC

28.9b

HORIZ SCALE: 500Hz/DIV
VERT SCALE: 10dB/DIV
f_{CUTOFF} = 1kHz
V_{IN} = 1V_{RMS} AT 700Hz
V_S = ±7.5V

CUTOFF FREQUENCY

28.9c

HORIZ SCALE. 200Hz/DIV
VERT SCALE: 10dB/DIV
f_{CUTOFF} = 1kHz
V_{IN} = 1V_{RMS} AT 330Hz
V_S = ±7.5V
$$\frac{1}{2\pi RC} = \frac{f_C}{1.93}$$

3RD HARMONIC = CUTOFF FREQUENCY

28.9d

Figure 28.9

fundamental frequency equals 1/2 or 1/3 of the filter cutoff frequency. This causes the 2nd or 3rd harmonics of the output to fall into the filter's passband edge, Figures 28.9a and 28.9b.

Figure 28.9c shows an input frequency of 700Hz and the filter's dynamic range under this condition is in excess of 80dB. This is true because the harmonics of the 700Hz input fall into the filter's stopband. With ±7.5V supplies

(or single 15V), the THD of the LTC1062 lies between 76dB and 83dB, depending on where the harmonics occur with respect to the circuit's band edge. A slight improvement, Figure 28.9d, can be achieved by increasing the value of the input resistor, R, such as $(1/2\pi RC) = f_C/1.93$. Under this condition, the filter no longer approximates a max flat ideal response since it becomes "droopy" above 30% of its cutoff frequency, as shown in Figure 28.3.

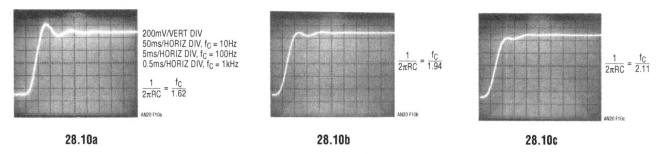

200mV/VERT DIV
50ms/HORIZ DIV, f_C = 10Hz
5ms/HORIZ DIV, f_C = 100Hz
0.5ms/HORIZ DIV, f_C = 1kHz

$$\frac{1}{2\pi RC} = \frac{f_C}{1.62}$$

28.10a

$$\frac{1}{2\pi RC} = \frac{f_C}{1.94}$$

28.10b

$$\frac{1}{2\pi RC} = \frac{f_C}{2.11}$$

28.10c

Figure 28.10 • Step Response to a 1V Peak Input Step

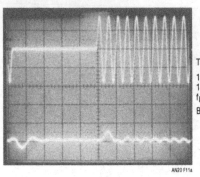

28.11a

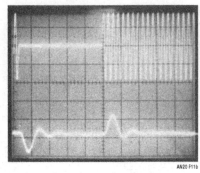

28.11b

Figure 28.11 ● LTC1062 Response to a 2V_P-P Sinewave Burst

Step response and burst response

The LTC1062 response to an input step approximates that of an ideal 5th order Butterworth lowpass filter. Butterworth filters are "ringy", Figure 28.10a, even though their passband is maximally flat. Figures 28.10b and 28.10c show a more damped step response which can be obtained by increasing the input (R • C) product and thereby sacrificing the maximum flatness of the filter's amplitude response. Figures 28.11a and 28.11b show the response of the LTC1062 to a 2V peak-to-peak sinewave burst which frequency is respectively equal to $2 • f_C$ and $4 • f_C$. It is interesting to compare Figure 28.11b to Figure 28.10a: In both figures the overshoots and the settling times are about equal since, from the filter's point of view, a high frequency burst looks like a step input.

LTC1062 shows little aliasing

Aliasing is a common phenomenon in sampled data circuits especially when signals approaching the sampling frequency are applied as inputs. Generally speaking, when an input signal of frequency (f_{IN}) is applied, an alias frequency equal to ($f_{CLK} \pm f_{IN}$) appears at the filter's output. If f_{IN} is less the ($f_{CLK}/2$), then the amplitude of the alias frequency equals the magnitude of f_{IN} multiplied by the gain of the filter at f_{IN}, times the (sinx/x) function of the circuit. For a lowpass filter, the gain around ($f_{CLK}/2$) is essentially limited by the attenuation floor of the filter and the ($f_{CLK} \pm f_{IN}$) alias signal is buried into the filter noise floor The problem arises when the input signal's frequency is greater than ($f_{CLK}/2$) and especially when it approaches f_{CLK}. Under these conditions ($f_{CLK}–f_{IN}$) falls either into the filter's passband or into the attenuation slope, and then aliasing occurs. If for instance a 5th order Butterworth switched-capacitor ladder filter has a 1kHz corner frequency and

operates with a 50kHz clock, a 49kHz, $1V_{RMS}$ input signal will cause an alias (1kHz, $0.7V_{RMS}$) signal to appear at the output of the circuit; a 48kHz input will appear as a 2kHz output attenuated by 30dB.

The LTC1062 internal circuitry has a 4th order sampled data network so, in theory, it will be subject to the above aliasing phenomenon. In practice, however; the input (R, C) band limits the incoming, clock-approaching signals, and aliasing is nearly eliminated. Experimental work shows the following data:

f_{IN}, 0dB LEVEL	LTC1062 V_{OUT} AT ($f_{CLK} - f_{IN}$)	STANDARD 6th ORDER SWITCHED CAPACITOR LOWPASS FILTER V_{OUT} AT ($f_{CLK} - f_{IN}$)
$0.97 • f_{CLK}$	−77dB	−22.0dB
$0.98 • f_{CLK}$	−64dB	−3.5dB
$0.99 • f_{CLK}$	−43dB	0dB
$0.995 • f_{CLK}$	−45dB	0dB
$0.999 • f_{CLK}$	−60dB	0dB

Cascading the LTC1062

Two LTC1062s can be cascaded with or without intermediate buffers. Figure 28.12 shows a DC accurate 10th order lowpass filter where the second LTC1062 input is taken directly from the DC accurate output of the first one. Because loading the junction of the input (R, C) causes passband error, the second resistor, R' should be much larger than R. The recommended ratio of (R'/R) is about

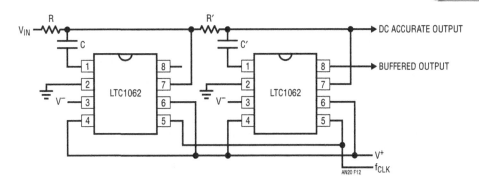

Figure 28.12 • Cascading Two LTC1062s

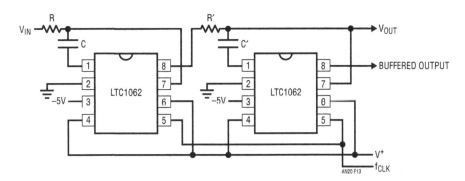

Figure 28.13 • Cascading Two LTC1062s. The 2nd Stage is Driven by the Buffered Output of the First Stage

117/1; beyond this, the passband error improvement is not worth the large value of R'. Also, under this condition $(1/2\pi RC) = f_C/1.57$ and $(1/2\pi R'C') = f_C/1.6$. For instance, a 10th order filter was designed with a cutoff frequency, f_C, of 4.16kHz, $f_{CLK} = 416$kHz and $R = 909\Omega$, $R' = 107$k, $C = 0.066\mu F$ and $C' = 574$pF. The maximum passband error was -0.6dB occurring around $0.5 \cdot f_C$. Figure 28.13 repeats the circuit of the previous figure but the second LTC1062 is fed from the buffered output of the first one. The filter's offset is the offset of the first LTC1062 buffer (which is typically

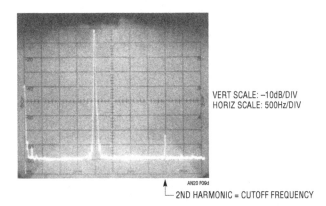

VERT SCALE: –10dB/DIV
HORIZ SCALE: 500Hz/DIV

AN20 F09d
└─ 2ND HARMONIC = CUTOFF FREQUENCY

Figure 28.14 • Response of the Filter of Figure 28.13 to a 2kHz 1V_{RMS} Input Sinewave. The 2nd Harmonic (Worst Case) Occurs at the Filter's Cutoff Frequency

under ±5mV and guaranteed 20mV over the full temperature range of the device). By using this intermediate buffer, impedance scaling is no longer required and the values of R and R' can be similar: With this approach the passband gain error is reduced to -0.2dB. The recommended equation of the two (R, C)s are the following: $(1/2\pi RC) = f_C/1.59$ and $(1/2\pi R'C') = f_C/1.64$ or vice versa.

A 4kHz lowpass filter was tested with the circuit of Figure 28.13. The measured component values were $R = 97.6$k and $C = 676$pF, $R' = 124$k and $C' = 508$pF. The wideband noise of the filter was $140\mu V_{RMS}$ and the worst-case second harmonic distortion occurred with $f_{IN} = 0.5 \cdot f_C$ as shown in Figure 28.14. With $1V_{RMS}$ input levels, the signal-to-noise ratio is 77dB and the worst-case dynamic range is 73dB.

Figure 28.15 illustrates a 12th order filter using two LTC1062s and a precision dual op amp. The first op amp is used as a precision buffer and the second op amp is used as a simple 2nd order noninverting lowpass filter to provide the remaining two poles and to eliminate any clock noise.

The output filter is tuned at the cutoff frequency of the LTC1062s and has a Q = 1 to improve the passband error around the cutoff frequency. For gain and Q equal to unity, the design equation for the center frequency, f_0, is simple: let C1/4C2 and R1 = R2, then $f_0 = 1/(\pi R1C1)$. The filter's overall frequency response is shown in Figure 28.16 with a 438kHz clock. The measured DC output offset of the filter was $100\mu V$, although the maximum guaranteed

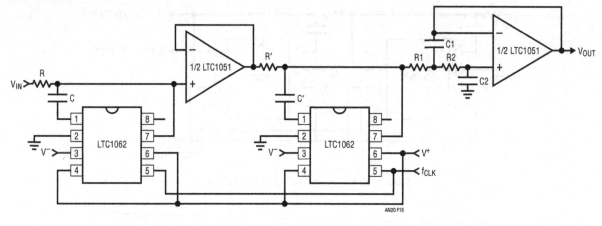

AN20 F15

Figure 28.15 • A Very Low Offset, 12th Order, Max Flat Lowpass Filter
R = 59k, C = 0.001µF, R1 = 5.7k, C1 = 0.01µF
R1 = R2 = 39.8k, C1 = 2000pF, C2 = 500pF, f_{CLK} = 438kHz, f_C = 4kHz

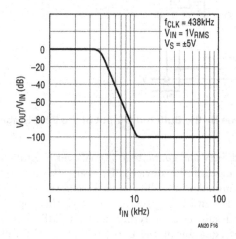

AN20 F16

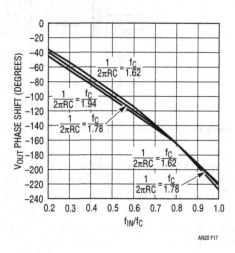

AN20 F17

Figure 28.16 • Frequency Response of the 12th Order Filter of Figure 28.15

Figure 28.17 • Phase Response of the LTC1062 for Various Input (R, C)s

offset of each op amp over temperature would be 400µV. Because the active (R, C) output filter is driven directly from the DC accurate output of the second LTC1062, impedance scaling is used with the resistor R'. The noise and distortion performance of this circuit is very similar to the one described for Figure 28.13.

Using the LTC1062 to create a notch

Filters with notches are generally difficult to design and they require tuning. Universal switched-capacitor filters can make very precise notches, but their useful bandwidth should be limited well below half the clock frequency; otherwise, aliasing will severely limit the filter's dynamic range.

The LTC1062 can be used to create a notch because the frequency where it exhibits −180° phase shift is inside its

passband as shown in Figure 28.17. It is repeatable and predictable from part to part. An input signal can be summed with the output of the LTC1062 to form a notch as shown in Figure 28.18. The 180° phase shift of the LTC1062 occurs at f_{CLK}/118.3 or 0.85 times the lowpass cutoff frequency. For instance, to obtain a 60Hz notch,

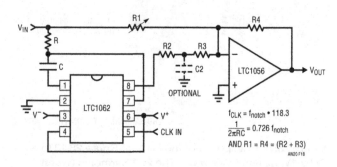

AN20 F18

$f_{CLK} = f_{notch} \cdot 118.3$

$\frac{1}{2\pi RC} = 0.726\ f_{notch}$

AND R1 = R4 = (R2 + R3)

Figure 28.18 • Using the LTC1062 to Create a Notch

the clock frequency should be 7.098kHz and the input $1/(2\pi RC)$ should be approximately 70.98Hz/1.63. The optional (R2C2) at the output of the LTC1062 filters the clock feedthrough. The $1/(2\pi R2C2)$ should be 12 to 15 times the notch frequency. *The major advantage of this notch is its wide bandwidth. The input frequency range is not limited by the clock frequency because the LTC1062 by itself does not alias.*

The frequency response of the notch circuit is shown in Figure 28.19 for a 25Hz notch. From part to part, the notch depth varies from 32dB to 50dB but it can be optimized by tuning resistor R1. Figure 28.20 shows an example of the wideband operation of the circuit. These pictures were taken with the filter operating with a 3kHz clock frequency and forming a 25Hz notch. Figure 28.20a shows the circuit's response to an input 1kHz, $1V_{RMS}$ sinewave; Figure 28.20b shows the response to a 10kHz, $1V_{RMS}$ sinewave. The high frequency distortion of the filter will depend on the quality

of the external op amp and not on the filter The measured wideband noise from DC to 20kHz was $138V_{RMS}$ and the measured noise from 50Hz to 20kHz was $30\mu V_{RMS}$.

The circuit of Figure 28.21 is an extension of the previous notch filter The input signal is summed with the

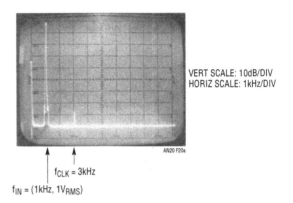

VERT SCALE: 10dB/DIV
HORIZ SCALE: 1kHz/DIV

f_{CLK} = 3kHz

f_{IN} = (1kHz, $1V_{RMS}$)

Figure 28.20a • Response of a 25Hz Notch Filter to a 1kHz, $1V_{RMS}$ Input Sinewave

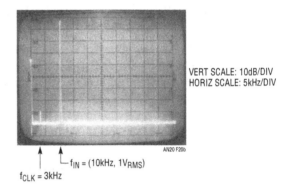

VERT SCALE: 10dB/DIV
HORIZ SCALE: 5kHz/DIV

f_{IN} = (10kHz, $1V_{RMS}$)

f_{CLK} = 3kHz

Figure 28.20b • Same as Above but the Input is a 10kHz Sinewave

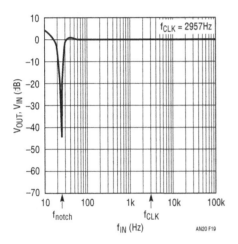

Figure 28.19

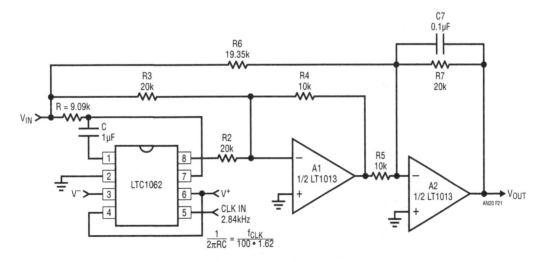

Figure 28.21 • A Lowpass Filter with a 60Hz Notch

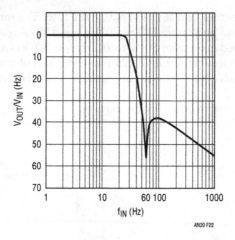

Figure 28.22 • Amplitude Response of the filter in Figure 28.21

lowpass filter output through A1, as previously described; then, the output of A1 is again summed with the input voltage through A2.

If R6 = R2 = R3 = R7 and R4 = R5 = 0.5R7, the output of A2 at least theoretically, should look like the output of LTC1062 Pin 8. If the ratio of (R6/R5) is slightly less than 2, a notch is introduced in the stopband of the LTC1062 as shown in Figure 28.22. The overall filter response looks pseudoelliptic lowpass. The frequency of the notch is at $f_{CLK}/47.3$ and the value of the resistor ratio (R6/R5) should be equal to 1.935.

Comments on capacitor types

Experimental work, done in a laboratory environment, shows that the passband gain error is the same when mylar, polystyrene and polypropylene capacitors are used. All the experiments done for this application note used mylar capacitors for 0.001µF and up and silver mica for less then 1000pF

Solid tantalum capacitors connected back to back, as shown in Figure 28.23, introduce an additional passband error of 0.05dB to 0.1dB. For cutoff frequencies well below

10Hz and for limited temperature range, the back to back solid tantalum capacitor approach offers an economical and board saving solution provided that their leakage and tolerances are acceptable. When disc ceramic capacitors were used as part of the required input (R, C) of the LTC1062, the passband accuracy of the filter was similar to that obtained with solid tantalum capacitors. Ceramic capacitors should be avoided primarily because of their large and unpredictable temperature coefficient. NPO ceramic capacitors, however, are highly recommended especially for military temperature range. Their maximum available value is of the order of 0.1µF, their physical size is reasonable and they are available with ±20ppm/°C tempco.

Clock circuits

Application Note 12 describes in detail various clock generation techniques which can be applied for the LTC1062 requirements. Two basic circuits are repeated and explained below:

1. Low frequency oscillators: A simple (R, C) oscillator is shown in Figure 28.24. It uses a medium speed comparator with hysteresis and a feedback (R1, C1) as timing elements. The capacitor, C1, charges and discharges to $2V^+/3$ and $V^+/3$ respectively. Because of this, the frequency of oscillation is, at least theoretically, independent from the power supply voltage. If the comparator swings to the supply rails, if the pull-up resistor is much smaller than the resistors R_h and if the propagation delay is negligible compared to the RC time constant, the oscillation frequency is:

$$f'_{osc} = \frac{0.72}{R1C1}$$

For LT111 or LT1011 type comparators, this holds for $f_{OSC} \leq 3kHz$. The circuit of Figure 28.24 is adequate to drive an LTC1062 tuned in the vicinity of 10Hz to 30Hz

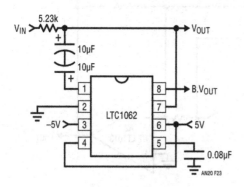

Figure 28.23 • A Low Frequency, 5Hz Filter Using Back to Back Solid Tantalum Capacitors

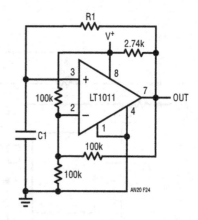

Figure 28.24 • A Low Frequency, Precision (R, C) Oscillator. For Bipolar ±5V Output Swing Refer the Ground Connection to –5V

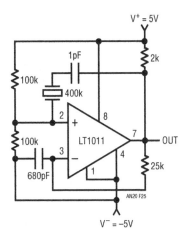

Figure 28.25 • Crystal Oscillator with 50% Duty Cycle

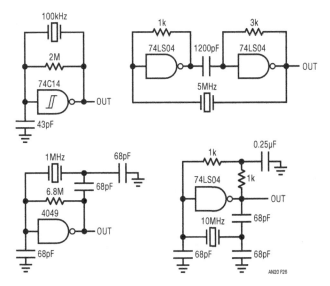

Figure 28.26 • Typical Gate Oscillators

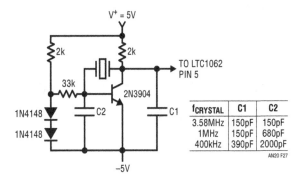

Figure 28.27 • Discrete, Low Cost Oscillator Using Parallel Type AT-CUT Crystal

$f_{CRYSTAL}$	C1	C2
3.58MHz	150pF	150pF
1MHz	150pF	680pF
400kHz	390pF	2000pF

2. The RC oscillator of Figure 28.24 can also be used up to 110kHz but the frequency of oscillation is about equal to 0.66/R1C1 and the duty cycle 60%. Again the major frequency drift component will be due to the drift of the R1C1. If the cutoff frequency of the filter should be made as temperature independent as possible, the (R • C) and (R1 • C1) products should also be made temperature independent. This can be achieved by choosing resistors and capacitors of nearly opposite temperature coefficients. For instance, TRW MTR-5/+120ppm/°C resistors can be used with −120ppm/°C±30ppm WESCO type 32-P capacitors.

3. Crystal oscillators: Figure 28.25 shows an LT1011 comparator biased in its linear mode and using a crystal to establish its resonant frequency. With this circuit we can achieve a few hundred kHz, temperature independent clock frequency with nearly 50% duty cycle. Many systems already have a crystal oscillator using digital gates as active elements, Figure 28.26. Their frequency, however, is usually above 1MHz and should be divided down before being applied to the LTC1062. Figure 28.27 shows an inexpensive discrete crystal oscillator using a single transistor as gain element. Its output can directly drive Pin 5 of the LTC1062 and its Pin 4, should they be converted to analog ground or negative supply to activate the internal divide by 2 or 4 of the circuit. This is necessary because the duty cycle at the collector of the crystal oscillator is not 50%.

Acknowledgement

For this chapter, the laboratory work was done by Guy Hoover whose meticulousness and contributions are greatly appreciated.

cutoff frequency. Also, when the input (R, C) of the LTC1062 and the (R1, C1) of the comparator have the same temperature coefficient, the cutoff frequency will drift but the passband error will be temperature independent since:

$$\frac{1}{2\pi RC} \cong \frac{f_C}{1.63} = \frac{0.72}{163 \bullet R1C1} \text{ or } (R1C1/RC) = 1/36$$

For C = 10C1, then R = 3.6R1, which yields a reasonable resistor and capacitor value spread.

Micropower circuits for signal conditioning

29

Jim Williams

Low power operation of electronic apparatus has become increasingly desirable. Medical, remote data acquisition, power monitoring and other applications are good candidates for battery driven, low power operation. Micropower analog circuits for transducer-based signal conditioning present a special class of problems. Although micropower ICs are available, the interconnection of these devices to form a functioning micropower circuit requires care. (See Box Sections A and C, "Some guidelines for micropower design and an example" and "Parasitic effects of test equipment on micropower circuits.") In particular, trade-offs between signal levels and power dissipation become painful when performance in the 10-bit to 12-bit area is desirable. Additionally, many transducers and analog signals produce inherently small outputs, making micropower

requirements complicate an already difficult situation. Despite the problems, design of such circuits is possible by combining high performance micropower ICs with appropriate circuit techniques.

Platinum RTD signal conditioner

Figure 29.1 shows a simple circuit for signal conditioning a platinum RTD. Correction for the platinum sensor's non-linear response is included. Accuracy is 0.25°C over a 2°C to 400°C sensed range. One side of the sensor is grounded, highly desirable for noise considerations. For a 2°C sensed temperature, current consumption is 250µA, increasing to 335µA for a 400°C sensed temperature.

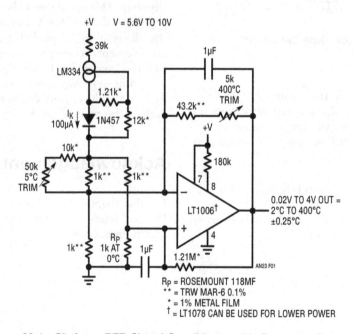

Figure 29.1 • Platinum RTD Signal Conditioner with Curvature Correction

Analog Circuit and System Design: A Tutorial Guide to Applications and Solutions. DOI: 10.1016/B978-0-12-385185-7.00029-9

The platinum sensor is placed in a current driven bridge with the 1k resistors. The LM334 current source drives the bridge and its associated resistors set a 100μA operating level. The diode provides temperature compensation (see LM334 data sheet). The 39k resistor deliberately sustains voltage drop, minimizing LM334 die temperature rise to ensure good temperature tracking with the diode. The 100μA current is split by the bridge. This light current saves power, but restricts the platinum sensor's output to about 200μV/°C. The circuit's 0.25°C accuracy specification requires the LT1006 low power precision op amp for stable gain. The LT1006 takes the signal differentially from the bridge to provide the circuit's output. Normally, the platinum sensor's slightly nonlinear response would cause several degrees error over the sensed temperature range. The 1.2M resistor gives slight positive feedback to correct for this. The amplifier's negative feedback path dominates, and the configuration is stable. The 1μF capacitors give a high frequency roll-off and the 180k resistor programs the LT1006 for 80μA quiescent current.

To calibrate this circuit, substitute a precision decade box (e.g., General Radio 1432) for R_P. Set the box to the 5°C value (1019.9Ω) and adjust the "5°C trim" for 0.05V output at the LT1006. Next, set the box for the 400°C value (2499.8Ω) and adjust the "400°C trim" for 4.000V output. Repeat this sequence until both points are fixed. The resistance values given are for a nominal 1000.0Ω (0°C) sensor. Sensors deviating from this nominal value can be used by factoring in the deviation from 1000.0Ω. This deviation, which is manufacturer-specified for each individual sensor, is an offset term due to winding tolerances during fabrication of the RTD. The gain slope of the platinum is primarily fixed by the purity of the material and is a very small error term.

Thermocouple signal conditioner

Figure 29.2 is another temperature sensing circuit, except the transducer is a thermocouple. Accuracy is within 1.5°C over a 0°C to 60°C sensed temperature and current consumption is about 125μA.

Thermocouples are inexpensive, have low impedances and feature self-generating outputs. They also produce low level outputs and require cold junction compensation, complicating signal conditioning. The bridge network, composed of the thermistor and R1-R4, provides cold junction compensation with the LT1004 acting as a voltage reference. The lithium battery noted allows the bridge to float and the thermocouple to be ground referred, eliminating the requirement for a differential amplifier For the battery specified, life will approach 10 years. This is a good way to avoid the additional power drain of a multi-amplifier differential stage. The LT1006 is set up with a gain scaled to produce the output shown and the 270k resistor programs it for low current drain. Note that this circuit requires no trims.

Sampled strain gauge signal conditioner

Strain gauge bridge-based transducers present a challenge where low power operation is needed. The 350Ω impedance combined with low signal outputs (typically 1mV to 3mV output per volt of drive) presents problems. Even with only 1V of drive, bridge consumption still approaches 3mA. Dropping drive to 100mV reduces current to acceptable levels, but precludes high accuracy operation due to the miniscule output available. In many situations, continuous transducer information is unnecessary and sampled operation is viable. Short sampling duty cycle permits high current bridge drive while maintaining low power operation. Figure 29.3 uses such a scheme to achieve dramatic power saving in a strain gauge bridge application (for a discussion of sampled operation considerations, see Box Section B, "Sampling techniques and components for micropower circuits").

In this circuit, Q2 is off when the "sample command" is low. Under these conditions only A4 and the CD4016

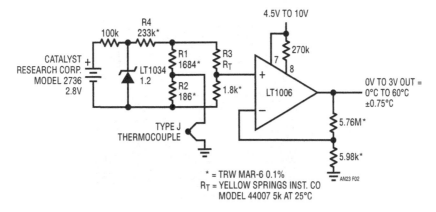

Figure 29.2 • Thermocouple Signal Conditioner with Cold Junction Compensation

receive power, and current drain is inside 125µA. When the sample command is pulsed high, Q2's collector (Trace A, Figure 29.4) goes high, providing power to all other circuit elements. The 10Ω-1µF RC at the LT1021 prevents the strain bridge from seeing a fast rise pulse which could cause long-term transducer degradation. The LT1021-5 reference output (Trace B) drives the strain bridge, and differential amplifier A1-A3's output appears at A2

(Trace C). Simultaneously, S1's switch control input (Trace D) ramps toward Q2's collector. At about one-half Q2's collector voltage (in this case just before mid-screen) S1 turns on, and A2's output is stored in C1. When the sample command drops low, Q2's collector falls, the bridge and its associated circuity shut down and S1 goes off. C1's stored value appears at gain scaled A4's output. The RC delay at S1's control input ensures glitch-free operation by

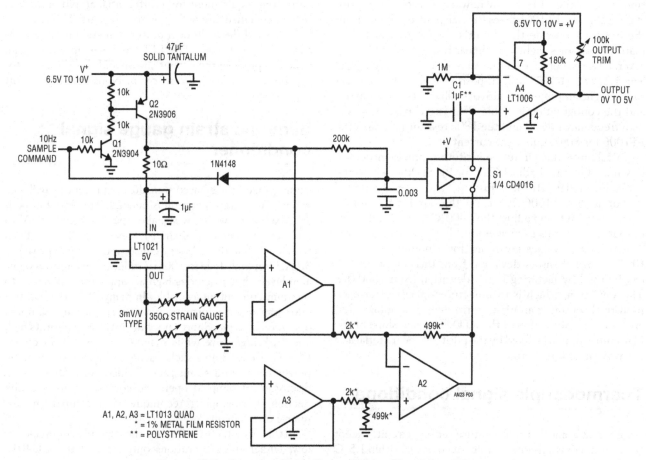

Figure 29.3 • Sampled Strain Gauge Bridge Signal Conditioner

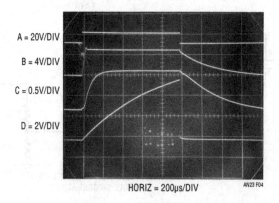

HORIZ = 200µs/DIV

Figure 29.4 • Waveforms for the Sampled Strain Gauge Signal Conditioner preventing C1 from updating until A2 has settled. During the 1ms sampling phase, supply current approaches 20mA, but a 10Hz sampling rate cuts effective drain below 200µA. Slower sampling rates will further reduce drain, but C1's droop rate (about 1mV/100ms) sets an accuracy constraint. The 10Hz rate provides adequate bandwidth for most transducers. For 3mV/V slope factor transducers, the gain trim shown allows calibration. It should be rescaled for other types. This circuit's effective current drain is about 300µA and A4's output is accurate enough for 12-bit systems.

Strobed operation strain gauge bridge signal conditioner

Figure 29.5's circuit also switches power to minimize strain bridge caused losses, but is not intended for continuously sampled operation. This circuit is designed to sit in the quiescent state for long periods with relatively brief on times. A typical application would be remote weight information in storage tanks where weekly readings are sufficient. This circuit has the advantage of not requiring a differential amplifier, despite the strain bridge's floating output. Additionally, it provides almost full rated drive to the strain bridge, enhancing accuracy. Quiescent current is about 150µA with on-state current typically 50mA.

With Q1's base unbiased, all circuitry is off except the LT1054 plus-to-minus voltage converter, which draws a 150µA quiescent current. When Q1's base is pulled low, its collector supplies power to A1 and A2. A1's output goes high, turning on the LT1054. The LT1054's output (Pin 5)

heads toward −5V and Q2 comes on, permitting bridge current to flow. To balance its inputs, A1 servo controls the LT1054 to force the bridge's midpoint to 0V

The bridge ends up with about 8V across it, requiring the 100mA capability LT1054 to sink about 24mA. The 0.02µF capacitor stabilizes the loop. The A1-LT1054 loop negative output sets the bridge's common mode voltage to zero, allowing A2 to take a simple single-ended measurement. The "output trim" scales the circuit for 3mV/V type strain bridge transducers, and the 100k-0.1µF combination provides noise filtering.

Thermistor signal conditioner for current loop application

4mA to 20mA "current loop" control is common in industrial environments. Circuitry used to modulate transducer data into this loop must operate well below the 4mA minimum current.

Figure 29.6 shows a complete 2-wire thermistor temperature transducer interface with a 4mA to 20mA output. Over a 0°C to 100°C range, accuracy is ±0.3°C and the circuit is current loop powered. No external supply is required. The LM134 current source absorbs the 40v input, preventing the LTC1040 from seeing too high a supply potential. It does this by fixing the current well below the 4mA loop minimum. The LTC1040 (detailed data on this device appears in Box Section B, "Sampling techniques and components for micropower circuits") senses the YSI thermistor network output and forces this voltage across the output resistor to set total circuit

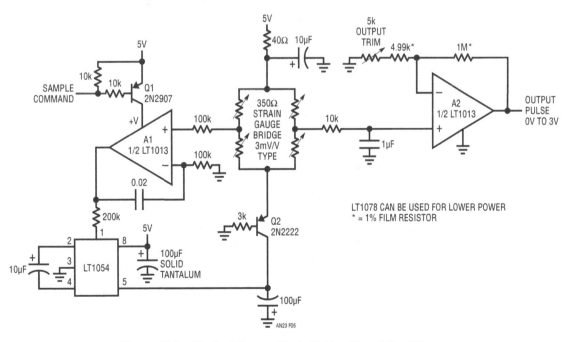

Figure 29.5 • Strobed Power Strain Bridge Signal Conditioner

current. Current is adjusted by varying the gate voltage on the 2N6657 FET. Note that the comparator output operates in pulse-width modulation mode, with the FET gate voltage filtered to DC by the 1M-1µF combination. An important LTC1040 feature is that very little current, on the order of nanoamperes, flows from the V^- supply. This allows the V^- supply to be connected to ground with negligible current error in the output sensing resistor. The differential input of the LTC1040 can sense the current through R_{OUT} because its common mode range includes the V^- supply. Trims shown are for 0°C and 100°C and are made by exposing the thermistor to those temperatures or by electrically simulating the conditions (see manufacturer's data sheet).

Microampere drain wall thermostat

Figure 29.7 shows a battery-powered thermostat using the LTC1041 (see Box Section B for details on this device). Temperature is sensed using a thermistor connected in a bridge with a potentiometer to set the desired temperature.

The bridge is not driven from the battery but from Pin 7 on the LTC1041. Pin 7 is the pulsed power (V_{P-P}) output and turns on only while the LTC1041 is sampling the inputs. With this pulse technique, average system power consumption is quite small. In this application the total

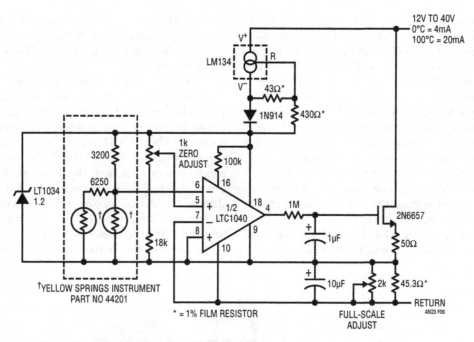

Figure 29.6 • Thermistor-Based Current Loop Signal Conditioner

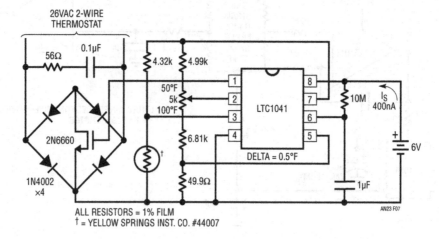

Figure 29.7 • Wall Type Thermostat

system current is below 1μA! This is far less than the self discharge rate of the battery, meaning battery life is shelf life limited. A lithium battery will run this circuit for 10 to 20 years.

An external RC network sets the sampling frequency. When an internal sampling cycle is initiated, power is turned on to the comparators and to the V_{P-P} output. The analog inputs are sampled and the resultant outputs are stored in CMOS latches. Power is then switched off although the outputs are maintained. The unclocked CMOS logic consumes almost no DC current. The sampling process takes approximately 80μs. During this 80μs interval, the LTC1041 draws typically 1.7mA of current at V^+=6V Because the sample rate is low, average power is extremely small.

The low sample rate is adequate for a thermostat because of the low rate of change normally associated with temperature.

A power MOSFET in a diode bridge switches 26VAC to the heater control circuitry. The MOSFET is a voltage controlled device with no DC current required from the battery.

The voltage from DELTA (Pin 5) to GND (Pin 4) sets the dead-band. Dead-band is desirable to prevent excessive heater cycling. The dead-band equals two times DELTA and is independent of both VIN (Pin 3) and SET POINT (Pin 2). This means that as the SET POINT is varied, the dead-band is fixed at two times DELTA. Conversely, as dead-band is varied, SET POINT does not move.

Freezer alarm

Figure 29.8 shows a very simple configuration for a freezer alarm. Such circuits are used in industrial and home freezers as well as refrigerated trucks and rail cars. The LTC1042 is a sampled operation window comparator (for details on this device see Box Section B). The 10M-0.05μF combination sets a sample rate of 1Hz, and the bridge values program the internal window comparator for the outputs shown. For normal freezer operation, Pin 1 is high and Pin 6 is low. Overtemperature reverses this state and can trigger an alarm. Circuit current consumption is about 80μA.

12-Bit A/D converter

Integrating A/D converters with low power consumption are available. Although capable of 12-bit measurements, they are quite slow, typically in the 100ms range. Higher speeds require a successive approximation (SAR) approach. No commercially produced 12-bit SAR converter features micropower (e.g., below 1mA) capability at the time of writing. Figure 29.9's design converts in 300μs, while consuming only 890μA.

Conceptually, this design is a straightforward SAR type converter, although some special measures are needed to achieve low power operation. The SAR chip and the DAC are arranged in the standard fashion, with C1 closing a loop. Normally, CMOS DACs are not used for SAR applications because their output capacitance slows operation. In this case, the CMOS DAC's low power consumption is attractive and speed is traded away. This is not too great a penalty, because micropower comparator C1 is a good speed match for the DAC specified. A limitation with CMOS DACs is that their outputs must terminate into 0V. This mandates a current summing comparison, meaning the reference must be of opposite polarity to the input. Since most micropower systems run from single-sided positive rails, it is unrealistic to expect the user to supply the A/D with a negative input. To be readily usable, the converter should accept positive inputs and derive a negative reference internally. This issue is addressed by C2 and the LTC1044 plus-to-minus voltage converter, which form a negative reference.

C2, compensated as an op amp, servo controls the LTC1044 via the boost transistor. The LTC1044's negative output is fed back to C2's input, closing a regulation loop.

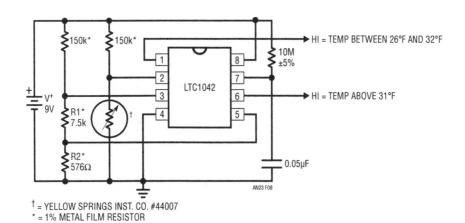

† = YELLOW SPRINGS INST. CO. #44007
* = 1% METAL FILM RESISTOR

Figure 29.8 • Freezer Alarm

Scaled current summing from the output and the LT1034 forces a 5.000V output. The Schottky diode prevents possible summing point negative overdrive during start-up. The choice of 5V for a reference maintains reasonable LSB overdrive for C1, but accounts for over half the circuit's current requirement. This limitation is set by the DAC's relatively low input impedance. Dropping the reference voltage would save significant power, but would also reduce LSB size below a millivolt. This would cause comparator offset and gain to become significant error sources.

Although the DAC has no negative supply, it can accept the negative reference because its thin film resistors are not intrinsic to the monolithic structure. Ground referred C1

cannot accept any negative voltages, however, and is Schottky clamped.

Performance includes a typical tempco of 30ppm/°C, 300μs conversion time, 890μA current consumption and an accuracy of ±2 LSBs. Trimming involves adjusting the 100k potentiometer for exactly −5V at V_{REF}. The DAC's internal feedback resistor serves as the input. Figure 29.10 shows operating waveforms. Trace A is the clock. Trace B is the convert command. The SAR is cleared on Trace B's falling edge and conversion commences on the rise. During conversion, C1's input (Trace C) sequentially converges towards zero. When conversion is complete, the status line (Trace D) drops low.

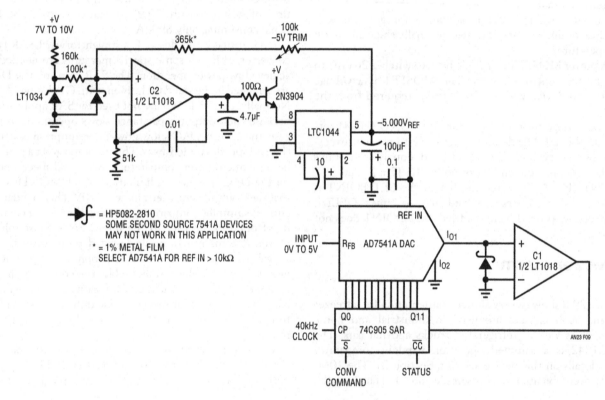

Figure 29.9 • 12-Bit, 300μs A/D

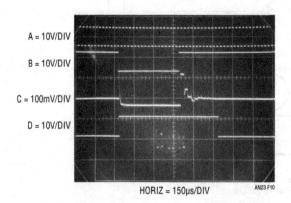

Figure 29.10 • Micropower SAR A/D Waveforms

10-Bit, 100μA A/D converter

Figure 29.11's A/D has less resolution than the previous circuit, but requires only 100μA. The design consists of a current source, an integrating capacitor, a comparator and some logic elements. When a pulse is applied to the convert command input (Trace A, Figure 29.12), the paralleled 74C906 sections reset the 0.075μF capacitor to zero (Trace B). Simultaneously, 74C14 inverter A goes low, biasing the 2N3809 current source on. During this interval the current source stabilizes, delivering its output to ground via the paralleled 74C906 sections. On the falling edge of the convert command pulse the 0.075μF capacitor begins to charge linearly. When the ramp voltage equals the input, C1 switches. Inverter A goes high, shutting off the current source. A small current is bled through the 10M diode connection to keep the ramp charging, but at a greatly reduced rate. This ensures overdrive for C1, but minimizes current source on-time, saving power. C1's output, a pulse (Trace C) width, is directly dependent on the value of Ex. This pulse width gates C2's clock output via the 74C00 configuration. The 74C00s also gate out the portion of C1's output due to the convert command pulse. Thus, the clock pulse bursts appearing at the output (Trace D) are proportional to Ex. For the arrangement shown, 1024 pulses appear for a 5V full-scale input. The current source scaling resistor and ramp capacitor specified provide good temperature compensation because of their opposing thermal coefficients. The circuit will typically hold ±1LSB accuracy over 0°C to 70°C with an additional ±1LSB due to the asynchronous relationship between the clock and the conversion sequence. If the conversion sequence is synchronized to the clock, the ±1LSB asynchronous limitation is removed, and total error falls to ±1LSB over 0°C to 70°C. The flop-flop shown in dashed lines permits such

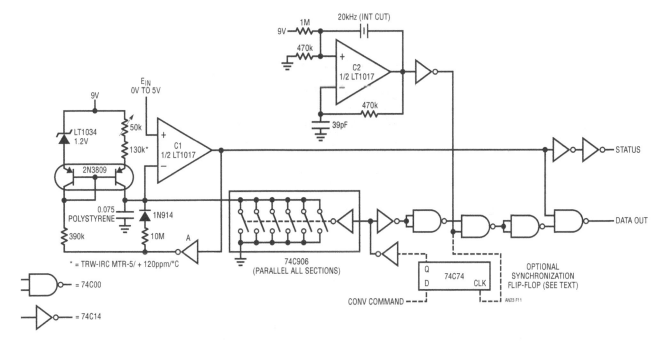

Figure 29.11 • 10-Bit, 100μA A/D

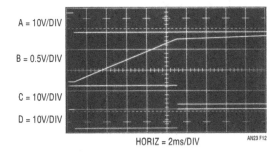

Figure 29.12 • Waveforms for the 100μA A/D

synchronization. Conversion rate varies with input. At tenth-scale 150Hz is possible, decreasing to 20Hz at full-scale.

Power consumption of the A/D is extremely low, due to the CMOS logic elements and the LT1017 comparator. Quiescent (E_{IN} = 0V) current is 100µA at V_{SUPPLY} = 9V decreasing to 80µA for V_{SUPPLY} = 7V. Because current source on-time varies with input, power consumption also varies. For E_{IN} = 5V current consumption rises to 125µA for E_{SUPPLY} = 9V and 105µA at E_{SUPPLY} = 7V. Additional power savings are possible by shutting off the current source during capacitor reset, but accuracy suffers due to current source settling time requirements. The 0.075µF capacitor's accumulated charge is thrown away at each reset. A smaller capacitor would help, but C1's bias currents would introduce significant error

Turning off the current source after C1 switches saves significant power. Figure 29.13, taken at a 25mV input, shows the ramp zero reset and the clean switching. When the current source switches off, the ramp slope decreases

but continues to move upward, ensuring overdrive. The 10M diode pair provides the charge, but less than a micro-ampere is lost.

20µs sample-hold

Figure 29.14 is a companion sample-hold for the SAR A/D. Acquisition time is 20µs, with low power operation (see Figure 29.14 table). This circuit takes full advantage of the programming pin on the LT1006 op amp to maximize speed-power performance. When the sample command (Trace A, Figure 29.15) is given, the CD4066 switches close. S1 and S2 allow A1's output (Trace B) to charge the capacitor (Trace C is capacitor current). Simulta-neously S3 and S4 close, raising the op amp's internal bias network. This puts both amplifiers into hyperdrive, boosting slew rate to speed acquisition time. A2 (Trace D) is seen to settle cleanly to ImV in 20µs.

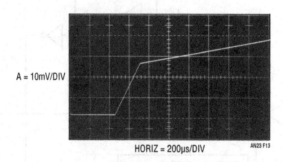

A = 10mV/DIV

HORIZ = 200µs/DIV   AN23 F13

Figure 29.13 • Detail of the Switched Slope Capacitor Charging

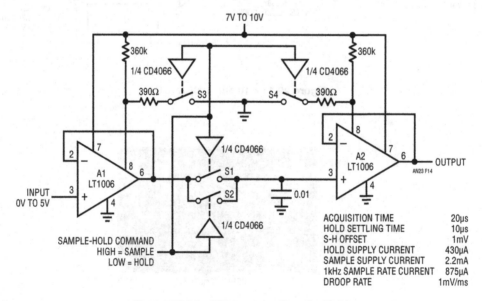

Figure 29.14 • Micropower Sample-Hold

When the sample command goes low, all switches go off, A2 follows the voltage stored on the capacitor, and supply current drops by a factor of five (see Figure 29.14 table). In normal operation, sample time is short compared to hold, and current consumption is low. The 360k resistors set the circuit's hold mode quiescent current at the value noted in the table.

10kHz voltage-to-frequency converter

Figure 29.16, another data converter, is a voltage-to-frequency converter. A 0V to 5V input produces a 0kHz to 10kHz output, with a linearity of 0.02%. Gain drift is

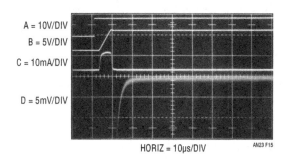

Figure 29.15 • Figure 29.14's Waveforms

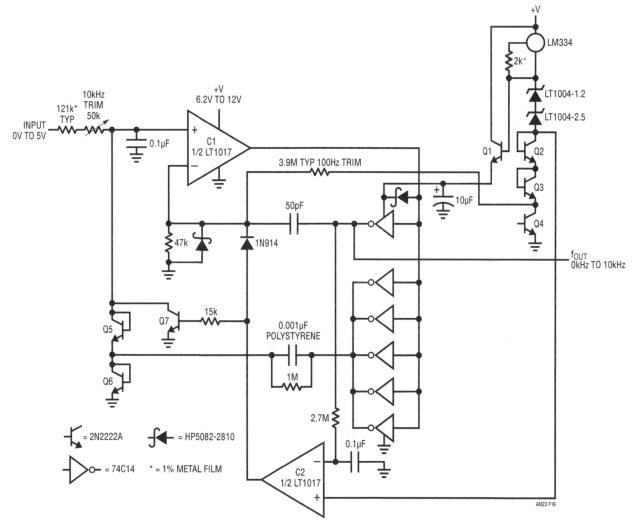

Figure 29.16 • Micropower 10kHz V→F Converter

40ppm/°C. Maximum current consumption is only 145μA, far below currently available units.

The evolution of this circuit is described in Box Section A, "Some guidelines for micropower design and an example". To understand circuit operation, assume C1's positive input is slightly below its negative input (C2's output is low). The input voltage causes a positive going ramp at C1's positive input (Trace A, Figure 29.17). C1's output is low, biasing the CMOS inverter outputs high. This allows current to flow from Q1's emitter, through the inverter supply pin to the 0.001μF capacitor. The 10μF capacitor provides high frequency bypass, maintaining low impedance at Q1's emitter. Diode connected Q6 provides a path to ground. The voltage to which the 0.001μF unit charges is a function of Q1's emitter potential and Q6's drop. When the ramp at C1's positive input goes high enough, C1's output goes high (Trace B) and the inverters switch low (Trace C). The Schottky clamp prevents CMOS inverter input overdrive. This action pulls current from C1's positive input capacitor via the Q5-0.001μF route (Trace D). This current removal resets C1's positive input ramp to a potential slightly below ground, forcing C1's output to go low. The 50pF capacitor connected to the circuit output furnishes AC positive feedback, ensuring that C1's output remains positive long enough for a complete discharge of the 0.001μF capacitor. The Schottky diode prevents C1's input from being driven outside its negative common mode limit. When the 50pF unit's feedback decays, C1 again switches low and the entire cycle repeats. The oscillation frequency depends directly on the input voltage derived current.

Q1's emitter voltage must be carefully controlled to get low drift. Q3 and Q4 temperature compensate Q5 and Q6 while Q2 compensates Q1's V_{BE}. The two LT1004s are the actual voltage reference and the LM334 current source provides 35μA bias to the stack. The current drive provides excellent supply immunity (better than 40ppm/V) and also aids circuit temperature coefficient. It does this by utilizing the LM334's 0.3%/°C tempco to slightly temperature modulate the voltage drop in the Q2-Q4 trio. This correction's sign and magnitude directly oppose that of the −120ppm/°C 0.001μF polystyrene capacitor, aiding overall circuit stability.

The Q1 emitter-follower delivers charge to the 0.001μF capacitor efficiently. Both base and collector current end up in the capacitor. The paralleled CMOS inverters provide low loss SPDT reference switching without significant drive losses. The 0.001μF capacitor, as small as accuracy permits, draws only small transient currents during its charge and discharge cycles. The 50pF-47k positive feedback combination draws insignificantly small switching currents. Figure 29.18, a plot of supply current vs operating frequency, reflects the low power design. At zero frequency, the LT1017's quiescent current and the 35μA reference stack bias accounts for all current drain. There are no other paths for loss. As frequency scales up, the charge-discharge cycle of the 0.001μF capacitor introduces the 7μA/kHz increase shown. A smaller value capacitor would cut power, but the effects of stray capacitance, charge imbalance in the 74C14, and LT1017 bias currents would introduce accuracy errors.

Circuit start-up or overdrive can cause the circuit's AC-coupled feedback to latch. If this occurs, C1's output goes high. C2, detecting this via the inverters and the 2.7M-0.1μF lag, also goes high. This lifts C1's negative input and grounds the positive input with Q7, initiating normal circuit action.

Because the charge pump is directly coupled to C1's output, response is fast. Figure 29.19 shows the output (Trace B) settling within one cycle for a fast input step (Trace A).

To calibrate this circuit, apply 50mV and select the value at C1's input for a 100Hz output. Then, apply 5V and trim the input potentiometer for a 10kHz output.

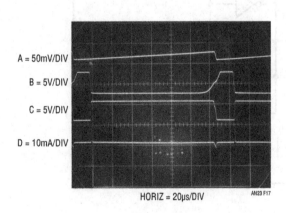

A = 50mV/DIV

B = 5V/DIV

C = 5V/DIV

D = 10mA/DIV

HORIZ = 20μs/DIV   AN23 F17

Figure 29.17 • Figure 29.16's Waveforms

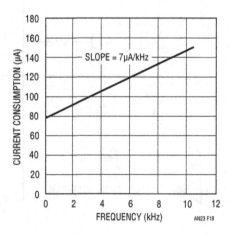

SLOPE = 7μA/kHz

AN23 F18

Figure 29.18 • Current Consumption vs Frequency for Figure 29.16

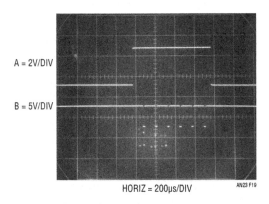

A = 2V/DIV

B = 5V/DIV

HORIZ = 200μs/DIV   AN23 F19

Figure 29.19 • Figure 29.16's Step Response

An evolutionary history of this design appears in Box Section A, "Some guidelines for micropower design and an example".

A nice day at the San Francisco Zoo with Celia Moreno M.D., instrumental in arriving at the final configuration, is happily acknowledged.

1MHz voltage-to-frequency converter

Figure 29.20 is also a V→F converter, but runs at 1MHz full-scale. Quiescent current is 90μA, ascending linearly to 360μA at 1MHz output. Obtaining higher operating

frequency requires trade-offs in power consumption and step response performance. Linearity is 0.02% over a 100Hz to 1MHz range, drift about 50ppm/°C and step response inside 350ms to full-scale.

This circuit has similarities to Figure 29.16, although operation is somewhat different. An input causes A1 to swing towards ground, biasing Q8. Q8's collector ramps (Trace A, Figure 29.21) as it charges the 3pF capacitor plus stray capacitance associated with Q7 and the 74C14 Schmitt input connected to the node. When the ramp hits the Schmitt's threshold its output (Trace B) goes low, turning on diode connected Q7. Q7's path discharges the node capacitances, forcing ramp reset. The 74C14 returns high, and oscillation commences. The 74C14 also drives the CD4024 divider, and serves as the circuit's output. The divider's ÷128 output (Trace C) controls a reference charge pump arrangement similar to Figure 29.16's. A2 furnishes a buffered reference. The 500pF capacitor is alternately charged and discharged by the LTC201 switch sections. The charge increments pulled through S1 continually force A1's 2μF capacitor to zero (Trace D), balancing the input derived current. The 0.022μF capacitor at the D1-D2 LTC201 node eliminates excessive differentiated response, preventing spurious modes. This action closes a loop around A1, and it servo controls the Q7, Q8, 74C14 oscillator to run at whatever frequency is required to maintain its negative input at zero. This servo behavior eliminates oscillator drift and nonlinearity as error terms, allowing the performance specifications noted. The 0.33μF capacitor at A1 stabilizes

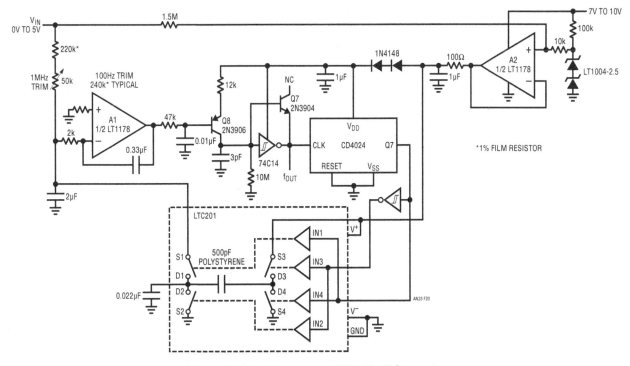

Figure 29.20 • Micropower 1MHz V→F Converter

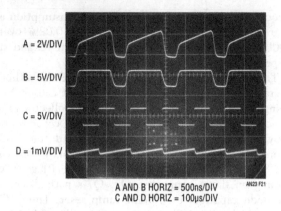

A = 2V/DIV

B = 5V/DIV

C = 5V/DIV

D = 1mV/DIV

A AND B HORIZ = 500ns/DIV
C AND D HORIZ = 100μs/DIV

AN23 F21

Figure 29.21 • Figure 29.20's Waveforms

the loop. This capacitor accounts for the circuit's 350ms settling time.

The resistor from the input to A2 sums a small input related voltage to the reference, improving linearity. The 10M resistor at Q8's collector deliberately introduces leakage to ground, dominating all node leakages. This ensures low frequency operation by forcing Q8 to source current to maintain oscillations.

The circuit's current drain, while low, is larger than Figure 29.16's. The increase is primarily due to high frequency oscillator and divider operation. The series diodes in the oscillator-divider supply line lower supply voltage, decreasing current consumption. Oscillator current is also heavily influenced by the capacitance and swing at Q8's collector. The swing is fixed by the 74C14 thresholds. Capacitance has been chosen at the lowest possible value commensurate with desired low frequency operation.

To trim this circuit, put in 500μV and select the indicated value at A1's positive input for 100Hz out. Then, put in 5V and trim the 50k potentiometer for 1MHz out. Repeat this procedure until both points are fixed.

Switching regulator

No discussion of micropower circuitry is complete without mention of switching regulators. Often, battery voltages must be efficiently converted to different potentials to meet circuit requirements. Figure 29.22 shows a micropower buck type switching regulator with a quiescent drain of 70μA and 20mA output current capability. When the output voltage drops (Trace A, Figure 29.23) C1's negative input also falls, causing its output (Trace B) to rise. This

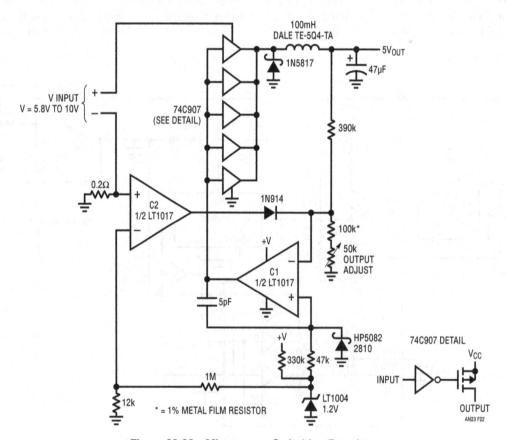

Figure 29.22 • Micropower Switching Regulator

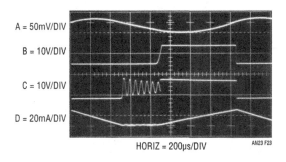

A = 50mV/DIV

B = 10V/DIV

C = 10V/DIV

D = 20mA/DIV

HORIZ = 200µs/DIV   AN23 F23

Figure 29.23 • Figure 29.22's Waveforms

turns on the paralleled 74C907 open-source buffers, and their outputs (Trace C) go high. Current ramps up through the inductor, maintaining the regulator output. When output voltage rises a small amount, C1's output returns low and the cycle repeats. This action maintains regulator output despite line and load changes. The LT1004 serves as a reference and the 5pF capacitor ensures clean switching at C1. The 2810 Schottky diode prevents negative overdrives due to the 5pF unit's differentiated response; the 1N5817 is a catch diode, preventing excessive inductor caused negative voltages.

This circuit's low quiescent drain is due to the LT1017's small operating currents and the 74C907's low input drive requirements. Circuit resistor values are kept high to save current. C2 shuts down the regulator when output current exceeds 50mA. It does this by comparing the potential across the 0.2Ω shunt to a resistively divided portion of the LT1004 reference. Excessive current drain trips C2 high, forcing C1's negative input high. This removes drive from the 74C907 buffers, shutting down the regulator

Utilization of a CMOS buffer as a pass switch for a switching regulator is somewhat unusual, but performance is quite good. Figure 29.24 plots efficiency vs output current at two input voltages. Efficiencies above 90% are possible, with output current to 20mA depending on input.

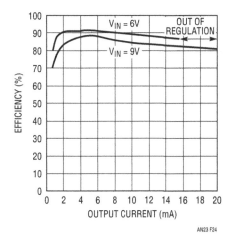

AN23 F24

Figure 29.24 • Figure 29.22's Efficiency vs Output Current

Post regulated micropower switching regulator

Figure 29.25 is another buck type switching regulator, but features a low loss linear post regulator, quiescent current of 40µA and 50mA output capacity. The LT1020 linear regulator provides lower noise than a straight switching approach. Additionally, it offers internal current limiting and contains an auxiliary comparator which is used to form the switching regulator.

The switching loop is similar to Figure 29.22's circuit. A drop at the switching regulator's output (Pin 3 of the LT1020 regulator; Trace A, Figure 29.26) causes the LT1020's comparator to go high. The 74C04 inverter chain switches, biasing the P-channel MOSFET switch's grid (Trace B). The MOSFET comes on (Trace C), delivering current to the inductor (Trace D). When the voltage at the inductor-220µF junction goes high enough (Trace A), the comparator switches high, turning off MOSFET current flow. This switching loop regulates the LT1020's input pin at a value set by the resistor divider in the comparator's negative input and the LT1020's 2.5V reference. The 680pF capacitor stabilizes the loop and the 1N5817 is the catch diode. The 270pF capacitor aids comparator switching and the 2810 diode prevents negative overdrives.

The low dropout LT1020 linear regulator smooths the switched output. Output voltage is set with the feedback pin associated divider. A potential problem with this circuit is start-up. The switching loop supplies the LT1020's input but relies on the LT1020's internal comparator to function. Because of this, the circuit needs a start-up mechanism. The 74C04 inverters serve this function. When power is applied, the LT1020 sees no input, but the inverters do. The 220k path lifts the first inverter high, causing the chain to switch, biasing the MOSFET and starting the circuit. The inverter's rail-to-rail swing also provides ideal MOSFET grid drive.

Even though this circuit's 40µA quiescent current is lower than Figure 29.22's, it can source more current. The extremely small quiescent current is due to the low LT1020 drain and the MOS elements. Figure 29.27 plots efficiency vs output current for two LT1020 input-output differential voltages. Efficiency exceeding 80% is possible, with outputs to 50mA available.

Figures 29.28 and 29.29 show two other LT1020 micropower regulator-based circuits. In many processor-based systems it is desirable to monitor or control the power-down sequence. Figure 29.28 produces a logical "1" output when the regulator output begins to drop out (e.g., battery is low). Here, the regulator is programmed for a 5V output with the 1M feedback resistors. The 0.001µF capacitor provides frequency compensation. The LT1020's internal comparator senses the difference between the chip's 2.5V reference and a small portion of the IC's pass transistor current (supplied at Pin 13). At the edge of dropout, the LT1020's pass transistor goes towards

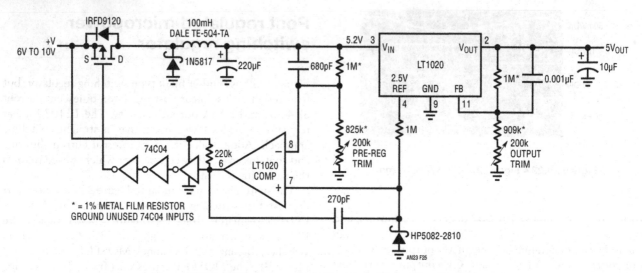

Figure 29.25 • Micropower Post-Regulated Switching Regulator

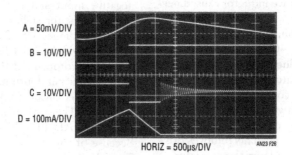

Figure 29.26 • Figure 29.25's Waveforms

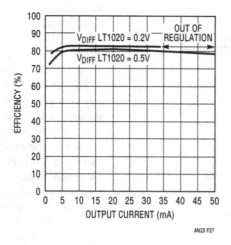

Figure 29.27 • Figure 29.25's Efficiency vs Output Current

saturation, raising Pin 13's voltage. This trips the comparator, and its output goes high. This signal can be used to alert a processor that power is about to go down.

Figure 29.29 is similar, except that power is turned completely off when dropout begins to occur, preventing unregulated supply conditions. The comparator feedback is arranged for a hysteresis type response. Although the output turns off at dropout, it will not turn on until:

$$\text{Turn on} = \frac{V_{IN} \cdot R2}{R1 + R2} = 2.5V$$

This prevents battery "creep back" from causing oscillation.

Figure 29.30 shows a simple way to shut the LT1020 down. In this state it draws only 40µA. The logic signal forces the feedback pin above the internal 2.5V reference, and all drive is removed from the output transistor.

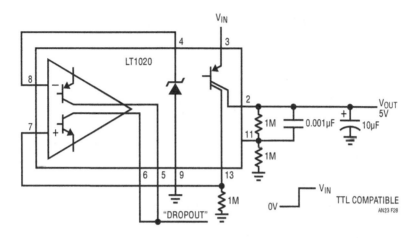

Figure 29.28 • Regulator with Logic Output on Dropout

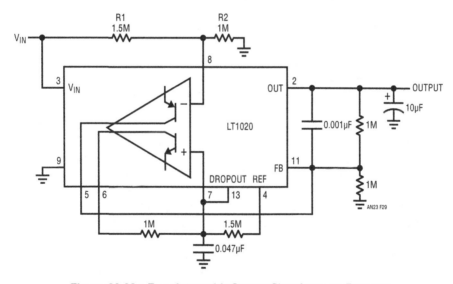

Figure 29.29 • Regulator with Output Shutdown on Dropout

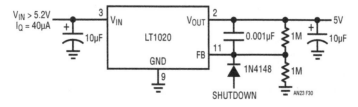

Figure 29.30 • LT1020 Shutdown

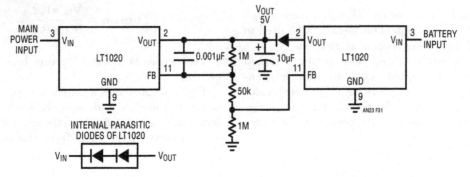

Figure 29.31 • Battery Backup Regulator

Figure 29.31 shows a low loss way to implement a "glitchless" memory battery backup. During line-powered operation, the right LT1020 does the work. The feedback string is arranged so that the left LT1020 does not conduct under line-powered conditions. When the line goes down, the associated LT1020 begins to go off, allowing the battery-driven regulator to turn on, maintaining the load.

Box Section A
Some guidelines for micropower design and an example

As with all engineering, micropower circuitry requires attention to detail, awareness of trade-offs and an opportunistic bent towards achieving the design goal.

The most obvious way to save power is to choose components which require little energy. Additional savings require more effort.

Circuits should be examined in terms of current flow. Consider such flow in all DC and AC paths. For example, do DC base currents go where they can do some useful work, or are they thrown away? Try to keep AC signal swings down, particularly if capacitors (parasitic or intended) must be continually charged and discharged.

Examine the circuit for areas where power strobing may be allowable.

Consider quiescent vs dynamic power requirements of components to avoid unpleasant surprises. Data sheets usually specify quiescent power because the manufacturer doesn't know what the user's circuit conditions are. For example, everyone "knows" that "MOS devices draw no current." Unfortunately, Mother Nature dictates that as frequency and signal swings go up, the capacitances associated with MOS devices begin to require more power. It is often a mistake to automatically associate low power operation with a process technology. While it's likely that CMOS will provide lower power

operation for a given function than 12AX7s, a bipolar approach may be even better. Consider individual situations on the basis of their specific requirements before committing to a technology. Very often, circuits require several technologies (e.g., CMOS, bipolar and discrete) for best results.

Usually, achieving low power operation requires performance trade-offs. Minimizing signal swings and current saves power, but moves circuit operation closer to the noise floor. Offsets, drift, bias currents and noise become increasingly significant error factors as signal amplitudes are constricted to save power. This is a fundamental trade-off and must be carefully considered. Circuits employing power strobing can sometimes get around this problem by utilizing low duty cycles. Text Figure 29.3 uses this technique to achieve dramatic power savings in a circuit with an on-state drain approaching 20mA (see also Box Section B, "Sampling techniques and components for micropower circuits").

Text Figure 29.16, a voltage-to-frequency converter, furnishes an example of the evolution of a low power design. Design goals included a 10kHz maximum output, fast step response, linearity inside 0.05% and a maximum supply current of 150µA. Other specifications appear in the text.

Figure A1 shows an early version of this circuit. Operation is similar to the text described for Figure 29.16, but a brief description follows. When the input current-derived ramp at C1's negative input crosses zero, C1's output drops low, pulling charge through C1. This forces the negative input below zero. C2 provides positive feedback, allowing a complete discharge for C1. When C2 decays, C1A's output goes high, clamping at the level set by D1, D2 and V_{REF}. C1 receives charge and recycling occurs when C1A's negative input again arrives at zero. The frequency of this action is related to the input voltage. Diodes D3 and D4 provide steering, and are temperature compensated by D1 and D2.

C1A's sink saturation voltage is uncompensated, but small. C1B is a start-up loop.

Although the LT1017 and LT1034 have low operating currents, this circuit pulls almost 400μA. The AC current paths include C1's charge-discharge cycle, and C2's branch. The DC path through D2 and V_{REF} is particularly costly. C1's charging must occur quickly enough for 10kHz operation, meaning the clamp seen by C1A's output must have low impedance at this frequency. C3 helps, but significant current still must come from somewhere to keep impedance low. C1A's current-limited output cannot do the job unaided, and the resistor from the supply is required. Even if C1A could supply the necessary current, V_{REF}'s settling time would be an issue. Dropping C1's value will reduce impedance

requirements proportionally, and would seem to solve the problem. Unfortunately, such reduction magnifies the effects of stray capacitance at the D3-D4 junction. It also mandates increasing R_{IN}'s value to keep scale factor constant. This lowers operating currents at C1A's negative input, making bias current and offset more significant error sources.

Figure A2 shows an initial attempt at dealing with these issues. This scheme is similar to Figure A1, except that Q1 and Q2 appear. V_{REF} receives switched bias via Q1, instead of being on all the time. Q2 provides the sink path for C1. These transistors invert C1A's output, so its input pin assignments are exchanged. R1 provides a light current from the supply, improving reference settling time. This arrangement decreases supply current to about 300μA, a significant improvement. Several problems do exist, however. Q1's switched operation is really effective only at higher frequencies. In the lower ranges, C1A's output is low most of the time, biasing Q1 on and wasting power. Additionally, when C1A's output switches, Q1 and Q2 simultaneously conduct during the transition, effectively shunting R2 across the supply. Finally, the base currents of both transistors flow to ground and are lost. The basic temperature compensation is as before, except that Q2's saturation term replaces the comparator's.

Figure A3 is better. Q1 is gone, Q2 remains but Q3, Q4 and Q5 have been added. V_{REF} and its associated diodes are biased from R1. Q3, an emitter-follower, is used to source current to C1. Q4 temperature compensates Q3's V_{BE}, and Q5 switches Q3.

This method has some distinct advantages. The V_{REF} string can operate at greatly reduced current because of Q3's current gain. Also, Figure A2's simultaneous conduction problem is largely alleviated because Q5 and Q2 are switched at the same

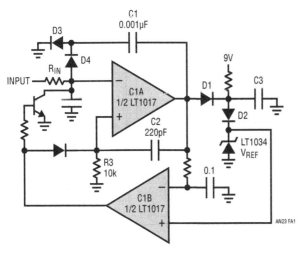

Figure A1

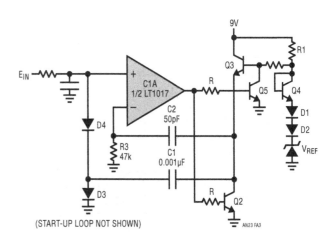

Figure A2   **Figure A3**

voltage threshold out of C1A. Q3's base and emitter currents are delivered to C1. Q5's currents are wasted, although they are much smaller than Q3's. Q2's small base current is also lost. The values for C2 and R3 have been changed. The time constant is the same, but some current reduction occurs due to R3's increase.

If C1 cannot be reduced for performance reasons, then its AC currents cannot be avoided. This leaves only the aforementioned Q5 and Q2 currents as significant wasted terms, along with R3's now smaller loss. Current drain for this circuit is about 200µA maximum. Text Figure 29.16's circuit is very similar; but eliminates Q5 and Q2's losses to achieve maximum operating current below 150µA with quiescent current under 80µA. Some other refinements are included, but the circuit is the final iteration of the three versions shown here. A complete description of Figure 29.16 appears in the text.

Box Section B
Sampling techniques and components for micropower circuits

The best way to get low power circuit characteristics is to turn off the power. While there are some obvious problems with this approach, it does point a way towards minimizing power consumption. In many applications continuous circuit power is not necessary. If bandwidth requirements are low, sampling techniques offer a simple way to save power. With low duty cycles, instantaneous current can be relatively high while average drain remains low. When considering a sampled approach some issues should be examined. The required circuit bandwidth dictates the minimum sampling frequency in accordance with Nyquist criteria. The sampling interval's duration is determined by circuit settling time to the required accuracy. This settling time should be considered for all circuit elements (transducers, ICs and discrete components) singularly and together. Additionally, effects of sampled operation on component life and operating characteristics should be examined. This is particularly the case for transducers, which may be designed and tested under DC operating conditions.

Once these issues have been addressed, components can be selected. The LTC1040, LTC1041 and LTC1042 have been specifically designed for sampled operation. Figure B1 details the LTC1040, dual micropower comparator. Its programmable internal oscillator sets the sampling rate with a sampling interval lasting 80µs. The V_{P-P} output supplies power during the sampling interval, allowing drive for external circuitry or transducers. Note that the input

common mode range includes both rails. Figure B2 plots supply current vs sampling frequency.

A related device is similar, but dedicated to "bang-bang" on-off type servo loops. The LTC1041 appears in Figure B3. Servo SET POINT and DELTA are controllable from the inputs. The associated diagram (Figure B4) graphically defines operation. Operating current is similar to the LTC1040.

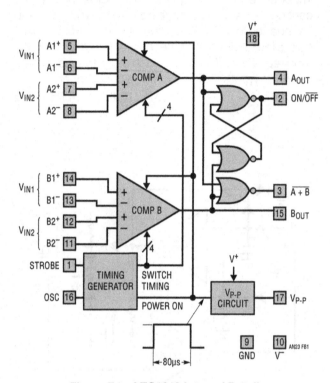

Figure B1 • LTC1040 Internal Details

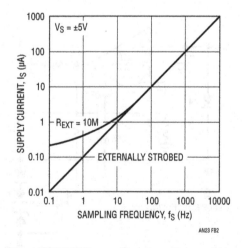

Figure B2 • LTC1040 Power Consumption vs Sampling Frequency

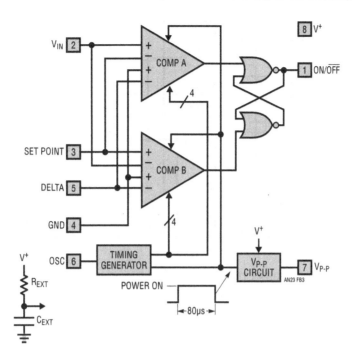

Figure B3 • LTC1041 Details

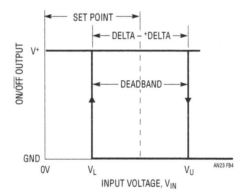

Figure B4 • LTC1041 Operation Diagram

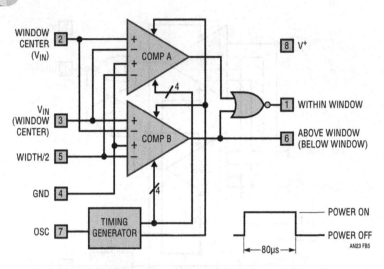

Figure B5 • LTC1042 "Window" Comparator Details

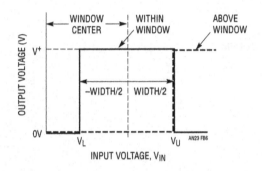

Figure B6 • LTC1042 Operating Diagram

A final device, the LTC1042, is also similar but is set up as a window comparator. Its internals appear in Figure B5 and the graphic operation description is shown in Figure B6. Operating current, input range and sampling characteristics are similar to the LTC1040 and LTC1041.

Box Section C
Parasitic effects of test equipment on micropower circuits

The energy absorbed by test equipment connections to micropower circuits can be significant. Under normal circumstances test equipment and probes have negligible power drain, but microampere level operating currents mandate care. Test instrumentation should be regarded as an integral part of the circuit. DC and AC loading and parasitic effects must be kept in mind to avoid unpleasant surprises. Such instrument connection errors can make the circuit under test look unfairly bad or good.

The DC resistance of oscilloscope probes varies from hundreds of ohms (1x types) to 10M (10x), with some 10x types as low as 1M. Contrary to some expectations, FET probes do not have high input resistance—some types are as low as 100k, although most are about 10M. The DC loading of a 10x 1M probe could introduce as much as 9μA of loss, almost 10% of Figure 29.11's total! The AC loading of a 10pF probe looking at Figure 29.11's 20kHz clock will cause apparent circuit consumption of 5μA, a significant loss in a low power circuit. 1x type probes present about 50pF of loading, with 1M DC resistance when connected to the 'scope. This kind of probe loading can cause large errors in micropower circuits, while virtually disabling some. Such a probe, introduced at Pin 6 of text Figure 29.7, would stop the circuit's oscillator. If placed across the supply of the same circuit it would consume 15 times the circuit's operating current. Similarly, the probe's 50pF input capacitance connected to Figure 29.20 (Q8's collector) results in a 25% apparent increase in circuit current at 1MHz output.

Probe AC and DC loading are not the only effects. Some DVMs produce "charge spitting" at their inputs. Such parasitic charge, introduced into high impedance nodes, can cause substantial errors. It's also worth remembering that DVM DC loading may change with range. Lower ranges may have very high input impedance, but higher ranges are typically 10MΩ. A 10MΩ DVM reading Figure 29.7's supply consumes 1 1/2 times the circuit current.

Figure C1 shows a way test equipment can make the circuit look too good, instead of too bad. If the pulse generator is adjusted more than a diode drop above the regulator's output, the bypass capacitor peak detects the charge delivered through the IC's internal diode. The regulator can't sink current, and with its output forced high it won't source anything.

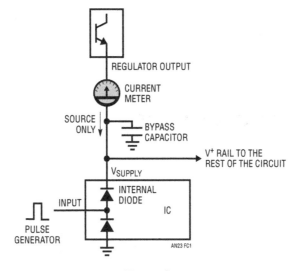

Figure C1

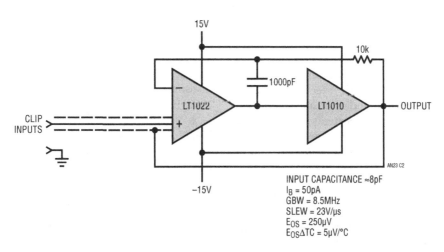

Figure C2

Under these conditions, the circuit functions while the current meter reads zero... a very low power circuit indeed!*

Figure C2 shows a very simple, but useful, circuit which greatly aids probe loading problems in micropower circuits. The LT1022 high speed FET op amp drives an LT1010 buffer. The LT1010's output allows DVM cable and probe driving and also biases the circuit's input shield. This bootstraps the input capacitance, reducing its effect. DC and AC errors of this circuit are low enough for almost all work, with enough bandwidth for just about any low power circuit. Built into a small enclosure with its own power supply, it can be used ahead of a 'scope or DVM with good results. Pertinent specifications appear in the diagram.

---

*Practically speaking, most regulators and power supplies can sink small amounts of current. Because of this, the current meter may actually read negative.

Thermocouple measurement

30

Jim Williams

Introduction

In 1822, Thomas Seebeck, an Estonian physician, accidentally joined semicircular pieces of bismuth and copper (Figure 30.1) while studying thermal effects on galvanic arrangements. A nearby compass indicated a magnetic disturbance. Seebeck experimented repeatedly with different metal combinations at various temperatures, noting relative magnetic field strengths. Curiously, he did not believe that electric current was flowing, and preferred to describe the effect as "thermo-magnetism." He published his results in a paper, "Magnetische Polarisation der Metalle und Erze durch Temperatur-Differenz" (see references).

Subsequent investigation has shown the "Seebeck Effect" to be fundamentally electrical in nature, repeatable, and quite useful. Thermocouples, by far the most common transducer, are Seebeck's descendants.

Thermocouples in perspective

Temperature is easily the most commonly measured physical parameter. A number of transducers serve temperature measuring needs and each has advantages and considerations. Before discussing thermocouple-based measurement it is worthwhile putting these sensors in perspective. Figure 30.2's chart shows some common contact temperature sensors and lists characteristics. Study reveals thermocouple strengths and weaknesses compared to other sensors. In general, thermocouples are inexpensive, wide range sensors. Their small size makes them fast and their low output impedance is a benefit. The inherent voltage output eliminates the need for excitation.

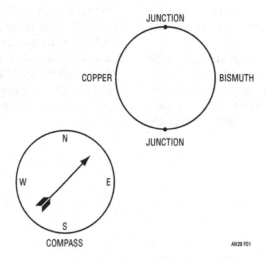

Figure 30.1 • The Arrangement for Dr. Seebeck's Accidental Discovery of "Thermo-Magnetism"

Analog Circuit and System Design: A Tutorial Guide to Applications and Solutions. DOI: 10.1016/B978-0-12-385185-7.00030-5

TYPE	RANGE OF OPERATION	SENSITIVITY AT 25°C	ACCURACY	LINEARITY	SPEED IN STIRRED OIL	SIZE	PACKAGE	COST	COMMENTS
Thermocouples (All Types)	−270°C to 1800°C	Typically Less Than 50μV/°C	±0.5°C with Reference	Poor Over Wide Range, Better Over 100°C	Typically 1 Sec. Some Types are Faster	0.02 In. Bead Typical. 0.0005 In. Units are Available	Metallic Bead, Variety of Probes Available	$1 to $50 Depending On Type, Specifications and Package	Requires Reference. Low Level Output Requires Stable Signal Conditioning Components
Thermistors and Thermistor Composites	−100°C to 450°C	5%/°C for Thermistors. ≈−0.5%/°C for Linearized Units	±0.1°C Standard from −40°C to 100°C; ±0.01°C from 0°C to 60°C Available	±0.2°C for Linearized Composite Units Over 100°C Ranges	1 to 10 Sec. is Standard; 3ms to 100ms Types are Available	Beads Can be as Small as 0.005 In., But 0.04 to 0.1 In. is Typical. "Flake" Types are Only 0.001 In. Thick	Glass, Epoxy, Teflon Encapsulated, Metal Housing, Etc.	$2 to $10 for Standard Units. $10 to $350 for High Precision Types and Specials	Highest Temperature Sensitivity of Any Common Sensor Special Units Required for Long-Term Stability Above 100°C
Platinum Resistance Wire	−250°C to 900°C	Approximately 0.5%/°C	±0.1°C Readily Available. ±0.01°C in Precision Standards—Lab Units	Nearly Linear Over Large Spans; Typically Within 1° Over 200°C Ranges	Typically Several Seconds	1/8 to 1/4 In. Typical. Smaller Sizes Available	Glass, Epoxy, Ceramic, Teflon, Metal, Etc.	$25 to $1000 Depending On Specs; Most Industrial Types Below $100	Sets Standard for Stability Over Long Term. Has Wider Temperature Range Than Thermistor but Lower Sensitivity
Diodes and Transistors	−270°C to 175°C	−2.2mV/°C (Approx. 0.33%/°C)	±2°C to ±5°C Over −55°C to 125°C	Within 2° Over Operating Range	1 to 10 Sec. is Standard. Small Diode Packages Permit Speeds in ms Range	Standard Diode and Transistor Case Sizes. Glass Passivated Chips Permit Extremely Small Sizes	Glass, Metal	Below 50¢. Cryogenic Units More Expensive	Require Individual Calibration. Must be Driven from Current Source for Optimum Performance. Extremely Inexpensive. Calibrated Cryogenic Types Available
Integrated Circuit	−85°C to 125°C Typical	0.4%/°C Typical	Over −55°C to 125°C	Within 1° (0.2° from 0°C to 70°C) Typical	Several Seconds	TO-18 Transistor Package Size. Also MiniDIP	Metal, Plastic	$1 to $10	Current and Voltage Outputs Available

Figure 30.2 • Characteristics of Some Contact Temperature Sensors (Chart Adapted from Reference 2)

an28f

JUNCTION MATERIALS	APPROXIMATE SENSITIVITY IN μV/°C AT 25°C	USEFUL TEMPERATURE RANGE (°C)	APPROXIMATE VOLTAGE SWING OVER RANGE	LETTER DESIGNATION
Copper—Constantan	40.6	–270 to 600	25.0mV	T
Iron—Constantan	51.70	–270 to 1000	60.0mV	J
Chromel—Alumel	40.6	–270 to 1300	55.0mV	K
Chromel—Constantan	60.9	–270 to 1000	75.0mV	E
Platinum 10%—Rhodium/Platinum	6.0	0 to 1550	16.0mV	S
Platinum 13%—Rhodium/Platinum	6.0	0 to 1600	19.0mV	R

Figure 30.3 • Temperature vs Output for Some Thermocouple Types

Signal conditioning issues

Potential problems with thermocouples include low level outputs, poor sensitivity and nonlinearity (see Figures 30.3 and 30.4). The low level output requires stable signal conditioning components and makes system accuracy difficult to achieve. Connections (see Appendix A) in thermocouple systems must be made with great care to get good accuracy. Unintended thermocouple effects (e.g., solder and copper create a 36μV/°C thermocouple) in system connections make "end-to-end" system accuracies better than 0.5°C difficult to achieve.

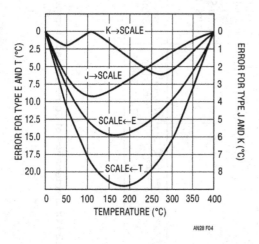

Figure 30.4 • Thermocouple Nonlinearity for Types J, K, E and T Over 0°C to 400°C. Error Increases Over Wider Temperature Ranges

Cold junction compensation

The unintended, unwanted and unavoidable parasitic thermocouples require some form of temperature reference for absolute accuracy. (See Appendix A for a discussion on minimizing these effects). In a typical system, a "cold junction" is used to provide a temperature reference (Figure 30.5). The term "cold junction" derives from

the historical practice of maintaining the reference junction at 0°C in an ice bath. Ice baths, while inherently accurate, are impractical in most applications. Another approach servo controls a Peltier cooler, usually at 0°C, to electronically simulate the ice bath (Figure 30.6). This approach[1] eliminates ice bath maintenance, but is too complex and bulky for most applications.

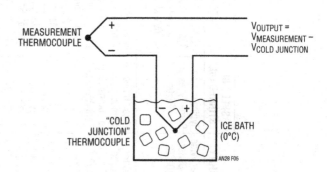

Figure 30.5 • Ice Bath Based Cold Junction Compensator

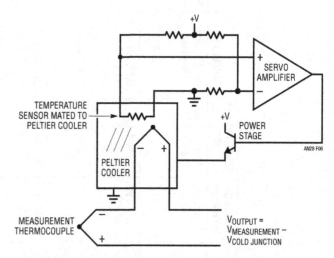

Figure 30.6 • A 0°C Reference Based on Feedback Control of a Peltier Cooler (Sensor is Typically a Platinum RTD)

Note 1: A practical example of this technique appears in LTC Application Note AN-25, "Switching Regulators for Poets."

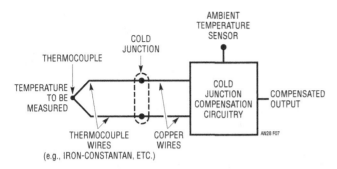

Figure 30.7 • Typical Cold Junction Compensation Arrangement. Cold Junction and Compensation Circuitry Must be Isothermal

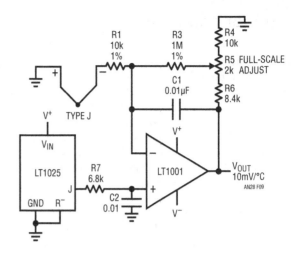

Figure 30.9 • LT1025 Cold Junction Compensates a Type J Thermocouple. The Op Amp Provides the Amplified Difference Between the Thermocouple and the LT1025 Cold Junction Output

Figure 30.7 conveniently deals with the cold junction requirement. Here, the cold junction compensator circuitry does not maintain a stable temperature but tracks the cold junction. This temperature tracking, subtractive term has the same effect as maintaining the cold junction at constant temperature, but is simpler to implement. It is designed to produce 0V output at 0°C and have a slope equal to the thermocouple output (Seebeck coefficient) over the expected range of cold junction temperatures. For proper operation, the compensator must be at the same temperature as the cold junction.

Figure 30.8 shows a monolithic cold junction compensator IC, the LT® 1025. This device measures ambient (e.g., cold junction) temperature and puts out a voltage scaled for use with the desired thermocouple. The low supply current minimizes self-heating, ensuring isothermal operation with the cold junction. It also permits battery or low power operation. The 0.5°C accuracy is compatible with overall achievable thermocouple system performance. Various compensated outputs allow one part to be used with many thermocouple types. Figure 30.9 uses an LT1025 and an amplifier to provide a scaled, cold

junction compensated output. The amplifier provides gain for the difference between the LT1025 output and the type J thermocouple. C1 and C2 provide filtering, and R5 trims gain. R6 is a typical value, and may require selection to accommodate R5's trim range. Alternately, R6 may be re-scaled, and R5 enlarged, at some penalty in trim resolution. Figure 30.10 is similar, except that the type K thermocouple subtracts from the LT1025 in series-opposed fashion, with the residue fed to the amplifier. The optional pull-down resistor allows readings below 0°C.

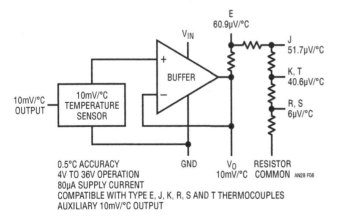

Figure 30.8 • LT1025 Thermocouple Cold Junction Compensator

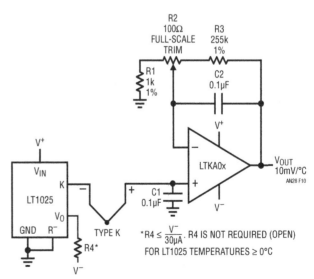

Figure 30.10 • LT1025 Compensates a Type K Thermocouple. The Amplifier Provides Gain for the LT1025-Thermocouple Difference

Amplifier selection

The operation of these circuits is fairly straightforward, although amplifier selection requires care.

Thermocouple amplifiers need very low offset voltage and drift, and fairly low bias current if an input filter is used. The best precision bipolar amplifiers should be used for type J, K, E and T thermocouples which have Seebeck coefficients to 40V/°C to 60μV/°C. In particularly critical applications, or for R and S thermocouples (6μV/°C to 15μV/°C), a chopper-stabilized amplifier is required. Linear Technology offers two amplifiers specifically tailored for thermocouple applications. The LTKA0x is a bipolar design with extremely low offset (30μV), low drift (1.5μV/°C), very low bias current (1nA), and almost negligible warm-up drift (supply current is 400μA).

For the most demanding applications, the LTC® 1052 CMOS chopper-stabilized amplifier offers 5μV offset and 0.05μV/°C drift. Input bias current is 30pA, and gain is typically 30 million. This amplifier should be used for R and S thermocouples, especially if no offset adjustments can be tolerated, or where a large ambient temperature swing is expected. Alternatively, the LTC1050, which has similar drift and slightly higher noise can be used. If board space is at a premium, the LTC1050 has the capacitors internally.

Regardless of amplifier type, for best possible performance dual-in-line (DIP) packages should be used to avoid thermocouple effects in the kovar leads of TO-5 metal can packages. This is particularly true if amplifier supply current exceeds 500μA. These leads can generate both DC and AC offset terms in the presence of thermal gradients in the package and/or external air motion.

In many situations, thermocouples are used in high noise environments, and some sort of input filter is required. To reject 60Hz pick-up with reasonable capacitor values, input resistors in the 10k to 100k range are needed. Under these conditions, bias current for the amplifier needs to be less than 1nA to avoid offset and drift effects.

To avoid gain error, high open-loop gain is necessary for single-stage thermocouple amplifiers with 10mV/°C or higher outputs. A type K amplifier; for instance, with 100mV/°C output, needs a closed-loop gain of 2,500. An ordinary op amp with a minimum loop of 50,000 would have an initial gain error of (2,500)/(50,000) = 5%! Although closed-loop gain is commonly trimmed, temperature drift of open-loop gain will have a deleterious effect on output accuracy. Minimum suggested loop gain for type E, J, K and T thermocouples is 250,000. This gain is adequate for type R and S if output scaling is 10mV/°C or less.

Additional circuit considerations

Other circuit considerations involve protection and common mode voltage and noise. Thermocouple lines are often exposed to static and accidental high voltages, necessitating circuit protection. Figure 30.11 shows two suggested approaches. These examples are designed to prevent excessive overloads from damaging circuitry. The added series resistance can serve as part of a filter. Effects of the added components on overall accuracy should be evaluated. Diode clamping to supply lines is effective, but leakage should be noted, particularly when large current limiting resistors are used. Similarly, IC bias currents combined with high value protection resistors can generate apparent measurement errors. Usually, a favorable compromise is possible, but sometimes the circuit configuration will be dictated by protection or noise rejection requirements.

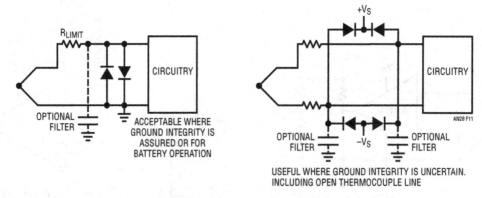

Figure 30.11 • Input Protection Schemes

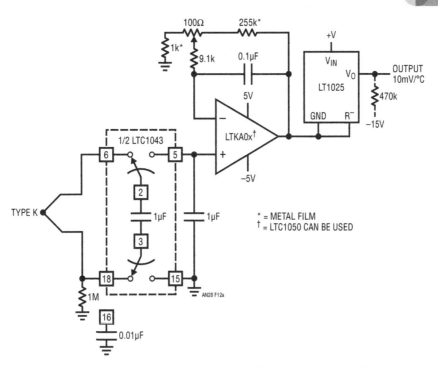

Figure 30.12a • Full Differential Input Thermocouple Amplifiers

Differential thermocouple amplifiers

Figure 30.12a shows a way to combine filtering and full differential sensing. This circuit features 120dB DC common mode rejection if all signals remain within the LTC1043 supply voltage range. The LTC1043, a switched-capacitor building block, transfers charge between the input "flying" capacitor and the output

capacitor. The LTC1043's commutating frequency, which is settable, controls rate of charge transfer, and hence overall bandwidth. The differential inputs reject noise and common mode voltages inside the LTC1043's supply rails. Excursions outside these limits require protection networks, as previously discussed. As in Figure 30.9, an optional resistor pull-down permits negative readings. The 1M resistor provides a bias path for the LTC1043's floating inputs. Figure 30.12b, for use with grounded thermocouples, subtracts sensor output from the LT1025.

Isolated thermocouple amplifiers

In many cases, protection networks and differential operation are inadequate. Some applications require continuous operation at high common mode voltages with severe noise problems. This is particularly true in industrial environments, where ground potential differences of 100V are common. Under these conditions the thermocouple and signal conditioning circuitry must be completely galvanically isolated from ground. This requires a fully isolated power source and an isolated signal transmission path to the ground referred output. Thermocouple work allows bandwidth to be traded for DC accuracy. With careful design, a single path can transfer floating power and isolated signals. The output may be either analog or digital, depending on requirements.

Figure 30.13 shows an isolated thermocouple signal conditioner which provides 0.25% accuracy at 175V common mode. A single transformer transmits isolated power and

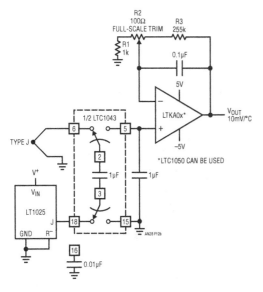

Figure 30.12b

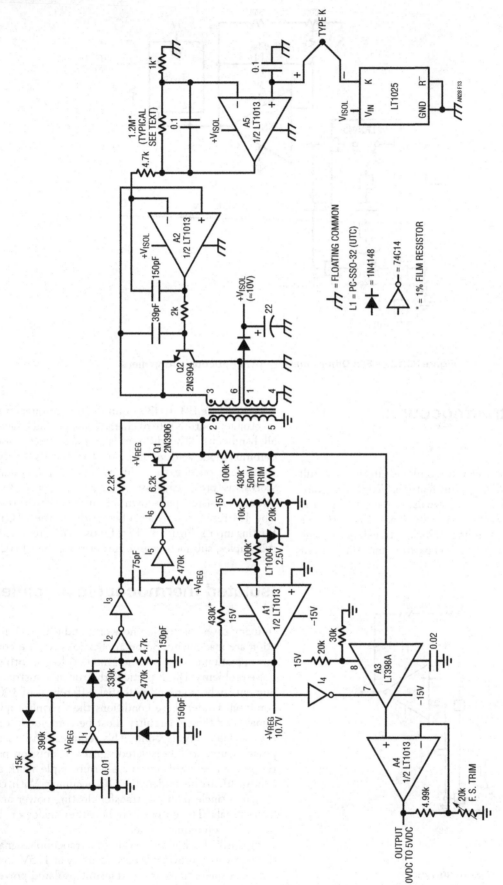

Figure 30.13 ● 0.25% Thermocouple Isolation Amplifier

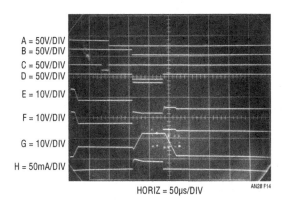

A = 50V/DIV
B = 50V/DIV
C = 50V/DIV
D = 50V/DIV
E = 10V/DIV
F = 10V/DIV
G = 10V/DIV
H = 50mA/DIV

HORIZ = 50µs/DIV   AN28 F14

Figure 30.14 • Waveforms for Figure 30.13's Thermocouple Isolation Amplifier

data. 74C14 inverter I_1 forms a clock (Trace A, Figure 30.14). I_2, I_3 and associated components deliver a stretched pulse to the 2.2k resistor (Trace B). The amplitude of this pulse is stabilized because A1's fixed output supplies 74C14 power. The resultant current through the 2.2k resistor drives L1's primary (Trace E). A pulse appears at L1's secondary (Trace F, Q2's emitter). A2 compares this amplitude with A5's signal conditioned thermocouple voltage. To close its loop, A2's output (Trace G) drives Q2's base to force L1's secondary (Pins 3 to 6) to clamp at A5's output value. Q2 operates in inverted mode, permitting clamping action even for very low A5 outputs. When L1's secondary (Trace F) clamps, its primary (Trace E) also clamps. After A2 settles, the clamp value is stable. This stable clamp value represents A5's thermocouple related information. Inverter I_4 generates a clock delayed pulse (Trace C) which is fed to A3, a sample-hold amplifier. A3 samples L1's primary winding clamp value. A4 provides gain scaling and the LT1004 and associated components adjust offset. When the clock pulse (Trace A) goes low, sampling ceases. When Trace B's stretched clock pulse goes low, the I_5-I_6 inverter chain output (Trace D) is forced low by the 470k-75pF differentiator's action. This turns on Q1, forcing substantial energy into L1's primary (Trace E). L1's secondary (Trace F) sees large magnetic flux. A2's output (Trace G) moves as it attempts to maintain its loop. The energy is far too great, however, and A2 rails. The excess energy is dumped into the Pin 1-Pin 4 winding, placing a large current pulse (Trace H) into the 22µF capacitor. This current pulse occurs with each clock pulse, and the capacitor charges to a DC voltage, furnishing the circuit's isolated supply. When the 470k-75pF differentiator times out, the I_5-I_6 output goes high, shutting off Q1. At the next clock pulse the entire cycle repeats.

Proper operation of this circuit relies on several considerations. Achievable accuracy is primarily limited by transformer characteristics. Current during the clamp interval is kept extremely low relative to transformer core capacity. Additionally, the clamp period must also be short relative

to core capacity. The clamping scheme relies on avoiding core saturation. This is why the power refresh pulse occurs immediately after data transfer, and not before. The transformer must completely reset before the next data transfer. A low clock frequency (350Hz) ensures adequate transformer reset time. This low clock frequency limits bandwidth, but the thermocouple data does not require any speed.

Gain slope is trimmed at A5, and will vary depending upon the desired maximum temperature and thermocouple type. The "50mV" trim should be adjusted with A5's output at 50mV. The circuit cannot read A5 outputs below 20mV (0.5% of scale) due to Q2's saturation limitations.

Drift is primarily due to the temperature dependence of L1's primary winding copper. This effect is swamped by the 2.2k series value with the 60ppm/°C residue partially compensated by I_3's saturation resistance tempco. Overall tempco, including the LT1004, is about 100ppm/°C. Increased isolation voltages are possible with higher transformer breakdown ratings.

Figure 30.15's thermocouple isolation amplifier is somewhat more complex, but offers 0.01% accuracy and typical drift of 10ppm/°C. This level of performance is useful in servo systems or high resolution applications. As in Figure 30.13, a single transformer provides isolated data and power transfer. In this case the thermocouple information is width modulated across the transformer and then demodulated back to DC. I_1 generates a clock pulse (Trace A, Figure 30.16). This pulse sets the 74C74 flip-flop (Trace B) after a small delay generated by I_2, I_3 and associated components. Simultaneously, I_4, I_5 and Q1 drive L1's primary (Trace C). This energy, received by L1's secondary (Trace H), is stored in the 47µF capacitor and serves as the circuit's isolated supply. L1's secondary pulse also clocks a closed-loop pulse width modulator composed of C1, C2, A3 and A4. A4's positive input receives A5's LT1025-based thermocouple signal. A4 servo-biases C2 to produce a pulse width each time C1 allows the 0.003µF capacitor (Trace E) to receive charge via the 430k resistor. C2's output width is inverted by I_6 (Trace F), integrated to DC by the 47k-0.68µF filter and fed back to A4's negative input. The 0.68µF capacitor compensates A4's feedback loop. A4 servo controls C2 to produce a pulse width that is a function of A5's thermocouple related output. I_6's low loss MOS switching characteristics combined with A3's supply stabilization ensure precise control of pulse width by A4. Operating frequency, set by the I_1 oscillator on L1's primary side, is normally a stability concern, but ratios out because it is common to the demodulation scheme, as will be shown.

I_6's output width's (Trace F) negative-going edge is differentiated and fed to I_7. I_7's output (Trace G) drives Q3. Q3 puts a fast spike into L1's secondary (Trace H). "Sing around" behavior by C1 is gated out by the diode at C2's positive input. Q3's spike is received at L1's primary, Pins 7 and 3. Q2 serves as a clocked synchronous demodulator pulling its collector low (Trace D) only when its base is high

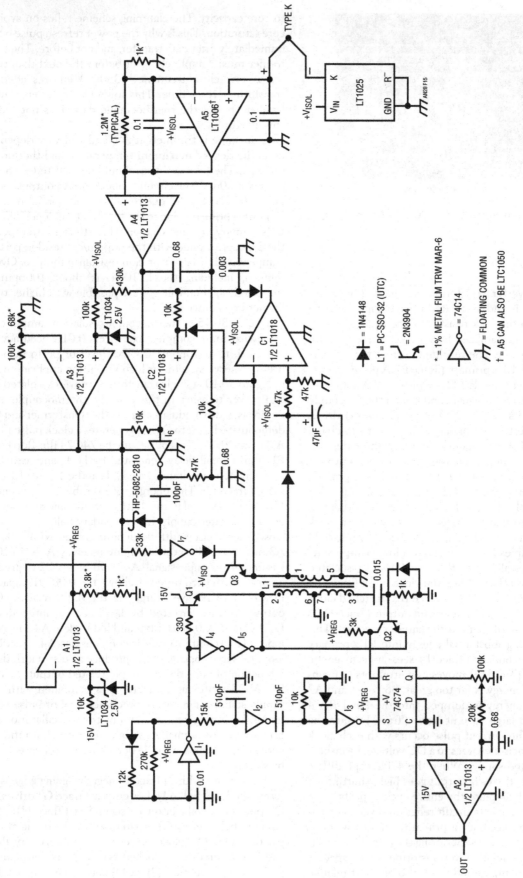

Figure 30.15 • 0.01% Thermocouple Isolation Amplifier

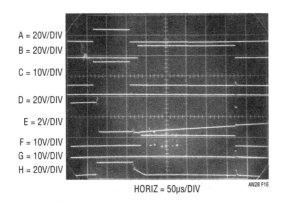

Figure 30.16 • Pulse-Width-Modulation Based Thermocouple Isolation Amplifier Waveforms

depends on A5's output. Variations with supply, temperature and I_1 oscillator frequency have no effect. A2 and its associated components extract the DC average by simple filtering. The 100k potentiometer permits desired gain scaling. Because this scheme depends on edge timing at the flip-flop, the delay in resetting the $0.003\mu F$ capacitor causes a small offset error. This term is eliminated by matching this delay in the 74C74 "set" line with the previously mentioned I_2-I_3 delay network. This delay is set so that the rising edge of the flip-flop output (Trace B) corresponds to I_6's rising edge. No such compensation is required for falling edge data because circuit elements in this path (I_7, Q3, L1 and Q2) are wideband. With drift matched LT1034s and the specified resistors, overall drift is typically 10ppm/°C with 0.01% linearity.

Digital output thermocouple isolator

and its emitter is low (e.g., when L1 is transferring data, not power). Q2's collector spike resets the 74C74 flip-flop. The MOS flip-flop is driven from a stable source (A1) and it is also clocked at the same frequency as the pulse-width modulator. Because of this, the DC average of its Q output

Figure 30.17 shows another isolated thermocouple signal conditioner. This circuit has 0.25% accuracy and features a

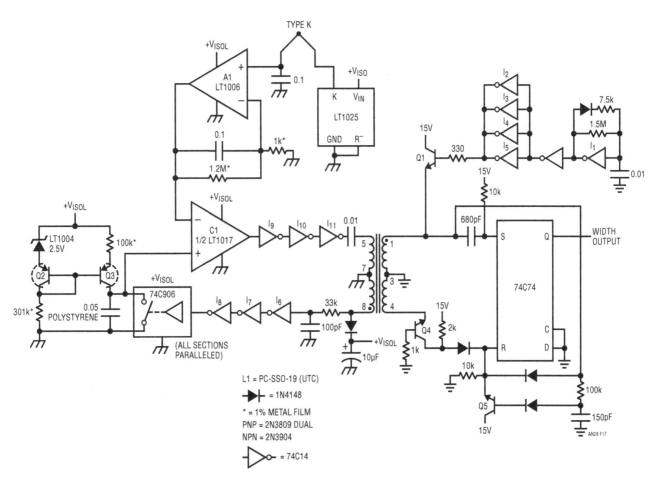

Figure 30.17 • Digital Output Thermocouple Isolator

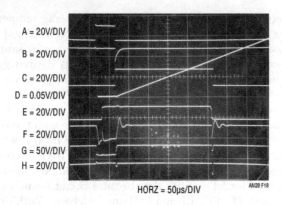

Figure 30.18 ● Waveforms for Digital-Output Thermocouple Isolator

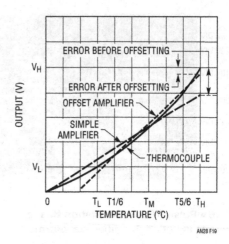

Figure 30.19 ● Offset Curve Fitting

digital (pulse width) output. I_1 produces a clock pulse (Trace A, Figure 30.18). I_2-I_5 buffers this pulse and biases Q1 to drive L1. Concurrently, the 680pF-10k values provide a differentiated spike (Trace B), setting the 74C74 flip-flop (Trace C). L1's primary drive is received at the secondary.

The 10µF capacitor charges to DC, supplying isolated power. The pulse received at L1's secondary also resets the 0.05µF capacitor (Trace D) via the inverters (I_6, I_7, I_8) and the 74C906 open-drain buffer. When the received pulse ends, the 0.05µF capacitor charges from the Q2-Q3 current source. When the resultant ramp crosses C1's threshold (A1's thermocouple related output voltage) C1 switches high, tripping the I_9-I_{11} inverter chain. I_{11} (Trace E) drives L1's secondary via the 0.01µF capacitor (Trace F). The 33k-100pF filter prevents regenerative "sing around". The resultant negative-going spike at L1's primary biases Q4, causing its collector (Trace G) to go low. Q4 and Q5 form a clocked synchronous demodulator which can pull the 74C74 reset pin low only when the clock is low. This condition occurs during data transfer, but not during power transfer The demodulated output (Trace H) contains a single negative spike synchronous with C1's (e.g., I_{11}'s) output transition. This spike resets the flip-flop, providing the circuit output. The 74C74's width output thus varies with thermocouple temperature.

Linearization techniques

It is often desirable to linearize a thermocouple-based signal. Thermocouples' significant nonlinear response requires design effort to get good accuracy. Four techniques are useful. They include offset addition, breakpoints, analog computation, and digital correction. Offset addition schemes rely on biasing the nonlinear "bow" with a constant term. This results in the output being high at low scale and low at high scale with decreased errors between these extremes (Figure 30.19). This compromise reduces overall error. Typically, this approach is limited to slightly

nonlinear behavior over wide ranges or larger nonlinearity over narrow ranges.

Figure 30.20 shows a circuit utilizing offset linearization for a type S thermocouple. The LT1025 provides cold junction compensation and the LTC1052 chopper-stabilized amplifier is used for low drift. The type S thermocouple output slope varies greatly with temperature. At 25°C it is 6µV/°C, with an 11µV/°C slope at 1000°C. This circuit gives 3°C accuracy over the indicated output range. The circuit, similar to Figure 30.10, is not particularly unusual except for the offset term derived from the LT1009 and applied through R4. To calibrate, trim R5 for $V_{OUT} = 1.669$ at $V_{IN} = 0.000$mV. Then, trim R2 for $V_{OUT} = 9.998$V at T = 1000°C or for V_{IN} (+ input) = 9.585mV.

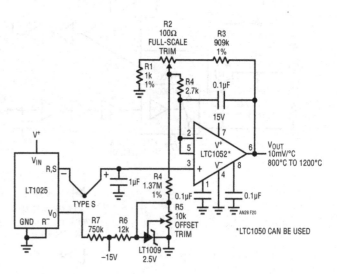

Figure 30.20 ● Offset-Based Linearization

Figure 30.21, an adaption of a configuration shown by Sheingold (Reference 3), uses breakpoints to change circuit gain as input varies. This method relies on scaling of the input and feedback resistors associated with A2-A6 and A7's reference output. Current summation at A8 is linear with the thermocouple's temperature. A3-A6 are the breakpoints, with the diodes providing switching when the respective summing point requires positive bias. As shown, typical accuracy of 1°C is possible over a 0°C to 650°C sensed range.

Figure 30.22, derived from Villanucci (Reference 8), yields similar performance but uses continuous function analog computing to replace breakpoints, minimizing amplifiers and resistors. The AD538 combines with appropriate scaling to linearize response. The causality of this circuit is similar to Figure 30.22; the curve fit mechanism (breakpoint vs continuous function) is the primary difference.

Digital techniques for thermocouple linearization have become quite popular. Figure 30.23, developed by Guy M. Hoover and William C. Rempfer, uses a microprocessor fed

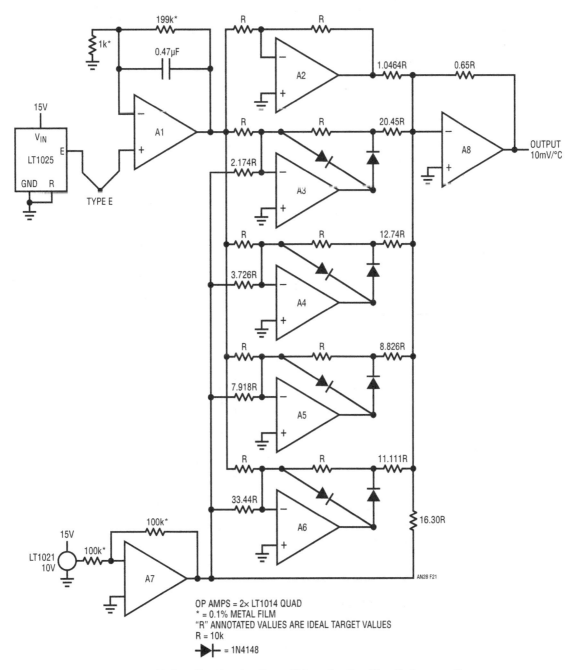

Figure 30.21 • Breakpoint-Based Linearization (See Reference 3)

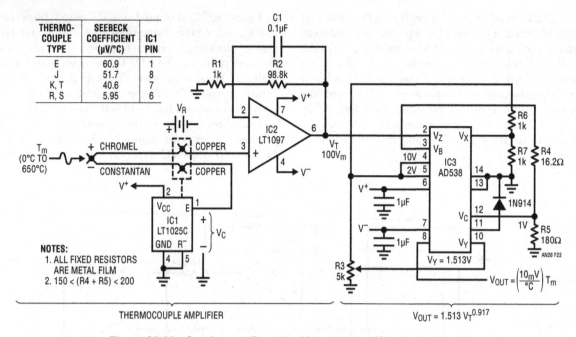

THERMO-COUPLE TYPE	SEEBECK COEFFICIENT (µV/°C)	IC1 PIN
E	60.9	1
J	51.7	8
K, T	40.6	7
R, S	5.95	6

NOTES:
1. ALL FIXED RESISTORS ARE METAL FILM
2. 150 < (R4 + R5) < 200

THERMOCOUPLE AMPLIFIER

$V_{OUT} = 1.513\ V_T{}^{0.917}$

$V_{OUT} = \left(\dfrac{10\,mV}{°C}\right) T_m$

Figure 30.22 • Continuous Function Linearization (See Reference 8)

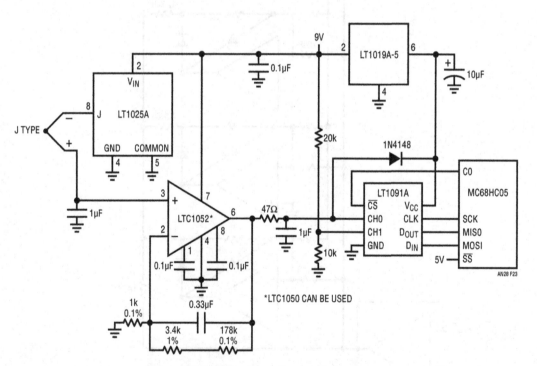

*LTC1050 CAN BE USED

Figure 30.23 • Processor-Based Linearization

```
                    *              TYPE J THERMOCOUPLE LINEARIZATION PROGRAM
                    *              WRITTEN BY GUY HOOVER LINEAR TECHNOLOGY CORPORATION
                    *              REV 1 10/4/87
                    *              N IS NUMBER OF SEGMENTS THAT THERMOCOUPLE RESPONSE IS DIVIDED INTO
                    *              TEMPERATURE (°C)=M•X+B
                    *              M IS SLOPE OF THERMOCOUPLE RESPONSE FOR A GIVEN SEGMENT
                    *              X IS A/D OUTPUT—SEGMENT END POINT
                    *              B IS SEGMENT START POINT IN DEGREES (°C • 2)
          ORG       $1000
          FDB       $00,$39,$74,$B0,$EE,$12B,$193,$262,$330,$397        TABLE FOR X
          ORG       $1020
          FDB       $85DD,$823A,$7FB4,$7DD4,$7CAE$7BC3,$7B8A,$7C24,$7C1E$7B3A    TABLE FOR M
          ORG       $1040
          FDB       $00,$3C,$78,$B4,$F0,$12C,$190,$258,$320,$384        TABLE FOR B
          ORG       $10FF
          FCB       $13                                                 N • 2
          ORG       $0100
          OPT       ]
          STA       $0A           LOAD CONFIGURATION DATA INTO $0A
          LDA       #$00          CONFIGURATION DATA FOR PORT A DDR
          STA       $04           LOAD CONFIGURATION DATA INTO PORT A
          LDA       #$FF          CONFIGURATION DATA FOR PORT B DDR
          STA       $05           LOAD CONFIGURATION DATA INTO PORT B
          LDA       #$F7          CONFIGURATION DATA FOR PORT C DDR
          STA       $06           LOAD CONFIGURATION DATA INTO PORT C
          JSR       HOUSEKP       INITIALIZE ASSORTED REGISTERS
MES92L    NPO
          JSR       CHECK
          LDA       #$0F          DIN WORD FOR LTC1091 CH0, W/RESPECT TO GND, MSB FIRST
          STA       $50           STORE IN DIN BUFFER
          JSR       READ91        READ LTC1091
LINEAR    LDX       $10FF         LOAD SEGMENT COUNTER INTO X
DOAGAIN   LDA       $1000,X       LOAD LSBs OF SEGMENT N
          STA       $55           STORE LSBs IN $55
          DECX                    DECREMENT X
          LDA       $1000,X       LOAD MSBs OF SEGMENT N
          STA       $54           STORE MSBs IN $54
          JSR       SUBTRCT
          BPL       SEGMENT
          JSR       ADDB
          DECX                    DECREMENT X
          JMP       DOAGAIN
SEGMENT   LDA       $1020,X       LOAD MSBs OF SLOPE
          STA       $54           STORE MSBs IN $54
          INCX                    INCREMENT X
          LDA       $1020,X       LOAD LSBs OF SLOPE
          STA       $55           STORE LSBs IN $55
          JSR       TBMULT        RETURNS RESULT IN $61 AND $62
          LDA       $1040,X       LOAD LSBs OF BASE TEMP
          STA       $55           STORE LSBs IN $55
          DECX                    DECREMENT X
          LDA       $1040,X       LOAD MSBs OF BASE TEMP
          STA       $54
          JSR       ADDB
CHECK     LDA       #$7F          DIN WORD FOR CH1
          STA       $50           LOAD DIN WORD INTO $50
          JSR       READ91        READ BATTERY VOLTAGE
          LDA       #$02          LOAD MSB OF MIN BATT VOLTAGE
          STA       $54           PUT IN MSB OF SUBTRACT BUFFER
          LDA       #$CC          LOAD LSB OF MIN BATT VOLTAGE
          STA       $55           PUT IN LSB OF SUBTRACT BUFFER
          JSR       SUBTRCT       COMPARE BATT VOLTAGE WITH MINIMUM
          BPL       NOPROB        IF BATT OK GOTO NOPROB
```

Figure 30.24 • Code for Processor-Based Linearization

```
                JSR     ADDB
                LDA     #$01
                STA     $56            SET BATTERY LOW FLAG
                RTS
NOPROB          JSR     ADDB
                CLR     $56            CLEAR LOW BATTERY FLAG
                RTS
READ91          LDA     #$50           CONFIGURATION DATA FOR SPCR
                STA     $0A            LOAD CONFIGURATION DATA
                LDA     $50
                BCLR    2,$02          BIT 0 PORT C GOES LOW (CS GOES LOW)
                STA     $0C            LOAD D_IN INTO SP1 DATA REG. START TRANSFER
BACK91          TST     $0B            TEST STATUS OF SPIF
                BPL     BACK91         LOOP TO PREVIOUS INSTRUCTION IF NOT DONE
                LDA     $0C            LOAD CONTENTS OF SPI DATA REG. INTO ACC
                STA     $0C            START NEXT CYCLE
                AND     #$03           CLEAR 6 MSBs OF FIRST D_OUT
                STA     $61            STORE MSBs IN $61
BACK92          TST     $0B            TEST STATUS OF SPIF
                BPL     BACK92         LOOP TO PREVIOUS INSTRUCTION IF NOT DONE
                BSET    2,$02          SET BIT 0 PORT C (CS GOES HIGH)
                LDA     $0C            LOAD CONTENTS OF SPI DATA INTO ACC
                STA     $62            STORE LSBs IN $62
                RTS


SUBTRCT         LDA     $62            LOAD LSBs
                SUB     $55            SUBTRACT LSBs
                STA     $62            STORE REMAINDER
                LDA     $61            LOAD MSBs
                SBC     $54            SUBTRACT W/CARRY MSBs
                STA     $61            STORE REMAINDER
                RTS
ADDB            LDA     $62            LOAD LSBs
                ADD     $55            ADD LSBs
                STA     $62            STORE SUM
                LDA     $61            LOAD MSBs
                ADC     $54            ADD W/CARRY MSBs
                STA     $61            STORE SUM
                RTS
TBMULT          CLR     $68
                CLR     $69
                CLR     $6A
                CLR     $6B
                STX     $58            STORE CONTENTS OF X IN $58
                LSL     $62            MULTIPLY LSBs BY 2
                ROL     $61            MULTIPLY MSBs BY 2
                LDA     $62            LOAD LSBs OF LTC1091 INTO ACC
                LDX     $55            LOAD LSBs OF M INTO X
                MUL                    MULTIPLY LSBs
                STA     $6B            STORE LSBs IN $6B
                STX     $6A            STORE IN $6A
                LDA     $62            LOAD LSBs OF LTC1091 INTO ACC
                LDX     $54            LOAD MSBs OF M INTO X
                MUL
                ADD     $6A            ADD NEXT BYTE
                STA     $6A            STORE BYTE
                TXA                    TRANSFER X TO ACC
                ADC     $69            ADD NEXT BYTE
                STA     $69            STORE BYTE
                LDA     $61            LOAD MSBs OF LTC1091 INTO ACC
                LDX     $55            LOAD LSBs OF M INTO X
```

Figure 30.24 • Code for Processor-Based Linearization (Continued)

```
            MUL
            ADD      $6A            ADD NEXT BYTE
            STA      $6A            STORE BYTE
            TXA                     TRANSFER X TO ACC
            ADC      $69            ADD NEXT BYTE
            STA      $69            STORE BYTE
            LDA      $61            LOAD MSBs OF LTC1091 INTO ACC
            LDX      $54            LOAD MSBs OF M INTO X
            MUL
            ADD      $69            ADD NEXT BYTE
            STA      $69            STORE BYTE
            TXA                     TRANSFER X TO ACC
            ADC      $68            ADD NEXT BYTE
            STA      $68            STORE BYTE
            LDA      $6A            LOAD CONTENTS OF $6A INTO ACC
            BPL      NNN
            LDA      $69            LOAD CONTENTS OF $69 INTO ACC
            ADD      #$01           ADD 1 TO ACC
            STA      $69            STORE IN $69
            LDA      $68            LOAD CONTENTS OF $68 INTO ACC
            ADC      #$00           FLOW THROUGH CARRY
            STA      $68            STORE IN $68
NNN         LDA      $68            LOAD CONTENTS OF $68 INTO ACC
            STA      $61            STORE MSBs IN $61
            LDA      $69            LOAD CONTENTS OF $69 INTO ACC
            STA      $62            STORE IN $62
            LDX      $58            RESTORE X REGISTER
            RTS                     RETURN
HOUSEKP     BSET     0,$02          SET B0 PORT C
            BSET     2,$02          SET B2 PORT C
            RTS
```

Figure 30.24 • Code for Processor-Based Linearization

from a digitized thermocouple output to achieve linearization. The great advantage of digital techniques is elimination of trimming. In this scheme a large number of breakpoints are implemented in software.

The 10-bit LTC1091A A/D gives 0.5°C resolution over a 0°C to 500°C range. The LTC1052 amplifies and filters the thermocouple signal, the LT1025A provides cold junction compensation and the LT1019A provides an accurate reference. The J type thermocouple characteristic is linearized digitally inside the processor Linear interpolation between known temperature points spaced 30°C apart introduces less than 0.1°C error. The 1024 steps provided by the LTC1091 (24 more that the required 1000) ensure 0.5°C resolution even with the thermocouple curvature.

Offset error is dominated by the LT1025 cold junction compensator which introduces 0.5°C maximum. Gain error is 0.75°C max because of the 0.1% gain resistors and, to a lesser extent, the output voltage tolerance of the LT1019A and the gain error of the LTC1091A. It may be reduced by trimming the LT1019A or gain resistors. The LTC1091A keeps linearity better than 0.15°C. The LTC1052's 5μV offset contributes negligible error (0.1°C or less). Combined errors are typically inside 0.5°C. These errors don't include the thermocouple itself. In practice, connection and wire errors of 0.5°C to 1°C are not uncommon. With care, these errors can be kept below 0.5°C.

The 20k-10k divider on CH1 of the LTC1091 provides low supply voltage detection (the LT1019A reference requires a minimum supply of 6.5V to maintain accuracy). Remote location is possible with data transferred from the MCU to the LTC1091 via the 3-wire serial port.

Figure 30.24 is a complete software listing[2] of the code required for the 68HC05 processor. Preparing the circuit involves loading the software and applying power. No trimming is required.

Note 2: Including of a software-based circuit was not without attendant conscience searching and pain on the author's part. Hopefully, the Analog Faithful will tolerate this transgression ...I'm sorry everybody, it just works too well!

References

1. Seebeck, Thomas Dr., "Magnetische Polarisation der Metalle und Erze durch Temperatur-Differenz", Abhaand-lungen der Preussischen Akademic der Wissenschaften (1822–1823), pg. 265–373.
2. Williams, J., "Designer's Guide to Temperature Sensors", EDN, May 5, 1977.
3. Sheingold, D.H., "Nonlinear Circuits Handbook", Analog Devices, Inc., pg. 92–97.

4. "Omega Temperature Measurement Handbook", Omega Engineering, Stamford Connecticut.
5. "Practical Temperature Measurements", Hewlett-Packard and Applications Note #290, Hewlett-Packard.
6. Thermocouple Reference Tables, NBS Monograph 125, National Bureau of Standards.

7. Manual on the Use of Thermocouples in Temperature Measurement, ASTM Special Publication 470A.
8. Villanucci, Robert S., "Calculator and IC Simplify Linearization", EDN, January 21, 1991.

Appendix A
Error sources in thermocouple systems

Obtaining good accuracy in thermocouple systems mandates care. The small thermocouple signal voltages require careful consideration to avoid error terms when signal processing. In general, thermocouple *system* accuracy better than 0.5°C is difficult to achieve. Major error sources include connection wires, cold junction uncertainties, amplifier error and sensor placement.

Connecting wires between the thermocouple and conditioning circuitry introduce undesired junctions. These junctions form unintended thermocouples. The number of junctions and their effects should be minimized, and kept isothermal. A variety of connecting wires and accessories are available from manufacturers and their literature should be consulted (Reference 4).

Thermocouple voltages are generated whenever dissimilar materials are joined. This includes the leads of IC packages, which may be kovar in TO-5 cans, alloy 42 or copper in dual-in-line packages, and a variety of other materials in plating finishes and solders. The net effect of these thermocouples is "zero" if all are at exactly the same temperature, but temperature gradients exist within IC packages and across PC boards whenever power is dissipated. For this reason, extreme care must be used to ensure that no temperature gradients exist in the vicinity of the thermocouple terminations, the cold junction compensator (e.g., LT1025) or the thermocouple amplifier. If a gradient cannot be eliminated, leads should be positioned isothermally, especially the LT1025 R⁻ and appropriate output pins, the amplifier input pins, and the gain setting resistor leads. An effect to watch for is amplifier offset voltage warm-up drift caused by mismatched thermocouple materials in the wire-bond/lead system of the IC package. This effect can be as high as tens of microvolts in TO-5 cans with kovar leads. It has nothing to do with the actual offset drift specification of the amplifier and can occur in amplifiers with measured "zero" drift. Warm-up drift is directly proportional to amplifier power dissipation. It can be minimized by avoiding TO-5 cans, using low supply current amplifiers, and by using the lowest possible supply voltages. Finally, it can be accommodated by calibrating and specifying the system after a five minute warm-up period.

A significant error source is the cold junction. The error takes two forms. The subtractive voltage produced by the cold junction must be correct. In a true cold junction (e.g., ice point reference) this voltage will vary with inability to maintain the desired temperature, introducing error. In a cold junction compensator like the LT1025, error occurs with inability to sense and track ambient temperature. Minimizing sensing error is the manufacturer's responsibility (we do our best!), but tracking requires user care. Every effort should be made to keep the LT1025 isothermal with the cold junction. Thermal shrouds, high thermal capacity blocks and other methods are commonly employed to ensure that the cold junction and the compensation are at the same temperature.

Amplifier offset uncertainties and, to a lesser degree, bias currents and open-loop gain should be considered. Amplifier selection criteria is discussed in the text under "Amplifier selection."

A final source of error is thermocouple placement. Remember that the thermocouple measures its own temperature. In flowing or fluid systems, remarkably large errors can be generated due to effects of laminar flow or eddy currents around the thermocouple. Even a "simple" surface measurement can be wildly inaccurate due to thermal conductivity problems. Silicone thermal grease can reduce this, but attention to sensor mounting is usually required. As much of the sensor surface as possible should be mated to the measured surface. Ideally, the sensor should be tightly mounted in a drilled recess in the surface. Keep in mind that the thermocouple leads act as heat pipes, providing a direct thermal path to the sensor. With high thermal capacity surfaces this may not be a problem, but other situations may require some thought. Often, thermally mating the lead wire to the surface or coiling the wire in the environment of interest will minimize heat piping effects.

As a general rule, skepticism is warranted, even in the most "obviously simple" situations. Experiment with several sensor positions and mounting options. If measured results agree, you're probably on the right track. If not, rethink and try again.

Section 2

Signal Conditioning

Applications for a switched-capacitor instrumentation building block (25)

This application note describes a wide range of useful applications for the LTC1043 dual precision instrumentation switched-capacitor building block. Some of the applications described are ultra high performance instrumentation amplifier, lock-in amplifier, wide range digitally controlled variable gain amplifier, relative humidity sensor signal conditioner, LVDT signal conditioner, charge pump F/V and V/F converters, 12-bit A/D converter and more.

Application considerations and circuits for a new chopper-stabilized op amp (26)

A discussion of circuit, layout and construction considerations for low level DC circuits includes error analysis of solder, wire and connector junctions. Applications include sub-microvolt instrumentation and isolation amplifiers, stabilized buffers and comparators and precision data converters.

Designing linear circuits for 5V operation (27)

This note covers the considerations for designing precision linear circuits which must operate from a single 5V supply. Applications include various transducer signal conditioners, instrumentation amplifiers, controllers and isolated data converters.

Applications for a DC accurate lowpass switched-capacitor filter (28)

Discusses the principles of operation of the LTC1062 and helpful hints for its application. Various application circuits are explained in detail with focus on how to cascade two LTC1062s and how to obtain notches. Noise and distortion performance are fully illustrated.

Micropower circuits for signal conditioning (29)

Low power operation of electronic apparatus has become increasingly desirable. This application note describes a variety of low power circuits for transducer signal conditioning. Also included are designs for data converters and switching regulators. Three appended sections discuss guidelines for micropower design, strobed power operation and effects of test equipment on micropower circuits.

Thermocouple measurement (30)

Considerations for thermocouple-based temperature measurement are discussed. A tutorial on temperature sensors summarizes performance of various types, establishing a perspective on thermocouples. Thermocouples are then focused on. Included are sections covering cold-junction compensation, amplifier selection, differential/isolation techniques, protection, and linearization. Complete schematics are given for all circuits. Processor-based linearization is also presented with the necessary software detailed.

Take the mystery out of the switched-capacitor filter: the system designer's filter compendium (31)

This note presents guidelines for circuits utilizing Linear's switched-capacitor filters. The discussion focuses on how to optimize filter performance by optimizing the printed wiring board, the power supply, and the output buffering of the filter. Many additional topics are discussed such as how to select the proper filter response for the application and how to characterize a filter's THD for DSP applications.

Bridge circuits (32)

Subtitled "Marrying gain and balance," this note covers signal conditioning circuits for various types of bridges. Included are transducer bridges, AC bridges, Wien bridge oscillators, Schottky bridges, and others. Special attention is given to amplifier selection criteria. Appended sections cover strain gauge transducers, understanding distortion measurements, and historical perspectives on bridge read-out mechanisms and Wein bridge oscillators.

High speed amplifier techniques (33)

This application note, subtitled "A designer's companion for wideband circuitry," is intended as a reference source for designing with fast amplifiers. Approximately 150 pages and 300 figures cover frequently encountered problems and their possible causes. Circuits include a wide range of amplifiers, filters, oscillators, data converters and signal conditioners. Eleven appended sections discuss related topics including oscilloscopes, probe selection, measurement and equipment considerations, and breadboarding techniques.

A seven nanosecond comparator for single supply operation (34)

This note is an extensive discussion of the causes and cures of problems in very high speed comparator circuits. A separate applications section uses the 7ns LT1394 in V-to-F converters, crystal oscillators, clock skew generators, triggers, sampling configurations and a nanosecond pulse stretcher. Appendices cover related topics.

Understanding and applying voltage references (35)

Just how do bandgaps and buried Zeners stack up against Weston cells? Did you know your circuit board may induce more drift in a reference than time and temperature? Learn the answers to these and other commonly asked reference questions ranging from burn-in recommendations to ΔV_{BE} generation in this application note.

Instrumentation applications for a monolithic oscillator (36)

Instrumentation applications for a monolithic programmable oscillator are presented in this publication. Circuits include platinum and thermistor based thermometers, an isolated thermometer and three relative humidity signal conditioners. Bipolar and FET input chopper stabilized amplifiers with noise below 45nV (0.1Hz to 10Hz) are detailed. Two clock tuneable sine wave generators with settable amplitude appear, as well as a tuneable notch filter, an interval generator and an A to D converter. The oscillator's performance is contrasted with other approaches and its interval operation discussed.

Slew rate verification for wideband amplifiers (37)

Wideband amplifiers achieve slew rates beyond 2500V/µs. Verifying slew rates at this speed requires special techniques. In particular, a subnanosecond rise time input step is necessary for accurate slew rate measurement. A pulse generator with a 360 picosecond rise time is shown, and its construction detailed. Slew rate test results using this generator are presented and compared to data taken with slower rise time generators. Appendices cover high speed measurement technique, generator output level shifting and picosecond signal path construction considerations.

Instrumentation circuitry using RMS → DC converters (38)

It is widely acknowledged that RMS measurement of waveforms furnishes the most accurate amplitude information. Rectify-and-average schemes, usually calibrated to a sine wave, are only accurate for one waveshape. Departures from this waveshape result in pronounced errors. Although accurate, RMS conversion often entails limited bandwidth, restricted range, complexity and difficult to characterize dynamic and static errors. The LTC1966/67/68 RMS converter family addresses these issues, making instrument grade applications practical. A variety of instrumentation oriented applications are presented. Included are basic circuits, a fully isolated AC line monitor, a distortionless AC line voltage regulator, wideband ×1000 pre-amplifiers, a quartz crystal RMS current meter, a crystal stabilized AC voltage reference, an RMS amplitude leveled random noise generator and an RMS amplitude level controller. Appended sections cover

RMS theory and converter operation, AC measurement and signal handling practice, test equipment recommendations, noise theory and noise diodes.

775 nanovolt noise measurement for a low noise voltage reference (39)

Frequently, voltage reference stability and noise define measurement limits in instrumentation systems. In particular, reference noise often sets stable resolution limits. Reference voltages have decreased with the continuing drop in system power supply voltages, making reference noise increasingly important. The compressed signal processing range mandates a commensurate reduction in reference noise to maintain resolution. Noise ultimately translates into quantization uncertainty in A to D converters, introducing jitter in applications such as scales, inertial navigation systems, infrared thermography, DVMs and medical imaging apparatus. A new low voltage reference, the LTC6655, has only 0.3ppm (775nV) noise at $2.5V_{OUT}$. Determining this figure constitutes a high order difficulty measurement, requiring an extremely low noise test fixture. Design details of this test fixture are presented, along with a thorough performance evaluation. Appended sections provide supporting material.

Take the mystery out of the switched-capacitor filter

The system designer's filter compendium

31

Richard Markell

Introduction

Overview

This chapter presents guidelines for circuits utilizing Linear Technology's switched-capacitor filter family. Although the switched-capacitor filter has been designed into "telecom" circuits for over 20 years, the newer devices are faster, quieter and lower in distortion. These filters now achieve total harmonic distortion (THD) below -76dB (LTC®1064-2), wideband noise below $55\mu V_{RMS}$ (LTC1064-3), high frequency of operation (LTC1064-2, LTC1064-3 and LTC1064-4 to 100kHz) and steep roll-offs from passband to stopband (LTC1064-1: -72dB at $1.5\times f_{CUTOFF}$). These specifications make the new generation of switched-capacitor filters from LTC candidates to replace all but the most esoteric of active RC filter designs.

Chapter 40 takes the mystery from the design of high performance active filters using switched-capacitor filter integrated circuits. To help the designer get the highest performance available, this chapter covers most of the problems prevalent in system level switched-capacitor filter design. The chapter covers both tutorial filter material and direct operating criteria for LTC's filter parts. Special attention is given to proper breadboarding techniques, proper power supply selection and design, filter response selection, aliasing, and optimization of dynamic range, noise and THD. These issues are presented after a short introduction to the switched-capacitor filter

The switched-capacitor filter

Why use switched-capacitor filters? One reason is that sampled data techniques economically and accurately imitate continuous time functions. Switched-capacitor filters can be made to model their active RC counterparts in the continuous time domain. The advantages of the switched-capacitor approach lie in the fact that a MOS integrated capacitor with a few switches replaces the resistor in the active RC biquad filter allowing full filter implementation on a chip. The building block for most filter designs, the integrator, appears in Figure 31.1. When implemented with resistors, capacitors and op amps it is expensive and sensitive to component tolerances. The switched-capacitor integrator, as seen in Figure 31.2, eliminates the resistors and replaces them with switched capacitors. The dependency on component tolerances is virtually eliminated because the switched-capacitor filter integrator depends on capacitor value ratios, and not on absolute values. This provides very good accuracy in setting center frequencies

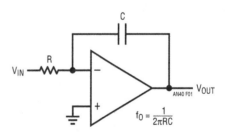

$$f_0 = \frac{1}{2\pi RC}$$

Figure 31.1 • Active RC Inverting Integrator

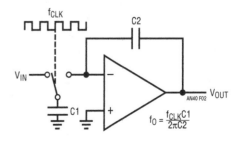

$$f_0 = \frac{f_{CLK}C1}{2\pi C2}$$

Figure 31.2 • Inverting Switched-Capacitor Integrator

Analog Circuit and System Design: A Tutorial Guide to Applications and Solutions. DOI: 10.1016/B978-0-12-385185-7.00031-7

and Q values. Additionally, this implementation allows the effective resistor value to be varied by the clock operating the switches. This allows the resistors (and therefore the filter center frequencies) to be varied over a wide range (typically 10,000:1 or more). Since existing integrated circuit technology can implement capacitor ratios much more accurately than resistor ratios (see Figure 31.2), the switched-capacitor filter can provide filters with inherent accuracy and repeatability. Active RC configurations are limited by resistor and capacitor tolerances (and to a secondary extent, the accuracy and bandwidth of the op amps) and usually require trimming. Switched-capacitor filters do have disadvantages compared to their active RC competitors. These include somewhat more noise in some circuit configurations and a phenomenon called clock feedthru. Clock feedthru is circuit clock artifacts feeding through to the filter output. It is present in virtually all sampled data systems to some degree. LTC has greatly improved this parameter over the past few years to the point where clock feedthru in the LTC1064 series is rarely a problem.

Present day switched-capacitor filters include one to four 2nd order sections per packaged integrated circuit. Not long ago the switched-capacitor filter was only considered for low frequency work where signal-to-noise ratio was non-critical. This has changed for the better in recent years. Recent switched-capacitor filter designs allow up to 8th order switched-capacitor filters to be implemented that compete successfully with the typical multiple operational amplifier designs in critical areas, including signal-to-noise ratios, stopband attenuation and maximum cutoff frequencies.[1]

Circuit board layout considerations

The most critical aspect of the testing and evaluation of any switched-capacitor filter is a proper circuit board layout. Switched capacitor filters are a peculiar mix of analog and digital circuitry that require the user to take time to lay out the circuit board, be it a breadboard or a board for the space station. All this hoopla is standard "do as I say, not as I do" in most data sheets. To give this issue its proper credibility, two breadboards were built of the same circuit. The first was built on a "protoboard." No bypass capacitors were used to isolate the power supply from the circuitry and, of course, there was no ground plane. The clock line consisted of flying wires as opposed to the coax used on the second or "recommended" breadboard. Additionally, when tests were run on the first breadboard, in most cases, no buffer operational amplifier at the filter output was used. A photograph of the first protoboard breadboard is shown as Figure 31.3. It should be emphasized here that the

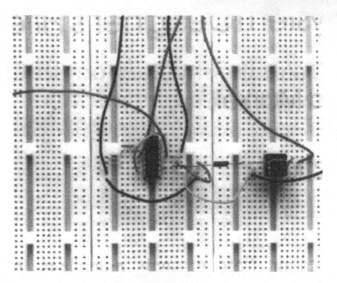

Figure 31.3 • Improperly Constructed "Poor" Switched-Capacitor Filter Breadboard

protoboard is a very good board for some types of breadboarding scenarios; it is just not very useful for the type of layout a switched-capacitor filter circuit requires.

Figure 31.5 shows a photograph of a properly constructed switched-capacitor filter breadboard. The circuit is built on a copper clad board which acts as a ground plane. The switched-capacitor filter and buffer operational amplifier are well bypassed. All leads are kept as short as possible and the switched-capacitor filter clock input is through a shielded cable. The second breadboard uses a single point ground in the form of the whole board. (Figure 31.4 shows the schematic for both breadboards.) Single point grounding techniques become more and more critical as switched-capacitor filter parts are incorporated into large boards containing many analog and digital devices. In these large boards, and in larger multi-board systems, lots of design time should be spent on single point grounding techniques and noise abatement.[2-6] It will be worthwhile in the long run.

Test results comparing the same circuit built on each breadboard are very interesting. Figure 31.6 shows the passband response of an 8th order Elliptic lowpass filter,

Note 1: Sevastopoulos, Nello and Markell, Richard, "Four-Section Switched-Cap Filter Chips Take on Discretes." Electronic Products, September 1, 1988.

Note 2: Brokaw, A. Paul, "An IC Amplifier User's Guide to Decoupling, Grounding, and Making Things Go Right for a Change." Analog Devices Data-Acquisition Databook 1984, Volume I, Pgs. 20-13 to 20-20.
Note 3: Morrison, Ralph, "Grounding and Shielding Techniques in Instrumentation." Second Edition, New York, NY: John Wiley and Sons, Inc., 1977.
Note 4: Motchenbacher, C.D., and Fitchen, F.C., "Low-Noise Electronic Design." New York, NY: John Wiley and Sons, Inc., 1973.
Note 5: Rich, Alan, "Shielding and Guarding." Analog Dialogue, 17, No. 1 (1983), 8-13. Also published as an Application Note in Analog Devices Data-Acquisition Databook 1984, Volume I, Pgs. 20-85 to 20-90.
Note 6: Rich, Alan, "Understanding Interference-Type Noise." Analog Dialogue, 16, No. 3 (1982), 16–19. Also published as an Application Note in Analog Devices Data-Acquisition Databook 1984, Volume I, Pgs. 20-81 to 20-84.

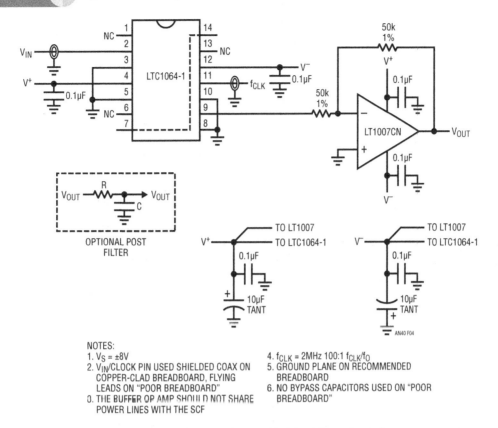

NOTES:
1. $V_S = \pm 8V$
2. V_{IN}/CLOCK PIN USED SHIELDED COAX ON COPPER-CLAD BREADBOARD, FLYING LEADS ON "POOR BREADBOARD"
3. THE BUFFER OP AMP SHOULD NOT SHARE POWER LINES WITH THE SCF

4. $f_{CLK} = 2MHz$ 100:1 f_{CLK}/f_O
5. GROUND PLANE ON RECOMMENDED BREADBOARD
6. NO BYPASS CAPACITORS USED ON "POOR BREADBOARD"

Figure 31.4 • Schematic of LTC1064-1 Breadboards

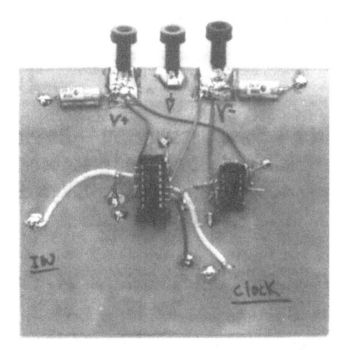

Figure 31.5 • Properly Constructed "Good" Switched-Capacitor Filter Breadboard

the LTC1064-1. The figure clearly illustrates the ripple in the passband that can be attributed not to the filter, but to the breadboarding technique. The good breadboard is a factor of 5 to 10 times better than the inadequate breadboard.

Figure 31.7 shows the two breadboard circuits measured in the filter stopband. The top trace is the inadequate breadboard while the bottom is the well constructed model. Note the loss of between 10dB and 20dB of attenuation. Also notice the notch, almost 80dB down from the input signal, which is clearly shown when using a good breadboard and which cannot be seen using the poorly constructed breadboard.

Figures 31.8 and 31.9 show the noise for the two different breadboards. Two plots were necessary as the noise was more than an order of magnitude greater when the poor breadboard was used. These tests were run using the identical switched-capacitor filter part (LTC1064-1CN) which was moved from one breadboard to the next to ensure exactly the same measurement conditions, except for the breadboards themselves.

Not shown graphically, but measured, was the offset voltage of the switched-capacitor filter part in each breadboard. In the poorly constructed board the offset was a whopping 266mV, while in the other circuit board it was 40.7mV, almost a factor of seven times less.

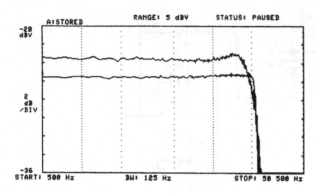

Figure 31.6 • Passband Response—8th Order Elliptic Lowpass Filter (LTC1064-1). Top Trace: Improperly Constructed Breadboard, No Buffer. Bottom Trace: Good Breadboard, with Buffer. Both: f_{CLK} = 2MHz, f_{CUTOFF} = 20kHz, V_S = ±8V. Note: Top Trace Was Offset to Increase Clarity

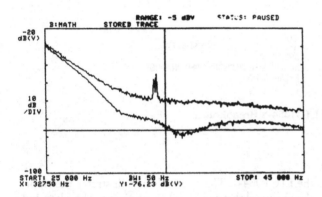

Figure 31.7 • Stopband Response—8th Order Elliptic Lowpass Filter (LTC1064-1). Top Trace: Improperly Constructed Breadboard, No Buffer. Bottom Trace: Good Breadboard, with Buffer. Both: f_{CLK} = 2MHz, f_{CUTOFF} = 20kHz, V_S = ±8V

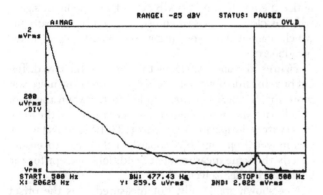

Figure 31.8 • Wideband Noise Measurement—8th Order Elliptic Lowpass Filter (LTC1064-1) Using Improper Breadboarding Techniques. f_{CLK} = 2MHz, f_{CUTOFF} = 20kHz, V_S = ±8V, Buffered with LT1007

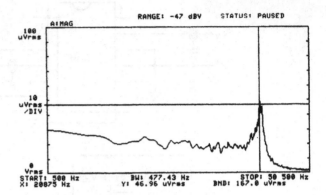

Figure 31.9 • Wideband Noise Measurement—8th Order Elliptic Lowpass Filter (LTC1064-1) Using Good Breadboarding Techniques. f_{CLK} = 2MHz, f_{CUTOFF} = 20kHz, V_S = ±8V, Buffered with LT1007

Carefully note that all measurements in this Application Note (with a few noted exceptions) were performed with a good 10×, low capacitance, scope probe (i.e., Tektronix P6133) at the output of the buffer operational amplifier. The probe ground lead length was kept below 1″ in length.

Moral of this section: Beware the breadboard that sits on your bench. It may not suffice to resurrect your old 741 breadboard to test today's switched-capacitor filters. They require fast clocks, buffering and good breadboarding techniques to deliver their optimum specifications.

Power supplies

Power supplies, proper bypassing of these supplies, and supply noise are usually given little attention in the board spectrum of chaos called system design. Beware! Sampled data devices such as switched-capacitor filters, A-to-D and D-to-A converters, chopper-stabilized operational amplifiers and sampled data comparators require careful power supply design. Poorly chosen or poorly designed power supplies may induce noise into the system. Similarly, improper bypassing may impair even the most ideal of power supplies. Common complaints range from noise in the passband of a switched-capacitor filter to spurious A-to-D outputs due to high frequency noise being aliased back into the signal bandwidth of interest. These effects are, at best, difficult to find and, at worst, worth a call to the local goblin extermination crew (LTC's Application Group!).

Figure 31.10 shows an example of how a switcher can cause problems. The figure was generated by using the breadboard in Figure 31.3. An industry standard 5V to ±15V switcher module was used to power the board. The switcher was unbypassed to better illustrate the potential problems.

Figure 31.10 can be compared directly with Figure 31.7 to see the switcher noise. The poor breadboard causes the stopband attenuation to be well above where it should be

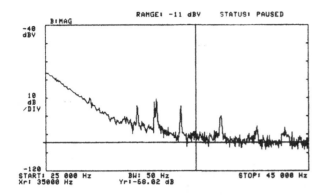

Figure 31.10 • Stopband Response—8th Order Elliptic Lowpass Filter (LTC1064-1). Power Supply is Unbypassed Industry Standard Modular 5V to ±15V Switcher. The Improperly Constructed Breadboard Was Used for This Test. f_{CLK} = 2MHz, f_{CUTOFF} = 20kHz, V_S = ±7.5V (Switched Zenered to ±7.5V)

when proper breadboarding techniques are used but switcher harmonics are also evident. These appear in Figure 31.10. Some of these peaks are only −45dB down from the signal of interest in the passband and could be confused with a legitimate signal. Clearly, if a filter is designed to be 80dB down in the stopband, using a noisy switcher will not do. Figure 31.11 shows the schematic diagram of a good, low noise switcher for system use. It produces ±7.5V with 200µV of noise. The switcher is an excellent example of good, low noise design techniques.[7]

All power supplies in a good system design should be properly bypassed. There are as many techniques for bypassing as voodoo curses, so we will not overly dwell on the subject. For switched-capacitor filters, we recommend good, low ESR bypass capacitors (0.1µF minimum, 0.22µF better) as close to the power supply pins of the part as possible. High quality capacitors are recommended for bypassing. For more details on how to identify an adequate capacitor see Appendix B. We recommend separate digital and analog grounds with the two only being tied together as close to supply common as practical. The ground lead to the bypass capacitors should go to the analog ground plane.

The prudent layout includes bypass capacitors on the pins of the switched-capacitor filter which are tied to a circuit potential for programming, such as the 50/100 pins. Should spikes and/or transients appear on this pin, trouble will ensue. The summing junction pins SA, SB, etc., are also candidates for bypassing if they are tied to analog ground in a single supply system where lots of noise is present. This last issue is probably icing on the cake in most cases, but it will help lower the noise in some cases.

Last, but not least, the power supplies used to power the switched-capacitor filters limit the maximum input and

Note 7: Williams, Jim, and Huffman, Brian, "Some Thoughts on DC/DC Converters," Linear Technology AN29.

output signals to and from the filters. Thus, power supplies have a direct effect on the system dynamic range. For the LTC1064 type filter ±7.5V supplies provide ±5V output swing, while ±5V supplies provide ±3.3V of swing. For a filter with 450mV$_{P-P}$ of output noise these numbers translate to 87dB and 83dB of dynamic range respectively.

Input considerations

This section considers some aspects of switched-capacitor filter design which are related to the input signal, the filter's input structure and the overall filter response.

Offset voltage nulling

Typical offset voltages for an LTC1064 or an LTC1064-X through the four sections may range up to 40mV or 50mV. While this may not be of concern for AC-coupled systems, it becomes important in DC-coupled applications. The anti-aliasing filter used before an A/D converter is a typical application where this is an important concern. For a 5V$_{O-P}$ input signal, the least significant bit of an 8-bit A/D converter is approximately 20mV, and one-half the LSB is approximately 10mV. This implies that use of a filter (or for that matter, any type of device other than a straight wire) before an 8-bit A/D converter requires offset voltages below 10mV. For a 12-bit converter this provision mandates stringent 600µV of offset at the A/D's input.

Several methods of offset cancellation are common. The usual method seen with operational amplifiers utilizes a potentiometer to inject a correction voltage. Figure 31.12 shows this arrangement with an LTC1064-1 Elliptic filter. This method can correct the initial offset, but both the adjustment circuit and the CMOS operational amplifiers in the switched-capacitor filters have temperature coefficients that are not zero. Thus, if this circuit is used to correct the offset at 25°C, it will not fully correct the offset at another temperature. An advantage of this type of offset nulling is that the filter's frequency response is affected very little. Figure 31.13 shows the time domain response of Figure 31.12's filter circuitry. The rising and falling edge overshoot is typical of high Q filters be it switched capacitor or active RC. (This time domain performance parameter, the rise time, is treated separately in a later section.) Compare Figure 31.13 with Figure 31.15 to observe the time domain response of an open-loop offset correction scheme (the potentiometer) versus Figure 31.14's closed-loop servo.

Figure 31.14 shows the circuit of an LTC1064-1 Elliptic filter with the same LT1007 output buffer amplifier used in Figure 31.13. An addition is the LT1012 operational amplifier used as a servo to zero the offset of the filter. This arrangement can provide offsets of less than 100µV which is quite acceptable for a 12-bit system. The servo generates a low frequency pole at about 0.16Hz which can interact with some signals of interest to some system users.

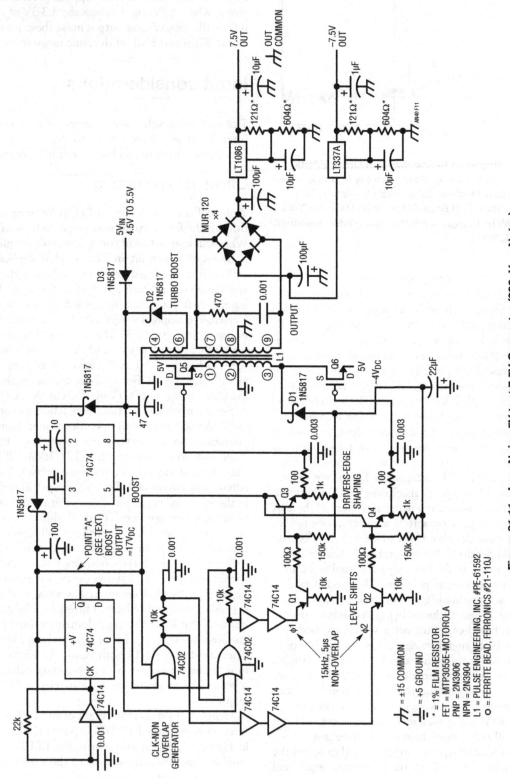

Figure 31.11 • Low Noise 5V to ±7.5V Converter (200μV$_{P-P}$ Noise)

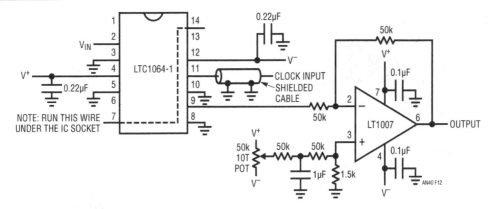

Figure 31.12 • Elliptic Filter with Offset Adjustment Potentiometer

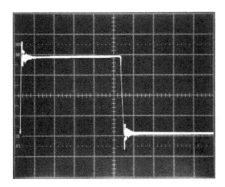

Figure 31.13 • Time Domain Response of Elliptic Filter (LTC1064-1) *Without Servo* **in Figure 31.12. Input Signal 1Hz Square Wave. LTC1064-1 Cutoff (f_O) = 100Hz. Horiz = 0.1s/ Div., Vert = 0.5V/Div (Photograph Reproduction of Original)**

Figure 31.15 shows the low frequency square wave response. Figure 31.16 shows the distortion introduced *for large-signal inputs* at a frequency near the servo pole. Figure 31.17 shows a sine wave input to the servo system, but at lower amplitude and higher frequency. The small distortion introduced at this higher frequency is probably traceable to the servo's high frequency cutoff. Figure 31.19 shows the servo response to the *small-signal input* at 0.092Hz shown in Figure 31.18. The servo tracks this input, but at a lower amplitude. The servo thus looks like a highpass filter to the input signal at input frequencies below the servo pole frequency. Servo offset nulling can be extremely useful in systems if the limitations described are tolerable.

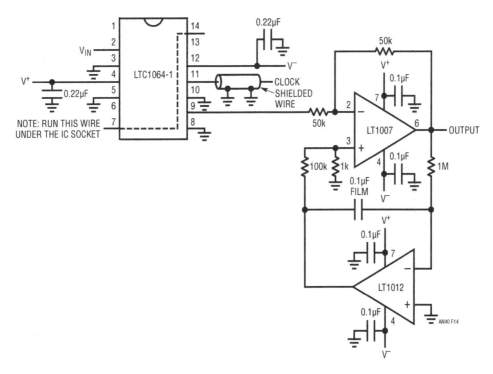

Figure 31.14 • Elliptic Filter (LTC1064-1) with Servo Offset Adjustment

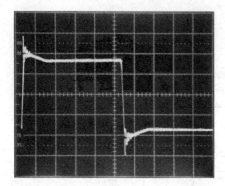

Figure 31.15 • Time Domain Response of Elliptic Filter (LTC1064-1) *with Servo* per Figure 31.14. Input Signal 1Hz Square Wave. LTC1064-1 Cutoff (f_O) = 100Hz. Horiz = 0.1s/Div., Vert = 0.5V/Div (Photograph Reproduction of Original)

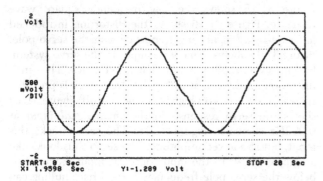

Figure 31.16 • Large-Signal Response—Output of Elliptic Filter (LTC1064-1) Plus Servo (Figure 31.14). Input: 1V$_{RMS}$, 0.092Hz Sine Wave. Filter f_{CUTOFF} = 100Hz

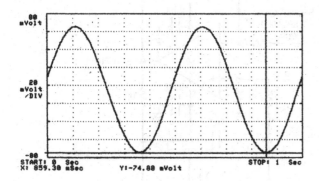

Figure 31.17 • Small-Signal Response—Output of Elliptic Filter (LTC1064-1) Plus Servo (Figure 31.14). Input: 50m V$_{RMS}$, 2Hz Sine Wave. Filter f_{CUTOFF} = 100Hz

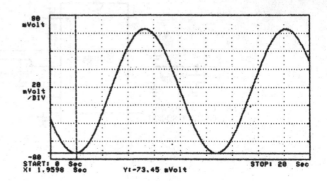

Figure 31.18 • Input Signal (50mV$_{RMS}$, 0.092Hz) to Elliptic Filter (LTC1064-1) as Shown in Figure 31.14

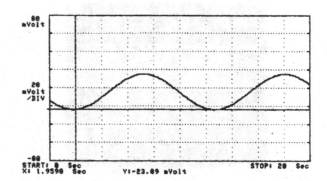

Figure 31.19 • Small-Signal Output Response of Elliptic Filter (LTC1064-1) Plus Servo to Input Shown as Figure 31.18. Filter f_{CUTOFF} = 100Hz

Slew limiting

The input stage operational amplifiers in active RC or switched-capacitor filters can be driven into slew limiting if the input signal frequency is too high. Slew limiting is usually caused by a capacitance load drive limitation in the op amps internal circuitry. Contemporary switched-capacitor filter devices are designed to avoid slew limiting in almost all cases.

As an example, the LTC1064 filter has a typical slew rate of 10V/μs. Since slew rate is a large-signal parameter, it also defines what is called the power bandwidth, given as:

$$f_P = \frac{SR}{2\pi E_{OP}} \text{(See Note 8)}$$

where f_P is the full power frequency and E_{OP} is the peak amplifier output voltage.

For the LTC1064 operating at V_S = ± 7.5V the device can swing ±5V or 10V peak. Based on the op amp slew

---

Note 8: Jung, Walter, "IC Op Amp Cookbook." Howard W. Sams, 1988.

rate performance *only*, the full power frequency is calculated as:

$$f_P = \frac{10V}{10^{-6}s2\pi10V} = 159kHz$$

This f_P is sufficient for all but the most stringent switched-capacitor filter applications of the LTC1064.

Aliasing

Since the switched-capacitor filter is based around a "switching capacitor" to generate variable filter parameters, it is by definition a sampled data device. Like all other such devices it is subject to aliasing. Aliasing is a complex subject with lots of mathematics involved. As such, its derivation is the subject of many paragraphs in textbooks.[9] The system designer can get a meaningful handle on the subject by a series of spectrum analyzer views.

Figure 31.20 through Figure 31.23 each show three spectrum segments while the switched-capacitor filter's input signal varies from 100Hz to 49.9kHz. Each figure shows a different input frequency and spectrum plots of the filter's passband (10Hz to 510Hz), the frequency spectrum around $f_{CLK}/2$ (25kHz) and the frequency spectrum around f_{CLK} (50kHz). The switched-capacitor filter is a lowpass elliptic filter (LTC1064-1) with a cutoff frequency set to 500Hz ($f_{CLK}/f_{CUTOFF} = 100:1$). The clock used for this cutoff is 50kHz. Thus, from sampled data theory, aliasing begins when the input signal passes the $f_{CLK}/2$ threshold (25kHz).

Figure 31.20 shows the LTC1064-1 in its normal mode of operation with the signal (100Hz) within the passband of the filter. Note the small second harmonic content (−80dB) component which also appears in the passband.

No signals are seen around $f_{CLK}/2$, however, the original signal (100Hz) appears attenuated (sin x/x envelope response) at 49.9kHz and 50.1kHz. This is consistent with sampled data theory and is a very important anomaly to be taken into consideration in some systems. Thus, for any signal input there will be side lobes at $f_{CLK} \pm f_{IN}$. These side lobes will be attenuated by the sin x/x envelope familiar to those who work with sampled data systems.

Figure 31.21 shows the same series of spectral photos for an input signal of 24.9kHz. This signal is outside of the filter passband, so the signals are much attenuated. The signal seen at 200Hz in this series of plots is actually the alias of the 49.8kHz second harmonic of the input signal. The signals seen around 25kHz are the 24.9kHz input and the $f_{CLK} - f_{IN}$ signals (see your textbook)!! The signals around and at 50kHz are clock feedthru, and the second harmonic images of the 24.9kHz input signal around the clock frequency.

Similarly, Figure 31.22 shows spectral plots for an input signal of 25.1kHz. The signals appear in the same locations as

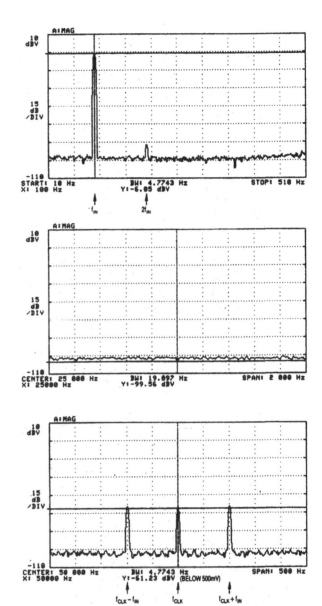

Figure 31.20 • Elliptic Switched-Capacitor Filter (LTC1064-1)
Aliasing Study. $f_O = 500Hz$, $f_{CLK} = 50kHz$, $V_S = \pm8V$, $V_{IN} = 100Hz$ at $500mV_{RMS}$. **Note Dynamic Range Limitation**

Figure 31.21, but for different reasons. The 200Hz signal in this figure is the alias of the 50.2kHz. The signals around 25kHz are the input signal at 25.1kHz and its alias at 24.9kHz.

Figure 31.23 details aliasing at its very worst. The input signal here is 49.9kHz which aliases back, in the filter passband, to 100Hz and appears almost identical to Figure 31.20's 100Hz signal. The input signal appears at 49.9kHz with its 50.1kHz mirror image.

Of additional note to the system designer is the LTC1062 5th order dedicated Butterworth lowpass filter. This device makes use of a continuous time input stage (an R and C) to prevent aliasing, making it attractive in some applications.

Note 9: Schwartz, Mischa, "Information Transmission, Modulation and Noise." New York, NY: McGraw Hill Book Co., 1980.

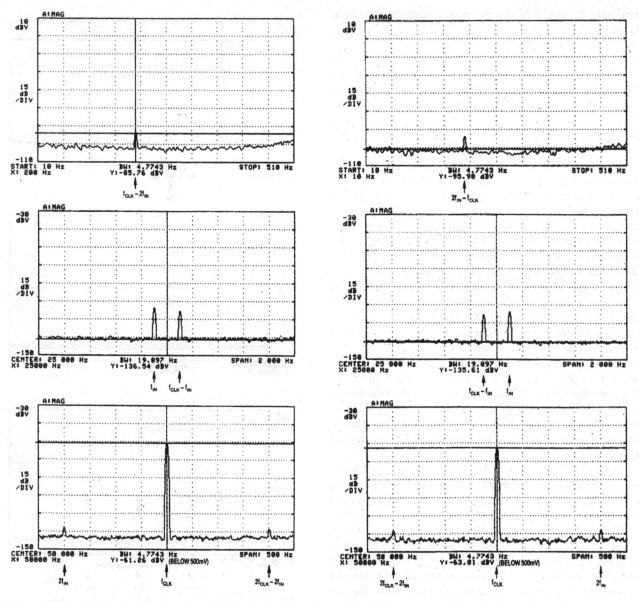

Figure 31.21 • Elliptic Switched-Capacitor Filter (LTC1064-1) *Aliasing Study.* f_O = 500Hz, f_{CLK} = 50kHz, V_S = ±8V, V_{IN} = 24.9kHz at 500mV$_{RMS}$

Figure 31.22 • Elliptic Switched-Capacitor Filter (LTC1064-1) *Aliasing Study.* f_O = 500Hz, f_{CLK} = 50kHz, V_S = ±8V, V_{IN} = 25.1kHz at 500mV$_{RMS}$

Filter response

What kind of filter do I use? Butterworth, Chebyshev, Bessel or Elliptic

One of the questions most asked among system designers at the shopping malls of America is "how do I choose the proper filter for my application?" Aside from the often used retort, "call Linear Technology," a discussion of the types of filters and when to use them is appropriate.

A typical application could involve lowpass filtering of pulses. This filter might be in the IF section of a digital radio receiver. The received pulses must pass through the filter without large amounts of overshoot or ringing. A Bessel filter is most likely the filter of choice.

Another application may be insensitive to filter pulse response, but requires as steep a cutoff slope as possible. This might occur in the detection of a series of continuous tones as in an EEG system. If filter ringing is irrelevant, an Elliptic filter may be a good choice.

The trade-offs involved in filter design are critical to the system designer's understanding of this topic. This means understanding what happens in both the time domain and in the frequency domain. At the risk of sounding pedantic, a discussion of this topic is appropriate. A time domain

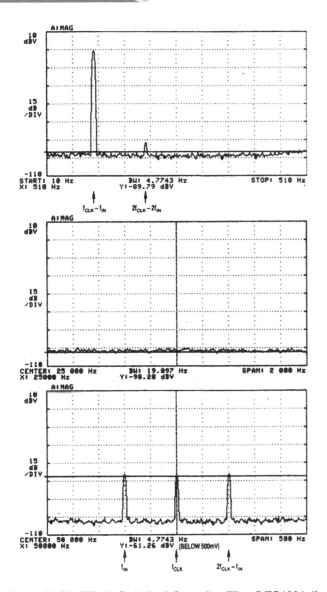

Figure 31.23 • Elliptic Switched-Capacitor Filter (LTC1064-1) *Aliasing Study*. f_O = 500Hz, f_{CLK} = 50kHz, V_S = ±8V, V_{IN} = 49.9kHz at 500mV$_{RMS}$. Note Dynamic Range Limitation

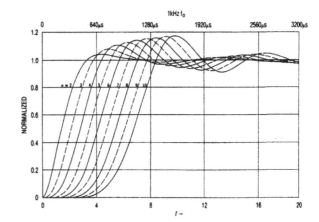

Figure 31.24 • Step Response for Butterworth Filters*

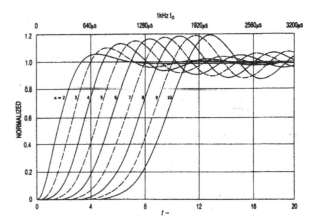

Figure 31.25 • Step Response for Chebyshev Filters with 0.1dB Ripple*

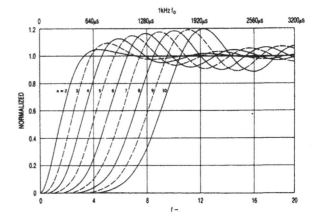

Figure 31.26 • Step Response for Chebyshev Filters with 0.01dB Ripple*

response can be viewed on an oscilloscope as amplitude versus time. It is in the time domain that pulse overshoot, ringing and distortion appear. The frequency domain (amplitude versus frequency) is traditionally where the designer looks at the filter's response (on a spectrum analyzer). A wonderful response in the frequency domain often appears ugly in the time domain. The converse may also be true. It is crucial to examine both responses when designing a filter!

Figures 31.24 through 31.28 reprinted here from Zverev[10] show the time domain step response for the

Note 10: Zverev, "Handbook of Filter Synthesis." New York, NY: John Wiley and Sons, Inc., Copyright 1967.

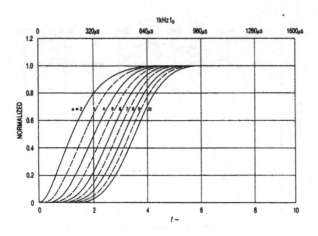

Figure 31.27 • Step Response for Maximally Flat Delay (Bessel) Filters*

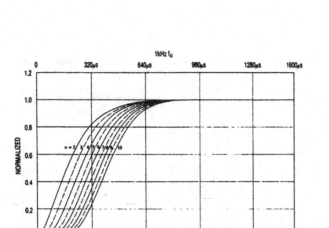

Figure 31.28 • Step Response for Synchronously Tuned Filters*

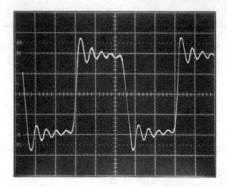

Figure 31.29 • LTC1064-1 Time Domain Response. Filter f_{CUTOFF} = 100Hz. Input: 10Hz Square Wave. Horiz = 20ms/ Div., Vert = 0.5V/Div (Photograph Reproduction of Original)

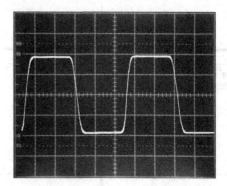

Figure 31.30 • LTC1064-3 Time Domain Response. Filter f_{CUTOFF} = 100Hz. Input: 10Hz Square Wave. Horiz = 20ms/ Div., Vert = 0.5V/Div (Photograph Reproduction of Original)

Butterworth, two types of Chebyshev, a Bessel and a synchronously tuned filter. (The synchronously tuned filter is one that is made up of multiple identical stages, but contains no notches. See AM27A for more discussion on this topic.) Figure 31.29 shows the time domain response of the LTC1064-1 8th order Elliptic LP filter to a 10Hz input square wave. The photo shows that a system requiring good pulse fidelity cannot use this filter. Figure 31.30 shows a better filter for this application. It is the LTC1064-3, an 8th order Bessel LPF. The response has a nice, rounded off rise time with no overshoot. The trade-off appears in the frequency domain.

Figures 31.31 through 31.34 show the frequency domain response of four 8th order filters, the LTC1064-1 through LTC1064-4. (Figure 31.35 shows an expansion of the comparison of the LTC1064-1 and LTC1064-4 roll

*From Anatol I. Zverev, "Handbook of Filter Synthesis," Copyright © 1967, John Wiley and Sons, Inc., reprinted by permission of John Wiley and Sons, Inc.

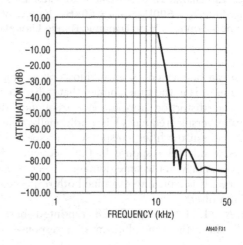

Figure 31.31 • Frequency Domain Response of LTC1064-1 Elliptic Filter. f_{CUTOFF} = 10kHz, V_S = ±7.5V, LT1007 Output Buffer, 100:1

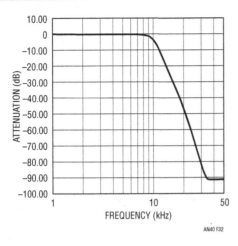

Figure 31.32 • Frequency Domain Response of LTC1064-2 Butterworth Filter. f_{CUTOFF} = 10kHz, V_S = ±7.5V, LT1007 Output Buffer, 50:1

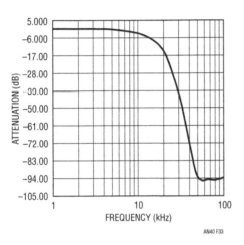

Figure 31.33 • Frequency Domain Response of LTC1064-3 Bessel Filter. f_{CUTOFF} = 10kHz, V_S = ±7.5V, LT1007 Output Buffer, 75:1

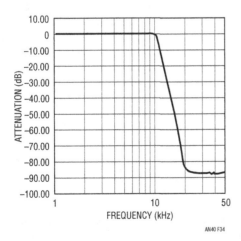

Figure 31.34 • Frequency Domain Response of LTC1064-4 Elliptic Filter. f_{CUTOFF} = 10kHz, V_S = ±7.5V, LT1007 Output Buffer, 50:1

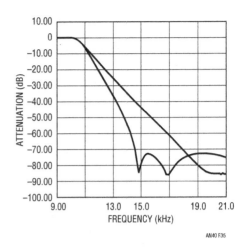

Figure 31.35 • Frequency Domain Response of LTC1064-1 and LTC1064-4 Transition Region Blow-Up. f_{CUTOFF} = 10kHz, V_S = ±7.5V, LT1007 Output Buffer

off from the passband to the stopband.) The LTC1064-1 and LTC1064-4 are 8th order Elliptic filters, while the LTC1064-2 is a Butterworth and the LTC1064-3 is a Bessel. The roll-off from the passband to the stopband is least steep for the LTC1064-3 Bessel filter. This is the price paid for the linear phase response which enable the filter to pass a square wave with good fidelity. Similarly, the LTC1064-2 Butterworth trades slightly worse transient response for steeper roll-off.

The system designer must carefully consider a potential filtering solution in the time and frequency domains. Figures 31.24 through 31.34 are intended as examples to help the designer with this process. There are an infinite variety of filters that can be implemented with switched-capacitor filters to obtain the precise response required for one's system.

Call Linear Technology Applications at (408) 432-1900 for additional help in choosing and/or defining a particularly difficult filtering problem.

The perennial question, "how fast can I sweep?" a filter from one frequency to another can be answered by looking at the transient response curves and renormalizing them to the desired cutoff frequency. Then the settling time can be read off the curve. A frequency sweep is in many aspects like the settling time performance to a pulse point.

Table 31.1 details the four filters mentioned previously and some of their key parameters. Note the wide variation in the stopband attenuation specification. This specification is a measure of the filters steepness of attenuation. This is a key specification for anti-aliasing filters found at an A/D converter's input. Note that for the LTC1064-1 in the figure (corner frequency set to 10kHz) the attenuation to a 15kHz signal would be about 72dB. The Butterworth gives approximately 27dB, while the Bessel only about 7dB attenuation. These latter two filters trade frequency domain roll off for good time domain response.

626

Table 1 Filter Selection Guide

Part number	Type	Passband ripple	Stopband attenuation	Wideband noise 1Hz – 1MHz	SNR		THD (1kHz)*	Supply voltage
LTC1064-1	Elliptic	±0.15dB	72dB at 1.5f_C	150μV_{RMS}	1V_{RMS} Input = 76dB		−76dB	V_S = ±5V
				165μV_{RMS}	3V_{RMS} Input = 85dB		−70dB	V_S = ±7.5V
LTC1064-2	Butterworth	3dB	90dB at 4f_C	80μV_{RMS}	1V_{RMS} Input = 82dB		−76dB	V_S = ±5V
				90μV_{RMS}	3V_{RMS} Input = 90dB		−70dB	V_S = ±7.5V
LTC1064-3	Bessel	3dB	60dB at 5f_C	55μV_{RMS}	1V_{RMS} Input = 85dB		−76dB	V_S = ±5V
				60μV_{RMS}	3V_{RMS} Input = 94dB		−70dB	V_S = ±7.5V
LTC1064-4	Elliptic	±0.1dB	80dB at 2f_C	120μV_{RMS}	1V_{RMS} Input = 78dB		−76dB	V_S = ±5V
				130μV_{RMS}	3V_{RMS} Input = 87dB		−70dB	V_S = ±7.5V

*These specifications from LTC data sheets represent typical values. Optimization may result in significantly better specifications. Call LTC for more details.

Filter sensitivity

How stable is my filter?

One of the great advantages of the switched-capacitor filter is the lack of discrete capacitors with their inherent tolerance and stability limitations. The active RC filter designed with theoretical capacitor values has problems with repeatability and stability when real world capacitors are used. The switched-capacitor filter has small errors in both the cutoff frequency, f_O and Q, but they are easier to deal with than those of the active RC filter.

Most universal switched-capacitor filters are arranged in the so called State-Variable-Biquad circuit configuration.[11] This configuration not only allows realization of all filter functions, LP, BP, HP, AP and notch, but also allows high Q filters to be realized with low sensitivity to component tolerances. (For a strict mathematical analysis of the sensitivity of the State-Variable-Biquad see reference 11, chapter 10.)

Manufacturing realities of the semiconductor business also affect the switched-capacitor filter design. Though this inaccuracy is much less than the active RC design (do the math in Daryanani) it does exist. Thus, switched-capacitor filters are available from manufacturers such as LTC with center frequency tolerances of generally 0.4% to 0.7%. This presumes an accurate stable clock. Operating the universal switched-capacitor filter in Mode 3 tends to make the center frequency error depend on the resistors since the equation for f_O is:

$$f_O = \frac{f_{CLK}\sqrt{R2/R4}}{50 \text{ or } 100}$$

Thus the manufacturing inaccuracy of the switched-capacitor filter is multiplied (and generally swamped) by the resistor inaccuracy. Mode 2 guarantees a filter designed with switched-capacitor filters has lower f_O sensitivity than Mode 3 by changing the equation for f_O to:

$$f_O = \frac{f_{CLK}}{50 \text{ or } 100}\sqrt{1 + R2/R4}$$

Here resistor sensitivity is mitigated by the one under the radical, and thus the inaccuracy is, in most cases, only caused by the manufacturing tolerances of the switched-capacitor filter. (See discussion of switched-capacitor filter modes in the LTC1060 and/or LTC1064 data sheets).

An excellent method of minimizing the filter dependence on resistor tolerances is to integrate the resistors directly on the semiconductor device. This can be done with the LTC1064 family of devices using silicon chrome resistors placed directly on the die.

Actual resistor values are generally better than 0.5%, but the important feature is that all thin film resistors track each other in both resistance and temperature coefficient. Thus, filters such as the LTC1064-1 through LTC1064-4 have virtually indistinguishable characteristics for each and every part.

The small tolerances in f_O using the switched-capacitor filter are trivial when compared to an active RC filter. An Elliptic filter like the LTC1064-1 requires no small amount of trimming when built with resistors, capacitors and op amps. Changing the f_O is even more impractical.

Note 11: Daryanani, Gobind, "Principles of Active Network Synthesis and Design." New York, NY: John Wiley and Sons, Inc., Copyright 1976.

Output considerations

THD and dynamic range

Presently, one of the biggest uses of filters is before A/D converters for antialiasing. The filter band limits the signal at the input to a Digital Signal Processing system. A critical concern is the filter's signal-to-noise ratio (SNR). Thus, if a filter has a maximum output swing of $2V_{RMS}$ with noise of $100\mu V_{RMS}$ it can be said to have an SNR (signal-to-noise ratio) of 86dB. This certainly seems to make it a candidate for antialiasing applications before a 14-bit A/D (required SNR approximately 84dB). But, *is this the only consideration??* What is missing in this analysis is a discussion of total harmonic distortion. This is a frequently ignored subtlety of system design.

This filter example has an SNR or 86dB. But suppose its THD is only −47dB. What this means can be better understood by applying a 1kHz signal to the system. What is desired is to digitize this 1kHz to 14-bits of accuracy. What happens is quite different. The 1kHz signal, along with its harmonics, will be digitized. THD (total harmonic distortion) is a measure of the unwanted harmonics that are introduced by nonlinearities in the system. Thus, the 1kHz pure tone will come out looking like 1kHz + 2kHz + 3kHz, etc. The A/D converter will digitize these signals adding errors to the data acquisition process.

Figure 31.36 illustrates a good way to characterize this potential problem. This figure shows a THD plot of an LTC1064-2 8-pole Butterworth LPF (circuit as shown in Figure 31.37). The graph shows THD versus input amplitude. A second horizontal scale labels SNR. The graph clearly shows, for instance, that for a $1.5V_{RMS}$ input the THD is below −70dB and the SNR is below −85dB. Thus, all the harmonics of the input signal (in this case 4kHz) are

below −70dB. Figure 31.38 shows a THD + N versus frequency curve for the same filter. At an input frequency of 2kHz the THD + N is approximately 0.018% or −74.9dB. Figure 31.39 shows good correlation with a spectrum analyzer in the frequency domain. Of course, there are an unlimited number of these plots that can be taken, for an almost unlimited number of cutoff frequencies and input frequencies. What must be considered as the most important issue is that *THD generally limits digitization accuracy,* **not SNR.** Figure 31.40 shows four units of the LTC1064-2 with a 1kHz input frequency. Here, as before, the −70dB THD specification is preserved up to $2.5V_{RMS}$ input.

Total harmonic distortion is a complicated phenomenon and is difficult to analyze all its potential causes. Some of these causes in the switched-capacitor filter are thought to be the charge transfer inherent in the SC process, the output drive and the swing internal to the switched-capacitor filter state variable filter.

THD in active RC filters

THD in RC active filters is generally assumed to be superior to switched-capacitor filters. Traditional filter textbooks seem to lack data on THD, either in a theoretical or a practical sense. The data presented here shows the RC active to be somewhat better than the switched-capacitor filter, but at a tremendous cost in terms of board space, non-tunability and cost.

Figure 31.41 shows a THD versus amplitude plot of the RC active equivalent of the LTC1064-1 8th order Elliptic filter. This filter requires 16 operational amplifiers, 31 resistors and eight capacitors on a board approximately 2-1/2 × 6 inches in size. An equivalent THD plot for the LTC1064-1 is shown in Figure 31.42.

The Elliptic filters are the worst choice for good THD specifications because of their high Q sections. Butterworth and Bessel filters have very good THD specifications.

Linear Technology has done extensive research in comparing the THD and SNR aspects of our switched-capacitor filters to those of active RC filters. In many cases, a filter may be optimized for THD by adjusting its design parameters. This process is specialized and thus data sheet THD specifications may not reflect the best achievable. Call us for more details.

Noise in switched-capacitor filters

Noise in switched-capacitor filters has been on the decline since the invention of the device. At this time many devices, like the LTC1064 family, have noise which competes with active RC filters. What is not immediately obvious is that the noise of the switched-capacitor filter is *constant independent of bandwidth.* The LTC1064-2 Butterworth filter has approximately $80\mu V_{RMS}$ noise from 1Hz to 50kHz (f_O equal to 50kHz), it also has $50\mu V_{RMS}$

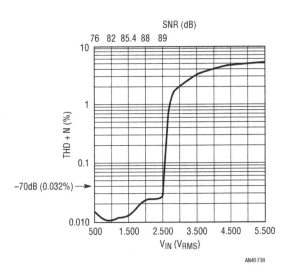

Figure 31.36 • LTC1064-2 THD + Noise vs Input Amplitude and Signal-to-Noise Ratio. Filter f_{CUTOFF} = 8kHz, f_{IN} = 4kHz, f_{CLK} = 800kHz, V_S = ±5V. Inverting Buffer LT1006

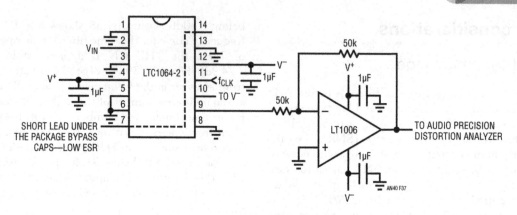

Figure 31.37 • LTC1064-2 Circuit Used for THD Testing. V$_S$ = ±5V

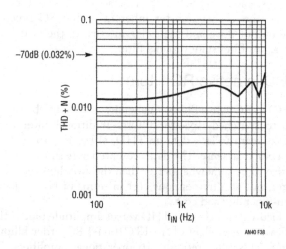

AN40 F38

Figure 31.38 • LTC1064-2 THD + Noise vs Input Frequency.
Filter f$_{CUTOFF}$ = 8kHz, f$_{CLK}$ = 800kHz, V$_S$ = ±5V, V$_{IN}$ =
1.5V$_{RMS}$, Buffered Output Using LT1006

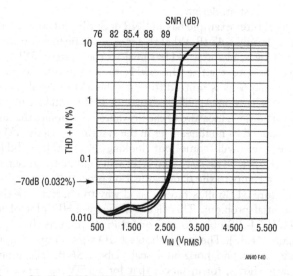

AN40 F40

Figure 31.40 • LTC1064-2 THD + Noise vs Input Amplitude
and Signal-to-Noise Ratio. Filter f$_{CUTOFF}$ = 8kHz, f$_{CLK}$ =
800kHz, f$_{IN}$ = 1kHz, V$_S$ = ±5V, Inverting Buffer Using
LT1006. Four Devices Superimposed

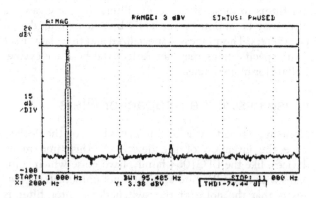

Figure 31.39 • LTC1064-2 Spectrum Analyzer Plot for V$_{IN}$ =
1.5V$_{RMS}$, f$_{IN}$ = 2kHz, V$_S$ = ±5V, Buffered Output Using
LT1006

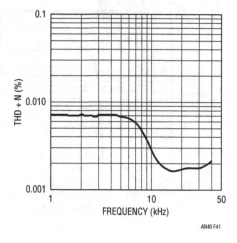

AN40 F41

Figure 31.41 • Active RC Implementation of LTC1064-1 8th
Order Elliptic Filter. V$_{IN}$ = 3V$_{RMS}$, V$_S$ = ±7.5V, Op Amps =
TL084, f$_C$ = 40kHz

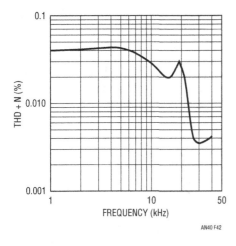

Figure 31.42 • LTC1064-1 8th Order Elliptic Filter, V_IN = 3V_RMS, V_S = ±7.5V, f_C = 40kHz, Buffered Output

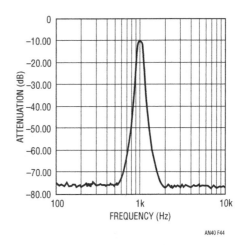

Figure 31.44 • 8th Order Bessel Using LTC1064. V_S = ±8V, V_IN = 1V_RMS, f_CLK = 50kHz, f_CUTOFF = 1kHz, Buffered Output

noise from 1Hz to 10kHz (f_O equal to 10kHz). Since the traditional RC active filter has noise specifications based on so many nV per Hz[12], the switched-capacitor filter is a better competitor to the active RC as the filter cutoff frequency increases. Future LTC devices will have even better noise specifications than the LTC1064-2.

Figure 31.43 compares noise between an RC active equivalent of the LTC1064-1 and the switched-capacitor filter. Both curves show typical peaking at the corner frequency. The illuminating feature seen in this figure is that the TL084 active filter noise is only slightly better than the LTC1064-1.

Bandpass filters and noise—an illustration

Figure 31.44 is the frequency response of an 8th order Bessel bandpass filter implemented with an LTC1064 as shown in Figure 31.45. This filter has a Q of approximately

nine and a very linear phase response (in the passband) as shown in Figure 31.47. As previously discussed, the Bessel response is very useful when signal phase is important.

Of particular interest to the present discussion is the noise of this bandpass filter (Figure 31.46). Note that the noise bandshape is identical to Figure 31.44. This is not unusual since the bandpass filter is letting only the noise at a particular f_C through the filter. This is **not** clock feedthru and it is **not** peculiar to the switched-capacitor filter. In an active RC, or even an LC passive bandpass filter with these characteristics, noise appears "like a signal" at the center frequency of the BPF.

Clock circuitry

Jitter

Clocking of switched-capacitor filters cannot be taken for granted. A good, stable clock is required to obtain device performance commensurate with the data sheet specifications. This implies that 555 type oscillators are *verboten*. Often, what appears as insufficient stopband attenuation or excessive passband ripple is, in fact, caused by poor clocking of the switched-capacitor filter device.

Figure 31.48 shows the LTC1064-1 set up to provide cutoff frequency of 500Hz. The clock was modulated (in the top curve measurement) to simulate approximately 50% clock jitter. The stopband attenuation at 750Hz is seen to be approximately 42dB instead of the specified (in the LTC1064-1 data sheet at 1.5× the cutoff

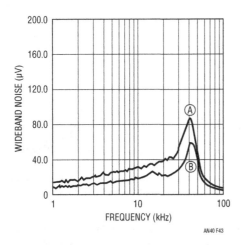

Figure 31.43 • Noise Comparison Between Figure 31.41 Active RC Elliptic Ⓑ and LTC1064-1 Ⓐ, Figure 31.42

Note 12: Ghausi, M.S., and K.R. Laker, "Modern Filter Design, Active RC and Switched Capacitor." Englewood Cliffs, New Jersey: Prentice-Hall, Inc., 1981.

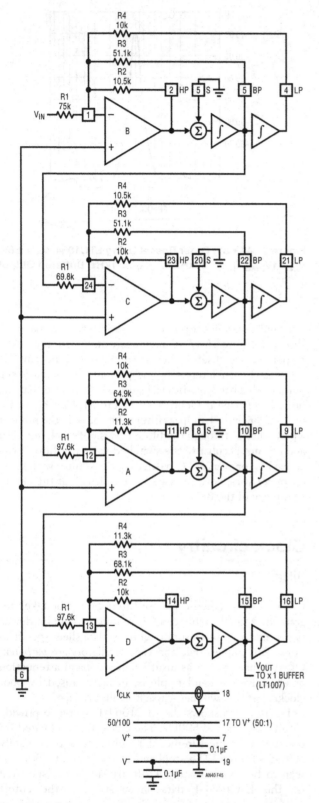

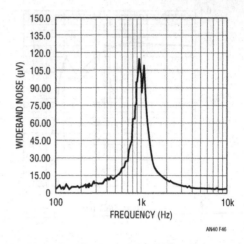

Figure 31.46 • Noise Spectrum of Bessel BPF as Shown in Figure 31.43. Input Grounded. Total Wideband Noise = 130µVRMS

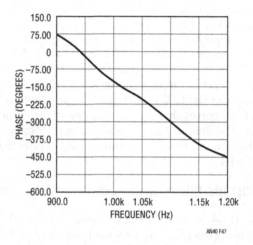

Figure 31.47 • Passband Phase Response of Bessel BPF as Shown in Figure 31.43. Note the Linear Phase Through the Passband

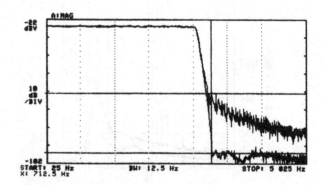

Figure 31.45 • Implementation, Section by Section, of Bessel BPF as per Figure 31.43 Using LTC1064. For f_CUTOFF = 1kHz, f_CLK = 50kHz

Figure 31.48 • Clock Jitter of Approximately 50% (Top Curve in Stopband) and Jitter-Free Clock (Bottom Curve in Stopband) Showing the Difference in Response. LTC1064-1, f_CLK = 50kHz, f_O = 500Hz, Buffered Output

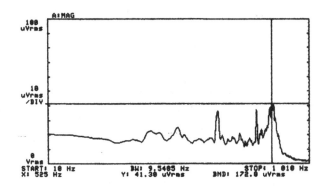

Figure 31.49 • Noise of LTC1064-1 as per Figure 31.48 with ≈50% Clock Jitter. V$_S$ = ±7.5V, Input Grounded

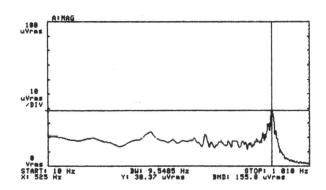

Figure 31.50 • Noise of LTC1064-1 as per Figure 31.48 with Low Jitter (<1ns). V$_S$ = ±7.5V, Input Grounded

frequency) 68dB. The second curve on the graph shows the situation when a good stable clock is used.

Similar graphs of the noise in Figures 31.49 and 31.50 show the effect of clock jitter on the noise. The wideband noise from 10Hz to 1kHz rises when a jittery clock is used from 156µV$_{RMS}$ to 173µV$_{RMS}$. This is an increase of approximately 11% due only to a poor clocking strategy.

LTC's Application Note 12 provides some good clock sources if none are available in the system.

Clock synchronization with A/D sample clock

Synchronizing the A/D and switched-capacitor filter clocks is highly recommended. This allows the A/D to receive filtered data at a constant time and to ensure that the system has settled to its desired accuracy.

Clock feedthru

While clock feedthru has been greatly improved in the recent generation of switched-capacitor filters, some users still want to further limit this anomaly.

The higher the clock-to-f$_{CUTOFF}$ ratio, the easier it is to filter out clock feedthru.

Figure 31.20 in the aliasing study shows the clock feedthru at 50kHz to be −61dB. This is below 0dB, which in this case is 500mV. Clock feedthru here is approximately 400uV. Inserting a simple RC filter (well outside the passband of the filter) at the output of the filter can reduce this by a factor of ten.

Figure 31.51 shows a similar set of curves with a simple RC on the output of the buffer amplifier (see Figure 31.4). The RC values were 9.64k and 3300pF. Figure 31.51 shows that clock feedthru has been reduced to −82dB below 500mV (40µV) when this post filter is used.

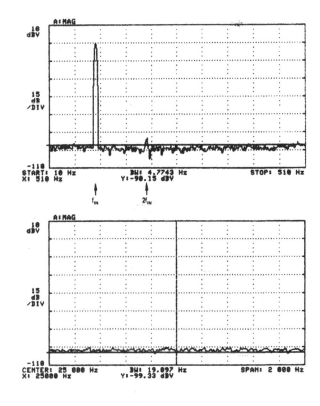

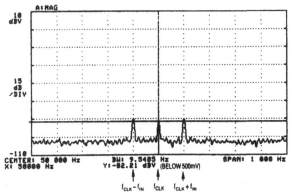

Figure 31.51 • Elliptic Switched-Capacitor Filter (LTC1064-1) Clock Feedthru Suppression. f$_O$ = 500Hz, f$_{CLK}$ = 50kHz, V$_S$ = ±8V, V$_{IN}$ = 100Hz at 500mV$_{RMS}$, RC Filter (9.64k, 3300pF) at Buffer Output, f$_{RC FILTER}$ = 1/2πRC = 5kHz

Post filtering is often unnecessary, as often times the clock feedthru is out of the band of interest.

Conclusions

Switched-capacitor filters are an evolving technology which continues to improve. As this evolution progresses, the switched-capacitor filter will replace greater numbers of active RC filters because of the switched-capacitor filter's inherent smaller size, better accuracy and tunability.

To best take advantage of this evolving technology the system designer must observe good engineering practices as described in this Application Note. More specifically, to properly evaluate and use the current crop of switched-capacitor filters as well as parts on the drawing boards one must observe certain precautions:

1) Utilize good breadboarding techniques.
2) Use a linear power supply. If this is impossible use a clean switcher. Properly bypass the supply.
3) Be aware that sampled data systems can alias and be prepared to deal with this limitation. Bandlimit!
4) Be aware that the ultimate response in the frequency domain is not the ultimate response in the time domain and vice versa. Look at both responses on the bench before committing a filter to the PCB or silicon.
5) Understand THD and signal-to-noise ratio and where one limits the other.
6) Provide a good clean clock to the switched-capacitor filter to avoid problems caused by too much clock jitter.

Appendix A

Square wave to sine wave conversion graphically illustrates the frequency domain, time domain and aliasing aspects of switched-capacitor filters

Figures A1 through A12 illustrate yet another use of the versatile switched-capacitor filter. In the past it has been difficult to obtain a good clean sine wave locked to a square wave input, if the square wave varies in frequency. In this example a square wave is filtered by an Elliptic filter (the LTC1064-1) to produce an excellent sine wave which is phase locked to the input. A Butterworth filter (the LTC1064-2) will also work for this application, but the sine wave will not be quite as pure.

The series of figures illustrates a varied input square wave (from 1kHz to 7500Hz) to an LTC1064-1 Elliptic filter. Recall that a square wave

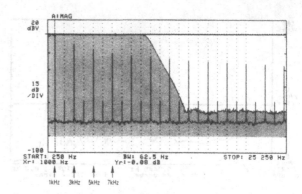

Figure A1 • Frequency Domain (Spectrum Analyzer) Plot of 1kHz Square Wave (5V$_{P-P}$) *Input* to LTC1064-1 as Configured in Figure 31.4. f$_{CUTOFF}$ = 10kHz. *Shaded Area* is the Frequency Response of the Filter

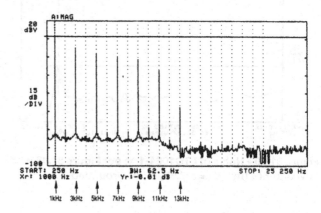

Figure A2 • Filter Output (Shown in Frequency Domain) for 1kHz Input. Note Raised Noise Floor Due to LTC1064-1 Approx. 75dB Noise Level

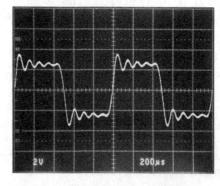

Figure A3 • Filter Time Domain Output for Input 1kHz Square Wave (Photograph Reproduction of Original)

consists of odd harmonics of the fundamental; that is, a 1kHz square wave should contain (in the mathematical work) 1kHz, 3kHz, 5kHz, 7kHz, 9kHz,

11kHz. . . spectral lines, *that is sine waves*, all added together to produce the 1kHz square wave.

Figures A2 and A3 show the Elliptic filter passing the 1kHz square wave with poor fidelity. As the input frequency increases, the fidelity decreases as fewer and fewer of the square waves' spectral lines are passed through the filter.

Figures A5 and A6 show the spectrum analyzer and oscilloscope response from the filter's output for an input square wave of 2.5kHz.

Figures A8 and A9 show the frequency and time domain responses from the filter's output for an input square wave of 5kHz. The 10kHz cutoff frequency of the filter passes only the first harmonic (5kHz) and the 10kHz second harmonic (which is a signal generator problem and should not be there at all). The output

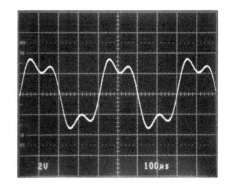

Figure A6 • Filter Time Domain Output for Input 2500Hz Square Wave (Photograph Reproduction of Original)

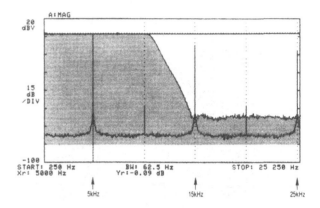

Figure A7 • Frequency Domain (Spectrum Analyzer) Plot of 5kHz Square Wave (5V$_{P-P}$) *Input* **to LTC1064-1 as Configured in Figure 31.4. f$_{CUTOFF}$ = 10kHz.** *Shaded Area* **is the Frequency Response of the Filter**

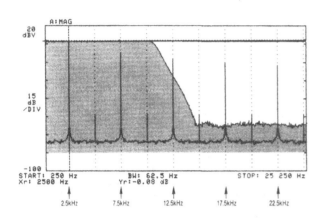

Figure A4 • Frequency Domain (Spectrum Analyzer) Plot of 2500Hz Square Wave (5V$_{P-P}$) *Input* **to LTC1064-1 as Configured in Figure 31.4. f$_{CUTOFF}$ = 10kHz.** *Shaded Area* **is the Frequency Response of the Filter**

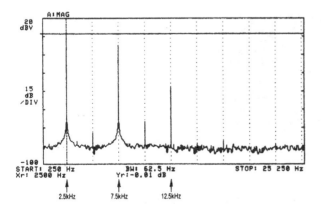

Figure A5 • Filter Output (Shown in Frequency Domain) for 2500Hz Input. Note Raised Noise Floor Due to LTC1064-1 Approx. 75dB Noise Level

appears to be a nice clean sine wave as seen in Figure A9.

Figure A10 shows the input spectrum of a 7.5kHz square wave. Figure A11 is the output from the filter in the frequency domain. In addition to the 7.5kHz spectral line producing the sine wave shown in Figure A12, there are other lines. These lines at 2.5, 5, 10 and 12.5kHz are the result of the 132nd, 133rd, 134th and 135th (!!!!!!) harmonics of the 7.5kHz fundamental frequency of the input square wave aliasing back into the passband of the filter. The perfect square wave should not contain even harmonics but the 133rd and 135th harmonic would remain. Again, WARNING: Bandlimit your input signal or risk aliasing!![1]

Note 1: Thanks to Lew Cronis for the inspiration to do this Appendix.

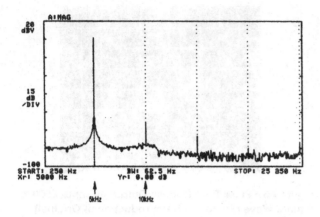

Figure A8 • Filter Output (Shown in Frequency Domain) for 5kHz Input. Note Raised Noise Floor Due to LTC1064-1 Approx. 75dB Noise Level

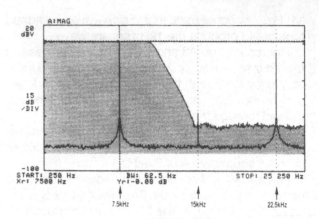

Figure A10 • Frequency Domain (Spectrum Analyzer) Plot of 7500Hz Square Wave (5V$_{P-P}$) *Input* to LTC1064-1 as Configured in Figure 31.4. f$_{CUTOFF}$ = 10kHz. *Shaded Area* is the Frequency Response of the Filter

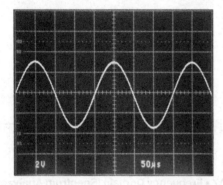

Figure A9 • Filter Time Domain Output for Input 5kHz Square Wave (Photograph Reproduction of Original)

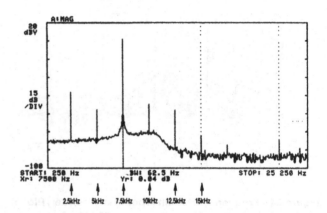

Figure A11 • Filter Output (Shown in Frequency Domain) for 7500Hz Input. Note Raised Noise Floor to LTC1064-1 Approx, 75dB Noise Level

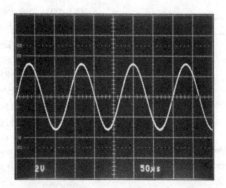

Figure A12 • Filter Time Domain Output for Input 7.5kHz Square Wave (Photograph Reproduction of Original)

Appendix B
About bypass capacitors

Bypass capacitors are used to maintain low power supply impedance at the point of load. Parasitic resistance and inductance in supply lines mean that the power supply impedance can be quite high. As frequency goes up, the inductive parasitic becomes particularly troublesome. Even if these parasitic terms did not exist, or if local regulation is used, bypassing is still necessary because no power supply or regulator has zero output impedance at 100MHz. What type of bypass capacitor to use is determined by the application, frequency domain of the circuit, cost, board space and many other considerations. Some useful generalizations can be made.

All capacitors contain parasitic terms, some of which appear in Figure B1. In bypass applications, leakage and dielectric absorption are 2nd order terms but series R and L are not. These latter terms limit the capacitor's ability to damp transients and maintain low supply impedance. Bypass capacitors must often be large values so they can absorb long transients, necessitating electrolytic types which have large series R and L.

Different types of electrolytics and electrolytic-non-polar combinations have markedly different characteristics. Which type(s) to use is a matter of passionate debate in some circles and the test circuit (Figure B2) and accompanying photos are useful. The photos show the response of five bypassing methods to the transient generated by the test circuit. Figure B3 shows an unbypassed line which sags and ripples badly at large amplitudes. Figure B4 uses an aluminum 10µF electrolytic to considerably cut the disturbance, but there is still plenty of potential trouble. A tantalum 10µF unit offers cleaner response in B5 and the 10µF aluminum combined with a 0.01 µF ceramic type is even better in B6. Combining electrolytics with non-polarized capacitors is a popular way to get good response but beware of picking the wrong duo. The right (wrong) combination of supply line parasitics and paralleled dissimilar capacitors can produce a resonant, ringing response, as in B7. Caveat!

Thanks to Nello Sevastopoulos, Philip Karantzalis and Kevin Vasconcelos for their generous assistance with this Application Note. Thanks to Lew Cronis for the inspiration to do Appendix A.

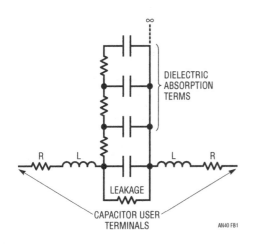

Figure B1 ● Parasitic Terms of a Capacitor

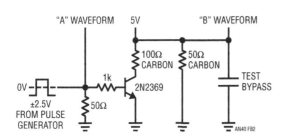

Figure B2 ● Bypass Capacitor Test Circuit

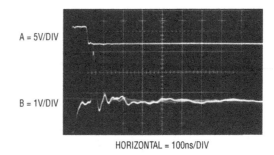

Figure B3 ● Response of Unbypassed Line

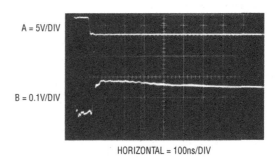

Figure B4 ● Response of 10µF Aluminum Capacitor

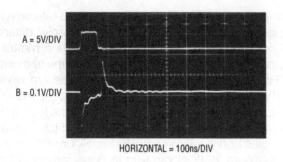

Figure B5 • Response of 10μF Tantalum Capacitor

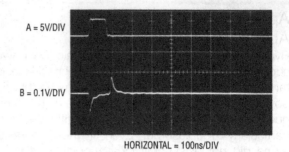

Figure B6 • Response of 10μF Aluminum Paralleled by 0.01μF Ceramic

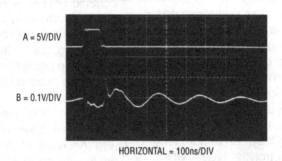

Figure B7 • Some Paralleled Combinations Can Ring. Try Before Specifying!

Bibliography

1. Sevastopoulos, Nello, and Markell, Richard, "Four-Section Switched-Cap Filter Chips Take on Discretes." Electronic Products, September 1, 1988.

2. Brokaw, A. Paul, "An IC Amplifier User's Guide to Decoupling, Grounding, and Making Things Go Right for a Change." Analog Devices Data-Acquisition Databook 1984, Volume I, Pgs. 20-13 to 20-20.

3. Morrison, Ralph, "Grounding and Shielding Techniques in Instrumentation." Second Edition, New York, NY: John Wiley and Sons, Inc., 1977.

4. Motchenbacher, C. D., and Fitchen, F. C., "Low-Noise Electronic Design." New York, NY: John Wiley and Sons, Inc., 1973.

5. Rich, Alan, "Shielding and Guarding." Analog Dialogue, 17, No. 1 (1983), 8-13. Also published as an Application Note in Analog Devices Data-Acquisition Databook 1984, Volume I, Pgs. 20-85 to 20-90.

6. Rich, Alan, "Understanding Interference-Type Noise." Analog Dialogue, 16, No. 3 (1982), 16-19. Also published as an Application Note in Analog Devices Data-Acquisition Databook 1984, Volume I, Pgs. 20-81 to 20-84.

7. Williams, Jim, and Huffman, Brian, "Some Thoughts on DC-DC Conver-ters," Linear Technology AN29.

8. Jung, Walter, "IC Op Amp Cookbook." Howard W. Sams, 1988.

9. Schwartz, Mischa, "Information, Transmission, Modulation and Noise." New York, NY: McGraw Hill Book Co., 1980.

10. Zverev, "Handbook of Filter Synthesis." New York, NY: John Wiley and Son, Inc., Copyright 1967.

11. Daryanani, Gobind, "Principles of Active Network Synthesis and Design." New York, NY: John Wiley and Sons, Inc., Copyright 1976.

12. Ghausi, M. S., and K. R. Laker, "Modern Filter Design, Active RC and Switched Capacitor." Englewood Cliffs, New Jersey: Prentice-Hall, Inc., 1981.

32

Bridge circuits

Marrying gain and balance

Jim Williams

Bridge circuits are among the most elemental and powerful electrical tools. They are found in measurement, switching, oscillator and transducer circuits. Additionally, bridge techniques are broadband, serving from DC to bandwidths well into the GHz range. The electrical analog of the mechanical beam balance, they are also the progenitor of all electrical differential techniques.

Resistance bridges

Figure 32.1 shows a basic resistor bridge. The circuit is usually credited to Charles Wheatstone, although S. H. Christie, who demonstrated it in 1833, almost certainly preceded him.[1] If all resistor values are equal (or the two sides *ratios* are equal) the differential voltage is zero. The

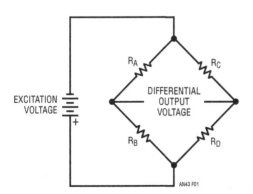

Figure 32.1 • The Basic Wheatstone Bridge, Invented by S. H. Christie

excitation voltage does not alter this, as it affects both sides equally. When the bridge is operating off null, the excitation's magnitude sets output sensitivity. The bridge output is nonlinear for a single variable resistor. Similarly, two variable arms (e.g., R_C and R_B both variable) produce nonlinear output, although sensitivity doubles. Linear outputs are possible by complementary resistance swings in one or both sides of the bridge.

A great deal of attention has been directed towards this circuit. An almost uncountable number of tricks and techniques have been applied to enhance linearity, sensitivity and stability of the basic configuration. In particular, transducer manufacturers are quite adept at adapting the bridge to their needs (see Appendix A, "Strain gauge bridges"). Careful matching of the transducer's mechanical characteristics to the bridge's electrical response can provide a trimmed, calibrated output. Similarly, circuit designers have altered performance by adding active elements (e.g., amplifiers) to the bridge, excitation source or both.

Bridge output amplifiers

A primary concern is the accurate determination of the differential output voltage. In bridges operating at null the absolute scale factor of the readout device is normally less important than its sensitivity and zero point stability. An off-null bridge measurement usually requires a well calibrated scale factor readout in addition to zero point stability. Because of their importance, bridge readout mechanisms have a long and glorious history (see Appendix B, "Bridge readout—then and now"). Today's investigator has a variety of powerful electronic techniques available to obtain highly accurate bridge readouts. Bridge amplifiers are designed to accurately extract the bridge's differential output from its common mode level. The ability to reject common mode signal is quite critical. A typical 10V powered strain gauge transducer produces only 30mV of signal "riding" on 5V of

Note 1: Wheatstone had a better public relations agency, namely himself. For Fascinating details, see Reference 19.

Analog Circuit and System Design: A Tutorial Guide to Applications and Solutions. DOI: 10.1016/B978-0-12-385185-7.00032-9

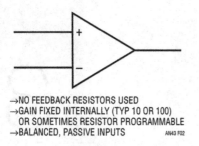

→NO FEEDBACK RESISTORS USED
→GAIN FIXED INTERNALLY (TYP 10 OR 100)
  OR SOMETIMES RESISTOR PROGRAMMABLE
→BALANCED, PASSIVE INPUTS   AN43 F02

Figure 32.2 • Conceptual Instrumentation Amplifier

common mode level. 12-bit readout resolution calls for an LSB of only 7.3µV... almost 120dB below the common mode signal! Other significant error terms include offset voltage, and its shift with temperature and time, bias current and gain stability. Figure 32.2 shows an "instrumentation amplifier," which makes a very good bridge amplifier. These devices are usually the first choice for bridge measurement, and bring adequate performance to most applications.

In general, instrumentation amps feature fully differential inputs and internally determined stable gain. The absence of a feedback network means the inputs are essentially passive, and no significant bridge loading occurs. Instrumentation amplifiers meet most bridge requirements. Figure 32.3 lists performance data for some specific instrumentation amplifiers. Figure 32.4's table summarizes some options for DC bridge signal conditioning. Various approaches are presented, with pertinent characteristics noted. The constraints, freedoms and performance requirements of any particular application define the best approach.

DC bridge circuit applications

Figure 32.5, a typical bridge application, details signal conditioning for a 350Ω transducer bridge. The specified strain gauge pressure transducer produces 3mV output per volt of

bridge excitation (various types of strain-based transducers are reviewed in Appendix A, "Strain gauge bridges"). The LT1021 reference, buffered by A1A and A2, drives the bridge. This potential also supplies the circuit's ratio output, permitting ratiometric operation of a monitoring A/D converter. Instrumentation amplifier A3 extracts the bridge's differential output at a gain of 100, with additional trimmed gain supplied by A1B. The configuration shown may be adjusted for a precise 10V output at full-scale pressure. The trim at the bridge sets the zero pressure scale point. The RC combination at A1B's input filters noise. The time constant should be selected for the system's desired lowpass cutoff. "Noise" may originate as residual RF/line pick-up or true transducer responses to pressure variations. In cases where noise is relatively high it may be desirable to filter ahead of A3. This prevents any possible signal infidelity due to nonlinear A3 operation. Such undesirable outputs can be produced by saturation, slew rate components, or rectification effects. When filtering ahead of the circuit's gain blocks remember to allow for the effects of bias current induced errors caused by the filter's series resistance. This can be a significant consideration because large value capacitors, particularly electrolytics, are not practical. If bias current induced errors rise to appreciable levels FET or MOS input amplifiers may be required (see Figure 32.3).

To trim this circuit apply zero pressure to the transducer and adjust the 10k potentiometer until the output *just* comes off 0V. Next, apply full-scale pressure and trim the 1k adjustment. Repeat this procedure until both points are fixed.

Common mode suppression techniques

Figure 32.6 shows a way to reduce errors due to the bridge's common mode output voltage. A1 biases Q1 to

PARAMETER	LTC1100	LT1101	LT1102	LTC1043 (USING LTC1050 AMPLIFIER)
Offset	10µV	160µV	500µV	0.5µV
Offset Drift	100nV/°C	2µV/°C	2.5µV/°C	50nV/°C
Bias Current	50pA	8nA	50pA	10pA
Noise (0.1Hz to 10Hz)	2µVP-P	0.9µV	2.8µV	1.6µV
Gain	100	10,100	10,100	Resistor Programmable
Gain Error	0.03%	0.03%	0.05%	Resistor Limited 0.001% Possible
Gain Drift	4ppm/°C	4ppm/°C	5ppm/°C	Resistor Limited <1ppm/°C Possible
Gain Nonlinearity	8ppm	8ppm	10ppm	Resistor Limited 1ppm Possible
CMRR	104dB	100dB	100dB	160dB
Power Supply	Single or Dual, 16V Max	Single or Dual, 44V Max	Dual, 44V Max	Single, Dual 18V Max
Supply Current	2.2mA	105µA	5mA	2mA
Slew Rate	1.5V/µs	0.07V/µs	25V/µs	1mV/ms
Bandwidth	8kHz	33kHz	220kHz	10Hz

Figure 32.3 • Comparison of Some IC Instrumentation Amplifiers

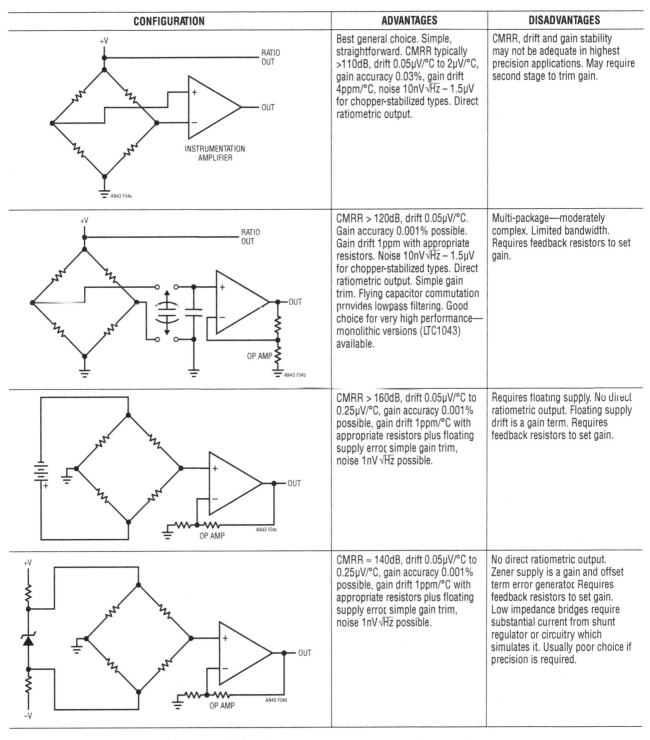

CONFIGURATION	ADVANTAGES	DISADVANTAGES
	Best general choice. Simple, straightforward. CMRR typically >110dB, drift 0.05µV/°C to 2µV/°C, gain accuracy 0.03%, gain drift 4ppm/°C, noise 10nV√Hz – 1.5µV for chopper-stabilized types. Direct ratiometric output.	CMRR, drift and gain stability may not be adequate in highest precision applications. May require second stage to trim gain.
	CMRR > 120dB, drift 0.05µV/°C. Gain accuracy 0.001% possible. Gain drift 1ppm with appropriate resistors. Noise 10nV√Hz – 1.5µV for chopper-stabilized types. Direct ratiometric output. Simple gain trim. Flying capacitor commutation provides lowpass filtering. Good choice for very high performance—monolithic versions (LTC1043) available.	Multi-package—moderately complex. Limited bandwidth. Requires feedback resistors to set gain.
	CMRR > 160dB, drift 0.05µV/°C to 0.25µV/°C, gain accuracy 0.001% possible, gain drift 1ppm/°C with appropriate resistors plus floating supply error, simple gain trim, noise 1nV √Hz possible.	Requires floating supply. No direct ratiometric output. Floating supply drift is a gain term. Requires feedback resistors to set gain.
	CMRR ≈ 140dB, drift 0.05µV/°C to 0.25µV/°C, gain accuracy 0.001% possible, gain drift 1ppm/°C with appropriate resistors plus floating supply error, simple gain trim, noise 1nV √Hz possible.	No direct ratiometric output. Zener supply is a gain and offset term error generator. Requires feedback resistors to set gain. Low impedance bridges require substantial current from shunt regulator or circuitry which simulates it. Usually poor choice if precision is required.

Figure 32.4 • Some Signal Conditioning Methods for Bridges

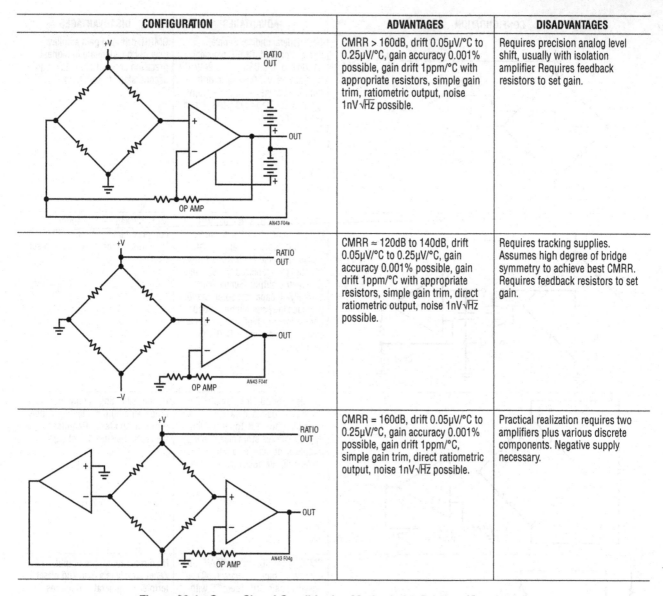

CONFIGURATION	ADVANTAGES	DISADVANTAGES
	CMRR > 160dB, drift 0.05μV/°C to 0.25μV/°C, gain accuracy 0.001% possible, gain drift 1ppm/°C with appropriate resistors, simple gain trim, ratiometric output, noise 1nV√Hz possible.	Requires precision analog level shift, usually with isolation amplifier Requires feedback resistors to set gain.
	CMRR ≈ 120dB to 140dB, drift 0.05μV/°C to 0.25μV/°C, gain accuracy 0.001% possible, gain drift 1ppm/°C with appropriate resistors, simple gain trim, direct ratiometric output, noise 1nV√Hz possible.	Requires tracking supplies. Assumes high degree of bridge symmetry to achieve best CMRR. Requires feedback resistors to set gain.
	CMRR = 160dB, drift 0.05μV/°C to 0.25μV/°C, gain accuracy 0.001% possible, gain drift 1ppm/°C, simple gain trim, direct ratiometric output, noise 1nV√Hz possible.	Practical realization requires two amplifiers plus various discrete components. Negative supply necessary.

Figure 32.4 • Some Signal Conditioning Methods for Bridges (Continued)

servo the bridge's left mid-point to zero under all operating conditions. The 350Ω resistor ensures that A1 will find a stable operating point with 10V of drive delivered to the bridge. This allows A2 to take a single-ended measurement, eliminating all common mode voltage errors. This approach works well, and is often a good choice in high precision work. The amplifiers in this example, CMOS chopper-stabilized units, essentially eliminate offset drift with time and temperature. Trade-offs compared to an instrumentation amplifier approach include complexity and the requirement for a negative supply. Figure 32.7 is similar, except that low noise bipolar amplifiers are used. This circuit trades slightly higher DC offset drift for lower

noise and is a good candidate for stable resolution of small, slowly varying measurands. Figure 32.8 employs chopper-stabilized A1 to reduce Figure 32.7's already small offset error. A1 measures the DC error at A2's inputs and biases A1's offset pins to force offset to a few microvolts. The offset pin biasing at A2 is arranged so A1 will always be able to find the servo point. The 0.01μF capacitor rolls off A1 at low frequency, with A2 handling high frequency signals. Returning A2's feedback string to the bridge's mid-point eliminates A4's offset contribution. If this was not done A4 would require a similar offset correction loop. Although complex, this approach achieves less than 0.05μV/°C drift, 1nV√Hz noise and CMRR exceeding 160dB.

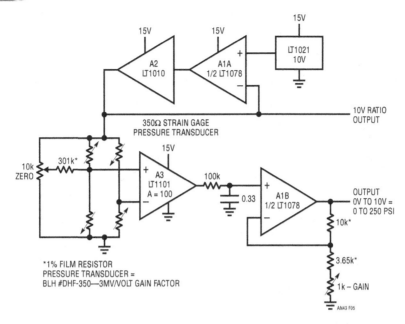

Figure 32.5 • A Practical Instrumentation Amplifier-Based Bridge Circuit

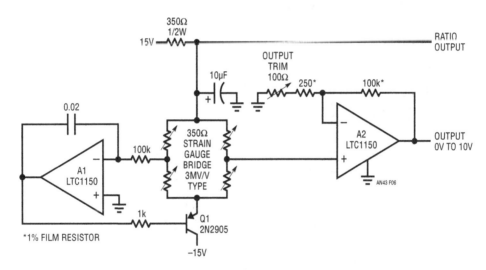

Figure 32.6 • Servo Controlling Bridge Drive Eliminates Common Mode Voltage

Single supply common mode suppression circuits

The common mode suppression circuits shown require a negative power supply. Often, such circuits must function in systems where only a positive rail is available. Figure 32.9 shows a way to do this. A2 biases the LTC®1044 positive-to-negative converter. The LTC1044's output pulls the bridge's output negative, causing A1's input to balance at

0V. This local loop permits a single-ended amplifier (A2) to extract the bridge's output signal. The 100k-0.33μF RC filters noise and A2's gain is set to provide the desired output scale factor. Because bridge drive is derived from the LT1034 reference, A2's output is not affected by supply shifts. The LT1034's output is available for ratio operation. Although this circuit works nicely from a single 5V rail the transducer sees only 2.4V of drive. This reduced drive results in lower transducer outputs for a given measurand value, effectively magnifying amplifier offset drift

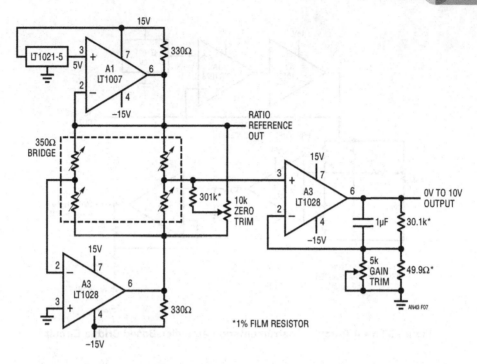

Figure 32.7 • Low Noise Bridge Amplifier with Common Mode Suppression

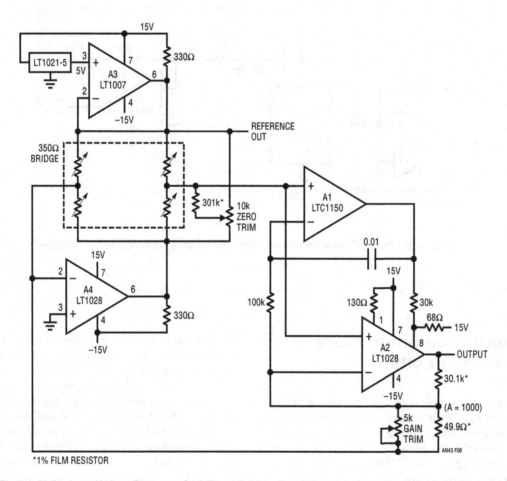

Figure 32.8 • Low Noise, Chopper-Stabilized Bridge Amplifier with Common Mode Suppression

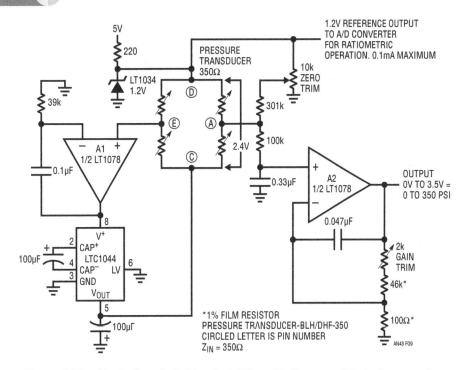

Figure 32.9 • Single Supply Bridge Amplifier with Common Mode Suppression

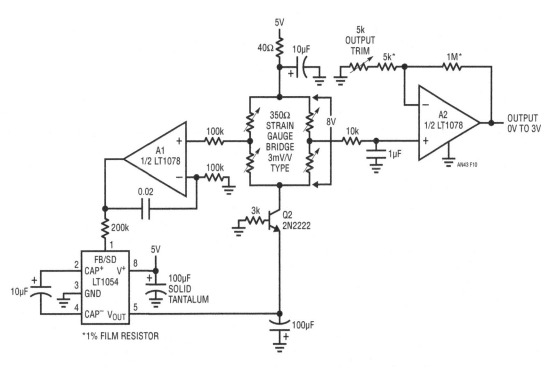

Figure 32.10 • High Resolution Version of Figure 32.9. Bipolar Voltage Converter Gives Greater Bridge Drive, Increasing Output Signal

terms. The limit on available bridge drive is set by the CMOS LTC1044's output impedance. Figure 32.10's circuit employs a bipolar positive-to-negative converter which has much lower output impedance. The biasing used permits 8V to appear across the bridge, requiring the 100mA capability LT1054 to sink about 24mA. This increased drive results in a more favorable transducer gain slope, increasing signal-to-noise ratio.

Switched-capacitor based instrumentation amplifiers

Switched-capacitor methods are another way to signal condition bridge outputs. Figure 32.11 uses a flying capacitor configuration in a very high precision-scale application. This design, intended for weighing human subjects, will resolve 0.01 pounds at 300.00 pounds full scale. The strain gauge based transducer platform is excited at 10V by the LT1021 reference, A1 and A2. The LTC1043 switched-capacitor building block combines with A3, forming a differential input chopper-stabilized amplifier. The LTC1043 alternately connects the 1μF flying capacitor between the strain gauge bridge output and A3's input. A second 1μF unit stores the LTC1043 output, maintaining A3's input at DC. The LTC1043's low charge injection maintains differential to single-ended transfer accuracy of about 1ppm at DC and low frequency. The commutation rate, set by the 0.01μF capacitor, is about 400Hz. A3 takes scaled gain, providing 3.0000V for 300.00 pounds full-scale output.

The extremely high resolution of this scale requires filtering to produce useful results. Very slight body movement acting on the platform can cause significant noise in A3's output. This is dramatically apparent in Figure 32.12's tracings. The total *force* on the platform is equal to gravity pulling on the body (the "weight") plus any additional accelerations within or acting upon the body. Figure 32.12 (Trace B) clearly shows that each time the heart pumps, the acceleration due to the blood (mass) moving in the arteries shows up as "weight". To prove this, the subject gets off the scale and runs in place for 15 seconds. When the subject returns to the platform the heart should work harder. Trace A confirms this nicely. The exercise causes the heart to work harder forcing a greater acceleration-per-stroke.[2]

Another source of noise is due to body motion. As the body moves around, its mass doesn't change but the instantaneous accelerations are picked up by the platform and read as "weight" shifts.

All this seems to make a 0.01 pound measurement meaningless. However, filtering the noise out gives a time

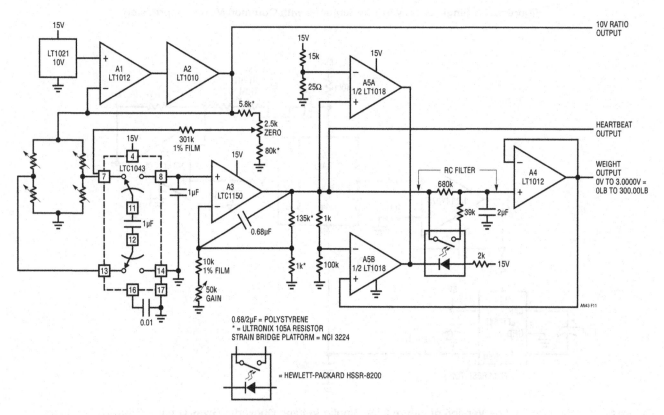

Figure 32.11 • High Precision Scale for Human Subjects

Note 2: Cardiology aficionados will recognize this as a form of *Ballistocardiograph* (from the Greek "ballein"—to throw, hurl or eject and "kardia," heart). A significant amount of effort was expended in attempts to reliably characterize heart conditions via acceleration detection methods. These efforts were largely unsuccessful when compared against the reliability of EKG produced data. See references for further discussion.

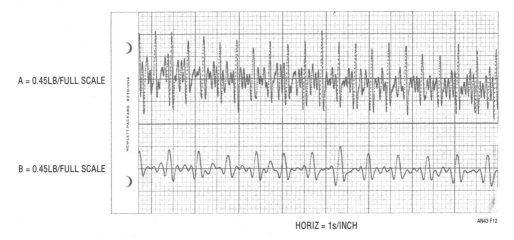

A = 0.45LB/FULL SCALE

B = 0.45LB/FULL SCALE

HEWLETT-PACKARD 9270-1004

HORIZ = 1s/INCH

AN43 F12

Figure 32.12 • High Precision Scale's Heartbeat Output. Trace B Shows Subject at Rest; Trace A After Exercise. Discontinuous Components in Waveforms Leading Edges Are Due to XY Recorder Slew Limitations

averaged value. A simple RC lowpass will work, but requires excessively long settling times to filter noise fundamentals in the 1Hz region. Another approach is needed.

A4, A5 and associated components form a filter which switches its time constant from short to long when the output has nearly arrived at the final value. With no weight on the platform A3's output is zero. A4's output is also zero, A5B's output is indeterminate and A5A's output is low. The MOSFET opto-coupler's LED comes on, putting the RC filter into short time constant mode. When someone gets on the scale A3's output rises rapidly. A5A goes high, but A5B trips low, maintaining the RC filter in its short time constant mode. The 2μF capacitor charges rapidly, and A4 quickly settles to final value ± body motion and heartbeat noise. A5B's negative input sees 1% attenuation from A3; its positive input does not. This causes A5B to switch high when A4's output arrives within 1% of final value. The opto-coupler goes off and the filter switches into long time constant mode, eliminating noise in A4's output. The 39k resistor prevents overshoot, ensuring monotonic A4 outputs. When the subject steps off the scale A3 quickly returns to zero. A5A goes immediately low, turning on the opto-coupler. This quickly discharges the 2μF capacitor, returning A4's output rapidly to zero. The bias string at A5A's input maintains the scale in fast time constant mode for weights below 0.50 pounds. This permits rapid response when small objects (or persons) are placed on the platform. To trim this circuit, adjust the zero potentiometer for 0V out with no weight on the platform. Next, set the gain adjustment for 3.0000V out for a 300.00 pound platform weight. Repeat this procedure until both points are fixed.

Optically coupled switched-capacitor instrumentation amplifier

Figure 32.13 also uses optical techniques for performance enhancement. This switched-capacitor based instrumentation amplifier is applicable to transducer signal conditioning where high common mode voltages exist. The circuit has the low offset and drift of the LTC1150 but also incorporates a novel switched-capacitor "front end" to achieve some specifications not available in a conventional instrumentation amplifier.

Common mode rejection ratio at DC for the front end exceeds 160dB. The amplifier will operate over a ±200V common mode range and gain accuracy and stability are limited only by external resistors. A1, a chopper stabilized unit, sets offset drift at 0.05μV/°C. The high common mode voltage capability of the design allows it to withstand transient and fault conditions often encountered in industrial environments.

The circuit's inputs are fed to LED-driven optically-coupled MOSFET switches, S1 and S2. Two similar switches, S3 and S4, are in series with S1 and S2. CMOS logic functions, clocked from A1's internal oscillator, generate non-overlapping clock outputs which drive the switch's LEDs. When the "acquire pulse" is high, S1 and S2 are on and C2 acquires the differential voltage at the bridge's output. During this interval, S3 and S4 are off. When the acquire pulse falls, S1 and S2 begin to go off. After a delay to allow S1 and S2 to fully open, the "read pulse" goes high, turning on S3 and S4. Now C1 appears as

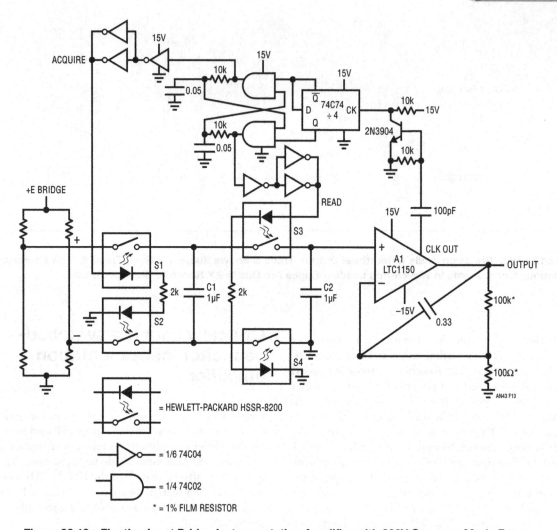

Figure 32.13 • Floating Input Bridge Instrumentation Amplifier with 200V Common Mode Range

a ground-referred voltage source which is read by A1. C2 allows A1's input to retain C1's value when the circuit returns to the acquire mode. A1 provides the circuit's output. Its gain is set in normal fashion by feedback resistors. The 0.33µF feedback capacitor sets roll-off. The differential-to-single-ended transition performed by the switches and capacitors means that A1 never sees the input's common mode signal. The breakdown specification of the optically-driven MOSFET switch allows the circuit to withstand and operate at common mode levels of ±200V. In addition, the optical drive to the MOSFETs eliminates the charge injection problems common to FET switched-capacitive networks.

Platinum RTD resistance bridge circuits

Platinum RTDs are frequently used in bridge configurations for temperature measurement. Figure 32.14's circuit

is highly accurate and features a ground referred RTD. The ground connection is highly desirable for noise rejection. The bridge's RTD leg is driven by a current source while the opposing bridge branch is voltage biased. The current drive allows the voltage across the RTD to vary directly with its temperature induced resistance shift. The difference between this potential and that of the opposing bridge leg forms the bridge's output.

A1A and instrumentation amplifier A2 form a voltage-controlled current source. A1A, biased by the LT1009 reference, drives current through the 88.7Ω resistor and the RTD. A2, sensing differentially across the 88.7Ω resistor, closes a loop back to A1A. the 2k-0.1µF combination sets amplifier roll-off, and the configuration is stable. Because A1A's loop forces a fixed voltage across the 88.7Ω resistor the current through R_P is constant. A1's operating point is primarily fixed by the 2.5V LT1009 voltage reference.

The RTD's constant current forces the voltage across it to vary with its resistance, which has a nearly linear positive

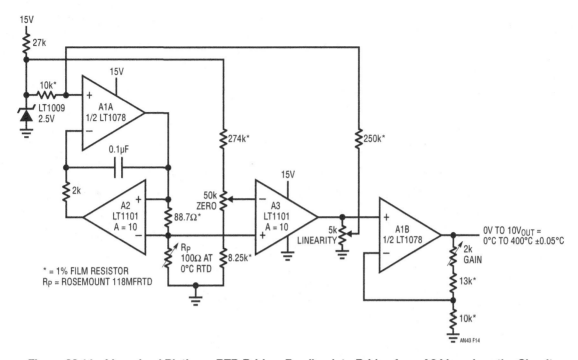

Figure 32.14 • Linearized Platinum RTD Bridge. Feedback to Bridge from A3 Linearizes the Circuit

temperature coefficient. The nonlinearity could cause several degrees of error over the circuit's 0°C to 400°C operating range. The bridge's output is fed to instrumentation amplifier A3, which provides differential gain while simultaneously supplying nonlinearity correction. The correction is implemented by feeding a portion of A3's output back to A1's input via the 10k-250k divider. This causes the current supplied to R_P to slightly shift with its operating point, compensating sensor nonlinearity to within ±0.05°C. A1B, providing additional scaled gain, furnishes the circuit output.

To calibrate this circuit, substitute a precision decade box (e.g., General Radio 1432k) for R_P. Set the box to the 0°C value (100.00Ω) and adjust the offset trim for a 0.00V output. Next, set the decade box for a 140°C output (154.26Ω) and adjust the gain trim for a 3.500V output reading. Finally, set the box to 249.0Ω (400.00°C) and trim the linearity adjustment for a 10.000V output. Repeat this sequence until all three points are fixed. Total error over the entire range will be within ±0.05°C. The resistance values given are for a nominal 100.00Ω (0°C) sensor. Sensors deviating from this nominal value can be used by factoring in the deviation from 100.00Ω. This deviation, which is manufacturer specified for each individual sensor, is an offset term due to winding tolerances during fabrication of the RTD. The gain slope of the platinum is primarily fixed by the purity of the material and has a very small error term.

Figure 32.15 is functionally identical to Figure 32.14, except that A2 and A3 are replaced with an LTC1043

switched-capacitor building block. The LTC1043 performs the differential-to-single-ended transitions in the current source and bridge output amplifier. Value shifts in the current source and output stage reflect the LTC1043's lack of gain. The primary trade-off between the two circuits is component count versus cost.

Digitally corrected platinum resistance bridge

The previous examples rely on analog techniques to achieve a precise, linear output from the platinum RTD bridge. Figure 32.16 uses digital corrections to obtain similar results. A processor is used to correct residual RTD nonlinearities. The bridge's inherent nonlinear output is also accommodated by the processor.

The LT1027 drives the bridge with 5V. The bridge differential output is extracted by instrumentation amplifier A1. A1's output, via gain scaling stage A2, is fed to the LTC1290 12-bit A/D. The LTC1290's raw output codes reflect the bridge's nonlinear output versus temperature. The processor corrects the A/D output and presents linearized, calibrated data out. RTD and resistor tolerances mandate zero and full-scale trims, but no linearity correction is necessary. A2's analog output is available for feedback control applications. The complete software code for the 68HC05 processor, developed by Guy M. Hoover, appears in Figure 32.17.

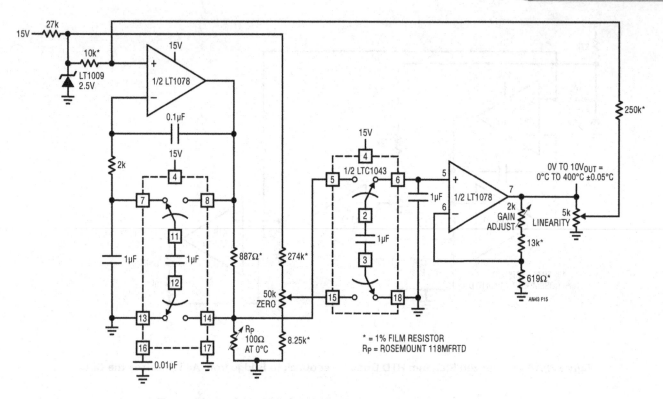

Figure 32.15 • Switched-Capacitor-Based Version of Figure 32.14

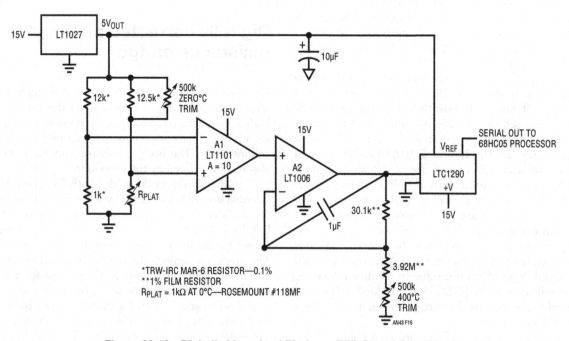

Figure 32.16 • Digitally Linearized Platinum RTD Signal Conditioner

```
*               PLATINUM RTD LINEARIZATION PROGRAM (0.0 TO 400.0 DEGREES C)
*                  WRITTEN BY GUY HOOVER LINEAR TECHNOLOGY CORPORATION
*                                      3/14/90
*             N IS THE NUMBER OF SEGMENTS THAT RTD RESPONSE IS DIVIDED INTO
*                             TEMPERATURE (DEG. C*10)=M*X+B
*                     M IS SLOPE OF RTD RESPONSE FOR A GIVEN SEGMENT
*                        X IS A/D OUTPUT MINUS SEGMENT END POINT
*                      B IS SEGMENT START POINT IN DEGREES C *10.
*
******************************************************************************************
*                               LOOK UP TABLES
*

        ORG  $1000
*                     TABLE FOR SEGMENT END POINTS IN DECIMAL
* X IS FORMED BY SUBTRACTING PROPER SEGMENT END POINT FROM A/D OUTPUT
        FDB  60,296,527,753,976,1195,1410,1621,1829,2032
        FDB  2233,2430,2623,2813,3000,3184,3365,3543,3718,3890
        ORG  $1030
*                             TABLE FOR M IN DECIMAL
*          M IS SLOPE OF RTD OVER A GIVEN TEMPERATURE RANGE
        FDB  3486,3535,3505,3605,3735,3784,3884,3934,3984,4083
        FDB  4133,4232,4282,4382,4432,4531,4581,4681,4730,4830
        ORG  $1060
*                             TABLE FOR B IN DECIMAL
*                          B IS DEGREES C TIMES TEN
        FDB  0,200,400,600,800,1000,1200,1400,1600,1800
        FDB  2000,2200,2400,2600,2800,3000,3200,3400,3600,3800
        ORG  $10FF
        FCB  39 (N*2)-1 IN DECIMAL
*
*                             END LOOK UP TABLES
******************************************************************************************
*                             BEGIN MAIN PROGRAM
*

        ORG  $0100
        LDA  #$F7            CONFIGURATION DATA FOR PORT C DDR
        STA  $06             LOAD CONFIGURATION DATA INTO PORT C
        BSET 0,$02           INITIALIZE B0 PORT C
MES90L  NOP
        LDA  #$2F            DIN WORD FOR 1290 CH4 WITH RESPECT
*                            TO CH5, MSB FIRST, UNIPOLAR, 16 BITS
        STA  $50             STORE DIN WORD IN DIN BUFFER
        JSR  READ90          CALL READ90 SUBROUTINE (DUMMY READ)
        JSR  READ90          CALL READ90 SUBROUTINE (MSBS IN $61 LSBS IN $62)
        LDX  $10FF           LOAD SEGMENT COUNTER INTO X      \ FOR N=20 TO 1
DOAGAIN LDA  $1000,X         LOAD LSBS OF SEGMENT N           \
        STA  $55             STORE LSBS IN $55                 \
        DECX                 DECREMENT X                        \
        LDA  $1000,X         LOAD MSBS OF SEGMENT N              \
        STA  $54             STORE MSBS IN $54                   \ FIND B
        JSR  SUBTRCT         CALL SUBTRCT SUBROUTINE            /
        BPL  SEGMENT         IF RESULT IS PLUS GOTO SEGMENT   /
        JSR  ADDB            CALL ADDB SUBROUTINE            /
        DECX                 DECREMENT X                    /
        JMP  DOAGAIN         GOTO CODE AT LABEL DOAGAIN      / NEXT N
```

Figure 32.17 • Software Code for 68HC05 Processor-Based RTD Linearization

```
*
*
*
*
*
*
SEGMENT  LDA   $1030,X        LOAD MSBS OF SLOPE                    \
         STA   $54            STORE MSBS IN $54                     \
         INCX                 INCREMENT X                           \ M*X
         LDA   $1030,X        LOAD LSBS OF SLOPE                    /
         STA   $55            STORE LSBS IN $55                     /
         JSR   TBMULT         CALL TBMULT SUBROUTINE                /
         LDA   $1060,X        LOAD LSBS OF BASE TEMP                \
         STA   $55            STORE LSBS IN $55                      \
         DECX                 DECREMENT X                              > B ADDED TO M*X
         LDA   $1060,X        LOAD MSBS OF BASE TEMP                /
         STA   $54            STORE MSBS IN $54                     /
         JSR   ADDB           CALL ADDB SUBROUTINE
*                            TEMPERATURE IN DEGREES C * 10 IS IN $61 AND $62
*                                 END MAIN PROGRAM
*****************************************************************************************
*
*
         JMP   MES90L         RUN MAIN PROGRAM IN CONTINUOUS LOOP
*
*****************************************************************************************
*                            SUBROUTINES BEGIN HERE
*
*****************************************************************************************
*        READ90 READS THE LTC1290 AND STORES THE RESULT IN $61 AND $62
*
READ90   LDA   #$50           CONFIGURATION DATA FOR SPCR \
         STA   $0A            LOAD CONFIGURATION DATA       > CONFIGURE PROCESSOR
         LDA   $50            LOAD DIN WORD INTO THE ACC   /
         BCLR  0,$02          BIT 0 PORT C GOES LOW (CS GOES LOW)         \
         STA   $0C            LOAD DIN INTO SPI DATA REG. START TRANSFER. |
BACK90   TST   $0B            TEST STATUS OF SPIF                         |
         BPL   BACK90         LOOP TO PREVIOUS INSTRUCTION IF NOT DONE    |
         LDA   $0C            LOAD CONTENTS OF SPI DATA REG. INTO ACC     |
         STA   $0C            START NEXT CYCLE                            |
         STA   $61            STORE MSBS IN $61                           | XFER
BACK92   TST   $0B            TEST STATUS OF SPIF                         | DATA
         BPL   BACK92         LOOP TO PREVIOUS INSTRUCTION IF NOT DONE    |
         BSET  0,$02          SET BIT 0 PORT C (CS GOES HIGH)             |
         LDA   $0C            LOAD CONTENTS OF SPI DATA REG INTO ACC      |
         STA   $62            STORE LSBS IN $62                           /
         LDA   #$04           LOAD COUNTER WITH NUMBER OF SHIFTS     \
SHIFT    CLC                  CLEAR CARRY                            \
         ROR   $61            ROTATE MSBS RIGHT THROUGH CARRY         \      RIGHT
         ROR   $62            ROTATE LSBS RIGHT THROUGH CARRY         /      JUSTIFY
         DECA                 DECREMENT COUNTER                      /      DATA
         BNE   SHIFT          IF NOT DONE SHIFTING THEN REPEAT LOOP  /
         RTS                  RETURN TO MAIN PROGRAM
*
*                                 END READ90
*****************************************************************************************
```

Figure 32.17 • Software Code for 68HC05 Processor-Based RTD Linearization (Continued)

```
**************************************************************************************
*
*         SUBTRCT SUBTRACTS $54 AND $55 FROM $61 AND $62. RESULTS IN $61 AND $62
*
SUBTRCT LDA   $62              LOAD LSBS
        SUB   $55              SUBTRACT LSBS
        STA   $62              STORE REMAINDER
        LDA   $61              LOAD MSBS
        SBC   $54              SUBTRACT W/CARRY MSBS
        STA   $61              STORE REMAINDER
        RTS                    RETURN TO MAIN PROGRAM
*
*                             END SUBTRCT
**************************************************************************************
**************************************************************************************
*
*ADDB RESTORES $61 AND $62 TO ORIGINAL VALUES AFTER SUBTRCT HAS BEEN PERFORMED
*
ADDB    LDA   $62              LOAD LSBS
        ADD   $55              ADD LSBS
        STA   $62              STORE SUM
        LDA   $61              LOAD MSBS
        ADC   $54              ADD W/CARRY MSBS
        STA   $61              STORE SUM
        RTS                    RETURN TO MAIN PROGRAM
*
*                             END ADDB
**************************************************************************************
**************************************************************************************
*
*TBMULT MULTIPLIES CONTENTS OF $61 AND $62 BY CONTENTS OF $54 AND $55.
*16 MSBS OF RESULT ARE PLACED IN $61 AND $62
*
TBMULT  CLR   $68              CLEAR CONTENTS OF $68    \
        CLR   $69              CLEAR CONTENTS OF $69     \ RESET TEMPORARY
        CLR   $6A              CLEAR CONTENTS OF $6A     / RESULT REGISTERS
        CLR   $6B              CLEAR CONTENTS OF $6B    /
        STX   $58              STORE CONTENTS OF X IN $58. TEMPORARY HOLD REG. FOR X
        LSL   $62              MULTIPLY LSBS BY 2 \
        ROL   $61              MULTIPLY MSBS BY 2  \
        LSL   $62              MULTIPLY LSBS BY 2   \
        ROL   $61              MULTIPLY MSBS BY 2    \ MULTIPLY $61 AND $62 BY 16
        LSL   $62              MULTIPLY LSBS BY 2    / FOR SCALING PURPOSES
        ROL   $61              MULTIPLY MSBS BY 2   /
        LSL   $62              MULTIPLY LSBS BY 2  /
        ROL   $61              MULTIPLY MSBS BY 2 /
        LDA   $62              LOAD LSBS OF 1290 INTO ACC
        LDX   $55              LOAD LSBS OF M INTO X
        MUL                    MULTIPLY CONTENTS OF $55 BY CONTENTS OF $62
        STA   $6B              STORE LSBS IN $6B
        STX   $6A              STORE MSBS IN $6A
        LDA   $62              LOAD LSBS OF 1290 INTO ACC
        LDX   $54              LOAD MSBS OF M INTO X
        MUL                    MULTIPLY CONTENTS OF $54 BY CONTENTS OF $62
        ADD   $6A              LSBS OF MULTIPLY ADDED TO $6A
        STA   $6A              STORE BYTE
        TXA                    TRANSFER X TO ACC
```

Figure 32.17 • Software Code for 68HC05 Processor-Based RTD Linearization (Continued)

```
            ADC   $69              ADD NEXT BYTE
            STA   $69              STORE BYTE
            LDA   $61              LOAD MSBS OF 1290 INTO ACC
            LDX   $55              LOAD LSBS OF M INTO X
            MUL                    MULTIPLY CONTENTS OF $55 BY CONTENTS OF $61
            ADD   $6A              ADD NEXT BYTE
            STA   $6A              STORE BYTE
            TXA                    TRANSFER X TO ACC
            ADC   $69              ADD NEXT BYTE
            STA   $69              STORE BYTE
            LDA   $61              LOAD MSBS OF 1290 INTO ACC
            LDX   $54              LOAD MSBS OF M INTO X
            MUL                    MULTIPLY CONTENTS OF $54 BY CONTENTS OF $61
            ADD   $69              ADD NEXT BYTE
            STA   $69              STORE BYTE
            TXA                    TRANSFER X TO ACC
            ADC   $68              ADD NEXT BYTE
            STA   $68              STORE BYTE
            LDA   $6A              LOAD CONTENTS OF $6A INTO ACC
            BPL   NNN              IF NO CARRY FROM $6A GOTO LABEL NNN
            LDA   $69              LOAD CONTENTS OF $69 INTO ACC
            ADD   #$01             ADD 1 TO ACC
            STA   $69              STORE IN $69
            LDA   $68              LOAD CONTENTS OF $68 INTO ACC
            ADC   #$00             FLOW THROUGH CARRY
            STA   $68              STORE IN $68
NNN         LDA   $68              LOAD CONTENTS OF $68 INTO ACC
            STA   $61              STORE MSBS IN $61
            LDA   $69              LOAD CONTENTS OF $69 INTO ACC
            STA   $62              STORE IN $62
            LDX   $58              RESTORE X REGISTER FROM $58
            RTS                    RETURN TO MAIN PROGRAM
*
*                                 END TBMULT
*********************************************************************************
*
*                                 END
*********************************************************************************
```

Figure 32.17 • Software Code for 68HC05 Processor-Based RTD Linearization (Continued)

Thermistor bridge

Figure 32.18, another temperature measuring bridge, uses a thermistor as a sensor. The LT1034 furnishes bridge excitation. The 3.2k and 6250Ω resistors are supplied with the thermistor sensor. The network's overall response is linearly related to the thermistor's sensed temperature. The network forms one leg of a bridge with resistors furnishing the opposing leg. A trim in this opposing leg sets bridge output to zero at 0°C. Instrumentation amplifier A1 takes gain with A2 providing additional trimmed gain to furnish a calibrated output. Calibration is accomplished in similar fashion to the platinum RTD circuits, with the linearity trim deleted.

Low power bridge circuits

Low power operation of bridge circuits is becoming increasingly common. Many bridge-based transducers are low impedance devices, complicating low power design. The most obvious way to minimize bridge power consumption is to restrict drive to the bridge. Figure 32.19a is identical to Figure 32.5, except that the bridge excitation has been reduced to 1.2V. This cuts bridge current from nearly 30mA to about 3.5mA. The remaining circuit elements consume negligible power compared to this amount. The trade-off is the sacrifice in bridge output signal. The reduced drive causes commensurately lowered bridge outputs, making the noise and drift floor a greater percentage

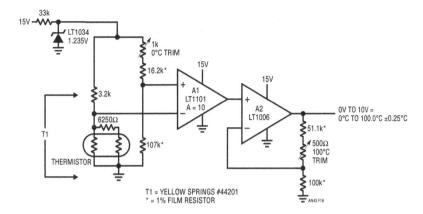

Figure 32.18 ● Linear Output Thermistor Bridge. Thermistor Network Provides Linear Bridge Output

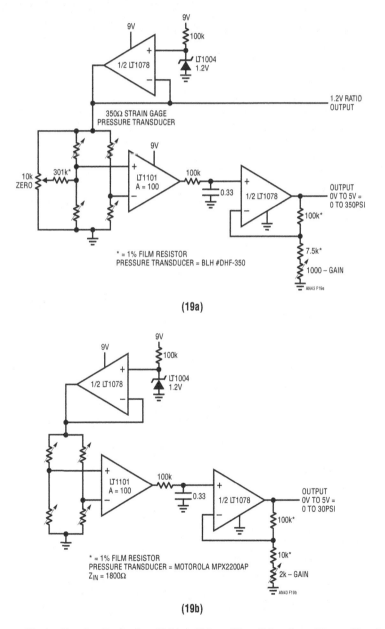

(19a)

(19b)

Figure 32.19 ● Power Reduction by Reducing Bridge Drive. Circuit is a Low Power Version of Figure 32.5

of the signal. More specifically, an 0.01% reading of a 10V powered 350Ω strain gauge bridge requires 3μV of stable resolution. At 1.2V drive, this number shrinks to a scary 360μV.

Figure 32.19b is similar, although bridge current is reduced below 700μA. This is accomplished by using a semiconductor-based bridge transducer. These devices have significantly higher input resistance, minimizing power dissipation. Semiconductor-based pressure transducers have major cost advantages over bonded strain gauge types, although accuracy and stability are reduced. Appendix A, "Strain gauge bridges," discusses trade-offs and theory of both technologies.

Strobed power bridge drive

Figure 32.20, derived directly from Figure 32.10, is a simple way to reduce power without sacrificing bridge signal output level. The technique is applicable where continuous output is not a requirement. This circuit is designed to sit in the quiescent state for long periods with relatively brief on-times. A typical application would be remote weight information in storage tanks where weekly readings are sufficient. Quiescent current is about 150μA with on-state current typically 50mA. Bridge power is conserved by simply turning it off.

With Q1's base unbiased, all circuitry is off except the LT1054 plus-to-minus voltage converter, which draws a 150μA quiescent current. When Q1's base is pulled low,

its collector supplies power to A1 and A2. A1's output goes high, turning on the LT1054. the LT1054's output (Pin 5) heads toward −5V and Q2 comes on, permitting bridge current to flow. To balance its inputs, A1 servo controls the LT1054 to force the bridge's midpoint to 0V. The bridge ends up with about 8V across it, requiring the 100mA capability LT1054 to sink about 24mA. The 0.02μF capacitor stabilizes the loop. The A1-LT1054 loop's negative output sets the bridge's common mode voltage to zero, allowing A2 to take a simple single-ended measurement. The "output trim" scales the circuit for 3mV/V type strain bridge transducers, and the 100k-0.1μF combination provides noise filtering.

Sampled output bridge signal conditioner

Figure 32.21, an obvious extension of Figure 32.20, automates the strobing into a clocked sequence. Circuit on-time is restricted to 250μs, at a clock rate of about 2Hz. This keeps average power consumption down to about 200μA. Oscillator A1A produces a 250μs clock pulse every 500ms (Trace A, Figure 32.22). A filtered version of this pulse is fed to Q1, whose emitter (Trace B) provides slew limited bridge drive. A1A's output also triggers a delayed pulse produced by the 74C221 one-shot output (Trace C). The timing is arranged so the pulse occurs well after the A1B-A2 bridge amplifier output (Trace D) settles. A

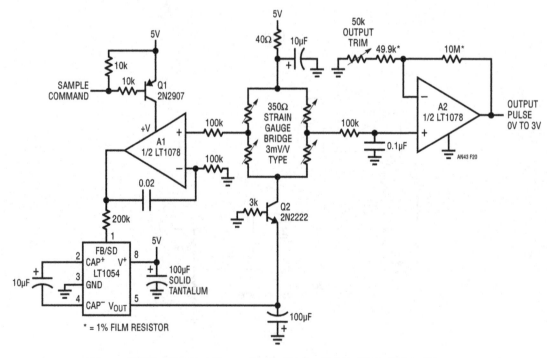

Figure 32.20 • Strobed Power Strain Gauge Bridge Signal Conditioner

monitoring A/D converter, triggered by this pulse, can acquire A1B's output.

The slew limited bridge drive prevents the strain gauge bridge from seeing a fast rise pulse, which could cause long term transducer degradation. To calibrate this circuit trim zero and gain for appropriate outputs.

Continuous output sampled bridge signal conditioner

Figure 32.23 extends the sampling approach to include a continuous output. This is accomplished by adding a sample-hold stage at the circuit output. In this circuit, Q2 is off when the "sample command" is low. Under these conditions only A2 and S1 receive power, and current drain is inside 60μA. When the sample command is pulsed high, Q2's collector (Trace A, Figure 32.24) goes high, providing power to all other circuit elements. The 10Ω-1μF RC at the LT1021 prevents the strain bridge from seeing a fast rise pulse, which could cause long term transducer degradation. The LT1021-5 reference output (Trace B) drives the strain bridge, and instrumentation amplifier A1 output responds (Trace C). Simultaneously, S1's switch control input (Trace D) ramps toward Q2's collector.

At about one-half Q2's collector voltage (in this case just before mid-screen) S1 turns on, and A1's output is stored in C1. When the sample command drops low, Q2's collector falls, the bridge and its associated circuitry shut down and S1 goes off. C1's stored value appears at gain scaled

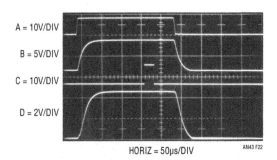

A = 10V/DIV

B = 5V/DIV

C = 10V/DIV

D = 2V/DIV

HORIZ = 50μs/DIV   AN43 F22

Figure 32.22 • Figure 32.21's Waveforms. Trace C's Delayed Pulse Ensures A/D Converter Sees Settled Output Waveform (Trace D)

A2's output. The RC delay at S1's control input ensures glitch-free operation by preventing C1 from updating until A1 has settled. During the 1ms sampling phase, supply current approaches 20mA but a 10Hz sampling rate cuts effective drain below 250μA. Slower sampling rates will further reduce drain, but C1's droop rate (about 1mV/100ms) sets an accuracy constraint. The 10Hz rate provides adequate bandwidth for most transducers. For 3mV/V slope factor transducers the gain trim shown allows calibration. It should be rescaled for other types. This circuit's effective current drain is about 250μA, and A2's output is accurate enough for 12-bit systems.

It is important to remember that this circuit is a sampled system. Although the output is continuous, information is

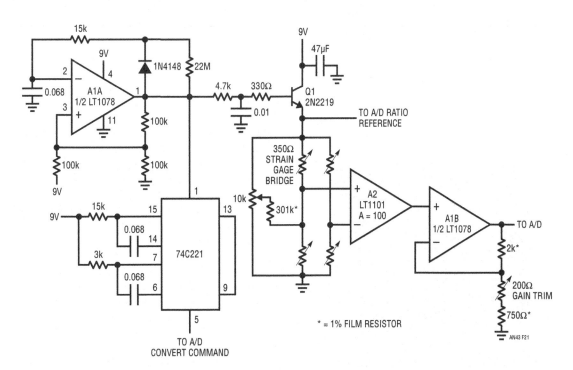

Figure 32.21 • Sampled Output Bridge Signal Conditioner Uses Pulsed Excitation to Save Power

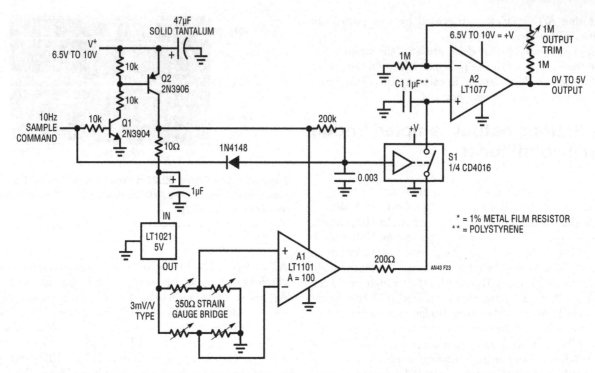

Figure 32.23 • Pulsed Excitation Bridge Signal Conditioner. Sample-Hold Stage Gives DC Output

being collected at a 10Hz rate. As such, the Nyquist limit applies, and must be kept in mind when interpreting results.

High resolution continuous output sampled bridge signal conditioner

Figure 32.25 is a special case of sampled bridge drive. It is intended for applications requiring extremely high resolution outputs from a bridge transducer. This circuit puts 100V across a 10V, 350Ω strain gauge bridge for short periods of time. The high pulsed voltage drive increases bridge output proportionally, without forcing excessive dissipation. In fact, although this circuit is not intended for power reduction, average bridge power is far below the normal 29mA obtained with $10V_{DC}$ excitation.

Combining the 10× higher bridge gain (300mV full scale versus the normal 30mV) with a chopper-stabilized amplifier in the sample-hold output stage is the key to the high resolution obtainable with this circuit.

When oscillator A1A's output is high Q6 is turned on and A2's negative input is pulled above ground. A2's output goes negative, turning on Q1. Q1's collector goes low, robbing Q3's base drive and cutting it off. Simultaneously, A3 enforces it's loop by biasing Q2 into conduction, softly turning on Q4. Under these conditions the voltage across the bridge is essentially zero. When A1A oscillates low (Trace A, Figure 32.26) RC filter driven Q6 responds by cutting off slowly. Now, A2's negative input sees current only through the 3.6k resistor. The input begins to head negative, causing A2's output to rise. Q1 comes out of saturation, and Q3's emitter (Trace B) rises. Initially this action is rapid (fast rise slewing is just visible at the start of Q3's ascent), but feedback to A2's negative input closes a control loop, with the 1000pF capacitor restricting rise time. The 72k resistor sets A2's gain at 20 with respect to the LT1004 2.5V reference, and Q3's emitter servo controls to 50V.

Simultaneously, A3 responds to the bridges biasing by moving its output negatively. Q2 tends towards cut-off,

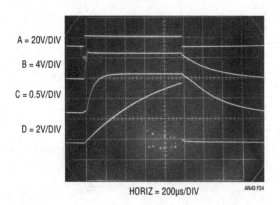

Figure 32.24 • Waveforms for Figure 32.23's Sampled Strain Gauge Signal Conditioner

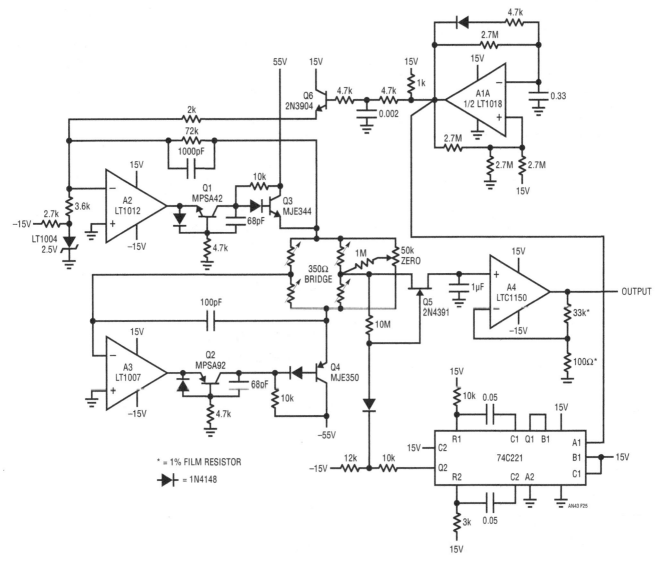

Figure 32.25 • High Resolution Pulsed Excitation Bridge Signal Conditioner. Complementary 50V Drive Increases Bridge Output Signal

increasing Q4's conduction. A3 biases its loop to maintain the bridge midpoint at zero. To do this, it must produce a complimentary output to A2's loop, which Trace C shows to be the case. Note that A3's loop roll-off is considerably faster than A2's, ensuring that it will faithfully track A2's loop action. Similarly, A3's loop is slaved to A2's loop output, and produces no other outputs.

Under these conditions the bridge sees 100V drive across it for the 1ms duration of the clock pulse.

A1A's clock output also triggers the 74C221 one-shot. The one-shot delivers a delayed pulse (Trace D) to Q5. Q5 comes on, charging the 1μF capacitor to the bridge's output voltage. With A3 forcing the bridge's left side midpoint to zero, Q5, the 1μF capacitor and A4 see a single-ended, low voltage signal. High transient common mode voltages

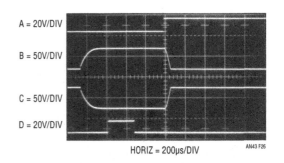

A = 20V/DIV

B = 50V/DIV

C = 50V/DIV

D = 20V/DIV

HORIZ = 200μs/DIV   AN43 F26

Figure 32.26 • Figure 32.25's Waveforms. Drive Shaping Results in Controlled, Complementary Bridge Drive Waveforms. Bridge Power is Low Despite 100V Excitation

658

are avoided by the control loop's complimentary controlled rise times. A4 takes gain and provides the circuit output. The 74C221's pulse width ends during the bridge's on-time, preserving sampled data integrity. When the A1A oscillator goes high the control loops remove bridge drive, returning the circuit to quiescence. A4's output is maintained at DC by the 1μF capacitor. A1A's 1Hz clock rate is adequate to prevent deleterious droop of the 1μF capacitor, but slow enough to limit bridge power dissipation. The controlled rise and fall times across the bridge prevent possible long-term transducer degradation by eliminating high $\Delta V / \Delta T$ induced effects.

When using this circuit it is important to remember that it is a sampled system. Although the output is continuous, information is being collected at a 1Hz rate. As such, the Nyquist limit applies, and must be kept in mind when interpreting results.

AC driven bridge/synchronous demodulator

Figure 32.27, an extension of pulse excited bridges, uses synchronous demodulation to obtain very high noise rejection capability. An AC carrier excites the bridge and synchronizes the gain stage demodulator. In this application, the signal source is a thermistor bridge which detects

extremely small temperature shifts in a biochemical microcalorimetry reaction chamber

The 500Hz carrier is applied at T1's input (Trace A, Figure 32.28). T1's floating output drives the thermistor bridge, which presents a single-ended output to A1. A1 operates at an AC gain of 1000. A 60Hz broadband noise source is also deliberately injected into A1's input (Trace B). The carrier's zero crossings are detected by C1. C1's output clocks the LTC1043 (Trace C). A1's output (Trace D) shows the desired 500Hz signal buried within the 60Hz noise source. The LTC1043's zero-cross-synchronized switching at A2's positive input (Trace E) causes A2's gain to alternate between plus and minus one. As a result, A1's output is synchronously demodulated by A2. A2's output (Trace F) consists of demodulated carrier signal and noncoherent components. The desired carrier amplitude and polarity information is discernible in A2's output and is extracted by filter-averaging at A3. To trim this circuit, adjust the phase potentiometer so that C1 switches when the carrier crosses through zero.

AC driven bridge for level transduction

Level transducers which measure angle from ideal level are employed in road construction, machine tools, inertial navigation systems and other applications requiring a gravity

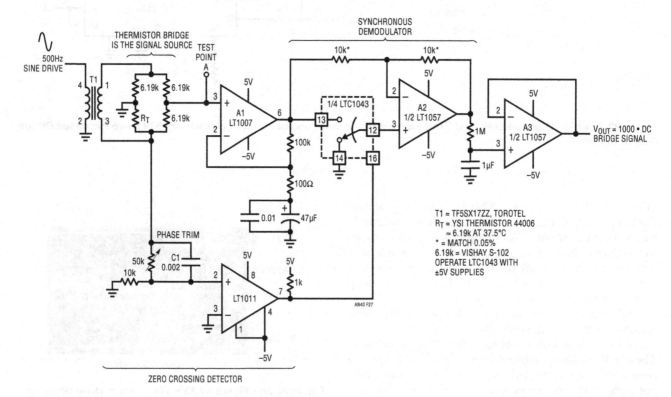

Figure 32.27 • "Lock-In" Bridge Amplifier. Synchronous Detection Achieves Extremely Narrow Band Gain, Providing Very High Noise Rejection

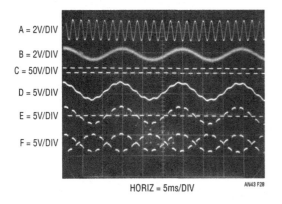

A = 2V/DIV
B = 2V/DIV
C = 50V/DIV
D = 5V/DIV
E = 5V/DIV
F = 5V/DIV

HORIZ = 5ms/DIV   AN43 F28

Figure 32.28 • Details of Lock-In Amplifier Operation. Narrowband Synchronous Detection Permits Extraction of Coherent Signals Over 120dB Down

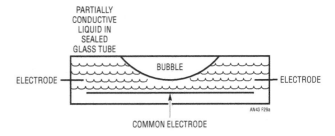

Figure 32.29a • Bubble-Based Level Transducer

reference. One of the most elegantly simple level transducers is a small tube nearly filled with a partially conductive liquid. Figure 32.29a shows such a device. If the tube is level with respect to gravity, the bubble resides in the tube's center and the electrode resistances to common are identical. As the tube shifts away from level, the resistances increase and decrease proportionally. By controlling the tube's shape at manufacture it is possible to obtain a linear output signal when the transducer is incorporated in a bridge circuit.

Transducers of this type must be excited with an AC waveform to avoid damage to the partially conductive liquid inside the tube. Signal conditioning involves generating this excitation as well as extracting angle information and polarity determination (e.g., which side of level the tube is on). Figure 32.29b shows a circuit which does this, directly producing a calibrated frequency output corresponding to level. A sign bit, also supplied at the output, gives polarity information.

The level transducer is configured with a pair of 2kΩ resistors to form a bridge. The required AC bridge excitation is developed at C1A, which is configured as a multivibrator. C1A biases Q1, which switches the LT1009's 2.5V potential through the 100μF capacitor to provide the AC bridge drive. The bridge differential output AC signal is converted to a current by A1, operating as a Howland current pump. This current, whose polarity reverses as bridge drive polarity switches, is rectified by the diode bridge. Thus, the 0.03μF capacitor receives unipolar charge. Instrumentation amplifier A2 measures the voltage across the capacitor and presents its single-ended output to C1B. When the voltage across the 0.03μF capacitor becomes high enough, C1B's output goes high, turning on the LTC201A switch. This discharges the capacitor. When C1B's AC positive feedback ceases, C1B's output goes low and the switch goes off. The 0.03μF unit again receives constant current charging and the entire cycle repeats. The frequency of this oscillation is determined by the magnitude of the constant current delivered to

the bridge-capacitor configuration. This current's magnitude is set by the transducer bridge's offset, which is level related.

Figure 32.30 shows circuit waveforms. Trace A is the AC bridge drive, while Trace B is A1's output. Observe that when the bridge drive changes polarity, A1's output flips sign rapidly to maintain a constant current into the bridge-capacitor configuration. A2's output (Trace C) is a unipolar ground-referred ramp. Trace D is C1B's output pulse and the circuit's output. The diodes at C1B's positive input provide temperature compensation for the sensor's positive tempco, allowing C1B's trip voltage to ratiometrically track bridge output over temperature.

A3, operating open loop, determines polarity by comparing the rectified and filtered bridge output signals with respect to ground.

To calibrate this circuit, place the level transducer at a known 40 arc-minute angle and adjust the 5kΩ trimmer at C1B for a 400Hz output. Circuit accuracy is limited by the transducer to about 2.5%.

Time domain bridge

Figure 32.31 is another AC-based bridge, but works in the time domain. This circuit is particularly applicable to capacitance measurement. Operation is straightforward. With S1 closed the comparators output is high. When S1 opens, capacitor C_X charges. When C_X's potential crosses the voltage established by the bridge's left side resistors the comparator trips low. The elapsed time between the switch opening and the comparator going low is proportionate to C_X's value. This circuit is insensitive to supply and repetition rate variations and can provide good accuracy if time constants are kept much larger than comparator and switch delays. For example, the LT1011's delay is about 200ns and the LTC201A contributes 450ns. To ensure 1% accuracy the bridge's right side time constant should not drop below 65μs. Extremely low values of capacitance may be

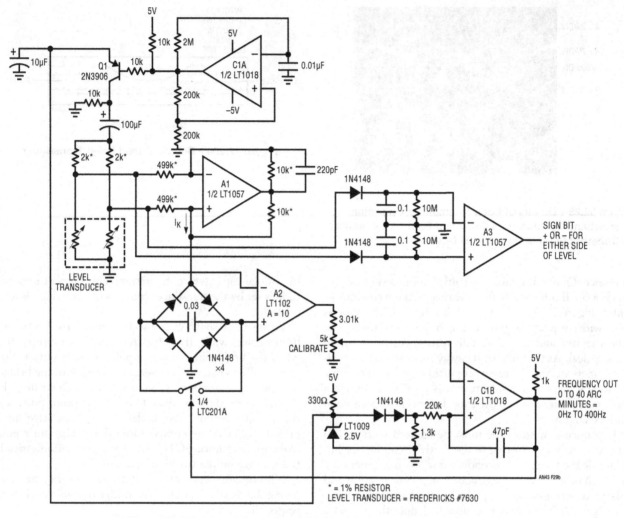

Figure 32.29b • Level Transducer Digitizer Uses AC Bridge Technique

influenced by switch charge injection. In such cases switching should be implemented by alternating the bridge drive between ground and +15.

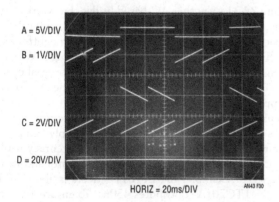

A = 5V/DIV

B = 1V/DIV

C = 2V/DIV

D = 20V/DIV

HORIZ = 20ms/DIV   AN43 F30

Figure 32.30 • Level Transducer Bridge Circuit's Waveforms

Bridge oscillator—square wave output

Only an inattentive outlook could resist folding Figure 32.31's bridge back upon itself to make an oscillator. Figure 32.32 does this, forming a bridge oscillator. This circuit will also be recognized as the classic op amp multi-vibrator. In this version the 10k to 20k bridge leg provides switching point hysteresis with C_X charged via the remaining 10k resistor When C_X reaches the switching point the amplifier's output changes state, abruptly reversing the sign of its positive input voltage. Cx's charging direction also reverses, and oscillations continue. At frequencies that are low compared to amplifier delays output frequency is almost entirely dependent on the bridge components. Amplifier input errors tend to ratiometrically cancel, and supply shifts are similarly rejected. The duty cycle is influenced by output saturation and supply asymmetrys.

661

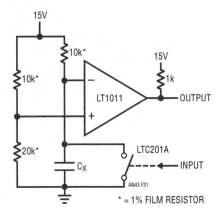

Figure 32.31 • Time Domain Bridge

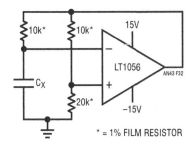

Figure 32.32 • "Bridge Oscillator" (Good Old Op Amp Multivibrator with a Fancy Name)

Quartz stabilized bridge oscillator

Figure 32.33, generically similar to Figure 32.32, replaces one of the bridge arms with a resonant element. With the crystal removed the circuit is a familiar noninverting gain of 2 with a grounded input. Inserting the crystal closes a positive feedback path at the crystal's resonant frequency. The amplifier output (Trace A, Figure 32.34) swings in an attempt to maintain input balance. Excessive circuit gain prevents linear operation, and oscillations commence as

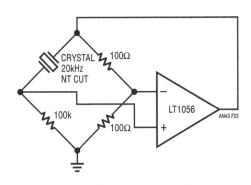

Figure 32.33 • Bridge-Based Crystal Oscillator

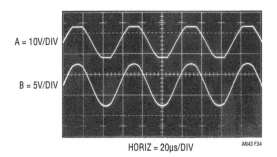

Figure 32.34 • Bridge-Based Crystal Oscillator's Waveforms. Excessive Gain Causes Output Saturation Limiting

the amplifier repeatedly overshoots in its attempts to null the bridge. The crystal's high Q is evident in the filtered waveform (Trace B) at the amplifier's positive input.

Sine wave output quartz stabilized bridge oscillator

Figure 32.35 takes the previous circuit into the linear region to produce a sine wave output. It does this by continuously controlling the gain to maintain linear operation. This arrangement uses a classic technique first described by Meacham in 1938 (see References).

In any oscillator it is necessary to control the gain as well as the phase shift at the frequency of interest. If gain is too low, oscillation will not occur. Conversely, too much gain produces saturation limiting, as in Figure 32.33. Here, gain control comes from the positive temperature coefficient of the lamp. When power is applied, the lamp is at a low resistance value, gain is high and oscillation amplitude builds. As amplitude builds, the lamp current increases, heating occurs and its resistance goes up. This causes a reduction in amplifier gain and the circuit finds a stable operating point. The 15pF capacitor suppresses spurious oscillation.

Operating waveforms appear in Figure 32.36. The amplifier's output (Trace A, Figure 32.36) is a sine wave, with about 1.5% distortion (Trace B). The relatively high distortion content is almost entirely due to the common

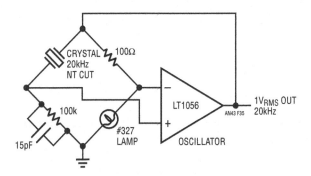

Figure 32.35 • Figure 32.33 with Lamp Added for Gain Stabilization

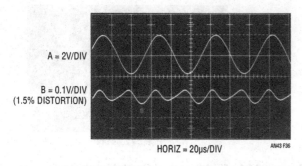

A = 2V/DIV

B = 0.1V/DIV
(1.5% DISTORTION)

HORIZ = 20µs/DIV

AN43 F36

Figure 32.36 • Lamp-Based Amplitude Stabilization Produces Sine Wave Output

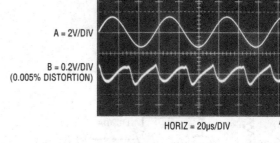

A = 2V/DIV

B = 0.2V/DIV
(0.005% DISTORTION)

HORIZ = 20µs/DIV

AN43 F38

Figure 32.38 • Distortion Measurements for Figure 32.37. Common Mode Suppression Permits 0.005% Distortion

mode swing seen by the amplifier. Op amp common mode rejection suffers at high frequency, producing output distortion. Figure 32.37 eliminates the common mode swing by using a second amplifier to force the bridge's midpoint to virtual ground.[3] It does this by measuring the midpoint value, comparing it to ground and controlling the formerly grounded end of the bridge to maintain its inputs at zero. Because the bridge drive is complementary the oscillator amplifier now sees no common mode swing, dramatically reducing distortion. Figure 32.38 shows less than 0.005% distortion (Trace B) in the output (Trace A) waveform.

Wien bridge-based oscillators

Crystals are not the only resonant elements that can be stabilized in a gain-controlled bridge. Figure 32.39 is a Wien bridge (see References) based oscillator. The configuration shown was originally developed for telephony

applications. The circuit is a modern adaptation of one described by a Stanford University student, William R. Hewlett,[4] in his 1939 Master's thesis (see Appendix C, "The Wien bridge and Mr. Hewlett").

The Wien network provides phase shift governed by the equation listed, and the lamp regulates amplitude in accordance with Figure 32.35's description. Figure 32.40 is a variable frequency version of the basic circuit. Output frequency range spans 20Hz to 20kHz in three decade ranges, with 0.25dB amplitude flatness.

The smooth, limiting nature of the lamp's operation, in combination with its simplicity, gives good results. Trace A, Figure 32.41, shows circuit output at 10kHz. Harmonic distortion, shown in Trace B, is below 0.003%. The trace shows that most of the distortion is due to second harmonic content and some crossover disturbance is noticeable. The low resistance values in the Wien network and the $3.8\text{nV}\sqrt{\text{Hz}}$ noise specification of the LT1037 eliminate amplifier noise as an error term.

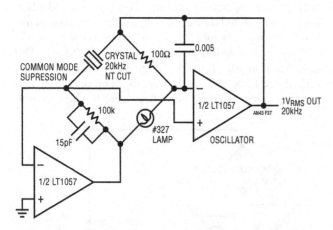

Figure 32.37 • Common Mode Suppression for Quartz Oscillator Lowers Distortion

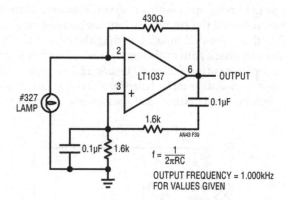

$$f = \frac{1}{2\pi RC}$$

OUTPUT FREQUENCY = 1.000kHz
FOR VALUES GIVEN

Figure 32.39 • Wien Bridge-Based Sine Wave Oscillator. Simple, Modern Version of an Old Circuit Has 0.0025% Distortion

Note 3: Sharp-eyed readers will recognize this as an AC version of the DC common mode suppression technique introduced back in Figure 32.6.

Note 4: History records that Hewlett and his friend David Packard made a number of these type oscillators. Then they built some other kinds of instruments.

At low frequencies, the thermal time constant of the small normal mode lamp begins to introduce distortion levels above 0.01%. This is due to "hunting" as the oscillator's frequency approaches the lamp thermal time constant. This effect can be eliminated, at the expense of reduced output amplitude and longer amplitude settling time, by switching to the low frequency, low distortion mode. The four large lamps give a longer thermal time constant and distortion is reduced. Figure 32.42 plots distortion versus frequency for the circuit.

Figure 32.43's version replaces the lamp with an electronic amplitude stabilization loop. The LT1055 compares the oscillators positive output peaks with a DC reference. The diode in series with the LT1004 reference provides temperature compensation for the rectifier diode. The op amp biases Q1, controlling its channel resistance. This influences loop gain, which is reflected in oscillator output amplitude. Loop closure around the LT1055 occurs, stabilizing oscillator amplitude. The 15μF capacitor stabilizes the loop, with the 22k resistor settling its gain.

Distortion performance for this circuit is quite disappointing. Figure 32.44 shows 0.15% 2f distortion (Trace B) in the output (Trace A), a huge increase over the lamp-based approach.[5] This distortion does not correlate with the rectifier peaking residue present at Q1's gate (Trace C). Where is the villain in this scheme?

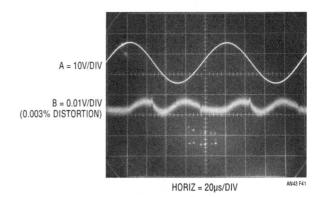

A = 10V/DIV

B = 0.01V/DIV
(0.003% DISTORTION)

HORIZ = 20μs/DIV   AN43 F41

Figure 32.41 • Figure 32.40's Distortion Characteristic at 10kHz

The culprit turns out to be Q1. In a FET gate voltage theoretically sets channel resistance. In fact, channel voltage also slightly modulates channel resistance. In this circuit Q1's channel sees large swings at the fundamental. This swing combines with the channel voltage-resistance modulation effect, producing distortion.

The cure for this difficulty is local feedback around Q1. Properly scaled, this feedback nicely cancels out the

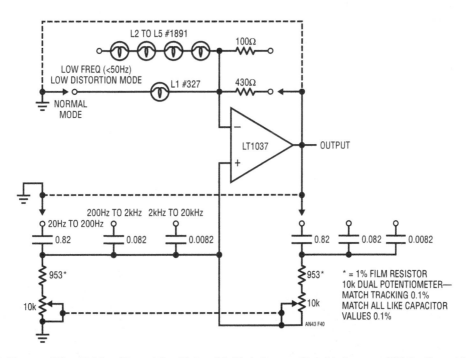

Figure 32.40 • Multirange Wien Bridge-Based Oscillator. Multiple Lamps Provide Lowered Distortion at Low Frequencies

Note 5: What else should be expected when trying to replace a single light bulb with a bunch of electronic components? I can hear Figure 32.39's #327 lamp laughing.

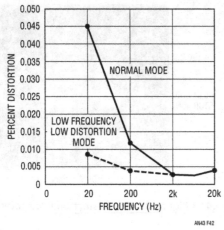

Figure 32.42 • Figure 32.40's Distortion vs Frequency

parasitic. Figure 32.45 shows the circuit redrawn with the inclusion of Q1's local loop. The 20k trimmer allows adjustment to optimize distortion performance. Figure 32.46 shows results. Distortion (Trace B) drops to 0.0018% and is composed of 2f, some gain loop rectification artifacts and noise. For reference the circuit's output (Trace A) and the LT1055 output (Trace C) are shown.

Figure 32.47 eliminates the trim, provides increased voltage and current output, and slightly reduces distortion. Q1 is replaced with an optically driven CdS photocell. This device has no parasitic resistance modulation effects. The LT1055 has been replaced with a ground sensing op amp running in single supply mode. This permits true integrator operation and eliminates any possibility of reverse biasing the (downsized) feedback capacitor. Additional feedback

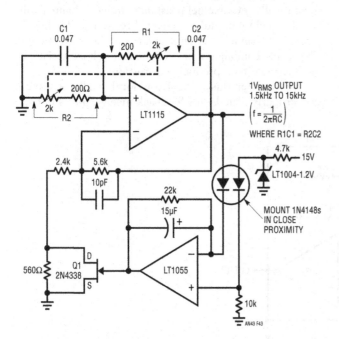

Figure 32.43 • Replacing the Lamp with an Electronic Equivalent

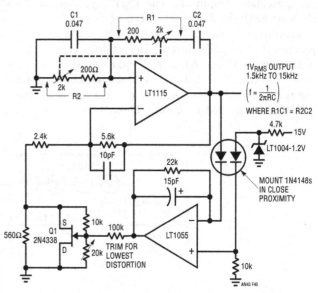

Figure 32.45 • Local Feedback Around Q1 Cures Channel Resistance Modulation, Reducing Distortion to 0.0018%

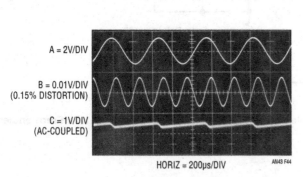

Figure 32.44 • Figure 32.43 Produces Excessive Distortion Due to Q1's Channel Resistance Modulation

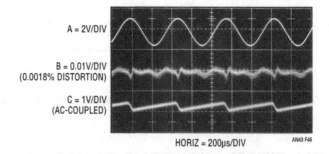

Figure 32.46 • Figure 32.45's 0.0018% Distortion Characteristic

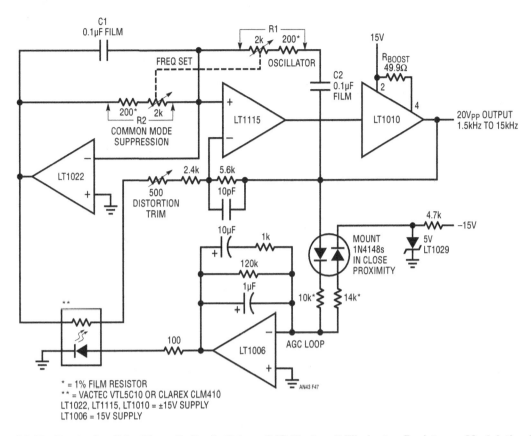

C1
0.1µF FILM

FREQ SET

R1
2k 200*

OSCILLATOR

15V

R_BOOST
49.9Ω

C2
0.1µF
FILM

LT1010

20V_PP OUTPUT
1.5kHz TO 15kHz

200* 2k
R2
COMMON MODE
SUPPRESSION

LT1115

LT1022

2.4k 5.6k

500
DISTORTION
TRIM

10pF

4.7k −15V

5V
LT1029

10µF 1k

120k

MOUNT
1N4148s
IN CLOSE
PROXIMITY

1µF

10k* 14k*

100

LT1006

AGC LOOP

* = 1% FILM RESISTOR
** = VACTEC VTL5C10 OR CLAREX CLM410
LT1022, LT1115, LT1010 = ±15V SUPPLY
LT1006 = 15V SUPPLY

AN43 F47

Figure 32.47 • Replacing Q1 with an Optically Driven CdS Photocell Eliminates Resistance Modulation Trim

components aid step response.[6] Distortion performance improves slightly to 0.0015%.

The last Wien bridge-based circuit borrows Figure 32.37's common mode suppression technique (which is simply an AC version of Figure 32.6's DC common mode suppression loop) to reduce distortion to vanishingly small levels. The LT1022 amplifier appears in Figure 32.48. This amplifier forces the midpoint of the bridge to virtual ground by servo biasing the formerly grounded bridge legs. As in Figure 32.37, common mode swing is eliminated, reducing distortion. The circuit's output (Trace A, Figure 32.49) contains less than 0.0003% (3ppm) distortion (Trace B), with no visible correlation to gain loop ripple residue (Trace C). This level of distortion is below the uncertainty floor of most distortion analyzers, requiring specialized equipment for meaningful measurement. (See Appendix D, guest written by Bruce Hofer of Audio Precision, Inc., for a discussion on distortion measurement considerations.)

Note 6: A much better scheme for a low ripple, fast response gain control loop is nicely detailed in the operating and service manual supplied with the Hewlett-Packard HP339A Distortion Analyzer.

Diode bridge-based 2.5MHz precision rectifier/AC voltmeter

A final circuit shows a way to achieve low AC error switching with diode bridge techniques. Diode bridges provide faster cleaner signal switching than any other technique.

Most precision rectifier circuits rely on operational amplifiers to correct for diode drops. Although this scheme works well, bandwidth limitations usually restrict these circuits to operation below 100kHz. Figure 32.50 shows the LT1016 comparator in an open-loop, synchronous rectifier configuration which has high accuracy out to 2.5MHz. An input 1MHz sine wave (Trace A, Figure 32.51) is zero cross detected by C1. Both of C1's outputs drive identical level shifters with fast (delay = 2ns to 3ns), ±5V outputs. These outputs bias a Schottky diode switching bridge (Traces B and C are the switched corners of the bridge). The input signal is fed to the left midsection of the bridge. Because C1 drives the bridge synchronously with the input signal, a half-wave rectified sine appears at the AC output (Trace D). The RMS value appears at the

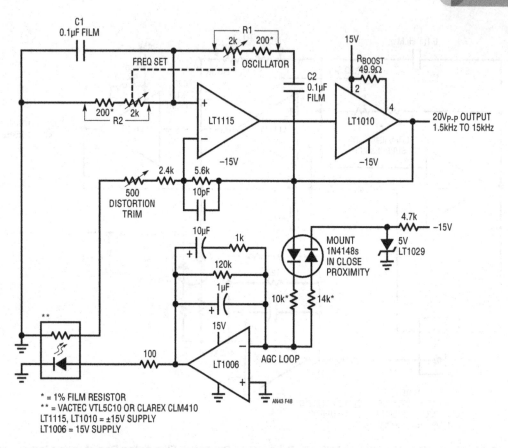

Figure 32.48 • Adding Common Mode Suppression Lowers Distortion to 0.0003%

DC output. The Schottky bridge gives fast switching without charge pump-through. This is evident in Trace E, which is an expanded version of Trace D. The waveform is clean

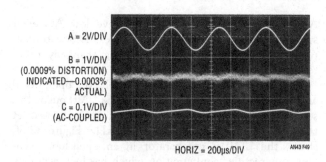

A = 2V/DIV

B = 1V/DIV
(0.0009% DISTORTION)
INDICATED—0.0003%
ACTUAL)

C = 0.1V/DIV
(AC-COUPLED)

HORIZ = 200μs/DIV   AN43 F49

Figure 32.49 • Figure 32.48's 3ppm Distortion is Below the Noise Floor of Most Analyzers

with the exception of very small disturbances where bridge switching occurs. To calibrate this circuit, apply a 1MHz to 2MHz $1V_{P-P}$ sine wave and adjust the delay compensation so bridge switching occurs when the sine crosses zero. The adjustment corrects for the small delays through the LT1016 and the level shifters. Next, adjust the skew compensation potentiometers for minimum aberrations in the AC output signal. These trims slightly shift the phase of the rising output edge of their respective level shifter. This allows skew in the complementary bridge drive signals to be kept within 1ns to 2ns, minimizing output disturbances when switching occurs. A 100mV sine input will produce a clean output with a DC output accuracy of better than 0.25%.

Note: *This application note was derived from a manuscript originally prepared for publication in EDN magazine.*

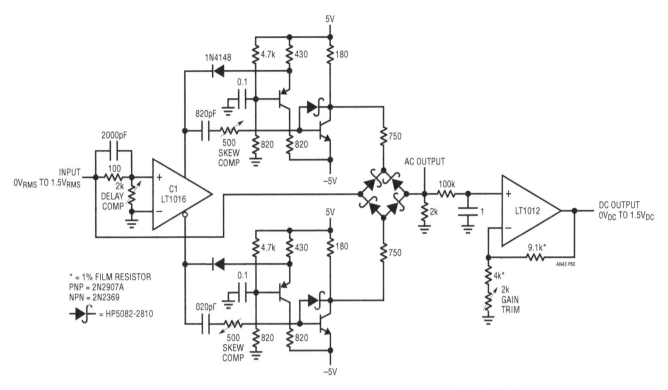

Figure 32.50 • Fast, Bridge-Switched Synchronous Rectifier-Based AC/DC Converter

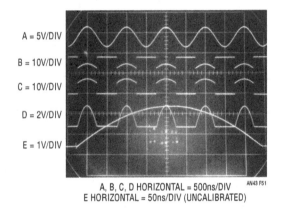

A, B, C, D HORIZONTAL = 500ns/DIV   AN43 F51
E HORIZONTAL = 50ns/DIV (UNCALIBRATED)

Figure 32.51 • Fast AC/DC Converter Operating at 1MHz. Clean Switching is Due to Bridge Symmetry and Compensations for Delay and Switching Skew

References

1. Sheingold, D.H., "Transducer Interfacing Handbook," Analog Devices Inc., 1980.

2. Arthur, K., "Transducer Measurements," Tektronix Inc., Concept Series, 1971.

3. Parry, C.H., "Diseases of the Heart," Vol. 2, London, Underwoods, p. 111, 1825.

4. Gordon, J.W., "On Certain Motor Movements of the Human Body Produced by the Circulation," J. Anatomy Physiology, 11:553–559, 1877.

5. Starr, I. and Noordegraff. A., "Ballistocardiography in Cardiovascular Research," Lippincott, 1967.

6. Weissler, A.M., "Non-Invasive Cardiology," Grune and Stratton, 1974.

7. Meade, M.L., "Lock-In Amplifiers and Applications," London: P. Peregrinus, Ltd.

8. Wien, Max, "Measung der induction constanten mit dern Optischen Telephon, " Ann. der Phys., Vol. 44, 1891, p. 704–7.

9. Meacham, L.A., "The Bridge Stabilized Oscillator," Bell System Technical Journal, Vol. 17, p. 574, Oct. 1938.

10. Hewlett, William R., "A New Type Resistance-Capacity Oscillator," M.S. Thesis, Stanford University, Palo Alto, California 1939.

11. Hewlett, William R., U.S. Patent No. 2,768,872, Jan. 6, 1942.

12. Bauer, Brunton, "Design Notes on the Resistance-Capacity Oscillator Circuit," Parts I and II, Hewlett-Packard Journal, Nov., Dec., 1949. Hewlett-Packard Company.

13. Williams, Jim, "Thermal Techniques in Measurement and Control Circuitry," Linear Technology Corporation Application Note 5, Linear Technology Corporation, Milpitas, California 1984.

14. Mattheys, R.L., "Crystal Oscillator Circuits," Wiley, New York, 1983.

15. Hewlett-Packard, "Schottky Diodes for High-Volume, Low Cost Applications,"

Application Note 942, Hewlett-Packard Company, 1973.

16. Tektronix, Inc., "Type 1S1 Sampling Plug-In Operating and Service Manual," Tektronix, Inc., 1965.

17. Mulvey, J., "Sampling Oscilloscope Circuits," Tektronix, Inc., Concept Series, 1970.

18. Hill, W. and Horowitz, P., "The Art of Electronics," Cambridge University Press, Cambridge, England 1989.

19. Bowers, B., "Sir Charles Wheatstone," Science Museum, London, England 1975.

20. Nahin, Paul J., "Oliver Heaviside: Sage in Solitude," IEEE Press, 1988.

21. Wilkinson, D.H., "A Stable Ninety-Nine Channel Pulse Amplitude Analyser for Slow Counting," Proceedings of the Cambridge Philosophical Society, Cambridge England, 46,508. 1950.

Appendix A
Strain gauge bridges

In 1856 Lord Kelvin discovered that applying strain to a wire shifted its resistance. This effect is repeatable, and is the basis for electrical output strain measurement. Early devices were simply wires suspended between two insulated points (Figure A1). The force to be measured mechanically biased the wire, changing its resistance. Modern devices utilize foil-based designs. The conductive material is deposited on an insulated carrier (Figure A2). Physically they take many forms, allowing for a variety of applications. The gages[1]

are usually configured in a bridge and mounted on a beam (Figure A3), forming a transducer.

A useful transducer must be trimmed for zero and gain, and compensated for temperature sensitivity. Figure A4 shows a typical arrangement. Zero is set with a parallel trim, with similar treatment used to set

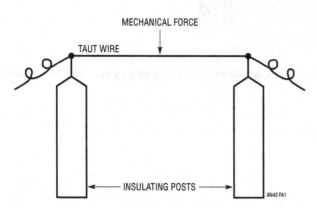

Figure A1 • A Very Basic Strain Gage

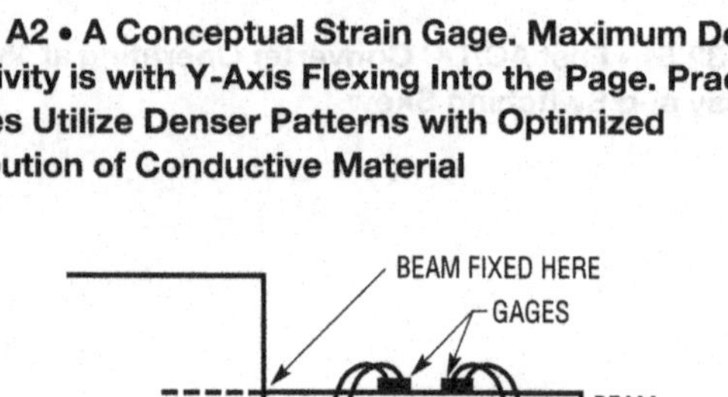

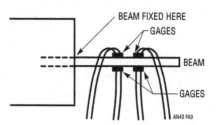

Figure A2 • A Conceptual Strain Gage. Maximum Device Sensitivity is with Y-Axis Flexing Into the Page. Practical Devices Utilize Denser Patterns with Optimized Distribution of Conductive Material

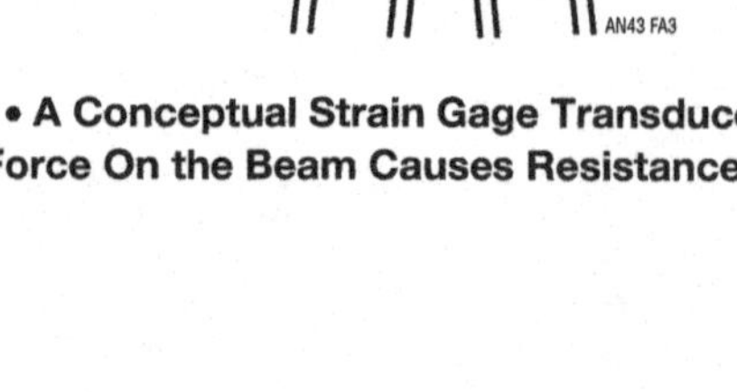

Figure A3 • A Conceptual Strain Gage Transducer. Bending Force On the Beam Causes Resistance Shifts

---

Note 1: The correct spelling is gauge, but prolonged grammatical assaults have assassinated the "u." Hence, "gage" assumes a claim to legitimacy.

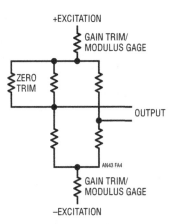
+EXCITATION

GAIN TRIM/
MODULUS GAGE

ZERO
TRIM

OUTPUT

AN43 FA4

GAIN TRIM/
MODULUS GAGE

−EXCITATION

Figure A4 • Simplified Strain Gage Transducer Schematic

gain. The gain trims include modulus gages to compensate beam material temperature sensitivity. Arranging these trims and completing the mechanical integration involves a fair amount of artistry, and is usually best left to specialists.[2]

Semiconductor based strain gage transducers utilize resistive shift in semiconducting materials. These devices, built in monolithic IC form, are considerably less expensive than manually assembled foil-based strain gage transducers. They have over ten times the sensitivity of foil-based devices, but are more sensitive to temperature and other effects. As such, they are best suited to somewhat less demanding applications than foil-based gages. Their monolithic construction and small size offer price and convenience advantages in many applications. Electrical form is similar to foil-based designs (e.g., a bridge configuration), although impedance levels are about ten times higher. The following guest written section details their characteristics.

Semiconductor based strain gages

Daniel A. Artusi, Randy K. Frank
Motorola Semiconductor Products Sector Discrete and Special Technologies Group

Strain gage technology, while based on a phenomena which dates back to the nineteenth century, has been of major importance in the areas of stress analysis, structural testing and transducer fabrication for more than 40 years.

First reports on semiconductor piezoresistive technology dates back to the observation by C.S. Smith[3] in the early 1950s of large piezoresistive coefficients in silicon and germanium.

There are several advantages to implementing strain gages using semiconductor technology. The immediate one is the very high gage factors of approximately two orders of magnitude higher than metallic gages. These higher gage factors allow improved signal-to-noise ratios for the measurement of small dynamic stresses and simplifies the signal conditioning circuitry.

Another advantage is the precise control of the piezoresistive coefficients including magnitude, sign, and the possibility of transverse and shear responses. Additional advantages are low cost, small size, and compatibility with semiconductor processing technology which allows for integration of additional circuit elements (i.e., operational amplifiers) on the same chip. The first phase of integration for silicon pressure sensors occurred when the strain gage and the diaphragm were combined into one monolithic structure. This was accomplished using the piezoresistive effect in semiconductors. A strain gage can be diffused or ion-implanted into a thin silicon diaphragm which has been chemically etched into a silicon substrate.

Piezoresistivity

In order to understand the implementation in silicon of strain gages, it is necessary to review the piezoresistive effect in silicon.

The analytic description of the piezoresistive effect in cubic silicon can be reduced to two equations which demonstrate the first order effects.

$$\Delta E_1 = P_0 \, I_1 (\pi_{11} X_1 + \pi_{12} X_2) \qquad (1)$$

$$\Delta E_2 = P_0 I_2 \pi_{44} X_6 \qquad (2)$$

where ΔE_1 and ΔE_2 are electric field flux density, P_0 is the unstressed bulk resistivity of silicon, Is are the excitation current density, πs are piezoresistive coefficients and Xs are stress tensors due to the applied force.

The effect described by equation (1) is that utilized in a pressure transducer of the Wheatstone bridge type. Regardless of whether the designer chooses N-type or P-type layers for the diffused sensing element, the piezoresistive coefficients π_{11} and π_{12} equation (1) will be oppose in sign.

Note 2: Those finding their sense of engineering prowess unalterably offended are referred to "SR-4 Strain Gage Handbook," available from BLH Electronics, Canton, Massachusetts. Have fun.

Note 3: Smith C.S., "Piezoresistance Effect in Germanium and Silicon," Physical Review, Volume 94, November 1, 1954 Pages 42–49.

This implies that through careful placement, and orientation with respect to the crystallographic axis, as well as a sufficiently large aspect ratio for the resistors themselves, it is possible to fabricate resistors on the same diaphragm which both increase and decrease respectively from their nominal values with the application of stress.

The effect described by equation (2) is typically neglected as a parasitic in the design of a Wheatstone bridge device. A closer look at its form, however, reveals that the incremental electrical field flux density, ΔE_2, due to the applied stress, X_6, is monotonically increasing for increasing X_6.

In fact, equation (2) predicts an extremely linear output since it depends on only one piezoresistive coefficient and one applied stress. Furthermore, the incremental electric field can be measured by a single stress sensitive element. This forms the theoretical basis for the design of the transverse voltage or shear stress piezoresistive strain gage.

Shear stress strain gage

Figure A5 shows the construction of a device which optimizes the piezoresistive effect of equation (2).[4] The diaphragm is anisotropically etched from a silicon substrate. The piezoresistive element is a single, 4-terminal strain gage that is located at the midpoint of the edge of the square diaphragm at an angle of 45 degrees as shown in Figure A5. The orientation of 45 degrees and location at the center of the edge of the diaphragm maximizes the sensitivity to shear stress, X_6, and the shear stress being sensed by the transducer by maximizing the piezoresistive coefficient, π_{44}.

Excitation current is passed longitudinally through the resistor (Pins 1 and 3) and the pressure that stresses the diaphragm is applied at a right angle to the current flow. The stress establishes a transverse electric field in the resistor that is sensed as an output voltage at Pins 2 and 4, which are the taps located at the midpoint of the resistor The single element shear-stress strain gage can be viewed as the mechanical analog of a Hall effect device. Figure A6 shows a cross section of a pressure transducer implemented in silicon and using the technique described. A differential pressure sensor chip is accomplished by opening the back side of the wafer.

Temperature compensation and calibration

The transverse voltage shear stress piezoresistive pressure transducer has been shown to present certain advantages over the Wheatstone bridge configuration. Specifically, improved linearity, and a more consistent reproducible offset (since it is defined by a single photolithographic step), as well as the added advantage of integrating stresses over a smaller percentage of the flexural element.

Very predictably, the transducer exhibits a negative temperature coefficient of span with a nominal value of 0.19%/°C, as well as a temperature coefficient of offset that can be in the range of $\pm15\mu V/°C$ or slightly larger before compensation. TC of span is due to the decrease of the piezoresistive coefficients with temperature due to increased thermal scattering in the lattice structure.

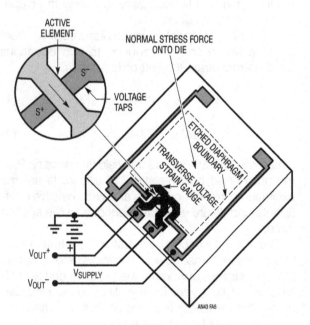

Figure A5 • Basic Sensor Element—Top View

Note 4: J.E. Gragg, U.S. Patent 4,317,126.

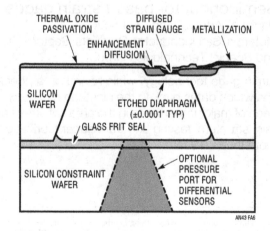

Figure A6 • Cross Section of Pressure Transducer

First let's consider the relationship of output voltage, ΔV_O, with excitation voltage, V_{EX}, as predicted by equation (2).

$$\Delta V_0 = w/l\,(\pi_{44}X_6)V_{EX} \qquad (3)$$

It is apparent that the output voltage varies directly with excitation, by a factor $w/l(\pi_{44}X_6)$, or conversely that the output is ratiometric to the excitation, V_{EX}.

A typical output characteristic for an uncompensated transducer with a constant V_{EX} applied is shown in Figure A7. Hence, it is apparent that by increasing the supply voltage at the same rate that the full-scale span is decreasing, the undesired temperature dependence of span may be eliminated. This is accomplished by means of a very low TCR resistor placed in series with the transducer excitation legs which, by design, have a TCR of 0.24%/°C (Figure A8). If the value of the zero-TCR span resistor is appropriately chosen, it will decrease the "net" TCR of the combination to the ideal +0.19%/°C required to exactly compensate the negative TC of SPAN. This technique is known as "self-compensation," and can be utilized in the described manner or with a constant current excitation and a parallel TC span compensation resistor.

The passive circuit utilized to achieve calibration and temperature compensation is shown in Figure A8. Since the single element design uses only one resistor for both the input and the output, a self-compensation scheme can be employed. This technique utilizes the temperature coefficient of the input resistance (TCR) to generate a temperature dependent voltage. The TCR of the strain gage has been specifically designed to be greater in absolute value

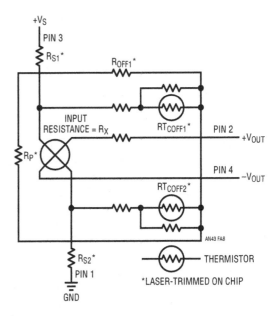

Figure A8 • On-Chip Temperature Compensation and Calibration

than the temperature coefficient of the span, so placing additional passive resistive elements in series with the strain gage modifies the effective TCR and allows temperature compensation based on the input resistance value at room temperature. A constant voltage source is all that is necessary external to the device to ensure accurate operation over a wide temperature range.

The self-compensation technique eliminates the requirement for thermistors which are used in most externally compensated Wheatstone bridge pressure sensors. In addition to the cost and nonlinearity characteristics of thermistors, their negative temperature coefficient precludes their integration on silicon. Thin film resistors, on the other hand, are easily deposited on the strain gage substrate using techniques similar to those required for the metallization of wire bond pads used to make connection to external leads. The laser trimming technique is similar to that used in the manufacturing of high accuracy, monolithic, 16-bit analog-to-digital and digital-to-analog data converters, except that in the case of a pressure transducer, the silicon diaphragm is exercised over the pressure range during the trimming procedure.

Four separate functions are accomplished by the laser trimming operation:
1) Zero calibration
2) Zero temperature compensation
3) Full-scale span temperature compensation
4) Full-scale span calibration

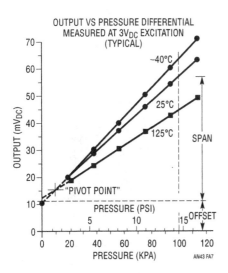

Figure A7 • Output Span for Uncompensated Transducer

The sequence in which the trimming operation is performed is important to avoid interaction of components and the addition of several iterations to the trimming process. The main factor that allows high volume manufacturing techniques, however, is the ability to achieve temperature compensation in the single element sensor without the necessity to change the temperature during the trim operation. Measurements of the sensor parameters are made prior to the laser trim operation. Computer calculations determine which resistors must be trimmed and the amount of trimming required. Resistor R_{OFF1} and R_{OFF2} act as a part of a voltage divider used to calibrate the offset. The output voltage is set to zero with zero pressure applied by trimming either offset resistor R_{OFF1} or R_{OFF2}.

To temperature compensate the offset, thermistors RTC_{OFF1} and RT_{COFF2}, a series of diffused silicon resistors with positive temperature coefficient and different values, are added as required to the circuit by cutting aluminum shorting links.

Full-scale span temperature compensation is accomplished by utilizing self temperature compensation—the addition of a single, series resistor to the input circuit when a constant voltage supply is used. The resistor is adjusted to compensate for changes in span with temperature by adjusting the magnitude of the excitation voltage applied to the active element. In order to minimize common mode errors, the "resistor" is actually split between the supply and ground side of the input so that RS1 = RS2. The span is adjusted to meet the specification by trimming resistor R_P, which is in parallel with the input resistance of the active element. The parallel resistor actually interacts with the series self-compensation network to provide a series-parallel temperature compensation which enhances the performance over the temperature range.

Performance of compensated sensors

The specification for key parameters of a 30PSI on-chip temperature compensated pressure sensor is shown in Figure A9. The excellent linearity is a result of the small active area of the single element strain gage—essentially a point condition. The temperature compensation which is achieved over 0°C to 85°C can be compared to commonly available alternatives.

Appendix B
Bridge readout—then and now

The contemporary monolithic components used to read bridge signals are the beneficiaries of almost 150 years of dedicated work in bridge readout mechanisms. Some early schemes made fiendishly ingenious use of available technology to achieve remarkable performance. Figure B1 shows a light beam galvanometer. This device easily resolved currents in the nanoampere range. The unknown current passed through a coil, producing a magnetic field. The coil is mounted within a static magnetic field. The two field's interactions mechanically biased a small mirror, which was centrally mounted on a tautly suspended wire. The mirror may be thought of as the elastically constrained shaft of a DC motor. The amplitude and sign of the coil current produced

PARAMETER	MIN	TYP	MAX
Pressure Range (in kPA)	—	—	100
Full-Scale Span (in mV)	38.5	40	41.5
Zero Pressure Offset (in mV)	—	+0.05	+1.0
Sensitivity (mV/PSI)	—	1.38	—
Linearity (% FS)	—	+0.1	+0.25
TEMPERATURE EFFECT FOR 0°C TO 85°C			
Full-Scale Span (% FS)	—	+0.5	+1.0
Offset (in mV)	—	+0.5	+1.0

Figure A9 • Specifications for a Typical Pressure Transducer

Figure B1 • The Light Beam Galvanometer is Essentially a Sensitive Meter Movement. It Takes Gain in the Optical Angle of a Mirror Reflected, Collimated Light Source (Courtesy the J. M. Williams Collection)

corresponding torque-like mirror movements. A collimated light source was bounced off the mirror, and its reflection collected on a surface equipped with calibrated markings. The instrument's high inherent sensitivity, combined with the gain in the optical angle, provided excellent results.

The tangent galvanometer (Figure B2) achieved similar nanoampere resolution. The actual meter movement was a compass, centrally mounted within a circular coil. Coil current is measured by noting compass deflection from the earth's magnetic north. Current flow is proportional to the tangent of the measured deflection angle.

These and similar devices were referred to as "null detectors." This nomenclature was well chosen, and reflected the fact that bridges were almost always read at null. This was so because the only technology available to accurately digitize electrical measurements was passive. "Bridge balances," including variable resistors, resistance decade boxes and Kelvin-Varley dividers, were cornerstones of absolute measurements. No source of stable, calibrated gain was available; although the null detectors provided high sensitivity. As such, bridge measurement depended on highly accurate balancing technology and sensitive null detectors.

Lee DeForest's triode (1908) began the era of electronic gain. Harold S. Black attempted to patent negative feedback in 1928, but the U.S. Patent Office, in their governmental wisdom, treated him as a crackpot. Black published in the 1930s, and the notion of

feedback stabilized gain was immediately utilized by more enlightened types. The technology of the day did not permit development of feedback-based amplifiers which could challenge conventional bridge techniques. While Hewlett could use feedback to build a dandy sinewave oscillator, it simply was not good enough to replace Kelvin-Varley dividers and null detectors. Doing so required amplifiers with very high open-loop gains and low zero drift. The second requirement was notably difficult and elusive.

E.A. Goldberg invented the chopper-stabilized amplifier in 1948, finally making stable zero performance practical. Electronic analog computers quickly followed, and historic George A. Philbrick Researches produced the first commercially available general purpose op amps in the 1950s.

Null detectors were the first bridge components to feel the impact of all this. A number of notable chopper-stabilized bridge null detectors were produced during the 1950s and 1960s. All of these were essentially chopper-based operational amplifiers configured as complete instruments. Notable among these was the Julie Research Laboratories sub-microvolt sensitivity ND-103, which featured a 93Hz mechanical chopper (to avoid any interaction with 60Hz noise components). The Hewlett-Packard HP-425 had similar sensitivity, and used a small synchronous clock motor, photocells and incandescent lamps[1] in an elegantly simple photo-chopping scheme. Latter versions of this instrument (the HP-419A) were completely solid state, although retaining a neon lamp-photocell chopping arrangement. Battery operation permitted floating the instrument across the bridge.

Concurrent to all this was the development of rackmounting-based devices called "instrumentation amplifiers." These devices, designed to be applied at the system level, featured settable gain and bandwidth, differential inputs, and good zero point stability. Some were chopper stabilized while others utilized transistorized differential connections. Sold by a number of concerns, they were quite popular for transducer signal conditioning. These devices were the forerunners of modern IC instrumentation amplifiers. Their ability to supply low errors at zero and stable gain made accurate off-null bridge measurement possible.

The development of analog-digital converters during the 1960s[2] provided the last ingredient necessary

Figure B2 • A Tangent Galvanometer Measures Small Currents by Indicating the Interaction Between Applied Current and the Earth's Magnetic Field. Absolute Current Value is Proportional to the Tangent of the Compass Deflection Angle (Courtesy the J. M. Williams Collection)

Note 1: The Hewlett-Packard Company and light bulbs have had a long and successful association.
Note 2: The first fully electronic analog-digital converter was developed by D.H. Wilkinson in 1949 (see References). The first analog-digital converters available as standard product were probably those produced by Pastoriza Electronics in the late 1960s.

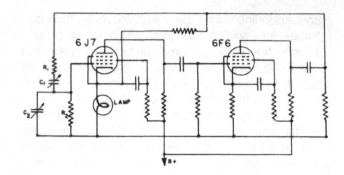

for practical digitized output off-null bridge measurement. It had required over 100 years of technological progress to replace the null detectors and bridge balances. This is something to think about when soldering in IC instrumentation amps and A/D converters. What Lord Kelvin would have given for a single mini-DIP!

Appendix C
The Wien bridge and Mr. Hewlett

The Wien bridge is easily the most popular basis for constructing sinewave oscillators. Circuits constructed around the Wien network offer wide dynamic range, ease of tuning, amplitude stability, low distortion and simplicity. Wien described his network (Figure C1) in 1891. Unfortunately, he had no source of electronic gain available, and couldn't have made it oscillate even if he wanted to. Wien developed the network for AC bridge measurement, and went off and used it for that.

Forty-eight years later William R. Hewlett combined Wien's network with controlled electronic gain in his Masters thesis. The results were the now familiar ''Wien bridge oscillator'' architecture and the Hewlett-Packard Company. Hewlett's circuit (Figure C2) utilized the relatively new tools of feedback theory (see References) to support stable oscillation. Two loops were required. A positive feedback loop from the amplifier's output (6F6 plate) back to its positive input (6J7 first grid) via the Wien bridge provided oscillation. Oscillation amplitude was stabilized by a second, negative, feedback loop. This loop was closed from the output (again, the 6F6 plate) back to the amplifier's negative input (the 6J7 cathode). The now famous lamp supplied a slight positive temperature coefficient to maintain gain at the proper value. For reference in interpreting the vacuum tube[1]

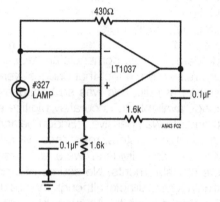

Figure C2 • A Copy of Hewlett's Thesis ''Figure 3'' Showing His Original Circuit. Modern Version Shown for Reference (Hewlett's Figure Courtesy Stanford University Archives)

configuration, a modern version (text Figure 32.39) of Hewlett's circuit appears as an insert.

Contemporary oscillators usually replace the lamps action with electronic equivalents to control loop settling time (see text).

Appendix D
Understanding distortion measurements
Bruce E. Hofer, Audio Precision, Inc.

Introduction
Analog signal distortion is unavoidable in the real world. It can be defined as any effect or process that causes the signal to deviate from ideal. Because ''distortion'' means significantly different things to different people let us distinguish between two general categories based upon frequency domain effect.

A *linear* distortion changes the amplitude and phase relationship between the existing spectral components of a signal without adding new ones. Frequency and phase response errors are the most common examples. Both can cause significant alteration of the time domain waveform.

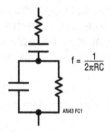

$$f = \frac{1}{2\pi RC}$$

AN43 FC1

Figure C1 • Wiens Network

Note 1: For those tender in years, ''vacuum tubes'' are thermionically activated FETs, descended from Lee DeForest.

A *nonlinear* distortion adds frequency components to the signal that were never there, nor should be to begin with. Nonlinear distortion alters both the time and frequency domain representations of a signal. Noise can be considered a form of nonlinear distortion in some applications.

Nonlinear distortion is generally considered to be more serious than linear distortion because it is impossible to determine if a specific frequency component in the output signal was present in the input. This brief discussion will focus on the measurement and meaning of nonlinear distortion only. The word "distortion" shall hereinafter be used accordingly.

Measures of distortion

One of the best and oldest methods of quantifying distortion is to excite a circuit or system with a relatively pure sinewave and analyze the output for the presence of signal components at frequencies other than the input sinewave. The sinewave is an ideal test signal for measuring nonlinear distortion because it is virtually immune to linear forms of distortions. With the exception of a perfectly tuned notch filter, the output of any linear distortion process will still be a sinewave!

"N-th" harmonic distortion is defined as the amplitude of any output signal at exactly N times the sinewave fundamental frequency. If the input sinewave is 400Hz any second harmonic distortion will show up at 800Hz, third harmonic at 1200Hz, etc. Spectrum analyzers, wave analyzers, and FFT analyzers are the typical instruments used to measure harmonic distortion. These instruments function by acting as highly selective voltmeters measuring the signal amplitude over a very narrow bandwidth centered at a specific frequency.

"THD" or total harmonic distortion is defined as the RMS summation of the amplitudes of *all* possible harmonics, although it is often simplified to include only the second through the fifth (or somewhat higher) harmonics. The assumption that higher order harmonic content is insignificant in the computation of THD can be quite invalid. The sinewave distortion of many function generators is usually dominated by high order harmonic products with only relatively small amounts of products below the fifth harmonic. The crossover distortion characteristic of class AB and B amplifiers can often exhibit significantly high harmonic content above the fifth order.

A far better definition of THD is to include *all* harmonics up to some prescribed frequency limit. Usually the specific application will suggest a relevant upper harmonic frequency limit. In audio circuits a justifiable upper frequency limit might be 20kHz to 25kHz because few people can perceive signals above that range. In practice it has proven desirable to use a somewhat higher limit (typically 80kHz) because nonlinear distortion products above 20kHz can provoke intermodulation problems in subsequent audio stages. In the world of FM and TV broadcast measurements it is common practice to use a 30kHz bandwidth limit even though the signals are inherently limited to 15kHz.

"THD+N" or total harmonic distortion plus noise is defined as the RMS summation of *all signal components*, excluding the fundamental, over some prescribed bandwidth. Distortion analyzers perform this measurement by removing the fundamental sinewave with a notch filter and measuring the leftover signal. Unfortunately some popular analyzers have excessive measurement bandwidth (>1MHz) with no provision for limiting. For the vast majority of applications a measurement bandwidth of >500kHz serves little purpose other than to increase noise contribution and sensitivity to AM radio stations. Today's better distortion analyzers offer a selection of measurement bandwidths typically including 20kHz to 22kHz, 30kHz, 80kHz, and wideband (300kHz to 500kHz).

At first glance it might appear that THD+N measurements are inferior to THD only measurements because of the sensitivity to wideband noise. Even with their noise contribution today's distortion analyzers offer the lowest residual distortion, hence the most accuracy in making ultralow distortion measurements. The typical residual contribution of spectrum analyzers is usually limited by their internal mixer stages to about 0.003% (−90dB). FFT analyzers do not fare much better due to A/D converter nonlinearities. The very best 16-bit converters available today do not guarantee residual distortion below about 0.002% although future developments promise to improve this situation. Distortion analyzers offer the lowest residual performance with at least one manufacturer claiming 0.0001% (typical).

"IMD" or intermodulation distortion is yet another technique for quantifying nonlinearity. It is a much more specialized form of testing requiring a multitone test signal. IMD tests can be more sensitive than THD or THD+N tests because the specific test frequencies, ratios, and analyzer measurement technique can be chosen to optimize response to only certain forms of nonlinearity. Unfortunately this is also one of the biggest disadvantages of IMD testing because there are so many tests that have been suggested: SMPTE, CCIF TIM, DIM, MTM, to name a few.

Distortion measurement accuracy

Nonlinear distortion is not a traceable characteristic in the sense that an unbroken chain of comparisons can be made to a truly distortion-*less* standard. Such a standard does not exist! Real world distortion measurements will always include the non-zero contributions from both the sinewave source and the analyzer.

It is a truly challenging task to *accurately* measure distortion below about 0.01% (−80dB). Indeed, distortion measurement errors can become quite large near residual levels. Harmonic contributions from the original sinewave and the analyzer can add algebraically, vectorially, or even cancel depending upon their relative phase. There are no general assumptions that can be made regarding how two residual contributions will add or subtract.

In the following equation let "M" be the measured value of the N-th harmonic, let "X" be the magnitude of the distortion contributed by the analyzer, and let "D" be the true distortion magnitude of some signal. The measured distortion will be influenced by the residual analyzer contribution:

$$M \cdot \sin(2\pi Nft + \phi) = D \cdot \sin(2\pi Nft) + x \cdot \sin(2\pi Nft + \theta)$$

$$M = \begin{cases} (D + X) & \text{if } \theta = 0° \\ (D^2 + X^2)^{1/2} & \text{if } \theta = \pm 90° \\ (D - X) & \text{if } \theta = 180° \end{cases}$$

$$(4)$$

Depending upon the relative phase between the distortion components (θ) a true distortion factor (D) of 0.0040% could be read as anything between 0.0025% to 0.0055% if the analyzer's internal distortion contribution (X) was 0.0015%. Conversely a 0.0040% reading could have resulted from a true distortion factor of anything from 0.0025% to 0.0055% with the same 0.0015% analyzer contribution.

It is very important to understand this concept when making distortion readings near the specified residual levels of the test equipment. A lower reading may not always signify lower distortion. A low reading could be the result of a fortuitous cancellation of two larger contributions. It is also illogical to conclude that the true value of distortion is always less than the reading because the non-zero residual contributions of the analyzer and sinewave. The service manual of one test equipment manufacturer incredibly states that a 0.0040% reading *verifies* their residual distortion guarantee of 0.0020% for both oscillator and analyzer!

All of the distortion measurement techniques give 0.5dB to 1.0dB (5% to 10%) reading accuracies at higher reading levels. Some distortion analyzers additionally provide average versus true RMS detection. Average detection is a carryover from the past and should be avoided because it will give erroneously low readings when multiple harmonics are present.

The ultimate meaning of THD and THD+N measurements

Both THD and THD+N are measures of signal *impurity*. Distortion analyzers measure THD+N, not THD. Spectrum, wave, and FFT analyzers measure individual harmonic distortion from which THD can be calculated, but not THD+N. Is one better than the other?

For most applications THD+N is the more meaningful measurement because it quantifies total signal impurity. Particularly as we enter the age of A/D and D/A based systems (for example, digital audio) the engineer is increasingly confronted with effects and imperfections that introduce non-harmonic components to a signal. Wideband noise itself can be viewed as an imperfection to be minimized. It is truly myopic to exclude other potentially serious and undesirable signal components in the determination of signal quality just because they do not happen to be a harmonic of the test signal. Why should a 60Hz component be acceptable in the calculation of 20Hz THD but be excluded when testing with a 1kHz fundamental?

On the other hand THD measurements are distinctly better than THD+N measurements if the application is to quantify a simple transfer function nonlinearity. Noise, hum, and other interference products are not introduced by these simple forms of nonlinearity and should not influence the measurement. Examples include the distortion due to component voltage coefficient effects and non-ohmic contact behavior.

Given that all real signals contain some distortion, how much THD or THD+N is acceptable? Only the designer can make that determination.

Appendix E
Some practical considerations for bridge interfaces

It is often desirable to route bridge outputs over considerable cable lengths. Cable driving should always be approached with caution. Even shielded cables are susceptible to noise pick-up, and input protection is often in order. Figure E1 shows some options. Simple RC filters often suffice for filtering. The upper

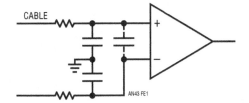

Figure E1 • RC Filter Alternatives

CLAMP TYPE	FORWARD DROP	LEAKAGE AT 25°C (15V REVERSE)
▶├ 1N4148	≈0.6V	≈10^{-9}A
▶∫ HP5082-2810	≈400mV	≈10^{-7}A
2N2222 / 2N4393	≈0.6V	≈10^{-11}A

Figure E2 • Various Devices Offer Different Clamp Characteristics

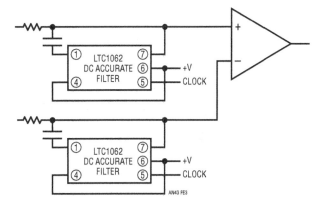

Figure E3 • Switched-Capacitor Techniques Permit a DC Accurate 5th Order Lowpass Filter

dates, and the largest practical non-electrolytic devices are about 1μF Often, a single capacitor (dashed lines) is all that is required. Diode clamps prevent high voltage spikes or faults (common in industrial environments) from damaging the amplifier. Figure E2 summarizes some clamp alternatives.

Figure E3 shows a high order switched-capacitor based filter. The LTC1062 has no DC error, and offers much better roll-off characteristics than the simple RC types. LTC Application Note 20, "Application Considerations for an Instrumentation Lowpass Filter," presents details.

Figure E4 shows a pre-amplifier used ahead of the remotely located instrumentation amplifier. The pre-amp raises cable signal level while lowering drive impedance. The asymmetrical bridge loading should be evaluated when using this circuit. Usually, the amplifier's input resistor can be made large enough to minimize its effect.

limit on resistor value is set by amplifier bias current. FET input amplifiers allow large values, useful for minimizing capacitance size and input protection. Leakage eliminates electrolytic capacitors as candi-

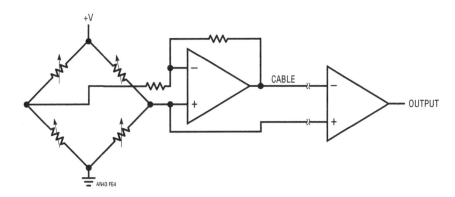

Figure E4 • Pre-Amplifer Provides Gain and Low Impedance Drive to Cable

33

High speed amplifier techniques

A designer's companion for wideband circuitry

Jim Williams

Preface

This publication represents the largest LTC commitment to an application note to date. No other application note absorbed as much effort, took so long or cost so much. This level of activity is justified by our belief that high speed monolithic amplifiers greatly interest users.

Historically, monolithic amplifiers have represented packets of inexpensive, precise and controllable gain. They have partially freed engineers from the constraints and frustrations of device level design. Monolithic operational amplifiers have been the key to practical implementation of high level analog functions. As good as they are, one missing element in these devices has been speed.

Devices presently coming to market are addressing monolithic amplifiers' lack of speed. They bring with them the ease of use and inherent flexibility of op amps. When Philbrick Researches introduced the first mass produced op amp in the 1950s (K2-W) they knew it would be used. What they couldn't possibly know was just how widely, and how many different types of applications there were. As good a deal as the K2-W was (I paid $24.00 for mine–or rather, my father did), monolithic devices are far better. The combination of ease of use, economy, precision and versatility makes modern op amps just too good to be believed.

Considering all this, adding speed to op amps' attractions seems almost certain to open up new application areas. We intend to supply useful high speed products and the level of support necessary for their successful application (such high minded community spirit is, of course, capitalism's deputy). We hope you are pleased with our initial efforts and look forward to working together.

Analog Circuit and System Design: A Tutorial Guide to Applications and Solutions. DOI: 10.1016/B978-0-12-385185-7.00033-0

Introduction

Most monolithic amplifiers have been relatively slow devices. Wideband operation has been the province of discrete and hybrid technologies. Some fast monolithic amplifiers have been available, but the exotic and expensive processing required has inflated costs, precluding widespread acceptance. Additionally, many of the previous monolithic designs were incapable of high precision and prone to oscillation or untoward dynamics, making them unattractive.

Recent processing and design advances have made inexpensive, precision wideband amplifiers practical. Figure 33.1 lists some amplifiers, along with a summary of their characteristics. Reviewing this information reveals extraordinarily wideband devices, with surprisingly good DC characteristics. All of these amplifiers utilize standard op amp architecture, except the LT1223 and LT1228, which are so-called current mode feedback types (see Appendix H, "About current mode feedback"). It is clear that the raw speed capabilities of these devices, combined with their inherent flexibility as op amps, permit a wide range of applications. What is required of the user is a familiarity with the devices and respect for the requirements of high speed circuitry.

This effort's initial sections are devoted to familiarizing the reader with the realities and difficulties of high speed circuit work. The mechanics and subtleties of achieving precision circuit operation at DC and low frequency have been well documented. Relatively little has appeared which discusses, in practical terms, how to get fast circuitry to work. In developing such circuits, even veteran designers sometimes feel that Nature is conspiring against them. In some measure this is true. Like all engineering endeavors, high speed circuits can only work if negotiated compromises with nature are arranged. Ignorance of, or contempt for, physical law is a direct route to frustration. Mother Nature laughs at dilettantism and crushes arrogance without even knowing she did it. Even without Einstein's revelations, the world of high speed is full of surprises. Working with events measured in nanoseconds requires the greatest caution, prudence and respect for Mother Nature. Absolutely nothing should be taken for granted, because nothing is. Circuit design is very much the art of compromise with parasitic effects. The "hidden schematic" (this descriptive was originated by Charly Gullet of Intel Corporation) usually dominates the circuit's form, particularly at high speed.

In this regard, much of the text and appendices are directed at developing awareness of, and respect for, circuit parasitics and fundamental limitations. This approach is maintained in the applications section, where the notion of negotiated compromises is expressed in terms of resistor values and compensation techniques. Many of the application circuits use the amplifier's speed to improve on a standard circuit. Some utilize the speed to implement a traditional function in a non-traditional way, with attendant advantages. A (very) few operate at or near the state-of-the-art for a given circuit type, regardless of approach. Substantial effort has been expended in developing these examples and documenting their operation. The resultant level of detail is justified in the hope that it will be catalytic. The circuits should stimulate new ideas to suit particular needs, while demonstrating fast amplifiers' capabilities in an instructive manner.

Perspectives on high speed design

A substantial amount of design effort has made Figure 33.1's amplifiers relatively easy to use. They are less prone to oscillation and other vagaries than some much slower amplifiers. Unfortunately, laws of physics dictate that the circuit's *environment* must be properly prepared. The performance limits of high speed circuitry are often determined by parasitics such as stray capacitance, ground impedance and layout. Some of these considerations are present in digital systems where designers are comfortable describing bit patterns, delays and memory access times in terms of nanoseconds. Figure 33.2's test circuit provides valuable perspective on just how fast these amplifiers are. Here, the pulse generator (Trace A, Figure 33.3) drives a 74S04 Schottky TTL inverter (Trace B), an LT1223 op amp connected as an inverter (Trace C), and a 74HC04 high speed CMOS inverter (Trace D). The LT1223 doesn't fare too badly. Its delay and fall times are about 2ns slower than the 74S04, but significantly faster than the 74HC04. In fact, the LT1223 has completely finished its transition before the 74HC04 even begins to move! Linear circuits operating with this kind of speed make many engineers justifiably wary. Nanosecond domain linear circuits are widely associated with oscillations, mysterious shifts in circuit characteristics, unintended modes of operation and outright failure to function.

Other common problems include different measurement results using various pieces of test equipment, inability to make measurement connections to the circuit without inducing spurious responses, and dissimilar operation between two identical circuits. If the components used in the circuit are good and the design is sound, all of the above problems can usually be traced to failure to provide a proper circuit environment. To learn how to do this requires studying the causes of the aforementioned difficulties.

The following segments, "Mr. Murphy's gallery of high speed amplifier problems" and the "Tutorial section", address this. The "Problems" section alerts the reader to trouble areas, while the "Tutorial" highlights theory and techniques which may be applied towards solving the problems shown. The tutorials are arranged in roughly the same order as the problems are presented.

Mr. Murphy's gallery of high speed amplifier problems

It sometimes seems that Murphy's Law dominates all physical law. For a complete treatise on Murphy's Law, see Appendix J, "The contributions of Edsel Murphy to the understanding of

PARAMETER	LT1122	LT1190	LT1191	LT1192	LT1193 DIFFERENTIAL	LT1194 DIFFERENTIAL	LT1220	LT1221	LT1222	LT1223	LT1224
Slew Rate	60V/µs	450V/µs	450V/µs	450V/µs	450V/µs	450V/µs	250V/µs	250V/µs	200V/µs	1000V/µs	300V/µs
Bandwidth	14MHz	50MHz	90MHz	400MHz	70MHz	70MHz	45MHz	150MHz	350MHz	100MHz	45MHz
Delay-Rise Time	15ns-65ns	4ns-7ns	3.5ns-1.6ns	5ns-7ns	4ns-7ns	4ns-7ns	4ns-4ns	5ns-5ns	5ns-5ns	3.5ns-3.5ns	4ns-4ns
Settling Time	340ns-0.01%	100ns-0.1%	100ns-0.1%	80ns-0.1%	100ns-0.1%	100ns-0.1%	90ns-0.1%	90ns-0.1%	90ns-0.1%	75ns-0.1%	90ns-0.1%
Output Current	6mA	50mA	50mA	50mA	50mA	50mA	24mA	24mA	24mA	50mA	40mA
Offset	600µV	4mV	2mV	2mV	3mV	3mV	2mV	1mV	1mV	3mV	1mV
Drift	6µV/°C						20µV/°C	15µV/°C	10µV/°C		20µV/°C
Bias Current	75pA	500nA	500nA	500nA	500nA	500nA	300nA	300nA	300nA	3µA	6µA
Gain	500,000	22,000	45,000	200,000	Adjustable	10	20,000	50,000	100,000	90dB	80dB
Gain Error (Minimum Gain)			A$_{VMIN}$ = 10	A$_{VMIN}$ = 10	0.1%	0.1%		A$_{VMIN}$ = 4	A$_{VMIN}$ = 10		A$_{VMIN}$ = 1
Gain Drift											
Power Supply	40V	18V$_{MAX}$	18V$_{MAX}$	18V$_{MAX}$	18V$_{MAX}$	18V$_{MAX}$	36V	36V	36V	36V$_{MAX}$	36V

Figure 33.1 • Characteristics of Some Different Fast IC Amplifiers

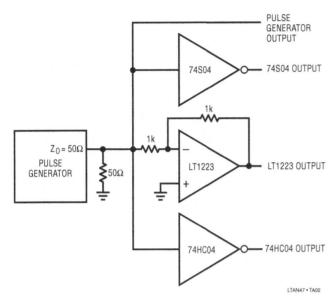

Figure 33.2 • A Race Between the LT1223 Amplifier and Some Fast Logic Inverters

the behavior of inanimate objects", by D.L. Klipstein. The law's consequences weigh heavily in high speed design. As such, a number of examples are given in the following discussion. The average number of phone calls we receive per month due to each "Murphy" example appears at the end of each figure caption.

Problems can start even before power is applied to the amplifier. Figure 33.4 shows severe ringing on the pulse edges at the output of an unterminated pulse generator cable. This is due to reflections and may be eliminated by terminating the cable. *Always terminate the source in its characteristic impedance when looking into cable or long PC traces. Any path over 1 inch long is suspect.*

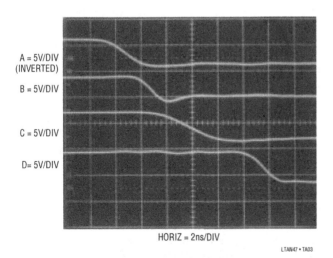

Figure 33.3 • The Amplifier (Trace C) is 3ns Slower than 74S Logic (Trace B), but 5ns Faster than High Speed HCMOS (Trace D)!

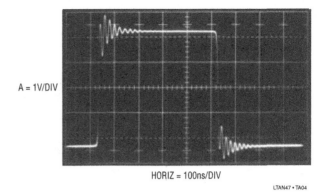

Figure 33.4 • An Unterminated Pulse Generator Cable Produces Ringing Due to Reflections – 3 ☎

In Figure 33.5 the cable is terminated, but ripple and aberration are still noticeable following the high speed edge transitions. In this instance the terminating resistor's leads are lengthy (≈3/4"), preventing a high integrity wideband termination.

The best termination for 50Ω cable is the BNC coaxial type. These devices should not simply be resistors in an enclosure. Good grade 50Ω terminators maintain true coaxial form. They use a carefully designed 50Ω resistor with significant effort devoted to connections to the actual resistive element. In particular, the largest possible connection surface area is utilized to minimize high speed losses. While these type terminators are practical on the test bench, they are rarely used as board level components. In general, the best termination resistors for PC board use are carbon or metal film types with the shortest possible lead lengths. These resistor's end-cap connections provide better high speed characteristics than the rod-connected composition types. Wirewound resistors, because of their inherent and pronounced inductive characteristics, are completely unsuitable for high speed work. This includes so-called non-inductive types.

Another termination consideration is disposal of the current flowing through the terminator. The terminating resistor's grounded end should be placed so that the high speed

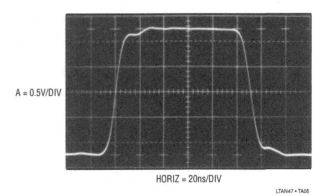

Figure 33.5 • Poor Quality Termination Results in Pulse Corner Aberrations – 1 ☎

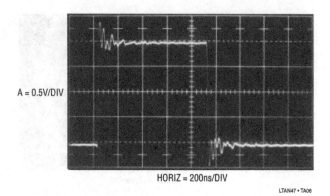

A = 0.5V/DIV

HORIZ = 200ns/DIV

LTAN47 • TA06

Figure 33.6 • Poor Probe Grounding Badly Corrupts the Observed Waveform – 53

currents flowing from it do not disrupt circuit operation. For example, it would be unwise to return terminator current to ground near the grounded positive input of an inverting op amp. The high speed, high density (5V pulses through a 50Ω termination generates 100mA current spikes) current flow could cause serious corruption of the desired zero volt op amp reference. This is another reason why, for bench testing, the coaxial BNC terminators are far preferable to discrete, breadboard mounted resistors. With BNC types in use the termination current returns directly to the source generator and never flows in the breadboard. (For more information see the tutorial section.) *Select terminations carefully and evaluate the effects of their placement in the test set-up.*

Figure 33.6 shows an amplifier output which rings and distorts badly after rapid movement. In this case, the probe ground lead is too long. For general purpose work, most probes come with ground leads about 6 inches long. At low frequencies this is fine. At high speed, the long ground lead looks inductive, causing the ringing shown. High quality probes are always supplied with some short ground straps to deal with this problem. Some come with very short spring clips which fix directly to the probe tip to facilitate a low impedance ground connection. For fast work, the ground connection to the probe should not exceed 1 inch in length. (Probes are covered in the tutorial section; also see Appendix

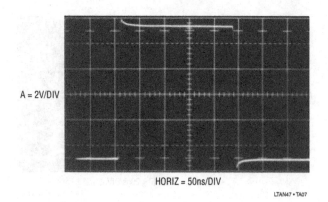

A = 2V/DIV

HORIZ = 50ns/DIV

LTAN47 • TA07

Figure 33.7 • Improper Probe Compensation Causes Seemingly Unexplainable Amplitude Error – 12

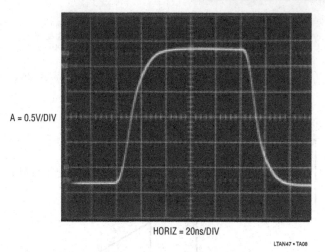

A = 0.5V/DIV

HORIZ = 20ns/DIV

LTAN47 • TA08

Figure 33.8 • Overcompensated or Slow Probes Make Edges Look Too Slow – 2

A, "ABC's of probes", guest written by the engineering staff of Tektronix, Inc.). *Keep the probe ground connection as short as possible. The ideal probe ground connection is purely coaxial. This is why probes mated directly to board mounted coaxial connectors give the best results.*

In Figure 33.7 the probe is properly grounded, but a new problem pops up. This photo shows an amplifier output excursion of 11V—quite a trick from an amplifier running from ±5V rails. This is a commonly reported problem in high speed circuits and can be quite confusing. It is not due to suspension of natural law, but is traceable to a grossly miscompensated or improperly selected oscilloscope probe. *Use probes which match your oscilloscope's input characteristics and compensate them properly.* (For discussions on probes, see Appendix A, "ABC's of probes", guest written by the engineering staff of Tektronix, Inc. and the tutorial section.) Figure 33.8 shows another probe-induced problem. Here the amplitude seems correct but the amplifier appears slow with pronounced edge rounding. In this case, the probe used is too heavily compensated or slow for the oscilloscope. Never use 1X or straight probes. Their bandwidth is 20MHz or less and capacitive loading is high. *Check probe bandwidth to ensure it is adequate for the measurement. Similarly, use an oscilloscope with adequate bandwidth.*

Mismatched probes account for the apparent excessive amplifier delay in Figure 33.9. Delay of almost 12ns (Trace A is the input, Trace B the output) is displayed for an amplifier specified at 6ns. Always keep in mind that various types of probes have different signal transit delay times. At high sweep speeds, this effect shows up in multi-trace displays as time skewing between individual channels. Using similar probes will eliminate this problem, but measurement requirements often dictate dissimilar probes. In such cases the differential delay should be measured and then mentally factored in to reduce error when interpreting the display. It is worth noting that active probes, such as FET and current probes, have signal transit times as long as 25ns. A fast 10X; or 50Ω probe delay

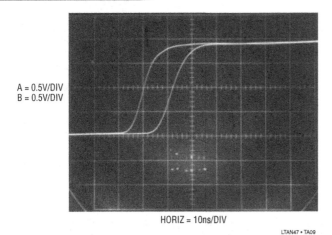

A = 0.5V/DIV
B = 0.5V/DIV

HORIZ = 10ns/DIV

LTAN47 • TA09

Figure 33.9 • Probes with Mismatched Delays Produce Apparent Time Skewing in the Display – 4 ☎

can be inside 3ns. *Account for probe delays in interpreting oscilloscope displays.*

The difficulty shown in Figure 33.10 is a wildly distorted amplifier output. The output slews quickly, but the pulse top and bottom recoveries have lengthy, tailing responses. Additionally, the amplifier output seems to clip well below its nominal rated output swing. A common oversight is responsible for these conditions. A FET probe monitors the amplifier output in this example. The probe's common-mode input range has been exceeded, causing it to overload, clip and distort badly. When the pulse rises, the probe is driven deeply into saturation, forcing internal circuitry away from normal operating points. Under these conditions the displayed pulse top is illegitimate. When the output falls, the probe's overload recovery is lengthy and uneven, causing the tailing. More subtle forms of FET probe overdrive may show up as extended delays with no obvious signal distortion. *Know your FET probe. Account for the delay of its active circuitry. Avoid saturation effects due to common-mode input limitations (typically ±1V). Use 10X and 100X attenuator heads when required.*

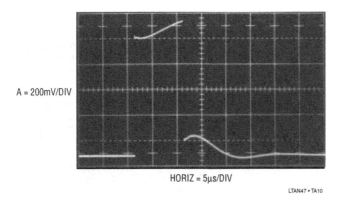

A = 200mV/DIV

HORIZ = 5µs/DIV

LTAN47 • TA10

Figure 33.10 • Overdriven FET Probe Produces Excessive Waveform Distortion and Tailing. Saturation Effects can Also Cause Delayed Response – 1 ☎

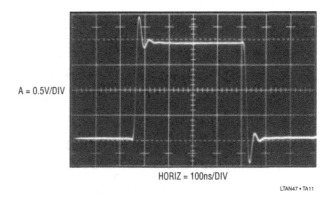

A = 0.5V/DIV

HORIZ = 100ns/DIV

LTAN47 • TA11

Figure 33.11 • Effect of a 10X, 10pF 'Scope Probe at the Summing Point – 2 ☎

Figure 33.11's probe-caused problem results in amplifier output peaking and ringing. In other respects the display is acceptable. This output peaking characteristic is caused by a second 10X probe connected to the amplifier's summing junction. Because the summing point is so central to analyzing op amp operation, it is often monitored. At high speed the 10pF probe input capacitance causes a significant lag in feedback action, forcing the amplifier to overshoot and hunt as it seeks the null point. Minimizing this effect calls for the lowest possible probe input capacitance, mandating FET types or special passive probes. (Probes are covered in the tutorial section; also see Appendix A, "ABC's of probes", guest written by the engineering staff of Tektronix, Inc.). *Account for the effects of probe capacitance, which often dominates its impedance characteristics at high speeds. A standard 10pF 10X probe forms a 10ns lag with a 1KΩ source resistance.*

A peaked, tailing response is Figure 33.12's characteristic. The photo shows the final 40mV of a 2.5V amplifier excursion. Instead of a sharp corner which settles cleanly, peaking occurs, followed by a lengthy tailing decay. This waveform was recorded with an inexpensive off-brand 10X probe. Such probes are often poorly designed, and constructed from materials inappropriate for high speed work. The selection and integration of materials for wideband probes is a specialized and difficult art. Substantial design effort is required to get good fidelity at high speeds. *Never use probes unless they are fully specified for wideband operation. Obtain probes from a vendor you trust.*

Figure 33.13 shows the final movements of an amplifier output excursion. At only 1mV per division the objective is to view the settling residue to high resolution. This response is characterized by multiple time constants, nonlinear slew recovery and tailing. Note also the high speed event just before the waveform begins its negative going transition. What is actually being seen is the oscilloscope recovering from excessive overdrive. Any observation that requires off-screen positioning of parts of the waveform should be approached with the greatest caution. Oscilloscopes vary widely in their response to overdrive, bringing displayed results into question. Complete treatment of high resolution settling time measurements and oscilloscope overload characteristics is

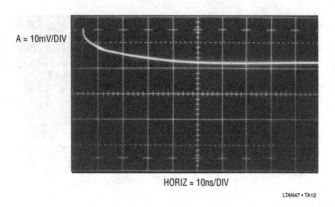

A = 10mV/DIV

HORIZ = 10ns/DIV

LTAN47 • TA12

Figure 33.12 • Poor Quality 10X Probe Introduces Tailing – 2

given in the tutorial section, "About oscilloscopes" and Appendix B "Measuring amplifier settling time". *Approach all oscilloscope measurements which require off-screen activity with caution. Know your instrument's capabilities and limitations.*

Sharp eyed readers will observe that Figure 33.14 is a duplicate of Figure 33.6. Such lazy authorship is excusable because almost precisely the same waveform results when no ground plane is in use. A ground plane is formed by using a continuous conductive plane over the surface of the circuit board. (The theory behind ground planes is discussed in the tutorial section). The only breaks in this plane are for the circuit's necessary current paths. The ground plane serves two functions. Because it is flat (AC currents travel along the surface of a conductor) and covers the entire area of the board, it provides a way to access a low inductance ground from anywhere on the board. Also, it minimizes the effects of stray capacitance in the circuit by referring them to ground. This breaks up potential unintended and harmful feedback paths. *Always use a ground plane with high speed circuitry.*

By far the most common error involves power supply bypassing. Bypassing is necessary to maintain low supply impedance. DC resistance and inductance in supply wires and PC traces can quickly build up to unacceptable levels.

A = 1V/DIV

HORIZ = 200ns/DIV

LTAN47 • TA14

Figure 33.14 • Instabilities Due to No Ground Plane Produce a Display Similar to a Poorly Grounded Probe – 62

This allows the supply line to move as internal current levels of the devices connected to it change. This will almost always cause unruly operation. In addition, several devices connected to an unbypassed supply can "communicate" through the finite supply impedances, causing erratic modes. Bypass capacitors furnish a simple way to eliminate this problem by providing a local reservoir of energy at the device. The bypass capacitor acts as an electrical flywheel to keep supply impedance low at high frequencies. The choice of what type of capacitors to use for bypassing is a critical issue and should be approached carefully (see tutorial, "About bypass capacitors"). An unbypassed amplifier with a 100Ω load is shown in Figure 33.15. The power supply the amplifier sees at its terminals has high impedance at high frequency. This impedance forms a voltage divider with the amplifier and its load, allowing the supply to move as internal conditions in the comparator change. This causes local feedback and oscillation occurs. *Always use bypass capacitors.*

In Figure 33.16 the 100Ω load is removed, and a pulse output is displayed. The unbypassed amplifier responds surprisingly well, but overshoot and ringing dominate. *Always use bypass capacitors.*

Figure 33.17's settling is noticeably better, but some ringing remains. This response is typical of lossy bypass capacitors,

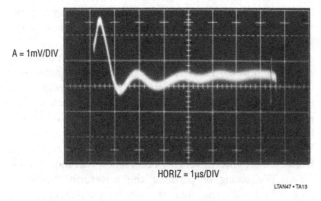

A = 1mV/DIV

HORIZ = 1µs/DIV

LTAN47 • TA13

Figure 33.13 • Overdriven Oscilloscope Display Says More About the Oscilloscope than the Circuit it's Connected to – 6

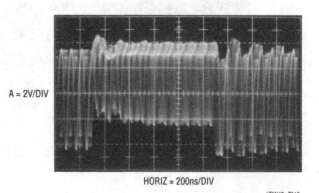

A = 2V/DIV

HORIZ = 200ns/DIV

LTAN47 • TA15

Figure 33.15 • Output of an Unbypassed Amplifier Driving a 100Ω Load Without Bypass Capacitors – 58

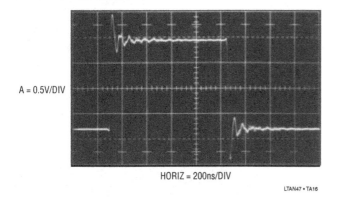

A = 0.5V/DIV

HORIZ = 200ns/DIV

LTAN47 • TA16

Figure 33.16 • An Unbypassed Amplifier Driving No Load is Surprisingly Stable . . . at the Moment – 49 📞

or good ones placed too far away from the amplifier. *Use good quality, low loss bypass capacitors, and place them as close to the amplifier as possible.*

The multiple time constant ringing in Figure 33.18 often indicates poor grade paralleled bypassing capacitors or excessive trace length between the capacitors. While paralleling capacitors of different characteristics is a good way to get wideband bypassing, it should be carefully considered.

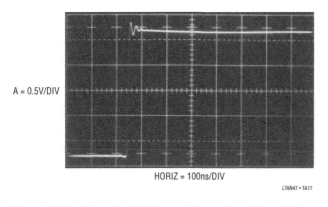

A = 0.5V/DIV

HORIZ = 100ns/DIV

LTAN47 • TA17

Figure 33.17 • Poor Quality Bypass Capacitor Allows Some Ringing – 28 📞

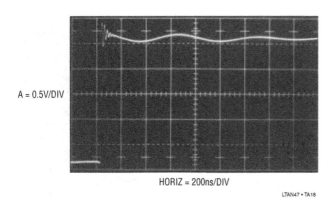

A = 0.5V/DIV

HORIZ = 200ns/DIV

LTAN47 • TA18

Figure 33.18 • Paralleled Bypass Capacitors Form a Resonant Network and Ring – 2 📞

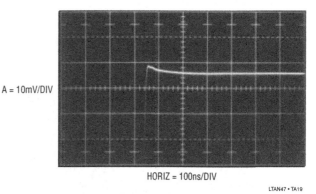

A = 10mV/DIV

HORIZ = 100ns/DIV

LTAN47 • TA19

Figure 33.19 • A More Subtle Bypassing Problem. Not-Quite-Good- Enough Bypassing Causes a Few Millivolts of Peaking – 1 📞

Resonant interaction between the capacitors can cause a waveform like this after a step.

This type response is often aggravated by heavy amplifier loading. *When paralleling bypass capacitors, plan the layout and breadboard with the units you plan to use in production.*

Figure 33.19 addresses a more subtle bypassing problem. The trace shows the last 40mV excursion of a 5V step almost settling cleanly in 300ns. The slight overshoot is due to a loaded (500Ω) amplifier without quite enough bypassing. Increasing the total supply bypassing from 0.1μF to 1μF cured this problem. *Use large value paralleled bypass capacitors when very fast settling is required, particularly if the amplifier is heavily loaded or sees fast load steps.*

The problem shown in Figure 33.20, peaking on the leading and trailing corners, is typical of poor layout practice (see tutorial section on "Breadboarding techniques"). This unity gain inverter suffers from excessive trace area at the summing point. Only 2pF of stray capacitance caused the peaking and ring shown. *Minimize trace area and stray capacitance at critical nodes. Consider layout as an integral part of the circuit and plan it accordingly.*

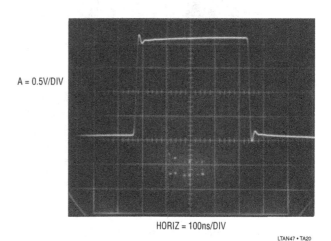

A = 0.5V/DIV

HORIZ = 100ns/DIV

LTAN47 • TA20

Figure 33.20 • 2pF Stray Capacitance at the Summing Point Introduces Peaking – 4 📞

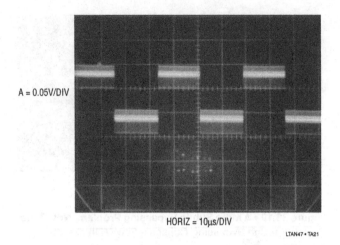

A = 0.05V/DIV

HORIZ = 10µs/DIV

LTAN47 • TA21

Figure 33.21 • Clock or Switching Regulator Noise Corrupts Output Due to Poor Layout – 3 ☎

Figure 33.21's low level square wave output appears to suffer from some form of parasitic oscillation. In actuality, the disturbance is typical of that caused by fast digital clocking or switching regulator originated noise getting into critical circuit nodes. *Plan for parasitic radiative or conductive paths and eliminate them with appropriate layout and shielding.*

Figure 33.22 underscores the previous statement. This output was taken from a gain-of-ten inverter with 1kΩ input resistance. It shows severe peaking induced by *only 1pF* of parasitic capacitance across the 1k resistor. The 50Ω terminated input source provides only 20mV of drive via a divider, but that's more than enough to cause problems, even with only 1pF stray coupling. In this case the solution was a ground referred shield at a right angle to, and encircling, the 1kΩ resistor. *Plan for parasitic radiative paths and eliminate them with appropriate shielding.*

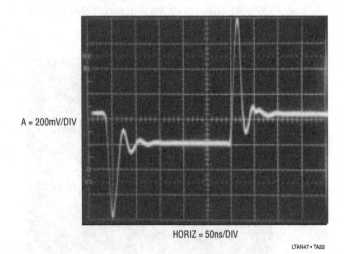

A = 200mV/DIV

HORIZ = 50ns/DIV

LTAN47 • TA22

Figure 33.22 • Output of an X10 Amplifier with 1pF Coupling from the Summing Point to the Input. Careful Shielding of the Input Resistor Will Eliminate the Peaked Edges and Ringing – 2 ☎

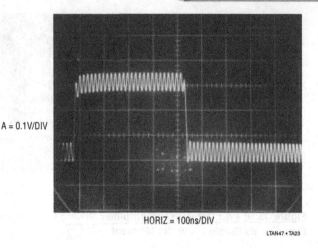

A = 0.1V/DIV

HORIZ = 100ns/DIV

LTAN47 • TA23

Figure 33.23 • Decompensated Amplifier Running at Too Low a Gain – 22 ☎

A decompensated amplifier running at too low a gain produced Figure 33.23's trace. The price for decompensated amplifiers' increased speed is restrictions on minimum allowable gain. Decompensated amplifiers are simply not stable below some (specified) minimum gain, and no amount of ignorance or wishing will change this. This is a common applications oversight with these devices, although the amplifier never fails to remind the user. *Observe gain restrictions when using decompensated amplifiers.*

Oscillation is also the problem in Figure 33.24, and it is due to excessive capacitive loading (see tutorial section on "Oscillation"). Capacitive loading to ground introduces lag in the feedback signal's return to the input. If enough lag is introduced (e.g., a large capacitive load) the amplifier may oscillate. Even if a capacitively loaded amplifier doesn't oscillate, it's always a good idea to check its response with step testing. It's amazing how close to the edge of the cliff you can get without falling off, except when you build 10,000 production units. *Avoid capacitive loading. If such loading is necessary, check performance margins and isolate or buffer the load if necessary.*

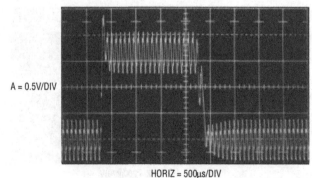

A = 0.5V/DIV

HORIZ = 500µs/DIV

LTAN47 • TA24

Figure 33.24 • Excessive Capacitive Load Upsets the Amplifier – 165 ☎

687

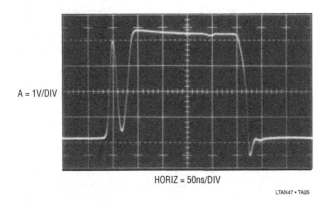

A = 1V/DIV

HORIZ = 50ns/DIV

LTAN47 • TA25

Figure 33.25 • Input Common Mode Overdrive Generates Odd Outputs – 3 ☎

Figure 33.25 appears to contain one cycle of oscillation. The output waveform initially responds, but abruptly reverses direction, overshoots and then heads positive again. Some overshoot again occurs, with a long tail and a small dip well before a non-linear slew returns the waveform to zero. Ugly overshoot and tailing completes the cycle. This is certainly strange behavior. What is going on here? The input pulse is responsible for all these anomalies. Its amplitude takes the amplifier outside its common-mode limits, inducing the bizarre effects shown. *Keep inputs inside specified common-mode limits at all times.*

Figure 33.26 shows an oscillation laden output (Trace B) trying to unity gain invert the input (Trace A). The input's form is distinguishable in the output, but corrupted with very high frequency oscillation and overshoot. In this case the amplifier includes a booster within its loop to provide increased output current. The disturbances noted are traceable to local instabilities within the booster circuit. (See Appendix C, "The oscillation problem—frequency compensation without tears"). *When using output booster stages, insure they are inherently stable before placing them inside an amplifier's feedback loop. Wideband booster stages are particularly prone to device level parasitic high frequency oscillation.*

Figure 33.27's booster augmented unity gain inverting op amp also oscillates, but at a much lower frequency.

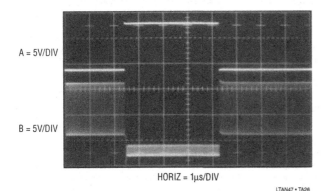

A = 5V/DIV

B = 5V/DIV

HORIZ = 1µs/DIV

LTAN47 • TA26

Figure 33.26 • Local Oscillations in a Booster Stage. Frequency is Typically High – 12 ☎

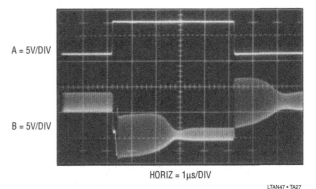

A = 5V/DIV

B = 5V/DIV

HORIZ = 1µs/DIV

LTAN47 • TA27

Figure 33.27 • Loop Oscillations in a Booster Stage. Note Lower Frequency than Local Oscillations in Previous Example – 28 ☎

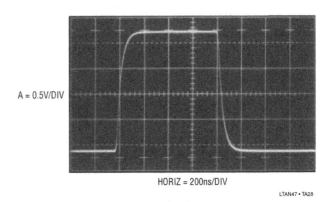

A = 0.5V/DIV

HORIZ = 200ns/DIV

LTAN47 • TA28

Figure 33.28 • Excessive Source Impedance Gives Serene But Undesired Response – 6 ☎

Additionally, overshoot and non-linear recovery dominate the waveform's envelope. Unlike the previous example, this behavior is not due to local oscillations within the booster stage. Instead, the booster is simply too slow to be included in the op amp's feedback loop. It introduces enough lag to force oscillation, even as it hopelessly tries to maintain loop closure. *Insure booster stages are fast enough to maintain stability when placed in the amplifier's feedback loop.*

The serene rise and fall of Figure 33.28's pulse is a welcome relief from the oscillatory screaming of the previous photos. Unfortunately, such tranquilized behavior is simply too slow. This waveform, reminiscent of Figure 33.8's band-limited response, is due to excessive source impedance. The high source impedance combines with amplifier input capacitance to band limit the input and the output reflects this action. *Minimize source impedance to levels which maintain desired bandwidth. Keep stray capacitance at inputs down.*

Tutorial section

An implied responsibility in raising the aforementioned issues is their solution or elimination. What good is all the rabble-rousing without suggestions for fixes? It is in this spirit that

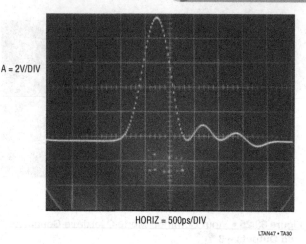

HORIZ = 500ps/DIV

LTAN47 • TA30

Figure 33.30 • Poor Grade Termination Produces Pronounced Ringing and Tailing in the GHz Range

this tutorial section is presented. Theory, techniques, prejudice and just plain gossip are offered as tools which may help avoid or deal with difficulties. As previously mentioned, the tutorials appear in roughly the same order as the problems were presented.

About cables, connectors and terminations

Routing of high speed signals to and from the circuit board should always be done with good quality coaxial cable. The cable should be driven and terminated in the system's characteristic impedance at the drive and load points. The driven end is usually an instrument (e.g., pulse or signal generator), presumably endowed with proper characteristics by its manufacturer. It is the cable and its termination, selected by the experimenter, that often cause problems.

All coaxial cable is not the same. Use cable appropriate to the system's characteristic impedance and of good quality. Poorly chosen cable materials or construction methods can introduce odd effects at very high speeds, resulting in observed waveform distortion. A poor cable choice can adversely effect 0.01% settling in the 100ns-200ns region. Similarly, poor cable can preclude maintenance of even the cleanest pulse generator's 1ns rise time or purity. Typically, inappropriate cable can introduce tailing, rise time degradation, aberrations following transitions, nonlinear impedance and other undesirable characteristics.

Termination choice is equally important. Good quality BNC coaxial type terminators are usually the best choice for breadboarding. Their impedance vs frequency is flat into the GHz range. Additionally, their construction insures that the (often substantial) drive current returns directly to the source, instead of being dumped into the breadboard's ground system. As previously discussed, BNC coaxial terminators are not simply resistors in a can. Special construction techniques insure optimum wideband response. Figures 33.29 and 33.30 demonstrate this nicely. In Figure 33.29 a

1ns pulse with 350ps rise and fall times[1] is monitored on a 1GHz sampling 'scope (Tektronix 556 with 1S1 sampling plug-in and P6032 probe). The waveform is clean, with only a slight hint of ring after the falling edge. This photo was taken with a high grade BNC coaxial type terminator in use. Figure 33.30 does not share these attributes. Here, the generator is terminated with a 50Ω carbon composition resistor with lead lengths of about 1/8 inch. The waveform rings and tails badly on turn-off before finally settling. Note that the sweep speed required a 2.5X reduction to capture these unwanted events.

Connectors, such as BNC barrel extensions and tee-type adaptors, are convenient and frequently employed. Remember that these devices represent a discontinuity in the cable, and can introduce small but undesirable effects. In general it is best to employ them as close as possible to a terminated point in the system. Use in the middle of a cable run provides minimal absorption of their mismatch and reflections. The worst offenders among connectors are adapters. This is unfortunate, as these devices are necessitated by the lack of connection standardization in wideband instrumentation. The mismatch caused by a BNC-to-GR874 adaptor transition at the input of a wideband sampling 'scope is small, but clearly discernible in the display. Similarly, mismatches in almost all adaptors, and even in "identical" adaptors of different manufacture, are readily measured on a high-frequency network analyzer such as the Hewlett-Packard 4195A[2] (for additional wisdom and terror along these lines see Reference 1).

BNC connections are easily the most common, but not necessarily the most desirable, wideband connection mechanism. The ingenious GR874 connector has notably superior

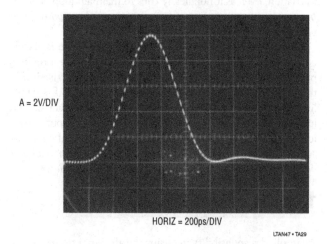

HORIZ = 200ps/DIV

LTAN47 • TA29

Figure 33.29 • 350ps Rise and Fall Times are Preserved by a Good Quality Termination

Note 1: The ability to generate such a pulse proves useful for a variety of tasks, including testing terminators, cables, probes and oscilloscopes for response. The requirements for this pulse generator are surprisingly convenient and inexpensive. For a discussion and construction details see Appendix D "Measuring probe-oscilloscope response"

Note 2: Almost no one believes any of this until they see it for themselves. I didn't. Photos of the network analyzer's display aren't included in the text because no one would believe it. I wouldn't.

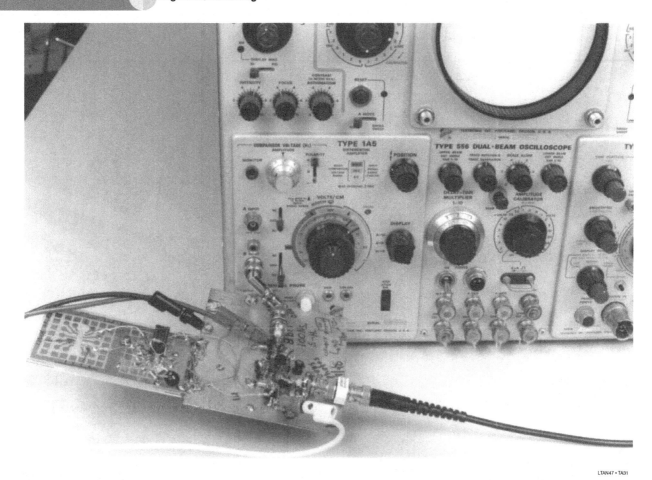

LTAN47 • TA31

Figure 33.31 • Sometimes the Best Probe is No Probe. Direct Connection to the Oscilloscope Eliminates a 10X Probe's Attenuation and Possible Grounding Problems in a Sample-Hold (Figure 33.124) Settling Time Measurement

high frequency characteristics, as does the type N. Unfortunately, it's a BNC world out there.

About probes and probing techniques

The choice of which oscilloscope probe to use in a measurement is absolutely crucial. The probe must be considered as an inherent part of the circuit under test. Rise time, bandwidth, resistive and capacitive loading, delay and other limitations must be kept in mind.

Sometimes, the best probe is no probe at all. In some circumstances it is possible and preferable to connect critical breadboard points directly to the oscilloscope (see Figure 33.31). This arrangement provides the highest possible grounding integrity, eliminates probe attenuation, and maintains bandwidth. In most cases this is mechanically inconvenient, and often the oscilloscope's electrical characteristics (particularly input capacitance) will not permit it. This is why oscilloscope probes were developed, and why so much effort has been put into their development (Reference 42 is excellent). In addition to the material presented here, an in-depth treatment of probes appears in Appendix A, "ABC's of probes", guest written by the engineering staff of Tektronix, Inc.

Probes are the most overlooked cause of oscilloscope mismeasurement. All probes have some effect on the point they are measuring. The most obvious is input resistance, but input capacitance usually dominates in a high speed measurement. Much time can be lost chasing circuit events which are actually due to improperly selected or applied probes. An 8pF probe looking at a 1kΩ source impedance forms an 8ns lag — substantially longer than a fast amplifier's delay time! Pay particular attention to the probe's input capacitance. Standard 10MΩ, 10X probes typically have 8pF-10pF of input capacitance, with 1X types being much higher. In general, 1X probes are not suitable for fast work because their bandwidth is limited to about 20MHz. Remember that all 10X probes cannot be used with all oscilloscopes; the probe's compensation range must match the oscilloscope's input capacitance. Low impedance probes (with 500Ω to 1kΩ resistance), designed for 50Ω inputs, usually have input capacitance of 1pF or 2pF. They are a very good choice if you can stand the low resistance. FET probes maintain high input resistance and keep capacitance at the 1pF level but have substantially more delay than passive probes. FET probes also have limitations on input common-mode range which must be adhered to or serious measurement errors will result. Contrary to popular belief, FET probes do not have extremely high input resistance — some types

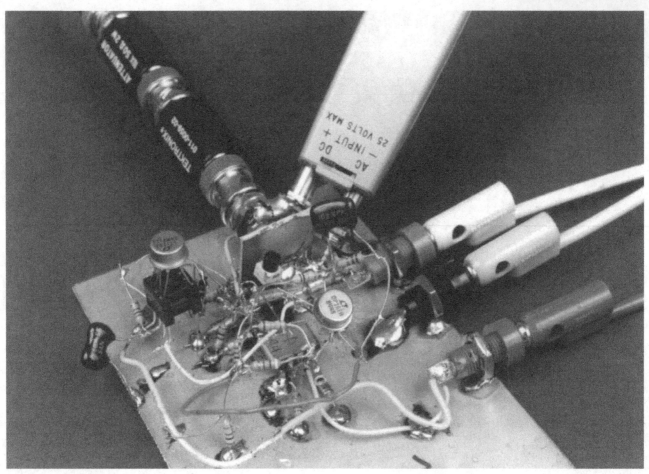

LTAN47 • TA32

Figure 33.32 • Using a Differential Probe to Verify the Integrity of a 2.5mV High Speed Input Pulse (Figure 33.76's X1000 Amplifier)

are as low as 100kΩ. It is possible to construct a wideband FET probe with very high input impedance, although input capacitance is somewhat higher than standard FET probes. For measurements requiring these characteristics, such a probe is useful. See Appendix E, "An ultra fast high impedance probe".

Regardless of which type probe is selected remember that they all have bandwidth and rise time restrictions. The displayed rise time on the oscilloscope is the vector sum of source, probe and 'scope rise times:

$$t_{RISE} = \sqrt{(t_{RISE}\,Source)^2 + (t_{RISE}\,Probe)^2 + (t_{RISE}\,Oscilloscope)^2}$$

This equation warns that some rise time degradation must occur in a cascaded system. In particular, if probe and oscilloscope are rated at the same rise time, the system response will be slower than either.

Current probes are useful and convenient.[3] The passive transformer-based types are fast and have less delay than the Hall effect-based versions. The Hall types, however, respond

at DC and low frequency and the transformer types typically roll off around 100Hz to 1kHz. Both types have saturation limitations which, when exceeded, cause odd results on the CRT which will confuse the unwary. The Tektronix type CT-1 current probe, although not nearly as versatile as the clip-on probes, bears mention. Although this is not a clip-on device, it may be the least electrically intrusive way of extracting wideband signal information. Rated at 1GHz bandwidth, it produces 5mV/mA output with only 0.6pF loading. Decay time constant of this AC current probe is ≈1%/50ns, resulting in a low frequency limit of 35kHz.

A very special probe is the differential probe. A differential probe may be thought of as two matched FET probes contained within a common probe housing. This probe literally brings the advantage of a differential input oscilloscope to the circuit board. The probes' matched, active circuitry provides greatly improved high frequency common mode rejection over single ended probing or even matched passive probes used with a differential amplifier. The resultant ability to reject common-mode signals and ground noise at high frequency allows this probe to deliver exceptionally clean results when monitoring

---

Note 3: A more thorough discussion of current probes is given in LTC Application Note 35, "Step Down Switching Regulators". See Reference 2.

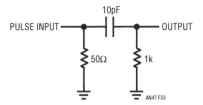

Figure 33.33 • Probe Test Circuit

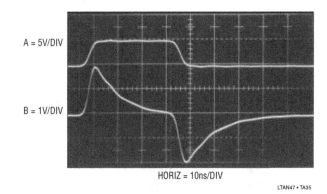

Figure 33.35 • Test Circuit Output with 9pF Probe and 0.25 Inch Ground Strap

small, fast signals. Figure 33.32 shows a differential probe being used to verify the waveshape of a 2.5mV input to a wideband, high gain amplifier (Figure 33.76 of the Applications section).

When using different probes, remember that they all have different delay times, meaning that apparent timing errors will occur on the CRT. Know what the individual probe delays are and account for them in interpreting the CRT display.

By far the greatest source of error in probe use is grounding. Poor probe grounding can cause ripples and discontinuities in the waveform observed. In some cases the choice and placement of a probe's ground strap will affect waveforms on another channel. In the worst case, connecting the probe's ground wire will virtually disable the circuit being measured. The cause of these problems is due to parasitic inductance in the probe's ground connection. In most oscilloscope measurements this is not a problem, but at nanosecond speeds it becomes critical. Fast probes are always supplied with a variety of spring clips and accessories designed to aid in making the lowest possible inductive connection to ground. Most of these attachments assume a ground plane is in use, which it should be. Always try to make the shortest possible connection to ground—anything longer than 1 inch may cause trouble. Sometimes it's difficult to determine if probe grounding is the cause of observed waveform aberrations. One good test is to disturb the grounding set-up and see if changes occur. Nominally, touching the ground plane or jiggling probe ground connectors or wires should have no effect. If a ground strap wire is in use try changing its orientation or simply squeezing it together to change and minimize its loop area. *If any waveform change occurs while doing this the probe*

grounding is unacceptable, rendering the oscilloscope display unreliable.

The simple network of Figure 33.33 shows just how easy it is for poorly chosen or used probes to cause bad results. A 9pF input capacitance probe with a 4 inch long ground strap monitors the output (Trace B, Figure 33.34). Although the input (Trace A) is clean, the output contains ringing. Using the same probe with a 1/4 inch spring tip ground connection accessory seemingly cleans up everything (Figure 33.35). However, substituting a 1pf FET probe (Figure 33.36) reveals a 50% output amplitude error in Figure 33.35! The FET probe's low input capacitance allows a more accurate version of circuit action. The FET probe does, however, contribute its own form of error. Note that the probe's response is tardy by 5ns due to delay in its active circuitry. Hence, separate measurements with each probe are required to determine the output's amplitude and timing parameters.

A final form of probe is the human finger. Probing the circuit with a finger can accentuate desired or undesired effects, giving clues that may be useful. The finger can be used to introduce stray capacitance to a suspected circuit node while observing results on the CRT. Two fingers, lightly moistened, can be used to provide an experimental resistance path. Some high speed engineers are particularly adept at these techniques and can estimate the capacitive and resistive effects created with surprising accuracy.

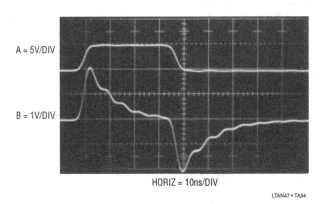

Figure 33.34 • Test Circuit Output with 9pF Probe and 4 Inch Ground Strap

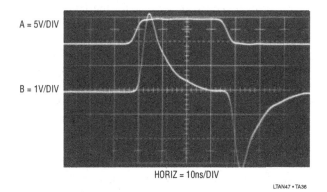

Figure 33.36 • Test Circuit Output with FET Probe

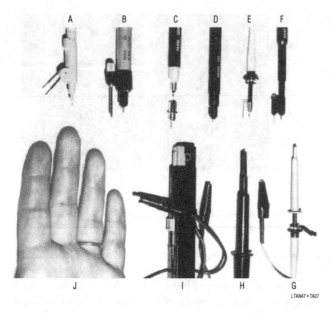

Figure 33.37 • Various Probe-Ground Strap Configurations

Examples of some of the probes discussed, along with different forms of grounding implements, are shown in Figure 33.37. Probes A, B, E, and F are standard types equipped with various forms of low impedance grounding attachments. The conventional ground lead used on G is more convenient to work with but will cause ringing and other effects at high frequencies, rendering it useless. H has a very short ground lead. This is better, but can still cause trouble at high speeds. D is a FET probe. The active circuitry in the probe and a very short ground connector ensure low parasitic capacitance and inductance. C is a separated FET probe attenuator head. Such heads allow the probe to be used at higher voltage levels (e.g., ±10V or ±100V). The miniature coaxial connector shown can be mounted on the circuit board and the probe mated with it. This technique provides the lowest possible parasitic inductance in the ground path and is especially recommended. I is a current probe. A ground connection is not usually required. However, at high speeds the ground connection may result in a cleaner CRT presentation. Because no current flows in the ground lead of these probes, a long strap is usually permissible. J is typical of the finger probes described in the text. Note the ground strap on the third finger.

The low inductance ground connectors shown are available from probe manufacturers and are always supplied with good quality, high frequency probes. Because most oscilloscope measurements do not require them, they invariably become lost. There is no substitute for these devices when they are needed, so it is prudent to take care of them. This is especially applicable to the ground strap on the finger probe.

About oscilloscopes

The modern oscilloscope is one of the most remarkable instruments ever constructed. The protracted and intense development effort put toward these machines is perhaps equaled only by the fanaticism devoted to timekeeping.[4] It is a tribute to oscilloscope designers that instruments manufactured over 25 years ago still suffice for over 90% of today's measurements. The oscilloscope-probe combination used in high speed work is the most important equipment decision the designer must make. Ideally, the oscilloscope should have at least 150MHz bandwidth, but slower instruments are acceptable if their limitations are well understood. Be certain of the characteristics of the probe-oscilloscope combination. Rise time, bandwidth, resistive and capacitive loading, delay, noise, channel-to-channel feedthrough, overdrive recovery, sweep nonlinearity, triggering, accuracy and other limitations must be kept in mind. High speed linear circuitry demands a great deal from test equipment and countless hours can be saved if the characteristics of the instruments used are well known. Obscene amounts of time have been lost pursuing "circuit problems" which in reality are caused by misunderstood, misapplied or out-of-spec equipment. Intimate familiarity with your oscilloscope is invaluable in getting the best possible results with it. In fact, it is possible to use seemingly inadequate equipment to get good results if the equipment's limitations are well known and respected. All of the circuits in the Applications section involve rise times and delays well above the 100MHz–200MHz region, but 90% of the development work was done with a 50MHz oscilloscope. Familiarity with equipment and thoughtful measurement technique permit useful measurements seemingly beyond instrument specifications. A 50MHz oscilloscope cannot track a 5ns rise time pulse, but it can measure a 2ns delay between two such events. Using such techniques, it is often possible to deduce the desired information. There are situations where no amount of cleverness will work and the right equipment (e.g., a faster oscilloscope) must be used. Sometimes, "sanity-checking" a limited bandwidth instrument with a higher bandwidth oscilloscope is all that is required. For high speed work, brute force bandwidth is indispensable when needed, and no amount of features or computational sophistication will substitute. Most high speed circuitry does not require more than two traces to get where you are going. Versatility and many channels are desirable, but if the budget is limited, spend for bandwidth!

Dramatic differences in displayed results are produced by probe-oscilloscope combinations of varying bandwidths. Figure 33.38 shows the output of a very fast pulse[5] monitored with a 1GHz sampling 'scope (Tektronix 556 with 1S1 sampling plug-in). At this bandwidth the 10V amplitude appears clean, with just a small hint of ringing after the falling edge. The rise and fall times of 350ps are suspicious, as the sampling oscilloscope's rise time is also specified at 350ps.[6]

Figure 33.39 shows the same pulse observed on a 350MHz instrument with a direct connection to the input (Tektronix

---

Note 4: In particular, the marine chronometer received ferocious and abundant amounts of attention. See References 4, 5, and 6. For an enjoyable stroll through the history of oscilloscope vertical amplifiers, see Reference 3. See also Reference 41.

Note 5: See Appendix D, "Measuring probe - oscilloscope response", for complete details on this pulse generator.

Note 6: This sequence of photos was shot in my home lab. I'm sorry, but 1GHz is the fastest 'scope in my house.

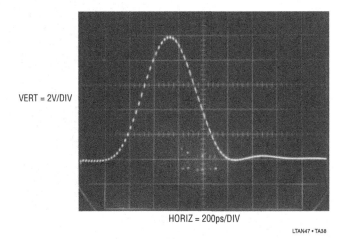

HORIZ = 200ps/DIV

LTAN47 • TA38

Figure 33.38 • A 350ps Rise/Fall Time 10V Pulse Monitored on 1GHz Sampling Oscilloscope. Direct 50Ω Input Connection is Used

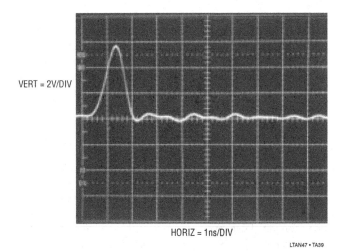

HORIZ = 1ns/DIV

LTAN47 • TA39

Figure 33.39 • The Test Pulse Appears Smaller and Slower On a 350MHz Instrument (t_{RISE} = 1ns). Deliberate Poor Grounding Creates Rippling After the Pulse Falls. Direct 50Ω Connection is Used

485/50Ω input). Indicated rise time balloons to 1ns, while displayed amplitude shrinks to 6V, reflecting this instrument's lesser bandwidth. To underscore earlier discussion, poor grounding technique (1½″ of ground lead to the ground plane) created the prolonged rippling after the pulse fall.

Figure 33.40 shows the same 350MHz (50Ω input) oscilloscope with a 3GHz 10× probe (Tektronix P6056). Displayed results are nearly identical, as the probe's high bandwidth contributes no degradation. Again, deliberate poor grounding causes overshoot and rippling on the pulse fall.

Figure 33.41 equips the same oscilloscope with a 10X probe specified at 290MHz bandwidth (Tektronix P6047). Additionally, the oscilloscope has been switched to its 1MΩ input mode, reducing bandwidth to a specified 250MHz. Amplitude degrades to less than 4V and edge times similarly increase. The deliberate poor grounding contributes the undershoot and underdamped recovery on pulse fall.

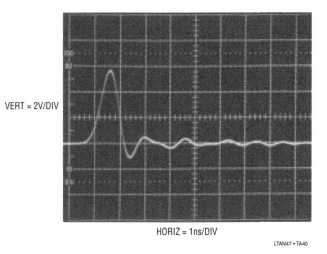

HORIZ = 1ns/DIV

LTAN47 • TA40

Figure 33.40 • Test Pulse on the Same 350MHz Oscilloscope Using a 3GHz 10X Probe. Deliberate Poor Grounding Maintains Rippling Residue

In Figure 33.42 a 100MHz 10X probe (Hewlett-Packard Model 10040A) has been substituted for the 290MHz unit. The oscilloscope and its set-up remain the same. Amplitude shrinks below 2V, with commensurate rise and fall times. Cleaned up grounding eliminates aberrations.

A Tektronix 454A (150MHz) produced Figure 33.43's trace. The pulse generator was directly connected to the input. Displayed amplitude is about 2V, with appropriate 2ns edges. Finally, a 50MHz instrument (Tektronix 556 with 1A4 plug-in) just barely grunts in response to the pulse (Figure 33.44). Indicated amplitude is 0.5V, with edges reading about 7ns. That's a long way from the 10V and 350ps that's really there!

A final oscilloscope characteristic is overload performance. It is often desirable to view a small amplitude portion of a large waveform. In many cases the oscilloscope is required to supply an accurate waveform after the display has been driven off

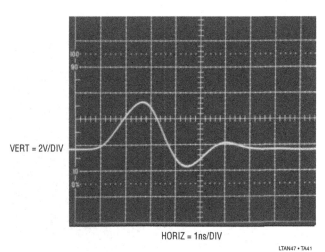

HORIZ = 1ns/DIV

LTAN47 • TA41

Figure 33.41 • Test Pulse Measures Only 3V High on a 250MHz 'Scope with Significant Waveform Distortion. 250MHz 10X Probe Used

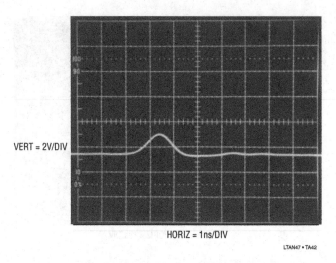

VERT = 2V/DIV

HORIZ = 1ns/DIV

LTAN47 • TA42

Figure 33.42 • Test Pulse Measures Under 2V High Using 250MHz 'Scope and a 100MHz Probe

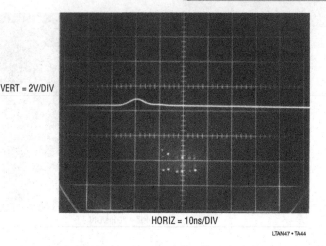

VERT = 2V/DIV

HORIZ = 10ns/DIV

LTAN47 • TA44

Figure 33.44 • A 50MHz Instrument Barely Grunts. 10V, 350ps Test Pulse Measures Only 0.5V High with 7ns Rise and Fall Times!

screen. How long must one wait after an overload before the display can be taken seriously? The answer to this question is quite complex. Factors involved include the degree of overload, its duty cycle, its magnitude in time and amplitude, and other considerations. Oscilloscope response to overload varies widely between types and markedly different behavior can be observed in any individual instrument. For example, the recovery time for a 100X overload at 0.005V/division may be very different than at 0.1V/division. The recovery characteristic may also vary with waveform shape, DC content and repetition rate. With so many variables, it is clear that measurements involving oscilloscope overload must be approached with caution. Nevertheless, a simple test can indicate when the oscilloscope is being deleteriously affected by overdrive.

The waveform to be expanded is placed on the screen at a vertical sensitivity which eliminates all off-screen activity. Figure 33.45 shows the display. The lower right hand portion is

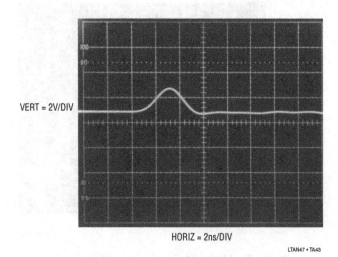

VERT = 2V/DIV

HORIZ = 2ns/DIV

LTAN47 • TA43

Figure 33.43 • 150MHz Oscilloscope (t_RISE = 2.4ns) with Direct Connection Responds to the Test Pulse

to be expanded. Increasing the vertical sensitivity by a factor of two (Figure 33.46) drives the waveform off-screen, but the remaining display appears reasonable. Amplitude has doubled and waveshape is consistent with the original display. Looking carefully, it is possible to see small amplitude information presented as a dip in the waveform at about the third vertical division. Some small disturbances are also visible. This observed expansion of the original waveform is believable. In Figure 33.47, gain has been further increased and all the features of Figure 33.46 are amplified accordingly. The basic waveshape appears clearer and the dip and small disturbances are also easier to see. No new waveform characteristics are observed. Figure 33.48 brings some unpleasant surprises. This increase in gain causes definite distortion. The initial negative-going peak, although larger, has a different shape. Its bottom appears less broad than in Figure 33.47. Additionally, the peak's positive recovery is shaped slightly differently. A new rippling disturbance is visible in the center of the screen. This kind of change indicates that the oscilloscope is having trouble. A further test can confirm that this waveform is being influenced by overloading. In Figure 33.49 the gain remains the same, but the vertical position knob has been used to reposition the display at the screen's bottom. This shifts the oscilloscope's DC operating point which, under normal circumstances, should not affect the displayed waveform. Instead, a marked shift in waveform amplitude and outline occurs. Repositioning the waveform to the screen's top produces a differently distorted waveform (Figure 33.50). It is obvious that for this particular waveform, accurate results cannot be obtained at this gain.

Differential plug-ins can address some of the issues associated with excessive overdrive, although they cannot solve all problems. Two differential plug-in types merit special mention. At low level, a high sensitivity differential plug-in is indispensable. The Tektronix 1A7, 1A7A and 7A22 feature 10μV sensitivity, although bandwidth is limited to 1MHz. The units also have selectable high and low pass filters and good high frequency common-mode rejection. Tektronix type 1A5, W and 7A13 are differential comparators. They

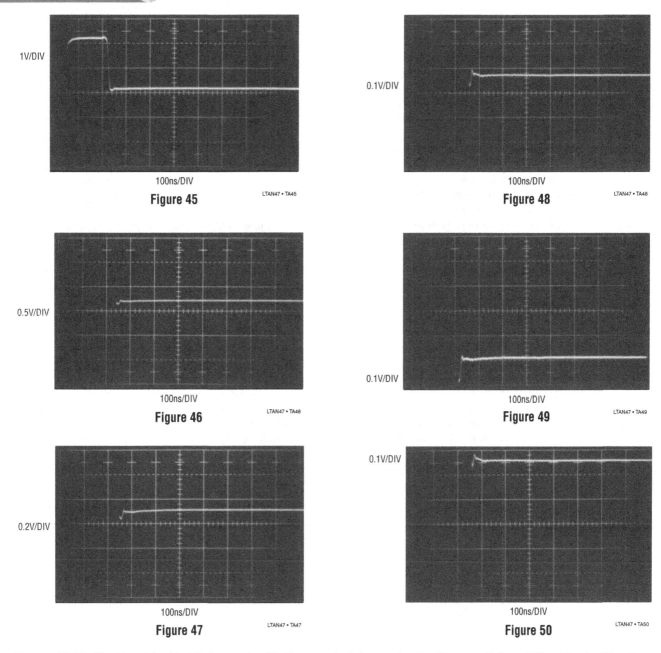

1V/DIV

100ns/DIV
Figure 45   LTAN47 • TA45

0.5V/DIV

100ns/DIV
Figure 46   LTAN47 • TA46

0.2V/DIV

100ns/DIV
Figure 47   LTAN47 • TA47

0.1V/DIV

100ns/DIV
Figure 48   LTAN47 • TA48

0.1V/DIV

100ns/DIV
Figure 49   LTAN47 • TA49

0.1V/DIV

100ns/DIV
Figure 50   LTAN47 • TA50

Figures 45-50 • The Overdrive Limit is Determined by Progressively Increasing Oscilloscope Gain and Watching for Waveform Aberrations

have calibrated DC nulling (slideback) sources, allowing observation of small, slowly moving events on top of common-mode DC or fast events riding on a waveform.

A special case is the sampling oscilloscope. By nature of its operation, a sampling 'scope in proper working order is inherently immune to input overload, providing essentially instantaneous recovery between samples. Appendix B, "Measuring amplifier settling time", utilizes this capability. See Reference 8 for additional details.

The best approach to measuring small portions of large waveforms, however, is to eliminate the large signal swing seen by the oscilloscope. Appendix B, "Measuring amplifier settling time" shows ways to do this when measuring DAC-amplifier settling time to very high accuracy at high speed.

In summary, while the oscilloscope provides remarkable capability, its limitations must be well understood when interpreting results.[7]

Note 7: Additional discourse on oscilloscopes will be found in References 1 and 7 through 11.

About ground planes

Many times in high frequency circuit layout, the term "ground plane" is used, most often as a mystical and ill-defined cure to spurious circuit operation. In fact, there is little mystery to the usefulness and operation of ground planes, and like many phenomena, their fundamental operating principle is surprisingly simple.

Ground planes are primarily useful for minimizing circuit inductance. They do this by utilizing basic magnetic theory. Current flowing in a wire produces an associated magnetic field. The field's strength is proportional to the current and inversely related to the distance from the conductor. Thus, we can visualize a wire carrying current (Figure 33.51) surrounded by radii of magnetic field. The unbounded field becomes smaller with distance. A wire's inductance is defined as the energy stored in the field set up by the wire's current. To compute the wire's inductance requires integrating the field over the wire's length and the total radial area of the field. This implies integrating on the radius from $R=R_W$ to infinity, a very large number. However, consider the case where we have two wires in space carrying the same current in either direction (Figure 33.52). The fields produced cancel.

In this case, the inductance is much smaller than in the simple wire case and can be made arbitrarily smaller by reducing the distance between the two wires. This reduction of inductance between current carrying conductors is the underlying reason for ground planes. In a normal circuit, the current path from the signal source through its conductor and back to ground includes a large loop area. This produces a large inductance for this conductor which can cause ringing due to LRC effects. It is worth noting that 10nH at 100MHz has an impedance of 6Ω. At 10mA a 60mV drop results.

A ground plane provides a return path directly under the signal carrying conductor through which return current can

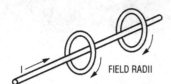

Figure 33.51 • Single Wire Case

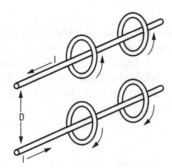

Figure 33.52 • Two Wire Case

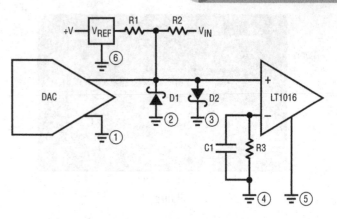

Figure 33.53 • Typical Grounding Scheme

flow. The conductor's small physical separation means the inductance is low. Return current has a direct path to ground, regardless of the number of branches associated with the conductor. Currents will always flow through the return path of lowest impedance. In a properly designed ground plane, this path is directly under the signal conductor. In a practical circuit, it is desirable to ground plane one whole side of the PC card (usually the component side for wave solder considerations) and run the signal conductors on the other side. This will give a low inductance path for all the return currents.

Aside from minimizing parasitic inductance, ground planes have additional benefits. Their flat surface minimizes resistive losses due to AC skin effect (AC currents travel along a conductor's surface). Additionally, they aid the circuit's high frequency stability by referring stray capacitances to ground.

Some practical hints for ground planes are:

1. Ground plane as much area as possible on the component side of the board, especially under traces that operate at high frequency.

2. Mount components that conduct substantial fast rise currents (termination resistors, ICs, transistors, decoupling capacitors) as close to the board as possible.

3. Where common ground potential is important (i.e., at comparator inputs), try to single point the critical components into the ground plane to avoid voltage drops.

   For example, in Figure 33.53's common A/D circuit, good practice would dictate that grounds 2, 3, 4 and 6 be as close to single point as possible. Fast, large currents must flow through R1, R2, D1 and D2 during the DAC settle time. Therefore, D1, D2, R1 and R2 should be mounted close to the ground plane to minimize their inductance. R3 and C1 don't carry any current, so their inductance is less important; they could be vertically inserted to save space and to allow point 4 to be single point common with 2, 3 and 6. In critical circuits, the designer must often trade off the beneficial effects of lowered inductance versus the loss of single point ground.

4. Keep trace length short. Inductance varies directly with length and no ground plane will achieve perfect cancellation.

About bypass capacitors

Bypass capacitors are used to maintain low power supply impedance at the point of load. Parasitic resistance and inductance in supply lines mean that the power supply impedance can be quite high. As frequency goes up, the inductive parasitic becomes particularly troublesome. Even if these parasitic terms did not exist, or if local regulation is used, bypassing is still necessary because no power supply or regulator has zero output impedance at 100MHz. What type of bypass capacitor to use is determined by the application, frequency domain of the circuit, cost, board space and many other considerations. Some useful generalizations can be made.

All capacitors contain parasitic terms, some of which appear in Figure 33.54. In bypass applications, leakage and dielectric absorption are second order terms but series R and L are not. These latter terms limit the capacitor's ability to damp transients and maintain low supply impedance. Bypass capacitors must often be large values so they can absorb long transients, necessitating electrolytic types which have large series R and L.

Different types of electrolytics and electrolytic-non-polar combinations have markedly different characteristics. Which type(s) to use is a matter of passionate debate in some circles

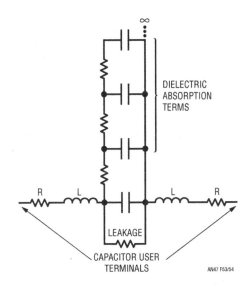

Figure 33.54 • Parasitic Terms of a Capacitor

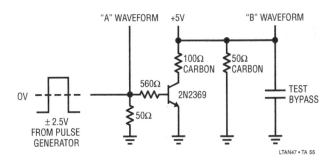

Figure 33.55 • Bypass Capacitor Test Circuit

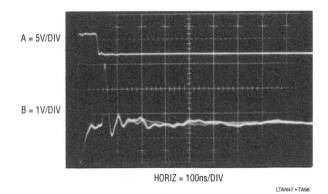

HORIZ = 100ns/DIV

LTAN47 • TA56

Figure 33.56 • Response of Unbypassed Line

and the test circuit (Figure 33.55) and accompanying photos are useful. The photos show the response of 5 bypassing methods to the transient generated by the test circuit. Figure 33.56 shows an unbypassed line which sags and ripples badly at large amplitudes. Figure 33.57 uses an aluminum 10μF electrolytic to considerably cut the disturbance, but there is still plenty of potential trouble. A tantalum 10μF unit offers cleaner response in Figure 33.58 and the 10μF aluminum combined with a 0.01μF ceramic type is even better in Figure 33.59. Combining electrolytics with non-polarized capacitors is a popular way to get good response but beware of picking the wrong duo. The right (wrong) combination of supply line parasitics and paralleled dissimilar capacitors can produce a resonant, ringing response, as in Figure 33.60. *Caveat!*

Breadboarding techniques

The breadboard is both the designer's playground and proving ground. It is there that Reality resides, and paper (or computer) designs meet their ruler. More than anything else, breadboarding is an iterative procedure, an odd amalgam of experience guiding an innocent, ignorant, explorative spirit. A key is to be willing to try things out, sometimes for not very good reasons. Invent problems and solutions, guess carefully and wildly, throw rocks and see what comes loose. Invent and design experiments, and follow them wherever they lead.

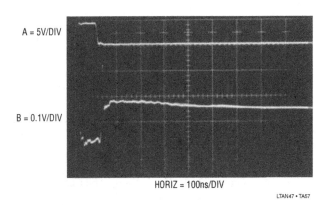

HORIZ = 100ns/DIV

LTAN47 • TA57

Figure 33.57 • Response of 10μF Aluminum Capacitor

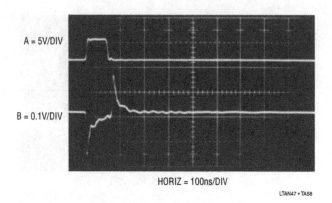

A = 5V/DIV

B = 0.1V/DIV

HORIZ = 100ns/DIV

LTAN47 • TA58

Figure 33.58 • Response of 10μF Tantalum Capacitor

Reticence to try things is probably the number one cause of breadboards that "don't work".[8] Implementing the above approach to life begins with the physical construction methods used to build the breadboard.

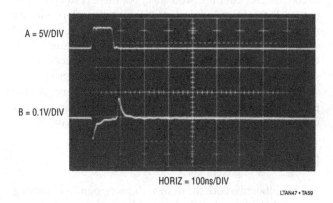

A = 5V/DIV

B = 0.1V/DIV

HORIZ = 100ns/DIV

LTAN47 • TA59

Figure 33.59 • Response of 10μF Aluminum Paralleled by 0.01μF Ceramic

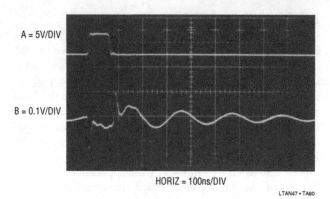

A = 5V/DIV

B = 0.1V/DIV

HORIZ = 100ns/DIV

LTAN47 • TA60

Figure 33.60 • Some Paralleled Combinations can Ring. Try before Specifying!

A high speed breadboard must start with a ground plane. Additionally, bypassing, component layout and connections should be consistent with high speed operations. Because of these considerations there is a common misconception that breadboarding high speed circuits is time consuming and difficult. This is simply not true. For high speed circuits of moderate complexity a complete and electrically correct breadboard can be assembled in 10 minutes if all necessary components are on hand. The key to rapid breadboarding is to identify critical circuit nodes and design the layout to suit them. This permits most of the breadboard's construction to be fairly sloppy, saving time and effort. Additionally, use all degrees of freedom in making connections and mounting components. Don't be bashful about bending IC pins to suit desired low capacitance connections, or air wiring components to achieve rapid or electrically optimum layout. Save time by using components, such as bypass capacitors, as mechanical supports for other components, such as amplifiers. It is true that eventual printed circuit construction is required, but when initially breadboarding forget about PC and

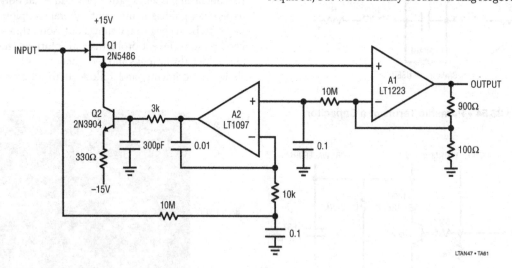

Figure 33.61 • The Stabilized FET Input Amplifier (Applications Figure 33.73) to be Breadboarded

Note 8: A much more eloquently stated version of this approach is found in Reference 12.

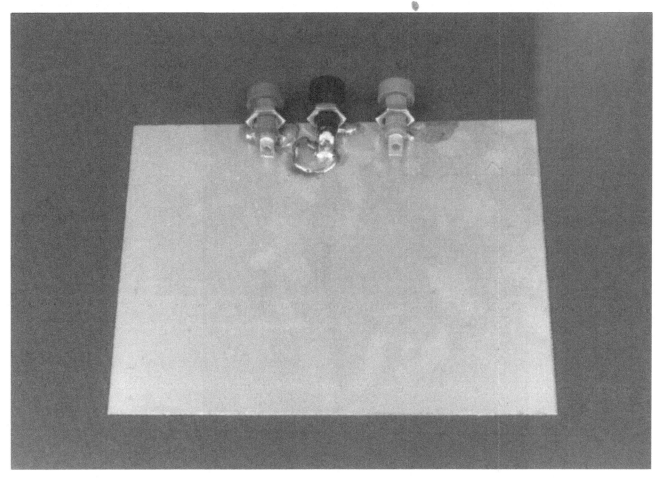

LTAN47 • TA62

Figure 33.62 • The Banana Jacks are Soldered to the Copper Clad Board

production constraints. Later, when the circuit works, and is well understood, PC adaptations can be taken care of.

Figure 33.61's amplifier circuit is a good working example. This circuit, excerpted from the Applications section (where its electrical operation is more fully explained) is a high impedance, wideband amplifier with low input capacitance. Q1 and A1 form the high frequency path, with the 900Ω-100Ω feedback divider setting gain. A2 and Q2 close a DC stabilization loop, minimizing DC offset between the circuit's input and output. Critical nodes in this circuit include Q1's gate (because of the desired low input capacitance) and A1's input related connections (because of their high speed operation). Note that the connections associated with A2 serve at DC and are much less sensitive to layout. These determinations dominate the breadboard's construction.

Figure 33.62 shows initial breadboard construction. The copperclad board is equipped with banana type connectors. The connector's mounting nuts are simply soldered to the clad board, securing the connectors. Figure 33.63 adds A1 and the bypass capacitors. Observe that A1's leads have been bent out, permitting the amplifier to sit down on the ground plane, minimizing parasitic capacitance. Also, the bypass capacitors

are soldered to the amplifier power pins right at the capacitor's body. The capacitor's lead lengths are returned to the banana power jacks. This connection method provides good amplifier bypassing while mechanically supporting the amplifier. It also eliminates separate wire runs to the power pins.

Figure 33.64 adds the discrete components in the high speed path. Q1's gate is connected directly to the input BNC, as is the 10MΩ resistor associated with A2's negative input. Note that the end of this resistor that sees high frequency is cut very short, while the other end is left uncut. The 900Ω-100Ω divider is installed at A1, with very short connections to A1's negative input. A1's 10MΩ resistor receives similar treatment to the BNC connected 10MΩ unit; the high frequency end is cut short, while the end destined for connection to A2 remains uncut. Q2's collector and Q1's source, high speed points, are tied closely together with A1's positive input.

Finally, DC amplifier A2 and its associated components are air wired into the breadboard (Figure 33.65). Their DC operation permits this, while the construction technique makes connections to the previously wired nodes easy. The previously uncommitted ends of the 10MΩ resistors may be bent in any way necessary to make connections. All other

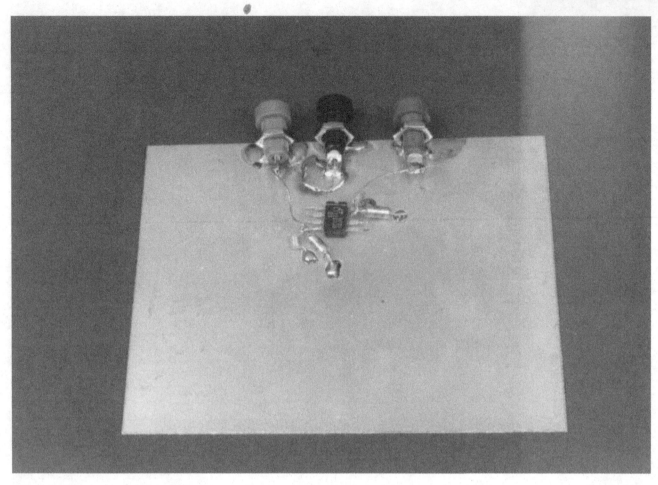

LTAN47 • TA63

Figure 33.63 • High Speed Amplifier A1 is Connected to Power. Bypass Capacitors Provide Support. Bending Amplifier Pins Eases Connections and Minimizes Distance to the Ground Plane

components associated with A2 receive similar treatment and the circuit is ready for experimentation.

Despite the breadboard's seemingly haphazard construction, the circuit worked well. Input capacitance measured a few pF (including BNC connector) with bias current of about 100pA. Slew rate was 1000V/µs, with bandwidth approaching 100MHz. Output, even with 50mA loading, was clean, with no sign of oscillation or other instabilities. Full details on this circuit appear in the Applications section. Additional examples of breadboard construction techniques appear in Appendix F, "Additional comments on breadboarding".

Once the breadboard seems to work, it's useful to begin thinking about PC layout and component choice for production. Experiment with the existing layout to determine just how sensitive nominally critical points are. Add controlled parasitic terms (e.g., resistors, capacitors and physical layout changes) to test for sensitivity. Gentle touching of suspect points with a finger can yield preliminary indication of sensitivity, giving clues that can be quite valuable.

In conclusion, when breadboarding, design the breadboard to be quick and easy to build, work with and modify. Observe the circuit and listen to what it is telling you before trying to get it to some desired state. Finally, don't hesitate to try just about anything; that's what the breadboard is for. Almost anything you do will cause some result — whether it's good or bad is almost irrelevant. Anything you do that enhances your ability to correlate events occurring on the breadboard can only be beneficial.

Oscillation

The forte of the operational amplifier is negative feedback. It is feedback which stabilizes the operating point and fixes the gain. However, positive feedback or delayed negative feedback can cause oscillation. Thus, a properly functioning amplifier constantly lives in the shadow of oscillation.

When oscillation occurs, several major candidates for blame are present. Power supply impedance must be low. If the supply is unbypassed, the impedance the amplifier sees at its power terminals is high, particularly at high frequency. This impedance forms a voltage divider with the amplifier, allowing the supply to move as internal conditions in the amplifier

LTAN47 • TA64

Figure 33.64 • Additional High Speed Discrete Components and Connectors are Added. Note Short Connections at Amplifier Input Pins (Left Side of Package). 10M Resistors Uncommitted Ends are Just Visible

change. This can cause local feedback and oscillation occurs. The obvious cure is to bypass the amplifier.

A second common cause of oscillation is positive feedback. In most amplifier circuits feedback is negative, although controlled amounts of positive feedback may be used. In a circuit that nominally has only negative feedback unintended positive feedback may occur with poor layout. Check for possible parasitic feedback paths and unwanted or overlooked feedback action. Always minimize (to the extent possible) impedances seen by amplifier inputs. This helps attenuate the effects of parasitic feedback paths to the inputs. Similarly, minimize exposed input trace area. Route amplifier outputs and other signals well away from sensitive nodes. Sometimes no amount of layout finesse will work and shielding is required. Use shielding only when required — extensive shielding is a sloppy substitute for good layout practice.

A final cause of oscillation is negative feedback arriving well delayed in time. Under these conditions the amplifier hopelessly tries to servo a feedback signal which consistently arrives too late. The servo action takes the form of an electronic tail chase, with oscillation centered around the ideal

servo point. The most common causes of this problem are reactive loading of the amplifier (most notably capacitive loads such as cable) and circuitry, such as power amplifiers, placed within the amplifier's feedback path. Reactive loads should be isolated from the amplifier's output (and feedback path) with a resistor or power amplifier. Sometimes rolling off the amplifier's frequency response will fix the problem, but in high speed circuits this may not be an option.

Placing power gain or other type stages within the amplifier's feedback path adds time delay to the stabilizing feedback. If the delay is significant, oscillation commences. Stages operating within the amplifier's loop must contribute minimum time lag compared to the amplifier's speed capability. At lower speeds this is not too difficult, but something destined for operation within a 100MHz amplifier's loop must be *fast*. As mentioned before, rolling off the amplifier's frequency response eases the job, but is usually undesirable in a wideband circuit. Every effort should be expended to maximize the added stages bandwidth before resorting to roll-off of the amplifier. In this way the fastest overall bandwidth is achieved while maintaining stability. Appendix C, "The oscillation

LTAN47 • TA65

Figure 33.65 • DC Servo Amplifier is Wired In and Connections to 10M Resistors Completed. This Part of the Circuit is Not Layout Sensitive

problem — frequency compensation without tears", discusses considerations surrounding operating power gain and other type stages within amplifier loops.

This completes the tutorial section. Hopefully, several notions have been imparted. First, in any measurement situation, test equipment characteristics are an integral part of the circuit. At high speed and high precision this is particularly the case. As such, it is imperative to know your equipment and how it works. There is no substitute for intimate familiarity with your tool's capabilities and limitations.[9]

In general, use equipment you trust and measurement techniques you understand. Keep asking questions and don't be satisfied until everything you see on the oscilloscope is accounted for and makes sense.

Fast monolithic amplifiers, combined with the precautionary notes listed above, permit fast linear circuit functions which are difficult or impractical using other approaches. Some of the applications presented represent the state-of-the-art for a particular circuit function. Others show simplified and/or improved ways to implement standard functions by utilizing the amplifier's easily accessed speed. All have been carefully (and painfully) worked out and should serve as good idea sources for potential users of the device. Have fun. I did.

Applications Section I — Amplifiers

Fast 12-bit digital-to-analog converter (DAC) amplifier

One of the most common applications for a high speed amplifier, transforming a 12-bit DAC's current output into a voltage, is also one of the most difficult. Although an op amp can easily do this, care is required to obtain good dynamic performance. A fast DAC can settle to 0.01% in 200ns or less, but its output also includes a parasitic capacitance term, making the amplifier's job more difficult. Normally, the DAC's current output is unloaded directly into the amplifier's summing junction, placing the parasitic capacitance from ground to the amplifier's input. The capacitance introduces feedback phase shift at high

---

Note 9: Further exposition and *kvetching* on this point is given in Reference 13.

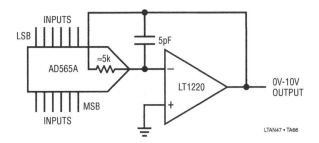

Figure 33.66 • Typical Output Amplifier Configuration for a 12-Bit D-to-A Converter

frequencies, forcing the amplifier to hunt and ring about the final value before settling. Different DACs have different values of output capacitance. CMOS DACs have the highest output capacitance, in the 100pF-150pF range, and it varies with code. Bipolar DACs typically have 20pF-30pF of capacitance, stable over all codes. As such, bipolar DACs are almost always used where high speed is required. Figure 33.66 shows the popular AD565A 12-bit DAC with an LT1220 output op amp. Figure 33.67 shows clean 0.01% settling in 280ns (Trace B) to an all-bits-on input step (Trace A). The requirements for obtaining Trace B's display are not trivial, and are fully detailed in Appendix B, "Measuring amplifier settling time".

2-Channel video amplifier

Figure 33.68 shows a simple way to multiplex two video amplifiers onto a single 75Ω cable. The appropriate amplifier is activated in accordance with the truth table in the figure[10]. Amplifier performance includes 0.02% differential gain error and 0.1° differential phase error. The 75Ω back termination looking into the cable means the amplifiers must swing $2V_{P-P}$ to produce $1V_{P-P}$ at the cable output, but this is easily handled.

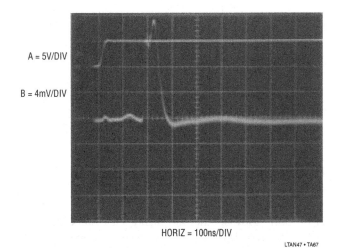

Figure 33.67 • Settling Residue (Trace B) for All Bits Switched On (Trace A). Output is Fully Settled in 280ns

Note 10: A truth table in an op amp circuit! *Et tu*, LTC!!

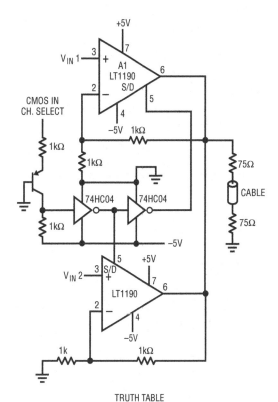

INPUT SELECT	A1 OUTPUT	A2 OUTPUT
5V	ACTIVE	INACTIVE
0V	INACTIVE	ACTIVE

Figure 33.68 • 2-Channel Multiplexed Video Amplifier

Simple video amplifier

Figure 33.69 is a simpler version of Figure 33.68. This is a single channel video amplifier, arranged (in this case) for a gain of ten. The double cable termination is retained and the circuit delivers a bandwidth of 55MHz.

Loop through cable receivers

Figure 33.70 is another cable related circuit. Here, the LT1193 differential amplifier simply hangs across a distribution cable, extracting the signal. The amplifier's true differential inputs reject common-mode signals. As in the previous circuit, differential gain and phase errors measure 0.2% and 0.1°, respectively. A separate input permits DC level adjustment.

DC stabilization — summing point technique

Often it is desirable to obtain the precision offset of a DC amplifier with the bandwidth of a fast device. There are a variety of techniques for doing this. Which method is best is heavily application dependent, so several configurations are presented.

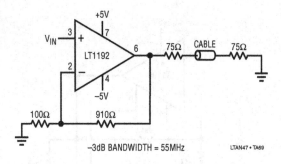

Figure 33.69 • Double Terminated Cable Driver

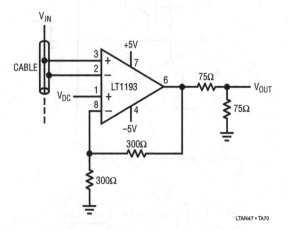

Figure 33.70 • Cable Sense Amplifier for Loop Through Connections with DC Adjust

Figure 33.71 shows a composite made up of an LT1097 low drift device and an LT1191 high speed amplifier. The overall circuit is a unity gain inverter with the summing node located at the junction of the two 1k resistors. The LT1097 monitors this summing node, compares it to ground and drives the LT1191's positive input, completing a DC stabilizing loop around the LT1191. The 100kΩ-0.01μF time constant at the LT1097 limits its response to low frequency signals. The LT1191 handles high frequency inputs while the LT1097 stabilizes the DC operating point. The 4.7k-220Ω divider at the LT1191 prevents excessive input overdrive during start-up.

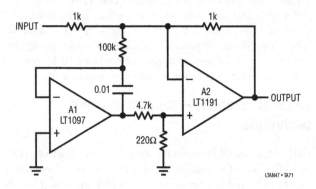

Figure 33.71 • A1 DC Stabilizes A2 by Forcing the Summing Point to Zero

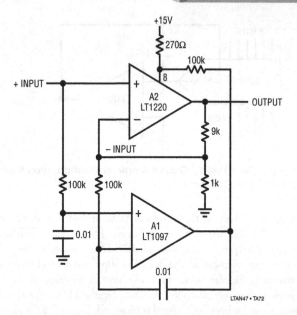

Figure 33.72 • A1 DC Stabilizes A2 by Forcing the Offset Pins to Produce a 0V Difference at A2's Inputs

This circuit combines the LT1097's 35μV offset and 1.5V/°C drift with the LT1191's 450V/μs slew rate and 90MHz bandwidth. Bias current, dominated by the LT1191, is about 500nA.

DC stabilization — differentially sensed technique

Figure 33.72 is similar to Figure 33.71, except that the sensing is done differentially, preserving access to both fast amplifier inputs. The LT1097 measures the DC error at the LT1220's input terminals and biases its offset pins to force offset within 50μV. The offset pin biasing at the LT1220 is arranged so the LT1097 will always be able to find the servo point. The 0.01μF capacitor rolls off the LT1097 at low frequency and the LT1220 handles high frequency signals. The combined characteristics of these amplifiers yield the following performance:

Offset voltage 50μV
Offset drift ....1μV/°C
Slew rate 250V/μs
Gain-bandwidth 45MHz

DC stabilization — servo controlled FET input stage

Figure 33.73 shows a wideband, highly stable gain-of-ten with high input impedance. Input capacitance is about 3pF. Q1 and Q2 constitute a simple, high speed FET input buffer. Q1 functions as a source follower, with the Q2 current source load setting the drain-source channel current. The LT1223 provides a 100MHz bandwidth gain of ten. Normally, this open loop configuration would be quite drifty because there

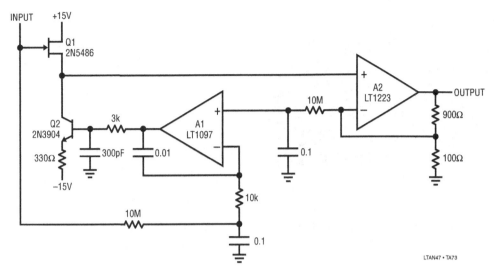

Figure 33.73 • A1 DC Stabilizes the Circuit by Controlling Q1's Channel Current

is no DC feedback. The LT1097 contributes this function to stabilize the circuit. It does this by comparing the filtered circuit output to a similarly filtered version of the input signal. The amplified difference between these signals is used to set Q2's bias, and hence Q1's channel current. This forces Q1's V_{GS} to whatever voltage is required to match the circuit's input and output potentials. The capacitor at A1 provides stable loop compensation. The RC network in A1's output prevents it from seeing high speed edges coupled through Q2's collector-base junction.

This circuit constitutes an extremely wideband (Q1 does not degrade A2's 100MHz performance), high input impedance amplifier. With an input capacitance of 3pF and bias current of 100pA, it is well suited for probing or as an ATE pin amplifier. As shown, gain is ten, but other gains are possible by varying the feedback ratio.

DC stabilization — full differential inputs with parallel paths

Figure 33.74 shows a way to get full differential inputs with DC stabilized operation. This circuit combines the output of two differential input amplifiers for overall DC corrected wideband operation. A1 and A2 both differentially sense the input at gains of ten. Wideband A1 feeds output amplifier A3 via a highpass network, while the slower A2 contributes DC and low frequency information to A3. A2 does not see high frequency inputs, because they are filtered by the 2k-200pF lowpass networks at its inputs. If the gain and bandwidth of the high and low frequency paths complement each other, A3's output should be an undistorted, amplified version (in this case ×10) of the input. Figure 33.75 shows this to be the case. Trace A is one side of a differential input applied to the circuit.

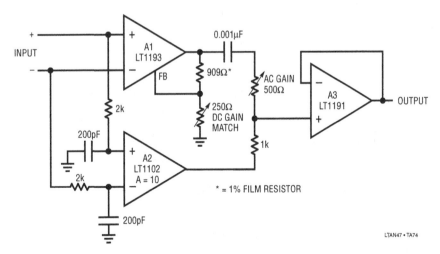

Figure 33.74 • A Parallel Path DC Stabilized Differential Amplifier. High Frequency Signals Go through A1, while A2 Handles DC and Low Frequency. A3 Sums Both Paths

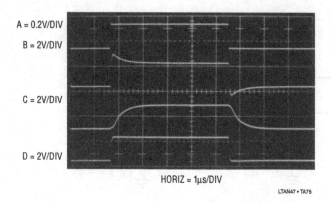

A = 0.2V/DIV

B = 2V/DIV

C = 2V/DIV

D = 2V/DIV

HORIZ = 1µs/DIV

LTAN47 • TA75

Figure 33.75 • Waveforms for the Parallel Path Differential Amplifier. Trace A is the Input; B, C and D are the High Pass, Low Pass and Output Nodes, Respectively

Trace B is A1's output taken at the 500Ω potentiometer −0.001µF junction. Trace C is A2's output. With the AC gain and DC gain match trims properly adjusted, the two paths' contributions match up and Trace D is singularly clean, with no residue. The adjustments are optimized by trimming the AC gain for the squarest corners and the DC gain match for a flat top. Bandwidth for this circuit exceeds 35MHz, slew rate is 450V/µs and DC offset about 200µV.

DC stabilization — full differential inputs, gain-of-1000 with parallel paths

Figure 33.76 is a very powerful extension of the previous circuit. Operation is similar, but gain is increased to 1000. Bandwidth is about 35MHz, rise time equals 7ns and delay is inside 7.5ns. Full power response is available to 10MHz, with

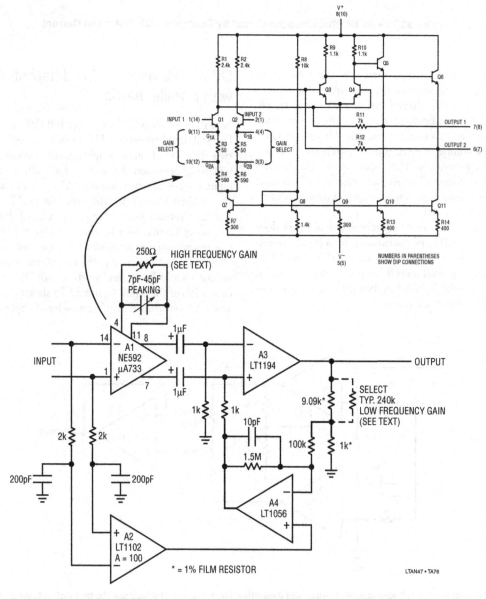

Figure 33.76 • A Full Differential, Parallel Path Amplifier. Gain is 1000, with 38MHz Bandwidth. Delay is Inside 7.5ns and Rise Time Under 7ns

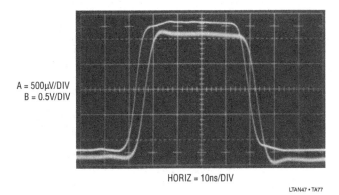

A = 500μV/DIV
B = 0.5V/DIV

HORIZ = 10ns/DIV

LTAN47 • TA77

Figure 33.77 • Pulse Response for the X1000 Differential Amplifier. Fidelity is Quite Good, with Only Slight Output Peaking (Trace B)

input noise about 15μV broadband. This kind of speed, coupled with full differential inputs, the gain of 1000, DC stability, and low cost make the circuit broadly applicable in wideband instrumentation. As before, two differential amplifiers, A1 and A2, simultaneously sense the inputs. In this case A1 is the popular and economical 592-733 type, operating at a gain of 100.[11] A1's differential outputs feed output amplifier A3 via 1μF-1kΩ high pass networks which strip off A1's DC content. A2, a precision DC differential type, operates in similar fashion to the previous circuit, supplying DC and low frequency information to A3 at a trimmed gain of 100. In this case output amplifier A3 is a differential gain block with a nominal committed gain of 10. This change is necessitated by A1's differential output, which must be single-ended to obtain the circuit's output. As such A2 does not directly apply its low frequency information to A3 as it did before. Instead, A4 measures the difference between A2's output and a divided down portion of A3's output. A4's output, biasing A3's positive input via the 1kΩ resistor, closes a loop around the circuit's DC-low frequency path. The divider feeding A4's negative input is adjusted so that the circuit's DC gain is known and equal to its AC gain.

Figure 33.77 shows the circuit's response to a 60ns, 2.5mV amplitude pulse (Trace A). The X1000 output (Trace B) responds cleanly, with delay and rise time in the 5ns-7ns range. Some small amount of peaking is evident, although it may be trimmed with the peaking adjustment at A1. Figure 33.78 plots the circuit's gain vs frequency. Gain is flat within 1/2dB to 20MHz, with the −3dB point at 38MHz. Figure 33.77's edge peaking shows up here as a very slight gain increase starting around 1MHz and continuing out to about 15MHz. The peaking trim will eliminate this effect.

To use this circuit, put in a low frequency or DC signal of known amplitude and adjust the low frequency gain for a X1000 output after the output has settled. Next, adjust the high frequency gain so that the signal's front and rear corners have amplitudes identical to the settled portion. Finally, trim

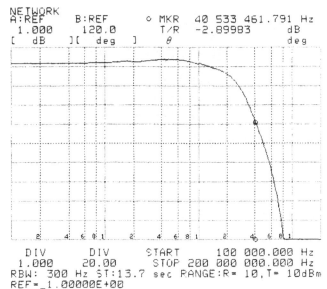

NETWORK
A:REF   B:REF   ○ MKR   40 533 461.791 Hz
  1.000   120.0   T/R   -2.89983   dB
[dB   ][deg   ]   θ   deg

DIV   DIV   START   100 000.000 Hz
1.000   20.00   STOP 200 000 000.000 Hz
RBW: 300 Hz ST:13.7 sec RANGE:R= 10,T= 10dBm
REF=_1.00000E+00

Figure 33.78 • Gain vs Bandwidth for the X1000 Differential Amplifier. Peaking Noted in Figure 33.77 Shows up as 0.25dB Peak at 5MHz, Which Could be Trimmed Out

the peaking adjustment for best settling of the output pulse's front and rear corners.

Figure 33.79 shows input (Trace A) and output (Trace B) waveforms with all adjustments properly set. Fidelity is excellent, with no aberrations or other artifacts of the parallel path operation evident. Figure 33.80 shows the effects of too much AC gain; excessive peaking on the edges with proper amplitude indicated only after the AC channel transitions through its highpass cut off. Similarly, excessive DC gain produces Figure 33.81's traces. The AC gain path provides proper initial response, but too much DC gain forces a long, tailing response to an incorrect amplitude.

High speed differential line receiver

High speed analog signals transmitted on a line often pick up substantial common-mode noise. Figure 33.82 shows a

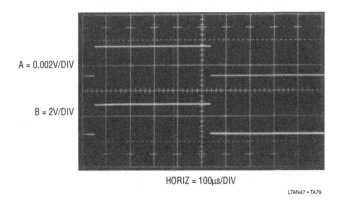

A = 0.002V/DIV

B = 2V/DIV

HORIZ = 100μs/DIV

LTAN47 • TA79

Figure 33.79 • Response of X1000 Amplifier with Bandwidth Crossover Points Properly Adjusted. A = Input; B = Output

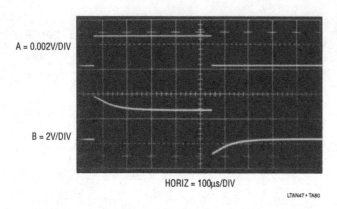

A = 0.002V/DIV

B = 2V/DIV

HORIZ = 100µs/DIV

LTAN47 • TA80

Figure 33.80 • Response of X1000 Amplifier with Excessive AC Gain. A = Input; B = Output

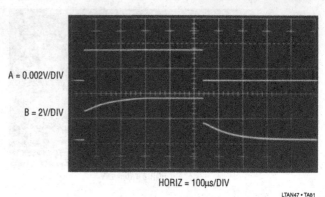

A = 0.002V/DIV

B = 2V/DIV

HORIZ = 100µs/DIV

LTAN47 • TA81

Figure 33.81 • Response of X1000 Amplifier with Too Much DC Gain. A = Input; B = Output

simple, fast differential line receiver using the LT1194 gain-of-ten differential amplifier. The differential line is fed to A1. The resistor-diode networks prevent overload and insure input bias for A1 under all conditions. A1's output represents the difference of the two line input times a gain-of-ten. In theory, all common-mode noise should be rejected. The test circuit shown in the figure confirms this. The sinewave oscillator drives T1 (Trace A, Figure 33.83), producing a differential line output at its secondary. T1's secondary is returned to ground through a broadband noise generator, flooding the line inputs with common-mode noise (traces B and C are A1's inputs). Trace D, A1's X10 version of the differential

signal at its inputs, is clean with no visible noise or disturbances. This circuit will easily provide a clean output with DC-5MHz noise dominating signal by a 100:1 ratio.

Transformer coupled amplifier

Figure 33.84 shows another way to achieve high common-mode rejection. Additionally, this circuit has the advantage of true 3 port isolation. The input, gain stage, and output are all galvanically isolated from each other. As such, this configuration is useful where large common-mode differences are encountered or where ground integrity is uncertain. A1 is set

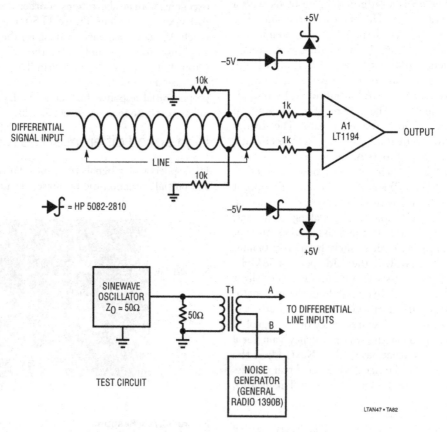

Figure 33.82 • Simple, Full Differential Line Receiver

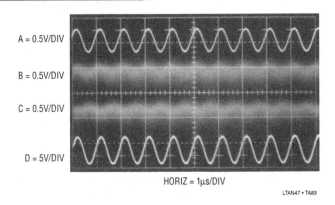

A = 0.5V/DIV

B = 0.5V/DIV

C = 0.5V/DIV

D = 5V/DIV

HORIZ = 1µs/DIV

LTAN47 • TA83

Figure 33.83 • Differential Line Receiver Easily Pulls Out a Signal Buried in Common-Mode Noise. Output is Clean, Despite 100:1 Noise-to-Signal Ratio

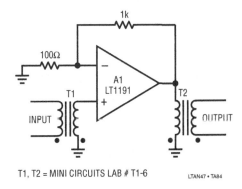

1k

100Ω

A1
LT1191

T1   T2

INPUT   OUTPUT

T1, T2 = MINI CIRCUITS LAB # T1-6   LTAN47 • TA84

Figure 33.84 • Transformer Coupled Amplifier. Note That A1 is Galvanically Isolated From Input and Output Nodes

up in a simple gain of 11. T1 feeds its input, and the output is taken from T2. Figure 33.85 shows results for a 4MHz input, with all "•" designated transformer leads referred to ground. The input (Trace A, Figure 33.85) is applied to T1, whose output (Trace B) feeds A1. A1 takes gain, and its output (Trace C) feeds T2. T2's output (Trace D) is the

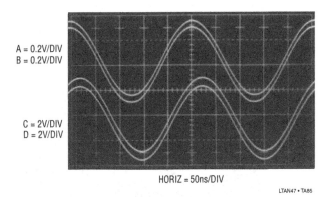

A = 0.2V/DIV
B = 0.2V/DIV

C = 2V/DIV
D = 2V/DIV

HORIZ = 50ns/DIV

LTAN47 • TA85

Figure 33.85 • Transformer Coupled Amplifier Responds to an Input (Trace A) with A Slightly Phase Shifted Output (Trace D). Traces B and C are T1 Secondary and T2 Primary, Respectively

circuit's output. Phase shift is evident, although tolerable. T1 and T2 are very wideband devices, with low phase shift. Note the negligible phase difference between the A-B and C-D trace pairs. A1 contributes essentially the entire phase error. Using the transformers specified, the circuit's low frequency cut-off is about 10kHz.

Differential comparator amplifier with adjustable offset

It is often desirable to examine or amplify one particular portion of a signal while rejecting all other portions. At high speed this can be difficult, because the amplifier may see fast, large common-mode swings. Recovery from such activity usually is dominated by saturation effects, making the amplifier's output questionable. The LT1193's differential amplifier's fast overload recovery permits this function, maintaining output fidelity to the input signal. Additionally, the input level amplitude at which amplification begins is settable, allowing any amplitude defined point to be selected. In Figure 33.86, A1, the LT1019 reference and associated

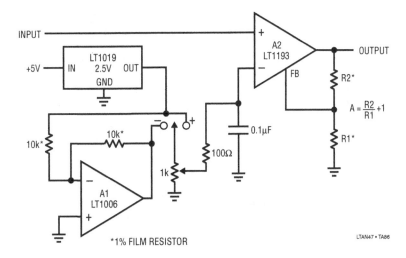

INPUT

A2
LT1193

OUTPUT

+5V

LT1019
2.5V
IN   OUT
GND

FB

R2*

$$A = \frac{R2}{R1} + 1$$

10k*   10k*

0.1µF

R1*

A1
LT1006

1k

100Ω

*1% FILM RESISTOR

LTAN47 • TA86

Figure 33.86 • Fast Differential Comparator Amplifier with Settable Offset

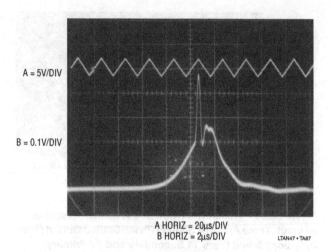

A = 5V/DIV

B = 0.1V/DIV

A HORIZ = 20µs/DIV
B HORIZ = 2µs/DIV

LTAN47 • TA87

Figure 33.87 • The Differential Comparator Amplifier Extracting Signal Detail From a Triangle Waveform's Peak. Triangle Generator's Switching Artifacts are Clearly Evident

Input signals below A2's negative input levels maintain A2's output in saturation, and no signal is seen at the output. When the positive input rises above the negative input's bias point A2 becomes active, providing an amplified version of the instantaneous difference between its inputs. Figure 33.87 shows what happens when the output of a triangle wave generator (Trace A) is applied to the circuit. Setting the bias level just below the triangle peak permits high gain, detailed operation of the turnaround at the peak. Switching residue in the generator's output is clearly observable in Trace B. Appropriate variations in the voltage source setting would permit more of the triangle slopes to be observed, with attendant loss of resolution due to oscilloscope overload limitations. Similarly, increasing A2's gain allows more amplitude detail while placing restrictions on how much of the waveform can be displayed. It is worth noting that this circuit performs the same function as differential plug-in units for oscilloscopes. This circuit's output is accurate and settled to 0.1% 100ns after it enters its linear region.

Differential comparator amplifier with settable automatic limiting and offset

Figure 33.88 extends the previous circuit's operation, allowing amplified observation of information between two settable, amplitude defined points. The amplitude setpoints are

components form an adjustable, bipolar voltage source which is coupled to differential amplifier A2's negative input. The input signal biases A2's positive input with A2's gain set by R1 and R2, in accordance with the equation given.

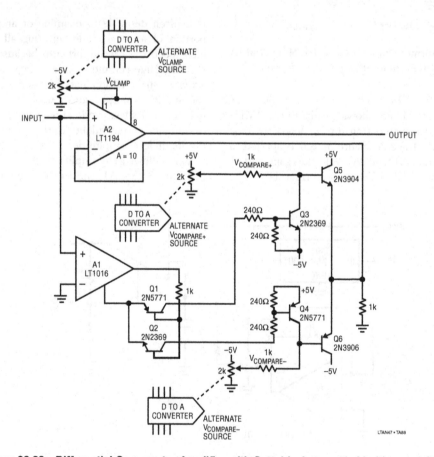

Figure 33.88 • Differential Comparator Amplifier with Settable Automatic Limiting and Offset

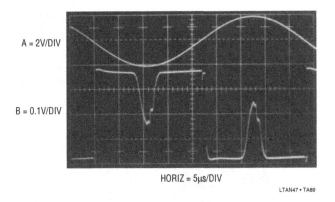

A = 2V/DIV

B = 0.1V/DIV

HORIZ = 5µs/DIV

LTAN47 • TA89

Figure 33.89 • The Automatic Differential Comparator Amplifier Finds Triangle Wave and Switching Residuals (Trace B) in Trace A's Peaks

settable in both magnitude and sign. In this circuit the polarity of the offset applied to A2's negative input is determined by comparator A1's output state. A1 compares the circuit's input to ground, generating polarity information at its outputs. Level shifters Q1-Q3 and Q2-Q4 bias followers Q5 and Q6. Positive circuit inputs result in Q5 supplying the "$V_{COMPARE+}$" potential to A2, while negative inputs route "$V_{COMPARE-}$" to A2. This eliminates the previous circuit's manual polarity switch, permitting automatic selection of the differencing polarity and amplitude. Additionally, this circuit takes advantage of A2's input clamp feature. This feature (see LT1194 data sheet) limits the dynamic range of the input, clamping the amplifier's input operating range. Signals inside the clamp limit are processed normally, while signals outside the limit are precluded from influencing the amplifier. This

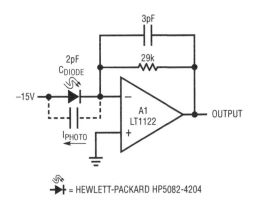

= HEWLETT-PACKARD HP5082-4204

RESPONSE DATA

LIGHT (900nM)	DIODE CURRENT	CIRCUIT OUTPUT
1mW	350µA	10.0V
100µW	35µA	1V
10µW	3.5µA	0.1V
1µW	350nA	0.01V
100nW	35nA	0.001V

LTAN47 • TA90

Figure 33.90 • A Simple Photodiode Amplifier

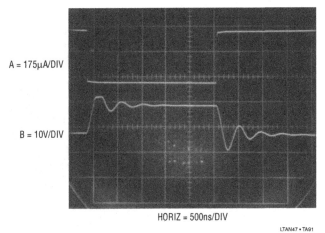

A = 175µA/DIV

B = 10V/DIV

HORIZ = 500ns/DIV

LTAN47 • TA91

Figure 33.91 • Response of Figure 33.90 Without a Feedback Capacitor

combination of circuit controls allows very tightly defined windows on a waveform to be selected for accurate amplification without overload restrictions.

Figure 33.89 shows the circuit output for a sine input (Trace A) from the same function generator used to test the previous circuit. The V^+ and V^- compare voltages are set just below the sinewave peaks, with "V_{clamp}" programmed to restrict amplification to the peak's excursion. Trace B, the circuit's output, simultaneously shows amplitude detail of *both* sine peaks. The observed distortion is directly traceable to this generator's imperfect internal triangle waveform (see Figure 33.87), as well as its sine shaper characteristics.

Photodiode amplifier

Amplification of fast photodiode signals over a wide range of intensity is a common requirement. Figure 33.90's fast FET amplifier serves well, giving wideband operation with 5 decades of photocurrent. The photodiode is set up in the conventional manner. Photocurrent is fed directly to A1's summing point, causing A1's output to move to the level required to maintain virtual ground at the negative input. The −15V diode bias aids diode response. The table in the figure details circuit operating characteristics with the diode specified.

Some care in frequency compensating this configuration is required. The diode has about 2pF of parasitic capacitance, forming a significant lag at A1's summing point. If no feedback capacitor is used, high speed dynamics are poor. Figure 33.91 shows circuit response to a photo input (Trace A) with the indicated 3pF feedback capacitor removed. A1's output overshoots and saturates before finally ringing down to final value. In contrast, replacing the 3pf capacitor provides Figure 33.92's results. The same input pulse (Trace A, Figure 33.92) produces a cleanly damped output (Trace B). The capacitor imposes a 50% speed penalty (note faster horizontal scale for Figure 33.92). This is unavoidable because suppressing the parasitic ringing's relatively low frequency mandates significant roll-off.

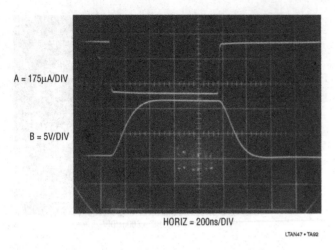

HORIZ = 200ns/DIV

LTAN47 • TA92

Figure 33.92 • Figure 33.90 Responding with a Feedback Capacitor

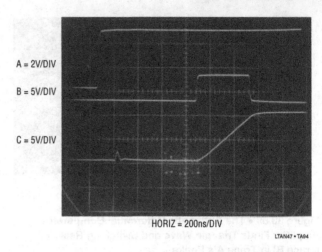

HORIZ = 200ns/DIV

LTAN47 • TA94

Figure 33.94 • The Photo Integrator Acquires (Trace C) an Input Light Pulse (Trace B) with the Control Line (Trace A) in the Run Mode. Charge Cancellation Action is Evident at Trace C's 400ns Point

Fast photo integrator

A related circuit to the photodiode amplifier is Figure 33.93's photo integrator. Here, the output represents the integral of the diode's photocurrent over some period of time. This circuit is particularly applicable in situations where the total energy in a light pulse (or pulses) must be measured. The

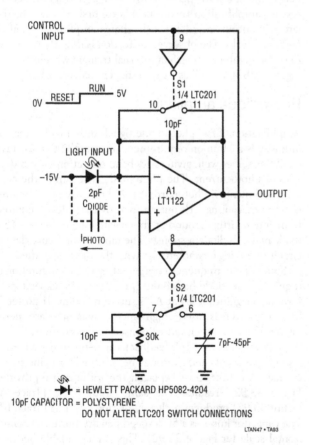

= HEWLETT PACKARD HP5082-4204
10pF CAPACITOR = POLYSTYRENE
DO NOT ALTER LTC201 SWITCH CONNECTIONS

LTAN47 • TA93

Figure 33.93A • Very Fast Photo Integrator. S2 Compensates Reset Switch S1's Small Charge Injection

circuit is a very fast integrator, with S1 used as a reset switch. S2, switched simultaneously with S1, compensates S1's charge injection error. With the control input (Trace A, Figure 33.94) low and no photocurrent, S1 is closed and A1 looks like a grounded follower. Under these conditions A1's output (Trace C) sits at 0V. When the control input goes high, A1 becomes an integrator as soon as S1 opens. Due to switch delay, this occurs about 150ns after the control input goes high. When S1 opens it delivers some parasitic charge to A1's summing point. S2 provides a compensatory charge based pulse at A1's positive terminal to cancel the effects of S1's charge error. This action shows up as a fast, small amplitude event in A1's output which settles rapidly back to 0V.

At this point in time the integrator is ready to receive and record a photo pulse. When light falls on the photodiode (Trace B triggers a light pulse seen by the photodiode) A1 responds by integrating. In this case A1's output integrates rapidly until the light pulse ceases. A1's voltage after the light event is over is related to the total energy seen by the diode during the event. A monitoring A-D converter can acquire A1's output. In typical operation the control line returns low, resetting A1 until the next event is to be integrated.

With only 10pF of integration capacitor, the circuit has an output droop rate of about 0.2V/µs. This can be increased, although integration speed will suffer accordingly. Integration times of nanoseconds to milliseconds and photocurrents ranging from nanoamperes to hundreds of microamperes are accommodated by the circuit as shown. Thus, light intensities spanning microwatts to milliwatts over wide ranges of duration are practical inputs. The primary accuracy restrictions are A1's 75pA bias current, its 12V output swing and the effectiveness of the charge cancellation network. Typically, full-scale accuracy of several percent is achievable if the charge cancellation network is trimmed. To do this, assure that the diode sees no light while repetitively pulsing the control line. Adjust the trimmer capacitor for 0V output at

A1 immediately after the disturbance associated with the S1-S2 switching settles.

Fiber optic receiver

A simple high speed fiber optic receiver appears in Figure 33.95. A1, a photocurrent-to-voltage converter similar to Figure 33.90, feeds comparator A2. A2 compares A1's output to a DC level established by the threshold adjust setting, producing a logic compatible output. Figure 33.96 shows typical waveforms. Trace A is a pulse associated with a photo input. Trace B is A1's response and Trace C is A2's output. The phase shift between the photo input and A2's output is due to A1's delay in reaching the threshold level. Reducing the threshold level will help, but moves operation closer to the noise floor. Additionally, the fixed threshold level cannot account for response changes in the emitter and detector diodes and fiber optic line over time and temperature.

40MHz fiber optic receiver with adaptive trigger

Receiving high speed fiber optic data with wide input amplitude variations is not easy. The high speed data and uncertain intensity of the light level can cause erroneous results unless the receiver is carefully designed. Figure 33.97 addresses the previous circuit's limitations, offering significant performance advantages. This receiver will reliably condition fiber optic inputs of up to 40MHz with input amplitude varying by >40dB. Its digital output features an adaptive threshold trigger which accommodates varying signal intensities due to component aging and other causes. An analog output is also available to monitor the detector output. The optical signal is detected by the PIN photodiode and amplified by A1. A second stage, A2, gives further amplification. The output of this stage biases a 2-way peak detector (Q1-Q4). The maximum peak is stored in Q2's emitter capacitor, while the minimum excursion is retained in Q4's emitter capacitor.

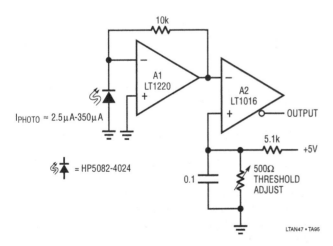

Figure 33.95 • A Simple Fiber Optic Receiver

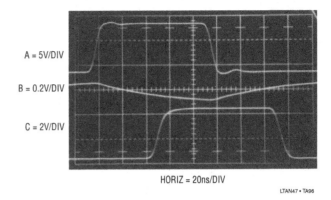

Figure 33.96 • Waveforms for the Simple Fiber Optic Receiver. A1 (Trace B) Lags the Input (Trace A), but Output (Trace C) is Clean

The DC value of A2's output signal's mid-point appears at the junction of the 500pF capacitor and the 22MΩ units. This point will always sit midway between the signal's excursions, regardless of absolute amplitude. This signal-adaptive voltage is buffered by the low bias LT1097 to set the trigger voltage at the LT1016's positive input. The LT1016's negative input is biased directly from A2's output. Figure 33.98 shows the results using the test circuit indicated in Figure 33.97. The pulse generator's output is Trace A, while A2's output (analog output monitor) appears in Trace B. The LT1016 output is Trace C. The waveforms were recorded with a 5µA photocurrent at about 20MHz. Note that A4's output transitions correspond with the midpoint of A2's output (plus A4's 10ns propagation delay) in accordance with the adaptive trigger's operation.

50MHz high accuracy analog multiplier

Although highly accurate, very wideband analog multipliers are available, their output takes a differential form. These differential outputs, which have substantial common mode content, are frequently inconvenient to work with. RF transformers can be used to single end the outputs, but DC and low frequency information is lost.

Figure 33.99 uses the LT1193 differential amplifier to accomplish the differential-to-single ended transition. The AD834 is set up in the recommended configuration (see Analog Devices AD834 Data Sheet, Reference 26). The LT1193 takes the differential signal from the AD834's 50Ω terminated output and provides a single ended output. The gain of two yields a ±1V output at full-scale.

The AD834 outputs come out riding on a common-mode level very close to the devices' positive supply. This common-mode level falls outside the LT1193's input common-mode range. The diodes in the 7.5V supply rails drop the supply at the AD834, biasing its outputs within the LT1193's input range. This scheme avoids the attenuation and matching problems presented by placing a level shift between the multiplier and amplifier. The ferrite beads combine with the diode's impedance to ensure adequate bypassing for the multiplier, a very wideband device.

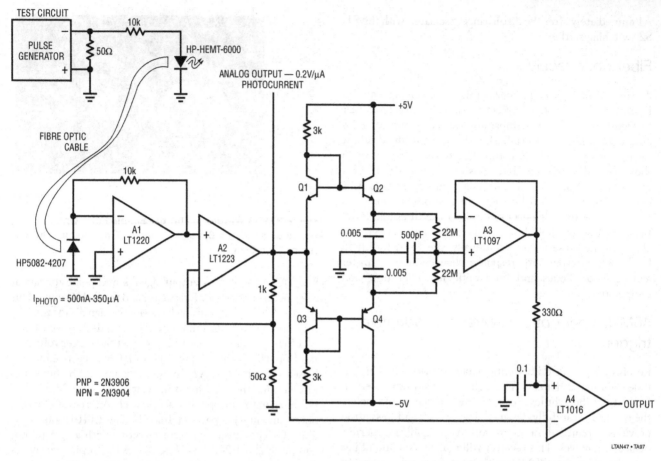

Figure 33.97 • Adaptively Triggered 40MHz Fiber Optic Receiver is Immune to Shifts in Operating Point

Performance for this circuit is quite impressive. Error remains within 2% over DC-50MHz, with feedthrough below −50dB. Trimming the circuit involves adjusting the variable capacitor at the amplifier for minimal output square wave peaking. Figure 33.100 shows performance when a 20MHz sine input is multiplied by Trace A's waveform. The output (Trace B) is a singularly clean instantaneous representation of the X•Y input products, with strict fidelity to their components.

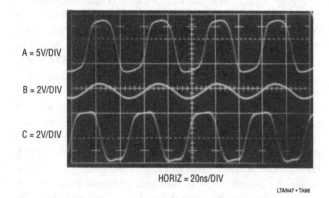

Figure 33.98 • Adaptively Triggered Fiber Optic Receiver's Waveforms at 20MHz with 5µA Diode Current

Power booster stage

Occasionally, it is necessary to supply larger output currents than an amplifier is capable of delivering. The power gain stage, sometimes called a booster, is usually placed within the monolithic amplifier's feedback loop, preserving the IC's low drift and stable gain characteristics.

Because the output stage resides in the amplifier's feedback path, loop stability is a concern. This is particularly the case with high speed amplifiers. The output stage's gain and AC characteristics must be considered if good dynamic performance is to be achieved. Overall circuit phase shift, frequency response and dynamic load handling capabilities are issues that cannot be ignored when designing a power gain stage for a monolithic amplifier. The output stage's added gain and phase shift can cause poor AC response or outright oscillation. Judicious application of frequency compensation methods is needed for good results (see Appendix C, "The oscillation problem—frequency compensation without tears", for discussion and details on compensation methods).

Figure 33.101 shows a 200mA power booster used with an LT1220 amplifier. Complementary emitter followers Q1-Q5 provide current gain for positive signals, with Q2 and Q6 handling negative excursions. Q3 and Q4 are V_{BE} based

715

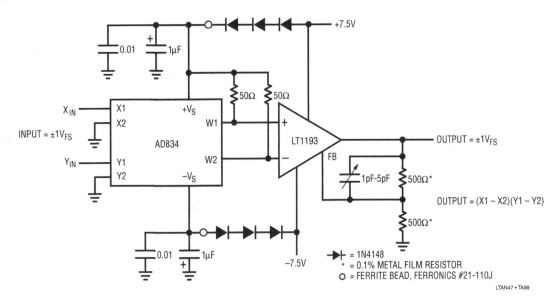

Figure 33.99 • Analog Multiplier with 2% Accuracy Over DC to 50MHz Has a Single Ended Output

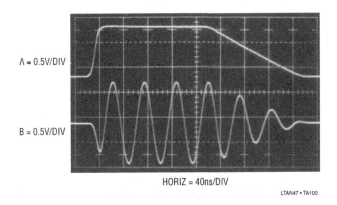

HORIZ = 40ns/DIV

Figure 33.100 • The Multiplier Produces a Modulated Sine Output (Trace B) in Accordance with Trace A's Envelope

current limits, coming on and robbing drive from the appropriate output transistor when current exceeds about 300mA. The diodes prevent Q1 and Q2 from seeing reverse V_{BE} during current limit. The 100Ω resistor and ferrite beads prevent the low impedance amplifier output from causing oscillation in Q1 and Q2 (see Appendix C).

To be effective, the booster must be exceptionally fast. A slow design will obviate the AC performance of the amplifier controlling it, or in the worst case, cause oscillation (again, see Appendix C). Figure 33.102 shows booster performance with the LT1220 removed from the circuit. The input pulse (Trace A) is applied to the booster input, with the output (Trace B) taken at the indicated spot. Evaluation of the photograph shows that booster rise and fall times are limited by the input pulse generator. Additionally, delay is in the 1ns

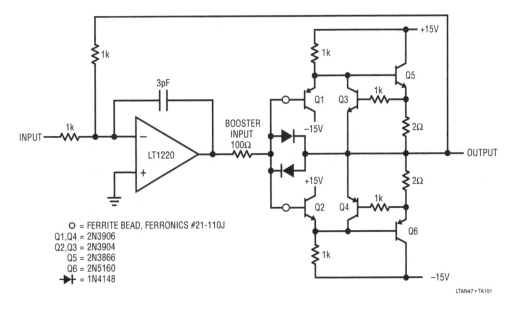

Figure 33.101 • A 200mA Output Wideband Booster Stage

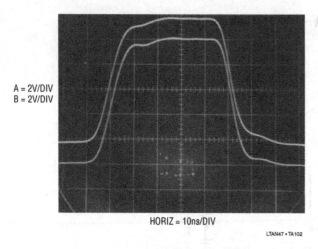

A = 2V/DIV
B = 2V/DIV

HORIZ = 10ns/DIV

LTAN47 • TA102

Figure 33.102 • Response of Figure 33.101's Booster Stage

A = 2V/DIV
B = 2V/DIV
(INVERTED)

HORIZ = 20ns/DIV

LTAN47 • TA103

Figure 33.103 • The Booster's Response When Inside an Amplifier's Loop

range. This kind of speed makes the circuit a good candidate for acceptable AC performance within a fast amplifier's loop.

Figure 33.103 shows pulse response with the LT1220 installed in the circuit with a 50Ω load. The booster's high speed contributes negligible delay and overall response is clean and predictable. The local 3pF roll-off at the LT1220 optimizes response, but is not absolutely necessary in this circuit. The input (Trace A) produces a nicely shaped LT1220 slew-limited output (Trace B).

High power booster stage

In theory, higher power booster stages should be achievable by utilizing bigger devices. This is partly the case, but lack of availability of wideband PNP power transistors is an issue. Figure 33.104 shows a way around this problem.

The circuit is essentially a 1A output version of Figure 33.101, with several differences. In the positive signal path output transistor Q4 is an RF power type, driven by Darlington connected

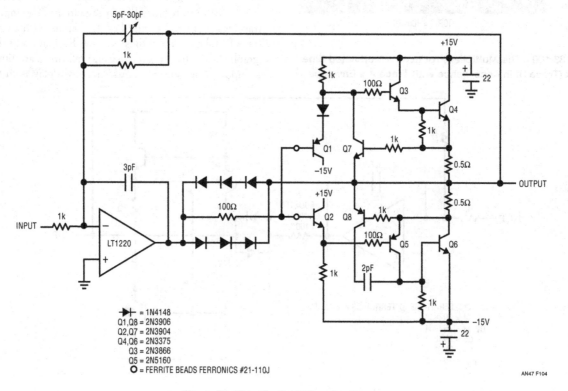

+ = 1N4148
Q1,Q8 = 2N3906
Q2,Q7 = 2N3904
Q4,Q6 = 2N3375
Q3 = 2N3866
Q5 = 2N5160
O = FERRITE BEADS FERRONICS #21-110J

AN47 F104

Figure 33.104 • Fast, 1A Booster Stage

717

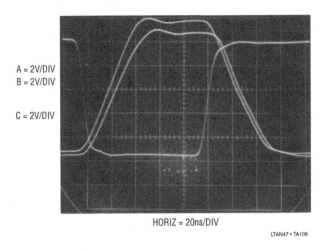

A = 2V/DIV
B = 2V/DIV

C = 2V/DIV

HORIZ = 20ns/DIV

LTAN47 • TA105

Figure 33.105 • The Boosted Op Amp Drives a 1A Load to 10V in 50ns

Q3. The diode in Q1's emitter compensates the additional V_{BE} introduced by Q3, preventing crossover distortion.

The negative signal path substitutes the Q5-Q6 connection to simulate a fast PNP power transistor. Although this configuration acts like a fast PNP follower, it has voltage gain and tends to oscillate. The local 2pF feedback capacitor suppresses these parasitic oscillations and the composite transistor is stable.

This circuit also includes a feedback capacitor trim to optimize AC response. This difference from the previous circuit is necessitated by this circuit's slightly slower characteristics and much heavier loading. Current limit operation and other characteristics are similar to the lower power circuit.

Figure 33.105 shows waveforms for a 10V negative input step (Trace A) with a 10Ω load. The amplifier responds (Trace B), driving the booster to the voltage required to close the loop. For this positive step, the amplifier provides about 1.5V overdrive to overcome Q3 and Q4's V_{BE} drops. The booster output, lagging by a few nanoseconds (Trace C), drives the load cleanly, with only minor peaking. This peaking may be minimized with the feedback capacitance trimmer.

Ceramic bandpass filters

Figure 33.106 is a highly selective bandpass filter using a resonant ceramic element and a single amplifier. The ceramic

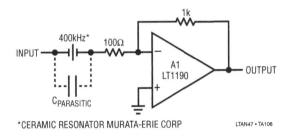

INPUT — 400kHz* — 100Ω — A1 LT1190 — OUTPUT

$C_{PARASITIC}$

1k

*CERAMIC RESONATOR MURATA-ERIE CORP

LTAN47 • TA106

Figure 33.106 • A Piezo-Ceramic Based Filter

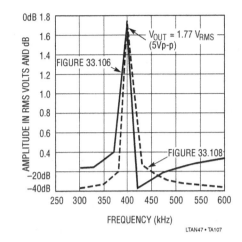

OdB 1.8

$V_{OUT} = 1.77 V_{RMS}$
(5Vp-p)

FIGURE 33.106

FIGURE 33.108

−20dB
−40dB

250 300 350 400 450 500 550 600

FREQUENCY (kHz)

LTAN47 • TA107

AMPLITUDE IN RMS VOLTS AND dB

Figure 33.107 • Response of Both Piezo-Ceramic Filters. Differential Network's Activity is Evident in Figure 33.108's Performance

element nominally looks like a high impedance off its resonant frequency, in this case 400kHz. For off resonance inputs, A1 acts like a grounded follower, producing no output. At resonance, the ceramic element has a low impedance and A1 responds as an inverter with gain. The 100Ω resistor isolates the ceramic element's capacitance from A1's summing point. This capacitance is quite substantial and limits the circuits' out of band rejection capability. Figure 33.107 shows this. This plot shows very steep rejection, with A1's output down almost 20dB at 300kHz and 40dB at 425kHz. The device's stray parasitic capacitance causes the gentle rise in output at higher frequencies and also sets the −20dB floor at 300kHz.

Figure 33.108 partially corrects this problem with a nulling technique. This circuit is similar to the previous one, except that a portion of the input is fed to A1's positive input. The RC network at this input is scaled to look like the ceramic resonator's off null impedance. As such, A1's inputs see similar signals for out of band components, resulting in attenuation via A1's common-mode rejection. At resonance, the added RC network appears as a much higher impedance than the ceramic element and filter response is similar to Figure 33.106's circuit. Figure 33.107 shows that

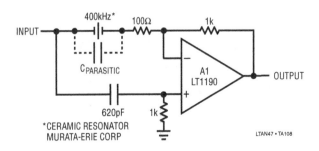

INPUT — 400kHz* — 100Ω — 1k

$C_{PARASITIC}$

A1 LT1190 — OUTPUT

620pF 1k

*CERAMIC RESONATOR MURATA-ERIE CORP

LTAN47 • TA108

Figure 33.108 • Differential Network Nulls Parasitic Capacitance of Ceramic Element

this circuit has much better out of band rejection than Figure 33.106. The high frequency roll-off is smooth, and over 20dB deeper than Figure 33.106 at 475kHz. The low frequency side of resonance has similar characteristics at 375kHz and below.

Crystal filter

Quartz crystals can also be used to make even higher selectivity filters at higher frequencies. Figure 33.109 replaces Figure 33.106's ceramic element with a 3.57MHz quartz crystal. Figure 33.110 shows almost 30dB attenuation only a few kHz on either side of resonance! The differential nulling technique used with the ceramic elements is less effective with quartz crystals. Crystals have significantly lower parasitic terms, making the cancellation less effective.

Applications Section II — Oscillators

Sine wave output quartz stabilized oscillator

Figure 33.111 places a crystal within the amplifier's feedback path, creating an oscillator. With the crystal removed, the circuit is a familiar non-inverting amplifier with a grounded input. Gain is set by the impedance ratio of the elements associated with A1's negative input. Inserting the crystal closes a positive feedback path at the crystal's resonant frequency and oscillations commence.

In any oscillator it is necessary to control the gain as well as the phase shift at the frequency of interest. If gain is too low, oscillation will not occur. Conversely, too much gain produces saturation limiting. Here, gain control comes from the positive temperature coefficient of the lamp at A1's negative input. When power is applied, the lamp is at a low resistance value, gain is high and oscillation amplitude builds. As amplitude builds, lamp current increases, heating occurs, and the lamp's resistance goes up. This causes a reduction in amplifier gain and the circuit finds a stable operating point. This circuit's sine wave output has all the stability advantages associated with quartz crystals. Although shown at 10MHz, it works well with a wide variety of crystal types over a 100kHz-20MHz range. The use of the lamp to control

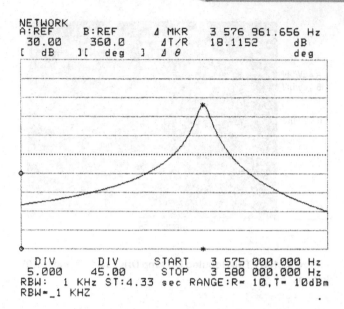

```
   DIV        DIV        START     3 575 000.000 Hz
  5.000      45.00       STOP      3 580 000.000 Hz
RBW:   1 KHz ST:4.33 sec RANGE:R= 10,T= 10dBm
RBW=_1 KHZ
```

Figure 33.110 • The Crystal Filter's Response

amplifier gain is a classic technique, first described by Meacham in 1938.[12] Electronic gain control, while more complex, offers more precise control of amplitude.

Sine wave output quartz stabilized oscillator with electronic gain control

Figure 33.112's quartz stabilized oscillator replaces the lamp with an electronic amplitude stabilization loop. A2 compares the A1 oscillator's positive output peaks with a DC reference. The diode in the DC reference path temperature compensates the rectifier diode. A2 biases Q1, controlling its channel resistance. This influences loop gain, which is reflected in

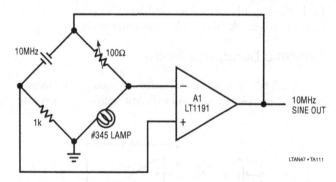

Figure 33.111 • 10MHz Quartz Stabilized Sine Wave Oscillator

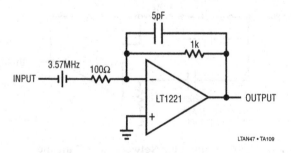

Figure 33.109 • Crystal Filter

Note 12: See Reference 20, as well as References 19 and 21 for supplemental information.

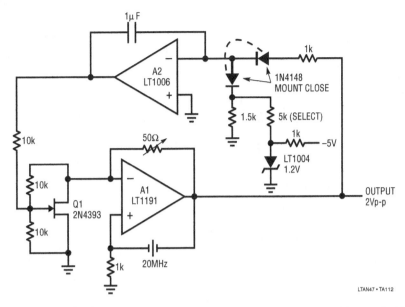

Figure 33.112 • 20MHz Quartz Stabilized Sine Wave Oscillator with Electronic AGC

oscillator output amplitude. Loop closure around A1 occurs, stabilizing oscillator amplitude. The 1μF capacitor compensates the gain control loop.

The DC reference network is set up to provide optimum temperature compensation for the rectifier diode, which sees a $2V_{P-P}$ 20MHz waveform out of A1. A1's small amplitude swing minimizes distortion introduced by channel resistance modulation in Q1. To use this circuit, adjust the 50Ω trimmer until $2V_{P-P}$ oscillations appear at A1's output.

Figure 33.113 is a spectrum analysis of the oscillator's output. The fundamental sits at 20MHz, with the second harmonic 47dB down at 40MHz. A third harmonic, 50dB down, occurs at 60MHz. Resolution bandwidth for the spectrum analysis is 1kHz.

DC tuned 1mhz-10mhz Wien bridge oscillator

In Figure 33.114 the quartz crystal is replaced with a Wien network at A2's positive input. A1 controls Q1 to amplitude stabilize A2's oscillations in identical fashion to the previous figure. Although the Wien network is not nearly as stable as a quartz crystal, it has the advantage of a variable frequency output. Normally, this is facilitated by varying either R, C, or both. Usually, manually adjustable elements such as dual potentiometers and two section variable capacitors are used. Here, the Wien network resistors are fixed at 360Ω, while the capacitive elements are realized with varactor diodes. The varactor diodes voltage-variable-capacitance characteristic allows DC tuning of the oscillator. DC inputs of 0V–10V to the varactors result in a 1MHz to 10MHz shift in oscillation frequency. The 0.1μF capacitor blocks the DC bias from A2's positive input while permitting the Wien network to function

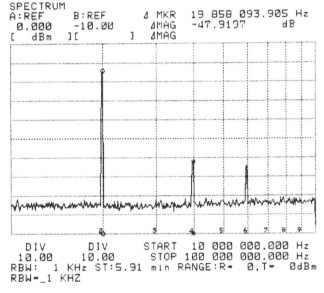

Figure 33.113 • Spectrum Analysis of the 20MHz Quartz Oscillator. Harmonic Content is at Least 47dB Down

normally. A2's $2V_{P-P}$ output minimizes the varactor's junction effects, aiding distortion.

This ±5V powered circuit requires voltage step-up to develop adequate varactor drive. A3 and the LT1172 switching regulator form a simple voltage step-up regulator. A3 controls the LT1172 to produce the output voltage required to close a loop at A3's negative input. L1's high voltage inductive flyback events, rectified by the diode and zener connected Q2, are stored in the 22μF output capacitor. The 7.5k-2.5k divider provides a sample of the output's value to

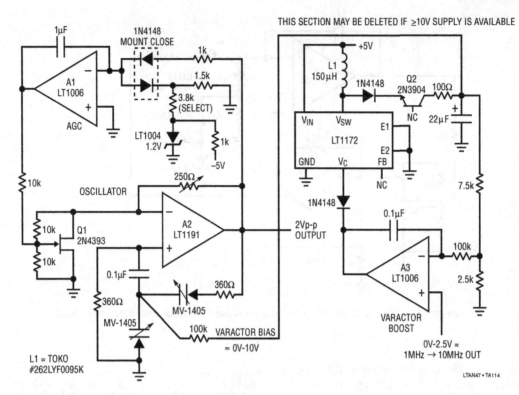

Figure 33.114 • Varactor Tuned 1MHz-10MHz Wien Bridge Oscillator

A3's negative input, closing the loop. The 0.1μF capacitor stabilizes this feedback action. Q2's zener drop allows the circuit to produce controlled outputs all the way down to 0V. This arrangement permits a 0V-2.5V input at A3 to produce a corresponding 0V-10V varactor bias. Figure 33.115, a spectral plot of the circuit running at 7.6MHz, shows the second harmonic down 35dB, with the third harmonic down almost 60dB. Resolution bandwidth is 3kHz.

Complete AM radio station

A complete microphone-to-antenna AM radio station appears in Figure 33.116.[13] The carrier is generated by A1, set up as a quartz stabilized oscillator similar to the one described in Figure 33.111. A1's output feeds A2, functioning as a modulated RF power output stage. A2's input signal range is restricted by the bias applied to offset pins 1 and 8 (see LT1194 data sheet for details). A3, a microphone amplifier, supplies bias to these pins, resulting in an amplitude modulated RF carrier at A2's output. The DC term summed with the microphone biases A3's output to the appropriate level for good quality modulation characteristics. Calibration of this circuit involves trimming the 100Ω potentiometer in the oscillator for a stable 1V_{P-P} 1MHz A1 output.[14]

Note 13: The construction and operation of this apparatus may require Federal Communications Commission review and/or licensing. See Appendix G for FCC licensing and application information.
Note 14: Operating frequency subject to FCC approval and assignment. See Footnote 13 and Appendix G.

Figure 33.117 shows typical AM carrier output at the antenna. In this case the modulation is supplied by Mr. Chuck Berry, singing "Johnny B. Goode".

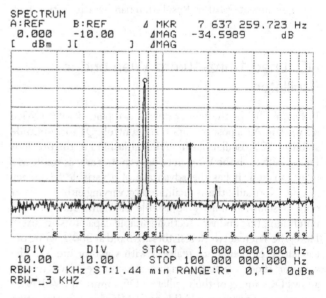

Figure 33.115 • Spectrum Analysis for the Varactor Tuned Wien Bridge. Harmonics are at Least 34dB Down From Fundamental

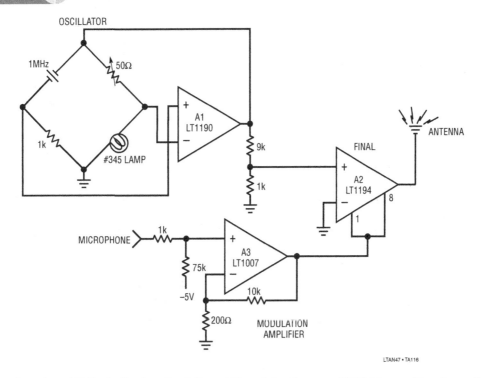

Figure 33.116 • A Complete AM Radio Station. Don't Forget Your Advertisers and FCC License (See Appendix G)

Applications section III — Data conversion

1Hz–1MHz voltage-controlled sine wave oscillator

The oscillators presented to this point have limited frequency tuning range. Although Figure 33.118 is not a true oscillator, it produces a synthesized sine wave output over a wide dynamic range. Many applications such as audio, shaker table driving and automatic test equipment require voltage-controlled oscillators (VCO) with a sine wave output. This circuit meets this need, spanning a 1Hz–1MHz range (120dB or 6 decades) for a 0V to 10V input. It maintains 0.25% frequency linearity and 0.40% distortion specifications.[15] To understand the circuit, assume Q5 is on and its collector (Trace A, Figure 33.119) is at −15V, cutting off Q1. The positive input voltage is inverted by A3, which biases the summing node of integrator A1 through the 3.6k resistor and the self-biased FETs. A current, −I, is pulled from the summing point. A2, a precision op amp, DC stabilizes A1. A1's output (Trace B, Figure 33.119) integrates positive until C1's input (Trace C) crosses 0V. When this happens, C1's

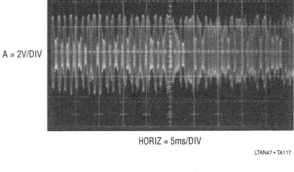

A = 2V/DIV

HORIZ = 5ms/DIV

Figure 33.117 • Chuck Berry Lays a Little Modulation on the 1MHz Carrier

inverting output goes negative, the Q4-Q5 level shifter turns off, and Q5's collector goes to +15V. This allows Q1 to come on. The resistors in Q1's path are scaled to produce a current, +2I, exactly twice the absolute magnitude of the current, −I, being removed from the summing node. As a result, the net current into the junction becomes +I and A1 integrates negatively at the same rate as its positive excursion.

When A1 integrates far enough in the negative direction, C1's "+" input crosses zero and its outputs reverse. This switches the Q4-Q5 level shifter's state. Q1 goes off and the entire cycle repeats. The result is a triangle waveform at A1's output. The frequency of this triangle is dependent on the circuit's input voltage and varies from 1Hz to 1MHz with

Note 14: Operating frequency subject to FCC approval and assignment. See Footnote 13 and Appendix G.

Note 15: Seasoned readers of LTC literature, a hardened corps, may recognize this and other circuits in this publication as updated versions of previous LTC applications. The partial repetition is justified based on improved specifications and/or simplification of the original circuit.

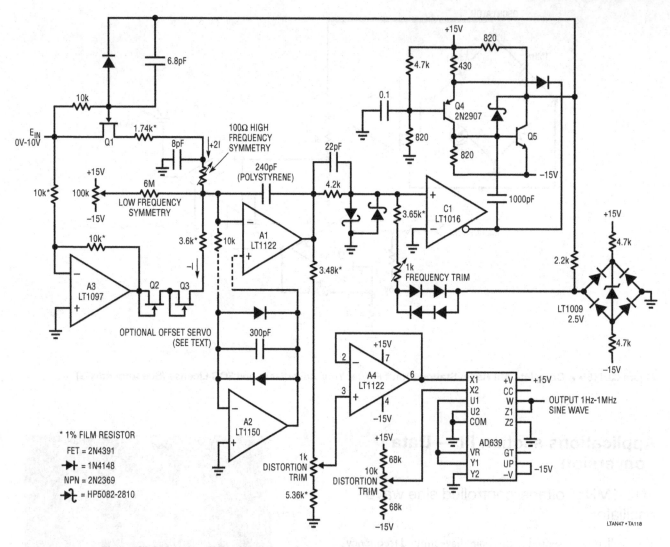

Figure 33.118 • 1Hz-1MHz Sine Wave Output VCO Has 0.25% Linearity and 0.4% Distortion

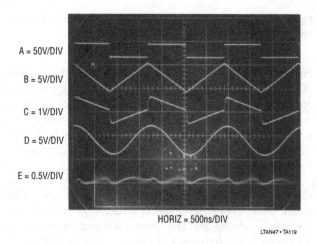

Figure 33.119 • Sine Wave VCO Waveforms

a 0V-10V input. The LT1009 diode bridge and the series-parallel diodes provide a stable bipolar reference which always opposes the sign of A1's output ramp. The Schottky diodes bound C1's "+" input, assuring it clean recovery from overdrive.

The AD639 trigonometric function generator, biased via A4, converts A1's triangle output into a sine wave (Trace D).

The AD639 must be supplied with a triangle wave which does not vary in amplitude or output distortion will result. At higher frequencies, delays in the A1 integrator switching loop result in late turn on and turn off of Q1. If these delays are not minimized, triangle amplitude will increase with frequency, causing distortion level to also increase with frequency. The total delay generated by the LT1016, the Q4-Q5 level shifter, and Q1 is 14ns. This small delay, combined with the 22pF feedforward network at the LT1016's input, keeps distortion to just 0.40% over the entire 1MHz range. At 100kHz,

723

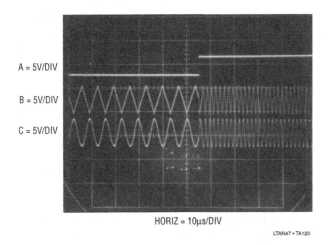

A = 5V/DIV

B = 5V/DIV

C = 5V/DIV

HORIZ = 10µs/DIV

LTAN47 • TA120

Figure 33.120 • Sine Wave Output VCO Step Response is Quick and Clean

distortion is typically inside 0.2%. The effects of gate-source charge transfer, which happens whenever Q1 switches, are minimized by the 8pF unit in Q1's source line. Without this capacitor, a sharp spike would occur at the triangle peaks, increasing distortion. The Q2-Q3 FETs compensate the temperature-dependent on-resistance of Q1, keeping the $+2I/-I$ relationship constant with temperature.

This circuit features extremely fast response to input changes, something most sine wave circuits cannot do.

Figure 33.120 shows what happens when the input switches between two levels (Trace A). A1's triangle output (Trace B) shifts frequency immediately, with no glitching or poor dynamics. The sine output (Trace C), reflecting this action, is similarly clean. To adjust this circuit, put in 10.00V and trim the 100Ω pot for a symmetrical triangle output at A1.

Next, put in 100µV and trim the 100k pot for triangle symmetry. Then, put in 10.00V again and trim the 1k frequency trim adjustment for a 1MHz output frequency. Finally, adjust the distortion trim potentiometers for minimum distortion as measured on a distortion analyzer (Trace E, Figure 33.119). Slight readjustment of the other potentiometers may be required to get lowest possible distortion. If operation below 100Hz is not required, the A2 based DC stabilization stage may be deleted. If this is done, A1's positive input should be grounded.

1Hz–10MHz V→F Converter

The LT1016 and the LT1122 high speed FET amplifier combine to form a high speed V→F converter in Figure 33.121. A variety of circuit techniques are used to achieve a 1Hz to 10MHz output. Overrange to 12MHz ($V_{IN} = 12$) is

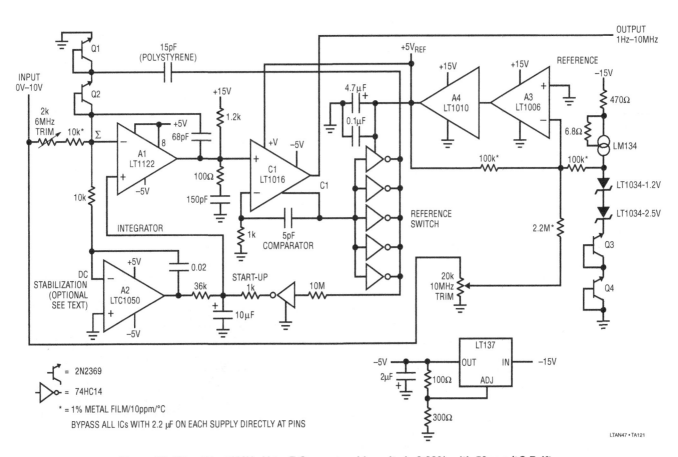

Figure 33.121 • 1Hz-10MHz V-to-F Converter. Linearity is 0.03% with 50ppm/°C Drift

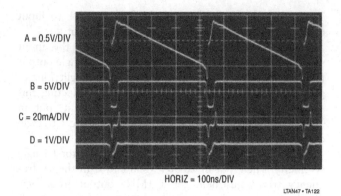

A = 0.5V/DIV

B = 5V/DIV

C = 20mA/DIV

D = 1V/DIV

HORIZ = 100ns/DIV

LTAN47 • TA122

Figure 33.122 • 10MHz V-to-F's Operating Waveforms. LT1122 Integrator is Completely Reset in 60ns

provided. This circuit has a wider dynamic range (140dB or 7 decades) than any commercially available unit. The 10MHz full-scale frequency is 10 times faster than currently available monolithic V→Fs. The theory of operation is based on the identity Q=CV.

Each time the circuit produces an output pulse, it feeds back a fixed quantity of charge (Q) to a summing node (Σ). The circuit's input furnishes a comparison current at the summing node. The difference signal at the node is integrated in a monitoring amplifier's feedback capacitor. The amplifier controls the circuit's output pulse generator, completing a feedback loop around the integrating amplifier. To maintain the summing node at zero, the pulse generator runs at a frequency which permits enough charge pumping to offset the input signal. Thus, the output frequency will be linearly related to the input voltage. A1 is the integrating amplifier.

0.05µV/°C offset drift performance is obtained by stabilizing A1 with A2, a chopper stabilized op amp. A2 measures the DC value of the negative input, compares it to ground, and forces the positive input to maintain offset balance in A1. Note that A2 is configured as an integrator and cannot see high frequency signals. It functions only at DC and low frequency.

A1 is arranged as an integrator with a 68pF feedback capacitor. When a positive voltage is applied to the input, A1's output integrates in a negative direction (Trace A, Figure 33.122). During this period, C1's inverting output is low. The paralleled HCMOS inverters form a reference voltage switch. The reference voltage is established by the LM134 current source driven LT1034's and the Q3-Q4 combination. Additionally, a small input voltage related term is summed into the reference, improving overall circuit linearity. A3-A4 provides low drift buffering, presenting a low impedance reference to the paralleled inverter's supply pin. The HCMOS outputs give low resistance, essentially errorless switching. The reference switch's output charges the 15pF capacitor via Q1's path.

When A1's output crosses zero, C1's inverting output goes high and the reference switch (Trace B) goes to ground. This causes the 15pF unit to dispense charge into the summing node via Q2's V_{BE}. The amount of charge dispensed is a direct

function of the voltage the 15pF unit was charged to (Q = CV). Q1 and Q2 are temperature compensated by Q3 and Q4 in the reference string. The current through the 15pF unit (Trace C) reflects the charge pumping action. The removal of current from A1's summing junction (Trace D) causes the junction to be driven very quickly negative. The initial negative-going 15ns transient at A1's output is due to amplifier delay. The input signal feeds directly through the feedback capacitor and appears at the output. When the amplifier finally responds, its output (Trace A) slew limits as it attempts to regain control of the summing node. The class A 1.2kΩ pull-up and the RC damper at A1's output minimizes erroneous output movement, enhancing this slew recovery. The amount of time the reference switch remains at ground depends on how long it takes A1 to recover and the 5pF-1000Ω hysteresis network at C1. This 60ns interval is long enough for the 15pF unit to fully discharge. After this, C1 changes state, the reference switch swings positive, the capacitor is recharged and the entire cycle repeats. The frequency at which this oscillation occurs is directly related to the voltage-input-derived current into the summing junction. Any input current will require a corresponding oscillation frequency to hold the summing point at an average value of 0V.

Maintaining this relationship at megahertz frequencies places severe restrictions on circuit timing. The key to achieving 10MHz full-scale operating frequency is the ability to transmit information around the loop as quickly as possible. The discharge-reset sequence is particularly critical and is detailed in Figure 33.123. Trace A is the A1 integrator output. Its ramp output crosses 0V at the first left vertical graticule division. A few nanoseconds later, C1's inverting output begins to rise (Trace B), switching the reference switch to ground (Trace C). The reference switch begins to head towards ground about 16ns after A1's output crosses 0V. 2ns later, the summing point (Trace D) begins to go negative as current is pulled from it through the 15pF capacitor. At 25ns, C1's inverting output is fully up, the reference switch is at ground, and the summing point has been pulled to its negative extreme. Now, A1 begins to take control. Its output (Trace A) slews rapidly in the positive direction, restoring the summing point. At 60ns, A1 is in control of the summing node and the integration ramp begins again.

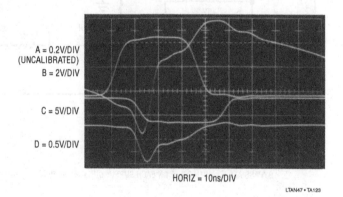

A = 0.2V/DIV
(UNCALIBRATED)

B = 2V/DIV

C = 5V/DIV

D = 0.5V/DIV

HORIZ = 10ns/DIV

LTAN47 • TA123

Figure 33.123 • Detail of 60ns Reset Sequence (Whoosh!)

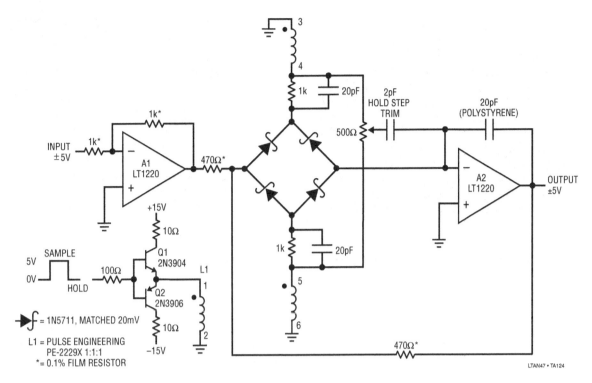

Figuro 33.124 • 8-Bit, 100ns Sample-Hold

Start-up and overdrive conditions could force A1's output to go to the negative rail and stay there. The AC-coupled nature of the charge dispensing loop can preclude normal operation and the circuit may latch. The remaining HCMOS inverter provides a watchdog function for this condition. If A1's output rails negative the reference switch tries to stay at ground. The remaining inverter goes high, lifting A1's positive input. This causes A1's output to slew positive, initiating normal circuit action. The 1k-10µF combination and the 10M-inverter input capacitance limit start-up loop bandwidth, preventing unwanted outputs.

The LM134 current source driving the reference string has a built in 0.33%/°C thermal coefficient, causing slight voltage modulation in the Q3-Q4 pair over temperature. This small change ($\approx$+120ppm/°C) opposes the −120ppm/°C drift in the 15pF polystyrene capacitor, aiding overall circuit tempco.

To trim this circuit, apply exactly 6V at the input and adjust the 2kΩ potentiometer for 6.000MHz output. Next, put in exactly 10V and trim the 20k unit for 10.000MHz output. Repeat these adjustments until both points are fixed. A2's low drift eliminates a zero adjustment. If operation below 600Hz is not required, A2 and its associated components may be deleted.

Linearity of this circuit is 0.03% with full-scale drift of 50ppm/°C. Zero point error, controlled by A2, is 0.05Hz/°C.

8-bit, 100ns sample-hold

Figure 33.124 shows a simple, very fast sample-hold circuit. This circuit will acquire a ±5V input to 8-bit accuracy in

100ns. Hold step is inside 1/4LSB with hold settling inside 25ns. Aperture time is 4ns and droop rate about 1/2LSB/µs.

The input is fed to a Schottky switching bridge via inverting buffer A1. The Schottky bridge, similar to types used in sampling oscilloscopes[16], gives 1ns switching and eliminates the charge pump-through that a FET switch would contribute. The switching bridge's output feeds output amplifier A2. A2, configured as an integrator, is the actual hold amplifier. Its output is fed back to the switching bridge's input, forming a summing point with A1's output resistor. This feedback loop places the bridge within a loop, enhancing accuracy.

The bridge is switched by driving the sample-hold input line. Q1 and Q2 drive L1's primary. L1's secondaries provide complementary drive to the bridge with almost no time skewing.

Figure 33.125 shows the circuit acquiring a full scale step. Trace A is the input command while Trace B is A2's output. The aberration visible in A2's output when switching into hold (hold step) is due to minute residual AC imbalances in the bridge. Figure 33.126 studies this effect in high resolution detail, with the hold step trim deliberately disconnected. After A2's output nominally settles at final value, the circuit is switched into hold. The bridge imbalance allows a small parasitic charge to be displaced into A2's summing point, causing A2 to step 10mV higher (in this case). If the trim is connected and properly adjusted, it supplies a small compensatory charge during switching. Figure 33.127 shows the effect of this on the output. The settled hold output is the

Note 16: See References 7, 8 and 28.

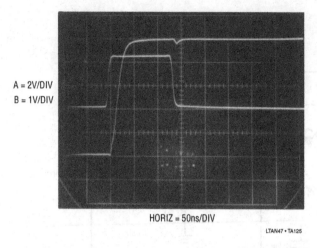

A = 2V/DIV
B = 1V/DIV

HORIZ = 50ns/DIV

LTAN47 • TA125

Figure 33.125 • Fast Sample-Hold Acquiring a Full-Scale Input

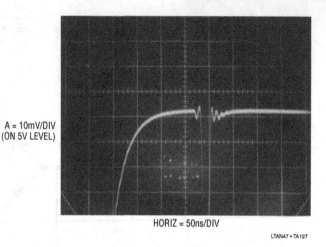

A = 10mV/DIV
(ON 5V LEVEL)

HORIZ = 50ns/DIV

LTAN47 • TA127

Figure 33.127 • Hold Step with Properly Adjusted Compensation

same as the acquired value. To trim this circuit, ground the input while pulsing the sample-hold control line. Next, adjust the trim for minimal amplitude step between the sample and hold states.

In contrast to low frequency sample-hold circuits this design cannot pass signal if left in the sample mode. The transformer's inherent AC coupling precludes such operation. Similarly, extended sample mode duration (e.g., >500ns) will cause transformer saturation, resulting in erroneous outputs and excessive Q1-Q2 dissipation. If extended logic high durations are possible at the control input, it should be AC coupled.

15ns current summing comparator

Figure 33.128 shows a way to build a high speed current comparator with resolution in the 12-bit range. Current comparison, the fastest way to compare D→A outputs and analog values, is commonly used in high speed A→D converters and instrumentation. A1 is set up as a Schottky bounded

amplifier. The bound diodes prevent A1 from saturating due to excessive summing point overdrive, aiding response time. The 3pF capacitor, a typical value, compensates DAC output capacitance and is selected for best amplifier damping. The 10k feedback resistor, also typical, is chosen for best gain-bandwidth performance. Voltage gains of 4 to 10 are common. Figure 33.129 shows performance. Trace A, a test input, causes A1's output (Trace B) to slew through zero (screen center horizontal line). When A1 crosses zero, C1's input biases negative and it responds (C1's output is Trace C) 10ns later with a TTL output. Total elapsed time from the test input arriving at a TTL high until the comparator output achieves a TTL high is inside 15ns.

50MHz adaptive threshold trigger circuit

Figure 33.130 is an extremely versatile trigger circuit. Designing a fast, stable trigger is not easy and often entails a considerable amount of discrete circuitry. This circuit reliably

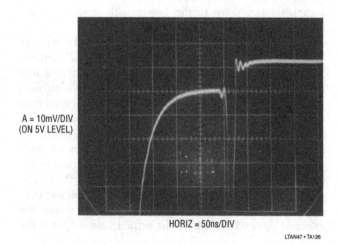

A = 10mV/DIV
(ON 5V LEVEL)

HORIZ = 50ns/DIV

LTAN47 • TA126

Figure 33.126 • Hold Step with Misadjusted Compensation

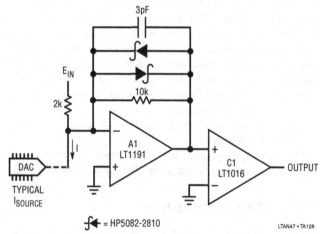

Figure 33.128 • Fast Summing Comparator

727

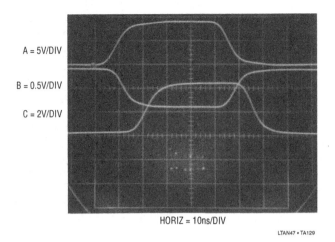

A = 5V/DIV

B = 0.5V/DIV

C = 2V/DIV

HORIZ = 10ns/DIV

LTAN47 • TA129

Figure 33.129 • Fast Summing Comparator's Waveforms. Total Delay is 15ns

triggers from DC-50MHz over a 2mV–300mV input range with no level adjustment required.

A1, a gain of ten preamplifier, feeds an adaptive trigger configuration identical to the one described in Figure 33.97's fiber optic receiver. The adaptive trigger maintains the A3 output comparator's trip point at 1/2 input signal amplitude, regardless of its magnitude. This insures reliable automatic triggering over a wide input amplitude range, even for very low level inputs. As an option, the network shown in dashed lines permits changing the trip threshold. This allows any point on the input waveform edge to be selected as the actual trigger point.[17]

Figure 33.131 shows performance for a 40MHz input sine wave (Trace A). A1's output (Trace B) takes gain and the A3 comparator gives a clean logic output (Trace C). At the highest frequencies, any bandwidth limiting in A1 is irrelevant; the adaptive trigger threshold will simply vary ratiometrically to maintain circuit output.

Fast time-to-height (pulsewidth-to-voltage) converter

The circuit of Figure 33.132 allows very short pulsewidths (in this case 250ns full-scale) to be determined to a typical accuracy of 1%. Digital methods of achieving similar results dictate clock speeds of 1GHz, which is cumbersome. In addition, processor based approaches using averaging techniques require repetitive pulses which this circuit does not. Circuits of this type are frequently required in automatic test equipment and nuclear and high energy physics work where determination of short pulsewidths is a common requirement.

The circuit functions by charging a capacitor during the period of a pulsewidth. When the pulse ends, charging ceases and the voltage across the capacitor is proportional to the width of the pulse.

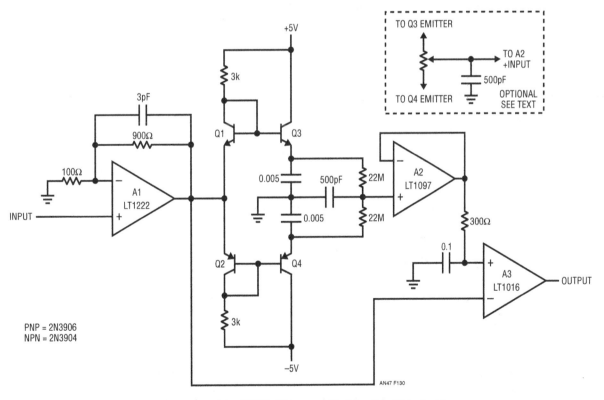

Figure 33.130 • 50MHz Trigger with Adaptive Threshold

Note 17: This technique is borrowed from oscilloscope trigger circuitry. See Reference 29.

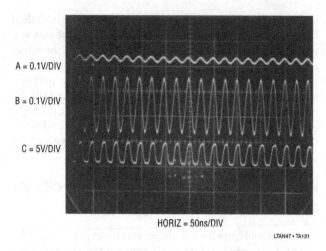

HORIZ = 50ns/DIV

LTAN47 • TA131

Figure 33.131 • The Trigger Responds to a 40MHz Input. Input Amplitude Variations from 2mV-300mV Have No Effect

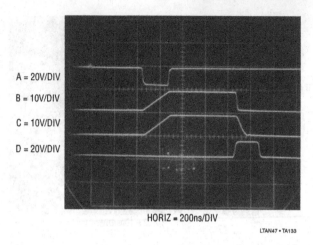

HORIZ = 200ns/DIV

LTAN47 • TA133

Figure 33.133 • Time-to-Height Converter Acquires a 250ns Pulse

The input pulse to be measured (Trace A, Figure 33.133) simultaneously biases the 74C221 dual one shot and Q3. Q3, aided by Baker[18] clamping, capacitive feedforward and optimized DC base biasing, turns off in a few nanoseconds. Current source Q2's emitter forward biases and Q2 supplies constant current to the 100pF integrating capacitor. Q1 supplies temperature compensation for Q2, with the 2.5V LT1009 referencing the current source. Q2's collector (e.g.,

the 100pF capacitor) charges in ramp fashion (Trace B). A1 supplies a buffered output (Trace C). When the input pulse ends, Q3 rapidly turns on, reverse biasing Q2's emitter and turning off the current source. A1's voltage is directly proportional to the input pulse width. A monitoring A→D converter can acquire this data.

After a time set by the 74C221's RC programmed delay, a pulse appears at its Q2 output (Trace D). This pulse turns on Q4, discharging the 100pF capacitor to zero and readying the circuit for the next input pulse.

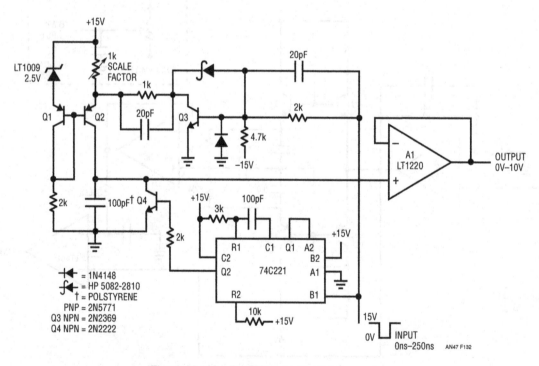

Figure 33.132 • Fast Time-to-Height Converter

Note 18: See Reference 45.

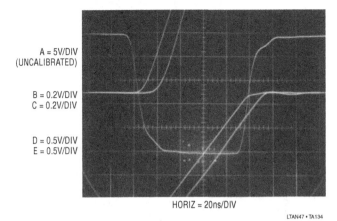

A = 5V/DIV
(UNCALIBRATED)

B = 0.2V/DIV
C = 0.2V/DIV

D = 0.5V/DIV
E = 0.5V/DIV

HORIZ = 20ns/DIV

LTAN47 • TA134

Figure 33.134 • Detail of Time-to-Height Converter's Ramp Switching

This circuit's accuracy and resolution are crucially dependent on minimizing delay in switching the Q1-Q2 current source. Figure 33.134 provides amplitude and time expanded versions of critical circuit waveforms. Trace A is the input pulse and Trace B is A1's input, showing the beginning of the ramp's ascent. Trace C, A1's output, shows about 13ns delay from A1's input. Traces D and E, A1's input and output respectively, record similar A1 delays for ramp turn-off. The photo reflects the extremely fast current source switching; the vast majority of delay is due to A1's delay. A1's delay is far less critical than current source switching delays; A1 will always settle to the correct value well before the one shot resets the circuit. In practice, a monitoring A→D converter should not be triggered until about 50ns after the circuit's input pulse has ceased. This gives A1 plenty of time to catch up to the 100pF capacitor's settled value.

As mentioned, current source switching speed is essential for good results. Figure 33.135 details current source turn off. Trace A is the circuit's input pulse rising edge and Trace B shows the top of the ramp. Turn off occurs in a few nanoseconds. Similar speed is characteristic of the input's falling

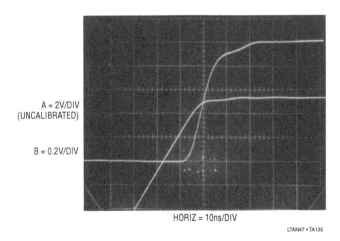

A = 2V/DIV
(UNCALIBRATED)

B = 0.2V/DIV

HORIZ = 10ns/DIV

LTAN47 • TA135

Figure 33.135 • Current Source Turn-Off Detail for the Time-to-Height Converter

edge (current source turn on). Additionally, it is noteworthy that circuit accuracy and resolution limits are set by the *difference* in current source turn on and off delays. As such, the *effective* overall delay is extremely small.

To calibrate this circuit, put in a 250ns width pulse and trim the 1kΩ potentiometer for 10V output. The circuit will convert pulse widths from 20ns to 250ns to a typical accuracy of 1%. The 20ns minimum measurable width is due to inability to fully discharge the 100pF capacitor. If this is objectionable, Q4 can be replaced with a lower saturation device or A1's output can be offset.

True RMS wideband voltmeter

Most AC RMS measurements use logarithmic techniques to compute the waveform's RMS value. This method limits bandwidth to below 1MHz and crest factor performance to about 10:1. Practically speaking, a waveform's RMS value is defined as its heating value in the load. Specialized instruments employ thermally based assemblies that compute the RMS value of the input. The thermal method provides substantially improved bandwidth and crest factor capability compared to logarithmically based converters.

Thermal RMS-DC converters are direct acting, thermoelectronic analog computers. The thermal technique is explicit, relying on first principles. The simple operation permits wideband performance unattainable with implicit, indirect methods based on logarithmic computing.

Figure 33.136 shows a classic scheme for implementing a thermally based RMS-DC converter. Here, the DC amplifier forces a second, identical, heater-sensor pair to the same thermal conditions as the input driven pair. This differentially sensed, feedback enforced loop makes ambient temperature shifts a common-mode term, eliminating their effect. Also, although the voltage and thermal interaction is non-linear, the input-output voltage relationship is linear with unity gain. The ability of this arrangement to reject ambient temperature shifts depends on the heater-sensor pairs being isothermal. This is achievable by thermally insulating them with a time constant well below that of ambient shifts. If the time

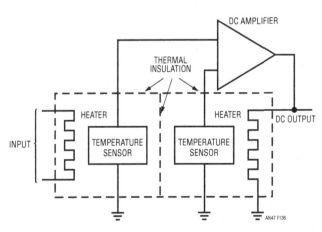

Figure 33.136 • Conceptual Thermal RMS-DC Converter

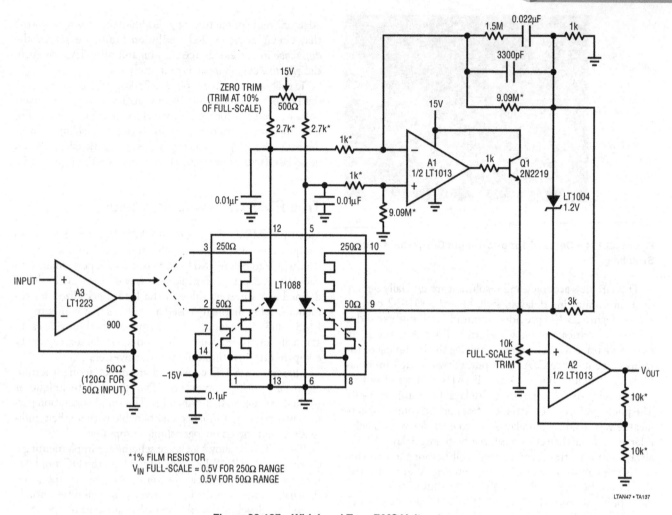

Figure 33.137 • Wideband True RMS Voltmeter

constants to the heater-sensor pairs are matched, ambient temperature terms will affect the pairs equally in phase and amplitude. The DC amplifier will reject this common-mode term. Note that, although the pairs are isothermal, they are insulated from each other. Any thermal interaction between the pairs reduces the system's thermally based gain terms. This would cause unfavorable signal-to-noise performance, limiting dynamic operating range. Figure 33.136's output is linear because the matched thermal pair's non-linear voltage-temperature relationships cancel each other.

The advantages of this approach have made its use popular in thermally based RMS-DC measurements. Typically, the assembly is composed of matched heater resistors, sensors and thermal insulation. These assemblies are relatively large and expensive to produce. Figure 33.137's economical wide-band thermally based voltmeter is based on a monolithic thermal converter. The LT1223 provides gain, and drives the LT1088 RMS-DC thermal converter.[19] The LT1088's temperature sensing diodes are biased from the supply. A1,

set up as a differential servo amplifier with a gain of 9000, extracts the diode's difference signal and biases Q1. Q1 drives one of the LT1088's heaters, completing a loop. The 3300pF capacitor gives a stable roll-off.

The 1.5M-0.022µF combination improves settling by reducing gain during output slew. The LT1088's square-law thermal gain means overall loop gain is lower for small inputs. Normally, this would result in slow settling for values below about 10%–20% of scale. The LT1004 1k–3k network is a simple breakpoint, boosting amplifier gain in this region to improve settling. A2, a gain trimmable output stage, serves to compensate for gain variations in the two sides of the LT1088. To trim the circuit, put in about a 10% scale DC signal (e.g., 0.05V). Adjust the zero trim so that $V_{OUT} = V_{IN}$. Next, apply a full-scale DC input and set the full-scale trim to that value at the output. Repeat the trims until both are fixed well within 1% of full-scale. An alternate trim scheme involves applying no input, grounding Q1's base and setting the zero trim until A1's output is active. Then, unground Q1's base, apply a full-scale input and trim the full-scale adjustment for that value at the output.

Note 19: Complete details on this device and a discussion on thermal conversion considerations are found in Reference 40.

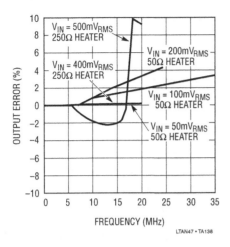

Figure 33.138 • Accuracy Plot for the RMS Voltmeter

Figure 33.138 is a plot of error vs input frequency. The LT1088 is specified at 2% to 100MHz (50Ω heater) or 1% to 20MHz (250Ω heater). As such, most of the error shown is due to bandwidth restrictions in A3, but performance is still impressive. The plots include data taken at various input levels into both heaters. A 500mV input into 250Ω dips to 1% at 8MHz and 2.5% at 14MHz before peaking badly beyond 17MHz. This input level forces a 9.5 V_{RMS} output at A3, introducing large signal bandwidth limitations. The 400mV input to the 250Ω heater shows essentially flat results to 20MHz, the LT1088's 250Ω heater specification limit.

The 50Ω heater provides significantly wider bandwidth, although A3's 50mA output limits maximum input to about 100mV_{RMS} (1.76V_{RMS} at the LT1088).

Applications section IV — Miscellaneous circuits

RF leveling loop

Figure 33.137's wideband AC conversion can be applied in other areas. A common RF requirement is to stabilize the amplitude of a waveform against variations in input, time and temperature. Instruments and transmitters frequently require this function, which is not easy if waveform purity must be maintained. Figure 33.139A shows a 25MHz RF leveling loop. The RF input is applied to the AD539 wideband multiplier. The multiplier's output drives A1. A1's output is converted to DC by the LT1088 based RMS-DC converter (see previous circuit). A servo amplifier compares this output with a settable DC reference and biases the multiplier's control channel, completing a loop. The 0.33μF capacitor provides frequency compensation by rolling off gain at a frequency well below the response of the LT1088 servo. The loop maintains the output's 25MHz RMS amplitude at the DC reference's value. Changes in load, input, power supply and other variables are rejected.

Figure 33.139B, a similar circuit, offers significantly lower cost although performance is not quite as good. The RF input is applied to LT1228 A1, an operational transconductance amplifier. A1's output feeds LT1228 A2, a current feedback amplifier. A2's output, the circuit's output, is sampled by the A3 based gain control configuration. This arrangement, similar to the gain control loops described in Figures 33.112 and 33.114, closes a gain control loop back at A1. The 4pF capacitor compensates rectifier diode capacitance, enhancing

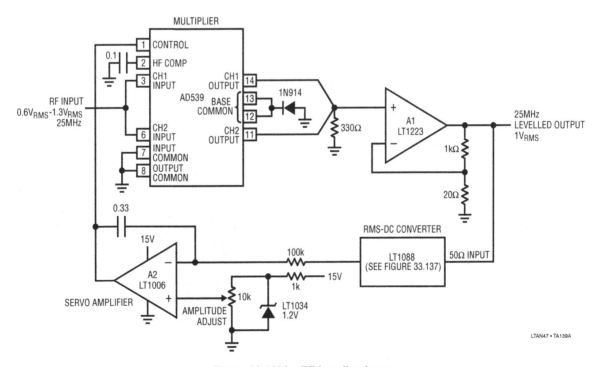

Figure 33.139A • RF Leveling Loop

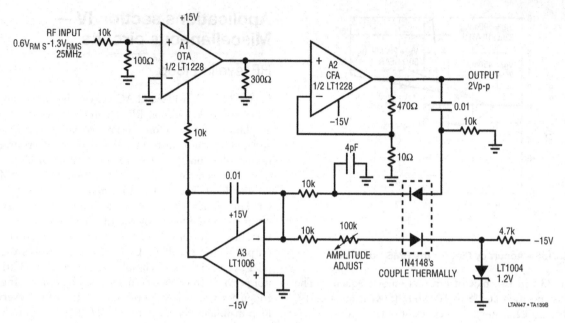

Figure 33.139B • Simple RF Leveling Loop

output flatness vs frequency. A1's I_{SET} input current controls its gain, allowing overall output level control. This approach to RF leveling is simple and inexpensive, although output drift, distortion and regulation are somewhat higher than in the previous circuit.

Voltage controlled current source

Figure 33.140 shows a voltage controlled current source with load and control voltage referred to ground. This simple, powerful circuit produces output current in accordance with the sign and magnitude of the control voltage. The circuit's scale factor is set by resistor R. A1, biased by V_{IN}, drives current through R (in this case 10Ω) and the load. A2, sensing differentially across R, closes a loop back to A1. The load

current is constant because A1's loop forces a fixed voltage across R. The 2k-100pF combination sets roll off and the configuration is stable. Figure 33.141 shows dynamic response. Trace A is the voltage control input while Trace B is the output current. Response is quick and clean, with delay of 5ns and no slew residue or aberration.

High power voltage controlled current source

Figure 33.142 is identical to the basic current source, except that it adds a 1A booster stage (adapted from Figure 33.104) for increased output power. Including the booster inside A1's feedback loop eliminates its DC errors. Note that the booster's current limiting features have been removed, because of this circuit's inherent current limiting nature of operation. Figure 33.143 shows this circuit's response to be as clean as the lower power version, although delay is about 20ns

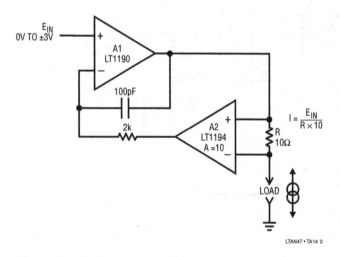

Figure 33.140 • Fast, Precise, Voltage Controlled Current Source with Grounded Load

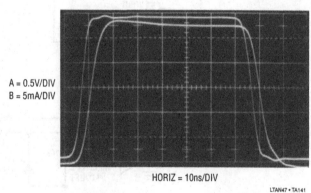

A = 0.5V/DIV
B = 5mA/DIV

HORIZ = 10ns/DIV

Figure 33.141 • Dynamic Response of the Current Source. Delay is 4ns, with Clean Settling

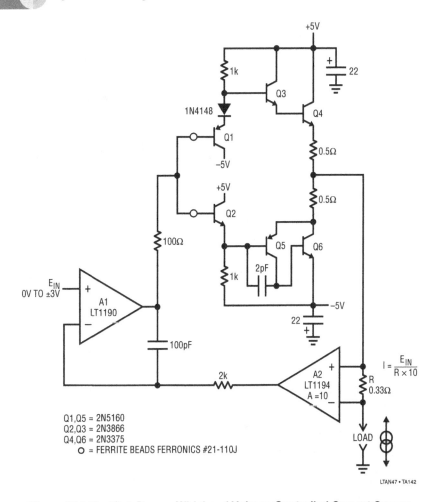

Figure 33.142 • High Power, Wideband Voltage Controlled Current Source

Q1,Q5 = 2N5160
Q2,Q3 = 2N3866
Q4,Q6 = 2N3375
O = FERRITE BEADS FERRONICS #21-110J

$$I = \frac{E_{IN}}{R \times 10}$$

LTAN47 • TA142

slower. It is worth mentioning that the loop stability considerations involved in placing A2 *and* the booster in A1's feedback path are significant. This circuit receives treatment in Appendix C, "The oscillation problem — frequency compensation without tears".

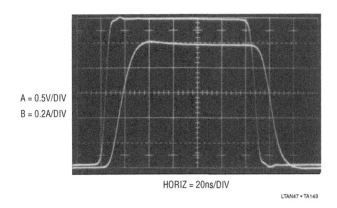

A = 0.5V/DIV
B = 0.2A/DIV

HORIZ = 20ns/DIV

LTAN47 • TA143

Figure 33.143 • 1A Pulse Response of the High Power Current Source

18ns circuit breaker

Figure 33.144 shows a simple circuit which will turn off current in a load 18ns after it exceeds a preset value. This circuit has been used to protect integrated circuits during developmental probing and is also useful for protecting expensive loads during trimming and calibration. The circuit's versatility is enhanced because one side of the load is grounded. Under normal conditions, Q1's emitter (Trace A, Figure 33.145, is Q1's current, and Trace C is its voltage) is biased on, supplying power to the load via the 10Ω current shunt. Differential amplifier A1's output resides below comparator A2's voltage programmed trip point and Q2 is off. When an overload occurs, Q1's emitter current begins to increase (Trace A, just prior to the third vertical division). A1's output (Trace B) begins to rise as it tracks the increase in the 10Ω shunt's voltage. The 9k-1k dividers keep A1 inputs inside their common-mode range. Simultaneously, Q1's emitter voltage (Trace C) begins to drop as it beta limits. When A1's version of the load current exceeds A2's trip point, A2 (Trace D) goes high, turning on Q2. Q2's turn on steals Q1's base drive, turning off the load current. Local positive feedback at A2's latch pin causes it to latch in this

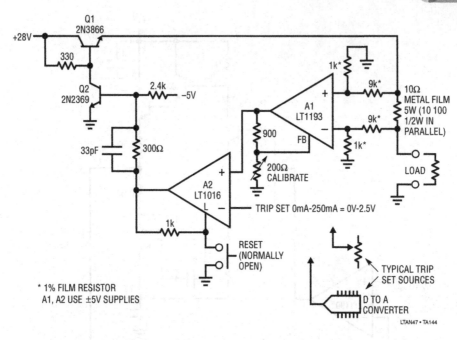

Figure 33.144 • 18ns Circuit Breaker with Voltage Programmable Trip Point

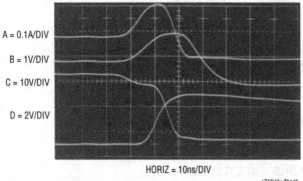

A = 0.1A/DIV

B = 1V/DIV

C = 10V/DIV

D = 2V/DIV

HORIZ = 10ns/DIV

LTAN47 • TA145

Figure 33.145 • Operating Waveforms for the 18ns Circuit Breaker. Circuit Output (Trace C) is Shut Down 18ns After Output Current (Trace A) Begins to Rise

off state. When the load fault has been cleared, the pushbutton can be used to reset the circuit. The delay from the onset of excessive load current to complete shutdown is inside 18ns. The 4ns delay of Trace A's current probe should be factored in when interpreting waveforms. To calibrate this circuit, ground Q2's base and install a 250mA load. Adjust the 200Ω trim for a 2.5V output at A1. Next, remove the load, unground Q2's base and press the reset button. Finally, put in the desired trip set voltage and the circuit is ready for use.

References

1. Orwiler, Bob, "Oscilloscope Vertical Amplifiers", Tektronix, Inc., Concept Series. 1969.

2. Williams, Jim, "Techniques and Equipment for Current Measurement", Appendix C, in "Step Down Switching Regulators", Linear Technology Corporation, Application Note 35 (August 1989).

3. Addis, John, "Fast Vertical Amplifiers and Good Engineering", in *Analog Circuit Design; Art, Science and Personalities*, Butterworths. (1991).

4. Brown, Lloyd A., "The Story of Maps" pp. 226-240. Little, Brown. Boston, Massachusetts. (1949).

5. Willsberger, Johann, "Clocks and Watches", Dial Press. (1975).

6. Mercer, Vaudrey, "John Arnold and Son – Chronometer Makers 1762-1843", Antiquarian Horological Society, London. (1972).

7. Tektronix, Inc., "Type 1S1 Sampling Plug-In Operating and Service Manual", Tektronix, Inc. (1965).

8. Mulvey, J. "Sampling Oscilloscope Circuits". Tektronix, Inc Concept Series. (1970).

9. Williams, Jim, "About Probes and Oscilloscopes", Appendix B, in "High Speed Comparator Techniques", Linear

Technology Corporation, Application Note 13 (April 1985).

10. Williams, Jim, "Evaluating Oscilloscope Overload Performance", Box Section A, in "Methods for Measuring Op Amp Settling Time", Linear Technology Corporation, Application Note 10 (July 1985).

11. Williams, Jim and Huffman, Brian, "Instrumentation for Converter Design", Appendix F, in "Some Thoughts on DC-DC Converters", Linear Technology Corporation, Application Note 29 (October 1988).

12. Gilbert, Barrie, "Where Do Little Circuits Come From" in *Analog Circuit Design; Art, Science and Personalities*, Butterworths. (1991).

13. Williams, Jim, "Should Ohm's Law Be Repealed?", in *Analog Circuit Design; Art, Science and Personalities*, Butterworths. (1991).

14. Williams, Jim, "Methods for Measuring Op Amp Settling Time", Linear Technology Corporation, Application Note 10 (July 1985).

15. Demerow, R., "Settling Time of Operational Amplifiers", *Analog Dialogue*, Volume 4-1, Analog Devices, Inc. (1970).

16. Pease, R.A., "The Subtleties of Settling Time", *The New Lightning Empiricist*, Teledyne Philbrick. (June 1971).

17. Harvey, Barry, "Take the Guesswork Out of Settling Time Measurements", EDN. (September 19, 1985).

18. Addis, John, "Sampling Oscilloscopes", Private Communication. (February, 1991).

19. Williams, Jim, "Bridge Circuits — Marrying Gain and Balance", Linear Technology Corporation, Application Note 43 (June 1990).

20. Meacham, L.A., "The Bridge Stabilized Oscillator", Bell System Technical Journal, Vol. 17, p. 574. (October 1938).

21. Hewlett, William R., "A New Type Resistance-Capacity Oscillator", M.S. Thesis, Stanford University, Palo Alto, California (1939).

22. Hewlett, William R., U.S. Patent No. 2,768,872 (January 6, 1942).

23. Bauer, Brunton, "Design Notes on the Resistance-Capacity Oscillator Circuit", Parts I and II, *Hewlett-Packard Journal*, Hewlett-Packard (November, December 1949).

24. Black, H.S., "Stabilized Feedback Amplifier", *Bell System Technical Journal*, Vol. 13, p. 1. (January 1934).

25. Tektronix, Inc., Type 111 Pretrigger Pulse Generator Operating and Service Manual, Tektronix, Inc. (1960).

26. Analog Devices, Inc, "Linear Products Databook", AD834 Datasheet, pp. 6-43 to 6-48 (1988).

27. Analog Devices, Inc., "Linear Products Databook", AD834 Datasheet, pp. 6-43 to 6-48. (1988).

28. Hewlett-Packard, "Schottky Diodes for High-Volume, Low Cost Applications", Application Note 942, Hewlett-Packard Company (1973).

29. Tektronix, Inc., "Trigger Circuit – Peak-Peak Automatic Operation", Model 2235 Oscilloscope Service Manual, Tektronix, Inc. (1983).

30. Wien, Max, "Measung der induction constanten mit dern Optischen Telephon", Ann. der Phys., Vol. 44, pp. 704-707. (1891).

31. Dostal, J., "Operational Amplifiers", Elsevier. (1981).

32. Philbrick Researches, "Applications Manual for Operational Amplifiers", Philbrick Researches. (1965).

33. Sheingold, D.H., "Analog-Digital Conversion Handbook", Prentice-Hall. (1986).

34. Bunze, V., "Matching Oscilloscope and Probe for Better Measurements", *Electronics*, pp. 88-93. (March 1, 1973).

35. Williams, Jim, "High Speed Comparator Techniques", Linear Technology Corporation, Application Note 13 (April 1985).

36. Morrison, Ralph, "Grounding and Shielding Techniques in Instrumentation", 2nd Edition, Wiley Interscience. (1977).

37. Hewlett-Packard, "Threshold Detection of Visible and Infra-Red Radiation with PIN Photodiodes", Application Note 915, Hewlett-Packard Company.

38. Roberge, J.K., "Operational Amplifiers: Theory and Practice", Wiley Interscience. (1975).

39. Ott, Henry W., "Noise Reduction Techniques in Electronic Systems", Wiley Interscience. (1976).

40. Williams, Jim, "A Monolithic IC for 100MHz RMS-DC Conversion". Linear Technology Corporation, Application Note 22 (September 1987).

41. Lee, Marshall M., "Winning With People: The First 40 Years of Tektronix", Tektronix, Inc. (1986).

42. Weber, Joe, "Oscilloscope Probe Circuits", Tektronix, Inc., Concept Series. (1969).

43. Chessman, M. and Sokol, N., "Prevent Emitter-Follower Oscillation", *Electronic Design* 13, pp. 110-113. (21 June 1976).

44. DeBella, G.B., "Stability of Capacitively-Loaded Emitter Followers – a Simplified Approach", *Hewlett-Packard Journal* 17, pp. 15-16. (April 1966).

45. Baker, R.H., "Boosting Transistor Switching Speed", *Electronics*, Vol. 31, pp. 190-193. (1957).

46. Williams, Jim, "Max Wien, Mr. Hewlett and a Rainy Sunday Afternoon", in Analog Circuit Design: Art, Science and Personalities, Butterworths. (1991).

Appendix A

ABC's of probes – Tektronix, Inc

This appendix, guest written by the engineering staff of Tektronix, Inc., is a distillation of their booklet, "ABC's of Probes". The complete booklet is available, at no charge, through any Tektronix sales office or call 800-835-9433 ext. 170. For excellent technical background on probe theory see Reference 42.

PART I: Understanding probes

The vital link in your measurement system

Probes connect the measurement test points in a DUT (device under test) to the inputs of an oscilloscope. Achieving optimized system performance depends on selecting the proper probe for your measurement needs.

Though you could connect a scope and DUT with just a wire, this simplest of connections would not let you realize the full capabilities of your scope. By the same token, a probe that is not right for your application can mean a significant loss in measurement results, plus costly delays and errors.

Why not use a piece of wire?

Good question: There are legitimate reasons for using a piece of wire or, more correctly, two pieces of wire; some low bandwidth scopes and special purpose plug-in amplifiers only provide binding post input terminals, so they offer a convenient means of attaching wires of various lengths.

DC levels associated with battery operated equipment could be measured. Low frequency (audio) signals from the same equipment could also be examined. Some high output transducers could also be monitored. However, this type of connection should be kept away from line-operated equipment for two basic reasons, safety and risk of equipment damage.

Safety. Attachment of hookup wires to line-operated equipment could impose a health hazard, either because the "hot" side of the line itself could be accessed, or because internally generated high voltages could be contacted. In both cases, the hookup wire offers virtually no operator protection, either at the equipment source or at the scope's binding posts.

Risk of equipment damage. Two unidentified hookup wires, one signal lead and one ground, could cause havoc in line-operated equipment. If the "ground" wire is attached to **any** elevated signal in line-operated equipment, various degrees of damage will result simply because both the scope and the equipment are (or should be) on the same three-wire outlet system, and short-circuit continuity is completed through one common ground.

Performance considerations. In addition to the hazards just mentioned, there are two major performance limitations associated with using hookup wires to transfer the signal to the scope: circuit loading and susceptibility to external pickup.

Circuit loading. This subject will be discussed in detail later, but circuit loading by the test equipment (scope-probe) is a combination of resistance and capacitance. Without the benefit of using an attenuator (10X) probe, the loading on the device under test (DUT) will be 1M ohm (the scope input resistance) and more than 15 picofarad (15pF), which is the typical scope input capacitance plus the stray capacitance of the hookup wire.

Figure 1-1 shows what a "real world" signal from a 500 ohm impedance source looks like when loaded by a 10M ohm, 10 pF probe: the scope-probe system is 300MHz. Observed risetime is 6 nSec.

Figure 1-2 shows what happens to the same signal when it is accessed by two 2-meter lengths of hookup wire: loading is 1MΩ (the scope input resistance) and about 20 pF (the scope input capacitance, plus the stray capacitance of the wires). Observed risetime has slowed to 10 nSec and the transient response of the system has become unusable.

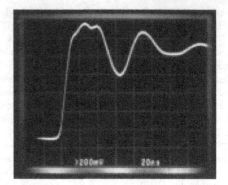

Figure 1-2

Susceptibility to external pickup. An unshielded piece of wire acts as an antenna for the pickup of external fields, such as line frequency interference, electrical noise from fluorescent lamps, radio stations and signals from nearby equipment. These signals are not only injected into the scope along with the wanted signal, but can also be injected into the device under test (DUT) itself.

The source impedance of the DUT has a major effect on the level of interference signals developed in the wire. A very low source impedance would tend to shunt any induced voltages to ground, but high frequency signals could still appear at the scope input and mask the wanted signal. The answer, of course, is to use a probe which, in addition to its other features, provides coaxial shielding of the center conductor and virtual elimination of external field pickup.

Figure 1-3 shows what a low level signal from a high impedance source (100mV from 100K ohm) looks like when accessed by a 300MHz scope-probe system. Loading is 10M ohm and 10pF. This is a true representation of the signal, except that probe resistive loading has reduced the amplitude by about 1%: the observed high frequency noise is part of the signal at the high impedance test point and would normally be removed

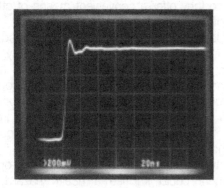

Figure 1-1

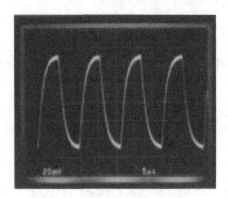

Figure 1-3

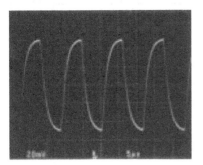

Figure 1-4

by using the BW (bandwidth) limit button on the scope. (See Figure 1-4.)

If we look at the same test point with our pieces of wire, two things happen. The amplitude drops due to the increased resistive and capacitive loading, and noise is added to the signal because the hookup wire is completely unshielded. (See Figure 1-5)

Most of the observed noise is line frequency interference from fluorescent lamps in the test area.

Probably the most annoying effect of using hookup wire to observe high frequency signals is its unpredictability. Any touching or rearrangement of the leads can produce different and nonrepeatable effects on the observed display.

Benefits of using probes

Not all probes are alike and, for any specific application, there is no one ideal probe; but they share common features and functions that are often taken for granted.

Probes are convenient. They bring a scope's vertical amplifier to a circuit. Without a probe, you would either need to pick up a scope and attach it to a circuit, or pick up the circuit and attach it to the scope. Properly used, probes are convenient, flexible and safe extensions of a scope.

Probes provide a solid mechanical connection. A probe tip, whether it's a clip or a fine solid point, makes contact at just the place you want to examine.

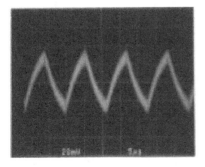

Figure 1-5

Probes help minimize loading. To a certain extent, all probes load the DUT—the source of the signal you are measuring. Still, probes offer the best means of making the connections needed. A simple piece of wire, as we have just seen, would severely load the DUT; in fact, the DUT might stop functioning altogether.

Probes are designed to minimize loading. Passive, non-attenuating 1X probes offer the highest capacitive loading of any probe type—even these, however, are designed to keep loading as low as possible.

Probes protect a signal from external interference. A wire connection, as described earlier, in addition to loading the circuit, would act as an antenna and pick up stray signals such as 60Hz power, CBers, radio and TV stations. The scope would display these stray signals as well as the signal of interest from the DUT.

Probes extend a scope's signal amplitude-handling ability. Besides reducing capacitive and resistive loading, a standard passive 10X (ten times attenuation) probe extends the on-screen viewability of signal amplitudes by a factor of ten.

A typical scope minimum sensitivity is 5V/division. Assuming an eight-division vertical graticule, a 1X probe (or a direct connection) would allow on-screen viewing of $40V_{P-P}$ maximum. The standard 10X passive probe provides $400V_{P-P}$ viewing. Following the same line, a 100X probe should allow 4kV on-screen viewing. However, most 100X probes are rated at 1.5kV to limit power dissipation in the probe itself.

Check the specs. Bandwidth is the probe specification most users look at first, but plenty of other features also help to determine which probe is right for your application. Circuit loading, signal aberrations, probe dynamic range, probe dimensions, environmental degradation and ground-path effects will all impact the probe selection process, as discussed in the pages that follow.

By giving due consideration to probe characteristics that your application requires, you will achieve successful measurements and derive full benefit from the instrument capabilities you have at hand.

How probes affect your measurements

Probes affect your measurements by loading the circuit you are examining. The loading effect is generally stated in terms of impedance at some specific frequency, and is made up of a combination of resistance and capacitance.

Source impedance. Obviously, source impedance will have a large impact on the net affect of any specific probe loading. For example, a device under test with a near zero output impedance would not be affected in

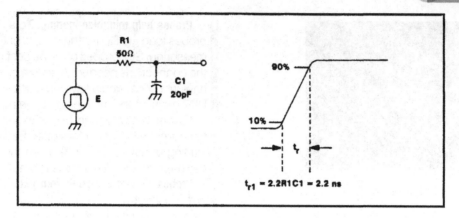

Figure 1-6

terms of amplitude or risetime to any significant degree by the use of a typical 10X passive probe. However, the same probe connected to a high impedance test point, such as the collector of a transistor, could affect the signal in terms of risetime and amplitude.

Capacitive loading. To illustrate this effect, let's take a pulse generator with a very fast risetime. If the initial risetime was assumed to be zero (tr = 0), the output tr of the generator would be limited by the associated resistance and capacitance of the generator. This integration network produces an output risetime equal to 2.2 RC. This limitation is derived from the universal time-constant curve of a capacitor.

Figure 1-6 shows the effect of internal source resistance and capacitance on the equivalent circuit. At no time can the output risetime be faster than 2.2 RC or 2.2 nSec.

If a typical probe is used to measure this signal, the probe's specified input capacitance and resistance is added to the circuit as shown in Figure 1-7.

Because the probe's 10M ohm resistance is much greater than the generator's 50 ohm output resistance, it can be ignored.

Figure 1-8 shows the equivalent circuit of the generator and probe, applying the 2.2 RC formula again. The actual risetime has slowed from 2.2 nSec. to 3.4 nSec.

Percentage change in risetime due to the added probe tip capacitance:

$$\% \text{ change} \frac{\text{tr}_2 - \text{tr}_1}{\text{tr}_1} \times 100 = \frac{3.4 - 2.2}{2.2} \times 100 = 55\%$$

Another way of estimating the effect of probe tip capacitance on a source is to take the ratio of probe tip capacitance (marked on the probe compensation box) to the known or estimated source capacitance.

Using the same values:

$$\frac{C_{\text{probe tip}}}{C_1} \times 100 = \frac{11\text{pF}}{20\text{pF}} \times 100 = 55\%$$

To summarize, any added capacitance slows the source risetime when using high impedance passive

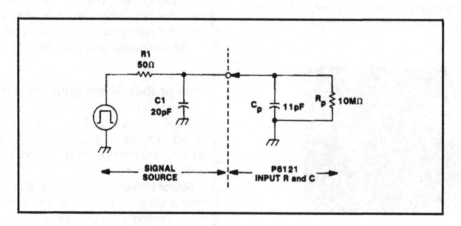

Figure 1-7

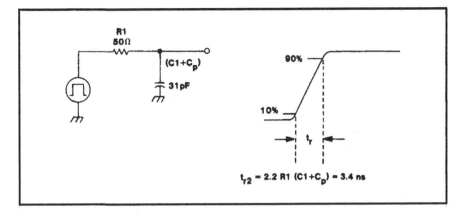

Figure 1-8

probes. In general, the greater the attenuation ratio, the lower the tip capacitance. Here are some examples:

Probe	Attenuation	Tip Capacitance
Tektronix P6101A	X1	54 pF
Tektronix P6105A	X10	11.2 pF
Tektronix P6007	X100	2 pF

Capacitive loading: Sinewave.
When probing continuous wave (CW) signals, the probe's capacitive reactance at the operating frequency must be taken into account.

The total impedance, as seen at the probe tip, is designated Rp and is a function of frequency. In addition to the capacitive and resistive elements, designed-in inductive elements serve to offset the pure capacitive loading to some degree.

Curves showing typical input impedance vs frequency, or typical Xp and Rp vs frequency are included in most Tektronix probe instruction manuals. Figure 1-9A shows the typical input impedance and phase relationship vs frequency of the Tektronix P6203

Active Probe. Note that the 10KΩ input impedance is maintained to almost 10 MHz by careful design of the associated resistive, capacitive and inductive elements.

Figure 1-9B shows a plot of Xp and Rp vs frequency for a typical 10MΩ passive probe. The dotted line (Xp) shows capacitive reactance vs frequency. The total loading is again offset by careful design of the associated R, C and L elements.

If you do not have ready access to the information and need a worst-case guide to probe loading, use the following formula:

$$Xp = \frac{1}{2\pi FC}$$

Xp = Capacitive reactance (ohms)
F = Operating frequency
C = Probe tip capacitance (marked on the probe body or compensation box)

For example, a standard passive 10M ohm probe with a tip capacitance of 11pF will have a capacitive reactance (Xp) of about 290 ohm at 50MHz.

Depending, of course, on the source impedance, this loading could have a major effect on the signal amplitude (by simple divider action), and even on the operation of the circuit itself.

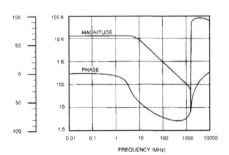

Figure 1-9A • Typical Input Impedance vs Frequency for the Tektronix P6203 Active Probe

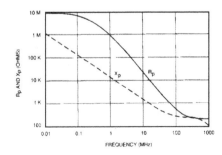

Figure 1-9B • Xp and Rp vs Frequency for a Typical 10 MΩ Passive Probe

Resistive loading. For all practical purposes, a 10X, 10M ohm passive probe has little effect on today's circuitry in terms of resistive loading; however, they do carry a trade-off in terms of relatively high capacitive loading as we have previously discussed.

Low Z passive probes. A "Low Z" passive probe offers very low tip capacitance at the expense of relatively high resistive loading. A typical 10X "50 ohm" probe has an input C of about 1pF and a resistive loading of 500 ohm: Figure 1-10 shows the circuit and equivalent model of this type of probe.

This configuration forms a high frequency 10X voltage divider because, from transmission line theory, all that the 450 ohm tip resistor "sees" looking into the cable is a pure 50 ohm resistance, no C or L component. No low frequency compensation is necessary because it is not a capacitive divider. Low Z probes are typically high bandwidth (up to 3.5GHz and rise-times to 100 pS) and are best suited for making rise-time and transit-time measurements. They can, however, affect the pulse amplitude by simple resistive divider action between the source and the load (probe). Because of its resistive loading effects, this type of probe performs best on 50 ohm or lower impedance circuits under test.

Note also that these probes operate into 50 ohm scope inputs only. They are typically teamed up with fast (500MHz to 1GHz) real time scopes or with scopes employing the sampling principle.

Bias-offset probes. A Bias/Offset probe is a special kind of Low Z design with the capability of providing a variable bias or offset voltage at the probe tip.

Bias/Offset probes like the Tektronix P6230 or P6231 are useful for probing high speed ECL circuitry, where resistive loading could upset the operating point. These special probes are fully described in Part 3: Advanced probing techniques.

The best of both worlds. From the foregoing, it can be seen that the totally "non-invasive" probe does not exist. However, one type of probe comes close—the active probe.

Active probes are discussed in the tutorial section, but in general, they provide low resistance loading (10M ohm) with very low capacitive loading (1 to 2pF). They do have trade-offs in terms of limited dynamic range, but under the right conditions, do indeed offer the best of both worlds.

Bandwidth. Bandwidth is the point on an amplitude versus frequency curve where the measurement system is down 3dB from a starting (reference) level. Figure 1-11

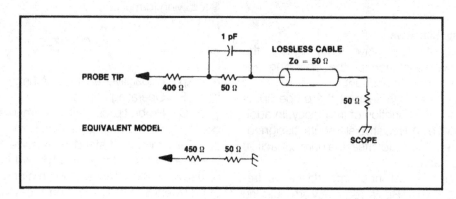

Figure 1-10

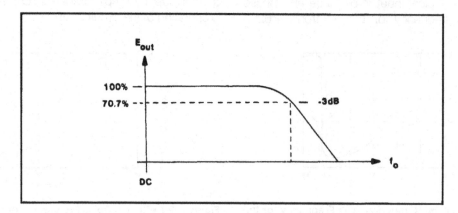

Figure 1-11

shows a typical response curve of an oscilloscope system.

Scope vertical amplifiers are designed for a Gaussian roll-off at the high end (a discussion of Gaussian response is beyond the scope of this primer). With this type of response, risetime is approximately related to bandwidth by the following equation:

$$Tr = \frac{.35}{BW}$$

or, for convenience:

$$\text{Risetime (nanoseconds)} = \frac{350}{\text{Bandwidth (MHz)}}$$

It is important to note that the measurement system is −3dB (30%) down in amplitude at the specified bandwidth limit.

Figure 1-12 shows an expanded portion of the −3dB area. The horizontal scale shows the input frequency derating factor necessary to obtain accuracies better than 30% for a specific bandwidth scope. For example, with no derating, a "100MHz" scope will have up to a 30% amplitude error at 100MHz (1.0 on the graph). If this scope is to have an amplitude accuracy better than 3%, the input frequency must be limited to about 30MHz (100MHz X.3).

For making amplitude measurements within 3% at a specific frequency, choose a scope with at least four times the specified bandwidth as a general rule of thumb.

Probe bandwidth. All probes are ranked by bandwidth. In this respect, they are like scopes or other amplifiers that are ranked by bandwidth. In these cases

we apply the square root of the sum of the squares formula to obtain the "system risetime." This formula states that:

$$\text{Risetime system} = \sqrt{Tr^2 \text{ displayed} - Tr^2 \text{ source}}$$

Passive probes do not follow this rule and should not be included in the square root of the sum of the squares formula.

Tektronix provides a probe bandwidth ranking system that specifies "the bandwidth (frequency range) in which the probe performs within its specified limits. These limits include: total abberrations, risetime and swept bandwidth."

Both the source and the measurement system shall be specified when checking probe specifications (see Test Methods, this page).

In general, a Tektronix "100MHz" probe provides 100MHz performance (−3dB) when used on a compatible 100MHz scope. In other words, it provides full scope bandwidth **at the probe tip**.

However, not all probe/scope systems can follow this general rule. "Scope bandwidth at the probe tip?."

Figure 1-13 shows examples of Tektronix scopes and their recommended passive probes.

Test methods. As with all specifications, matching test methods must be employed to obtain specified performance. In the case of bandwidth and risetime measurements, it is essential to connect the probe to a properly terminated source. Tektronix specifies a 50 ohm source terminated in 50 ohm, making this a 25 ohm source impedance. Furthermore, the probe

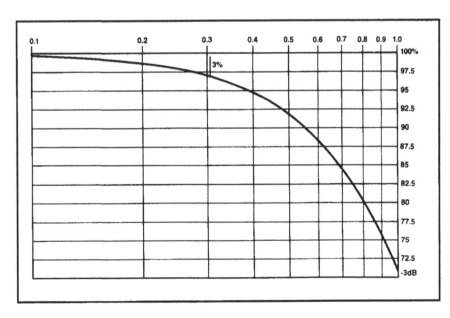

Figure 1-12

SCOPE	BW (1 MΩ input)	PROBE	BW	SYSTEM
2235	100	P6109	150	100
2245A	100	P6109	150	100
2246A	100	P6109	150	100
2445B	150	P6133 Opt 25	150	150
485	350	P6106A	250	250
2465B	400	P6137	400	400
2467B	400	P6137	400	400

Figure 1-13

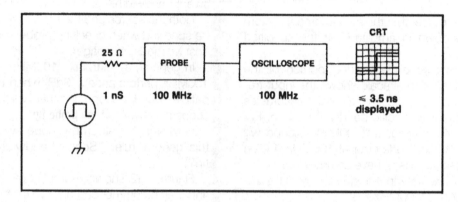

Figure 1-14

must be connected to the source via a proper probe tip to BNC adaptor. (Figure 1-14).

Figure 1-14 shows an equivalent circuit of a typical setup. The displayed risetime should be a 3.5 nsec or faster.

Figure 1-15 shows an equivalent circuit of a typical passive probe connected to a source.

Scope bandwidth at the probe tip?

Most manufacturers of general-purpose oscilloscopes that include standard accessory probes in the package, promise and deliver the advertised scope bandwidth **at the probe tip**.

For example, the Tektronix 2465B 400 MHz Portable Oscilloscope and its standard accessory P6137 Passive Probes deliver 400 MHz (−3db) at the probe tip.

However, not all high performance scopes can offer this feature, even when used with their recommended passive probes. For example, the Tektronix 11A32 400 MHz plug-in has a system bandwidth of 300 MHz when used with its recommended P6134

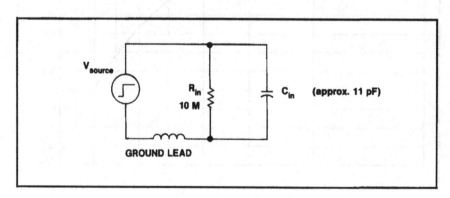

Figure 1-15

passive probe. This is simply because even the highest impedance passive probes are limited to about 300 to 350MHz, while still meeting their other specifications.

It is important to note that the above performance is only obtainable under strictly controlled, and industry recognized conditions; which state that the signal must originate from a 50Ω back-terminated source (25Ω), and that the probe must be connected to the source by means of a probe tip to BNC (or other) adaptor.

This method ensures the shortest ground path and necessary low impedance to drive the probe's input capacitance, and to provide the specified bandwidth at the signal acquisition point, the probe tip.

Real-world signals rarely originate from 25Ω sources, so less than optimum transient response and bandwidth should be expected when measuring higher impedance circuits.

How ground leads affect measurements

A ground lead is a wire that provides a local ground-return path when you are measuring any signal. An inadequate ground lead (one that is too long or too high in inductance) can reduce the fidelity of the high frequency portion of the displayed signal.

What grounding system to use. When making **any** measurement, some form of ground path is required to make a basic two-terminal connection to the DUT. If you want to check the presence or absence of signals from low-frequency equipment, **and** if the equipment is line-powered and plugged into the same outlet system as the scope, then the common 3-wire ground system provides the signal ground return. However, this indirect route adds inductance in the signal path—it can also produce ringing and noise on the displayed signal and is not recommended.

When making any kind of absolute measurement, such as amplitude, risetime or time delay measurements, you should use the shortest grounding path possible, consistent with the need to move the probe among adjacent test points. The ultimate grounding system is an in-circuit ECB (etched circuit board) to probe tip adaptor. Tektronix can supply these for either miniature, compact or subminiature probe configurations.

Figure 1-15 shows an equivalent circuit of a typical passive probe connected to a source. The ground lead, L and C_{in}, form a series resonant circuit with only 10M ohm for damping. When hit with a pulse, it will ring. Also, excessive L in the ground lead will limit the changing current to C_{in}, limiting the risetime.

Without going into the mathematics, an 11pF passive probe with a 6-inch ground lead will ring at about 140MHz when excited by a fast pulse. As the ring frequency increases, it tends to get outside the pass-band of the scope and is greatly attenuated. So to increase the ring frequency, use the shortest ground lead possible and use a probe with the lowest input C.

Probe ground lead effects. The effect of inappropriate grounding methods can be demonstrated several ways. Figs 1-16A, B and C show the effect of a 12-inch ground lead when used on various bandwidth scopes.

In Figure 1-16A, the display on the 15MHz scope looks OK because the ringing aberrations are beyond the passband of the instrument and are greatly attenuated. Figs 1-16B and C show what the same signal looks like on 50MHz and 100MHz scopes.

Even with the shortest ground lead, the probe-DUT interface has the **potential** to ring. The potential to ring depends on the **speed** of the step function. The ability to **see** the resultant ringing oscillation depends on the scope system bandwidth.

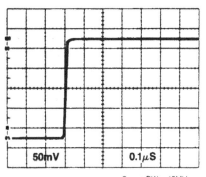

| 50mV | | | 0.1μS |

Scope BW = 15MHz
Ground lead 12 inches

Figure 1-16A

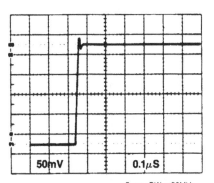

| 50mV | | | 0.1μS |

Scope BW = 50MHz
Ground lead 12 inches

Figure 1-16B

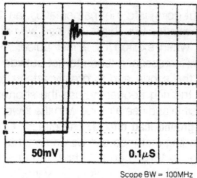

50mV   0.1μS

Scope BW = 100MHz
Ground lead 12 inches

Figure 1-16C

Figs. 1-17A through F show the effects of various grounding methods and ground lead lengths on the display of a very fast pulse. This is the most critical way of looking at ground lead effects: we used a fast pulse, with a risetime of about 70 picoseconds and a fast (400 MHz) scope with a matching P6137 probe.

Fig. 1-17A shows the input pulse under the most optimum conditions when using 50 ohm coax cable. Scope: the Tektronix 2465B with 50 ohm input and 50 ohm cable from a 50 ohm source. Displayed risetime is <1nSec.

Fig. 1-17B shows the same signal when using the scope-probe combination under the most optimum conditions. A BNC to probe adaptor or an in-circuit test jack provides a coaxial ground that surrounds the probe ground ring. This sytem provides the shortest probe ground connection available. Displayed risetime is <1nSec.

Figures 1-17C through E show the effects of longer ground leads on the displayed signal. Fig. 1-17C shows the effect of a short semi-flexible ground connection, called a "Z" lead. Finally, Fig. 1-17F shows what happens when no probe ground lead is used.

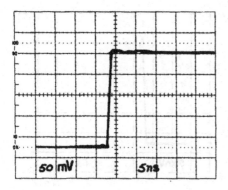

50 mV   5ns

50 ohm Source/Cable/2465B/50 ohm input

Figure 1-17A

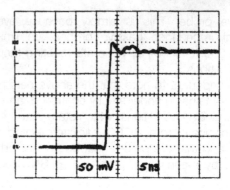

50 mV   5ns

P6137-BNC/Probe Adaptor Tr = < 1nS

Figure 1-17B

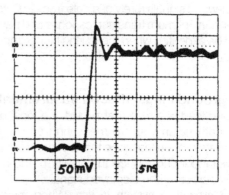

50mV   5ns

P6137 - Probe/Z Ground Tr = 1.5 nS

Figure 1-17C

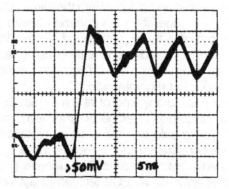

>50mV   5ns

P6137 - Probe/3" Gnd Lead Tr = 4 nS

Figure 1-17D

How probe design affects your measurements

Probes are available in a variety of sizes, shapes and functions, but they do share several main features: a probe head, coaxial cable and either a compensation box or a termination.

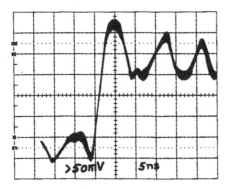

P6137 - Probe/6" Gnd Lead Tr = 4 nS

Figure 1-17E

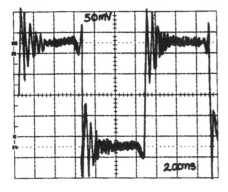

No Ground Lead

Figure 1-17F

The probe head contains the signal-sensing circuitry. This circuitry may be passive (such as a 9-M ohm resistor shunted by an 11 pF capacitor in a passive voltage probe or a 125-turn transformer secondary in a current probe); or active (such as a source follower or Hall generator) in a current probe or active voltage probe.

The coaxial cable couples the probe head output to the termination. Cable types vary with probe types. The termination has two functions:

- to terminate the cable in its characteristic impedance
- to match the input impedance of the scope.

The termination may be passive or active circuitry. For easy connection to various test points, many probes feature interchangeable tips and ground leads.

A unique feature of most Tektronix probes is the Tektronix-patented coaxial cable that has a resistance-wire center conductor. This distributed resistance suppresses ringing caused by impedance mismatches between the cable and its terminations when you're viewing fast pulses on wideband scopes.

PART II: Effects of probe compensation—understanding probes

Tips on using probes

Compensating the probe. The most common mistake in making scope measurements is forgetting to compensate the probe. Improperly compensated probes can distort the waveforms displayed on the scope. The probe should be compensated as it will be used when you make the measurement.

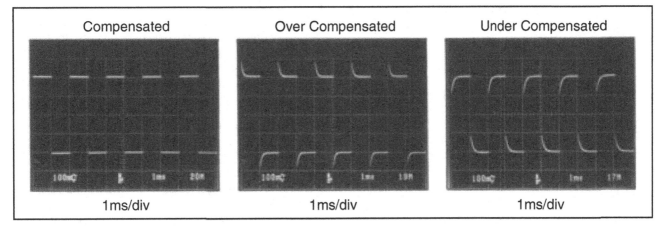

Figure 2-1 • Shows the Display Associated With Correctly And Incorrectly Compensated Probes.

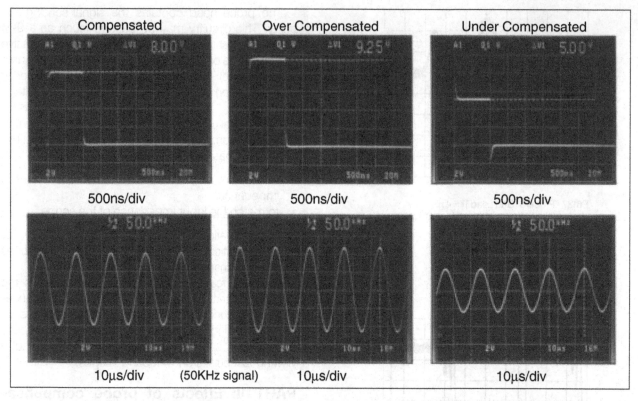

Compensated	Over Compensated	Under Compensated
500ns/div	500ns/div	500ns/div
10µs/div (50KHz signal)	10µs/div	10µs/div

Figure 2-2 • Shows The Effects on Faster Pulses And Sinewaves When An Incorrectly Compensated Probe is Used. Note That The Much Faster Sweep Rates Used To Correctly View These Waveforms Does Not Warn The User of An Adjustment Problem

The basic low frequency compensation (L.F. comp.) procedure is simple:

- Connect the probe tip to the scope CALIBRATOR (refer to Scope Calibrator Outputs)
- Switch the channel 1 input coupling to dc
- Turn on the scope and move the CH1 VOLTS/DIV switch to produce about four divisions of vertical display
- Set the sweep rate to 1mSec/div. (for line-driven calibrators see Scope Calibrators Outputs)
- Use a non-metalic alignment tool to turn the compensation adjust until the tops and bottoms of the square-wave are flat.

PART III: Advanced probing techniques

Introduction:

In Part III we will examine some of the more advanced probing techniques associated with accessing high frequency and complex signals, such as fast ECL, waveforms offset from ground, and true differential signals.

Most of the techniques to be described follow recommended practices outlined throughout this Booklet, and to a large extent involve proper grounding techniques.

Workers in the audio and relatively low frequency fields may wonder what all the fuss is about, and may comment "I don't have any of these problems", or "I can't see any difference when I use different ground lead lengths, or even when I leave the ground lead completely off?"

In order to see abberations caused by poor grounding techniques, two conditions must exist:

1. The scope system bandwidth must be great enough to handle the high frequency content existing at the probe tip.
2. The input signal must contain enough high frequency information (fast risetime) in order to cause ringing and aberrations due to poor grounding techniques.

To illustrate these points, a 20 MHz scope was used to access a 1.7nS pulse by using a standard passive probe with a 6" ground lead.

NOTE: A fast scope can be made into a slow scope simply by pushing the Bandwidth Limit (B/W Limit) button?.

We used a 350MHz scope with a 20MHz B/W Limit Function.

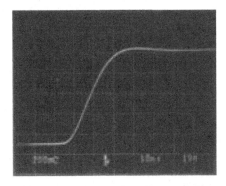

Figure 3-1 • Resultant Clean, But Incorrect Display Caused By Inadequate Scope System Bandwidth.

Figure 3-1 shows the resultant clean displayed pulse with a risetime of about 20nS (17.5MHz).

This display does not represent conditions actually existing at the probe tip, because the 20MHz measurement system cannot "see" what's really happening.

Figure 3-2 shows what the probe tip signal really looks like when a 350MHz scope is used under the same conditions (B/W Limit off).

The observed risetime has improved to about 2nS, but we have serious problems with ringing and aberrations, caused by incorrect grounding techniques.

The problem can now be seen because the scope system bandwidth is great enough to pass and display all the frequency content existing at the probe tip.

To further stress the points about high frequency content and scope system bandwidth, let's assume an input pulse with a risetime of about 20nS. If the signal is accessed by the same probe/6″ ground lead/350MHz system, it would look very much like the display in Figure 3-1.

There would be no frequency content higher than 17.5MHz (20nS Tr). The 6″ ground lead would not ring, and would therefore be the correct choice for accessing this relatively slow signal.

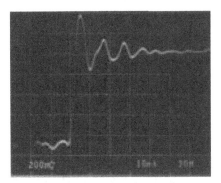

Figure 3-2 • The Same Input Signal As Shown In Figure 3-1, But Accessed By A 350MHz System Bandwidth Scope (Same 6″ Ground Lead)

In the following sections we discuss how to recognize signal acquisition problems, and how to avoid them.

Techniques for probing ECL, high speed 50Ω environments, and accessing true differential signals are also discussed.

Probe ground lead effects. In Part I we discussed the basic need for probe grounding, and showed several different ways of looking at the effects of correct, and incorrect probe grounding.

In this section, we will expand upon these techniques and show how to identify problem areas.

When a probe (high Z, low Z, passive or active) is connected to the circuit under test via an ECB to Probe Tip Adaptor (test point), the coaxial environment existing at the probe tip is extended through the adaptor to the signal pick-off point, and to the ECB ground plane (or device ground).

Figure 3-3 shows what a typical 1nS Tr pulse looks like when a suitable probe is connected to the circuit via an ECB to Probe Tip Adaptor.

Figure 3-4 shows a typical ECB to Probe Tip Adaptor (test point) installation.

These test points are available in three sizes to accept miniature, compact or sub-miniature series probes.

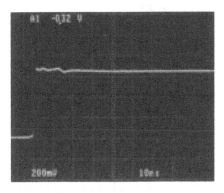

Figure 3-3 • 1nS Tr Pulse Accessed Via an ECB to Probe Tip Adaptor (test point)

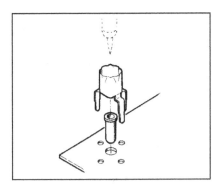

Figure 3-4 • Typical ECB to Probe Tip Adaptor installation

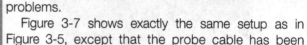

If a flexible ground lead is used in place of the ECB to Probe Tip Adaptor, the 1ns Tr input step (with high frequency content up to 350MHz) will cause the ground lead to ring at a frequency determined by the ground lead inductance and the probe tip and source capacitance.

Figure 3-5 shows the effect of using a 6″ ground lead to make the ground connection.

The ring frequency for the 6″ ground lead/probe tip C combination is 87.5MHz. This signal is injected in series with the wanted signal and appears at the probe tip, as shown in Figure 3-6.

Unfortunately, the problem is not this simple.

The probe's coaxial environment has been disrupted at the signal acquisition point by ground lead inductance, and is no longer correctly terminated (for high speed signal acquisition).

This abrupt transition leaves the probe's outer shield susceptible to ring frequency injection (the ground lead inductance is in series with the outer braid)

The now unterminated probe cable system develops reflections, which intermix with the ring frequency and the signal to produce a multitude of problems and unpredictable results.

Herein lies the key to the identification of ground lead problems.

Figure 3-7 shows exactly the same setup as in Figure 3-5, except that the probe cable has been moved, and a hand has been placed over part of the probe cable.

KEY: If touching or moving the probe cable produces changes in the display, you have a probe grounding problem.

A correctly grounded (terminated) probe should be completely insensitive to cable positioning or touch.

Ground lead length. All things being equal, the shortest ground lead produces the highest ring frequency.

If the lead is very short, the ring frequency might be high enough to be outside the passband of the scope, and/or the input frequency content may not be high enough to stimulate the ground lead's resonant circuit.

In all cases, the shortest ground lead should be used, consistent with the need for probe mobility.

If possible, use 3″ or shorter ground leads, such as the Low Impedance Contact (Z Lead). These are

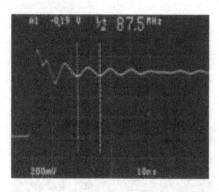

Figure 3-5 • Effect of a 6″ Ground Lead on a 1nS Tr Input Step.

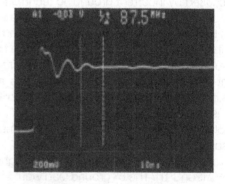

Figure 3-7 • The Same Setup As in Figure 3-5, Except That The Probe Cable Has Been Repositioned, And a Hand Has Been Placed Over Part of The Probe Cable.

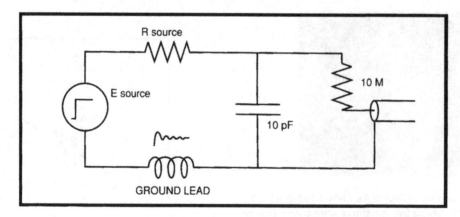

Figure 3-6 • Equivalent Circuit, Ground Lead Inductance (Excess Inductance).

supplied with the Tektronix P613X and P623X family or probes.

One final note. The correct probe grounding method depends on the signal's high frequency content, the scope system bandwidth, and the need for mobility between test points.

A 12" ground lead may be perfect for many lower frequency applications. It will provide you with extra mobility, and nothing will be gained by using shorter leads.

If in doubt, apply the cable touch test outlined previously.

Ground loop noise injection. Another form of signal distortion can be caused by signal injection into the grounding system.

This can be caused by unwanted current flow in the ground loop existing between the common scope and test circuit power line grounds, and the probe ground lead and cable.

Normally, all these points are, or should be at zero volts, and no ground current will flow.

However, if the scope and test circuit are on different building system grounds, there could be small voltage differences, or noise on one of the building ground systems.

The resulting current flow (at line frequency or noise frequency) will develop a voltage drop across the probe cable's outer shield, and be injected into the scope in series with the desired signal.

Inductive pickup in ground loops. Noise can enter a common ground system by induction into long 50Ω signal acquisition cables, or into standard probe cables.

Proximity to power lines or other current-carrying conductors can induce current flow in the probe's outer cable, or in standard 50Ω coax. The circuit is completed through the building system common ground.

Prevention of ground loop noise problems. Keep all signal acquisition probes and/or cables away from sources of potential interference.

Verify the integrity of the building system ground.

If the problem persists, open the ground loop:

1. By using a Ground Isolation Monitor like the Tektronix A6901.
2. By using a power line isolation transformer on either the test circuit or on the scope.
3. By using an Isolation Amplifier like the Tektronix A6902B,
4. By using differential probes (see Differential Measurements).

NOTE: Never defeat the safety 3-wire ground system on either the scope or on the test circuit.

Do not "float" the scope, except by using an approved isolation transformer, or preferably, by using the Tektronix A6901 Ground Isolation Monitor.

The A6901 automatically reconnects the ground if scope ground voltages exceed ±40 V.

Induced noise in probe ground leads. The typical probe ground lead resembles a single-turn loop antenna when it is connected to the test circuit.

The relatively low impedance of the test circuit can couple any induced voltages into the probe, as shown in Figure 3-8.

High speed logic circuits can produce significant electromagnetic (radiated) noise at close quarters.

If the probe ground lead is positioned too close to certain areas on the board, interference signals could be picked up by the loop antenna formed by the probe ground lead, and mix with the probe tip signal.

Question: Is this what my signal really looks like? Moving the probe ground lead around will help identify the problem.

If the **noise level** changes, you have a ground lead induced noise problem.

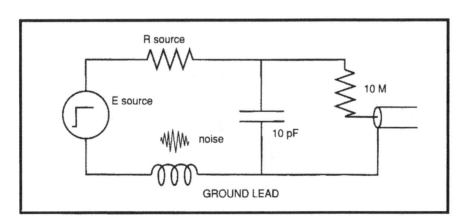

Figure 3-8 • Equivalent Circuit. Ground Lead Induced Noise

A more positive way of identification is to disconnect the probe from the signal source and clip the ground lead to the probe tip.

Now use the probe/ground lead as a loop antenna and search the board for radiated noise.

Figure 3-9 shows what can be found on a logic board, **with the probe tip shorted to the ground lead**.

This is radiated noise, induced in the single-turn loop and fed to the probe tip.

The significance of any induced or injected noise increases with reduced working signal levels, because the signal to noise ratio will be degraded. This is especially true with ECL, where signal levels are 1V or less.

Prevention: If possible, use an ECB to Probe Tip Adaptor (test point). If not, use a Z Lead or short flexible ground lead.

Also, bunch the ground leads together to make the loop area as small as possible.

Bias offset probes. A Bias/Offset probe is a special kind of Low Z design with the capability of providing a variable bias or offset voltage at the probe tip.

Bias/Offset probes like the Tektronix P6230 and P6231 are useful for probing high speed ECL circuitry, where resistive loading could upset the operating point.

They are also useful for probing higher amplitude signals (up to ±5 V), where resistive loading could affect the DC level at some point on the waveform.

Bias/Offset probes are designed with a tip resistance of 450Ω (10X). When these probes are connected into a 50Ω environment, this loading results in a 10% reduction in peak to peak source amplitude. This round-figure loading is more convenient to handle than that produced by a standard 500Ω (10X) Low Z probe, which would work out at 9.09% under the same conditions.

It is important to note that bias/offset probes always present a 450Ω resistive load to the source, regardless of the bias/offset voltage selected.

The difference between bias/offset and standard Low Z probes lies in their ability to null current flow **at some specific and selectable point** on the input waveform (within ± 5V).

To see how bias/offset probes work, let's take a typical 10×500Ω Low Z probe and connect it in the circuit shown in Figure 3-10.

By taking a current flow approach we find that at one point on the waveform the source voltage is −4V, therefore the load current will be:

$$1 = ER = 4/Rs + Rp + R\,scope = 4/550 = 7.27\,mA$$

Therefore the voltage drop across the 50Ω source resistance (Rs) will be;

$$E = IR = 7.27 \times 10^{-3} \times 50 = 0.363V$$

And the measured pulse amplitude will be −4 −0.363 = 3.637V (E dut), or about 9% down from its unloaded state.

If we substitute the 500Ω Low Z probe with a 450Ω bias/offset probe, the circuit will look like Figure 3-11.

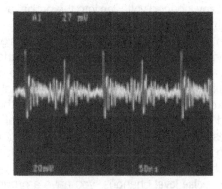

Figure 3-9 • Induced Noise In The Probe Ground Loop (Tip Shorted To The Ground Clip)

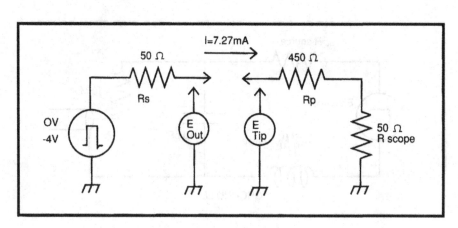

Figure 3-10 • Low Z 10X 500Ω Probe Connected To A 50Ω Source

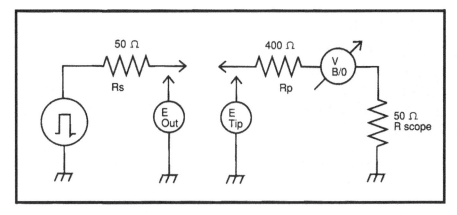

Figure 3-11 ● A 450Ω Bias/Offset Probe Connected To A 50Ω Source

With the bias/offset adjusted for 0V, the effect on the circuit will be similiar to a 500Ω Low Z probe, except for the small resistive change.

Figure 3-12 shows the source waveform acquired by a 10MΩ probe.

Figure 3-13 shows the effect on the waveform when the 450Ω probe is added.

As expected, the pulse amplitude has reduced from −4V to 3.60V, or exactly 10% down.

Figure 3-14 shows the effect of adjusting the offset to −4V. The −4V bias opposes the signal at the −4V level and results in zero current flow, and the source is effectively unloaded **at this point**.

However, when the signal returns to ground level, there is a 4V differential between the top of the pulse and the bias/offset source. Current will flow, and Ohms Law will dictate that the top of the pulse will go negative by −40mV (10%).

Sometimes it is desirable to adjust the offset midway between the peak to peak excursions. This distributes the effect of resistive loading between the two voltage swings.

Figure 3-15 shows the effect of adjusting the bias/offset to −2V. Current flow will be the same for both

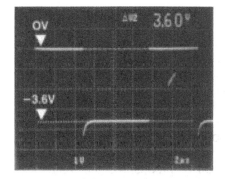

Figure 3-13 ● Effect of Connecting a 450Ω Bias/Offset Probe (Offset = 0V). Minus Level Has Been Reduced by 10%

signal swings, and they will be equally down by 5%, for a total of 10%.

Summary:
1. Bias/Offset probes can be adjusted (within ± 5V) to provide zero resistive (effective) loading at one selected point on the input waveform.
2. Bias/Offset probes can be used to simulate the effect of pull-up or pull-down voltages (within

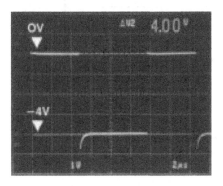

Figure 3-12 ● Unloaded Negative-Going 4V Pulse Acquired By a 10 MΩ Probe

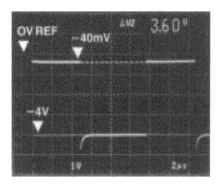

Figure 3-14 ● Bias/Offset Adjusted For − 4 V. Signal Current At The −4V Level is Zero. Current Flow At Ground Level Is Maximum. Peak to Peak Amplitude Remains The Same (10% Down)

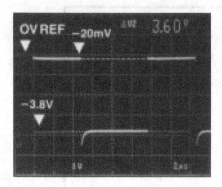

Figure 3-15 • Bias/Offset Adjusted for −2 V. Load Current Distributed Between The Negative And Positive-Going Swings. Peak to Peak Amplitude Remains the Same (10% Down).

± 5V) on the circuit under test (voltage source impedance is 450Ω).

3. Bias/Offset probes always present a total resistive load of 450Ω, and reduce the peak to peak amplitude of 50Ω sources by 10%.

4. For simplicity, we have ignored the effects of capacitive loading. Typically, Bias/Offset probes have less than 2pF tip C.

Bias/Offset probes like the Tektronix P6230 or P6231 have bandwidths to 1.5GHz, 450Ω, input R, and 1.3pF (P6230), or 1.6pF (P6231) input C.

They provide offset voltages of ± 5 V DC, and function with 1 MΩ or 50Ω input systems (P6231, 50Ω only).

The P6230 obtains operating power either from the scope itself, or from the Tektronix 1101A or 1102 Power Supply.

The P6231 is designed to operate with the Tektronix 11000 Series scopes, and obtains operating power and bias/offset from the scope. Offset is selectable from the mainframe touch screen.

Differential probing techniques. Accessing small signals elevated from ground, either at an AC level or a combination of AC and DC, requires the use of differential probes and a differential amplifier system.

One of the problems associated with differential measurements is the maintenance of high common mode rejection ratio (CMRR) at high common mode frequencies.

Poor common mode performance allows a significant portion of the common (elevated) voltage to appear across the differential probe's inputs. If the common mode voltage is pure DC, the result may only be a displayed baseline shift. However, if the common mode voltage is AC, or a combination of AC and DC, a significant portion may appear across the differential input and will mix with the desired signal.

Figure 3-16 shows the basic items necessary to make a differential measurement.

In this example two similar but unmatched passive probes are used. The probe ground leads are usually either removed or clipped together. They are **never** connected to the elevated DUT (device under test).

CMRR depends upon accurate matching of the probe-pair's electrical characteristics, including cable length. System CMRR can be no better than the differential amplifier's specifications, and in all cases, CMRR degrades as a function of frequency.

Figure 3-17 shows a simplified diagram of a DUT with a pulsed output of 1V$_{P-P}$ floating on a 5 V$_{P-P}$ 20MHz sinewave.

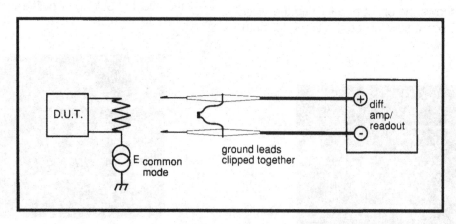

Figure 3-16 • Basic Connections To A Device Under Test To Make A Differential Measurement

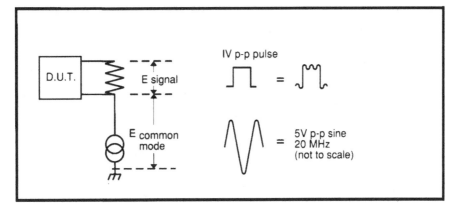

Figure 3-17 • Simplified Diagram. Elevated DUT. Common Mode Rejection Is 10:1 At 20MHz

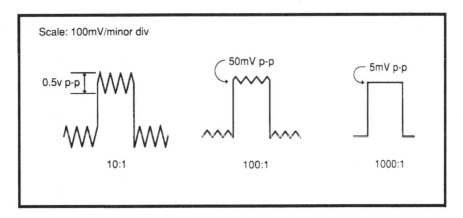

Figure 3-18 a, b and c • Displayed Waveforms From The Circuit Shown In Figure 3-17 At CMRR's of 10:1,100:1 and 1000:1

CMRR at 20MHz is a poor 10:1 because of the un-matched probes.

Observed signal, (referred to probe input) = $1V_{P-P}$ pulse + ($5V_{P-P}$ sine/10) = $1V_{P-P}$ pulse + $0.5V_{P-P}$ sine.

Figure 3-18a shows what the displayed wave-form might look like under the conditions shown in Figure 3-17.

In comparison, Figures 3-18b and 3-18c show what the displayed signal might look like at CMRR's of 100:1 and 1000:1.

Appendix B
Measuring amplifier settling time

High resolution measurement of amplifier settling at high speed is often necessary. Frequently, the amplifier is driven from a digital-to-analog converter (DAC). In particular, the time required for the DAC-amplifier com-bination to settle to final value after an input step is especially important. This specification allows setting a circuit's timing margins with confidence that the data produced is accurate. The settling time is the total length of time from input step application until the amplifier output remains within a specified error band around the final value.

Figure B1 shows one way to measure DAC amplifier settling time. The circuit uses the false sum node tech-nique. The resistors and amplifier form a summing net-work. The amplifier output will step positive when the DAC moves. During slew, the oscilloscope probe is bounded by the diodes, limiting voltage excursion. When settling occurs, the summing node is arranged so the oscilloscope probe voltage should be zero. Note that the resistor divider's attenuation means the probe's output will be one-half the actual settled voltage.

In theory, this circuit allows settling to be observed to small amplitudes. In practice, it cannot be relied upon to produce useful measurements. Several flaws exist. The oscilloscope connection presents problems. As probe capacitance rises, AC loading of the resistor junction will influence observed settling waveforms. The 20pF probe shown alleviates this problem but its 10X attenuation sacrifices oscilloscope gain. 1X

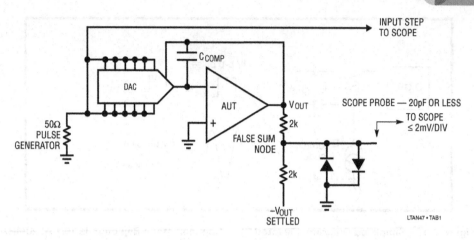

Figure B1 • One (Not Very Good) Way to Measure DAC-Op Amp Settling Time

probes are not suitable because of their excessive input capacitance. An active 1X FET probe might work, but another issue remains.

The clamp diodes at the probe point are intended to reduce swing during amplifier slewing, preventing excessive oscilloscope overdrive. Unfortunately, osci-llo-scope overdrive recovery characteristics vary widely among different types and are not usually specified. The diodes' 600mV drop means the oscilloscope may see an unacceptable overload, bringing displayed results into question (for a discussion of oscilloscope overdrive considerations, see the tutorial section on oscilloscopes). With the oscilloscope set at 1mV per division, the diode bound allows a 600:1 overdrive. Schottky diodes can cut this in half, but this is still much more than any real-time vertical amplifier is designed to accommodate.[1] The oscilloscope's overload recovery will completely dominate the observed waveform and all measurements will be meaningless.

One way to achieve reliable settling time measure-ments is to clip the incoming waveform in *time*, as well as amplitude. If the oscilloscope is prevented from see-ing the waveform until settling is nearly complete, over-load is avoided. Doing this requires placing a switch at the settle circuit's output and controlling it with an input-triggered, variable delay. FET switches are not suitable because of their gate-source capacitance. This capacitance will allow gate drive artifacts to cor-rupt the oscilloscope display, producing confusing readings. In the worst case, gate drive transients will be large enough to induce overload, defeating the switch's purpose.

Figure B2 shows a way to implement the switch which largely eliminates these problems. This circuit allows settling within 1mV to be observed. The Schottky sampling bridge is the actual switch. The

bridge's inherent balance, combined with matched diodes and very high speed complementary bridge switching, yields a clean switched output. An output buffer stage unloads the settle node and drives the diode bridge.

The operation of the DAC-amplifier is as before. The additional circuitry provides the delayed switching function, eliminating oscilloscope overdrive. The settle node is buffered by A1, a unity gain broadband FET input buffer with 3pF input capacitance and 350MHz bandwidth. A1 drives the Schottky bridge. The pulse generator's output fires the 74123 one shot. The one shot is arranged to produce a delayed (controllable by the 20k potentiometer) pulse whose width (controllable by the 5k potentiometer) sets diode bridge on-time. If the delay is set appropriately, the oscilloscope will not see any input until settling is nearly complete, eliminat-ing overdrive. The sample window width is adjusted so that all remaining settling activity is observable. In this way the oscilloscope's output is reliable and meaning-ful data may be taken. The one shot's output is level shifted by the Q1-Q4 transistors, providing comple-mentary switching drive to the bridge. The actual switching transistors, Q1-Q2, are UHF types, permit-ting true differential bridge switching with less than 1ns of time skew.[2] The bridge's output may be observed directly (by oscilloscopes with adequate sensitivity) or A2 provides a times 10 amplified version. A2's gain of 20 (and the direct output's ÷2 scaling) derives from the 2k-2k settle node divider's attenuation. A third output, taken directly from A1, is also available. This output, which bypasses the entire switching circuitry, is designed to be monitored by a sampling oscilloscope. Sampling oscilloscopes are inherently immune to over-load.[3] As such, a good test of this settling time test

---

Note 1: See Reference 3 for history and wisdom about vertical amplifiers.

Note 2: The Q1-Q4 bridge switching scheme, a variant of one described in Reference 14, was developed at LTC by George Feliz.
Note 3: See References 7, 8 and 18.

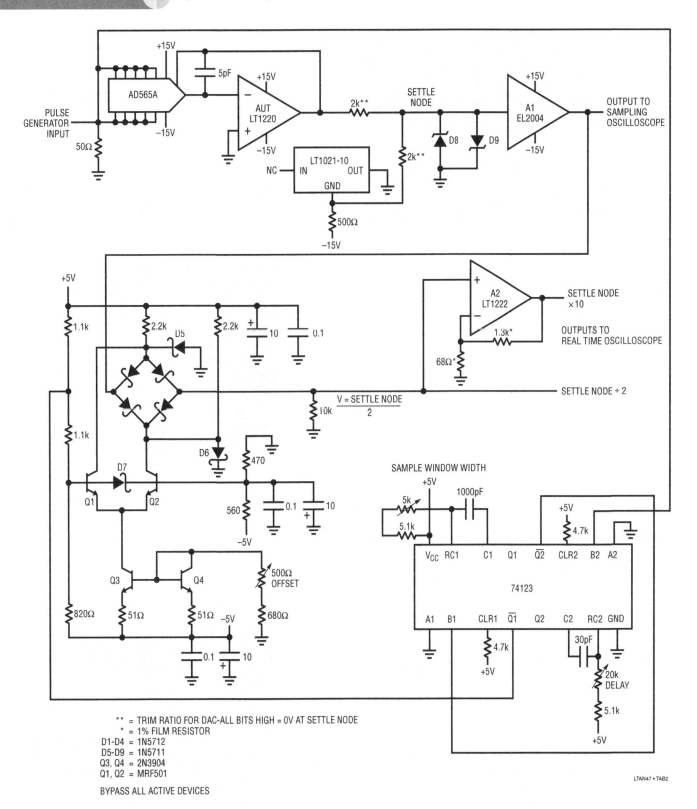

** = TRIM RATIO FOR DAC-ALL BITS HIGH = 0V AT SETTLE NODE
 * = 1% FILM RESISTOR
D1-D4 = 1N5712
D5-D9 = 1N5711
Q3, Q4 = 2N3904
Q1, Q2 = MRF501

BYPASS ALL ACTIVE DEVICES

LTAN47 • TAB2

Figure B2 • Settling Time Test Circuit Using a Sampling Bridge Eliminates Oscilloscope Overdrive

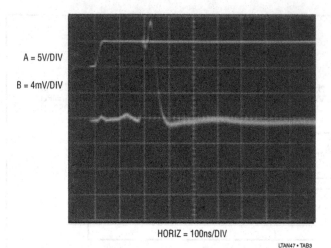

A = 5V/DIV

B = 4mV/DIV

HORIZ = 100ns/DIV

LTAN47 • TAB3

Figure B3 • 280ns Settling Time as Measured by Figure B2's Circuit. Sampling Switch Closes Just Before Third Vertical Division, Allowing Settling Detail to be Observed Without Overdriving the Oscilloscope

fixture (and the above statement) is to compare the signals displayed by the sampling 'scope and the Schottky bridge-aided real time 'scope. As an *additional* test, a completely different method of measuring settling time (albeit considerably more complex) described by Harvey[4] was also employed. If all three app-roaches represent good measurement technique and are constructed properly, results should be identical.[5] If this is the case, the identical data produced by the three methods has a high probability of being valid. Figures B3, B4 and B5 show settling time details of an AD565A DAC and an LT1220 op amp. The photos represent the sampling bridge, sampling 'scope and Harvey methods, respectively. Photos B3 and B5 display the input step for convenience in ascertaining elapsed time. Photo B4, taken with a single trace sampling oscilloscope (Tektronix 1S1 with P6032 cathode follower probe in a 556 mainframe) uses the leftmost vertical graticule line as its zero time reference. All methods agree on 280ns to 0.01% settling (1mV on a 10V step). Note that Harvey's method inherently adds 30ns, which must be subtracted from the displayed 310ns to get the real number.[6] Additionally, the shape of the settling waveform, in every detail, is identical in all three photographs. This kind of agreement provides a high degree of credibility to the measured results.

Some poorly designed amplifiers exhibit a substantial thermal tail after responding to an input step. This

phenomenon, due to die heating, can cause the output to wander outside desired limits long after settling has apparently occurred. After checking settling at high speed, it is always a good idea to slow the oscilloscope sweep down and look for thermal tails. Often the thermal tail's effect can be accentuated by loading the amplifier's output. Such a tail can make an amplifier appear to have settled in a much shorter time than it actually has.

To get the best possible settling time from any amplifier, the feedback capacitor, C_F, should be carefully chosen. C_F's purpose is to roll-off amplifier gain at

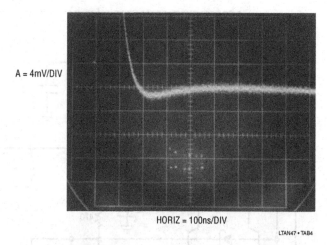

A = 4mV/DIV

HORIZ = 100ns/DIV

LTAN47 • TAB4

Figure B4 • 280ns Settling Time as Measured at Figure B2's Sampling Oscilloscope Output by a Sampling 'Scope. Settling Time and Waveform Shape is Identical to Figure B3

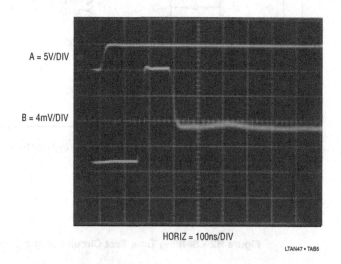

A = 5V/DIV

B = 4mV/DIV

HORIZ = 100ns/DIV

LTAN47 • TAB5

Figure B5 • 280ns Settling Time as Measured by Harvey's Method. After Subtraction of this Method's Inherent 30ns Delay, Settling Time and Waveform Shape are Identical to Figures B3 and B4

Note 4: See Reference 17.
Note 5: Construction details of the settling time fixture discussed here appear (literally) in Appendix F, "Additional comments on breadboarding". Also see the tutorial section on breadboarding techniques.
Note 6: See Reference 17.

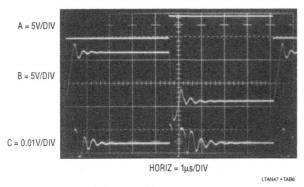

Figure B6 • No Feedback Capacitor

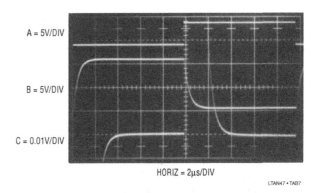

Figure B7 • Relatively Large Feedback Capacitor

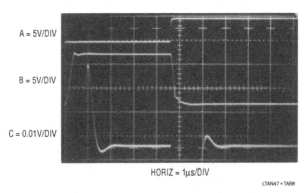

Figure B8 • Reduced Feedback Capacitor Gives Fastest Settling

Figures B6-B8 • Effects of Different Feedback Capacitors on a DAC-Op Amp Combination

the frequency which permits best dynamic response. The optimum value for C_F will depend on the feedback resistor's value and the characteristics of the source. DAC's are one of the most common sources and also one of the most difficult. DAC's current outputs must often be converted to a voltage. Although an op amp can easily do this, care is required to obtain good dynamic performance. A fast DAC can settle to 0.01% in 200ns or less but its output also includes a parasitic capacitance term, making the amplifier's job more difficult. Normally, the DAC's current output is unloaded directly into the amplifier's summing junction, placing the parasitic capacitance from ground to the amplifier's input. The capacitance introduces feedback phase shift at high frequencies, forcing the amplifier to hunt and ring about the final value before settling. Different DACs have different values of output capacitance. CMOS DACs have the highest output capacitance and it varies with code. Bipolar DACs typically have 20pF-30pF of capacitance, stable over all codes. Because of their output capacitance, DACs furnish an instructive example in amplifier compensation. Figure B6 shows the response of an industry standard DAC-80 type and a relatively slow (for this publication)

op amp. Trace A is the input, while Traces B and C are the amplifier and settle outputs, respectively. In this example no compensation capacitor is used and the amplifier rings badly before settling. In Figure B7, an 82pF unit stops the ringing and settling time goes down to 4μs. The overdamped response means that C_F dominates the capacitance at the AUT's input and stability is assured. If fastest response is desired, C_F must be reduced. Figure B8 shows critically damped behavior obtained with a 22pF unit. The settling time of 2μs is the best obtainable for this DAC-amplifier combination. Higher speed is possible with faster amplifiers and DACs but the compensation issues remain the same.

Appendix C
The oscillation problem — frequency compensation without tears

All feedback systems have the propensity to oscillate. Basic theory tells us that gain and phase shift are required to build an oscillator. Unfortunately, feedback systems, such as operational amplifiers, have gain and phase shift. The close relationship between oscillators and feedback amplifiers requires careful attention when an op amp is designed. In particular, excessive input-to-output phase shift can cause the amplifier to oscillate when feedback is applied. Further, any time delay placed in the amplifier's feedback path introduces additional phase shift, increasing the likelihood of oscillation. This is why feedback loop enclosed stages can cause oscillation.

A large body of complex mathematics is available which describes stability criteria and can be used to predict stability characteristics of feedback amplifiers. For the most sophisticated applications, this approach is required to achieve optimum performance.

However, little has appeared which discusses, in practical terms, how to understand and address the issues of compensating feedback amplifiers. Specifically, a practical approach to stabilizing amplifier-power gain stage combinations is discussed here, although the considerations can be generalized to other feedback systems.

Oscillation problems in amplifier-power booster stage combinations fall into two broad categories; local and loop oscillations. *Local* oscillations can occur in the boost stage, but should not appear in the IC op amp, which presumably was debugged prior to sale. These oscillations are due to transistor parasitics, layout and circuit configuration-caused instabilities. They are usually relatively high in frequency, typically in the 0.5MHz to 100MHz range. Usually, local booster stage oscillations do not cause loop disruption. The major loop continues to function, but contains artifacts of the local oscillation. Text Figure 33.101, repeated here as Figure C1 for reader convenience, furnishes an instructive example. The Q1, Q2 emitter follower pair has reasonably high f_t. These devices will oscillate if driven

from a low impedance source (see insert, Figure C1 and References 43 and 44). The 100Ω resistor and the ferrite beads are included to make the op amps output look like a higher impedance to prevent problems. Q5 and Q6, also followers, have even higher f_t, but are driven from 330Ω sources, eliminating the problem. The photo in Figure C2 shows Figure C1 following an input with the 100Ω resistor and ferrite beads removed. Trace A is the input, while Trace B is the output. The resultant high frequency oscillation is typical of locally caused disturbances. Note that the major loop is functional, but the local oscillation corrupts the waveform.

Eliminating such local oscillations starts with device selection. Avoid high f_t transistors unless they are needed. When high frequency devices are in use, plan layout carefully. In very stubborn cases, it may be necessary to lightly bypass transistor junctions with small capacitors or RC networks. Circuits which use local

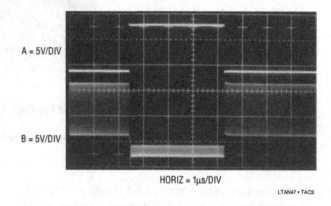

A = 5V/DIV

B = 5V/DIV

HORIZ = 1µs/DIV

LTAN47 • TAC2

Figure C2 • Local Oscillations Due to Booster Stage Instabilities

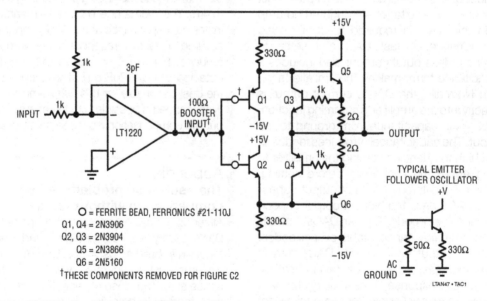

O = FERRITE BEAD, FERRONICS #21-110J
Q1, Q4 = 2N3906
Q2, Q3 = 2N3904
Q5 = 2N3866
Q6 = 2N5160
†THESE COMPONENTS REMOVED FOR FIGURE C2

LTAN47 • TAC1

Figure C1 • Figure 33.101's Booster Circuit with a Few Components Removed Begins Our Study of Loop Stability

feedback can sometimes require careful transistor selection and use. For example, transistors operating in a local loop may require different f_ts to achieve stability. Emitter followers are notorious sources of oscillation and should never be directly driven from low impedance sources (again, see References 43 and 44).

Loop oscillations are caused when the added gain stage supplies enough delay to force substantial phase shift. This causes the control amplifier to run too far out of phase with the gain stage. The control amplifier's gain combined with the added delay causes oscillation. Loop oscillations are usually relatively low in frequency, typically 10Hz–1MHz.

A good way to eliminate loop-caused oscillations is to limit the gain-bandwidth of the control amplifier. If the booster stage has higher gain-bandwidth than the control amplifier, its phase delay is easily accommodated in the loop. When control amplifier gain-bandwidth dominates, oscillation is assured. Under these conditions, the control amplifier hopelessly tries to servo a feedback signal which consistently arrives too late. The servo action takes the form of an electronic tail chase with oscillation centered around the ideal servo point.

Frequency response roll-off of the control amplifier will almost always cure loop oscillations. In many situations it is preferable to brute force compensation using large capacitors in the major feedback loop. As a general rule, it is wise to stabilize the loop by rolling off control amplifier gain-bandwidth. The feedback capacitor serves only to trim step response and should not be relied on to stop outright oscillation.

Figures C3 and C4 illustrate these issues. The LT1006 amplifier used with the LT1010 current buffer produces the output shown in Figure C4. As before, Trace A is the input and Trace B the output. The LT1006 has less than 1MHz gain-bandwidth. The LT1010's 20MHz gain-bandwidth introduces negligible loop delay, and dynamics are clean. In this case, the LT1006's internal roll-off is well below that of the output stage and stability is achieved with no external compensation components. Figure C5 uses a 100MHz bandwidth LT1223 as the control amplifier. The associated photo (Figure C6) shows the results. Here, the control amplifier's roll-off is well beyond the output stage's, causing problems. The phase shift through the LT1010 is now appreciable and oscillations occur. Stabilizing this circuit requires degenerating the control amplifier's gain-bandwidth.

The fact that the slower op amp circuit doesn't oscillate is a key to understanding how to compensate booster loops. With the slow device, compensation is free.

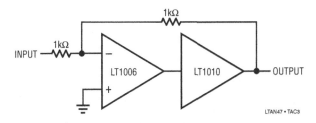

Figure C3 • A Slow Op Amp and a Medium Speed Booster

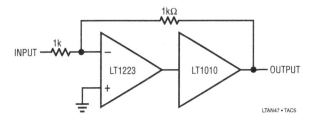

Figure C5 • A Fast Op Amp and a Medium Speed Booster

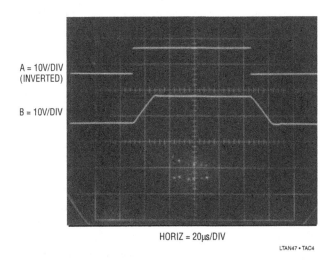

Figure C4 • Loop Stability is "Free" When the Op Amp is Much Slower than the Booster

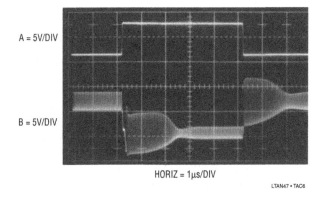

Figure C6 • Loop Oscillation is "Free" When the Op Amp is Much Faster than the Booster

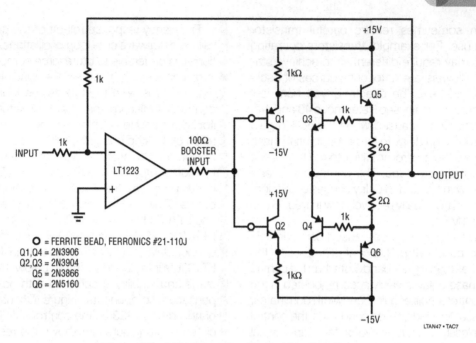

Figure C7 • A Very Fast Amplifier with a Fast Booster

The faster amplifier makes the AC characteristics of the output stage become significant and requires roll-off components for stability. Practically, the LT1223's speed is simply too much for the LT1010. A somewhat slower amplifier is the way to go. Alternately, a faster booster may be employed. Figure C7 attempts this, but doesn't quite make it. Photo C8 is less corrupted, but 100MHz oscillation indicates the booster stage (borrowed from text Figure 33.101) is still too slow for the LT1223. Attempts to use another booster design in Figure C9 (similarly purloined from text Figure 33.104) fail for the same reason. Figure C10 shows 40MHz oscillation, indicative of this high power booster's slower speed.

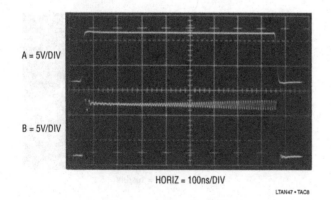

HORIZ = 100ns/DIV

LTAN47 • TAC8

Figure C8 • Figure C7's Booster is Not Quite Quick Enough to Prevent Loop Oscillation

Figure C11 has much more pleasant results. Here a 45MHz gain-bandwidth LT1220 has been substituted for the 100MHz LT1223 in Figure C7's circuit. The slower amplifier, combined with light local compensation, works well with the booster stage in its loop. Figure C12 shows a well controlled high speed output, nicely damped, with no sign of oscillations.

Power boosters are not the only things that can be placed within an amplifier's feedback loop. Text Figure 33.140's current source, reproduced here as Figure C13, is an interesting variation. There is no power booster in the loop, but rather a 40MHz differential amplifier with a gain of 10. To stabilize the circuit the slowest amplifier in the 1190 family, the 50MHz LT1190, is chosen. The local 100pF feedback slows it down a bit more and the loop is fast and stable (Figure C14). What happens if we remove the 100pF feedback path? Figure C15 shows that the loop is no longer stable under this condition because the LT1190 control amplifier cannot servo the phase shifted feedback at higher frequency. Put that 100pF capacitor back in!

It's worth mentioning that similar results to those obtained back in Figure C3 are obtainable by substituting a very slow control amplifier (e.g., an LT1006 which has less than 1MHz gain-bandwidth). The slower amplifier would give "free" compensation, eliminating the necessity for the 100pF unit. However, the circuit's frequency response would be severely degraded.

Text Figure 33.142's high power current source furnishes further instruction. This loop contains the

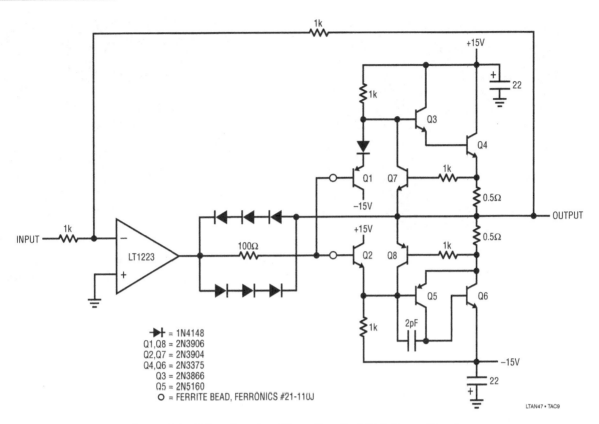

Figure C9 • A Very Fast Amplifier with a Fast High Power Booster

differential amplifier *and* a booster, seemingly making things even more difficult. Figure C16, recognizable as text Figure 33.142's high power current source with the 100pF local compensation removed, oscillates above 10MHz. Replacing the compensation restores proper response. Figure C17 shows the loop has no oscillations. What this tells us is that the control amplifier doesn't care just what generates the causal feedback between its input and output, so long as there isn't excessive delay. This circuit has a fairly busy feedback loop, but the control amplifier is oblivious to its bustling nature... unless you leave that 100pF feedback capacitor out!

When compensating loops like these, remember to investigate the effects of various loads and operating conditions. Sometimes a compensation scheme which appears fine gives bad results for some conditions. For this reason, check the completed circuit over as wide a variety of operating conditions as possible.

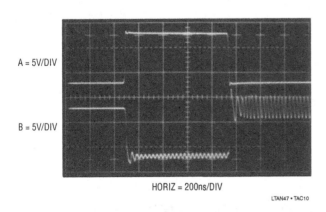

A = 5V/DIV

B = 5V/DIV

HORIZ = 200ns/DIV

Figure C10 • C9's High Power Booster is Fast But Causes Loop Oscillations

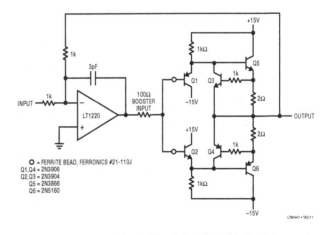

Figure C11 • Figure 33.101 (Again) with 100Ω Resistor and Beads Reinstalled

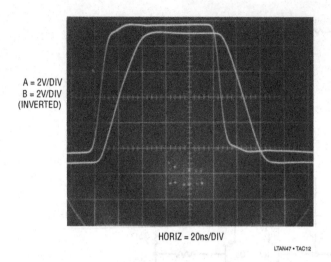

A = 2V/DIV
B = 2V/DIV
(INVERTED)

HORIZ = 20ns/DIV

LTAN47 • TAC12

Figure C12 • Lovely!

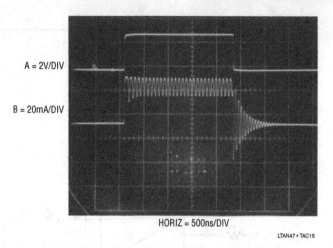

A = 2V/DIV

B = 20mA/DIV

HORIZ = 500ns/DIV

LTAN47 • TAC15

Figure C15 • Removing the 100pF Capacitor Allows the Op Amp to See Phase Shifted Feedback, Causing Oscillation. Put that 100pF Back In!

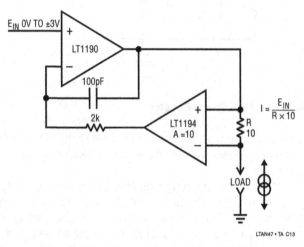

E_{IN} 0V TO ±3V

LT1190

100pF

2k

LT1194
A =10

+

R
10

$I = \dfrac{E_{IN}}{R \times 10}$

LOAD

LTAN47 • TA C13

Figure C13 • Figure 33.140's Current Source. What Do the RC Components Do?

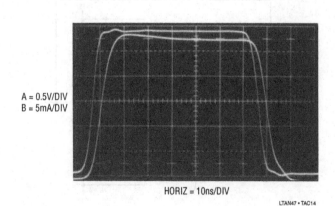

A = 0.5V/DIV
B = 5mA/DIV

HORIZ = 10ns/DIV

LTAN47 • TAC14

Figure C14 • Response of the Current Source with the RC Components in Place

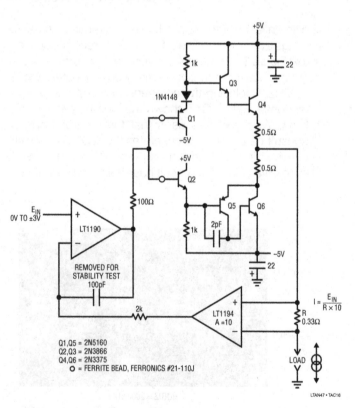

+5V

1k

Q3

1N4148

Q1

Q4

0.5Ω

−5V

+5V

Q2

0.5Ω

Q5 Q6

E_{IN}
0V TO ±3V

LT1190

100Ω

1k

2pF

−5V

22

REMOVED FOR
STABILITY TEST
100pF

2k

LT1194
A =10

+

R
0.33Ω

$I = \dfrac{E_{IN}}{R \times 10}$

LOAD

Q1,Q5 = 2N5160
Q2,Q3 = 2N3866
Q4,Q6 = 2N3375
O = FERRITE BEAD, FERRONICS #21-110J

LTAN47 • TAC16

Figure C16 • Text Figure 33.142's High Power Current Source. When the 100pF Capacitor is Removed, 10MHz Loop Oscillations Result

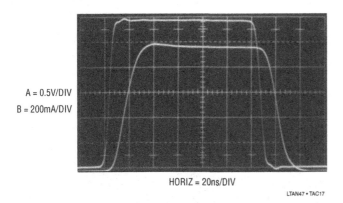

A = 0.5V/DIV
B = 200mA/DIV

HORIZ = 20ns/DIV

LTAN47 • TAC17

Figure C17 • Much Better. Leave that 100pF Capacitor in There!

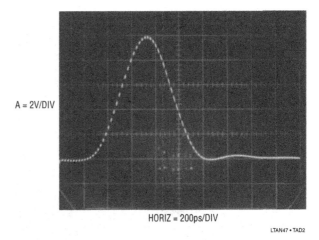

A = 2V/DIV

HORIZ = 200ps/DIV

LTAN47 • TAD2

Figure D2 • The Avalanche Pulse Generator's Output Monitored on a 1GHz Sampling Oscilloscope

Appendix D
Measuring probe-oscilloscope response

Verifying the rise time limit of wideband test equipment set-ups is a difficult task. In particular, the end-to-end rise time of oscilloscope-probe combinations is often required to assure measurement integrity. Conceptually, a pulse generator with rise times substantially faster than the oscilloscope-probe combination can provide this information. Figure D1's circuit does this, providing a 1ns pulse with rise and fall times inside 350ps. Pulse amplitude is 10V with a 50Ω source impedance. This circuit, built into a small box and powered by a 1.5V battery, provides a simple, convenient way to verify the rise time capability of almost any oscilloscope-probe combination.

The LT1073 switching regulator and associated components supply the necessary high voltage. The LT1073 forms a flyback voltage boost regulator. Further voltage step-up is obtained from a diode-capacitor voltage step-up network. L1 periodically receives charge and its flyback discharge delivers high voltage events to the step-up network. A portion of the step-up network's DC output is fed back to the LT1073 via the 10M, 24k divider, closing a control loop.

The regulator's 90V output is applied to Q1 via the 1M- 2pF combination. Q1, a 40V breakdown device,

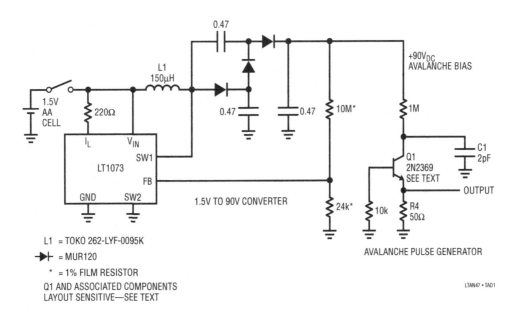

L1 = TOKO 262-LYF-0095K
➤⊢ = MUR120
 * = 1% FILM RESISTOR
Q1 AND ASSOCIATED COMPONENTS
LAYOUT SENSITIVE—SEE TEXT

LTAN47 • TAD1

Figure D1 • 350ps Rise/Fall Time Avalanche Pulse Generator

LTAN47 • TAD3

Figure D3 • Details of the Avalanche Pulse Generator's Head. 90V_DC Enters at Lower Right BNC, Pulse Exits at Top Left BNC. Note Short Lead Lengths Associated with Output

non-destructively avalanches when C1 charges high enough. The result is a quickly rising, very fast pulse across R4. C1 discharges, Q1's collector voltage falls and breakdown ceases. C1 then recharges until breakdown again occurs. This action causes free running oscillation at about 200kHz.[1,2] Figure D2 shows the output pulse. A 1GHz sampling oscilloscope (Tektronix 556 with 1S1 sampling plug-in) measures the pulse at 10V high with about a 1ns base. Rise time is 350ps, with fall time also indicating 350ps. There is a slight hint of ring after the falling edge, but it is well controlled. The figures may actually be faster, as the 1S1 is specified with a 350ps limit.[3]

Q1 may require selection to get avalanche behavior. Such behavior, while characteristic of the device specified, is not guaranteed by the manufacturer. A sample of 50 Motorola 2N2369s, spread over a 12 year date code span, yielded 82%. All good devices switched in less than 650ps. C1 is selected for a 10V amplitude output. Value spread is typically 2pF-4pF. Ground plane type construction with high speed layout techniques are essential for good. results from this circuit. Current drain from the 1.5V battery is about 5mA.

Figure D3 shows the physical construction of the actual generator. Power, supplied from a separate box, is fed into the generator's enclosure via a BNC connector. Q1 is mounted *directly* at the output BNC connector, with grounding and layout appropriate for wideband operation. Lead lengths, particularly Q1's and C1's, should be experimented with to get best output pulse purity. Figure D4 is the complete unit.

Note 1: This method of generating fast pulses borrows heavily from the Tektronix type 111 Pretrigger Pulse Generator. See References 8 and 25.
Note 2: This circuit replaces the tunnel diode based arrangement shown in AN13, Appendix D. While AN13's circuit works well, it generates a smaller, more irregularly shaped pulse and the tunnel diodes have become quite expensive.
Note 3: Just before going to press the pulse was measured at Hewlett-Packard Laboratories with a HP-54120B 12GHz sampling oscilloscope. Rise and fall times were 216ps and 232ps, respectively. Photo available on request.

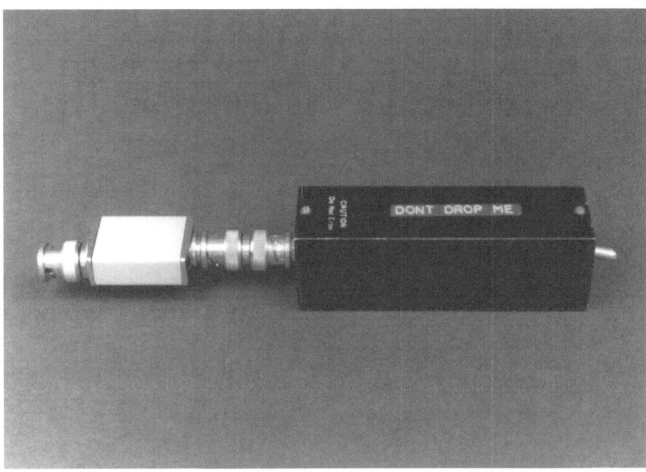

LTAN47 • TAD4

Figure D4 • The Packaged Avalanche Pulser. 1.5V-90V Converter is in the Black Box. Avalanche Head is at Left

Appendix E
An ultra-fast high impedance probe

Under most circumstances the 1pF-2pF input capacitance and 10MΩ resistance of FET probes is more than adequate for difficult probing situations. Occasionally, however, very high input resistance with high speed is needed. At some sacrifice in speed and input capacitance compared to commercial probes, it is possible to construct such a probe. Figure E1 shows schematic details. A1, a 350MHz hybrid FET buffer, forms the electrical core of the probe. This device is a low input capacitance, wideband FET source follower driving a fast bipolar output stage. The input of the probe goes to this device via a 51Ω resistor, reducing the possibility of oscillations in the follower input stage when the probe sees low AC impedance. A1's output drives a guard shield around the probe's input line, reducing effective input capacitance to about 4pF. A ground referred shield encircles the guard shield,

reducing pickup and making high quality ground connections to the circuit under test easy. A1 drives the output BNC cable to feed the oscilloscope. Normally, it is undesirable to back terminate the cable at A1 because the oscilloscope will see only half of A1's output. While a back termination provides the best signal dynamics, the resulting attenuation is a heavy penalty. The RC damper shown can be trimmed for best edge response while maintaining an unattenuated output.

What can't be seen in the schematic is the probe's physical construction. Very careful construction is required to maintain low input capacitance, low bias current and wide bandwidth. The probe head is particularly critical. Every effort should be made to minimize the length of wire between A1's input and the probe tip. In our lab, we have found that discarded pieces of broken 10× probes, particularly attenuator boxes and probe heads, provide an excellent packaging basis for

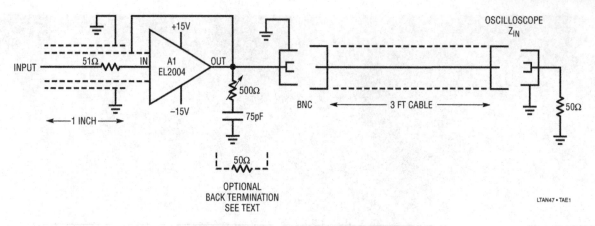

Figure E1 • Ultra Fast Buffer Probe Schematic

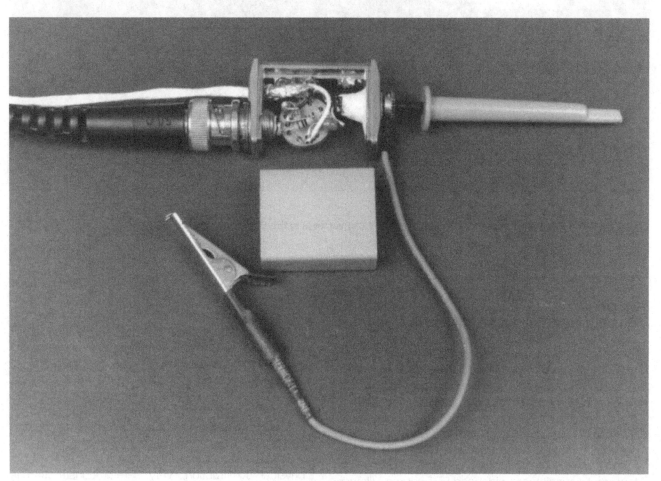

Figure E2 • Physical Layout of Ultra Fast Buffer Probe

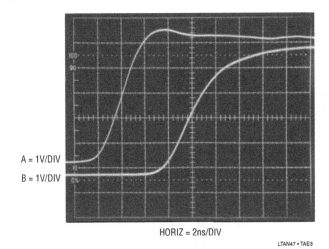

A = 1V/DIV

B = 1V/DIV

HORIZ = 2ns/DIV

LTAN47 • TAE3

Figure E3 • Probe Response (Trace B) to Input Pulse (Trace A)

this probe.[1] Figure E2 shows the probe head. Note the compact packaging. Additionally, A1's package is arranged so that it's (not insubstantial) dissipated heat is transferred to the probe case body when the snap-on cover (shown in photo) is in place. This reduces A1's

---

Note 1: This is not to encourage or even accept the breakage of probes. The author regards the breakage of oscilloscope probes as the lowest possible human activity. The sole exception to this condemnation is poor quality probes, which should be destroyed as soon as their deficiencies are discovered.

substrate temperature, keeping bias current down. A1's input is directly connected to the probe head to minimize parasitic capacitance. The power supply for A1, located in a separate enclosure, is fed in through separate wires. A1's output is delivered to the oscilloscope via conventional BNC hardware.

Figure E3 shows the probe output (Trace B) responding to an input (Trace A) as monitored on a 350MHz oscilloscope (Tektronix 485). Measured specifications for our version of this probe include a rise time of 6ns, 6ns delay and 58MHz bandwidth. The delay time contribution is about evenly split between the amplifier and cable. Input capacitance is about 4pF without the probe hook tip and 7pF with the hook tip. Input bias current measured 400pA and gain error about 5%. (A1 is an open loop device.)

Appendix F
Additional comments on breadboarding
This section contains, in visual form, commentary on some of the breadboards of the circuits described in the text. The breadboards appear in roughly corresponding order to their text presentation and comments are brief but hopefully helpful. The bit pushers have commented software; why not commented hardware?

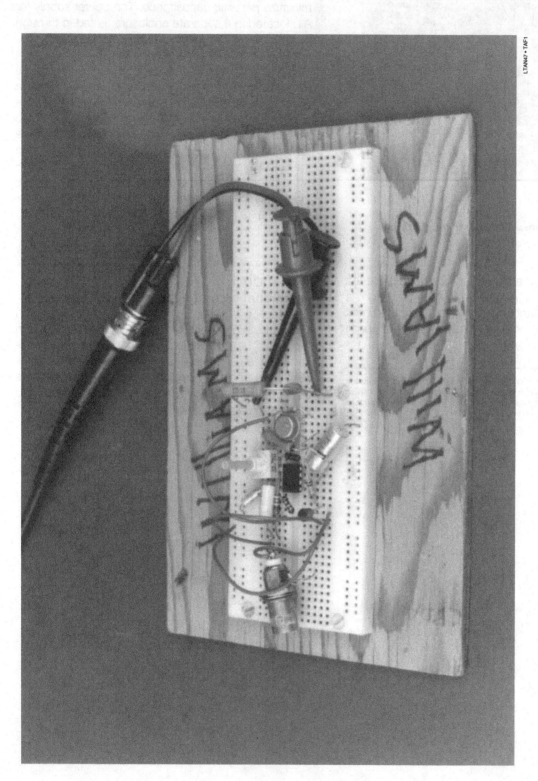

Figure F1 • No

Figure F2 • No

LTAN47 • TAF2

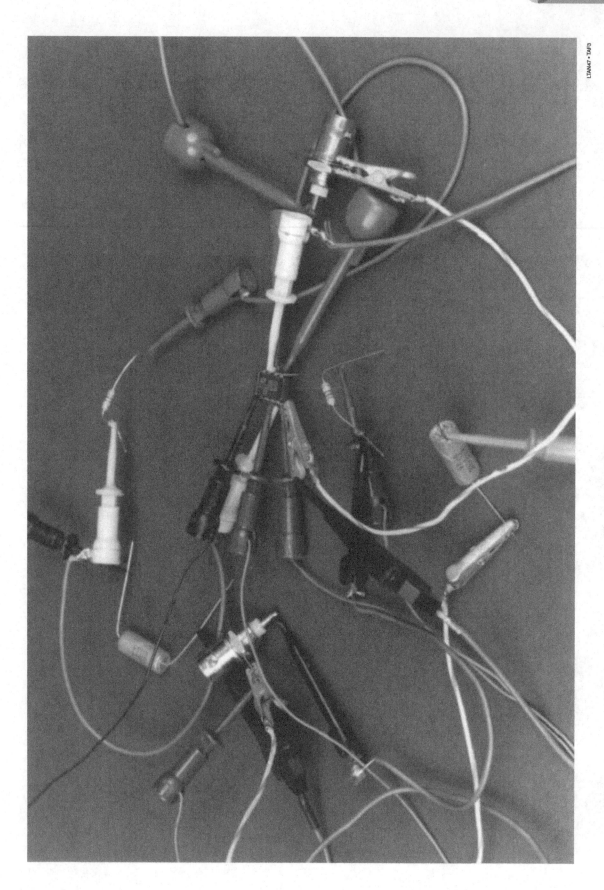

Figure F3 • No

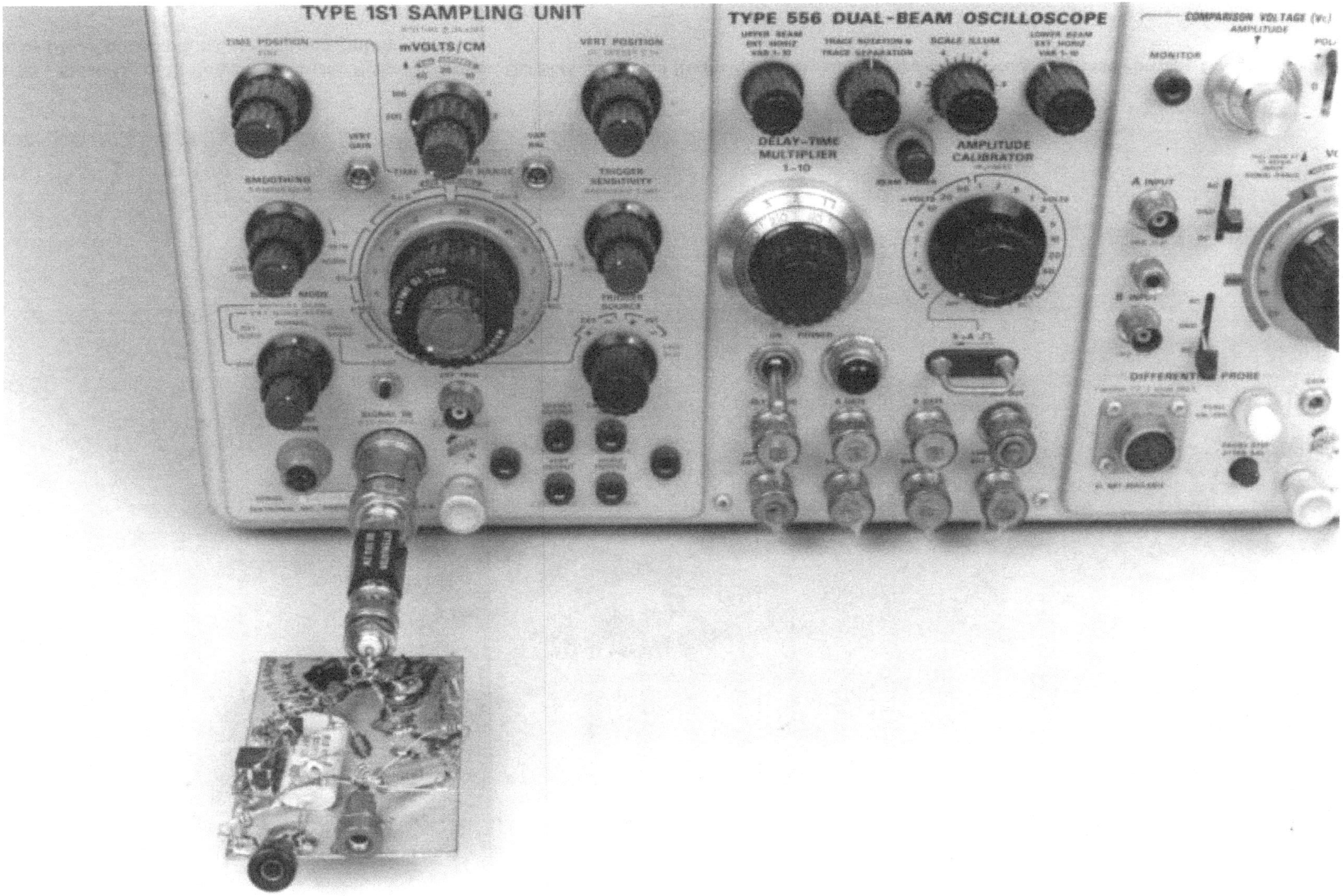

LTAN47 • TAF4

Figure F4 • Prototype Avalanche Pulser Under Test. Direct Connection to Oscilloscope Eliminates Cable or Probe Effects

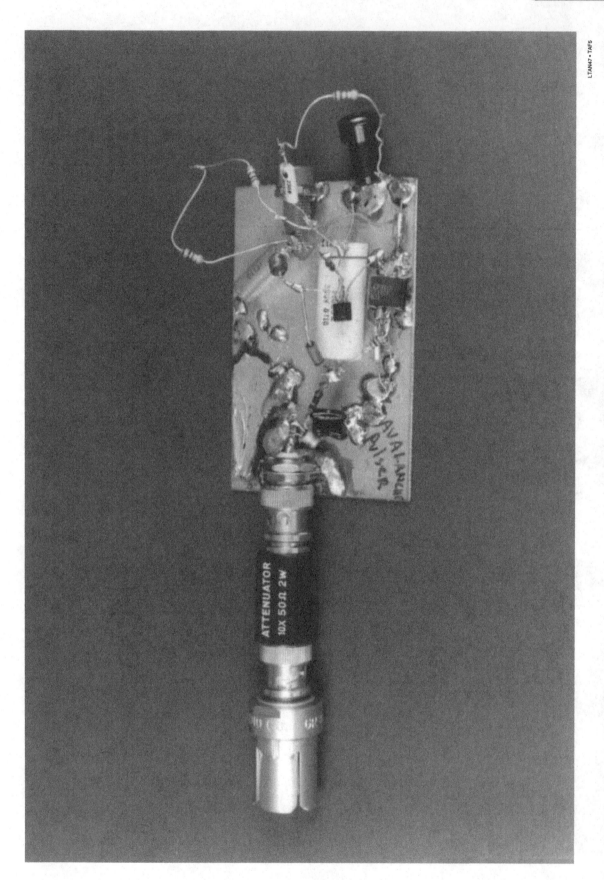

Figure F5 • Close-Up of Prototype Avalanche Pulse Generator. DC Bias Generator (Right Side of Board) is Carelessly Wired, but Pulse Forming Circuitry (Left Side of Board, by BNC Connector) is Carefully and Tightly Wired

Figure F6 ● The Settling Time Test Fixture Described in Appendix B. DAC and Amplifier are in Center Right of Photo. Note Break in Clad Separating Analog and Digital Grounds and Attention to Layout in Switching Bridge (Lower Left). Switching Bridge is Returned Separately to Ground — Its Board is Mechanically Stood-Off From Main Board by 10 MΩ Resistors. Output Section, Driving the Large P6032 Follower Probe, is at Lower Right

LTAN47 • TAF7

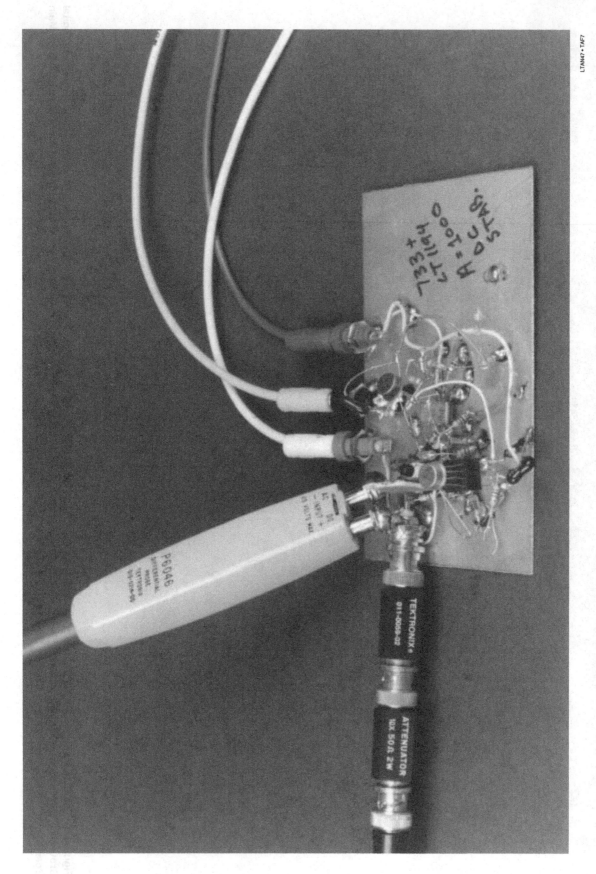

Figure F7 • The X1000 38MHz Differential Amplifier. DC and Low Frequency Electronics use Sockets (Foreground) and Air Wire Techniques (Center Right) for Easy and Fast Breadboarding. Wideband Circuitry Hugs the Ground Plane, and is Clustered Near the Input BNC

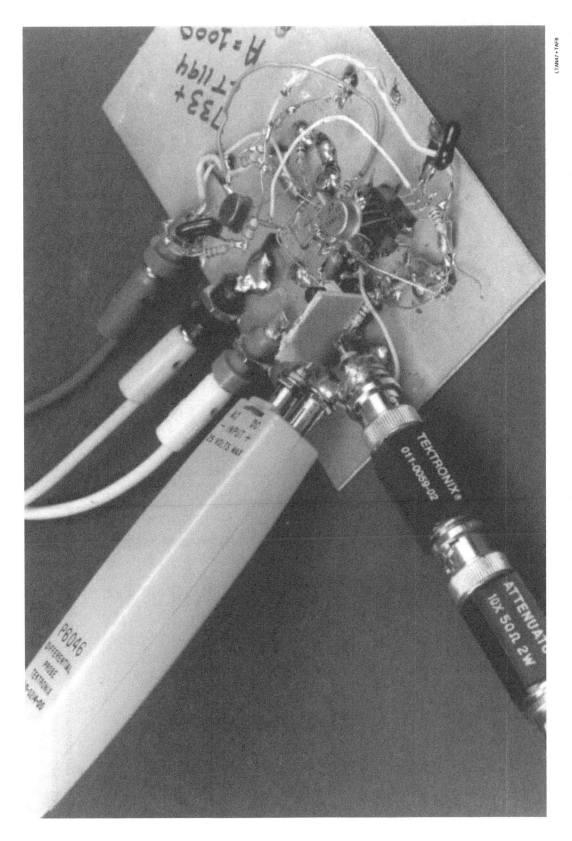

LTAN47 • TAF8

Figure F8 • Input Detail of X1000 Differential Amplifier. Clad Shield (Center Right) Prevents BNC Radiation from Corrupting Low Level Circuitry. Differential Probe Verifies Fidelity of 2.5mV Pulse Out of the X100 Attenuator Stacked Sections. Note DIP Packages Hugging Ground Plane, While Cans Operating at Low Frequency are Carelessly Wired

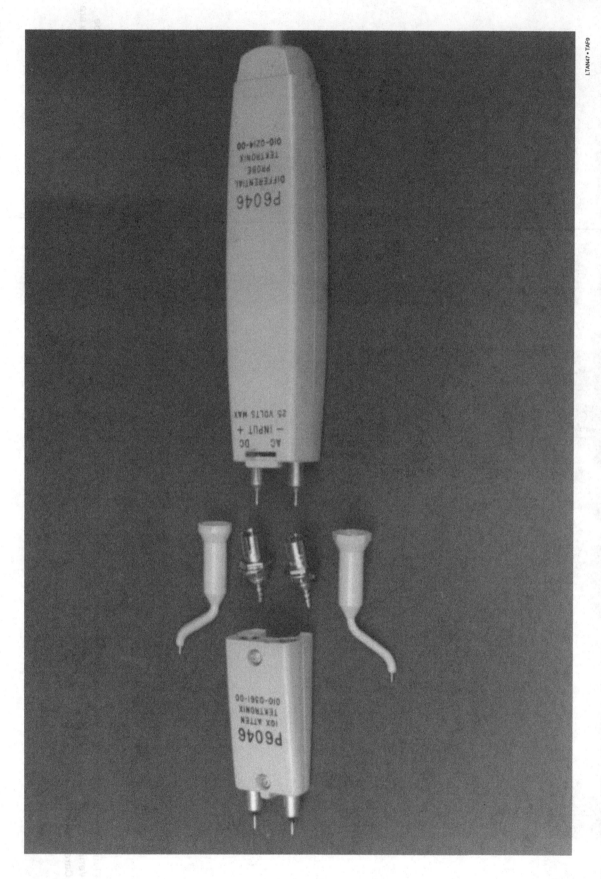

Figure F9 • The Differential Probe and Its 10X Attenuator. Offset Probe Tips are Convenient for Making Differential Connections, but Sockets Maintain a True Coaxial Environment and are Preferred

LTAN47 • TAF9

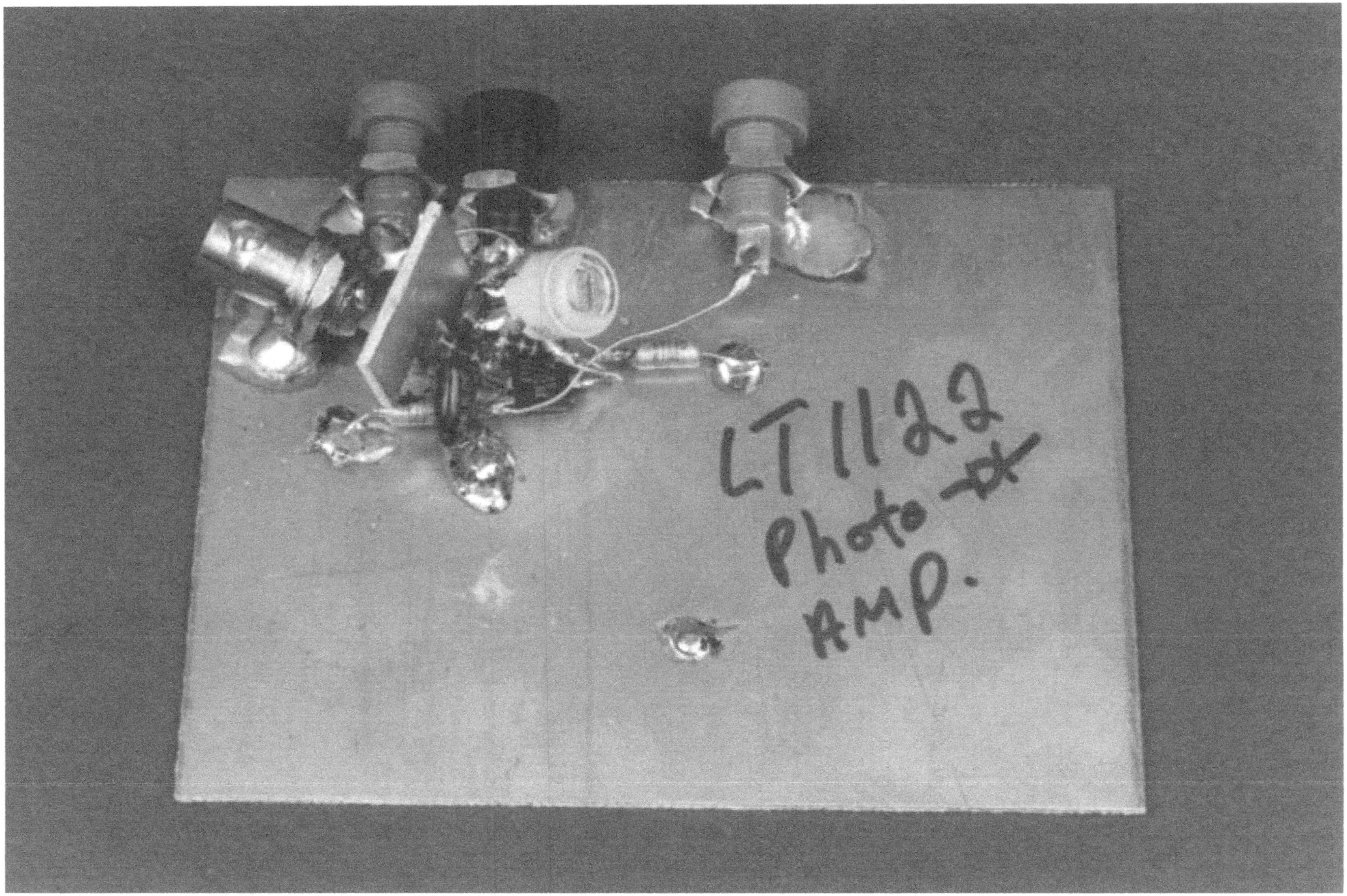

Figure F10 • The Photodiode Amplifier Layout Emphasizes Low Capacitance at Amplifier (Located Below Trimmer Capacitor, Photo Center Upper Left). Vertical Guard Shield Breaks Up BNC Radiation; was Used When Photo Input was Simulated with a Pulse Generator

LTAN47 • TAF10

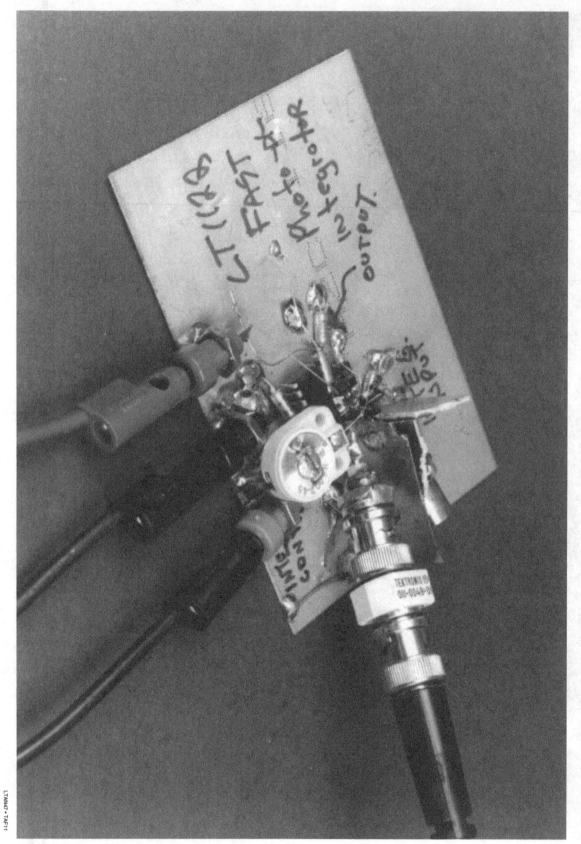

Figure F11 • Fast Photo Integrator Under Test with Pulse Generator Simulating a Photo Input. BNC Radiation is Controlled with Extensive Shielding at Integrator Input (Just Visible Upper Center). Control Input (Cable Connected BNC) is Less Critical; Does Not Require Shielding

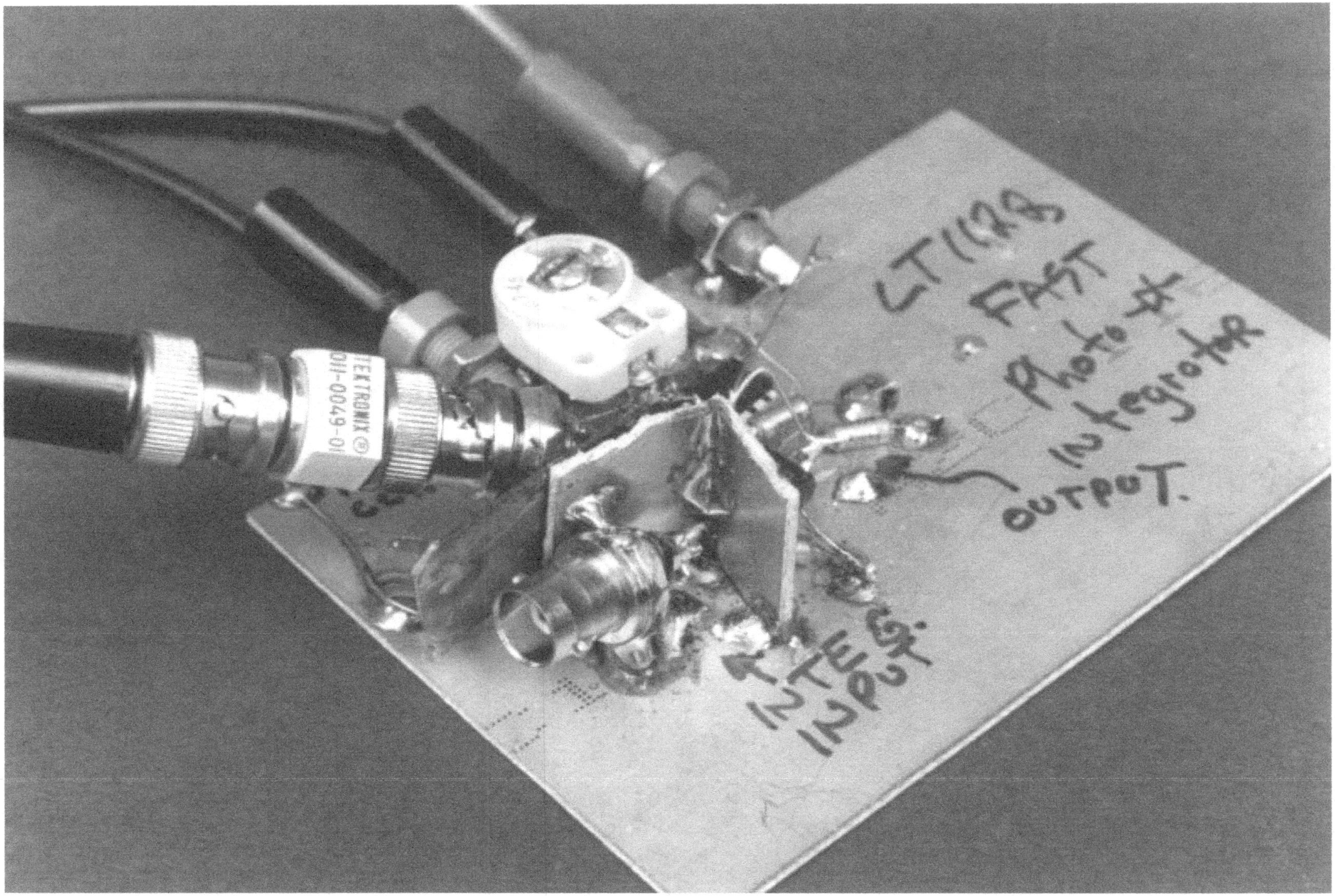

Figure F12 • Photo Integrator Details. Integrator Input BNC is Fully Shielded From Integrator Amp — 1pF Coupling From BNC Output to Summing Point will Cause Excessive Peaking. Amplifier and Switch ICs are Just Visible

LTAN47 • TAF12

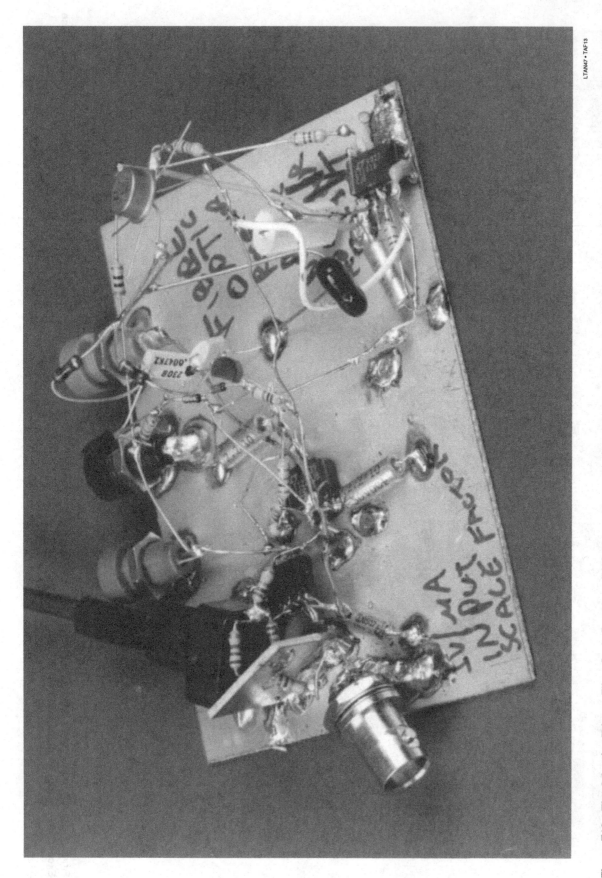

LTAN47 • TAF13

Figure F13 • The Adaptive Trigger Fiber Optic Receiver. BNC Photo-Simulation Input and Fiber Optic Line Both Connected. Low Frequency Wiring is Haphazardly Constructed While High Frequency Sections are Tight and Hug the Ground Plane. Note Vertical Shield at Photo-Simulation Input BNC

Figure F14 • Detail of the Fiber Optic Receiver's Photo-Simulation BNC Input. Resistor From BNC is Routed Through a Small Hole in Vertical Shield, Minimizing Capacitance. Another Resistor on the Shield's Other Side Divides Effects of Residual Capacitance to Keep Summing Point Clean

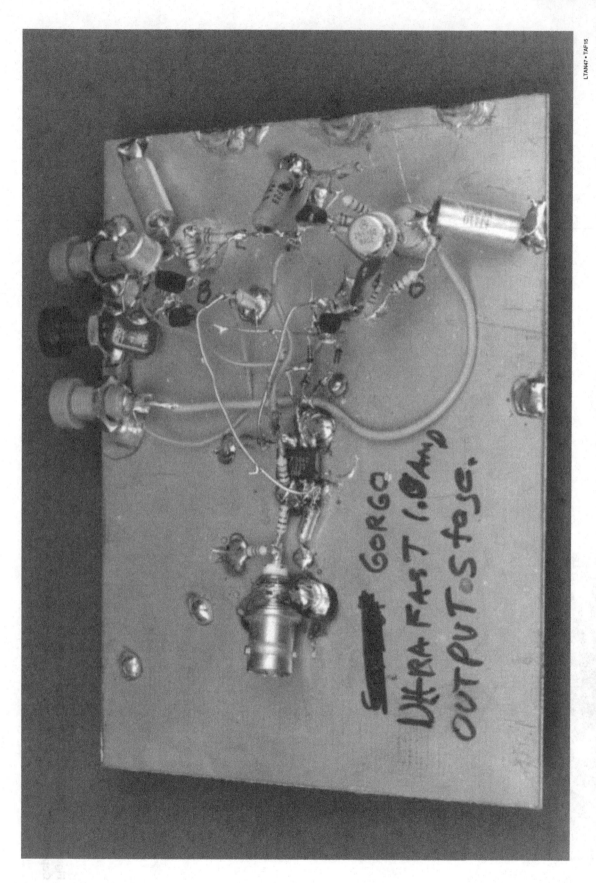

LTAN47 • TAF15

Figure F15 • The 1A Booster. Note Heavy Bypassing Right at the Output Power Transistors (Both Stud-Mounted to Clad). Local Compensation Capacitors (Right Side Upper and Lower) Have Short Leads

LTAN47 • TAF16

Figure F16 • 20MHz Sine Wave Crystal Oscillator. DC-AGC Section is at Lower Left, Oscillator is in Center. Control FET is Located at Oscillator Amplifier. Slow Gate Control Signal Arrives via Long-Leaded Resistor (Photo Center Upper Left)

LTAN67 • TAF17

Figure F17 • The Varactor Tuned Wien Bridge. DC-AGC Section is on Vertical Board — High Frequency Section Hugs the Ground Plane. Control FET (Center Left), Located at Oscillator, is Biased From a Long Line Originating on AGC Board. Note FET Gate Resistor is Located at FET, Not DC Board. Oscillator Output Receives Reverse Treatment

Figure F18 • The 1Hz-10MHz V to F Breadboard with Probes Attached. DC Servo Amplifier is Socketed, with Long Leads. Reference Section, Starting at Breadboard (Upper Right), Works Toward Reference Switch (Large DIP at Board Center). Note Very Tight Layout in Amplifier-Comparator Region (Board Left Center)

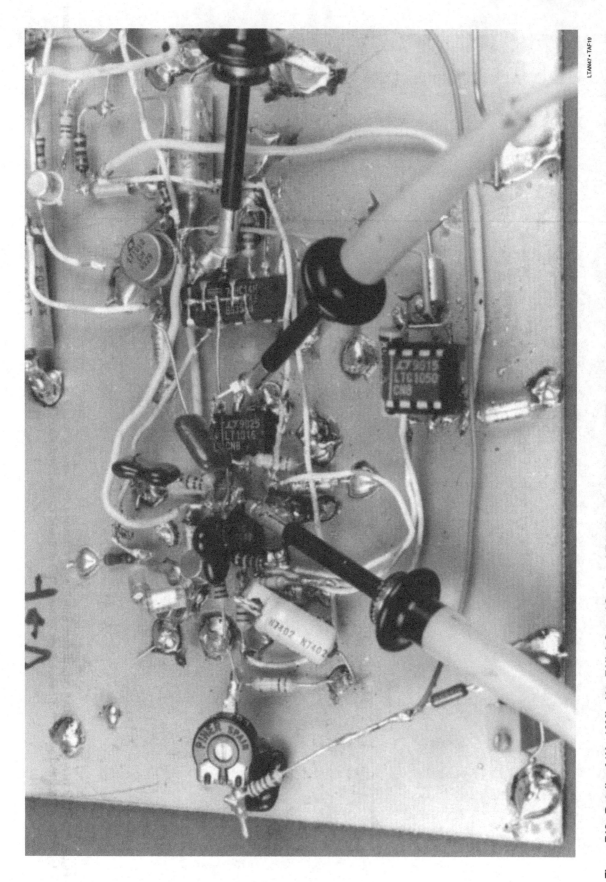

Figure F19 • Details of 1Hz-10Mhz V to F High Speed Section. LT1122 Integrator is Just Visible Under its Associated Discrete Components. Summing Point (Left Side of Amplifier) is Layout's Electrical Center. LT1016 Wiring is Also Very Tight Except for its Output Which Goes to Reference Switch. DC Servo Amplifier Sleeps in its Socket. Note Probe Tip Connectors

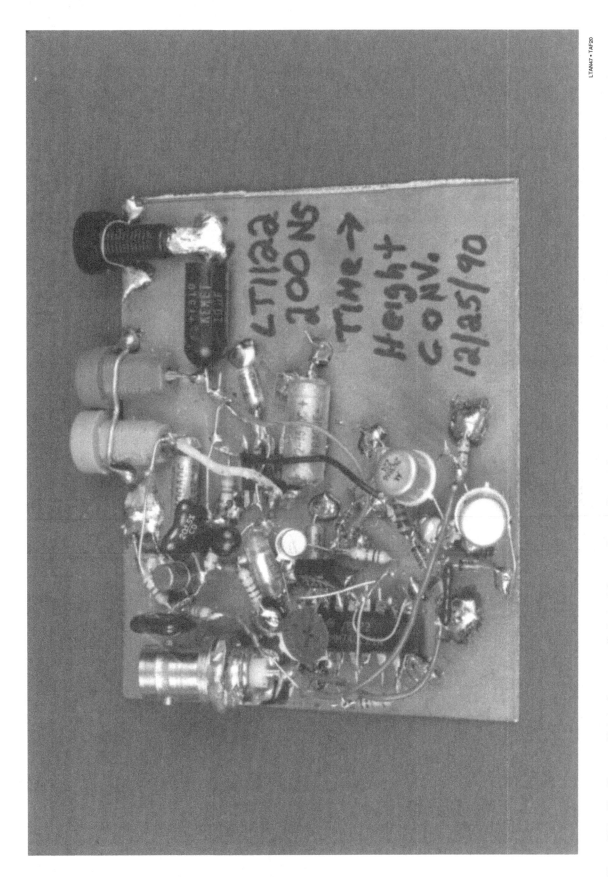

Figure F20 • The Time-to-Height Converter. Switched Current Source (Board Center Left), has Very Tight Layout. Follower Amplifier is at Board Upper Center. Major Components (in Order from Top to Bottom), Include Current Source Switch Transistor, Current Source Transistor (Black Case) Integrator Capacitor (Black Case) Integrator Capacitor (Silver) and Reset Transistor. Note Short Connection to Amplifier Input Pin

Figure F21 • The Automatic Trigger. Low Frequency Automatic Level Section is Spread Out (Right Side of Board). Wideband Circuitry Hugs Ground Plane and is Located Near Input BNC. Amplifier's Low Impedance, Fast Output Feeds LT1016 Output Comparator Over a Relatively Long Wire Run, Routed Through Insensitive Section of DC Circuitry

LTAN47 • TAF21

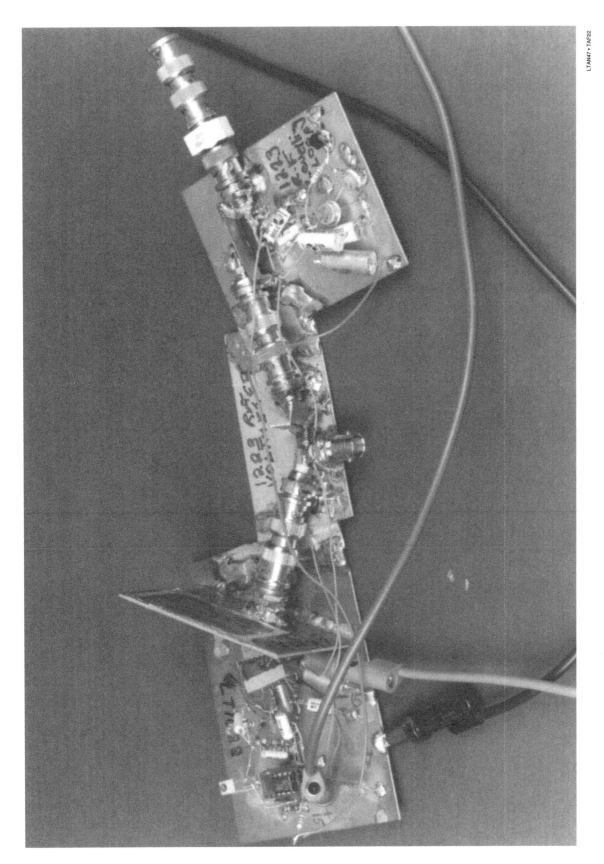

Figure F22 • The RF Leveling Loop. The RMS to DC Converter (Left Board) was Built First, Then the RF Pre-Amp (Center Board) and Finally the Multiplier-Servo Board. Each Board's Performance was Verified Before Joining Them. Note Copper Tape Maintaining Ground Plane Integrity Between Boards

LTAN47 • TAF22

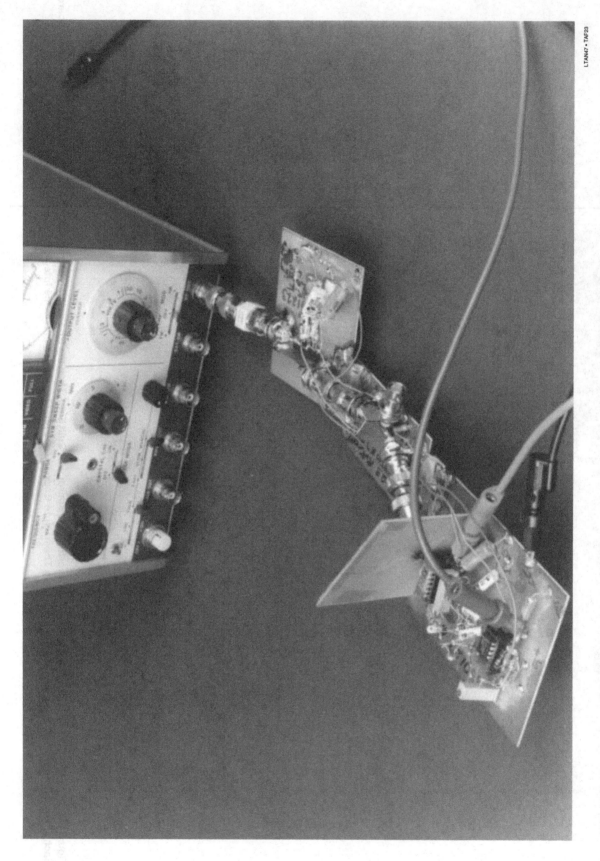

Figure F23 • RF Leveling Loop Attached Directly to the Test Generator. Cable Uncertainties are Eliminated Because There is No Cable

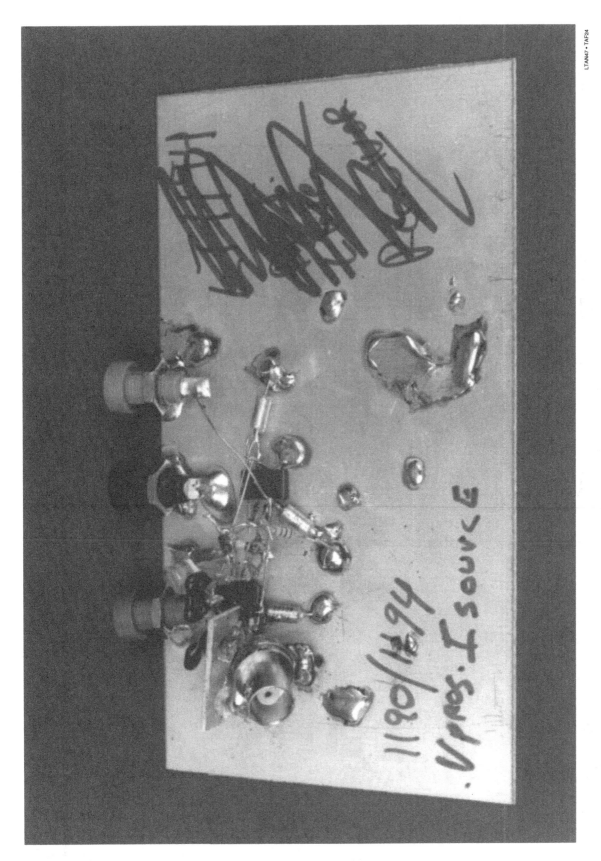

LTAN47 • TAF24

Figure F24 • The Voltage Controlled Current Source. Vertical Shield (Upper Left) Absorbs Input BNC Radiation. Amplifier-Loop Compensation Components are Located Directly at Amplifier (Behind Shield)

Figure F25 • The Good Life. The High Frequency Amplifier Demonstration Board Discussed in Appendix I. Sockets are a Compromise Between Best Performance and Flexibility

Appendix G

FCC licensing and construction permit applications for commerical AM broadcasting stations

In accordance with the application for Figure 33.116's circuit, and our law-abiding nature, find facsimiles of the appropriate FCC applications below. The complete forms are available by writing to:

Federal Communications Commission
Washington, D. C. 20554

Federal Communications Commission
Washington, D. C. 20554

FCC 301

Approved by OMB
3060-0027
Expires 2/28/92
See Page 25 for information regarding public burden estimate

APPLICATION FOR CONSTRUCTION PERMIT FOR COMMERCIAL BROADCAST STATION

For COMMISSION Fee Use Only	FEE NO:	For APPLICANT Fee Use Only
	FEE TYPE:	Is a fee submitted with this application? ☐ Yes ☐ No
	FEE AMT:	If fee exempt (see 47 C.F.R. Section 1.1112), indicate reason therefor (check one box): ☐ Noncommercial educational licensee ☐ Governmental entity
	ID SEQ:	FOR COMMISSION USE ONLY FILE NO.

Section I - GENERAL INFORMATION

1. Name of Applicant	Send notices and communications to the following person at the address below: Name
Street Address or P.O. Box	Street Address or P.O. Box
City / State / ZIP Code	City / State / ZIP Code
Telephone No. (Include Area Code)	Telephone No. (Include Area Code)

2. This application is for: ☐ AM ☐ FM ☐ TV

(a) Channel No. or Frequency

(b) Principal Community — City / State

(c) Check one of the following boxes:

☐ Application for NEW station

☐ MAJOR change in licensed facilities; call sign: ______

☐ MINOR change in licensed facilities; call sign: ______

☐ MAJOR modification of construction permit; call sign: ______

File No. of construction permit: ______

☐ MINOR modification of construction permit; call sign: ______

File No. of construction permit: ______

☐ AMENDMENT to pending application; Application file number: ______

NOTE: It is not necessary to use this form to amend a previously filed application. Should you do so, however, please submit only Section I and those other portions of the form that contain the amended information.

3. Is this application mutually exclusive with a renewal application? ☐ Yes ☐ No

If Yes, state:

Call letters	Community of License	
	City	State

G1

Federal Communications Commission
Washington, D.C. 20554

APPLICATION FOR NEW BROADCAST STATION LICENSE
(Carefully read instructions before filing out Form)
RETURN ONLY FORM TO FCC

Approved by OMB
3060-0029
Expires 9/30/90

For Commission Fee Use Only	FEE NO:	For Applicant Fee Use Only
	FEE TYPE:	Is a fee submitted with this application? ☐ Yes ☐ No
	FEE AMT:	If No, indicate reason therefor (check one box): ☐ Nonfeeable application
	ID SEQ:	Fee Exempt (See 47 C.F.R. Section 1.1112) ☐ Noncommercial educational licensee ☐ Governmental entity

SECTION I - GENERAL DATA

For Commission Use Only
File No.

Legal Name of Applicant	Mailing Address
	City / State / Zip Code
	Telephone No. (include area code)

1. Facilities authorized by construction permit

This application is for: ☐ Commercial ☐ Noncommercial

☐ AM Directional ☐ AM Non-Directional ☐ FM Directional ☐ FM Non-Directional ☐ TV

Call Letters	Community of License	Construction Permit File No.	Modification of Construction Permit File No(s).	Expiration Date of Last Construction Permit

2. Is the station now operating pursuant to automatic program test authority in accordance with 47 C.F.R. Section 73.1620? ☐ Yes ☐ No

If No, explain.

3. Have all the terms, conditions, and obligations set forth in the above described construction permit been fully met? ☐ Yes ☐ No

If No, state exceptions.

4. Apart from the changes already reported, has any cause or circumstance arisen since the grant of the underlying construction permit which would result in any statement or representation contained in the construction permit application to be now incorrect? ☐ Yes ☐ No

If Yes, explain.

5. Has the permittee filed its Ownership Report (FCC Form 323) or ownership certification in accordance with 47 C.F.R. Section 73.3615(b)? ☐ Yes ☐ No ☐ Does not apply

If No, explain.

G2

Figure G1-G2 • The FCC Forms Appropriate for Figure 33.116's Circuit

Appendix H
About current feedback

Contrary to some enthusiastic marketing claims, current feedback isn't new. In fact, it is much older than "normal" voltage feedback, which has been so popularized by op amps. The current feedback connection is at *least* 50 years old, and probably much older. William R. Hewlett used it in 1939 to construct his now famous sine wave oscillator.[1] "Cathode feedback" was widely applied in RF and wideband instrument design throughout the 30s, 40s and 50s. It was a favorite form of feedback, if for no other reason than there wasn't any place else left to feed back to!

In the early 1950s G.A. Philbrick Researches introduced the K2-W, the first commercially available packaged operational amplifier. This device, with its high impedance differential inputs, permitted the voltage type feedback so common today. Although low frequency instrumentation types were quick to utilize the increased utility afforded by high impedance feedback nodes, RF and wideband designers hardly noticed. They continued to use cathode feedback, called (what else?) emitter feedback in the new transistor form.

Numerous examples of the continued use of current feedback in RF and wideband instruments are found in designs dating from the 1950s to the present.[2] With

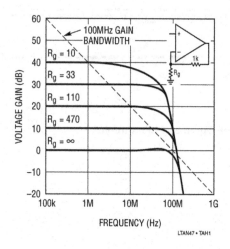

Figure H1 • Voltage Gain vs Frequency for Current Feedback Amplifier (Family of Curves) and a Conventional Voltage Amplifier (Straight Line)

Note 1: See Appendix C, "The Wien bridge and Mr. Hewlett", in Reference 19. See also References 20 through 24 and 46.
Note 2: See the "General Electric Transistor Manual", published by G.E. in 1964. See also operating and service manuals for the Hewlett-Packard 3400A RMS Voltmeter, 1120A FET probe, and the Tektronix P6042 current probe.

ostensibly easier to use voltage type feedback a reality during this period, particularly as monolithic devices became cheaper, why did discrete current feedback continue to be used? The reason for the continued popularity of current techniques was (and is) bandwidth. Current feedback is simply much faster. Additionally, within limits, a current feedback based amplifier's bandwidth does not degrade as closed loop gain is increased. This is a significant advantage over voltage feedback amplifiers, where bandwidth falls as closed loop gain is increased.

Relatively recently, current based designs have become available as general purpose, easy to use monolithic and hybrid devices. This brings high speed capability to a much wider audience, hopefully opening up new applications. So, while the technique is not new, marketing claims notwithstanding, the opportunity is. Although current based designs have poorer DC performance than voltage amplifiers, their bandwidth advantage is undeniable. What's the magic?

Current feedback basics
William H. Gross

The distinctions of how current feedback amplifiers differ from voltage feedback amplifiers are not obvious at first, because, from the outside, the differences can be subtle. Both amplifier types use a similar symbol, and can be applied on a first order basis using the same equations. However, their behavior in terms of gain bandwidth tradeoffs and large signal response is another story.

Unlike voltage feedback amplifiers, small signal bandwidth in a current feedback amplifier isn't a straight inverse function of closed loop gain, and large signal response is closer to ideal. Both benefits are because the feedback resistors determine the amount of current driving the amplifier's internal compensation capacitor. In fact, the amplifier's feedback resistor (R_f) from output to inverting input works with internal junction capacitances to set the closed loop bandwidth. Even though the gain set resistor (R_g) from inverting input to ground works with the R_f to set the voltage gain, just as in a voltage feedback op amp, the closed loop bandwidth does not change. The explanation of this is fairly straightforward. The equivalent gain bandwidth product of the CFA is set by the Thevenin equivalent resistance at the inverting input and the internal compensation capacitor. If R_f is held constant and gain changed with R_g, the Thevenin resistance changes by the same amount as the gain. From an overall loop standpoint, this change in feedback attenuation will produce a change in noise gain, and a proportionate reduction of open loop bandwidth (as in a conventional op amp). With current feedback, however, the key point is that changes in Thevenin resistance also produce compensatory changes in open loop bandwidth, unlike a conventional fixed gain bandwidth amplifier. As a result, the net closed loop bandwidth of a current-fed-back amplifier remains the same for various closed loop gains.

Figure H1 shows the LT1223 voltage gain vs frequency for five gain settings driving 100Ω. Shown for comparison is a plot of the fixed 100MHz gain bandwidth limitation that a voltage feedback amplifier would have. It is obvious that for gains greater than one, the LT1223 provides 3–20 times more bandwidth.

Because the feedback resistor determines the compensation of the LT1223, bandwidth and transient response can be optimized for almost every application. When operating on ±15V supplies, R_f should be

1kΩ or more for stability, but on ±5V, the minimum value is 680Ω, because the junction capacitors increase with lower voltage. For either case, larger feedback resistors can also be used, but will slow down the LT1223 (which may be desirable in some applications).

The LT1223 delivers excellent slew rate and bandwidth with better DC performance than previous current feedback amplifiers (CFAs). On ±15V supplies with a 1k feedback resistor, the small signal bandwidth is 100MHz into a 400Ω load and 75MHz into 100Ω. The input will follow slew rates of 250V/μs with the output generating over 500V/μs, and output slew rate is well over 1000V/μs for large input overdrive. Input offset voltage is 3mV (max), and input bias current is 3μA (max). A 10kΩ pot, connected to pins 1 and 5 with wiper to V^+, provides optional offset trimming. This trim

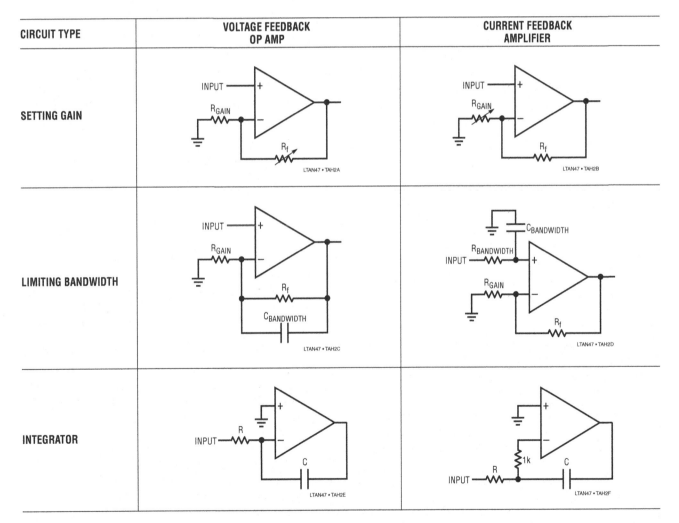

Figure H2 • Some Practical Differences in Applying Current and Voltage Amplifiers

shifts inverting input current about $\pm10\mu A$, effectively producing input voltage offset.

The LT1223 also has shutdown control, available at pin 8. Pulling more than $200\mu A$ from pin 8 drops the supply current to less than 3mA, and puts the output into a high impedance state. The easy way to force shutdown is to ground pin 8 using an open collector (drain) logic stage. An internal resistor limits current, allowing direct interfacing with no additional parts. When pin 8 is open, the LT1223 operates normally.

The difference in operating characteristics between op amps and CFAs result in slight differences in common circuit configurations. Figure H2 summarizes some popular circuit types, showing differences between op amps and CFAs. Gain can be set with either R_{IN} or R_f in an op amp, while a CFA's feedback resistor (R_f) is fixed. Op amp bandwidth is controllable with a feedback capacitor; for a CFA, bandwidth must be limited at the input. A feedback capacitor is never used. In an integrator, the 1k resistor must be included in the CFA so its negative input sees the optimal impedance. Finally, (not shown) there is no correlation between bias currents of a CFA's inputs. Because of this, source impedance matching will not improve DC accuracy. Matching input source impedances aids offset performance in op amps that do not have internal bias current cancellation.

Appendix I

High frequency amplifier evaluation board

LTC demo board 009 (photo, Figure I1, schematic, Figure I2) is designed to simplify the evaluation of high speed operational amplifiers. It includes both an inverting and non-inverting circuit, and extra holes are provided to allow the use of board-mounted BNC or SMA connectors. The two circuits are independent with the exception of shared power supply and ground connections.

Layout is a primary contributor to the performance of any high speed amplifier. Poor layout techniques adversely affect the behavior of a finished circuit. Several important layout techniques, all used in demo board 009, are described below:

1. Top side ground plane. The primary task of a ground plane is to lower the impedance of ground connections. The inductance between any two points on a uniform sheet of copper is less than the inductance of a thin, straight trace of copper connecting the same two points. The ground plane approximates the characteristics of a copper sheet

and lowers the impedance at key points in the circuit, such as the grounds of connectors and supply bypass capacitors.

2. Ground plane voids. Certain components and circuit nodes are very sensitive to stray capacitance. Two good examples are the summing node of the op amp and the feedback resistor. Voids are put in the ground plane in these areas to reduce stray ground capacitance.

3. Input/output matching. The width of the input and output traces is adjusted to a stripline impedance of 50Ω. Note that the terminating resistors (R3 and R7) are connected to the end of the input lines, not at the connector. While stripline techniques aren't absolutely necessary for the demo board, they are important on larger layouts where line lengths are longer. The short lines on the demo board can be terminated in 50Ω, 75Ω, or 93Ω without adversely affecting performance.

4. Separation of input and output grounds. Even though the ground plane exhibits a low impedance, input and output grounds are still separated. For example, the termination resistors (R3 and R7) and the gain-setting resistor (R1) are grounded in the vicinity of the input connector. Supply bypass capacitors (C1, C2, C4, C5, C7, C8, C9 and C10) are returned to ground in the vicinity of the output connectors.

The circuit board is designed to accommodate standard 8-pin miniDIP, single operational amplifiers such as the LT1190 and LT1220 families. Both voltage and current feedback types can be used. Pins 1, 5 and 8 are outfitted with extra holes for use in adjusting DC offsets, compensation, or, in the case of the LT1223 and LT1190/1/2, for shutting down the amplifier.

If a current feedback amplifier such as the LT1223 is being evaluated, omit C3/C6. R4 and R6 are included for impedance matching when driving low impedance lines. If the amplifier is supposed to drive the line directly, or if the load impedance is high, R4 and R8 can be replaced by jumpers. Similarly R10 and R12 can be used to establish a load at the output of the amplifier.

Low profile sockets may be used for the op amps to facilitate changing parts, but performance may be affected above 100MHz.

High speed operational amplifiers work best when their supply pins are bypassed with R_f-quality capacitors. C1, C5, C8 and C10 should be 10nF disc ceramic or other capacitors with a self-resonant frequency greater than 10MHz. The polarized capacitors (C2,

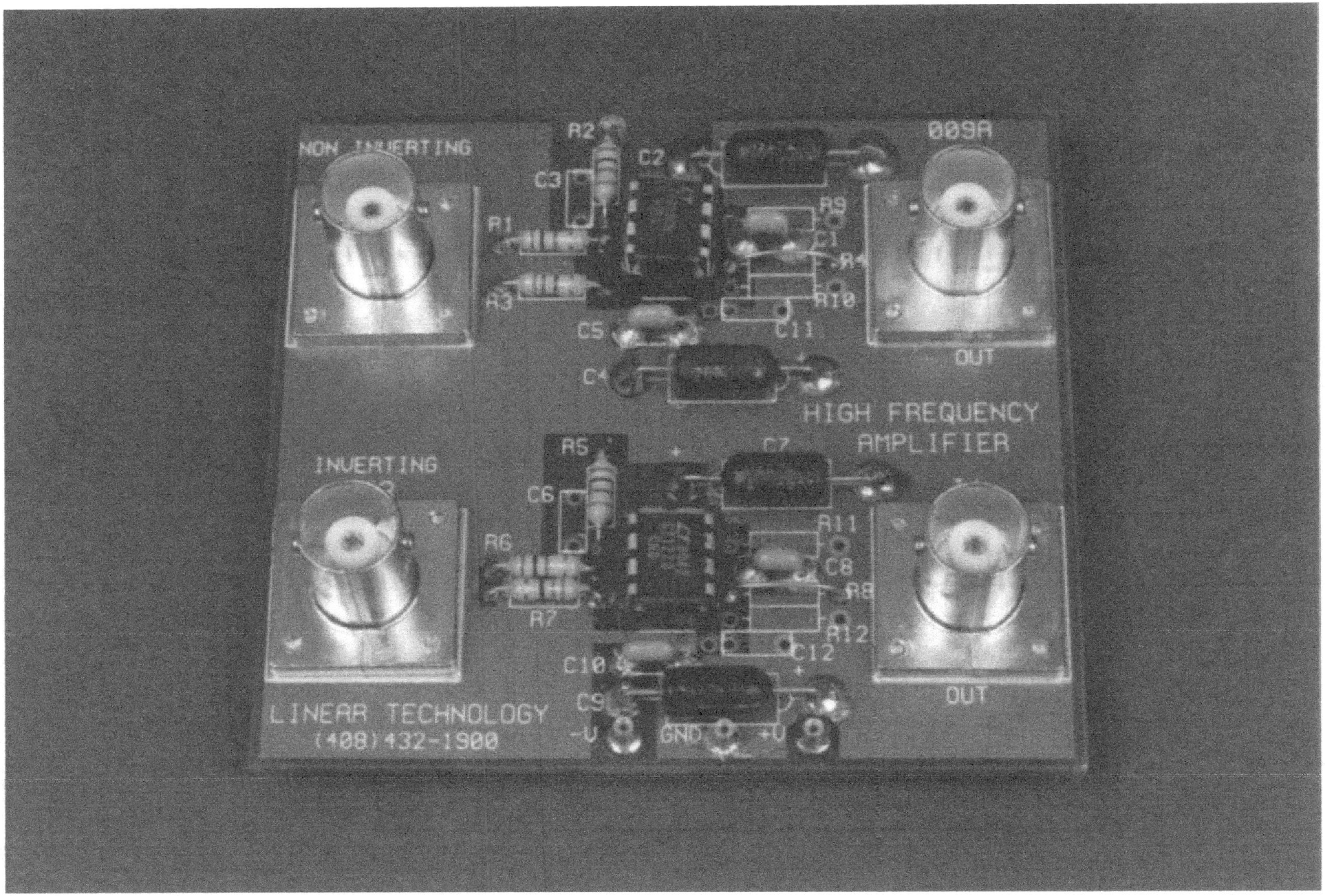

LTAN47 • TAI1

Figure I1 • The Enticing LTC High Frequency Amplifier Demonstration Board. Sockets are Not Optimal, but Allow Trying Different Amplifiers

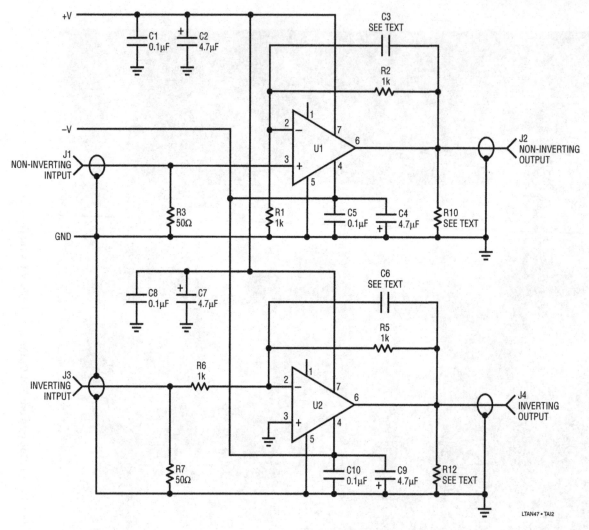

Figure I2 • High Frequency Amplifier Demonstration Board Schematic

LTAN47 • TAI2

C4, C7, and C9) should be 1µF to 10µF tantalums. Most 10nF ceramics are self-resonant well above 10MHz and 4.7µF solid tantalums (axial leaded) are self-resonant at 1MHz or below. Lead lengths are critical; the self-resonant frequency of a 4.7µF tantalum drops by a factor of 2 when measured through 2″ leads. Although a capacitor may become inductive at high frequencies, it is still an effective bypass component above resonance because the impedance is low.

Appendix J

Reprinted, with permission from Cahners Publishing Co., from EEE Magazine, Vol. 15, No. 8, August 1967.

The contributions of Edsel Murphy to the understanding of the behavior of inanimate objects

Abstract—Consideration is given to the effects of the contributions of Edsel Murphy to the discipline of electronics engineering. His law is stated in both general and special form. Examples are presented to corroborate the author's thesis that the law is universally applicable.

I. Introduction

IT HAS LONG BEEN the consideration of the author that the contributions of Edsel Murphy, specifically his general and special laws delineating the behavior of inanimate objects, have not been fully appreciated. It is deemed that this is, in large part, due to the inherent simplicity of the law itself.

It is the intent of the author to show, by references drawn from the literature, that the law of Murphy has produced numerous corollaries. It is hoped that by noting these examples, the reader may obtain a greater appreciation of Edsel Murphy, his law, and its ramifications in engineering and science.

As is well known to those versed in the state-of-the-art, Murphy's Law states that "If anything can go wrong, it will." Or, to state it in more exact mathematical form:

$$1 + 1 > = < 2 \qquad (1)$$

where $>=<$ is the mathematical symbol for hardly ever.

Some authorities have held that Murphy's Law was first expounded by H. Cohen[1] when he stated that "If anything can go wrong, it will — during the demonstration." However, Cohen has made it clear that the broader scope of Murphy's general law obviously takes precedence.

To show the all-pervasive nature of Murphy's work, the author offers a small sample of the application of the law in electronics engineering.

II. General Engineering

II.1.   A patent application will be preceded by one week by a similar application made by an independent worker.

II.2.   The more innocuous a design change appears, the further its influence will extend.

II.3.   All warranty and guarantee clauses become void upon payment of invoice.

II.4.   The necessity of making a major design change increases as the fabrication of the system approaches completion.

II.5.   Firmness of delivery dates is inversely proportional to the tightness of the schedule.

II.6.   Dimensions will always be expressed in the least usable term. Velocity, for example, will be expressed in furlongs per fortnight.[2]

II.7.   An important Instruction Manual or Operating Manual will have been discarded by the Receiving Department.

II.8.   Suggestions made by the Value Analysis group will increase costs and reduce capabilities.

II.9.   Original drawings will be mangled by the copying machine.[3]

III. Mathematics

III.1.   In any given miscalculation, the fault will never be placed if more than one person is involved.

III.2.   Any error that can creep in, will. It will be in the direction that will do the most damage to the calculation.

III.3.   All constants are variables.

III.4.   In any given computation, the figure that is most obviously correct will be the source of error.

III.5.   A decimal will always be misplaced.

III.6.   In a complex calculation, one factor from the numerator will always move into the denominator.

IV. Prototyping and Production

IV.1.   Any wire cut to length will be too short.

IV.2.   Tolerances will accumulate unidirectionally toward maximum difficulty of assembly.

IV.3.   Identical units tested under identical conditions will not be identical in the field.

IV.4.   The availability of a component is inversely proportional to the need for that component.

IV.5.   If a project requires n components, there will be n-1 units in stock.[4]

IV.6.   If a particular resistance is needed, that value will not be available. Further, it cannot be developed with any available series or parallel combination.[5]

IV.7.   A dropped tool will land where it can do the most damage. (Also known as the law of selective gravitation.)

IV.8.   A device selected at random from a group having 99% reliability, will be a member of the 1% group.

IV.9.   When one connects a 3-phase line, the phase sequence will be wrong.[6]

IV.10.   A motor will rotate in the wrong direction.[7]

IV.11.   The probability of a dimension being omitted from a plan or drawing is directly proportional to its importance.

IV.12.   Interchangeable parts won't.

IV.13.   Probability of failure of a component, assembly, subsystem or system is inversely proportional to ease of repair or replacement.

IV.14.   If a protoype functions perfectly, subsequent production units will malfunction.

IV.15.   Components that must not and cannot be assembled improperly will be.

IV.16.   A dc meter will be used on an overly sensitive range and will be wired in backwards.[8]

IV.17.   The most delicate component will drop.[9]

IV.18. Graphic recorders will deposit more ink on humans than on paper.[10]

IV.19. If a circuit cannot fail, it will.[11]

IV.20. A fail-safe circuit will destroy others.[12]

IV.21. An instantaneous power-supply crowbar circuit will operate too late.[13]

IV.22. A transistor protected by a fast-acting fuse will protect the fuse by blowing first.[14]

IV.23. A self-starting oscillator won't.

IV.24. A crystal oscillator will oscillate at the wrong frequency — if it oscillates.

IV.25. A pnp transistor will be an npn.[15]

IV.26. A zero-temperature-coefficient capacitor used in a critical circuit will have a TC of -750ppm/°C.

IV.27. A failure will not appear till a unit has passed Final Inspection.[16]

IV.28. A purchased component or instrument will meet its specs long enough, and only long enough, to pass Incoming Inspection.[17]

IV.29. If an obviously defective component is replaced in an instrument with an intermittent fault, the fault will reappear after the instrument is returned to service.[18]

IV.30. After the last of 16 mounting screws has been removed from an access cover, it will be discovered that the wrong access cover has been removed.[19]

IV.31. After an access cover has been secured by 16 hold-down screws, it will be discovered that the gasket has been omitted.[20]

IV.32. After an instrument has been fully assembled, extra components will be found on the bench.

IV.33. Hermetic seals will leak.

V. Specifying

V.1. Specified environmental conditions will always be exceeded.

V.2. Any safety factor set as a result of practical experience will be exceeded.

V.3. Manufacturers' spec sheets will be incorrect by a factor of 0.5 or 2.0, depending on which multiplier gives the most optimistic value. For salesmen's claims these factors will be 0.1 or 10.0.

V.4. In an instrument or device characterized by a number of plus-or-minus errors, the total error will be the sum of all errors adding in the same direction.

V.5. In any given price estimate, cost of equipment will exceed estimate by a factor of 3.[21]

V.6. In specifications, Murphy's Law supersedes Ohm's.

References*

[1] H. Cohen, Roundhill Associates, private communication.

[2] P. Birman, Kepco, private communication.

[3] T. Emma, Western Union, private communication.

[4] K. Sueker, Westinghouse Semiconductor, private communication.

[5] ———, *loc cit.*.

[6] ———, *loc cit.*.

[7] ———, *loc cit.*

[8] P. Muchnick, Sorensen, private communication.

[9] A. Rosenfeld, Micro-Power, private communication.

[10] P. Muchnick, *loc cit.*

[11] R. Cushman, McCann/ITSM, private communication.

[12] ———, *loc cit.*

[13] ———, *loc cit.*

[14] S. Froud, Industrial Communications Associates, private communication.

[15] L. LeVieux, Texas Instruments, private communication.

[16] G. Toner, Sylvania, private communication.

[17] H. Roth, Power Designs, private communication.

[18] W. Buck, Marconi Instruments, private communication.

[19] A. de la Lastra, SBD Systems, private communication.

[20] ———, *loc cit.*

[21] P. Dietz, Data Technology, private communication.

The man who developed one of the most profound concepts of the twentieth century is practically unknown to most engineers. He is a victim of his own law. Destined for a secure place in the engineering hall of fame, something went wrong.

His real contribution lay not merely in the discovery of the law but more in its universality and in its impact. The law itself, though inherently simple, has formed a foundation on which future generations will build.

*In some cases where no reference is given, the source material was misplaced during preparation of this paper (another example of Murphy's Law). In accordance with the law, these misplaced documents will turn up on the date of publication of this paper.

A seven-nanosecond comparator
for single supply operation

Guidance for putting civilized speed to work

Jim Williams

Introduction

In 1985 Linear Technology Corporation introduced the LT1016 Comparator. This device was the first readily usable, high speed TTL comparator. Previous ICs were either too slow or unstable, preventing widespread acceptance. The LT1016 was, and is, a highly successful product.

Recent technology trends have emphasized low power, single supply operation. The LT1016, although capable of such operation, does not include ground in its input range. As such, it must be biased into its operating common mode range for practical single supply use. A new device, the LT1394, maintains the speed and application civility of its predecessor while including ground in its input operating range. Additionally, the new comparator is faster and pulls significantly lower operating current than the LT1016.

This publication borrows shamelessly from earlier LTC efforts, while introducing new material.[1] It approximates, affixes, appends, abridges, amends, abbreviates, abrogates, ameliorates and augments the previous work.[2] More specifically, the applications section has been almost entirely refurbished, reflecting the LT1394's single supply agility. Additionally, tutorial content has been expanded beyond previous efforts. This approach is necessitated by the continuing need for tutorial guidance in the application of high speed linear devices. The rules of the game are never obviated by new components; rather, they become even more significant as performance increases.

Comparators may be the most underrated and underutilized monolithic linear component. This is unfortunate because comparators are one of the most flexible and universally applicable components available. In large measure the lack of recognition is due to the IC op amp, whose versatility allows it to dominate the analog design world. Comparators are frequently perceived as devices that crudely express analog signals in digital form—a 1-bit A/D converter. Strictly speaking, this viewpoint is correct. It is also wastefully constrictive in its outlook. Comparators don't "just compare" in the same way that op amps don't "just amplify."

Comparators, in particular high speed comparators, can be used to implement linear circuit functions which are as sophisticated as any op amp-based circuit. Judiciously combining a fast comparator with op amps is a key to achieving high performance results. In general, op amp-based circuits capitalize on their ability to close a feedback loop with precision. Ideally, such loops are maintained continuously over time. Conversely, comparator circuits are often based on speed and have a discontinuous output over time. While each approach has its merits, a fusion of both yields the best circuits.

This effort's initial sections are devoted to familiarizing the reader with the realities and difficulties of high speed comparator circuit work. The mechanics and subtleties of achieving precision circuit operation at DC and low frequency have been well documented. Relatively little has appeared that discusses, in practical terms, how to get fast circuitry to work. In developing such circuits, even the most veteran designers sometimes feel that nature is conspiring against them. In some measure this is true. Like all engineering endeavors, high speed circuits can only work if negotiated compromises with nature are arranged. Ignorance of, or contempt for, physical law is a direct route to frustration. In this regard, much of the text and appendices are directed at developing awareness of, and respect for, circuit parasitics and fundamental limitations. This approach is maintained in the applications section, where the notion of "negotiated compromises" is expressed in terms of resistor values and compensation techniques. Many of the application circuits use the LT1394's speed to improve on a standard circuit. Some utilize the speed to

Note 1: In particular LTC Application Note 13, "High Speed Comparator Techniques." Additional text has been similarly purloined from other LTC sources. See the References section following the main text for specifics.
Note 2: An alliterative amalgamated assemblage.

Analog Circuit and System Design: A Tutorial Guide to Applications and Solutions. DOI: 10.1016/B978-0-12-385185-7.00034-2

implement a traditional function in a nontraditional way, with attendant advantages. A (very) few operate at or near the state-of-the-art for a given circuit type, regardless of approach. Substantial effort has been expended in developing these examples and documenting their operation. The resultant level of detail is justified in the hope that it will be catalytic. The circuits should stimulate new ideas to suit particular needs, while demonstrating the LT1394's capabilities in an instructive manner.

The LT1394 — an overview

A new ultrahigh speed comparator, the LT1394, features TTL-compatible complementary outputs and 7ns response time. Other capabilities include a latch pin and good DC input characteristics (see Figure 34.1). The LT1394's outputs directly drive all 5V families, including the higher speed ASTTL, FAST and HC parts. Additionally, TTL

outputs make the device easier to use in linear circuit applications where ECL output levels are often inconvenient.

A substantial amount of design effort has made the LT1394 relatively easy to use. It is much less prone to oscillation and other vagaries than some slower comparators, even with slow input signals. In particular, the LT1394 is stable in its linear region. Additionally, output stage switching does not appreciably change power supply current, further enhancing stability. Finally, current consumption is far lower than previous devices. These features make the 200GHz gain bandwidth LT1394 considerably easier to apply than other fast comparators. Unfortunately, laws of physics dictate that the circuit *environment* the LT1394 works in must be properly prepared. The performance limits of high speed circuitry are often determined by parasitics such as stray capacitance, ground impedance and layout. Some of these considerations are present in digital systems, where designers are comfortable describing bit patterns and memory access times in terms of nanoseconds. The LT1394 can be used in such fast digital systems and Figure 34.2 shows just how fast the device is. The simple test circuit allows us to see that the LT1394's (Trace B) response to the pulse generator (Trace A) is faster than a TTL inverter (Trace C)! Linear circuits operating with this kind of speed make many engineers justifiably wary. Nanosecond domain linear circuits are widely associated with oscillations, mysterious shifts in circuit characteristics, unintended modes of operation and outright failure to function.

Other common problems include different measurement results using various pieces of test equipment, inability to make measurement connections to the circuit without inducing spurious responses and dissimilar operation between two "identical" circuits. If the components used in the circuit are good and the design is sound, all of the above problems can usually be traced to failure to provide a proper circuit "environment." To learn how to do this requires studying the causes of the aforementioned difficulties.

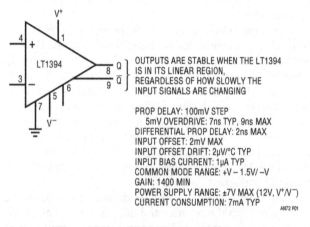

OUTPUTS ARE STABLE WHEN THE LT1394 IS IN ITS LINEAR REGION, REGARDLESS OF HOW SLOWLY THE INPUT SIGNALS ARE CHANGING

PROP DELAY: 100mV STEP
   5mV OVERDRIVE: 7ns TYP, 9ns MAX
DIFFERENTIAL PROP DELAY: 2ns MAX
INPUT OFFSET: 2mV MAX
INPUT OFFSET DRIFT: 2µV/°C TYP
INPUT BIAS CURRENT: 1µA TYP
COMMON MODE RANGE: +V – 1.5V/ –V
GAIN: 1400 MIN
POWER SUPPLY RANGE: ±7V MAX (12V, V$^+$/V$^-$)
CURRENT CONSUMPTION: 7mA TYP

AN72 F01

Figure 34.1 • The LT1394 at a Glance

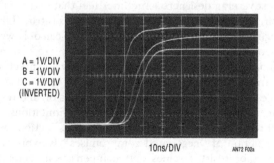

A = 1V/DIV
B = 1V/DIV
C = 1V/DIV
(INVERTED)

10ns/DIV   AN72 F02a

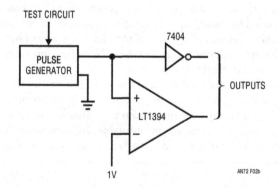

TEST CIRCUIT

PULSE GENERATOR

7404

LT1394

OUTPUTS

1V

AN72 F02b

Figure 34.2 • LT1394 vs a TTL Gate

The rogue's gallery of high speed comparator problems

By far the most common error involves power supply bypassing. Bypassing is necessary to maintain low supply impedance. DC resistance and inductance in supply wires and PC traces can quickly build up to unacceptable levels. This allows the supply line to move as internal current levels of the devices connected to it change. This will almost always cause unruly operation. In addition, several devices connected to an unbypassed supply can "communicate" through the finite supply impedances, causing erratic modes. Bypass capacitors furnish a simple way to eliminate this problem by providing a local reservoir of energy at the device. The bypass capacitor acts like an electrical flywheel to keep supply impedance low at high frequencies. The choice of what type of capacitors to use for bypassing is a critical issue and should be approached carefully (see "About bypass capacitors" in the Tutorial section). An unbypassed LT1394 is shown responding to a pulse input in Figure 34.3. The power supply the LT1394 sees at its terminals has high impedance at high frequency. This impedance forms a voltage divider with the LT1394, allowing the supply to move as internal conditions in the comparator change. This causes local feedback and oscillation occurs. Although the LT1394 responds to the input pulse, its output is a blur of 100MHz oscillation. *Always use bypass capacitors.*

In Figure 34.4 the LT1394's supplies are bypassed, but it still oscillates. In this case, the bypass units are either too far from the device or are lossy capacitors. *Use capacitors with good high frequency characteristics and mount them as close as possible to the LT1394. An inch of wire between the capacitor and the LT1394 can cause problems.*

In Figure 34.5 the device is properly bypassed but a new problem pops up. This photo shows both outputs of the comparator. Trace A appears normal, but Trace B shows an excursion of almost 8V—quite a trick for a device running from a 5V supply. This is a commonly reported problem in high speed circuits and can be quite confusing. It is not due to suspension of natural law, but is traceable to a grossly miscompensated or improperly selected oscilloscope probe. *Use probes that match your oscilloscope's input*

characteristics and compensate them properly (for a discussion on probes, see "About probes and probing techniques" in the Tutorial section). Figure 34.6 shows another probe-induced problem. Here, the amplitude seems correct but the 7ns response time LT1394 appears to have 50ns edges! In this case, the probe used is too heavily compensated or slow for the oscilloscope.

Never use 1× or "straight" probes. Their bandwidth is 20MHz or less and capacitive loading is high. *Check probe bandwidth to ensure that it is adequate for the measurement. Similarly, use an oscilloscope with adequate bandwidth.*

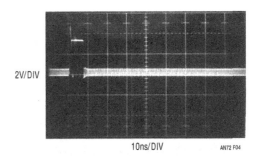

Figure 34.4 • LT1394 Response with Poor Bypassing

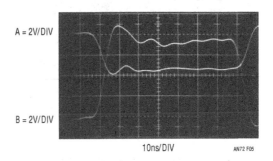

Figure 34.5 • Improper Probe Compensation Causes Seemingly Unexplainable Amplitude Error

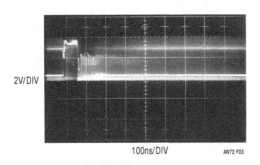

Figure 34.3 • Unbypassed LT1394 Response

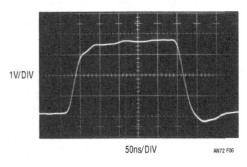

Figure 34.6 • Overcompensated or Slow Probes Make Edges Look Too Slow

In Figure 34.7 the probes are properly selected and applied but the LT1394's output rings and distorts badly. In this case, the probe ground lead is too long. For general purpose work most probes come with ground leads about six inches long. At low frequencies this is fine. At high speed, the long ground lead looks inductive, causing the ringing shown. High quality probes are always supplied with some short ground straps to deal with this problem. Some come with very short spring clips that fix directly to the probe tip to facilitate a low impedance ground connection. For fast work, the ground connection to the probe should not exceed one inch in length. *Keep the probe ground connection as short as possible.*

The difficulty in Figure 34.8 is delay and inadequate amplitude (Trace B). A small delay on the leading edge is followed by a large delay before the falling edge begins. Additionally, a lengthy, tailing response stretches 70ns before finally settling out. The amplitude only rises to 1.5V. A common oversight is responsible for these conditions.

A FET probe monitors the LT1394 output in this example. The probe's common mode input range has been exceeded, causing it to overload and clip the output badly. The small delay on the rising edge is characteristic of active probes and is legitimate. During the time the output is high, the probe is driven deeply into saturation. When the output falls, the probe's overload recovery is lengthy and uneven, causing the delay and tailing.

Know your FET probe. Account for the delay of its active circuitry. Avoid saturation effects due to common mode input limitations (typically ±1V). Use 10× and 100× attenuator heads when required.

Figure 34.9 shows the LT1394's output (Trace B) oscillating near 40MHz as it responds to an input (Trace A). Note that the input signal shows artifacts of the oscillation. This example is caused by improper grounding of the comparator. In this case, the LT1394's ground pin connection is one inch long. The ground lead of the LT1394 must be as short as possible and connected directly to a low impedance ground point. Any substantial impedance in the LT1394's ground path will generate effects like this. The reason for this is related to the necessity of bypassing the power supplies. The inductance created by a long device ground lead permits mixing of ground currents, causing undesired effects in the device. The solution here is simple. *Keep the LT1394's ground pin connection as short (typically 1/4 inch) as possible and run it directly to a low impedance ground. Do not use sockets.*

Figure 34.10 addresses the issue of the "low impedance ground," referred to previously. In this example, the output is clean except for chattering around the edges. This photograph was generated by running the LT1394 without

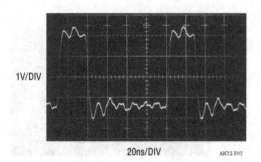

Figure 34.7 • Typical Results Due to Poor Probe Grounding

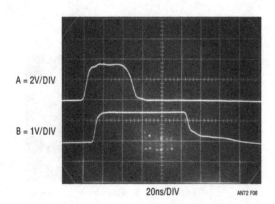

Figure 34.8 • Overdriven FET Probe Causes Delayed, Tailing Response

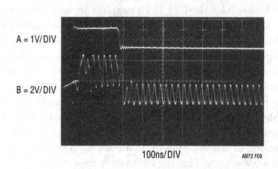

Figure 34.9 • Excessive LT1394 Ground Path Resistance Causes Oscillation

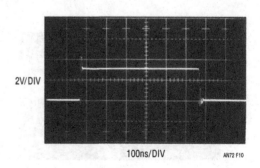

Figure 34.10 • Transition Instabilities Due to No Ground Plane

a "ground plane." A ground plane is formed by using a continuous conductive plane over the surface of the circuit board (ground plane theory is discussed in the Tutorial section). The only breaks in this plane are for the circuit's necessary current paths. The ground plane serves two functions. Because it is flat (AC currents travel along the surface of a conductor) and covers the entire area of the board, it provides a way to access a low inductance ground from anywhere on the board. Also, it minimizes the effects of stray capacitance in the circuit by referring them to ground. This breaks up potential unintended and harmful feedback paths. *Always use a ground plane with the LT1394.*

"Fuzz" on the edges is the difficulty in Figure 34.11. This condition appears similar to Figure 34.10, but the oscillation is more stubborn and persists well after the output has gone low. This condition is due to stray capacitive feedback from the outputs to the inputs. A 3kΩ input source impedance and 3pF of stray feedback allowed this oscillation. The solution for this condition is not too difficult. *Keep source impedance as low as possible, preferably 1kΩ or less. Route output and input pins and components away from each other.*

The opposite of stray-caused oscillations appears in Figure 34.12. Here, the output response (Trace B) badly lags the input (Trace A). This is due to some combination of high source impedance and stray capacitance to ground at the input. The resulting RC forces a lagged response at the input and output delay occurs. An RC combination of 2kΩ source resistance and 10pF to ground gives a 20ns time constant—significantly longer than the LT1394's response time. *Keep source impedance low and minimize stray input capacitance to ground.*

Figure 34.13 shows another capacitance-related problem. Here the output does not oscillate, but the transitions are discontinuous and relatively slow. The villain of this situation is a large output load capacitance. This could be caused by cable driving, excessive output lead length or the input characteristics of the circuit being driven. In most situations this is undesirable and may be eliminated by buffering heavy capacitive loads. In a few circumstances it may not affect overall circuit operation and is tolerable. *Consider the comparator's output load characteristics and their potential effect on the circuit. If necessary, buffer the load.*

Another output-caused fault is shown in Figure 34.14. The output transitions are initially correct but end in a ringing condition. The key to the solution here is the ringing. What is happening is caused by an output lead that is too long. The output lead looks like an unterminated transmission line at high frequencies and reflections occur. This

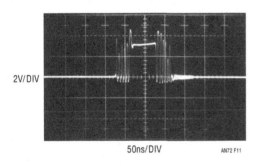

2V/DIV

50ns/DIV   AN72 F11

Figure 34.11 • 3pF Stray Capacitive Feedback with 3kΩ Source Can Cause Oscillation

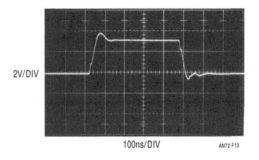

2V/DIV

100ns/DIV   AN72 F13

Figure 34.13 • Excessive Load Capacitance Forces Edge Distortion

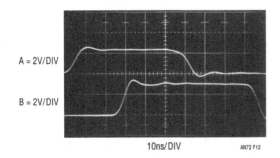

A = 2V/DIV

B = 2V/DIV

10ns/DIV   AN72 F12

Figure 34.12 • Stray 5pF Capacitance from Input to Ground Causes Delay

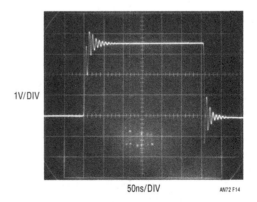

1V/DIV

50ns/DIV   AN72 F14

Figure 34.14 • Lengthy, Unterminated Output Lines Ring from Reflections

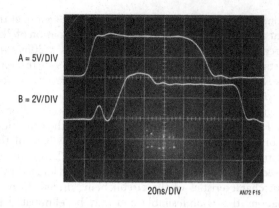

A = 5V/DIV

B = 2V/DIV

20ns/DIV   AN72 F15

Figure 34.15 • Input Common Mode Overdrive Generates Odd Outputs

accounts for the abrupt reversal of direction on the leading edge and the ringing. If the comparator is driving TTL this may be acceptable, but other loads may not tolerate it. In this instance, the direction reversal on the leading edge might cause trouble in a fast TTL load. Similarly, outputs and inputs can see excursions outside supply bounds due to poorly terminated lines, causing device misfunction or failure. *Keep output lead lengths short. If they get much longer than a few inches, terminate with a resistor (typi-cally 250Ω to 400Ω). Ensure that device terminals remain inside supply limits at all times.*

A final malady is presented in Figure 34.15. These waveforms are reminiscent of Figure 34.12's input RC-induced delay. The output waveform initially responds to the input's leading edge, but then returns to zero before going high again. When it does go high, it slews slowly. Additional odd characteristics include pronounced overshoot and pulse top aberration. The fall time is also slow and well delayed from the input. This is certainly strange behavior from a TTL output. What is going on here? The input pulse is responsible for all these anomalies. Its 10V amplitude is well outside the 5V-powered LT1394's common mode input range. Internal input clamps prevent this pulse from damaging the LT1394, but an overdrive of this magnitude results in poor response. *Keep input signals inside the LT1394's common mode range at all times.*

Tutorial section

An implied responsibility in raising the aforementioned issues is their solution or elimination. What good is all the rabble-rousing without suggestions for fixes? It is in this spirit that this tutorial section is presented. Theory, techniques, prejudice and just plain gossip are offered as tools that may help avoid or deal with difficulties.

About pulse generators

A significant consideration for fast comparator development work is the pulse generator. Features such as variable rise and fall time, output DC biasing capability, triggering facilities and amplitude range are highly desirable. General purpose pulse generators usually provide some or all of these capabilities, and little editorial comment is required. Less common, however, particularly at any reasonable price, are really fast pulse generators suitable for LT1394 work. Relatively few generators have transition times below 2.5ns. This kind of speed is highly desirable and some noteworthy instruments bear mention.

The current production Hewlett-Packard 8110A has a fairly complete set of features and 2ns transition times, and is typical of modern, high speed instruments. The older HP-8082A is more versatile, has clean sub-nanosecond transitions and the "knob driven" panel controls are intuitively easy to use. The Phillips PM-5771, also well endowed with features, has 2.4ns transitions and is quite inexpensive. Finally, the HP215A, long out of manufacture, is a special case. This instrument has a restricted 0 to 100ns width range and no rise time control, but other features make it uniquely useful. The output has sub-nanosecond transitions with extraordinarily well-controlled and specified pulse shape parameters. The trigger is very agile, permitting continuous time phase adjustment from before to after the main output. External trigger impedance, polarity and sensitivity are also variable. The output, controlled by a stepped attenuator, will put ±10V into 50Ω in 800ps.

In general, select a pulse generator with the features needed for the circuit of interest. Also, take the time to acquaint yourself with output pulse characteristics, particularly at the highest speeds. Finally, (this is unadulterated author prejudice), instruments with knobs remain easier and faster to set up and modify than "menu-driven" types. This is important in bench work where the ability to quickly and easily change instrument operating point is paramount. In this regard knobs have no equal. Menus belong in restaurants.

About cables, connectors and terminations

High speed signals should always be routed to and from the circuit board with good quality coaxial cable. The cable should be driven and terminated in the system's characteristic impedance at the drive and load points. The driven end is usually an instrument (e.g., pulse or signal generator), presumably endowed with proper characteristics by its manufacturer. It is the cable and its termination, selected by the experimenter, that often cause problems.

All coaxial cable is not the same. Use cable appropriate to the system's characteristic impedance and of good quality. Poorly chosen cable materials or construction methods can introduce odd effects at very high speeds, resulting in observed waveform distortion. A poor cable choice can adversely affect 0.01% settling in the 100ns to 200ns region. Similarly, poor cable can preclude maintenance of even the cleanest pulse generator's 1ns rise time or purity. Typically, inappropriate cable can introduce tailing, rise time degradation, aberrations following transitions, nonlinear impedance and other undesirable characteristics.

Termination choice is equally important. Good quality BNC coaxial type terminators are usually the best choice for breadboarding. Their impedance vs frequency is flat into the GHz range. Additionally, their construction ensures that the (often substantial) drive current returns directly to the source, instead of being dumped into the breadboard's ground system. BNC coaxial terminators are not simply resistors in a can. Good grade 50Ω terminators maintain true coaxial form. They use a carefully designed 50Ω resistor with significant effort devoted to connections to the actual resistive element. In particular, the largest possible connection surface area is utilized to minimize high speed losses. These construction techniques ensure optimum wideband response. Figures 34.16 and 34.17

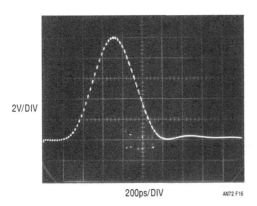

2V/DIV

200ps/DIV   AN72 F16

Figure 34.16 • 350ps Rise and Fall Times Are Preserved by a Good Quality Termination

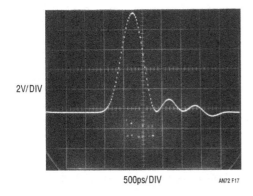

2V/DIV

500ps/DIV   AN72 F17

Figure 34.17 • Poor Grade Termination Produces Pronounced Ringing and Tailing in the GHz Range

demonstrate this nicely. In Figure 34.16 a 1ns pulse with 350ps rise and fall times[3] is monitored on a 1GHz sampling 'scope (Tektronix 556 with 1S1 sampling plug-in and P6032 probe). The waveform is clean, with only a slight hint of ring after the falling edge. This photo was taken with a high grade BNC coaxial type terminator in use. Figure 34.17 does not share these attributes. Here, the generator is terminated with a 50Ω carbon composition resistor with lead lengths of about 1/8 inch. The waveform rings and tails badly on turn-off before finally settling. Note that the sweep speed required a 2.5× reduction to capture these unwanted events. Variable attenuators must provide performance similar to fixed types for meaningful results. The HP-355 series are excellent units, with high fidelity response to 1GHz.

Connectors, such as BNC barrel extensions and tee-type adaptors, are convenient and frequently employed. Remember that these devices represent a discontinuity in the cable, and can introduce small but undesirable effects. In general it is best to employ them as close as possible to a terminated point in the system. Use in the middle of a cable run provides minimal absorption of their mismatch and reflections. The worst offenders among connectors are adapters. This is unfortunate, as these devices are necessitated by the lack of connection standardization in wideband instrumentation. The mismatch caused by a BNC-to-GR874 adaptor transition at the input of a wideband sampling 'scope is small, but clearly discernible in the display. Similarly, mismatches in almost all adaptors, and even in "identical" adaptors of different manufacture, are readily measured on a high frequency network analyzer such as the Hewlett-Packard 4195A[4] (for additional wisdom and terror along these lines see Reference 1).

BNC connections are easily the most common, but not necessarily the most desirable, wideband connection mechanism. The ingenious GR874 connector has notably superior high frequency characteristics, as does the type N. Unfortunately, it's a BNC world out there.

About probes and probing techniques

The choice of which oscilloscope probe to use in a measurement is absolutely crucial. The probe must be considered as an inherent part of the circuit under test. Rise time, bandwidth, resistive and capacitive loading, delay and other limitations must be kept in mind.

Sometimes, the best probe is no probe at all. In some circumstances it is possible and preferable to connect critical breadboard points *directly* to the oscilloscope (see Figure 34.18). This arrangement provides the highest possible grounding integrity, eliminates probe attenuation, and

Note 3: The ability to generate such a pulse proves useful for a variety of tasks, including testing terminators, cables, probes and oscilloscopes for response. The requirements for this pulse generator are surprisingly convenient and inexpensive. For a discussion and construction details see Appendix B "Measuring probe-oscilloscope response."
Note 4: Almost no one believes any of this until they see it for themselves. I didn't. Photos of the network analyzer's display aren't included in the text because no one would believe them. I wouldn't.

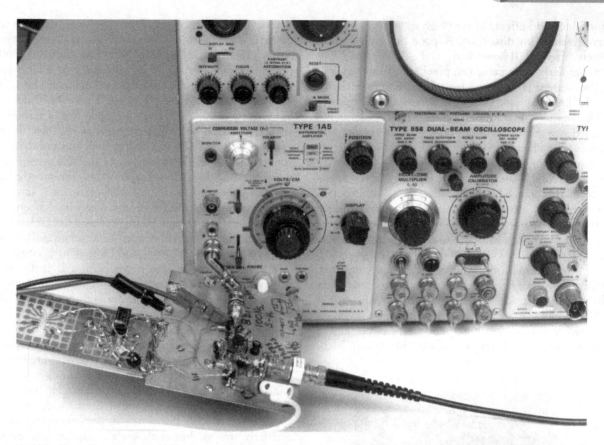

Figure 34.18 • Sometimes the Best Probe Is No Probe. Direct Connection to the Oscilloscope Eliminates a 10× Probe's Attenuation and Possible Grounding Problems

maintains bandwidth. In most cases this is mechanically inconvenient, and often the oscilloscope's electrical characteristics (particularly input capacitance) will not permit it. This is why oscilloscope probes were developed, and why so much effort has been put into their development (Reference 11 is excellent).

Probes are the most overlooked cause of oscilloscope mismeasurement. All probes have some effect on the point they are measuring. The most obvious is input resistance, but input capacitance usually dominates in a high speed measurement. Much time can be lost chasing circuit events that are actually due to improperly selected or applied probes. An 8pF probe looking at a 1kΩ source impedance forms an 8ns lag— longer than the LT1394's delay time! Pay particular attention to the probe's input capacitance. Standard 10MΩ, 10× probes typically have 8pF to 10pF of input capacitance, with 1× types being much higher. In general, 1× probes are not suitable for fast work because their bandwidth is limited to about 20MHz. Remember that all 10× probes cannot be used with all oscilloscopes; the probe's compensation range must match the oscilloscope's input capacitance. Low impedance probes (with 500Ω to 5kΩ resistance) designed for 50Ω inputs, usually have input capacitance of 1pF or 2pF. They are a very good choice if you can stand

the low resistance. FET probes maintain high input resistance and keep capacitance at the 1pF level but have substantially more delay than passive probes. FET probes also have limitations on input common mode range which must be adhered to or serious measurement errors will result. Contrary to popular belief, FET probes *do not* have extremely high input resistance—some types are as low as 100kΩ.

Regardless of which type probe is selected, remember that they all have bandwidth and rise time restrictions. The displayed rise time on the oscilloscope is the vector sum of source, probe and 'scope rise times.

$$t_{RISE}=\sqrt{\left(t_{RISE}\ Source\right)^2+\left(t_{RISE}\ Probe\right)^2+\left(t_{RISE}\ Oscilloscope\right)^2}$$

This equation warns that some rise time degradation must occur in a cascaded system. In particular, if probe and oscilloscope are rated at the same rise time, the system response will be slower than either.

Current probes are useful and convenient.[5] The passive transformer-based types are fast and have less delay than the Hall effect stabilized versions. The Hall types,

---

Note 5: A more thorough discussion of current probes is given in LTC Application Note 35, "Step-Down Switching Regulators." See Reference 2.

however, respond at DC and low frequency and the transformer types typically roll off around 100Hz to 1kHz. Both types have saturation limitations, which, when exceeded, cause odd results on the CRT, confusing the unwary. The Tektronix type CT-1 current probe, although not nearly as versatile as the clip-on probes, bears mention. Although this is not a clip-on device, it may be the least electrically intrusive way of extracting wideband signal information. Rated at 1GHz bandwidth, it produces 5mV/mA output with only 0.6pF loading. Decay time constant of this AC current probe is ≈1%/50ns, resulting in a low frequency limit of 35kHz.

A very special probe is the differential probe. A differential probe may be thought of as two matched FET probes contained within a common probe housing. This probe literally brings the advantage of a differential input oscilloscope to the circuit board. The probes matched, active circuitry provides greatly improved high frequency common mode rejection over single-ended probing or even matched passive probes used with a differential amplifier. The resultant ability to reject common mode signals and ground noise at high frequency allows this probe to deliver exceptionally clean results when monitoring small, fast signals. Figure 34.19 shows a differential probe being used to verify the waveshape of a 2.5mV circuit input.

When using different probes, remember that they all have different delay times, meaning that apparent timing errors will occur on the CRT. Know what the individual probe delays are and account for them in interpreting the CRT display.

By far the greatest source of error in probe use is grounding. Poor probe grounding can cause ripples and discontinuities in the observed waveform. In some cases the choice and placement of a probe's ground strap will affect waveforms on another channel. In the worst case, connecting the probe's ground wire will virtually disable the circuit being measured. The cause of these problems is parasitic inductance in the probe's ground connection. In most oscilloscope measurements this is not a problem, but at nanosecond speeds it becomes critical. Fast probes are always supplied with a variety of spring clips and accessories designed to aid in making the lowest possible inductive connection to ground. Most of these attachments assume a ground plane is in use, which it should be. Always try to make the shortest possible connection to ground—anything longer than one inch may cause trouble. Sometimes it's difficult to determine if probe grounding is the cause of observed waveform aberrations. One good test is to disturb the grounding setup and see if changes occur. Nominally, touching the ground plane or jiggling probe ground

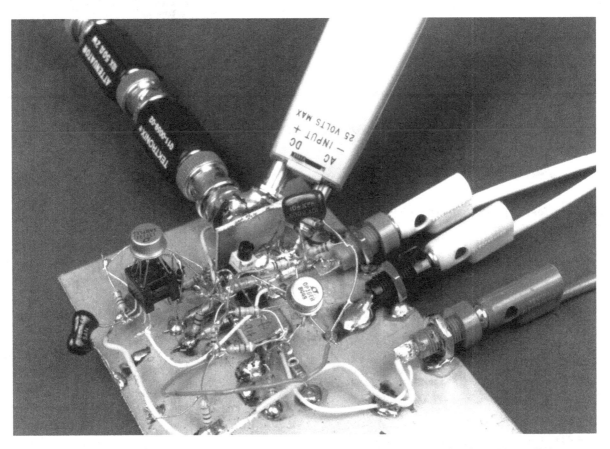

Figure 34.19 • Using a Differential Probe to Verify the Integrity of a 2.5mV High Speed Input Pulse

connectors or wires should have no effect. If a ground strap wire is in use try changing its orientation or simply squeezing it together to change and minimize its loop area. *If any waveform change occurs while doing this the probe grounding is unacceptable, rendering the oscilloscope display unreliable.*

The simple network of Figure 34.20 shows just how easy it is for poorly chosen or used probes to cause bad results. A 9pF input capacitance probe with a 4-inch long ground strap monitors the output (Trace B, Figure 34.21). Although the input (Trace A) is clean, the output contains ringing. Using the same probe with a 1/4-inch spring tip ground connection accessory seemingly cleans up everything (Figure 34.22). However, substituting a 1pF FET probe (Figure 34.23) reveals a 50% output amplitude error in Figure 34.22! The FET probe's low input capacitance allows a more accurate version of circuit action. The FET probe does, however, contribute its own form of error.

Note that the probe's response is tardy by 5ns due to delay in its active circuitry. Hence, separate measurements with each probe are required to determine the output's amplitude and timing parameters. An alternative would employ two matched FET probes to minimize delay uncertainty.

A final form of probe is the human finger. Probing the circuit with a finger can accentuate desired or undesired effects, giving clues that may be useful. The finger can be used to introduce stray capacitance to a suspected circuit node while observing results on the CRT. Two fingers, lightly moistened, can be used to provide an experimental resistance path. Some high speed engineers are particularly adept at these techniques and can estimate the capacitive and resistive effects created with surprising accuracy.

Examples of some of the probes discussed, along with different forms of grounding implements, are shown in Figure 34.24. Probes A, B, E and F are standard types equipped with various forms of low impedance grounding attachments. The conventional ground lead used on G is more convenient to work with but will cause ringing and

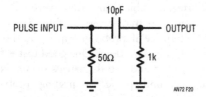

Figure 34.20 • Probe Test Circuit

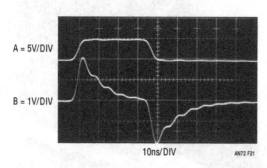

Figure 34.21 • Test Circuit Output with 9pF Probe and 4-Inch Ground Strap

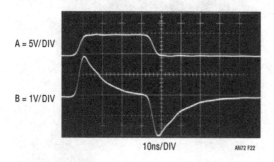

Figure 34.22 • Test Circuit Output with 9pF Probe and 1/4-Inch Ground Strap

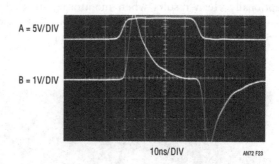

Figure 34.23 • Test Circuit Output with FET Probe

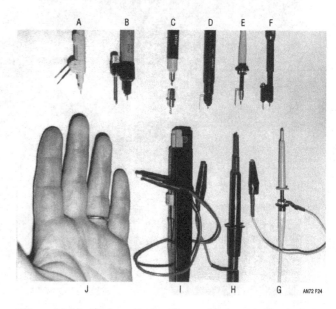

Figure 34.24 • Various Probe-Ground Strap Configurations

other effects at high frequencies, rendering it useless. H has a very short ground lead. This is better, but can still cause trouble at high speeds. D is a FET probe. The active circuitry in the probe and a very short ground connector ensure low parasitic capacitance and inductance. C is a separated FET probe attenuator head. Such heads allow the probe to be used at higher voltage levels (e.g., ±10V or ±100V). The miniature coaxial connector shown can be mounted on the circuit board and the probe mated with it. This technique provides the lowest possible parasitic inductance in the ground path and is especially recommended. I is a current probe. A ground connection is not usually required. However, at high speeds the ground connection may result in a cleaner CRT presentation. Because no current flows in the ground lead of these probes, a long strap is usually permissible. J is typical of the finger probes described in the text. Note the ground strap on the third finger.

The low inductance ground connectors shown are available from probe manufacturers and are always supplied with good quality, high frequency probes. Because most oscilloscope measurements do not require them, they invariably become lost. There is no substitute for these devices when they are needed, so it is prudent to take care of them. This is especially applicable to the ground strap on the finger probe.

About oscilloscopes

The modern oscilloscope is one of the most remarkable instruments ever constructed. The protracted and intense development effort put toward these machines is perhaps equaled only by the fanaticism devoted to timekeeping.[6] It is a tribute to oscilloscope designers that instruments manufactured over 30 years ago still suffice for over 90% of today's measurements. The oscilloscope-probe combination used in high speed work is the most important equipment decision the designer must make. Ideally, the oscilloscope should have at least 150MHz bandwidth, but slower instruments are acceptable if their limitations are well understood. Be certain of the characteristics of the probe-oscilloscope combination. Rise time, bandwidth, resistive and capacitive loading, delay, noise, channel-to-channel feedthrough, overdrive recovery, sweep nonlinearity, triggering, accuracy and other limitations must be kept in mind. High speed linear circuitry demands a great deal from test equipment and countless hours can be saved if the characteristics of the instruments used are well known. Obscene amounts of time have been lost pursuing "circuit problems" that in reality are caused by misunderstood, misapplied or out-of-spec equipment. Intimate familiarity with your oscilloscope is invaluable in getting the best possible results with it. In fact, it is possible to use seemingly inadequate equipment to get good results if the equipment's limitations are well known and respected. All

of the circuits in the Applications section involve rise times and delays well above the 100MHz to 200MHz region, but 90% of the development work was done with a 50MHz oscilloscope. Familiarity with equipment and thoughtful measurement technique permit useful measurements seemingly beyond instrument specifications. A 50MHz oscilloscope cannot track a 5ns rise time pulse, but it can measure a 2ns delay between two such events. Using such techniques, it is often possible to deduce the desired information. There are situations where no amount of cleverness will work and the right equipment (e.g., a faster oscilloscope) must be used. Sometimes, "sanity-checking" a limited bandwidth instrument with a higher bandwidth oscilloscope is all that is required. For high speed work, brute force bandwidth is indispensable when needed, and no amount of features or computational sophistication will substitute. Most high speed circuitry does not require more than two traces to get where you are going. Versatility and many channels are desirable, but if the budget is limited, spend for bandwidth!

Dramatic differences in displayed results are produced by probe-oscilloscope combinations of varying bandwidths. Figure 34.25 shows the output of a very fast pulse[7] monitored with a 1GHz sampling 'scope (Tektronix 556 with 1S1 sampling plug-in). At this bandwidth the 10V amplitude appears clean, with just a small hint of ringing after the falling edge. The rise and fall times of 350ps are suspicious, as the sampling oscilloscope's rise time is also specified at 350ps.[8]

Figure 34.26 shows the same pulse observed on a 350MHz instrument with a direct connection to the input (Tektronix 485/50Ω input). Indicated rise time balloons to 1ns, while displayed amplitude shrinks to 6V, reflecting

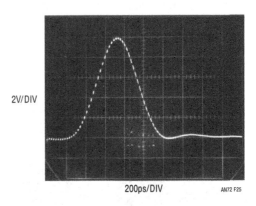

2V/DIV

200ps/DIV   AN72 F25

Figure 34.25 • A 350ps Rise/Fall Time 10V Pulse Monitored on 1GHz Sampling Oscilloscope. Direct 50Ω Input Connection Is Used

---

Note 6: In particular, the marine chronometer received ferocious and abundant amounts of attention. See References 4, 5 and 6. For an enjoyable stroll through the history of oscilloscope vertical amplifiers, see Reference 3.

Note 7: See Appendix B "Measuring probe-oscilloscope response," for complete details on this pulse generator.
Note 8: This sequence of photos was shot in my home lab. I'm sorry, but 1GHz was the fastest 'scope in my house at the time. See Appendix B for a higher speed representation of this pulse.

this instrument's lesser bandwidth. To underscore earlier discussion, poor grounding technique (1 1/2″ of ground lead to the ground plane) created the prolonged rippling after the pulse fall.

Figure 34.27 shows the same 350MHz (50Ω input) oscilloscope with a 3GHz 10× probe (Tektronix P6056). Displayed results are nearly identical, as the probe's high bandwidth contributes no degradation. Again, deliberate poor grounding causes overshoot and rippling on the pulse fall.

Figure 34.28 equips the same oscilloscope with a 10× probe specified at 290MHz bandwidth (Tektronix P6047). Additionally, the oscilloscope has been switched to its 1MΩ input mode, reducing bandwidth to a specified 250MHz. Amplitude degrades to less than 4V and edge times similarly increase. The deliberate poor grounding contributes the undershoot and underdamped recovery on pulse fall.

In Figure 34.29, a 100MHz 10× probe (Hewlett-Packard Model 10040A) has been substituted for the 290MHz unit. The oscilloscope and its setup remain the same. Amplitude shrinks below 2V, with commensurate rise and fall times. Cleaned-up grounding eliminates aberrations.

A Tektronix 454A (150MHz) produced Figure 34.30's trace. The pulse generator was directly connected to the input. Displayed amplitude is about 2V, with appropriate 2ns edges. Finally, a 50MHz instrument (Tektronix 556 with 1A4 plug-in) just barely grunts in response to the pulse (Figure 34.31). Indicated amplitude is 0.5V, with edges reading about 7ns. That's a long way from the 10V and 350ps that's really there!

A final oscilloscope characteristic is overload performance. It is often desirable to view a small amplitude portion of a large waveform. In many cases the oscilloscope is required to supply an accurate waveform after the display has been driven off screen. How long must one wait after an overload before the display can be taken seriously? The answer to this question is quite complex. Factors

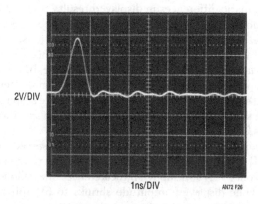

Figure 34.26 • The Test Pulse Appears Smaller and Slower on a 350MHz Instrument (t_{RISE} = 1ns). Deliberate Poor Grounding Creates Rippling After the Pulse Falls. Direct 50Ω Connection Is Used

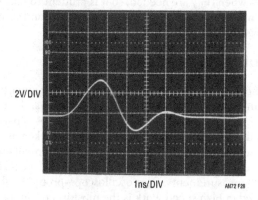

Figure 34.28 • Test Pulse Measures Only 3V High on a 250MHz 'Scope with Significant Waveform Distortion. 290MHz 10× Probe Used

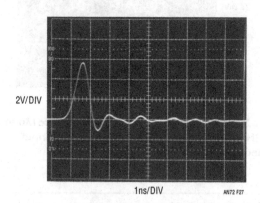

Figure 34.27 • Test Pulse on the Same 350MHz Oscilloscope Using a 3GHz 10× Probe. Deliberate Poor Grounding Maintains Rippling Residue

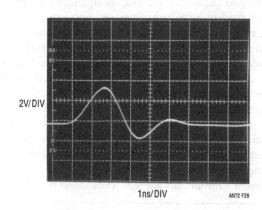

Figure 34.29 • Test Pulse Measures Under 2V High Using 250MHz 'Scope and a 100MHz Probe

involved include the degree of overload, its duty cycle, its magnitude in time and amplitude and other considerations. Oscilloscope response to overload varies widely between types and markedly different behavior can be observed in any individual instrument. For example, the recovery time for a 100× overload at 0.005V/division may be very different than at 0.1V/division. The recovery characteristic may also vary with waveform shape, DC content and repetition rate. With so many variables, it is clear that measurements involving oscilloscope overload must be approached with caution. Nevertheless, a simple test can indicate when the oscilloscope is being deleteriously affected by overdrive.

The waveform to be expanded is placed on the screen at a vertical sensitivity which eliminates all off-screen activity. Figure 34.32 shows the display. The lower right hand portion is to be expanded. Increasing the vertical sensitivity by a factor of two (Figure 34.33) drives the waveform off-screen, but the remaining display appears reasonable. Amplitude has doubled and waveshape is consistent with the original display. Looking carefully, it is possible to see

small amplitude information presented as a dip in the waveform at about the third vertical division. Some small disturbances are also visible. This observed expansion of the original waveform is believable. In Figure 34.34, gain has been further increased and all the features of Figure 34.33 are amplified accordingly. The basic waveshape appears clearer and the dip and small disturbances are also easier to see.

No new waveform characteristics are observed. Figure 34.35 brings some unpleasant surprises. This increase in gain causes definite distortion. The initial negative-going peak, although larger, has a different shape. Its bottom appears less broad than in Figure 34.34. Additionally, the peak's positive recovery is shaped slightly differently. A new rippling disturbance is visible in the center of the screen. This kind of change indicates that the oscilloscope is having trouble. A further test can confirm that this waveform is being influenced by overloading. In Figure 34.36 the gain remains the same, but the vertical position knob has been used to reposition the display at the screen's bottom. This shifts the oscilloscope's DC operating point, which, under normal circumstances, should not affect the displayed waveform. Instead, a marked shift in waveform amplitude and outline occurs. Repositioning the waveform to the screen's top produces a differently distorted waveform (Figure 34.37). It is obvious that for this particular waveform, accurate results cannot be obtained at this gain.

Differential plug-ins can address some of the issues associated with excessive overdrive, although they cannot solve all problems. Two differential plug-in types merit special mention. At low level, a high sensitivity differential plug-in is indispensable. The Tektronix 1A7, 1A7A and 7A22 feature 10μV sensitivity, although bandwidth is limited to 1MHz. The units also have selectable highpass and lowpass filters and good high frequency common mode rejection. Tektronix types 1A5, W and 7A13 are differential comparators. They have calibrated DC nulling (slide back) sources, allowing observation of small, slowly moving events on top of common mode DC or fast events riding on a waveform.

A special case is the sampling oscilloscope. By nature of its operation, a sampling 'scope in proper working order is inherently immune to input overload, providing essentially instantaneous recovery between samples. See Reference 8 for additional details.

The best approach to measuring small portions of large waveforms, however, is to eliminate the large signal swing seen by the oscilloscope. Reference 17 discusses applicable techniques in detail.

In summary, although the oscilloscope provides remarkable capability, its limitations must be well understood when interpreting results.[9]

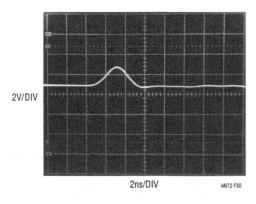

Figure 34.30 • 150MHz Oscilloscope (t_{RISE} = 2.4ns) with Direct Connection Responds to the Test Pulse

2V/DIV

2ns/DIV   AN72 F30

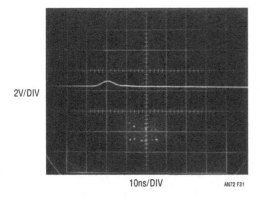

2V/DIV

10ns/DIV   AN72 F31

Figure 34.31 • A 50MHz Instrument Barely Grunts. 10V, 350ps Test Pulse Measures Only 0.5V High with 7ns Rise and Fall Times!

Note 9: Additional discourse on oscilloscopes will be found in References 1 and 7 through 10.

814

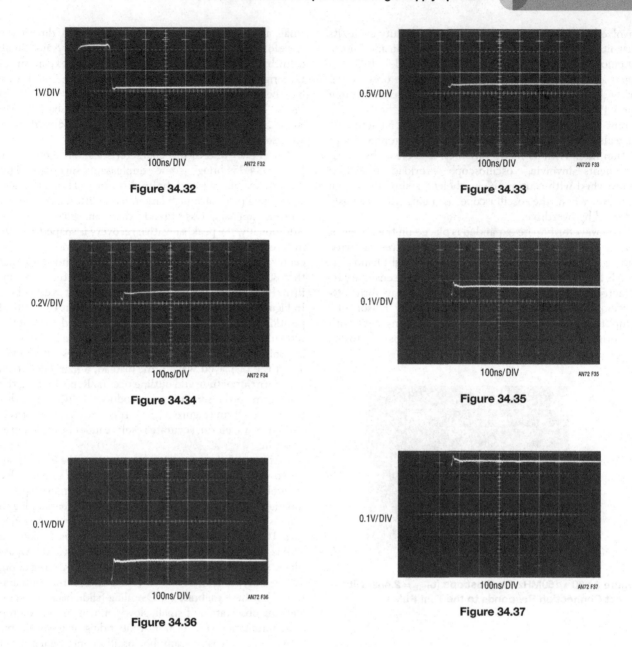

1V/DIV

100ns/DIV   AN72 F32

Figure 34.32

0.5V/DIV

100ns/DIV   AN720 F33

Figure 34.33

0.2V/DIV

100ns/DIV   AN72 F34

Figure 34.34

0.1V/DIV

100ns/DIV   AN72 F35

Figure 34.35

0.1V/DIV

100ns/DIV   AN72 F36

Figure 34.36

0.1V/DIV

100ns/DIV   AN72 F37

Figure 34.37

Figures 34.32 to 34.37 The Overdrive Limit Is Determined by Progressively Increasing Oscilloscope Gain and Watching for Waveform Aberrations

About ground planes

Many times in high frequency circuit layout, the term "ground plane" is used, most often as a mystical and ill-defined cure to spurious circuit operation. In fact, there is little mystery to the usefulness and operation of ground planes, and like many phenomena, their fundamental operating principle is surprisingly simple.

Ground planes are primarily useful for minimizing circuit inductance. They do this by utilizing basic magnetic theory. Current flowing in a wire produces an associated magnetic field. The field's strength is proportional to the current and inversely related to the distance from the conductor. Thus, we can visualize a wire carrying current (Figure 34.38) surrounded by radii of magnetic field. The unbounded field becomes smaller with distance. A wire's inductance is defined as the energy stored in the field set up by the wire's current. To compute the wire's inductance requires integrating the field over the wire's length and the total radial area of the field. This implies integrating on the radius from $R=R_W$ to infinity, a very large number.

However, consider the case where we have two wires in space carrying the same current in either direction (Figure 34.39). The fields produced cancel.

In this case, the inductance is much smaller than in the simple wire case and can be made arbitrarily smaller by reducing the distance between the two wires. This reduction of inductance between current carrying conductors is the underlying reason for ground planes. In a normal circuit, the current path from the signal source through its conductor and back to ground includes a large loop area. This produces a large inductance for this conductor which can cause ringing due to LRC effects. It is worth noting that 10nH at 100MHz has an impedance of 6Ω. At 10mA a 60mV drop results.

A ground plane provides a return path directly under the signal carrying conductor through which return current can flow. The conductor's small physical separation means the inductance is low. Return current has a direct path to ground, regardless of the number of branches associated with the conductor. Currents will always flow through the return path of lowest impedance. In a properly designed ground plane, this path is directly under the signal conductor. In a practical circuit, it is desirable to use one whole side of the PC card (usually the component side for wave solder considerations) as a ground plane and run the signal conductors on the other side. This will give a low inductance path for all the return currents.

Aside from minimizing parasitic inductance, ground planes have additional benefits. Their flat surface minimizes resistive losses due to AC skin effect (AC currents travel along a conductor's surface). Additionally, they aid the circuit's high frequency stability by referring stray capacitances to ground.

Some practical hints for ground planes are:

1. Utilize a ground plane over as much area as possible on the component side of the board, especially under traces that operate at high frequency.
2. Mount components that conduct substantial fast rise currents (termination resistors, ICs, transistors, decoupling capacitors) as close to the board as possible.
3. Where common ground potential is important (i.e., at comparator inputs), try to single point the critical components into the ground plane to avoid voltage drops.
4. Keep trace length short. Inductance varies directly with length and no ground plane will achieve perfect cancellation.

About bypass capacitors

Bypass capacitors are used to maintain low power supply impedance at the point of load. Parasitic resistance and inductance in supply lines mean that the power supply impedance can be quite high. As frequency goes up, the inductive parasitic becomes particularly troublesome. Even if these parasitic terms did not exist, or if local regulation were used, bypassing is still necessary because no power supply or regulator has zero output impedance at 100MHz. What type of bypass capacitor to use is determined by the application, frequency domain of the circuit, cost, board space and many other considerations. Some useful generalizations can be made.

All capacitors contain parasitic terms, some of which appear in Figure 34.40. In bypass applications, leakage and dielectric absorption are second order terms but series

Figure 34.38 ● Single Wire Case

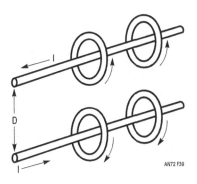

Figure 34.39 ● Two Wire Case

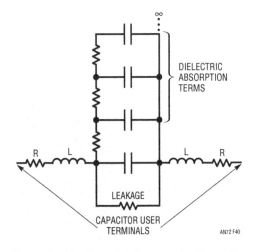

Figure 34.40 ● Parasitic Terms of a Capacitor

R and L are not. These latter terms limit the capacitor's ability to damp transients and maintain low supply impedance. Bypass capacitors must often be large values so they can absorb long transients, necessitating electrolytic types which have large series R and L.

Different types of electrolytics and electrolytic-nonpolar combinations have markedly different characteristics. Which type(s) to use is a matter of passionate debate in some circles and the test circuit (Figure 34.41) and accommpanying photos are useful. The photos show the response of five bypassing methods to the transient generated by the test circuit. Figure 34.42 shows an unbypassed line which sags and ripples badly at large amplitudes. Figure 34.43 uses an aluminum 10μF electrolytic to considerably cut the disturbance, but there is still plenty

of potential trouble. A tantalum 10μF unit offers cleaner response in Figure 34.44 and the 10μF aluminum combined with a 0.01μF ceramic type is even better in Figure 34.45. Combining electrolytics with nonpolarized capacitors is a popular way to get good response but beware of picking the wrong duo. The right (wrong) combination of supply line parasitics and paralleled dissimilar capacitors can produce a resonant, ringing response, as in Figure 34.46. *Caveat!*

Breadboarding techniques

The breadboard is both the designer's playground and proving ground. It is there that Reality resides, and paper (or

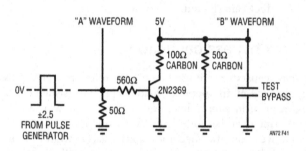

Figure 34.41 • Bypass Capacitor Test Circuit

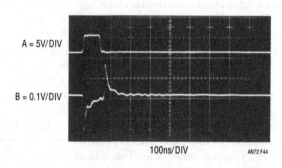

Figure 34.44 • Response of 10μF Tantalum Capacitor

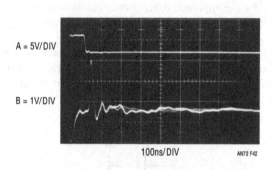

Figure 34.42 • Response of Unbypassed Line

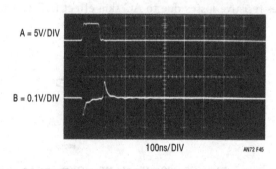

Figure 34.45 • Response of 10μF Aluminum Paralleled by 0.01μF Ceramic

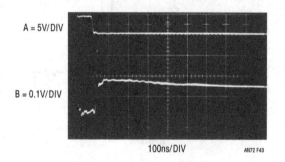

Figure 34.43 • Response of 10μF Aluminum Capacitor

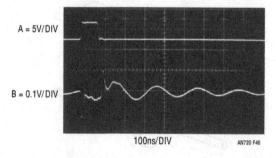

Figure 34.46 • Some Paralleled Combinations Can Ring. Try Before Specifying!

computer) designs meet their ruler. More than anything else, breadboarding is an iterative procedure, an odd amalgam of experience guiding an innocent, ignorant, explorative spirit. A key is to be willing to try things out, sometimes for not very good reasons. Invent problems and solutions, guess carefully and wildly, throw rocks and see what comes loose. Invent and design experiments, and follow them wherever they lead. Reticence to try things is probably the number one cause of breadboards that "don't work."[10] Implementing the above approach to life begins with the physical construction methods used to build the breadboard.

A high speed breadboard must start with a ground plane. Additionally, bypassing, component layout and connections should be consistent with high speed operations. Because of these considerations there is a common misconception that breadboarding high speed circuits is time consuming and difficult. This is simply not true. For high speed circuits of moderate complexity a complete and electrically correct breadboard can be assembled in 10 minutes if all necessary components are on hand. The key to rapid breadboarding is to identify critical circuit nodes and design the layout to suit them. This permits most of the breadboard's construction to be fairly sloppy, saving time and effort. Additionally, use all degrees of freedom in making connections and mounting components. Don't be bashful about bending IC pins to suit desired low capacitance connections, or air wiring components to achieve rapid or electrically optimum layout. Save time by using components, such as bypass capacitors, as mechanical supports for other components, such as amplifiers. It is true that eventual printed circuit construction is required, but when initially breadboarding forget about PC and production constraints. Later, when the circuit works, and is well understood, PC adaptations can be taken care of.[11]

Once the breadboard seems to work, it's useful to begin thinking about PC layout and component choice for production. Experiment with the existing layout to determine just how sensitive nominally critical points are. Add controlled parasitic terms (e.g., resistors, capacitors and physical layout changes) to test for sensitivity. Gentle touching of suspect points with a finger can yield preliminary indication of sensitivity, giving clues that can be quite valuable.

In conclusion, when breadboarding, design the breadboard to be quick and easy to build, work with and modify. Observe the circuit and listen to what it is telling you before trying to get it to some desired state. Finally, don't hesitate to try just about anything; that's what the breadboard is for. Almost anything you do will cause some result—whether it's good or bad is almost irrelevant. Anything you do that enhances your ability to correlate events occurring on the breadboard can only be beneficial.

This completes the tutorial section. Hopefully, several notions have been imparted. First, in any measurement situation, test equipment characteristics are an integral part of the circuit. At high speed and high precision this is particularly the case. As such, it is imperative to know your equipment and how it works. There is no substitute for intimate familiarity with your tool's capabilities and limitations.[12]

In general, use equipment you trust and measurement techniques you understand. Keep asking questions and don't be satisfied until everything you see on the oscilloscope is accounted for and makes sense.

The LT1394, combined with the precautionary notes listed above, permits fast linear circuit functions that are difficult or impractical using other approaches. Some of the applications presented in the following section represent the state-of-the-art for a particular circuit function. Others show simplified and/or improved ways to implement standard functions by utilizing the comparator's easily accessed speed. All have been carefully (and painfully) worked out and should serve as good idea sources for potential users of the device. Have fun. I did.

Applications

Crystal oscillators

Figure 34.47's circuits are crystal oscillators. In the circuit (a) the resistors at the LT1394's positive input set a DC

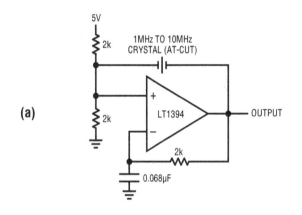

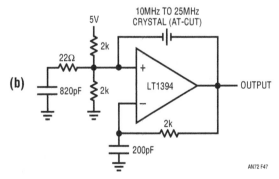

Figure 34.47 • Crystal Oscillators for Outputs to 30MHz. Circuit (b)'s Damper Network Suppresses Overtone Crystal's Harmonic Modes

---

Note 10: A much more eloquently stated version of this approach is found in Reference 12.

Note 11: See Reference 17 for a pictorially enhanced version of this discussion.

Note 12: Further exposition and *kvetching* on this point is given in Reference 13.

bias point. The 2k-0.068µF path sets up phase shifted feedback and the circuit looks like a wideband unity-gain follower at DC. The crystal's path provides resonant positive feedback and stable oscillation occurs. The circuit (b) is similar, but supports oscillation frequencies to 30MHz. Above 10MHz, AT-cut crystals operate in overtone mode. Because of this, oscillation can occur at multiples of the desired frequency. The damper network rolls off gain at high frequency, ensuring proper operation.

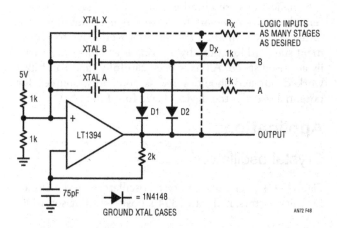

Figure 48 • Switchable Output Crystal Oscillator. Biasing A or B High Places Associated Crystal in Feedback Path. Additional Crystal Branches Are Permissible

Switchable output crystal oscillator

Figure 34.48 permits crystals to be electronically switched by logic commands. This circuit is similar to the previous examples, except that oscillation is only possible when one of the logic inputs is biased high.

Temperature-compensated crystal oscillator (TXCO)

Figure 34.49 is a temperature-compensated crystal oscillator (TXCO). This circuit reduces oscillator temperature drift by inserting a temperature-dependent compensatory correction into the crystal's frequency trimming network. This open-loop correction technique relies on matching the oscillator's frequency versus temperature characteristic, which is quite repeatable.

The LT1394 and associated components form the crystal oscillator, operating similarly to Figure 34.47's examples. The LM134, a temperature-dependent current source, biases A1. A1 takes gain referred to the LM134's output and the negative offset supplied via the 470kΩ-LT1004 reference path. Note that the LT1004's negative voltage bias is bootstrapped from the oscillator's output, maintaining single supply operation. This arrangement delivers temperature-dependent bias to the varactor diode, causing a scaled variation in the crystal's resonance versus ambient temperature. The varactor's bias-dependent capacitance shift pulls crystal frequency to complement the circuit's temperature drift. The simple first

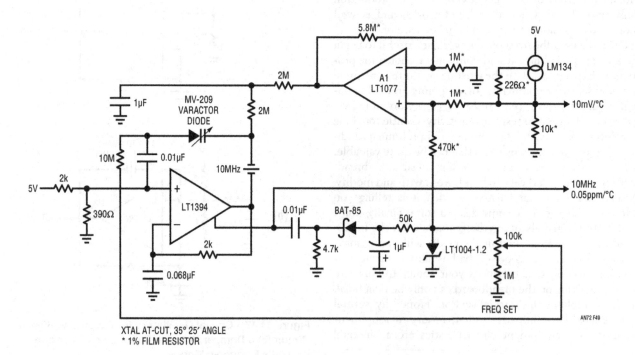

Figure 34.49 • Temperature-Compensated 10MHz Crystal Oscillator. Temperature-Dependent Varactor Bias Reduces Drift by 20:1

order fit provided by the compensation is very effective. Figure 34.50 shows results. The −70ppm frequency shift over 0°C to 70°C is corrected within a few ppm. The "FREQ SET" trim also biases the varactor, allowing

accurate output frequency setting. It is worth noting that better compensation is possible by including higher order terms in the temperature-to-voltage conversion.

Voltage-controlled crystal oscillator (VCXO)

Figure 34.51, also a variant of the basic crystal oscillator, permits voltage tuning the output frequency. Such voltage-controlled crystal oscillators (VCXO) are often employed where slight variation of a stable carrier is required. This example is specifically intended to provide a 4× NTSC sub-carrier tunable oscillator suitable for phase locking.

The LT1394 is set up as a crystal oscillator, operating similarly to Figure 34.47(a). The varactor diode is biased from the tuning input. The tuning network is arranged so a 0V to 5V drive provides a reasonably symmetric, broad tuning range around the 14.31818MHz center frequency. The indicated selected capacitor sets tuning bandwidth. It should be picked to complement loop response in phase locking applications. Figure 34.52 is a plot of tuning input voltage versus frequency deviation. Tuning deviation from the 4× NTSC 14.31818MHz center frequency exceeds ±240ppm for a 0V to 5V input.

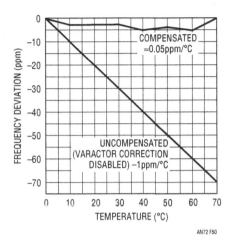

Figure 34.50 • Figure 34.49's Compensated vs Uncompensated Temperature Dependence. First Order Compensation Reduces Oscillator Drift to 0.05ppm/°C

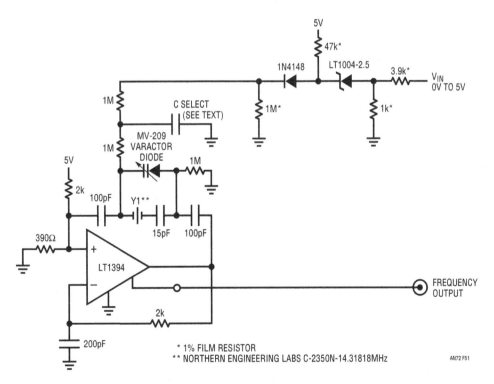

Figure 34.51 • A 4× NTSC Sub-Carrier Voltage-Tunable Crystal Oscillator. Tuning Range and Bandwidth Accommodate Variety of Phase Locked Loops

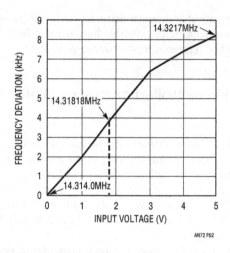

AN72 F52

Figure 34.52 • Control Voltage vs Output Frequency for Figure 34.51. Tuning Deviation from Center Frequency Exceeds ±240ppm

Voltage-tunable clock skew generator

It is sometimes necessary to generate pairs of identical clock signals that are phase skewed in time. Further, it is desirable to be able to set the amount of time skew via a tuning voltage. Figure 34.53's circuit does this by utilizing comparators to digitize phase information from a varactor-tuned time domain bridge. A 0V to 2V control signal provides ≈ ±10ns of output skew. The input is applied to the CMOS inverters, which deliver noninverting drive to the bridge network (Trace A, Figure 34.54). The bridge, essentially composed of two RC sections, responds in ramp fashion at both of its outputs (Trace B is "fixed" output, Trace C is "skewed" output). The "skewed" bridge half's capacitance is tuned by a varactor diode, biased from A1, and hence the control input. The comparators, referenced to 1/2 supply voltage, trigger (Traces D and E are C1 and C2 output, respectively) when their positive inputs exceed

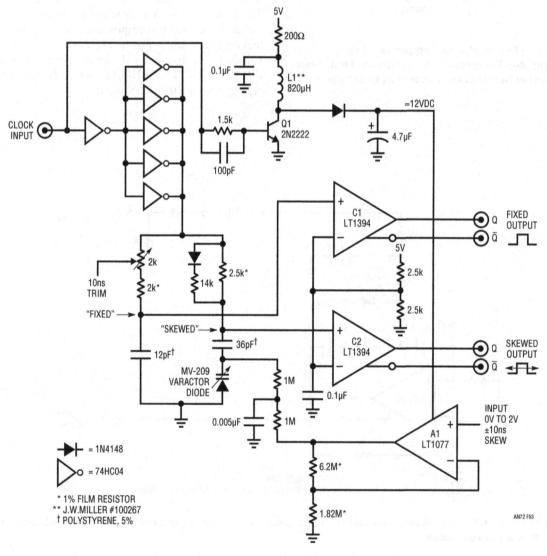

Figure 34.53 • Comparators Extract Phase Difference from Varactor-Tuned Bridge, Permitting Controllable Clock Skew

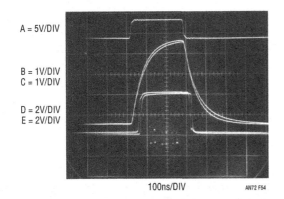

A = 5V/DIV

B = 1V/DIV
C = 1V/DIV

D = 2V/DIV
E = 2V/DIV

100ns/DIV   AN72 F54

Figure 34.54 • Clocked (Trace A), Varactor-Tuned Bridge Has Phase Shifted Outputs (Traces B and C). Comparators (Traces D and E) Digitize Information, Providing Output

the reference point. The time skew of this response is determined by imbalance in the bridge's RC time constants, which is controlled via the voltage input. The diode-resistor network across the 2.5k bridge resistor compensates for ramp-induced variation of varactor capacitance, enhancing control symmetry. Q1 and associated components form a simple voltage boost stage, enabling A1 to supply adequate varactor bias. The bridge's ratiometric operation permits almost 100:1 power supply rejection ratio over a $4.5V_{IN}$ to $5.5V_{IN}$ range. To trim this circuit, put in 2V and adjust the 2k potentiometer for 10ns

skew in the outputs. Over a 0V to 2V range, output skew will continuously vary from −10ns through 0, to 10ns.

Simple 10MHz voltage-to-frequency converter

Figure 34.55 is a voltage-to-frequency converter. A 0V to 2.5V input produces a 0Hz to 10MHz output with 40dB of dynamic range, 1% linearity and 400ppm/°C gain drift. Power supply rejection is 0.5% for 4.75V to 5.25V supply excursions.

To understand circuit operation, assume C1's positive input is slightly below its negative input. The input voltage causes a positive-going ramp at C1's positive input (Trace A, Figure 34.56). C1's output is low, biasing the CMOS inverters high. This allows current flow from diode Q1's collector, through the CMOS inverter supply pin to the 10pF capacitor. The 4.7μF capacitor provides high frequency bypass, maintaining low impedance at Q1's collector. Diode connected Q3 provides a path to ground. The voltage the 10pF capacitor charges to is a function of Q1's collector potential and Q3's drop. When the ramp at C1's positive input goes high enough, C1's output goes high and the paralleled inverters switch low (Trace B). This action pulls current from C1's positive input capacitor via the Q4-10pF route (Trace D). This current removal resets C1's positive input ramp to a potential slightly below ground, forcing C1's output low and the paralleled inverters high. The 8pF capacitor at C1's inverting output furnishes AC

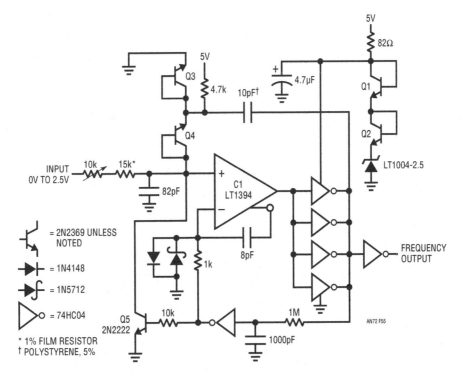

Figure 34.55 • Simple Charge Pump-Based 10MHz Voltage-to-Frequency Converter Has 40dB Dynamic Range, Operates from 5V Supply

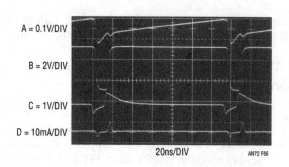

A = 0.1V/DIV

B = 2V/DIV

C = 1V/DIV

D = 10mA/DIV

20ns/DIV   AN72 F56

Figure 34.56 • Waveforms for the 10MHz Voltage-to-Frequency Converter. Charge Pump-Based Feedback Provides Linearity and Fast Response to Input

positive feedback to C1's negative input (Trace C). This ensures that C1's output remains high long enough for a complete discharge of the 10pF unit. The Schottky diode prevents C1's input from being driven outside its negative common mode limit. When the 8pF capacitor's feedback decays, C1 again switches high and the entire cycle repeats. The oscillation frequency depends directly on the input-derived current.

The LT1004 is the circuit's voltage reference, with Q1 and Q2 temperature compensating Q3 and Q4.

Start-up or overdrive can cause the circuit's AC-coupled feedback to latch. If this occurs, C1's output goes high, causing the paralleled inverters to go low. After a time

determined by the 1M-1000pF RC the associated lone inverter goes high. This lifts C1's negative input and grounds the positive input with Q5, initiating normal circuit action.

To calibrate this circuit, apply 2.5V and adjust the 10k potentiometer for a 10MHz output.

Precision 1Hz to 10MHz voltage-to-frequency converter

Significant performance improvements over the previous circuit are achievable if increased complexity is tolerable. The LT1394 and the LT1122 high speed FET amplifier combine to form a high speed V/F converter in Figure 34.57. A variety of circuit techniques are used to achieve a 1Hz to 10MHz output. Overrange to 12MHz ($V_{IN} = 12V$) is provided. This circuit has a wider dynamic range (140dB or 7 decades) than any commercially available unit. The 10MHz full-scale frequency is ten times faster than currently available monolithic V/Fs. The theory of operation is based on the identity $Q = CV$.

Each time the circuit produces an output pulse, it feeds back a fixed quantity of charge (Q) to a summing node (Σ). The circuit's input furnishes a comparison current at the summing node. The difference signal at the node is integrated in a monitoring amplifier's feedback capacitor. The amplifier controls the circuit's output pulse generator, completing a feedback loop around the integrating

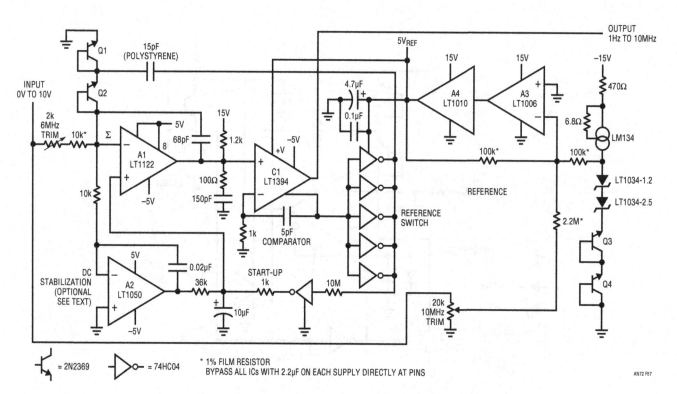

Figure 34.57 • A Very High Performance 1Hz to 10MHz Voltage-to-Frequency Converter. Linearity is 0.03% with 50ppm/°C Drift

amplifier. To maintain the summing node at zero, the pulse generator runs at a frequency that permits enough charge pumping to offset the input signal. Thus, the output frequency will be linearly related to the input voltage. A1 is the integrating amplifier.

0.05μV/°C offset drift performance is obtained by stabilizing A1 with A2, a chopper-stabilized op amp. A2 measures the DC value of the negative input, compares it to ground, and forces the positive input to maintain offset balance in A1. Note that A2 is configured as an integrator and cannot see high frequency signals. It functions only at DC and low frequency.

A1 is arranged as an integrator with a 68pF feedback capacitor. When a positive voltage is applied to the input, A1's output integrates in a negative direction (Trace A, Figure 34.58). During this period, C1's inverting output is low. The paralleled HCMOS inverters form a reference voltage switch. The reference voltage is established by the LM134 current source driven LT1034's and the Q3-Q4 combination. Additionally, a small input voltage-related term is summed into the reference, improving overall circuit linearity. A3-A4 provides low drift buffering, presenting a low impedance reference to the paralleled inverter's supply pin. The HCMOS outputs give low resistance, essentially errorless switching. The reference switch's output charges the 15pF capacitor via Q1's path.

When A1's output crosses zero, C1's inverting output goes high and the reference switch (Trace B) goes to ground. This causes the 15pF unit to dispense charge into the summing node via Q2's V_{BE}. The amount of charge dispensed is a direct function of the voltage the 15pF unit was charged to (Q = CV). Q1 and Q2 are temperature compensated by Q3 and Q4 in the reference string. The current through the 15pF unit (Trace C) reflects the charge pumping action. The removal of current from A1's summing junction (Trace D) causes the junction to be driven very quickly negative. The initial negative-going 15ns transient at A1's output is due to amplifier delay. The input signal feeds directly through the feedback capacitor and appears at the output. When the amplifier finally responds, its output (Trace A) slew limits as it attempts to regain

control of the summing node. The class A 1.2kΩ pull-up and the RC damper at A1's output minimize erroneous output movement, enhancing this slew recovery. The amount of time the reference switch remains at ground depends on how long it takes A1 to recover and the 5pF-1000Ω time constant at C1. This 60ns interval is long enough for the 15pF unit to fully discharge. After this, C1 changes state, the reference switch swings positive, the capacitor is recharged and the entire cycle repeats. The frequency at which this oscillation occurs is directly related to the voltage input-derived current into the summing junction. Any input current will require a corresponding oscillation frequency to hold the summing point at an average value of 0V.

Maintaining this relationship at megahertz frequencies places severe restrictions on circuit timing. The key to achieving 10MHz full-scale operating frequency is the ability to transmit information around the loop as quickly as possible. The discharge-reset sequence is particularly critical and is detailed in Figure 34.59. Trace A is the A1 integrator output. Its ramp output crosses 0V at the first left vertical graticule division. A few nanoseconds later, C1's inverting output begins to rise (Trace B), switching the reference switch to ground (Trace C). The reference switch begins to head towards ground about 16ns after A1's output crosses 0V. 2ns later, the summing point (Trace D) begins to go negative as current is pulled from it through the 15pF capacitor. At 25ns, C1's inverting output is fully up, the reference switch is at ground, and the summing point has been pulled to its negative extreme. Now, A1 begins to take control. Its output (Trace A) slews rapidly in the positive direction, restoring the summing point. At 60ns, A1 is in control of the summing node and the integration ramp begins again.

Start-up and overdrive conditions could force A1's output to go to the negative rail and stay there. The AC-coupled nature of the charge dispensing loop can preclude normal operation and the circuit may latch. The remaining HCMOS inverter provides a watchdog function for this condition. If A1's output remains negative the reference switch tries to stay at ground. The remaining inverter goes high, lifting A1's positive input. This causes A1's output to slew positive, initiating normal circuit action. The 1k-10μF

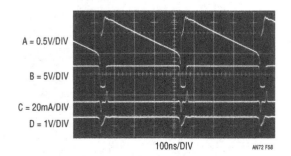

A = 0.5V/DIV

B = 5V/DIV

C = 20mA/DIV

D = 1V/DIV

100ns/DIV   AN72 F58

Figure 34.58 • Precision 10MHz Voltage-to-Frequency's Operating Waveforms. LT1122 Integrator Is Completely Reset in 60ns

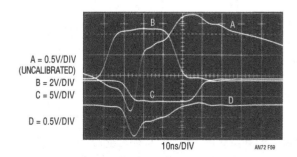

A = 0.5V/DIV
(UNCALIBRATED)

B = 2V/DIV

C = 5V/DIV

D = 0.5V/DIV

10ns/DIV   AN72 F59

Figure 34.59 • Detail of 60ns Reset Sequence (Whoosh!)

combination and the 10M-inverter input capacitance limit start-up loop bandwidth, preventing unwanted outputs.

The LM134 current source driving the reference string has a built in 0.33%/°C thermal coefficient, causing slight voltage modulation in the Q3-Q4 pair over temperature. This small change ($\approx$120ppm/°C) opposes the −120ppm/°C drift in the 15pF polystyrene capacitor, aiding overall circuit tempco.

To trim this circuit, apply exactly 6V at the input and adjust the 2kΩ potentiometer for 6.000MHz output. Next, put in exactly 10V and trim the 20k unit for 10.000MHz output. Repeat these adjustments until both points are fixed. A2's low drift eliminates a zero adjustment. If operation below 600Hz is not required, A2 and its associated components may be deleted.

Linearity of this circuit is 0.03% with full-scale drift of 50ppm/°C. Zero point error, controlled by A2, is 0.05Hz/°C.

Fast, high impedance, variable threshold trigger

A frequent requirement in instrumentation is a fast trigger with a variable threshold. Often, a high impedance input is also required. Figure 34.60 meets these requirements. Comparator C1 is the basic trigger, with threshold voltage set at its negative input. Source follower Q1 provides high impedance with about 2pF input capacitance and 50pA bias current. Normally, Q1's source bias point would be uncertain and drifty, but stabilization techniques eliminate this concern. A1 measures filtered versions of Q1's gate and source voltages. A1's output biases Q2, forcing Q1's channel current to whatever value is required to equalize A1's inputs, and hence Q1's gate and source voltages. A1's

input filtering and roll-off are far slower than input frequencies of interest; its action does not interfere with the circuit's main signal path. The 330pF capacitor prevents fast edges coupled through Q2's collector base junction from influencing A1's operation.

Q1 should contribute negligible timing error to minimize overall delay. Figure 34.61's photo verifies Q1's wideband operation. Trace B, Q1's source, lags the input (Trace A) by only 300ps. Input, FET buffer output and C1 output appear as Traces A, B and C, respectively in Figure 34.62. As before, the FET buffer is seen to contribute small timing error, and C1's output is about 8ns delayed from the input.

High speed adaptive trigger circuit

Line and fibre-optic receivers often require an adaptive trigger to compensate for variations in signal amplitude and DC offsets. The circuit in Figure 34.63 triggers on 2mV to 175mV signals from 100Hz to 45MHz while oper-

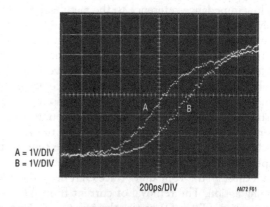

A = 1V/DIV
B = 1V/DIV

200ps/DIV   AN72 F61

Figure 34.61 • Trigger Buffer's 300ps Delay Minimizes Timing Error. 4GHz Sampling Oscilloscope's Output Is a Series of Dots

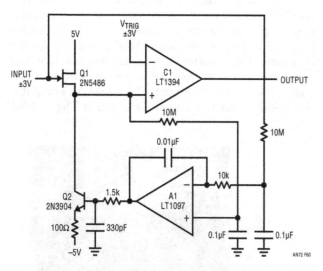

Figure 34.60 • Buffer Provides 2pF, 50pA Input Characteristics for Fast Trigger. Amplifier-Stabilized Biasing Eliminates FET Offset

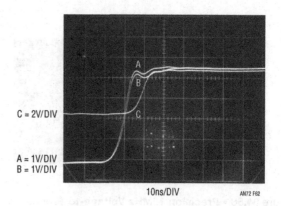

C = 2V/DIV

A = 1V/DIV
B = 1V/DIV

10ns/DIV   AN72 F62

Figure 34.62 • Input (Trace A), FET Source (Trace B) and Output (Trace C) Waveforms for the Trigger. Total Delay Is 8ns

ating from a single 5V rail. A1, operating at a gain of 15, provides wideband AC gain. The output of this stage biases a 2-way peak detector (Q1 through Q4). The maximum peak is stored in Q2's emitter capacitor, while the minimum excursion is retained in Q4's emitter capacitor. The DC value of the midpoint of A1's output signal appears at the junction of the 500pF capacitor and the 3MΩ units. This point always sits midway between the signal's excursions, regardless of absolute amplitude. This signal-adaptive voltage is buffered by A2 to set the trigger voltage at the LT1394's positive input. The LT1394's negative input is biased directly from A1's output. The LT1394's output, the circuit's output, is unaffected by >85:1 signal amplitude variations. Bandwidth limiting in A1 does not affect triggering because the adaptive trigger threshold varies ratiometrically to maintain circuit output.

Figure 34.64 shows operating waveforms at 45MHz. Trace A's input produces Trace B's amplified output at A1. The comparator's output is Trace C.

Split supply versions of this circuit can achieve bandwidths to 50MHz with wider input operating range (see Reference 17).

18ns, 500µV sensitivity comparator

The ultimate limitation on comparator sensitivity is available gain. Unfortunately, increasing gain invariably involves giving up speed. The gain vs speed trade-off in a fast comparator is usually a practical compromise designed to satisfy most applications. Some situations, however, require more sensitivity (e.g., higher gain) with minimal impact on

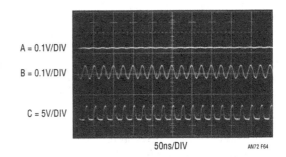

A = 0.1V/DIV

B = 0.1V/DIV

C = 5V/DIV

50ns/DIV   AN72 F64

Figure 34.64 • Adaptive Trigger Responding to a 40MHz, 5mV Input. Input Amplitude Variations from 2mV to 175mV Are Accommodated

speed. Figure 34.65's circuit adds a differential preamplifier ahead of the LT1394, increasing gain. This permits 500µV comparisons in 18ns. A parallel path DC stabilization approach eliminates preamplifier drift as an error source. A1 is the differential preamplifier, operating at a gain of 100. Its output is AC-coupled to the LT1394. A1 has poorly defined DC characteristics, necessitating some form of DC correction. A2 and A3, operating at a differential gain of 100, provide this function. They differentially sense a band limited version of A1's inputs and feed DC and low frequency amplified information to the comparator. The low frequency roll-off of A1's signal path complements A2-A3's high frequency roll-off. The summation of these two signal channels at the LT1394 inputs results in flat response from DC to high frequency.

Figure 34.66 shows waveforms for the high gain comparator. Trace A is a 500µV overdrive on a 1mV step

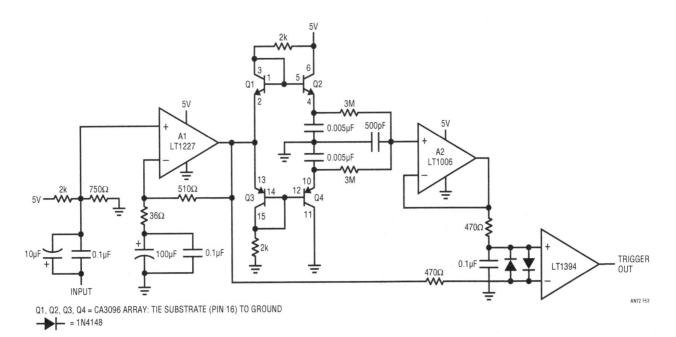

Q1, Q2, Q3, Q4 = CA3096 ARRAY: TIE SUBSTRATE (PIN 16) TO GROUND
▶️├ = 1N4148

Figure 34.63 • 45MHz Single Supply Adaptive Trigger. Output Comparator's Threshold Varies Ratiometrically with Input Amplitude, Maintaining Data Integrity over >85:1 Input Amplitude Range

applied to the circuit's positive input (negative input grounded). Trace B shows the resulting amplified step at A1's positive output. Trace C is A2's band limited output. A1's wideband output combines with A2's DC corrected information to yield the correct, amplified composite signal at the LT1394's positive input in Trace D. The LT1394's output is Trace E. Figure 34.67 details circuit propagation delay. The output responds in 18ns to a 500μV overdrive on a 1mV step. Figure 34.68 plots response time versus overdrive. As might be expected,

propagation delay decreases at higher overdrives. A1's noise limits usable sensitivity.

Voltage-controlled delay

The ability to set a precise, predictable delay has broad application in pulse circuitry. Figure 34.69's configuration sets a 0 to 300ns delay from a corresponding 0V to 3V control voltage. It takes advantage of the LT1394's speed and the clean dynamics of an emitter switched current source.

Q1 and Q2 form a current source that charges the 1000pF capacitor. When the trigger input is high (Trace A, Figure 34.70) both Q3 and Q4 are on. The current source is off and Q2's collector (Trace B) is at ground. The latch input at the LT1394 prevents it from responding and its output remains high. When the trigger input goes low, the LT1394's latch input is disabled and its output drops low. Q4's collector (Trace C) lifts and Q2 comes on,

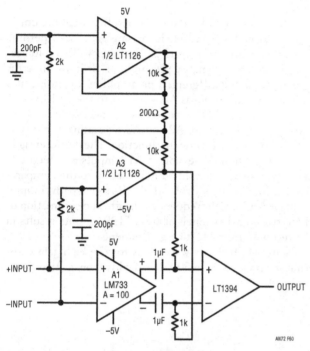

Figure 34.65 • Parallel Preamplified Paths Allow 18ns Comparator Response to 500μV Overdrive

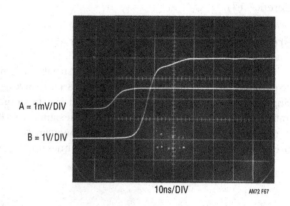

Figure 34.67 • Parallel Path Comparator Shows 18ns Response (Trace B) to 500μV Overdrive (Trace A)

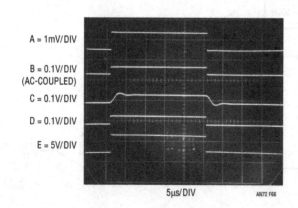

Figure 34.66 • 500μV Input (Trace A) Is Split into Wideband and Low Frequency Gain Paths (Traces B and C) and Recombined (Trace D). Comparator Output Is Trace E

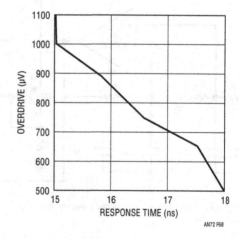

Figure 34.68 • Response Time vs Overdrive for the Composite Comparator

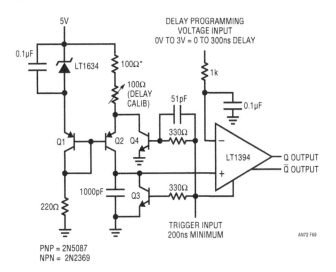

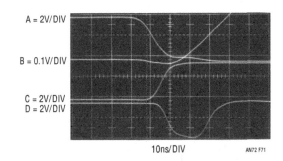

Figure 34.69 • Fast, Precise, Voltage-Controlled Delay. Emitter Switched Current Source Has Clean, Predictable Dynamics

Figure 34.71 • High Speed Expansion of Figure 34.70. Ramp (Trace B) Begins When Trigger (Trace A) Falls and Current Source Turns On (Trace C). Trace D is Output

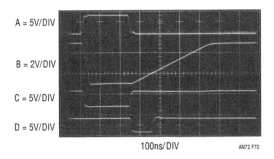

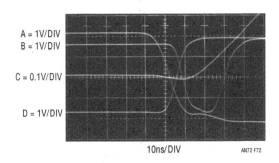

Figure 34.70 • Voltage-Controlled Delay's Waveforms. Programming Voltage Determines Delay Between Input (Trace A) Falling Edge and Output (Trace D) Rising Edge. High Linearity Timing Ramp (Trace B) Permits 1ns Accuracy and 100ps Repeatability

Figure 34.72 • Delay's Output Switching Begins with Trigger Falling Low (Trace A). Ramp (Trace C) Starts 3ns After Current Source Turn-On (Trace D). Output (Trace B) Begins 4ns Later

delivering constant current to the 1000pF capacitor (Trace B). The resulting linear ramp at the LT1394's positive input is compared to the delay programming voltage input. When a crossing occurs, the comparator goes high (Trace D). The length of time the comparator was low is directly proportional to the delay programming voltage. The fast switching and ramp linearity permits 1ns accuracy and 100ps repeatability. Figure 34.71, a high speed expansion of the current source turn-on, details the clean switching. Q4 goes off within 2ns of the trigger input (Trace A) dropping low, enabling the current source (Q2's emitter is Trace C). Concurrently, the 1000pF capacitor's ramp (Trace B) begins. The LT1394's output (Trace D) drops low about 7ns later, returning high after crossing (in this case) a relatively low programming voltage. Figure 34.72 juxtaposes the waveforms differently, permitting enhanced study of circuit timing. Switching begins with the input trigger

falling low (Trace A). The ramp (Trace C) begins 3ns after the current source turns on (Q2 emitter is Trace D). The output pulse (Trace B) begins about 4ns later.

To calibrate this circuit apply a trigger input and 3V to the programming input. Adjust the 100Ω trim for a 300ns width at the LT1394's output.

10ns sample-and-hold

Figure 34.73's 10ns sample-and-hold applies the previous circuit. This sample-hold circuit is extremely fast, although it can only be used with repetitive signals. Here, C1 drives differential integrator A1's input. Feedback from the integrator back to C1 closes a loop around the circuit. Figure 34.74 shows what happens when a waveform (Trace A) is applied to the input. C2 generates a trigger signal for a programmable delay generator identical to the previously described circuit. The 74121 one-shot is triggered from the delay's output. It's Q output produces a 30ns pulse which is fed into a logic network with its $\bar{Q}$ signal. The two inverter delays in Q's path give its associated gate a shorter duration output (Trace C) than $\bar{Q}$'s gate (Trace B). The last gate subtracts these two signals and generates a 10ns spike. This is inverted (Trace D) and fed

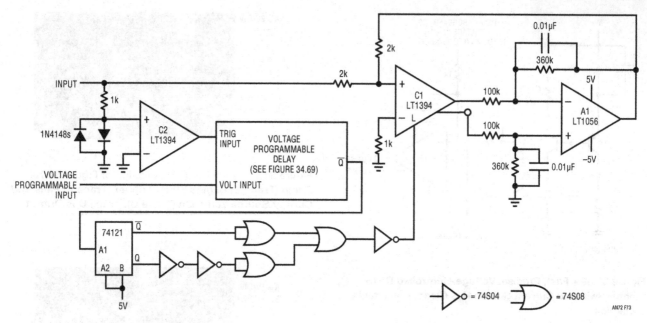

Figure 34.73 • 10ns Sample-and-Hold for Repetitive Signals. Feedback Loop Around Comparator and Programmable Delay Allow Controllable Sampling of Input

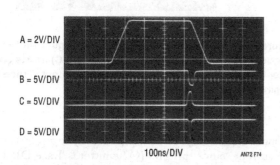

A = 2V/DIV

B = 5V/DIV

C = 5V/DIV

D = 5V/DIV

100ns/DIV   AN72 F74

Figure 34.74 • Sampling Pulse (Trace D) May Be Positioned at Desired Point on Input Waveform (Trace A)

to C1's latch pin. Each time the latch is enabled the comparator responds to the condition of the summing junction at its "+" input. If summing error is positive, A1 pulls current. If the error is negative, A1 sources current to the junction. After a number of input cycles, A1's output settles at a DC value that is the same as the level sampled during the time the latch is enabled. The delay's voltage programming allows the 10ns sampling "window" to be positioned anywhere on the input waveform.

Programmable, sub-nanosecond delayed pulse generator

The preceding circuit's 10ns wide sampling window limits sampling speed. Faster sampling requires narrower pulses.

This circuit uses an avalanche pulse generator[13] to create extremely short duration events. The combination of a controllable, calibrated delay and a very fast pulse generator has broad applicability in fast sampling circuitry.

In Figure 34.75, C1 and Q1 through Q4 form a voltage programmable delay identical to the one described in Figure 34.69. Q5, the LT1082 switching regulator and associated components comprise the avalanche pulse generator. The generator provides an 800ps pulse with rise and fall times inside 250ps. Pulse amplitude is 10V with a 50Ω source impedance.

The pulse generator requires high voltage bias for operation. The LT1082 switching regulator forms a high voltage switched mode control loop. The LT1082 pulse width modulates at its 40kHz clock rate. L1's inductive events are rectified and stored in the 2μF output capacitor. The adjustable resistor divider provides feedback to the LT1082. The 10k-1μF RC provides noise filtering.

The high voltage is applied to Q5, a 40V breakdown device, via the R3-C1 combination. The high voltage "bias adjust" control should be set at the point where free running pulses across R4 *just* disappear. This puts Q5 slightly below its avalanche point. When C1's output pulse is applied to Q5's base, it avalanches. The result is a quickly rising, very fast pulse across R4. C1 discharges, Q1's collector voltage falls and breakdown ceases. C1 then recharges to just below the avalanche point. At C1's next pulse this action repeats[14].

Note 13: See References 17, 20, 22, 27 and 28 for background on avalanche pulse generator theory and practice.
Note 14: This circuit is based on the operation of the Tektronix Type 111 pulse generator. See Reference 20.

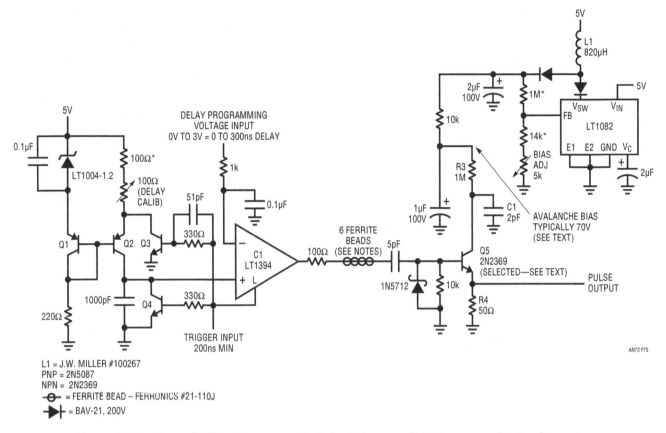

Figure 34.75 • Figure 34.69's Programmable Delay Triggers a Sub-Nanosecond Pulse Generator

Figure 34.76 shows the circuit input trigger (Trace A) that initiates the delay. After a time set by the programming input voltage, C1 goes high (Trace B). The avalanche pulse output is indicated in Trace C, but probe and oscilloscope bandwidth limitations prevent an accurate representation.

Figure 34.77, taken with a 3.9GHz bandpass instrument (Tektronix 661 with 4S2 sampling plug-in), shows more detail. Trace A is C1's output, and Trace B is the avalanche pulse. When avalanche occurs, Q5's reverse base current

rises so abruptly that C1's output cannot directly absorb it. The 100Ω resistor and the ferrite beads present impedance at frequency, allowing C1 to handle the load. Without this network, C1's positive-going output will completely reverse direction and ring severely before completing its transition, corrupting avalanche behavior. Even with these

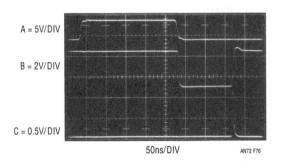

Figure 34.76 • Input Trigger (Trace A) Initiates Delay (Trace B) with Resultant Output Pulse (Trace C). Oscilloscope Bandwidth Limitations Prevent Accurate Output Pulse Representation

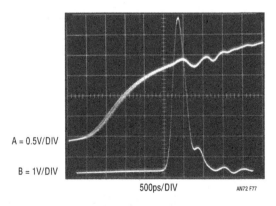

Figure 34.77 • 3.9GHz Sampling Oscilloscope Measures Delay Output and Avalanche Pulse. Pulse-Induced Loading Is Buffered by Ferrite Beads, but Artifacts Appear in Delay Output

components, artifacts of the avalanche induced base current are clearly visible in C1's trace.

The avalanche pulse measures 8V high with a 1.2ns base. Rise time is 250ps, with fall time indicating 200ps. The times are probably slightly faster, as the oscilloscope's 90ps rise time influences the measurement.[15]

Q5 may require selection to get avalanche behavior. Such behavior, while characteristic of the device specified, is not guaranteed by the manufacturer. A sample of 50 Motorola 2N2369s, spread over a 12-year date code span, yielded 82%. All "good" devices switched in less than 600ps. C1 is selected for a 10V amplitude output. Value spread is typically 2pF to 4pF. Ground plane type construction with high speed layout, connection and termination techniques is essential for good results from this circuit.

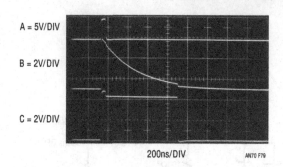

A = 5V/DIV

B = 2V/DIV

C = 2V/DIV

200ns/DIV   AN70 F79

Figure 34.79 • Waveforms for the Pulse Stretcher. Input (Trace A) Triggers Ramp Decay (Trace B), Resulting in Stretched Output (Trace C). Output Is a Composite of Input and Comparator-Based Response

Fast pulse stretcher

The minimum input pulse width required to operate a pulse stretcher is usually in the 5ns to 10ns range. Additionally, the rise and delay times are of the same order. Figure 34.78's circuit is considerably faster. It produces a stretched pulse from a 2ns width input with rise and delay times of 650ps.

The input pulse (Trace A, Figure 34.79) causes Q1 to conduct, charging the timing capacitor, C_T (Trace B). The

input pulse is also fed forward around C1, via D1, to the output (Trace C). Additionally, C_T's potential, buffered by Q3, is similarly fed forward to the output. C1 responds to C_T's charging by going high. Its output turns Q2 on, augmenting the outputs high state. C1's 7ns delay does not affect output delay or waveshape because the feedforward paths "fill in" the dead time before the comparator responds. The output pulse is a composite of the input and comparator-based response. The small change in output amplitude when the input ceases is related to this, but is not deleterious. When the input pulse falls, C1's output,

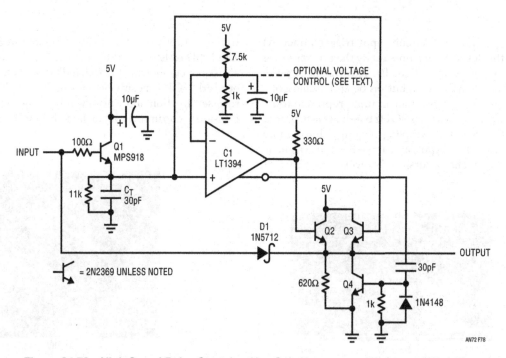

Figure 34.78 • High Speed Pulse Stretcher Has Sub-Nanosecond Delay and Rise Time

---

Note 15: I'm sorry, but 3.9GHz is the fastest 'scope in my house (as of November 1996).

and hence the circuits output, remains high until C_T discharges below C1's negative input. When C1 goes low its inverting output goes high, pulsing Q4 to pull the output down in 5ns.

The feedforward paths are crucial to circuit operation. The effect of D1's path is readily understood, but the C_T-originated route is less obvious. A good way to see the effect of C_T's path is to eliminate it. Figure 34.80's photo, taken with Q3's base open, is quite revealing. Trace A is the input pulse, Trace B the output and Trace C is C1's output. The absence of the C_T-based feedforward path is clearly evident. The output (again, Trace B) sags for 8ns before the comparator responds, restoring output amplitude.

Evaluating circuit operation requires a fast pulse generator and a wideband oscilloscope. Figure 34.81's photo, taken at $100\times$ Figure 34.79's sweep rate, shows the pulse stretcher's input-output relationship in a 3.9GHz sampled bandpass. Trace A is the input pulse and Trace B the output. As in Figure 34.79, output amplitude drops slightly when

the input ceases, but the logical high state is maintained. Also visible on the input's leading edge is a 0.5V amplitude 500ps aberration which occurs about 3V into the transition.

Figure 34.82 further increases sweep rate to examine the input (Trace A) and output (Trace B) leading edges. The output is delayed from the input by only 650ps, with rise time also about 650ps. The input transition aberration, now clearly visible, is due to the circuit's nonlinear input impedance. It occurs above a logical high level, and is acceptable.

Output pulse width is approximately equal to the input pulse width added to 25ns/pF of C_T. The ratiometric biasing of C1's inputs provides supply variation immunity from 5V $\pm$5%. The output width can be voltage controlled by biasing C1's negative input, but supply immunity will be compromised. The minimum input trigger width to maintain programmed output width within 1% is 2ns.

20ns response overvoltage protection circuit

It is often desirable to protect an expensive load from supply overvoltage. Overvoltage events may derive from supply failure or poor transient response. In Figure 34.83, Q1, a source follower, receives gate overdrive bias from the 12V bias supply and is saturated. The regulator driving Q1's drain takes feedback from the source, eliminating Q1's saturation resistance as an output impedance term.

C1 monitors the 3V output feeding the protected load. Under normal conditions C1's positive input is below its negative input, and its output is low. Q2 through Q5 are off and the load receives drive via Q1. Figure 34.84 shows what happens when an overvoltage event occurs. The 3V output (Trace A) begins to rise (note upward excursion beginning about center screen). This is detected at C1,

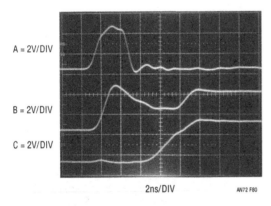

A = 2V/DIV

B = 2V/DIV

C = 2V/DIV

2ns/DIV   AN72 F80

Figure 34.80 • Results of Disconnecting C_T-Originated Feedforward Path. Output (Trace B) Sags for 8ns Before C1 (Trace C) Can Restore Its Amplitude

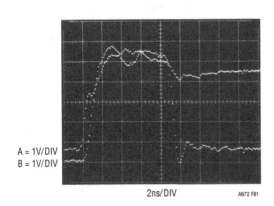

A = 1V/DIV
B = 1V/DIV

2ns/DIV   AN72 F81

Figure 34.81 • Pulse Stretchers Input-Output Relationship in a 3.9GHz Bandpass. Output Amplitude Drops When Input Decays, but Logic Level Is Maintained. Sampling Oscilloscope Display Is a Series of Dots

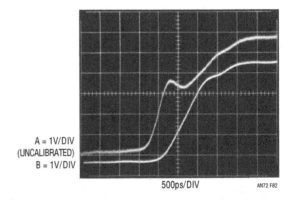

A = 1V/DIV
(UNCALIBRATED)
B = 1V/DIV

500ps/DIV   AN72 F82

Figure 34.82 • Pulse Stretcher Waveforms in 3.9GHz Bandpass Show 650ps Output Rise and Delay Times (Trace B). Nonlinear Loading Causes Input Transition Aberration (Trace A), but Is Not Deleterious. Trace Granularity Derives from Sampling Oscilloscope Operation

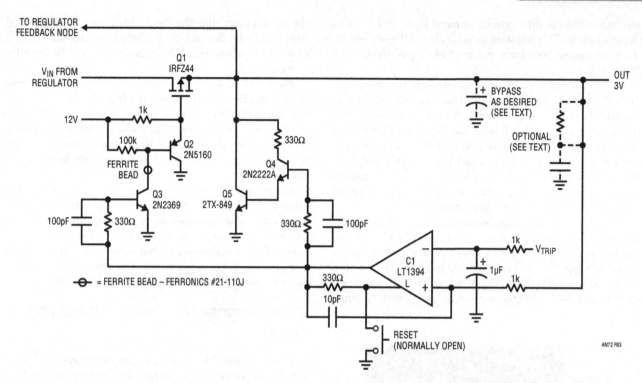

Figure 34.83 • A 20ns Response Time Overvoltage Protection Circuit. Latching Comparator Drives a Turn-Off Optimized Series—Shunt Switch

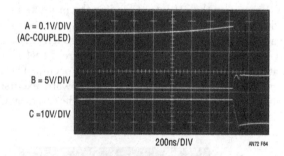

Figure 34.84 • Overvoltage Event (Note Upward Excursion, Trace A) Triggers Comparator (Trace B), Resulting in Gate Bias Collapse (Trace C)

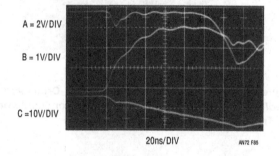

Figure 34.85 • Detail of Protection Circuit's Behavior. Output Amplitude Excursion (Trace A) Triggers Comparator (Trace B), Resulting Gate Drive Removal (Trace C). Overvoltage is Arrested in 20ns, Complete Shutdown Requires 150ns

and its output (Trace B) goes high. Q2 and Q3 come on very quickly, pulling down Q1's gate (Trace C). Q4 and Q5, slower devices, turn on after Q2-Q3, and shunt Q1's residual output to ground without experiencing excessive current. C1's output is fed via a 330Ω resistor to its latch pin. This causes C1 to latch high, preventing any output until the overvoltage cause is corrected. Reset is accomplished by breaking the latch with the normally open reset switch.

The switching is arranged to optimize turn-off time; Figure 34.85 shows just how fast the circuit is. As before, Trace A is the 3V output; Trace B, C1's output and Trace C, Q1's gate. The output's amplitude (Trace A) excursion begins just prior to the second vertical division. C1

responds (Trace B) by going high, turning on Q2 and Q3. This initial turn-on pulls Q1's gate downwards (Trace C), arresting the output excursion in 20ns. As Q2 pulls charge out of Q1, gate bias decays. When Q4 and Q5 come on, Q1 is out of saturation and the output drops rapidly. The overvoltage event is arrested in 20ns with total shutdown taking 150ns. Bypassing of Q1's source is optional—it will slow down the overvoltage rise time, but also restricts turn-off time. Similarly, the optional RC filter will eliminate noise-induced nuisance tripping at the expense of response time.

References

1. Orwiler, Bob, "Oscilloscope Vertical Amplifiers," Tektronix, Inc., Concept Series (1969).
2. Williams, Jim, "Techniques and Equipment for Current Measurement," Appendix C, in "Step-Down Switching Regulators," Linear Technology Corporation, Application Note 35 (August 1989).
3. Addis, John, "Fast Vertical Amplifiers and Good Engineering," in *Analog Circuit Design; Art, Science and Personalities*, Butterworths (1991).
4. Brown, Lloyd A., "The Story of Maps" pp. 226–240. Little, Brown. Boston, Massachusetts (1949).
5. Willsberger, Johann, "Clocks and Watches," Dial Press (1975).
6. Mercer, Vaudrey, "John Arnold and Son – Chronometer Makers 1762–1843," Antiquarian Horological Society, London (1972).
7. Tektronix, Inc., "Type 1S1 Sampling Plug-In Operating and Service Manual," Tektronix, Inc. (1965).
8. Mulvey, J., "Sampling Oscilloscope Circuits," Tektronix, Inc., Concept Series (1970).
9. Williams, Jim, "About Probes and Oscilloscopes," Appendix B, in "High Speed Comparator Techniques," Linear Technology Corporation, Application Note 13 (April 1985).
10. Williams, Jim, "Evaluating Oscilloscope Overload Performance," Box Section A, in "Methods for Measuring Op Amp Settling Time," Linear Technology Corporation, Application Note 10 (July 1985).
11. Weber, Joe, "Oscilloscope Probe Circuits," Tektronix, Inc., Concept Series (1969).
12. Gilbert, Barrie, "Where Do Little Circuits Come From" in *Analog Circuit Design; Art, Science and Personalities*, Butterworths (1991).
13. Williams, Jim, "Should Ohm's Law Be Repealed?" in *Analog Circuit Design; Art, Science and Personalities*, Butterworths (1991).
14. Williams, J., "Simple Techniques Fine-Tune Sample-Hold Performance," *Electronic Design*, page 235. (November 12, 1981).
15. Bunze, V., "Matching Oscilloscope and Probe for Better Measurements," *Electronics*, pp. 88–93 (March 1, 1973).
16. Williams, Jim, "High Speed Comparator Techniques," Linear Technology Corporation, Application Note 13 (April 1985).
17. Williams, J. "High Speed Amplifier Techniques," Linear Technology Corporation, Application Note 47 (August 1991).
18. Dendinger, S., "One IC Makes Precision Sample and Hold," *EDN*. (May 20, 1977).
19. Pease, R. A., "Amplitude to Frequency Converter," U.S. patent #3,746,968. (Filed September 1972).
20. Tektronix, Inc., Type 111 Pretrigger Pulse Generator Operating and Service Manual, Tektronix, Inc. (1960).
21. Baker, R. H. "Boosting Transistor Switching Speed," *Electronics*, Vol. 31, pp. 190–193 (1957).
22. Williams, J., "Practical Circuitry for Measurement and Control Problems," Linear Technology Corporation, Application Note 61 (August 1994).
23. Hewlett-Packard, "Schottky Diodes for High Volume, Low Cost Applications," Hewlett-Packard Company, Application Note 942 (1973).
24. Tektronix, Inc., "Trigger Circuit—Peak-Peak Automatic Operation," Model 2235 Oscilloscope Service Manual, Tektronix, Inc. (1983).
25. Williams, J., "Circuit Techniques for Clock Sources," Linear Technology Corporation, Application Note 12 (October 1985).
26. Williams, J., "Designs for High Performance Voltage-to-Frequency Converters," Linear Technology Corporation, Application Note 14 (March 1986).
27. Haas, Isy, "Millimicrosecond Avalanche Switching Circuits Utilizing Double-Diffused Silicon Transistors," Fairchild Semiconductor, Application Note 8/2 (December 1961).
28. Beeson, R. H., Haas, I., Grinich, V. H., "Thermal Response of Transistors in the Avalanche Mode," Fairchild Semiconductor, Technical Paper 6 (October 1959).
29. Tektronix, Inc., "ABC's of Probes," Tektronix, Inc. (1991).
30. Mattheys, R. L., "Crystal Oscillator Circuits," Wiley, New York (1983).
31. Frerking, M. E., "Crystal Oscillator Design and Temperature Compensation," Van Nostrand Reinhold, New York (1978).

Appendix A
About level shifts

The LT1394's logic output will interface with many circuits directly. Many applications, however, require some form of level shifting of the output swing. With LT1394-based circuits this is not trivial because it is desirable to maintain very low delay in the level shifting stage. When designing level shifters, keep in mind that the TTL output of the LT1394 is a sink-source pair (Figure A1) with good ability to drive capacitance (such as feedforward capacitors). Figure A2 shows a noninverting voltage gain stage with a 15V output. When the LT1394 switches, the base-emitter voltages at the 2N2369 reverse, causing it to switch very quickly. The 2N3866 emitter-follower gives a low impedance output and the Schottky diode aids current sink capability.

Figure A3 is a very versatile stage. It features a bipolar swing that is set by the output transistor's supplies. This 3ns delay stage is ideal for driving FET

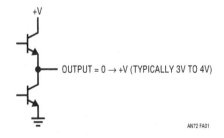

OUTPUT = 0 → +V (TYPICALLY 3V TO 4V)

AN72 FA01

Figure A1 • Simplified LT1394 Output Stage

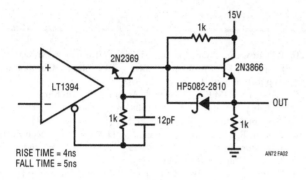

RISE TIME = 4ns
FALL TIME = 5ns

AN72 FA02

Figure A2 • Level Shift Has Noninverting Voltage Gain

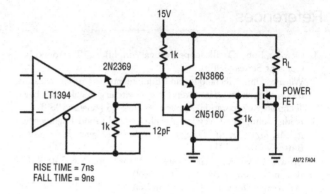

RISE TIME = 7ns
FALL TIME = 9ns

AN72 FA04

Figure A4 • Noninverting Voltage Gain Level Shift

switch gates. Q1, a gated current source, switches the Baker-clamped output transistor, Q2. The heavy feedforward capacitor from the LT1394 is the key to low delay, providing Q2's base with nearly ideal drive. This capacitor loads the LT1394's output transition (Trace A, Figure A5), but Q2's switching is clean (Trace B, Figure A5) with 3ns delay on the rise and fall of the pulse. Figure A4 is similar to A2 except that a sink transistor has replaced the Schottky diode. The two emitter-followers drive a power MOSFET that switches 1A at 15V. Most of the 7ns to 9ns delay in this stage occurs in the MOSFET and the 2N2369.

When designing level shifters, remember to use transistors with fast switching times and high f_Ts. To get the kind of results shown, switching times in the nanosecond range and f_Ts approaching 1GHz are required.

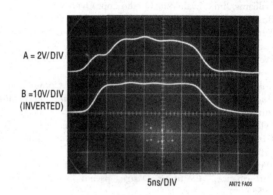

A = 2V/DIV

B = 10V/DIV
(INVERTED)

5ns/DIV   AN72 FA05

Figure A5 • Figure A3's Waveforms

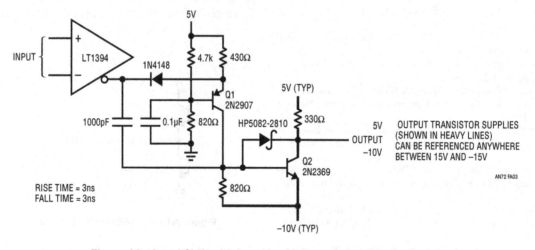

RISE TIME = 3ns
FALL TIME = 3ns

OUTPUT TRANSISTOR SUPPLIES
(SHOWN IN HEAVY LINES)
CAN BE REFERENCED ANYWHERE
BETWEEN 15V AND −15V

AN72 FA03

Figure A3 • Level Shift with Inverting Voltage Gain—Bipolar Swing

Appendix B
Measuring probe-oscilloscope response

The LT1394's 7ns response time and the circuitry it is used in will challenge the best test equipment. Many of the measurements made utilize equipment near the limit of its capabilities. It is a good idea to verify parameters such as probe and scope rise time and differences in delays between probes and even oscilloscope channels. Verifying the limits of wideband test equipment setups is a difficult task. In particular, the end-to-end rise time of oscilloscope-probe combinations is often required to assure measurement integrity. Conceptually, a pulse generator with rise times substantially faster than the oscilloscope-probe combination can provide this information. Figure B1's circuit does this, providing a 1ns pulse with rise and fall times inside 250ps. Pulse amplitude is 10V with a 50Ω source impedance. This circuit, built into a small box and powered by a 1.5V battery, provides a simple, convenient way to verify the rise time capability of almost any oscilloscope-probe combination.

The LT1073 switching regulator and associated components supply the necessary high voltage. The LT1073 forms a flyback voltage boost regulator. Further voltage step-up is obtained from a diode-capacitor voltage step-up network. L1 periodically receives charge and its flyback discharge delivers high voltage events to the step-up network. A portion of the step-up network's DC output is fed back to the LT1073 via the 10M-24k divider, closing a control loop.

The regulator's 90V output is applied to Q1 via the 1M-2pF combination. Q1, a 40V breakdown device, nondestructively avalanches when C1 charges high enough. The result is a quickly rising, very fast pulse across R4. C1 discharges, Q1's collector voltage falls and breakdown ceases. C1 then recharges until breakdown again occurs. This action causes free running oscillation at about 200kHz.[1,2] Figure B2 shows the output pulse. A 12.4GHz sampling oscilloscope measures the double-terminated pulse at 4.8V high with about a 700ps base. Rise time is 216ps, with fall time 232ps. There is a slight hint of ring after the falling edge, but it is well controlled.

Q1 may require selection to get avalanche behavior. Such behavior, while characteristic of the device specified, is not guaranteed by the manufacturer. A sample of 50 Motorola 2N2369s, spread over a 12-year date code span, yielded 82%. All good devices switched in less than 650ps. C1 is selected for a 10V amplitude output. Value spread is typically 2pF to 4pF. Ground plane type construction with high speed layout techniques is essential for good results from this circuit. Current drain from the 1.5V battery is about 5mA.

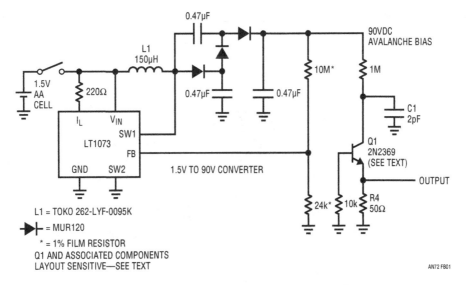

Figure B1 • 250ps Rise/Fall Time Avalanche Pulse Generator

Note 1: This method of generating fast pulses borrows heavily from the Tektronix type 111 Pretrigger Pulse Generator. See References 17, 20, 22, 27 and 28.
Note 2: If desired, the avalanche pulse generator may be externally triggered. See Figure 34.75 and associated text. See also References 20 and 22.

Figure B3 shows the physical construction of the actual generator. Power, supplied from a separate box, is fed into the generator's enclosure via a BNC connector. Q1 is mounted *directly* at the output BNC connector, with grounding and layout appropriate for wideband operation. Lead length, particularly Q1's and C1's, should be experimented with to get best output pulse purity. Figure B4 is the complete unit.

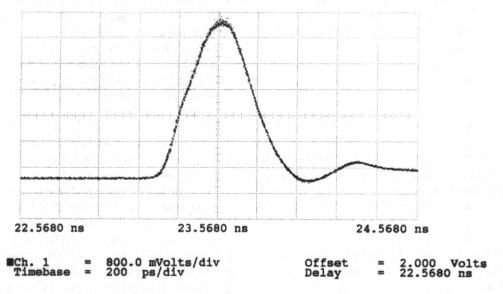

22.5680 ns	23.5680 ns	24.5680 ns

■Ch. 1	=	800.0 mVolts/div	Offset	=	2.000 Volts
Timebase	=	200 ps/div	Delay	=	22.5680 ns

Figure B2 • The Avalanche Pulse Generator's Output Monitored on a Hewlett-Packard 54120B 12GHz Sampling Oscilloscope. Double-Terminated Output Reduces Pulse Amplitude

(Courtesy of T. Hornak, Hewlett-Packard Laboratories)

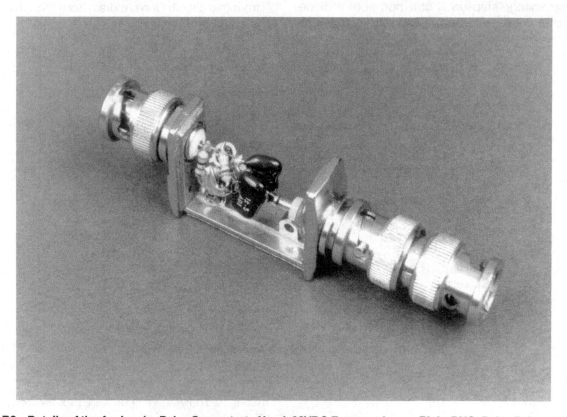

Figure B3 • Details of the Avalanche Pulse Generator's Head. 90VDC Enters at Lower Right BNC, Pulse Exits at Top Left BNC. Note Short Lead Lengths Associated with Output

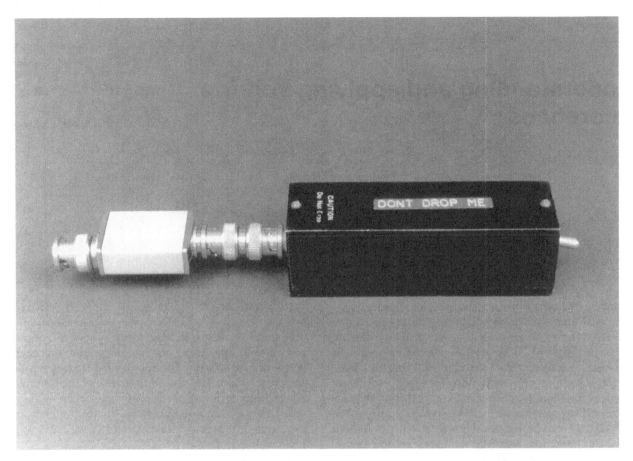

Figure B4 • The Packaged Avalanche Pulser. 1.5V-to-90V Converter Is in the Black Box. Avalanche Head Is at Left

Understanding and applying voltage references

35

Mitchell Lee

Specifying the right reference and applying it correctly is a more difficult task than one might first surmise, considering that references are only 2- or 3-terminal devices. Although the word "accuracy" is most often spoken in reference to references, it is dangerous to use this word too freely because it can mean different things to different people. Even more perplexing is the fact that a reference classified as a dog in one application is a panacea in another. This chapter will familiarize the reader with the various aspects of reference "accuracy" and present some tips on extracting maximum performance from any reference.

As with other specialized electronic fields, the field of monolithic references has its own vocabulary. We've already learned the first word in our reference vocabulary, "accuracy." This is the yardstick with which references are graded and compared. Unfortunately, there are at least five or six good units for gauging accuracy. To keep you from reaching a full understanding of the topic, industry pundits use a special technique called "unit-hopping" to confuse and confound everyone from newcomer to seasoned veteran. You mention an accuracy figure and the pundit quickly hops to a new unit so that you cannot follow his line of reasoning. Figure 35.1 neutralizes the pundits' callous intentions and allows its possessor to unit-hop with equal ease and full comprehension. Refer to Figure 35.1 as you read this chapter.

Today's IC reference technology is divided along two lines: bandgap references, which balance the temperature coefficient of a forward-biased diode junction against that of a ΔV_{BE} (see Appendix B); and buried Zeners (see Appendix A), which use subsurface breakdown to achieve outstanding long-term stability and low noise. With few exceptions, both reference types use additional on-chip circuitry to further minimize temperature drift and trim output voltage to an exact value. Bandgap references are generally used in systems of up to 12 bits; buried Zeners take over from there in higher accuracy systems.

In circuits and systems, monolithic references face competition from discrete Zener diodes and 3-terminal voltage regulators only where accuracy is not a concern. 5% Zeners and 3% voltage regulators are commonplace; these represent 4- or 5-bit accuracy. At the other end of the spectrum—laboratory standards—the performance of the best monolithic references is exceeded only by saturated Weston cells and Josephson arrays, leaving monolithic references in command of every conceivable circuit and system application.

Reference accuracy comprises multiple electrical specifications. These are summarized in Table 35.1. Most commonly specified by circuit designers is *initial accuracy*. This is a measure of the output voltage error expressed in percent or in volts. Initial accuracy is specified at room temperature (25°C), with a fixed input voltage and zero load current, or for shunt references, a fixed bias current.

Tight initial accuracy is a concern in systems where calibration is either inconvenient or impossible. More commonly, absolute accuracy is only a secondary concern, as a final trim is performed on the finished product to reconcile the summation of all system inaccuracies. A final trim effects considerable cost savings by eliminating the need for tight initial accuracy in every reference, DAC, ADC, amplifier and transducer in the system.

Monolithic reference initial accuracy ranges from 0.02% to 1%, representing 1LSB error in 6-bit to 12-bit systems. Weston cells and Josephson arrays clock in at 1ppm to 10ppm and 0.02ppm initial accuracy, respectively (0.02ppm is less than 1LSB error in a 25-bit system).

Temperature-induced changes in reference output voltage can quickly overshadow a tight initial accuracy specification. Considerable effort is therefore expended to minimize the *temperature coefficient* (tempco) of a reference. Most references are guaranteed in the range of 2ppm/°C to 40ppm/°C, with a few devices falling outside this range. A properly applied LTZ1000 temperature stabilized reference can demonstrate 0.05ppm/°C.

Analog Circuit and System Design: A Tutorial Guide to Applications and Solutions. DOI: 10.1016/B978-0-12-385185-7.00035-4

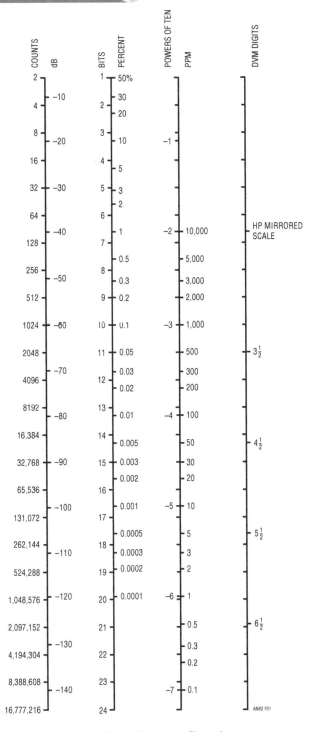

Figure 35.1 • Accuracy Translator

Table 35.1 Reference Accuracy Specifications		
PARAMETER	**DESCRIPTION**	**PREFERRED UNIT(S)**
Initial Accuracy	Initial Output Voltage at 25°C	V, %
Temperature Coefficient	$\dfrac{V_{MAX} - V_{IN}}{\text{Total Temperature Range}}$	ppm/°C
Long-Term Stability	Change in Output vs Time Measured Over 1000 Hours or More	ppm√kh
Noise	0.1Hz to 10Hz	μV_{P-P}, ppm_{P-P}
	10Hz to 1kHz	μV_{RMS}, ppm_{RMS}

are applied to the equation shown, resulting in an average temperature coefficient expressed in V/°C. This is further manipulated to find ppm/°C, as used in the data sheet. The tempco is an average over the operating range, rather than an incremental slope measured at any specific point. In the case of the LT1021 and LT1236, the incremental slope at 25°C is also guaranteed.

A data sheet figure for tempco can be used to directly calculate the output voltage tolerance over the entire operating temperature range. A device with a tempco of 10ppm/°C, specified for 0°C to 70°C, could drift up to 700ppm from the initial value (about 3 counts in a 12-bit system). A 0.1% reference with 700ppm tempco error is guaranteed 0.17% accurate over its entire operating temperature range.

Two exceptions to this rule are the LT1004 and LT1034, which simply guarantee absolute output voltage accuracy over the entire operating temperature range. The LT1009

Tempco is specified as an average over the operating temperature range in units of ppm/°C or mV/°C. This average is calculated in what is called the "box" method. Figure 35.2 shows how box method tempco figures are defined and calculated. The reference in question (LT®1019 bandgap) is tested over the specified operating temperature range. The minimum and maximum recorded output voltages

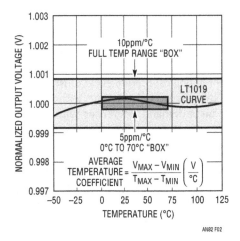

Figure 35.2 • The Box Method Expresses Absolute Output Accuracy Over Temperature as a Drift Term

and LT1029 use a combination of the two, called the "bow tie" or "butterfly" method (see the LT1009 data sheet for a detailed explanation).

Neither the bandgap nor the buried Zener, in their basic form, are inherently low drift. Special on-chip circuitry is used to improve the tempco of the reference core. A buried Zener is first-order compensated against temperature changes by adding a P-N junction diode. The Zener itself measures $+2mV/°C$ and the diode $-2mV/°C$. The combination of the two in series cancel to about $0.2mV/°C$ ($\approx 30ppm/°C$) out of a total of 7V. Interestingly, this is very close to the tempco of a saturated Weston cell, which measures $-40\,\mu V/°C$, or $-39ppm/°C$. Weston cells are held in a temperature-controlled bath; monolithic buried Zener references are further compensated against temperature changes by carefully adding fractional V_{BE} and/or ΔV_{BE} terms to the output. Post-manufacturing trims are used on both bandgap and buried Zener products to further minimize tempco of the finished reference.

Another detractor from accuracy is *long-term stability*. The output of a reference changes, usually in one direction, as it ages. The effect is logarithmic; that is, the output changes less and less as time progresses. The units of long-term stability, $ppm/\sqrt{kh}$ (kh=1000 hours), reflect the logarithmic decline of the output change vs time. Because long-term changes in the output are small and occur over the course of months or years, it is impossible to devise an affordable manufacturing test to guarantee the true stability of all references. Instead, this parameter is characterized by aging dozens of units in a temperature-controlled chamber at 25°C to 30°C for 1000 hours or more. Note that the absolute temperature is unimportant, but it must remain invariant during the course of the test. Mathematically extrapolating long-term stability data from high temperature, accelerated life tests leads to erroneously optimistic room temperature results.

When long-term stability is guaranteed, it is done by means of a 4-week burn-in, during which multiple output voltage measurements are made. Even with this elaborate, costly procedure, the guaranteed limit is about three to four times the typical drift.

Unless the product is designed for frequent calibration or is relatively low performance, long-term stability may be an important aspect of reference performance. Products designed for a long calibration cycle must hold their accuracy for extended periods of time without intervention. These products demand references with good long-term stability. You can expect buried Zeners to perform better than $20ppm/\sqrt{kh}$, and bandgaps between 20ppm and $50ppm/\sqrt{kh}$. Some of this drift is attributed to the trim and compensation circuitry wrapped around the reference core. The LTZ1000 dispenses with trim and compensation overhead in favor of an on-chip heater. The remaining Zener/diode core drifts $0.5ppm/\sqrt{kh}$ in the first year of operation, approaching the stability of a Weston cell.

Most of the long-term stability figures shown in LTC reference data sheets are for devices in metal can packages, where assembly and package stresses are minimized. You can expect somewhat less performance for the same reference in a plastic package.

One last factor that affects accuracy is short-term variation of output voltage, otherwise known as noise. Reference noise is typically characterized over two frequency ranges: 0.1Hz to 10Hz for short-term, peak-to-peak drift, and 10Hz to 1kHz for total "wideband" RMS noise. Noise voltage is usually proportional to output voltage, so the output noise expressed in ppm is constant for all voltage options of any given reference. Wideband noise ranges from 4ppm to 16ppm RMS for bandgap references, to 0.17ppm to 0.5ppm RMS for buried Zeners. Noise improves with increased reference current, regardless of reference type. But since the reference core operating current is set internally, the noise characteristics cannot be changed except by external filtering (the LT1027 features a noise filtering pin). The LT1034 and LTZ1000 buried Zeners are externally accessible, allowing the user to increase the bias current and reduce noise.

Adding output bypassing or external compensation will affect the character of a reference's noise. In particular, if the compensation is "peaky," the spot noise will likely rise to a peak somewhere in the 100Hz to 10kHz range. Critical damping will eliminate this noise peak.

Reference noise can affect the dynamic range of a high resolution system, obscuring small signals. Low frequency noise also complicates the measurement of output voltage. Modern, high accuracy digital voltmeters can average many readings to help filter low frequency noise effects and provide a stable reading of a reference's true output voltage.

Essential features

There are two styles of references: shunt, functionally equivalent to a Zener diode; and series, not unlike a 3-terminal regulator. Bandgaps and buried Zeners are available in both configurations (see Figure 35.3). Some series references are designed to also operate in shunt mode by simply biasing the output pin and leaving the input pin

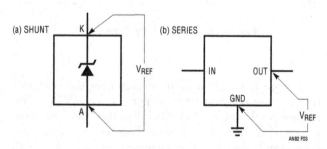

Figure 35.3 • References Are Supplied in Either 2-Terminal Zener Style (a) or 3-Terminal Voltage Regulator Style (b)

open circuit. Series-mode references have the advantage that they draw only load and quiescent current from the input supply, whereas shunt references must be biased with a current that exceeds the sum of the maximum quiescent and maximum expected load currents. Since they are biased by a resistor, shunt references can operate on a very wide range of input voltages.

About half of LTC's reference offerings include a pin for external (customer) trimming. Some are designed for precision trimming of the reference output, whereas others have a wide trim range, allowing the output voltage to be adjusted several percent above or below the intended operating point.

If load current steps must be handled, transient response is important. Transient response varies widely from reference to reference and comprises three distinct qualities: turn-on characteristics, small-signal output impedance at high frequency and settling behavior when subjected to a fast, transient load. References exhibit these qualities because almost all contain an amplifier to buffer and/or scale the output.

The LT1009 is optimized for fast start-up characteristics, and it settles in a little over 1µs, as shown in Figure 35.4. For some references, optimum settling is obtained with an external compensation network. As shown in Figure 35.5, a 2µF/2Ω damper optimizes the settling and high frequency output impedance of an LT1019 reference. Fastest settling is obtained with an LT1027, which settles to 13 bits accuracy in 2µs. This impressive feat is illustrated by the oscillograph of Figure 35.6, which clearly shows the output recovering from a 10mA load step.

Reference pitfalls

References look deceptively simple to use, but like any other precision product, maximum performance is not necessarily easy to achieve. Here are a few common pitfalls reference users face, and ways to beat them.

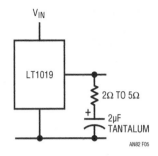

Figure 35.5 • . Optimum Settling Realized with RC Compensation at Output

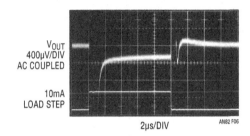

2µs/DIV

Figure 35.6 • The LT1027 is Optimized for Fast Settling in Response to Load Steps

Current-hungry loads

Most references are specified for maximum load currents (or shunt currents) of 10mA to 20mA. Nevertheless, best performance is not obtained by running the reference at maximum current. A number of effects, including thermal gradients across the die and thermocouples formed between the leads and external circuit connections, may limit the short-term stability of the output voltage. Adding an external pass transistor, as shown in Figure 35.7, removes the load current from the reference. For loads greater than 300µA, the pass transistor carries almost all

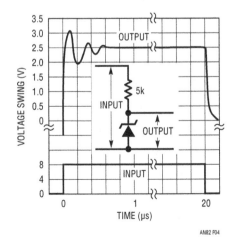

Figure 35.4 • The LT1009 is Optimized for Rapid Settling at Power-Up

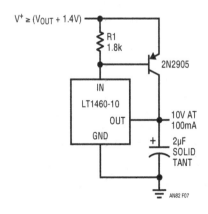

Figure 35.7 • An External Transistor is Useful for Boosting Output Current as Well as for Removing Load Current from the Reference. This Trick Works On All 3-Terminal References

of the current and eliminates short-term thermal drift. This circuit is also useful for applications requiring more than 20mA, and easily supports up to 100mA, limited only by transistor beta and dissipation.

"NC" pins

If references need only two or three external connections, why are they supplied in 8-pin packages? There are several reasons, but the one we'll cover here is post-package trimming. To guarantee tight output tolerances, some factory trimming is necessary after the device has been packaged. In packaged form we no longer have direct access to the die, so the extra pins on an 8-pin package are used to effect post-package trimming.

For some ICs, "NC" means "this pin is floating, you can hook it up to whatever you want." In the case of a reference, it means "don't connect anything to this pin." That includes ESD and board leakage, as well as intentional connections. External connections will, at best, cause output voltage shifts and, at worst, permanently shift the output voltage out of spec.

A similar caution applies to the TRIM pin on references with adjustable outputs. The TRIM pin is akin to an amplifier's summing node; do not inject current into a TRIM pin unless you want to trim the output, of course. Here board leakage or capacitive coupling to noise sources are pitfalls to avoid.

Board leakage

A new specter has entered the field of references: board leakage caused by the residues of water-soluble flux. The effect is not unlike that produced by the sticky juice extravasated from a ruptured electrolytic capacitor. Leakage from ground, supply rails and other circuit potentials into NC, trim and other sensitive pins through conductive flux residues will cause output voltage shifts. Even if the leakage paths do not shift the reference out of spec, external leakage can manifest itself as long-term output voltage drift, as the resistance of the flux residue changes with shifts in relative humidity and the diffusion of external contaminants. Water-soluble flux residues must be removed from the board and package surfaces, or completely avoided. In one case, the author observed an LT1009 shifted out of spec by a gross leakage path of approximately 80kΩ between the trim pin and a nearby power supply trace. The leakage was traced to water-soluble flux.

Figure 35.8 shows how a good reference can go bad with only a very small leakage. A hypothetical industrial control board contains an LT1027A producing 5V for various data acquisition circuits. A nearby trace carries 24V. Just 147MΩ leakage into the noise filtering pin (NR) causes a typical device to shift +200ppm, and out of spec. Clearly, a 24V circuit trace doesn't belong anywhere near a 0.02% reference. This example is oversimplified but clearly demonstrates the potential for disaster.

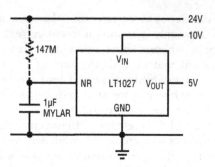

Figure 35.8 • Board Leakage Can Wreak Havoc with a Precision Reference. Here, a 147MΩ Leakage Path to 24V Pushes the 5V Output Out of Spec

A tightly packed circuit board may leave no choice but to agglomerate incompatible traces. In this case, use a guard ring to eliminate reference shift (see Figure 35.9). The output of the reference is divided down to 4.4V, equal to the potential on the NR pin, and used to bias a guard ring encircling the trace connecting NR to the noise filter capacitor. This reduces the effect of board leakage paths by more than two orders of magnitude, shunting the errant leakage away from the guarded traces.

Trim-induced temperature drift

About half of LTC's reference offerings include a pin for external (customer) trimming. Trimming may be necessary to calibrate the system, but it can also adversely affect the tempco of the reference. For example, in the LT1019 bandgap reference, external trim resistors won't match the tempco of the internal resistors. The mismatch causes a small (1ppm/°C) worst-case shift in the output voltage tempco, as explained on the data sheet. The LT1021-5 and LT1236-5 standard trim circuit can be modified, as shown in Figure 35.10, to prevent upsetting the references' inherently low temperature coefficients. Trimming the LT1027 has little effect on the output voltage tempco, and it needs no special consideration. Always check the reference data sheet for specific recommendations.

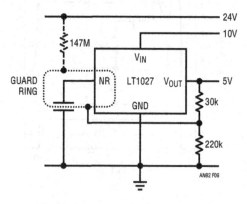

Figure 35.9 • Adding a Guard Ring Protects Against Errant Leakage Paths

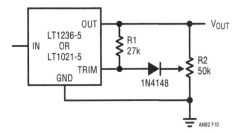

Figure 35.10 • The LT1021 or LT1236 Output Trim is Made Temperature Insensitive by the Addition of a Diode and a Resistor

Burn-in

Most manufacturers of high-accuracy systems run their products through a burn-in procedure. Burn-in solves two problems at once: it relieves stresses built into the reference and circuit board during assembly and it ages the reference beyond the highest long-term drift region, which occurs when power is first applied to the part. A typical burn-in procedure calls for operating the board at 125°C ambient for 168 hours. If the main concern is stress relief, a shorter, unpowered burn-in cycle can be used.

Board stress

Burn-in can help "relax" a stuffed board, but additional mechanical stress may be introduced when the board is mounted into the product. Stress has a directly measurable effect on reference output. If the stress changes over a period of time, it may manifest itself as unacceptable long-term drift. Circuit boards are not perfectly elastic, so bending forces may cause permanent deformation and a permanent step-change in reference output voltage. Devices in metal (TO-5 and TO-46) packages are largely immune to board stress, owing to the rigidity of the package and the flexibility of the leads. Plastic and surface mount packages are another matter.

Board stress effects are easily observed by monitoring the output of a reference while applying a bending force to the board. A controlled experiment was performed to measure the effect of board stress on an LT1460CS8-2.5 surface mount reference. Devices were mounted in the center of 7″×9″ rectangular boards, as shown in Figure 35.11. The boards were then deflected out-of-plane 18 mils per inch, as shown in steps 1 through 4. Figure 35.12 shows the net effect on the output of one representative sample measured over eight cycles of flexure.

The original board showed about 60ppm peak-to-peak shift. The board was then slotted on a vertical mill, forming a 0.5″×0.5″ tab with the reference located in its center (also illustrated in Figure 35.11). The test continued with the slotted configuration, and the output voltage variations were reduced to ±1 count (10μV) on the meter, or approximately 4ppm peak-to-peak. This represents a ten-fold improvement in stress-induced output voltage shift.

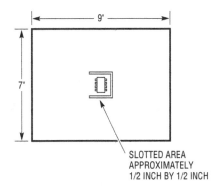

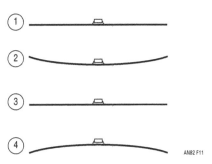

Figure 35.11 • Reference Sensitivity to Stress Was Evaluated by Assembling Devices On a 7″×9″ Circuit Board and Flexing, as Shown in Steps 1 Through 4

Several other techniques can be employed to minimize this effect, without resorting to a milled board. Anything that can be done to restrict the board from bending is helpful. A small, thick board is better than a large, thin board. Stiffeners help immunize the board against flexure. Mount the circuit board with grommets, flexible standoffs or card-cage style so that minimal force is applied to the mounting holes and board.

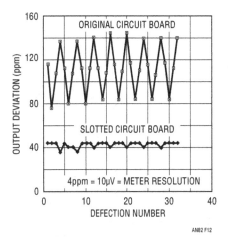

Figure 35.12 • Isolating Stress by Slotting the Circuit Board Reduces Reference Variations by More Than an Order of Magnitude (LTC1460S8-2.5)

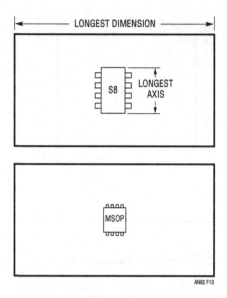

Figure 35.13 • Arranging the Longest Axes of the Board and Package in Perpendicularity Minimizes Stress-Induced Output Changes

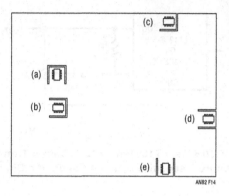

Figure 35.14 • Slotting the Area Around the Reference Can Help Isolate it from Board Stress if Properly Applied (See Text)

Part placement and orientation are just as important. If a board is squeezed from opposite edges, the bending force tends to concentrate in a line down the center. Locate the reference away from the middle of the board. Since the longer side of a board is more flexible than the shorter, locate the reference along the shorter edge. These recommendations are generalities; the placement, mounting method and orientation of other components and assemblies on the circuit board will influence the mechanical strengths and weaknesses of the circuit board.

Bench tests indicate that the strongest axis for plastic packages is along the shorter dimension of the body of the plastic. Figure 35.13 shows the correct orientation for surface mount parts. Note that the part's longest axis is placed perpendicular to that of the circuit board. The devices in Figure 35.13 are shown in the center of the board for illustrative purposes only; comments about placement still apply.

In spite of all precautions, extraneous effects may adversely affect the reference's resistance to board stress. Watch out for adhesives and solder and flux debris under the package. These will create pressure points and induce unpredictable stresses in the package. If a board has been subjected to a high bending force, some of the glass fibers and layers may break or shear apart, permanently weakening the board. Subsequent bending forces will concentrate their stress at points thus weakened.

Figure 35.14 shows various schemes for routing stress-relief slots on a circuit board, along with optimum package orientation. Note that the longest axis of the reference is aligned with the tab, not the shortest axis of the circuit board. This is in anticipation of flexing forces transmitted into the tab. The best orientation for the tab is in line with the longest axis of the board as in (b), (c) and (d). Bending forces along the weaker (longer) axis of the board could be coupled into (a) and (e). Note that the ICs are aligned to resist this force. Use configuration (c) when the part is located along the longer edge of the board, and (d) when it is located along the shorter edge. Use (b) when the part is not located along any edge.

Temperature-induced noise

Even though references operate on very meager supply currents, dissipation in the reference is enough to cause small temperature gradients in the package leads. Variations in thermal resistance, caused by uneven air flow, lead to differential lead temperatures, thereby causing thermoelectric voltage noise at the output of the reference. Figure 35.15 dramatically demonstrates this effect. The first half of the plot was made with an LT1021H-7 buried Zener reference, which was shielded from ambient air with a small foam cup (Dart Container Corporation Stock No. 8J8 or similar). The cup was removed at six minutes elapsed time for the second half of the test. Ambient in both cases was a lab bench-top with no excessive turbulence from air conditioners, opening/closing doors, foot

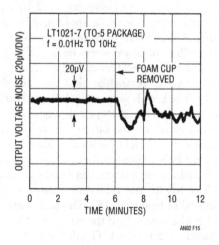

Figure 35.15 • Air Turbulence Induces Low Frequency Noise and Compromises Reference Accuracy

traffic or 547 exhaust. Removing the foam cup increased the output noise by almost an order of magnitude in the 0.01Hz to 10Hz band.

The Kovar leads of the TO-5 working against copper circuit traces are the primary culprit. Copper lead frames used on DIP and surface mount packages are not nearly as sensitive to air turbulence because they are intrinsically matched. Still, external components create thermocouples of their own with potentials of $10\mu V/^\circ C$ or more per junction. In a LT1021-7 reference, this represents more than $1ppm/^\circ C$ shift from each thermoelectric generator. Temperature gradients across the circuit board and dissipation within external components can lead to the same kind of noise as shown in Figure 35.15.

Temperature gradients may arise from heat generators on the board. Position the reference and its associated external components far from heat sources and, if necessary, use routing techniques to create an isothermal island around the reference circuitry. Minimize air movement either by adding a small enclosure around the reference circuitry, or by encapsulating the reference circuitry in self-expanding polyurethane foam.

Reference applications

The unique pocket reference shown in Figure 35.16 is a good match for a pair of AAA alkaline cells, because the circuit draws less than $16\mu A$ supply current. Two outputs are provided: a buffered, 1.5V voltage output, and a regulated $1\mu A$ current source. The current source compliance ranges from approximately 1V to $-43V$.

The reference is self-biased, completely eliminating line regulation as a concern. Start-up is guaranteed by the LT1495 op amp, whose output saturates at 11mV from the negative rail. Once powered, there is no reason to turn the circuit off. One AAA alkaline contains 1200mAH capacity, enough to power the circuit throughout the 5-year shelf life of the battery. Voltage output accuracy is about 0.17% and current output accuracy is about 1.2%.

Trim R1 to calibrate the voltage ($1k\Omega$ per 0.1%), and R3 to calibrate the output current (250Ω per 0.1%).

Low noise synthesizers need quiet power supplies for their VCOs and other critical circuitry. 3-terminal regulators exhibit far too much noise for this application, calling instead for a regulator constructed from a reference. A practical example is shown in Figure 35.17. Current through the LT1021-5 reference is used to drive the base of a PNP pass device, resulting in an available output current of at least 1A. In this example, the current is intentionally limited to 200mA by the addition of emitter degeneration and base clamping. The low noise of the reference is preserved, giving a 100-fold improvement over the noise of an equivalent 5V, 3-terminal regulator, not to mention improved initial accuracy and long-term stability. Typical output noise is $7\mu V_{P-P}$ over a 10kHz bandwidth.

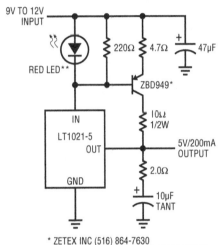

* ZETEX INC (516) 864-7630
** GLOWS IN CURRENT LIMIT. DO NOT OMIT.

Figure 35.17 • Ultralow Noise 5V, 200mA Supply Output Noise is $7\mu V_{RMS}$ Over a 10Hz to 10kHz Bandwidth. Reference Noise is Guaranteed to be Less Than $11\mu V_{RMS}$. Standard 3-Terminal Regulators Have One Hundred Times the Noise and No Guarantees

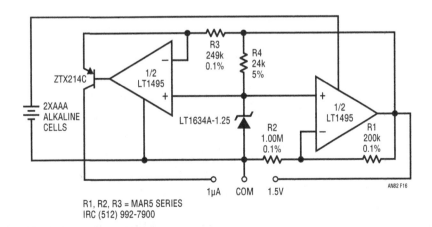

R1, R2, R3 = MAR5 SERIES
IRC (512) 992-7900

Figure 35.16 • This Pocket Reference Operates for Five Years on One Set of AAA Cells

Conclusion

When specifying a reference, keep in mind that initial accuracy, temperature coefficient and long-term stability all play a role in overall accuracy of the finished product. By taking some care in applying the reference, and by avoiding some key pitfalls, the reference's inherent accuracy can be preserved.

For further reading

1. Spreadbury, Peter J. "The Ultra-Zener-A Portable Replacement for the Weston Cell?" *by IEEE Transactions on Instrumentation and Measuremen*, Vol.40(No. 2), April 1991, pp. 343–346.
2. Huffman, Brian. Application Note 42: Voltage Reference Circuit Collection. Linear Technology Corporation June 1991.
3. Lee, Albert. "4.5µA Li-Ion Battery Protection Circuit" *Linear Technology*, Volume 9, Number 2, June 1999, p. 36.

Appendix A
Buried Zener: low longterm drift and noise

The Zener diode has long been used in reference service in many noncritical applications. Integrated circuit designers sometimes use an NPN emitter-base junction operating in reverse breakdown as a Zener reference. Breakdown occurs at the surface of the die, where the effects of contamination and oxide charge are most pronounced. These junctions are noisy and suffer from unpredictable short- and long-term drift.

The buried Zener, developed as a precision IC reference, places the junction below the surface of the silicon, well away from contamination and oxide effects. The result is a Zener with excellent long-term stability, low noise, and relatively accurate initial tolerance.

Figure A1 shows the first steps in fabricating a buried Zener. A region of n^+ buried layer is located beneath the Zener structure so as to shield subsequent diffusions from contact with the substrate.

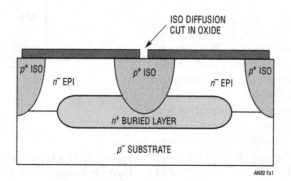

Figure A1 • Iso is Diffused to Form the Anode. Highest Dopant Concentration Occurs Directly Under the Mask Opening

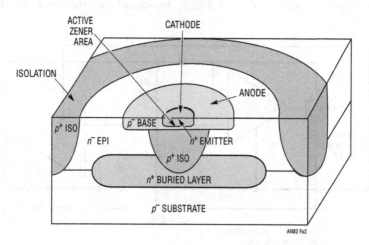

Figure A2 • An Emitter Diffusion Forms the Cathode. Breakdown Occurs Under the Center of the Emitter, Where Both Emitter and Iso + Base Dopant Concentrations are Highest

After growth of the n^- epitaxial layer, p^+ isolation is diffused through a small opening at the center of the Zener. At the same time, isolation is diffused around the periphery to form a separate tub containing the entire Zener structure.

Isolation diffuses both downward and laterally. The central diffusion is shielded from contact with the substrate by the buried layer, while the isolation walls are allowed to reach substrate and form an isolated tub. It is important to note that the highest concentration of p^+ occurs directly under the mask opening and that the dopant concentration is weakest at the fringes of a diffusion.

The last steps include a p^- base diffusion and an emitter diffusion, located at the center of the Zener (see Figure A2). The emitter becomes the cathode, whereas the combined isolation and base diffusion serve as the anode.

Breakdown occurs at the bottom of the of the cathode, where the emitter and isolation + base dopant concentrations are richest. Lighter doping concentrations result in a higher breakdown voltage at the iso-buried layer, base-epi and iso-epi junctions, and at the outer fringes of the emitter diffusion, ensuring that these areas are not active when the buried junction is biased into breakdown. The result is an extremely stable subsurface breakdown mechanism that has near-theoretical noise and is unaffected by surface contamination or oxide effects

Appendix B
ΔV_{BE}: integrated circuit workhorse

It is, perhaps, a cruel fate for IC designers that no single IC device or structure is invariant with changes in temperature. Various combinations of devices have been devised to stabilize circuits against changes in temperature. As explained in the text, Zener-based references use a Zener and a forward-biased diode connected in series to achieve near-zero temperature coefficient and a bandgap relies on a ΔV_{BE} in series with a forward-biased diode.

An indispensable technique in integrated circuit design, the ΔV_{BE} is not widely known in other fields. Before explaining the theory of ΔV_{BE}, let's skip ahead

to the two most important results: two identical diode (or base emitter) junctions running different currents produce different voltage drops. The ratio of the currents controls the absolute value of the offset voltage. Further, this offset has a predictable, positive temperature coefficient of approximately 3.4µV/°C for each room-temperature millivolt of offset. By combining the positive TC of a ΔV_{BE} with the negative TC of a diode drop, a zero TC bandgap reference is formed. As we shall soon see, it takes a ΔV_{BE} offset of 650mV to cancel the −2.18mV/°C TC of a hypothetical diode.[1]

Two transistors (or diodes) produce an offset given by the following equation:

$$\Delta V_{BE} = V_{BE1} - V_{BE2} = (kT/q)\ln(J_{E1}/J_{E2}) \qquad (1)$$

where ΔV_{BE} = offset voltage, k = Boltzmann's constant ($1.381 \cdot 10^{-23}$ Joules/K), T = absolute temperature (298K at room), q = charge of an electron ($1.6 \cdot 10^{-19}$ Coulombs), and J_E = emitter current density. The actual units of area used to calculate J_{E1} and J_{E2} cancel each other, so that only the area ratio is important. Similarly, only the current ratio is important. If we restrict ourselves to using two identical transistors, Equation (1) reduces to

$$\Delta V_{BE} = V_{BE1} - V_{BE2} = (kT/q)\ln(I_{C1}/I_{C2}) \qquad (2)$$

where I_c = collector current (see Figure B1). The temperature coefficient is given by

$$TC = d\Delta V_{BE}/dT = (k/q)\ln(I_{C1}/I_{C2}) \qquad (3)$$

where k/q = 86.3µV/°C.

Calculating the current ratio required to produce +2.18mV/°C (corresponding to 650mV offset) we find that it is unmanageably large, about $9.44 \cdot 10^{10}$:1. In practice, a much smaller offset is generated by a ΔV_{BE} cell and then amplified to 650mV. As an example, see Figure B2. Using a 10:1 current ratio,[2] we find a room temperature offset from Equation (2) of 59.2mV, and a temperature coefficient of 199µV/°C. Applying a gain of slightly less than eleven brings us to 650mV and +2.18mV/°C.

Adding a PNP emitter follower to the output of this circuit forms a crude "bandgap" reference, with an output voltage equal to the sum of 650mV and the PNP's V_{BE}. Assuming V_{BE} = 600mV, the output would be 1.25V. The reference could be further improved by trimming the gain of eleven so that the ΔV_{BE} exactly canceled the PNP's base-emitter temperature coefficient. IC bandgap references are constructed in a similar way.

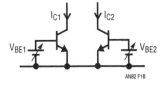

AN82 F1B

Figure B1 ● The Current Ratio Required to Produce a Certain V_{BE} Offset is Defined by Equations (1) and (2)

Note 1: The numbers have been massaged for those who want to reproduce the calculations.
Note 2: or a combination of current and area scaling to achieve a 10:1 current density ratio in Equation (1).

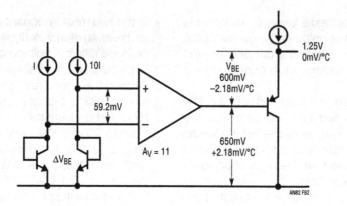

Figure B2 • A Bandgap Reference is Formed by Stacking a ΔV_{BE} Generator and a V_{BE}

Instrumentation applications for a monolithic oscillator

A clock for all reasons

36

Jim Williams

Introduction

Oscillators are fundamental circuit building blocks. A substantial percentage of electronic apparatus utilizes oscillators, either as timekeeping references, clock sources, for excitation or other tasks. The most obvious oscillator application is a clock source in digital systems.[1] A second area is instrumentation. Transducer circuitry, carrier based amplifiers, sine wave formation, filters, interval generators and data converters all utilize different forms of oscillators. Although various techniques are common, a simply applied, broadly tuneable oscillator with good accuracy has not been available.

Clock types

Commonly employed oscillators are resonant element based or RC types.[2] Figure 36.1 shows two of each. Quartz crystals and ceramic resonators offer high initial accuracy and low drift (particularly quartz) but are essentially untuneable over any significant range. Typical RC types have lower initial accuracy and increased drift but are easily tuned over broad ranges. A problem with conventional RC oscillators is that considerable design effort is required to achieve good specifications. A new device, the LTC1799, is also an RC type but fills the need for a simply applied, broadly tuneable, accurate oscillator. Its accuracy and drift specifications fit between resonator based types

CLOCK TYPE	TYPICAL FREQUENCY ACCURACY	TYPICAL FREQUENCY RANGE	TUNEABILITY	TEMPERATURE COEFFICIENT	POWER SUPPLY REJECTION RATIO	COMMENTS
Quartz	0.005%	10kHz to 200MHz	Poor	0.5ppm/°C Easily Achieved. See Comments	1ppm/V	High Stability and Initial Accuracy at Expense of Tuneability. Essentially No Tuneability. $1 \cdot 10^{-9}$ Stability Achievable with Compensation Techniques
Ceramic Resonator	0.5%	250kHz to 60MHz	Poor	30ppm/°C	20ppm/V	Lower Performance and cost than Quartz. Essentially Untuneable
LTC1799	1.5%	1kHz to 33MHz	Good	40ppm/°C Plus Resistor Temperature Coefficient	500ppm/V	Add 10 to 50ppm/°C Temperature Coefficient, Depending on Resistor Type. Extremely Small Footprint— SOT-23 and 1 Resistor
Typical RC Based Clock	10%	1Hz to 25MHz	Good	200ppm/°C	2500ppm/V	Requires Careful Design and Component Selection for Best Results

Figure 36.1 • LTC1799 Compared to Other Oscillators. Quartz and Ceramic Based Types Offer Higher Frequency Accuracy and Lower Drift but Lack Tuneability. RC Designs are Tuneable but Accuracy, Temperature Coefficient and PSRR are Poor

Note 1: Strictly speaking, an oscillator (from the Latin verb, "*oscillo*," to swing) produces sinusoids; a clock has rectangular or square wave output. The terms have come to be used interchangeably and this publication bends to that convention.

Note 2: This forum excludes such exotica as rubidium and cesium based atomic resonance devices, nor does it admit mundane but dated approaches such as tuning forks.

Analog Circuit and System Design: A Tutorial Guide to Applications and Solutions. DOI: 10.1016/B978-0-12-385185-7.00036-6

and typical RC oscillators. Additionally, its board footprint, a 5-pin SOT-23 package and a single resistor, is notably small. Note that no external timing capacitor is required.

A (very) simple, high performance oscillator

Figure 36.2 shows how simple to use the LTC1799 is. A single resistor (R_{SET}) programs the device's internal clock and pin-settable decade dividers scale output frequency. Various combinations of resistor value and divider choice permit outputs from 1kHz to 33MHz.[3] Figure 36.3 shows R_{SET} vs output frequency for the three divider pin states and the governing equation. The inverse relationship between resistance and frequency means that LTC1799 period vs resistance is linear.

Figure 36.4 reveals that the LTC1799 has speciated into a family. There are two additional devices. The LTC6900, quite similar, cuts supply current to 500μA but gives up some frequency range. The LTC6902, designed for noise smoothed, multiphase power applications, has multiphase outputs and spread spectrum capability. Spread spectrum clocking distributes power switching over a settable frequency range, preventing significant noise peaking at any given point. This greatly reduces EMI concerns.

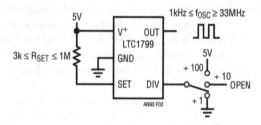

Figure 36.2 • LTC1799 Oscillator Frequency Is Determined by R_{SET} and Divider Pin (DIV). Tunable Range Spans 1kHz to 33MHz

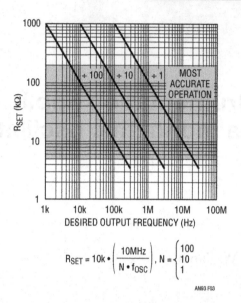

$$R_{SET} = 10k \cdot \left(\frac{10MHz}{N \cdot f_{OSC}} \right), \; N = \begin{cases} 100 \\ 10 \\ 1 \end{cases}$$

Figure 36.3 • R_{SET} vs Output Frequency for the Three Divider Pin States and Governing Equation. Relationship between R_{SET} and Frequency Is Inverse; R_{SET} vs Period has Linear Characteristic

The LTC1799's combination of simplicity, broad tuneability and good accuracy invites use in instrumentation circuitry. The following text utilizes the device's attributes in a variety of such applications.

Platinum RTD digitizer

A platinum RTD, used for R_{SET} in Figure 36.5, results in a highly predictable O1 output period vs temperature. O1's output, scaled via counters, is presented to a clocked, period determining logic network which delivers digital output data. Over a 0°C to 100°C sensed temperature, 1000 counts are delivered, with accuracy inside 1°C.

DEVICE TYPE	FREQUENCY RANGE	FREQUENCY ACCURACY	TEMPERATURE COEFFICIENT	PSRR	COMMENTS
LTC1799	1kHz to 33MHz	1.5%	40ppm/°C + Resistor Drift	0.05%/V	I_{SUPPLY} = 1mA
LTC6900	1kHz to 20MHz	1.5%	40ppm/°C + Resistor Drift	0.04%/V	Low Power (I_{SUPPLY} = 500μA) Version of LTC1799
LTC6902	5kHz to 20MHz	1.5%	40ppm/°C + Resistor Drift	0.04%/V	2-, 3- or 4-Phase Outputs. Programmable Width Spread Spectrum Frequency Modulation. Intended for Multiphase Power Supply Applications

Figure 36.4 • Oscillator Family Details. LTC6900 Is Low Power Version of LTC1799. LTC6902, Intended for Noise Sensitive, High Power Switching Regulator Applications, Has Multiphase, Spread Spectrum Outputs. All Types Have Excellent Tunability, Good Frequency Accuracy, Low Temperature Coefficient and High PSRR

Note 3: This deceptively simple operation derives from noteworthy internal cleverness. See Appendix A, "LTC1799 internal operation" for a description.

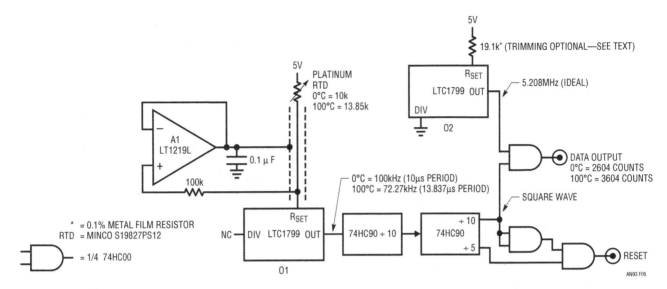

Figure 36.5 • Platinum RTD Digitizer Accurate within 1°C Over 0°C to 100°C. Platinum RTD Value Is Linearly Converted to Period by LTC1799. Logic and Second LTC1799 Clock Digitize Period into Output Data Bursts. A1 Drives RTD Shield at R_{SET} Potential, Bootstrapping Pin Capacitance to Permit Remotely Located Sensor

Extended range (sensor limits are −50°C to 400°C) is possible by using a monitoring processor to implement linearity correction in accordance with sensor characteristics.[4]

If the RTD is at the end of a cable, the cable shield should be driven by A1 as shown. This bootstraps the cable shield to the same potential as R_{SET}, eliminating jitter inducing capacitive loading effects at the R_{SET} node.[5]

Figure 36.6 shows operating waveforms. The RTD determines O1's output (Trace A), which is divided by 100 and assumes square wave form (Trace B). The logic network combines with O2's fixed frequency to digitize period measurement, which appears as output data bursts (Trace C). The logic also produces a reset output (Trace D), facilitating synchronization of monitoring logic.

As shown, accuracy is about 1.5°C, primarily due to LTC1799 initial error. Obtaining accuracy inside 1°C involves simulating a 100°C temperature (13,850Ω) at the sensor terminals and trimming R_{SET} for appropriate output. A precision resistor decade box (e.g., ESI DB62) allows convenient calibration.

Thermistor-to-frequency converter

Figure 36.7's circuit also directly converts temperature to digital data. In this case, a thermistor sensor biases the R_{SET} pin. The LTC1799 frequency output is predictable, although nonlinear. The inverse R_{SET} vs frequency relationship combines with the thermistor's nonlinear characteristic to give Figure 36.8's data. The curve is nonlinear, although tightly controlled.

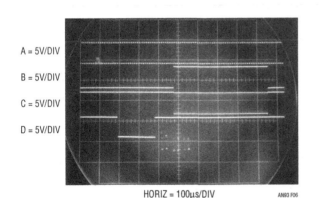

HORIZ = 100µs/DIV   AN93 F06

Figure 36.6 • Platinum RTD Biased LTC1799 Produces Output (Trace A) which Is Divided by 100 (Trace B) and Gated with 5.2MHz Clock. Resultant Data Bursts (Trace C) Correspond to Temperature. Reset Pulse (Trace D), Preceding Each Data Burst, Permits Synchronization of Monitoring Logic

Note 4: Linearity deviation over −50°C to 400°C is several degrees. See Reference 1.
Note 5: The R_{SET} node, while not unduly sensitive, requires management of stray capacitance. See Appendix B, "R_{SET} node considerations" for detail.

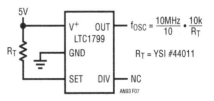

Figure 36.7 • Simple Temperature-to-Frequency Converter Biases R_{SET} with Thermistor. Frequency Output Is Predictable, Although Nonlinear

852

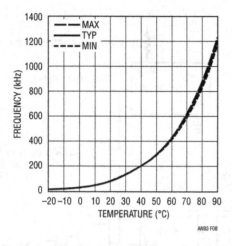

Figure 36.8 • LTC1799 Inverse Resistance vs Frequency Relationship and Nonlinear Thermistor Characteristic Result in above Data. Curve Is Nonlinear, Although Tightly Controlled

Isolated, 3500V breakdown, thermistor-to-frequency converter

This circuit, building on the previous approach, galvanically isolates the thermistor from the circuit's power and data output ports. The 3500V breakdown barrier between the thermistor and power/data output ports permits operation at high common mode voltages. Such conditions are often encountered in industrial measurement situations.

Figure 36.9's pulse generator, C1, running around 10kHz, produces a 2.5µs wide output (Trace A, Figure 36.10). Q1-Q2 provide power gain, driving T1 (Trace B is Q2's collector). T1's secondary responds, charging the 100mF capacitor to a DC level via the 1N5817 rectifier. The capacitor powers O1, which oscillates at the sensor determined frequency. O1's output, differentiated to conserve power, switches Q4. Q4, in turn, drives T1's secondary, T1's primary receives Q4's signal and Q3 amplifies it, producing the circuit's data output (Trace C). Q3's collector also lightly modulates C1's negative input (Trace D), synchronizing T1's primary drive to the data output. C2 prevents erratic circuit operation below 4.5V by removing Q1's drive.

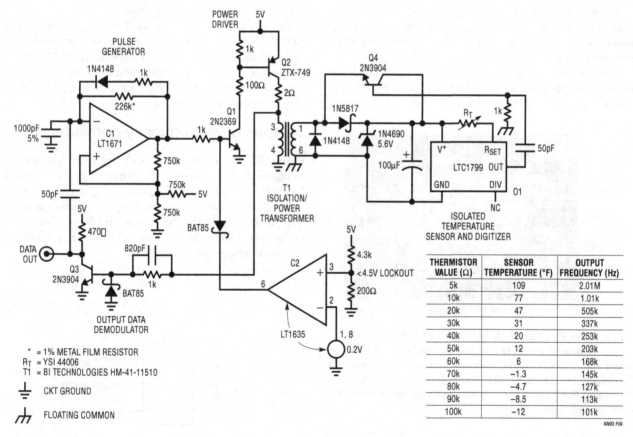

THERMISTOR VALUE (Ω)	SENSOR TEMPERATURE (°F)	OUTPUT FREQUENCY (Hz)
5k	109	2.01M
10k	77	1.01k
20k	47	505k
30k	31	337k
40k	20	253k
50k	12	203k
60k	6	168k
70k	−1.3	145k
80k	−4.7	127k
90k	−8.5	113k
100k	−12	101k

Figure 36.9 • A Galvanically Isolated Thermistor Digitizer. C1 Sources Pulsed Power to Thermistor Biased LTC1799 via Q1, Q2 and T1. LTC1799 Output Modulates T1 through Q4. Q3 Extracts Data, Presents Ouput. T1's 3500V Breakdown Sets Isolation Limit

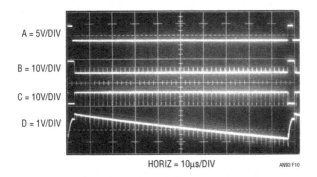

Figure 36.10 • Isolated Thermistor Digitizer's Waveforms Include C1's Output (Trace A), Q2's Collector Drive to T1 (Trace B), Data Output (Trace C) and C1's Negative Input (Trace D). C1's Negative Input (Trace D) Is Lightly Modulated by Q3, Synchronizing Transformer Power Drive to Data Output

C1's continuous clocking, while maintaining O1's isolated DC power supply, generates periodic cessations in the frequency coded output. These interruptions can be used as markers to control operation of monitoring logic. Output frequency vs thermistor characteristics are included in Figure 36.9.

Relative humidity sensor digitizer-hetrodyne based

Figure 36.11 converts the varying capacitance of a linearly responding relative humidity sensor to a frequency output.

The 0Hz to 1kHz output corresponds to 0% to 100% sensed relative humidity (RH). Circuit accuracy is 2%, plus an additional tolerance dictated by the selected sensor grade. Circuit temperature coefficient is ≈400ppm/°C and power supply rejection ratio is <1% over 4.5V to 5.5V. Additionally, one sensor terminal is grounded, often beneficial for noise rejection.

This is basically a hetrodyne circuit. Two oscillators, one variable, one fixed, are mixed, producing sum and difference frequencies. The variable oscillator is controlled by the capacitive humidity sensor. The demodulated difference frequency is the output.[6] The hetrodyne frequency subtraction approach permits a sensed 0% RH to give a 0Hz output, even though sensor capacitance is not zero at RH = 0%.

C1, the sensor controlled variable oscillator, runs between the indicated output frequencies for the RH sensor excursion noted. The RH sensor is AC coupled, in accordance with its manufacturer's data sheet.[7] Reference oscillator O1 is tuned to C1's nominal 25% RH dictated frequency.

The two oscillators are mixed at Q1's base (Figure 36.12, Trace A). Q1 amplifies the mixed frequency components, although collector filtering attenuates the sum frequency. The RH determined difference frequency, appearing as a sine wave at Q1's collector (Trace B), remains. This waveform is filtered and AC coupled to zero crossing detector C2. AC hysteretic feedback at C2's input (Trace C) produces clean C2 output (Trace D). Counter based scaling at C2's output combines with slight sensor padding (note 2pF value across the sensor) to provide

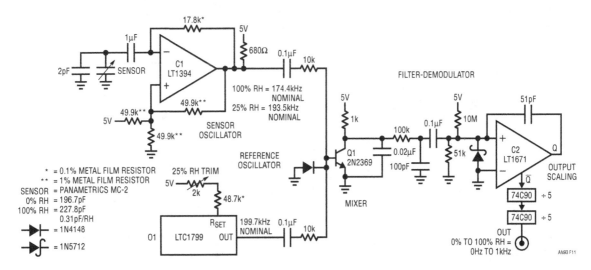

Figure 36.11 • Hetrodyne Based Humidity Transducer Digitizer Has Grounded Sensor, 2% Accuracy. Capacitively Sensed Hygrometer Beats Humidity Dependent Oscillator (C1) Against Stable Oscillator O1. Difference Frequency Is Demodulated by Q1, Converted to Pulse Form at C2. Counters Scale Output for 0kHz to 1kHz = 0% to 100% Relative Humidity

Note 6: Hetrodyne techniques, usually associated with communications circuitry, have previously been applied to instrumentation. This circuit's operation was adapted from approaches described in References 2, 3 and 4.
Note 7: DC coupling introduces destructive electromigration effects. See Reference 6.

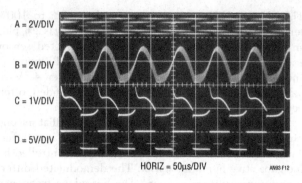

Figure 36.12 • Sensor and Stable Oscillators are Mixed at Q1's Base (Trace A); Difference Frequency Appears at Q1's Collector (Trace B). Filtering and AC Hysteresis at C2's + Input (Trace C) Produce Clean Response at C2's Output (Trace D)

numeric output frequency correspondence to RH. Calibration involves simulating the RH sensor's 25% value and trimming 01 for a 250Hz output. The simulated value may be built up from known discrete capacitors or simply dialed out on a precision variable air capacitor (General Radio 1422D).

When evaluating circuit operation, it is useful to consider that C1's frequency changes inversely with sensor capacitance; its *period* is linear vs sensor capacitance. This would normally corrupt the desired linear output relationship between frequency and RH. Practically, because the sensor's excursion range is small compared to its 0% RH value, the error is similarly small. This term almost entirely accounts for the circuit's stated 2% accuracy.

Relative humidity sensor digitizer—charge pump based

Figure 36.13 also digitizes the capacitive humidity sensor's output but has better specifications than the previous circuit. Circuit accuracy is 0.3%, plus the selected sensor grade's tolerance. Temperature coefficient is about 300ppm/°C and power supply rejection ratio is 0.25% for 5V ±0.5V. Compromises include a floating sensor and somewhat more complex circuitry.

01 (Trace A, Figure 36.14) clocks an LTC1043 switch array based charge pump. This configuration alternately connects the AC coupled RH sensor to a 4V reference derived potential and then discharges it into A1's summing point. A1, an integrator, responds with a ramping output, Trace B of Figure 36.14. When A1's output exceeds C1's negative input voltage, C1's Q output (Trace C, Figure 36.14) goes high, triggering Q1 and resetting the

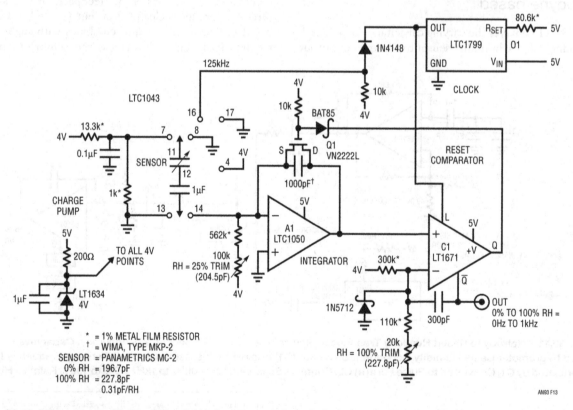

Figure 36.13 • Hygrometer Digitizer Has 0.3% Accuracy, Although Sensor Must Float Off-Ground. Humidity Sensor Determines Charge Delivered to A1 Integrator During Each Charge Pump Cycle. Resultant A1 Output Ramp Is Reset by Level Triggered C1 via Q1. Output Frequency, Taken at C1, Varies with Humidity

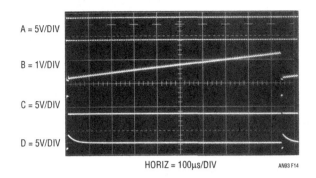

A = 5V/DIV

B = 1V/DIV

C = 5V/DIV

D = 5V/DIV

HORIZ = 100µs/DIV   AN93 F14

Figure 36.14 • LTC1799 Clock (Trace A) Drives Humidity Sensor Based Charge Pump, Producing A1 Output Ramp (Trace B). C1 Q Output, Trace C, Biases Q1, Resetting Ramp. AC Feedback at C1 (Trace D) Permits Complete Ramp Reset, Sets Output Pulse Width

ramp. AC feedback to C1's negative input (Trace D) ensures long enough Q1 on-time for complete ramp reset. This action's repetition rate depends on RH sensor value. The A1-C1 loop is synchronized to the charge pump's clocking by 01's output path to C1's latch input. In theory, if the charge pump, offset term (25% trim current) and ramp amplitude are tied to the same potential, this circuit does not require a voltage reference. In practice, the sensor's extremely small capacitance shifts magnify the effect of charge pump errors vs supply, necessitating powering the LTC1043 from the 4V reference. Once this is done, the mentioned points are tied to the 4V reference. Note that the 5V powered 01's output must be level shifted to drive the LTC1043.

A trimmed DC offset current (100k potentiometer) into A1's summing junction compensates the RH sensor's offset term (e.g., 0% RH ≠ 0pF). Output frequency is scaled by the 20kΩ trim at C1 so 0% to 100% RH = 0Hz to 1kHz. Trimming involves substituting capacitance for the sensor's known 100% and 25% values and trimming the appropriate adjustments. The adjustments are somewhat interactive, necessitating repetition until convergence occurs. A precision variable capacitor (General Radio type 1422D) is invaluable in this regard, although acceptable results are possible with built-up calibrated discrete capacitors.

Relative humidity sensor digitizer—time domain bridge based

Figure 36.15, also a relative humidity (RH) digitizer, features 1% accuracy, PSRR of 1% over 4.5V to 5.5V, temperature coefficient of 350ppm/°C and a ground referred sensor. Additionally, the circuit's trim scheme accommodates wide tolerance grade RH sensors. The circuit is basically a time domain bridge; it subtracts time intervals representing sensor and sensor offset values to determine

sensor value extrapolated to RH = 0%. This measurement is digitized and scaled so zero to 100 counts equals 0% to 100% RH at the output.

01's nominal 12.77MHz output, conditioned by a counter chain and an inverter configured gate, presents a 12.4kHz, 2.5µs pulse (Trace A, Figure 36.16) to Q1A and Q1B. The transistors' collectors fall (Trace B = Q1A collector, Trace C = Q1B collector) to zero volts. When the base drive ceases, both collectors ramp towards 5V. Trace B's ramp slope varies with the RH sensor's capacitance; Trace C's ramp slope represents the sensor's offset value (0% RH ≠ 0pF). C1 and C2 switch when their associated ramp inputs cross the comparators' common DC input potential. The comparator outputs (Trace D = C2, Trace E = C1) define a "both high" time region proportional to the ramp slopes' difference and, hence, an offset corrected version of sensor value. This time interval is gated with 01's output, providing Trace F's data output.

Circuit operation is fairly straightforward, although some details bear mention. Q1, a dual transistor, promotes cancellation of the individual transistors' V_{CE} vs temperature terms, minimizing their error contribution. The unit specified, a 2-die type, minimizes crosstalk; monolithic types should not be substituted. Similarly, a dual comparator should not be substituted for the single types specified for C1 and C2. Also, the comparators operate at high source impedance relative to their input characteristics but symmetry provides adequate error cancellation. Finally, the 5.6k resistor combines with the output gates' input capacitance, forming a ≈20ns lag. This delay prevents false output data transients when the ramps are resetting.

Trimming procedure is similar to the previous RH circuit. It involves substituting capacitance for the sensor's known 100% and 25% values and trimming the indicated adjustments. The adjustments are somewhat interactive, necessitating repetition until convergence occurs. A precision variable capacitor (General Radio type 1422D) is invaluable for this work, although acceptable results are possible with calibrated discrete capacitor assemblies.

40nV noise, 0.05µV/°C drift, chopped bipolar amplifier

Figure 36.17's circuit, adapted from Reference 7, combines the low noise of an LT1028 with a chopper based carrier modulation scheme to achieve an extraordinarily low noise, low drift DC amplifier. DC drift and noise performance exceed any currently available monolithic amplifier. Offset is inside 1µV, with drift less than 0.05µV/°C. Noise in a 10Hz bandwidth is less than 40nV, far below monolithic chopper stabilized amplifiers. Bias current, set by the bipolar LT1028 input, is about 25nA. The circuit is powered by a single 5V supply, although its output will swing ±2.5V. Additionally, a carefully selected chopping frequency prevents deleterious

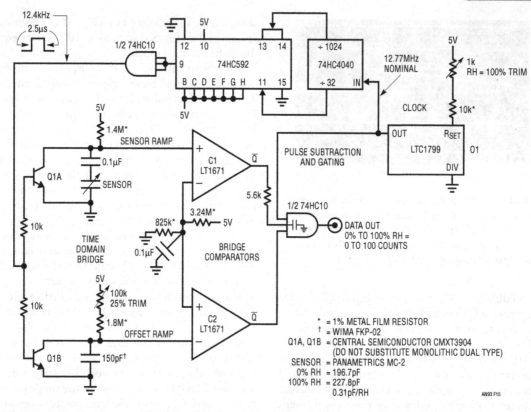

Figure 36.15 • Humidity Transducer Digitizer Has Grounded Sensor, 1% Accuracy; Trim Scheme Allows Low Tolerance Sensors. Clocked Q1A-Q1B Configurations Produce Ramp Outputs. Q1A Ramp Slope Varies with Humidity Sensor Value, Q1B Ramp Represents Sensor's Offset (0% RH ≠ 0pF). C1, C2 Digitize Ramp Times. Gate Extracts Time Difference, Presents 0 to 100 Counts Out for 0% to 100% Relative Humidity

interaction with 60Hz related components at the amplifier's input. These specifications suit demanding transducer signal conditioning situations such as high resolution scales and magnetic search coils.

01's 37kHz output is divided down to form a 2-phase 925Hz square wave clock. This frequency, harmonically

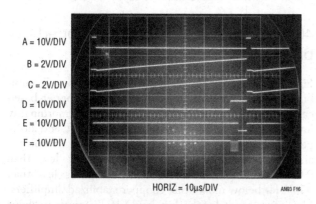

HORIZ = 10μs/DIV

Figure 36.16 • Humidity Sensor Time Domain Bridge Waveforms. Gate (Figure 36.15, Upper Left) Clocks (Trace A) Q1A and Q1B. Sensor and Offset Ramps Are Traces B and C. C1 and C2 Outputs are Traces D and E. Gate Extracts C1-C2 Time Difference, Presents Trace F's Digitized Output

unrelated to 60Hz, provides excellent immunity to harmonic beating or mixing effects which could cause instabilities. S1 and S2 receive complementary drive, causing A1 to see a chopped version of the input voltage. A1 amplifies this AC signal. A1's square wave output is synchronously demodulated by S3 and S4. Because these switches are synchronously driven with the input chopper, proper amplitude and polarity information is presented to A2, the DC output amplifier. This stage integrates the square wave into a DC voltage, providing the output. The output is divided down (R2 and R1) and fed back to the input chopper where it serves as a zero signal reference. Gain, in this case 1000, is set by the R1-R2 ratio. Because A1 is AC coupled, its DC offset and drift do not affect overall circuit offset, resulting in the extremely low offset and drift noted. A1's input damper minimizes offset voltage contribution due to nonideal switch behavior.

Normally, this single supply amplifier's output would be unable to swing to ground. This restriction is eliminated by powering the circuit's negative rail from a charge pump. 01's 37kHz output excites the charge pump, comprised of paralleled logic inverters and discrete components. Deliberate 10Ω loss terms combine with the specified 47μF capacitors to form a very low noise power source. These precautions eliminate charge pump noise which might otherwise degrade amplifier noise performance.

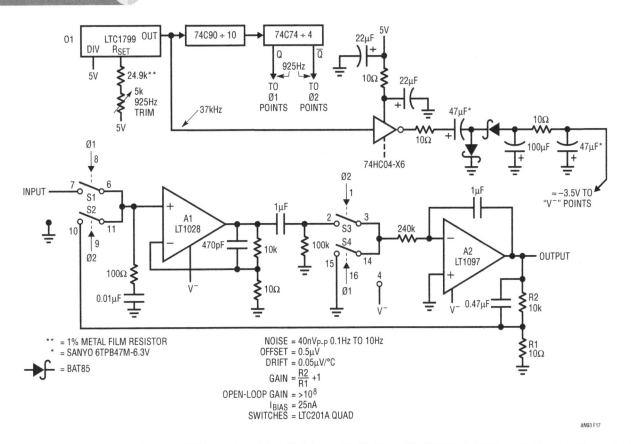

Figure 36.17 • 5V Powered, Chopped Bipolar Amplifier. Noise Is ≈40nV with 0.05µV/°C Drift. DC Input Is Carrier Modulated, Amplified by A1, Demodulated to DC and Fed Back from A2. 925Hz Carrier Clock Prevents Interaction with 60Hz Line Originated Components. Negative Supply, Derived via Charge Pump, Allows Zero Volt Output Swing

Figure 36.18, a noise plot of the amplifier in a 0.1Hz to 10Hz bandwidth, shows about 40nV of peak-to-peak noise. A1 and the 60Ω resistance of S1-S2 contribute about equally to form this noise. When using this amplifier, it is important to realize that A1's bias current flowing through the input source impedance causes additional noise. In general, to maintain low noise performance, source resistance should be kept below 500Ω. Fortunately, transducers such as strain gauge bridges, RTDs and magnetic detectors are well below this figure.

45nV noise, 0.05µV/°C drift, chopped FET amplifier

Figure 36.19 replaces the previous circuit's input stage with a pair of extremely low noise J-FETs. In most other

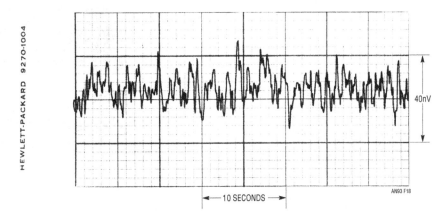

Figure 36.18 • Noise in a 0.1Hz-10Hz Bandwidth Is about 40nV with 0.05µV/°C Drift

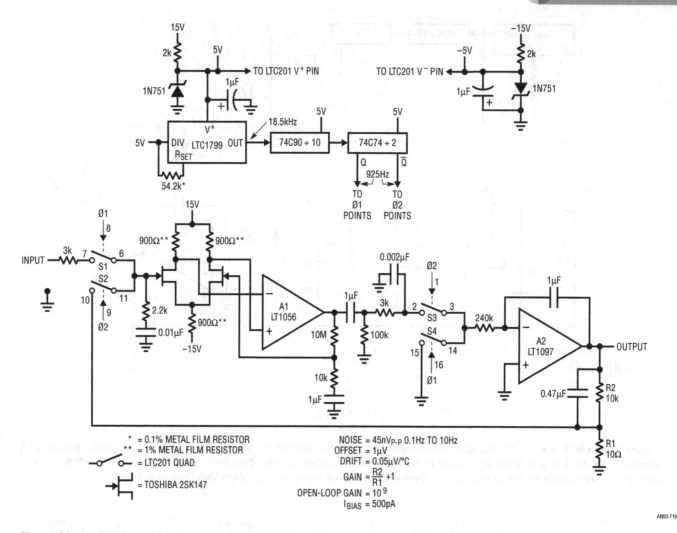

Figure 36.19 • FET Input Version of Figure 36.17 Has 500pA Bias Current. 925Hz Clock Is Retained, Noise Increases Slightly to ≈45nV

respects, circuit operation is similar. Noise increases very slightly, to ≈45nV, but bias current decreases to only 500pA—50 times lower than the previous circuit. The noise performance is especially noteworthy—it is almost 17 times better than currently available monolithic chopper stabilized amplifiers and nearly equals the best bipolar designs. Other performance specifications, appearing in the figure, are similar to Figure 36.17.

The 925Hz clock is retained, although this ±15V powered design uses Zeners to derive internal ±5V points. The clock and logic run from 5V and the LTC201 switches use ±5V. The switches' low voltage rails reduce charge injection, minimizing its effect on offset voltage. RC damper networks further attenuate parasitic switch behavior effects, resulting in the 1μV offset specification.

Noise measured over Figure 36.20's 50 second interval is about 45nV in a 0.1Hz to 10Hz bandwidth. This is

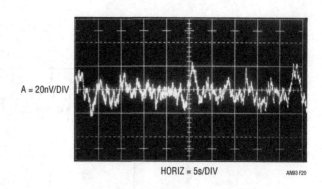

A = 20nV/DIV

HORIZ = 5s/DIV   AN93 F20

Figure 36.20 • Chopped FET Input Amplifier Noise Is ≈45nV in 0.1Hz to 10Hz Bandwidth

spectacularly low noise for a J-FET based design and is directly attributable to the input pairs' die size and current density.[8]

Clock tunable, filter based sine wave generator

A feedback loop enclosed resonator can be made to oscillate. Figure 36.21's sine wave generator takes advantage of this and eliminates the need for an amplitude control loop. This circuit, a mildly modified form of the Regan resonant bandpass loop, is clock tuneable and produces sine and cosine outputs.[9]

The LTC1060 switched capacitor filter is set up as a clock tuneable bandpass filter with a Q of 10. 01 clocks the filter at 100kHz, resulting in a 1kHz bandpass. C1, switched by the sine output, supplies square wave drive to the filter input in regenerative fashion. The loop is self-sustaining, resulting in continuous sine wave outputs at the indicated points. Zener bridge clamping of C1's output stabilizes square wave amplitude applied to the filter and, hence, the sine wave outputs. This form of amplitude control eliminates AGC loop settling times and potential instabilities. Changes in 01's clock frequency permit bandpass tuning, with no amplitude shifts during or after tuning.

Figure 36.22 shows operating waveforms. The bandpass filter, responding to C1's clamped output (Trace A), produces sine (Trace C) and cosine (Trace B) outputs. Distortion, Trace D, dominated by filter clock residue, is 2%.

Clock tunable, memory based sine wave generator

This circuit generates a variable frequency sine wave by continuously clocking a sine coded lookup table memory. The memory's state is converted to an analog output by a DAC. A strength of this technique is its rapid, high fidelity response to frequency and amplitude change commands.

01, set to one of three output frequencies dictated by its digital control inputs, clocks the 74HC191 counters. These counters parallel load a 2716 EPROM programmed to produce an 8-bit (256 states) digitally coded sine wave. The program, developed by Sean Gold and Guy M. Hoover, appears in Figure 36.24.[10] The 2716's parallel output is fed to a DAC, producing the analog output.

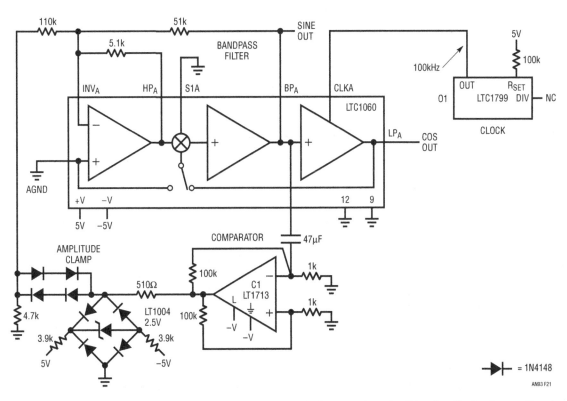

Figure 36.21 • The Regan Resonant Bandpass Loop. A Bandpass Filter, Driven by C1's Oscillation Loop, Continuously Rings at Resonance. Clock Controls Output Frequency. Zener Bridge Clamp Sets Sine and Cosine Output Amplitude

Note 8: See References 8 and 9.
Note 9: This circuit draws heavily on a scheme originated by Tim Regan. See Reference 10.

Note 10: See Reference 11.

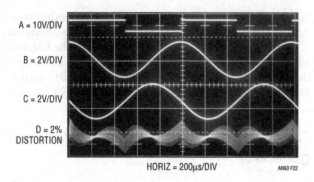

A = 10V/DIV

B = 2V/DIV

C = 2V/DIV

D = 2%
DISTORTION

HORIZ = 200µs/DIV   AN93 F22

Figure 36.22 • Bandpass Filter, Responding to C1's Loop Enforced Excitation (Zener Clamp Output, Trace A), Produces Sine (Trace C) and Cosine (Trace B) Outputs. Distortion (Trace D), Dominated by Switched Capacitor Filter Clock Residue, Is 2%

Trace A in Figure 36.25 is the sine wave output, in this case tuned to 60Hz. Distortion, appearing as Trace B, is mostly composed of clock residue and measures about 0.75%. In Figure 36.26, the digital inputs abruptly change output frequency to 400Hz and then promptly return it to 60Hz. These frequency shifts occur crisply, with no alien components or untoward behavior. Amplitude shifts, accomplished by driving the DAC's reference input (see LTC1450 data sheet), are similarly well behaved. Figure 36.27 shows Trace B's amplitude faithfully responding to Trace A's DAC reference input step. As before, the lack of control loop time constants promotes uncorrupted response.

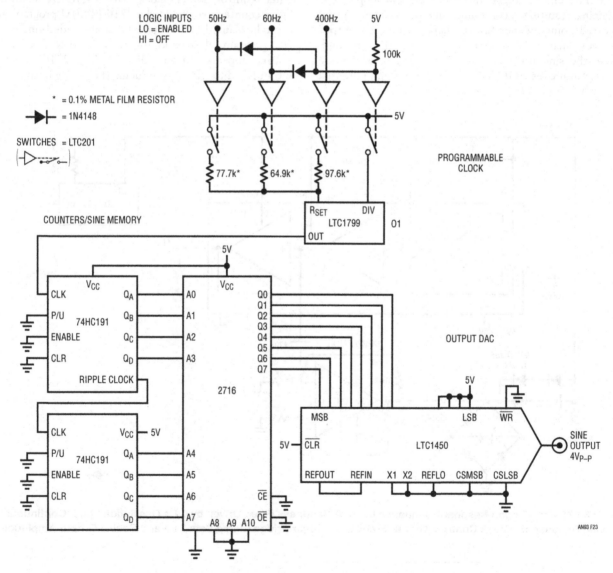

Figure 36.23 • Counter Driven, Sine Encoded Memory Produces 0.75% Distortion Sinewave via D/A Converter. LTC1799 Oscillator Frequency, Controlled by Digital Inputs, Sets Output Frequency

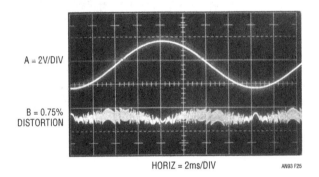

```
Line 10736    Column    Wrap              APL2/PC
      GENCODES
FF FF FF FF FE FE FE FD FD FC FB FA F9 F9 F7 F6
F5 F4 F3 F1 F0 EE ED EB E9 E8 E6 E4 E2 E0 DE DC
D9 D7 D5 D2 D0 CE CB C9 C6 C3 C1 BE B9 B8 B6 B3
B0 AD AA A7 A4 A1 9E 9B 98 95 92 8E 8B 88 85 82
7F 7C 78 75 72 6F 6C 69 66 63 60 5D 5A 57 54 51
4E 4B 48 45 42 40 3D 3A 38 35 33 30 2E 2B 29 27
25 22 20 1E 1C 1A 18 17 15 13 11 10 0E 0D 0C 0A
09 08 07 06 05 04 03 03 02 02 01 01 00 00 00 00
00 00 00 00 01 01 02 02 03 03 04 05 06 07 08 09
0A 0C 0D 0E 10 11 13 15 17 18 1A 1C 1E 20 22 25
27 29 2B 2E 30 33 35 38 3A 3D 40 42 45 48 4B 4E
51 54 57 5A 5D 60 63 66 69 6C 6F 72 75 78 7C 7F
82 85 88 8B 8E 92 95 98 9B 9E A1 A4 A7 AA AD B0
B3 B6 B8 BB BE C1 C3 C6 C9 CB CE D0 D2 D5 D7 D9
DC DE E0 E2 E4 E6 E8 E9 EB ED EE F0 F1 F3 F4 F5
F6 F7 F9 F9 FA FB FC FD FD FE FE FE FF FF FF FF
```

Figure 36.24 • Sinewave Generation Code for the Memory

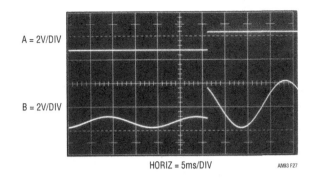

Figure 36.27 • Trace B's Sinewave Amplitude Instantaneously and Faithfully Responds to DAC Reference Input Step, Trace A

Clock tunable notch filter

Figure 36.28 shows a quick, clean way to tune a notch filter's center frequency by varying a single resistor, which could be switched. The LTC1062 switched capacitor filter and A1 form a clock tunable notch (see LTC1062 data sheet). 01, running from the 5V supply, furnishes the clock, which is level shifted by Q1 to drive the ±5V powered

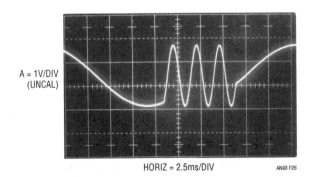

Figure 36.25 • Sinewave Output (Trace A) and Its Distortion (Trace B). Clock Related Products Are Evident in Distortion Presentation

R	CLOCK FREQUENCY	NOTCH CENTER FREQUENCY
210k	4.75kHz	60Hz
249k	3.96kHz	50Hz
31.5k	31.72kHz	400Hz

$$\frac{R1}{R2} = 1.234, \quad \frac{f_{CLOCK}}{f_{NOTCH}} = \frac{79.3}{1}$$

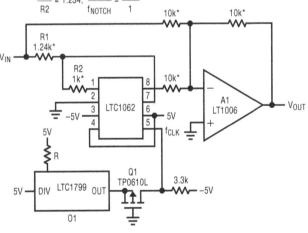

* = 1% METAL FILM RESISTOR

Figure 36.28 • A Clock Tuned, Highly Selective Notch Filter. LTC1799 Oscillator Sets Notch Center Frequency According to Table. R Value Could be Switched Under Digital Control

Figure 36.26 • Fast Oscillator Frequency Shifting Permits Crisp Sinewave Output Frequency Change

LTC1062. In this case, three common notch frequencies are listed; others are selectable by tuning 01 in accordance with the equivalency listed.

Figure 36.29 shows notch performance at a 60Hz center frequency. Response is down over 45dB at 60Hz, with steep slopes on either side of the notch. This characteristic is maintained as center frequency is clock-tuned over broad ranges.

Clock tunable interval generator with 20×10^6:1 dynamic range

An accurate interval generator with large dynamic range appears in Figure 36.30.[11] The circuit is made up of a clock, a counter and a dual flip-flop. Clock frequency and counter modulo are programmable. A trigger input is passed to flip-flop 1's $\bar{Q}$ output (Trace A, Figure 36.31) synchronously with O1's clock (Trace C). This output going low sets flip-flop 2's $\bar{Q}$ output, a circuit output, high (Trace B).

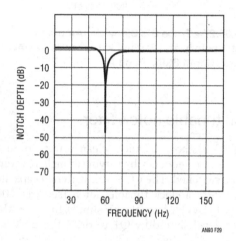

Figure 36.29 • Notch Characteristic at 60Hz Center Frequency. Response Is Essentially Identical as Center Frequency Is Tuned over Broad Range

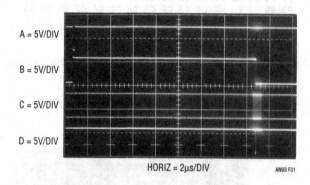

Figure 36.31 • Trace A's Trigger Pulse Sets Circuit Output (Trace B) High. LTC1799 Oscillator (Trace C) Clocks Counter until Selected Counter Output Biases Inverter Low (Trace D), Resetting Circuit Output (Trace B). Reset Sequence Intensified for Photographic Clarity

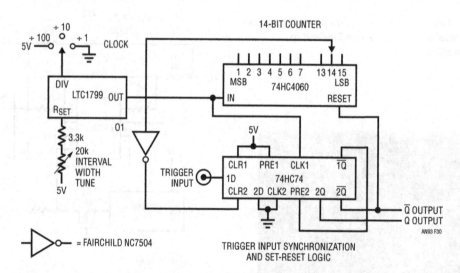

Figure 36.30 • 1% Accurate Interval Generator with 20×10^6:1 Dynamic Range. Flip-Flop Output, Set by Trigger Input, Resets when Counter Times Out. LTC1799 Oscillator Controls Timing Sequence

---

Note 11: Pedestrian laboratory argot for interval generator is "one shot."

Simultaneously, flip-flop 2's $\bar{Q}$ output resets the 4060 counter, allowing it to accumulate clock pulses (again, Trace C). When enough clock pulses occur to set the selected 4060 output high, flip-flop 2's clear input (Trace D) is pulled low, ending the circuit's output width. The output width is settable by O1's frequency and the counter's modulo, both variable over many decades. As shown, the interval is programmable over 800 nanoseconds to 16 seconds, although other counters can extend this range. Interval accuracy and stability is almost entirely dependent on O1's programming resistor.

8-bit, 80μs, passive input, A/D converter

In general, monolithic A/D converters have replaced discrete types. Occasionally, specific desirable circuit characteristics dictate a discrete design. Examples of such special cases include the need for a passive analog input, output data format, control protocol or economic constraints. Figure 36.32's 8-bit design has 90ppm/°C drift (<1LSB 0°C to 70°C) and converts in 80μs. The circuit consists of a current source, an integrating capacitor, a comparator, logic and a clock.[12]

Applying a pulse to the convert command input causes flip-flop $\bar{Q1}$ output to go high (Trace A, Figure 36.33) when the CLK1 input is clocked by O1. This turns on Q3, resetting the 0.01μF capacitor (Trace B). Simultaneously, $\bar{Q1}$ goes low, pulling the CLK2-CLR2 input down. C1's $\bar{Q}$ output, the circuit's status output (Trace C), also goes low and C1's $\bar{Q}$ output rises high. This logic

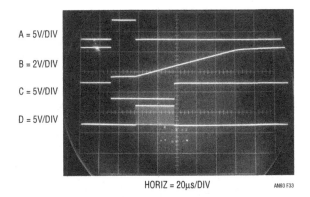

A = 5V/DIV
B = 2V/DIV
C = 5V/DIV
D = 5V/DIV

HORIZ = 20μs/DIV   AN93 F33

Figure 36.33 • 8-Bit A/D Converter Waveforms (E_{IN} = 1V) Include Synchronized Convert Command (Trace A), Reference Ramp (Trace B), Status Output (Trace C) and Data Output (Trace D). Conversion, Initiable when Status Output Is High, Begins when Command Line Goes Low

state prevents any of O1's clock pulses from being transmitted to the circuit's data output (Trace D). When the convert command falls, $\bar{Q1}$ goes low, Q3 turns off and the 0.01μF capacitor begins to ramp. Concurrently, $\bar{Q1}$ goes high, allowing clock pulses to appear at the data output. When the ramp crosses E_{IN}'s voltage, C1's outputs exchange state, pulling the CLK2-CLR2 line low and data output pulses cease. Thus, the O1 originated clock burst appearing at the data output is directly and solely

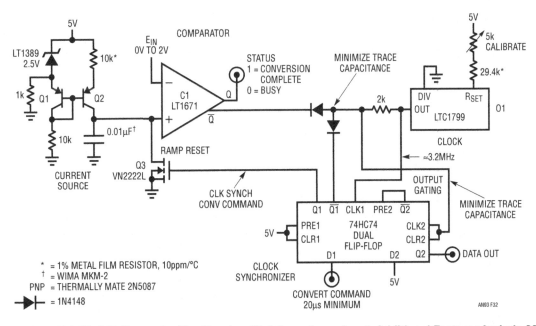

Figure 36.32 • Simple 8-Bit A/D Converter Has Passive, High Impedance Input. Additional Features Include 80μs Conversion Time, Accurate 0°C to 70°C Operation

Note 12: This circuit is a modern incarnation of the earliest electronic A/D known to the author. See Reference 13.

proportional to E_{IN}. For the arrangement shown, 256 pulses appear for a 2V full-scale input. Conversion time decreases with the time required for the ramp to cross E_{IN}. A full-scale conversion requires 80μs, linearly descending to 8μs at 0.1 scale.

Flip-flop 2, connected as a logic buffer, duplicates the high impedance diode-2kΩ node's logic state. As such, this node's trace capacitance should be minimized. This is facilitated by locating the diodes and 2k resistor adjacent to the CLK2-CLR2 inputs. Circuit trimming is accomplished by applying a 2V input and adjusting 01's frequency output ("calibrate") for 256 data output pulses per conversion.

Note: This Application Note was derived from a manuscript originally prepared for publication in EDN magazine.

References

1. Minco Products, Inc., Bulletin TS-102 (N), Minco Products, Inc., 2002.
2. Benjaminson, Albert, "The Linear Quartz Thermometer—A New Tool for Measuring Absolute and Differential Temperatures," Hewlett-Packard Journal, March 1965.
3. Hewlett-Packard Company, "Model 2801A Quartz Thermometer Operating and Service Manual," Hewlett-Packard Company, 1969.
4. Tektronix, Inc., "Type 130 LC Meter Operating and Service Manual," Tektronix, Inc., 1959.
5. Yellow Springs Instrument Company, "YSI Precision Thermistors," Yellow Springs Instrument Company.
6. Panametrics, Inc., "MiniCap 2—Relative Humidity Sensor," Panametrics, Inc., 2000.
7. Williams, J., "Measurement and Control Circuit Collection," Linear Technology Corporation, Application Note 45, June 1991.
8. Toshiba, "2SK147 Datasheet," Toshiba Corporation, Tokyo, Japan.
9. Williams, J., "Practical Circuitry for Measurement and Control Problems," Linear Technology Corporation, Application Note 61, August 1994.
10. Regan, Tim, "Introducing the MF-10: A Versatile Monolithic Active Filter Building Block," National Semiconductor Corporation, Application Note 307, August 1982.
11. Williams, J., "Step-Down Switching Regulators," Linear Technology Corporation, Application Note 35, August 1989.
12. Linear Technology Corporation, LTC1799 Data Sheet, Linear Technology Corporation, August 2001.
13. Wilkinson, D.H., "A Stable Ninety-Nine Channel Pulse Amplitude Analyzer for Slow Counting," Proceedings of the Cambridge Philosophical Society, Cambridge, England, 46,508, 1950.

Appendix A
LTC1799 internal operation

As shown in Figure A1, the LTC1799's master oscillator is controlled by the ratio of the voltage between the V+ and SET pins and the current entering the SET pin (I_{RES}). The voltage on the SET pin is forced to approximately 1.13V below V+ by the PMOS transistor and its gate bias voltage. This voltage is accurate to ±7% at a particular input current and supply voltage (see Figure A2). The effective input resistance is approximately 2k.

A resistor R_{SET}, connected between the V+ and SET pins, "locks together" the voltage (V+ – V_{SET}) and current, I_{RES}, variation. This provides the LTC1799's high precision. The master oscillation frequency reduces to:

$$f_{MO} = 10MHz \cdot \left(\frac{10k\Omega}{R_{SET}}\right)$$

The LTC1799 is optimized for use with resistors between 10k and 200k, corresponding to master oscillator frequencies between 0.5MHz and 10MHz. Accurate frequencies up to 20MHz (R_{SET} = 5k) are attainable if the supply voltage is greater than 4V.

To extend the output frequency range, the master oscillator signal may be divided by 1, 10 or 100 before driving OUT (Pin 5). The divide-by value is determined by the state of the DIV input (Pin 4). Tie DIV to GND or drive it below 0.5V to select ÷1. This is the highest frequency range, with the master output frequency passed directly to OUT. The DIV pin may be floated or driven to midsupply to select ÷10, the intermediate frequency range. The lowest frequency range, ÷100, is selected by tying DIV to V+ or driving it to within 0.4V of V+. Figure A3 shows the relationship between R_{SET}, divider setting and output frequency, including the overlapping frequency ranges near 100kHz and 1MHz.

The CMOS output driver has an on resistance that is typically less than 100Ω. In the ÷1 (high frequency) mode, the rise and fall times are typically 7ns with a 5V supply and 11ns with a 3V supply. These times maintain a clean square wave at 10MHz (20MHz at 5V supply). In the ÷10 and ÷100 modes, where the output frequency is much lower, slew rate control circuitry in the output driver increases the rise/fall times to typically 14ns for a 5V supply and 19ns for a 3V supply. The reduced slew rate lowers EMI (electromagnetic interference) and supply bounce.

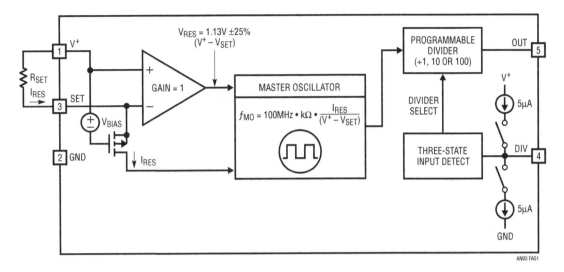

Figure A1 • LTC1799 Master Oscillator Frequency Is Controlled by Ratio of Voltage Between V⁺ and SET and Current Entering SET. Pin-Programmable Frequency Divider Permits Output Frequency Ranging

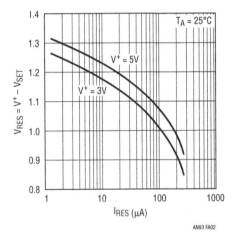

Figure A2 • V⁺ – V$_{SET}$ Variation with I$_{RES}$

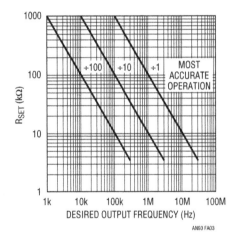

Figure A3 • R$_{SET}$ vs Desired Output Frequency for Three Output Divider Settings

Appendix B
R_{SET} node considerations

The R_{SET} node is the LTC1799's sole analog input. Figure B1, a partial LTC1799 block diagram (see Appendix A for more detail) shows that the node is a MOSFET source and an amplifier input. Equivalent input resistance is about 2kΩ and the point sits approximately 1.13V below the LTC1799 V$^+$ pin. Excessive stray capacitance or noise at R_{SET} will influence amplifier operation, causing master oscillator jitter. Stray capacitance at R_{SET} should be limited to <10pF and signal lines, particularly those operat-

ing at high speed, should be routed away from R_{SET}. A simple guideline is to place the programming resistor directly at R_{SET}. In cases where R_{SET} is a transducer (e.g., a temperature sensor), it may be desirable to locate the transducer at the end of cable. Maintaining low effective capacitance at the R_{SET} node requires "bootstrap" driving the cable shield at the R_{SET} potential (Figure B2). This negates the effect of shield capacitance, because charge cannot transfer between it and R_{SET}. An amplifier capable of driving the shield is required but this is accommodatable. Text Figure 36.5 is a practical incarnation of this technique.

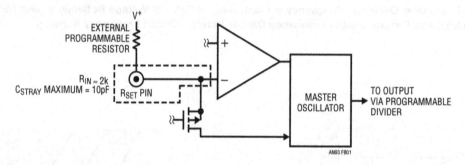

Figure B1 • R_{SET} Pin Has Effective Input Resistance of ≈2k. Stray Pin Capacitance Must be <10pF to Avoid Output Frequency Jitter

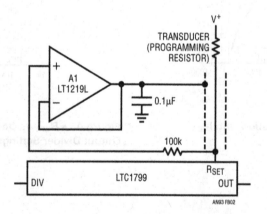

Figure B2 • A1 Senses R_{SET} Voltage, Bootstraps Cable Drive Potential. Arrangement Prevents Cable Capacitance from Influencing R_{SET} Node Because Charge Cannot Transfer. 100k Resistor Isolates A1's Input and Trace Capacitance

Slew rate verification for wideband amplifiers

The taming of the slew

37

Jim Williams

Introduction

Slew rate defines an amplifier's maximum rate of output excursion. This specification sets limits on undistorted bandwidth, an important capability in A/D driver applications. Slew rate also influences achievable performance in D/A output stages, filters, video amplification and data acquisition. Because of its importance, amplifier slew rate must be verified by measurement. Deriving a measurement approach requires understanding slew rate's relationship to amplifier dynamics.

Amplifier dynamic response

Figure 37.1 shows that amplifier dynamic response components include delay, slew and ring times. The *delay time* is small and is almost entirely due to amplifier propagation delay. During this interval there is no output movement. During *slew time* the amplifier moves at its highest possible speed towards the final value. *Ring time* defines the region where the amplifier recovers from slewing and ceases movement within some defined error band. The total

elapsed time from input application until the output arrives at and remains within a specified error band around the final value is the *settling time*.[1]

Slew rate, normally measured during the middle 2/3 of output movement at A = +1, is expressed in volts/ microsecond. Discounting the initial and final movement intervals ensures that amplifier gain-bandwidth limitations during partial input overdrive do not influence the measurement.

Historically, slew rate measurement has been relatively simple.[2] Early amplifiers had slew rates of typically 1V/µs, with later versions sometimes reaching hundreds of volts/µs. Standard laboratory pulse generators easily supplied rise times well beyond amplifier speeds. As slew rates have crossed 1000V/µs, the pulse generator's finite rise time has become a concern. A recent device, the LT1818 (see Box Section, "A 2500V/µs slew rate amplifier with −85dBc

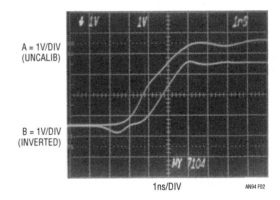

A = 1V/DIV
(UNCALIB)

B = 1V/DIV
(INVERTED)

1ns/DIV   AN94 F02

Figure 37.2 • LT1818 Slew Rate (Upper Trace) is Comparable to Schottky TTL Transition Time (Lower Trace)

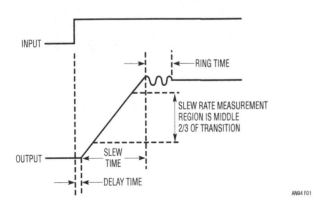

INPUT

RING TIME

SLEW RATE MEASUREMENT REGION IS MIDDLE 2/3 OF TRANSITION

OUTPUT

SLEW TIME

DELAY TIME

AN94 F01

Figure 37.1 • Amplifier Response Components Include Delay, Slew and Ring Times. Slew Rate is Typically Measured During Middle 2/3 of Slew Time

Note 1. Although not considered here, settling time determination is a high order measurement challenge. It is treated in considerable detail in References 2, 3, and 4.
Note 2. The term "slew rate" has a clouded origin. Although used for many years in amplifier literature, there is no mention of it on the Philbrick K2-W (the first standard product op amp, introduced in January 1953) data sheet, dated 1964. Rather, the somewhat more dignified "maximum rate of output swing" is specified.

Analog Circuit and System Design: A Tutorial Guide to Applications and Solutions. DOI: 10.1016/B978-0-12-385185-7.00037-8

distortion at 5MHz"), has a 2500V/µs slew rate, or 2.5V/ *nanosecond*. Figure 37.2 puts this transition rate in perspective. The LT1818's slew rate (Trace A) is comparable to a Schottky TTL gate's (Trace B) transition time. Such speed eliminates almost all pulse generators as candidates for putting the amplifier into slew rate limiting.

A 2500V/µS slew rate amplifier with −85dBc distortion at 5MHz

A/D driving, D/A output stages, data acquisition, video amplification and high frequency filters require low distortion, wideband amplifiers. The LT1818 amplifier (LT1819 is a dual version), with 2500V/µs slew rate, 400MHz GBW and −85dBc distortion, is designed for these applications. Additionally, only 9mA supply current is required. The table provides short form specifications.

LT1818 Short form specifications

CHARACTERISTIC	SPECIFICATION
Gain – Bandwidth Product	400MHz (Typ) 270MHz (Min)
Full Power Bandwidth	95MHz (Typ)
Slew Rate	2500V/µs (Typ), A = +1
Delay	1ns (Typ)
Settling Time	10ns to 0.1% (Typ)
Distortion	−85dBc at 5MHz (Typ)
Input Noise Voltage	6nV/√Hz (Typ)
DC Gain	2500 (Typ) 1500 (Min)
Output Current	±70mA (Typ) ±40mA (Min)
Input Voltage Range	±3.5V at ±5V Supplies (Min)
Input Bias Current	8µA (Max)

Pulse generator rise time effects on measurement

Pulse generator rise time limitations are a significant concern when attempting to accurately determine slew rate. Figures 37.3 through 37.6 demonstrate this by recording amplifier (at A = +1) response to progressively faster pulse generator rise times. Figure 37.3's apparent slew rate limit is ≈385V/µs when driven by a 10ns rise time pulse generator. Figure 37.4 indicates 800V/µs using a 5ns rise time generator. A 3.5ns rise time generator prompts Figure 37.5's 1400V/µs response and a 1ns rise time

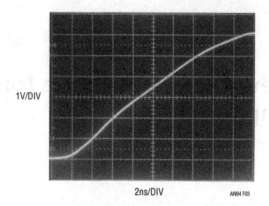

Figure 37.3 • LT1818 Slew Rate Measures ≈385V/µs When Driven By Ten Nanosecond Rise Time Pulse Generator

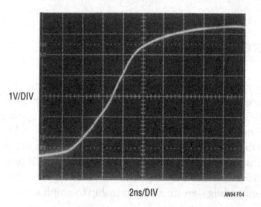

Figure 37.4 • Five Nanosecond Rise Time Pulse Generator Indicates 800V/µs Slew Rate

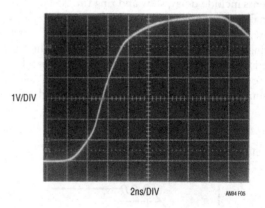

Figure 37.5 • 1400V/µs Slew Rate is Observed with Faster (t$_{RISE}$ = 3.5ns) Pulse Generator

instrument results in Figure 37.6's 2500V/µs observed slew rate. Figure 37.7's plot summarizes results. The data shows a nonlinear slew rate increase as pulse generator rise time decreases. The continuous slew rate increase with decreasing generator rise time, although approaching a zero rise time enforced bound, hints that slew rate limit has not

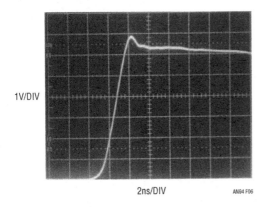

1V/DIV

2ns/DIV   AN94 F06

Figure 37.6 • One Nanosecond Rise Time Pulse Generator Results in 2500V/μs Slew Rate. Verifying Slew Rate Limiting Occurrence Requires Repeating Measurement with Subnanosecond Rise Time Pulse Generator

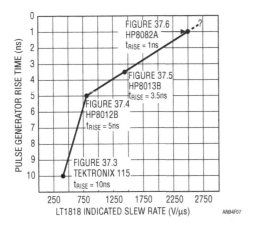

Figure 37.7 • Summarized Data for the Pulse Generators. Decreasing Rise Time Promotes Higher Observed Slew Rate. Verifying Slew Rate Limiting Occurrence Requires Subnanosecond Rise Time Pulse Generator

been reached. Determining if this is so requires a faster pulse generator than Figure 37.6's 1ns rise time unit.

Subnanosecond rise time pulse generators

The majority of general purpose pulse generators have rise times in the 2.5ns to 10ns range. Instrument rise times below 2.5ns are relatively rare, with only a select few types getting down to 1ns.[3] The ranks of subnanosecond rise time generators are even thinner. Subnanosecond rise time generation, particularly if relatively large swings (e.g. 5V to 10V) are desired, employs arcane technologies and exotic

Note 3: See Reference 3 for further discussion and recommendations.

construction techniques (see References 5-16 and 20). Available instruments in this class work well, but can easily cost $10,000 with prices rising towards $30,000 depending on features. For slew rate testing in a laboratory or production environment there is a substantially less expensive alternative.

360ps rise time pulse generator

Figure 37.8 shows a circuit for producing subnanosecond rise time pulses. Rise time is 360ps, with adjustable pulse amplitude. Output pulse occurrence is settable from before-to-after a trigger output. This circuit uses an avalanche pulse generator to create extremely fast rise time pulses.[4]

Q1 and Q2 form a current source that charges the 1000pF capacitor. When the LTC1799 clock is high (trace A, Figure 37.9) both Q3 and Q4 are on. The current source is off and Q2's collector (trace B) is at ground. C1's latch input prevents it from responding and its output remains high. When the clock goes low, C1's latch input is disabled and its output drops low. The Q3 and Q4 collectors lift and Q2 comes on, delivering constant current to the 1000pF capacitor (trace B). The resulting linear ramp is applied to C1 and C2's positive inputs. C2, biased from a potential derived from the 5V supply, goes high 30 nanoseconds after the ramp begins, providing the "trigger output" (trace C) via its output network. C1 goes high when the ramp crosses the potentiometer programmed delay at its negative input, in this case about 170ns. C1 going high triggers the avalanche-based output pulse (trace D), which will be described. This arrangement permits the delay programming control to vary output pulse occurrence from 30 nanoseconds before to 300 nanoseconds after the trigger output. Figure 37.10 shows the output pulse (trace D) occurring 25ns before the trigger output. All other waveforms are identical to Figure 37.9.

When C1's output pulse is applied to Q5's base, it avalanches. The result is a quickly rising pulse across Q5's emitter termination resistor. The 10pF collector capacitor and the charge line discharge, Q5's collector voltage falls and breakdown ceases. The 10pF collector capacitor and the charge line then recharge. At C1's next pulse, this action repeats. The 10pF capacitor supplies the initial pulse response, with the charge lines prolonged discharge contributing the pulse body. The 40″ charge line length forms an output pulse width about 12ns in duration.

Avalanche operation requires high voltage bias. The LT1533 low noise switching regulator and associated components supply this high voltage. The LT1533 is a "push-pull" output switching regulator with controllable transition times.

Note 4: Additional examples of avalanche pulse generators and theoretical discussion appear in Reference 3 and References 5 through 16. The circuit detailed here produces positive going pulses referred to a zero volt baseline. Level shifting options are presented in Appendix B, "Pulse generator output level shifting."

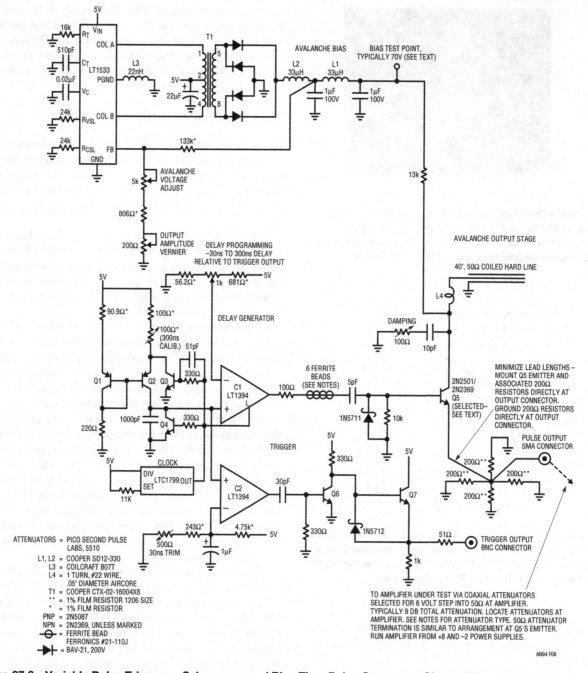

Figure 37.8 • Variable Delay Triggers a Subnanosecond Rise Time Pulse Generator. Charge Line at Q5's Collector Determines ≈10 Nanosecond Output Width. Output Pulse Occurrence is Settable from Before-to-After Trigger Output

Output harmonic content ("noise") is notably reduced with slower switch transition times.[5] Switch current and voltage transition times are controlled by resistors at the R_{CSL} and R_{VSL} pins, respectively. In all other respects the circuit behaves as a classical push-pull, step-up converter.

Circuit optimization

Circuit optimization begins by setting the "Output Amplitude Vernier" to maximum and grounding Q4's collector. Next, set the "Avalanche Voltage Adjust" so free running pulses *just* appear at Q5's emitter, noting the bias test points voltage. Readjust the "Avalanche Voltage Adjust" five volts below this voltage and unground Q4's collector. Set the "30ns Trim" so the trigger output goes

Note 5: The LT1533's low noise performance and its measurement are discussed in Reference 17.

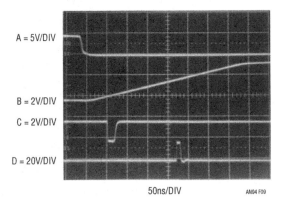

A = 5V/DIV

B = 2V/DIV

C = 2V/DIV

D = 20V/DIV

50ns/DIV   AN94 F09

Figure 37.9 • Pulse Generator's Waveforms Include Clock (Trace A), Q2's Collector Ramp (Trace B), Trigger Output (Trace C) and Pulse Output (Trace D). Delay Sets Output Pulse ≈170ns After Trigger Output

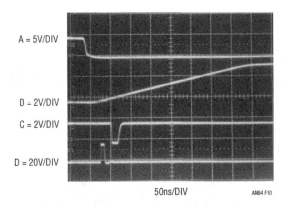

A = 5V/DIV

D = 2V/DIV

C = 2V/DIV

D = 20V/DIV

50ns/DIV   AN94 F10

Figure 37.10 • Pulse Generator's Waveforms with Delay Adjusted for Output Pulse Occurrence (Trace D) 25ns Before Trigger Output (Trace C). All Other Activity is Identical to Previous Figure

low 30ns after the clock goes low. Adjust the delay programming control to maximum and set the "300ns Calib." so C1 goes high 300ns after the clock goes low. Slight interaction between the 30ns and 300ns trims may require repeating their adjustments until both points are calibrated.

Q5 requires selection for optimal avalanche behavior. Such behavior, while characteristic of the device specified, is not guaranteed by the manufacturer. A sample of 30 2N2501s, spread over a 17-year date code span, yielded ≈90%. All "good" devices switched in less than 475ps with some below 300ps.[6] In practice, Q5 should be selected for "in-circuit" rise time under 400 picoseconds. Once this is done, output pulse shape is optimized for slew rate testing by adjusting Q5's collector damping trim. The optimization procedure takes full advantage of the

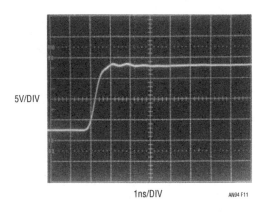

5V/DIV

1ns/DIV   AN94 F11

Figure 37.11 • Excessive Damping is Characterized by Front Corner Rounding and Minimal Pulse-Top Aberrations. Trade Off is Relatively Slow Rise Time

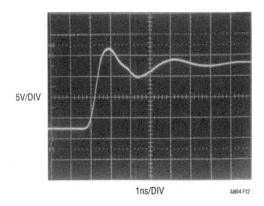

5V/DIV

1ns/DIV   AN94 F12

Figure 37.12 • Minimal Damping Accentuates Rise Time, Although Pulse-Top Ringing is Excessive

freedom that pulse purity is *not* required for slew rate testing. Normally, the pulse edge is carefully adjusted so that maximum transition speed is attained with minimal sacrifice of pulse purity. Slew rate testing does not require this, considerably simplifying optimization.[7]

Slew rate testing permits overshoot and post-transition aberrations if they do not influence amplifier response in the measurement region. Figures 37.11 through 37.13 detail the optimization procedure. In Figure 37.11, the damping trim is set for significant effect, resulting in a reasonably clean pulse but sacrificing rise time.[8] Figure 37.12 represents the opposite extreme. Minimal damping accentuates rise time, but pronounced post-transition ring may influence amplifier operation during slew testing. Figure 37.13's compromise damping is more realistic. Edge rate is only slightly reduced, but post-transition ring is significantly attenuated. The damping photographs were taken with a 1GHz real time oscilloscope (Tektronix 7104/7A29/7B15) with a

---

Note 6: 2N2501s are available from Semelab plc. Sales@semelab.co.uk; Tel. 44-0-1455-556565. A more common transistor, the 2N2369, may also be used but switching times are rarely less than 450ps. See also Footnotes 10 and 11.

Note 7: Optimization procedures for obtaining high degrees of pulse purity while preserving rise time appear in References 3, 5 and 6.
Note 8: The strata is becoming rarefied when a subnanosecond rise time is described as "sacrificed."

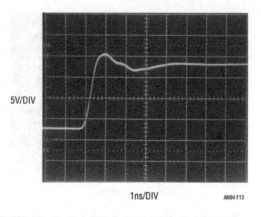

Figure 37.13 • **Optimal Damping Retards Pulse-Top Ringing; Preserves Rise Time in Slew Rate Measurement Region**

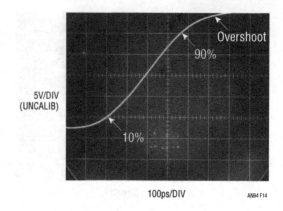

Figure 37.14 • **Figure 37.13's Rise Time Measures 360 Picoseconds in 3.9GHz Sampled Bandpass**

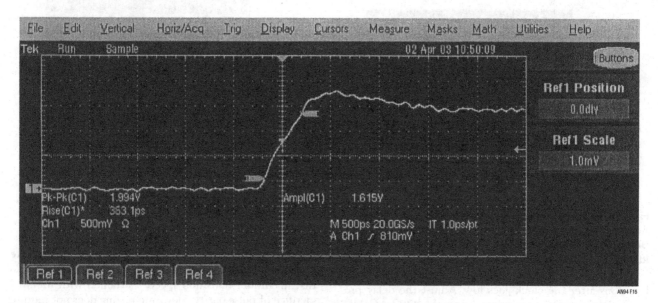

Figure 37.15 • **360 Picosecond Rise Time Monitored in 6GHz Sampled Bandwidth Assures Measurement Integrity** (Courtesy of Michael J. Martin, Tektronix, Inc.)

350ps rise time limit. Accurately determining Figure 37.13's rise time requires more bandwidth.[9] Figure 37.14, taken with a 3.9GHz (t_{RISE} = 90ps) bandwidth oscilloscope (Tektronix 556 with 1S2 sampling plug-in) indicates 360 picosecond output rise time.[10] Figure 37.15 aids measurement confidence by verifying 360 picosecond rise time in a 6GHz (t_{RISE} = 60 picoseconds) oscilloscope bandwidth (Tektronix TDS 6604). The 360 picosecond rise time is almost three times faster than Figure 37.6's 1

nanosecond rise time pulse generator, which promoted a 2500V/µs slew rate. Figure 37.16 puts this kind of speed into perspective. Trace A's 360ps rise time has completed its transition before trace B's 400MHz LT1818 amplifier begins to move! Trace A's rise time is actually faster than depicted, as the 1GHz real time measurement bandwidth limits observed response. Applying this faster rise time pulse should add useful information to Figure 37.7's data.

Refining slew rate measurement

Figure 37.17 shows amplifier (A = +1) response to the 360 picosecond rise time pulse in a 1GHz real time

---

Note 9: Accurate rise time determination at these speeds mandates verifying measurement signal path (cables, attenuators, probes, oscilloscope) integrity. See Appendix A, "verifying rise time measurement integrity" and Appendix C, "Connections, cables, adapters, attenuators, probes and picoseconds."
Note 10: Experimental adjustment, iterated towards favorable results, of Q5's lead lengths, impedances and layout may be required for fastest rise time.

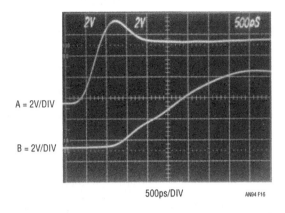

Figure 37.16 • Trace A's 360 Picosecond Rise Time Pulse Completes Transition Before Amplifier Output (Trace B) Begins Movement. Trace A's Rise Time is Actually ≈150ps Faster than Depiction, as 1GHz Measurement Bandwidth Limits Observed Response

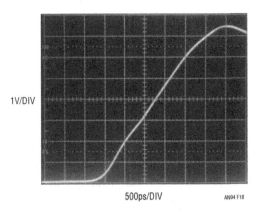

Figure 37.18 • Time Expansion of Figure 37.17 Shows ≈2800V/μs Slew Rate, Revealing 11% Error in Figure 37.6's 1 Nanosecond Rise Time Driven 2500V/μs Response

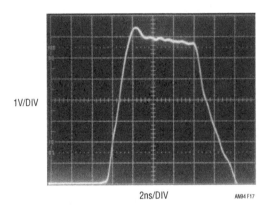

Figure 37.17 • LT1818 Slew Response When Driven from Avalanche Pulse Generator Appears Faster Than Figure 37.6's 2500V/μs

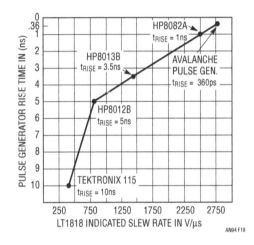

Figure 37.19 • Figure 37.7's Data Restated to Include Avalanche Pulse Generator Results. Significant Slew Rate Increase is Unlikely Because Required Input Step Rise Time Approaches Zero

bandpass. The middle 2/3 of the positive transaction, the slew rate measurement region, appears faster than Figure 37.6. Figure 37.18 increases sweep speed to 500 picoseconds/division. The photograph shows a measurement region slew rate of ≈2800V/μs, revealing an 11% error in Figure 37.6's determination. Applying these findings to Figure 37.7's plot produces Figure 37.19. The new data suggests that, while slew rate "hard" limiting may not be occurring, little practical improvement is possible because rise time is approaching zero. A faster rise time pulse generator could confirm this, but any slew rate improvement would likely be academic.[11] Realistically,

the large signal, 360 picosecond rise time input required to promote 2800V/μs slew rate is rarely encountered in practical circuitry.

Note: This chapter was derived from a manuscript originally prepared for publication in EDN magazine.

Note 11: Faster rise times are possible, although considerable finesse is required in Q5's selection, layout, mounting, terminal impedance choice and triggering. The 360ps rise time quoted in the text represents readily reproducible results. Rise times below 300ps have been achieved, but require considerable and tedious effort. See References 5 and 6.

References

1. Shakespeare, William, "The Taming of the Shrew," 1593-94.
2. Williams, Jim, "Component and Measurement Advances Ensure 16-Bit DAC Settling Time," Linear Technology Corporation, Application Note 74, July 1998.
3. Williams, Jim, "30 Nanosecond Settling Time Measurement for a Precision Wideband Amplifier," Linear Technology Corporation, Application Note 79, September 1999.
4. Williams, Jim, "A Standards Lab Grade 20-Bit DAC with 0.1ppm/°C Drift," Linear Technology Corporation, Application Note 86, January 2001.
5. Braatz, Dennis, "Avalanche Pulse Generators," Private Communication, Tektronix, Inc. 2003.
6. Tektronix, Inc., Type 111 Pretrigger Pulse Generator Operating and Service Manual, Tektronix, Inc. 1960.
7. Haas, Isy, "Millimicrosecond Avalanche Switching Circuit Utilizing Double-Diffused Silicon Transistors," Fairchild Semiconductor, Application Note 8/2, December 1961.
8. Beeson, R. H., Haas, I., Grinich, V. H., "Thermal Response of Transistors in Avalanche Mode," Fairchild Semi-

conductor, Technical Paper 6, October 1959.
9. Chaplin, G. B. B. "A Method of Designing Transistor Avalanche Circuits with Applications to a Sensitive Transistor Oscilloscope," paper presented at the 1958 IRE-AIEE Solid State Circuits Conference, Philadelphia, PA., February 1958.
10. Motorola, Inc., "Avalanche Mode Switching," Chapter 9, 285–304. Motorola Transistor Handbook, 1963.
11. Williams, Jim, "A Seven-Nanosecond Comparator for Single Supply Operation," "Programmable, Sub-nanosecond Delayed Pulse Generator," 32-34, Linear Technology Corporation, Application Note 72, May 1998.
12. Hamilton, D. J. Shaver, F. H. and Griffith P. G. "Avalanche Transistor Circuits for Generating Rectangular Pulses," Electronic Engineering, December, 1962.
13. Seeds, R. B. "Triggering of Avalanche Transistor Pulse Circuits," Technical Report No. 1653-1, August 5, 1960, Solid-State Electronics Laboratory, Stanford Electronics Laboratories, Stanford University, Stanford, California.

14. Williams, Jim, "Measurement and Control Circuit Collection," Linear Technology Corporation, Application Note 45, June 1991.
15. Williams, Jim, "High Speed Amplifier Techniques," Linear Technology Corporation, Application Note 47, August 1991.
16. Williams, Jim, "Practical Circuitry for Measurement and Control Problems," Linear Technology Corporation, Application Note 61, August 1994.
17 Williams, Jim, "A Monolithic Switching Regulator with 100µV Output Noise," Linear Technology Corporation, Application Note 70, October 1997.
18. Andrews, James R. "Pulse Measurements in the Picosecond Domain," Picosecond Pulse Labs, Application Note AN-3a, 1988.
19. Martin, Michael J., "Fast Rise Time Oscilloscope Measurement," Private Communication, Tektronix, Inc. 2003.
20. Madden, C. J. Rodwell, M. J. W. Marsland, R. A. Bloom, D. M. and Pao, Y. C. "Generation of 3.5ps fall-time shock waves on a monolithic nonlinear transmission line," *IEEE Electron Device Lett.* 9, 303–305.

Appendix A
Verifying rise time measurement integrity

Any measurement requires the experimenter to insure measurement confidence. Some form of calibration check is always in order. High speed time domain measurement is particularly prone to error, and various techniques can promote measurement integrity.

Figure A1's battery-powered 200MHz crystal oscillator produces 5ns markers, useful for verifying oscilloscope time base accuracy. A single 1.5V AA cell supplies the LTC3400 boost regulator, which produces 5 volts to run the oscillator. Oscillator output is delivered to the 50Ω load via a peaked attenuation network. This provides well defined 5ns markers (Figure A2) and prevents overdriving low level sampling oscilloscope inputs.

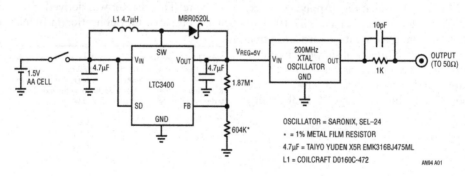

Figure A1 • 1.5V Powered, 200MHz Crystal Oscillator Provides 5 Nanosecond Time Markers. Switching Regulator Converts 1.5V to 5V to Power Oscillator

MANUFACTURER	MODEL NUMBER	RISE TIME	AMPLITUDE	AVAILABILITY	COMMENTS
Avtech	AVP2S	40ps	0V to 2V	Current Production	Free Running or Triggered Operation, 0MHz to 1MHz
Hewlett-Packard	213B	100ps	≈175mV	Secondary Market	Free Running or Triggered Operation to 100kHz
Hewlett-Packard	1105A/1108A	60ps	≈200mV	Secondary Market	Free Running or Triggered Operation to 100kHz
Hewlett-Packard	1105A/1106A	20ps	≈200mV	Secondary Market	Free Running or Triggered Operation to 100kHz
Picosecond Pulse Labs	TD1110C/TD1107C	20ps	≈230mV	Current Production	Similar to Discontinued HP1105/1106/8A. See above
Stanford Research Systems	DG535 OPT 04A	100ps	0.5V to 2V	Current Production	Must be Driven with Stand-alone Pulse Generator
Tektronix	284	70ps	≈200mV	Secondary Market	50kHz Repetition Rate. Pre-trigger 75ns to 150ns Before Main Output. Calibrated 100MHz and 1GHz Sine Wave Auxilary Outputs
Tektronix	111	500ps	≈±10V	Secondary Market	10kHz to 100kHz Repetition Rate. Positive or Negative Outputs. 30ns to 250ns Pre-trigger Output. External Trigger Input. Pulse Width Set with Charge Lines
Tektronix	067-0513-00	30ps	≈400mV	Secondary Market	60ns Pre-trigger Output. 100kHz Repetition Rate
Tektronix	109	250ps	0V to ±55V	Secondary Market	≈600Hz Repetition Rate (High Pressure Hg Reed Relay Based). Positive or Negative Outputs. Pulse Width Set by Charge Lines

Figure A3 ● Picosecond Edge Generators Suitable for Rise Time Verification. Considerations Include Speed, Features and Availability

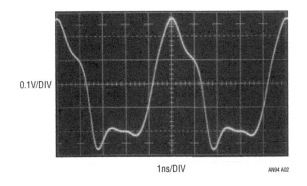

0.1V/DIV

1ns/DIV

AN94 A02

Figure A2 ● Time Mark Generator Output Terminated into 50Ω. Peaked Waveform is Optimal for Verifying Time Base Calibration

Once time base accuracy is confirmed it is necessary to check rise time. The lumped signal path rise time, including attenuators, connections, cables, oscilloscope and anything else, should be included in this measurement. Such "end-to-end" rise time checking is an effective way to promote meaningful results. A guideline for insuring accuracy is to have 4x faster measurement path rise time than the rise time of interest. Thus, text Figure 37.14's 360 picosecond rise time measurement requires a verified 90 picosecond measurement path rise time to support it. Verifying the 90 picosecond measurement path rise time, in turn, necessitates a ≤ 22.5 picosecond rise time test step. Figure A3 lists some very fast edge generators for rise time checking.[1]

Note 1: This is a fairly exotic group, but equipment of this caliber really is necessary for rise time verification.

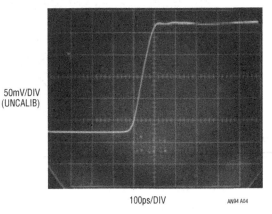

50mV/DIV
(UNCALIB)

100ps/DIV

AN94 A04

Figure A4 ● 20 Picosecond Step Produces ≈90 Picosecond Oscilloscope Rise Time, Verifying Text Figure 37.14's Measurement Path Fidelity

The Hewlett-Packard 1105A/1106A, specified at 20 picoseconds rise time, was used to verify text Figure 37.14's measurement signal path. Figure A4 indicates a 90 picosecond rise time, promoting measurement confidence.

Appendix B
Pulse generator output level shifting

The text's avalanche pulse generator produces a positive 15V to 20V output. This is not suitable for most amplifiers. Various amplifier configurations require different forms of level shifting. A difficulty is

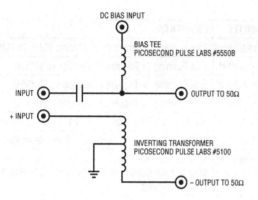

Figure B1 • Output Level Shifters Include Bias Tee for DC Offsetting and Inverting Transformer for Negative Outputs. Practical Realization of Conceptually Simple Networks Requires Care to Maintain Picosecond Speed Fidelity

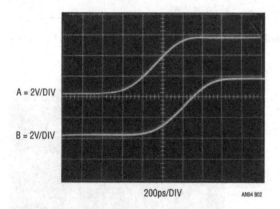

Figure B2 • Bias Tee's Level Shifted Output (Trace B) Faithfully Reproduces Input (Trace A) in 3.9GHz Sampled Bandpass. 300 Picosecond Timing Skew Derives from Bias Tee and Measurement Fixture Delays

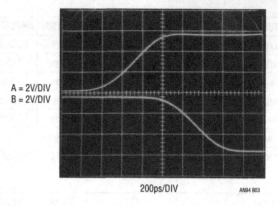

Figure B3 • Wideband Pulse Transformers Output (Trace B) Inverts Input (Trace A) with Uncompromised Fidelity and 600 Picosecond Delay

that whatever level shift mechanism is employed must not degrade pulse rise time.

The simplest level shift is pure attenuation, facilitated by the coaxial attenuators listed in text Figure 37.8's notes. These devices, well specified in the time domain, have 8 picosecond rise time. Combining these attenuators with amplifier power supplies of 8V and −2V permitted a 6V pulse to bias the unity gain follower used in the text's test.

In some cases the pulse must be negatively biased or inverted. Figure B1 shows ways to do this. The "bias tee" network capacitively strips the input's DC component, reestablishing it with the potential present at the DC bias input. The inverting network, an autotransformer, magnetically accomplishes pulse inversion at unity gain. These conceptually straightforward networks are deceptively simple in appearance. Maintaining pulse fidelity and rise time at

picosecond speeds involves numerous construction subtleties. The commercially available units noted in the figure are recommended.

Figure B2 shows bias tee response (trace B) to a fast input step (trace A). The output, in this case biased from −3V, faithfully reproduces the input with only 300 picoseconds skew, primarily due to uncompensated measurement fixture delays.

Figure B3 records inverting transformer response (trace B) to trace A's input. Rise time and fidelity are uncompromised, with about 600 picoseconds propagation delay.

Appendix C
Connections, cables, adapters, attenuators, probes and picoseconds

Subnanosecond rise time signal paths must be considered as transmission lines. Connections, cables, adapters, attenuators and probes represent discontinuities in this transmission line, deleteriously effecting its ability to faithfully transmit desired signal. The degree of signal corruption contributed by a given element varies with its deviation from the transmission lines nominal impedance. The practical result of such introduced aberrations is degradation of pulse rise time, fidelity, or both. Accordingly, introduction of elements or connections to the signal path should be minimized and necessary connections and elements must be high grade components. Any form of connector, cable, attenuator or probe must be fully specified for high frequency use. Familiar BNC hardware becomes lossy at rise times much faster than 350ps. SMA components are preferred for the rise times described in the text. Additionally, the avalanche pulse generator output cable should be 50Ω "hard line" or, at least, Teflon-based coaxial cable fully specified for high frequency operation. Optimal connection practice eliminates any cable by coupling the generator output (via the necessary coaxial attenuators—see Figure 37.8) *directly* to the amplifier under test input. For example, replacing 18″ of output cable with a direct connection improved generator rise time from 380 picoseconds to 360 picoseconds.

Mixing signal path hardware types via adapters (e.g. BNC/SMA) should be avoided. Adapters introduce significant parasitics, resulting in reflections, rise time degradation, resonances and other degrading behavior. Similarly, oscilloscope connections should be made directly to the instrument's 50Ω inputs, avoiding probes. If probes must be used, their introduction to the signal path mandates attention to their connection mechanism and high frequency compensation. Passive "Z_0" types, commercially available in 500Ω (10×) and 5kΩ (100×) impedances, have input capacitance below 1pf. Any such probe must be carefully frequency compensated before use or misrepresented measurement will result. Inserting the probe into the signal path necessitates some form of signal pick-off which nominally does not influence signal transmission. In practice, some amount of disturbance must be tolerated and its effect on measurement results evaluated. High quality signal pick-offs always specify insertion loss, corruption factors and probe output scale factor.

The preceding emphasizes vigilance in designing and maintaining a signal path. Skepticism, tempered by enlightenment, is a useful tool when constructing a signal path and no amount of hope is as effective as preparation and directed experimentation.

Instrumentation circuitry using RMS-to-DC converters

RMS converters rectify average results

38

Jim Williams

Introduction

It is widely acknowledged that RMS (root of the mean of the square) measurement of waveforms furnishes the most accurate amplitude information.[1] Rectify-and-average schemes, usually calibrated to a sine wave, are only accurate for one waveshape. Departures from this waveshape result in pronounced errors. Although accurate, RMS conversion often entails limited bandwidth, restricted range, complexity and difficult to characterize dynamic and static errors. Recent developments address these issues while simultaneously improving accuracy. Figure 38.1 shows the LTC®1966/LTC1967/LTC1968 device family. Low frequency accuracy, including linearity and gain error, is inside 0.5% with 1% error at bandwidths extending to 500kHz. These converters employ a sigma-delta based computational scheme to achieve their performance.[2]

Figure 38.2's pinout descriptions and basic circuits reveal an easily applied device. An output filter capacitor is all that is required to form a functional RMS-to-DC converter. Split and single supply powered variants are shown. Such ease of implementation invites a broad range of application; examples begin with Figure 38.3.

Isolated power line monitor

> BEFORE PROCEEDING ANY FURTHER, THE READER IS WARNED THAT CAUTION MUST BE USED IN THE CONSTRUCTION, TESTING AND USE OF THIS CIRCUIT. HIGH VOLTAGE, LETHAL POTENTIALS ARE PRESENT IN THIS CIRCUIT. EXTREME CAUTION MUST BE USED IN WORKING WITH, AND MAKING CONNECTIONS TO, THIS CIRCUIT. REPEAT: THIS CIRCUIT CONTAINS DANGEROUS, HIGH VOLTAGE POTENTIALS. USE CAUTION.

Figure 38.3's AC power line monitor has 0.5% accuracy over a sensed 90VAC to 130VAC input and provides a safe, fully isolated output. RMS conversion provides accurate reporting of AC line voltage regardless of waveform distortion, which is common.

PART NUMBER	LINEARITY ERROR TYP/MAX (%)	CONVERSION GAIN ERROR TYP/MAX (%)	1% ERROR BANDWIDTH (kHz)	3dB ERROR BANDWIDTH (kHz)	SUPPLY VOLTAGE		I SUPPLY MAX (µA)
					MIN(V)	MAX(V)	
LTC1966	0.02/0.15	0.1/0.3	6	800	2.7	±5	170
LTC1967	0.02/0.15	0.1/0.3	200	4MHz	4.5	5.5	390
LTC1968	0.02/0.15	0.1/0.3	500	15MHz	4.5	5.5	2.3mA

Figure 38.1 • Primary Differences in RMS to DC Converter Family are Bandwidth and Supply Requirements. All Devices Have Rail-to-Rail Differential Inputs and Output

Note 1: See Appendix A, "RMS-to-DC Conversion" for complete discussion of RMS measurement.
Note 2: Appendix A details sigma-delta based RMS-to-DC converter operation.

Analog Circuit and System Design: A Tutorial Guide to Applications and Solutions. DOI: 10.1016/B978-0-12-385185-7.00038-X

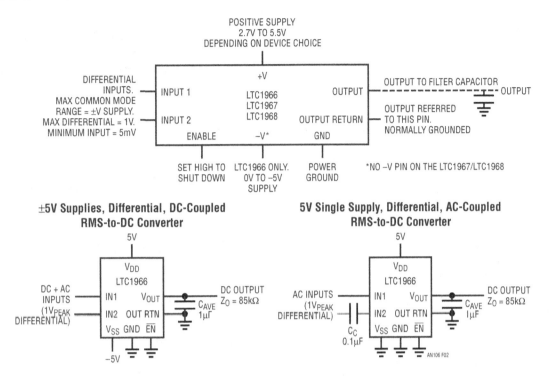

Figure 38.2 • RMS Converter Pin Functions (Top) and Basic Circuits (Bottom). Pin Descriptions are Common to All Devices, with Minor Differences

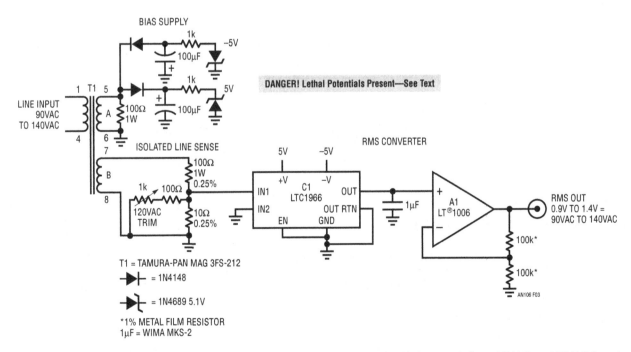

Figure 38.3 • Isolated Power Line Monitor Senses Via Transformer with 0.5% Accuracy Over 90VAC to 130VAC Input. Secondary Loading Optimizes Transformer Voltage Conversion Linearity

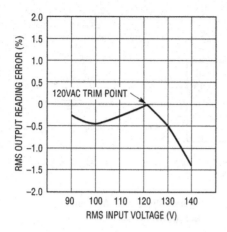

Figure 38.4 • Error Plot for Isolated Line Monitor Shows 0.5% Accuracy from 90VAC to 130VAC, Degrading to 1.4% at 140VAC. Transformer Parasitics Account for Almost All Error

The AC line voltage is divided down by T1's ratio. An isolated and reduced potential appears across T1's secondary B, where it is resistively scaled and presented to C1's input. Power for C1 comes from T1's secondary A, which is rectified, filtered and Zener regulated to DC. A1 takes gain and provides a numerically convenient output. Accuracy is increased by biasing T1 to an optimal loading point, facilitated by the relatively low resistance divider values. Similarly, although C1 and A1 are capable of single supply operation, split supplies maintain symmetrical T1 loading. The circuit is calibrated by adjusting the 1k trim for 1.20V output with the AC line set at 120VAC. This adjustment is made using a variable AC line transformer and a well floated (use a line isolation transformer) RMS voltmeter.[3]

Figure 38.4's error plot shows 0.5% accuracy from 90VAC to 130VAC, degrading to 1.4% at 140VAC. The beneficial effect of trimming at 120VAC is clearly evident; trimming at full scale would result in larger overall error, primarily due to non-ideal transformer behavior. Note that the data is specific to the transformer specified. Substitution for T1 necessitates circuit value changes and recharacterization.

Fully isolated 2500V breakdown, wideband RMS-to-DC converter

NOTE: BEFORE PROCEEDING ANY FURTHER, THE READER IS WARNED THAT CAUTION MUST BE USED IN THE CONSTRUCTION, TESTING AND USE OF THIS CIRCUIT. HIGH VOLTAGE, LETHAL POTENTIALS ARE PRESENT IN THIS CIRCUIT.

Note 3: See Appendix B, "AC Measurement and Signal Handling Practice," for recommendations on RMS voltmeters and other AC measurement related gossip.

EXTREME CAUTION MUST BE USED IN WORKING WITH, AND MAKING CONNECTIONS TO, THIS CIRCUIT. REPEAT: THIS CIRCUIT CONTAINS DANGEROUS, HIGH VOLTAGE POTENTIALS. USE CAUTION.

Accurate RMS amplitude measurement of SCR chopped AC line related waveforms is a common requirement. This measurement is complicated by the SCR's fast switching of a sine wave, introducing odd waveshapes with high frequency harmonic content. Figure 38.5's conceptual SCR-based AC/DC converter is typical. The SCRs alternately chop the 220VAC line, responding to a loop enforced, phase modulated trigger to maintain a DC output. Figure 38.6's waveforms are representative of operation. Trace A is one AC line phase, trace B the SCR cathodes. The SCR's irregularly shaped waveform contains DC and high frequency harmonic, requiring wideband RMS conversion for measurement. Additionally, for safety and system interface considerations, the measurement must be fully isolated.

Figure 38.7 provides isolated power and data output paths to an RMS-to-DC converter, permitting safe, wideband, digital output RMS measurement. A pulse generator configured comparator combines with Q1 and Q2 to drive T1, resulting in isolated 5V power at T1's rectified, filtered and Zener regulated output. The RMS-to-DC converter senses either 135VAC or 270VAC full-scale inputs via a resistive divider. The converter's DC output feeds a self-clocked, serially interfaced A/D converter; optocouplers convey output data across the isolation barrier. The LTC6650 provides a 1V reference to the A/D and biases the RMS-to-DC converter's inputs to accommodate the voltage divider's AC swing. Calibration is accomplished by adjusting the 20k trim while noting output data agreement with the input AC voltage. Circuit accuracy is within 1% in a 200kHz bandwidth.

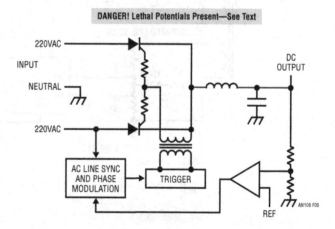

Figure 38.5 • Conceptual AC/DC Converter is Typical of SCR-Based Configurations. Feedback Directed, AC Line Synchronized Trigger Phase Modulates SCR Turn-On, Controlling DC Output

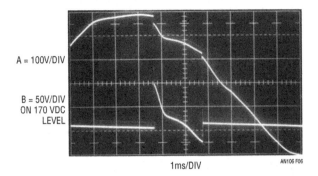

A = 100V/DIV

B = 50V/DIV
ON 170 VDC
LEVEL

1ms/DIV

AN106 F06

Figure 38.6 • Typical SCR-Based Converter Waveforms Taken at AC Line (Trace A) and SCR Cathodes (Trace B). SCR's Irregularly Shaped Waveform Contains DC and High Frequency Harmonic, Requiring Wideband RMS Converter for Measurement

Low distortion AC line RMS voltage regulator

NOTE: BEFORE PROCEEDING ANY FURTHER, THE READER IS WARNED THAT CAUTION MUST BE USED IN THE CONSTRUCTION, TESTING AND USE OF THIS CIRCUIT. HIGH VOLTAGE, LETHAL POTENTIALS ARE PRESENT IN THIS CIRCUIT. EXTREME CAUTION MUST BE USED IN WORKING WITH, AND MAKING CONNECTIONS TO, THIS CIRCUIT. REPEAT: THIS CIRCUIT CONTAINS DANGEROUS, HIGH VOLTAGE POTENTIALS. USE CAUTION.

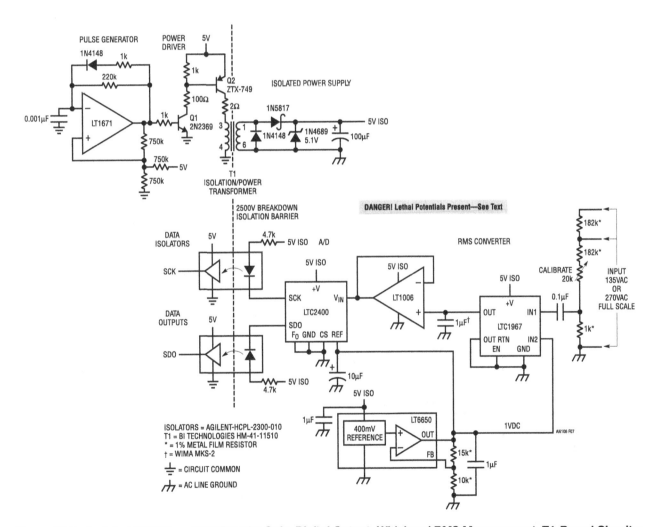

Figure 38.7 • Isolated RMS Converter Permits Safe, Digital Output, Wideband RMS Measurement. T1-Based Circuitry Supplies Isolated Power. RMS-to-DC Converter Senses High Voltage Input via Resistive Divider. A/D Converter Provides Digital Output Through Optoisolators. Accuracy is 1% in 200kHz Bandwidth

Almost all AC line voltage regulators rely on some form of waveform chopping, clipping or interruption to function. This is efficient, but introduces waveform distortion, which is unacceptable in some applications. Figure 38.8 regulates the AC line's RMS value within 0.25% over wide input swings and does not introduce distortion. It does this by continuously controlling the conductivity of a series pass MOSFET in the AC line's path. Enclosing the MOSFET in a diode bridge permits it to operate during both AC line polarities.

The AC line voltage is applied to the Q2-diode bridge. The Q2-diode bridge output is sensed by a calibrated variable voltage divider which feeds C1. C1's output, representing the regulated line's RMS value, is routed to control amplifier A1 and compared to a reference. A1's output biases Q1, controlling drive to a photovoltaic optoisolator. The optoisolator's output voltage provides level-shifted bias to diode bridge enclosed Q2, closing a control loop which regulates the output's RMS voltage against AC line

and load shifts. RC components in A1's local feedback path stabilize the control loop. The loop operates Q2 in its linear region, much like a common low voltage DC linear regulator. The result is absence of introduced distortion at the expense of lost power. Available output power is constrained by heat dissipation. For example, with the output adjustment set to regulate 10V below the normal input, Q2 dissipates about 10W at 100W output. This figure can be improved upon. The circuit regulates for $V_{IN} \geq 2V$ above V_{OUT}, but operation in this region risks regulation dropout as V_{IN} varies.

Circuit details include JFET Q5 and associated components. The passive components associated with Q5's gate form a slow turn-on negative supply for C1. They also provide gate bias for Q5. Q5, a soft-start, prevents abrupt AC power application to the output at start-up. When power is off, Q5 conducts, holding A1's "+" input low. When power is applied, A1 initially has a zero volt reference, causing the control loop to set the output at zero. As

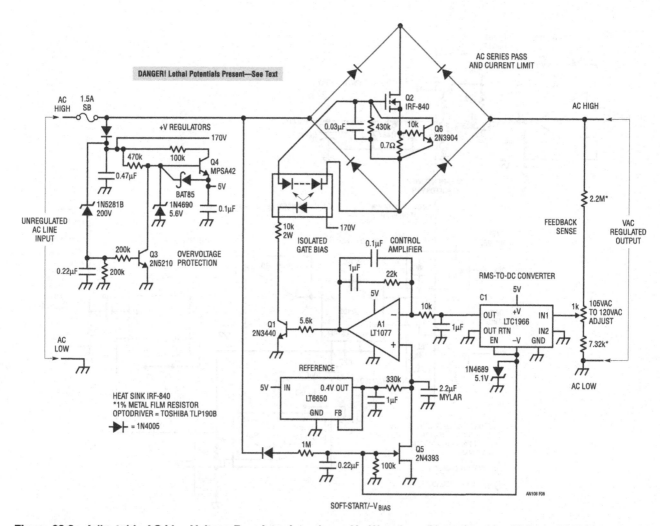

Figure 38.8 • Adjustable AC Line Voltage Regulator Introduces No Waveform Distortion. Line Voltage RMS Value is Sensed and Compared to a Reference by A1. A1 Biases Photovoltaic Optocoupler via Q1, Setting Q2-Diode Bridge Conductivity and Closing a Control Loop. V_{IN} Must be ≥ 2 V Above V_{OUT} to Maintain Regulation

the 1MΩ 0.22μF combination charges, Q5's gate moves negative, causing its channel conductivity to gradually decay. Q5 ramps off, A1's positive input moves smoothly towards the LT6650's 400mV reference, and the AC output similarly ascends towards its regulation point. Current sensor Q6, measuring across the 0.7Ω shunt, limits output current to about 1A. At normal line inputs (90VAC to 135VAC) Q4 supplies 5V operating bias to the circuit. If line voltage rises beyond this point, Q3 comes on, turning off Q4 and shutting down the circuit.

X1000 DC stabilized millivolt preamplifier

The preceding circuits furnish high level inputs to the RMS converter. Many applications lack this advantage and some form of preamplifier is required. High gain pre-amplification for the RMS converter requires more attention than might be supposed. The preamplifier must have low offset error because the RMS converter (desirably) processes DC as legitimate input. More subtly, the preamplifier must have far more bandwidth than is immediately apparent. The amplifier's −3db bandwidth is of interest, but its closed loop 1% amplitude error bandwidth must be high enough to maintain accuracy over the RMS converter's 1% error passband. This is not trivial, as very high open-loop gain at the maximum frequency of interest is required to avoid inaccurate closed-loop gain.

Figure 38.9 shows an ×1000 preamplifier which preserves the LTC1966's DC-6kHz 1% accuracy. The amplifier may be either AC or DC coupled to the RMS converter. The 1mV full-scale input is split into high and low frequency paths. AC coupled A1 and A2 take a cascaded, high frequency gain of 1000. DC coupled, chopper stabilized A3 also has ×1000 gain, but is restricted to DC and low frequency by its RC input filter. Assuming the switch is set to "DC+AC", high and low frequency path

information recombine at the RMS converter. The high frequency path's 650kHz −3db response combines with the low frequency section's microvolt level offset to preserve the RMS converter's DC-6kHz 1% error. If only AC response is desired, the switch is set to the appropriate position. The minimum processable input, set by the circuits noise floor, is 15μV.

Wideband decade ranged ×1000 preamplifier

The LTC1968, with a 500kHz, 1% error bandwidth, poses a significant challenge for an accurate preamplifier, but Figure 38.10 meets the requirement. This design features decade ranged gain to ×1000 with a 1% error bandwidth beyond 500kHz, preserving the RMS converter's 1% error bandwidth. Its 20μV noise floor maintains wideband performance at microvolt level inputs.

Q1A and Q1B form a low noise buffer, permitting high impedance inputs. A1 and A2, both gain switchable, take cascaded gain in accordance with the figure's table. The gains are settable via reed relays controlled by a 2-bit code. A2's output feeds the RMS converter and the converter's output is smoothed by a Sallen-Keys active filter. The circuit maintains 1% error over a 10Hz to 500kHz bandwidth at all gains due to the preamplifier's −3db, 10MHz bandwidth. The 10Hz low frequency restriction could be eliminated with a DC stabilization path similar to Figure 38.9's but its gain would have to be switched in concert with the A1-A2 path.

Figure 38.11 shows preamplifier response to a 1mV input step at a gain of ×1000. A2's output is singularly clean, with trace thickening in the pulse flat portions due to the 20μV noise floor. The 35ns risetime indicates a 10MHz bandwidth.

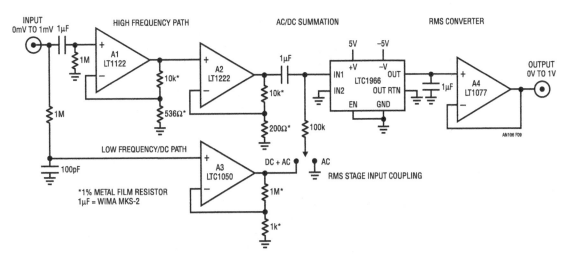

Figure 38.9 • ×1000 Preamplifier Allows 1mV Full-Scale Sensitivity RMS-to-DC Conversion. Input Splits Into High and Low Frequency Amplifier Paths, Recombining at RMS Converter. Amplifier's −3dB, 650kHz Bandwidth Preserves RMS-to-DC Converter's 6kHz, 1% Error Bandwidth. Noise Floor is 15μV

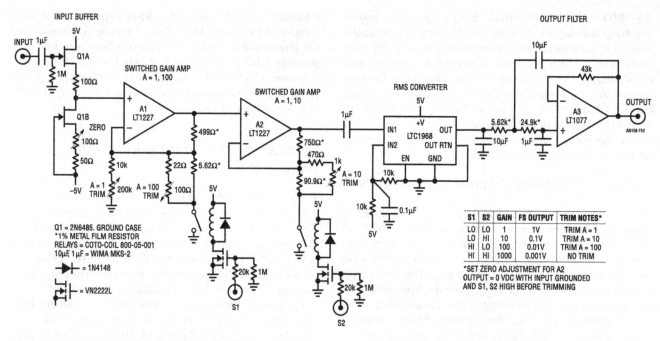

Figure 38.10 • Switched Gain 10MHz (−3dB) Preamplifier Preserves LTC1968's 500kHz, 1% Error Bandwidth. Decade Ranged Gains (See Table) Allow 1mV Full Scale with 20μV Noise Floor. JFET Input Stage Presents High Input Impedance. AC Coupling, 3rd Order Sallen-Key Filter Maintains 1% Accuracy Down to 10Hz

S1	S2	GAIN	FS OUTPUT	TRIM NOTES*
LO	LO	1	1V	TRIM A = 1
LO	HI	10	0.1V	TRIM A = 10
HI	LO	100	0.01V	TRIM A = 100
HI	HI	1000	0.001V	NO TRIM

*SET ZERO ADJUSTMENT FOR A2
OUTPUT = 0 VDC WITH INPUT GROUNDED
AND S1, S2 HIGH BEFORE TRIMMING

Q1 = 2N6485. GROUND CASE
*1% METAL FILM RESISTOR
RELAYS = COTO-COIL 800-05-001
10μF, 1μF = WIMA MKS-2

⊦⊣— = 1N4148

⊦⊣— = VN2222L

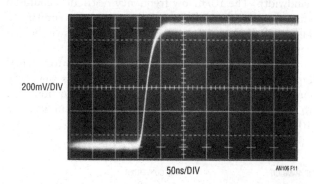

200mV/DIV

50ns/DIV

Figure 38.11 • Figure 38.10's A2 Output Responds to a 1mV Input Step at ×1000 Gain. 35ns Risetime Indicates 10MHz Bandwidth. Trace Thickening in Pulse Flat Portions Represents Noise Floor

To calibrate this circuit first set S1 and S2 high, ground the input and trim the "zero" adjustment for zero VDC at A2's output. Next, set S1 and S2 low, apply a 1V, 100kHz input, and trim "A=1" for unity gain, measured at the circuit output, in accordance with the table in the figure. Continue this procedure for the remaining three gains given in the table. A good way to generate the accurate low level inputs required is to set a 1.00VAC level and divide it down with a high grade 50Ω attenuator such as the Hewlett-Packard 350D or the Tektronix 2701. It is prudent to verify the attenuator's output with a precision RMS voltmeter.[4]

Wideband, isolated, quartz crystal RMS current measurement

Quartz crystal RMS operating current is critical to long-term stability, temperature coefficient and reliability. Accurate determination of RMS crystal current, especially in low power types, is complicated by the necessity to minimize introduced parasitics, particularly capacitance, which corrupt crystal operation. Figure 38.12, a form of Figure 38.10's wideband amplifier, combines with a commercially available closed core current probe to permit the measurement. An RMS-to-DC converter supplies the RMS value. The quartz crystal test circuit shown in dashed lines exemplifies a typical measurement situation. The Tektronix CT-2 current probe monitors crystal current while introducing minimal parasitic loading (see Figure 38.14). The probe's 50Ω termination allows direct connection to A1—Figure 38.10's FET buffer is deleted. Additionally, because quartz crystals are not common below 4kHz, A1's gain does not extend to low frequency.

Figure 38.13 shows results. Crystal drive, taken at Q1's collector (trace A), causes a 25μA RMS crystal current which is represented at the RMS-to-DC converter input

Note 4: See Appendix B for recommendations on RMS voltmeters.

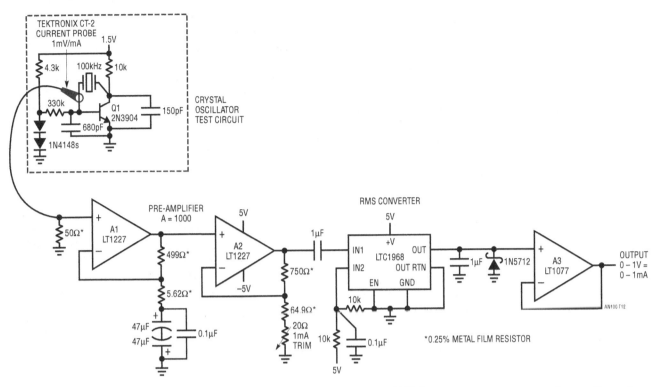

Figure 38.12 • Figure 38.10's Wideband Amplifier Adapted for Isolated RMS Current Measurement of Quartz Crystal Current. FET Input Buffer is Deleted; Current Probe's 50Ω Impedance Allows Direct Connection to A1. Current Probe Provides Minimal Crystal Loading in Oscillator Test Circuit

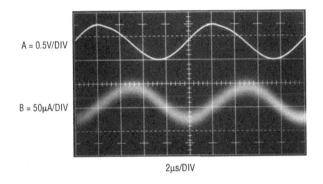

Figure 38.13 • Crystal Voltage (Trace A) and Current (Trace B) for Figure 38.12's Test Circuit. 25µA RMS Crystal Current Measurement Includes Preamplifier 5µA RMS Noise Floor Contribution

PARAMETER	CT-1	CT-2
Sensitivity	5mV/mA	1mV/mA
Accuracy	3%	3%
Low Frequency Additional 1% Error BW*	98kHz	6.4kHz
–3dB Bandwidth	25kHz to 1GHz	1.2kHz to 200MHz
Noise Floor with Amplifier Shown*	1µA RMS	5µA RMS
Capacitive Loading	1.5pF	1.8pF
Insertion Impedance at 10MHz	1Ω	0.1Ω

*As measured. Not vendor specified

Figure 38.14 • Relevant Specifications of Two Tektronix Current Probes. Primary Trade-Off is Low Frequency Error and Sensitivity. Noise Floor is Due to Amplifier Limitations

(trace B). The trace enlargement is due to the preamplifier's 5µA RMS equivalent noise contribution.

Figure 38.14 details characteristics of two Tektronix closed core current probes. The primary trade-off is low frequency error versus sensitivity. There is essentially no probe noise contribution and capacitive loading is notably low. Circuit calibration is achieved by putting 1mA RMS

current through the probe and adjusting the indicated trim for a 1V circuit output. To generate the 1mA, drive a 1k, 0.1% resistor with 1V$_{RMS}$.[5]

Note 5: This measurement technique has been extended to monitor 32.768kHz "watch crystal" sub-microampere operating currents. Contact the author for details.

AC voltage standard with stable frequency and low distortion

Figure 38.15 utilizes the RMS-to-DC converter's stability in an AC voltage standard. Initial circuit accuracy is 0.1% and long-term (6 months at 20°C to 30°C) drift remains within that figure. Additionally, the 4kHz operating frequency is within 0.01% and distortion inside 30ppm.

A1 and its power buffer A3 sense across a bridge composed of a 4kHz quartz crystal and an RC impedance in one arm; resistors and an LED driven photocell comprise the other arm. A1 sees positive feedback at the crystal's 4kHz resonance, promoting oscillation. Negative feedback, stabilizing oscillation amplitude, occurs via a control path which includes an RMS-to-DC converter and amplitude control amplifier, A5. A5 acts on the difference between A3's RMS converted output and the LT1009 voltage reference. Its output controls the LED driven photocell to set A1's negative feedback. RC components in A5's feedback path stabilize the control loop. The 50k trim sets the optically driven resistor's value to the point where lowest A3 output distortion occurs while maintaining adequate loop stability.

Normally the bridge's "bottom" would be grounded. While this connection will work, it subjects A1 to common mode swings, increasing distortion due to A1's finite common mode rejection versus frequency. A2 eliminates this concern by forcing the bridge's mid-points, and hence common mode voltage, to zero while not influencing desired circuit operation. It does this by driving the bridge "bottom" to force its input differential to zero. A2's output swing is 180° out of phase with A3's circuit output. This action eliminates common mode swing at A1, reducing circuit output distortion by more than an order of magnitude. Figure 38.16 shows the circuit's $1.414V_{RMS}$ ($2.000V_{PEAK}$) output in trace A while trace B's distortion constituents include noise, fundamental related residue and 2F components.

The 4kHz crystal is a relatively large structure with very high Q factor. Normally, it would require more than

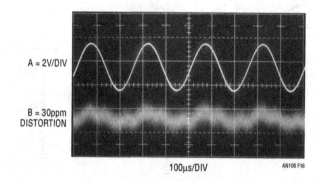

A = 2V/DIV

B = 30ppm DISTORTION

100µs/DIV   AN106 F16

Figure 38.16 • A3's 1.414V$_{RMS}$ (2.000V$_{PEAK}$), 4kHz Reference Output (Trace A) Shows 30ppm Distortion in Trace B. Distortion Constituents Include Noise, Fundamental Related Residue and 2F Components

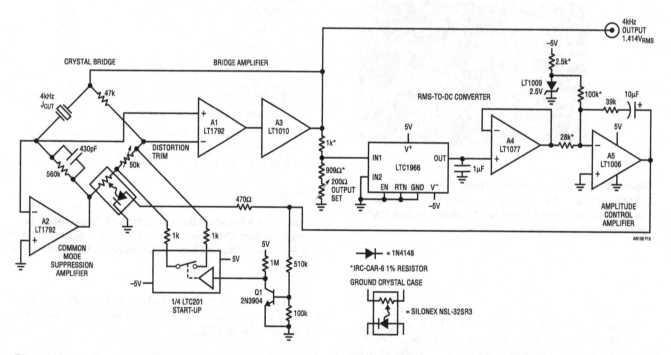

Figure 38.15 • Quartz Stabilized Sine Wave Output AC Reference Has 0.1% Long-Term Amplitude Stability. Frequency Accuracy is 0.01% with <30ppm Distortion. Positive Feedback Around A1 Causes Oscillation at Crystal's Resonance. A5, Acting on A3's RMS Amplitude, Supplies Negative Feedback to A1 via Bridge Network, Stabilizing RMS Output Amplitude. Optocoupler Minimizes Feedback Induced Distortion. Q1 Closes Switch During Start-Up, Ensuring Rapid Oscillation Build-Up

30 seconds to start and arrive at full regulated amplitude. This is avoided by inclusion of the Q1-LTC201 switch circuitry. At start-up A5's output goes high, biasing Q1. Q1's collector goes low, turning on the LTC201. This sets A1's gain abnormally high, increasing bridge drive and accelerating crystal start-up. When the bridge arrives at its operating point A5's output drops to a lower value, Q1, and the LTC201 switch go off, and the circuit transitions into normal operation. Start-up time is several seconds.

The circuit requires trimming for amplitude accuracy and lowest distortion. The distortion trim is made first. Adjust the trim for minimal output distortion as measured on a distortion analyzer. Note that the absolute lowest level of distortion coincides with the point where control loop gain is

just adequate to maintain oscillation. As such, find this point and retreat from it into the control loop's active region. This necessitates giving up about 5ppm distortion, but 30ppm is achievable with good control loop stability. Output amplitude is trimmed with the indicated adjustment for exactly $1.414V_{RMS}$ ($2.000V_{PEAK}$) at the circuit output.

RMS leveled output random noise generator

Figure 38.17 uses the RMS-to-DC converter in a leveled output random noise generator. Noise diode D1 AC biases A1, operating at a gain of 2.[6] A1's output feeds a 1kHz to

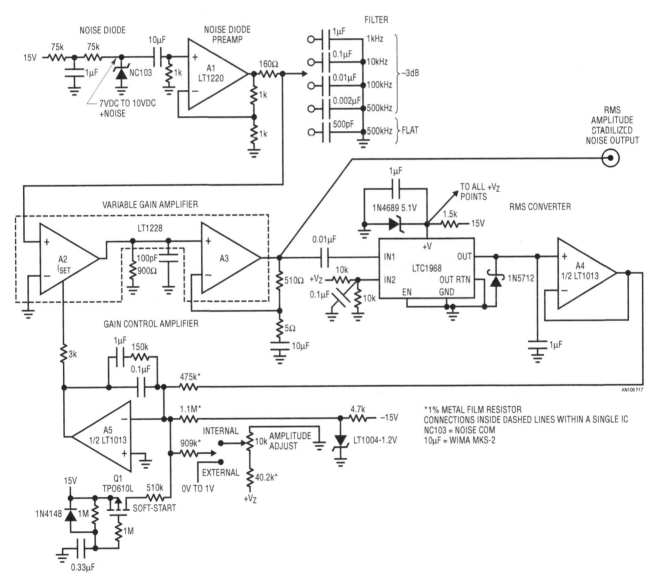

Figure 38.17 • An RMS Levelled Output Random Noise Generator. Amplified (A1) Diode Noise Is Filtered, Variable Gain Amplified (A2-A3) and RMS Converted. Converter Output Feeds Back to A5 Gain Control Amplifier, Closing RMS Stabilized Loop. Output Amplitude, Taken at A3, is Settable

500kHz switch selectable lowpass filter. The filter output biases the variable gain amplifier, A2-A3. A2-A3, contained on one chip, include a current controlled transconductance amplifier (A2) and an output amplifier (A3). This stage takes AC gain, biases the LTC1968 RMS-to-DC converter and is the circuit's output. The RMS converter output at A4, feeds back to gain control amplifier A5, which compares the RMS value to a variable portion of the 5.1V Zener potential. A5's output sets A2's gain via the 3k resistor, completing a control loop to stabilize noise RMS output amplitude. The RC components in A5's local feedback path stabilize this loop. Output amplitude is variable by the 10k potentiometer; a switch permits external voltage control. Q1 and associated components, a soft-start circuit, prevent output overshoot at power turn-on.

Figure 38.18 shows circuit output noise in the 10kHz filter position; Figure 38.19's spectral plot reveals essentially flat RMS noise amplitude over a 500kHz bandwidth.

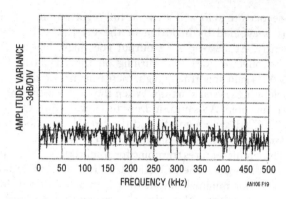

Figure 38.19 • Amplitude vs Frequency for the Random Noise Generator is Essentially Flat to 500kHz. NC103 Diode Contributes Even Noise Spectrum Distribution; RMS Converter and Loop Stabilize Amplitude. Sweep Time is 2.8 Minutes, Resolution Bandwidth, 100Hz

RMS amplitude stabilized level controller

Figure 38.20 borrows the previous circuit's gain control loop to stabilize the RMS amplitude of an arbitrary input waveform. The unregulated input is applied to variable gain amplifier A1-A2 which feeds A3. DC coupling at A1-A2 permits passage of low frequency inputs. A3's output is

taken by RMS-to-DC converter C1-A4, which feeds the A5 gain control amplifier. A5 compares the RMS value to a variable reference and biases A1, closing a gain control loop. The 0.15μF feedback capacitor stabilizes this loop, even for waveforms below 100Hz. This feedback action stabilizes output RMS amplitude despite large variations in input amplitude while maintaining waveshape. Desired output level is settable with the indicated potentiometer or an external control voltage may be switched in.

Figure 38.21 shows output response (trace B) to abrupt reference level set point changes (trace A). The output settles within 60 milliseconds for ascending and descending transitions. Faster response is possible by decreasing A5's compensation capacitor, but low frequency waveforms would not be processable. Similar considerations apply to Figure 38.22's response to an input waveform step change. Trace A is the circuit's input and trace B its output. The output settles in 60 milliseconds due to A5's compensation. Reducing compensation value speeds response at the expense of low frequency waveform processing capability. Specifications include 0.1% output amplitude stability for inputs varying from $0.4V_{RMS}$ to $5V_{RMS}$, 1% set point accuracy, 0.1kHz to 500kHz passband and 0.1% stability for 20% power supply deviation.

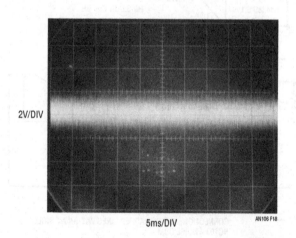

2V/DIV

5ms/DIV   AN106 F18

Figure 38.18 • Figure 38.17's Output in the 10kHz Filter Position

---

Note 6: See Appendix C "Symmetrical White Gaussian Noise," guest written by Ben Hessen-Schmidt of Noise Com, Inc. for tutorial on noise and noise diodes.

Note: This chapter was derived from a manuscript originally prepared for publication In EDN magazine.

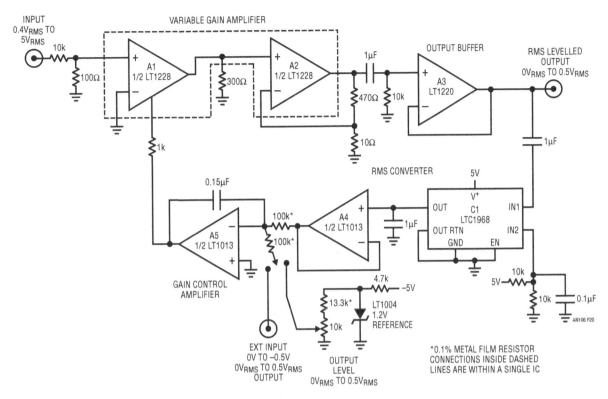

Figure 38.20 • RMS Amplitude Level Control Uses Figure 38.17's Gain Control Loop. A1-A3 Provide Variable Gain to Input. RMS Converter Feeds Back to A5 Gain Control Amplifier, Closing Amplitude Stabilization Loop. Variable Reference Permits Settable, Calibrated RMS Output Amplitude Independent of Input Waveshape

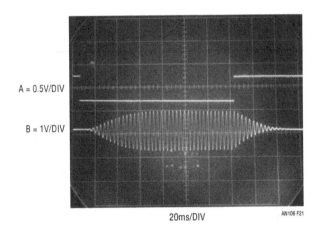

20ms/DIV   AN106 F21

Figure 38.21 • Amplitude Level Control Response (Trace B) to Abrupt Reference Changes (Trace A). Settling Time is Set by A5's Compensation Capacitor, Which Must be Large Enough to Stabilize Loop at Lowest Expected Input Frequency

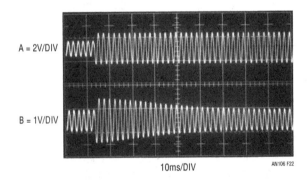

10ms/DIV   AN106 F22

Figure 38.22 • Amplitude Level Control Output Reacts (Trace B) to Input Step Change (Trace A). Slow Loop Compensation Allows Overshoot But Output Settles Cleanly

References

1. Hewlett-Packard Company. "1968 Instrumentation. Electronic—Analytical—Medical," AC Voltage Measurement. Hewlett-Packard Company 1968, pp. 197-198.
2. Sheingold, D. H. (editor), "Nonlinear Circuits Handbook," 2nd Edition. Analog Devices, Inc., 1976
3. Lambda Electronics, Model LK-343A-FM Manual.
4. Grafham, D. R., "Using Low Current SCRs," General Electric AN200.19. Jan. 1967.
5. Williams, J. "Performance Enhancement Techniques for Three-Terminal Regulators," Linear Technology Corp. AN-2. (August, 1984). "SCR Preregulator," pp. 3–6.
6. Williams, J., "High Efficiency Linear Regulators," Linear Technology Corporation, Application Note 32, "SCR Pre-regulator." March 1989, pp. 3–4.
7. Williams, J., "High Speed Amplifier Techniques," Linear Technology Corporation, Application Note 47, "Parallel Path Amplifiers," August 1991, pp. 35–37.
8. Williams, J., "Practical Circuitry for Measurement and Control Problems," Broadband Random Noise Generator," "Symmetrical White Gaussian Noise," Appendix B, Linear Technology Corporation, Application Note 61, August 1994, pp.24–26, pp. 38–39.
9. Williams, J., "A Fourth Generation of LCD Backlight Technology," "RMS Voltmeters," Linear Technology Corporation, Application Note 65, November 1995, pp. 82–83.
10. Meacham, L. A., "The Bridge Stabilized Oscillator," Bell System Technical Journal, Vol. 17, p. 574, October 1938.
11. Williams, Jim, "Bridge Circuits—Marrying Gain and Balance," Linear Technology Corporation, Application Note 43, June, 1990.

Appendix A
RMS-to-DC conversion
Joseph Petrofsky

Definition of RMS
RMS amplitude is the consistent, fair and standard way to measure and compare dynamic signals of all shapes and sizes. Simply stated, the RMS amplitude is the heating potential of a dynamic waveform. A $1V_{RMS}$ AC waveform will generate the same heat in a resistive load as will 1V DC. See Figure A1.

Mathematically, RMS is the "root of the mean of the square":

$$V_{RMS} = \sqrt{\overline{V^2}}$$

Alternatives to RMS
Other ways to quantify dynamic waveforms include peak detection and average rectification. In both cases, an average (DC) value results, but the value is only accurate at the one chosen waveform type for which it is calibrated, typically sine waves. The errors with average rectification are shown in Table A1. Peak detection is worse in all cases and is rarely used.

The last two entries of Table A1 are chopped sine waves as is commonly created with thyristors such as SCRs and Triacs. Figure A2a shows a typical circuit and Figure A2b shows the resulting load voltage, switch voltage and load currents. The power delivered to the load depends on the firing angle, as

Table A1 Errors with Average Rectification vs True RMS

WAVEFORM	V_{RMS}	AVERAGE RECTIFIED (V)	ERROR*
Square wave	1.000	1.000	11%
Sine wave	1.000	0.900	*Calibrate for 0% error
Triangle wave	1.000	0.866	−3.8%
SCR at 1/2 power, $\Theta = 90°$	1.000	0.637	−29.3%
SCR at 1/4 power, $\Theta = 114°$	1.000	0.536	−40.4%

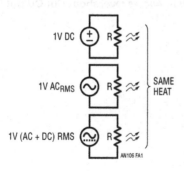

Figure A1

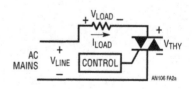

Figure A2a

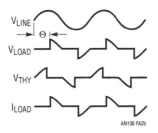

Figure A2b

well as any parasitic losses such as switch "ON" voltage drop. Real circuit waveforms will also typically have significant ringing at the switching transition, dependent on exact circuit parasitics. Here, "SCR Waveforms" refers to the ideal chopped sine wave, though the LTC1966/LTC1967/LTC1968 will do faithful RMS-to-DC conversion with real SCR waveforms as well.

The case shown is for $\Theta=90°$, which corresponds to 50% of available power being delivered to the load. As noted in Table A1, when $\Theta=114°$, only 25% of the available power is being delivered to the load and the power drops quickly as Θ approaches 180°.

With an average rectification scheme and the typical calibration to compensate for errors with sine waves, the RMS level of an input sine wave is properly reported; it is only with a non-sinusoidal waveform that errors occur. Because of this calibration, and the output reading in V_{RMS}, the term True-RMS got coined to denote the use of an actual RMS-to-DC converter as opposed to a calibrated average rectifier.

How an RMS-to-DC converter works

Monolithic RMS-to-DC converters use an implicit computation to calculate the RMS value of an input signal. The fundamental building block is an analog multiply/divide used as shown in Figure A3. Analysis of this topology is easy and starts by identifying the inputs and the output of the lowpass filter. The input to the LPF is the calculation from the multiplier/

divider; $(V_{IN})^2/V_{OUT}$. The lowpass filter will take the average of this to create the output, mathematically:

$$V_{OUT} = \overline{\left(\frac{(V_{IN})^2}{V_{OUT}}\right)},$$

Because V_{OUT} is DC,

$$\overline{\left(\frac{(V_{IN})^2}{V_{OUT}}\right)} = \frac{\overline{\left((V_{IN})^2\right)}}{V_{OUT}}, \text{ so}$$

$$V_{OUT} = \frac{\overline{\left((V_{IN})^2\right)}}{V_{OUT}}, \text{ and}$$

$$(V_{OUT})^2 = \overline{(V_{IN})^2}, \text{ or}$$

$$V_{OUT} = \sqrt{\overline{(V_{IN})^2}} = RMS(V_{IN})$$

Unlike the prior generation RMS-to-DC converters, the LTC1966/LTC1967/LTC1968 computation does NOT use log/antilog circuits, which have all the same problems, and more, of log/antilog multipliers/dividers, i.e., linearity is poor, the bandwidth changes with the signal amplitude and the gain drifts with temperature.

How the LTC1966/LTC1967/LTC1968 RMS-to-DC converters work

The LTC1966/LTC1967/LTC1968 use a completely new topology for RMS-to-DC conversion, in which a $\Delta\Sigma$ modulator acts as the divider, and a simple polarity switch is used as the multiplier as shown in Figure A4.

The $\Delta\Sigma$ modulator has a single-bit output whose average duty cycle ($\overline{D}$) will be proportional to the ratio of the input signal divided by the output. The $\Delta\Sigma$ is a 2nd order modulator with excellent linearity. The single-bit output is used to selectively buffer or invert the input signal. Again, this is a circuit with excellent linearity, because it operates at only two points: ±1 gain; the average effective multiplication over time will be on the straight line between these two points. The combination of these two elements again creates a

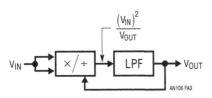

Figure A3 • RMS-to-DC Converter with Implicit Computation

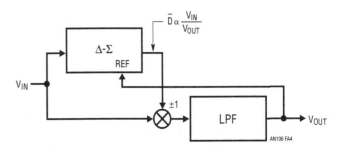

Figure A4 • Topology of the LTC1966/LTC1967/LTC1968

lowpass filter input signal equal to $(V_{IN})^2/V_{OUT}$, which, as shown above, results in RMS-to-DC conversion.

The lowpass filter performs the averaging of the RMS function and must be a lower corner frequency than the lowest frequency of interest. For line frequency measurements, this filter is simply too large to implement on-chip, but the LTC1966/LTC1967/LTC1968 need only one capacitor on the output to implement the lowpass filter. The user can select this capacitor depending on frequency range and settling time requirements.

This topology is inherently more stable and linear than log/antilog implementations primarily because all of the signal processing occurs in circuits with high gain op amps operating closed loop.

Note that the internal scalings are such that the $\Delta\Sigma$ output duty cycle is limited to 0% or 100% only when V_{IN} exceeds $\pm 4 \cdot V_{OUT}$.

Linearity of an RMS-to-DC converter

Linearity may seem like an odd property for a device that implements a function that includes two very nonlinear processes: squaring and square rooting.

However, an RMS-to-DC converter has a transfer function, RMS volts in to DC volts out, that should ideally have a 1:1 transfer function. To the extent that the input to output transfer function does not lie on a straight line, the part is nonlinear.

A more complete look at linearity uses the simple model shown in Figure A5. Here an ideal RMS core is corrupted by both input circuitry and output circuitry that have imperfect transfer functions. As noted, input offset is introduced in the input circuitry, while output offset is introduced in the output circuitry.

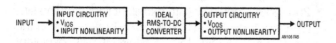

Figure A5 • Linearity Model of an RMS-to-DC Converter

Any nonlinearity that occurs in the output circuity will corrupt the RMS in to DC out transfer function. A nonlinearity in the input circuitry will typically corrupt that transfer function far less simply because with an AC input, the RMS-to-DC conversion will average the nonlinearity from a whole range of input values together.

But the input nonlinearity will still cause problems in an RMS-to-DC converter because it will corrupt the accuracy as the input signal shape changes. Although an RMS-to-DC converter will convert any input waveform to a DC output, the accuracy is not necessarily as good for all waveforms as it is with sine waves. A common way to describe dynamic signal wave shapes is Crest Factor. The crest factor is the ratio of the peak value relative to the RMS value of a waveform. A signal with a crest factor of 4, for instance, has a peak that is four times its RMS value. Because this peak has energy (proportional to voltage squared) that is 16 times (4^2) the energy of the RMS value, the peak is necessarily present for at most 6.25% (1/16) of the time.

The LTC1966/LTC1967/LTC1968 perform very well with crest factors of 4 or less and will respond with reduced accuracy to signals with higher crest factors. The high performance with crest factors less than 4 is directly attributable to the high linearity throughout the LTC1966/LTC1967/LTC1968.

Appendix B
AC measurement and signal handling practice

Accurate AC measurement requires trustworthy instrumentation, proper signal routing technique, parasitic minimization, attention to layout and care in component selection. The text circuits DC-500kHz, 1% error bandwidth seems benign, but unpleasant surprises await the unwary.

An accurate RMS voltmeter is required for serious AC work. Figure B1 lists types used in our laboratory. These are high grade, specialized instruments specifically intended for precise RMS measurement. All are thermally based.[1] The first three entries, general purpose instruments with many ranges and features, are easily used and meet almost all AC measurement needs. The last entry is more of a component than an instrument. The A55 series of "thermal converters" provide millivolt level outputs for various inputs. Typical input ranges are $0.5V_{RMS}$, $1V_{RMS}$, $2V_{RMS}$ and $5V_{RMS}$ and each converter is supplied with individual calibration data. They are somewhat cumbersome to use and easily destroyed but are highly accurate. Their primary use is as reference standards to check other instrument's performance.

AC signal handling for high accuracy is a broad topic, involving a considerable degree of depth. This forum must suffer brevity, but some gossip is possible.

Layout is critical. The most prevalent parasitic in AC measurement is stray capacitance. Keep signal path connections short and small area. A few

---

Note 1: See References 1 and 2 for details on thermally based RMS-to-DC conversion.

MODEL	MANUFACTURER	1V RANGE	INPUT	BANDWIDTH	COMMENTS
3400A/3400B	Hewlett-Packard	1%	AC	10MHz/20MHz	Metered Instrument. Most Common RMS Voltmeter
3403C	Hewlett-Packard	0.2%	AC, AC + DC	100MHz	Digital Display, 1μV Sensitivity (2MHz BW), dB Ranges, Relative dB
8920/8921A	Fluke	0.7%	AC, AC + DC	20MHz	Digital Display, 10μV Sensitivity (2MHz BW), dB Ranges, Relative dB
A55	Fluke	0.05%	AC + DC	50MHz	Set of Individually Calibrated Thermal Converters. Reference Standards. Not for General Purpose Measurement

Figure B1 ● Precision Wideband RMS Voltmeters Useful for AC Measurement. All are Thermally Based, Permitting High Accuracy and Wide Bandwidth Independent of Input Waveshape. A55 Reference Standards, Although Unsuitable for General Purpose Measurement, Have Best Accuracy

picofarads of coupling into a high impedance node can upset a 500kHz, 1% accuracy signal path. To the extent possible, keep impedances low to minimize parasitic capacitive effects. Consider individual component parasitics and plan to accommodate them. Examine effects of component placement and orientation on the circuit board. If a ground plane is in use it may be necessary to relieve it in the vicinity of critical circuit nodes or even individual components.

Passive components have parasitics that must be kept in mind. Resistors suffer shunt capacitance whose effects vary with frequency and resistor value. It is worth noting that different brands of resistors, although nominally similar, may exhibit markedly different parasitic behavior. Capacitors in the signal path should be used so that their outer foil is connected to the less sensitive node, affording some relief from pick-up and stray capacitance induced effects. Some capacitors are marked to indicate the outer foil terminal, others require consulting the data sheet or vendor contact. Avoid ceramic capacitors in the signal path. Their piezoelectric responses make them unsuitable for precision AC circuitry. In general, any component in the signal path should be examined in terms of its potential parasitic contribution.

Active components, such as amplifiers, must be treated as potential error sources. In particular, as stated in the text, ensure that there is enough open loop gain at the frequency of interest to assure needed closed loop gain accuracy. Margins of 100:1 are not unreasonable. Keep feedback values as low as possible to minimize parasitic effects.

Route signals to and from the circuit board coaxially and at low impedance, preferably 50Ω, for best results. In 50Ω systems, remember that terminators and attenuators have tolerances that can corrupt a 1% amplitude accuracy measurement. Verify such terminator and attenuator tolerances by measurement and account for them when interpreting

measurement results. Similarly, verify the accuracy of any associated instrument's 50Ω input or output impedance and account for deviations.

This all seems painful but is an essential part of achieving 1% accurate, 500kHz signal integrity. Failure to observe the precautions listed above risks degrading the RMS-to-DC converter's system level performance.

Appendix C
Symmetrical white Gaussian noise

Ben Hessen-Schmidt, NOISE COM, INC.

White noise provides instantaneous coverage of all frequencies within a band of interest with a very flat output spectrum. This makes it useful both as a broadband stimulus and as a power-level reference.

Symmetrical white Gaussian noise is naturally generated in resistors. The noise in resistors is due to vibrations of the conducting electrons and holes, as described by Johnson and Nyquist.[1] The distribution of the noise voltage is symmetrically Gaussian, and the average noise voltage is:

$$\overline{V}_n = 2\sqrt{kT \int R(f)\, p(f)\, df} \qquad (1)$$

where:
k=1.38E–23J/K (Boltzmann's constant)
T=temperature of the resistor in Kelvin
f=frequency in Hz
h=6.62E-34 Js (Planck's constant)
R(f)=resistance in ohms as a function of frequency

$$p(f) = \frac{hf}{kT\,[\exp{(hf/kT)} - 1]} \qquad (2)$$

Note 1: See "Additional Reading" at the end of this section.

p(f) is close to unity for frequencies below 40GHz when T is equal to 290°K. The resistance is often assumed to be independent of frequency, and Údf is equal to the noise bandwidth (B). The available noise power is obtained when the load is a conjugate match to the resistor, and it is:

$$N = \frac{\overline{V}_n^2}{4R} = kTB \qquad (3)$$

where the "4" results from the fact that only half of the noise voltage and hence only 1/4 of the noise power is delivered to a matched load.

Equation 3 shows that the available noise power is proportional to the temperature of the resistor; thus it is often called thermal noise power. Equation 3 also shows that white noise power is proportional to the bandwidth.

An important source of symmetrical white Gaussian noise is the noise diode. A good noise diode generates a high level of symmetrical white Gaussian noise. The level is often specified in terms of excess noise ratio (ENR):

$$\text{ENR (in dB)} = 10 \text{ Log} \frac{(Te - 290)}{290} \qquad (4)$$

Te is the physical temperature that a load (with the same impedance as the noise diode) must be at to generate the same amount of noise.

The ENR expresses how many times the effective noise power delivered to a non-emitting, nonreflecting load exceeds the noise power available from a load held at the reference temperature of 290°K (16.8°C or 62.3°F).

The importance of high ENR becomes obvious when the noise is amplified, because the noise contributions of the amplifier may be disregarded when the ENR is 17dB larger than the noise figure of the amplifier (the difference in total noise power is then less than 0.1dB). The ENR can easily be converted to noise spectral density in dBm/Hz or $\mu V/\sqrt{Hz}$ by use of the white noise conversion formulas in Table 1.

When amplifying noise it is important to remember that the noise voltage has a Gaussian distribution. The peak voltages of noise are therefore much larger than the average or RMS voltage. The ratio of peak voltage to RMS voltage is called crest factor, and a good crest factor for Gaussian noise is between 5:1 and 10:1 (14 to 20dB). An amplifier's 1dB gain-compression point should therefore be typically 20dB larger than the desired average noise-output power to avoid clipping of the noise.

For more information about noise diodes, please contact NOISE COM, INC. at (973) 386-9696.

Table 1 Useful White Noise Conversion

dBm	=	dBm/Hz+10log (BW)
dBm	=	$20\log (\overline{V}n) - 10\log (R) + 30dB$
dBm	=	$20\log (\overline{V}n) + 13dB$ for R = 50Ω
dBm/Hz	=	$20\log (\mu\overline{V}n\sqrt{Hz}) - 10\log (R) - 90dB$
dBm/Hz	=	$-174dBm/Hz+ENR$ for ENR>17dB

Additional reading

1. Johnson, J. B.. , "Thermal Agitation of Electricity in Conductors," Physical Review., July 1928, pp. 97–109.

2. Nyquist, H., "Thermal Agitation of Electric Charge in Conductors," Physical Review., July 1928, pp. 110–113.

775 nanovolt noise measurement for a low noise voltage reference

Quantifying silence

Jim Williams

Introduction

Frequently, voltage reference stability and noise define measurement limits in instrumentation systems. In particular, reference noise often sets stable resolution limits. Reference voltages have decreased with the continuing drop in system power supply voltages, making reference noise increasingly important. The compressed signal processing range mandates a commensurate reduction in reference noise to maintain resolution. Noise ultimately translates into quantization uncertainty in A to D converters, introducing jitter in applications such as scales, inertial navigation systems, infrared thermography, DVMs and medical imaging apparatus. A new low voltage reference, the LTC6655, has only 0.3ppm (775nV) noise at $2.5V_{OUT}$. Figure 39.1 lists salient specifications in tabular form. Accuracy and temperature coefficient are characteristic of high grade, low voltage references. 0.1Hz to 10Hz noise, particularly noteworthy, is unequalled by any low voltage electronic reference.

Noise measurement

Special techniques are required to verify the LTC6655's extremely low noise. Figure 39.2's approach appears innocently straightforward but practical implementation represents a high order difficulty measurement. This 0.1Hz to 10Hz noise testing scheme includes a low noise pre-amplifier, filters and a peak-to-peak noise detector. The pre-amplifiers 160nV noise floor, enabling accurate measurement, requires special design and layout techniques. A forward gain of 10^6 permits readout by conventional instruments.

Figure 39.3's detailed schematic reveals some considerations required to achieve the 160nV noise floor. The references DC potential is stripped by the 1300μF, 1.2k resistor combination; AC content is fed to Q1. Q1-Q2, extraordinarily low noise J-FET's, are DC stabilized by A1, with A2 providing a single-ended output. Resistive feedback from A2 stabilizes the configuration at a gain of 10,000. A2's output is routed to amplifier-filter A3-A4

LTC6655 Reference Tabular Specifications

SPECIFICATION	LIMITS
Output Voltages	1.250, 2.048, 2.500, 3.000, 3.300, 4.096, 5.000
Initial Accuracy	0.025%, 0.05%
Temperature Coefficient	2ppm/°C, 5ppm/°C
0.1Hz to 10Hz Noise	0.775μV at V_{OUT} = 2.500V Peak-to-Peak Noise is within this Figure in 90% of 1000 Ten Second Measurement Intervals
Additional Characteristics	5ppm/Volt Line Regulation, 500mV Dropout, Shutdown Pin, I_{SUPPLY} = 5mA, $V_{IN} = V_O + 0.5V$ to $13.2V_{MAX}$, $I_{OUT(SINK/SOURCE)}$ = ±5mA, I_{SHORT} Circuit = 15mA

Figure 39.1 • LTC6655 Accuracy and Temperature Coefficient Are Characteristic of High Grade, Low Voltage References. 0.1Hz to 10Hz Noise, Particularly Noteworthy, Is Unequalled by Any Low Voltage Electronic Reference

Analog Circuit and System Design: A Tutorial Guide to Applications and Solutions. DOI: 10.1016/B978-0-12-385185-7.00039-1

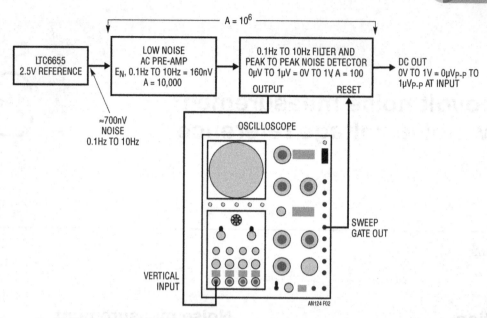

$$A = 10^6$$

| LTC6655 2.5V REFERENCE | → | LOW NOISE AC PRE-AMP E_N, 0.1Hz TO 10Hz = 160nV A = 10,000 | → | 0.1Hz TO 10Hz FILTER AND PEAK TO PEAK NOISE DETECTOR 0µV TO 1µV = 0V TO 1V A = 100 | → | DC OUT 0V TO 1V = 0µV$_{P-P}$ TO 1µV$_{P-P}$ AT INPUT |

OUTPUT   RESET

≈700nV NOISE 0.1Hz TO 10Hz

OSCILLOSCOPE

SWEEP GATE OUT

VERTICAL INPUT

AN124 F02

Figure 39.2 • Conceptual 0.1Hz to 10Hz Noise Testing Scheme Includes Low Noise Pre-Amplifier, Filter and Peak to Peak Noise Detector. Pre-Amplifier's 160nV Noise Floor, Enabling Accurate Measurement, Requires Special Design and Layout Techniques

which provides 0.1Hz to 10Hz response at a gain of 100. A5-A8 comprise a peak-to-peak noise detector read out by a DVM at a scale factor of 1 volt/microvolt. The peak-to-peak noise detector provides high accuracy measurement, eliminating tedious interpretation of an oscilloscope display. Instantaneous noise value is supplied by the indicated output to a monitoring oscilloscope. The 74C221 one-shot, triggered by the oscilloscope sweep gate, resets the peak-to-peak noise detector at the end of each oscilloscope 10-second sweep.

Numerous details contribute to the circuit's performance. The 1300µF capacitor, a highly specialized type, is selected for leakage in accordance with the procedure given in Appendix B. Further, it, and its associated low noise 1.2k resistor, are fully shielded against pick-up. FETs Q1 and Q2 differentially feed A2, forming a simple low noise op amp. Feedback, provided by the 100k-10Ω pair, sets closed loop gain at 10,000. Although Q1 and Q2 have extraordinarily low noise characteristics, their offset and drift are uncontrolled. A1 corrects these deficiencies by adjusting Q1's channel current via Q3 to minimize the Q1-Q2 input difference. Q1's skewed drain values ensure that A1 is able to capture the offset. A1 and Q3 supply whatever current is required into Q1's channel to force offset within about 30µV The FETs' V_{GS} can vary over a 4:1 range. Because of this, they must be selected for 10% V_{GS} matching. This matching allows A1 to capture the offset without introducing significant noise. Q1 and Q2 are thermally mated and lagged in epoxy at a time constant much greater than A1's DC stabilizing loop roll-off, preventing offset instability and hunting. The entire A1-Q1-Q2-A2 assembly and the reference under test are

completely enclosed within a shielded can.[1] The reference is powered by a 9V battery to minimize noise and insure freedom from ground loops.

Peak-to-peak detector design considerations include J-FETs used as peak trapping diodes to obtain lower leakage than afforded by conventional diodes. Diodes at the FET gates clamp reverse voltage, further minimizing leakage.[2] The peak storage capacitor's highly asymmetric charge-discharge profile necessitates the low dielectric absorption polypropelene capacitors specified.[3] Oscilloscope connections via galvanically isolated links prevent ground loop induced corruption. The oscilloscope input signal is supplied by an isolated probe; the sweep gate output is interfaced with an isolation pulse transformer. Details appear in Appendix C.

Noise measurement circuit performance

Circuit performance must be characterized prior to measuring LTC6655 noise. The pre-amplifier stage is verified

Note 1: The pre-amplifier structure must be carefully prepared. See Appendix A, "Mechanical and Layout Considerations", for detail on preamplifier construction.
Note 2: Diode connected J-FET's superior leakage derives from their extremely small area gate-channel junction. In general, J-FET's leak a few picoamperes (25°C) while common signal diodes (e.g. 1N4148) are about 1,000× worse (units of nanoamperes at 25°C).
Note 3: Teflon and polystyrene dielectrics are even better but the Real World intrudes. Teflon is expensive and excessively large at 1µF. Analog types mourn the imminent passing of the polystyrene era as the sole manufacturer of polystyrene film has ceased production.

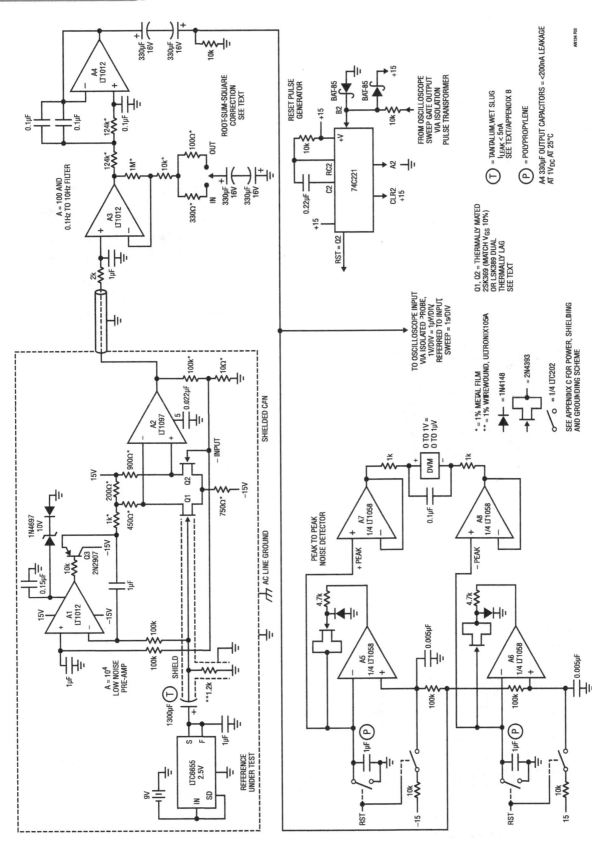

Figure 39.3 • Detailed Noise Test Circuitry. Thermally Lagged Q1-Q2 Low Noise J-FET Pair Is DC Stabilized by A1-Q3; A2 Delivers A = 10,000 Pre-Amplifier Output. A3-A4 form 0.1Hz to 10Hz, A = 100, Bandpass Filter; Total Gain Referred to Pre-Amplifier Input Is 10^6. Peak to Peak Noise Detector, Reset by Monitoring Oscilloscope Sweep Gate, Supplies DVM Output

for >10Hz bandwidth by applying a 1μV step at its input (reference disconnected) and monitoring A2's output. Figure 39.4's 10ms risetime indicates 35Hz response, insuring the entire 0.1Hz to 10Hz noise spectrum is supplied to the succeeding filter stage.

Figure 39.5 describes peak-to-peak noise detector operation. Waveforms include A3's input noise signal (Trace A), A7 (Trace B) positive/A8 (Trace C) negative peak detector outputs and DVM differential input (Trace D). Trace E's oscilloscope supplied reset pulse has been lengthened for photographic clarity.

Circuit noise floor is measured by replacing the LTC6655 with a 3V battery stack. Dielectric absorption effects in the large input capacitor require a 24-hour settling period before measurement. Figure 39.6, taken at the circuit's oscilloscope output, shows 160nV 0.1Hz to 10Hz noise in a 10 second sample window. Because noise adds in root-sum-square fashion, this represents about a 2% error in the LTC 6655's expected 775nV noise figure. This term is accounted for by placing Figure 39.3's "root-sum-square correction" switch in the appropriate position during reference testing. The resultant 2% gain attenuation first order corrects LTC6655 output noise reading for the circuit's 160nV noise floor contribution. Figure 39.7, a strip-chart recording of the peak-to-peak noise detector output over 6 minutes, shows less than 160nV test circuit noise.[4] Resets occur every 10 seconds. A 3V battery biases the input capacitor replacing the LTC6655 for this test.

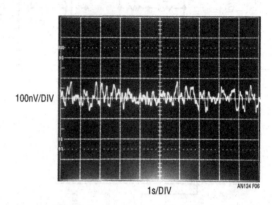

100nV/DIV

1s/DIV   AN124 F06

Figure 39.6 • Low Noise Circuit/Layout Techniques Yield 160nV 0.1Hz to 10Hz Noise Floor, Ensuring Accurate Measurement. Photograph Taken at Figure 39.3's Oscilloscope Output with 3V Battery Replacing LTC6655 Reference. Noise Floor Adds ≈2% Error to Expected LTC6655 Noise Figure Due to Root-Sum-Square Noise Addition Characteristic; Correction is Implemented at Figure 39.3's A3

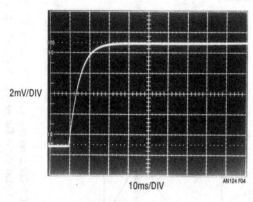

2mV/DIV

10ms/DIV   AN124 F04

Figure 39.4 • Pre-Amplifier Rise Time Measures 10ms; Indicated 35Hz Bandwidth Ensures Entire 0.1Hz to 10Hz Noise Spectrum Is Supplied to Succeeding Filter Stage

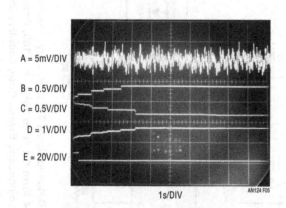

A = 5mV/DIV

B = 0.5V/DIV

C = 0.5V/DIV

D = 1V/DIV

E = 20V/DIV

1s/DIV   AN124 F05

Figure 39.5 • Waveforms for Peak to Peak Noise Detector Include A3 Input Noise Signal (Trace A), A7 (Trace B) Positive/A8 (Trace C) Negative Peak Detector Outputs and DVM Differential Input (Trace D). Trace E's Oscilloscope Supplied Reset Pulse Lengthened for Photographic Clarity

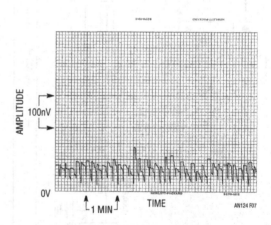

AMPLITUDE

100nV

0V

1 MIN   TIME   AN124 F07

Figure 39.7 • Peak to Peak Noise Detector Output Observed Over 6 Minutes Shows <160nV Test Circuit Noise. Resets Occur Every 10 seconds. 3V Battery Biases Input Capacitor, Replacing LTC6655 for This Test

Note 4: That's right, a *strip-chart recording*. Stubborn, locally based aberrants persist in their use of such archaic devices, forsaking more modern alternatives. Technical advantage could account for this choice, although deeply seated cultural bias may be indicated.

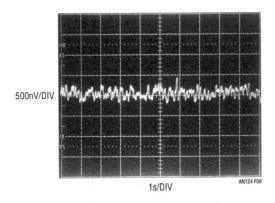

500nV/DIV

1s/DIV

AN124 F08

Figure 39.8 • LTC6655 0.1Hz to 10Hz Noise Measures 775nV in 10 Second Sample Time

Figure 39.8 is LTC6655 noise after the indicated 24-hour dielectric absorption soak time. Noise is within 775nV peak-to-peak in this 10 second sample window with the root-sum-square correction enabled. The verified, extremely low circuit noise floor makes it highly likely this data is valid. In closing, it is worth mention that the approach taken is applicable to measuring any 0.1Hz to 10Hz noise source, although the root-sum-square error correction coefficient should be re-established for any given noise level.

References

1. Morrison, Ralph, "Grounding and Shielding Techniques in Instrumentation," Wiley-Interscience, 1986.
2. Ott Henry W., "Noise Reduction Techniques in Electronic Systems," Wiley-Interscience, 1976.
3. LSK-389 Data Sheet, Linear Integrated Systems.
4. 2SK-369 Data Sheet, Toshiba.
5. LTC6655 Data Sheet, Linear Technology Corporation.
6. LT1533 Data Sheet, Linear Technology Corporation.
7. Williams, Jim, "Practical Circuitry for Measurement and Control Problems," Linear Technology Corporation, Application Note 61, August 1994.
8. Williams, Jim, "A Monolithic Switching Regulator with 100µV Output Noise," Linear Technology Corporation, Application Note 70, October 1997.
9. Williams, Jim and Owen, Todd, "Performance Verification of Low Noise, Low Dropout Regulators," Linear Technology Corporation, Application Note 83, March 2000.
10. Williams, Jim, "Low Noise Varactor Biasing with Switching Regulators," Linear Technology Corporation, Application Note 85, August 2000, pages 4–6.
11. Williams, Jim, "Minimizing Switching Regulator Residue in Linear Regulator Outputs," Linear Technology Corporation, Application Note 101, July 2005.
12. Williams, Jim, "Power Conversion, Measurement and Pulse Circuits," Linear Technology Corporation, Application Note 113, August 2007.
13. Williams, Jim, "High Voltage, Low Noise, DC-DC Converters," Linear Technology Corporation, Application Note 118, March 2008.
14. Tektronix, Inc., "Type 1A7 Plug-In Unit Operating and Service Manual," Tektronix, Inc., 1965.
15. Tektronix, Inc., "Type 1A7A Differential Amplifier Operating and Service Manual," Tektronix, Inc. 1968.
16. Tektronix, Inc. "Type 7A22 Differential Amplifier Operating and Service Manual," Tektronix, Inc., 1969.
17. Tektronix, Inc., "AM502 Differential Amplifier Operating and Service Manual," Tektronix, Inc., 1973.

Appendix A
Mechanical and layout considerations

The low noise ×10,000 preamplifier; crucial to the noise measurement, must be quite carefully prepared. Figure A1 shows board layout. The board is enclosed within a shielded can, visible in A1A. Additional shielding is provided to the input capacitor and resistor (A1A left); the resistor's wirewound construction has low noise but is particularly susceptible to stray fields. A1A also shows the socketed LTC6655 reference under test (below the large input capacitor shield) and the JFET input amplifier associated components. Q3 (A1A upper right), a heat source, is located away from the JFET printed circuit lands, preventing convection currents from introducing noise. Additionally, the JFET's are contained within an epoxy filled plastic cup (Figure A1B center), promoting thermal mating and lag.[1] This thermal management of the FETs prevents offset instability and hunting in A1's stabilizing loop from masquerading as low frequency noise. ±15V power enters the enclosure via banana jacks; the reference is supplied by a 9V battery (both visible in A1A). The A = 100 filter and peak-peak detector circuitry occupies a separate board outside the shielded can. No special commentary applies to this section although board leakage to the peak detecting capacitors should be minimized with guard rings or flying lead/Teflon stand-off construction.

Appendix B
Input capacitor selection procedure

The input capacitor, a highly specialized type, must be selected for leakage. If this is not done, resultant errors can saturate the input pre-amplifier or introduce noise. The highest grade wet slug 200°C rated tantalum capacitors are utilized. The capacitor operates at a small fraction of its rated voltage at room temperature, resulting in much lower leakage than its specification indicates.

The capacitor's dielectric absorption requires a 24-hour charge time to insure meaningful measurement. Capacitor leakage is determined by following

Figure A1A

Figure A1B

Figure A1 • Preamplifier Board Top (Figure A1A) and Bottom (A1B) Views. Board Top Includes Shielded Input Capacitor (Upper Left) and Input Resistor (Upper Center Left). Stabilized JFET Input Amplifier Occupies Board Upper Center Right; Output Stage Adjoins BNC Fitting. Reference Under Test Resides in Socket Below Input Capacitor. ±15 Power Enters Shielded Enclosure Via Banana Jacks (Extreme Right). 9V Battery (Lower) Supplies Reference Under Test. Board Bottom's Epoxy Filled Plastic Cup (A1B Center) Contains JFETs, Provides Thermal Mating and Lag

Note 1: The plastic cup, supplied by Martinelli and Company, also includes, at no charge, 10 ounces of apple juice.

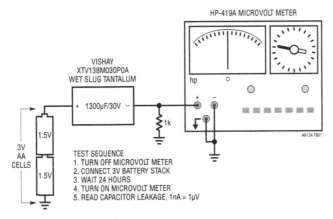

Figure B1 • Pre-Amplifier Input Capacitor Selected for <5nA Leakage to Minimize DC Error and Capacitor Introduced Noise. Capacitor Dielectric Absorption Requires 24 Hour Charge Time to Insure Meaningful Measurement. Highest Grade Wet Slug Tantalum Capacitors are Required to Pass This Test

the 5-step procedure given in the figure. Yield to required 5-nanoampere leakage exceeds 90%.[1]

Note 1: This high yield is most welcome because the specified capacitors are spectacularly priced at almost $400.00. There may be a more palatable alternative. Selected commercial grade aluminum electrolytics can approach the required DC leakage although their aperiodic noise bursts (mechanism not understood; reader comments invited) are a concern.

Appendix C
Power, grounding and shielding considerations

Figure 39.3's circuit requires great care in power distribution, grounding and shielding to achieve the reported results. Figure C1 depicts an appropriate

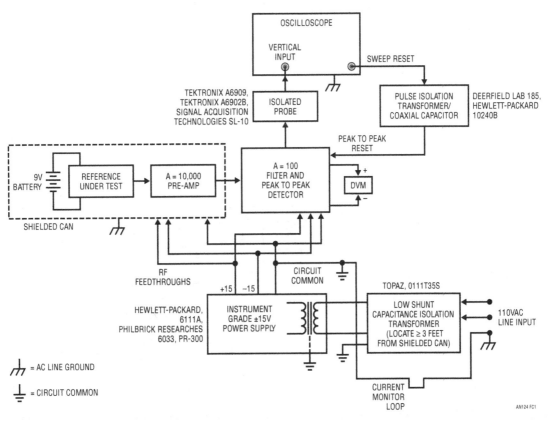

Figure C1 • Power/Grounding/Shielding Scheme for Low Noise Measurement Minimizes AC Line Originated Interference and Mixing of Circuit Return and AC Line Ground Current. No Current Should Flow in Current Monitor Loop

scheme. A low shunt capacitance line isolation transformer powers an instrument grade ±15V supply, furnishing clean, low noise power. The pre-amplifier's shielded can is tied to the 110V AC ground terminal, directing pick-up to earth ground. Filter/peak-to-peak detector oscilloscope connections are made via an isolated probe and a pulse isolation transformer, precluding error inducing ground

loops.[1] The indicated loop, included to verify no current flow between circuit common and earth ground, is monitored with a current probe. Figures C2 and C3, both optional, show battery powered supplies which replace the line isolation transformer

---

Note 1: An acceptable alternative to the isolated probe is monitoring Figure 39.3's A4 output current into a grounded 1k resistor with a DC stabilized current probe (e.g. Tektronix P6042, AM503). The resultant isolated 1V/μV oscilloscope presentation requires 10Hz lowpass filtering (see Appendix D) due to inherent current probe noise.

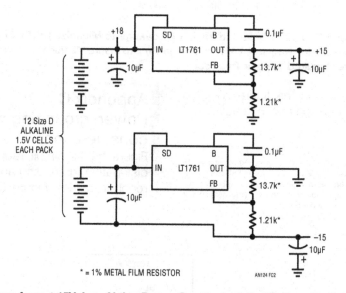

Figure C2 • LT1761 Regulators form ±15V, Low Noise Power Supply. Isolated Battery Packs Permit Positive Regulator to Supply Negative Output and Eliminate Possible AC Line Referred Ground Loops

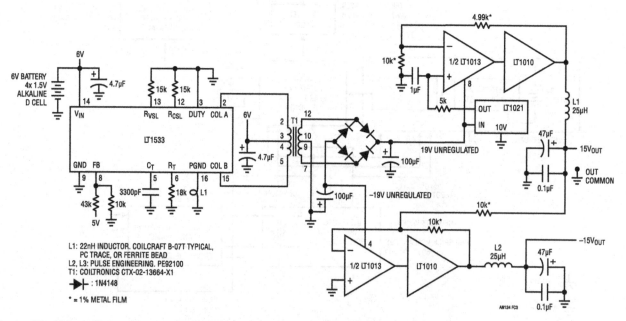

Figure C3 • A Low Noise, Bipolar, Floating Output Converter. Grounding LT1533 "DUTY" Pin and Biasing FB Puts Regulator into 50% Duty Cycle Mode. LT1533's Controlled Transition Times Permit <100μV Broadband Output Noise; Discrete Linear Regulators Maintain Low Noise, Provide Regulation

and instrumentation grade power supplies. C2 uses linear regulators to furnish low noise ±15V. Because the batteries float, positive regulators suffice for both positive and negative rails. In C3, a single battery stack supplies an extremely low noise DC-DC converter to furnish positive and negative rails via low noise discrete linear regulators.[2] Both of these battery supplied approaches are economical compared to the AC line powered version but require battery maintenance.

The indicated commercial products accompanying Figure C1's blocks represent typical applicable units which have been found to satisfy requirements. Other types may be employed but should be verified for necessary performance.

Appendix D
High sensitivity, low noise amplifiers

Figure D1 lists some useful low level amplifiers for setting up and troubleshooting the text's circuit. The table lists both oscilloscope plug-in amplifiers and stand-alone types. Two major restrictions apply. The filters in these units are single-pole types resulting in somewhat pessimistic bandwidth cut-offs. Additionally, the amplifiers listed do not include 10Hz lowpass frequency filters, although they are easily modified to provide this capability. Figure D2 lists four amplifiers with the necessary modification information.[1]

Note 2: References 6 and 8 detail the specialized DC DC converter used.

Note 1: See References 14-17.

INSTRUMENT TYPE	MANUFACTURER	MODEL NUMBER	–3dB BANDWIDTH	MAXIMUM SENSITIVITY/GAIN	AVAILABILITY	COMMENTS
Differential Amplifier	Tektronix	1A7/1A7A	1MHz	10μV/DIV	Secondary Market	Requires 500 Series Mainframe, Settable Bandstops
Differential Amplifier	Tektronix	7A22	1MHz	10μV/DIV	Secondary Market	Requires 7000 Series Mainframe, Settable Bandstops
Differential Amplifier	Tektronix	5A22	1MHz	10μV/DIV	Secondary Market	Requires 5000 Series Mainframe, Settable Bandstops
Differential Amplifier	Tektronix	ADA-400A	1MHz	10μV/DIV	Current Production	Stand-Alone with Optional Power Supply, Settable Bandstops
Differential Amplifier	Preamble	1822	10MHz	Gain = 1000	Current Production	Stand-Alone, Settable Bandstops
Differential Amplifier	Stanford Research Systems	SR-560	1MHz	Gain = 50000	Current Production	Stand-Alone, Settable Bandstops, Battery or Line Operation
Differential Amplifier	Tektronix	AM-502	1MHz	Gain = 100000	Secondary Market	Requires TM-500 Series Power Supply, Settable Bandstops

Figure D1 • Some Useful High Sensitivity, Low Noise Amplifiers. Trade-Offs Include Bandwidth, Sensitivity and Availability

MANUFACTURER	MODEL NUMBER	MODIFICATION
Tektronix	1A7	Parallel C370A with 1μF
Tektronix	1A7A	Parallel C445A with 1μF
Tektronix	7A22	Parallel C426H with 3μF
Tektronix	AM502	Parallel C449 with 3μF

Figure D2 • Modification Information for Various Tektronix Low Level Oscilloscope Plug-In's and Amplifiers Permits 10Hz High Frequency Filter Operation in 100Hz Panel Switch Position. All Cases Utilize 100V, Mylar Capacitors

Section 3

High Frequency/RF Design

WCDMA ACPR and AltCPR measurements (40)

ACPR (adjacent channel power ratio) and AltCPR (alternate channel power ratio) are both measures of spectral regrowth. They are important performance metrics for digital communication systems that use, for example, WCDMA (wideband code division multiple access) modulation. This publication highlights key considerations for accurate measurements of these parameters. In particular, highly linear direct I/Q modulators such as the LT5528 require high performance measurement equipment and careful techniques to characterize their spectral regrowth.

Measuring phase and delay errors accurately in I/Q modulators (41)

A large image rejection can be achieved in an RF transmitter system using an I/Q modulator after performing a phase and gain calibration of the I and Q signals. This is usually done by monitoring the RF output signal and using an optimization algorithm in the baseband processor. However, delay errors in the system prevent a good image rejection from extending over a large bandwidth. This application note helps to characterize the delay errors in the system accurately, using a three-step measurement approach. It can derive the corresponding phase errors for each block in the system and the most dominant delay error source(s) can be identified.

LT5528 WCDMA ACPR, AltCPR and noise measurements

Doug Stuetzle

Introduction

ACPR (adjacent channel power ratio), AltCPR (alternate channel power ratio), and noise are important performance metrics for digital communication systems that use, for example, WCDMA (wideband code division multiple access) modulation. ACPR and AltCPR are both measures of spectral regrowth. The power in the WCDMA carrier is measured using a 5MHz measurement bandwidth; see Figure 40.1. In the case of ACPR, the total power in a 3.84MHz bandwidth centered at 5MHz (the carrier spacing) away from the center of the outermost carrier is measured and compared to the carrier power. The result is expressed in dBc. For AltCPR, the procedure is the same, except we center the measurement 10MHz away from the center of the outermost carrier.

To measure ACPR and AltCPR, refer to the test setup shown in Figure 40.2. The DUT (device under test) is the LT5528, which is a high linearity direct I/Q modulator. It accepts WCDMA modulation at the baseband inputs, and generates a WCDMA modulated signal at the RF output. Note that a free running RF generator provides the LO signal. This type of generator is used because of its superior noise performance. This is critical, as a noisy LO signal may corrupt the modulator output, and consequently the

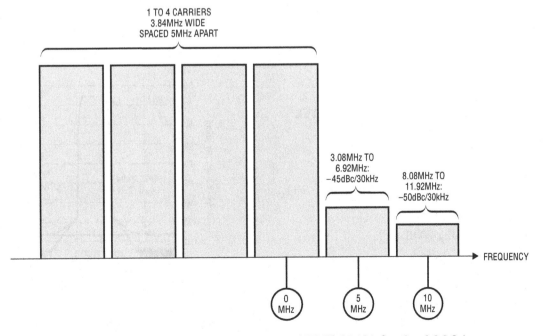

Figure 40.1 • WCDMA ACPR Limits, Per 3GPP TS 25.104, Section 6.6.2.2.1

Analog Circuit and System Design: A Tutorial Guide to Applications and Solutions. DOI: 10.1016/B978-0-12-385185-7.00040-8

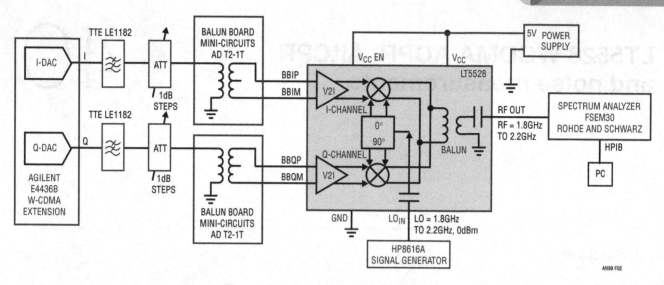

Figure 40.2 • ACPR Measurement Setup

ACPR measurement. For the generator shown, the automatic level control must be switched off to avoid degrading its broadband noise floor. Also, the operating frequency can drift slightly, so manual frequency correction could be needed.

The spectrum analyzer must have a wide dynamic range. That means a high input 3rd order intercept point, and a low noise floor. The analyzer shown in Figure 40.2 meets both of these requirements.

In general, the baseband source will not be ideal. It generates spectral regrowth and noise which may swamp the performance of the DUT. The lowpass filters shown at the baseband generator outputs reduce these impairments to a tolerable level. Filters suggested for this purpose are listed below:

- 1-channel ACPR measurement
  Filter part number LE1182 2.5M-50-720B, from TTE Engineering
  Rejection at 2.6MHz >20dB
  Rejection at 3.08MHz >80dB
- 2-channel ACPR measurement
  Filter part number LE1182 5M-50-720B, from TTE Engineering
  Rejection at 5.2MHz >20dB
  Rejection at 8.08MHz >80dB
- 4-channel ACPR measurement
  Filter part number LE1182 10M-50-720B, from TTE Engineering
  Rejection at 10.4MHz >20dB
  Rejection at 13.08MHz >80dB

An accurate measurement of the spectral regrowth of a highly linear device such as the LT5528 is difficult because its dynamic range may rival that of the measurement equipment. Because of this, it is important to account for the noise of the measurement system; i.e., the spectrum analyzer. Refer to Figure 40.3.

To do this, first measure the noise floor of the spectrum analyzer with a 50Ω input termination. The input attenuation of the analyzer should be set to 0dB. This will minimize the input 3rd order intercept point of the measurement system, as well as the noise figure. A 30kHz resolution bandwidth is used because the spectrum analyzer shown has the lowest noise figure (about 24dB) at that resolution bandwidth. This spectrum analyzer includes an RMS display detector mode, which is specifically designed to measure noise-like signals. For spectrum analyzers that do not offer this mode, it is important to set the video bandwidth to at least 3 times the resolution bandwidth. In this case, use a video bandwidth of 100kHz. If the ratio of video to resolution bandwidth is too low, the power measurement will be inaccurate. For example, if the ratio is 1:1, the measured power may be 0.35dB lower than the true power. Note that the sweep

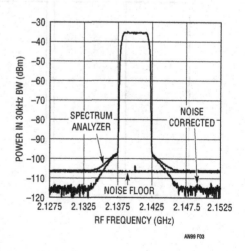

Figure 40.3 • ACPR Spectrum for a Single Carrier WCDMA Signal

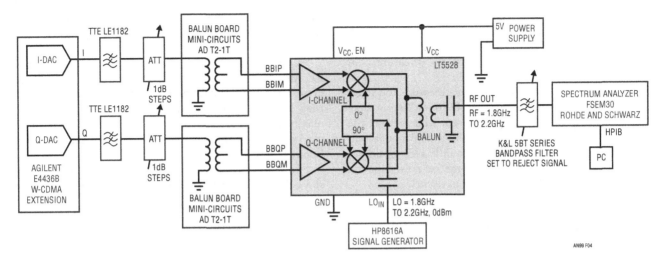

Figure 40.4 • 30MHz Offset Noise Measurement Setup

time must be increased by a factor of 10 or so in order to take advantage of the higher number of samples in this mode. Video averaging helps smooth the result; 100 averages gives good results. The channel power utility of the analyzer is used to find the total power within a 3.84MHz bandwidth.

Next measure the output spectrum of the DUT using the same settings. For ACPR, center the measurement band 5MHz above the center of the highest carrier. To find the true spectral regrowth power, convert the measured spectral power levels to mW and subtract the spectrum analyzer noise floor from the measured DUT power. Reconvert to dBm to get the true spectral regrowth. Do the same for the band 5MHz below the center of the lowest carrier. Take the average of the two dBc figures to arrive at the average spectral regrowth.

The ACPR/AltCPR is equal to the difference in dBc between the signal power and the spectral regrowth.

The spectrum analyzer shown also offers an ACPR measurement utility. This utility will not, however, give accurate results for highly linear devices, as it does not compensate for the measurement system noise floor.

To measure noise at 30MHz offset, the test setup is modified as shown in Figure 40.4. A tunable bandpass filter is added to the DUT output. This filter should be set to reject the main signal, but not attenuate the noise 30MHz above the outermost channel. Noise measurements that do not use this filter technique will produce degraded results for strong RF signal levels. The reason is that the dynamic range of the spectrum analyzer is not sufficient to accept the power of the main signal while accurately measuring the noise. So the main signal will

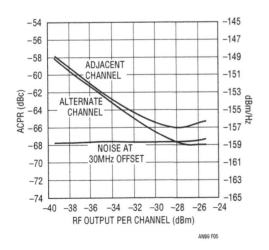

Figure 40.5 • LT5528 4-channel WCDMA Adjacent and Alternate CPR and 30MHz Noise Floor Measurement vs Channel Power

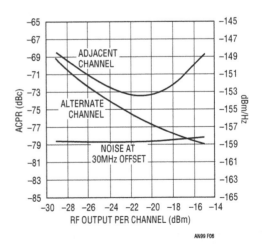

Figure 40.6 • LT5528 1-channel WCDMA Adjacent and Alternate CPR and 30MHz Noise Floor Measurement vs Channel Power

tend to overload the RF front end of the analyzer. The bandpass filter helps to reduce the amount of signal reaching the analyzer by approximately 20dB, while adding only about 1dB of noise figure to the measurement frequency.

Proceeding with this measurement, measure the noise floor of the spectrum analyzer with a 50Ω input termination. A narrow span of 100kHz can be used, while setting the resolution and video bandwidths to 30kHz. The input attenuation of the analyzer should be set to 0dB. Here again, set the detector mode for RMS. Use the marker noise function and video averaging to obtain a result in dBm/Hz.

Next, connect the test setup as shown in Figure 40.4. Measure the noise at 30MHz from the center of the signal frequency, using the same settings as above. To find the true noise level, convert the noise power and the noise floor to mW and subtract the spectrum analyzer noise floor from the DUT output noise power. Reconvert to dBm/Hz to get the true noise level.

ACPR and AltCPR vary with output signal level. For low RF output power levels, these are limited by the output noise floor of the DUT. At high RF output power levels, they are determined by the linearity of the DUT. The maximum ACPR/AltCPR are observed between these extremes, where the spectral regrowth is equal to the noise floor of the DUT.

Some sample results of these measurements are shown to illustrate this in Figures 40.5 and 40.6. Figure 40.5 is a plot of ACPR, AltCPR, and noise versus baseband drive level for a 4-carrier WCDMA signal. Figure 40.6 shows the same results for a single WCDMA carrier.

Measuring phase and delay errors accurately in I/Q modulators

41

P. Stroet

Introduction

This chapter describes a method to accurately measure internal and external phase and timing errors for a high performance direct I/Q modulator. A direct I/Q modulator, such as the LT5528, translates baseband I and Q signals to RF, and combines them to produce a modulated single sideband signal with (ideally) minimal residual carrier (LO feedthrough) and image signals (undesired sideband). In an ideal I/Q modulator, with perfect 90° phase shift between the I mixer and Q mixer local oscillators (LOI and LOQ), and with no other undesired phase and gain impairments, the modulator output will contain only the desired sideband. In practice, this is very difficult to accomplish. For example, with a requirement of −60dBc image suppression, the residual I-Q phase error is required to be below 0.16°. In practice, there are other sources of phase error, particularly in the baseband signal processing. Examples include baseband skew or other frequency dependent phase shifts in the modulator baseband circuitry; skew errors due to phase or delay mismatched baseband connection paths (e.g., cabling); and phase mismatch between the I and the Q paths in the baseband signal source (e.g., baseband DACs or signal generators.). These phase errors can cause RF output spectra to be shaped like those shown in Figures 41.1, 41.2 and 41.3.

For each plot, the (single) channel is chosen to be at −7.5MHz, −2.5MHz, 2.5MHz and 7.5MHz offset from the RF carrier by choosing the frequency offset function on the baseband generator. As can be seen, the residual sideband spectra are not flat vs RF frequency. Usually, the image rejection calibration is done using one (baseband) frequency, preferably in the center of the desired channel. However, if the uncalibrated residual sideband is not flat versus frequency, it causes the image rejection after calibration to degrade at the edges of the channel. This can be seen in Figure 41.4 where the image rejection is less than

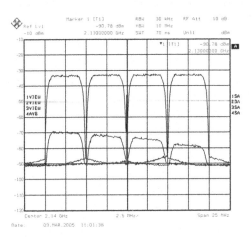

Figure 41.1 • Measurement Compilation of Four One-Channel W-CDMA I/Q Modulator RF Output Spectra Selected to be at −7.5MHz, −2.5MHz, 2.5MHz and 7.5MHz Offset Frequency from the 2.14GHz Carrier Using Baseband I/Q W-CDMA Channel Selection with Uncalibrated Image

60dBc and the image channel has a shape in the form of the letter "M". A delay difference between the I and the Q baseband paths can cause the image power to be falling vs RF frequency as in Figure 41.1, rising as in Figure 41.2 or to have a "V" shape as in Figure 41.3. The sign and magnitude of the quadrature phase error in the I/Q modulator, and the sign and magnitude of the I/Q baseband delay difference determine whether the situation is as in Figure 41.1, 41.2 or 41.3.

The residual sideband spectrum of Figure 41.4 can be improved by adding a compensating delay to the I or Q baseband paths. This is shown in Figure 41.5.

In order to achieve the best image rejection for a broadband communications channel (such as W-CDMA), it is important to understand what error source(s) causes the image response to be non-flat over frequency. This chapter

Analog Circuit and System Design: A Tutorial Guide to Applications and Solutions. DOI: 10.1016/B978-0-12-385185-7.00041-X

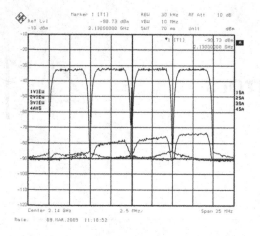

Figure 41.2 • Measurement Compilation of Four One-Channel W-CDMA I/Q Modulator RF Output Spectra Selected to be at −7.5MHz, −2.5MHz, 2.5MHz and 7.5MHz Offset Frequency from the 2.14GHz Carrier Using Baseband I/Q W-CDMA Channel Selection with Uncalibrated Image

provides a measurement method to determine the sources of both RF and baseband phase error, whether it comes from the baseband generator and/or the I/Q modulator. The method consists of three different measurements, each with a slightly different measurement setup. From these measurements, we can determine the quadrature error of the I/Q modulator φ_{LO}, the baseband phase error of the I/Q modulator φ_{MOD}, and the baseband phase error of the baseband signal generator φ_{DGEN}. It is very likely that φ_{DGEN} and φ_{MOD} result from internal skew or time delay errors (τ_{DQEN}

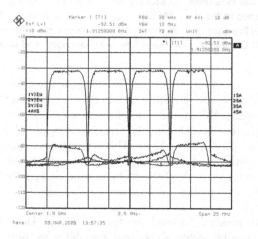

Figure 41.3 • Measurement Compilation of Four One-Channel W-CDMA I/Q Modulator RF Output Spectra Selected to be at −7.5MHz, −2.5MHz, 2.5MHz and 7.5MHz Offset Frequency from the 2.14GHz Carrier Using Baseband I/Q W-CDMA Channel Selection with Uncalibrated Image

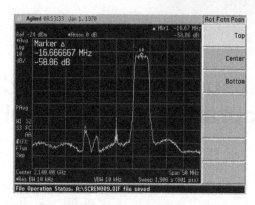

Figure 41.4 • Measurement of a One-Channel W-CDMA Spectrum at the I/Q Modulator Output After Image Nulling at 7.5MHz Baseband Calibration Frequency. The W-CDMA Channel is Located at an Offset of 7.5MHz and the Image is Located at an Offset of −7.5MHz with Respect to the Carrier

and τ_{MOD}, respectively). Therefore, we can write in a more general case for different baseband frequencies (ω_{BB}):

$$\varphi_{DGEN} = \varphi_{DGEN0} + \omega_{BB} \bullet \tau_{DGEN}$$

and

$$\varphi_{MOD} = \varphi_{MOD0} + \omega_{BB} \bullet \tau_{MOD}$$

In the analysis that follows, we disregard amplitude mismatches, because our measurements indicate that phase errors are dominant, and it greatly simplifies the math.

In order to resolve the uncontrolled, systematic phase errors, φ_{LO}, φ_{DGEN} and φ_{MOD}, our technique requires there to be a controllable, adjustable baseband phase offset, φ_{GEN}. This adjustable phase is used to null out the image signal under various measurement conditions. The nulling phases are used to calculate the individual system phase errors.

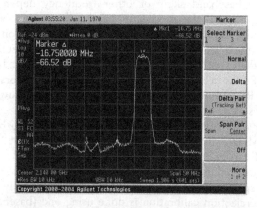

Figure 41.5 • Measurement of a One-Channel W-CDMA Spectrum at the I/Q Modulator Output After Image Nulling at 7.5MHz Baseband Calibration Frequency Using a Baseband I/Q Delay Correction. The W-CDMA Channel is Located at an Offset of 7.5MHz and the Image is Located at an Offset of −7.5MHz with Respect to the Carrier

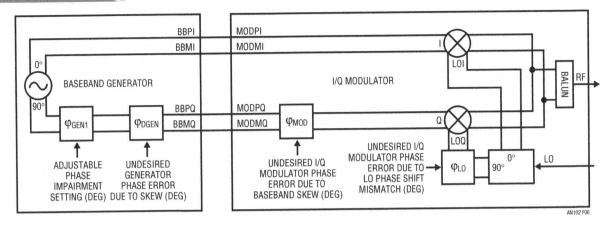

Figure 41.6 • Measurement Setup for Configuration 1

Measurements

First measurement—null out the I/Q modulator image signal with normal signal connections (Figure 41.6)

A phase error φ_{LO} exists between the quadrature signals LOI and LOQ in the modulator. We try to cancel this with an extra phase shift φ_{GEN1} between the baseband signals I and Q. However, as shown and defined in Figure 41.6, there can be delay differences between the I and the Q path for both the baseband generator (φ_{DGEN}) and within the I/Q modulator itself (φ_{MOD}). At a particular baseband frequency $\omega_{BB} = 2\pi f_{BB}$, the baseband signals at the modulator's I and Q mixers are given by:

$$I = \cos(\omega_{BB} \bullet t)$$
$$Q = \sin(\omega_{BB} \bullet t + \varphi_{GEN1} + \varphi_{DGEN} + \varphi_{MOD})$$

Note that the placement of the error terms φ_{DGEN}, φ_{MOD} and φ_{LO} in the I or Q paths is arbitrary and does not affect the final conclusions of this analysis.

Here φ_{GEN1} is a controllable phase offset that can be adjusted as needed to compensate for other phase errors in the system.

$$LOI = \cos(\omega_{LO} \bullet t)$$
$$LOQ = \sin(\omega_{LO} \bullet t + \varphi_{LO})$$

$$RF = \cos(\omega_{BB} \bullet t) \bullet \cos(\omega_{LO} \bullet t) + \sin(\omega_{BB} \bullet t + \varphi_{GEN1} + \varphi_{DGEN} + \varphi_{MOD}) \bullet \sin(\omega_{LO} \bullet t + \varphi_{LO})$$

$$\cos(\alpha) \bullet \cos(\beta) = 1/2 \cos(\alpha - \beta) + 1/2 \cos(\alpha + \beta)$$

$$\sin(\alpha) \bullet \sin(\beta) = 1/2 \cos(\alpha - \beta) - 1/2 \cos(\alpha + \beta)$$

$$RF = 1/2 \cos[(\omega_{LO} - \omega_{BB})t] + 1/2 \cos[(\omega_{LO} + \omega_{BB} \bullet t] + 1/2 \cos[(\omega_{LO} - \omega_{BB})t + \varphi_{LO} - \varphi_{GEN1} - \varphi_{DGEN} - \varphi_{MOD}] - 1/2 \cos[(\omega_{LO} - \omega_{BB})t + \varphi_{LO} + \varphi_{GEN1} + \varphi_{DGEN} + \varphi_{MOD}]$$

$$\cos(\alpha + \beta) = \cos(\alpha)\cos(\beta) - \sin(\alpha)\sin(\beta)$$

$$\begin{aligned}
RF = {} & 1/2 \cos[(\omega_{LO} - \omega_{BB})t] + 1/2 \cos[(\omega_{LO} + \omega_{BB}) \bullet t] \\
& + 1/2 \cos(\varphi_{LO} - \varphi_{GEN1} - \varphi_{DGEN} - \varphi_{MOD}) \\
& \bullet \cos[(\omega_{LO} - \omega_{BB})t] \\
& - 1/2 \sin(\varphi_{LO} - \varphi_{GEN1} - \varphi_{DGEN} - \varphi_{MOD}) \\
& \bullet \sin[(\omega_{LO} - \omega_{BB})t] \\
& - 1/2 \cos(\varphi_{LO} + \varphi_{GEN1} + \varphi_{DGEN} + \varphi_{MOD}) \\
& \bullet \cos[(\omega_{LO} + \omega_{BB})t] \\
& + 1/2 \sin(\varphi_{LO} + \varphi_{GEN1} + \varphi_{DGEN} + \varphi_{MOD}) \\
& \bullet \sin[(\omega_{LO} + \omega_{BB})t]
\end{aligned}$$

$$\cos(\varphi) = 1 - \frac{\varphi^2}{2} + \frac{\varphi^4}{24} - \frac{\varphi^6}{720} + \ldots \approx 1 - \frac{\varphi^2}{2} \approx 1$$

(Small angle approximation)

$$\sin(\varphi) = \varphi - \frac{\varphi^3}{6} + \frac{\varphi^5}{120} - \frac{\varphi^7}{5040} + \ldots \approx \varphi$$

(Small angle approximation)

$$\begin{aligned}
RF = {} & \cos[(\omega_{LO} - \omega_{BB}) \bullet t] - 1/2 (\varphi_{LO} - \varphi_{GEN1} - \varphi_{DGEN} \\
& - \varphi_{MOD}) \bullet \sin[(\omega_{LO} - \omega_{BB}) \bullet t] + 1/2 (\varphi_{LO} + \varphi_{GEN1} \\
& + \varphi_{DGEN} + \varphi_{MOD}) \bullet \sin[(\omega_{LO} + \omega_{BB})t]
\end{aligned}$$

In addition to the desired lower sideband signal at $(\omega_{LO} - \omega_{BB})$ we also see some upper sideband signal at $(\omega_{LO} + \omega_{BB})$.

For small phase errors, the upper sideband amplitude is approximately given by:

$$A_{USB} \approx 1/2 (\varphi_{LO} + \varphi_{GEN1} + \varphi_{DGEN} + \varphi_{MOD})$$

and the upper sideband suppression is given by:

$$\begin{aligned}
R_{SB}(dB) = {} & 20 \bullet \log[1/2 (\varphi_{LO} + \varphi_{GEN1} + \varphi_{DGEN1} + \varphi_{MOD})] \\
= {} & 20 \bullet \log(\varphi_{LO} + \varphi_{GEN1} + \varphi_{DGEN} + \varphi_{MOD}) \\
& - 6.02 \ (dB)
\end{aligned}$$

Note that the phases φ are in radians.

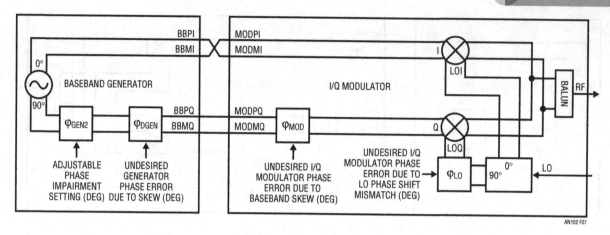

Figure 41.7 • Measurement Setup for Configuration 2

The image term can be minimized by adjusting the generator (impairment) phase setting to:

$$\varphi_{GEN1} = -\varphi_{LO} - \varphi_{DGEN} - \varphi_{MOD}$$

Second measurement—null out the I/Q modulator image signal with reversed differential baseband signals to the modulator's differential I-channel inputs (Figure 41.7)

This configuration differs from that of Figure 41.6 in that the differential baseband signals to the modulator's I inputs are reversed. In this configuration the image component of the RF output signal is measured and nulled by adjustment of the controllable signal generator phase, φ_{GEN2}. Note that the length of the I signal path is assumed not to change by flipping BBPI and BBMI; the connectors on the baseband generator are just flipped.

$$
\begin{aligned}
I &= -\cos(\omega_{BB} \bullet t), \; Q = \sin(\omega_{BB} \bullet t + \varphi_{GEN2} \\
&\quad + \varphi_{DGEN} + \varphi_{MOD}) \\
LOI &= \cos(\omega_{LO} \bullet t), \; LOQ = \sin(\omega_{LO} \bullet t + \varphi_{LO}) \\
RF &= -\cos(\omega_{BB} \bullet t) \bullet \cos(\omega_{LO} \bullet t) + \sin(\omega_{BB} \bullet t \\
&\quad + \varphi_{GEN2} + \varphi_{DGEN} + \varphi_{MOD}) \bullet \sin(\omega_{LO} \bullet t + \varphi_{LO})
\end{aligned}
$$

Using trigonometric identities, this can be expanded to:

$$
\begin{aligned}
RF &= -1/2 \cos[(\omega_{LO} - \omega_{BB})t] - 1/2 \cos[(\omega_{LO} + \omega_{BB}) \bullet t] \\
&\quad + 1/2 \cos(\varphi_{LO} - \varphi_{GEN2} - \varphi_{DGEN} - \varphi_{MOD}) \\
&\quad \bullet \cos[(\omega_{LO} - \omega_{BB})t] \\
&\quad - 1/2 \sin(\varphi_{LO} - \varphi_{GEN2} - \varphi_{DGEN} - \varphi_{MOD}) \\
&\quad \bullet \sin[(\omega_{LO} - \omega_{BB})t] \\
&\quad - 1/2 \cos(\varphi_{LO} + \varphi_{GEN2} + \varphi_{DGEN} + \varphi_{MOD}) \\
&\quad \bullet \cos[(\omega_{LO} + \omega_{BB})t] \\
&\quad + 1/2 \sin(\varphi_{LO} + \varphi_{GEN2} + \varphi_{DGEN} + \varphi_{MOD}) \\
&\quad \bullet \sin[(\omega_{LO} + \omega_{BB})t]
\end{aligned}
$$

Again using the small angle approximations, this becomes:

$$
\begin{aligned}
RF &= -\cos[(\omega_{LO} + \omega_{BB}) \bullet t] - 1/2\,(\varphi_{LO} - \varphi_{GEN2} - \varphi_{DGEN} \\
&\quad - \varphi_{MOD}) \bullet \sin[(\omega_{LO} - \omega_{BB}) \bullet t] + 1/2\,(\varphi_{LO} + \varphi_{GEN2} \\
&\quad + \varphi_{DGEN} + \varphi_{MOD}) \bullet \sin[(\omega_{LO} + \omega_{BB})t]
\end{aligned}
$$

Now, the desired signal is the upper sideband signal ($\omega_{LO} + \omega_{BB}$), and the image signal is at ($\omega_{LO} - \omega_{BB}$).

For small phase errors, the lower side band amplitude is given by:

$$A_{LSB} \approx 1/2\,(\varphi_{LO} - \varphi_{GEN2} - \varphi_{DGEN} - \varphi_{MOD})$$

The lower sideband suppression is given by:

$$
\begin{aligned}
R_{SB}(dB) &= 20 \bullet \log[1/2\,(\varphi_{LO} - \varphi_{GEN2} - \varphi_{DGEN} \\
&\quad - \varphi_{MOD})] \\
&= 20 \bullet \log(\varphi_{LO} - \varphi_{GEN2} - \varphi_{DGEN} \\
&\quad - \varphi_{MOD}) - 6.02(dB)
\end{aligned}
$$

In this configuration, the image is minimized by adjusting:

$$\varphi_{GEN2} = \varphi_{LO} - \varphi_{DGEN} - \varphi_{MOD}$$

Third measurement—null out the I/Q modulator image signal after reversing the I and Q inputs to the modulator (Figure 41.8)

This configuration differs from that of Figure 41.6 in that the I and Q differential inputs are exchanged. Note that the connection lengths in the I and Q path did not change by reversing BBI and BBQ, the connectors on the baseband generator are just flipped.

In this configuration, the image component of the RF output signal is measured and nulled by adjustment of the controllable signal generator phase, φ_{GEN3}.

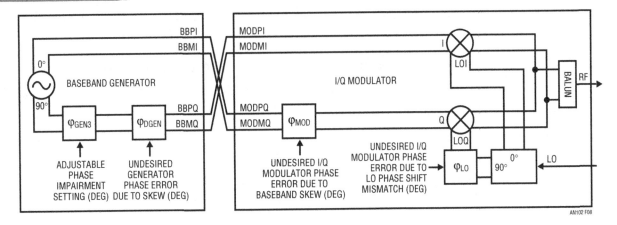

Figure 41.8 • Measurement Setup for Configuration 3

$$
\begin{aligned}
Q &= \cos(\omega_{BB} \bullet t + \varphi_{MOD}) \\
I &= \sin(\omega_{BB} \bullet t + \varphi_{GEN3} + \varphi_{DGEN}) \\
LOI &= \cos(\omega_{LO} \bullet t) \\
LOQ &= \sin(\omega_{LO} \bullet t + \varphi_{LO}) \\
RF &= \sin(\omega_{BB} \bullet t + \varphi_{GEN3} + \varphi_{DGEN}) \bullet \cos(\omega_{LO} \bullet t) \\
&\quad + \cos(\omega_{BB} \bullet t + \varphi_{MOD}) \bullet \sin(\omega_{LO} \bullet t + \varphi_{LO})
\end{aligned}
$$

Using trigonometric identities and small angle approximations, this can be expanded to:

$$
\begin{aligned}
RF &\approx 1/2 \, (\varphi_{GEN3} + \varphi_{DGEN} + \varphi_{LO} - \varphi_{MOD}) \bullet \cos[(\omega_{LO} \\
&\quad - \omega_{BB})t] + 1/2 \, (\varphi_{GEN3} + \varphi_{DGEN} + \varphi_{LO} + \varphi_{MOD}) \\
&\quad \bullet \cos[(\omega_{LO} + \omega_{BB})t] + \sin[(\omega_{LO} + \omega_{BB}) \bullet t]
\end{aligned}
$$

Now, the desired signal is the upper sideband frequency component $(\omega_{LO} + \omega_{BB})$ and the image is a lower sideband signal at $(\omega_{LO} - \omega_{BB})$.

For small phase errors, the lower sideband amplitude is given by:

$$
A_{LSB} \approx 1/2 \, (\varphi_{GEN3} + \varphi_{DGEN} + \varphi_{LO} - \varphi_{MOD})
$$

The lower sideband suppression is given by:

$$
\begin{aligned}
R_{SB}(dB) &= 20 \bullet \log[1/2 \, (\varphi_{GEN3} + \varphi_{DGEN} + \varphi_{LO} - \varphi_{MOD})] \\
&= 20 \bullet \log(\varphi_{GEN3} + \varphi_{DGEN} + \varphi_{LO} \\
&\quad - \varphi_{MOD}) - 6.02(dB)
\end{aligned}
$$

In this configuration, the image is minimized by adjusting:

$$
\varphi_{GEN3} = -\varphi_{LO} - \varphi_{DGEN} + \varphi_{MOD}
$$

Calculation of phase impairments

$$
\begin{aligned}
\text{From} \quad IIA &: \varphi_{GEN1} = -\varphi_{LO} - \varphi_{DGEN} - \varphi_{MOD} \\
IIB &: \varphi_{GEN2} = -\varphi_{LO} - \varphi_{DGEN} - \varphi_{MOD} \\
IIC &: \varphi_{GEN3} = -\varphi_{LO} - \varphi_{DGEN} + \varphi_{MOD}
\end{aligned}
$$

We can solve these three equations, with three unknowns, to give:

$$
\varphi_{LO} = (\varphi_{GEN2} - \varphi_{GEN1})/2 \tag{1}
$$

$$
\varphi_{DGEN} = -(\varphi_{GEN2} + \varphi_{GEN3})/2 \tag{2}
$$

$$
\varphi_{MOD} = (\varphi_{GEN3} - \varphi_{GEN1})/2 \tag{3}
$$

Note that we can express the phases φ in these equations in radians or degrees.

The equations above hold for an I/Q modulator with an output relationship given by:

$$
RF = I \bullet \cos(\omega_{LO} \bullet t) + Q \bullet \sin(\omega_{LO} \bullet t) \tag{2D_A}
$$

The I/Q modulators provided by Linear Technology will satisfy the above equation.

However, other I/Q modulators may have the following output characteristic:

$$
RF = I \bullet \cos(\omega_{LO} \bullet t) - Q \bullet \sin(\omega_{LO} \bullet t) \tag{2D_B}
$$

There is no international convention on which I/Q modulator equation is the "right" one.

Using an I/Q modulator with the latter relationship will affect the derivations somewhat. In this case, in configuration 1, the desired signal will be then at the *upper* sideband $(\omega_{LO} + \omega_{BB})$ and image nulling will be achieved for:

$$
\varphi_{GEN1} = \varphi_{LO} - \varphi_{DGEN} - \varphi_{MOD}
$$

In configuration 2, the desired signal will be at the *lower* sideband $(\omega_{LO} - \omega_{BB})$ and image nulling will be achieved for:

$$
\varphi_{GEN2} = -\varphi_{LO} - \varphi_{DGEN} - \varphi_{MOD}
$$

In configuration 3, the desired signal will be again at the *lower* sideband $(\omega_{LO} - \omega_{BB})$ and image nulling will be achieved for:

$$
\varphi_{GEN3} = \varphi_{LO} - \varphi_{DGEN} - \varphi_{MOD}
$$

We can again solve these three equations, with three unknowns, to give:

$$
\varphi_{LO} = (\varphi_{GEN1} - \varphi_{GEN2})/2 \tag{4}
$$

Table 41.1 Image Rejection Null Vectors for Configurations 1, 2 and 3 for 5MHz and 10MHz Baseband Frequency. The Amplitude Adjustment Required for Nulling (Not Shown) is <0.35% (Worst Case)

	BASEBAND FREQUENCY = 5MHz			BASEBAND FREQUENCY = 10MHz		
	config1	config2	config3	config1	config2	config3
	φ_{DGEN1}	φ_{DGEN2}	φ_{DGEN3}	φ_{DGEN1}	φ_{DGEN2}	φ_{DGEN3}
UNIT	DEGREE	DEGREE	DEGREE	DEGREE	DEGREE	DEGREE
1	−0.90	1.93	0.83	−0.48	2.41	−0.41
2	1.13	−0.07	1.24	1.60	0.30	1.74
3	0.32	0.81	0.37	0.73	1.30	0.84
4	0.36	0.74	0.44	0.80	1.20	0.92
5	0.51	0.60	0.62	−0.03	0.10	0.03

$$\varphi_{DGEN} = -(\varphi_{GEN2} + \varphi_{GEN3})/2 \qquad (5)$$

$$\varphi_{MOD} = (\varphi_{GEN3} - \varphi_{GEN1})/2 \qquad (6)$$

Note that the sign is different for the φ_{LO} calculation, and the equations for φ_{DGEN} and φ_{MOD} stay the same.

Applying the method

For five different LT®5528 boards the image rejection null-vectors for configurations 1, 2 and 3 described above are measured and logged in Table 41.1, for baseband frequencies 5MHz and 10MHz. A QPSK signal is programmed into a Rohde & Schwartz AMIQ baseband generator with a bit sequence of 00011011. The symbol rate is 40MHz with oversampling of 2 for the 10MHz baseband frequency, and the symbol rate is 20MHz with an oversampling of 4 for the

5MHz baseband frequency, both resulting in a sample rate of 80MHz. In all cases better than 75dBc image rejection is achieved after nulling.

The quadrature phase error of the LT5528 φ_{LO}, the baseband phase error of the LT5528 φ_{MOD} and the baseband phase error of the generator, φ_{DGEN} can be determined using Equations 1, 2 and 3. The amplitude mismatch results are discarded. The results for the phase errors in degrees are given in Table 41.2. Also equivalent delays are derived from the phase errors, assuming all phase error is caused by a delay.

Conclusion

The method described here is capable of accurately measuring various sources of phase error. The measured quadrature error φ_{LO} using 5MHz and 10MHz baseband frequencies are equal within 0.05 degrees, suggesting quadrature error can be measured quite accurately for

Table 41.2 Phase Error Measurement Results of the LT5528

BOARD	BASEBAND FREQUENCY = 5MHz					BASEBAND FREQUENCY = 10MHz				
	φ_{LO}	φ_{MOD}	τ_{MOD}	φ_{DGEN}	τ_{DGEN}	φ_{LO}	φ_{MOD}	τ_{MOD}	φ_{DGEN}	τ_{DGEN}
UNIT	DEGREE	DEGREE	ps	DEGREE	ps	DEGREE	DEGREE	ps	DEGREE	ps
1	1.415	0.035	19.4	−0.55	306	1.445	0.035	9.7	−1.0	278
2	−0.60	0.055	30.6	−0.585	325	−0.65	0.07	19.4	−1.02	283
3	0.245	0.025	13.9	−0.59	328	0.285	0.055	15.3	−0.785	218
4	0.19	0.04	22.2	−0.59	328	0.20	0.06	16.7	−0.86	239
5	0.045	0.055	30.6	−0.61	339	0.08	0.1	27.8	−1.08	300

relatively high baseband frequencies. It can be seen that the baseband signal generator phase error φ_{DGEN} is dominant in this setup. τ_{DGEN} is about 300ps, compared to the LT5528's baseband delay error τ_{MOD}, which is only about 25ps to 30ps.

A somewhat surprising result is the magnitude of the phase error in the baseband signal source φ_{DGEN}. This base-band signal source phase error may be a dominant error source in a direct I/Q modulation scheme. It should be carefully characterized and compensated. Otherwise, it may limit the extent of image suppression. This is especially important in a broadband application, such as W-CDMA, if the baseband source phase error is a skew (time delay) error, which results in a frequency dependent phase error.

Subject Index

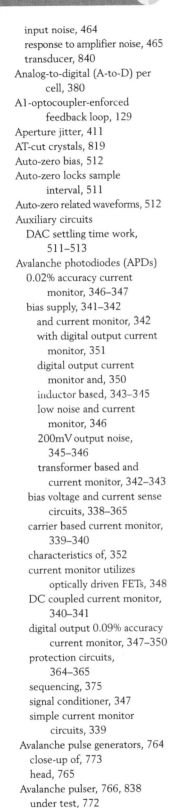

Printed in the United States
By Bookmasters